T0175912

Automotive Maintenance & Light Repair

Second Edition

Rob Thompson

Australia • Brazil • Canada • Mexico • Singapore • United Kingdom • United States

Automotive Maintenance & Light Repair,
Second Edition
Rob Thompson

SVP, GM Skills & Global Product Management: Jonathan Lau

Product Director: Matthew Seeley

Senior Product Manager: Katie McGuire

Senior Director, Development: Marah Bellegarde

Senior Product Development Manager: Larry Main

Senior Content Developer: Meaghan Tomaso

Product Assistant: Mara Ciacelli

Vice President, Marketing Services: Jennifer Ann Baker

Associate Marketing Manager: Andrew Ouimet

Senior Content Project Manager: Kenneth McGrath

Design Director: Jack Pendleton

Text Designer: Jay Purcell

Cover Designer: Dave Gink

Cover Credit: Boophuket/Shutterstock.com

Library of Congress Control Number: 2017952177

ISBN-13: 978-1-337-56439-7

Cengage
200 Pier 4 Boulevard
Boston, MA 02210
USA

Notice to the Reader

Publisher does not warrant or guarantee any of the products described herein or perform any independent analysis in connection with any of the product information contained herein. Publisher does not assume, and expressly disclaims, any obligation to obtain and include information other than that provided to it by the manufacturer. The reader is expressly warned to consider and adopt all safety precautions that might be indicated by the activities described herein and to avoid all potential hazards. By following the instructions contained herein, the reader willingly assumes all risks in connection with such instructions. The publisher makes no representations or warranties of any kind, including but not limited to, the warranties of fitness for particular purpose or merchantability, nor are any such representations implied with respect to the material set forth herein, and the publisher takes no responsibility with respect to such material. The publisher shall not be liable for any special, consequential, or exemplary damages resulting, in whole or part, from the readers' use of, or reliance upon, this material.

Printed in the United States of America
Print Number: 04 Print Year: 2022

DEDICATION

Without the encouragement, support, and especially the patience of my wife, Vondra, this would not have been possible. Thank you.

Thanks really need to go to my students. Over my years of teaching, through all the ups and downs that are part of being an educator, I've been lucky enough to have many great students who help inspire and motivate me to always try harder and do better the next day.

I want to thank a long-time friend, colleague, and mentor, Jack Erjavec. It was Jack who many years ago put the idea of teaching in my head. Jack also introduced me to Cengage, where I have been lucky to have worked on many different projects and meet a lot of great people. Without Jack's interventions in my life, I would not be where I am today.

CONTENTS

PREFACE

Automotive Maintenance & Light Repair (AM&LR) is designed to guide and prepare students enrolled in automotive maintenance and light repair automotive programs. The textbook and accompanying workbook cover the fundamental theories, real world examples, and practical applications for each of the 2017 ASE Education Foundation Maintenance and Light Repair (MLR) tasks.

The second edition has been thoroughly revised and updated. What started out as a book going in for some maintenance and light repair work turned into a complete overhaul of many areas. This completely updated edition is designed to meet the needs of MLR programs and to prepare students for the ASE Student Certification tests.

Written by a high school automotive technology instructor, this textbook provides in-depth detail about each task, including the underlying concepts necessary to understand how and why components and systems operate. In addition, real-world examples of inspecting and servicing these components and systems are provided in both the text and the workbook.

Today's automotive students face a challenging career—the technological changes taking place with modern cars and trucks are vast. Systems unheard of only ten years ago are now common on many vehicles. The rate of adoption of new technology can only be expected to increase as consumer expectations change and technology becomes more affordable. Keeping up with technology is one of the biggest challenges technicians face today. A theme of AM&LR is helping the student become a lifelong learner; to learn how to find information and how to use the information productively. As part of this theme is the inclusion of developing the "soft" skills, such as communication, which in the modern work place are as important as technical skills.

To help prepare students for MLR tasks, the text includes a chapter on reviewing and reinforcing fundamental academic and professional skills. Chapter 4 includes a review of basic math and science skills, computer use and information about job seeking, resumes, interviewing, and work ethic. These topics are addressed because having technical skills is not always enough to be able to get and keep a job.

The layout of AM&LR is designed to logically progress from basic industry and shop operations and shop safety to automotive systems operation, service and repair. Safety is emphasized throughout the text to reinforce safe work practices addressed in Chapters 2 and 3. The sequence of chapters is from basic systems and services to complex, although it is not necessary to follow the sequence of chapters as presented.

The workbook contains a corresponding question and answer section for each textbook chapter. These sections can be used to help guide the students' reading of the textbook by requiring the answering of questions directly from the text. The workbook also contains additional activities to reinforce concepts found in the core text, as well as selected lab activities and worksheets. The lab worksheets provided are meant to reinforce important fundamental skills that each student should master.

Whether used in a high school or post-secondary training program, AM&LR is designed to guide students through the MLR tasks and onto becoming automotive professionals.

This edition is also correlated to Precision Exams' Automotive Service, Introduction exam, part of the Transportation, Distribution & Logistics Career Cluster's Facility and Mobile Equipment Maintenance pathway.

Thank you,

Rob Thompson

ACKNOWLEDGMENTS

No book project is ever the work of a single person and this is no exception. This book would not have been possible without the help and support of many others. If there are errors in content, the fault is mine and not theirs.

Vondra Hoop-Thompson

Jack Erjavec

Laurie Sandall
South-Western City Schools

Derek Fitzer
South-Western City Schools

Ron Cross
South-Western City Schools

Bill Henning
South-Western City Schools

Robert MacConnell
South-Western City Schools

Jay Dimasso
South-Western City Schools

Tim Gilles
Delmar/Cengage Learning

Beriky Ouk
VW Service Technician, Hatfield Volkswagon

Kaylee Daw
Honda Service Technician, Hugh White Honda

Madeline Ginther
Chevrolet Service Technician, Byers Chevrolet

Brenda York
Chevrolet Service Technician, Byers Chevrolet

Jordan Thompson
Service Advisor, Byers Imports

Scott Barkow
Porsche Service Technician, Byers Imports

Danny Foor
Instructor, Columbus State Community College
Jaguar Service Technician, Byers Imports

REVIEWERS

The author and publisher would also like to thank the instructors who provided invaluable feedback during the development of the project:

Tim Campbell
Wenatchee Valley Technical Skills Center
Wenatchee, WA

Dave Kapitulik
Connecticut Technical High School
Middletown, CT

Brian LaCroix
Capital Region BOCES, Career & Technical School
Albany, NY

Arminio Lopes
Greater New Bedford Regional Vocational Technical High School
New Bedford, MA

Robert Wilson
California Department of Education
Sacramento, CA

ABOUT THE AUTHOR

An experienced automotive technician and educator, Rob Thompson is the author of multiple publications on automotive technology, repair, and service. In addition to teaching high school automotive courses at South-Western Career Academy in Grove City, Ohio, he has served as an adjunct faculty member at Columbus State Community College. He is a past board member and past President of the North American Council of Automotive Teachers (NACAT).

SUPPLEMENTS

INSTRUCTOR RESOURCES

Time-saving instructor resources are available on CD or at the Instructor Companion Website found on cengagebrain.com. Either delivery option offers the following components to help minimize instructor preparation and engage students:

- PowerPoint chapter presentations with selected images that present the highlights of each chapter
- An Instructor's Guide in electronic format
- Cengage Learning Testing Powered by Cognero® delivers hundreds of test questions in a flexible, on-line system. You can choose to author, edit, and manage test bank content from multiple Cengage Learning solutions and deliver tests from your LMS, or you can simply download editable Word documents from the Instructor Resource CD or Instructor Companion Website.
- An Image Gallery includes photos and illustrations from the text.
- A ASE Education Foundation Correlation Guide

WORKBOOK

The Workbook to accompany Automotive Maintenance & Light Repair, 2e is designed to work hand-in-hand with the textbook to offer additional opportunities for review and application of the chapter material. The *Workbook* includes theory-based **Activities**, procedure-based **Lab Worksheets**, and finally, **Review Questions** to help reinforce what was learned from studying the core text.

MINDTAP FOR AUTOMOTIVE MAINTENANCE & LIGHT REPAIR

MindTap is a personalized teaching experience with relevant assignments that guide students to analyze, apply, and improve thinking, allowing you to measure skills and outcomes with ease.

- Personalized Teaching: Becomes yours with a Learning Path that is built with key student objectives. Control what students see and when they see it. Use it as-is or match to your syllabus exactly –hide, rearrange, add and create your own content.
- Guide Students: A unique learning path of relevant readings, multimedia and activities that move students up the learning taxonomy from basic knowledge and comprehension to analysis and application.
- Promote Better Outcomes: Empower instructors and motivate students with analytics and reports that provide a snapshot of class progress, time in course, engagement and completion rates.

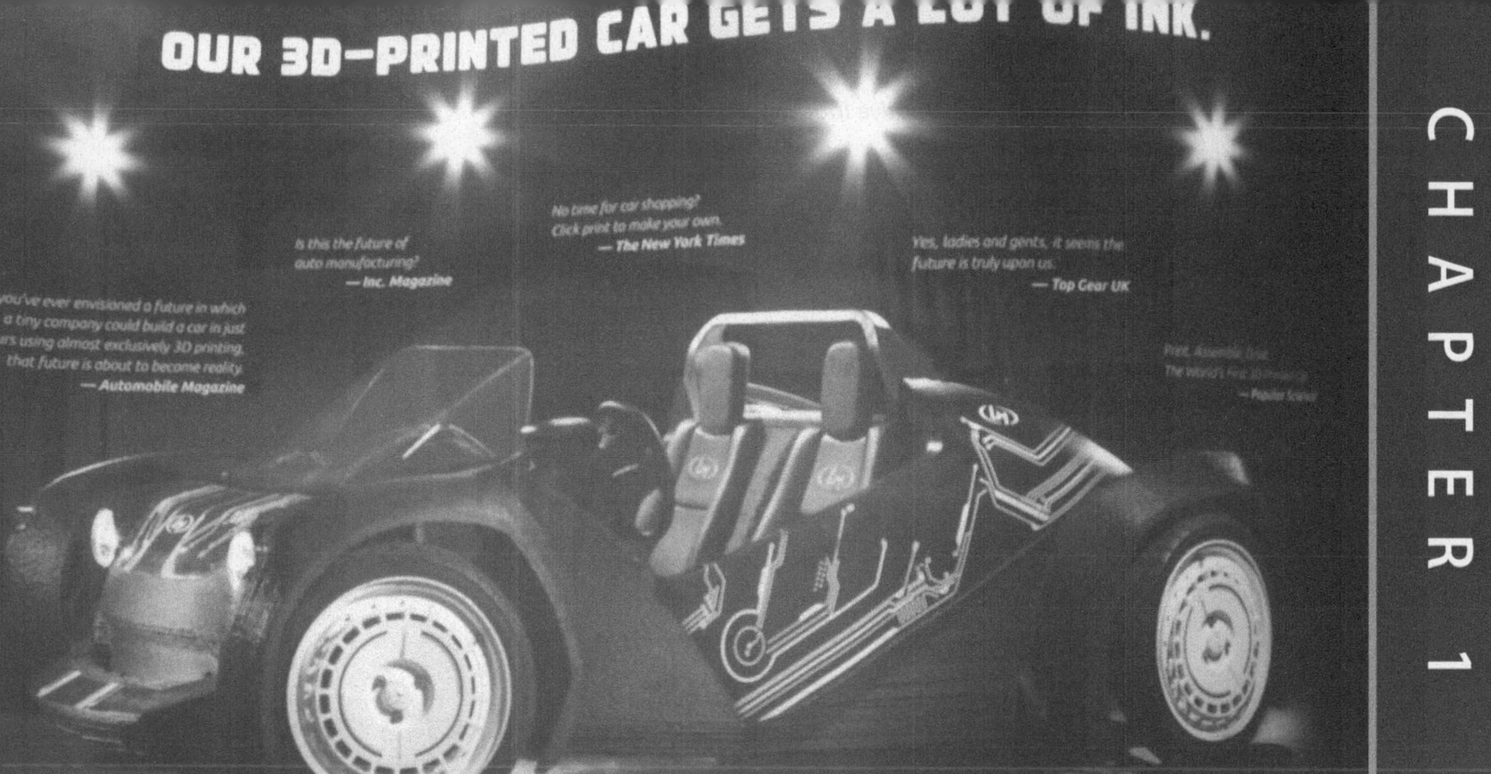

OUR 3D–PRINTED CAR GETS A LOT OF INK.
f you've ever envisioned a future in which a tiny company could build a car in just ours using almost exclusively 3D printing, that future is about to become reality.
— Automobile Magazine
Is this the future of auto manufacturing?
— Inc. Magazine
No time for car shopping? Click print to make your own.
— The New York Times
Yes, ladies and gents, it seems the future is truly upon us.
— Top Gear UK

CHAPTER 1

Introduction to the Automotive Industry

Chapter Objectives

At the conclusion of this chapter, you should be able to:

- Describe the types of jobs available in the automotive industry.
- Explain training and education options for technicians.
- Explain the areas of ASE certification.
- Describe the reasons for the changes in automotive design and construction.

KEY TERMS

ASE Education Foundation
collision technician
entry-level technician
lifelong learning
line technician
National Institute for Automotive Service Excellence (ASE)
parts technician
service advisor

The history of the modern automobile, a vehicle using a combustion engine to propel itself, can be traced back to the late 1800s. The first self-propelled vehicles were hand built in very limited quantities by pioneers such as Gottlieb Daimler, Wilhelm Maybach, and Karl Benz. An example of what is considered the first automobile is shown in **Figure 1-1**.

In the early days of motorized transportation, gasoline was not the only fuel source. Even in the late 1800s and early 1900s, many models of electric vehicles existed. Other vehicles had steam engines, or used types of alcohol or kerosene as their energy source. **Figure 1-2** shows a map of electric vehicle charging stations in New York City from 1923.

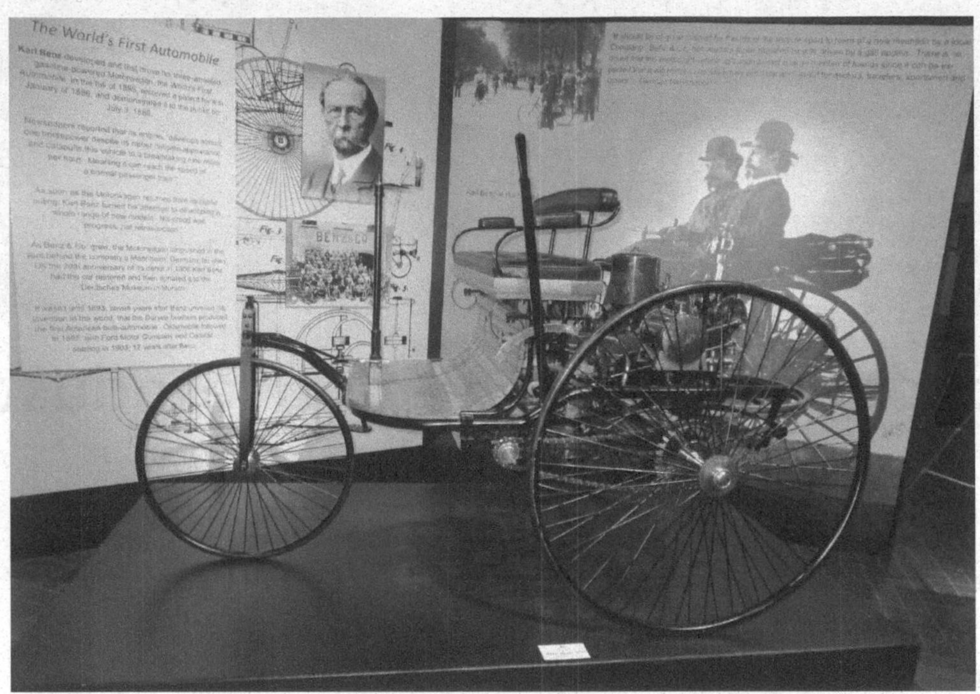

FIGURE 1-1 The earliest automobiles, such as this, were motorized horse buggies.

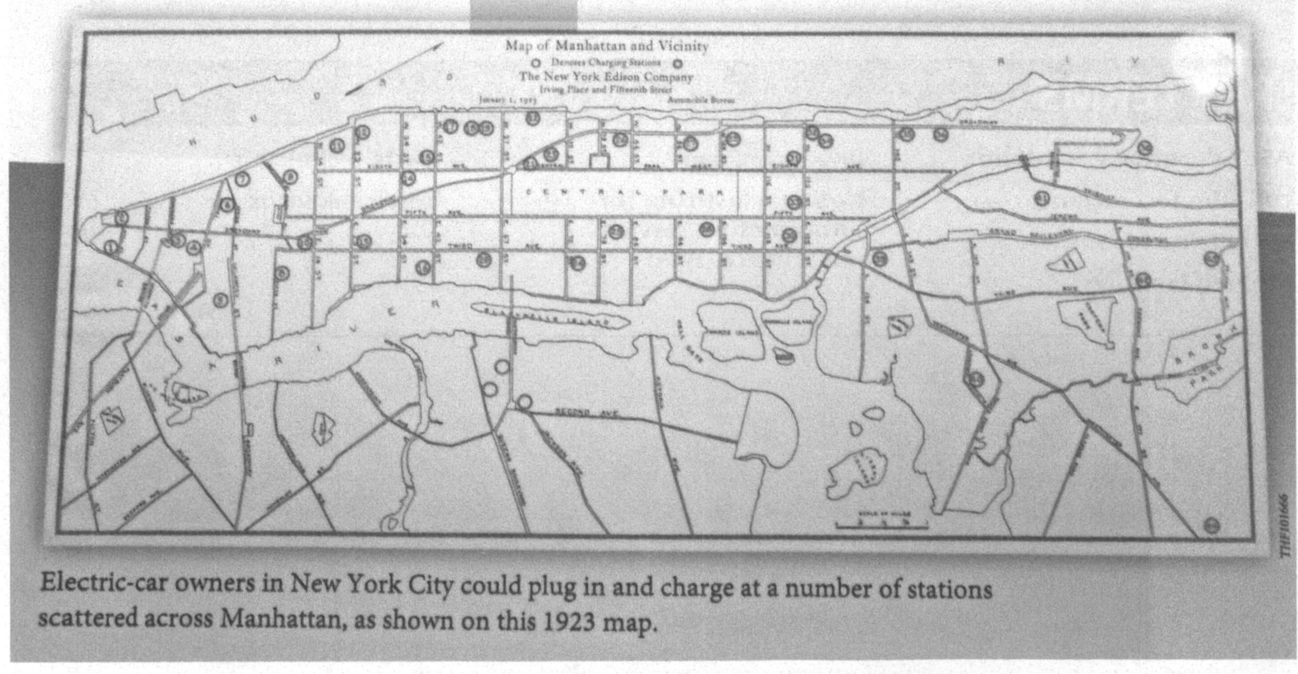

Map of Manhattan and Vicinity
The New York Edison Company
Electric-car owners in New York City could plug in and charge at a number of stations scattered across Manhattan, as shown on this 1923 map.

FIGURE 1-2 Electric cars were very popular and had many charging stations in the early 1900s.

In the early 1900s, Ransom E. Olds began mass production of the Oldsmobile. This process was greatly improved upon by Henry Ford in the 1910s with the Model T. Ford's use of standardized parts and the assembly line brought the cost of manufacturing the Model T down enough that it became affordable to many more Americans.

Decreased manufacturing costs created fierce competition among brands and helped to push design advancements. However, despite the advances in service life, safety, and efficiency, cars and trucks remained largely unchanged for decades. Not until electronics began to be integrated into new car technology in the 1970s did major changes come to the automotive industry.

The Automotive Industry

The automotive industry is part of the domestic (and international) transportation industry. In the United States, the number of jobs associated with automobiles is large; however, these jobs are just part of a bigger picture encompassing all transportation-related jobs. Approximately 1 in 7, or about 14%, of all jobs in the United States are transportation related. This includes indirect jobs that support the cars and light trucks on the country's roads, heavy-duty trucks and equipment, aviation, shipping, and off-road and recreational vehicles. All these industries have changed due to the growth or loss of markets; changes in technology; and changes in laws, regulations, and the economy. Regardless of the path you choose, if you remain in a transportation-related career, you will need to be able to adapt and grow as things change around you.

CHANGE IN THE AUTO INDUSTRY

There are many reasons for the changes and advancements made over the last 100 years. Improvements in manufacturing, materials, and electronics have played significant roles in the industry's evolution, and how these improvements came into being deserves some attention.

■ ***Emissions and the Environment.*** At the end of World War II, the American economy, booming due to the needs of war production, needed to change to consumer production. Factories that had been producing tanks, airplanes, and war supplies shifted to producing household goods and automobiles. General Motors, Ford, and Chrysler restarted new car production in 1946. Returning veterans and their families needed housing and transportation as suburban development began. As the number of vehicles on the road increased each year, noticeable changes occurred in the air around certain parts of the country.

As more vehicles were sold and more road miles traveled, more pollution was released into the atmosphere. In parts of California, the combination of pollution and weather patterns created a thick, heavy haze over cities, called *smog*, a combination of smoke and fog (**Figure 1-3**). The California government knew that the automobile was contributing to the pollution and began to take steps to decrease the amount of pollution produced by cars and trucks. The very first emission control

FIGURE 1-3 Automotive exhaust contributes to air pollution. Emission control has been a major contributor to automotive design since the 1970s.

device, the positive crankcase ventilation (PCV) valve, was introduced in 1957. Since then, passage of the Clean Air Act, Clean Water Act, and many more emission control laws have forced vehicle manufacturers to meet increasingly strict exhaust emission standards.

■ ***The Economy and Fuel Prices.*** Before the energy crisis in 1973, American cars and trucks were, in general, large, heavy, powerful, and not fuel efficient. The energy crisis of the 1970s caused a shift in consumer attitudes toward the cars the domestic auto makers produced. In 1975, Congress passed the corporate average fuel economy (CAFE) standards. These standards require auto makers to reach increasingly higher fuel economy ratings across all their vehicles sold in the United States. Currently, the National Highway Traffic Safety Administration (NHTSA) projects fleet average fuel economy will be between 40.3 and 41 mpg by 2021 (**Figure 1-4**). Changes in engines, vehicle construction, and other areas will be necessary as future vehicles will be required to achieve better fuel economy.

■ ***Market Share.*** Imported cars had a small percentage of the total automotive market share before the mid-1970s. When oil and gas prices rose, many car buyers started to look at the small, fuel-efficient models offered by Honda, Toyota, Datsun (later Nissan), VW, and others. In 1970, Americans bought about 313,000 Japanese-manufactured vehicles and approximately 750,000 vehicles from Germany. In comparison, sales by General Motors, Ford, and Chrysler exceeded 7.1 million vehicles—more than 85% of the market. In 2015, sales of imported cars in the United States accounted for about 55% of total sales. A chart of U.S. auto sales is shown in **Figure 1-5**.

One factor in market share was product quality. Even though domestic manufacturers began making smaller and more efficient vehicles, they were still being surpassed in quality by the imports. The cars imported from Japan and Germany were more efficient, were well built, and often lasted longer than domestically built vehicles. Consumers responded to this by buying more and more imported cars and trucks. While General Motors, Ford, and Chrysler made improvements, the availability of so many other makes and models reduced their market share steadily over the years.

■ ***The Electronic Revolution.*** As production of smaller and less expensive electronic components and magnets increased, more accessories, such as power windows, could be supplied at lower cost. Features such as power door locks, rear window defoggers, and air conditioning, once expensive options, are standard equipment on today's vehicles.

Most modern options such as navigation systems, Bluetooth phone integration, adaptive cruise control, occupant safety systems, antilock brakes, and vehicle stability control would not exist without low-cost electronics. In addition, electrical and electronic components are replacing items that have traditionally been either mechanical or hydraulic (powered by pressurized fluid). For example, some vehicles use electrically controlled torque-vectoring systems, designed to increase handling and performance by controlling the power to each drive wheel (**Figure 1-6**). Electric power steering is common on many cars, and Nissan has a steer-by-wire system available on the Infiniti Q50. Using electronic components gives design engineers more flexibility and often reduces manufacturing, maintenance, and repair costs.

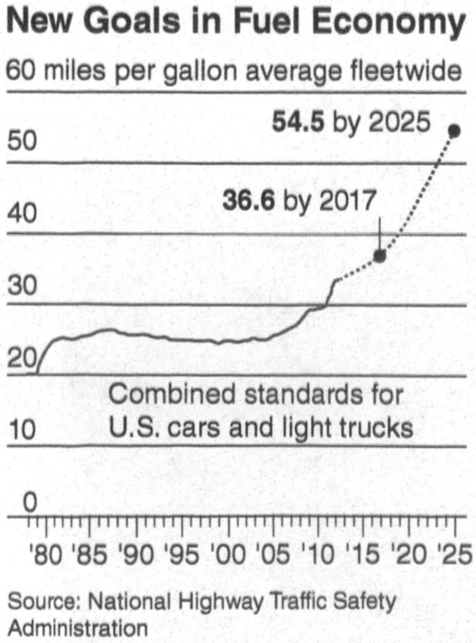

FIGURE 1-4 Reducing emissions and increasing fuel economy have been major factors in vehicle design.

United States Automotive Market Share 2015

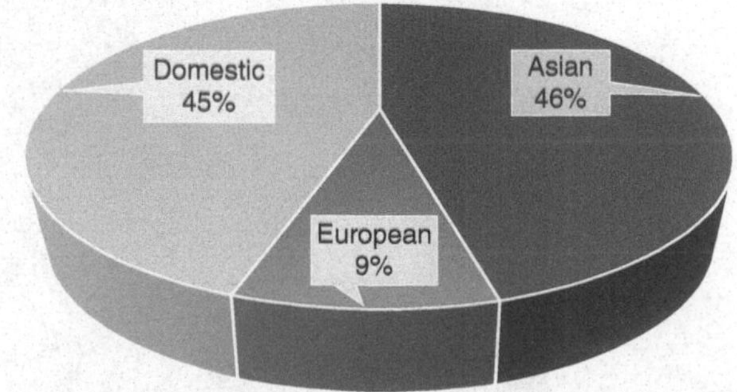

FIGURE 1-5 The share of new car sales among the U.S. production and foreign manufacturers.

FIGURE 1-6 Changes in technology continue to drive automotive design. Systems that were once only found on very high-end cars, such as ABS, are now standard on all new cars sold in the United States.

■ ***Safety.*** Founded in 1969, the Insurance Institute for Highway Safety (IIHS) has worked to improve vehicle safety by concentrating on three areas involved in motor vehicle crashes: human, vehicle, and environment factors. Prior to the work of the IIHS, there was little safety regulation regarding the design and construction of passenger vehicles. Seat belts did not become a standard feature until 1968, due in part to the way in which vehicle manufacturers evaluated vehicle safety and conducted tests. In fact, the federal government did not incorporate vehicle crash testing until the NHTSA began testing in 1979.

Since then, the United States government, the IIHS, and the auto makers themselves have conducted extensive testing to improve vehicle crashworthiness. This has led to improved vehicle designs that not only reduce injuries but also help prevent accidents from occurring. Because of the work of organizations such as the IIHS and NHTSA, vehicles now are designed from the start with safety as a primary concern. This has led to the standardization of many safety systems, such as:

- Antilock brake systems (ABS) and stability control
- Supplemental restraint systems
- Tire pressure monitoring systems
- Rear-view backup cameras

Also, due to crash test studies, vehicles are built with features—impact zones or crumple zones, collapsible steering columns, door reinforcements, dash bracing, and so on—that allow the vehicle to absorb and deflect impact forces away from the occupants and to reduce injuries (**Figure 1-7**). These factors have made modern cars and trucks much safer and have helped reduce traffic accident deaths over the years.

■ ***Passenger Comfort and Expectations.*** As the consumer electronics industry introduces new and more exciting products, customers demand more of these types of devices and conveniences for their vehicles. GPS-linked navigation, adaptive lighting and cruise control, wireless media integration, voice recognition, and intelligent computer systems are all standard or available on many of today's vehicles. Most new cars and trucks are equipped with display screens that are used for the driver and passengers to interface with the various onboard systems (**Figure 1-8**).

As these features become less expensive to integrate, they eventually trickle down to even the lowest-cost

FIGURE 1-7 Safety is another key aspect of the evolution of the modern automobile. As vehicles become smaller and lighter, manufacturers must develop new technologies to keep the passengers and pedestrians safe.

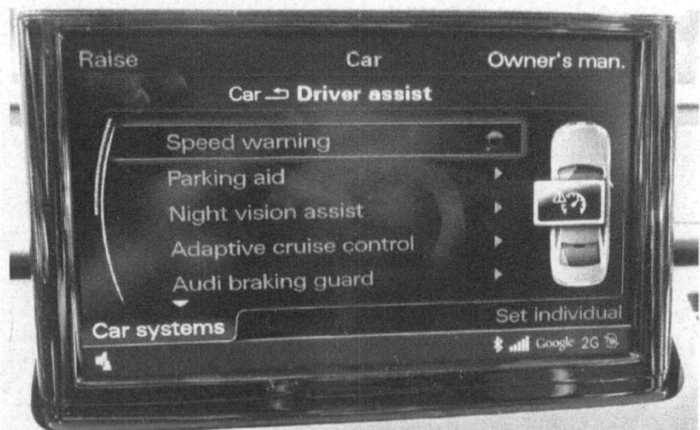

FIGURE 1-8 Customers expect voice-activated systems such as navigation, phone, entertainment, and customizable vehicle settings.

vehicles as options or standard equipment. However, as vehicles become more complex, the necessity of qualified technicians to service and repair these systems when they fail increases.

CAREERS

Without the transportation industry, the U.S. and world economies would stop. There are more than 11 million jobs in the United States directly related to transportation, including sales, marketing, engineering, and production. The career choices available are staggering. Beyond the servicing and repairing of automobiles, there is collision repair; diesel, heavy-duty, and agriculture equipment repair fields; small engine; marine and air transportation; and countless other jobs related to cars and trucks.

■ ***Auto Technicians.*** According to the Bureau of Labor Statistics, there are currently more than 630,000 automotive technicians in the United States servicing more than 241 million cars, SUVs, and light trucks. This does not include collision, heavy-duty, off-road, or power sports technicians. Most automotive technicians begin their careers in an entry-level position, then progress to jobs with more responsibility and higher pay.

An **entry-level technician** is expected to be able to perform basic inspections and maintenance services. Most will need to have a basic tool set consisting of socket sets, wrenches, pliers, screwdrivers, and an assortment of other basic tools (**Figure 1-9**). As a beginning technician, you should have at least a basic understanding of the various systems found on modern vehicles and the ability to make simple repairs. For example, it is expected that an entry-level technician should be able to:

- Perform an engine oil and filter change and reset the maintenance reminder system.
- Service tires and tire pressure monitoring systems.
- Inspect the brake system and perform basic brake repairs.
- Replace batteries and either maintain or restore memory functions.
- Replace transmission and differential fluids.
- Service the cooling system.
- Replace various lights, wipers, and other maintenance items.
- Perform some computer programming functions.

FIGURE 1-9 An example of a starter tool kit, suitable for an entry-level technician.

As important as these skills and equally important are the abilities to locate and interpret technical information, to work well with colleagues and customers, and to have a good work ethic. As cars and trucks become more complex, the ability to locate and correctly interpret technical information is of significant importance. Very little can be done to modern vehicles that do not require looking up information on a computer or connecting to the vehicle's onboard network with a scan tool.

Interpersonal skills for working in teams are important for the overall operation of the business. Customer service industries also require the ability to communicate well, to present a professional image, and to relate well to others. These qualities, plus having initiative, good attendance, and a positive attitude, are necessary to be successful in the modern workplace.

After an entry-level technician acquires additional skills and experience, he or she may become a line technician. A **line technician** is one who is certified and has experience with most of the systems on the vehicle. As a line technician, you will be expected to perform increasingly more difficult repairs quickly and profitably, and you may even assist in training new employees.

Many shops designate their technicians by categories, such as A, B, and C technicians. An A technician, also called a *lead technician*, has the most experience and certifications. He or she can perform repairs on all the vehicle systems and can generate a lot of income. A C-level technician is often young and has the least experience in the shop. He or she may have a couple of certifications and only a couple of years of experience. C-level technicians often work on specific areas such as brakes or suspension systems. B-level technicians generally fall in between C- and A-level technicians. An example of technician skill-level work is shown in **Figure 1-10**.

OTHER TYPES OF TECHNICIANS

As stated previously, there are many more career opportunities in the transportation industry than being an

	Skill Level	Warranty Time	Standard
(1) HUB & BEARING, R&R			
All Models (2WD)			
One Side	B	(0.8)	1.0
Both Sides	B	(1.5)	1.9
Parts			
		Mfg. Part No.	**Price (MSRP)***
HUB & BEARING			
All Models			
1/2 Ton			
w/Crew Cab		15946732	$509.50
w/o Crew Cab		15233111	$453.46

FIGURE 1-10 Skill levels shown in a parts and time guide. A shop may have one A-level or master technician, several B-level technicians, and many C-level technicians.

FIGURE 1-11 Engine machinists specialize in engine rebuilding.

FIGURE 1-12 Collision technicians repair body damage, straighten the frame, and restore the paint and finish.

automotive technician. Many people who love cars and trucks specialize in one of the many other fields related to the auto industry.

■ ***Engine Machinist.*** Automotive machinists are generally employed in specialty shops, called *machine shops*. Machinists are those men and women who repair and rebuild engines, service cylinder heads and blocks, and, in some cases, build high-performance racing engines. They perform work such as reboring cylinder blocks, replacing cylinder liners or sleeves, machining the block deck and cylinder head surfaces, fitting pistons and connecting rods, and performing various types of crack repairs (**Figure 1-11**).

■ ***Collision Technician.*** A **collision technician** repairs a vehicle after it has been involved in a collision or has suffered some type of body or structural damage. Depending on the type of repair facility, one technician may perform nonstructural body repairs while another technician is responsible for painting and refinishing (**Figure 1-12**). In some shops, one technician may perform the entire repair and paint the vehicle also. In the *aftermarket*, many collision technicians specialize in painting and refinishing or custom painting. Custom painting has seen an increase in demand as more people want to customize their cars, trucks, motorcycles, boats, and other vehicles.

Word Wall

Aftermarket—Aftermarket means the parts and service suppliers to the automotive industry not supplied by the vehicle manufacturers and their dealers.

■ ***Parts Technician.*** A **parts technician** may work for a dealer, an independent store, or a chain store. A parts person is knowledgeable about the parts industry and works with the public, helping to find the right parts for the customer (**Figure 1-13**). Some parts persons work only with commercial accounts, such as local auto and truck repair facilities.

Being a parts person requires good people skills and some basic automotive knowledge because many parts stores provide free services such as wiper blade and battery installation.

■ ***Service Advisor.*** An automobile service advisor specializes in communications between the customer and others in the shop (**Figure 1-14**). In addition to communication skills, a service advisor is knowledgeable about vehicle systems, has sales skills, and often can perform some basic shop operations.

The **service advisor** typically greets the customer upon entering the service department, completes the service order, then routes the service order to the appropriate technician. Once the technician has completed the diagnosis and estimate, the service advisor then contacts the customer to discuss the needed repairs and services and their costs. It is important for a service advisor to have a good understanding of the various automotive systems so that he or she can accurately communicate with the technicians and customers.

FIGURE 1-13 Parts technicians specialize in parts distribution and working with customers.

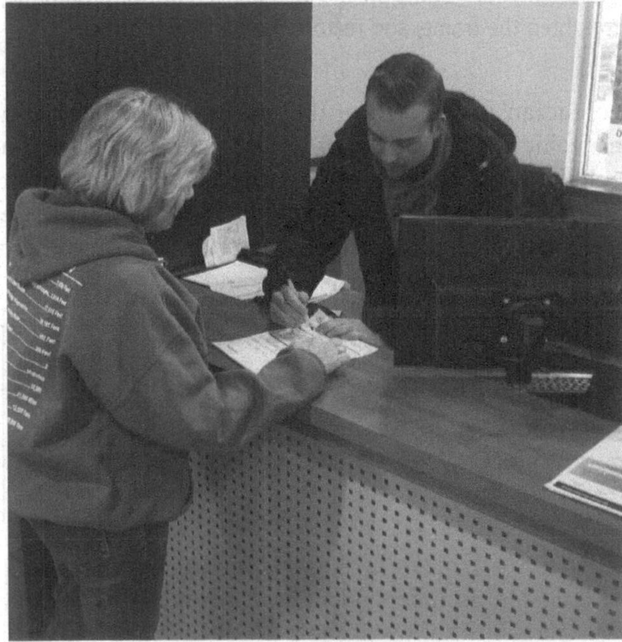

FIGURE 1-14 Service advisors work with customers when dropping the vehicle off, contacting them with the estimate, explaining what repairs and services are required, and handle the bill when the car is picked up.

■ ***Sales.*** Salespeople are often the most visible people in the automotive industry. A salesperson provides the expertise about the various makes and models of vehicles that helps the customer make a decision when buying a vehicle. Salespeople often work with the service department to ensure that new vehicles are properly prepped and ready for delivery.

■ ***Training and Education.*** Some technicians become trainers and educators. Vehicle manufacturers, equipment suppliers, tool companies, and even parts stores employ technical trainers to provide training on tools and repair procedures.

Some technicians become high school and college instructors. To teach, an instructor needs to be an experienced technician. In addition, he or she should have excellent communication and people skills and the desire to help others succeed in the automotive industry.

■ ***Diesel, Heavy-Duty, and Agriculture Technicians.*** One of the greatest needs of the transportation industry is for qualified diesel technicians. Every heavy-duty truck, piece of construction equipment, and large ship have a diesel engine (**Figure 1-15**). Even though the diesel industry was slower to adopt

FIGURE 1-15 Heavy-duty equipment technicians keep trucks and other equipment in operation.

the use of electronics for vehicle management than the auto industry, modern diesels often have as much or more electronic monitoring and controlling of engine operation than a gasoline-powered engine. This is due in part to the cost of a heavy-duty diesel engine and in part to increasingly strict diesel emission standards. More than 180,000 jobs existed in 2015 in the diesel, heavy-duty, and agriculture equipment repair fields.

■ ***Motorcycle and Other Outdoor Power Equipment.*** There are more motorcycles, ATVs, snowmobiles, and personal watercraft in use today than ever before, with the numbers expected to increase each year.

In 2015, there were more than 65,000 motorcycle, marine, and small engine technicians in the United States. With the expected increases in ownership of these types of vehicles, there is continued growth in the need for service technicians. Just as cars and trucks have become more complex in the last couple of decades, motorcycles and off-road machines also are becoming more complex (**Figure 1-16**). Many manufacturers now use electronic spark control and fuel injection systems.

Snowmobiles, ATVs, and lawn maintenance equipment may also be serviced by a motorcycle technician during the off-season.

■ ***Aircraft Technicians.*** There were about 124,000 aircraft and avionics technicians in 2015. Most of those jobs are located around major airports in large cities, though employment can be found at smaller commercial and private airports as well. Service and repair of general aviation aircraft, such as small propeller-driven planes, is very similar to automotive repair. Many of the same skills used in fixing cars and trucks are used for repairing aircraft. Propeller-driven aircraft use small gasoline engines, hydraulic brakes, and an electrical system for lighting and instrumentation.

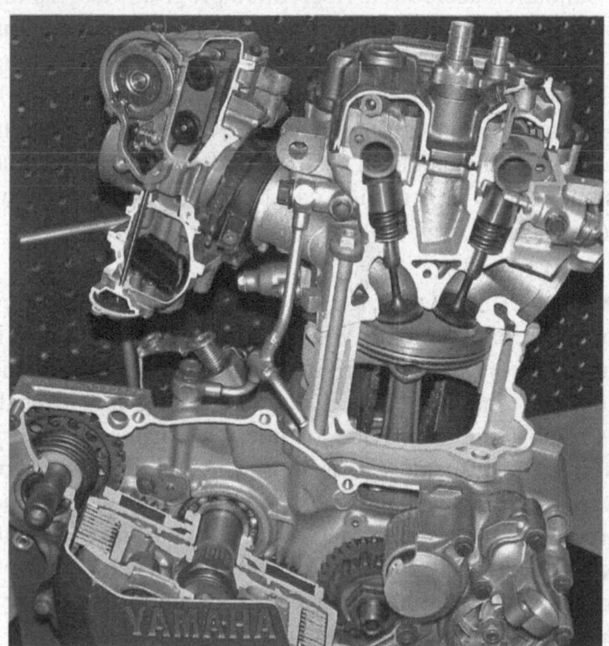

FIGURE 1-16 Many aspects of automotive repair are similar to working on motorcycles, ATVs, boats, and other forms of transportation.

EDUCATION AND TRAINING

The most important consideration for future auto technicians is education. Before 1975, the only electronic components likely to be found on a vehicle were in the radio. In 1975, the need to increase fuel economy and reduce exhaust emissions saw the implementation of electronics in automobiles and light trucks. Small, mysterious boxes began to replace mechanical ignition parts that had been in use for decades. What had been for years a repair-based industry started to move to a diagnose-and-replace industry. These changes required mechanics to learn new skills and adapt to the increasing amount of electronics and changing technology. Today's vehicles have dozens of electronic modules monitoring and controlling nearly every aspect of the automobile. Technology use will continue to increase as consumers expect more of their vehicles.

■ ***Secondary Schools.*** The best preparation for a future technician is participation in a formal training program. Many high schools and career centers throughout the United States provide training in automotive repair, collision repair, diesel repair, and even aviation repair. These programs offer training and experience in the automotive repair industry and are often linked with local dealerships and community colleges. Some schools participate in the Automotive Youth Education System (AYES), a partnership between several vehicle manufacturers and secondary schools, that provides work experience in addition to automotive training.

High school programs take many different forms, though one- and two-year courses are common. In this type of program, students can learn about the basic aspects of auto technology, collision repair, or diesel technology. Due to time constraints, it is difficult for high school programs to prepare graduates for more than entry-level positions. For those students who plan on attending post-secondary education, a high school program often can provide the student with advanced placement credit with cooperating colleges and universities.

■ ***Post-Secondary Schools.*** Automotive programs at community colleges, technical schools, and universities are another common way to obtain training as a technician (**Figure 1-17**). These programs are typically two-year associate degree programs, but some four-year degree programs exist. Many community college programs work cooperatively with a manufacturer, providing a training program that is specific to that manufacturer's vehicles. Programs such as the Ford ASSET, General Motors ASEP, BMW STEP, and Toyota T-TEN combine a two-year associate degree in automotive technology with working in a dealership as a technician (**Figure 1-18**).

■ ***Technical Schools.*** There are many different post-secondary technical schools in the United States. Schools such as Universal Technical Institute, Lincoln Technical Institute, and WyoTech, specialize in training auto technicians, collision technicians, and heavy-duty diesel technicians. These schools differ from community colleges, in that some are privately owned and operated, meaning that they do not receive any funding from taxes. In many cases, these schools offer both degree and diploma programs. A diploma program provides training in the technical skills but does not include any academic courses, such as communications, math, science, or the humanities. While this may sound ideal to a high school student, a diploma may not provide the same opportunities in the future that an associate degree can offer. For example, if a technician wants to someday move up into management, a degree may be needed.

FIGURE 1-17 An automotive lab at a community college. Community colleges and private schools offer automotive degree programs.

FIGURE 1-18 This school is a GM ASEP training school where students become factory-trained GM technicians.

Technical schools often have shorter program completion times than a community college because students spend between four and eight hours a day, five days a week, in their technical content courses. This means that a two-year program can be condensed into a 9- to 12-month course. It is important to note that with some private schools, credit earned may not transfer to an accredited college or university. Be sure to check on accreditation and transfer credit before signing on to attend a technical school.

■ ***ASE Education Foundation Program Accreditation.*** While each program is different, most auto tech programs share some similarities. Many high school programs and nearly all post-secondary schools are evaluated and accredited by the **ASE Education Foundation**. Each program has to provide documentation and pass an onsite evaluation to become ASE Education Foundation accredited. Accredited programs display the sign showing that the program has achieved accreditation (**Figure 1-19**). This ensures that the school is teaching the competencies and standards prescribed by ASE Education Foundation. Because of this standardization, all core skills taught in each accredited program are the same. More information can be found at http://www.aseeducation.org/.

■ ***Lifelong Learning.*** Regardless of what area of the repair industry you decide to pursue, you will need to continue to learn and acquire new skills as the industry changes. This is called **lifelong learning**. It is impossible to provide a student all of the necessary skills and information he or she needs during a two-year program, even at the college level. Today's cars and trucks are too complex and changes occur too rapidly to know everything there is to know. Because of this, you will need to continue your education and training even after you complete an associate's degree.

If working for a new car dealership, you will be expected to complete new product training, take web-based coursework, and attend training classes so that you are familiar with the changes for each model year. An example of a manufacturer's training program in shown in **Figure 1-20**. To be able to perform warranty repairs, technicians must first complete required training so that both the technician and the dealership are reimbursed for the repairs.

FIGURE 1-19 This sign shows that a school has met the ASE Education Foundation requirements for an automotive training program.

If you plan on working in the aftermarket or even owning your own shop, you will need to continue your training to remain current with technology. This may require taking night classes or other types of specialized training provided by a parts company such as NAPA or CarQuest or from a tool company such as Snap-On or Mac Tools.

Regardless of where you work, you will need to accept that change is the only constant in the automotive industry and that if you do not keep up, you will quickly get left behind.

■ ***Professional Organizations.*** Several national organizations exist for automotive technicians, including the Automotive Service Association (ASAshop.org), Society of Automotive Engineers (SAE.org), and the International Automotive Technicians Network (iATN.net).

ASA was founded more than 50 years ago and is dedicated to and governed by independent automotive service and repair professionals. ASA sponsors training events and provides members with information regarding legislation and news from the industry.

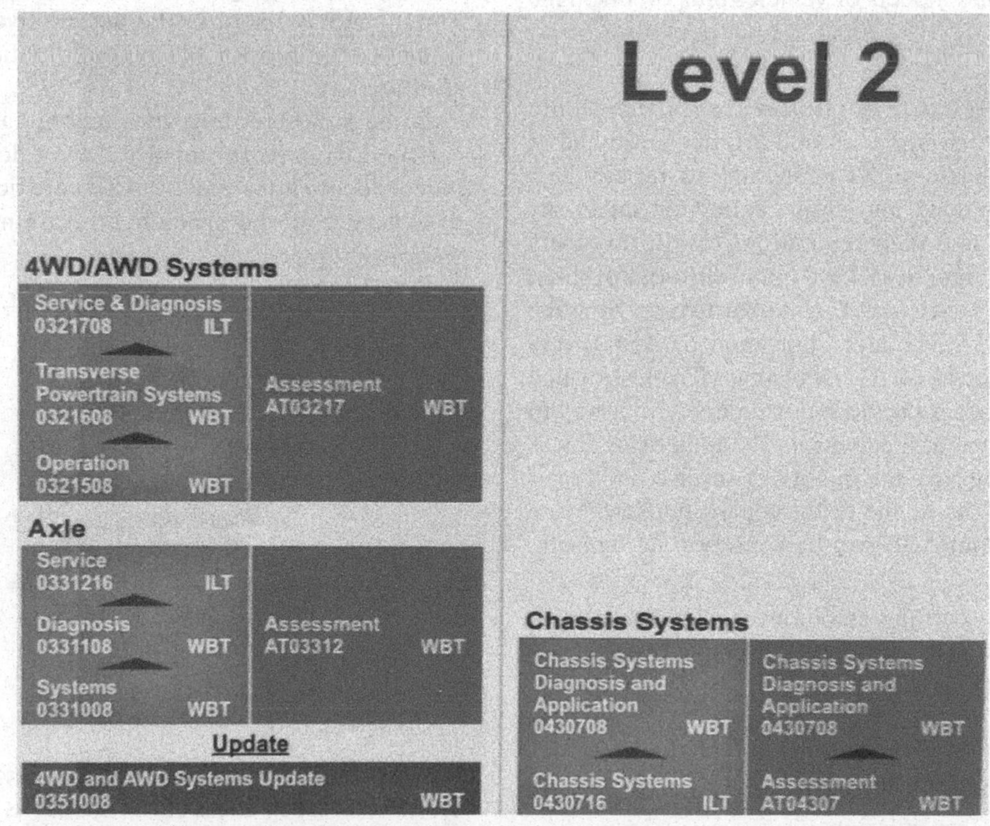

FIGURE 1-20 Every vehicle manufacturer has a training system in place for its technicians. ILT means instructor led training and WBT courses are web based training.

SAE is comprised of engineers, executives, educators, and others who share information and exchange ideas for advancing the engineering of all types of self-propelled vehicles. SAE develops standards for the automotive industry, such as motor oil viscosity ratings, and onboard data communication protocols.

The iATN is the largest online network of automotive repair technicians in the world. iATN is a community where technicians, automotive trainers and educators, shop owners, and others can discuss ideas, present problems for discussion, and post requests for help with a challenging problem.

■ ***Publications.*** Even though the Internet has become the dominant source of information for many, there still exist many excellent print publications for the professional technician. Among these are *Motor Magazine*, *Motor Age*, and the publications by Babcox such as *Brake & Front End*, *Import Car*, *Underhood Service*, and *Tomorrow's Technician*. In addition, these magazines offer online access to current and back issues.

Regardless of how you keep up with current trends in the industry, Internet or print, what is important is that you keep current. Publications like these and others provide information about what is happening with current makes and models of vehicles plus events in the auto industry.

■ ***ASE Certification.*** Before 1972, the public had no way to determine if an auto mechanic had at minimum the basic skills necessary to repair their vehicles. Mechanics and shops relied on customer word-of-mouth and their reputations to inform others about their abilities. In 1972, a nonprofit organization, the **National Institute for Automotive Service Excellence** (NIASE, later shortened to **ASE**), was formed and began to certify mechanics. The testing and certification process provides customers with a way to recognize competent mechanics. The goal of ASE was and still is to improve the quality of service and repair and to improve the image of the repair industry. More information about ASE can be found on its website, www.ase.com.

ASE currently certifies automotive technicians in nine areas and offers advanced certifications for engine performance specialists.

A1 Engine Repair

A2 Automatic Transmissions

A3 Manual Transmissions

A4 Suspension & Steering

A5 Brakes

A6 Electrical/Electronics

A7 Heating and Air Conditioning

A8 Engine Performance

A9 Light Duty Diesel

L1 Advanced Engine Performance Specialist

Once a technician has passed all eight of the automotive tests, A1 through A8, he or she is considered an ASE Master Automobile Technician. Of the nearly 300,000 certified automotive technicians, approximately one-third are master technicians. Once master certified, technicians earn the gold ASE Certified patch (**Figure 1-21**).

The ASE Advanced Engine Performance Specialist (L1) Test contains questions that test technicians' knowledge in the diagnosis and repair of computer-controlled engine systems. In addition, the test measures technicians' abilities in diagnosing the cause of emission failures. This is especially important in states where basic inspection/maintenance (I/M), enhanced I/M (ASM or IM240), and OBD testing programs are in effect.

The A9 Light Vehicle Diesel Engines Test was added in 2009 in response to the expected increase in the use of diesel engines in cars and light-duty trucks. Though A9 is listed in the automotive test section, it is not required for Master Automotive Technician certification.

Some states require auto technicians to be ASE certified to perform repairs. Many dealerships and independent shops require ASE certification in order to advance, attend specialized training, or perform warranty repairs.

FIGURE 1-21 A master technician has passed all eight areas of ASE certification and is often regarded as an A-level technician. ASE certification is a nationally recognized method of proving your skills.

In 2013, ASE began offering student certification tests. These tests do not require any job experience and are valid for two years. You need to pass one or more of the eight automobile tests to receive your certificate. Passing these tests and being able to present the certificate(s) to a potential employer will show you have taken the first steps to becoming a professional technician. In addition, having these certifications means you are more valuable as an employee because you already have the knowledge and skills necessary to pass the ASE Student Certification Exam. More information can be found at www.asestudentcertification.com.

■ ***Other ASE Certifications.*** In addition to those listed above, ASE also has certification tests for the following:

- Maintenance and Light Repair (G1)
- Compressed Natural Gas Vehicle Technician (F1)
- Exhaust Systems (X1)
- Collision Repair and Refinish Technicians (B2–B6)
- Medium/Heavy-Duty Truck Technicians (T1–T8 and L2)
- Transit Bus Technicians (B1–B8)
- School Bus Technicians (S1–S7)
- Automotive Service Consultant (C1)
- Parts Specialist (P1, P2, and P4)

■ ***Manufacturer Certifications.*** In addition to ASE certifications, most vehicle manufacturers have their own systems of training and certifying their technicians. These systems generally include self-paced courses via the Internet or product training manuals and a series of classes with hands-on training (**Figure 1-22**). Once a technician successfully completes a training series, he or she becomes certified in that area. This allows the technician to then perform warranty repairs in that area. As the technician acquires more training and certifications, he or she often progresses further up the pay scale and can earn more income by being able to perform more types of repairs.

FIGURE 1-22 An example of a portion of a manufacturer's training program certifications.

SUMMARY

The automobile has evolved from a custom-built, high-priced oddity into an integral component of modern society.

Today's cars and trucks are equipped with features and designs only dreamed of a few decades ago.

The modern automobile has created a need for highly qualified, critical thinking technicians to diagnose and repair complex systems on today's cars and trucks.

Regardless of career path, nearly countless opportunities are available for someone wanting to pursue a career in the transportation repair industry.

Approximately one in seven U.S. jobs are transportation related, creating a diverse and plentiful job market for those who wish to enter.

Both ASE and manufacturer certifications are used to show the skills acquired by technicians.

REVIEW QUESTIONS

1. A person who greets the customer, writes the repair order, and communicates with the technician and the customer is called a ________________ ________________.
2. Most technicians start as an ________________ ________________ technician before acquiring the skills necessary to become a line technician.
3. A ________________ is a technician who specializes in repairing and rebuilding engines.
4. For a technician to achieve ASE ________________ Automotive Technician certification, he or she must pass eight tests.
5. The acceptance of ________________ ________________ means that as a technician, you will need to continue your education and training over the years to stay current with changes in technology.
6. Which of the following would not be considered a responsibility of an entry-level technician?
   a. Inspect and maintain fluids.
   b. Check brake pad wear.
   c. Perform a tire rotation.
   d. Replace a transmission.
7. Which of the following statements about ASE certification is correct?
   a. You must attend an ASE Education Foundation-accredited school to become ASE certified.
   b. ASE certifies automotive, heavy-duty truck, and aviation technicians.
   c. A technician must pass eight automobile tests to achieve Master Technician certification.
   d. ASE certification is the same as being certified by the vehicle manufacturer.
8. *Technician A* says that fuel economy and emission requirements have led to the extensive use of electronics in modern vehicles. *Technician B* says that changes in safety requirements have caused the vehicle manufacturers to rely on electronics and computerization. Who is correct?
   a. Technician A
   b. Technician B
   c. Both A and B
   d. Neither A nor B
9. Which of the following automotive systems relies on electronic sensing and control?
   a. Restraint system
   b. Engine and transmission control
   c. Brake system
   d. All of the above
10. Which of the following are transportation jobs related to the auto industry?
   a. Aircraft technician
   b. Paint and refinish specialist
   c. Heavy-duty truck technician
   d. All of the above

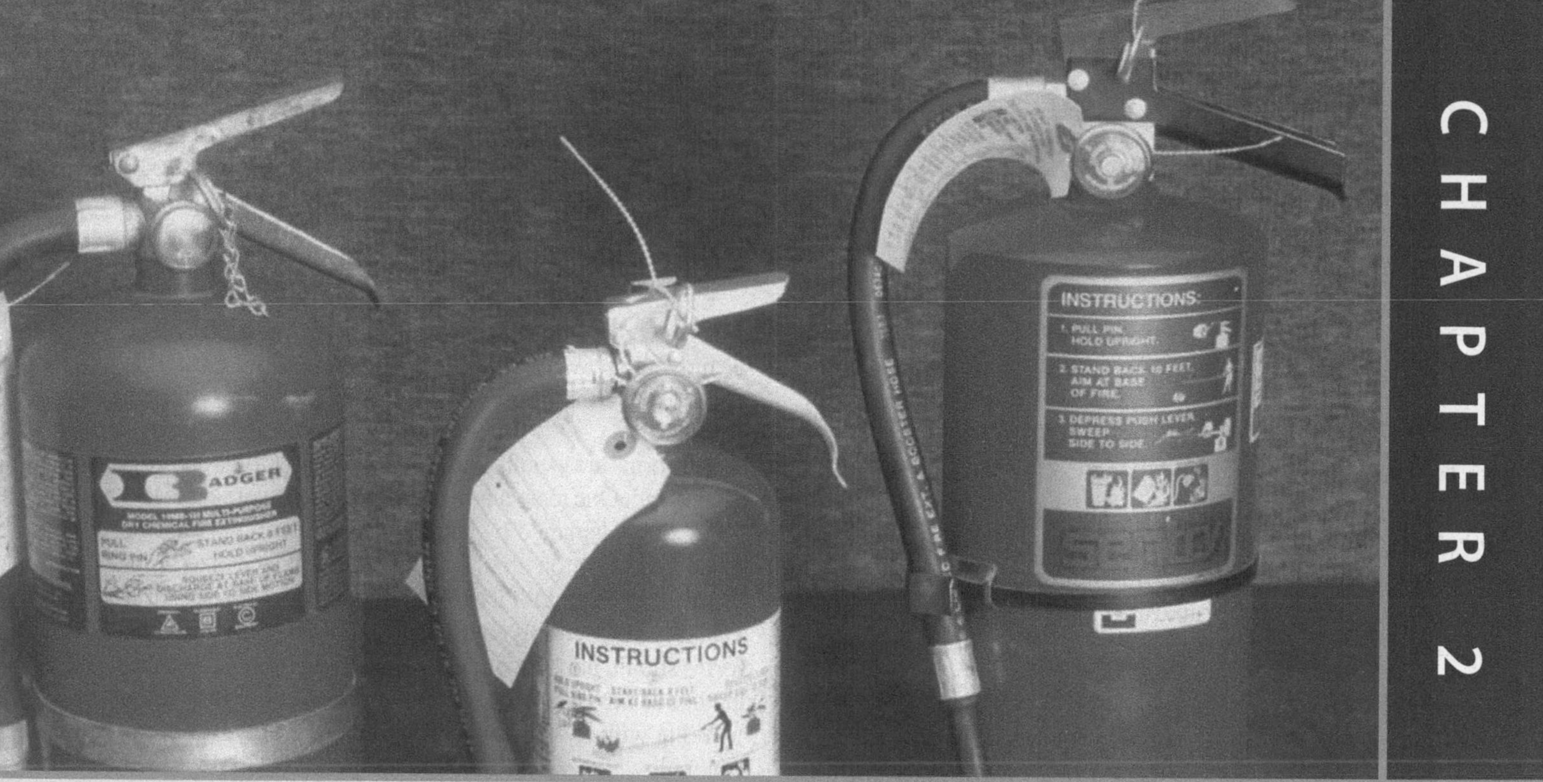

CHAPTER 2

Safety

Chapter Objectives

At the conclusion of this chapter, you should be able to:

- Identify and demonstrate the use of personal protective equipment. (ASE Education Foundation RST Safety #10)
- Identify and demonstrate correct behaviors for the automotive shop. (ASE Education Foundation RST Safety #1, 6, 7, 9, 11, & 12)
- Inspect and demonstrate the use of basic shop equipment. (ASE Education Foundation RST Safety #2)
- Identify and use emergency showers and eyewash stations. (ASE Education Foundation RST Safety #8)
- Locate, read, and interpret material safety data sheets. (ASE Education Foundation RST Safety #15)
- Demonstrate safe battery handling.
- Explain how to safely operate vehicles in the shop. (ASE Education Foundation RST Safety #5)
- Identify chemical hazards and demonstrate safe use and disposal of various automotive chemicals. (ASE Education Foundation RST Safety #2 & 5)
- Identify fire hazards and fire extinguishers and demonstrate fire extinguisher use. (ASE Education Foundation RST Safety #7)

KEY TERMS

air tools
asbestos
blow guns
carbon monoxide (CO)
creepers
engine hoists
eyewash stations
floor jack
hazardous wastes
material safety data sheet (MSDS)
mechanical safety latch
mechanic's gloves
nitrile gloves
personal hygiene
personal protective equipment (PPE)
respirators
safety data sheets (SDS)
work ethic
Z.87 standard

There is no aspect of automotive repair and service that is more critical than that of safety. Damaged vehicles can be repaired or replaced. Damage to the human body is not so easily fixed. Failure to follow safety rules or policies can endanger yourself and those around you as well.

Personal Safety and Behaviors

Personal safety is about taking care of yourself, your appearance, and your body. The most talented technicians are useless if they are dangerous to themselves or others. There are dangers inherent to automotive repair, but awareness and preparation will allow you to work safely and productively.

CLOTHING AND PERSONAL PROTECTIVE EQUIPMENT

Clothing and **personal protective equipment (PPE)** is your first line of defense in the automotive shop (**Figure 2-1**). Wearing the proper clothing helps keep you safe and shows a professional image to the public. Safe work practices and utilizing PPE correctly will benefit you and your coworkers as well.

■ ***Proper Automotive Clothing.*** If there is anywhere that fashion is not a priority, it is the auto repair lab or shop. Proper clothing is necessary to prevent many types of injuries. Clothing that is too big can be caught in moving machinery. Long, unbuttoned sleeves or untucked shirts can be pulled into rotating tools or engine belts. Improper clothing can be unsafe and looks unprofessional (**Figure 2-2**). While clothing styles change, some styles do not give a professional appearance to the public. This may result in customers not returning to a shop because of how the employees are dressed (Figure 2-2).

Clothing that is too tight can restrict movement. Shirts and pants should fit comfortably, allowing a full range of movement. Very tight clothing can be uncomfortable and cut off circulation while working in and around the automobile.

Uniforms are usually made of polyester or a cotton/polyester blend. One reason is because polyester clothing is less likely to be damaged by battery acid and other chemicals compared to cotton clothing. Polyester also tends to wrinkle less and maintain its color and appearance over time. When using welding equipment, however, polyester uniforms should not be worn without additional PPE. Welding leather or heavy canvas overclothes should be worn when working with welding or cutting torches as polyester alone can melt when exposed to high temperatures.

Regardless of the type of uniform worn, it is important to wear it properly and keep it as clean as possible. It is very likely that you will be responsible for moving customer vehicles as well as working inside the passenger compartment. Wearing a dirt- and grease-covered uniform inside a customer's car can result in an angry customer and supervisor. The use of seat covers, steering wheel covers, and floor mats will ensure that no grease gets onto the surfaces that the technician must touch. However, you will still need to check for grease on your clothing before moving the vehicle.

Most repair shops contract their uniforms from a company that maintains and launders them. Dirty uniforms are discarded at the end of the workday, and the company

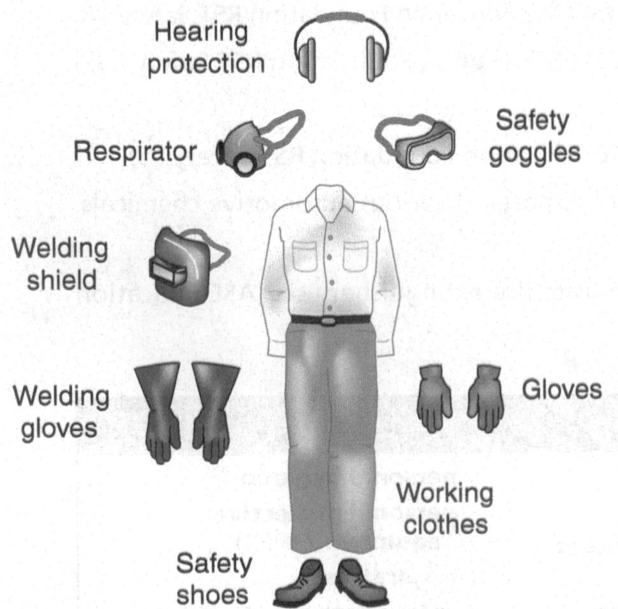

FIGURE 2-1 Personal protective equipment is designed to keep you safe while working in the shop.

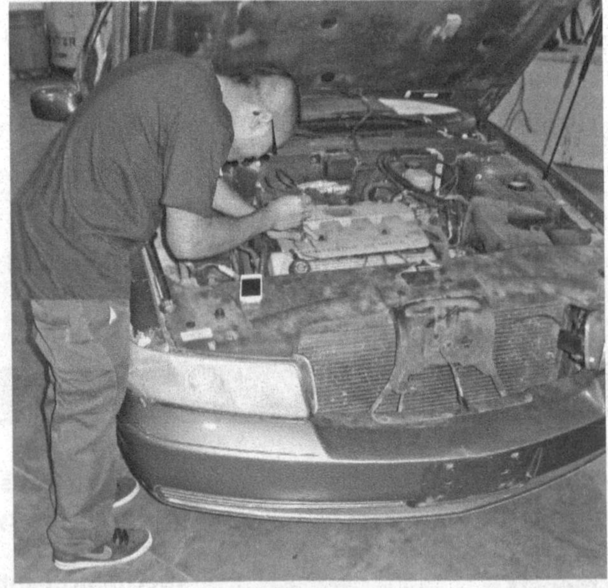

FIGURE 2-2 Baggy clothes are dangerous and look unprofessional. Wear proper fitting work clothes when in the shop.

picks up the dirty clothes and drops off clean uniforms on a weekly basis. Using a uniform company saves the technician time and ensures that the employees look professional and clean.

■ ***Proper Appearance.*** Proper appearance is more than a clean, well-fitting uniform. Hairstyle, facial hair, tattoos, piercings, and hygiene are very important to many people. Many businesses require employees to maintain short hair, neatly trimmed facial hair, remove piercings, and cover tattoos while at work. This is to provide the best possible appearance to the public. While appearance does not determine the abilities of a technician or anyone else, many people equate appearance with competence. There is an old expression, "perception is reality." A neat, clean, professional appearance often leads customers to think that the person is professional and competent. Examine the technician in **Figure 2-3** and think about how appearance plays a role in your opinion of his professionalism.

■ ***Safety Glasses.*** An eye injury can result in permanent loss of vision since there is no way to transplant an eye to restore lost sight. Safety glasses, safety goggles, or a face shield are required when workers can be exposed to the following:

- Flying particles, such as when using a bench or hand grinder.
- Molten metal, possible when using a torch to heat or cut something.
- Liquid chemicals, such as brake cleaner or penetrating oil.
- Acids, which can leak from a car battery.
- Chemical gases or vapor, such as refrigerant.
- Injurious light radiation from using a torch or welder.

These situations occur regularly in the automotive shop, so it is vital that proper eye protection is worn at all times (**Figure 2-4**).

According to government data, in 2005, there were more than 1,200 reported eye injuries to automotive technicians. In 2014, the number dropped to 600. Even though there were fewer reported injuries, many of these may have been preventable. Using protective eyeglasses or safety glasses will greatly reduce your risk of injury. Safety glasses for the shop need to meet the **ANSI Z.87 standard** (**Figure 2-5**). Wearing glasses that do not meet the safety standards may not adequately protect your eyes. Safety goggles should be worn when working with liquids, fumes, powders, or other materials that could splash or go around standard safety glasses. Full-face shields should be worn whenever working with grinders. Welding helmets provide protection for your entire face while using welding equipment, and they reduce the amount of light transmitted to your eyes.

Ensuring all machine guards and screens are installed will also reduce the chances of injury. All hand-held grinding equipment, bench grinders, and brake lathes

FIGURE 2-3 Uniforms or coveralls are the best apparel for the shop environment. Well-fitted uniforms, safety boots, and glasses protect you and present a professional image to the customers.

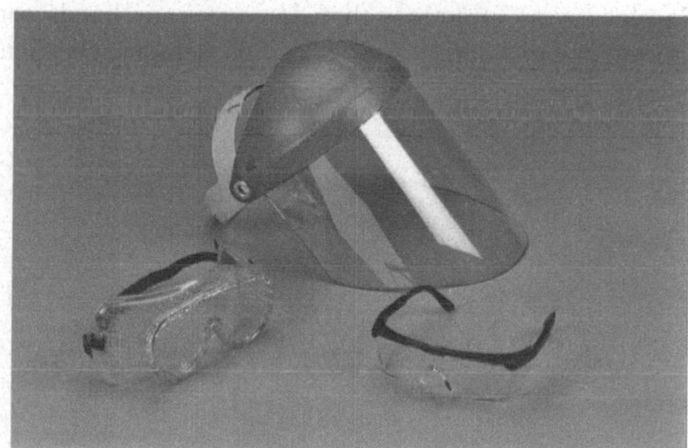

FIGURE 2-4 Safety glasses, goggles, and shields are necessary to protect your eyes. Goggles and shields are used with chemicals and grinding operations.

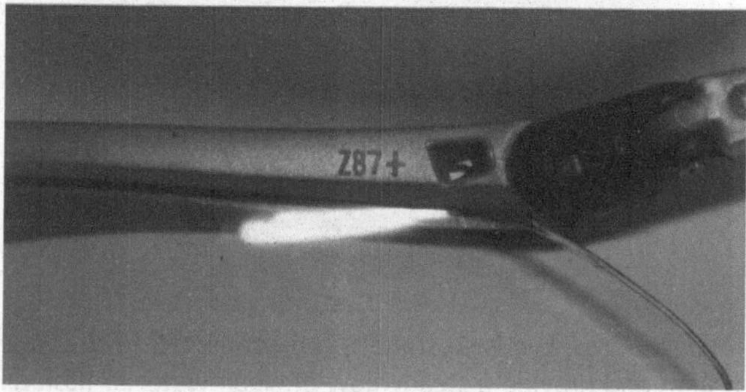

FIGURE 2-5 Safety glasses must have the ANSI Z.87 stamp.

must have their guards installed properly. Welding areas are required to have protective screens in place to prevent accidental exposure of other workers to the harmful ultra violet light produced by welding equipment.

Eyewash stations should be located within the work area and should be easily reached (**Figure 2-6**). If you do accidentally get something in your eyes, go immediately to the nearest eyewash station—do not rub your eyes! Rubbing may scratch or embed the material in your eye, causing further damage.

To use an eyewash station, first press the handle or step on the pedal to start the water flowing. Next, hold your eye open and place your face into the water so that it washes your eye socket. If necessary, have a coworker assist you. Continue to flush your eye until the debris is removed.

If you are exposed to a chemical, flush your eyes for at least 15 minutes. Have someone check the chemical's safety data sheets (SDS) for how to treat eye exposure. Safety data sheets are discussed in detail later in this chapter. If the foreign object will not dislodge or if eye pain or burning continues after flushing, seek medical care immediately. An injury or chemical contact may require you to go to a hospital as soon as possible. If necessary, have someone call an ambulance as you continue to flush your eyes.

■ ***Gloves.*** Technicians' hands are their main tools. Careless work practices that damage your hands can shorten or end a career. Everyday cuts, scrapes, chemical exposure, and stress on the bones, joints, tendons, and ligaments of your hands, over time, can cause pain and limit their use. The best way to keep your hands in the best possible condition is to use gloves appropriate for the task.

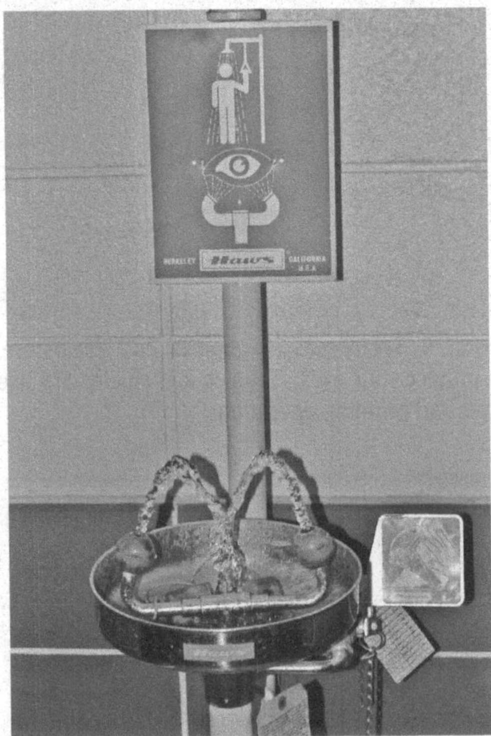

FIGURE 2-6 Eyewash stations are connected to the water system and provide a steady flow of clean water to flush dirt and irritants out of your eyes.

> **Life Skill**
> Remember that your hands are not hammers; use the proper tool for the job. Using your hands or palms to hammer on objects can result in carpal tunnel syndrome, which is extremely painful and requires surgery to repair.

Many technicians wear work gloves or **mechanic's gloves** (**Figure 2-7**). These gloves provide protection against minor cuts, scrapes, and burns. Keeping several pairs of inexpensive jersey gloves around for occasional use is a cheap and effective way to help protect your hands. When working around hot engines and exhaust systems, some technicians recycle old socks into arm protectors or purchase high-temperature sleeves, which are available from tool dealers.

Working on cars and trucks means exposure to many chemicals and waste products. These chemicals can leach into the bloodstream just through skin contact. Exposing cuts on your hands to chemicals such as used motor oil is an invitation to serious medical problems. Use light-duty **nitrile gloves** when you handle all fluids, especially when you come into contact with used motor oil, brake fluid, coolants, and cleaning solvents. **Figure 2-8** shows common chemical-handling gloves. Nitrile gloves are allergy safe and stronger than latex gloves and offer good protection against dirt and chemicals while still being light and comfortable to wear.

Waste motor oil contains hazardous wastes and *carcinogens*, and even limited exposure can cause skin irritation and allergic reactions in some people. Extensive exposure to petroleum products has been known to cause cancer in some cases. Used coolant contains heavy metals and other chemicals that can cause reactions when exposed to bare skin. Any chemical, whether new or used, should be handled carefully and with the proper equipment to prevent exposure.

FIGURE 2-7 Wearing gloves also keeps your hands clean and in good condition.

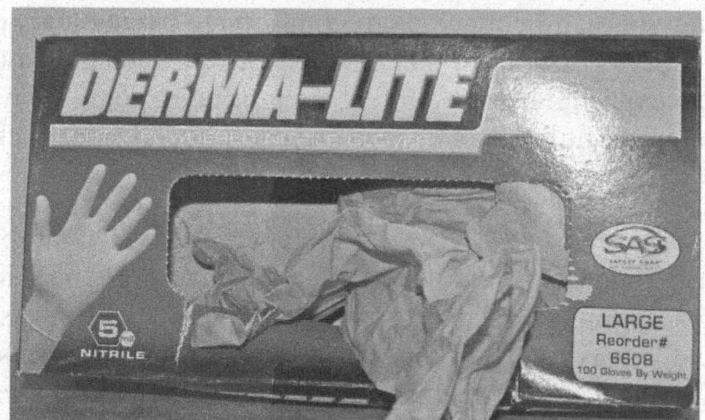

FIGURE 2-8 Nitrile gloves protect your skin when working with fuels, solvents, and other chemicals.

> **Word Wall**
> *Carcinogen*—A carcinogen is a substance that can cause cancer.

An additional incentive to protect your skin is to reduce the chances of becoming infected by contact with another person's blood or other bodily fluids. Many diseases, such as hepatitis and HIV, are caused by *bloodborne pathogens*, which can be transmitted by contact with blood. While your everyday job duties will not often place you into contact with someone's blood, you may need to provide first aid to a coworker. If this happens, protect yourself with gloves.

> **Word Wall**
> *Bloodborne pathogens*—Infectious microorganisms in human blood that transmit disease.

If you are working in a salvage yard or in a collision shop, you may work on a vehicle in which a person or persons were injured. Be careful when handling parts that may have blood on them. Some bloodborne pathogens can survive for an extended period of time outside the human body. Use of gloves will ensure your danger of exposure is small.

Regular use of hand sanitizer is also a good idea. As a technician, you will be in contact with customer cars, tools, replacement parts, phones, keyboards, and other items, all of which have been handled by other people. To reduce your exposure to common illnesses, use an antibacterial hand sanitizer often and avoid touching your face while working in the shop (**Figure 2-9**).

Many technicians also keep hand lotion in their toolbox. Hand lotion helps fight dry skin common in winter weather. These lotions can also help replace the natural moisture and oils that are often removed by harsh hand soaps. Washing your hands on a regular basis and then exposing partially wet hands to wind or extreme cold will chap the skin and can result in dermatitis, which can create painful cracking of the skin. The use of nitrile gloves will limit the number of times technicians must wash their hands in a day, further protecting skin from damage.

FIGURE 2-9 Since you will be handling tools, computers, steering wheels, and many other objects that can carry bacteria, using hand sanitizer can be an effective method of reducing your chances of illness.

■ ***Hearing Protection.*** An auto shop can be a very loud place to work. When you are using grinders, air hammers, air drills, and similar equipment, ear protection should always be used. By OSHA standards, when the sound level is at or above 85 db, hearing protection must be used. This is slightly less than the sound levels created when using a hair dryer or lawn mower, or by the noise of a passing motorcycle.

Losing your hearing is often a gradual process and is usually less noticeable than other types of workplace injuries. The type of hearing loss most often associated with the workplace, sensory hearing loss, results from damage to the inner ear and cannot be corrected medically or surgically, meaning the hearing loss is permanent.

To avoid damage to your hearing, hearing protection devices such as earplugs or earmuffs should be worn when the noise levels reach damaging levels. Earplugs, either PVC or foam, provide an inexpensive, lightweight choice, while earmuffs are better for use over a longer period of time and often are more comfortable (**Figure 2-10**).

■ ***Back Braces.*** Technicians often have to lift heavy objects, such as batteries, wheel/tire assemblies, and cylinder heads. Whenever you are lifting a heavy object, always use your leg muscles, which are much larger than those in your back. **Figure 2-11** shows the proper way to lift a

FIGURE 2-10 Ear protection should be used when working with air hammers, air drills, and other tools that are loud.

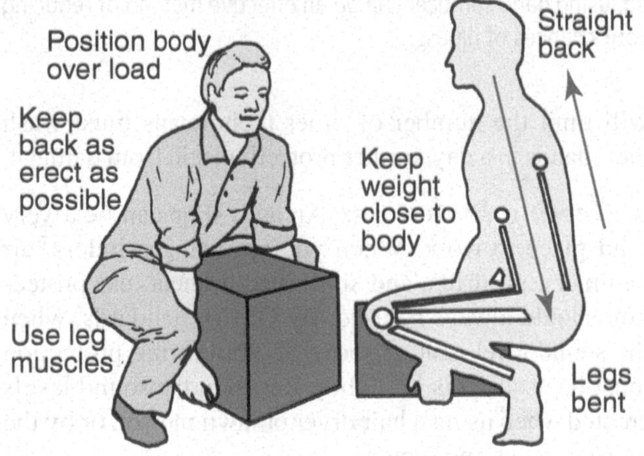

FIGURE 2-11 Back injuries can ruin a career and are usually preventable. Always use your legs to lift instead of your back.

heavy object. Back braces are used to help prevent back injuries by forcing the wearer to use the leg muscles instead of the back muscles. Back injuries can cause long-term disability and bring an early end to a technician's career.

■ ***Protective Footwear.*** Most automotive shops require the use of safety boots or shoes. This footwear should be hard-soled, nonslip, and oil and chemical resistant. Many labs and shops require steel-toed safety boots or shoes. Steel-toed sneakers are not acceptable as they do not have slip-resistant soles or offer much protection from oils and other chemicals.

It is a good idea to try several different shoe manufacturers to find a pair that fits well and is comfortable. You will be spending many hours in these boots or shoes, and you will be much more comfortable buying a good quality boot or shoe that fits well. As a technician, you can expect to be on your feet eight or more hours a day. A poorly fitting boot or shoe can cause foot, leg, or back pain, making your workday very uncomfortable.

■ ***Respirators.*** There are three ways in which a substance can enter your body, inhalation, ingestion, and skin absorption. Technicians may inhale dust, fumes, mists, and vapors from solvents, cleaners, and other chemicals. The inhalation of chemicals can damage the tiny air sacs in the lungs, called *alveoli*. Once damaged, your lungs are more susceptible to respiratory diseases. In some cases, these cannot be cured and could eventually lead to death.

Respirators, worn over the mouth and nose, trap contaminates before you inhale them and should be used any time there is a danger of airborne chemicals or dust being inhaled. In the automotive shop, brake and clutch system repairs often require a method of dust collection and filtration. Respirators should be worn when using aerosol cleaners or other airborne chemicals (**Figure 2-12**). There are several types of respirators that may be used depending on the work being performed. Auto technicians most often use air purifying respirators, which can be a simple dust mask or a full-face respirator. Selecting the correct respirator depends on the type of contaminant present, the form the contaminant is in, how toxic the contaminant is, and the concentration (how much is in the air) of the contaminant. Always check the condition of the respirator before use and ensure it provides a tight, leak-proof fit against your face. Because different respirator cartridges are used for certain types of dangers, always make sure you are using the correct filter for the application. A mask used for dust may not provide any protection against inhaling chemical vapors. Refer to the respirator's instructions for the correct filter to be used for your situation. Do not just assume that just any respirator will protect you when you are using chemicals and paint products, and read the packaging to ensure it is rated for the chemicals in use!

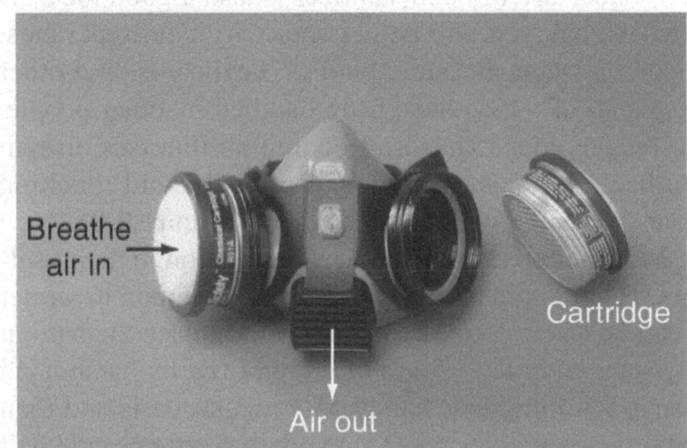

FIGURE 2-12 Respirators and dust masks are used to prevent inhalation of chemicals and asbestos.

> **Word Wall**
> *Contaminant*—Something that makes a place or substance no longer suitable for use.

■ ***Personal Hygiene.*** As discussed earlier, your hygiene is important in how the customer accepts you as a professional. Technicians, however talented, may not be able to develop the necessary dialog and relationship with customers if they have poor **personal hygiene**. The following are important guidelines regarding hygiene:

- Wash and comb your hair. Approaching a customer with a bad case of "bed head" is not a good idea. If you have long hair, keeping it pulled back into a ponytail or in a hat will help keep your hair clean and out of the way when working.
- Wash your face and trim any facial hair. Beards, mustaches, and goatees are fine, but remember that looking like you have been lost in the wilderness for the last five years may not appeal to your customer or employer.
- Take baths or showers often. Do not try to cover a missed shower with a lot of body spray or cologne. Many people find excessive application of colognes offensive, and it can even lead to allergic reactions. Once you are off work, take another bath or shower to remove the dirt and smells of the auto shop.
- Keep your hands clean and in good condition. Wear gloves to prevent cuts and scrapes. Keep your fingernails trimmed and clean. Use plenty of hand cleaner and hand lotion if your skin gets dry.
- Keep your uniform neat and presentable. If you get covered in dirt, grease, or a chemical, change into a fresh uniform, and do not get into a vehicle if you are dirty or covered in grease or other chemicals.
- Last, while not strictly a matter of hygiene, it is worth mentioning that piercings and tattoos, while more acceptable in today's culture than ever before, are not always well received by employers and customers. Keep visible piercings to a minimum and, if necessary, replace any questionable jewelry with a stud or another low-profile item. Cover tattoos as necessary. The jewelry and tattoos you exhibit to the world outside of your hours of employment may not be acceptable to your employer or your customers.

BEHAVIORS

■ ***Working in the Automotive Shop Environment.*** You will find that working in an automotive lab is very different from a traditional classroom environment. You will be working on real vehicles using tools and equipment found in modern auto repair facilities. Because of this, you will be expected to act as a mature young professional. Why is behavior so important, you might ask? Primarily, it is for safety reasons. Beyond safety, it is about professionalism and developing your work ethic.

Your safety and the safety of every other person in the class depend on your behavior. An auto shop is, by the nature of the tools and equipment, a potentially dangerous place. Improper use of a tool or piece of equipment is an invitation to an accident and injury. Intentionally using a tool or piece of equipment in an unsafe manner is foolish and can result in serious unexpected consequences, such as injury or death of yourself or a coworker. You should not accept unsafe behaviors from those who work near you, nor should they be placed at risk by any unsafe actions you take. There is never a time where playing around in a lab or shop is acceptable. Focusing on what you are working on is an important part of safe work practices. Letting your mind wander or becoming distracted by music, other students, or anything not related to your work can lead to an accident. If you find that your attention is drifting, take a moment to refocus as this is a dangerous situation for you and others. If necessary, take a short break when needed to help you refocus and continue your work.

To be a professional technician, your goal is to perform every service and repair 100% correctly every time. To accomplish this, there are several things to consider.

- **Safety:** Perform all work safely and by following the recommended service procedures. Do not take shortcuts, like lifting a vehicle with only a floor jack to slide under the car for a quick look. If the hydraulic jack fails, the vehicle will drop and you can be injured or killed.
- **Professional attitude:** This is more than just dressing as a professional technician. A professional takes time to ensure every procedure is done properly and always double-checks his or her work. Doing a job properly means not taking shortcuts and ensuring the quality of the work. If you are in doubt, double-check yourself or have a more experienced or more knowledgeable person double-check your work.
- **Admit what you do not know:** No one knows everything; there is to know about any vehicle. Most people will have more respect for technicians who can admit that they do not know something and are willing to find the answer than by those who bluff their way through an answer, only to be proven wrong in the future. In addition, you will not make your instructor or supervisor happy if you make claims about knowledge and abilities only to prove yourself wrong someday.

- **Work ethic:** Loosely defined, **work ethic** is the idea that personal accountability and responsibility have intrinsic or internal value. Generally, this means having initiative, being honest, and dependable. Showing up for work late every day and pretending you are there on time is not a good work ethic. Be polite, work your hardest, do your best, and strive to learn as much as possible each day from your instructors and more experienced technicians. These simple things will go a long way toward your success as a technician because others will respect you more and be more willing to help you when you need it.

■ ***Safe Working Behaviors.*** As stated earlier, safety is always of the utmost importance. Maintaining a safe work environment requires a constant awareness of your surroundings. Unfortunately, you may be the only person in the shop who is concerned with safety. Hopefully that is not the case, but if it is, do not allow the attitudes of others to have dangerous consequences for you. The following are some general safe working habits to incorporate into your own work ethic.

- Clean up after yourself. Keep your work areas and tools clean and organized.
- Use the correct tool for the job and keep your tools in good working condition.
- Clean any spills immediately. A wet floor is an accident waiting to happen.
- Check all electrical cords and electrical tools before using them. Inspect the ground lug on cords and check for any fraying or damaged insulation. Do not use anything with a damaged power cord or plug.
- Check air hoses and air tools regularly. Leaking hoses or fittings should be replaced immediately. Do not use faulty air tools or air hoses.
- Ensure the proper operation of floor jacks and stands before lifting a vehicle. Do not use any damaged equipment. Do not overload the floor jack or stands.
- Check the operation of vehicle lifts regularly. Check for hydraulic fluid loss and make sure the safety catches are working properly. Never use any lift that is not functioning exactly as designed.
- Never use flammable chemicals near hot surfaces.
- Do not use any piece of equipment unless you have been trained in its safe operation.
- Make sure all guards, guides, and safety disconnects are present and functioning on equipment you use.

While it is true that accidents do happen, many shop accidents can be prevented by maintaining an awareness of safety procedures. Being able to work safely and professionally will make you a valued employee.

Shop Safety

Shop safety is about being able to work safely in the automotive shop. A typical shop contains equipment from floor jacks to grinders and presses. Technicians must be able to understand and operate the types of equipment used in the day-to-day operation of the shop.

EQUIPMENT AND SHOP TOOL SAFETY

Proper care and use of shop tools is critical to maintaining a safe and productive shop. Misuse of a piece of equipment can cause damage to the equipment, to a vehicle, or to you and others. Make sure you understand how to properly use a piece of equipment before you attempt to use it. Also, you should keep in mind that all personal safety rules and PPE use continue to apply when you are using shop tools and equipment. Some tools and equipment may require additional specific pieces of PPE for safe operation.

In many shops, certain sections will be marked with safety or warning tape to indicate a hazardous area, such as around bench grinders and brake lathes. These safety zones are to warn you and others to use extra caution when working near one of these areas (**Figure 2-13**).

■ ***Floor Jacks and Jack Stands.*** Almost all auto shops have floor jacks and jack stands (**Figure 2-14**). These are used to lift a vehicle off the floor for inspections and repairs. A **floor jack** uses a hydraulic cylinder to convert the large up-and-down movements of the jack handle into a smaller upward movement of the jack.

Different styles of hydraulic jacks are available, but all require the use of jack stands to support the vehicle

FIGURE 2-13 Areas around brake lathes, solvent tanks, grinders, and other equipment should be marked. Only the person using the equipment should be within the marked area.

FIGURE 2-14 Floor jacks and jack stands are some of the most common pieces of equipment for an auto shop.

once it is raised. **Never work on a vehicle supported only with a hydraulic jack.** The internal seals could fail, allowing the jack to drop suddenly and causing serious injury or death to anyone trapped under the weight of the vehicle. Complete operation and safety for using floor jacks and jack stands are discussed in Chapter 3.

■ ***Lifts.*** Lifts, also called *hoists*, are used to raise the vehicle up to safe and comfortable working heights (**Figures 2-15** and **2-16**). The lifts shown in Figure 2-15 are called *inground swing arm lifts* because the four arms swing or pivot to allow placement under a vehicle. The lift in Figure 2-16 is an above ground swing arm lift.

Lifts are either pneumatic-hydraulic or electric-hydraulic. *Pneumatic-hydraulic* or air-over-hydraulic lifts use the shop's compressed air system to pressurize hydraulic oil, which is used to raise and lower the vehicle. Electric-hydraulic lifts use an electric motor pump to pressurize the oil for lift operation. Before attempting to use either type of lift, be sure you are properly trained and understand how the lift operates. Do not use any lift that has a leaking air or hydraulic system.

Word Wall

Pneumatic-hydraulic—Pneumatic means using air or compressed air to power a tool or perform work. Hydraulic means to use a liquid to apply force or motion to perform work.

All lifts have some type of **mechanical safety latch**. This safety should apply automatically as the lift is raised. When the lift is lowered, the safety will engage, preventing the lift from lowering. Always lower the vehicle onto the safety latch once you have reached the height you need. The safety latch is a mechanical lock that acts like a jack stand, keeping the vehicle from dropping in the event the hydraulics fail. To disengage the safety, you must raise the lift backup, disengage the safety, and then lower the lift.

FIGURE 2-15 Lifts come in a variety of sizes and configurations.

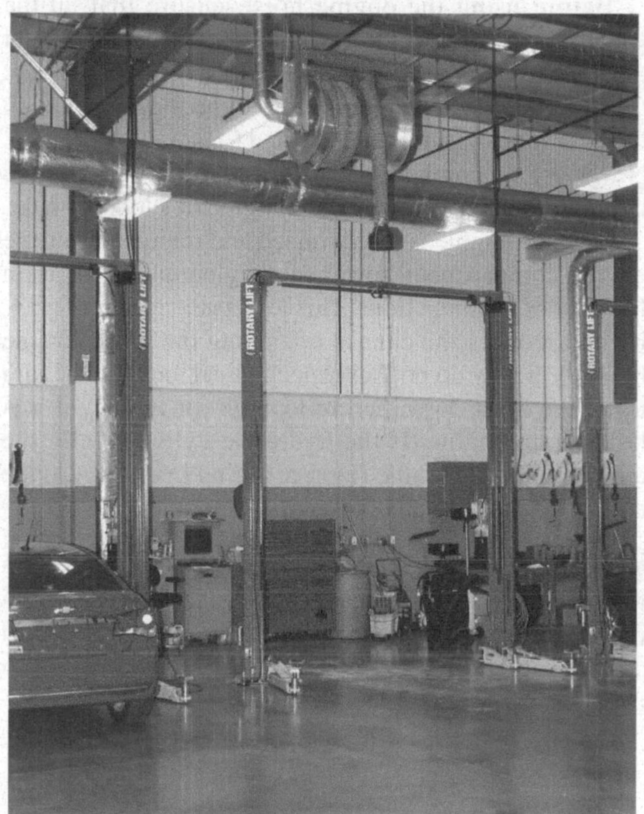

FIGURE 2-16 This type of unequal length swing arm lifts are common in auto shops.

Never use a lift if the safety latch does not operate. If the lift should lose pressure, it will come crashing down. The safety latch prevents this in the event of a loss of pressure.

■ ***Hoists and Cranes.*** Many auto shops have a crane or hoist for engine removal. Some shops may even have overhead cranes for lifting large, heavy components, although these are more often found in heavy-duty truck shops.

Engine hoists are used when removing an engine from a vehicle and when mounting an engine on an engine stand for repairs. **Figure 2-17** shows an example of a typical engine hoist. Most hoists have extendable booms and legs to allow for greater reach over a vehicle. If the boom is extended, the weight-carrying capacity of the hoist is reduced. This capacity should be indicated on the hoist. Never overload an engine hoist by attempting to lift more than the rated weight capacity. This can cause the hoist to fail, resulting in injury and vehicle damage. If the boom is extended, the legs of the hoist must also be extended to maintain balance. Moving the boom without extending the legs will place the weight of the engine too far from the hoist's center of gravity and can cause the hoist to tip forward. This can result in personal injury and damage to the engine attached to the hoist.

Before using the engine hoist, ensure that all the wheels and casters roll freely. A frozen caster can prevent the hoist from moving easily, which can result in the hoist tipping over when being moved around the shop. Check the hydraulic cylinder for leaks and for smooth operation. Never use an engine hoist with a hydraulic leak as the cylinder may fail, allowing the engine to suddenly fall.

Once an engine is safely raised and removed from the engine compartment, lower the engine until it is just a few inches from the floor. This lowers the center of gravity and reduces the chance of tipping the hoist. Engine hoists typically do not have mechanical locks, so be sure to let the hoist down before leaving for the day if it is holding an engine. If the hydraulic cylinder fails, the engine will crash to the floor, resulting in damage to the engine or possibly an unknowing bystander.

■ ***Bench Grinders.*** A bench grinder is used to dress tools and shape and prepare the edges of metal. Dressing a tool means to clean up or restore it to its original shape, such as regrinding the cutting edge of a chisel that has worn down with use. Many bench grinders also have a wire wheel. A typical bench grinder is shown in **Figure 2-18**. Used to remove rust and sharp metal edges, the wire wheel can shed wire bristles that can penetrate skin and eyes. Always wear a full-face shield when using a bench grinder due to the debris and wire bristles that can be ejected during use.

FIGURE 2-17 An engine hoist is used to remove the engine from the vehicle and support it until the engine is mounted on a work stand.

Before using a bench grinder or any electrically powered tool, check the power cord. Make sure the ground lug is intact and that the cord is not frayed or damaged. Inspect the guards and shields (**Figure 2-19**).

Inspect the guards on the grinder. The guard is placed in front of the grinder or wire wheel. This guard can be used as a tool rest when dressing a tool and also helps prevent tools and other objects from getting pulled into the wheel housing when in use. Be sure to consult the operator's manual for the proper clearance. Typically, no more than an 1/8-inch gap should be between the wheel and the tool rest.

FIGURE 2-18 Bench grinders must have protective shields and guards installed for safe operation. Many bench grinders have one grinding stone and a wire wheel.

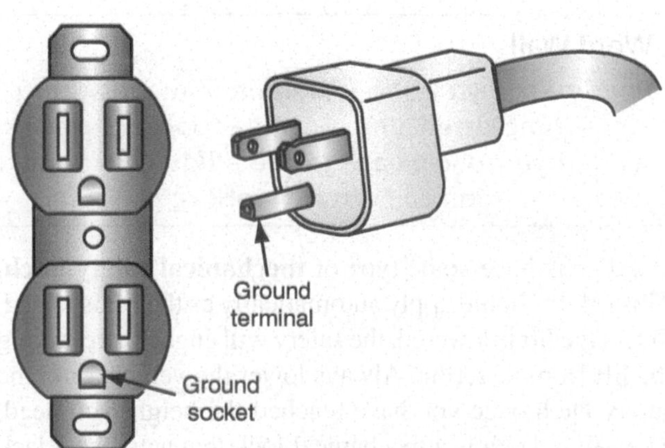

FIGURE 2-19 Before using any electrical device, check the power cord, plug, and outlet condition first. Do not use any equipment with a damaged cord or plug.

The shields are usually adjustable so the operator can adjust their angle for protection and access to the wheels. Bench grinders must have the appropriate shields in place to prevent debris from flying up at the operator. Do not use a bench grinder that shows signs of damage to the grinder wheel or the wire wheel, or has missing guards or shields. **Inspect the wheels for cracks and damage before using them.**

■ ***Air Tools.*** **Air tools** are common in the shop. Most technicians and shops have many types of air-powered tools. These include air impact guns, air ratchets, grinders, cutoff wheels, and sanders (**Figure 2-20**). Regardless of the type of air tool being used, always follow the same basic safety precautions:

- Check the air hose and fittings before using the air tool. Do not use leaking hoses or fittings. Damaged hoses and fittings must be replaced. Do not attempt to fix or patch the air hose.
- When connecting air tools to the shop's compressed air system, the air line should be off and there should not be any pressure in the hose. Connect the tool to the air hose and then open the valve. When finished with the tool, turn the air supply line off and operate the tool to bleed air pressure from the hose. Once the pressure is gone, disconnect the tool from the air hose.
- Be sure you are trained and understand how to use the air tool before you actually use it.
- Use only approved attachments for air chisels.
- When using a grinder or cutoff wheel, use only discs that are rated for the tool being used. Do not use a grinder disc or cutoff wheel with a lower revolutions per minute (RPM) rating than the tool you are using. Using an incorrect disc or wheel can cause the wheel to explode, resulting in serious injuries.
- Use only impact sockets with impact guns. Impact sockets are black and are designed to take the forces associated with the hammering action of an impact gun. Using standard chrome sockets on an impact gun can cause the socket to break and come apart while in use, which can cause serious injury.
- Do not trigger the impact gun and socket unless the socket is being used on a fastener. Spinning the socket in the air can cause the socket to fly off, leading to injury.
- Do not hold impact sockets or air hammer adaptors in your hands while using the tool. The socket or adaptor could break during use and can cause serious injury if you are holding on to it while in use.
- When using an air ratchet, be sure there is enough room to contain your hand and the ratchet where you are working. Air ratchets can force your hand around into a tight spot, trapping it there, leading to injured fingers and broken bones.

■ ***Blow Guns.*** **Blow guns** are used to blow away dirt and debris and to dry components, such as wheel bearings, that have been cleaned in a liquid solvent. Blow guns come in many shapes and sizes (**Figure 2-21**). Before using a

FIGURE 2-20 Air tools often make a technician's work easier and faster but only if used correctly.

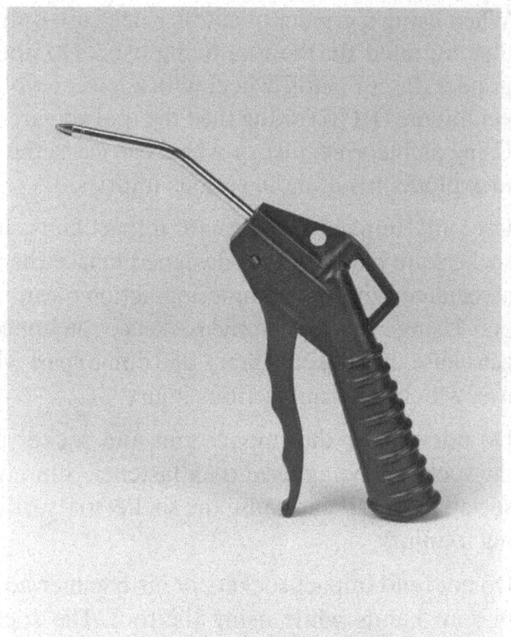

FIGURE 2-21 Blow guns should never be used to clean dirt from your uniform or body. Never point a blow gun at yourself or anyone else. Serious injury—even death—can result from using compressed air against a person.

FIGURE 2-22 Creepers are used when under a vehicle. When not in use, always stand them up and against a bench or wall, never leave a creeper lying on the floor.

blow gun, make sure the air hose you are using is in good condition with no leaks. Check that the air fitting on the gun is made to work with the air fitting on the air hose, as there are several types and sizes of air fittings available.

Blow guns should never be pointed directly at yourself or at others. Do not use a blow gun to blow dirt from your uniform. This can cause dirt and debris to lodge in your skin. Never point a blow gun at any part of your body. The air pressure exiting the blow gun is sufficient to cause serious injury or death if pointed toward your head or any body cavity. Because your heart is a hydraulic pump, if compressed air gets into the bloodstream, your heart will be unable to pump blood, which will result in death.

■ ***Creepers.*** **Creepers** are used by technicians to move around under a vehicle that is only raised off the floor a slight amount. Most creepers are padded, which provides a little bit of comfort while working (**Figure 2-22**). Some creepers have adjustable head rests for even greater comfort.

Before using a creeper, check that the casters operate smoothly. If you have long hair, you should secure it into a ponytail or hat before using a creeper. Long hair can get caught in the moving wheels, which can lead to an unexpected haircut!

When you are not using the creeper, stand it upright and lean it against a workbench or wall. Do not lean the creeper against the vehicle or leave it lying on the floor. A creeper left on the floor is an invitation to a serious fall.

FIRST AID, SHOWERS, AND EYEWASH

A good understanding of basic first-aid procedures is important so that in the event of an accident or injury, you will be prepared and will know what steps to take without hesitation. This section will cover some procedures for common injuries.

■ ***First-Aid Kits.*** Every shop should have at least a basic first-aid kit, containing bandages, gauze, and medical tape (**Figure 2-23**). First-aid kits should be in a visible, easy-to-reach location that is accessible to everyone. You should notify your instructor or supervisor anytime you have an injury, even a minor cut. Wearing mechanic's gloves can prevent many minor cuts and abrasions, but when they do happen, thoroughly clean and bandage the areas affected.

■ ***Basic First Aid.*** Knowing some basic first-aid procedures will help you be prepared and to know what steps to take in the event of an accident. Taking a course to become first aid certified is a good idea and may be paid for by your employer. In addition, having first-aid training may give you an edge when seeking employment. The following are some procedures for common injuries:

- **Cuts and scrapes:** First, stop the bleeding. Apply gentle pressure with a clean cloth or bandage, holding pressure for 20 to 30 minutes if necessary. Clean the wound with clean water; avoid using soap as this can irritate the wound. Apply an antibiotic such as Neosporin and cover the wound. Change the dressing at least once per day.
- **Burns:** First, determine the degree of the burn. The least serious burns are first-degree burns. This is when only the outer layer of skin is burned. Pain,

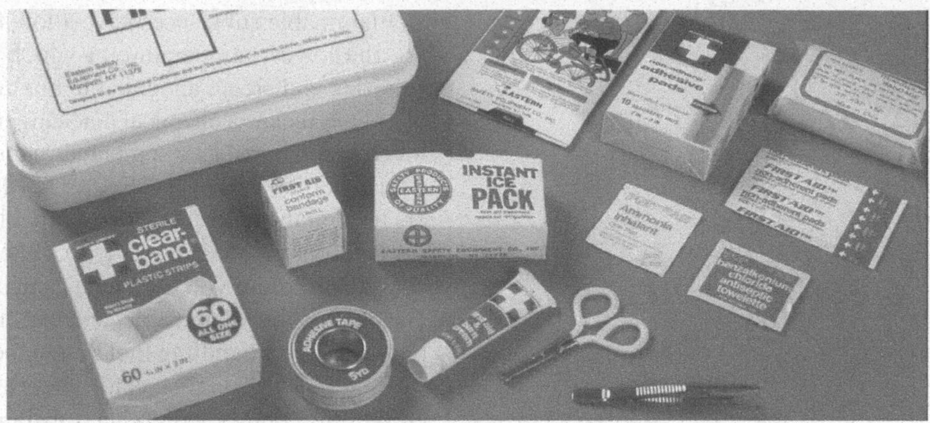

FIGURE 2-23 A basic first-aid kit should be well stocked and accessible in case of a minor injury.

swelling, and red skin are present. When the first layer of skin is burned through, it is considered a second-degree burn. Blisters and significant reddening of the skin are present.

Second-degree burns cause severe pain and swelling around the burn area. Second-degree burns larger than about three inches or over a joint should be considered serious, and medical help should be sought immediately. For a second-degree burn, cool the area by running it under cold water for at least five minutes or until the pain subsides. Loosely cover the burn with a sterile gauze bandage.

The most serious type of burn is a third-degree burn as it involves all of the layers of the skin and can cause permanent damage. For a serious burn, call 911. Until medical help arrives, follow these steps: make sure the skin is no longer in contact with smoldering materials but do not remove the burnt clothing, and elevate the burn areas above the level of the heart if possible.

- **Chemical burns:** First, remove the cause of the burn by flushing with water for 20 minutes or more. Remove any clothing or jewelry that is contaminated by the chemical. Apply a cool, wet cloth to help relieve the pain and wrap the area loosely with a dry sterile dressing. If the burn has penetrated the first layer of skin or any other reactions appear to be taking place, seek medical assistance.
- If the chemical has entered the eyes, flush with water for at least 20 minutes. Do not rub the eye. Seek medical assistance.
- **Foreign object in eye:** First, examine the eye and attempt to flush the object out. Do not rub the eye. Do not try to remove any object that is embedded in the eye. Seek medical help if the object will not come out, is embedded, vision is affected, or pain or redness persists after the object has been removed. It is a good idea to seek medical attention for any type of eye injury.
- **General seizures (grand mal):** Keep calm. Do not try to hold the person down or stop his or her movements. Remove any objects that the person may contact while the seizure is occurring. Try to place something soft, like a folded jacket, under the person's head. Turn the person onto their side to help keep the airway clear. Stay with the person until the seizure stops.

■ ***Emergency Showers.*** Emergency showers are used in the event that you or a coworker get splashed with a chemical, solvent, acid, gasoline, diesel fuel, or other hazardous liquid. **Figure 2-24** shows an example of an emergency shower. To use, stand under the shower and pull the handle down.

■ ***Emergency Eyewash.*** Eyewash stations provide a fountain of water to flush foreign objects from the eye. Even with proper eye protection, dirt, debris, and liquids can still get into your eyes. If this happens, do not rub your eyes. Get to the eyewash station; ask for help if you need it. For eyewash stations like that in Figure 2-24, hold your face over the station and press the lever to start the water flow. Hold your eye open and flush with water. If a chemical or other liquid has gotten into your eyes, flush for at least 15 minutes and have someone call 911. Chemical burns can cause serious eye damage and blindness. Have your eyes checked by a physician as soon as possible.

HAZARDOUS WASTES AND SAFETY DATA SHEETS

Hazardous wastes can be any substance that can affect public health or damage the environment. Nearly every liquid, chemical, and solvent used by the vehicle and for service and repair is considered a hazardous waste. Because of this, there are strict laws and regulations

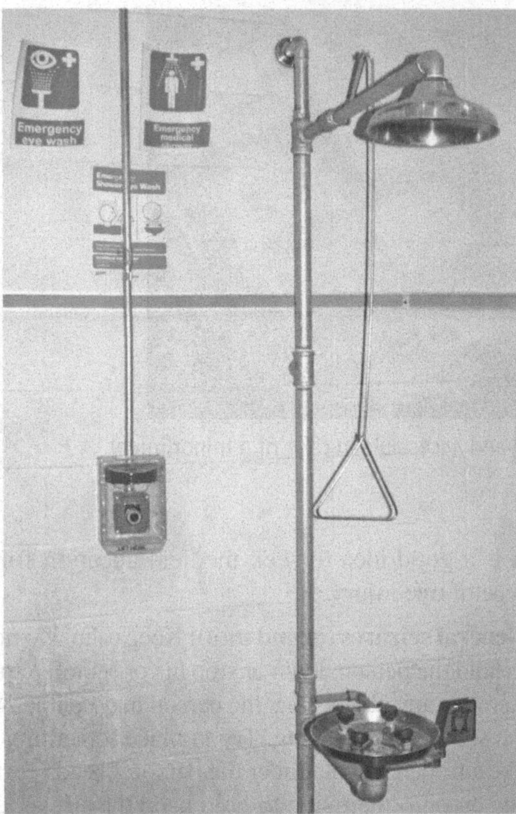

FIGURE 2-24 Emergency showers are plumbed into the water system and are often located near the eyewash station.

regarding the use, handling, and disposal of these substances. Information about each type of substance is contained in a safety data sheet (SDS).

■ ***Hazardous Wastes.*** Hazardous wastes not only can cause personal injury, but when handled or disposed of improperly can damage the environment.

Hazardous wastes are identified by the following characteristics:

- Corrosive
- Flammable
- Reactive
- Toxic

Corrosive substances can eat through metals, clothing, skin, and other items. A common corrosive found in auto service is battery acid, also called *electrolyte*. Electrolyte is a combination of water and sulfuric acid and is highly corrosive. Exposure to electrolyte can cause serious skin reactions, burning, and skin damage. Skin exposed to electrolyte should be immediately washed with water for 15 minutes. Exposure to battery acid can also be from corrosion on the battery. The acid, even though it is solidified and powdery, is still a danger and can burn your skin or eyes.

Flammable substances have a low flash point, meaning the lowest temperature at which the chemical can vaporize and create an ignitable mixture in the air. Simply put, it means a chemical that can ignite or catch fire at low temperatures. Low temperature means the liquid has a flash point below 199.4°F (93°C). Many aerosol cleaners used in the auto shop are flammable and can ignite if sprayed onto a hot surface. Check the labels on aerosol chemicals before you use them, as many are flammable. Gasoline is also flammable, and extreme caution should be taken when handling fuel system components.

Reactive substances can, when allowed to mix or come into contact with other substances, create harmful gases or start a fire. Shop rags, when they are soaked with chemicals and are allowed to come into contact with other shop rags with different chemicals soaked into them, can lead to combustion.

Toxic substances are poisonous to people and the environment. Antifreeze is a toxic substance, especially after it has been in the engine's cooling system. The antifreeze picks up traces of heavy metals from the engine. Antifreeze and other toxic chemicals must be captured and reclaimed from the vehicle and should not be allowed to enter the environment.

■ ***Safety Data Sheets.*** **Safety data sheets (SDS)** are the standardized and modernized version of material safety data sheets (MSDS). Safety data sheets are compliant with the global harmonized system (GHS) standards of classifying and labeling of chemicals.

The data sheets must be located at the worksite and be made available to every employee. Every automotive lab and shop is required to have SDSs available to employees and students as well. **Figure 2-25** shows an example of SDS location. An example of an SDS is shown in **Figure 2-26**. Information detailed in the SDS includes the following:

- Product name and chemical components
- Information about the substance's flammability, reactivity, toxicity, and corrosiveness
- The types of possible exposure and the health effects for exposure
- Correct handling, including the type of PPE required and proper disposal of the substance
- First aid for accidental exposure

■ ***Environmental Protection Agency.*** The Environmental Protection Agency (EPA) is a federal agency that enforces environmental laws passed by Congress. States often have their own version of the EPA. Both levels of the EPA oversee the enforcement of laws and regulations regarding the safe handling, storage,

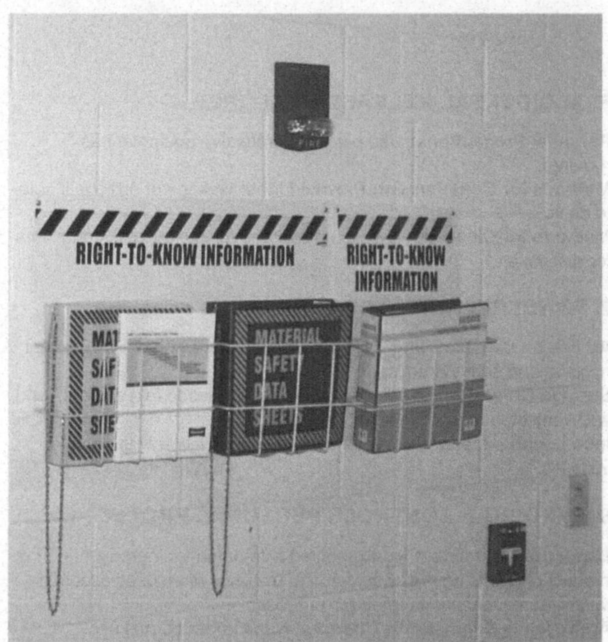

FIGURE 2-25 SDS and MSDS information must be accessible and kept current.

transportation, and disposal of hazardous wastes. As auto shops generate hazardous wastes, certain procedures must be followed for their safe storage and disposal. Failure to comply with EPA rules and regulations can result in fines, imprisonment, and closure of the business.

■ ***Occupational Safety and Health Administration.*** Occupational Safety and Health Administration (OSHA) is the government agency responsible for the safety and health of workers. OSHA regulations include many aspects of the automotive shop, from the training of employees to how much air pressure is allowed in the shop's compressed air system. Workplace accidents are investigated by OSHA, and shops found to be negligent in their compliance can be fined or even closed for business.

Employees must be trained regarding the hazardous materials that they may be exposed to on the job. OSHA has enacted the Hazardous Communication Standard, which includes the Right-to-Know standard. This standard mandates that employees be notified about workplace hazards and trained how to protect themselves.

BATTERY SAFETY

Every modern car and truck has at least one battery, and some have two. The battery provides the energy to start the engine and power the electrical system. The amount of electrical power stored in a single lead-acid battery is equivalent to about one-quarter of the power used in an average-sized home. This power comes from the chemical reaction inside the battery between the battery acid and the lead-plated cell components. Once the engine is started, the vehicle is powered by the generator. The generator also recharges the battery. When the battery charges, it releases hydrogen, which can, if exposed to a spark or flame, cause an explosion. An example of the battery warning label is shown in **Figure 2-27**.

The following are safety precautions specific to working around automotive batteries:

- Never smoke or create a spark or flame near a battery. This could cause the hydrogen gas to explode the battery.
- Never lay tools, parts, or equipment on or near the battery, battery terminals, or battery cables.
- When disconnecting a battery, always remove the negative cable first, then the positive. Connect the positive cable first, then the negative.
- Never attempt to charge or jump-start a frozen battery as this can cause the battery to explode.
- Only charge batteries in a well-ventilated area.
- Following simple battery safety rules and using common sense will prevent battery services from causing accidents and injuries.

■ ***Battery Acid.*** Before attempting to service a battery, inspect it for swelling, damage such as cracks, and leaking electrolyte (**Figure 2-28**). Look for corrosion on the top of the battery, around the cables, and in the battery tray. Do not handle batteries such as those shown in Figure 2-31 until the acid and corrosion have been removed. When handling, use a battery carrying strap to prevent acid contact with your skin and clothing.

■ ***Disconnecting and Reconnecting a Battery.*** **Before removing a battery, make sure the ignition and all electrical accessories are off.** Always remove the battery's negative cable first. This is because if the tool you are using accidently contacts the vehicle, the negative or ground side will not arc or conduct to other grounded parts. Once the negative cable is off, remove the positive cable.

Use a battery strap or other carrying device when moving a battery. Do not pick the battery up with your bare hands or allow the battery to come into contact with your skin or clothing. When reinstalling the battery, install the positive cable first. This is because if the tool you are using contacts metal or another grounded component, the positive will not arc or conduct because the battery is not grounded to the vehicle yet. Once the positive is installed, connect the negative cable.

Safety Data Sheet

Issue Date: 16-Aug-2013 **Revision Date:** 16-Aug-2013
Version: 1

1. IDENTIFICATION

Product Name: OzzyMat® FL-3 Fluid Activation Mat
Other means of identification:
SDS#: CF-010 **Product Code:** FL-3
Recommended Use: Mat and hydrocarbon degrader for the SmartWasher system.
SDS Supplier Address:
ChemFree Corporation
8 Meca Way
Norcross, GA 30093
Company Phone Number: 800-521-7182 or 770-564-5580
Company Fax: 770-564-5533
Emergency Telephone (24 hr): INFOTRAC 1-352-323-3500 (International) / 1-800-535-5053 (North America)

2. HAZARDS IDENTIFICATION

Classification:
This chemical does not meet the hazardous criteria set forth by the 2012 OSHA Hazard Communication Standard (29 CFR 1910.1200). However, this Safety Data Sheet (SDS) contains valuable information critical to the safe handling and proper use of this product. This SDS should be retained and available for employees and other users of this product.
Appearance: Solid filter with granular powder. Buff to brown color in the middle.
Physical State: Solid

3. COMPOSITION/INFORMATION ON INGREDIENTS

The manufacturer lists no ingredients as hazardous according to OSHA 29 CFR 1910.1200.

4. FIRST-AID MEASURES

General Advice: It is unlikely that emergency treatment will be required.
Eye Contact: If adverse effects occur, rinse eyes with large amounts of water until irritation subsides. Get medical attention if necessary.
Skin Contact: Wash with soap and water.
Inhalation: If adverse effects occur, remove to fresh air and observe. Get medical attention if necessary.
Ingestion: Do not ingest.
Most important symptoms and effects: Direct contact with eyes may cause temporary irritation.
Indication of any immediate medical attention and special treatment needed: NOTES TO PHYSICIAN: Treat symptomatically.

5. FIRE-FIGHTING MEASURES

Suitable Extinguishing Media: Use extinguishing measures that are appropriate to local circumstances and the surrounding environment.
Unsuitable Extinguishing Media: Not determined.
Specific Hazards Arising from the Chemical: None.
Protective equipment and precautions for firefighters: As in any fire, wear self-contained breathing apparatus pressure-demand, MSHA/NIOSH (approved or equivalent) and full protective gear.

6. ACCIDENTAL RELEASE MEASURES

Personal Precautions: Use personal protective equipment as required.
Methods for Containment: Prevent further leakage or spillage if safe to do so.
Methods for Clean-Up: Sweep up and shovel into suitable containers for disposal.

7. HANDLING AND STORAGE

Advice on safe handling: Handle in accordance with good industrial hygiene and safety practice.
Storage Conditions: Keep containers tightly closed in a dry, cool and well-ventilated place. Store at 32°F - 105°F.
Incompatible Materials: Water should be avoided until the mat is installed.

8. EXPOSURE CONTROLS/PERSONAL PROTECTION

Exposure Guidelines: No exposure limits noted for ingredient(s) This product presents no health hazards to the user when used according to label directions for its intended purposes.
Appropriate Engineering Controls: None under normal use conditions.
Appropriate Personal Protective Equipment:
Eye/Face Protection: Safety glasses should always be worn in an industrial operation.
Skin and Body Protection: Wear chemical resistant, impervious gloves for routine industrial use.
Respiratory Protection: No protective equipment is needed under normal use conditions.
General Hygiene: Handle in accordance with good industrial hygiene and safety practice.

9. PHYSICAL AND CHEMICAL PROPERTIES

Physical State: Solid	**Odor:** Not Determined	**Odor Threshold:** Not determined
Appearance: Solid filter with granular powder. Buff to brown color in the middle.		**Color:** Buff to brown

pH	Not determined	**Specific Gravity**	0.5-0.7
Melting Point / Freezing Point	Not determined	**Water Solubility**	Forms slurry
Boiling Point / Boiling Range	Not determined	**Solubility in other solvents**	Not determined
Flash Point	None	**Partition Coefficient**	Not determined
Evaporation Rate	Not determined	**Autoignition Temperature**	Not determined
Flammability (Solid, Gas)	None	**Decomposition Temperature**	Not determined
Upper Flammability Limits	Not applicable	**Kinematic Viscosity**	Not determined
Lower Flammability Limit	Not applicable	**Dynamic Viscosity**	Not determined
Vapor Pressure	Not determined	**Explosive Properties**	Not determined
Vapor Density	Not determined	**Oxidizing Properties**	Not determined

FIGURE 2-26 SDS sheets contain important information about chemical use, handling, and disposal.

■ ***Battery Chargers.*** Battery chargers are used to recharge a battery that has become weak or discharged (**Figure 2-29**). During the charging process, hydrogen gas is emitted from the battery. A spark or heat source near the battery could set off a hydrogen explosion, causing the battery to explode also. Follow these steps when charging a battery:

FIGURE 2-27 Batteries are explosive and contain acid, which is corrosive. Never work on a battery without using PPE.

1. Ensure the charger is turned off before hooking the cables to the battery.
2. Connect the positive charger lead to the battery's positive terminal.
3. Connect the negative charger lead to the battery's negative lead.
4. Plug the charger into the electrical outlet.
5. Set the proper charge rate and time on the charger.
6. Monitor battery voltage and temperature.
7. Once charging is complete, turn the charger off, unplug the electrical cord, and then remove the charging clamps from the battery.
8. Wrap the charger power cord and charging cables around the charger handle for storage. Do not attach the charging clamps to the handle.

■ ***Jump-Starting.*** Booster or battery jump-start boxes are often used to jump-start a discharged battery to get a vehicle into the shop. Caution must be taken when using a booster pack because any spark may ignite any hydrogen produced by the battery. Be careful not to hook up a battery backward (reversing the positive and negative leads) as this can cause the battery to explode and/or cause damage to the car's computers and electronics. This would be a very expensive mistake. To use a booster box, connect the positive cable to the battery's positive connection and the negative to a good, nonmoving engine ground. Start the vehicle and let it idle for a moment. Disconnect the negative cable, then the positive cable.

If you are using jumper cables, follow these steps.

- Make sure the vehicles are not touching.
- Connect one positive cable clamp to the booster battery and the other end to the discharged battery.

FIGURE 2-28 Battery corrosion is solidified battery acid, which is highly corrosive. Do not handle a battery without wearing nitrile or similar gloves.

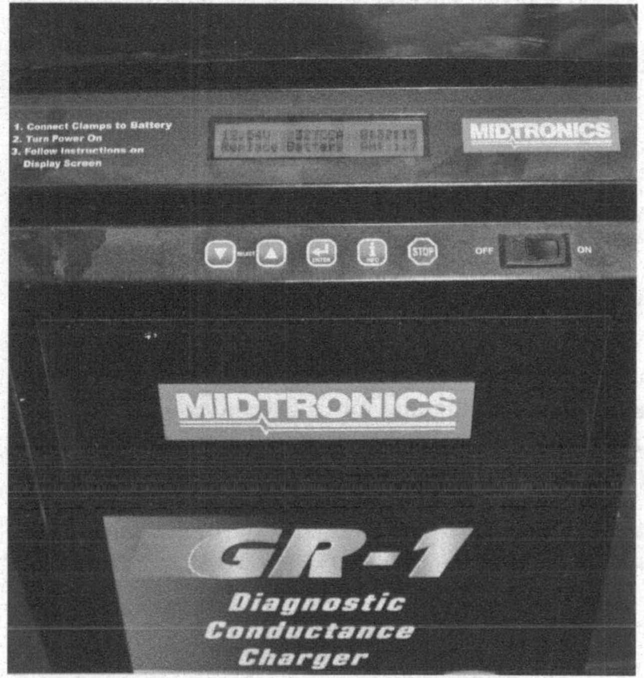

FIGURE 2-29 Battery charging requires special safety. Do not attempt to charge a battery until you are properly instructed.

- Then connect the negative cable clamp to the booster battery and the other end to a good, nonmoving ground on the vehicle being jump-started (**Figure 2-30**).
- Start the engine of the vehicle providing the jump-start, then start the vehicle being jump-started.
- Remove the negative jumper cable from the ground on the vehicle that was jump-started first, then the negative from the booster vehicle.
- Then remove the positive from the jump-started vehicle and then from the booster vehicle.

■ ***Electric Vehicle and Hybrid Vehicle High-Voltage Batteries.*** Electric vehicles (EVs) and hybrid-electric vehicles (HEVs) use high-voltage electric motors, some over 500 volts. EV and hybrid vehicles are discussed throughout this text, but the following are some general hybrid-vehicle safety precautions.

- Never touch or work on any orange-colored high-voltage wiring if the high-voltage system is powered up. An example of the high-voltage orange wiring is shown in **Figure 2-31**.
- Do not attempt to repair any system you have not been properly trained on; the voltages produced and stored in a hybrid can cause serious injury or death.
- Do not attempt to push start a hybrid.
- Do not push a hybrid vehicle or EV with the engine off.
- Follow all manufacturer service warnings and procedures exactly when working on an EV or hybrid vehicle.
- For some repairs, special high-voltage gloves are required. These gloves must be checked and certified every six months.

RUNNING AND MOVING VEHICLES

As part of your job as a technician, you will be working on running vehicles, performing test drives, and working near other running and moving vehicles. To avoid possible injury and damage, there are some safety precautions that must be followed.

■ ***Starting Vehicles.*** Usually starting a vehicle does not present any dangers, but when working in auto shop,

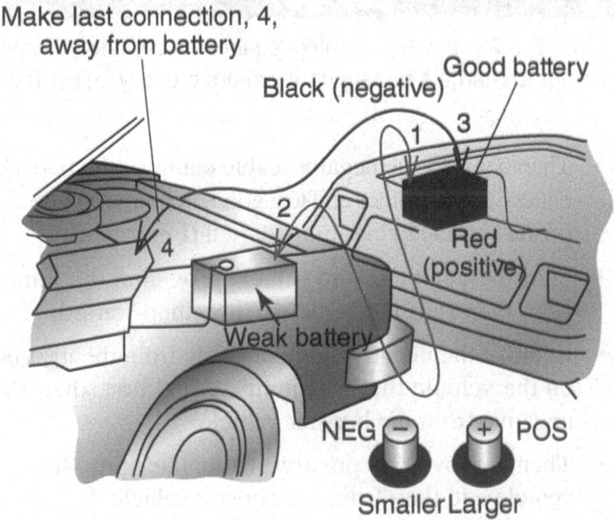

FIGURE 2-30 When jump-starting a dead battery, connect the cables in the correct order to prevent sparks from occurring at the battery.

FIGURE 2-31 The orange wiring signifies the high-voltage components on a hybrid vehicle.

several situations are common that require caution when cranking or starting an engine.

Do not try to crank or start a vehicle that someone else is working on without verifying the following:

- Ensure the fuel system is not open for service. Turning on the ignition causes the fuel pump on most vehicles to operate for two seconds to get the engine started. If the fuel system is open and the key is turned on, fuel can spray from the fuel lines. This can lead to injury or a fire.
- The engine is assembled enough to be cranked. When you are replacing the timing belts and chains, engine damage can occur if the engine is cranked over without the belt or chain installed.
- Tools and other equipment are not under the hood or loose in the engine compartment.

Many shops use signs and steering wheel covers to alert others in the shop not to attempt to start an engine (**Figure 2-32**).

■ ***Carbon Monoxide and Ventilation.*** Many times a vehicle needs to be running in the shop for inspection and service. When the engine is running, the exhaust must be ventilated properly. The exhaust contains **carbon monoxide (CO)**, a poisonous gas. CO is colorless, odorless, tasteless, and nonirritating, so it is impossible to detect it in the air without a CO detector.

When CO enters the body, it combines with hemoglobin. Hemoglobin is the primary oxygen-carrying compound in the blood. When the CO bonds to the hemoglobin, it reduces the oxygen-carrying capacity of the blood. In effect, the CO causes oxygen to stay attached to the hemoglobin instead of being released to the body. This can result in hypoxia, or being deprived of oxygen. Needless to say, oxygen deprivation is not good. Symptoms of CO poisoning include headaches, vertigo, and flu-like symptoms. Excessive exposure can cause death.

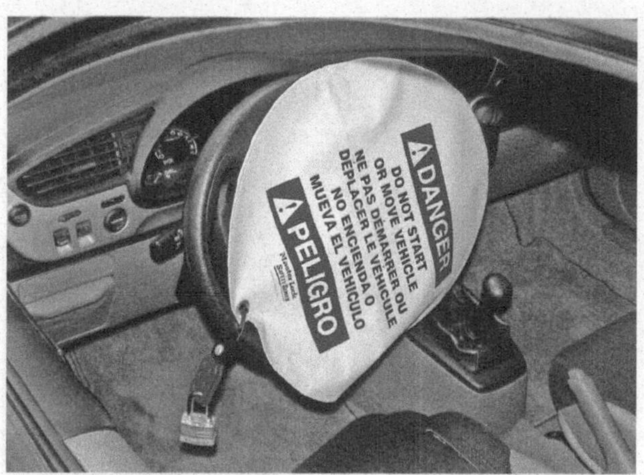

FIGURE 2-32 To prevent accidents, a vehicle lock-out procedure is used.

FIGURE 2-33 Anytime a vehicle is running in the shop, an exhaust hose must be connected to remove the exhaust gases.

To avoid exposure to CO, shops have some type of exhaust gas ventilation system. This is usually a network of hoses, ducts, and a fan to route the gases outside. A hose is attached to the tailpipe of a vehicle to remove the exhaust gas. A typical system is shown in **Figure 2-33**.

■ ***Moving Vehicles.*** There will be vehicles moving in, out, and around the auto shop during the workday. The following are a few simple safety precautions regarding moving vehicles:

- Only fully licensed and insured people should be moving any vehicle.
- Be sure you know how to safely operate the vehicle before attempting to drive. For example, do not attempt to start or move a vehicle with a manual transmission if you do not know how to drive a manual transmission vehicle.
- Make sure others know that you are going to be moving a vehicle before you start moving.
- Remove any wheel chocks and exhaust hoses before moving. Double-check that no tools or equipment are under or near the vehicle.
- Check the brake pedal feel and travel before starting to move. Do a brake check just as you start to move to ensure the brakes work properly. Do not attempt to move a vehicle if the brake pedal goes to the floor.
- If necessary, have a coworker act as a guide or spotter to maneuver around other vehicles, equipment, or other obstacles.
- Honk the horn when passing through garage doors and approaching intersections or areas where people cannot see if a vehicle is coming.
- Many modern vehicles are push-to-start. To operate these vehicles, press the brake pedal then the Start button. To shut down the vehicle, place the transmission in Park or Neutral on a manual transmission, take your foot off the brake, and press the Start button. Make sure the vehicle is off and the dash display shuts down before exiting the vehicle.
- Hybrid and electric vehicles make almost no noise when driven in a shop. Use extra caution when driving a "stealth" vehicle to help prevent injury to a coworker who might not hear you coming.

Chemical Safety

Technicians come into contact with different chemicals, many of which can be dangerous if they are handled or used improperly.

OILS, LUBRICANTS, AND COOLANTS

All vehicles need chemicals to operate. Engine oil is the lifeblood of the engine, and without it, the engine will quickly self-destruct. Lubricants are used in suspension and steering components to allow for smooth operation and to reduce wear. Coolants are used to maintain correct engine operating temperatures, prevent the engine from freezing in cold weather, and provide heat for the passenger compartment.

■ ***Common Automotive Fluids.*** Engine oil becomes contaminated with combustion by-products while protecting the internal engine's components. These by-products can, if allowed to come into contact with your skin, cause skin irritation, rashes, and other health concerns. Studies have shown that high levels of exposure to waste engine oil can cause an increased risk for skin cancer and rheumatoid arthritis. An example of an oil warning label is shown in **Figure 2-34**. Wearing the proper protective equipment, such as nitrile gloves, protects your skin and greatly reduces your exposure.

API Service SN & SM ILSAC GF-5 & GF-4

- *Formulated to Provide Outstanding Wear Protection for Gasoline Engines*
- *Engineered to Extend Engine Life by Reducing Wear and Preventing Sludge Build-Up*
- *Designed to Provide Easier Cold-Weather Starting*

Always consult your owner's manual for specific recommendations.

CAUTION: Avoid prolonged or repeated skin contact with used motor oil. Used motor oil has been shown to cause skin cancer in laboratory animals. Thoroughly wash exposed areas with soap and water. KEEP OUT OF REACH OF CHILDREN. DON'T POLLUTE. CONSERVE RESOURCES. RETURN USED FLUID TO COLLECTION CENTERS. PLEASE RECYCLE CONTAINER.

FIGURE 2-34 An example of a warning label on a bottle of oil. Wear the recommended PPE when handling any of the fluids used in the automobile.

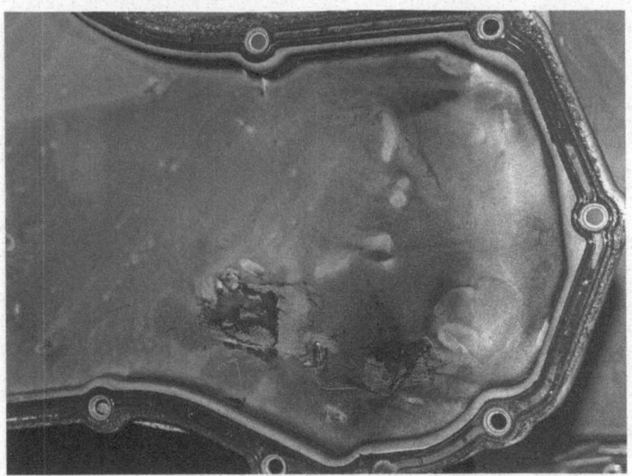

FIGURE 2-35 Automotive fluids pick up all sorts of contaminates. Wear protective gloves when handling all fluids.

Waste oil from diesel engines is especially dirty and can stain your skin. Protective gloves should always be worn when working with used diesel oil.

Transmission fluids, while not as harmful as waste engine oils, should also be handled with caution. Over time, transmission fluid becomes contaminated with very fine metal shavings and clutch disc friction material (**Figure 2-35**). The combination of the fluid and the contaminants can cause skin irritations, so protective gloves should be worn whenever handling these fluids.

Power steering fluid is another petroleum-based fluid. While it is not generally harmful due to contamination, chemical gloves should still be worn to protect your skin. Differential lubricant is another fluid that does not get contaminated like engine oil does. It can, however, have a smell that many find offensive. Wearing chemical gloves can help prevent your skin from smelling like used differential lubricant.

As with all liquids, clean any spills immediately. A small amount of oil on the floor can quickly lead to a fall and serious injury. In addition, waste oils can be flammable and, if left unchecked, can cause a fire if any sparks or heat sources find their way to the spill.

■ ***Lubricants, Cleaners, and Greases.*** Many lubricants and greases are found in the auto shop. Lubricants and penetrants, such as WD-40 and PB Blaster, are often used in day-to-day shop operation. Gloves should be worn when handling these chemicals, and special attention must be paid when using them near sparks or flames.

Many aerosol cleaners used in the auto shop are flammable and can easily catch fire if they are used incorrectly. Read and follow all usage warnings and instructions as printed on the product you are using (**Figure 2-36**).

Even though there are fewer components that require grease during routine services, many vehicles still have ball joints and other components that need an occasional shot of grease. Most automotive greases are safe when they are handled properly, but they can cause both skin and respiratory irritation for those who are sensitive to chemical exposure.

■ ***Coolant Handling.*** Used coolant is a hazardous waste and must be treated as such. Prevent exposure to your skin by using nitrile or chemical gloves when handling coolant. Over time, coolant becomes contaminated with lead, grit, metal particles, and combustion by-products. Handling coolant without the proper protection can cause skin irritation and breathing concerns.

■ ***Coolant Recycling.*** Because waste coolant is a hazardous waste, it is illegal to dispose of improperly. Waste coolant must be reclaimed and recycled. There are three ways waste coolant can be recycled; by an on-site unit, a mobile service, or an off-site service.

On-site recycling requires the use of a coolant reclamation and recycling unit (**Figure 2-37**). These machines, when connected to the vehicle, remove the old coolant and refill the system with fresh coolant. The used coolant is processed through a series of filters to remove the contaminants, then the addition of an additive package restores the coolant so that it can be used again.

A mobile recycling service is one that travels from shop to shop and recycles coolant. This reduces the cost to the shop because there is no recycling equipment to purchase.

Many shops use an off-site recycling service. In this situation, waste coolant is collected and removed by a licensed contractor. The shop pays the contractor to remove and recycle the coolant. This charge is often passed on to the customer as a hazardous waste fee.

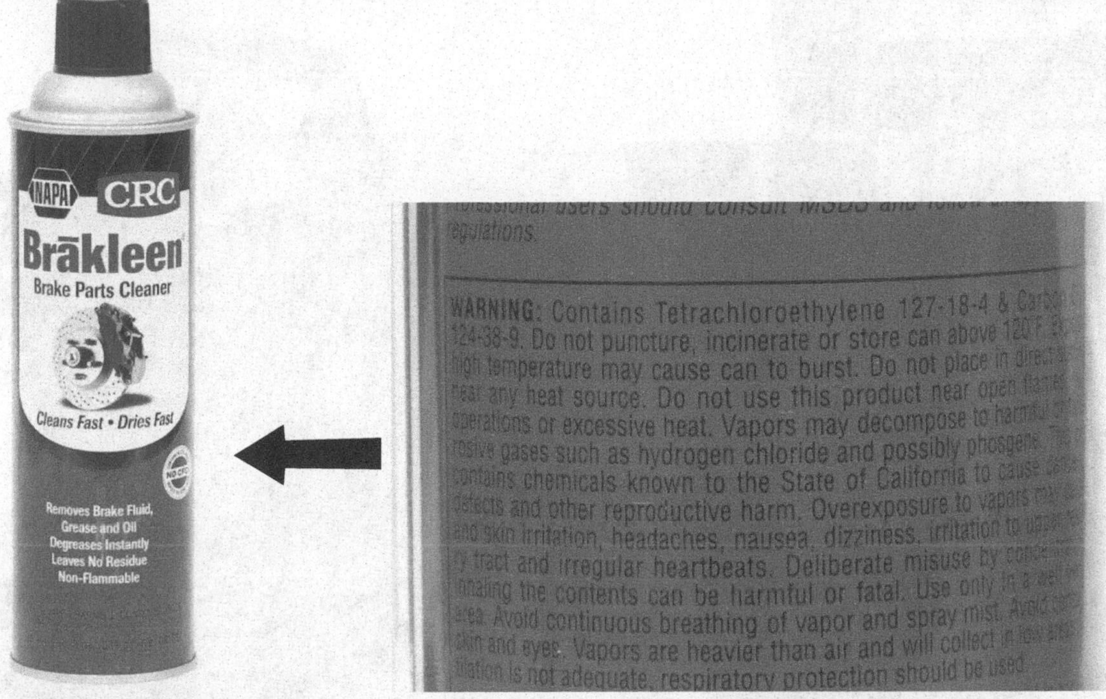

FIGURE 2-36 Always read the warning labels on chemicals before use. Using a flammable cleaning chemical on a hot component can cause a fire.

FIGURE 2-37 Used coolant is collected and recycled. Always follow local laws and regulations concerning used coolant.

■ ***Hazardous Waste Disposal.*** There are many forms of hazardous wastes found in the auto shop, from car batteries to refrigerants. To prevent injury and damage to the environment, each waste must be handled, stored, and disposed of properly. Proper disposal of hazardous waste is the responsibility of everyone in the shop.

Waste chemicals, such as coolant, oils, and refrigerants, must all be stored in the proper type of container. Two common automotive wastes are used oil and antifreeze. Used motor oil is contaminated by combustion byproducts and is considered toxic. In addition, waste oil can also be flammable. Shops have a waste oil storage tank. The storage tank may be connected to a system that allows the waste oil to be pumped from the oil drains to the storage tank (**Figure 2-38**). When the tank is nearly full, a company that handles hazardous waste is called to pump out the waste oil. This oil is then taken to a recycling center, where it can be treated and used for other applications.

Some auto shops have a waste oil heating system. The used oil is stored on-site and used as a fuel for heat during the cold months. Waste antifreeze is collected and recycled similarly to waste oil. Waste oil and antifreeze should never be allowed to mix and must be stored in separate containers. When you are disposing of hazardous wastes, refer to the SDS to determine the proper safe handling procedures for the waste.

Waste engine oil, transmission fluid, power steering fluid, and gear oils can be stored in the same container. This container must also be sealed and marked as waste oils. Under no circumstances can waste coolants and oils be mixed together. The shop hazardous waste contractor may

FIGURE 2-38 Waste oil is collected for shop heating or recycling.

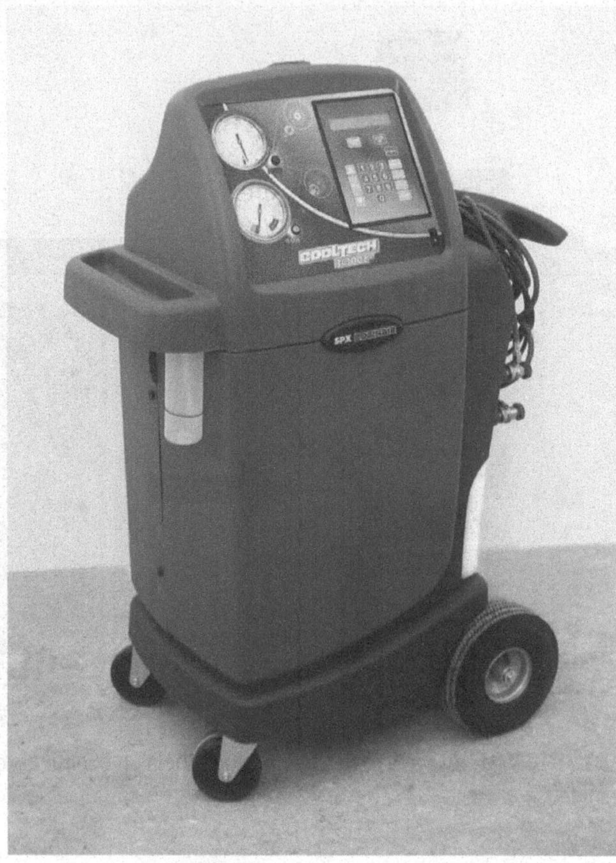

FIGURE 2-39 Refrigerants used in the air conditioning system must be recovered by special service equipment.

not accept any cross-contaminated wastes or may charge many times the standard fees to dispose of the mixed fluids.

Old batteries should be kept on a wood pallet in a well-ventilated area while awaiting pickup by a battery recycler. Do not stack batteries or allow them to tip over.

Recovered air conditioning refrigerants must be stored in approved containers. Many air conditioning service machines not only reclaim and recharge the A/C system but also recycle the refrigerant (**Figure 2-39**). This process removes any moisture and contaminants from the refrigerant, allowing it to be reused. Refrigerants must not be allowed to escape into the atmosphere.

FUELS

Technicians come into contact with gasoline, diesel, and ethanol fuels during many routine services and repairs. All fuels are hazardous, but proper handling and using PPE decrease the chance of injury when dealing with these materials.

■ ***Safety Precautions for Handling and Storing Fuels.*** Fuels present a unique danger for the technician; not only are they hazardous substances, they are flammable. Gasoline and diesel fuels should never come into contact with your skin or eyes, nor should you breathe their fumes. Gasoline can cause skin irritation, dry skin, and dermatitis—an inflammation of the skin as a result of an allergy or irritant. Diesel fuel can contain micro organisms, which can enter your system through open cuts or sores. Always wear protective gloves when handling either gasoline or diesel fuel. If gasoline or diesel gets into your eyes, flush with water immediately and seek medical attention.

Gasoline fumes are extremely flammable and can easily ignite if exposed to an ignition source. Clean any spilled gasoline immediately, and keep the area well ventilated to remove all fumes. Even though diesel fuel is not as easily ignited as gasoline, never expose diesel fuel to any sparks or flame.

Gasoline and other flammable liquids must be stored in a flammable storage cabinet (**Figure 2-40**). This is a special type of storage cabinet designed to contain the fire in the event of ignition.

CLEANING CHEMICALS

Technicians use a variety of cleaning chemicals in their day-to-day work. Always read and follow all instructions regarding the safe use and proper PPE required when you

FIGURE 2-40 Gasoline and other flammable materials are stored in special flammable storage cabinets.

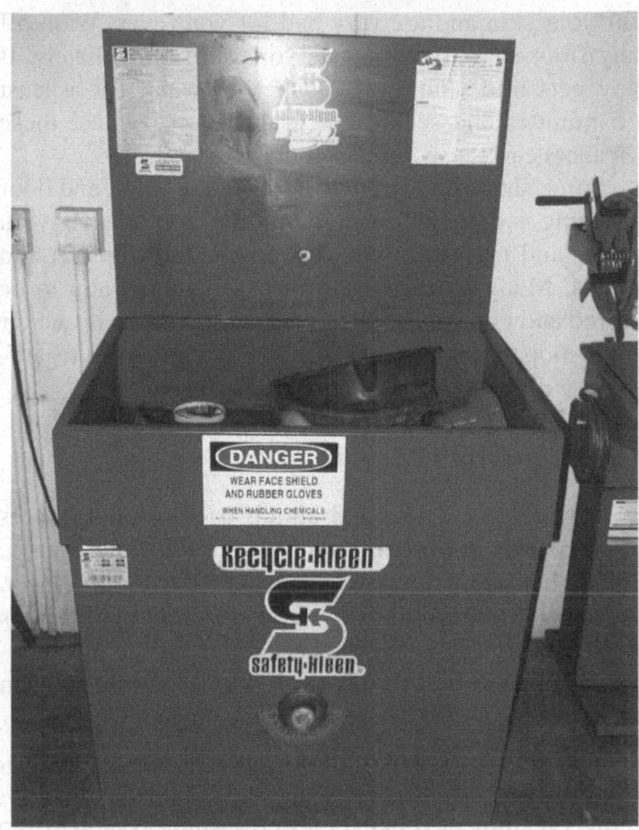

FIGURE 2-41 Cleaning solvents may be petroleum or water based and each had different use and disposal guidelines.

are using these chemicals. The following are examples of some of the more common chemicals you may use.

■ ***Solvents.*** Many shops use a solvent tank or parts washer to remove grease, oil, and other forms of dirt from vehicle parts. There are two basic types of solvent tanks, petroleum-based and water-based. Petroleum-based solvents have been in use for many years and are good at removing most forms of dirt and crud. **Figure 2-41** shows a typical solvent cleaning tank. This type of cleaning system requires periodic replacement of the solvent as it becomes less effective the dirtier it gets. This type of solvent is often picked up and replaced with fresh solvent by a contracted service.

Some systems recycle the solvent in the cleaning unit. Self-recycling systems eliminate the need to remove the old solvent that has become a hazardous waste. Once the recycling cycle has been performed, the contaminants are removed from the solvent. This waste is then drained from the parts washer and poured into the shop waste oil collection tank.

Newer cleaning solvents are aqueous or water-based solvents. These products are biodegradable, and nontoxic, nonflammable, and perform as well as the older petroleum-based systems. This allows a shop to use a more environmentally friendly cleaning system and reduces hazardous waste disposal fees.

■ ***Aerosols.*** Aerosol products, such as brake parts cleaners, lubricants, and penetrants, are commonly used shop supplies. Always read and follow all instructions and precautions listed on the product before using it. Some aerosol cleaners are flammable, while others are not. Using the wrong can of cleaner on a hot surface can result in a fire and injury. Others are made of chlorinated solvents, which, if allowed to burn, create harmful or even deadly fumes. Also, use caution when handling the aerosol cans as the contents are under pressure. Just dropping the can may cause it to rupture, spraying the contents all over you and the shop. When you are using any aerosol, always make sure there is adequate ventilation, not just for yourself but for those around you.

■ ***Floor Cleaners.*** All shops get dirty and need to be cleaned on a regular basis. The better you are at cleaning spills and messes when they occur, the less often you will need to clean the entire shop. But when the need does arise, there are some precautions you should be aware of when using floor cleaners.

Most floor soaps are heavy-duty degreasing agents. While these work well on the floor, they are not good for your skin and are very bad for your eyes. Wash off any floor soap that contacts your skin immediately. If any gets into your eyes, flush with water for at least 15 minutes and seek medical attention because these cleaners can cause burns and damage to the eye.

Some shops do not have drains in the floor, and floor cleaning is done with mops. Mix the floor cleaner with water and mop the floors to remove built-up dirt and stains. Many cities require the used mop water to be stored and removed as hazardous waste. If this is true for your shop, be sure to follow all local regulations regarding the disposal of the used water.

BRAKE AND CLUTCH DUST

■ ***Asbestos.*** Brake and clutch friction components may contain **asbestos**, a compound with microscopic fibers that with prolonged exposure causes lung cancer. Asbestos has been used in many different applications due to its high resistance to heat and its insulating qualities. This high heat resistance is why asbestos has been used in making brake pads and clutch discs.

When you are working on brake or clutch systems, use either a high-efficiency particulate air (HEPA) vacuum containment system or a wet washing system to trap and contain asbestos fibers. Never use compressed air to blow brake or clutch dust. Even though asbestos use has been greatly reduced, there is the possibility of asbestos being present in brake and clutch friction components.

■ ***PPE.*** When you are working on brakes or clutch systems, there is the possibility of encountering asbestos dust. Even if the parts do not contain asbestos, it is a good idea to protect yourself from breathing in the dust created by the brake and clutch friction components. To prevent exposure to this dust, follow these precautions:

- Wear a respirator approved for handling asbestos.
- Use either a negative-pressure HEPA vacuum enclosure or a wet cleaning system (**Figure 2-42**).
- Never have food or drinks in an area where brake or clutch repair is taking place.

■ ***Capturing and Disposal.*** If you are using a brake dust vacuum to capture the dust, the dust will remain in the vacuum until it is time to replace the filters. If you are using the wet cleaning method, place a pan under the brake parts being cleaned and spray the parts with the cleaning solvent or aerosol cleaner. The wet dust will drip into the pan and remain there.

Used vacuum HEPA filters are considered hazardous waste and must be disposed of by a licensed hauler. These filters, along with all collected asbestos dust, must be placed into sealed containers and removed by a licensed hauler.

FLAMMABLE MATERIALS AND FIRE SAFETY

Many of the products auto technicians use on a daily basis are flammable. Understanding the proper storage, use, and cleanup of these chemicals is vital.

■ ***Common Shop Flammables.*** Improper use of chemicals in the shop can result in a fire, injury, or even death. The following is a list of commonly used chemicals that present a fire hazard:

Gasoline	Petroleum-based cleaning solvents
Diesel fuel	Aerosol battery cleaners
Brake cleaner	Aerosol gasket remover
Carburetor and throttle body cleaner	Waste oil
Starting fluid/ether	WD-40

■ ***Fire Prevention and Proper Handling.*** Common sense is required when you are working with many of these flammable substances. For instance, do not spray a flammable chemical on a hot component or near an open flame.

- Before you use a can of spray brake cleaner on the brake parts, make sure the parts have cooled. Spraying brake cleaner on a set of hot brakes can easily result in a fire.
- Never use throttle body, starting fluid, carburetor, or choke cleaner on a hot engine.

FIGURE 2-42 Capturing brake dust with a wet sink.

- If you are using a torch to heat a part, do not spray any penetrating oil around the flame or onto the part being heated.

■ ***Fire Extinguishers.*** All shops are required to have working fire extinguishers, and all employees (and students) should be trained on their proper operation and use.

There are four types of fire extinguishers in use today. Refer to the fire extinguisher chart in **Figure 2-43** for the types and uses of each kind of extinguisher.

To use a typical pressurized extinguisher, follow these steps, called *the PASS method*:

- Pull the pin at the top of the extinguisher (**Figure 2-44**).
- Aim at the base of the fire to extinguish the fuel for the fire.
- Squeeze the lever to discharge the extinguishing agent.
- Sweep the nozzle from side to side until the fire is out.

■ ***Fire Blankets.*** A fire blanket is used in place of a fire extinguisher. The blanket is made of fire retardant materials that smother the fire. To use, remove the blanket from its package and place over the fire (**Figure 2-45**).

■ ***If a Fire Occurs.*** First, do not panic. Many times, a small fire can be safely and completely extinguished with a fire blanket or a standard fire extinguisher. Next, identify the cause of the fire; this is so you can correctly put the fire out. If the fire is a result of a flammable liquid, use a fire blanket or a B, C, or ABC extinguisher on the fuel and put the fire out. If the fire is from an electrical source, you need to remove the power source, which will remove the heat for the fire. If the fire is not extinguished by using the blanket or fire extinguishers, call 911 and report the fire.

	Class of Fire	Typical Fuel Involved	Type of Extinguisher
Class A Fires (green)	**For Ordinary Combustibles** Put out a Class A fire by lowering its temperature or by coating the burning combustibles.	Wood Paper Cloth Rubber Plastics Rubbish Upholstery	Water*[1] Foam* Multipurpose dry chemical[4]
Class B Fires (red)	**For Flammable Liquids** Put out a Class B fire by smothering it. Use an extinguisher that gives a blanketing flame-interrupting effect; cover whole flaming liquid surface.	Gasoline Oil Grease Paint Lighter fluid	Foam* Carbon dioxide[5] Halogenated agent[6] Standard dry chemical[2] Purple K dry chemical[3] Multipurpose dry chemical[4]
Class C Fires (blue)	**For Electrical Equipment** Put out a Class C fire by shutting off power as quickly as possible and by always using a nonconducting extinguishing agent to prevent electric shock.	Motor Appliances Wiring Fuse boxes Switchboards	Carbon dioxide[5] Halogenated agent[6] Standard dry chemical[2] Purple K dry chemical[3] Multipurpose dry chemical[4]
Class D Fires (yellow)	**For Combustible Metals** Put out a Class D fire of metal chips, turnings, or shaving by smothering or coating with a specially designed extinguishing agent.	Aluminum Magnesium Potassium Sodium Titanium Zirconium	Dry powder extinguishers and agents only

*Cartridge-operated water, foam, and soda-acid types of extinguishers are no longer manufactured. These extinguishers should be removed from service when they become due for their next hydrostatic pressure test.

Notes:

(1) Freeze in low temperatures unless treated with antifreeze solution, usually weighs over 20 pounds, and is heavier than any other extinguisher mentioned.

(2) Also called ordinary or regular dry chemical (solution bicarbonate).

(3) Has the greatest initial fire-stopping power of the extinguishers mentioned for class B fires. Be sure to clean residue immediately after using the extinguisher so sprayed surfaces will not be damaged (potassium bicarbonate).

(4) The only extinguishers that fight A, B, and C class fires. However, they should not be used on fires in liquified fat or oil of appreciable depth. Be sure to clean residue immediately after using the extinguisher so sprayed surfaces will not be damaged (ammonium phosphates).

(5) Use with caution in unventilated, confined spaces.

(6) May cause injury to the operator if the extinguishing agent (a gas) or the gases produced when the agent is applied to a fire is inhaled.

FIGURE 2-43 Common classes of fire extinguishers and the types of fires each is used for.

FIGURE 2-44 Be sure you know how to use a fire extinguisher before you actually have to put out a fire.

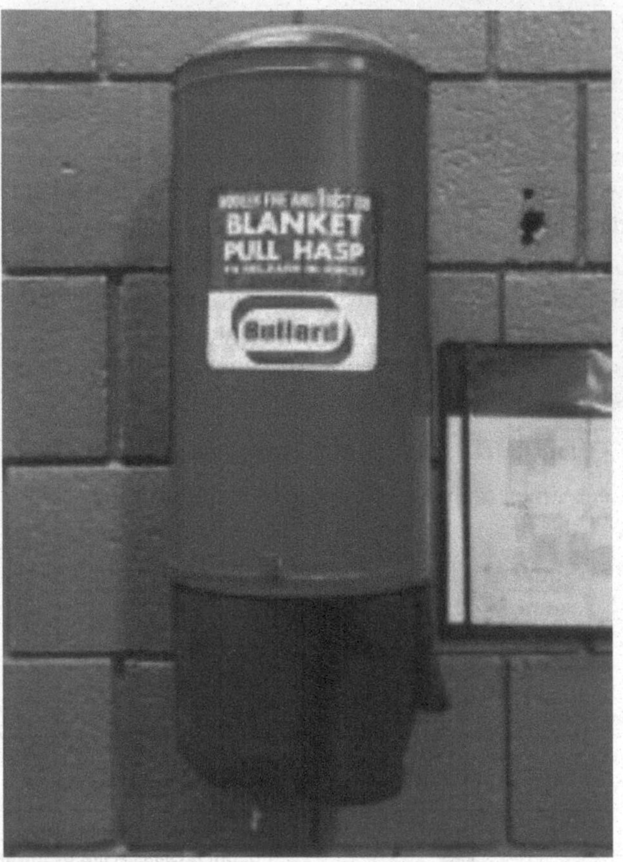

FIGURE 2-45 Fire blankets can be used to smother a small fire.

If you are able to put out the fire, make sure everyone is safe and unharmed, and then assess the damage. You may need to keep a watch on the area in which the fire occurred to make sure it does not restart.

In the event you need to evacuate the shop, leave by the nearest exit. To comply with fire department regulations, a fire evacuation route is typically posted in plain sight (**Figure 2-46**).

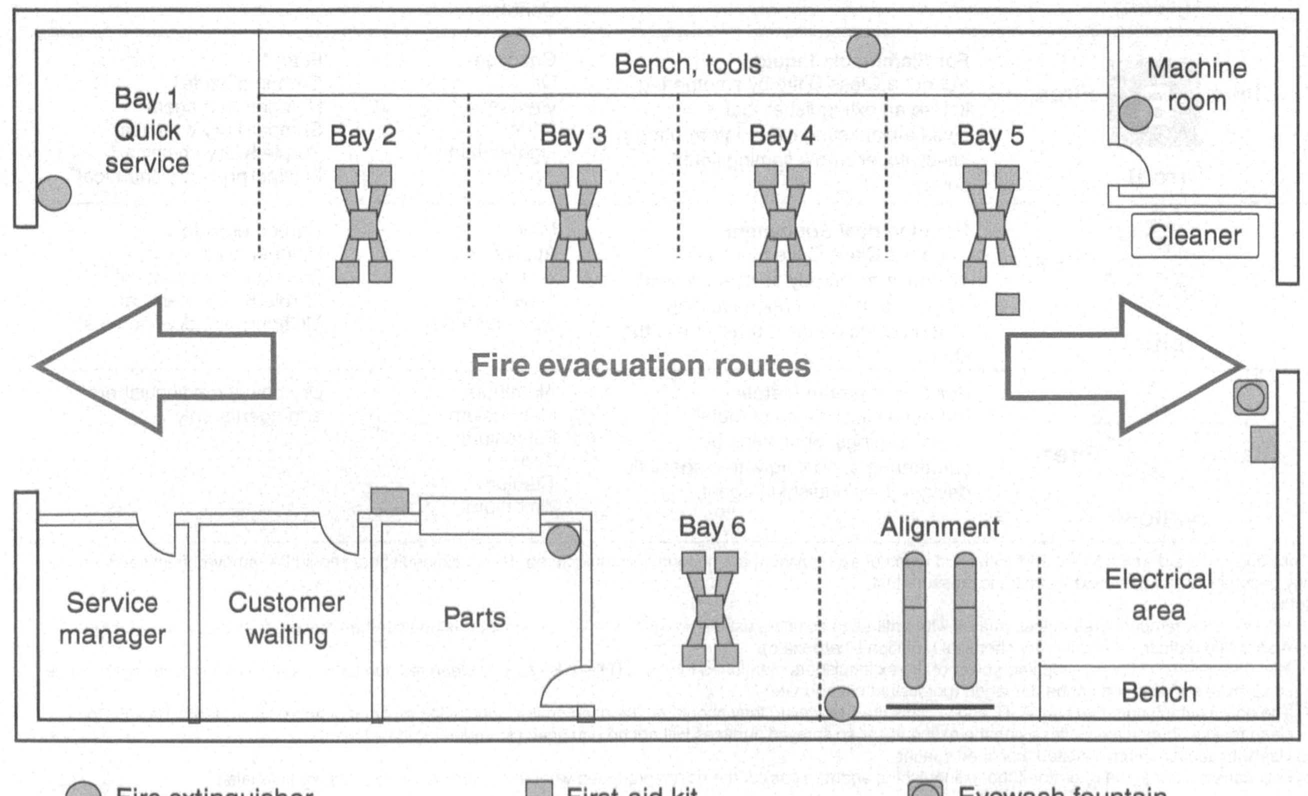

FIGURE 2-46 A fire evacuation route will show the floor plan of the building and the nearest exits.

SUMMARY

Safe working habits are developed by always following the rules and precautions for the auto shop. Over time these habits will become second nature to you, and you may begin to take safety for granted—don't! You will need to keep safety foremost in your mind to keep yourself and those around you safe. Remember these general rules for safe conduct in the auto shop, and you should have many years of accident-free work ahead of you.

Wear clothing that fits properly, neither too tight nor too loose.

Safety glasses should never be an option. Always properly protect your eyes.

Use gloves to protect your hands from cuts, burns, and chemicals.

The auto shop is not a playground, and you should never act in any manner that is dangerous to yourself or others around you.

Never support a vehicle with only a floor jack. Always use jack stands.

Check and double-check any vehicle raised on a lift before working underneath it.

Make sure all guards and shields are in place on bench grinders.

Do not use blow guns and compressed air to clean your clothes or your skin.

Know how to use the emergency shower, eyewash, and first-aid kit.

Do not attempt to handle any chemical without first reading the MSDS and understanding how to properly protect yourself.

Never lay tools on or near a battery.

Never expose a battery to a flame or spark as this can cause the battery to explode.

Never touch the orange high-voltage wiring found on hybrid vehicles. Never run a vehicle in the shop without connecting an exhaust hose.

Check the brakes before moving any vehicle.

Always locate and use the proper PPE whenever handling waste fluids or other chemicals.

Follow approved methods when working around brake or clutch dust.

Know the location, type, and proper use of the fire extinguishers in your shop.

REVIEW QUESTIONS

1. Carbon ______________ is a colorless and odorless gas formed during combustion that can cause illness and death.
2. High-voltage wiring is indicated by ______________ ______________ colored conduit and connectors.
3. Safety ______________ ______________ provide information about how a chemical should be stored, handled, and disposed of.
4. ______________ ______________ ______________ are the items you wear to prevent injury while working in the automotive shop.
5. When disconnecting a battery, disconnect the ______________ terminal first and then the ______________ terminal.
6. Two technicians are discussing shop chemicals PPE: *Technician A* says wearing mechanic's gloves prevents skin absorption of chemicals. *Technician B* says that a respirator should be worn when working with certain chemicals. Who is correct?
   a. Technician A
   b. Technician B
   c. Both A and B
   d. Neither A nor B
7. *Technician A* says that prescription eye glasses will protect you against shop hazards. *Technician B* says that work boots must meet the ANSI Z.87 safety standard. Who is correct?
   a. Technician A
   b. Technician B
   c. Both A and B
   d. Neither A nor B
8. All of the following statements are correct except:
   a. Long hair can be caught in moving machinery.
   b. Always use jack stands to support the vehicle when lifting with a floor jack.

c. Common aerosol chemicals are often flammable.

d. Waste oil and antifreeze can be disposed of in the same container.

9. *Technician A* says that corrosive wastes include items such as car batteries. *Technician B* says toxic wastes, such as used coolant, must be collected for recycling. Who is correct?

a. Technician A

b. Technician B

c. Both A and B

d. Neither A nor B

10. *Technician A* says that the SDS enforces the employee Right-to-Know standard regarding shop safety. *Technician B* says that OSHA is responsible for environmental safety regulations. Who is correct?

a. Technician A

b. Technician B

c. Both A and B

d. Neither A nor B

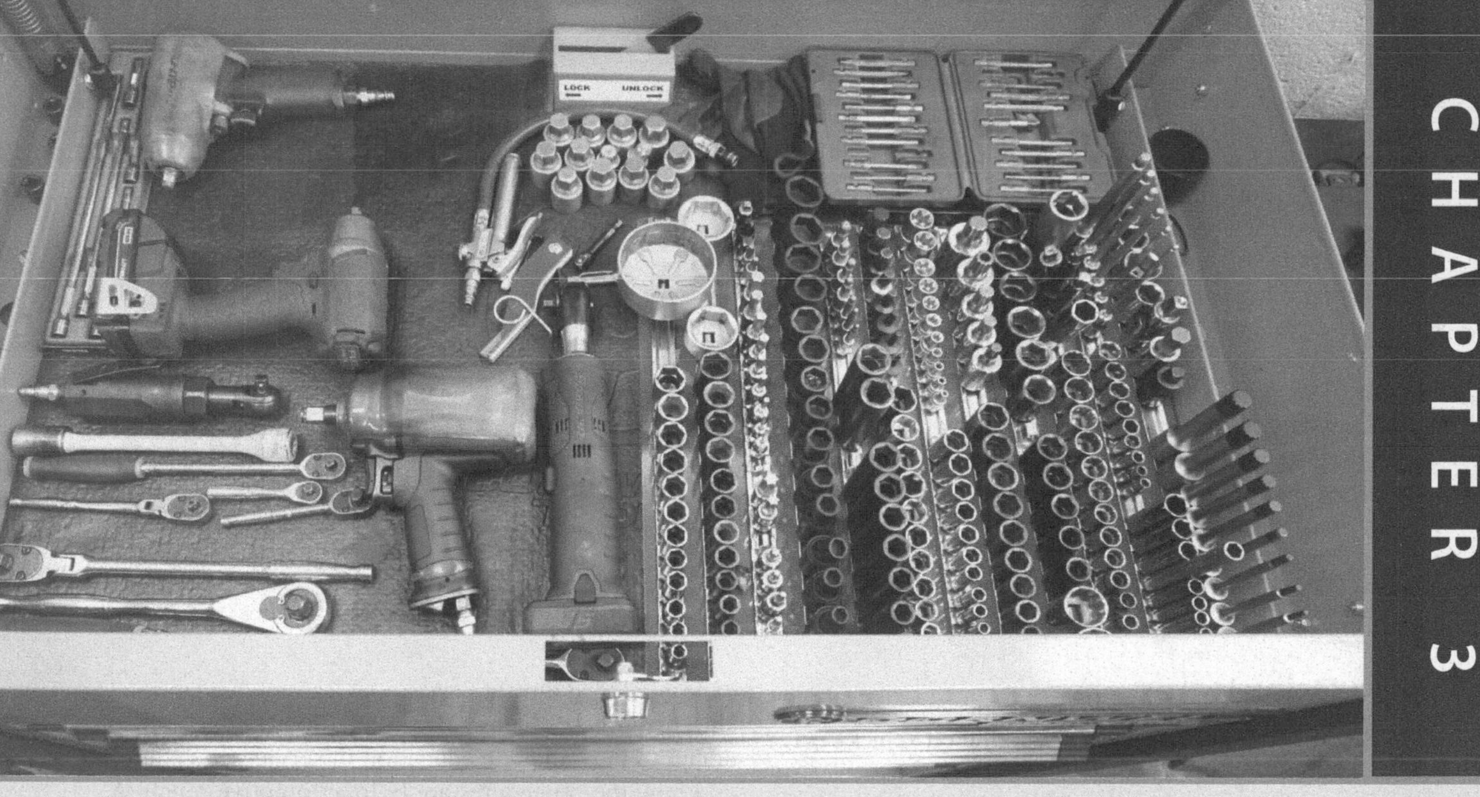

CHAPTER 3

Shop Orientation

Chapter Objectives

At the conclusion of this chapter, you should be able to:

- Identify the job requirements, routines, and housekeeping procedures for automotive technicians. (ASE Education Foundation RST Safety #1)
- Identify basic hand and power tools and the correct operation for these tools. (ASE Education Foundation RST Tools #1, 2, 3, & 4)
- Demonstrate safe operation of vehicle hoists and jacking equipment. (ASE Education Foundation RST Safety #3 & 4)
- Identify and locate vehicle identification information. (ASE Education Foundation RST Service #1)
- Identify and complete a repair order and maintain customer and technician records. (ASE Education Foundation RST Service #1 & 5)

KEY TERMS

above-ground lift
air hammer
battery charger
bench grinders
calibration decal
chain of command
combination wrench
dead blow hammer
emissions decal
entry level
impact wrench
in-ground lift
parts washer
pliers
punches
repair order (RO)
rubber mallet
screwdriver
service order
sockets
VIN
vise

While the automotive lab is often designed to mirror working in a real repair facility as much as possible, some differences are worth noting. Automotive training classes typically perform limited types of services on real customers' vehicles. This is because the program instructors seek to provide the right kind and amount of experience to help the students. This place limits on what actual repairs can be made by the students so that they are not overwhelmed or put into situations in which they are not confident in performing certain tasks. This is different from how many repair facilities operate, especially new car dealerships, where the shop and its technicians are expected to be able to accommodate whatever services and repairs are necessary.

FIGURE 3-1 An example of a college automotive classroom. This room is also used for GM ASEP students.

Shop Orientation

Even if you have prior experience working on vehicles or even working in a shop, there are going to be certain rules, procedures, and practices that you will need to learn for each shop in which you work. Orienting yourself to how a shop operates is necessary so that you become a productive and important part of the overall team.

Part of shop orientation is to help the instructor(s) learn what experiences the students have. While it is always a bonus to have students who already have some experience, it is important for those students to be open-minded about learning or relearning certain tasks. Just because you or other students in the class may have done automotive work in the past, does not mean that it was done correctly and to professional standards. One of the main goals of the program instructor is to teach you how to perform repairs safely and by the manufacturer's recommended service procedures. Even if you know how to perform a certain task more quickly by taking a shortcut, that does not mean it will always be the best way to do that particular job. In fact, shortcuts performed incorrectly can often lead to additional problems to fix.

LAB AND SHOP OPERATIONS

Working in an automotive lab is different from working in an actual full-service shop. This is because the automotive technology course focuses on training students safely and properly so that they may become productive technicians. In the full-service shop, the expectation is that the technicians will be able to make a profit for the business. Because of these different expectations, the lab and the shop, while sharing many similarities, will have some operating differences.

■ ***Lab Operation.*** Many factors determine how a lab operates. Among these are location, student population, physical facilities, equipment, and funding. Many automotive technology programs provide both classroom and lab-based instruction. Future technicians should expect to spend time in a classroom learning fundamental skills and theory. This "classroom" knowledge is then added to the hands-on education in service procedures and how the procedures apply to automotive repair. A classroom may be used to have large group discussions, demonstrations, and instruction (**Figure 3-1**). The lab is where students will learn to apply classroom concepts and acquire hands-on skills (**Figure 3-2**).

For training purposes, there should be a good supply of vehicles on which the students can practice their hands-on skills. Vehicles may be owned by the school or may consist of live work, meaning cars and trucks owned by customers who bring them in for service. Students should have adequate time to practice their skills before applying them to real-world vehicles, though.

An auto tech lab will typically consist of classroom discussion or instructor demonstration followed by student practice. It is important for the students to have the necessary time to be able to work with the parts and systems being taught so that students develop an understanding of how each works together in a system and within the scope of the entire vehicle.

Additionally, you will spend time each day in the care and cleanup of the lab and equipment. In most repair shops, technicians are responsible for maintaining and

FIGURE 3-2 An example of a college automotive lab.

cleaning their own areas. This means that floors, tools, and equipment are cleaned regularly, and workbenches are kept clean and orderly.

■ ***Shop Orientation.*** A real-world repair facility will operate differently than an auto tech class and lab. The primary objectives in a teaching environment are safety and education. In a repair shop, the primary goals are to service and repair vehicles as quickly and efficiently as possible and to make a profit. While there should be a real-world correlation between the auto course and the repair shop, there will be differences in how each operates.

In a teaching environment, the instructor and school staff are the people in charge. The students are similar to the shop employees and may have some input in the operation of the lab, but probably not a large amount. Students have to live by a code of conduct, school rules, and shop policies. Violation of those rules can result in discipline and loss of privileges.

In a shop, there may be the shop owner, store manager, general manager, or service manager who is in control of the operation. The employees may have varying degrees of input in the shop operation. One big difference between school and work is that violations of the shop rules and policies can result in the employee being fired.

Most students who are enrolled in or have completed an auto tech program often start in a repair shop as an entry-level employee. This often means that, since that person is an unknown quantity to the employer, the new employee must first complete a probationary period. During this time, which can last from 30 to 180 days, the employee is closely monitored for attendance, punctuality, dependability, and initiative, in addition to the performance of service and repair tasks. Once the employee proves he or she is responsible and an asset to the shop, an increase in pay may follow.

Do not expect to be hired immediately as an experienced technician, nor expect to be offered huge amounts of money when you are just entering the field. Even though you may have some experience, that usually does not mean much at the beginning. The owner or manager of the shop is taking a risk each time a new employee is hired. A technician may talk the talk, meaning that it sounds like he or she is experienced and knowledgeable, but in reality may not be. It takes time to demonstrate to the owner or manager that you can work safely, productively, accurately, and well with other employees.

DAILY ROUTINES

Most auto tech courses have some type of daily routine that students follow. These routines will vary depending on the time of day the class meets, how many students are in the class, and the facilities in which the learning takes place.

■ ***Starting the Day.*** Typically, the day will start with classroom time for attendance, assignments, lectures, demonstrations, and other items that may need to be addressed. Lab time may be structured by assignments, objectives, or projects. The end-of-the-class time usually includes time to clean up and restore the lab and equipment.

Daily shop routines may start with employee meetings, team meetings, and work assignments. Many shops require the technicians to be responsible for keeping their work bays cleaned and organized, so the day typically concludes with some cleaning and preparing for the next day.

■ ***Clean Up.*** A clean shop is much easier to maintain, is more productive, and is more professional than a cluttered and dirty shop. Regardless of whether you are in a 50-year-old auto tech lab or a new state-of-the-art repair facility, cleanliness is everyone's responsibility. Students and employees are responsible for maintaining their own areas (**Figure 3-3**).

Start by making sure your tools and the shop's tools are clean, organized, and returned to the proper locations. When you are working as a technician, your tools are used to earn your income. To be productive, keep your tools clean and organized. If you have to spend time looking around in a messy toolbox looking for a tool, you are losing money. Just placing the tool on top of the toolbox or bench in the tool room is not the same as returning it to its proper place. Each tool should be returned to the specific location where it is stored.

In addition to your tools, keep your immediate work area clean and organized. This not only helps your productivity, it presents a positive, professional image to those around you.

■ ***Chain of Command.*** In a repair shop, the **chain of command** may be as few as just two people or a half-dozen levels between the technician and the upper management. Regardless of the depth of the chain of command, it is important to understand who is responsible for what and to whom. A technician working in a new car dealership may have an immediate supervisor, such as a team leader, who reports to a shop supervisor, who reports to the service manager, who reports to the service director, who reports to the general manager. Even though you may only deal with your immediate supervisor on a day-to-day basis, it is important to know who is also in your chain of command in the event that your supervisor is absent for some reason (**Figure 3-4**).

After you have been working in a position for a while, you may be placed in a supervisory position over new employees or even be in charge of their training. If you

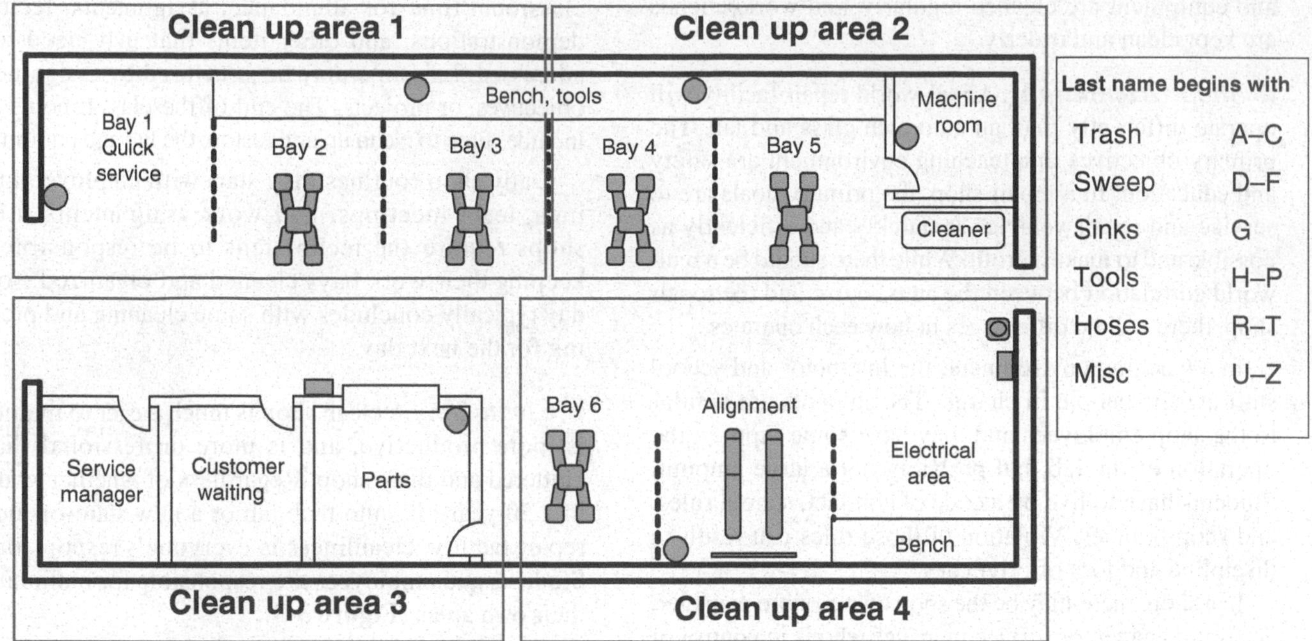

FIGURE 3-3 An example of a shop clean-up plan.

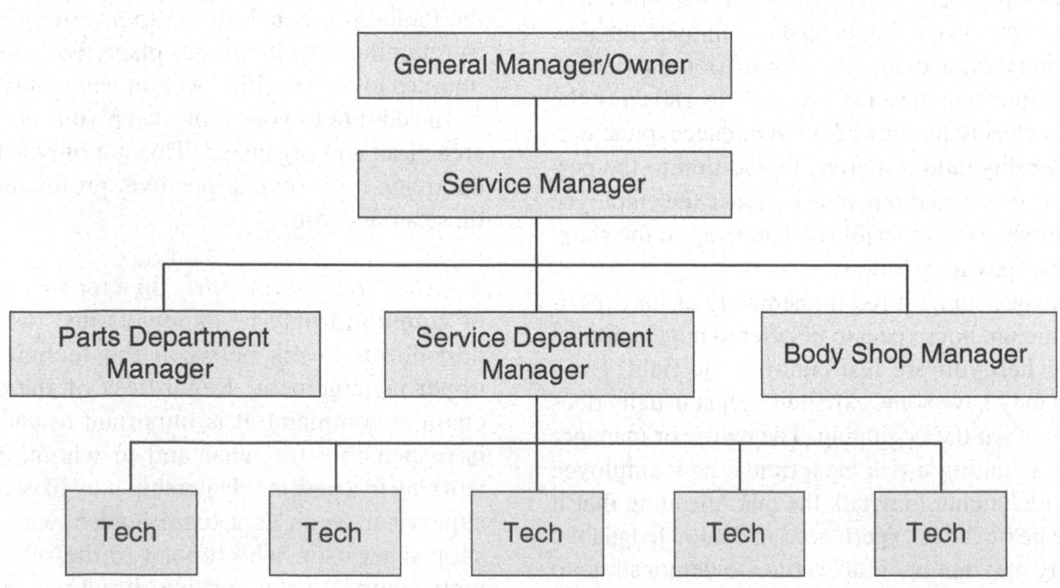

FIGURE 3-4 The chain of command may be simple or complex depending on your place of employment. What is important to know is who you report to and who to go to with any problems.

do supervise others, be sure to treat them with the same respect with which you treat your supervisor. Make sure that they understand the day-to-day operation of the facility and how to perform their jobs. You may be evaluated on how well you do on training new employees so do the best that you can.

JOB REQUIREMENTS

The requirements and expectations for entry-level jobs differ from those for an experienced technician.

■ ***Entry-Level Jobs.*** All technicians start out somewhere at the entry level. **Entry level** means starting in the lowest-skill and lowest-paid position. In the automotive service industry, this can mean starting work in several different ways, including as a technician's assistant, an express lube technician, or in new car preparation (**Figure 3-5**).

Some shops start their entry-level workers by assigning them to a more experienced technician who oversees their work and helps them learn to function in their new job. This mentoring is an excellent way to learn your new job and to learn from an experienced technician. Mentoring also provides a safety net of someone who watches over the work you perform.

In many shops, entry-level employees have very limited responsibilities for servicing vehicles; instead they are first used to help with getting parts, moving vehicles, shuttling customers to and from the shop, and performing housekeeping duties. This allows the employer to see how they perform doing mundane tasks and provides insight into their work ethic. An employee who shows up on time every day, does all that is asked, and shows initiative is considered an asset and will usually be given more and more responsibility.

Many students start in the shop as a lube tech. This job performs a lot of engine oil changes, tire rotations, and basic inspections. Even though this is an entry-level position, it requires skill, concentration, and the ability to multitask. Once a technician proves himself or herself as a dependable lube tech, he or she is often provided training and advancement opportunities within the shop.

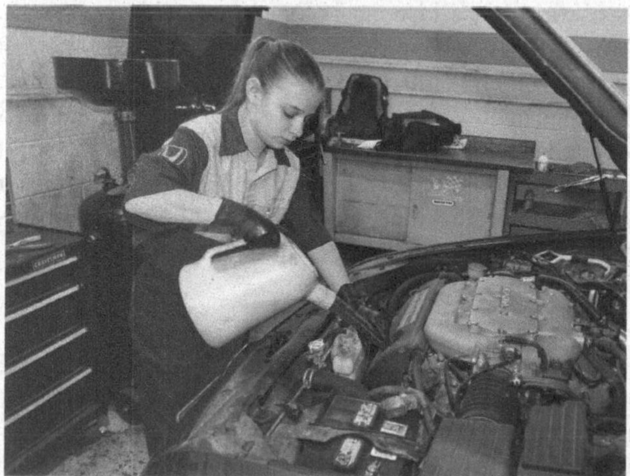

FIGURE 3-5 Entry-level technicians are often given basic tasks and have the lowest pay in the shop until they prove themselves as dependable employees. This technician works on an express lube rack at a Honda dealership.

At some shops, particularly at new car dealerships, a new employee may start by being assigned to new car preparation, often called *a predelivery inspection (PDI)*. This requires getting vehicles that have just arrived from the manufacturer ready for sale. Often the PDI technician is responsible for removing exterior and interior protective coverings, installing hubcaps, checking fluids and tires, washing the vehicle, and other similar jobs.

The lube tech and PDI positions are often used to weed-out those who are unable or unwilling to perform their jobs well and handle more responsibility. Employers use these entry-level jobs as a way to determine the character and work ethic of the employee. Once you prove you are a productive asset to the company, better opportunities may arise.

It is possible that a newly hired entry-level technician will have to earn a position in the shop. Every new employee is a risk for the employer since there are so many unknown qualities about the person. Until the employee demonstrates acceptable work behaviors over a period of time, many shops will not risk putting that person into a service position.

While it is important for the new employees to want to demonstrate their work ethic and skills, it is very important not to be boastful, making claims of knowledge and experience that are not true. The employer wants and expects honesty. If you cannot perform a certain skill, you need to make sure your employer knows that ahead of time. Do not exaggerate your abilities because when it comes time to show what you can or cannot do, your claims of skills that you have not acquired will likely cause your employer to reevaluate you and your employment.

■ ***Lifelong Learning.*** If you are serious about being an automotive technician, then you need to accept the necessity of lifelong learning. This means that you are not done with school upon graduation. Once you are working in the industry, you will need to update your skills as cars and trucks change from year to year. To do so will require that you attend training classes provided by your employer (if offered) or by locating and attending classes on your own. Technicians who are working in new car dealerships take online training and attend update classes at regional training centers (**Figure 3-6**). This training is necessary so that the technicians are familiar with the new models, systems, and components that appear each year. Technicians who work for independent and national corporations should also attend classes to remain current.

Training is like adding tools to your toolbox. As you accumulate more tools, you can perform more types of repairs, which allows you to earn more money. Training is basically the same. The more knowledge you have,

Training Path

My Status | NATEF - Maintenance and Light Repair Specialist

Required

Course	Status	Action	Tablet Compatible	Delivery Method	Test Out
Professional Skills Assessment	Passed			BWBT	Test Out
Time Management	Passed			BWBT	Test Out
Problem Solving		Launch		BWBT	Test Out
Continuous Improvement		Launch		BWBT	Test Out
HVAC Systems Operation		Launch		WBT	

FIGURE 3-6 Staying in the auto industry means that you will have to continually update your knowledge and skills as new vehicles and technology are released. This typically involves a lot of online training before being sent to a training center.

the more skilled you become and the more money you can make. Technicians who do not believe in investing in proper tools or training are fooling themselves and are placing serious limits on what they can accomplish.

Technician's Tools and Their Use

Technicians can accumulate thousands of dollars in tools, from the most basic screwdrivers and pliers to specialty tools with only one purpose. Regardless of quality or brand name, take proper care of your tools, cleaning and maintaining them on a daily or weekly basis. A tool that is broken or damaged can result in lost productivity and could even cause personal injury or damage to a vehicle.

BASIC HAND TOOLS

Wrenches, sockets, pliers, screwdrivers, and similar tools are called *hand tools*. This is because they require hand power to operate. Even though many of these tools have not changed in many years, improvements in design and quality have changed.

Once you know what the tools are and what each tool does, you then need to be able to use the tool properly. Using a tool improperly can cause tool damage or failure, damage to the vehicle, and personal injury. **When you are loosening or tightening a fastener, especially a very tight fastener, pull the tool toward yourself instead of pushing away.** If the tool slips off or the fastener breaks free, applying a lot of force while pushing away can cause you to follow the movement of the tool, possibly injuring yourself.

The following are basic tools found in a technician's toolbox (**Figure 3-7**).

■ ***Wrenches*** The most common type of wrenches are **combination wrenches** (**Figure 3-8**). These have an enclosed end called a *box end* for high-torque uses, such as when you are loosening or tightening a fastener. The opposite end is called *the open end.* The open end is useful once a fastener is loose to quickly remove it. Most wrenches have the box end at an angle to the rest of the wrench (**Figure 3-9**). This allows extra clearance for your hand when you are working. The box end of a wrench may have six points or 12 points. A six-point provides better grip, while the 12-point allows for easier access since the increased number of contact points allows for working in tighter spaces (**Figure 3-10**).

Tool Use.

Always use the box end of a wrench when you are attempting to loosen a tight fastener. This is because the box end has more contact with the fastener and is less likely to slip off. Trying to loosen a stuck bolt or nut with the open end of a wrench will often cause the wrench to slip off, rounding the head of the fastener and possibly hurting your knuckles. The same applies if you are using a wrench to tighten a fastener.

FIGURE 3-7 A technician's tools are as important as the knowledge and skills he or she has. Taking care of your tools is a priority as a technician.

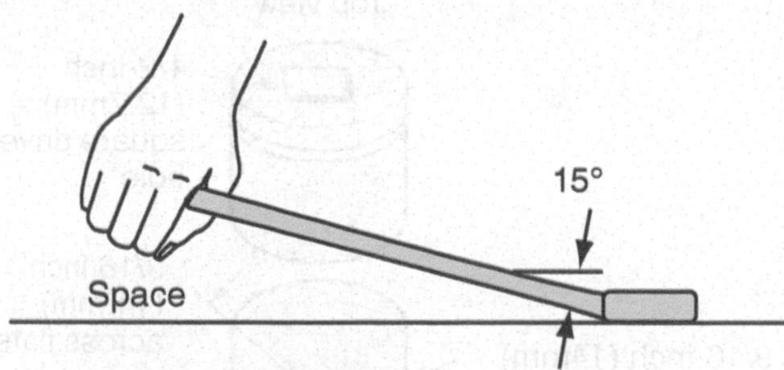

FIGURE 3-9 Most combination wrenches are offset to allow for clearance when working on a vehicle. When loosening or tightening a fastener, pull the wrench toward your body instead of pushing away. This can save you from injury if the tool slips or the fastener lets go.

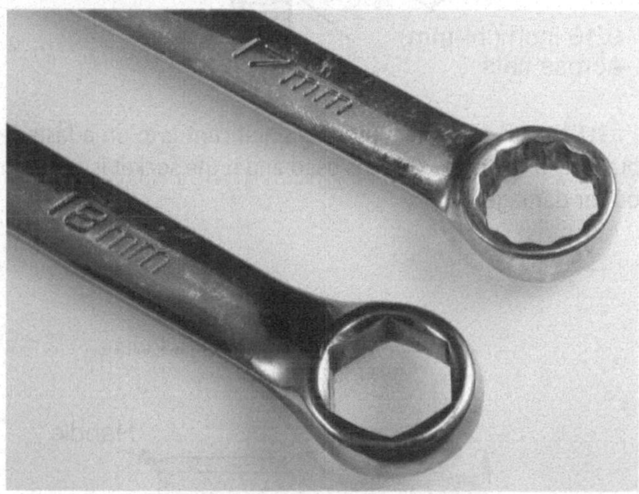

FIGURE 3-10 Examples of 6- and 12-point box end wrenches.

■ ***Sockets and Ratchets.*** **Sockets** are used with a ratchet to quickly and safely loosen and tighten fasteners. Sockets come in deep, mid, and shallow depths, 6 and 12 points, triple square, chrome and impact grades, and in various drive sizes. Sockets allow better contact with the fastener than open-end wrenches and can be combined with extensions and other tools to gain better access and/or leverage (**Figure 3-11**). The most common drive sizes for sockets are ¼, ⅜, and ½ inch. Technicians working on heavy duty trucks and commercial equipment often use ¾- and 1-inch drive socket sets. An example of a ½-inch drive is shown in **Figure 3-12**.

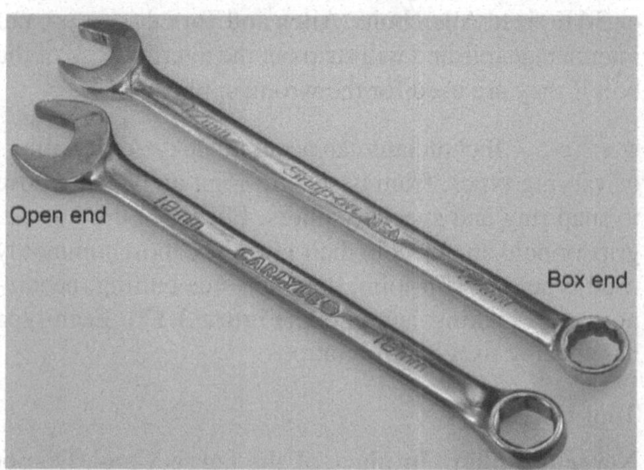

FIGURE 3-8 Combination wrenches are still common technician tools. Many different varieties of wrenches are available.

The ratchet allows for quick switching from one size socket to another and is a means to switch from tightening to loosening with just a click of the ratchet selector (**Figure 3-13**). Ratchets should be cleaned and lubricated regularly to stay in proper working condition. Sockets should be checked for cracks and worn teeth and replaced as needed. Socket sets usually include a variety of extensions in different lengths for getting additional working room or clearance.

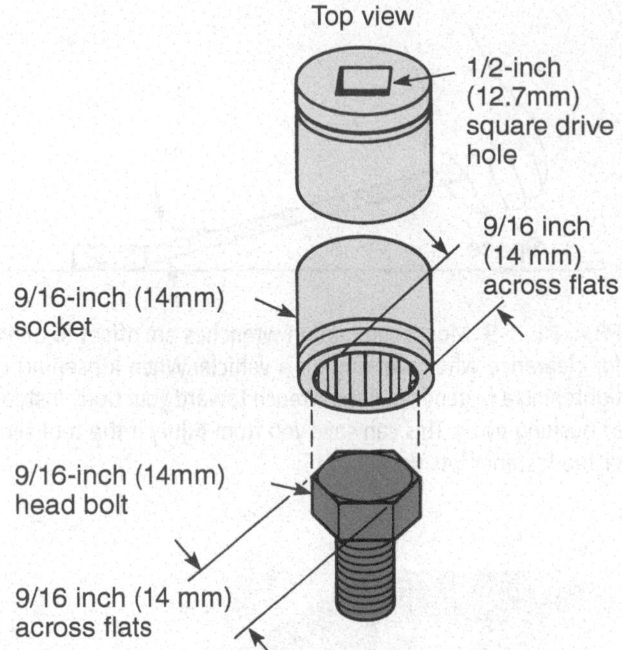

FIGURE 3-11 Sockets provide significant grip on a fastener but only if the correct size is used and if the socket is not worn out or damaged.

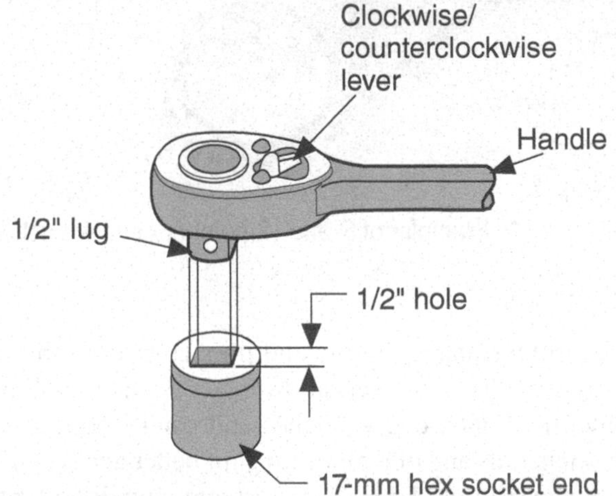

FIGURE 3-12 A ratchet is used to drive a socket either clockwise or counter clockwise.

Tool Use.

First, make sure you are using the correct size socket for the fastener. A socket should fit closely onto the fastener with very little or no slop or wiggle when attempting to rotate the socket on the fastener. Chose the correct size ratchet, ¼-inch, ⅜-inch, or ½-inch drive based on the size of the nut or bolt. Using a ¼-inch drive to try to remove a lug nut is not going to be effective and using the ½-inch drive to tighten very small bolts can cause you to overtighten and break them off. Use only enough of an extension as needed. Adding extensions and universal joints can change operating angles, causing you to slip off the nut or bolt and get injured.

■ ***Screwdrivers.*** The most common **screwdrivers** used in auto repair are the standard or straight, the Phillips, and Torx. Many technicians use cordless screwdrivers or drills with driver bits for work with screwdriver types of fasteners. **Figure 3-14** shows a Phillips screwdriver, and **Figure 3-15** shows a Torx screwdriver.

Tool Use.

Do not use a screwdriver as a pry bar, chisel, or punch. This can easily damage the screwdriver and the vehicle. Use the screwdriver that has the best fit to the screw head. There are three common Phillips head screwdrivers, the number one, number two, and number three. A number one Phillips is the smallest of the three, while the number two is the most commonly used size. The number three Phillips screws are not common automotive fasteners. When you are using a screwdriver to tighten fasteners and hose clamps, do not overtighten the screws. Screws are often used on plastic parts, and they can easily break if they are overtightened.

■ ***Drivers.*** Many fasteners use recessed or internal six-sided drivers or Allen wrenches, Torx drives, or triple square. These are commonly found on brake caliper bolts and engine fasteners. Allen drivers can be English or metric sizes, while Torx drives can be found in a tamper deterring safety Torx configuration. **Figure 3-16** shows examples of various driver bits.

Tool Use.

Use the correct type and size drivers. Most Allen fasteners are metric, though many older vehicles have English or SAE-sized Allen bolts. Allen and Torx drivers are not interchangeable and will strip out the internal head of the bolt if they are used for the wrong application.

■ ***Pliers.*** Technicians can accumulate dozens of pliers of varying types, from basic slip-joint and needle-nose to snap ring and specialty pliers. **Pliers** are used to cut, grip, or hold an object in their jaws. The most commonly used types are slip-joint, diagonal side cutting, needle-nose, and locking-jaw pliers (**Figure 3-17**). Each type has specific uses and limitations.

Tool Use.

Never use pliers in place of the correct tool. Do not loosen or tighten fasteners with pliers or use them to remove drum brake springs. Using pliers on nuts and

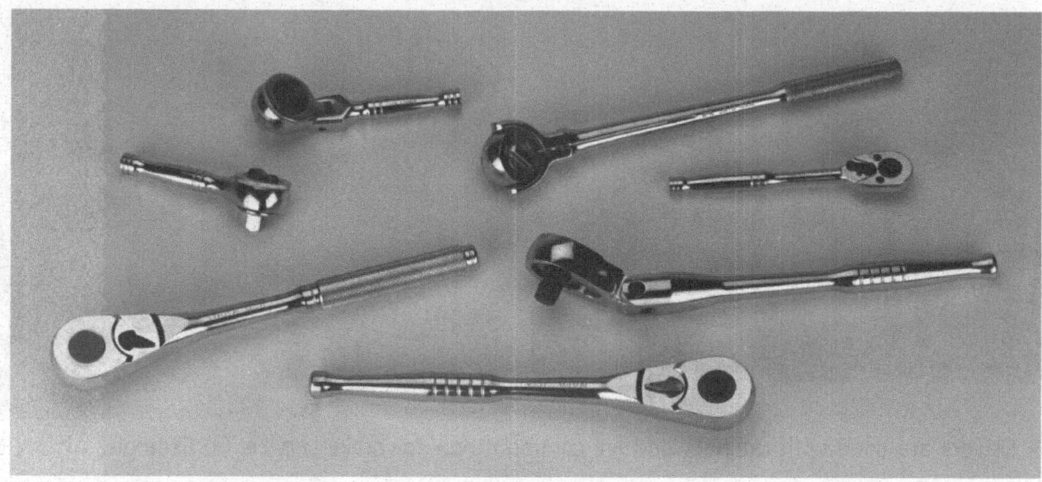

FIGURE 3-13 There are as many types of ratchets available as there are sockets to use them on.

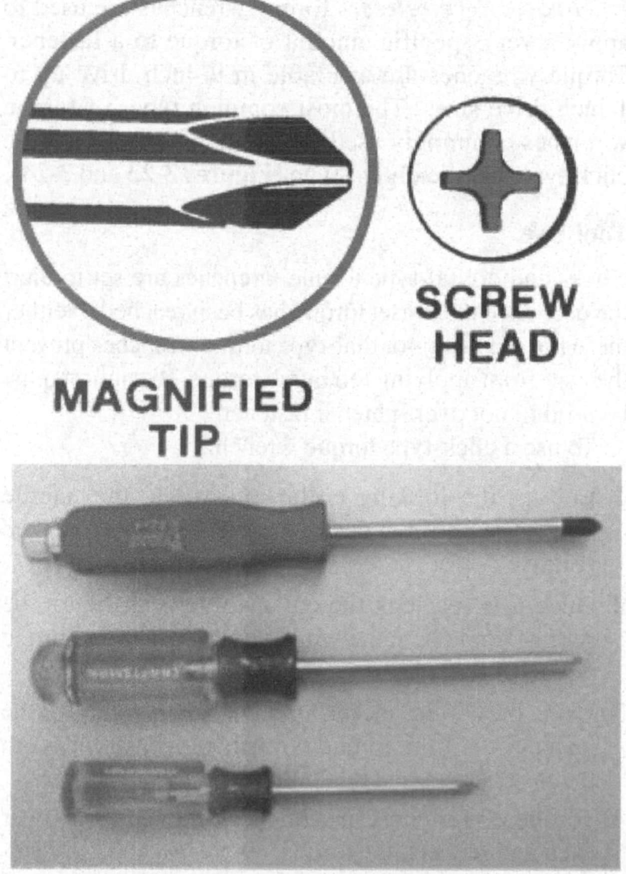

FIGURE 3-14 Phillips screwdrivers are used in interior trim and exterior light assemblies and are used instead of standard or flat head screws because they are less prone to slipping.

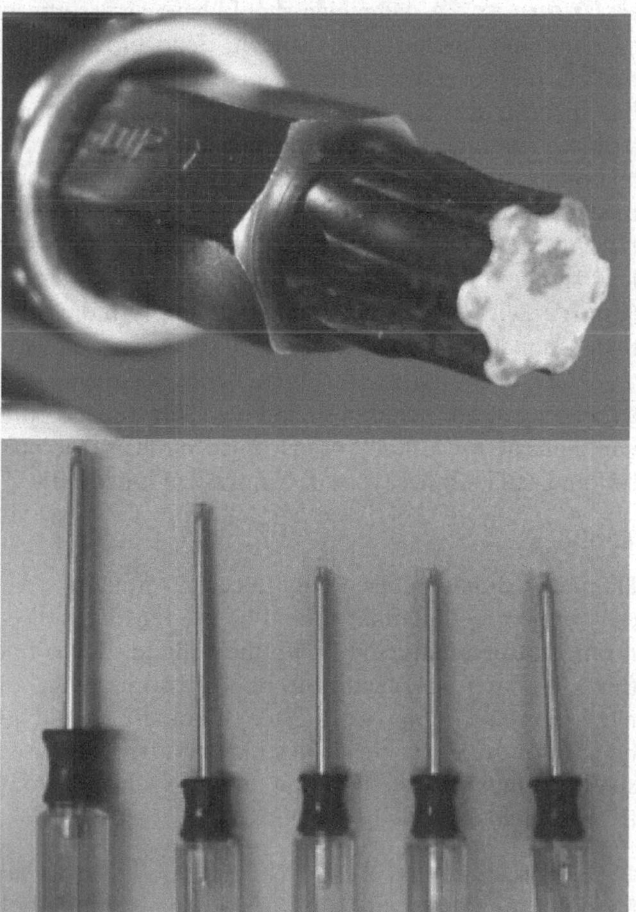

FIGURE 3-15 Torx heads use a six-sided star-shaped driver. Torx screws are common in trim, lighting assemblies, and other applications.

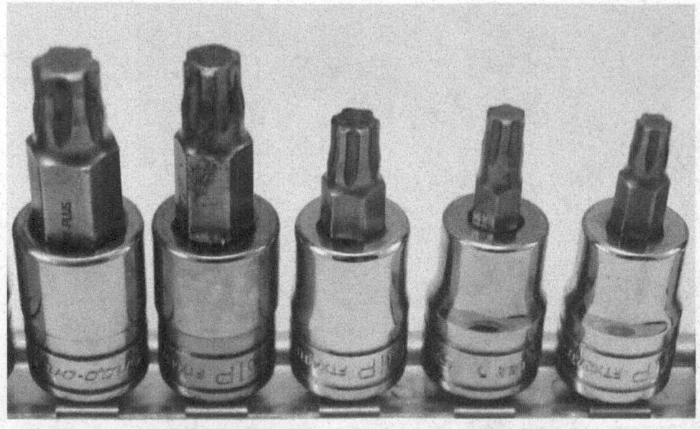

FIGURE 3-16 Drivers are used with ratchets and are common tools for brake service. (a) Examples of Torx drive bits and (b) Allen drive bits.

bolts will quickly round off the head of the fastener. Using pliers on brake springs can damage or cut the springs and can also cause the springs to fly off unpredictably. Using pliers in place of the correct tool can be dangerous and lead to an unhappy instructor.

■ ***Hammers.*** A technician should have an assortment of hammers, including ball-peen, dead blow, rubber mallets, plastic, and brass hammers. Each type of hammer has specific uses and precautions. A standard 16-ounce ball-peen hammer can be used while servicing ball joints and other suspension parts but should never be used on a machined surface or where the possibility of sparking or flaking can occur. A **dead blow hammer** is filled with lead shot and reduces the elastic rebound of the hammer (**Figure 3-18**). Dead blow hammers apply more force than a standard hammer. **Rubber mallets** are used when a surface could be marred by a harder hammer, such as when you are installing a hub cap. Plastic and brass hammers are used on soft or machined surfaces that could be damaged by other types of hammers (**Figure 3-19**).

Tool Use.

Hammers should be used only in certain situations, and only the correct hammer should be used. Most parts do not need to be hammered onto the vehicle. If you find that a hammer is necessary to reinstall something, you need to recheck your work. Steel hammers should not be used on steel, plastic, or machined parts. Never use a steel hammer on threaded components, such as tie-rods, as the threads are easily flattened.

■ ***Chisels, Punches, and Files.*** Chisels are used to remove rivets and spread apart pinch-bolt clamps; punches can be flat, centering, or drift types (**Figure 3-20**). **Punches** are used to remove hollow rolled pins, cotter pins, or to create an indentation in preparation for drilling. A drift or aligning punch is used as an alignment tool to center holes for installation of a fastener (**Figure 3-21**). Files are used to remove sharp edges and small metal fragments called *burrs*, to clean up the metal around a hole, or to file down slight imperfections. Files may also be used to dress the tips of screwdrivers, punches, and chisels (**Figure 3-22**).

■ ***Torque Wrenches.*** Torque wrenches are used to apply a very specific amount of torque to a fastener. Torque wrenches are available in ¼-inch drive up to 1-inch drive sizes. The most common types of torque wrenches commonly used in automotive repair are the click-type and the digital-type (**Figures 3-23** and **3-24**).

Tool Use.

Click- and digital-type torque wrenches are set to alert the user once the preset torque has been reached. Neither these nor the beam- or dial-type torque wrenches prevent the user from applying too much torque, so caution must be used to not overtighten a fastener.

To use a click-type torque wrench:

1. Loosen the locking collar and rotate the handle clockwise until the desired torque setting is reached (**Figure 3-25**).
2. Once it is set, lock the collar to prevent the torque setting from changing while you are using the torque wrench.
3. Using the correct socket, smoothly apply torque to the fastener until the torque wrench clicks. Do not turn the fastener beyond this point.
4. Torque all fasteners in sequence. Then go back over each and recheck the torque.
5. Once you are finished with all the fasteners, unlock the collar and turn the handle all the way back down counterclockwise. This takes the tension off the internal spring inside of the torque wrench.

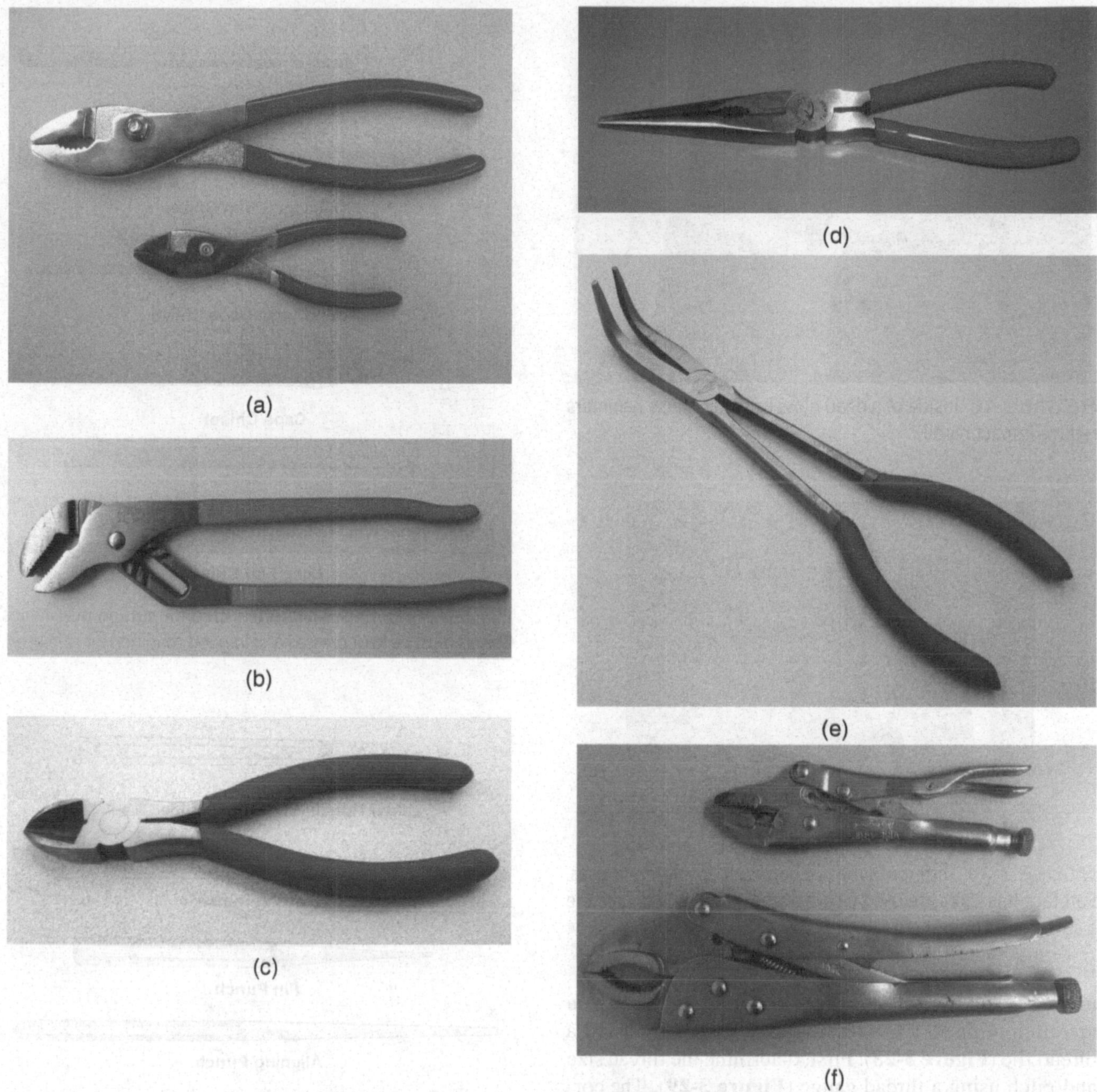

FIGURE 3-17 (a) Examples of slip-joint pliers. (b) Large slip-joint pliers, often called channel locks. (c) Side-cutting or diagonal-cutting pliers. (d) Needle-nose pliers. (e) Extended reach needle-nose pliers. (f) Locking pliers are often called by the common brand name of Vice Grips. Do not use pliers in place of a wrench or socket just because the pliers may be closer at hand. Pliers are not meant to be used to loosen or tighten nuts and bolts.

To set a digital torque wrench, simply press the up and down buttons until the desired setting is reached. Apply torque until the wrench buzzes or beeps and then turn the torque wrench off for storage.

■ ***Tap and Die Sets.*** These are used to repair damaged threads, both internal and external. Taps and dies are used to clean up (often called *chasing*) threads or to cut new threads (**Figure 3-26**). Taps and dies are made of hardened steel that can cut threads in a variety of metals. Taps are bolt shaped and are used on internal threads. Dies are round pieces of steel with the thread cutting edge in the center and are used to chase or cut threads on studs and bolts. Often thread files are contained in a tap and die set (**Figure 3-27**). This special type of file is used to straighten damaged threads on bolts and studs.

Tool Use.

Sooner or later you will encounter a damaged fastener or bolt hole. When this happens, you will likely need to

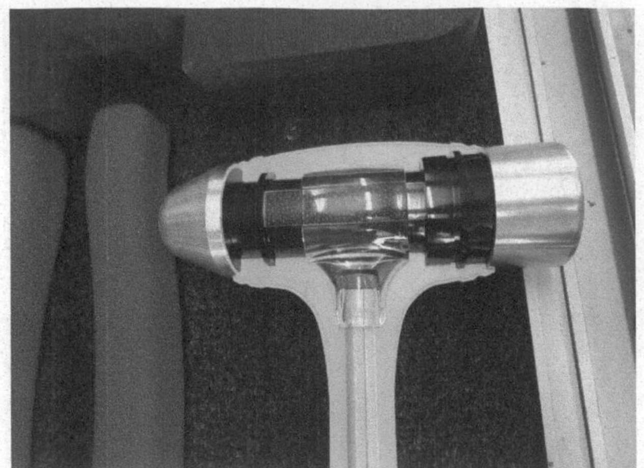

FIGURE 3-18 Inside of a dead blow hammer. These hammers reduce impact recoil.

FIGURE 3-19 Different types of hammers for different uses. Soft face hammers are used where a steel hammer will damage a surface or component.

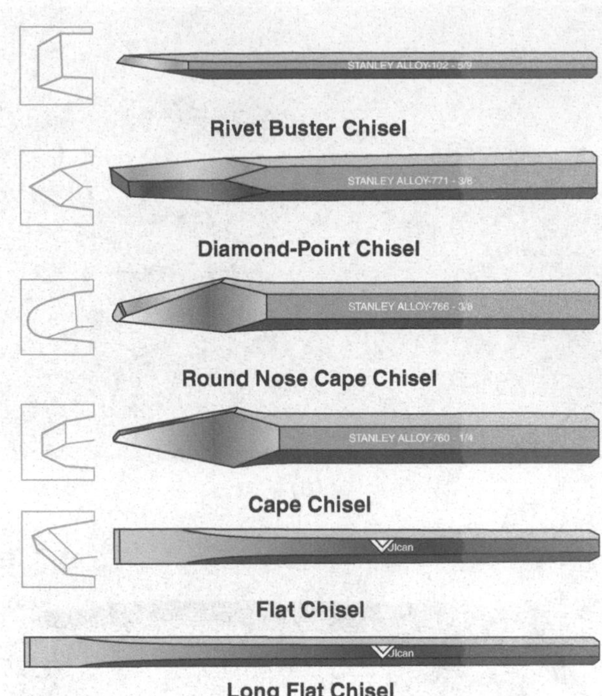

FIGURE 3-20 Chisels are often used for cutting rivet heads. Chisels must be kept dressed and in good condition for safe use.

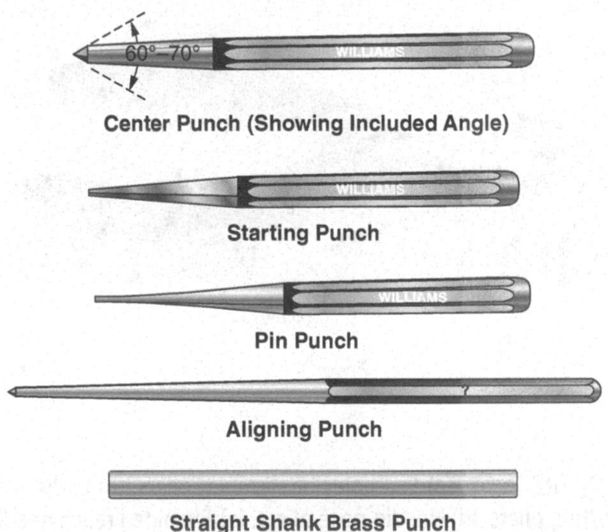

FIGURE 3-21 A selection of punches. Punches are used for driving out pins and aligning components.

use a tap and die set to correct the damage. When you are attempting to fix external threads, start by using a thread file (**Figure 3-28**). First, determine the thread size and pitch using a thread gauge (**Figure 3-29**). The correct pitch gauge will fit exactly with the threads being checked. If the gauge does not fit perfectly in the threads, select another gauge. Once the thread pattern is identified, select the correct thread file and start to work the file across the damaged threads while keeping the file perfectly in line with the threads (**Figure 3-30**).

If the threads are severely damaged, a die may be needed. Select the correct die that matches the thread pattern of the fastener. Next, slowly and carefully begin to thread the die onto the threads of the fastener (**Figure 3-31**). Rotate the die one-quarter turn at a time, and then back the die off slightly. Repeat this process until the die has corrected the damaged threads and moves easily on the fastener.

Using a tap is similar to using a die. First, select the correct size tap for the size of the threads needing repair. Carefully begin to thread the tap into the hole, working it slowly clockwise and then counterclockwise (**Figure 3-32**). Ensure the tap is perfectly aligned with the hole since the tap can easily destroy the old threads and cut new threads if misaligned. Once the threads are repaired, use compressed air to blow any metal shavings from the hole.

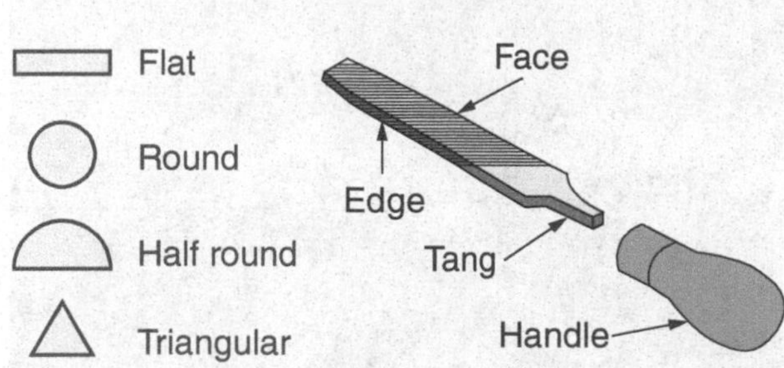

FIGURE 3-22 Files are often used to dress screw-drivers, chisels, and files. Files are also used to remove small slivers of metal, and to shape metal parts.

FIGURE 3-23 Torque wrenches are a necessity for any technician. Using a torque wrench ensures that a nut or bolt is not too tight or too loose.

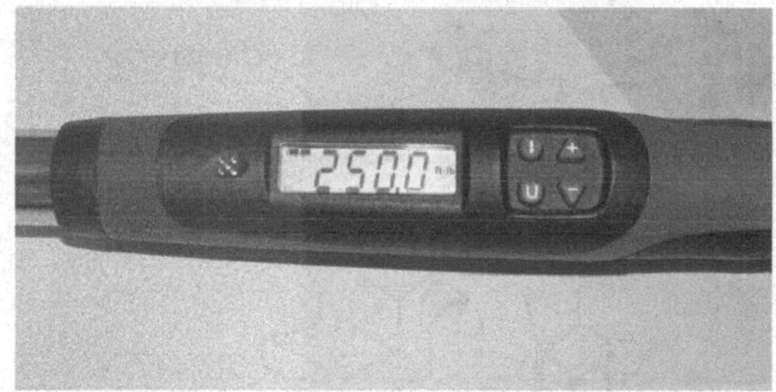

FIGURE 3-24 An example of a digital torque wrench.

It is common for the threads in a hole to become so damaged or stripped that they cannot be repaired. In this case, a threaded insert can often be used to fix the damage. A threaded insert kit contains an oversized tap, inserts, and insert installer (**Figure 3-33**). Some kits may contain the drill bit that is used to enlarge the hole in which the insert is installed (**Figure 3-34**).

To begin, start by drilling the hole slightly oversized. The size of the drill bit that is needed is indicated on the insert kit. Be sure the hole is drilled straight and to the correct depth. Next, use the provided tap to cut new threads into the hole. Once the threads are cut, clean the hole using compressed air and ensure that no metal shavings remain. Install the insert onto the installation tools and thread the insert down into the threads in the hole. Once the insert is seated, use a small chisel to remove the bottom tang of the insert to prevent interference with the fastener.

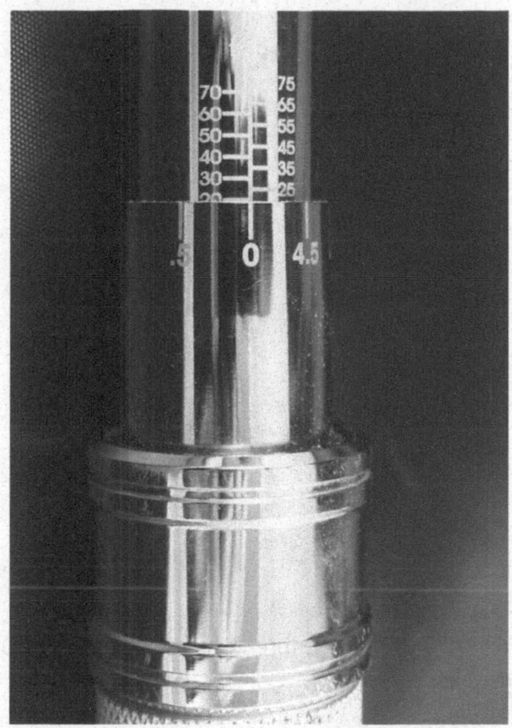

FIGURE 3-25 To set a click-type torque wrench, slide the sleeve down and turn the handle. Once the setting is made, release the sleeve to lock the handle in place. In this example, the wrench is set to 20 ft.lbs.

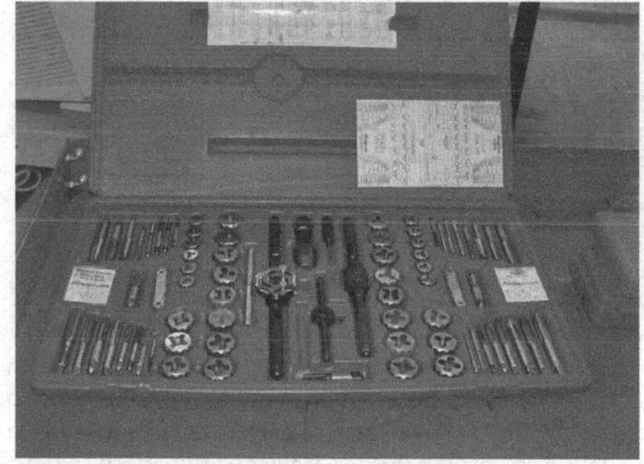

FIGURE 3-26 Repairing damaged threads is a common procedure for a technician. Investing in a good quality tap and die set will make thread repairs faster and easier.

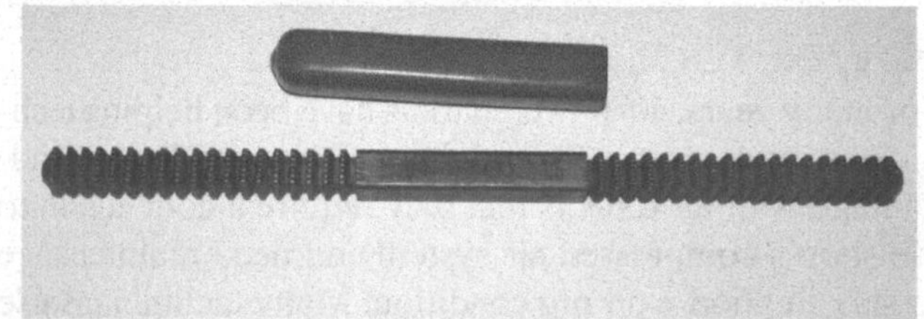

FIGURE 3-27 Damaged threads, studs, and bolts can often be repaired using a thread file. This type has eight different thread pitches.

FIGURE 3-28 Damaged threads are a common occurrence in the auto lab.

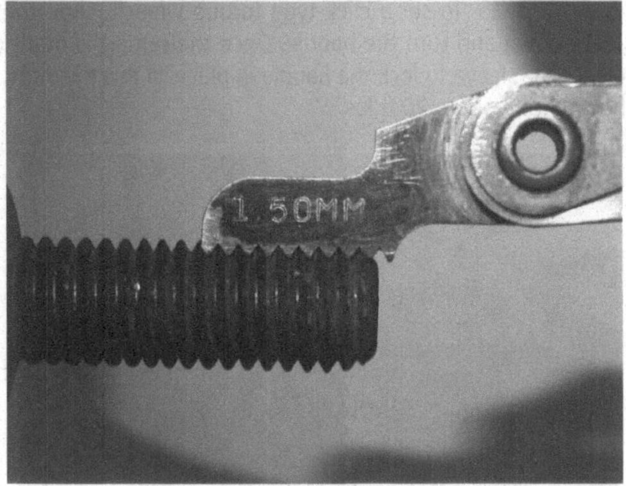

FIGURE 3-29 Measure thread pitch using a thread pitch gauge. Gauges are either English or metric and measure in threads per inch (TPI) or threads per millimeter (TPMM). Align the gauge against the threads. If the two fit perfectly against each other, you have found the correct pitch. If gaps remain between the gauge and threads, keep checking until the correct pitch is found.

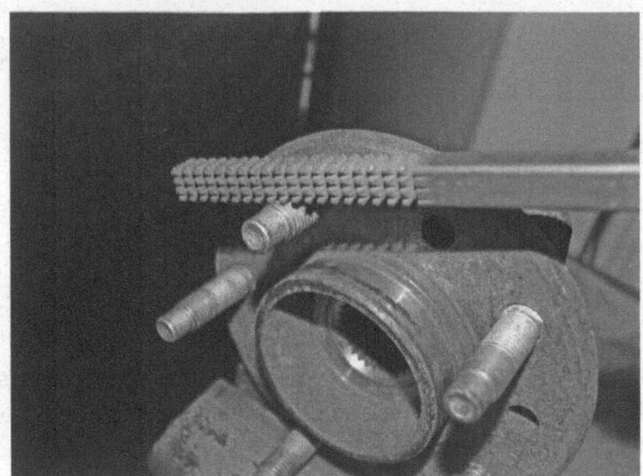

FIGURE 3-30 A thread file can be used to straighten damaged external threads. Select the correct file pitch and move the file back and forth across the damaged area to restore the threads.

Repairing Threads with Tap and Dies

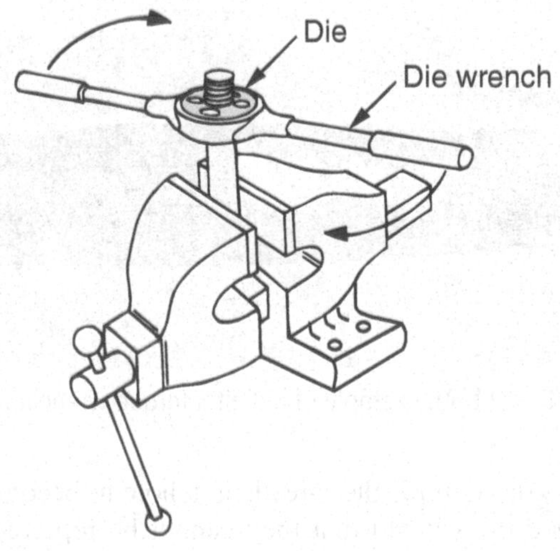

FIGURE 3-31 To repair external threads with a die, select the correct size and start the die onto the threads. Work the die slowly tightening and then loosening and retightening until the damaged threads are repaired.

POWER TOOLS

For many years, air-powered tools have been helping technicians work more efficiently and more quickly. A disadvantage with air tools is that they require a connection to the shop's compressed air system and need maintenance to stay in good working condition. Many technicians use battery-powered electric tools instead of air tools for many types of work. These tools are often, smaller, lighter, and easier to use and maintain than air-powered tools.

■ ***Impact Wrenches.*** **Impact wrenches** are used to remove lug nuts and other tight fasteners. Many technicians use electric impacts for all but the tightest of fasteners. Air-powered impact wrenches usually have more torque and speed than battery-powered impacts (**Figure 3-35**).

■ ***Air Hammers.*** **Air hammers** are used with chisels and cutting bits on exhaust repairs or to separate bushings or suspension components. Air tools should be kept clean

FIGURE 3-32 A tap is used to clean or cut new internal threads. After selecting the correct tap, start it into the hole and slowly work it down and back up. Blow metal debris out with a blow gun.

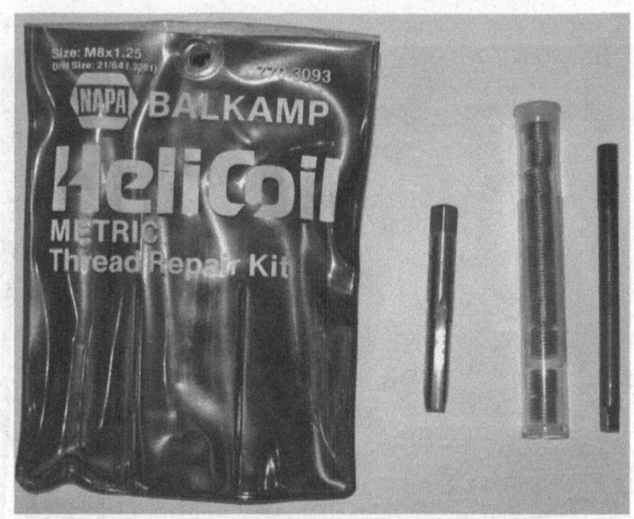
Size: M8x1.25
NAPA BALKAMP
HeliCoil
METRIC
Thread Repair Kit

FIGURE 3-33 A thread insert kit is used to replace damaged threads and restore the hole to its original thread size. This is used when tapping the hole oversize and rethreading is not an option.

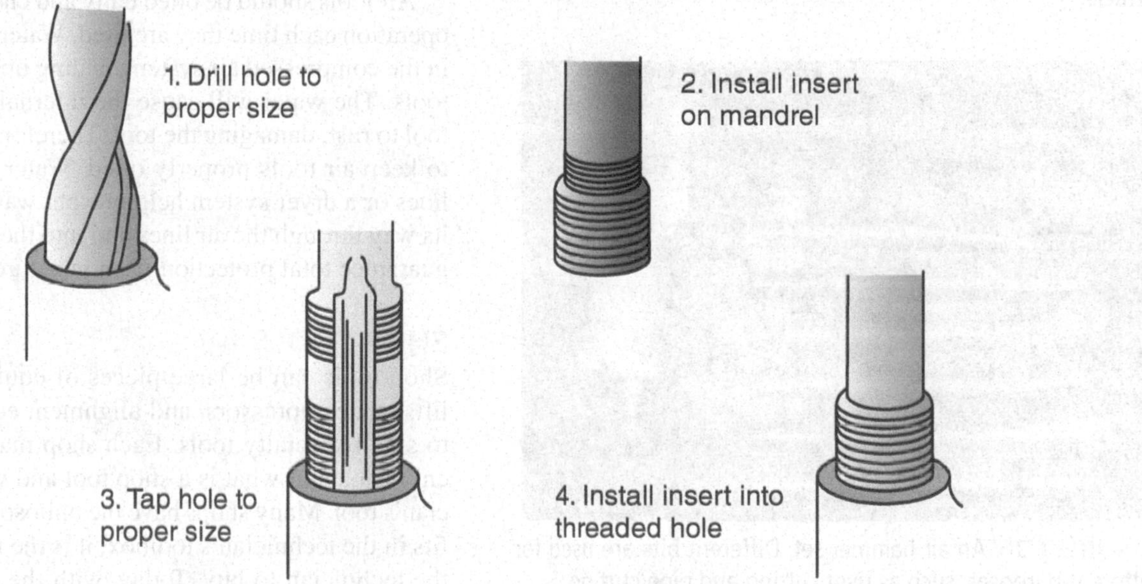
1. Drill hole to proper size
2. Install insert on mandrel
3. Tap hole to proper size
4. Install insert into threaded hole

FIGURE 3-34 The basic steps for installing a thread insert begin with drilling the hole oversize and tapping the hole to accept the insert. Next, thread the insert into the hole and secure the insert in place.

and should be lubricated before each use. Some common examples of air-powered tools are shown in **Figure 3-36**.

■ ***Air or Electric Ratchets.*** While typically larger than a standard ratchet, air or electric ratchets allow you to work with increased speed. However, do not start a nut or bolt with a powered ratchet since it is easy to get it cross-threaded. An air-powered ratchet is shown in **Figure 3-37**.

Tool Use.

Use only sockets, adaptors, and bits rated for impact tools. Using chrome sockets or other tools that are not rated for impact use can cause damage and injury if the tool comes apart during use. Many air tools are loud, and hearing protection should be worn during use.

■ ***Tool Maintenance.*** Shops and technicians invest thousands of dollars in tools, all for particular purposes.

FIGURE 3-35 Power tools save time and make many jobs much easier. However, it is important to properly maintain these tools and use them correctly to prevent damage to the vehicle.

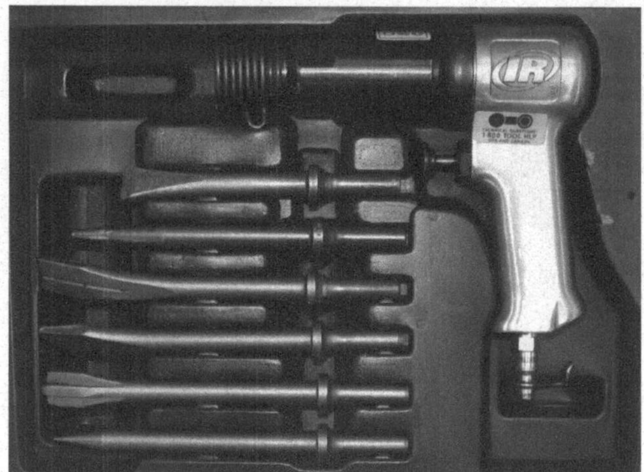

FIGURE 3-36 An air hammer set. Different bits are used for different purposes, such as rivet cutting and pipe cutting.

FIGURE 3-37 An air ratchet can make removing and installing nuts and bolts much faster. However, do not use the tool to start the nuts or bolts; always start fasteners by hand and finish with a torque wrench.

It is unwise to use tools inappropriately, since it can lead to damage to the tool, damage to a vehicle, and personal injury. All tools should be cleaned regularly and inspected for wear or damage. A tool that is worn excessively or is cracked or damaged can cause injury and damage to the vehicle on which the work is being performed.

Sockets should be kept clean and inspected for cracks. A socket that is worn at the contact point with the fastener could slip, resulting in a rounded fastener and a hurt hand. Ratchets should be cleaned and lubricated, and the ratcheting mechanism should be checked regularly. A damaged, dirty, or dry ratchet gear set can bind or break, resulting in hand injuries.

Screwdrivers, punches, and chisels should be clean and dressed as needed. Dressing means maintaining the tips and edges in their original condition. Tools with worn or damaged tips can slip, causing damage to the fastener and the operator. Pliers should be inspected regularly. Check the teeth and pivot for damage or wear. Dull diagonal-cutting pliers should be either redressed or replaced.

Air tools should be oiled daily and checked for proper operation each time they are used. Water can accumulate in the compressed air system, ending up inside of the air tools. The water will cause the internal parts of the air tool to rust, damaging the tool. Therefore, it is important to keep air tools properly oiled. Water traps on the air lines or a dryer system help prevent water from making its way through the air lines and into the tools but do not guarantee total protection from moisture.

SHOP TOOLS

Shop tools can be large pieces of equipment, such as lifts, air compressors, and alignment equipment, down to small specialty tools. Each shop may have a different idea about what is a shop tool and what is a technician's tool. Many shops have the philosophy that if a tool fits in the technician's toolbox, it is the responsibility of the technician to buy. Today, with the development of smaller, faster, and more complex and expensive diagnostic tools, that practice is changing in some shops.

Shops and technicians must be careful not to fall into the new tool trap. Just because a new tool is available does not mean it is a good idea to purchase it. Each purchase, especially of expensive diagnostic equipment, should be carefully considered. As a technician or shop owner, ask yourself these questions before purchasing a tool or piece of equipment:

- Will the tool increase productivity or sales?
- How long before the tool pays for itself?
- What can the new tool do that an existing tool does not do?

Questions like these should be answered honestly before any new major purchase. It is important to note here that many young technicians begin buying tools on credit, which can lead to financial problems. Making weekly payments on tools can consume a large amount of your paycheck. Do not be too eager to buy a lot of new tools, especially very expensive tools and toolboxes, just because you have started your first automotive job.

Most shops will have the following pieces of equipment: bench grinders, cleaning equipment, battery chargers, a compressed air system, and lifting and jacking equipment.

■ ***Bench Grinder.*** A **bench grinder** has one or two grinding stones or wire wheels driven by a powerful motor (**Figure 3-38**). Bench grinders are used to reshape metal and dress tools. The wire wheel can be used to remove rust and clean the threads on bolts.

Inspect the grinding or wire wheels and the guards before using a bench grinder. Do not use it if a dressing wheel is damaged or if the guards are not in place. Before you use a grinder, you should put on a pair of mechanic's gloves to help prevent injury from flying debris and because the parts being ground tend to get hot. However, do not use gloves that are too large and can get caught up in the wheels.

To use the grinding wheel, stand aside from the grinder, turn the power switch on, and let the motor get up to speed. Next, place the object you are grinding on the tool rest in front of the grinding wheel. Slowly and carefully move the workpiece up to the grinding wheel. When you are finished, remove the workpiece and turn off the motor.

■ ***Parts Washer.*** Many shops will have some type of solvent-based cleaning equipment called a **parts washer** (**Figure 3-39**). Solvents can be either petroleum or water based. Parts are placed into the parts cleaner, and the solvent and a brush are used to remove dirt, oil, grease, rust, or other substances. This type of cleaner should be used with chemical gloves to prevent skin reactions to the solvent.

To use, locate the power switch to turn on the parts washer pump. Place the item being cleaned into the cleaning tray and use the solvent and brushes to clean the component. When you are finished, allow the solvent to drip from the part and clean out any mess left from the cleaning.

■ ***Battery Chargers.*** **Battery chargers** are used to recharge a battery that has become discharged or to charge newly filled batteries. Chargers can be small, low-amperage float or trickle chargers, designed to deliver a small amount of current over many hours (**Figure 3-40**), or larger output booster-style chargers (**Figure 3-41**). Most have the option to select the charge rate and amount of time for charging to occur. Many shops have smart chargers that can be set and will then shut themselves off once the battery is fully charged (**Figure 3-42**).

Before using a battery charger, inspect the power cord and the charging cables and clamps. Do not use a charger with damaged or frayed cords or damaged clamps. Make sure the charger is off and unplugged before you connect the charging clamps to the battery. Connect the positive clamp to the positive battery terminal first, then connect the negative charging clamp to the negative battery terminal. Plug the charger into the outlet and set the charging rate and time as indicated on the charger. When you are finished charging the battery, turn the charger off and unplug the power cord. Remove the negative clamp and

FIGURE 3-38 Bench grinders are used to clean parts, shape metal, and dress tools. Never use a bench grinder without inspecting the wheels, guards, and shields and making sure all are in good working condition.

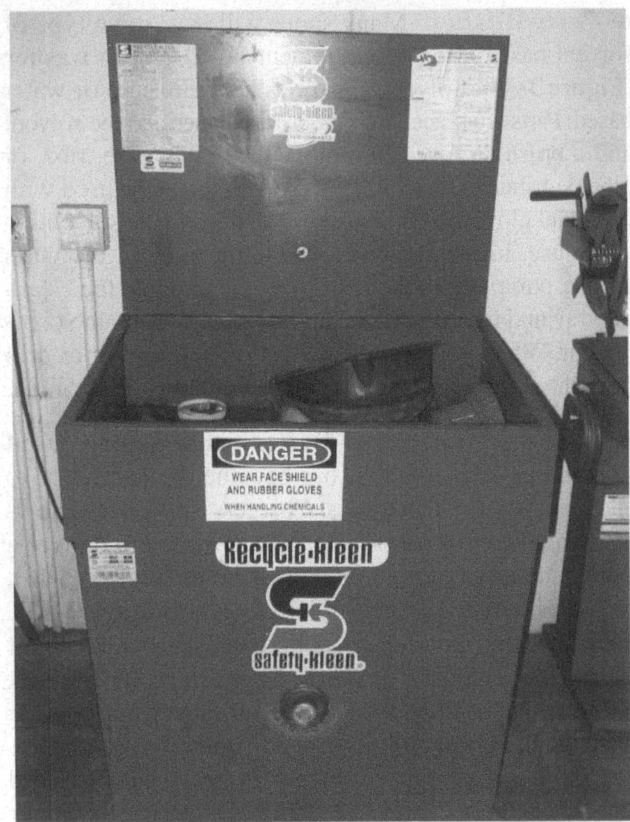

FIGURE 3-39 Solvent cleaning tanks are used for light-duty cleaning of dirt, grease, and oil.

FIGURE 3-41 This type of battery charger can perform slow, fast, and even jump-start boosting power. Never leave a battery charging at a high rate without monitoring its voltage and temperature.

FIGURE 3-40 A trickle or slow charger supplies a couple of amps over a long period of time to recharge a battery slowly. This charger does not have an on/off switch so caution must be used when connecting and disconnecting.

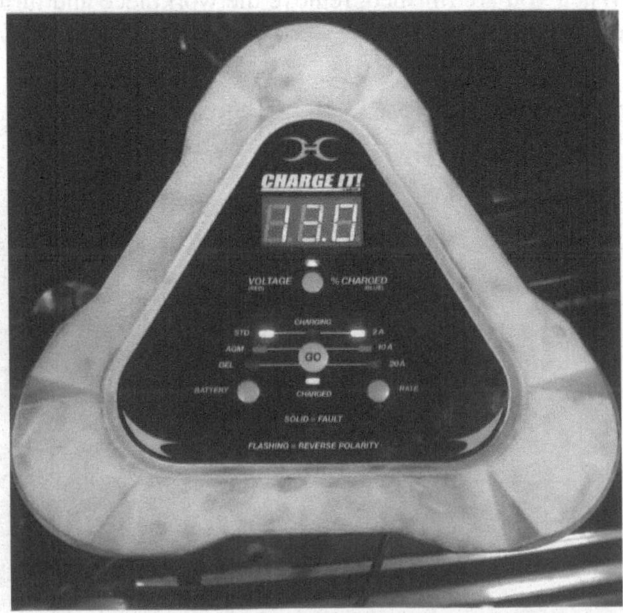

FIGURE 3-42 This type of battery charger, called a smart charger, can be used on regular lead-acid, AGM, and gel batteries.

FIGURE 3-43 A bench vise is used to hold objects securely in place while working. Do not overtighten the vise or close the jaws on machined components.

then the positive clamp. Carefully wrap the power cord and charging cables around the handle. Do not clamp the charging clamps to the handle.

■ ***Bench Vice.*** **Vises** are used to hold an object (**Figure 3-43**). The jaws of a vice may be steel, brass, wood, or aluminum depending on what it is designed to hold. Different jaws are used for different applications. Soft jaws, such as brass and aluminum, are used to hold something that can be damaged, such as a machined surface. Caution should be used to avoid accidentally pinching your fingers in a vice while clamping a work piece.

To open the vise, turn the handle counterclockwise; to close it, turn the handle clockwise. When it is not in use, leave the vise jaws slightly loose and the handle pointing down.

■ ***Compressed Air.*** **Compressed air** is used to power tools and hoists. The shop will be equipped with an air compressor and a network of air lines and hoses. The air compressor uses a large electric motor to draw in and compress air—similar to an internal combustion engine. The air compressor, however, compresses the air and then stores it in a large tank. The tank is connected to the shop's air lines, to which the technician then connects his or her air tools (**Figure 3-44**). To use an air tool, follow these steps:

1. Connect the air tool to the air hose. Pull the locking collar rearward and slide the connector over the fitting on the tool. Once fully connected, release the collar to lock the connector and fitting together.
2. Open the air line valve.
3. Listen and check for air leaks from the hose and connections.

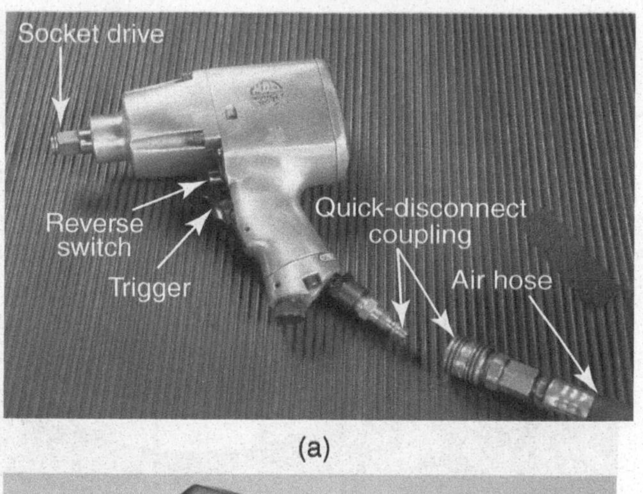

(a)

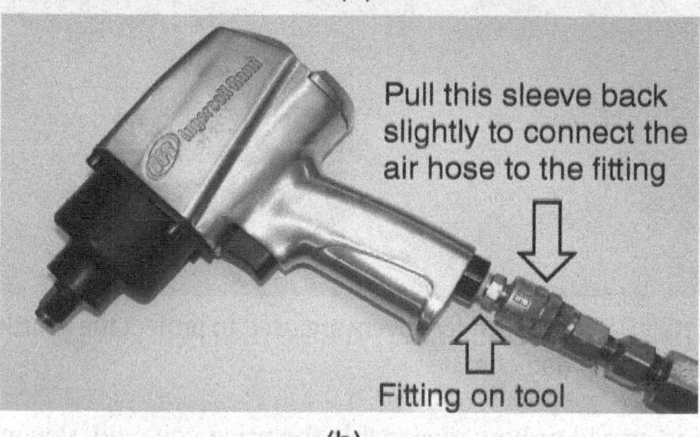

(b)

FIGURE 3-44 (a) The shop compressed air system has an air compressor, lines, and hoses. Pneumatic tools attach to the air hose using a special quick-disconnect fitting. (b) To attach the tool to the air hose, pull the sleeve away from the open end of the fitting and slide the fitting over the male fitting on the tool. Release the sleeve and the two fittings are locked together.

4. When finished with the tool, close the air line valve.
5. Purge the air from the hose by operating the tool.
6. Disconnect the air tool from the hose by pulling the locking collar rearward.

■ ***Compressed Air Warning.*** Shop air pressure should be regulated at 90 psi at connection points. This provides adequate pressure to operate most air tools. Compressed air and blow guns should never be used to blow dirt and debris from your clothing or skin. The air pressure can force dirt into your skin, and if it is used around an open wound, face, or head, it can cause serious, even deadly injury.

KEEPING THE VEHICLE CLEAN

Imagine you take your car or truck to a shop for repair. Once the work is done, you go to your vehicle and find that the door panel, seat, and steering wheel are greasy and dirty from the shop. No matter how well the work was

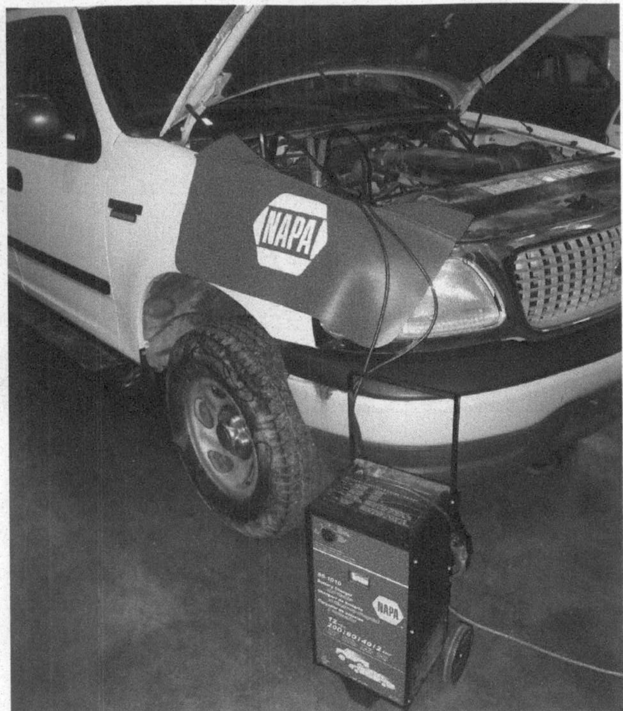

FIGURE 3-45 Fender covers are used to protect the vehicle's finish from damage.

performed or how reasonable the price you paid, you will not be happy about how your vehicle was returned to you.

Vehicle covers are some of the most important items in the shop. Fender covers are used to protect the vehicle's finish and underhood components during service (**Figure 3-45**). Fender covers should always be placed over areas where work is being performed to prevent damage from dropped tools or chemical spills.

Seat, steering wheel, and floor covers should be used whenever you are inside the customer's vehicle (**Figure 3-46**). Even though your uniform may look clean, dirt and oil may transfer from your clothing to the interior. Work boots tend to collect dirt and chemicals, so a floor mat should always be used to protect the vehicle's carpet. Customers appreciate having their vehicle returned to them as clean as (or even cleaner than) when it was dropped off for service.

Fasteners and Torque

Fasteners, the nuts, bolts, screws, and other hardware that hold the vehicle together are as important to the vehicle as any component or system. Cars and trucks contain hundreds of nuts and bolts, most are different from each other and have very specific purposes. Each fastener also has a torque specification, meaning a certain way in which it is correctly tightened.

FASTENERS

A fastener is anything that mechanically joins two or more pieces together and includes: nuts, bolts, screws, rivets, clips, pins, clamps, and more. On modern cars and trucks, each of these and more are used to hold components together.

FIGURE 3-46 Seat, steering wheel, and floor covers are used to protect the interior of the vehicle during service.

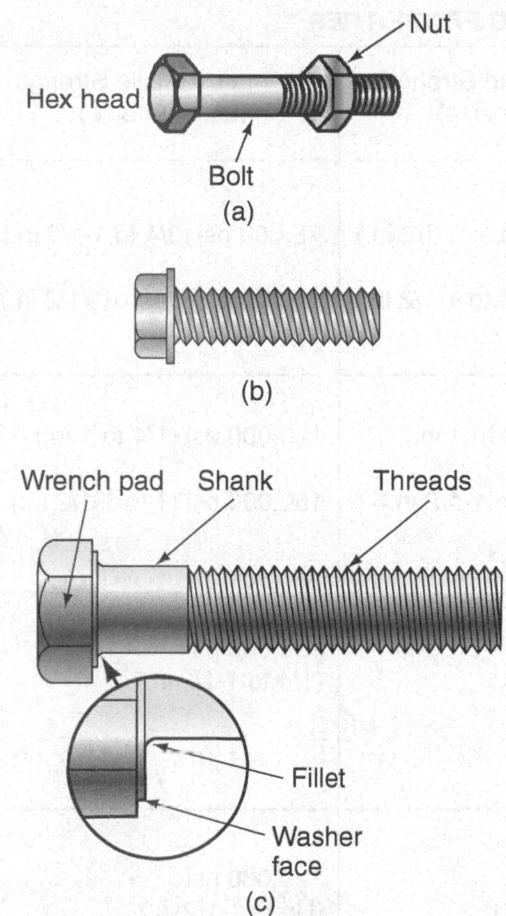

FIGURE 3-47 (a) Bolts are one-piece of metal formed into a threaded section and a drive section. Nuts are used on bolts and studs and are rated for the strength just as bolts are. (b) A bolt may have a washer or fillet under the head and the end is straight-cut. (c) Cap screws look like bolts but have a washer or fillet under the head and chamfered ends.

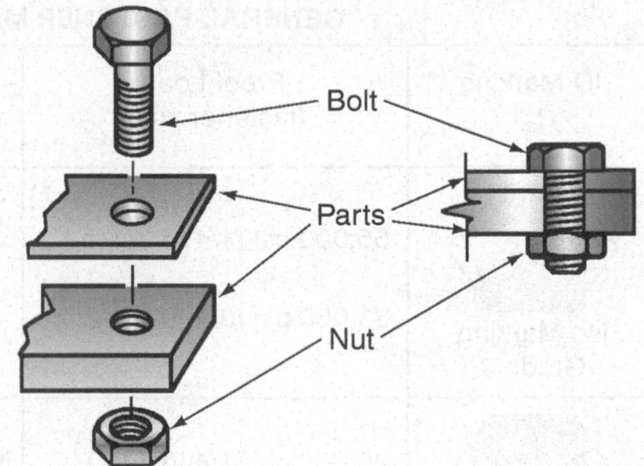

FIGURE 3-48 Bolts are used with a nut to hold parts together.

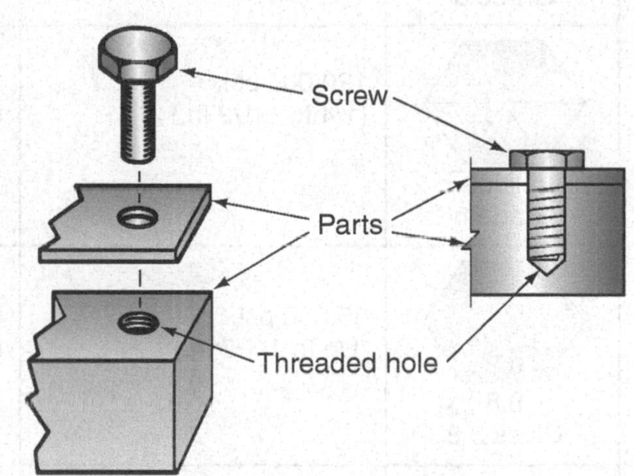

FIGURE 3-49 Cap screws thread into a blind hole to hold parts together.

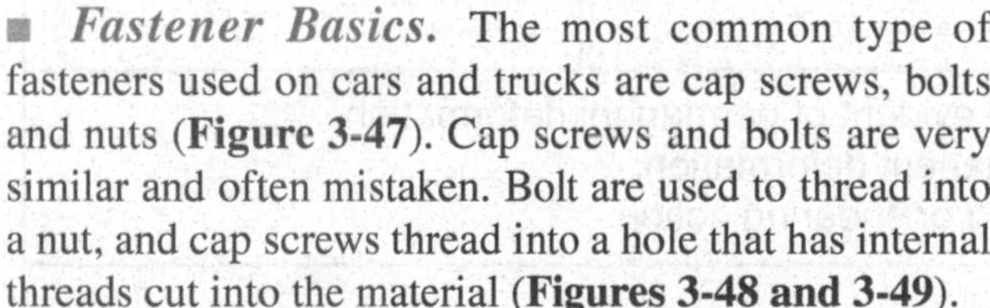

■ ***Fastener Basics.*** The most common type of fasteners used on cars and trucks are cap screws, bolts and nuts (**Figure 3-47**). Cap screws and bolts are very similar and often mistaken. Bolt are used to thread into a nut, and cap screws thread into a hole that has internal threads cut into the material (**Figures 3-48 and 3-49**).

Most bolts and nuts are made of steel and vary in strength. Bolts have a specific range in which they are tightened without damage. The lines on an English or SAE bolt indicate the strength of the bolt. More lines mean that the bolt or nut can take greater clamping and twisting loads, which means more torque can be applied (**Figure 3-50 a**). Metric bolts are marked in numbers, such as 8.8 and 10.9, the higher number indicating a higher-strength bolt. As the load on a bolt increases, the greater the stress, which can lead to the bolt failing (**Figure 3-50b**). Proof load is the amount of tension that can be applied to the fastener without any permanent damage. Over-tightening a bolt can cause the bolt to stretch before it breaks (**Figure 3-51**). When this happens, the bolt loses clamping force and is significantly weakened as the bolt shank is stretched thin.

Once a bolt reaches its breaking point, the bolt will break in two, releasing all its clamping force. When this happens, the parts being held together are no longer held tightly around the area of the broken bolt. This can cause problems from fluid leaks to destroyed engines, depending on the location and purpose of the bolt.

In addition to bolt grade markings, bolts are measured by their physical size (**Figure 3-52**). When you are selecting a replacement bolt, you must first determine the correct grade and then the size of the bolt. To find the size, begin by determining the thread pitch using a thread pitch gauge (**Figure 3-53**). Next, measure the diameter and the length of the bolt. Once you have determined the bolt grade, thread pitch, diameter, and length,

GENERAL FASTENER MARKINGS AND PROPERTIES

ID Marking	Proof Load (fastener size)	Minimum Yield Strength (fastener size)	Minimum Tensile Strength (fastener size)
No Marking Grade 2	55,000 psi (1/4 to 3/4 in.) 33,000 psi (3/4 to 1-1/2 in.)	57,000 psi (3/4 to 1-1/2 in.) 36,000 psi (3/4 to 1-1/2 in.)	75,000 psi (3/4 to 1-1/2 in.) 60,000 psi (3/4 to 1-1/2 in.)
3 Lines Grade 5	85,000 psi (1/4 to 1 in.) 74,000 psi (1 to 1-1/2 in.)	92,000 psi (1/4 to 1 in.) 81,000 psi (1 to 1-1/2 in.)	120,000 psi (1/4 to 1 in.) 150,000 psi (1 to 1-1/2 in.)
6 lines Grade 8	120,000 psi (1/4 to 1-1/2 in.)	130,000 psi (1/4 to 1-1/2 in.)	150,000 psi (1/4 to 1-1/2 in.)
8.8 8.8 Class 8.8	85,000 psi (Up to 1-1/2 in.)	92,000 psi (Up to 1-1/2 in.)	120,000 psi (Up to 1-1/2 in.)
10.9 10.9 Class 10.9	120,000 psi (Up to 1-1/2 in.)	130,000 psi (Up to 1-1/2 in.)	150,000 psi (Up to 1-1/2 in.)

Proof load: Axial tensile load without material shows no evident of permanent deformation.
Yield strength: Axial load before material shows a permanent deformation.
Tensile strength: Maximum load in tension during pulling or shearing action.

(a)

FIGURE 3-50 **(a) Common bold head grade markings. Do not use a lower grade bolt in place of the correct bolt. (b) This chart illustrates as the load on a bolt increases, the greater the stress, which can lead to the bolt failing.**

select the correct replacement bolt. Do not substitute a lower-grade bolt in place of a high-strength bolt—the weaker bolt may not be able to handle the torque applied to it and may break.

Because the threads in a nut must correspond to the threads on a bolt or stud, you often will need to determine the thread pitch and size for a nut. This can be done with a thread pitch gauge and ruler. The easiest way to find thread size is by matching the nut with a known bolt type. Find a bolt that matches the size and pitch of the threads in the nut and thread the nut onto the bolt. If the threads are correct, the nut should thread easily onto the bolt. Pay close attention if you are using a nut and bolt to find out the thread size as it is possible to install a nut that is too large over a bolt and incorrectly judge the size of the threads. Do not substitute a lower-grade nut in

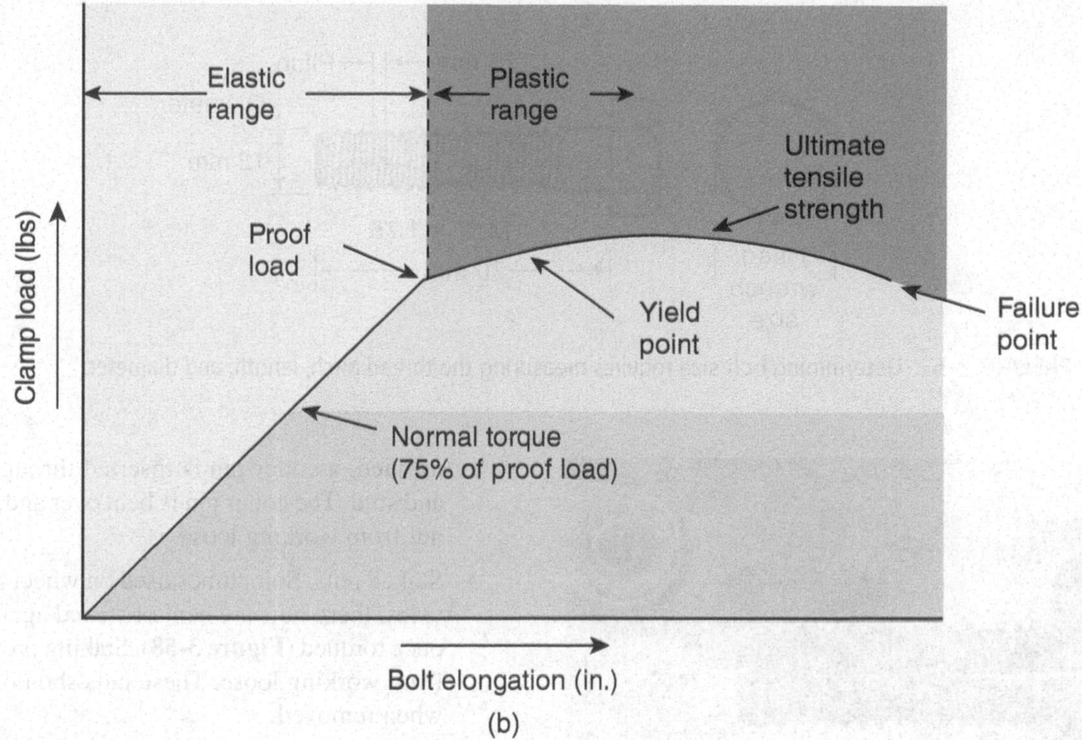

FIGURE 3-50 (continued)

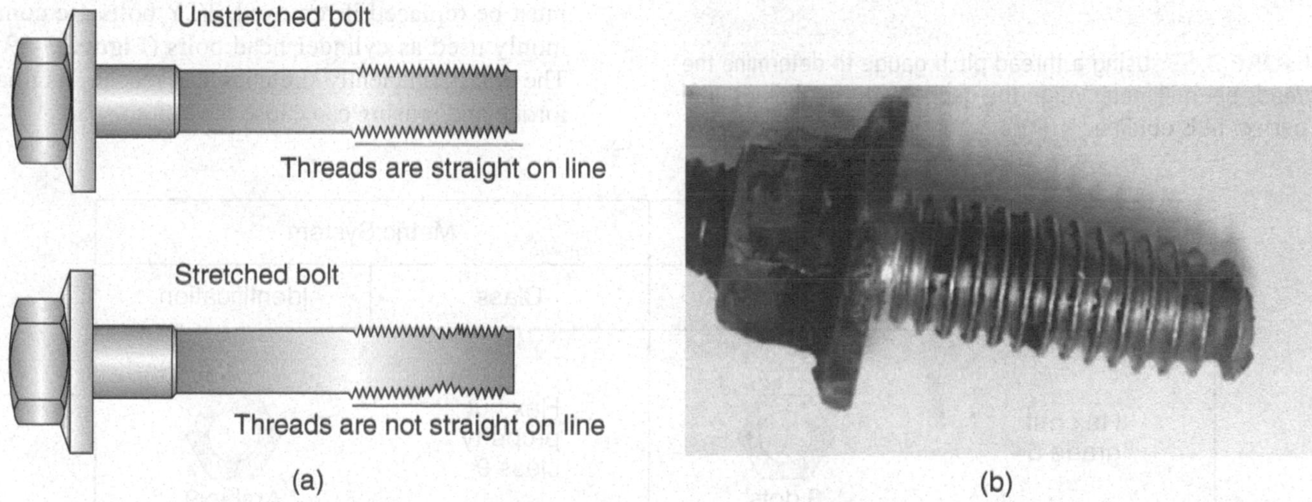

FIGURE 3-51 (a) Once a bolt is over-tightened and stretched, the bolt is weakened and must be replaced. (b) An over-torqued bolt. Note the stretching near the break.

place of a high-strength nut as this may cause the threads on the nut to strip as it is tightened (**Figure 3-54**).

■ ***Specialty and One-Time Use Fasteners.*** Many automotive fasteners are special, meaning specific for an application. Some are one-time use nuts and bolts. Once these have been torqued, and if removed, they must be replaced. Common examples include:

- Prevailing torque nuts. These types of nuts are designed to resist working loose and may have a slight distortion built into the nut. This causes the nut to have greater friction against the threads, helping to prevent it from working loose (**Figure 3-55**).
- Nylon lock nuts. Also called *Nyloc* or *Nylock nuts*, these fasteners have a nylon ring insert in the top of the nut (**Figure 3-56**). Once installed, the nylon ring provides friction against loosening. Once removed, they need to be replaced.
- Castle nuts. This type of nut is common on steering and suspension components (**Figure 3-57**). Once

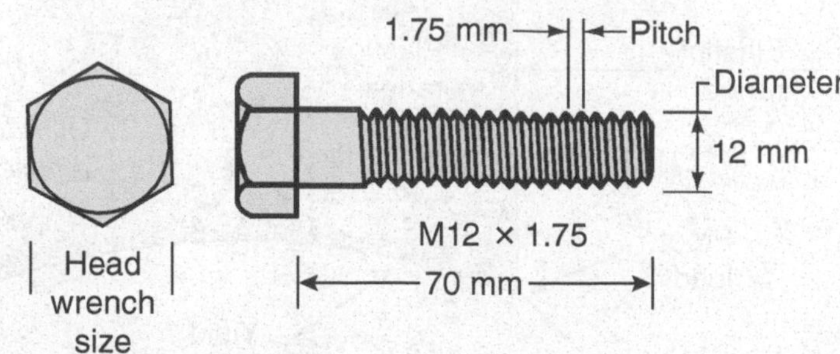

FIGURE 3-52 Determining bolt size requires measuring the thread pitch, length, and diameter.

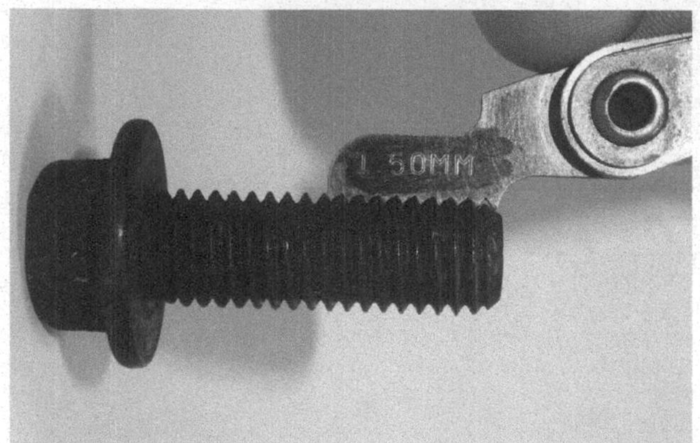

FIGURE 3-53 Using a thread pitch gauge to determine the threads per millimeter. Align the gauge with the threads until a perfect fit is obtained.

torqued, a cotter pin is inserted through a window and stud. The cotter pin is bent over and prevents the nut from working loose.

- Staked nuts. Sometimes used on wheel bearings and axles, these nuts are bent or staked against a groove once torqued (**Figure 3-58**). Staking prevents the nut from working loose. These nuts should be replaced when removed.
- Torque-to-yield bolts. Also called *TTY bolts*, this type of bolt is made to stretch when torqued and must be replaced if removed. TTY bolts are commonly used as cylinder head bolts (**Figure 3-59**). The bolt permanently stretches as it reaches its final torque and reusing can cause the bolt to snap.

Inch System		Metric System	
Grade	Identification	Class	Identification
Hex nut grade 5	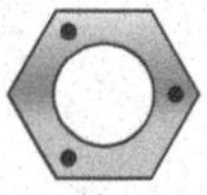 3 dots	Hex nut property class 9	 9 Arabic 9
Hex nut grade 8	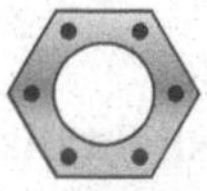 6 dots	Hex nut property class 10	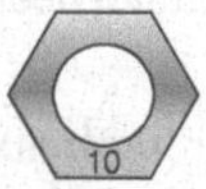 10 Arabic 10
Increasing dots represent increasing strength.		Can also have blue finish or paint dab on hex flat. Increasing numbers represent increasing strength.	

FIGURE 3-54 Nuts are rated for the strength just as bolts are. Using a lower-quality nut can cause the fastener to strip or work loose.

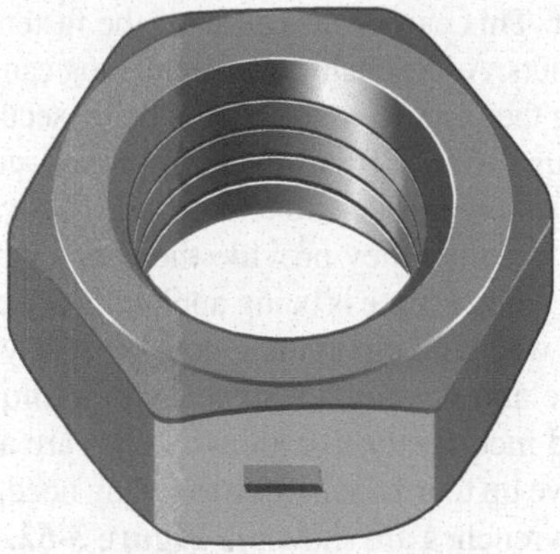

FIGURE 3-55 An example of a lock nut. Note the rectangular indentation marking this as a locking nut.

FIGURE 3-56 Nylon-insert locking nuts are common on steering and suspension parts. These nuts are one-time use and should be replaced when removed.

Screws are used either in situations where the screw cuts its own threads into a material or in the case of machine screws, have bolt-like threads but are much smaller in diameter, and have screw heads instead of bolt heads (**Figure 3-60**). Screws with screw heads, such as slotted, Phillips, Torx, and others, are used in low-torque applications, such as interior and exterior trim pieces.

TORQUE

Torque, as it is applied to tightening nuts and bolts, is the amount of twisting force applied over distance. The amount of torque applied to a fastener is based on the

FIGURE 3-57 Castle nuts have windows cut into them to align with the hole in the stud. Tighten the nut to specs and then tighten if necessary to align the cotter pin hole.

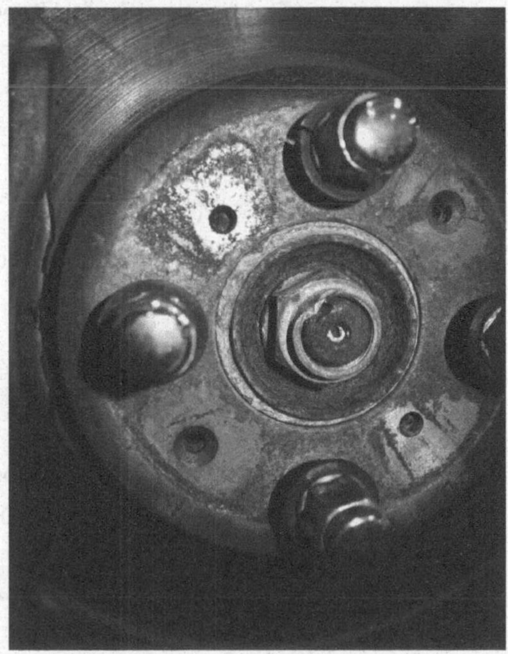

FIGURE 3-58 An example of a staked nut. These are common on axle shafts and some wheel bearing applications.

amount of force applied to the tool and the length of the tool (**Figure 3-61**). When removing a very tight bolt, increase the length of the tool you are using to apply more torque. However, do not use too big of a tool when tightening nuts and bolts so that you and damage or break the fastener.

■ ***Fastener Torquing.*** Because fasteners have limits to how tight they can be torqued and because many components require significant clamping force to remain tight, every fastener has a torque specification. Under-tightening does not supply enough clamping force and components can come apart. This can be disastrous if the fasteners are the wheel lug nuts. Additionally, over-tightening can easily lead to breaking the fastener and the parts being secured.

To ensure that a fastener is tightened correctly, a torque-indicating torque wrench is used. These are torque-indicating because they provide the user with feedback about how much torque is being applied. They do not limit the torque, which is important to note because you can still over-torque and destroy a fastener with a torque wrench if it is used incorrectly. Torque wrenches are available in ¼-inch drive up to a 1-inch drive to fit any need. Examples of torque wrenches are shown in **Figure 3-62**.

The most common torque wrenches used today are the click and digital type. Click-type wrenches have an adjustable sleeve that is used to set the desired torque. Once it is set, tighten the fastener and listen for the click. You should be able to feel the torque wrench click in addition to hearing the click, though the feedback is less pronounced at lower torque settings. You will need to pay careful attention to a click-type torque wrench when you are using it at low torque settings.

Digital wrenches have an LCD display and buttons to change the units and torque setting (**Figure 3-63**). This

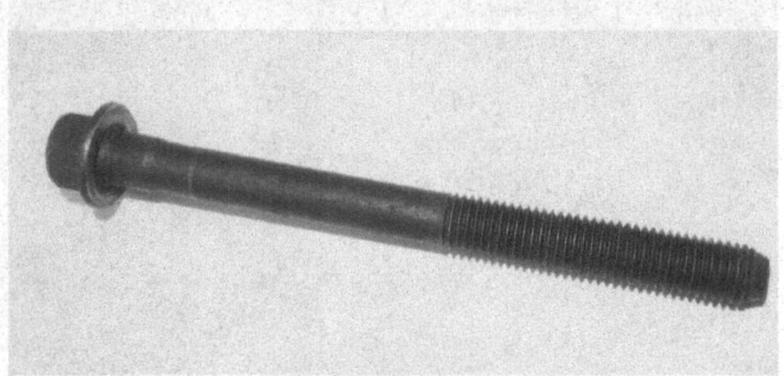

FIGURE 3-59 Torque-to-yield bolts stretch when torqued to spec and cannot be reused.

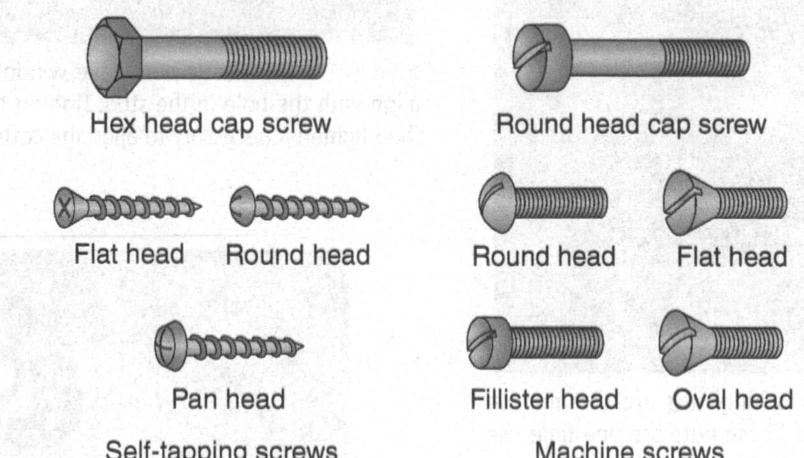

FIGURE 3-60 Examples of types of screws. Screws are used in interior and exterior trim.

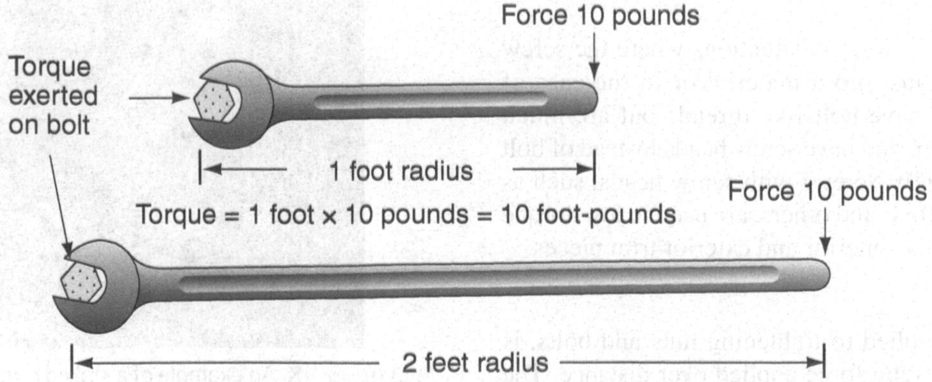

FIGURE 3-61 Torque is force applied over distance. Using torque to your advantage makes some work easier. Using torque correctly prevents damage to fasteners and components.

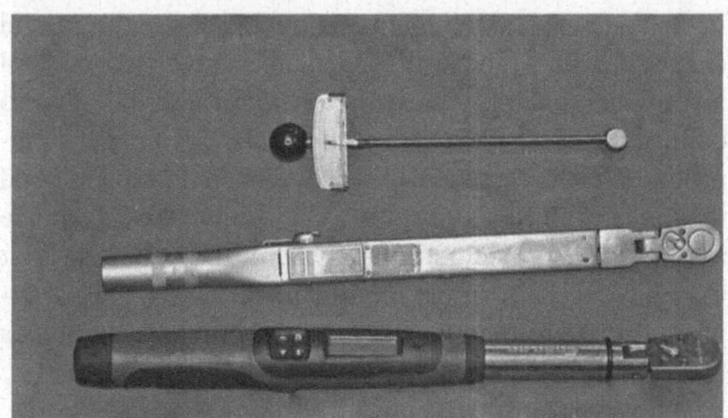

FIGURE 3-62 Examples of torque wrenches.

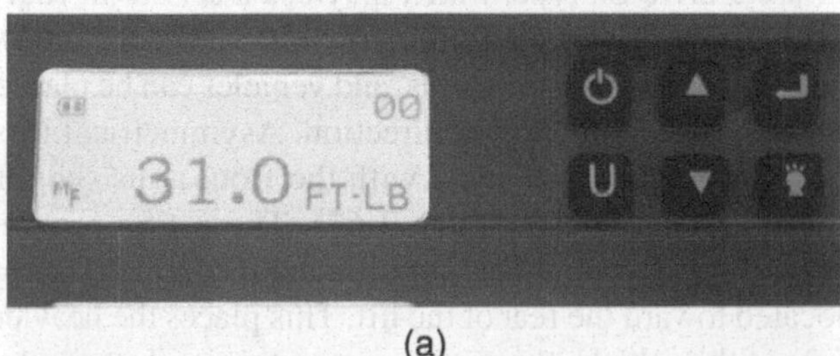

(a)

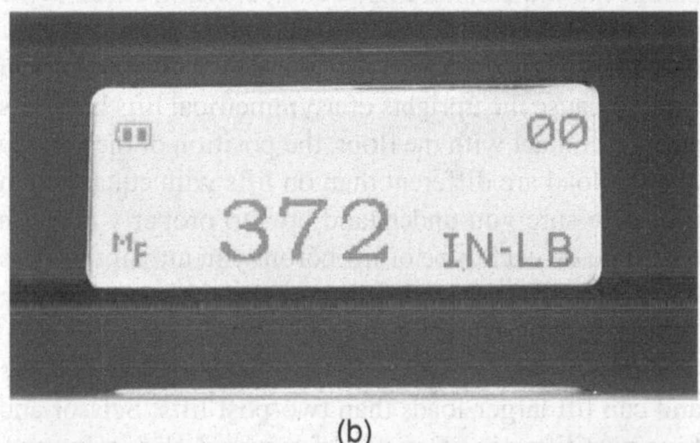

(b)

(c)

FIGURE 3-63 A digital torque wrench can be used for (a) foot-pounds, (b) inch-pounds, or (c) Newton-meters, making it a very versatile tool.

torque wrench can be set for foot-pounds, inch-pounds, and Newton-meters. Some digital torque wrenches can be used for angles (**Figure 3-64**).

■ ***Torque Wrench Care.*** As with any tool, proper care and maintenance are important for torque wrenches if they are to maintain their calibration.

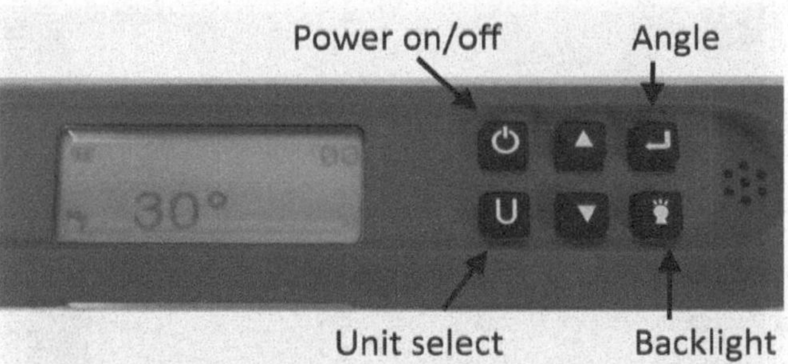

FIGURE 3-64 A torque wrench that can be set for angles in addition to standard torque settings.

FIGURE 3-65 In-ground lifts may have one or two posts and take up very little floor space since most of the components are located in the floor.

FIGURE 3-66 Symmetrical lifts have equal length arms.

- Be careful not to drop torque wrenches on the floor. Also, do not use a torque wrench as a pry bar or breaker bar.
- Inspect the wrench for proper operation of the ratcheting mechanism each time it is used.
- Make sure the adjustment sleeve on click-type wrenches moves smoothly and the locking collar locks and releases properly.
- Clean and oil the ratchet regularly.

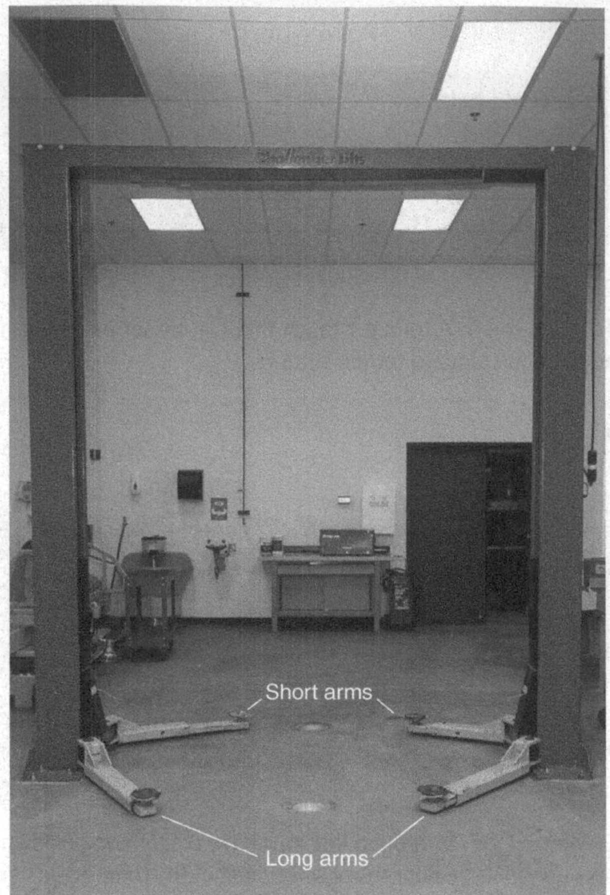

FIGURE 3-67 Asymmetrical lifts have shorter front arms and longer rear arms. This changes the way the vehicle is placed on the lift compared to symmetrical lifts.

- Always store click-type wrenches with the adjustment set at the lowest possible position.
- Store the wrench in its case when not in use.
- Have the wrench checked and calibrated at least yearly. Keep the certification papers supplied by the company that performs the service.

Keeping records of tool calibration is important for legal reasons if there is ever any question about how a repair was performed. Being able to prove that proper torque procedures were followed with a certified accurate tool may be your only defense in a legal dispute.

Vehicle Lifting and Jacking

One of the most frequent operations a technician will perform is the lifting and supporting of vehicles. For that reason, we will take a close look at the steps for safely lifting a vehicle.

VEHICLE LIFTS

Lifts or hoists take many shapes. In-ground lifts can be single-post, twin-post, or in some older shops, twin axle lifts. **In-ground lifts** have the hydraulic cylinders buried in the concrete and the controls mounted above the ground (**Figure 3-65**). These lifts typically have two control levers, one to supply pressure to the lift and the other to control the lift speed and operation. The advantages of in-ground lifts are that there are no overhead crossbeams and they take up little floor space. The primary disadvantage is that if a leak occurs in the system, the concrete floor may have to be torn out for repairs. Newer styles of in-ground lifts are manufactured in self-contained modules that require less extensive plumbing and reduced cost.

Above-ground lifts are of either a swing arm-type, such as two-post symmetrical and asymmetrical designs, or are a drive-on type, which may use a scissor or four-post design. Two-post symmetrical lifts (**Figure 3-66**), have equal length swing arms, and vehicles can be placed onto the lift facing either direction. Asymmetrical lifts have unequal-length arms, with the front arms shorter than the rear arms (**Figure 3-67**). These lifts require the vehicle to be pulled in with most the vehicle being located toward the rear of the lift. This places the heavier part of the vehicle, the engine compartment, closer to the uprights. Asymmetrical lifts have a smaller footprint, or less contact area to the floor, compared to symmetrical lifts. Because the uprights of asymmetrical lifts have less surface contact with the floor, the position of the vehicle and the load are different than on lifts with equal length arms. Be sure you understand how to properly position a vehicle on each type of lift before you attempt to raise the vehicle. Improper loading of the lift can cause the vehicle to fall off the lift or even lift failure.

Drive-on lifts are often for heavier-duty applications and can lift larger loads than two-post lifts. Scissor and four-post lifts are often used for wheel alignment purposes (**Figure 3-68**).

Regardless of lift type, several safety precautions apply when using a lift;

- Do not attempt to operate a lift without first receiving instruction on its safe operation.
- Familiarize yourself with the lift controls and the lift operation.
- Understand how to correctly load the lift, set the swing arms and contact points, and operate the lift controls before use. Improper loading can cause the vehicle to overload the lift. The vehicle could fall from the lift or damage the lift if too much weight is placed incorrectly.

To raise a vehicle, follow these steps.

1. Locate the lift contact points on the vehicle. Proper setting of the contact pads to the vehicle is critical. Improper placement can cause damage to the vehicle and could cause the vehicle to fall. Lift contact points are available from service information resources and from a lift guide available from the American Lift

FIGURE 3-68 Drive-on lifts are commonly used for wheel alignments and are available in many different types of configurations.

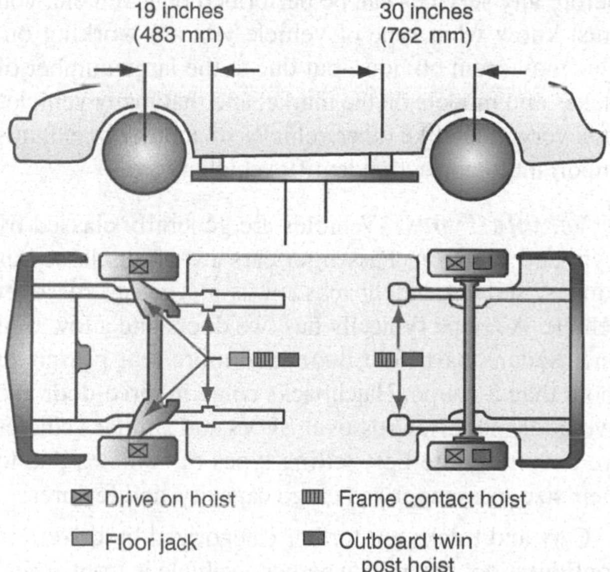

FIGURE 3-69 Identifying and using the correct lifting points on the vehicle is important to prevent damage to the vehicle, damage to the lift, and to prevent injury.

Institute. These references show the proper lifting and jacking points (**Figure 3-69**). Some vehicles prominently show where to place the pads of a swing arm lift (**Figure 3-70**).

2. Once the lift contact pads are set, raise the vehicle approximately 6 inches off the floor and stop.
3. Recheck the contact points and give the vehicle a bounce at the front or rear. If the contacts are correct and the vehicle is stable, continue to raise it to a working height.
4. When the vehicle is at the desired height, lower the lift onto the mechanical safety locks. This provides greater stability for the lift and prevents the vehicle from dropping unexpectedly should there be a hydraulic failure. **Never work on or under the vehicle until the lift is lowered onto the safety locks.**
5. Once the vehicle is secured and work begins, it is important to remember that some types of repairs will change the weight balance of the vehicle on the lift. For example, removing a rear axle assembly from a vehicle will significantly reduce the weight at the rear and may unbalance the load placed on the lift, allowing the car or truck to tilt forward. This can be extremely dangerous since the different weight distribution could cause the vehicle to fall off the lift.
6. When you are ready to lower the vehicle, first make sure nothing is under the vehicle, such as another student, a toolbox, or drain pan. Then raise the lift off the safety locks. Disengage the safety locks, and lower the vehicle to the floor. Make sure no one is standing near the vehicle and lift as it approaches the floor. Being too close to either could result in someone's foot getting trapped under a tire or swing arm.

(a)

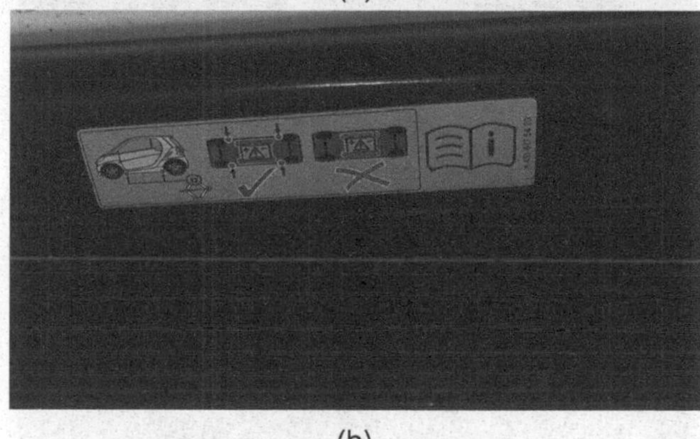

(b)

FIGURE 3-70 Locate the lifting point on the vehicle as specified in a lift manual or service information. Placing the lift contacts in the wrong places can cause serious damage to the vehicle.

■ ***Using Floor Jacks.*** Using a floor jack and jack stands is also common in instructional labs and some repair facilities. Examine the condition of the floor jack prior to its use. Make sure that the casters (wheels) move freely and

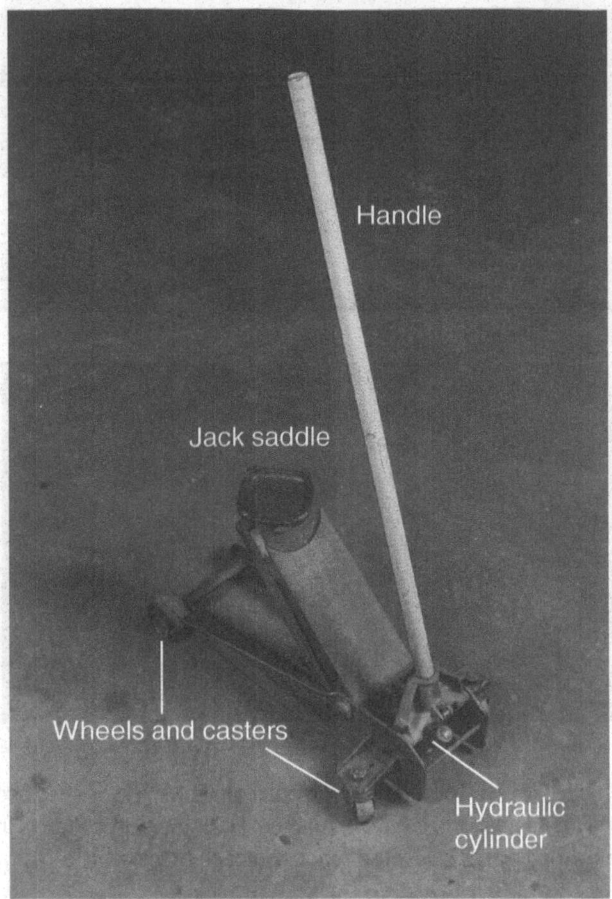

FIGURE 3-71 A floor jack is used to raise a vehicle up off the floor. Once in place, jack stands are used to support the vehicle. Never work under a car or truck supported only with a hydraulic floor jack.

provide a full range of motion. The jack must be able to roll slightly when raising the vehicle. If the casters are frozen, the jack could bind and be kicked out from under the car.

To raise the jack, turn the handle clockwise fully until it stops. Move the jack handle up and down to raise the jack, then turn the handle counterclockwise to lower the jack (**Figure 3-71**). The hydraulics should work smoothly and hold solidly. Any leak from the hydraulic cylinder can cause the jack to lose pressure and drop.

Check each jack stand before use. Examine the lock and release mechanisms. The lock should fully catch the teeth of the jack. You should ensure that the jack stand is not damaged or bent and that there are no cracks in the welds. Faulty jack stands should be discarded. Using a weak or broken jack stand could allow the vehicle to drop, causing serious injury or death.

Follow these steps, shown in **Photo Sequence 1**, to safely jack and support a vehicle.

Floor jacks require periodic maintenance to ensure proper operation and a long service life. Lubricate the casters and jack pivot with the specified lubricants, usually a light oil for the casters and chassis grease for the pivots. Ensure the jack is not leaking hydraulic oil from the cylinder. Do not use a jack that is leaking as it could fail completely, allowing a vehicle to drop suddenly.

Locating Vehicle Information

Today's technicians have access to millions of pages of service information. Being able to quickly and accurately find the correct information is an important skill. Vehicle identification starts with figuring out the manufacturer and model of the vehicle. Finding more specific details is done by locating the various tags and decals located throughout the vehicle. Every new vehicle has several identification tags and labels. Knowing the locations and the information that can be found on these labels will greatly decrease the amount of time you spend just trying to find basic information.

BASIC VEHICLE IDENTIFICATION

Before any services can be performed on a vehicle, you must know what type of vehicle you are working on. This may seem obvious, but due to the large number of makes and models on the market and that many vehicles look very much like other vehicles of another brand, it is important to correctly identify vehicle types.

■ ***Vehicle Types.*** Vehicles are generally classed by style and function. Passenger cars are often classed as coupes, sedans, hatchbacks, or as a type of crossover vehicle. A coupe typically has two doors and a low roof line. Sedans have four doors and more rear passenger room than a coupe. Hatchbacks come in three-door and five-door models. Crossover SUVs and similar vehicles are becoming the best-selling types of vehicles due to their size, passenger and cargo capacity, and features.

Cars and trucks are further categorized by drivetrain configuration, meaning whether a vehicle is front-wheel drive (FWD), rear-wheel drive (RWD), four-wheel drive (4WD), or all-wheel drive (AWD). Most passenger cars and some SUVs currently on the market are FWD vehicles, meaning that the engine and transmission sit in the front of the vehicle and drive the front wheels through short axles (**Figure 3-72**).

Larger SUVs and light trucks, as well as some passenger cars, especially sports cars, have RWD configurations. These vehicles most often have the engine and transmission mounted in the front of the vehicle and drive the rear wheels (**Figure 3-73**). Some sports cars use a mid- or rear-engine configuration and drive the rear wheels or all four wheels. Many SUVs and some passenger cars offer all-wheel drive. Often, these vehicles are based on FWD configurations and then add a method of applying power to the rear wheels for increased traction and safety.

PHOTO SEQUENCE 1

PS1-1 First, locate a suitable jacking point on the vehicle. For this vehicle, the body seam is used as a jacking point.

PS1-2 Next, position the jack saddle on the contact point. Raise the jack until it contacts the vehicle and recheck the contact.

PS1-3 Jack the vehicle up until the tire is off the ground. You may need to raise or lower the vehicle slightly to properly place the jack stand. Place the jack handle up so it is not a tripping hazard.

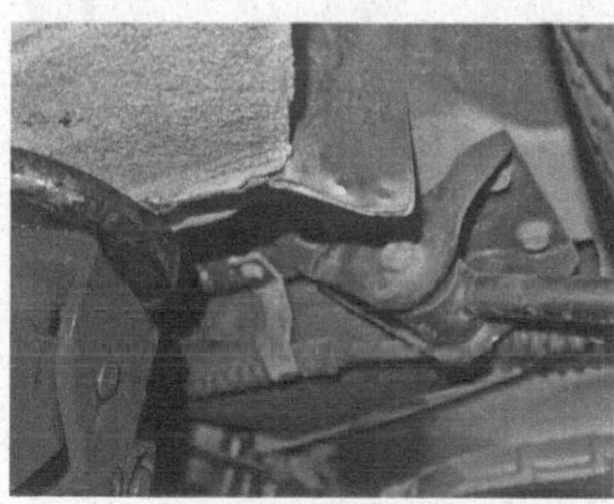

PS1-4 Locate a place to set the jack stand. These places will vary depending on the vehicle. Do not place the stand on components that will move, such as the rear suspension arm shown here.

PS1-5 Set the jack stand to the desired height and position it so it will make solid contact with the vehicle.

PS1-6 Slowly and carefully lower the vehicle onto the jack stand. Double check the contact between the jack and vehicle. Once you have verified, the jack is well placed and secure, remove the floor jack.

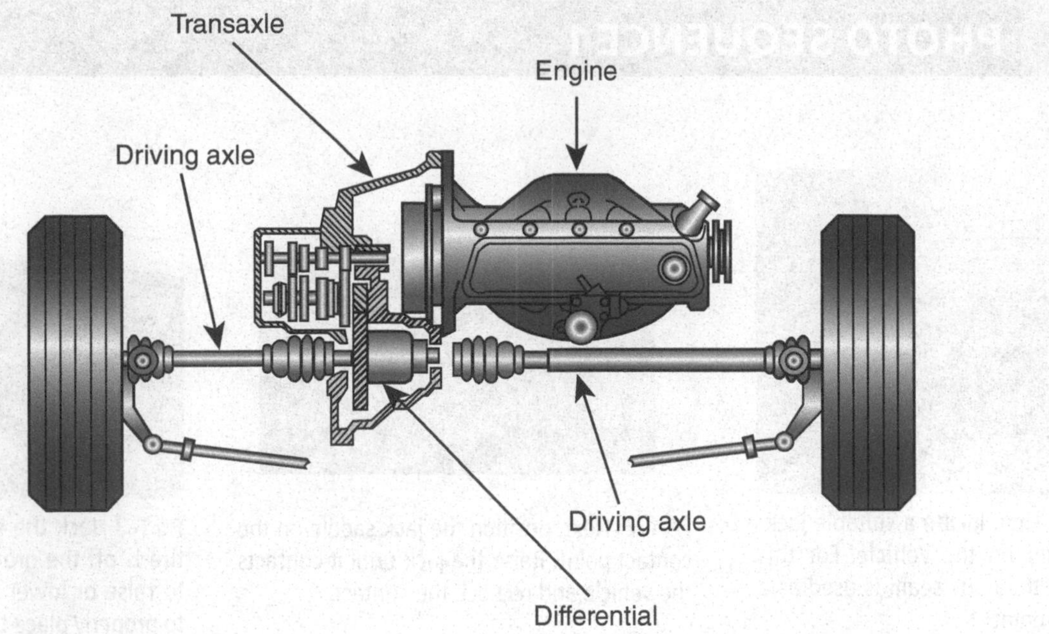

FIGURE 3-72 Most modern cars and small SUVs are front-wheel drive (FWD).

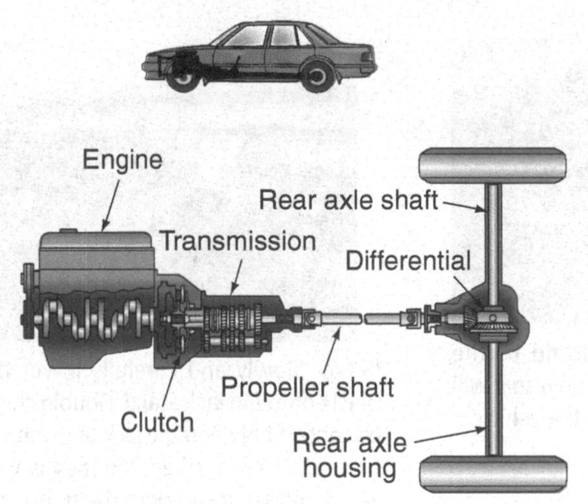

FIGURE 3-73 Larger SUVs, trucks, and some cars are rear-wheel drive (RWD).

FIGURE 3-74 The vehicle may display the name of the manufacturer or division as this truck does.

FIGURE 3-75 Many cars do not spell out the manufacturer, instead an emblem is used to show who made the vehicle.

■ ***Basic Vehicle Identification.*** Begin by looking at the vehicle and locating manufacturer badges (**Figure 3-74**). If there is no manufacturer's name plate, check for badges on the hood, trunk, fenders, or wheels (**Figure 3-75**). Next, determine the model and look for any additional model details, such as EX, LT, SE, GT, or other similar designations on the vehicle (**Figure 3-76**). These may be located on fenders or at the rear of the vehicle. Many times, these model types are important for correct parts ordering and labor calculations.

Finally, it is good to know if the vehicle is a product or division of a manufacturer; for example, Chevrolet, Buick, and Cadillac all are divisions of General Motors (GM). **Figure 3-77** shows current brands sold in the United States by manufacturer and market share. This information can be important because some vehicle models have an identical model sold by a different division.

VEHICLE IDENTIFICATION NUMBERS (VIN)

Just as many items, such as TVs and computers, have serial numbers, so do cars, trucks, motorcycles, and other vehicles. These serial numbers contain a lot of information about the vehicle. As the number of vehicles produced and sold over the years has increased, the

FIGURE 3-76 Check for model designations like the LTZ model. This information is often needed to correctly identify submodels and options.

identification method has also changed. Vehicles sold in the United States use a 17-digit identifier called the Vehicle Identification Number, or **VIN**.

■ ***What Is the VIN?*** Each vehicle sold in the United States has a unique vehicle identification number, called the VIN (**Figure 3-77**). The current VIN system has been in use since 1980 and contains 17 letters and numbers, which provide detailed information about each vehicle.

Decoding the information in the VIN requires a chart (**Figure 3-78**). The first three digits or letters denote the country of origin and manufacturer. The second digit is the manufacturer, the third is the division of the manufacturer, such as Buick is a division of GM. In some cases, the first two digits represent the country of manufacturer. The fourth and fifth digits designate the body or car type. The eighth digit is often the engine designation. The tenth represents the model year of the vehicle, and the last six digits are the actual number of the vehicle built of that type for that year.

■ ***Locations.*** The federal government mandates that the VIN be placed in the left corner of the dash, near the A pillar for easy access (**Figure 3-79**). The VIN also appears on stickers on body panels, etched into glass, on firewalls, and on engines and transmissions.

One of the reasons that the VIN is placed in so many different places on the vehicle is to help prevent the sale of parts removed from a stolen vehicle. The VIN provides a way to track down parts that have been removed from a stolen vehicle and resold.

OTHER VEHICLE ID TAGS

The VIN is just one of many different types of identification tags found on modern cars and trucks. There may be six or seven different types of tags or decals located

Manufacturer	Brands sold in the US	Country of Origin	US Market share
General Motors	Buick, Cadillac, Chevrolet, GMC (used to produce Saturn, Pontiac, Oldsmobile, Saab)	USA	17.4
Ford	Ford, Lincoln (used to produce Mercury)	USA	14.2
Toyota	Toyota, Lexus, Scion	Japan	13.7
Fiat Chrysler Automobiles	Fiat, Chrysler, Dodge, Ram, Alpha Romeo, Ferarri, Jeep (used to produce Plymouth)	Italy-USA	13.4
Honda	Honda, Acura	Japan	9.3
Renault-Nissan Alliance	Renault, Nissan, Infiniti	France -Japan	8.9
Hyundai	Hyundai, Kia	South Korea	8
Subaru	Subaru	Japan	3.8
Volkswagon	VW, Audi, Porsche, Bentley, Bugatti	Germany	3.2
Daimler AG	Mercedes Benz, AMG Smart	Germany	2.4
BMW	BMW, Roll Royce, Mini	Germany	2.1
Mazda	Mazda	Japan	1.7
Jaguar	Jaguar, Land Rover	England	0.6
Mitsubishi	Mitsubishi	Japan	0.5
Volvo	Volvo	Sweden	0.4
Tesla	Tesla	USA	0.3

FIGURE 3-77 This provides information about what brands are made by and sold under which vehicle manufacturer.

Vehicle Identification Number (VIN) System

Position	Definition	Character	Description
The vehicle identification number (VIN) plate (1) is the legal identifier of the vehicle. The VIN plate is located on the upper left corner of the instrument panel (I/P) and can be seen through the windshield from the outside of the vehicle.			
1	Country of Origin	1	USA
2	Manufacturer	G	General Motors
3	Make	6	Cadillac
4 - 5	Vehicle Line/Series	D/A	CTS Base
		D/B	CTS Base with Navigation
		D/C	CTS Base (AWD)
		D/D	CTS Base with Navigation (AWD)
		D/E	CTS Luxury Collection
		D/F	CTS Luxury Collection with Navigation
		D/G	CTS Luxury Collection (AWD)
		D/H	CTS Luxury Collection with Navigation (AWD)
		D/J	CTS Performance Collection
		D/K	CTS Performance Collection with Navigation
		D/L	CTS Performance Collection (AWD)
		D/M	CTS Performance Collection with Navigation (AWD)
		D/N	CTS Premium Collection
		D/P	CTS Premium Collection with Navigation
		D/R	CTS Premium Collection (AWD)
		D/S	CTS Premium Collection with Navigation (AWD)
		D/T	CTS - RHD
		D/V	CTS - V
		D/2	CTS Sport Appearance Package
		D/9	CTS (Export Only)
6	Body Style	1	47-Coupe, 2 Door, Notchback Special
		5	69-Sedan, 4 Door, 4 Window, Notchback
		8	35-Station Wagon, 4 Door
7	Restraint System	D	Active Manual Belts, Airbags-Driver and Passenger-Front (1st row), and Front Seat Side (1st row)
		E	Active Manual Belts, Airbags-Driver and Passenger-Front (1st row), Front Seat Side (1st row), Roof Side (all seating rows)
8	Engine Code	5	RPO LFW, Gas, 6 Cylinder, 3.0L, SIDI, DOHC, VVT, E85 Max, Aluminum, GM
		3	RPO LFX, Gas, 6 Cylinder, 3.6L, SIDI, DOHC, VVT, E85 Max, Aluminum, GM
		P	RPO LSA, Gas, 8 Cylinder, 6.2L, SFI, ER, Aluminum, INTR CLR SC, GM
9	Check Digit	Varies	Check Digit
10	Model Year	C	2012
11	Plant Location	0	Lansing - Grand River, MI USA
12-17	Plant Sequence Number	-	100001

FIGURE 3-78 The vehicle identification number (VIN) is one of the most important pieces of information on the car. Learning how to read a VIN is an important skill.

throughout the vehicle, each for a different purpose. Knowing what these tags are for and their common locations can save you a lot of time when you need to track down certain pieces of information about a vehicle.

■ ***Vehicle Emissions Control Information Decal.*** Located under the hood of each vehicle is the **emissions decal** (**Figure 3-80**). This decal contains such information as the emission year for which the vehicle is certified, engine size, installed emission control devices, and what emission standards the vehicle meets. Other information on the decal may include spark plug gap, valve lash settings, and a vacuum schematic. This decal may also tell how to adjust the idle and ignition timing on older vehicles.

Callout	Description
	Vehicle Certification Label
	The vehicle certification label is located on the driver side of the B pillar and displays the following assessments: • Gross Vehicle Weight Rating (GVWR) • Gross Axle Weight Rating (GAWR), front and rear • The gross vehicle weight (GVW) is the weight of the vehicle and everything it carries. The gross vehicle weight must not exceed the GVWR. Include the following items when figuring the GVW: o The base vehicle weight (factory weight) o The weight of all vehicle accessories o The weight of the driver and the passengers o The weight of the cargo
1	Name of Manufacturer
2	Gross Vehicle Weight Rating
3	Gross Axle Weight Rating (FRONT, REAR)
4	Canadian Safety Mark (w/RPO Z49)
5	Certification Statement
6	Vehicle Class Type (Pass Car, etc.)
7	Vehicle Identification Number
8	Date of Manufacture (Mo/Yr)
	Tire Placard
	The tire placard label is located on the driver side of the B pillar and displays the following assessments:
9	Specified Occupant Seating Positions
10	Maximum Vehicle Capacity Weight
11	Tire Pressure, Front, Rear, and Spare (Cold)
12	Original Equipment Tire Size
	Service Parts ID Label
	The vehicle service parts identification label is located in the rear compartment under the spare tire cover. The label is use to help identify the vehicle original parts and options.
13	Vehicle Identification Number
14	Engineering Model Number (Vehicle Division, Line and Body Style)
15	Interior Trim Level and Decor
16	Exterior (Paint Color) WA Number
17	Paint Technology
18	Special Order Paint Colors and Numbers
19	Vehicle Option Content
20	**Anti-Theft Label** The Federal law requires that General Motors label certain body parts on this vehicle with the VIN. The purpose of the law is to reduce the number of motor vehicle thefts by helping in the tracing and recovery of parts from stolen vehicles. Labels are permanently affixed to an interior surface of the part. The label on the replacement part contains the letter R, the manufacturer's logo, and the DOT symbol. The anti-theft label must be covered before any painting, and rustproofing procedures, and uncovered after the procedures. Failure to follow the precautionary steps may result in liability for violation of the Federal Vehicle Theft Prevention Standard and possible suspicion to the owner that the part was stolen.

FIGURE 3-78 (continued)

If the door decal containing the production date is missing from the vehicle, refer to the vehicle emissions control information (VECI) because it will show for which model year the vehicle is certified.

■ ***Door Decals.*** The stickers located on the door jambs can contain a lot of information, though each manufacturer will supply different details on one or more decals on one or more door jambs. An example of a Ford door decal is shown in **Figure 3-81** and a GM decal in **Figure 3-82**. Typically, the door decal will contain the vehicle production date and gross vehicle weight, and may contain information about wheelbase, transmission and differential identification, and paint codes.

The tire information decal has information about tire size, wheel size, and the recommended inflation pressures. This sticker may also list any optional tire sizes for that vehicle (**Figure 3-83**). The tire information may be located on a glove box door, console door, fuel door, or with the spare tire.

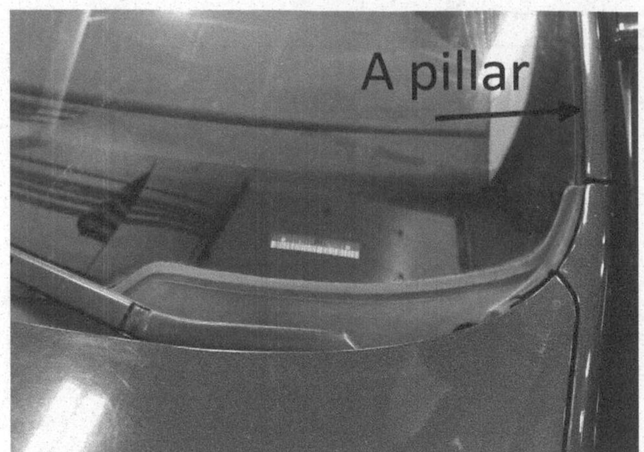

FIGURE 3-79 The VIN plate has to be located on the dash near the driver's A pillar on modern cars and trucks. It is also found on many other decals, body panels, and even glass to help deter theft.

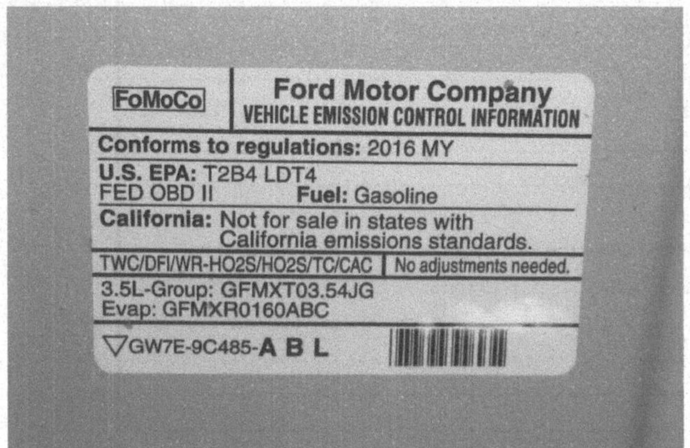

FIGURE 3-80 The vehicle emission control information (VECI) decal is located under the hood and contains information specific to the engine and emission control devices installed in the vehicle.

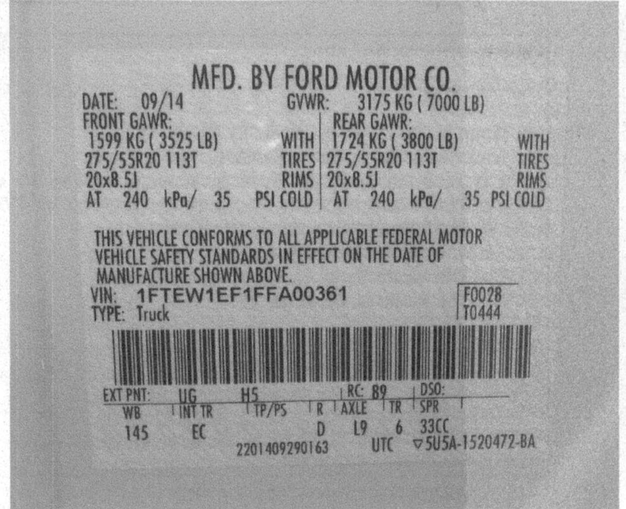

FIGURE 3-81 This door decal contains information such as production date, tire specifications, paint and trim colors, vehicle wheel base, rear axle ratio, and more.

FIGURE 3-82 This door decal gives build and weight information.

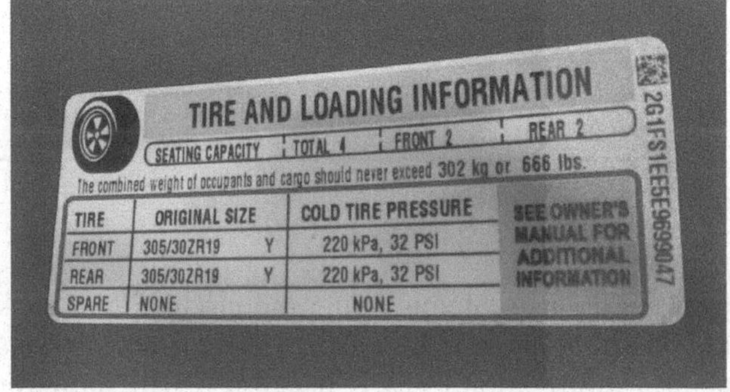

FIGURE 3-83 Tire decals show the tire and wheel size for the vehicle, the recommended tire pressure, and spare tire information.

Warning decals are not uncommon on modern cars and trucks (**Figure 3-84**). These warning decals are used to inform occupants and those who work on the vehicles of dangerous substances used in the construction of the vehicle.

Paint and trim code stickers may also be located on the door jamb (**Figure 3-85**).

■ ***Calibration and Production Decals.*** Some manufacturers have **calibration decals** on the engine or transmission (**Figure 3-86**). This decal provides specific information about emissions system devices and calibrations. A production tag is often used on the engine and transmission or transaxle (**Figure 3-87**). This sticker can indicate when the transmission was assembled, at which plant and shift, and gear ranges.

■ ***SPO Tags.*** Vehicles produced by GM have a Service Parts Organization (SPO) (**Figure 3-88**). This tag contains information about all of the options installed on the vehicle. This includes items such as interior styles and colors, radio options, and many others. As you can see, however, the tag contains a lot of three-digit codes, which unless you know exactly which one you are

FIGURE 3-84 If a vehicle contains hazardous substances, such as mercury, warning labels on the door will indicate which components.

FIGURE 3-85 An example of a paint code sticker.

FIGURE 3-86 The calibration decal has information about the engine and emission control system.

FIGURE 3-87 An example of a transmission ID tag.

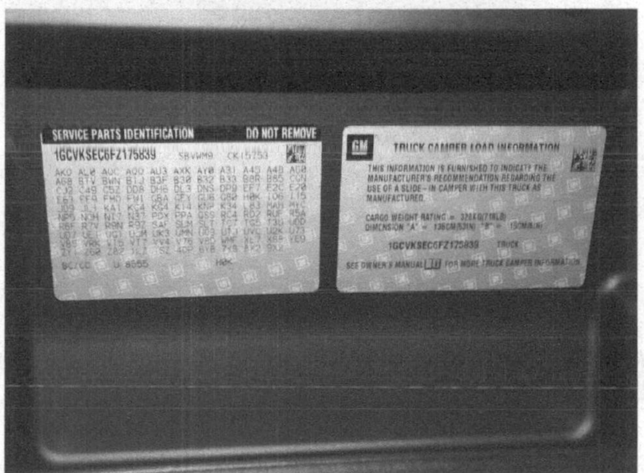

FIGURE 3-88 An example of a General Motors Service Parts Organization (SPO) decal. This contains information about installed options and colors used in the vehicle.

looking for, mean nothing. Fortunately, these codes can be deciphered by using the proper service information, from Internet sources, and by a GM parts department. Other vehicle manufacturers use a similar system for tracking the options installed on the vehicle.

■ ***Accessing Identification Labels under the Hood.*** Because of the number and varying locations of labels throughout a vehicle, it is worth noting how to access some of these labels.

VECI and engine calibration tags are located under the hood in the engine compartment. Also under the hood are decals with information about the air conditioning system, cooling system, accessory belt routing, and possibly others. To get to these labels, you must first open the hood. Locate the interior hood release mechanism (**Figure 3-89**). This releases the primary hood latch. Once the interior release is pulled, the hood should pop up slightly, allowing access to the secondary release. This release is used to prevent the hood from accidently

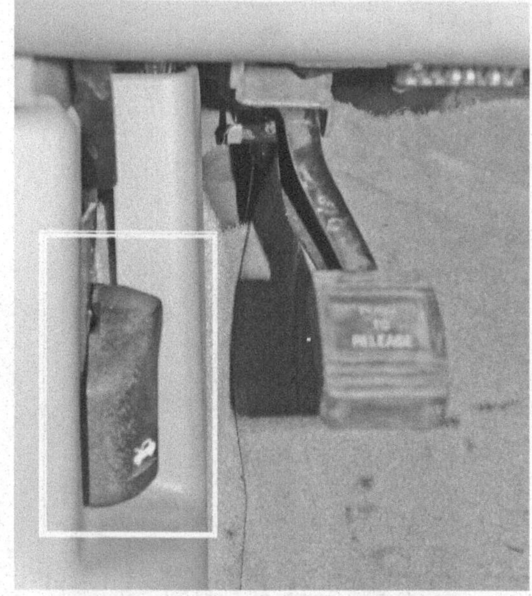

(a)

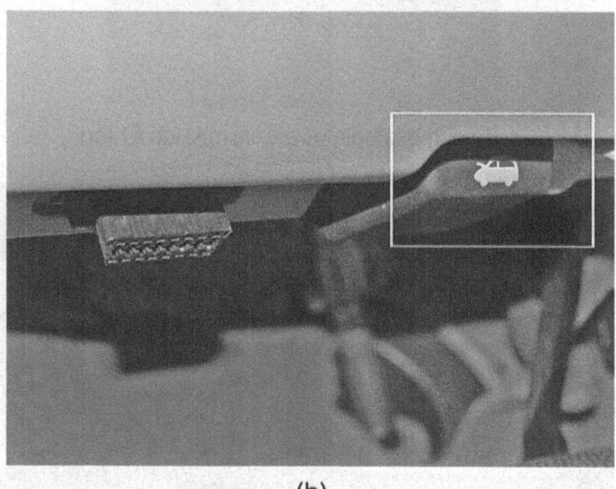

(b)

FIGURE 3-89 Hood release levers on the left (driver's) side kick panel.

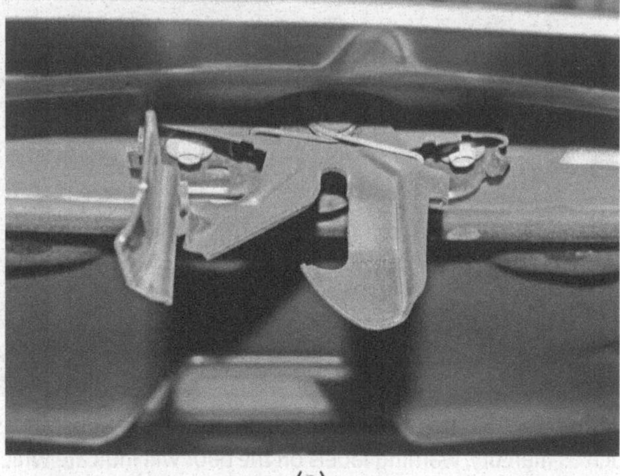

(a)

(b)

FIGURE 3-90 The hood latch has the secondary hood release. Locate the release lever to fully open the hood.

flying open if the driver pulls the primary release by accident while driving (**Figure 3-90**). The secondary release lever may be located under the hood in the center, or left or right of center, or may be tucked into or behind the grill. The locations and operation of the secondary release vary greatly among makes and models of vehicles.

Once the secondary latch is released, raise the hood. Many vehicles have a hood prop that must be set into a slot in the hood (**Figure 3-91**). Other vehicles use either gas-charged hood supports, similar to shock absorbers, to keep the hood raised or a set of springs that support the hood. Before getting under the hood, make sure that it is securely supported on the hood prop or that the hood struts or springs will support the weight. Hood struts tend to lose their ability to keep the hood up over time, resulting in a situation where the hood can suddenly drop.

Service Orders

A **service order** or **repair order (RO)** is a contract between the customer and the shop. It contains pertinent customer information, vehicle information, and a record of services and repairs. A service order is also a legal document, designed for both the shop and the customer.

CUSTOMER RECORDS

One of the main purposes of the service order is to maintain customer records. This database contains customers' contact information and the service histories for their vehicles. Many shops also use this database to send reminders to their customers when certain services are due, such as engine oil changes.

FIGURE 3-91 Many vehicles have a simple prop rod to hold the hood open. Locate the recess or hole in the hood for the prop rod. Do not forget to remove and secure the prop rod before trying to close the hood.

■ ***Components of a Service Order.*** When a vehicle is in for service, a service order should be completed (**Figure 3-92**). A service order allows the shop to provide a written estimate and to document what services are performed. It should contain the following information:

1. Customer name, address, and contact information. Include cell phone numbers and email addresses as needed to ensure your ability to contact the customer easily.
2. Vehicle information, such as the year, make, model, license number, and VIN.
3. A complete and accurate description of the complaint(s), if applicable.
4. A description of the services, repairs, or diagnostic procedures to be performed and their costs.
5. The method of approving additional repairs or services.
6. The customer's signature and receipt of an estimate.

Many states have laws requiring the shop to provide an estimate for work if the expected cost is over a certain amount, such as $25. The repair order should include adequate space for the documentation of communication with the customer about estimates and delivery times.

Most shops use a computerized repair order system. This allows for easy storage and retrieval of customer records without the necessity of storing large amounts of paper records. Electronic repair order generation also allows for maintaining a customer database, useful for quickly finding customer information, past repair histories, and comments and recommendations made during previous visits (**Figure 3-93**).

Computerized repair orders also allow for more precise job dispatching (assigning) to the technicians. While some shops may send repair orders out to the next available technician, many shops select what a technician will receive based on his or her abilities. Doing this ensures that the most qualified technicians receive the work, increasing productivity and customer satisfaction.

Once a technician receives the work order, notes regarding diagnosis, needed parts and labor, and other recommendations can be entered or written on the paper copy and sent back to the service advisor. The service advisor will then cost-out the order and contact the customer. Accurate and complete documentation of diagnostic steps, test results, and recommendations are critical. Technician and shop income can be reduced by incomplete documentation. Some states allow for a slight amount of difference between a repair estimate and the final cost; this difference is typically 10% or less of the total bill. If the expected cost of repairs does change due to unforeseen problems, the customer needs to be notified and given a revised estimate.

■ ***Technician Records.*** The service order also provides a method of determining how much technicians are paid. This is because many shops pay their technicians a commission, often based on how much labor is produced or in some cases, the total amount of parts and labor sales. Regardless of which method is used, if a technician's pay is based on a commission, it is important for the technician to correctly document all the work he or she performed.

When you are completing a service order, the technician or the service advisor fills in part and labor codes (**Figure 3-94**). These codes identify the parts used and labor operations for the vehicle. These codes are important, especially when performing warranty repairs, so that both the technician and the shop are properly reimbursed.

■ ***Legal Issues.*** Because the service order is a legal document, a type of contract between the customer and the shop, there is usually some type of legal disclaimer that the customer signs. An example of this may state that

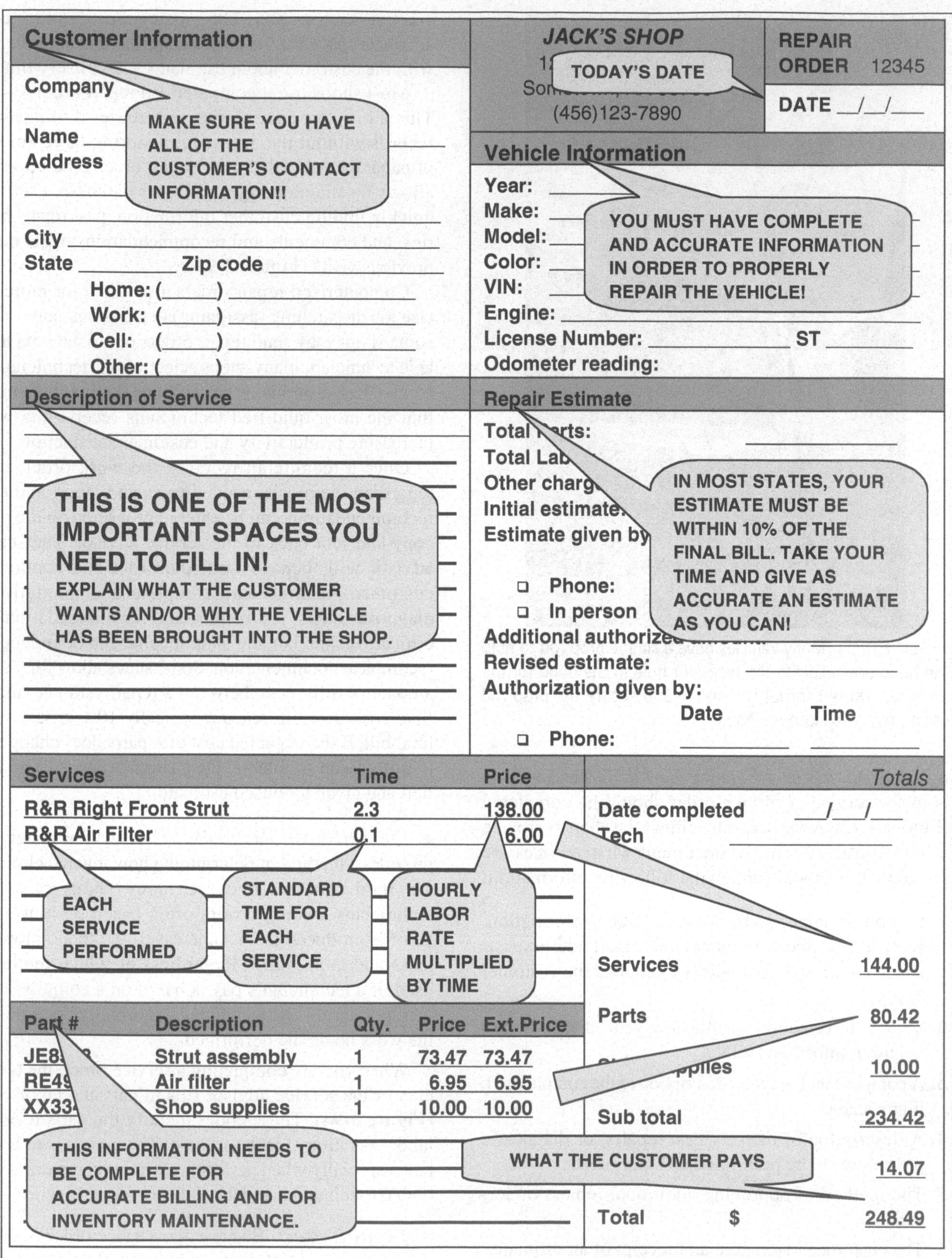

Customer Information

Company

Name

Address

City

State Zip code

Home: ( )

Work: ( )

Cell: ( )

Other: ( )

JACK'S SHOP

(456)123-7890

REPAIR ORDER 12345

DATE __/__/___

Vehicle Information

Year:

Make:

Model:

Color:

VIN:

Engine:

License Number: ST

Odometer reading:

Description of Service

Repair Estimate

Total Parts:

Total Lab

Other charg

Initial estimate:

Estimate given by

❑ Phone:

❑ In person

Additional authorized

Revised estimate:

Authorization given by:

Date Time

❑ Phone:

Services	Time	Price
R&R Right Front Strut	2.3	138.00
R&R Air Filter	0.1	6.00

Part #	Description	Qty.	Price	Ext.Price
JE8	Strut assembly	1	73.47	73.47
RE49	Air filter	1	6.95	6.95
XX33	Shop supplies	1	10.00	10.00

Totals

Date completed __/__/__

Tech

Services	144.00
Parts	80.42
pplies	10.00
Sub total	234.42
	14.07
Total $	248.49

FIGURE 3-92 A repair order is a legal contract between the shop and the customer. All relevant information should be included on the RO.

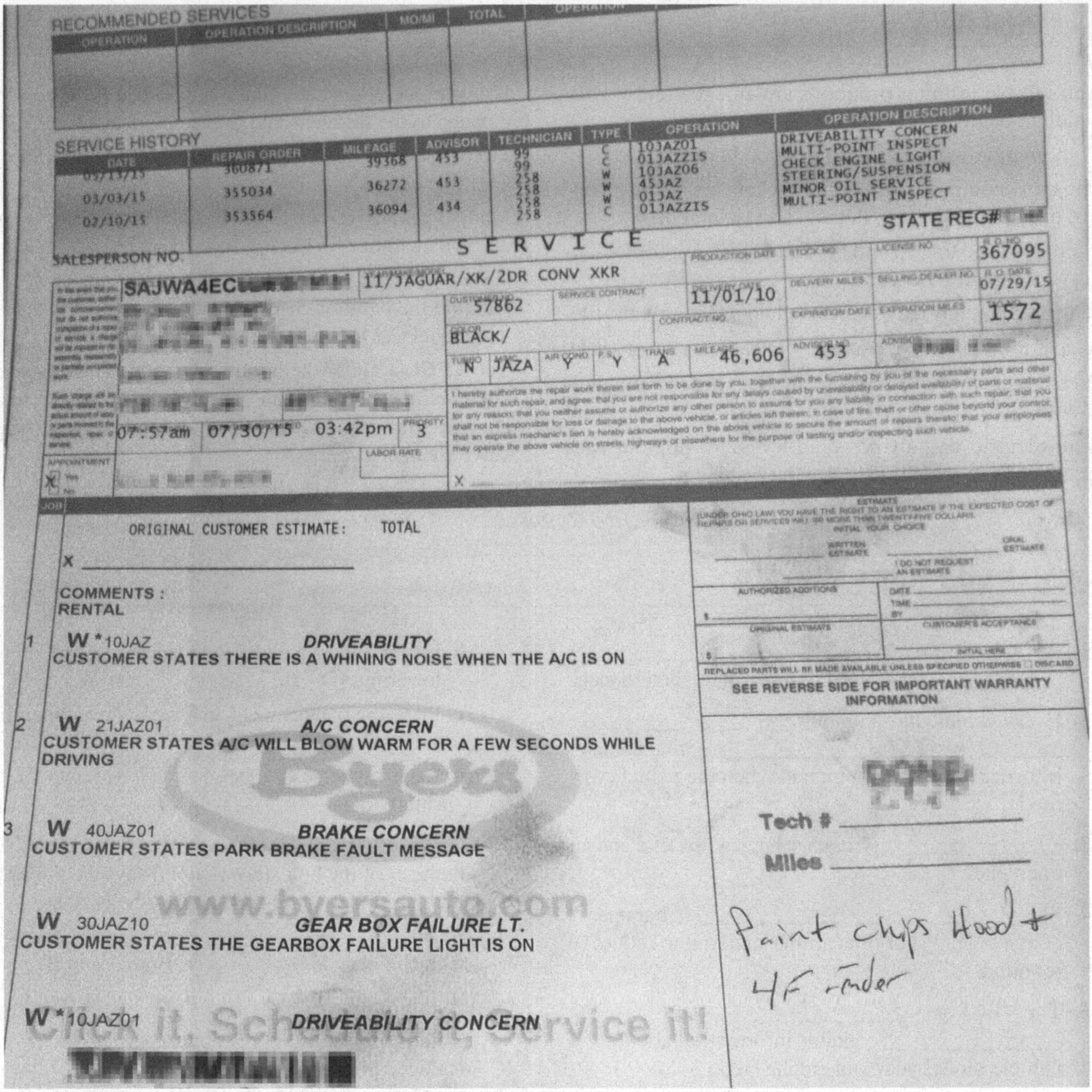

```
RECOMMENDED SERVICES
OPERATION | OPERATION DESCRIPTION | MO/MI | TOTAL

SERVICE HISTORY
DATE      REPAIR ORDER  MILEAGE  ADVISOR  TECHNICIAN  TYPE  OPERATION  OPERATION DESCRIPTION
                                           99          C     10JAZ01    DRIVEABILITY CONCERN
                                           99          C     01JAZZIS   MULTI-POINT INSPECT
03/03/15  355034        36272    453       258         W     10JAZ06    CHECK ENGINE LIGHT
                                           258         W     45JAZ      STEERING/SUSPENSION
02/10/15  353564        36094    434       258         W     01JAZ      MINOR OIL SERVICE
                                           258         C     01JAZZIS   MULTI-POINT INSPECT
                                                                         STATE REG#
SALESPERSON NO.        S E R V I C E                                    R.O. NO. 367095
11/JAGUAR/XK/2DR CONV XKR                                                R.O. DATE 07/29/15
CUSTOMER NO. 57862        DELIVERY DATE 11/01/10                         1572
COLOR BLACK/
TURBO N  JAZA  AIR COND. Y  P.S. Y  TRANS A  MILEAGE 46,606  ADVISOR NO. 453
07:57am  07/30/15  03:42pm  PRIORITY 3
APPOINTMENT  X Yes  No

JOB
ORIGINAL CUSTOMER ESTIMATE:   TOTAL
X ____________________
COMMENTS :
RENTAL
1  W *10JAZ        DRIVEABILITY
   CUSTOMER STATES THERE IS A WHINING NOISE WHEN THE A/C IS ON
2  W  21JAZ01      A/C CONCERN
   CUSTOMER STATES A/C WILL BLOW WARM FOR A FEW SECONDS WHILE
   DRIVING
3  W  40JAZ01      BRAKE CONCERN
   CUSTOMER STATES PARK BRAKE FAULT MESSAGE
   W  30JAZ10      GEAR BOX FAILURE LT.
   CUSTOMER STATES THE GEARBOX FAILURE LIGHT IS ON
   W *10JAZ01      DRIVEABILITY CONCERN

SEE REVERSE SIDE FOR IMPORTANT WARRANTY INFORMATION
Tech # ____________
Miles ____________
Paint chips Hood + LF fender
www.byersauto.com
Click it. Schedule it. Service it!
```

FIGURE 3-93 An example of an electronic RO. Note the service history and current concerns.

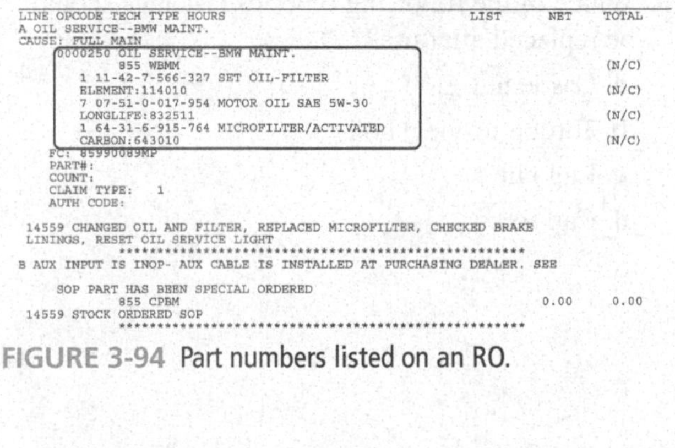

```
LINE OPCODE TECH TYPE HOURS                              LIST     NET     TOTAL
A OIL SERVICE--BMW MAINT.
CAUSE: FULL MAIN
    0000250 OIL SERVICE--BMW MAINT.
            855 WBMM                                                      (N/C)
        1 11-42-7-566-327 SET OIL-FILTER
        ELEMENT:114010                                                    (N/C)
        7 07-51-0-017-954 MOTOR OIL SAE 5W-30
        LONGLIFE:832511                                                   (N/C)
        1 64-31-6-915-764 MICROFILTER/ACTIVATED
        CARBON:643010                                                     (N/C)
    FC: 85990089MP
    PART#:
    COUNT:
    CLAIM TYPE:   1
    AUTH CODE:

 14559 CHANGED OIL AND FILTER, REPLACED MICROFILTER, CHECKED BRAKE
 LININGS, RESET OIL SERVICE LIGHT
            ****************************************************
B AUX INPUT IS INOP- AUX CABLE IS INSTALLED AT PURCHASING DEALER. SEE

     SOP PART HAS BEEN SPECIAL ORDERED
            855 CPBM                                              0.00    0.00
 14559 STOCK ORDERED SOP
            ****************************************************
```

FIGURE 3-94 Part numbers listed on an RO.

once the customer agrees to the estimate and authorizes the shop to begin repairs, labor charges will be incurred if the customer halts the repair before completion. Another example is that customers have the right to keep their old parts. If there is a core charge on a part, the customers must pay the core charge to obtain the old part.

SUMMARY

Automotive training programs are an excellent way in which to begin a career in the auto repair industry.

Combination wrenches have open and box ends.

The most common ratchet and socket sizes for automotive use are the ¼-inch, ⅜-inch, and ½-inch drive.

Pliers are used to grip and hold an object, not for loosening or tightening fasteners.

Damaged threads may be cleaned up using a thread file.

A tap is used to clean or create new threads in a nut or a hole.

Dies are used to clean or create new threads on a bolt or stud.

Both electric and air-powered tools are popular with technicians to save time loosening fasteners.

Use a torque wrench to tighten nuts and bolts to prevent damage to the fasteners and to the components.

Friction and nylon lock nuts are meant to be replaced after one use.

Many types are lifts are found in auto shops. Never operate a lift without first being trained on its proper setup and operation.

Get into the habit of checking the vehicle's year, make, and model (YMM) when you begin to work on it.

The VECI is under the hood and contains model year, engine size, and emission control information. It may contain other information, such as spark plug gap and valve clearance specs.

REVIEW QUESTIONS

1. A ______________ ______________________ __________set is used to repair damaged threads.
2. Every vehicle has a unique ______________ ______________ ______________ that contains information about where and when the vehicle was made.
3. ______________wrenches have a box end and an open end.
4. A ______________ ______________hammer contains lead shot or a similar material to reduce rebound.
5. The VECI or ______________ ______________contains information about vehicle model year and engine size.
6. *Technician A* says information commonly found on door decals includes vehicle build date, gross weight, and recommended tire inflation pressure. *Technician B* says that door decals may also contain paint codes and other installed option data. Who is correct?
   a. Technician A
   b. Technician B
   c. Both A and B
   d. Neither A nor B
7. Which of the following is not a safety precaution when using an above-ground lift?
   a. Read and understand the operating instructions of the lift before use.
   b. Ensure the area under the vehicle is clear before lowering.
   c. Place the pad contacts on any flat area under the vehicle.
   d. Raise to the desired height and lower onto the safety locks.
8. *Technician A* says bolt thread pitch is based on the diameter of the bolt. *Technician B* says the thread pitch on some metric and standard bolts is the same.
   a. Technician A **c.** Both A and B
   b. Technician B **d.** Neither A nor B
9. Which of the following is not used to repair damaged threads?
   a. Tap **c.** Thread file
   b. Thread insert **d.** Thread puller
10. Which of the following types of fasteners should be replaced after use??
    a. Castle nut
    b. Torque-to-yield bolt
    c. Lug nut
    d. Cap screw

CHAPTER 4

Basic Technician Skills

Chapter Objectives

At the conclusion of this chapter, you should be able to:

- Identify and demonstrate skills necessary for employment.
- Demonstrate proper communication skills.
- Identify and apply math skills related to automotive applications.
- Identify and apply science skills related to automotive applications.

KEY TERMS

aftermarket	**initiative**	**ratio**
body language	**linear measurement**	**résumé**
chemical reaction	**metric measurement**	**specific gravity**
employability skills	**motivation**	**torque**
employment plan	**nonverbal communication**	**vacuum**
flat-rate guide	**OE parts**	**warranty time**
hydraulics	**percentage**	**work ethic**

Long-term success in the automotive industry requires more than being able to diagnose and repair cars and trucks. Technicians need lifelong learning skills to adapt to constantly changing technology. Employers need employees who have a good work ethic, can work as part of a team, and have excellent communication and interpersonal skills.

Education for auto technology students should include science, language arts, math, and computers skills in addition to formal automotive training. The modern automotive industry needs technicians who are knowledgeable, can adapt, and learn new skills.

Years ago, one service manual could contain most of the service and repair information needed for most vehicles likely to be encountered. Today, each year of each model requires a complete service manual with hundreds of pages. **Figure 4-1** shows an example of an older repair manual, which covers multiple years and makes of vehicles. This type of manual has been replaced in automotive shops by computerized information systems. Online service information systems (e.g., AllData, Identifix, etc.) contain millions of pages of service data. Technicians who do not attend training, or at the very least read about new technology and service practices, will soon find themselves out-of-date as the amount of new data continues to increase.

Employability Skills

In general, regardless of what type of job you are in, employers agree that there are eight basic skills you should have, these are called **employability skills** and are: communication, teamwork, problem solving, initiative, organization, self-management, being willing to learn, and being adept with technology. Being employed as an automotive technician requires using more than just hands-on skills to service vehicles. In many shops, you will find that you are part of a group or team who are all working for common goals but each with individual responsibilities. To function as part of the group, you will need employability skills. Sometimes called *soft skills*, these and other skills are discussed in this chapter.

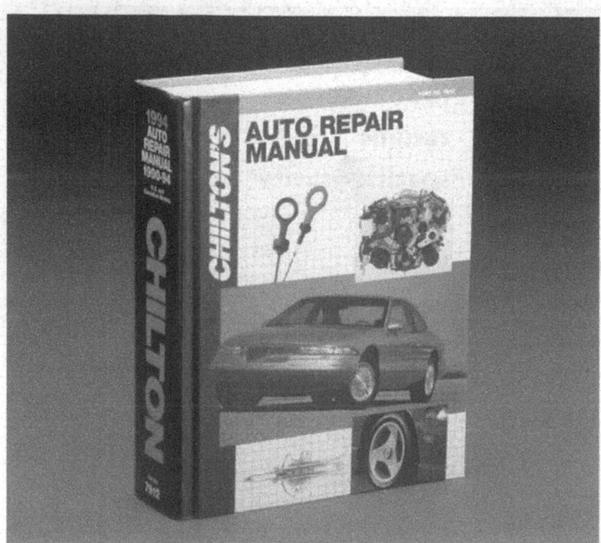

FIGURE 4-1 An example of a repair manual that covers several years and makes of vehicles. This type of manual has largely been replaced by electronic service information.

Working in an automotive lab is very different from a traditional classroom environment. You will be working on real vehicles using tools and equipment used in modern auto repair shops. Because of this, you will be expected to act accordingly. Why is behavior so important? Your behavior not only affects your safety but also it reflects your professionalism and work ethic.

PROFESSIONALISM

To be a professional technician, you should try to perform every service and repair correctly every time. To accomplish this, there are several things to consider:

- Perform all work safely and follow the service procedures.
- A professional takes his or her time to make sure each procedure is done properly, always double-checking his or her work. Doing a job properly means not taking shortcuts.
- A good **work ethic** means you are dependable, trustworthy, and show initiative.

■ ***Work Ethic.*** Of all the qualities employers want most in their employees, a good work ethic is at the top of the list. Countless employers have stated, "Give me employees who show up to work every day and have the right attitude, and I will train them." One of the biggest reasons for not hiring or retaining student-technicians is poor attendance. Poor attendance costs the employer money. If a shop employing five technicians has one person who calls off or is frequently late, the other technicians must pick up the workload. This creates problems for the customers and the shop as vehicles may not be completed on time. Many shops have been forced to let good technicians go simply because they missed too much work or were late to work too often.

The "right attitude" is harder to define or measure but often means having respect for people and property, being reliable, showing initiative, and exhibiting good social skills. A good work ethic and positive attitude are intrinsic values; things that a person internally believes have value.

Some shops operate with their technicians as part of a team, working and problem solving as a group. Being a team player is important even if you are not employed in one of those shops. Unless you work for yourself in a one-person shop (even a one-person shop has to deal with the public, the parts suppliers, and others), you will be working with other people. Being a team player means being cooperative, doing your fair share, being honest, and keeping commitments. Another part of being part of a team is keeping others informed. As a technician, you will have to keep customers, other employees, and your supervisor informed about your work.

■ ***Motivation.*** **Motivation** can take many forms and can be very different from person to person. You should understand your own motivations for wanting to be in the auto repair industry. For some, it is because they like the challenge of fixing things. Some simply enjoy working on and around cars and trucks. For others, it may be money.

Motivators are either internal or external. An internal motivator is something that you do because it makes you happy, while an external motivator comes from outside, such as earning a salary. Employers want workers who are motivated by more than just a paycheck. Studies have shown that money as a motivator tends to last for a limited amount of time. To offset this, other factors, such as respect, recognition, and a sense of belonging, tend to be much stronger motivators than money alone. Employees who are truly motivated tend to look for better ways to perform their jobs, produce better quality work, and are more productive than those who are not.

■ ***Initiative.*** **Initiative** is a behavior that causes a person to start to do something or to help out in some way without having been asked to do so. In practical terms, initiative is the act of moving from one task to the next without being told to do so, or if once a task is completed, engaging in some productive activity, such as cleaning, until the next task is assigned. Employers want employees with initiative, who can move through different tasks without being constantly reminded of what needs to be done next.

FINDING EMPLOYMENT

Before you are able to display your skills, you must first find a job. As discussed in Chapter 1, there are many different career options in the automotive industry. Regardless of what aspect of the industry interests you, there are basic skills and characteristics necessary to getting the job you want.

■ ***Developing an Employment Plan.*** Before you enrolled in an automotive technology course, hopefully you spent some time thinking about why you want to work in the automotive industry. For many students, the answer is as simple as wanting to work on cars and trucks, while for others, it may be an interest in the industry but not necessarily as a technician. Whatever your motivation is, it should relate to your interests in automobiles.

But just having an interest in cars and trucks may not be enough to be successful as a technician. To be successful as a technician, you also need some aptitude. Many students, especially at the high school level, do not really know if they have the skills to become a technician until they have spent some time, several weeks at least, in an auto tech class. Sometimes students who are really interested in the auto industry just do not have the mechanical aptitude to really be successful. That does not mean that they cannot be successful in the automotive industry, just that the basic skills to be successful technicians are lacking. One of the best aspects of the automotive industry is that there are many different types of career options available, as discussed in Chapter 1.

> **Word Wall**
> *Aptitude*—A natural or acquired ability or talent to do something.

For many students, developing an employment plan is a helpful tool. An **employment plan** often consists of a list of personal strengths and weaknesses, short- and long-term goals, and steps to accomplish the goals (**Figure 4-2**).

Goal 1: Get a job working as an auto technician before graduation.
Step 1: Maintain or improve grade point average in auto and other classes.
Step 2: Continue perfect attendance this school year.
Step 3: Get recommendation letters from my teachers.
Step 4: Fill out applications at local shops.
Goal 2: Start automotive degree program at Delmar Community College.
Step 1: Complete automotive course and graduate from high school.
Step 2: Apply at college and take placement tests.
Step 3: Meet with advisor about classes and advanced credit.
Step 4: Enroll in classes.
Strengths: good hands-on skills, excellent attendance, really want to work on cars.
Areas for improvement: math, not sure how to start college program.

FIGURE 4-2 List your goals and steps required to reach those goals as part of an employment plan. Review your plan, make changes, and update it often.

This form is only provided as a service and a guide. It may not be compliant with local laws and is not warranted as such. This form may need to be modified to fit local laws and regulations.

FOR OFFICE USE ONLY
EMP. NO. ______
W4
WORKING PAPER # ______

EMPLOYMENT APPLICATION

PERSONAL INFORMATION: *(please print clearly)*

NAME Rob (First) D (Middle Initial) Thompson (Last) SOC. SEC. # / TAX ID NO. 000-00-0000

ADDRESS 123 Maple St. CITY Anytown STATE/PROVINCE Ohio ZIP/POSTAL CODE 55555

TELEPHONE (555) 555-5555 Have you ever worked for Delmar's Automotive Shop before? ☐ Yes ☒ No If yes, when/where?

Are you 16 years of age or over? ☒ Yes ☐ No (Proof of age or a work permit may be required.)

In Case of Emergency Notify:

NAME Thompson (Last) Beth (First) R (Middle) TELEPHONE (555) 555-5555 (Area Code)

ADDRESS 123 Maple St CITY Anytown STATE/PROVINCE Ohio ZIP/POSTAL CODE 55555

AVAILABILITY :

Are you legally able to be employed in this country? ☒ Yes ☐ No (If hired, verification will be required by law)

What type of position are you seeking? ☒ Part time ☐ Full time ☐ Seasonal ☐ Temporary

Are you able to meet the attendance requirements of the position? ☒ Yes ☐ No

HOURS AVAILABLE		S	M	T	W	T	F	S
	From		2	2	2	2	2	8
	To		7	7	7	7	7	5

Total hours available per week 30
Date available to start work 4·18·17

SCHOOL MOST RECENTLY ATTENDED :

NAME Ohio High School ADDRESS 321 School St.

CITY Anytown STATE Ohio TELEPHONE (555) 555 1234

TEACHER OR COUNSELOR Smith LAST GRADE COMPLETED 11 GRADE AVERAGE B+

GRADUATED? ☐ Yes ☒ No NOW ENROLLED? ☒ Yes ☐ No

Sports or activities? Skills USA

MOST RECENT EMPLOYMENT :

Company N/A Address ______
City ______ State ______ Telephone ( )
Position ______ Supervisor ______ Dates worked: From ______ To ______
Wage ______ Reason for leaving ______
Mgmt. ref. ck. done by ______

Company N/A Address ______
City ______ State ______ Telephone ( )
Position ______ Supervisor ______ Dates worked: From ______ To ______
Wage ______ Reason for leaving ______
Mgmt. ref. ck. done by ______

Do we have your permission to contact your current employer? ☐ Yes ☐ No
If NO, please explain: ______

REFERENCES: (Please do not use family members)

Name: Mr. Smith Telephone: (555) 555 1234 Years Known 1½
Address Ohio High School, 321 School St. City Anytown State Ohio

Name: Mrs. Jones Telephone: (555) 555 1234 Years Known 3½
Address Ohio High School, 321 School St City Anytown State Oh

WE ARE AN EQUAL OPPORTUNITY EMPLOYER

FIGURE 4-3 A legible and complete application is the first step in getting a job.

This form is only provided as a service and a guide. It may not be compliant with local laws and is not warranted as such. This form may need to be modified to fit local laws and regulations.

FOR OFFICE USE ONLY

EMP. NO. ______
W4 ______
WORKING PAPER # ______

EMPLOYMENT APPLICATION

PERSONAL INFORMATION: *(please print clearly)*

NAME Rob (First) ______ (Middle Initial) Thompson (Last) SOC. SEC. # / TAX ID NO. ?

ADDRESS 123 Maple CITY Anytown STATE/PROVINCE OH ZIP/POSTAL CODE 55555

TELEPHONE (555) 555-5555 Have you ever worked for Delmar's Automotive Shop before? ☒ Yes ☒ No If yes, when/where?

Are you 16 years of age or over? ☒ Yes ☐ No (Proof of age or a work permit may be required.)

In Case of Emergency Notify:

NAME my Mom - Beth (Last) (First) (Middle) TELEPHONE ( ) same (Area Code)

ADDRESS same CITY ______ STATE/PROVINCE ______ ZIP/POSTAL CODE ______

AVAILABILITY :

Are you legally able to be employed in this country? ☐ Yes ☐ No (If hired, verification will be required by law)

What type of position are you seeking? ☒ Part time ☒ Full time ☒ Seasonal ☒ Temporary

Are you able to meet the attendance requirements of the position? ☒ Yes ☐ No

HOURS AVAILABLE		S	M	T	W	T	F	S
	From						[scribbled out]	
	To						[scribbled out]	

I don't want to work on Fridays or Sat. Please

Total hours available per week any

Date available to start work now

SCHOOL MOST RECENTLY ATTENDED :

NAME OHS ADDRESS school St.

CITY anytown STATE Ohio TELEPHONE ( )

TEACHER OR COUNSELOR Swartz LAST GRADE COMPLETED 11 GRADE AVERAGE yes

GRADUATED? ☐ Yes ☒ No NOW ENROLLED? ☒ Yes ☐ No

Sports or activities? went to football games

MOST RECENT EMPLOYMENT :

Company none Address ______
City ______ State ______ Telephone ( )
Position ______ Supervisor ______ Dates worked: From ______ To ______
Wage ______ Reason for leaving ______
Mgmt. ref. ck. done by ______

Company none Address ______
City ______ State ______ Telephone ( )
Position ______ Supervisor ______ Dates worked: From ______ To ______
Wage ______ Reason for leaving ______
Mgmt. ref. ck. done by ______

Do we have your permission to contact your current employer? ☐ Yes ☐ No

If NO, please explain: ______

REFERENCES: (Please do not use family members)

Name: Beth Thompson mom Telephone: ( ) Years Known ______
Address same as mine City ______ State ______
Name: Mr. Smith Telephone: ( ) Years Known 1½
Address OHS City ______ State ______

WE ARE AN EQUAL OPPORTUNITY EMPLOYER

Please complete reverse side

FIGURE 4-4 This application is nearly incomprehensible and is likely to be thrown away as soon as it is received.

■ ***Seeking and Applying for Employment.*** The first impression you make to a prospective employer is on a job application (**Figures 4-3** and **4-4**). Notice how the first application is easy to read and has all the sections completed, while the second application has some parts that are nearly illegible. On which application do you think an employer is more likely to follow up?

Many larger companies use electronic applications (**Figure 4-5**). Some electronic applications can be completed anywhere and some you may have to complete at the place of business. Online applications may allow or require submitting a resume and other documents, such as proof of certifications, when filling out the online form. Some online applications even include tests of your knowledge or personality tests to see what type of person you are.

When you are going to shops to complete applications, dress appropriately and be prepared to complete the application in person. Although some companies may let you take the application home and return it completed, you may find yourself asked for an on-the-spot interview.

If you are completing applications at a place of business, be sure to take a pen and all the information you will need to complete the application. This means having names, addresses, and phone numbers of previous employers and references. Do not list references without having asked their permission ahead of time.

Most applications contain the same basic components: the applicant's personal information, including previous employment experience, education, and references. Applications for automotive positions will often include spaces for the your driver's license information. Many companies and even some postsecondary schools require a driver's license check. Technicians or students with severe infractions or excessive points, sometimes just two points on their license, may find it difficult or impossible to find an automotive job.

The application may also include questions about salary expectations. Be realistic about your pay expectations. As an entry-level employee, you will not earn as much as an experienced technician. However, you should find out what the average pay for an entry-level person is in your area as part of your search for employment. Your first job is not going to start you out at $50,000 per year. Have a realistic idea of what to expect when discussing pay.

Do not leave sections of the application blank. If an item does not apply to you, write not applicable (N/A) in that box. Be as legible as you possibly can. A poorly written application may not even get past the first step in the evaluation process and may end up in the trash.

When being considered for employment, be prepared for a drug test and a background check. Technicians are entrusted with the safety of their customers and their customers' families every time a vehicle comes in for

NARROW SEARCH

CATEGORY

Other (6) x

COMPANY

West Subaru >>

Southern Volkswagon >>

Eastside Kia >>

Northland Dodge (2) >>

Northern Chevrolet >>

JOB TITLE	COMPANY	LOCATION	DATE POSTED
Express Lube Technician Apply	West Subaru	Columbus, OH	11/16/2016
Parts Specialist Apply	Northland Dodge	Columbus, OH	11/04/2016
Service Advisor Apply	Eastside Kia	Columbus, OH	10/28/2016
Cashier Apply	Southern Volkswagen	Columbus, OH	10/26/2016
Express Lube Technician Apply	Northern Chevrolet	Columbus, OH	10/25/2016
Technician Apply	Northland Dodge	Columbus, OH	10/19/2016

FIGURE 4-5 Many large companies use electronic applications that may include a form of personality test as part of the screening process.

service. Technicians are also accountable for thousands of dollars in tools, equipment, and inventory. It is uncommon for an employer to hire anyone without having them first submit to a drug test and background check. Once the technician is hired, many companies continue to conduct required periodic and random retests.

Many companies may request or require a résumé. A **résumé** showcases your experience, education, and career goals and allows potential employers to quickly learn about you. Included with a résumé should be a cover letter. A cover letter introduces you to the company, states what job you are seeking, and gives a very brief reason why you should be selected for the job (**Figure 4-6**). The actual résumé should contain a career objective, your personal information, a concise version of your education and employment histories, and at least two references. An organized and effective résumé will provide enough information for the company to be interested in having you return to interview in person (**Figure 4-7**). Fortunately, modern technology has made it easier for you to create cover letters and résumés. A computer with a word-processing program, such as Microsoft Word, has built-in templates for creating many different types of documents, including cover letters and résumés. In addition, many resume templates and examples are available on the Internet.

Many people are now using online résumés or personal digital portfolios. These are web pages specifically used to provide information as a résumé and to show evidence of learning, certifications, and examples of work performed.

Once an application is submitted, you may be called for an interview. You may even be asked for an interview when you are completing applications, so it is important to dress accordingly and be prepared. The interview process can be intimidating, but it can be made less so by preparing in advance.

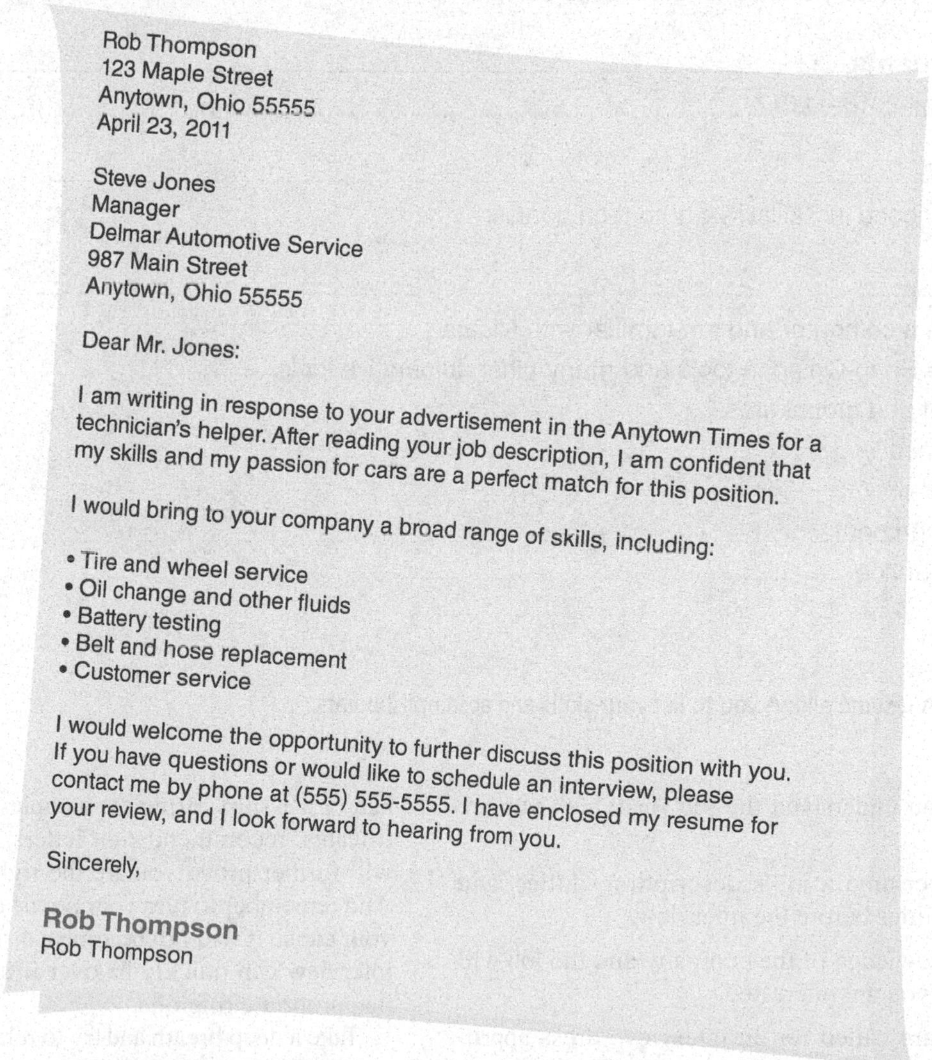

Rob Thompson
123 Maple Street
Anytown, Ohio 55555
April 23, 2011

Steve Jones
Manager
Delmar Automotive Service
987 Main Street
Anytown, Ohio 55555

Dear Mr. Jones:

I am writing in response to your advertisement in the Anytown Times for a technician's helper. After reading your job description, I am confident that my skills and my passion for cars are a perfect match for this position.

I would bring to your company a broad range of skills, including:

- Tire and wheel service
- Oil change and other fluids
- Battery testing
- Belt and hose replacement
- Customer service

I would welcome the opportunity to further discuss this position with you. If you have questions or would like to schedule an interview, please contact me by phone at (555) 555-5555. I have enclosed my resume for your review, and I look forward to hearing from you.

Sincerely,

Rob Thompson
Rob Thompson

FIGURE 4-6 A well-written cover letter should accompany your résumé. Be sure to get correct information about whom to send the letter and résumé to.

Rob Thompson

123 Maple Street
Anytown, Ohio 55555
(555) 555-5555

Education

Ohio High School – 321 School Street, Anytown, Ohio 55555
2013 – 2017 (expected graduation May, 2017)

Experience

Auto Tech class, 2015 –2017

- Attended auto tech class. Performed tire service, oil changes, battery service, replaced belts and hoses.
- Inspected and serviced brake systems
- Performed wheel alignments
- Performed battery and lighting systems service

Achievements

- Honor Roll 2016 – 2017

Activities

- Placed second in SkillsUSA auto tech contest

Skills

- I can use a computer and am familiar with Alldata.
- I can use Snap-On scan tools and many other automotive tools.
- Dismount and mount tires
- Tire balance
- Brake repairs
- Wheel alignment
- Battery service

FIGURE 4-7 A résumé allows you to list your skills and accomplishments.

- Be sure you understand the job for which you are applying.
- Try to determine a job's description, duties, and responsibilities before the interview.
- Having knowledge of the company and the job will also show you are interested.

When you are called for an interview, dress appropriately and arrive 10 minutes early. Appropriate dress includes a good pair of pants and a dress shirt that is tucked in. Do not wear baggy pants, pants with holes, a hat, or a T-shirt. Bring extra copies of your résumé, certificates, recommendation letters, and other items that will further prove you are the right person for the job. And remember to turn your phone off, or even leave it in your car so it will not become a distraction. A promising interview can quickly be over if it is interrupted by an inappropriate ringtone.

Take a deep breath and try to relax when the interview starts. The interview could be one-on-one, by a committee, or as a group. Regardless of the interview type, remember these pointers:

- Concentrate on your body language. Make and maintain eye contact. Have a firm handshake, but do not show off your strength. Sit straight and try not to look nervous.
- Look at the person who is talking to you, and maintain eye contact with him or her (and others if there is more than one interviewer).
- Often, applicants are asked, "Tell me (us) about yourself." As an opening question. Do not read from your résumé. Provide a few pieces of additional information not on the résumé, highlight any extracurricular or public service activities (or other academic or athletic accomplishments), include additional experience that pertains to the job you are seeking, and give a brief description of why you are the right person for the job. Do not give your life history or discuss family and friends; be concise and remain focused. Do not talk about things such as video game accomplishments, money, or topics that are not relevant to getting a job.
- When you are asked questions, keep your answers brief while providing enough detail to thoroughly answer the question. If you do not understand a question, ask the interviewer to repeat or to rephrase it. Do not guess at the intent of the question if you are unsure.
- Have a brief list of questions ready. Most interviewers will ask if you have any questions at the end of the interview. Not having any questions can give the impression that you are not very interested. Questions could include asking for more details about the job duties or requirements, what opportunities are available for advancement, if there are incentives for additional training or education, and when to expect to be notified of a decision. Asking for an estimate of the salary range is acceptable.
- At the end, take a moment to summarize your qualifications, work ethic, and other positive attributes that will close the interview on a positive note.
- A day or two after the interview, send a thank you note to the interviewer or contact person, thank the person for the opportunity to talk with him or her and mention that you are available for a follow-up interview if needed.

It is very important to follow up after completing an application or an interview. This shows the employer that you are serious about the job; you respect their time, and you appreciate the opportunity to meet with them. Successfully navigating the application and interview process will hopefully lead to a job offer. Resources for you to develop your interviewing skills may be available through your school's guidance department or through SkillsUSA.

■ ***Accepting Employment.*** Once you accept a job offer, be prepared to submit to a drug test and a background check.

Once these preliminary requirements are met, you will be required to complete paperwork for the employer, such as tax withholding forms, insurance forms, confidentiality statements, and possibly many other documents related to the company's policies.

Federal and state tax withholding forms, often referred to as W-4 forms, are required so that all federal, state, and local taxes can be deducted from your wages (**Figure 4-8**). You will need to declare the number of tax exemptions on the W-4. This, in part, determines how much tax is withheld. A single person, especially a teen still living at home, usually claims zero or no exemptions. Choosing no exemptions applies a standard tax rates so that enough tax is withheld and so that you should not have to pay additional taxes when you are filing your federal tax returns. If you claim exemptions, less tax will be withheld, which may sound good right now but may require you to have to pay taxes when filing your annual income taxes. If you are not sure what to claim on the W-4, you should consult a tax specialist to ensure you are not paying too much or too little in taxes.

Once you start receiving a paycheck, you may be surprised to see the difference between what you thought you were getting paid and what you actually get paid. The difference is called *gross pay* and *net pay*. Gross pay is your hourly wage times the number of hours you work. As shown in **Figure 4-9**, 32 hours at $12.00 per hour equals a gross wage of $384.00. However, after taxes are taken out, the net pay is $318.52. Expect to pay Federal, State, Local, Medicare and possibly additional taxes, and for medical insurance once employed.

Once you accept employment, it is time to live up to the expectations created during the application and interview process. While employers want technicians, even student technicians, who have competence in vehicle service and repair, there are other qualities that employers consider equally or even more important.

Communication Skills

Perhaps the most important of all nontechnical skills technicians must have are communication skills. The ability to communicate clearly to a customer, coworker, or supervisor is critical.

Often a customer does not want to talk to you. Do not take it personally. The customer has come to you because he or she is having trouble with his or her vehicle. The trouble may be something new or a reoccurrence

Cut here and give Form W-4 to your employer. Keep the top part for your records.

Form **W-4** Department of the Treasury Internal Revenue Service

Employee's Withholding Allowance Certificate

▶ **Whether you are entitled to claim a certain number of allowances or exemption from withholding is subject to review by the IRS. Your employer may be required to send a copy of this form to the IRS.**

OMB No. 1545-0074 **2010**

1 Type or print your first name and middle initial.	Last name	2 **Your social security number**
Home address (number and street or rural route)	3 ☐ Single ☐ Married ☐ Married, but withhold at higher Single rate. **Note.** If married, but legally separated, or spouse is a nonresident alien, check the "Single" box.	
City or town, state, and ZIP code	4 **If your last name differs from that shown on your social security card, check here. You must call 1-800-772-1213 for a replacement card.** ▶ ☐	

5 Total number of allowances you are claiming (from line **H** above **or** from the applicable worksheet on page 2) 5

6 Additional amount, if any, you want withheld from each paycheck . . . . 6 $

7 I claim exemption from withholding for 2010, and I certify that I meet **both** of the following conditions for exemption.

- Last year I had a right to a refund of **all** federal income tax withheld because I had **no** tax liability **and**
- This year I expect a refund of **all** federal income tax withheld because I expect to have **no** tax liability.

If you meet both conditions, write "Exempt" here . . . . ▶ 7

Under penalties of perjury, I declare that I have examined this certificate and to the best of my knowledge and belief, it is true, correct, and complete.

Employee's signature
(Form is not valid unless you sign it.) ▶ **Date** ▶

8 Employer's name and address (Employer: Complete lines 8 and 10 only if sending to the IRS.)	9 Office code (optional)	10 Employer identification number (EIN)

For Privacy Act and Paperwork Reduction Act Notice, see page 2. Cat. No. 10220Q Form **W-4** (2010)

FIGURE 4-8 Part of the hiring process includes completing paperwork. Make sure you fill out wage and tax forms correctly.

SMART Autmotive Repair

Period:	10/14/2016	Employee Name	Rob Thompson	Employee ID	0U812
Tax Status	1	Federal Allowance (From W-4)	0	Hours Worked	32
Hourly Rate	$12.00	Overtime Rate	$18.75	Sick Hours	0
Social Security Tax	$23.81	Federal Income Tax	$29.70	Vacation Hours	0
Medicare Tax	$3.20	State Tax	$5.57	Overtime Hours	0
Insurance Deduction	$0.00	Other Regular Deduction	$3.20	Gross Pay	$384.00
Total Taxes and Regular Deductions	$65.48	Other Deduction	$0.00	Total Taxes and Deductions	$67.85
				Net Pay	$318.52

SMART Automotive Repair, INC.
5150 Speed Way
Columbus, Ohio 43224

Advice number: 0000458852
Pay date: 10/21/2016

Deposited to the account of	account number	transit ABA	amount
Rob Thompson	xxxxxx5341	xxxx xxxx	$ 318.52

THIS IS NOT A CHECK

NON-NEGOTIABLE

FIGURE 4-9 An example of a pay check.

of a previous problem, and it may have made the customer late for work or caused some other issues about which the customer is unhappy. This is not the best way to begin or continue a relationship, especially if a reoccurring problem is one with which you or your shop is dealing. Regardless of the circumstance, it is important for you to treat the customer correctly, and that requires strong communication skills. Let us begin by examining various methods of communication.

NONVERBAL AND VERBAL COMMUNICATION

Nonverbal communication provides other people with as much or more information about you than what can often be determined by what you are saying. Because of this, it is important to understand nonverbal communication and what it says.

■ ***Appearance.*** You probably know the expression "You never get a second chance to make a good first impression." A customer may decide about you as a technician or your shop in the very first couple of seconds of interaction. What is your impression of the young technician shown in **Figure 4-10**?

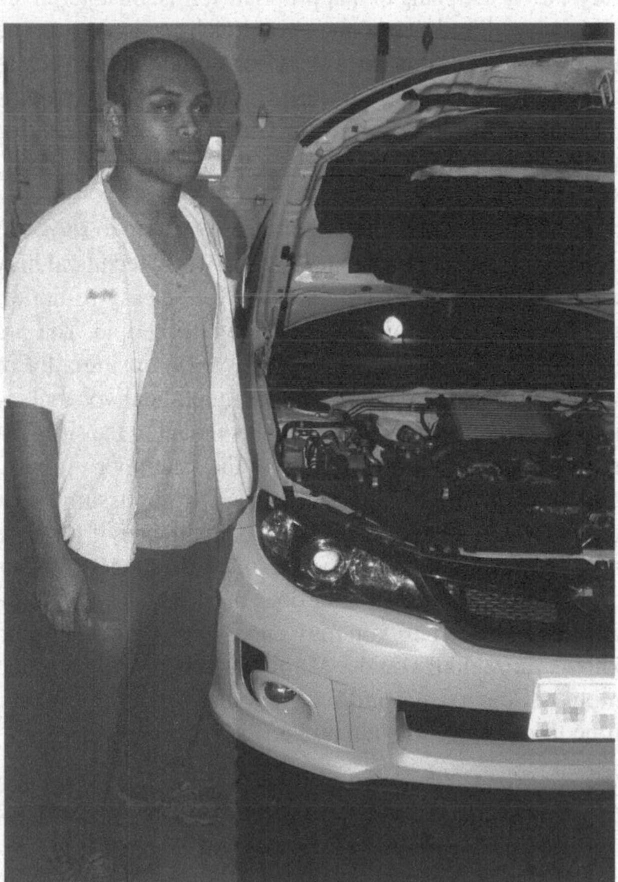

FIGURE 4-10 What does this technician say to you? His cloths are clean but his appearance is sloppy.

A long-term auto technician stereotype is of a dirt-covered guy who is trying to rip off the customer. What does your appearance say about you? Is your uniform neat and clean, or do you look like you have been living in puddles of oil and grease? Your shirt and pants should be as clean as possible, and your shirt should be tucked in. If your shirt buttons up the front, use the buttons. The public does not come to your shop to see your chest or underclothes under your unbuttoned shirt.

On the subject of pants, make sure yours are pulled up and are secured around your waist. The top of a pair of pants has a waistband, so pull your pants up to your waist. No one wants to see your underwear (or anything else) because your pants are down too low, especially when you bend over. Additionally, uniform providers can and do hem work pants to the proper length so that you do not have to try to walk on the bottoms of your pant legs.

Keep your work boots laced and clean. Dirt and grease on your boots and uniform can end up on the carpet and upholstery of the customer's interior. Do not create problems by leaving dirt and grease all over someone's car.

As discussed in Chapter 2, wash your hands and arms thoroughly and often. Wearing nitrile or mechanic's gloves makes keeping clean easier, is better for your skin, and is safer and healthier for you. Many technicians wear mechanic's gloves for protection while working, and they wear special chemical-resistant gloves when they are handling cleaners and solvents. All customers appreciate a technician who is clean and respectful of their vehicle. Regardless of gloves or uniform care, a clean, professional appearance is always more pleasing to the public.

■ ***Body Language.*** **Body language** is an important part of communicating, sometimes more important than what is actually being said. Body language can make or break a conversation, depending on what is interpreted by the parties involved. When you are dealing with others, especially with customers, it is important to note that the correct body language is used to display attentiveness, openness, and concern.

Possibly the biggest nonverbal sign is eye contact or the lack of eye contact. When you are speaking to another person, look that person in the eyes and maintain eye contact during the conversation. This shows attentiveness and respect for the person speaking.

In addition to eye contact, keep your body and feet pointed toward the person to whom you are talking. This indicates that you are attentive and not looking for a way to escape. Try to keep your arms uncrossed or from putting your hands in your pockets. Some hand gestures or movement is normal, but avoid using your hands excessively while speaking unless you are trying to explain some physical aspect about the vehicle.

■ ***What Does Body Language Say?*** **Figure 4-11** shows an example of a customer and a service advisor discussing a problem with a vehicle. Examine the photo and think about what the body language of both people say to you. Next, look at **Figure 4-12** and notice the differences in the body language of the technician.

What does the service advisor's body language say about the situation shown in the first photo? What can you determine about the customer in the first photo? If you were the customer in the second photo, how would you feel about the appearance of the technician? Think of your own experiences in which body language had either a positive or a negative impact on a situation.

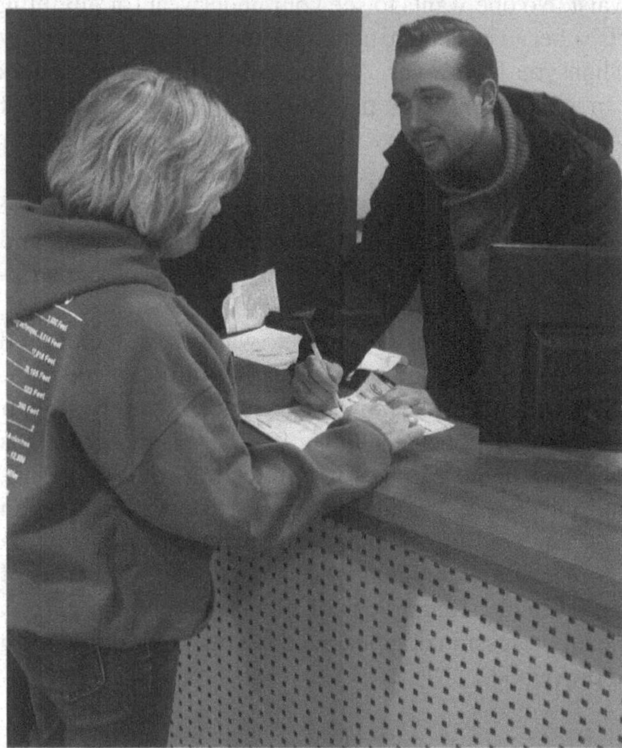

FIGURE 4-11 Eye contact and posture are important when interacting with others in a professional setting.

FIGURE 4-12 The technician shown here is not presenting the most professional image to the customer.

VERBAL COMMUNICATION

Technicians often must be able to interview the customer, interpret what the customer is saying, and share important information about what services or repairs need to be performed. Often customers will have only a vague idea about what may be wrong with their vehicle, or they may attempt to re-create the noise, movement, or condition that brought them to your shop. Careful listening and questioning can make the difference between a quick and correct diagnosis and hours of chasing false leads. If the interview process does not adequately provide an understanding of the problem, perform a test drive with the customer driving. This will allow you to experience the concern in person and start to form an opinion about its cause.

■ ***Standard English and Slang.*** Use proper English while talking with your instructors, customers, supervisors, and everyone else for whom professional language is necessary. Never use slang or inappropriate terms; misinterpreted slang may offend a customer. Avoid such terms as "screwed up," "bad" as in "this part is bad," or similar words, such as "shot," "toast," "dead," or others. What is a "bad" part? If a part is faulty, describe how or why. The idea of a leaking water pump is easy to convey, but try explaining a "bad" water pump.

But be aware that this is not the time to try to show off all your technical vocabulary. Being overly technical may cause you to lose a customer as well. Use easy-to-understand language but do not talk down to people. You are the expert; you do not have to prove it by using a lot of words customers may not know. If technical vocabulary is needed, be sure to explain what is meant in the simplest and clearest way you can. Show the customer the part, a picture, or diagram if necessary. Do not assume customers know what you are talking about even if they are nodding their heads in response to your statements. Also, do not assume the customer does not know what you are talking about either. You may encounter a customer who knows more than you do. Many customers will be much more willing to trust and accept your opinions if you take time to carefully explain and answer their questions.

■ ***Examples of Good Communication.*** If you are responsible for meeting and greeting customers, always greet them with a smile and a friendly face. Be courteous and attentive while gathering their information. While interviewing the customer, spend as much time as necessary to discuss and correctly document the concerns and what procedures are necessary to diagnose and correct the problem. Make sure all charges for inspections and services are covered and what procedures will be used to provide an estimate for any further repairs. Once complete, send the customer off with a smile and thank the customer for his or her business.

Even if you are not the person who handles all the customer communications, you may still be required to talk with customers. Always use formal names, such as Mr. (last name) or Ms. (last name), or address them as sir or miss. You should avoid calling customers by their first names unless they make it a point for you to do so. Be sure to greet customers with a smile and a "hello" as they enter.

READING AND WRITING SKILLS

Reading and writing skills are just as important for the automotive technician as are hands-on skills. Reading skills are necessary because of the amount of service information that must be read and understood for today's complex cars and trucks. Writing skills are needed to properly document the work performed. Failure to accurately document your work can cause you to not be properly paid. In an extreme example, proper documentation can be the difference in the outcome of a judge's decision in a legal dispute.

■ ***Why Reading Skills Are Important.*** The complexity of modern vehicles and the need to be able to read and interpret technical information to work on these vehicles requires good reading ability and reading comprehension. It is impossible to memorize everything there is to know about every car and truck that arrives for service, so important service information must be looked up and interpreted. Because of the vast amount of information available, just finding the correct service information can be a challenge. Once the information is located, careful reading of procedures and specs is necessary. Careless mistakes caused by inattention to reading and thinking about what the service information contains can lead to serious and expensive mistakes.

■ ***Reading for Technical Content.*** Technical service information is usually written in the form of a sequential flow chart (**Figure 4-13**), in narrative form, or as a list (**Figure 4-14**). In either example, the entire process may hinge on one key word or phrase. For example, in the flow chart shown in Figure 4-13, step 3 asks if a visual inspection was performed according to the procedure listed at the beginning of the symptoms section. If you do not take the time to go to the symptoms section and actually read what the inspection requires, you cannot accurately decide where to go from step 3.

Being able to record the correct repair information on the customer repair order (RO) is as important as being able to access, interpret, and utilize the repair information itself. Today's vehicles are complex and continue to become more so each year. The amount of information available for automotive repair today is vast and requires careful interpretation. The chances of making a mistake are increased if a technician has difficulty reading and understanding the information. Even simple procedures, such as an engine oil change, require careful reading and understanding of technical information more than ever before.

For example, vehicle manufacturers have very specific engine oil specifications for their engines. Not only are the oil viscosity ratings critical, but also many engines have very specific oil requirements to maintain the warrantees and prevent engine damage. Using an unapproved oil can result in oil sludge buildup, rapid component wear, and premature engine failure.

More importantly, technicians' safety and the safety of those around them is directly related to how seriously technicians take the use of information. Incorrect testing or service techniques can result in vehicle damage, a fire, personal or coworker injury, or even death. Many hybrid vehicles operate at 300–600 volts, more than enough to cause serious injury or death if the correct service procedures are not followed. Proper understanding of safety warnings is critical to reduce the risks associated with automotive repair. **Figure 4-15** shows a high-voltage warning decal from a hybrid vehicle. Is this an important piece of information?

The best way to understand service and repair information is to spend time reading and developing a familiarity with the language, content, and organization of the material. Written for professional technicians, information systems use language that is very technical and often contains tables, charts, and diagrams. The content usually assumes an implied understanding of the vehicle, the systems, and components. Some information services provide basic information, though it is assumed that the reader is familiar with certain tools and tests necessary to perform the procedures. Each type of repair information system is organized differently, so moving from one type or provider to another requires adjusting strategies. Many technicians will spend their down or slow time, lunch break, and even time after work reading service information, training materials, or professional magazines to improve their skills and understanding of systems and procedures. The more time spent reading the language, the better you will understand concepts

Hesitation, Sag, and Stumble

Step	Action	Values	Yes	No
Definition: Momentary lack of response as the accelerator is pushed down. Can occur at all vehicle speeds. Usually most severe when first trying to make the vehicle move, as from a stop sign. May cause engine to stall if severe enough.				
1	1. Was the Powertrain On-Board Diagnostic (OBD) System Check performed?	–	Go to Step 2	Go to Powertrain OBD System Check 2.4 L Powertrain OBD System Check 2.2 L
2	1. Perform a bulletin search. 2. If a bulletin that addresses the symptom is found, correct the condition per bulletin instructions. Was a bulletin found that addresses the symptom?	–	Go to Step 16	Go to Step 3
3	1. Perform the careful visual/physical checks as described at the beginning of the Symptoms section. Is the action complete?	–	Go to Step 4	–
4	1. Check the TP sensor for binding or sticking. Voltage should increase at a steady rate as the throttle is moved toward wide-open throttle. 2. Repair if found faulty. Was a problem found?	–	Go to Step 16	Go to Step 5
5	1. Check the MAP sensor for proper operation. Refer to MAP Sensor Output. Check 2.4 L or MAP Sensor Output. Check 2.2 L. 2. Repair if found faulty. Was a problem found?	–	Go to Step 16	Go to Step 6
6	1. Check for fouled spark plugs. 2. Replace the spark plugs if found faulty. Was a problem found?	–	Go to Step 16	Go to Step 7
7	1. Check the ignition control module for proper grounds. 2. Repair if grounds are found faulty. Was a problem found?	–	Go to Step 16	Go to Step 8

FIGURE 4-13 Service and repair information is often complex and requires careful reading. Missing one word can cause a completely incorrect diagnosis.

and procedures necessary to adapt to new technology and vehicles, and in turn, the more money you will make.

■ ***Written Communication.*** Proper written and/or electronic documentation is important in the auto repair industry. Records must be legible (if handwritten) and accurate, and spelling and grammar matter. It is a poor reflection on the technician and the facility if the person who is dealing with the public cannot spell (or uses text-messaging spelling and grammar), uses slang terms, or cannot write legibly. Technicians, at a minimum, should be able to complete repair orders with proper spelling, especially of automotive terms. Customers may associate your ability to repair their vehicle with your ability

3.3 L and 3.8 L Engines

1. Before servicing the vehicle, refer to the Precautions section.
2. Relieve the fuel system pressure.
3. Drain the cooling system.
4. Remove the negative battery cable.
5. Disconnect AFS (A) and breather hose (B).
6. Remove air cleaner upper cover (D) and intake hose (C).
7. Remove or disconnect the following:
   - The RH oxygen sensor connector
   - The RH injector connector and ignition coil connector
   - The Purge Control Solenoid Valve (PCSV) connector, Manifold Absolute Pressure (MAP) sensor connector, and PCSV hose
   - The Electronic Throttle Control (ETC) connector and knock sensor connector
   - The water hoses from ETC
   - The PCV hose
   - The brake vacuum hose
   - The surge tank stay
   - The connector bracket from surge tank
   - The surge tank
   - The breather Pipe assembly
   - The LH injector connector.
8. Remove intake manifold and gasket

FIGURE 4-14 Many service procedures are written out step-by-step. These steps often include specific procedures and specifications which must be followed.

FIGURE 4-15 Always read and follow all warning decals when working on a vehicle, especially if the vehicle is a hybrid.

to write and spell. Additionally, good documentation creates a record of the concerns, parts, labor, and any further recommendations made to the customer. The documentation process begins with the customer RO.

The RO serves several purposes. First, it contains the customer's personal and vehicle information. The customer's name, address, and contact numbers need to be included. Having home, work, cell phone numbers, and an e-mail address makes it possible to contact the customer quickly and easily. One cause of lost production in a shop is when a vehicle is inspected and an estimate is ready for approval, but the customer cannot be reached because someone did not get the contact information.

For the technician, the RO provides a record of why the customer brought the vehicle in for service (**Figure 4-16**). All concerns (complaints) should be clearly defined. Few things can be as aggravating to a technician as an RO that simply states, "Customer hears a noise while driving." Less time can be spent guessing if more time is invested in obtaining good information from the customer at the beginning. Tests and inspections should be suggested that coincide with the concerns.

Each shop will have its own policy regarding charges for testing and inspection. While you are talking with the customer, document the costs associated for the testing. The RO may also show times when an estimate or completed repairs are expected. The customer signs the completed RO and receives a copy.

You will need to be aware of local regulations about how authorization for additional costs can be obtained. Make sure that there is a process for obtaining and documenting additional charges before the

CUSTOMER #: 51479

238557

INVOICE

WILLIAMS

Phone: • Toll Free:

SERVICE · PARTS · SALES

Service with "Personal Attention"

PAGE 1

BUS: CELL: SERVICE ADVISOR: 124 JENNIFER

COLOR	YEAR	MAKE/MODEL	VIN	LICENSE	MILEAGE IN / OUT	TAG
	05	FORD FOCUS	1FAFP34N35W		90474/90474	T148

DEL. DATE	PROD. DATE	WARR. EXP.	PROMISED	PO NO.	RATE	PAYMENT	INV. DATE
01JAN05 DD			17:00 10FEB15		0.00	CASH	17FEB15

R.O. OPENED	READY	OPTIONS: DLR:02215 ENG:2.0_Liter_DOHC
08:56 10FEB15	12:32 17FEB15	

LINE OPCODE TECH TYPE HOURS LIST NET TOTAL

A CUSTOMER STATES CHECK ENGINE LIGHT ON -AND HEARING LOUD EXHUAST NOISE
1A FOUND FRONT OYXGEN SENSOR FAULTY -CUSTOMER
DECLINED OYXGEN SENSOR-REPAIRED MUFFLER AND
FLEX PIPE
2409 CP 105.97 105.97
SUBL ALL STAR REPLACED MUFFELER/FLEX PIPE
CP 225.00 225.00
PARTS: 0.00 LABOR: 105.97 OTHER: 225.00 TOTAL LINE A: 330.97
****

B CUSTOMER STATES WHEN USING ACCERSORIES -WILL HEAR A CLICKING NOSIE
FROM THE DASH AREA-AND FEELS LIKE IT IS LOOSE POWER -ABS LIGHT
AND BRAKE LIGHT COME ON WHEN THIS HAPPENS AS EL
1A FOUDN BOLT FROM LOWER MOUNT BROKEN -RELACED
BOLT-CASED FRONT AXLE SHAFTS TO PULL OUT AND
LEAK AND TRANSMISSIONN LINES PULLED AND
LEAKING -
2409 CP 105.97 105.97
PARTS: 0.00 LABOR: 105.97 OTHER: 0.00 TOTAL LINE B: 105.97
****

C CUSTOMER STATES WHEN SHIFTING INTO DRIVER OR REVS THERE IS A CLUNKING
NOISE
17000 SEE LINE B -ALSORACK AND PINION IS LEAKING
AND POSSIBLE FUEL PUMP ISSUE ALSO MULTIPLE
ABS CODES AND AIG BAG FRONT CRASH SENSOR
POSSIBLE CONN
2409 CP 0.00 0.00
PARTS: 0.00 LABOR: 0.00 OTHER: 0.00 TOTAL LINE C: 0.00
****

D THW WORKS PACKAGE CHANGE OIL AND FILTER,ROTATE TIRES, INSPECT BRAKES,
PERFORM COMPLEMENTARY MULTI POINT INSPECTION
NWP NO WORK PERFORMED
2409 CP 0.00 0.00
PARTS: 0.00 LABOR: 0.00 OTHER: 0.00 TOTAL LINE D: 0.00
****

CUSTOMER

MISC. SHOP SUPPLY CHARGE
A nominal charge is included to cover misc. lubricants, rags, washer solvent, top off fluid levels, cleaners, adhesives, seat covers, floor mats, certain misc. hardware, tape, etc.

SHOP SUPPLY COSTS: A charge equal of 10% of the total cost of labor, not to exceed $25.00 will be added to the repair order for shop supplies used in conjunction with the repair. A minimum charge of $2.00.

Repairs or Services performed have been explained or demonstrated for customer.

ALL PARTS INSTALLED ARE NEW UNLESS SPECIFIED OTHERWISE

THE DEALER HEREBY DISCLAIMS ALL WARRANTIES, EXPRESS OR IMPLIED, INCLUDING ANY IMPLIED WARRANTIES OF MERCHANTABILITY OR FITNESS FOR A PARTICULAR PURPOSE REGARDING ANY PRODUCTS OR SERVICES PROVIDED, UNLESS OTHERWISE INDICATED ON THE SERVICE REPAIR ORDER AND NEITHER ASSUMES NOR AUTHORIZES ANY OTHER PERSON TO ASSUME FOR IT ANY LIABILITY IN CONNECTION WITH THE SALE OF ANY PARTS OR THIS REPAIR. THIS DISCLAIMER IN NO WAY AFFECTS THE PROVISIONS OF ANY MANUFACTURER OF OTHER SUPPLIERS WARRANTIES.

CUSTOMER SIGNATURE

DESCRIPTION	TOTALS
LABOR AMOUNT	
PARTS AMOUNT	
OIL	
SUBLET AMOUNT	
MISC. CHARGES	
TOTAL CHARGES	
LESS DISCOUNT	
SALES TAX	
PLEASE PAY THIS AMOUNT	

CUSTOMER COPY

FIGURE 4-16 An example of a repair order.

customer leaves. Many states have laws requiring a written and signed estimate for repairs above a certain dollar amount, such as $25.00, and how additional approval can be granted and by whom. Never perform repairs on a customer's vehicle without proper and documented authorization. Doing so can result in lost customers, lost income, and possibly severe legal consequences.

The RO serves as the record of tests, inspections, and findings by the technician once the vehicle is in the service bay. This record keeping is important for both the shop and the technician. A written record of tests and their findings are important for further communication with the customer. Additional testing or repairs may not be approved if specific details cannot be provided. The documentation may also provide a record for the technician's pay. Many shops pay their technicians based on the amount of labor they produce. If the technician performs multiple tests and repairs, each will have a specific labor time and code to be documented. Improper documentation can mean incorrect or insufficient pay for the technician or incorrect billing of parts and labor to the customer. Instances of incorrect billing can result in technician dissatisfaction and disciplinary action by the shop against the technician.

APPLICATIONS OF COMMUNICATION SKILLS

This section provides examples of using correct communication skills.

■ ***Answering the Phone.*** Most shops have a standard greeting used when employees answer the phone, which includes identifying the facility and the person answering the phone. For example, "Hello, thank you for calling ABC Automotive. This is (insert your name here), how may I help you?" Be sure to speak clearly and loudly enough to be heard. Do not mumble; articulate what you are saying, and try to sound confident.

■ ***Talking with a Customer.*** Conversations with customers vary from pleasant to unpleasant depending on the circumstances. Regardless of the situation, remain calm and professional. Most your interactions with customers will be positive and productive. Always treat the customer with respect, look him or her in the eye, and keep your verbal and nonverbal language professional. If you do have to deal with an angry customer, it is best to let the person vent first while remaining calm, quiet, and professional. Do not argue, antagonize, or belittle the customer. This only shows your lack of professionalism and will do nothing to resolve the problem. Once the customer has had a chance to vent, things usually calm down and a rational conversation can take place. If the customer is angry about something that is the fault of the shop, accept the responsibility, apologize, and work to correct the situation. Most people are willing to forgive mistakes when the problem is resolved appropriately.

If the person who is angry at you is your instructor or your boss, follow the same advice. If you have acted incorrectly or made a mistake that has caused the problem, accept responsibility, apologize, and correct the situation. Be advised that if you repeatedly are the source of problems with customers or the boss, you will not be given an unlimited number of times to make things right. If you find yourself in this situation, you may need to reexamine your behavior or attitude. If the primary conflict is with a coworker or supervisor, make every attempt to correct the situation so that it does not cause further problems. If, however, the issues cannot be resolved, it may be necessary to look for a new employer.

Math Skills

Math is an integral component of being an automotive technician. It is important that the math used by automotive technicians be performed correctly. Inaccurate measurements or calculations can result in part failure, broken fasteners, personal injury, customer injury, or worse.

BASIC MATH FOR TECHNICIANS

There is no automobile without math, nor is there any way to correctly service or repair a vehicle without math. Technicians need to understand how math applies to the various aspects of the automobile, from estimates for repairs, to ratios, proportions, units of measurement, geometry, and many other areas.

■ ***Math for Estimates.*** Technicians are often responsible for preparing the repair estimate, which can include not just parts and labor costs, but also figuring parts markup, calculating shop supply and hazardous waste charges, and sales tax. Additionally, the technician's pay is often directly related to the labor charge or a combination of the parts and labor sales, so proper estimating and calculation should be a priority for all technicians.

Labor cost may be based on a labor estimating guide, also called a **flat-rate guide**. The flat-rate guide provides the shop and technician with information about original equipment (OE) parts cost, aftermarket parts cost (some software), and estimated labor times for specific repairs. An example of a parts and time guide is shown in **Figure 4-17**. When preparing an estimate, technicians refer to the labor guide to determine the amount of time specified to complete a repair. The labor guide may show the labor time, warranty time, and skill-level information.

The labor time is used as the basis for what the shop will charge a customer for labor. Labor operations are broken down into tenths of an hour, shown as 0.1 hour. **Figure 4-18** shows how an hour is broken down into tenths. According to the estimating guide, if the labor time is 1.0 hour, a professional technician with the proper tools and knowledge of the vehicle and system should be able to perform the labor operation in 1 hour. Technicians who perform the job in less than 1 hour

	Skill Level	Warranty Time	Standard
(1) HUB & BEARING, R&R			
All Models (2WD)			
One Side	B	(0.8)	1
Both Sides	B	(1.5)	1.9

Parts

	Mfg. Part No.	Price (MSRP)*
HUB & BEARING		
All Models		
1/2 Ton		
w/Crew Cab	15946732	$509.50
w/o Crew Cab	15233111	$453.46

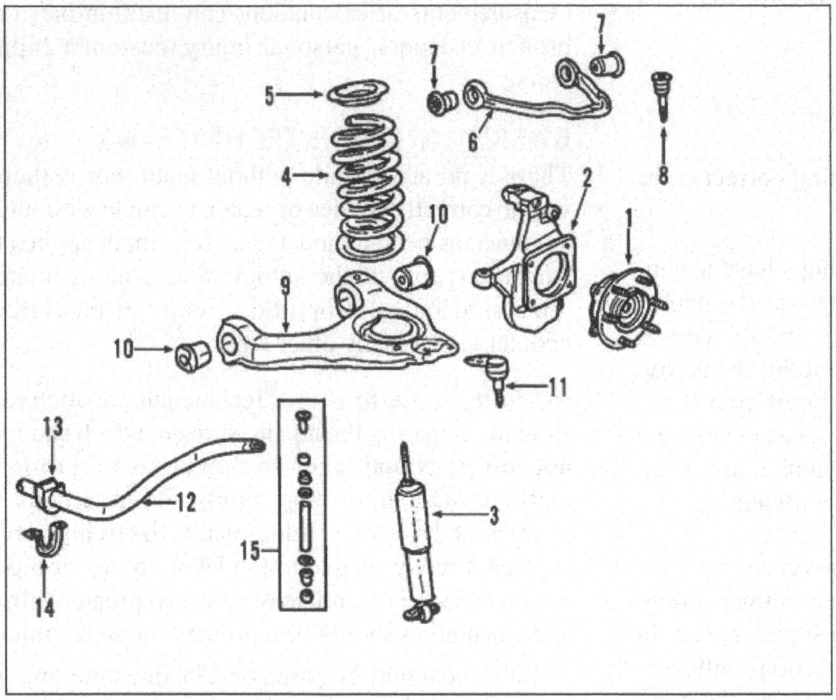

FIGURE 4-17 Services and repairs are typically charged by flat-rate time, which breaks an hour into tenths. Flat-rate time, warranty time, and skill-level codes are typically part of the labor guide. The work shown here is appropriate for a B-level technician.

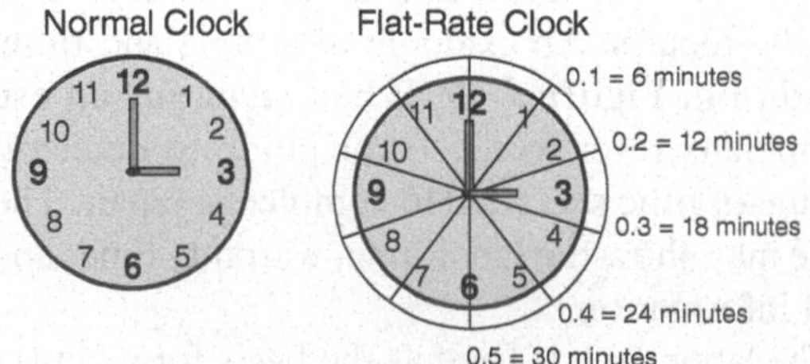

FIGURE 4-18 How flat-rate time is broken down into minutes. A repair that calls for 0.8 hours means it should take about 48 minutes to complete. Find the actual time by multiplying the indicated time by 6 minutes.

can move on to the next job while having been paid for 1 hour of labor but actually working less than the 1 hour specified. This is how experienced technicians can earn more, by beating the labor or clock time also known as flat rate. The reverse is also true, though. An inexperienced technician may spend an hour and a half completing a repair that pays one hour. In this case, the technician is losing money due to the repair taking longer to complete than what the labor time pays. Once a technician becomes experienced and knows what tools and procedures are needed, beating the flat rate can be quite profitable. Do not expect as an entry-level technician to be able to compete with more experienced technicians and beat flat rate. Only through experience, trial and error, and developing skills and routines will a technician become proficient and faster at the work.

Many shops pay their technicians solely on the flat-rate system. This means the technician's pay is based completely on how many labor hours the technician produces. The technician receives a percentage of the labor as commission. For example, if a shop labor charge is $100 per flat-rate hour and the technicians earn 15% of the labor they produce, the technicians would make $15 per flat-rate hour. This is figured by multiplying 100 by 15% (or 100 × 0.15). This may sound very good, but remember, technicians who take more time to complete repairs than the flat-rate guide specifies will not earn pay for each hour they actually worked. If Technician A, who is inexperienced, works 50 hours in a week but only produces 35 labor hours, Technician A will only receive pay for 35 hours. Across the shop, Technician B, who is experienced, works 40 hours but has produced 50 hours of labor in a week will receive pay for 50 hours.

Another factor in flat-rate pay is **warranty time**. Technicians employed at new car dealerships perform repairs on vehicles that are still covered under the manufacturer's warranty. Remember the two different labor times listed in the labor guide? The larger amount of time is for customer pay (CP) repairs; the lesser of the two is warranty time. Warranty time is generally less because one or all of the following usually apply:

1. The technician is trained, experienced, and skilled with the particular vehicle and repair procedures.
2. The technician has access to all the manufacturer-specific tools and information.
3. Because the vehicle is under warranty and (probably) still in better condition than an older vehicle, the repairs will be easier because fasteners and components have not had as much time to rust or seize.

Most manufacturers require their warranty technicians to be specially trained and certified to perform these repairs. For the dealership to receive reimbursement from the manufacturer for warranty repairs, the warranty documentation generally requires information about which technician performed the repair. If the technician is not trained or certified for the repair, the warranty claim can be denied.

When figuring an estimate, you probably need to calculate sales tax. Depending on your location, sales tax may be applied to parts or labor or both. **Figure 4-19** shows an example of an estimate and how to calculate sales tax. Many shops also charge a percentage of the total bill before tax for shop supplies. A common shop supplies surcharge is 5% of the parts and labor up to a maximum of $5. To calculate this amount, subtotal the parts and labor and multiply times 5% (0.05) to get the cost of shop supplies. The example in **Figure 4-20** shows shop supplies are calculated only on labor. This amount is added to the parts and labor subtotal and is generally subject to sales tax.

■ ***Parts.*** Original equipment manufacturer (OEM) parts are replacement parts that come from the original parts manufacturer. AC Delco is a parts supplier of original equipment for General Motors, and if you purchase a part from a GM dealer's parts department and the part was originally made by AC Delco, you should get an AC Delco replacement part. OEM, often just referred to as **OE parts**, applies to all parts purchased from the manufacturer and even parts from an aftermarket source, if the parts are the same brand and application as the original parts. **Aftermarket** means parts and services (and repair shops) supplied by companies other than the OEM.

Aftermarket parts suppliers, from sources such as NAPA, Carquest, or Advance Auto Parts, supply parts to shops and do-it-yourselfers (DIYs). The parts from parts stores are usually not OEM parts, although these stores may carry some OE parts, such as AC Delco, Motorcraft, and other products. These parts come from companies that have purchased the manufacturing rights to produce parts to replace the OE parts.

Much can be said about parts and parts pricing, but as a rule, name brand or OE parts are higher quality and cost than store brand parts. The expression "you get what you pay for" is good advice for purchasing automotive parts. The use of inferior quality parts just to have the lowest price can lead to repeated failures, customer and technician frustration, and ultimately lost income due to having to replace a defective "new" part.

Parts costs vary depending on the parts supplier (aftermarket or OE), geographic location, and even the volume of parts purchased by the shop. Once a part is purchased, shop policies on markup, based on many factors, dictate how much for which the part is marked up and resold to the customer. Markup is the difference in price between what a part costs and how much it is sold for; it is usually figured as a percent.

Shops that use aftermarket parts and/or OE parts usually receive that part at wholesale cost. Wholesale is what the parts supplier can sell the part for at a discount, and still make a profit. For example, a parts store may purchase a wheel bearing from their supplier for $42.15. The retail cost to the public may be $125.65, and the wholesale price to the shop may be $67.85. This way the parts store makes a profit, though not as much as at the retail price. The shop can purchase the part and sell it for the original list price (retail) of $125.65 and make money off the part. In some cases, the shop may mark-up the part over its actual retail price. Because shops cannot survive on labor alone, part markup is needed to make the repair profitable.

In this example, the shop paid $67.85 for a part being resold at $125.65. The labor to install the part was $104.00, then the total sale (excluding taxes and shop

SMART Automotive Repair

5150 Speed Way
Columbus, Ohio 43224

Estimate Creation Date: 11/19/2016 09:22

Estimate #:
Rob Thompson

2002 GMC Sierra 1500 4.8L, V8, Gas, VIN V, 16V, USA/Canada
Miles

Wheel Hub And Bearing Assembly R&R

Type	Description	Part#/OpCode/Tax Option	Qty/Hrs	Price	Total
Part	Wheel hub and bearing		1.00	$125.65	$125.65
Labor	Wheel Hub And Bearing Assembly R&R - One Side	Wheel Hub And Bearing Assembly R&R	1.30	$80.00	$104.00
					$229.65

Parts Total:	$125.65
Labor Total:	$104.00
Shop Supplies:	$0.00
Hazardous Materials:	$0.00
Fees:	$0.00
Subtotal:	$229.65
Sales Tax:	$17.22
Total:	**$246.87**

The subtotal of parts and labor, $229.65, is multiplied by the sales tax percentage, in this example, 7.5%. This can be done three ways:
1. 229.65 x 0.075 = 17.22 or
2. 229.65 x 7.5% = 17.22 or
3. 229.65 x 1.075 = 246.87

FIGURE 4-19 Sales tax varies greatly across the country, so be sure you understand how it is applied in your area.

fees) would be $229.65. The $229.65 may sound like a good return on a cost of $67.85, but the technician must be paid and so does the rent on the building, the utilities, insurances, taxes, tool and equipment costs, and many other items, all of which take a part of the profit. In this case, the customer paid $229.65; subtract the parts cost, $67.85, and the technician's pay, $15 per flat rate hour, or $19.50, so gross profit is $142.30. From that subtract for all the costs of operating the business, and soon the net profit (the amount left after all expenses are paid) will not be much. High operating costs, too low of a labor rate, and low parts markup can mean low net profits and can result in the shop either making very little money or even losing money on this job. An example of parts cost, markup, and shop expenses is shown in Figure 4-20.

Parts markup can vary widely depending on shop location, part cost, part availability, and other factors. Markup can often range from 40% to 200%. If this sounds excessive, consider that a soft drink at a fast food restaurant may cost the restaurant 12 cents per serving but sells retail for $1.49, a markup of about 12 times (1200%) the cost of the drink.

Once the labor and parts costs are determined, any additional charges such as shop supplies, hazardous waste fees, and sales tax can be determined. Charged by some shops, shop supplies fees help cover the costs of shop rags, miscellaneous fasteners or small parts (nuts and bolts), and chemicals used during service (**Figure 4-21**). Shop supplies fees may be charged as a flat percentage of the total parts and labor, typically up to

Part retail price from supplier (list)	Part cost from supplier (wholesale)	Price part sold to customer (retail)	Part profit
$125.65	$67.85	$125.65	$57.80
	Labor cost (technician pay)	**Labor sold to customer**	**Labor profit**
	$19.50	$104.00	$84.50
		Gross profit	**$142.30**

Expenses / month	
Building rent	$1,800
Electric	$550
Gas	$225
Insurance	$300
Phone/Internet	$400
Service Information	$175
Taxes	$425
	$3,875

Net profit

FIGURE 4-20 Markup is how much the selling price of a part or service increased over the cost of the part or service.

```
LABOR-----------------------------------------------------------------------------------
J# 1 01POZZIS         MULTI-POINT INSPECT                  TECH(S):415           200.00
            MULTI-POINT INSPECTION REPORT CARD
            CUSTOMER REQUESTS A PRE-PURCHASE INSPECTION
            SEE ATTACHED SHEET
            INSPECTION COMPLETE
            THE REAR TIRES ARE INTO THE WEAR BARS AND THE TIME FOR
            MAINT. LIGHT IS ON.
MISC------CODE-------DESCRIPTION---------------------------CONTROL NO---------
                SS1  CREDIT SHOP SUPPLIES                                         -4.50
                                                              TOTAL  MISC         -4.50
JOB#  1 TOTALS--------------------------------------------------------------------------
                                                              LABOR              200.00
                                                              MISC                -4.50

                                  JOB# 1 JOURNAL PREFIX  POCS  JOB#  1 TOTAL     195.50
MISC------CODE-------DESCRIPTION---------------------------CONTROL NO---------
JOB # A          SS  SHOP SUPPLIES/RECYCLING FEES                                 24.50
                                                              TOTAL - MISC        24.50
COMMENTS--------------------------------------------------------------------------------
DROP

TOTALS----------------------------------------------------------------------------------
**************************************************      TOTAL LABOR          200.00
*                                                *      TOTAL PARTS.....       0.00
*  [ ] CASH     [ ] CHECK   CK NO. [          ]  *      TOTAL SUBLET...        0.00
*                                                *      TOTAL G.O.G....        0.00
*  [ ] VISA     [ ] MASTERCARD     [ ] DISCOVER  *      TOTAL MISC CHG.       24.50
*                                                *      TOTAL MISC DISC       -4.50
*  [ ] AMER XPRESS    [ ] OTHER    [ ] CHARGE    *      TOTAL TAX.......      15.15
*                                                *
**************************************************      TOTAL INVOICE $     235.15
```

FIGURE 4-21 Shop supplies may be a percentage of parts, labor, or both. In this example, it is only calculated on labor. Practice your math skills if necessary to prevent over- or undercharging a customer.

a maximum amount of $5. Hazardous waste fees can be applied to cover the cost of used tire disposal and waste oil and antifreeze reclamation. These fees may be a set dollar amount instead of a percentage but are usually a couple of dollars. Sales tax can be figured on both parts and labor or just one. Some states tax both, while others may tax only parts.

Work that is subcontracted out, such as having a windshield replaced by a mobile glass company, is added into the RO as a sublet service and may or may not have taxes applied. Regardless of what charges may be included, technicians need to be able to figure percentages on different parts of the bill and on the entire bill. If sales tax is 6.5% and is charged to parts and labor, an RO subtotal of \$167.25 plus tax would be calculated 167.25 × 6.5% (167.25 × 0.065) = \$10.87, for a total of \$178.12.

Basic math skills, addition, subtraction, multiplication, and division, are basic tools of a technician's toolbox, but as is often the case, many technicians are not as proficient in math as they should be. Calculating parts and labor costs, shop supplies percentage, and sales tax for an RO may seem mundane, but they are very important to the customer and the shop. A simple mistake on the RO can cause embarrassment on the part of the technician, feelings of mistrust by the customer, or lost money for the shop because of a failure to charge the correct amount for repairs.

■ ***Basic Math Functions.*** There are four basic math functions: addition, subtraction, multiplication, and division. While most people rely on calculators for doing more than very basic math calculations, it is important to remember how to perform these operations in the event a calculator is not available.

When adding whole numbers, such as when completing an estimate, it is often easier to arrange the numbers from largest to smallest (**Figure 4-22**). When adding the rightmost column, remember to add any carryover digits to the next column to the left. Continue to add all columns of numbers until the total sum is found. If necessary, double-check your work to make sure the total is correct.

When subtracting, place the larger of the two numbers on top of the smaller number (**Figure 4-23**). When subtracting a bigger number from a smaller number you have to borrow from the next number to the left (Figure 4-22). When you borrow, remember that you have to decrease the number borrowed from by one, otherwise your total will be incorrect.

When multiplying, arrange the numbers with the larger number on top of the smaller number (**Figure 4-24**). Begin by multiplying the number on the bottom right by the number directly above, and write down the product. Continue to multiply and write the products (Figure 4-23). Once all of the numbers have been multiplied, add the two lines together to find the total.

Division uses addition, subtraction, and multiplication to solve for the quotient. **Figure 4-25** shows how to perform long division.

Step 1. Organize from lowest to highest.

18.78
56.90
76.55

Step 2. Add the numbers in the rightmost column.

8 + 0 + 5 = 13

Step 3. Since this number is larger than 10, place the 3 under the rightmost column and carry the 1 to the next column to the left as shown.

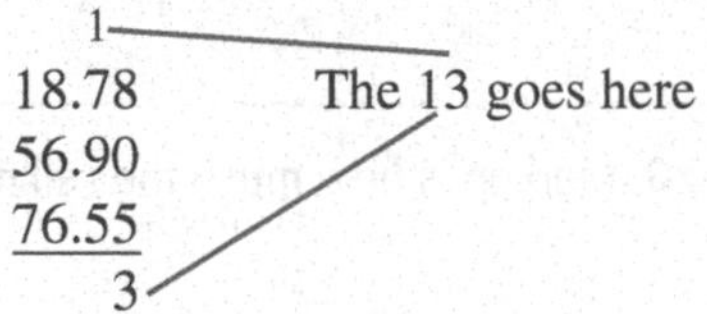

Step 4. Add the second column from the right.

1 + 7 + 9 + 5 = 22

Step 5. Place a 2 under the second 5 from the right and the other 2 over the third column from the right as shown. Place a decimal to the left of the 2 so that it aligns with the decimal above the line.

2
18.78 The 22 goes here
56.90
76.55
.23

Step 6. Add the third column as shown.

2 + 8 + 6 + 6 = 22

2
18.78
56.90
76.55
2.23

Step 7. Add the last column as shown.

2 + 1 +5 + 7 = 15

2
18.78
56.90
76.55
152.23

FIGURE 4-22 In the age of calculators, it is a good idea to remember how to do basic math functions without a calculator.

Step 1. To subtract, place the larger number on top of the smaller, as shown.

$$\begin{array}{r} 89.77 \\ -\underline{15.91} \end{array}$$

Step 2. Subtract 1 from 7 from the rightmost column.

$$\begin{array}{r} 7-1=6 \\ 89.77 \\ -\underline{15.91} \\ 6 \end{array}$$

Step 3. Subtract 9 from 7 in the second column from the right. To subtract 9 from 7, you must borrow from the 9 in the third column from the right as shown.

$$\begin{array}{r} 8\ 17 \\ 89.77 \\ -\underline{15.91} \\ 86 \end{array}$$

Borrowing from the 9 makes the 9 become 8 and adds 10 to the 7, making it 17. So now 9 is subtracted form 17, which equals 8.

Step 4. Subtract 5 from 8 in the third colum as shown.

$$\begin{array}{r} 88.77 \\ -\underline{15.91} \\ 3.86 \end{array} \qquad 8-5=3$$

Step 5. Subtract 1 from 8 in the last colum as shown.

$$\begin{array}{r} 88.77 \\ -\underline{15.91} \\ 73.86 \end{array} \qquad 8-1=7$$

FIGURE 4-23 More basic math functions.

To multiply 23 times 12, arrange the numbers as shown.

$$\begin{array}{r} 23 \\ \times\underline{12} \end{array}$$

Step 1. Multiply the 2 from 12 times the 3 as shown.

$$\begin{array}{r} 23 \\ \times\underline{12} \\ 6 \end{array} \qquad 2\times3=6$$

Step 2. Multiply the 2 from 12 times the 2 above.

$$\begin{array}{r} 23 \\ \times\underline{12} \\ 46 \end{array} \qquad 2\times2=4$$

Step 3. Multiply the 1 times 3 and place below the four. This is because the 3 is in the tens place and needs to remain in that place.

$$\begin{array}{r} 23 \\ \times\underline{12} \\ 46 \\ 3 \end{array} \qquad 1\times3=3$$

Step 4. Multiply the 1 times 2 and place the 2 at the left of the 3. This is the hundred column.

$$\begin{array}{r} 23 \\ \times\underline{12} \\ 46 \\ 23 \end{array} \qquad 1\times2=2$$

Step 5. Add the 46 and the 23, keeping the three columns as shown.

$$\begin{array}{r} 23 \\ \times\underline{12} \\ 46 \\ \underline{23} \\ 276 \end{array} \qquad \begin{array}{l} 6+0=6 \\ 4+3=7 \\ 2+0=2 \end{array}$$

FIGURE 4-24 Proper multiplication is important when figuring out parts and labor costs, as well as your paycheck.

■ ***Fractions and Decimals.*** Other common types of numbers used by technicians are fractions and decimals. A fraction is a method of expressing nonwhole numbers. Whole numbers are 0, 1, 2, 3, and so on. A fraction is a type of rational number, which may contain a whole number and a fraction. Just like whole numbers, fractions can be added, subtracted, multiplied, and divided. Before any functions can be performed on different fractions, they must first be converted to the lowest common denominator. Fractions have a numerator, the number on top, and a denominator, the number on the bottom. For these examples, simple numbers with common denominators are used. If a common denominator is not easily figured, you will need to factor each number until a common number is found.

Figure 4-26 shows how three basic fractions are added, subtracted, multiplied, and divided.

Decimals are another way to represent nonwhole numbers and they can be used in place of fractions, making calculations of larger numbers much easier. Decimals are compared to regular fractions (**Figure 4-27**). The first digit right of the decimal point is equivalent to 1/10, the next is 1/100, next is 1/1,000, and so on. **Figure 4-28** shows how to perform basic math functions on decimals.

Both fractions and decimals are used in automotive measurement. With the exception of fractional tool sizes, decimal numbers are more common than fractions. Fractions may be used for measurements such as ride height and wheelbase. Decimals are used extensively in specifications and angles.

Step 1. Determine which number is the divisor and which is the dividend. Since we are dividing 425 by 25, the divisor is 25 and 425 is the dividend. Arrange the numbers as shown.

$25\overline{)425}$

Step 2.

$$\begin{array}{r} 0 \\ 25\overline{)425} \\ \underline{0} \end{array}$$

Since 4 is not divisible by 25, 0 is placed above the 4. This 0 is then placed below the 4 as 25 × 0 = 0

Step 3.

$$\begin{array}{r} 0 \\ 25\overline{)425} \\ \underline{-0} \\ 4 \end{array}$$

Now subtract 0 from the 4 and carry the 4 down.

Step 4.

$$\begin{array}{r} 01 \\ 25\overline{)425} \\ \underline{0} \\ 42 \\ \underline{-25} \\ 17 \end{array}$$

Next, bring the 2 down. Divide 42 by 25. 42/25 = 1 with a remainder of 17.

Step 5.

$$\begin{array}{r} 017 \\ 25\overline{)425} \\ \underline{0} \\ 42 \\ \underline{-25} \\ 175 \\ \underline{-175} \\ 0 \end{array}$$

Now bring the 5 down with the 17 remaining from Step 4. Divide 175 by 25 to get 7, which goes above the division line as shown.

FIGURE 4-25 A refresher in how to solve long division problems.

■ ***Percentages.*** **Percentage** means the part of the whole and is used to calculate sales tax, fees, parts markup, and mixtures of antifreeze and water. Percentages are shown with the % symbol or written as 50%. Fifty percent is equal to 50/100 or 0.50. When you are calculating percentages, remember that the % symbol is treated as the constant of 1/100 or 0.01. This means that when you are calculating a percentage, such as 40% of 400, it can be written as 40/100 · 4 = 160. Once you remember that the number for a percentage is equal to the number divided by 100, it is quicker to turn the number into decimal form. In the example above, 40% of 400 can be calculated directly by using 400 × 0.4 to get the same result.

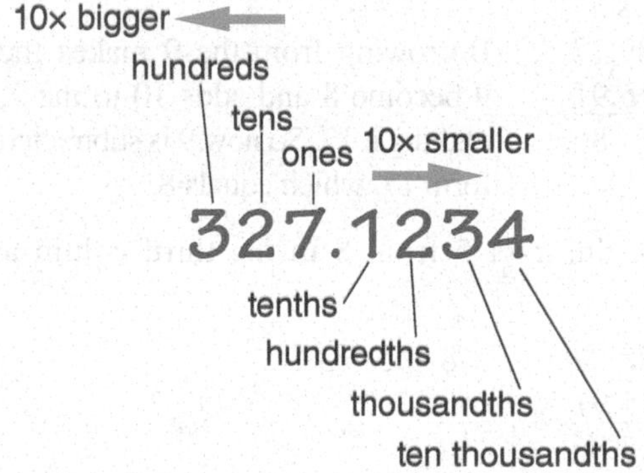

A fraction such as 1/4 is converted into a decimal by dividing the 1 by 4. So 1/4 is equal to 0.250 in decimal

FIGURE 4-27 Decimals are commonly used in auto repair, and you should have a good understanding of the value of each number on either side of the decimal.

$\frac{1}{2}+\frac{1}{4}= \frac{2}{4}+\frac{1}{4}=\frac{3}{4}$ To add fractions, find the lowest common denominator.

Convert $\frac{1}{2}$ to $\frac{2}{4}$ by multiplying each number by 2. Next, simply add across the numerator and keep the common denominator to find the answer.

$\frac{3}{4}-\frac{1}{8}=\frac{6}{8}-\frac{1}{8}=\frac{5}{8}$ To subtract fractions, find the lowest common denominator. Next, subtract across the numerator and keep the common denominator to find the answer.

$\frac{1}{2}\cdot\frac{1}{4}=\frac{1\cdot 4}{2\cdot 4}=\frac{1}{8}$ To multiply fractions, multiply across both the numerator and the denominator.

$\frac{3}{4}\div\frac{1}{8}=\frac{3}{4}\cdot\frac{8}{1}=\frac{24}{4}=\frac{6}{1}=6$ To divide fractions, use the reciprocal of the divisor (the number divided by) and multiply times the dividend and reduce as needed.

FIGURE 4-26 Working with fractions may be required when performing wheel alignments.

Adding decimals 3.25, .075, and 5.

First, line the decimals up:

```
  3.25
  0.075
+ 5.
```

Fill in empty places with zeros:

```
  3.250
  0.075
+ 5.000
```

Add:

```
  1
  3.250
  0.075
+ 5.000
  8.325
```

Subtracting decimals .55 from 7.005

Line the decimals up:

```
  7.005
- 0.55
```

Fill in empty place with zeros:

```
  7.005
- 0.550
```

Subtract:

```
  6 9 10
  7.005
- 0.550
  6.455
```

To multiply decimals such as 102 × .22:

First, multiply without decimal points:

102 × 22 = 2244

Next since 102 has 0 decimal places and 0.22 has 2 decimal places, the answer has 2 decimal places:

2244 becomes 22.44 after moving the decimal two places to the left.

To divide decimals use this example:

Divide 9.1 by 7. Ignore the decimal point and use long division so you set up the problem as shown.

```
   13
7)91
  9
  7
  21
  21
   0
```

Next, place the decimal point in the answer directly above the decimal point in the dividend 9.1 so 13 becomes 1.3

FIGURE 4-28 You will need to be skilled with decimals for writing estimates and measuring components.

Using percentages is necessary when you are figuring part costs, shop supplies and/or hazardous waste fees, and sales tax. For parts, the cost of the part is multiplied by the markup percentage. For example, a part costing $72.89 may be marked up 60% for resale. To determine the resale amount multiply 72.89 × 1.6 (72.89 × 160%). When you are figuring the sales tax, the parts and labor subtotals are multiplied by the sales tax percentage. For example, a sales tax of 6.5% is equal to $0.065. If the subtotal is $155.10, then $155.10 × 0.065 (155.10 × 6.5%) equals the sales tax, or $10.08.

MEASUREMENTS

Correct measurement is part math and part technical skill. Making a correct measurement can be the difference between not replacing an out-of-specification component or installing the wrong part and creating further problems. Taking correct measurements is more than obtaining a reading; it also means understanding what is being measured, how it is being measured, and what the reading means. The following section examines several types of measurements.

■ ***Linear Measurements.*** When you are using a ruler or tape measure, a **linear measurement** is being made. These measurements may be made from point A to point B, such as using a tape measure to find the wheelbase of a vehicle, or using a ruler to determine pedal travel or pulley diameter. It could also be taking two measurements and subtracting one from the other to obtain a reading such as determining if an object is out-of-round, such as a brake drum.

This type of measurement will be in either English or metric units. English measurements, which are still used in the United States, are based on miles, yards, feet, inches, and fractions of an inch. Inches are broken down into fractions. Common fractions found in automotive tools and specifications are in half an inch, a quarter inch, eighths of an inch, sixteenths, thirty-seconds, and even sixty-fourths of an inch; fractional units smaller than a sixty-fourth of an inch are not practical. While the U.S. auto makers have converted to the metric system (to align with the rest of the world), millions of vehicles still require English measurements and tools.

Metric measurement uses the meter and portions of the meter, all based on units of 10. Decameters (1/10 or 0.1 meter) divide the meter into 10 units; these are further divided into centimeters (1/100 or 0.01 meter), which are then divided into millimeters (1/1,000 or 0.001 meter). A meter multiplied by 1,000 equals a kilometer (kilo meaning 1,000). Automotive measurements are usually in millimeters (mm) or centimeters (cm). Since all measurements are factors of 10, changing from one metric unit to another is just a matter of moving the decimal place. For example, a 32-mm socket is the same as a 3.2-cm socket, or cylinder bore of 101.6 mm is equal to 10.16 cm. **Figure 4-29** shows a comparison of English and metric measurement and **Figure 4-30** shows an example of a meter and how it compares to the English yard.

Since both English and metric systems are used, it is helpful to be able to switch between the two. **Figure 4-31** shows common conversion factors:

■ ***Precision Measurement.*** Inspecting and testing brake, engine, and transmission components often requires technicians make measurements to thousandths (0.001) or even ten thousandths (0.0001) of an inch or hundredths of a millimeter (0.01 mm). To make these measurements, technicians use tools such as micrometers, dial calipers, and dial indicators (**Figures 4-32–4-34**). These tools make it possible to determine extremely small variations as well as to make very exact measurements. How to read these types of tools is covered in detail in the chapters in which these types of measurement are necessary.

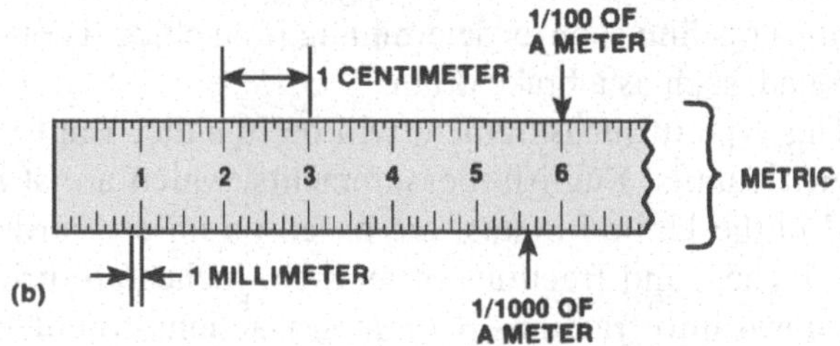

FIGURE 4-29 Both English and metric are used, and you will need to know how to work with both forms of measurement. While a 9/16 and 14 mm bolt head are very close in size, often allowing either size tool to be used on the bolt, English and metric threads never are close enough to allow you to use one in place of the other.

To Find		Multiply	x	Conversion Factors
millimeters	=	inches	x	25.40
centimeters	=	inches	x	2.540
centimeters	=	feet	x	32.81
meters	=	feet	x	0.3281
kilometers	=	feet	x	0.0003281
kilometers	=	miles	x	1.609
inches	=	millimeters	x	0.03937
inches	=	centimeters	x	0.3937
feet	=	centimeters	x	30.48
feet	=	meters	x	0.3048
feet	=	kilometers	x	3048.
yards	=	meters	x	1.094
miles	=	kilometers	x	0.6214

FIGURE 4-31 A conversion chart is very handy when working with both units of measurement. Conversion charts are typically part of repair information and available as apps for mobile devices.

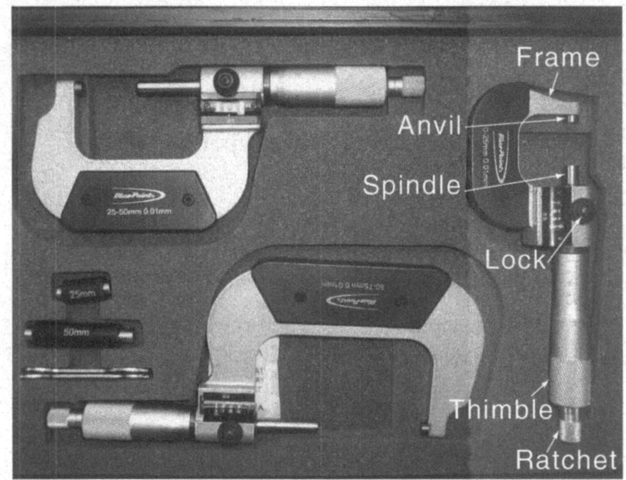

FIGURE 4-32 Micrometers are used to measure brake rotor thickness down to 0.01 mm. Outside micrometers measure the outside size of a component, such as pistons, bolts, washers, and many other items.

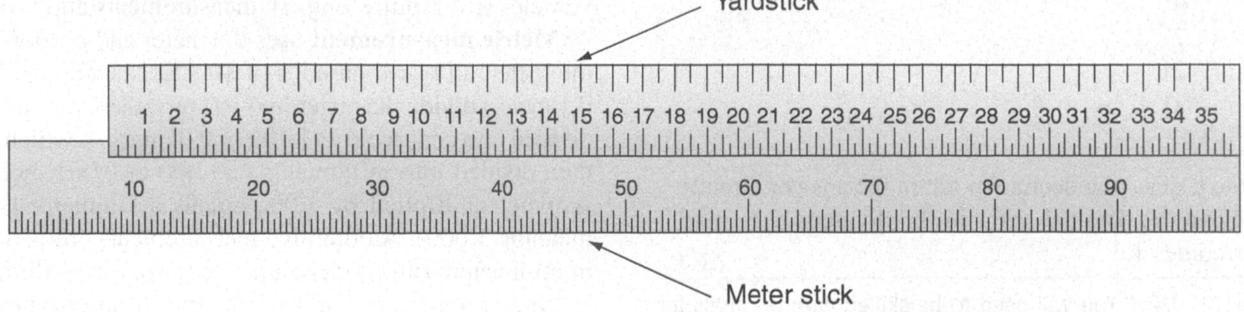

FIGURE 4-30 A meter is a little longer than 39 inches. There is no metric unit that corresponds to a foot or an inch.

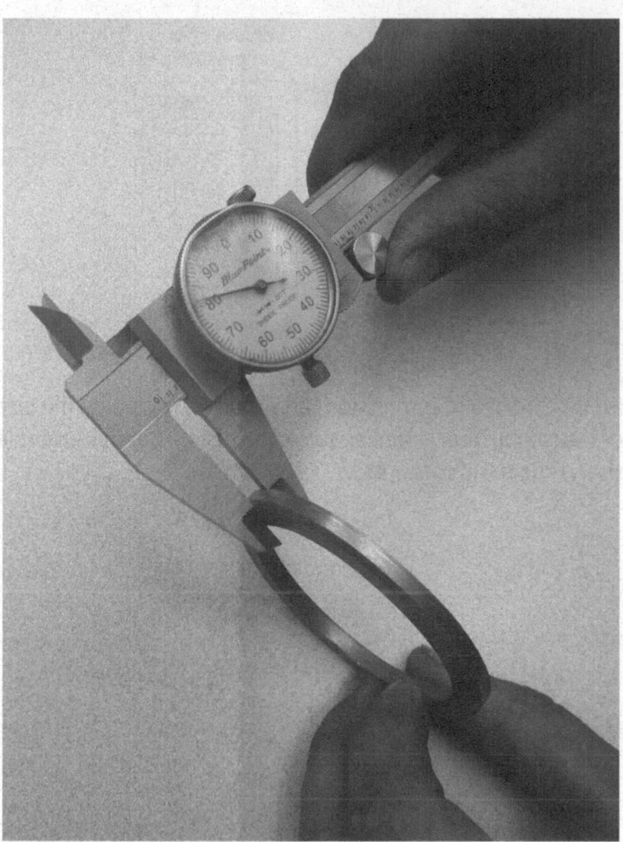

FIGURE 4-33 Dial calipers can measure a component's outside size, inside size, and the depth of a hole.

FIGURE 4-34 A dial indicator is used to measure very small amounts of movement or distortion, such as rotor runout, the amount a wheel is bent or the play in a crankshaft.

Other types of measurements include:

- Backlash; the space between two gear teeth in contact.
- Out-of-round; the readings of a circular area 90 degrees apart and finding the difference between them (see Chapter 13).
- Taper; the difference between a cylinder or another object's diameter from end-to-end.
- Lateral runout; the amount of side-to-side variation of an object like a wheel or tire (see Chapter 15).
- Radial runout; the amount of circumferential variation of a round object like a tire (see Chapter 5).
- Depth readings of a bolt hole or cylinder bore.
- Warpage; the amount of surface distortion from completely flat measured with a straightedge and feeler gauge.

Directions for using micrometers and other precision measuring tools and performing these types of measurements are discussed in the appropriate service chapters.

■ ***Volume.*** Linear measurements can be one- or two-dimensional, but volume measurement is three-dimensional. If you are measuring the square footage of a shop, the width and length are determined and multiplied together to find the area. To find the volume of an object, a third measurement is needed, height. If a room is 10 feet wide by 10 feet long, the total area is 100 square feet ($10 \times 10 = 100$). If the room is also 10 feet high, the total volume is calculated 10 feet × 10 feet × 10 feet, providing a result of 1,000 cubic feet.

To find the area of a circle, the formula πr^2 is used. A circle with a diameter of 3 inches (76.2 mm) will have a radius of 1.5 inches (38.1 mm). The area would be equal to 3.14 × 1.5 squared or 7.068 inches (3.14 × 38.1 squared = 1,454.61 mm).

In practice, a technician may need to calculate the volume of a cylinder in an engine. To find the volume of a cylinder, the formula $\pi r^2 \times$ length is used. Some technicians use diameter × 0.785 to find area and volume, because less calculating needs to be done. The number 0.785 is one-quarter of π, when multiplied times diameter^2, provides the same result as using πr^2. To calculate the volume of a cylinder with a 4-inch bore and 3.48-inch stroke, you can use $0.785 \times 4^2 \times 3.48 = 43.7088$ cubic inches. Bore applies to the diameter of an engine cylinder, and the stroke is the distance the piston travels up and down in the cylinder (**Figure 4-35**).

■ ***Ratios and Proportions.*** A **ratio** is a method of expressing a relationship of two values relative to one another. A common example is an engine's compression

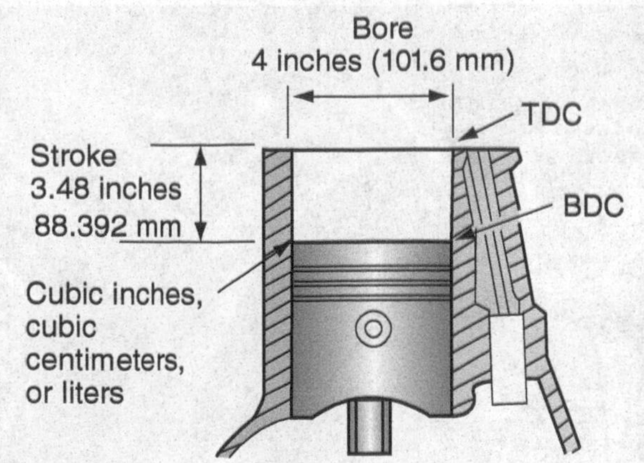

FIGURE 4-35 Factory engine sizes are easily found, but you may need to measure displacement on an engine that has been modified by boring or changing the stroke.

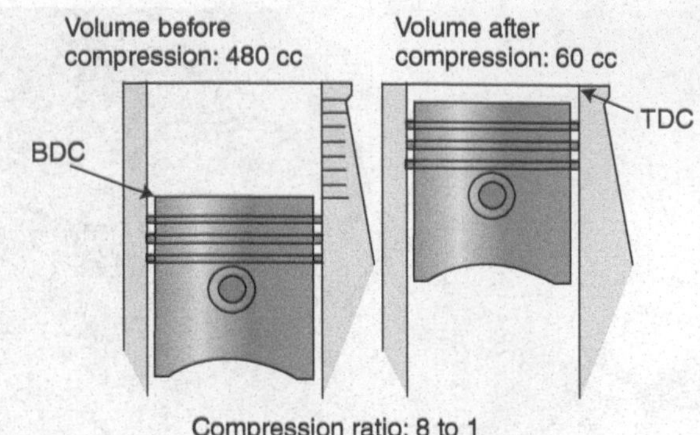

FIGURE 4-36 Compression ratio is also affected by the size of the combustion chamber above the piston, but this example shows the basic concept.

ratio. The compression ratio, expressed as two numbers such as 10:1, means the whole consists of 10 parts, which is reduced to 1 part in 10 or to one-tenth of the original volume (**Figure 4-36**). The volume of the cylinder shown in the figure is divided into eight equal parts when the piston is at the bottom. As the piston moves to the top of the cylinder, the volume decreases, eventually stopping with one-eighth of the original volume remaining. In this case, the air in the cylinder is compressed by a factor of 8, giving the ratio of 8:1.

Ratios are also used in defining gear interaction. When two different-size gears are meshed together, the smaller gear will turn faster than the larger gear. This is expressed as a mathematical relationship between the driving and driven gears (**Figure 4-37**). Meshed together, a drive gear with 12 teeth will spin twice as fast as a driven gear of 24 teeth, or at a 2:1 ratio. The larger gear, while turning at half the speed of the smaller gear, will turn with twice as much torque. This is because a gear can be thought of as a bunch of levers (**Figure 4-38**). As the size of the gear increases, the distance from the center to the teeth increases (**Figure 4-39**). This distance acts as leverage, and therefore the greater the leverage, the more force that can be applied. Used in the transmission and final drive gearing, various ratios achieve the different speeds and torque output. The different gear ratios in the transmission allow for an increase in engine torque or speed.

A variation of a ratio is proportion, such as a 50/50 solution of antifreeze and water. Antifreeze is mixed with water to obtain the best overall protection against freezing and boiling (**Figure 4-40**). In most cases, the antifreeze to water proportion is 50/50, meaning 50% antifreeze and 50% water.

FIGURE 4-37 Gear ratio is based on driving gear size to driven gear size.

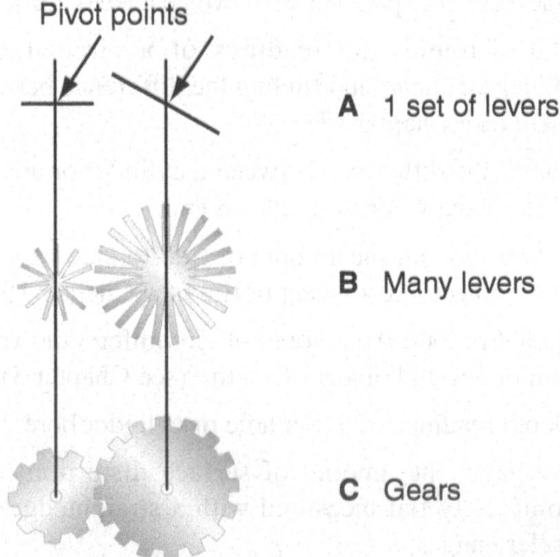

FIGURE 4-38 Arrangements of levers around a common center for gears.

■ ***Torque.*** **Torque** is a measurement of force applied over a distance (**Figure 4-41**). Typically, torque describes how much twisting force an engine or motor can generate. Readers of car and truck magazines often see engines rated for horsepower and torque at a given speed in revolutions per minute (rpm). But what do these numbers mean? If you attempted to lift a one-pound (0.45 kg) weight that is bolted to the floor, no matter

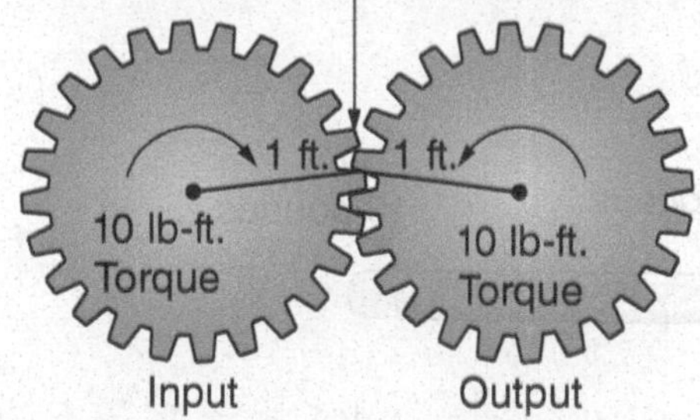

A 10 lb-ft. ÷ 1 ft. = 10 lb = 10 lb-ft.

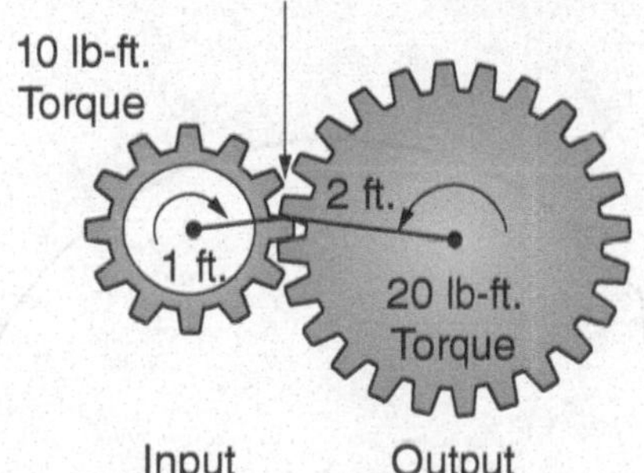

B 10 lb-ft. ÷ 1 ft. = 10 lb × 2 ft. = 20 lb-ft.

FIGURE 4-39 Each tooth of a gear is a lever. The longer the distance from center to the tooth, the more leverage produced and the more torque the gear applies.

FIGURE 4-40 Refilling a cooling system typically requires a 50/50 mix of antifreeze and water. Different proportions may be needed to correct for a system that has had too much antifreeze or water added.

how much force you apply, no work is done because the weight does not move. If the weight is unbolted, and you pick it up 1 foot (0.3 m) from the floor, you lift with 1 pound of force over a distance of 1 foot, or 0.45 kg over 0.3 m. Lifting the weight 1 foot equals 1 foot-pound of work. If you accomplish this task in 1 minute, you are working at the rate of 1 foot-pound per minute. If you accomplish the task in 1 second, the rate of work is 1 foot-pound per second or 60 foot-pounds per minute (1 lb. × 1 ft. × 60 seconds = 60 ft.-lb/min).

In the 1700s, James Watt (think of a 60-watt light bulb or 1,000-watt amplifier) defined the rate at which work is performed. Watt calculated that a horse could lift 550 pounds 1 foot in 1 second, or 33,000 foot-pounds per minute (550 lb × 1 ft. × 60 = 33,000 ft.-lb), which he equated to 1 horsepower (HP) (**Figure 4-42**).

Torque, however, is a measurement of twisting force. One ft.-lb of torque will support a 1-pound load at the end of a bar (a weightless bar for this example) 1 foot from the fulcrum. To determine HP, torque is what is measured and HP is calculated. Horsepower is not seen or felt; however, torque can be felt and measured. In **Figure 4-43**, if we can imagine a 1-pound weight on a 1-foot-long bar and rotate it one complete revolution, we get 6.2832 ft.-lbs. of work (π × 2 ft. circle = 6.2832 ft. × 1 lb = 6.2832 ft.-lb of work). Remember that 33,000 ft.-lbs. of work for 1 minute is equal to 1 HP. Divide 33,000 ft.-lb by 6.2832 ft.-lb per revolution, and you get 5,252. One foot pound of torque at 5,252 rpm is equal to 33,000 ft.-lb per minute of work, which is equal to 1 horsepower. Therefore, HP = torque × rpm/5,252, which is exactly how horsepower figures are determined for an engine. Engineers measure the highest torque output from an engine on a dynamometer, record the rpm at which it is reached, and divide by 5,252 to get horsepower. The metric system uses the Newton-meter, abbreviated N·m, which is equivalent to 0.737 ft.-lb of force.

Why is torque so important? As a technician, you will have to use torque every day. The correct application of torque (and leverage) affects how much work (physical exertion) is used when loosening or tightening a fastener. To tighten a wheel lug nut to 100 ft.-lb (135 N·m), a lot of force must be applied either over a short distance or less force over a longer distance. Which of these sounds easier? Which means more work from you, the short or long distance?

Two common situations young technicians face are not being able to loosen tight fasteners effectively and breaking fasteners from applying too much force to them. This is often a result of not understanding how to apply leverage and torque. To loosen tight fasteners, such as lug nuts, use a large ratchet or breaker bar to provide leverage and reduce the amount of force you need to apply to the tool. Conversely, when tightening small nuts and bolts, use smaller tools, such as a ¼-inch drive ratchet to help prevent

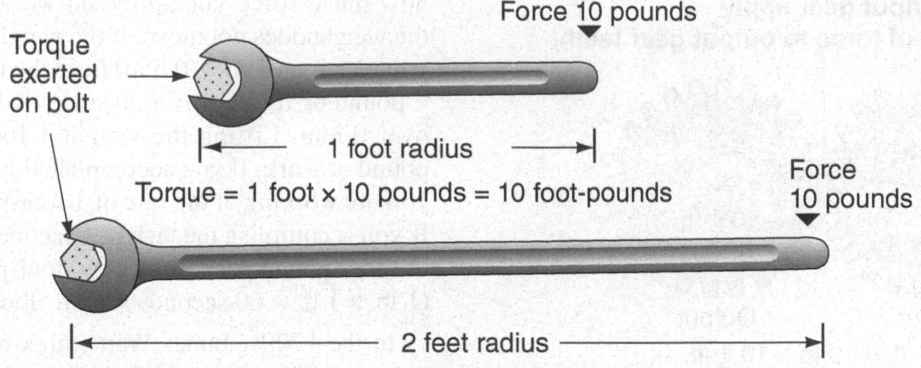

FIGURE 4-41 Torque is force multiplied by distance, an important application for all technicians.

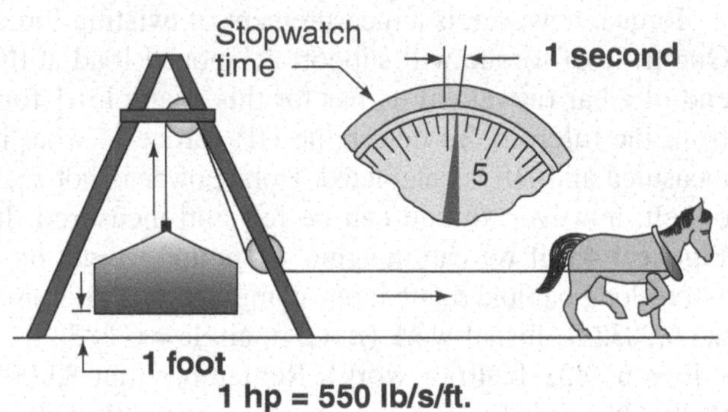

FIGURE 4-42 Torque is measured and felt, but horsepower is a calculation of how much work can be performed.

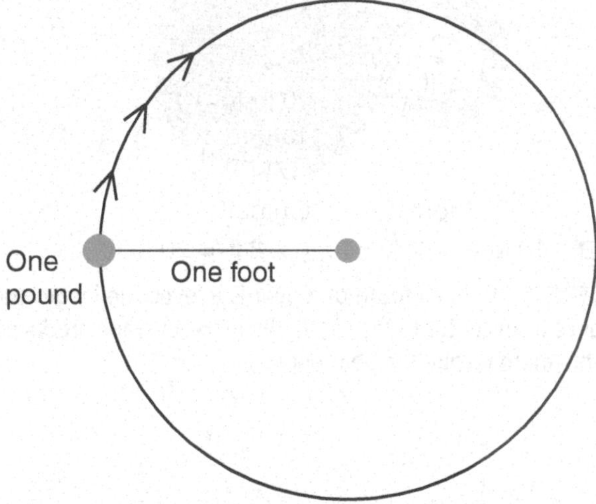

As the one pound weight moves in a circle with a radius of one foot, the amount of work is calculated as:

pi × r² = 3.1415 × 1² = 6.2832 feet x 1 pound which equals 6.2832 pounds of work per foot or 6.2832 pound-feet

FIGURE 4-43 One pound applied over a circle with a radius of one foot equals about 6.3 foot-pounds of work.

over-tightening and damage to the fastener. **Figure 4-44** shows common bolt and nut sizes, torque ranges, and which tool is often used to loosening and tightening. **The chart is not meant to replace torque specifications. Always look up tightening specifications in the service information.**

■ ***Other Measurements.*** Technicians must be able to read more than a ruler or a micrometer. Working with pressures, whether in tires or the air conditioning system, requires an understanding of what pressure readings mean. Air pressure is the result of the weight of the components of the atmosphere (78 % nitrogen, 21 % oxygen, and about 1 % other gases) pressing down from the edge of the atmosphere all the way to the ground. This force, known as atmospheric pressure, applies 14.7 pounds of pressure per square inch (psi) at sea level (**Figure 4-45**). This pressure increases if it is measured below sea level, such as in Death Valley, California, and decreases with a rise in elevation, approximately 1 psi for every 2000 feet above sea level. Atmospheric pressure in Denver, Colorado, which is at approximately 5,000 feet above sea level, is approximately 12.2 psi (**Figure 4-46**).

Pressure also refers to the amount of force exerted by a gas or liquid against an enclosed volume. Imagine two plastic bottles, one with water and the other with your favorite carbonated soda. Now imagine that instead of the normal bottle cap, you have installed a pressure gauge. The pressure within both bottles is zero pounds of pressure. Now imagine shaking the bottles vigorously. What would the gauges read now? The bottled water would remain at zero, but the soda bottle would show a slight pressure. Where did the pressure come from? By shaking the bottle, the carbon dioxide in the soda is released quickly, and because it is trapped in a sealed container, it causes an increase in pressure.

If you use a tire pressure gauge to measure a tire with no air in it, does it read 14.7 psi? No, because it is calibrated to read atmospheric pressure as zero pounds of pressure.

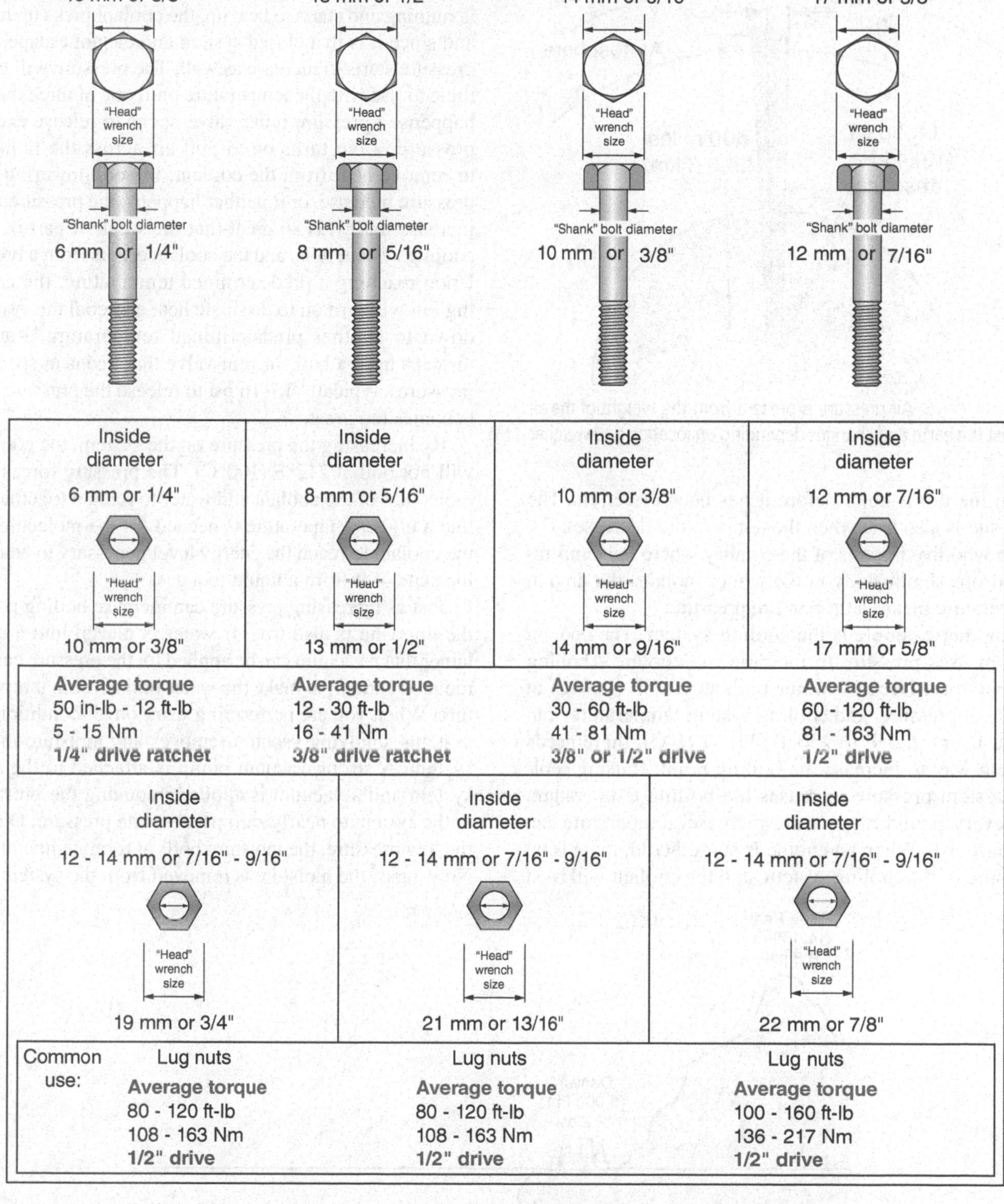

FIGURE 4-44 Examples of torque loads based on bolt size. NOT to be used as a torque specification chart.

There is atmospheric pressure present but it does not register, just like in our bottle of water. A flat tire has atmospheric pressure, but our tire gauge is made to read 14.7 psi air pressure as 0 psi. The gauge will read pressure when more air is forced into the tire. As a tire gets hot from rolling along the road, the tire pressure will increase. The rolling resistance of the tire generates heat; this heats the air inside the tire, and because temperature and pressure are related, the pressure will rise with the temperature. This is an example of Charles' Law. Because the air is trapped in a sealed container, as the temperature increases, the energy of the air molecules also increases. This causes the air to exert more pressure against the tire than before the temperature increase. This is why tire pressures should be checked

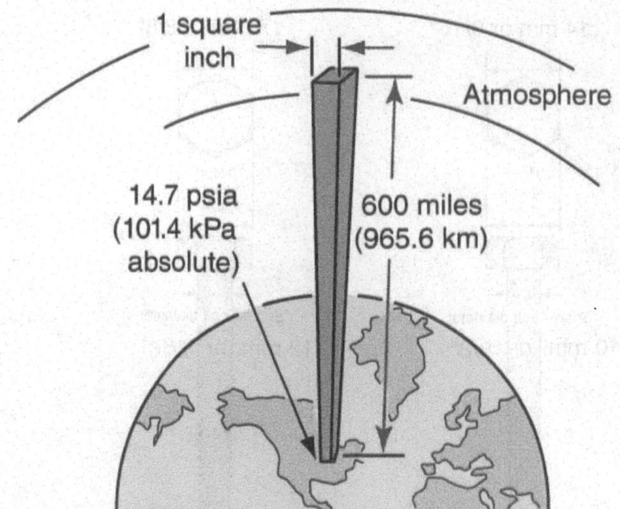

FIGURE 4-45 Air pressure is present from the weight of the air against the earth and changes depending on location and weather.

when the tire is cold, before it has been driven on. The opposite is also true when the temperature decreases. For those who live in parts of the country where the temperature drops significantly in the winter months, the drop in temperature means a drop in tire pressure.

Another example is the cooling system. The cooling system uses pressure to increase the coolant's boiling temperature. Because water boils at 212°F (100°C) at sea level pressure, and cooling system temperatures can range from 230°F to 250°F (110–121°C), there needs to be a way to increase the boiling point. Raising cooling system pressure increases the boiling temperature. For every pound of pressure increase, temperature can increase 3°F. When an engine is started cold, there is no pressure in the cooling system, and the coolant will be at ambient (surrounding air) temperature. Once the engine is running and starts to heat up, the coolant picks up heat, and since it is in a closed system and cannot escape, the pressure starts to increase as well. The pressure will continue to rise with the temperature until one of three things happens: a pressure relief valve opens to release excess pressure, a fan turns on to pull air across the radiator to remove heat from the coolant, thereby lowering the pressure increase, or if neither happens, the pressure will increase until it is so great that the weakest part of the cooling system fails, and the coolant erupts from a break. Upon reaching a predetermined temperature, the cooling fan will turn on to dissipate heat and cool the system down to another predetermined temperature. Radiator caps have a built-in vent valve that opens at specific pressures, typically 13–16 psi to release the pressure if it becomes too great.

By increasing the pressure on the system, the coolant will not boil at 212°F (100°C). The pressure forces the molecules of the coolant and water to remain together so that a higher temperature is needed for the molecules in the coolant to reach the energy level necessary to change the state of it from a liquid to a gas.

Just as increasing pressure can increase boiling point, the opposite is also true. If water is placed into a container that a vacuum can be applied to, the pressure can be reduced enough to make the water boil at room temperature. When you are performing work on air conditioning systems, applying vacuum removes any moisture in the system. A strong vacuum pump is attached to the A/C system and a vacuum is applied, reducing the pressure in the system to nearly zero psi absolute pressure. Due to the low pressure, the moisture boils at room temperature. Now a gas, the moisture is removed from the system.

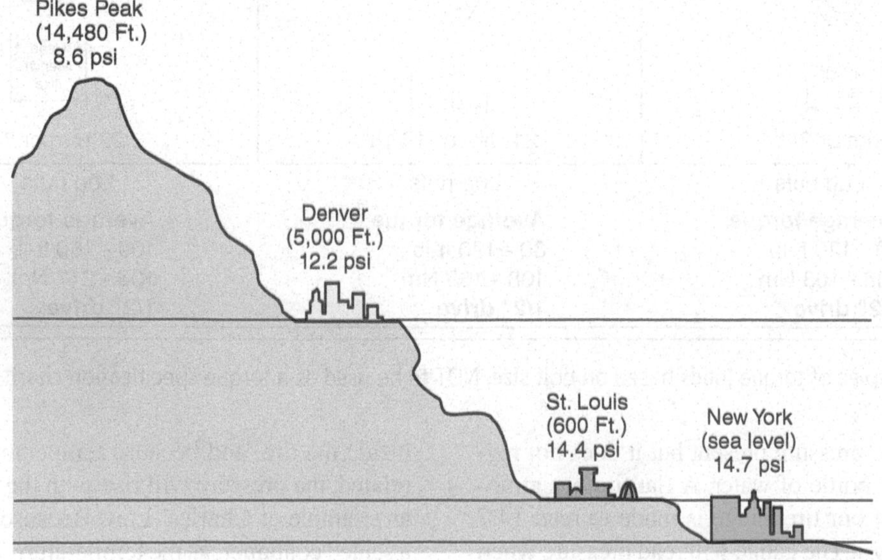

FIGURE 4-46 Air pressure decreases as elevation increases. This has a significant impact on engine operation as less air is drawn into the engine.

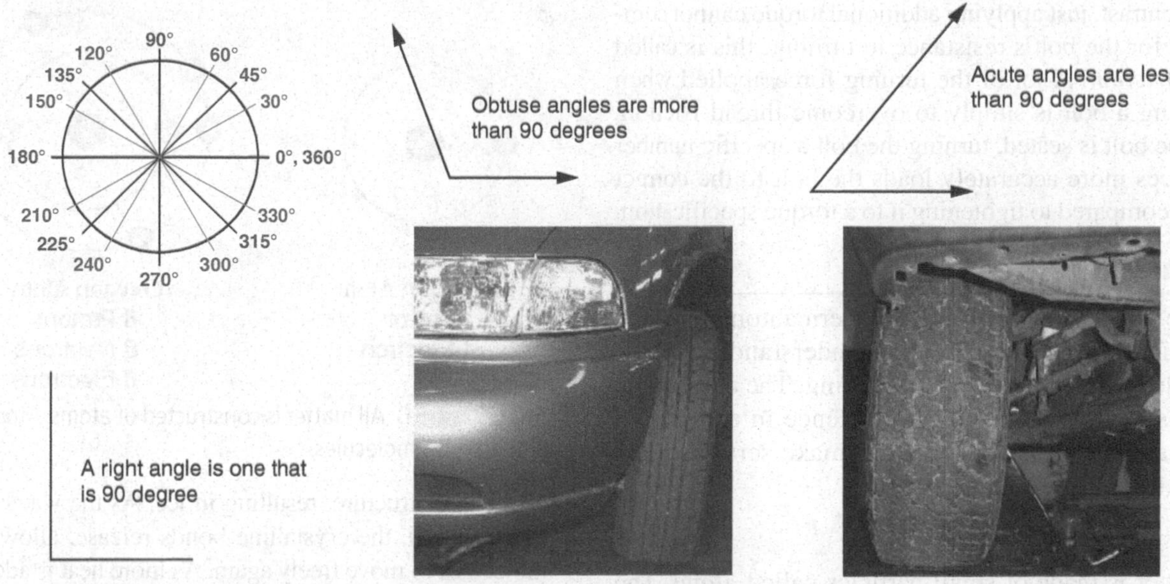

FIGURE 4-47 Angles are used in wheel alignment and torquing.

■ *Angles.* Often used when discussing the geometry of the vehicle's suspension system, angles are also used when tightening a fastener to specifications and require using a torque-angle gauge. An angle is a measurement of 1/360 of a circle or full rotation. Angles can be broken down into smaller units than a degree, called *minutes of arc* and *seconds of arc*. There are 60 minutes of arc per degree and 60 seconds of arc per minute of arc. Most automotive applications use degrees and fractions of a degree, such as 1.5 degrees or − 0.25 degrees. Angles can also be acute, less than 90 degrees, or obtuse, more than 90 degrees (**Figure 4-47**).

A perfectly vertical tire is 90 degrees from the ground and will have 0 degrees of tilt or lean away from vertical. If the tires are leaning in toward or away from the vehicle, they will have a measurable tilt. This angle, measured in degrees, is called camber and is one of the wheel alignment angles. Each wheel has angles of movement in which it travels while the vehicle is in operation. When you are performing steering and suspension repairs, specifically wheel alignments, these angles are measured and wheel position are adjusted.

To measure the torque applied to a fastener, a torque-angle gauge is used (**Figure 4-48**). Many engine designs use a standard ft.-lb setting for the initial torque and then specify an additional number of degrees to be turned. For example, a cylinder head bolt may have a torque specification of 45 ft.-lb plus 60 degrees. This means after the initial torque is applied, a torque-angle gauge is used to apply 60 degrees of additional twist to the head bolt to complete the tightening sequence. Newer torque wrenches include torque angle as a setting so an additional tool is not required (**Figure 4-49**).

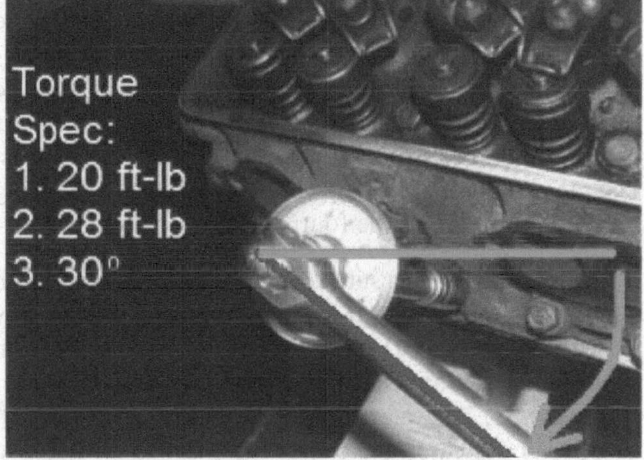

FIGURE 4-48 Torque specs often include angles in addition to standard fastener torque. A torque-angle gauge can be used to apply the correct amount of load on the fastener.

FIGURE 4-49 Digital torque wrenches can be set for inch-pounds, foot-pounds, Newton-meters, or degrees of angle, making them very versatile tools.

Engineers use the torque angle because it is a more accurate application of torque than can be achieved since the angle has a direct relationship to how tight the bolt will be. In contrast, just applying additional torque cannot compensate for the bolt's resistance to turning; this is called *thread friction*. A lot of the turning force applied when tightening a bolt is simply to overcome thread friction. Once the bolt is seated, turning the bolt a specific number of degrees more accurately loads the bolt to the correct tension compared to tightening it to a torque specification.

Science Skills

Science plays a key role in the modern automobile, and technicians should have a good understanding of the principles with which they are working. The automobile contains countless examples of science in action, and understanding these concepts can make service of the vehicle safer and easier.

MATTER AND ENERGY

All matter is made of small particles called *atoms*. The common view of an atom is similar to a tiny solar system where protons and neutrons cluster at the center and electrons orbit the nucleus (**Figure 4-50**). Atoms combine into structures that form the universe we inhabit. While atoms may all appear similar, by combining atoms differently we get a multitude of different substances. These substances then fall into one of three states or categories of how matter is grouped, as a solid, liquid, or a gas. Plasma, a high-energy state of energy such as a lightning bolt, is referred to as the fourth state of matter.

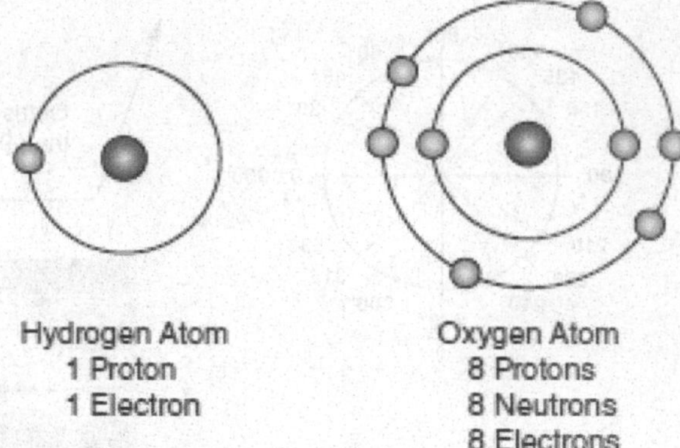

FIGURE 4-50 All matter is constructed of atoms. Atoms combine to form molecules.

■ ***Reactions.*** One substance, water, can exist in all three states in our day-to-day environment, though only in one state at a time. Heat, either added or removed from a substance, can force a change from one state to another (**Figure 4-51**). As water cools (heat removed), its molecular activity decreases and causes the water to form a solid crystalline structure, resulting in ice. As the water warms (heat added), the crystalline bonds release, allowing the molecules to move freely again. As more heat is added, the molecules are excited and move with more energy. Once the water reaches 212°F or 100°C, enough heat energy has been added to separate the hydrogen–oxygen bond, allowing for the evaporation of the water into the air. Once the steam cools, it will recombine into a liquid in a process called *condensation*.

Reactions, the changing of the state or the arrangement of molecules, can be spontaneous, meaning no additional energy is required for the reaction to take place. An example of this is when certain types of mint are introduced to a diet cola. A nonspontaneous reaction requires additional energy, such as light, heat, or electricity to take place. An example of this type of reaction is the release of the stored energy in gasoline and air during combustion. The air–fuel mixture is compressed and heated in the combustion chamber of the engine, and energy is released as heat and pressure against the piston.

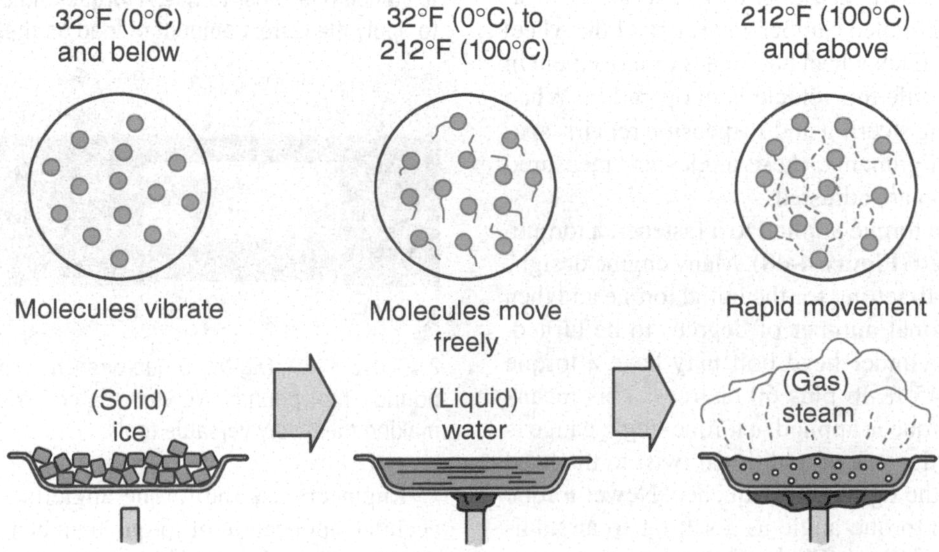

FIGURE 4-51 Matter can exist in three basic forms and can change from one form to another under the right conditions.

■ ***Energy.*** You may have heard that energy cannot be created or destroyed, just converted from one form to another. A technician should be familiar with the types of energy used in an automobile. For most vehicles, the energy to propel the vehicle comes from the internal combustion engine. Fossil fuels, burned in the engine's combustion chamber, generate large amounts of heat and pressure. This energy pushes on the pistons that turn the crankshaft, which ultimately turns the drive wheels (**Figure 4-52**). Unfortunately, the internal combustion engine (ICE) is not 100% efficient. Most of the heat energy generated by the combustion process is lost through the cooling system and exhaust. Only about one-third of the energy produced is used to power the vehicle; the rest is wasted (**Figure 4-53**).

Once the vehicle is moving, it has kinetic energy, or the energy of motion. Once it is in motion, the vehicle performs work. The rotating mechanical energy of the engine turns the AC generator (alternator), converting the mechanical energy into electrical energy used to power the vehicle's electrical system. When a technician starts the vehicle, the starter motor turns the crankshaft using the electrical energy stored in the battery by converting electrical energy into mechanical energy.

Hybrid and electric vehicles use what is known as regenerative braking. When applying the brakes on a regular vehicle, kinetic energy is converted into heat by the friction of the brake system components. On non-hybrid vehicles, as the heat is dissipated, the energy is wasted. On hybrid vehicles, the electric motor used to drive the wheels (or assist the ICE) also functions as a generator. When the brakes are applied, the on-board computer system switches the electrical motor operation into electrical energy generation. The generation of electrical energy in the motor/generator creates a physical load that acts as a resistance, slowing the vehicle down. In this way, the captured kinetic energy recharges the battery that supplies power for the motor/generator. This means reduced gasoline consumption and wear on the brake system.

GENERAL SCIENTIFIC PRINCIPLES

The automobile is a complex marriage of both ancient and modern science, combining ideas thousands of years old with modern expectations.

■ ***The Simple Machines.*** Thousands of years ago, people devised ways to make everyday work easier by applying the principles of the simple machines. A simple machine requires the application of one force to perform work. This is often to reduce the amount of effort needed to lift or move very heavy objects. By utilizing simple machines, the building of the pyramids of Egypt along with the structures of the Greeks and Romans

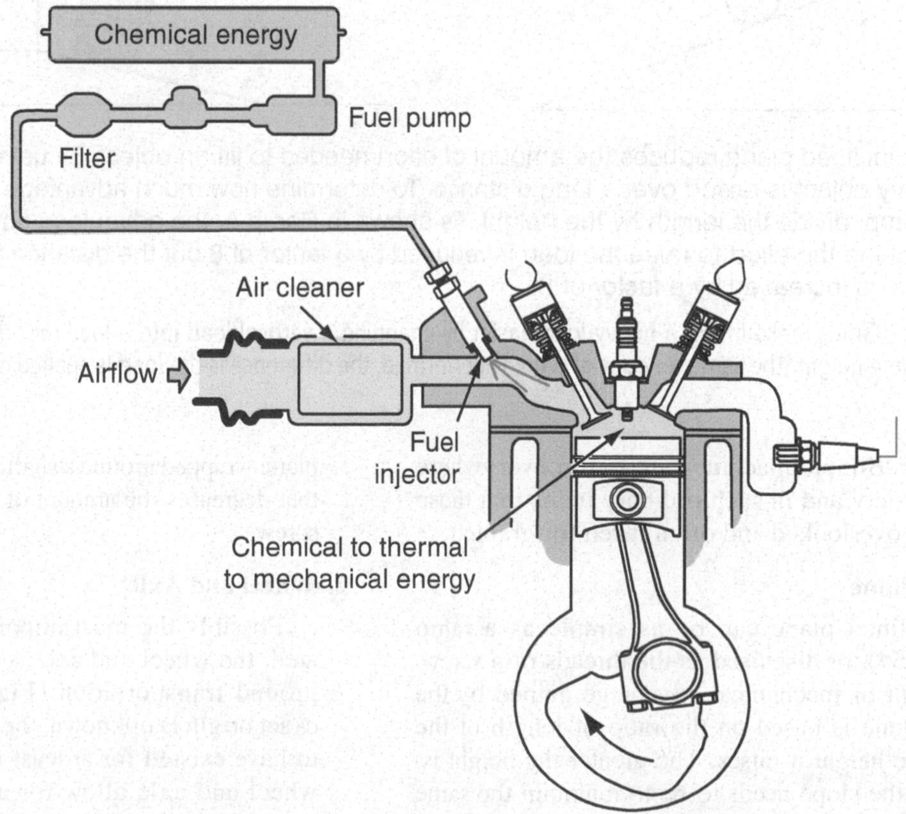

FIGURE 4-52 Energy conversion is found throughout the automobile; in the engine, brakes, and electrical systems.

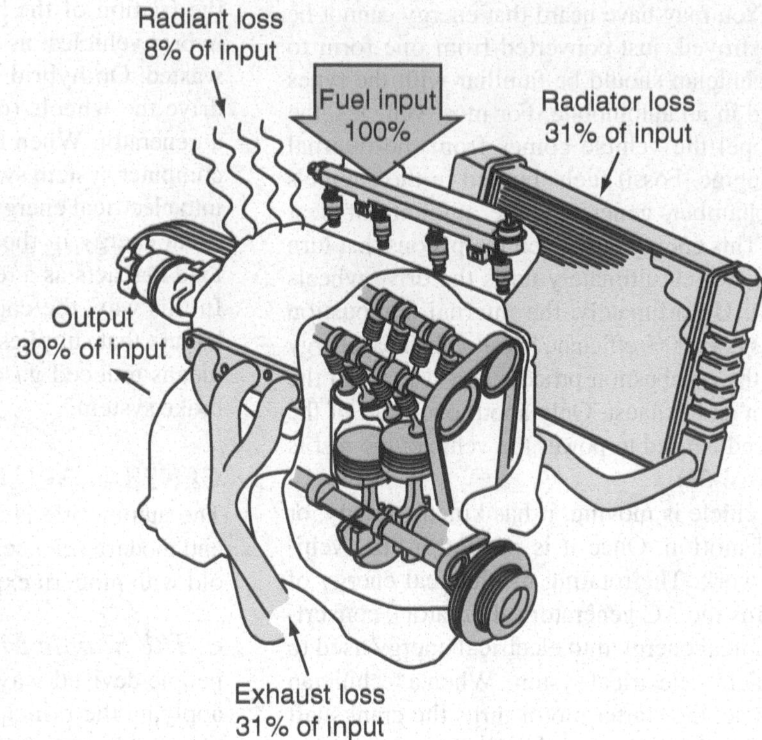

FIGURE 4-53 Unfortunately, most of the energy produced by the engine is wasted as heat.

The inclined plane reduces the amount of effort needed to lift an object. By using a ramp, a heavy object is raised over a long distance. To determine how much advantage is gained by a ramp, divide the length by the height. As shown in Ramp A, the advantage equals 8, meaning the effort to raise the load is reduced by a factor of 8 but the distance the load must move is increased by a factor of 8.

FIGURE 4-54 Ramps make lifting a heavy load easier by changing a vertical load into a load raised over a longer distance but reaching the same height. The same amount of work is performed, the difference is the load is applied over a much longer distance.

were possible. Simple machines are in use everywhere in modern society and in such ordinary items that these machines are overlooked and often taken for granted.

1. Inclined Plane

An inclined plane can be as simple as a ramp (**Figure 4-54**), or disguised as the threads of a screw. The amount of mechanical advantage gained by the inclined plane is based on the ratio of length of the slope to the height it raises. The greater the height is, the longer the slope needs to be to maintain the same amount of work being done. A screw uses the inclined plane wrapped around a shaft. The threads create a ramp that decreases the amount of work needed to insert the screw.

2. Wheel and Axle

Possibly the most important invention ever created, the wheel and axle is the basis for all modern ground transportation (**Figure 4-55**). Though the exact origin is unknown, the wheel and axle is known to have existed for at least five thousand years. The wheel and axle allow for a much lower resistance to motion compared to dragging an object. Other

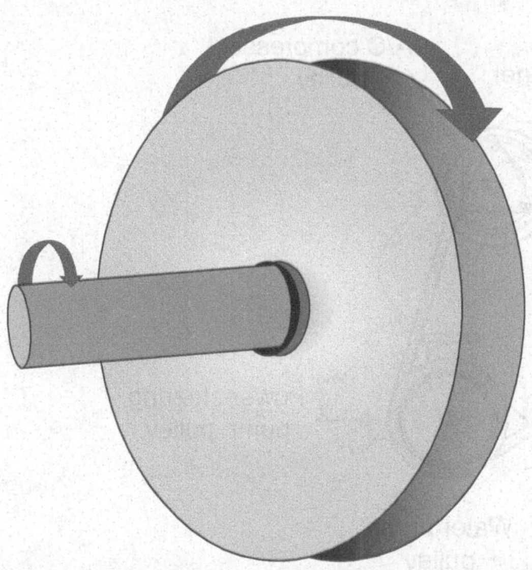

FIGURE 4-55 Possibly the most significant invention ever was the wheel and axle. It allows for much easier movement over ground.

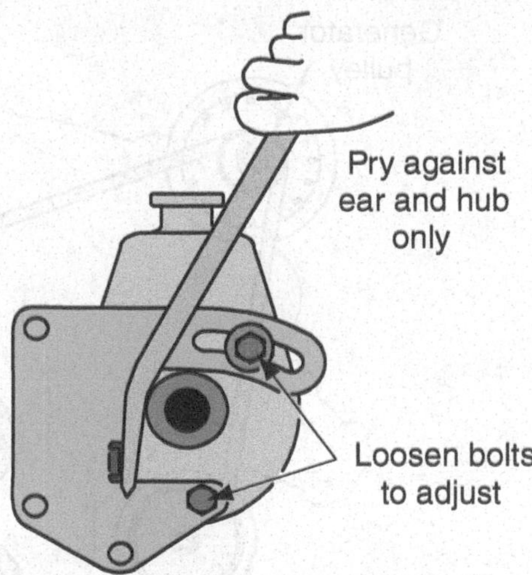

FIGURE 4-56 The lever is similar to the inclined plane in that it allows force to be applied over a longer distance.

applications of this machine are the rotating door knob, screwdriver, and a mechanical pencil sharpener.

3. The Lever

One of the greatest inventers and thinkers of all time, Archimedes, reportedly said that given a place to stand and a large enough lever, he could move the earth. Used with a pivot, known as a fulcrum, levers in their various forms as pry bars, wheelbarrows, teeter totters, and shovels multiply mechanical force to an object. Leverage, also called *mechanical advantage*, is important to technicians; too little leverage means more effort is required to perform a task, too much can cause broken fasteners or components. Technicians should be able to understand and apply leverage to make the work easier. **Figure 4-56** shows how a pry bar acts as a lever to tighten an accessory belt. To apply a large amount of force at the end of the lever, the force is applied over a long distance. To determine how much force is being applied, we can measure the distance from the fulcrum to each end of the pry bar and calculate how much force is being applied by the technician.

4. The Pulley

A pulley is simply a wheel with a groove or passage along the outside edge to hold a rope, cable, or in automobiles, a belt. As simple machines, pulleys are used to change the direction of loads. Several pulleys combined into a block and tackle arrangement will provide a mechanical advantage when a person is lifting a load. On the automobile, pulleys transmit the rotational motion of the crankshaft to belt-driven accessories, such as the power steering pump, generator, and air conditioning (**Figure 4-57**). By changing the diameter of the pulley on the accessory, the speed at which the accessory will spin is determined by the ratio of the diameter of the driving and driven pulleys. As you may know, generator drive pulleys are much smaller than the crankshaft pulley. This is because generator output is very low at slow speeds. Decreasing the pulley size increases the rotational speed and output.

5. The Wedge

A wedge is two inclined planes placed together. Axes, scissors, knives, forks, and nails are examples of uses of a wedge, as are chisels (**Figure 4-58**).

6. The Screw

A screw is an example of an elongated inclined plane. The threads of the screw convert rotational force (torque) into a linear force. The length of the plane (the threads) applies force over a much greater area when compared to a nail. Screws, while similar to bolts, are not exactly the same (**Figure 4-59**). A screw creates the opening and applies its force directly to the material; no internal threads are required. A bolt matches the threads of a hole or nut in which the same thread pattern already exists. Modern vehicles consist of thousands of screws and bolts, fastening together the engine, transmission, and countless other components.

OTHER PRINCIPLES

■ ***Hydraulics.*** **Hydraulics** is the science of the mechanical properties of fluids. Although formalized in the 1600s by Blaise Pascal, the principles of hydraulics have been in use for thousands of years. Pascal's law, simply stated, says that increasing the pressure on a fluid

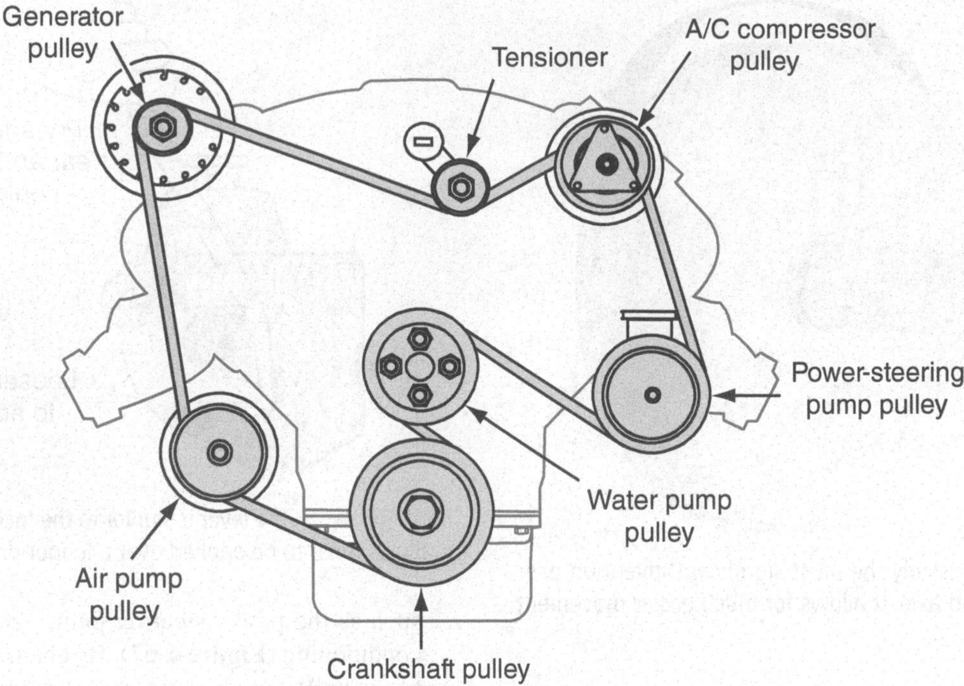

FIGURE 4-57 Pulleys are similar to gears and levers in that the size of the pulley determines the speed and force with which it rotates. Pulleys can be used to decrease loads and to change the direction in which loads are applied.

FIGURE 4-58 A chisel is a common use of a wedge.

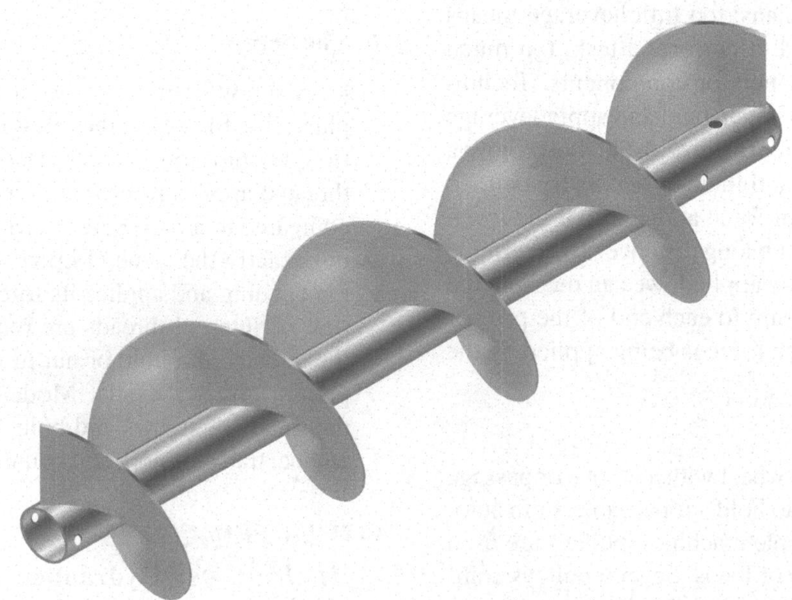

FIGURE 4-59 Screws, bolts, and nuts are examples of an inclined plane in a different form.

in a confined system will cause an equal increase in pressure at every other point in the system. Modern hydraulic power steering, brake systems, and automatic transmissions utilize hydraulics to perform work. The same principle applies to the hydraulic floor jacks, hoists, and presses found in automotive shops. Pressure is created on the fluid in the system by applying force to a piston. A piston on the output side responds to the fluid pressure and exerts an equal output force if the two pistons are the same diameter. Using pistons of different sizes can customize output forces.

Figure 4-60 shows how a floor jack uses hydraulics to lift a vehicle off the ground. As the technician moves the jack handle up and down, the handle, acting as a lever, applies force to the input hydraulic piston, labeled piston A, which has a surface area of one square inch. The force applied to the piston is based on the force applied to the handle times the length of the handle, which in this case equals 100 pounds of force. This force on the piston equals 100 psi of pressure on the fluid, and the pressure is transmitted to the output piston, which has five times the surface area of the input piston. The force generated by the output piston is equal to five times the pressure on the fluid, or 500 pounds of force. Because both leverage and hydraulics can increase forces by increasing the distance over which a force is applied, the output piston, while moving with much greater force, also moves much less distance than the input piston. If the input piston moves 1 inch, the output piston, because of the 5:1 ratio, will move 1/5 of an inch. This explains why when a floor jack is used to lift a vehicle, it takes many up and down movements of the jack handle over a very long arc to raise the vehicle just a couple of inches.

■ ***Density/Specific Gravity.*** **Specific gravity** is a measurement of the density of a substance compared to water. Density refers to how much mass is contained in a material in a given volume. For example, mercury compared to water, by volume, is much more dense. This means that while wood will float on water, iron will float on mercury. This is due to the greater concentration of mercury atoms than water molecules for the same volume. Specific gravity compares the density of a material to water, which has a specific gravity of 1.0. When added to water, a more dense substance, such as the sulfuric acid used in automotive batteries, increases the combined specific gravity (**Figure 4-61**). Antifreeze added to water also increases the overall specific gravity of the coolant, while motor oil, which has a lower specific gravity than water, will float on top of water.

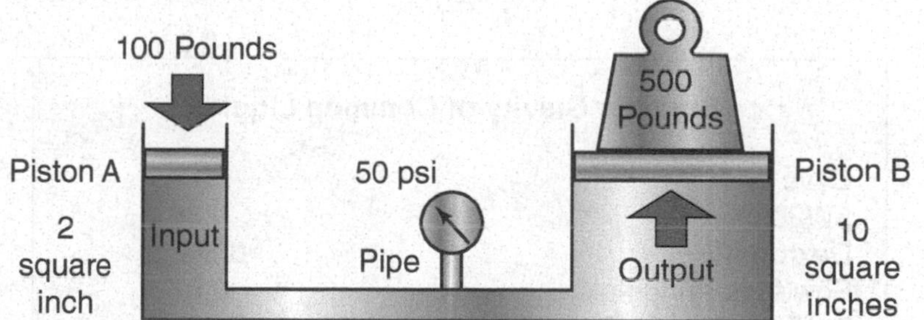

FIGURE 4-60 Hydraulics use liquids to transfer force and motion. These principles are used in the brake system, power steering, and automatic transmission.

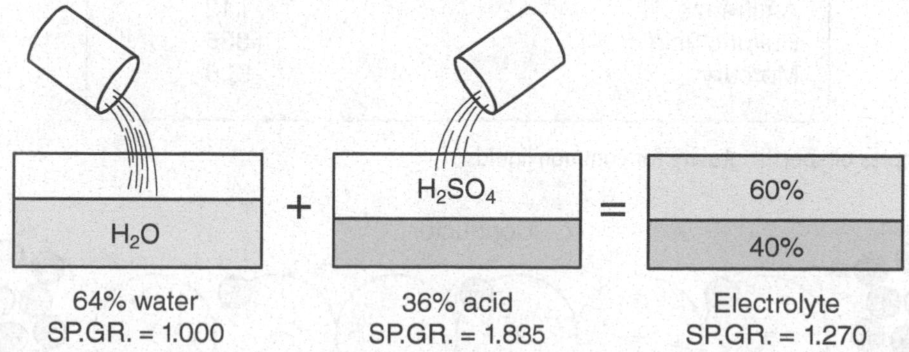

FIGURE 4-61 The principle that determines how heavy or light an object is comes from its density. Lead is much more dense than aluminum, and water is more dense than new motor oil.

Specific gravity measurements are used to determine both battery acid and coolant condition (**Figure 4-62**).

■ ***Electricity.*** One of the most misunderstood aspects of automotive repair, even by experienced technicians, is electricity. Until experiments were conducted by Benjamin Franklin, Alessandro Volta, Georg Simon Ohm, André-Marie Ampère, Michael Faraday, and Luigi Galvani in the 1700s and 1800s most people thought electricity and lightning were acts of magic or of the gods. The work of these scientists defined and quantified the properties of electricity and led to its eventual applications for everyday life. It is difficult to imagine our world without electricity as we are so dependent upon it.

Many people have difficulty understanding electricity since it is, most of the time, an unseen force. Often seen in a negative way, such as with lightning strikes or showers of sparks from a broken high voltage line, electricity is also commonly feared.

Electricity, at its most basic, can be expressed as the flow of electrons (**Figure 4-63**). Conductors such as copper and gold have four or fewer electrons in their outer or valance ring. These substances, given the proper conditions, will give up one of their outer electrons to another atom. Amperage, also called *current flow*, is the movement of electrons from atom to atom in an electrical circuit. Named after André Ampère, amperage is the measurement of electron flow through a circuit. To get the electrons to move, there must be an incentive. This incentive is a difference in potential between two points in a circuit. In a battery, there is an excess of electrons at one side and a similar excess of protons on the other side. The force of the electrons and protons to reunite is great, thus providing the incentive for the electrons to move or flow. Given the proper circumstances, the difference in potential, and a path for electron flow, the electrons will move to rejoin the protons.

When an electrical circuit is constructed, a resistance, called *a load*, is placed in the path of the electrons. A simple circuit is shown in **Figure 4-64**. Work can be performed when the electrons move along the path and through the load. This work may be the lighting of a light bulb or the release of sound from a speaker. Once the work has been performed, the electrons rejoin the protons. If the circuit is left on, the work by the electrons will eventually cause all the electrons to rejoin with all the protons. This situation, where there is no longer any incentive for electron flow, is evident as a dead battery. By charging the battery, separating the electrons and protons, the potential for work to take place is restored.

Modern vehicles use electricity in many ways. The battery supplies the power to turn the electric starter motor that turns the engine during startup. The AC generator changes mechanical movement into electrical

Specific Gravity of Common Liquids	
Olive oil	0.7
5W20 oil	0.86
Dexron ATF	0.862
Power steering fluid	0.87
DOT 5 brake fluid	0.958
DOT 3 brake fluid	1.0 to 1.07
Pure water	1
Salt water	1.028
Antifreeze	1.12
Sulfuric acid	1.835
Mercury	13.6

FIGURE 4-62 Examples of specific gravity for common liquids.

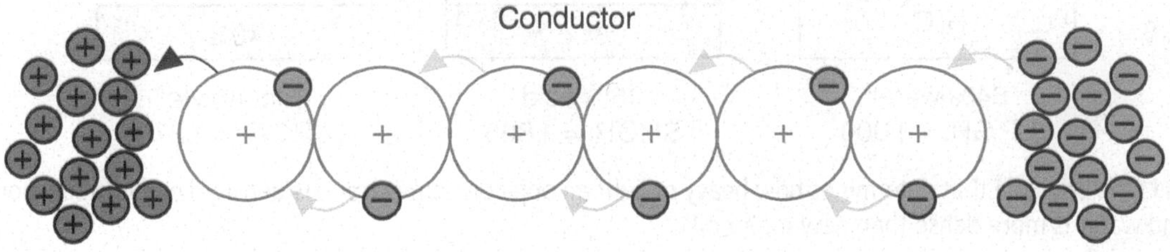

FIGURE 4-63 Electricity is the flow of electrons along a conductive path.

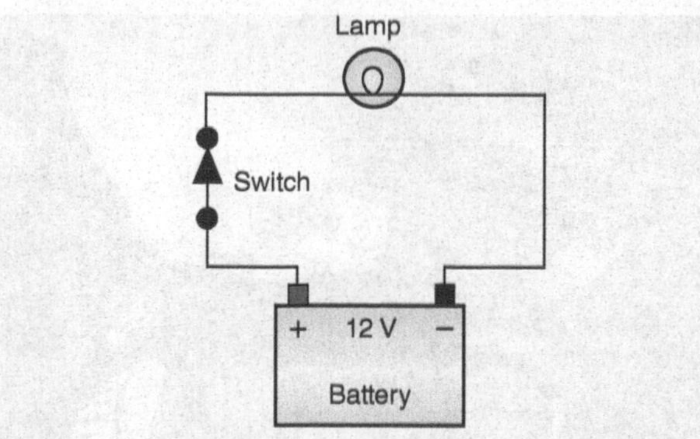

FIGURE 4-64 A circuit is a path for electrons to flow and perform work, such as illuminating a light bulb.

energy to power the electrical system and recharge the battery. The ignition system converts battery voltage into a high-voltage spark to ignite the air–fuel mixture in the engine. Electrical signals are used by the on-board computer systems to communicate and control functions such as transmission shifting and power window operation.

■ ***Vacuum.*** **Vacuum** is the term used to describe air pressure less than atmospheric pressure. The relationship between gauge pressure, absolute pressure, and vacuum are shown in **Figure 4-65**. Gauge pressure, also referred to as *per square inch gauge (psig)*, is pressure indicated on a gauge that is calibrated to read atmospheric pressure as zero pressure. Gauges of this type include tire pressure and cylinder compression gauges. Absolute pressure is read from zero pressure. Absolute pressure gauges are not commonly used in automotive repair. A vacuum gauge reads pressure below 14.7 psi. A reading of 30 inches of vacuum on a vacuum gauge equals zero absolute air pressure.

When the engine begins to turn, the pistons move in the cylinders (**Figure 4-66**). Downward movement causes a low pressure area, which is used to draw air from above into the cylinder. The shape of the throttle body speeds up the flow of air entering the engine, which reduces the pressure of the air (**Figure 4-67**). The restrictions of the throttle plates and around the valves cause the pressure in the engine to fall below the atmospheric pressure outside of the engine. This is called *pressure differential*, and the lower pressure is called *vacuum*. The difference in pressure outside compared to inside the engine causes the air to flow into the low pressure areas in the cylinders. The same principle is used whenever you use a straw to drink from a cup or container. Atmospheric pressure outside the container pushes the drink up through the straw to the lower pressure area in your mouth.

Vacuum is measured several ways, with the most common being in inches of mercury, shown as inches Hg or in. Hg. For many years, this unit was displayed on scan tools to show pressure in the intake manifold. Most vehicles now use kilopascals, abbreviated as kPa. At sea level,

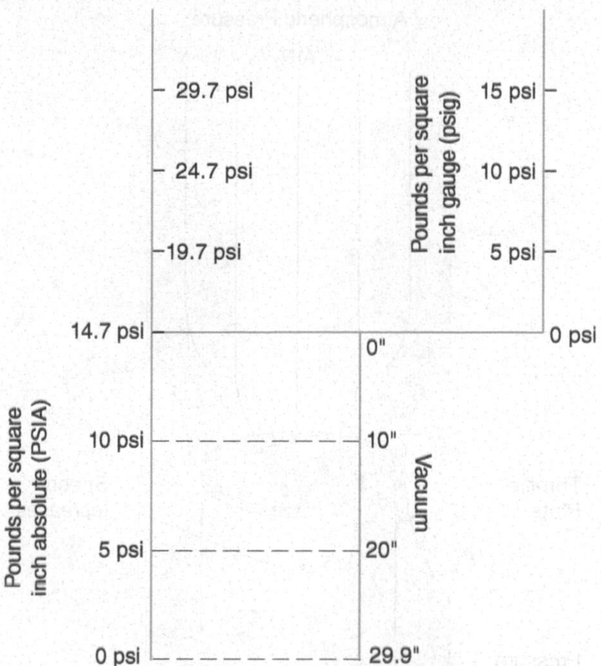

FIGURE 4-65 Pressure and vacuum are important for a technician to understand. What is measured as vacuum is just pressure lower than atmospheric pressure.

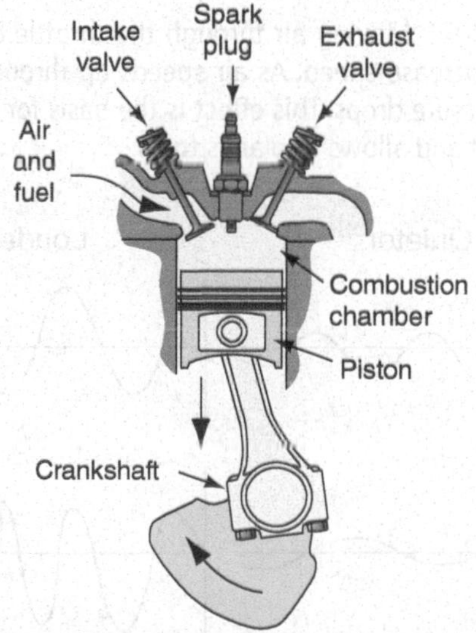

FIGURE 4-66 The motion of the piston in the cylinder creates a low-pressure area, which causes air to flow into the cylinder.

atmospheric pressure is approximately 101 kPa. With the engine running, pressure in the intake is reduced, and the scan tool displays that information in kPa.

■ ***Sound.*** The sounds we hear are the result of the vibrations of objects and the subsequent vibration of the air surrounding us. When an object moves, it disturbs the air around it, causing the air to move in waves away from the source. These waves carry the sound and travel in frequencies that we perceive as sound (**Figure 4-68**).

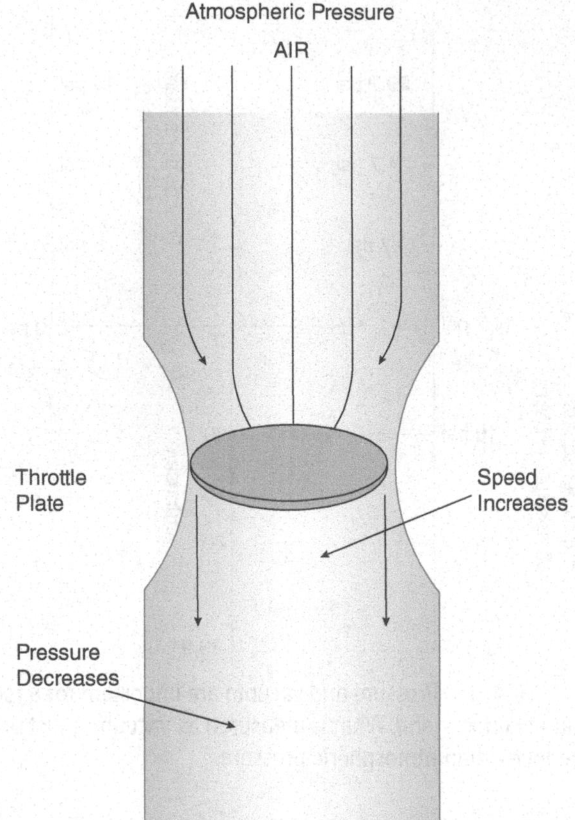

FIGURE 4-67 Moving air through the throttle body causes the air to increase speed. As air speeds up through a restriction, its pressure drops. This effect is the basis for how a wing provides lift and allows airplanes to fly.

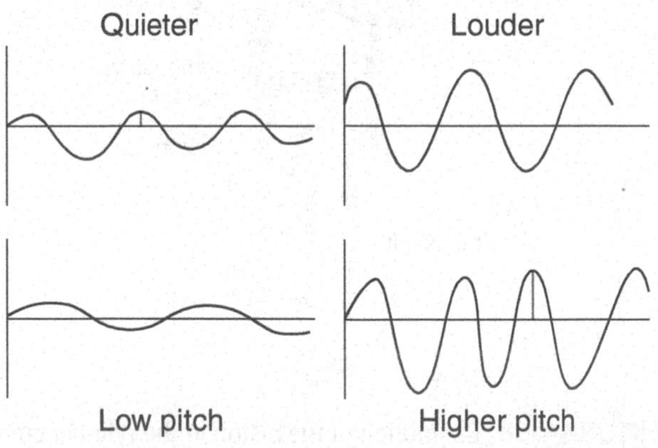

FIGURE 4-68 Sound waves create ripples in the air, which we interpret as sound. In this example, the taller the sound wave is, the louder the sound. The more times per second the sound wave oscillates, the higher the pitch of the sound.

Sound waves with low frequencies, such as from a bass guitar, have rather slow periods of oscillation while higher-frequency sounds, such as from a disc brake wear indicator, vibrate much more rapidly and produce higher-frequency sounds.

Sound travels faster through steel and other metals than it does through the air. This is important to consider when listening for noises. Mechanical noises can travel through things, like the engine or frame, making them difficult to isolate sometimes. You can also use this situation to your advantage by using a tool as a stethoscope if a mechanics stethoscope is not available. You can use a long screwdriver, prybar, or socket extension to amplify and help pinpoint where noises are coming from (**Figure 4-69**).

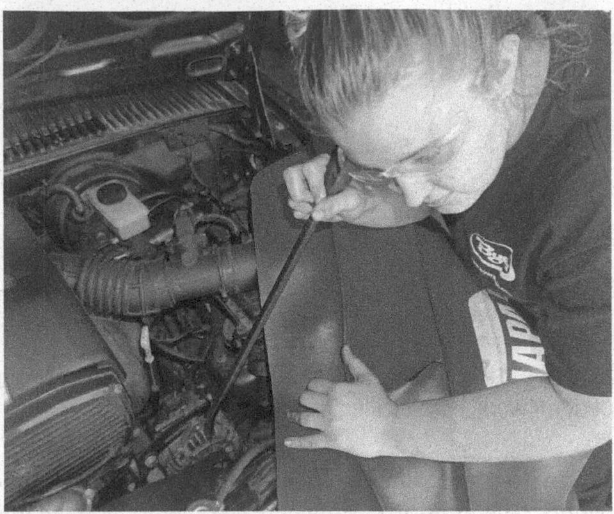

FIGURE 4-69 Sound travels easily through metal. Here a technician is using a prybar to locate a bearing noise. Use caution when working around running engines and moving parts. Do not allow the tool to contact moving parts.

BASIC CHEMISTRY

Chemistry is the science of matter and the changes that it undergoes, specifically regarding the composition, behavior, and properties of matter and the changes that take place during chemical reactions. Several chemical reactions take place in the automobile, such as in the battery to store and produce electricity, during combustion to release energy, and in forms of rust and corrosion.

■ ***Atoms and Molecules.*** In the common model of the atom, the electrons arrange themselves around the nucleus in a series of energy levels, called *rings* or *shells*. Each shell has room for an exact number of electrons (**Figure 4-70**). Some atoms have incomplete shells and tend to join with other atoms so that eventually every electron will be in a complete shell. This type of joining together is called *electron sharing*. For example, two hydrogen atoms can share their electrons to form a complete inner shell, which results in a molecule of hydrogen gas, represented as H_2. The symbol for hydrogen is H and the subscript 2 shows that two atoms of hydrogen have formed a molecule. When two hydrogen atoms and one oxygen atom join, a molecule of water H_2O, is formed. **Figure 4-71** shows how the hydrogen and oxygen bond together. Water is a molecule because it is made up of different elements joined together.

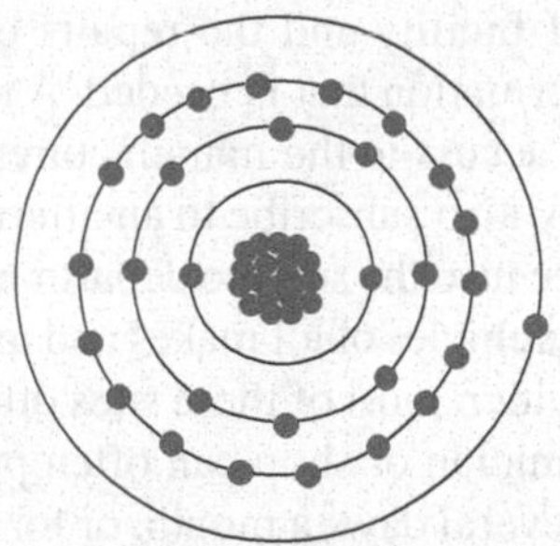

FIGURE 4-70 Electrons are arranged in rings or shells of different levels of energy. The electron in the outer shell is often able to leave or be joined by another electron. This allows atoms to bond with other atoms to form all of the substances we encounter.

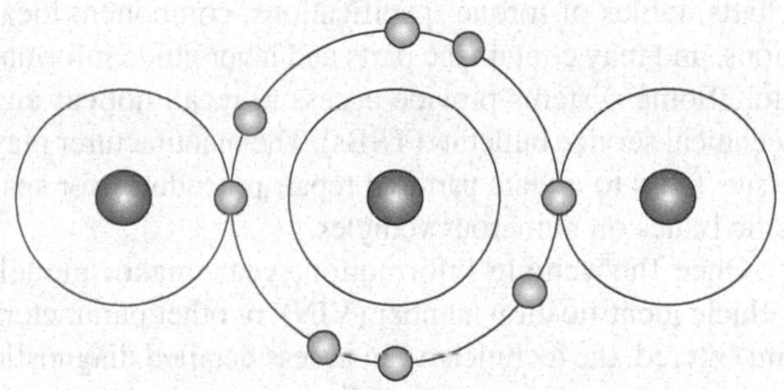

FIGURE 4-71 A form of chemical bonding where two different atoms share electrons.

How atoms join together is as important as which atoms join. Two compounds may contain the same number of atoms and molecules but differ in how they bond together, which results in the two being distinctly different substances. The use of hydrocarbons, an aspect of organic chemistry which focuses on organic molecules and is associated with carbon-based compounds, generally applies to living organisms. Inorganic chemistry and the study of inorganic molecules is based on nonliving compounds. Hydrocarbons form the basis of petroleum products, which provides gasoline, diesel, kerosene, lubricants, and innumerable other products used in day-to-day life. How the hydrocarbons react during chemical reactions is important because of the results of these reactions.

■ ***Chemical Reactions.*** As different elements combine with each other, the shifting of electrons form different compounds. Any change in the bonds between atoms constitutes a **chemical reaction**. Atoms may join together, molecules may join together or come apart, an atom can replace another atom in a molecule, and atoms in different molecules can trade places. Because atoms and molecules can combine, separate, and recombine, equations are used to express how chemical reactions take place.

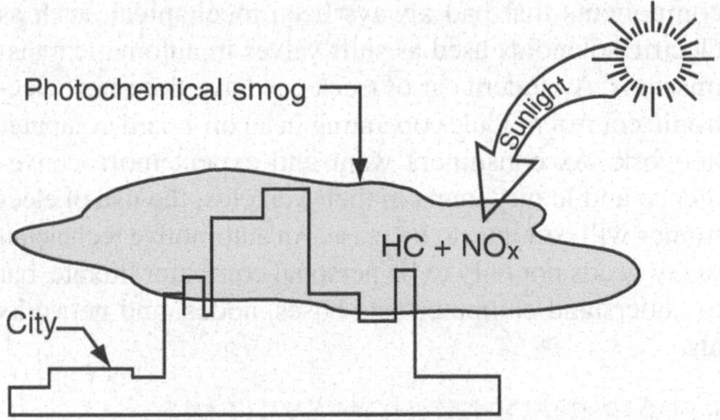

FIGURE 4-72 Smog, a form of air pollution, is caused by chemical reactions between exhaust emissions and sunlight.

For example, if complete combustion is taking place in an internal combustion engine, hydrocarbons (HC) combine with O_2 to form H_2O and CO_2. In this example, the hydrocarbons separate and combine with the oxygen to form water and carbon dioxide. Unfortunately, complete combustion does not occur. A reaction in which oxygen atoms are added to another substance is called *oxidation*. When oxygen is removed it is called *reduction*. The term redox is often used when talking about reactions involving either the addition or the loss of oxygen. In the previous example, oxygen combines with hydrogen from the hydrocarbons and forms water vapor. This is an example of oxidation. Rust is another very common example of the oxidation process. Both oxidation and reduction take place in the vehicle's catalytic converter to reduce the amount of harmful exhaust emissions.

Other reactions that take place during combustion cause the formation of carbon monoxide (CO) and oxides of nitrogen (NOx). When there is insufficient oxygen to fully convert all of the HCs with O_2 during combustion, CO is formed. CO is a colorless and odorless gas that in high concentrations leads to illness and death. When inhaled, CO bonds to hemoglobin, which is the primary carrier of oxygen in the blood. The CO decreases the ability of the hemoglobin to carry oxygen, which can lead to essentially being starved of oxygen.

Oxides of nitrogen are formed under the high temperatures of combustion. When NOx enters the atmosphere with hydrocarbons and is exposed to sunlight, the result leads to smog and ground-level ozone (**Figure 4-72**). Both smog and ground-level ozone contribute to health concerns.

Computer Literacy

Until the 1970s, all the systems on vehicles were mechanical. The only electronics were in the radios, and beginning in the mid-1970s, the ignition systems. Since the 1980s, more functions have become electrical or electronic. Current vehicles are utilizing electronics to replace components that had always been mechanical, such as electric solenoids used as shift valves in automatic transmissions. A modern car or truck can have dozens of electronic control modules operating in an on-board computer network. As consumers want and expect more convenience and luxury items in their vehicles, the use of electronics will continue to increase. An automotive technician today needs not only to be personal computer literate, but to understand computer data buses, nodes, and networks also.

REPAIR INFORMATION SYSTEMS

Repair information has evolved from large paper manuals to customizable digital content. Information can be stored and accessed from CD/DVDs or from web-based systems. Disk-based storage is useful where Internet access is limited. Information may be one disk or as many as a dozen or more. One drawback to CD/DVD-based information is that it is usually only updated quarterly compared to web-based information, which can be updated as often as necessary. A negative of web-based information is that if the provider's servers crash or the Internet connection is disrupted, there is no way to access information, which could be costly for a technician or shop. By using computer-based systems, a technician can locate the information he or she needs, print it, and take it back to the vehicle. When information was kept mostly in large repair manuals, only one technician could easily use it one time, and the book was subject to the hazards of the service bay, such as grease prints, chemical spills, physical damage from being dropped, or the book wearing out and falling apart.

Regardless of the access method, there is no question about the need for accurate information. One of the most important skills a technician can acquire is the ability to locate and use technical information. It is impossible to memorize all the information needed to repair any vehicle that may come in for repairs, and even if it was possible, service information changes with new parts, tools, and procedures.

Online service information is available from the auto manufacturer that is specific to a company, such as GM or Toyota, and from aftermarket companies. The information from the manufacturer (OE) usually is very thorough and contains the most up-to-date information for servicing their vehicles. Aftermarket information usually covers all the common make and model vehicles, but often sacrifices depth for broader coverage.

The type of facility and the repairs offered usually dictate the information that is needed. A new car dealership will have access to the manufacturer's information system but may also subscribe to an aftermarket company's service because the service department may have to repair trade-in vehicles of all makes and models. Regardless of the provider, most of these sites offer subscription service; a technician or shop can often purchase access for 24 hours, several days, a month, or for a year depending on needs. One company, Alldata, offers information access for professional technicians and a service for non-professionals or enthusiasts.

The information systems will usually contain information about theory and operation, diagnostics, diagrams and charts, tables of torque specifications, component locations, and may contain the parts and labor guide information. Some systems provide access to recall notices and technical service bulletins (TSBs). The manufacturer may issue TSBs to update part and repair procedures for specific issues on numerous vehicles.

Once the vehicle information, year, make, model, vehicle identification number (VIN), or other parameters are entered, the technician can access detailed diagnostic and repair procedures, wiring diagrams, and other details necessary for servicing the vehicle.

■ ***The Necessity for Information.*** The heavy use of on-board computers and electronics, while creating a more complicated electrical system, also provides better diagnostic capabilities, safer vehicle operation and increased fuel economy. On-board technology provides accessories that are more sophisticated, such as radar- and laser-based cruise control systems, collision avoidance systems, parking assist, and GPS tracking. Modern vehicles are equipped with an amazing amount of computer power, controlling various functions, and communicating module-to-module on a computer network, called a *data bus* (**Figure 4-73**). The modules communicate, pass on data, process data, and take actions based on their function and programming. An engine control module and an audio module (the radio), even though they are not on the same network, may need to communicate if the vehicle is equipped with automatic volume control. Used to compensate for the increased noise of traveling at higher speeds, data the engine control module is used to tell the audio module to increase the volume based on the vehicle speed. Volume then decreases when the vehicle

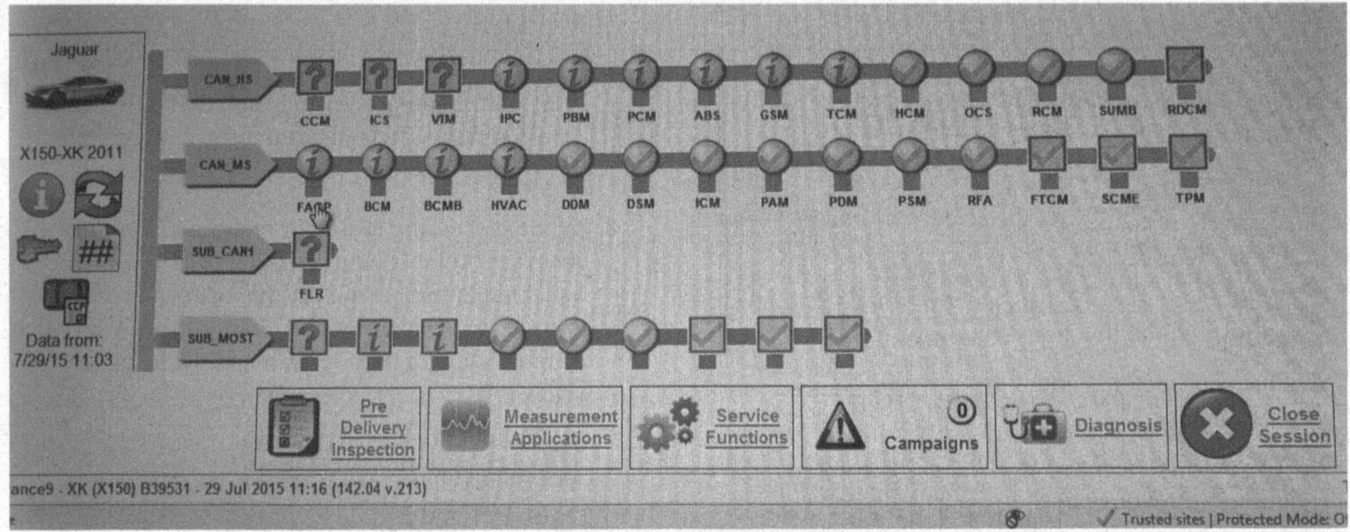

FIGURE 4-73 Modern cars and trucks have many different modules and several different types and speeds of computer networks. Data buses connect modules and networks together.

slows. It is this type of integration of technologies that, when not working properly, will require a skilled technician to diagnose and repair the cause of the problem.

Becoming an automotive technician requires more than a set of hand tools; it requires interpersonal skills, the ability to thinking critically, academic and mechanical skills, as well as the ability to adapt and learn new skills all the time. Vehicles are becoming more complex each year, and new technologies, such as hybrid and alternative-fuel vehicles, continue to push the design and service envelope. Who will be there to fix these cars and trucks in the future?

SUMMARY

Employability skills include: communication, teamwork, problem solving, initiative, organization, self-management, being willing to learn, and being adept with technology.

Nonverbal communication includes body language, eye contact, and appearance.

A good resume and cover letter can make the difference between getting a job or not.

Gross pay is the total amount of money you make and net pay is what you actually get to keep.

Repair orders are legal documents between the customer and the shop.

Flat-rate pay is when a technician earns his or her income based on the amount of labor time billed to the customer.

Technicians should be able to easily use both the English and metric systems.

Gears are a group of levers arranged in a circle. Each tooth acts as a lever against the tooth of another gear.

Torque is force applied over distance.

Atmospheric pressure decreases as you go up in altitude.

Atoms are the basic building blocks of matter.

Electricity is the movement of electrons along a path from a source of higher potential to lower potential.

Vacuum is pressure that is less than atmospheric pressure.

REVIEW QUESTIONS

1. The estimated labor time to complete a repair is found in a _______________ _______________ _______________.
2. _______________ is the science of using fluids to perform work.
3. Body _______________ is how you communicate with someone in a nonverbal way.
4. One centimeter is equal to _______________ millimeters.
5. The amount of money you actually bring home from your paycheck is called your _______________ pay.
6. *Technician A* says when searching for employment, you should always apply online so you do not need to submit a résumé. *Technician B* says a résumé should contain a summary of all your life experiences. Who is correct?
   a. Technician A
   b. Technician B
   c. Both A and B
   d. Neither A nor B
7. Which of the following statements about replacement parts is correct?
   a. OE parts are those parts that are supplied by the vehicle manufacturer.
   b. Aftermarket parts are usually made by independent companies and sold through parts stores such as NAPA or AutoZone.
   c. A part manufacturer may sell both OE and aftermarket parts under the same brand name.
   d. All of the above
8. While discussing customer relations: *Technician A* says customers may base their opinion to trust a technician based on how the technician looks. *Technician B* says that appearance can say as much or more about a technician than how he or she speaks. Who is correct?
   a. Technician A
   b. Technician B
   c. Both A and B
   d. Neither A nor B
9. *Technician A* says that a good understanding of math and science skills is important for a technician. *Technician B* says that good communication skills are important for a technician. Who is correct?
   a. Technician A
   b. Technician B
   c. Both A and B
   d. Neither A nor B
10. All of the following are important communication skills except:
    a. Maintaining a clean uniform and appearance
    b. Looking at the person you are talking to in the eyes
    c. Being polite
    d. Having a powerful handshake

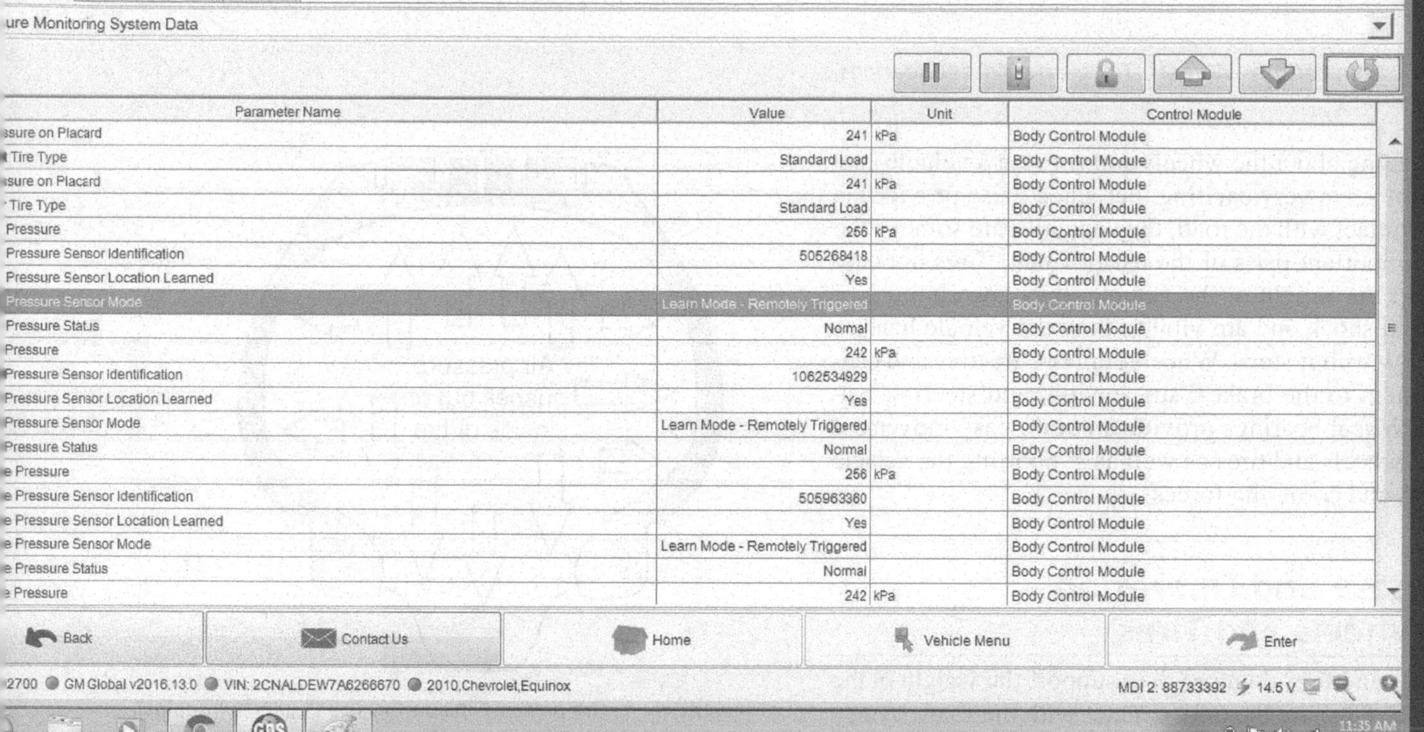

Parameter Name	Value	Unit	Control Module
ssure on Placard	241	kPa	Body Control Module
Tire Type	Standard Load		Body Control Module
ssure on Placard	241	kPa	Body Control Module
Tire Type	Standard Load		Body Control Module
Pressure	256	kPa	Body Control Module
Pressure Sensor Identification	505268418		Body Control Module
Pressure Sensor Location Learned	Yes		Body Control Module
Pressure Sensor Mode	Learn Mode - Remotely Triggered		Body Control Module
Pressure Status	Normal		Body Control Module
Pressure	242	kPa	Body Control Module
Pressure Sensor Identification	1062534929		Body Control Module
Pressure Sensor Location Learned	Yes		Body Control Module
Pressure Sensor Mode	Learn Mode - Remotely Triggered		Body Control Module
Pressure Status	Normal		Body Control Module
e Pressure	256	kPa	Body Control Module
e Pressure Sensor Identification	505963360		Body Control Module
e Pressure Sensor Location Learned	Yes		Body Control Module
e Pressure Sensor Mode	Learn Mode - Remotely Triggered		Body Control Module
e Pressure Status	Normal		Body Control Module
e Pressure	242	kPa	Body Control Module

CHAPTER 5

Wheels, Tires, and Wheel Bearings

Chapter Objectives

At the conclusion of this chapter, you should be able to:

- Identify different types of tires and their construction. (ASE Education Foundation MLR 4.D.1)
- Rotate tires per the manufacturer's specifications. (ASE Education Foundation MLR 4.D.2)
- Perform tire mounting and balancing. (ASE Education Foundation MLR 4.D.3)
- Inspect the wheel and tire for air loss and repair tire. (ASE Education Foundation MLR 4.D.5 & 4.D.6)
- Check for lateral and radial runouts. (ASE Education Foundation MLR 4.D.1)
- Check and adjust tire air pressure. (ASE Education Foundation MLR 4.D.1)
- Identify and service components of the tire pressure monitoring system. (ASE Education Foundation MLR 4.D.4, 4.D.7, & 4.D.8)
- Service wheel bearings. (ASE Education Foundation MLR 3.D.1 & 5.F.1)
- Inspect and replace wheel studs. (ASE Education Foundation MLR 5.F.6)

KEY TERMS

air pressure
all-season tires
asymmetrical tires
centerbore
contact patch
direct TPMS
directional tires
directional tread
dynamic wheel balance
hubcap
wheel offset
indirect TPMS
lateral runout
performance tires
pounds per square inch
radial runout
run-flat tires
shimmy
static wheel balance
temperature rating
tire rotation
TPMS reset
traction rating
treadwear rating
wheel bearing

Reading about the wheels and tires on a vehicle may not seem very exciting, but since these provide the only contact with the road, they actually are some of the most important parts of the car or truck. Tires not only support the weight of the vehicle, they absorb much of the road shock and are vital to how the vehicle handles and how well it stops. Wheels support the tires and connect them to the brakes, suspension, and steering systems. Wheel bearings provide smooth, easy movement of the wheels and tires as well as supporting the vehicle weight and cornering forces.

Purpose and Operation of Wheels and Tires

Wheels and tires do more than support the weight of the vehicle. Tires are the only contact with the road, so they affect how the car or truck steers, how it rides and handles bumps, and how effective the brakes are. Wheels, in addition to supporting the tires, are also part of the overall look and design of modern vehicles.

TIRE PRINCIPLES AND OPERATION

Tires use pressurized air to support weight and to provide some dampening of the bumps in the road. Tires that rely on air pressure are called *pneumatic tires*, just as air-powered tools are called *pneumatic tools*.

Beyond supporting the weight of the vehicle, the tire must be able to roll smoothly, provide good traction under different driving conditions, generate low amounts of noise, wear well, and allow for good handling and braking performance. Overall, tires have a very difficult and often conflicting job to do.

■ ***How 30 psi of Air Pressure Supports a 4,000-Pound Vehicle.*** Many people, technicians included, seldom think about the fact that an air-filled tire is supporting the weight of the vehicle. Did you ever wonder how that works? It is actually fairly simple.

The tire is mounted on the wheel (**Figure 5-1**). Both are made to be airtight so that when the air pressure in the tire is increased, the pressure will hold. The empty space inside the tire has a lot of surface area. As the air pressure in the tire increases, the amount of force the air exerts against the inner surface area increases. We read that force in pounds per square inch as (psi) or in kilopascals (kPa). When a tire pressure gauge reads 30 psi, it means the **air pressure** in the tire that you are checking is applying a force of 30 **pounds per square inch** for each inch of surface area inside the tire (**Figure 5-2**). When you add up all the square inches of surface area inside a tire, you can begin to see how much pressure is applied to the inside of the tire.

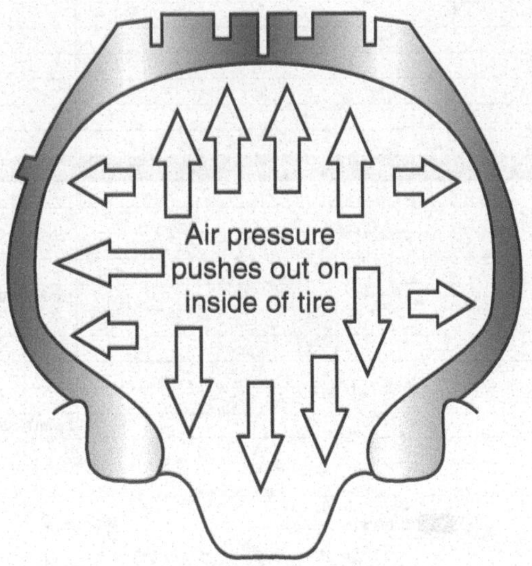

FIGURE 5-1 The pressure inside the tire is exerted against every square inch of internal surface area.

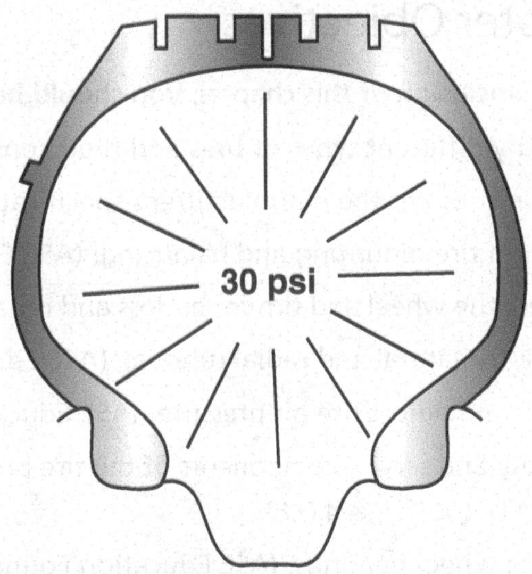

FIGURE 5-2 The pressure against the inside of the tire allows it to maintain its shape and to carry weight. As weight is placed on the tire, the pressure inside pushes back against the load.

When you are looking at a properly inflated tire that is not mounted on a car, it should look perfectly round. But when the weight of the vehicle is on the tire, the weight tends to make the bottom of the tire flatten out, so the tire is not perfectly round. This flattened area is called the **contact patch**. The contact patch is where the tire supports the weight of the car.

The contact patch area is the width times the length of the tread section that is contacting the road (**Figure 5-3**). To determine how many pounds a tire can support, you

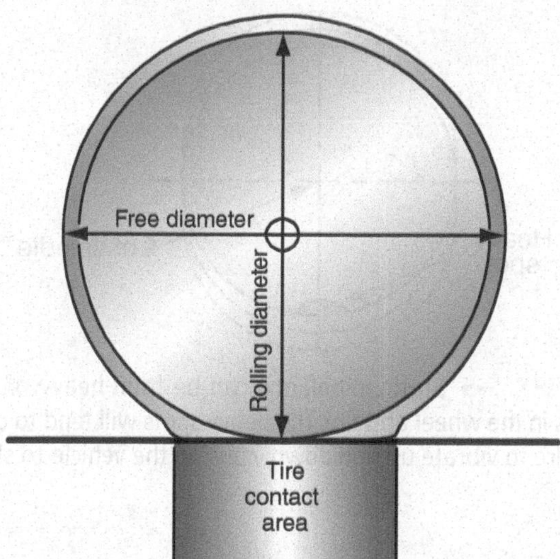

FIGURE 5-3 The contact patch is where the tire contacts the road surface. A typical passenger car tire has a contact patch of about 40 square inches. The larger the tire, the larger the contact patch.

multiply the size of the contact patch times the specified tire pressure. For example, a passenger car tire may have a contact patch 7 inches wide by 5 inches long. This equals an area of 35 square inches. The area times the inflation pressure, for example, 30 psi, equals 1,050 pounds of weight carrying capacity. If the standard tire is replaced by a temporary spare tire, the contact patch gets smaller. A temporary spare may have a contact patch of only 20 square inches.

As you can see, a small tire with a small contact patch will not be able to carry as much weight as a larger tire with a larger contact patch. As the contact patch decreases, the tire pressure needs to increase if the tire is to be able to carry the same amount of weight. This is why compact spare tires require 60 psi, while standard tires often require about 30 psi.

■ ***Forces Acting upon Wheels and Tires during Operation.*** Because wheels are made of steel, aluminum, or other strong materials, they do not flex or deform the way tires do during operation. However, the wheels are the connection between the tires and the brakes, steering, and suspension systems.

When the weight of the vehicle is on the tires, the tires tend to flatten slightly, which increases the size of the contact patch. While this may be good for increased traction and braking, it also means more rolling resistance since more of the tread is in contact with the road. The friction between the rolling tire and the road generates heat. The more contact there is between the two, the more friction that is generated. The ability of the tire to dissipate heat is one of the universal tire quality guidelines discussed later in this chapter.

As the tire temperature increases, the pressure inside of the tire also increases. This is because the pressure of a gas in a closed space will increase if the temperature increases, or conversely, the pressure will decrease when the temperature decreases. This is why tires gain pressure when driven and lose pressure in the winter as the air temperature drops. This is also why regular checks and adjustments of tire pressure are so important. Low tire pressure places more load on the tire sidewall, which affects ride quality, increases rolling resistance, and can cause tire damage and failure. Excessive tire pressure increases the tire temperature, causes a harsh ride, and can cause tire failure as well.

Beyond carrying the weight of the vehicle, tires should roll smoothly and quietly, absorb bumps and road shock, and provide good handling and braking performance. When the vehicle is cornering, additional loads are placed on the tire sidewall. This compresses the sidewall and flattens the tire slightly. The tires on the inside of the turn lose a little bit of weight load as the vehicle weight is shifted, and the tires lose a little grip with the road. At the same time, the tires on the outside of the turn are placed under more load and tend to compress or squat slightly. The constantly changing road conditions mean that the tire is always flexing, deforming, and trying to return to its normal shape.

■ ***Wheel and Tire Balance.*** Because each wheel and tire rotate several hundred times per mile, and around 15 times per second at freeway speeds, they need to be balanced to reduce vibrations. Even a small amount of imbalance will cause the wheel and tire to vibrate. This vibration can be felt in the passenger compartment as a shake in the steering wheel or in the seats (**Figure 5-4**). A vibration at freeway speeds is one of the most common customer complaints heard in an auto shop.

There are two types of tire balance, static and dynamic. A statically balanced wheel and tire means that there is an even distribution of weight around the axis of rotation (**Figure 5-5**). When the **static wheel balance** is incorrect, a heavy or light spot is present in the wheel or tire. When the tire is rolling down the road, these heavy and light spots will cause the tire to try to speed up and slow down as it rotates. An excessively heavy spot will try to lift the tire up off the ground as the weight moves toward the top of the circle of rotation and then slam the tire back down as the weight moves downward. It does not take very much weight to imbalance a wheel and tire.

An average wheel and tire assembly weighs between 30 and 40 pounds. Mounting a new tire on a new rim and checking the balance may show that the assembly is statically out-of-balance 1 ounce. This may not seem like

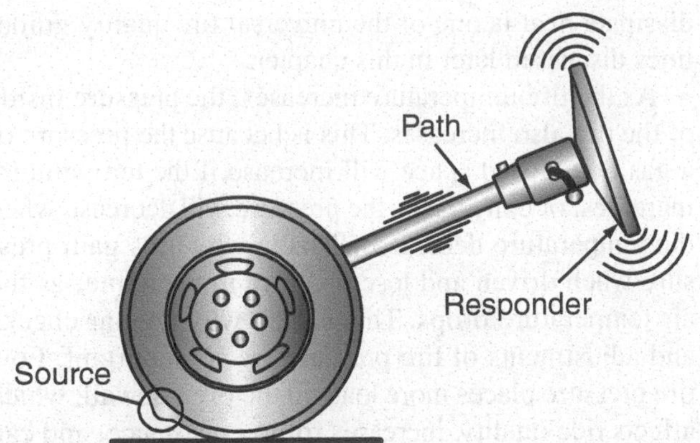

FIGURE 5-4 An out-of-balance tire or a tire with a defect will vibrate. The vibration will carry up to the steering wheel and be felt by the driver and passengers.

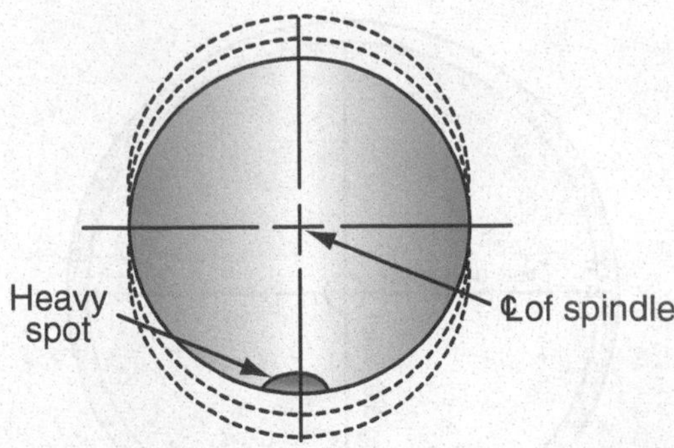

FIGURE 5-5 Static imbalance can be from heavy or light spots in the wheel and tire. The heavy spots will tend to cause the tire to vibrate up and down, causing the vehicle to shake.

very much considering the overall weight of the wheel and tire; however, the wheel and tire are rotating. The faster the wheel rotates, the more the impact the weight has on the balance.

Dynamic wheel balance ensures that there is equal weight distribution across the width of the wheel and tire. Uneven weight distribution from side-to-side pulls the wheel and tire back and forth, resulting in steering wheel shake or shimmy (**Figure 5-6**).

Both static and dynamic balance problems are corrected with modern tire balancing machines. The application of wheel weights, in the proper locations, will offset the heavy and light spots in the wheel and tire and provide a vibration-free driving experience.

TIRE CONSTRUCTION

Tires have evolved significantly since the wooden rims and solid tires found on very early cars. As the speed and load-carrying capacity of vehicles increased, the capabilities of tires had to change as well. Modern tires have to perform the same functions as tires did a hundred years ago, but because of the changes in vehicles and the need to adapt tires accordingly, the design and materials used today are almost completely different from those used then.

■ ***Radial Tires.*** By far the most common type of tire installed today is the radial tire. Radial tires are constructed of layers of belts or plies that are wrapped around the tire in circular bands (**Figure 5-7**). The belts are constructed of cords, often made from steel, polyester, and other materials, which when layered provide a strong tire carcass that is also flexible. These cords lie 90 degrees from the radial plies that run from bead to bead. This design reduces internal friction and heat buildup, which increases tire life.

The outside sections of the tire are the bead, sidewall, and tread. The bead contains thick steel wire and makes up the inside diameter of the tire. It needs to be very strong and rigid because it has to hold the tire firmly to the wheel. The tire sidewall is constructed of belts that run from bead to bead. The sidewall supports the tire and absorbs road shocks. The flexing of the sidewall is what allows the tire

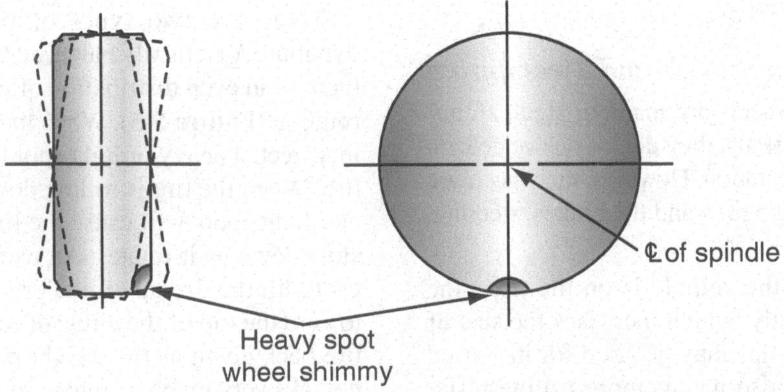

FIGURE 5-6 Dynamic imbalance causes the tire to move side-to-side as it rotates. This causes the steering wheel to shimmy back and forth.

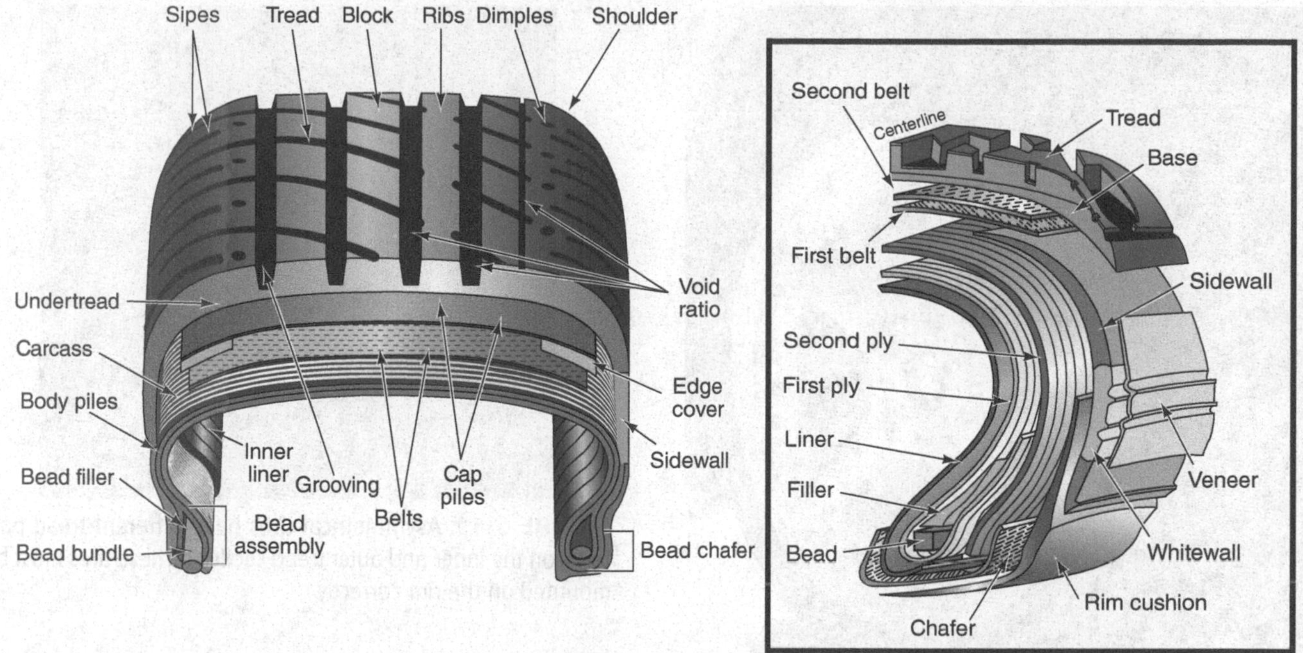

FIGURE 5-7 A tire is comprised of many different layers and pieces. Radial tires have overlapping belts around the circumference of the tire and are the most common type of tire construction.

to conform to road conditions and helps smooth the vehicle's ride. Generally, the shorter the sidewall is, the stiffer the tire will be, and consequently the firmer the ride quality. Many sports cars have very low profile sidewall tires. This reduces the amount of flex the tire has and improves cornering ability but at the expense of a firmer ride.

■ ***Tire Types.*** The outer tread is literally where the rubber meets the road. The design of the tread determines how the tire will perform under various conditions. Tires can be designed for all-season use, just for wet/dry performance, for mud and snow, or even just for snow conditions.

All-season tires are found on the majority of vehicles on the road today. This type of tire provides good traction under nearly all operating conditions. This all-purpose tire combines good wet and dry traction, ride quality, long life, and low noise in one tire. The trade-off is that all-season tires do not perform as well in heavy snow, nor do they provide superior handling qualities found in performance tires. **Figure 5-8** shows an example of a common all-season tire.

Many sports cars are equipped with **performance tires** that are suitable for most wet and dry driving conditions, but are not intended to be used in snowy conditions. These tires typically have a lower profile and a very firm ride. An example of a sport tire is shown in **Figure 5-9**. Sport tires often have very large tread blocks arranged in an aggressive tread pattern to maximize grip. These tires should not be used in snow as the snow fills the grooves and channels of the tread. Some high-performance sport tires do not perform well in cold temperatures, such as below 40°F (4°C) and should not be used in cold weather. This is because the tire will not

FIGURE 5-8 An all-season tire has many different-sized grooves and blocks and is designed to provide good overall performance in different driving conditions.

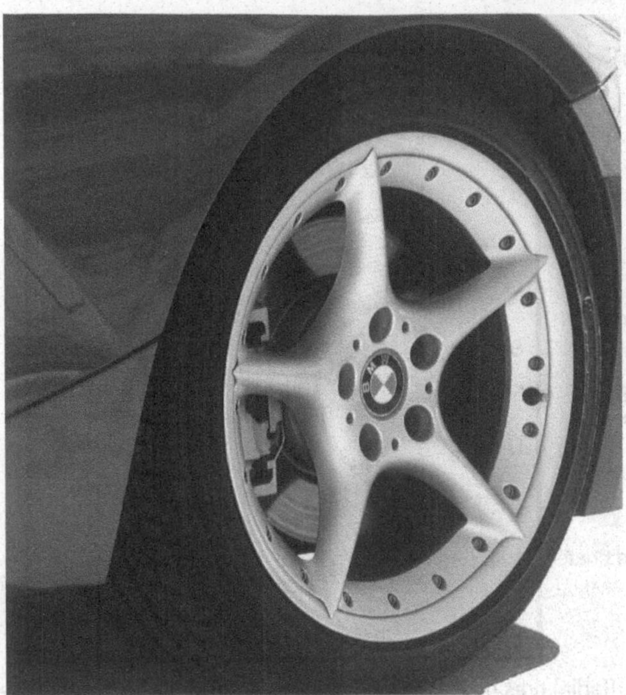

FIGURE 5-9 Sport tires are typically low-profile, meaning they have short sidewalls. Sport tires often sacrifice service life for improved handling.

generate sufficient heat to grip the road properly and handling will be compromised. An example of a sport tire door decal is shown in **Figure 5-10**.

Tire tread patterns can be symmetrical, asymmetrical, or directional. **Asymmetrical tires** have different tread patterns from the inside to the outside of the tread and require that a specific side of the tire be mounted toward the outside of the vehicle (**Figure 5-11**). **Directional tires** have a tread pattern that is designed to be used to rotate in one direction only, meaning that two of the tires must be mounted on the right side of the vehicle and two must be mounted specifically for the left side. **Figure 5-12** shows a directional tread pattern.

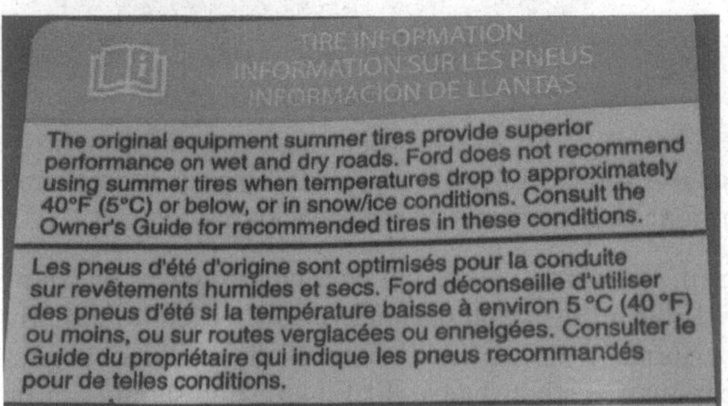

FIGURE 5-10 It is important to note and remind your customers that sport tires may not be suitable for cold weather or winter use.

FIGURE 5-11 Asymmetrical tires have different tread patterns on the inner and outer tread sections. These tires must be mounted on the rim correctly.

FIGURE 5-12 Directional tires have a clearly defined tread pattern that points in the direction of proper rotation.

Directional tires will also have markings on the sidewall to indicate how the tire must be mounted to the vehicle. Mud- and snow-rated tires have large grooves that run from the edge of the tread to the center. This type of tire is designed to provide better traction in adverse driving conditions than what can be achieved by an all-season tire. A mud- and snow-rated tire may have M&S, M+S, M/S, or MS stamped in the sidewall (**Figure 5-13**).

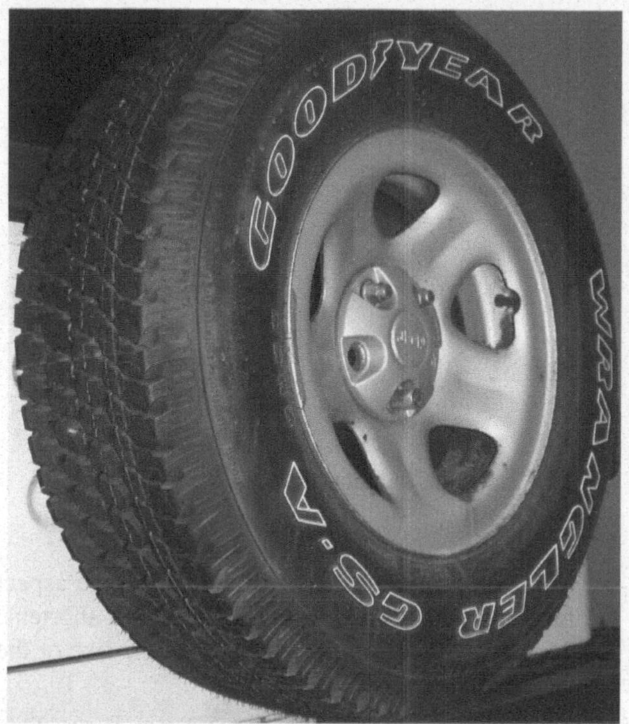

FIGURE 5-13 Light truck and mud and snow (M+S) tires tend to have aggressive tread blocks.

FIGURE 5-14 A compact spare tire is very small compared to the standard tires and has a higher inflation pressure due to the smaller size.

A newer version of mud and snow tires is the severe snow tires, which have a mountain and snowflake symbol along with the M+S designation. These symbols mean the tire has been tested for use and meets certain criteria for use in heavy snow.

■ ***Spare Tires.*** Many vehicles have temporary use spare tires that are significantly smaller than the rest of the tires on the vehicle (**Figure 5-14**). These temporary tires, often called *mini spares* or *doughnut spares*, save weight and space in the trunk or the underside of the vehicle where they are stored. Temporary spares typically are rated for use for 50 miles (81 km) and up to 50 mph or 81 kph. They also require much higher inflation pressures because they are much smaller than the standard tires. Most temporary tires require 60 psi air pressure.

Some vehicles are sold without a spare. Cars and trucks with factory-equipped run-flat tires do not have a spare tire. Other vehicles have an air compressor (**Figure 5-15**), used to inflate a low or flat tire. Eliminating the spare tire reduces vehicle weight and allows the space for the spare to be used for other purposes.

■ ***Run-Flat Tires.*** **Run-flat tires** and self-sealing tires are a relatively recent development to passenger car tires. The basic idea is that because tire blowouts are dangerous, making a tire that can either seal or support itself and the vehicle, even if air pressure is lost, the

FIGURE 5-15 A vehicle without a spare tire but with an air compressor.

driver will be able to maintain safe operation until the tire is replaced. There are a couple of different types of these tires; the most common are the self-sealing tire, the self-supporting tire, and the auxiliary supported tire.

A self-sealing tire contains an extra lining along the undertread. This lining contains a sealant that can immediately seal small punctures in the tire, up to about 3/16 of an inch or about 5 mm.

A self-supporting run-flat tire has very stiff sidewalls and tread, which allow the tire to support itself temporarily, even if all the pressure has been lost. This type of tire also has a stronger bead to ensure that the tire remains seated on the wheel when there is no air pressure to keep the bead forced against the bead area of the wheel. A run-flat tire has an RFT symbol on the sidewall to identify the tire as a run-flat tire (**Figure 5-16**). Even though the tire can support itself without air pressure, it cannot do so indefinitely. Most run-flat tires of this type are limited to 50 miles and no more than 55 mph without air pressure.

The auxiliary supported run-flat tire has a support ring attached to the wheel that the tire rests on if pressure is lost. The disadvantage of this system is that the unique wheels cannot be used with standard tires, so the total cost of the system is higher than that of the other run-flat systems.

Self-supporting run-flat tires require the use of tire pressure monitoring systems, also called *TPM systems.* This is because the driver may not be able to notice a difference in how the vehicle drives, even with the tire deflated. TPM systems are discussed in more detail later in this chapter.

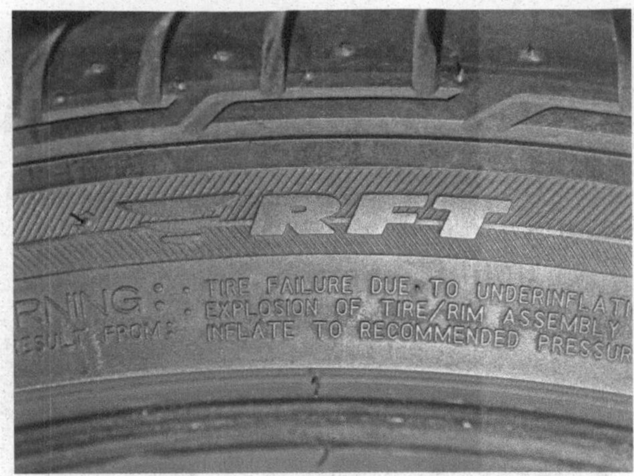

FIGURE 5-16 This indicates that this is a run-flat tire.

■ ***Tire Size and Sidewall Information.*** As you are probably aware, tires come in many different sizes and types, but how do you know which is the correct tire for a particular application? Information about the original equipment or OE tires is usually found on the tire decal located on the vehicle. In addition, you must be able to decipher the information that is found on the tire to be able to decide what is the correct and best tire for the vehicle.

All street-legal tires have a lot of information molded into the sidewalls (**Figure 5-17**). This information includes:

- Tire size: The tire size markings for most passenger car tires in the United States use a combination of letters and numbers and are called *the P-Metric size.* (**Figure 5-18**). The first three numbers of the tire size are the width of the tire, sidewall to sidewall, in millimeters. The second set of numbers is the aspect ratio, or the ratio of height to width. A tire that has a section width of 225 and an aspect ratio of 65 will have a sidewall height equal to 225 × 65 percent (225 × 0.65) = 146.25 millimeters. As the aspect ratio number decreases, the tire sidewall shortens. The last numbers in the tire size are the size of the rim in inches.
- Tire construction materials and tire type: This explains what the tire plies are constructed of and how many plies there are for the tread and sidewalls (**Figure 5-19**).
- Uniform tire quality grade or (UTQG) information: This information enables consumers to compare tires based on guidelines accepted by the tire manufacturers. The tires have a **treadwear rating, traction rating**, and **temperature rating** (**Figure 5-20**). In general, a higher treadwear number means the tire will wear better and last longer than a tire with a lower treadwear number. Traction is rated AA, A, B, or C and is based on the tire's traction during a wet skid test. This rating does not indicate the tire's dry braking or cornering abilities or its resistance to hydroplaning. The tire's resistance to heat and its ability to dissipate heat is graded as A, B, or C. All tires sold in the United States must meet the minimum C rating, which indicates the tire can withstand the heat generated while operating at 85 mph or about 138 kph.
- Maximum tire load capacity and maximum air pressure: All tires have load carrying weight limits (**Figure 5-21**). These limits are based on each tire's construction materials and the number of plies that are used in the tire. Maximum inflation pressure is usually listed with the maximum load capacity. Most passenger car tires have a maximum inflation of between 35 psi and 44 psi, while commercial truck tires may hold up to 120 psi. Tires should never be inflated beyond the maximum pressure shown on the tire.

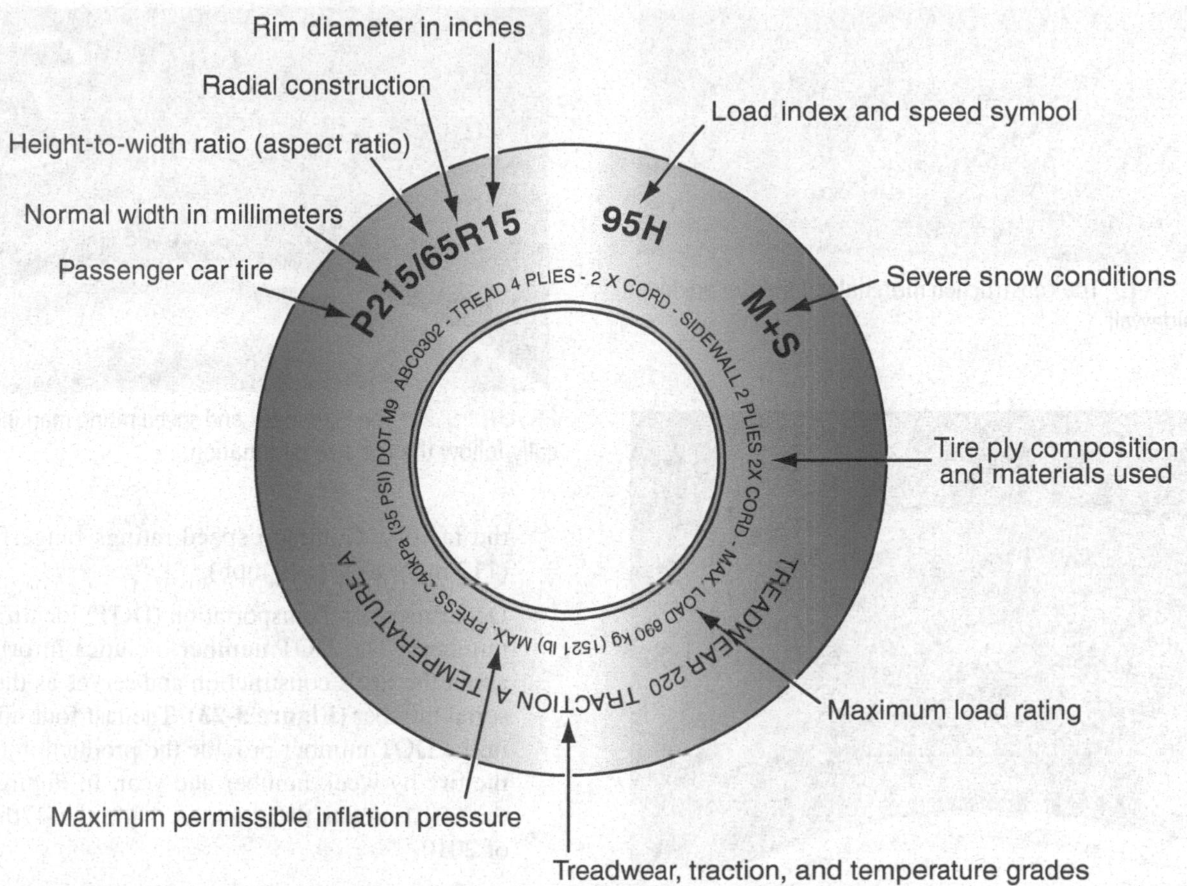

FIGURE 5-17 A lot of information is molded into the sidewall of the tire. This information is important when buying new tires or working in a shop and selling new tires.

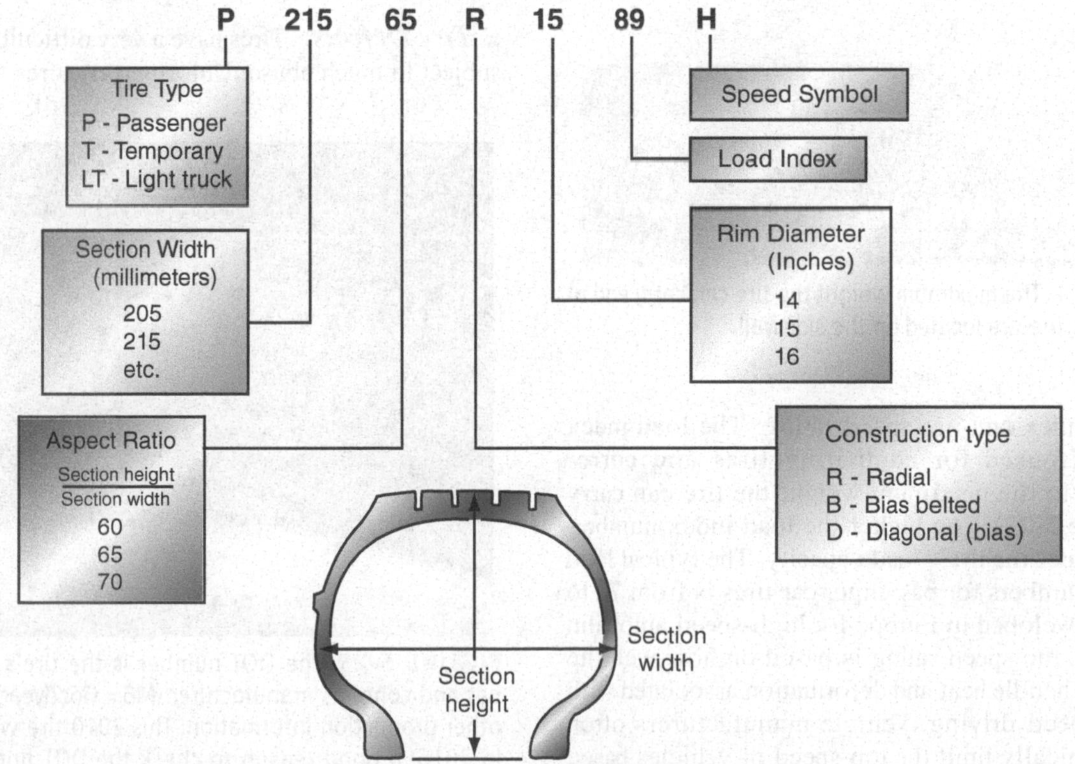

FIGURE 5-18 An explanation of the tire size information found on the sidewall.

FIGURE 5-19 The construction materials of the tire are found on the sidewall.

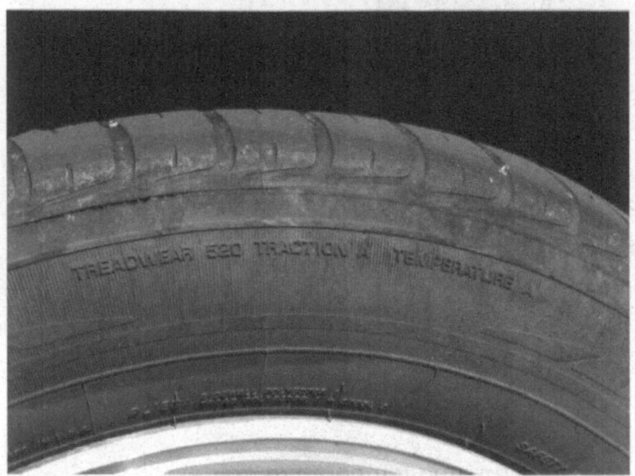

FIGURE 5-20 Treadwear, traction, and temperature ratings are part of the uniform tire quality gradings (UTQG) and are located on the sidewall.

FIGURE 5-21 The maximum weight the tire can carry and at what air pressure are located on the sidewall.

- Load index and tire speed rating: The load index value is used for comparing tires and corresponds to the maximum weight the tire can carry (**Figure 5-22**). The higher the load index number, the greater the tire's load capacity. The typical load index numbers for passenger car tires is from 70 to 110. Developed in Europe for high-speed autobahn driving, the speed rating is based on how well the tire can handle heat and deformation associated with high-speed driving. Vehicle manufacturers often electronically limit the top speed of vehicles based on the speed rating of the tires that are installed at

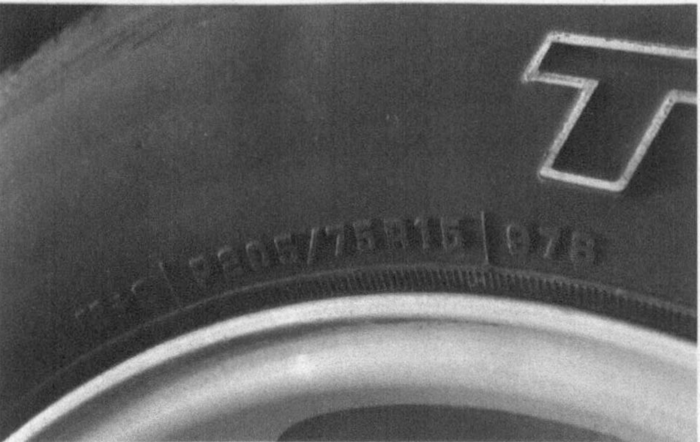

FIGURE 5-22 The load index and speed rating markings typically follow the tire size information.

the factory. Common speed ratings range from S (112 mph) to V (149 mph).

- Department of Transportation (DOT) identification numbers: The DOT number includes information about the tire's construction and serves as the tire's serial number (**Figure 5-23**). The last four numbers in the DOT number provide the production date of the tire by week number and year. In Figure 5-23, the 2010 means the tire was made the 47th week of 2010.

Other information on the tire's sidewall include the tire manufacturer, tire model, and symbols for tire use, such as all season, M+S, or severe snow.

■ ***Tire Defects.*** Tires have a very difficult job and are subject to much abuse. Unfortunately, tires may also be

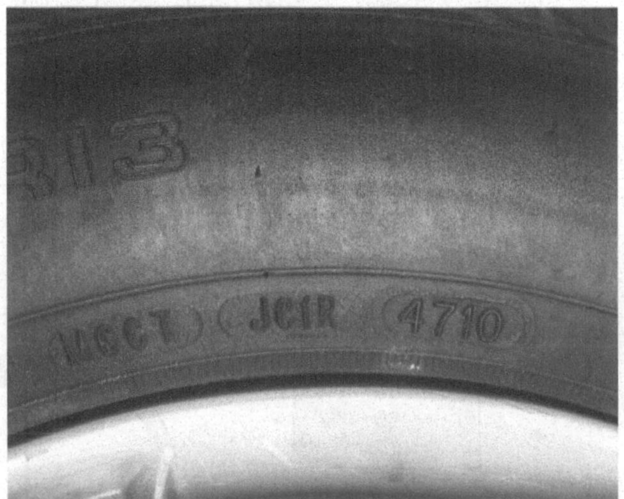

FIGURE 5-23 The DOT number is the tire's serial number and contains manufacturer (M6 = Goodyear), plant, and other production information. This 2010 tire was sold new in 2016; a good reason to check the DOT numbers when buying tires.

manufactured with defects that can cause problems that are difficult or impossible to correct.

The tire is constructed of layers of plies that are then heated and molded into the final tire form. Sometimes these layers do not mold into the exact position in which they are intended. This can cause the tire to roll incorrectly, resulting in tire pull, vibration, and wobble (**Figure 5-24**).

This problem can sometimes be less severe and may not cause any visible tire deformation, but it can cause the tire to lead or pull in one direction. This is called *tire pull* or *radial pull*. Tire pull can also be caused by conicity. Conicity is when the tire is slightly cone shaped. As such, the tire tries to roll in a circle, just as a cone or funnel rolls in a circle. The misalignment can also cause the tire to vibrate against the road, and it cannot be corrected by balancing the tire. These conditions are uncorrectable and can only be solved by replacing the tire.

WHEELS

Wheels connect the vehicle to and support the tires. Wheels also are one of the easiest ways to customize the look of a vehicle, which is why they are available in hundreds of sizes and designs.

■ ***Steel Wheels.*** Many vehicles are equipped with steel wheels (**Figure 5-25**). Steel wheels are usually two pieces of stamped steel, the inner hub section and the outer rim section welded together. Steel wheels are

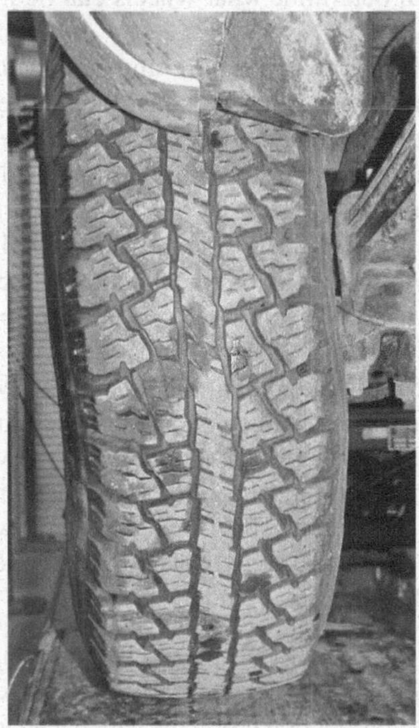

FIGURE 5-24 This tire has a broken belt, which is causing the curve in the tread. This type of defect is extremely dangerous as the tire can completely come apart while driving.

FIGURE 5-25 An example of a steel wheel.

strong but heavy. Most steel wheels have openings cut into the hub section to reduce weight. Some steel wheels are chromed to provide a nicer finished look, while others are not decorative and have a hubcap installed on the outside to dress up the vehicle's appearance.

A **hubcap** is usually secured to the wheel in one of three ways: by spring-type mounting clips around the circumference of the hubcap, by the lug nuts against the hubcap, or by false lug nuts that thread onto the actual lug nuts, keeping the hubcap tight against the rim. Always check to see how a hubcap is attached before you try to remove it from the wheel. Prying on a lug nut-secured hubcap can lead to damage that may require hubcap replacement.

To remove a spring-retained hubcap, locate a gap between the hubcap and rim. Next, carefully insert a screwdriver into the gap and gently pry the cap away from the rim (**Figure 5-26**). Once it is loose, pull the hubcap off the rim and set it aside on a workbench. To install the hub cab, align the notch in the hubcap and the valve stem and push the hubcap against the wheel. Start with one section of the hubcap and gently tap it into place along the circumference of the wheel. Do not use hammers or excessive force, as this will likely damage or destroy the hubcap.

Many vehicles have plastic lug nut caps that thread onto the lug nuts and hold the hubcap in place. To remove one, loosen the plastic lug nuts with a socket and pull the hubcap off the wheel. When you are reinstalling it, align the hubcap to the wheel and valve stem and start each lug nut cap by hand. Do not over-tighten the caps as they will crack and break.

Some cars use the lug nuts to hold the hubcaps on. This type of lug nut has a larger washer installed on the nut, which presses against the hubcap to hold it in place.

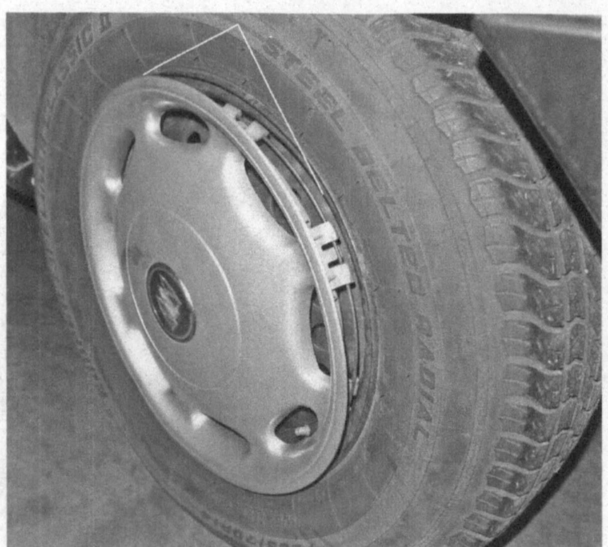

FIGURE 5-26 Be careful when removing and installing hubcaps to prevent damage. Many hubcaps have fake lug nuts, which will snap off if you try to remove them.

FIGURE 5-27 An aftermarket alloy wheel. Aluminum or alloy wheels reduce unsprung weight and can improve ride quality and handling.

■ ***Aluminum and Alloy Wheels.*** Many vehicles come from the factory with aluminum or alloy wheels (**Figure 5-27**). This type of wheel is usually lighter than a steel wheel of equal size and can be made into a nearly unlimited number of designs and colors. The term alloy wheel refers to a wheel made from a combination of materials, with aluminum being the largest component. These combinations are used to reduce weight or increase strength compared to a wheel made entirely of aluminum.

■ ***Wheel Design and Dimensions.*** Regardless of what a wheel is constructed of, all passenger vehicle wheels have the same basic layout and characteristics. (**Figure 5-28**).

The center of the wheel is the hub section. It has a large center hole, called a **centerbore**, to mount to the hub of the vehicle and four or more smaller holes for the lugs to fit through. The hub section is made to center the wheel onto the vehicle hub assembly. The lug holes often have a tapered seat (**Figure 5-29**), that allows the taper on the lug nuts or bolts to squarely seat against the wheel and provide a positive fit.

Most wheels are hub-centric, which means the hub is what centers the wheel to the hub. Original equipment or OE wheels generally have the centerbore machined to precisely match the vehicle. Installing an aftermarket wheel may require the use of a centering ring to position the wheel correctly on the hub (**Figure 5-30**) Always check for proper fit when replacing stock wheels with aftermarket wheels (**Figure 5-31**).

Some wheels are nonhub-centric; these are called *lug-centric*. When you are installing lug-centric wheels, you must torque the wheels with the wheel off the ground. This allows the lugs to center and seat the wheel properly before any weight is applied to the wheel.

As stated above, changing wheels may require the use of a centering ring for the hub. Another consideration is the spacing of the holes for the lug nuts. This is called *the bolt pattern* (**Figure 5-32**). The dimensions of the bolt pattern determine what wheels can be installed on a vehicle.

Aside from the centerbore and bolt pattern, other wheel dimensions must be considered when you are choosing a replacement wheel. Wheel diameter is simply how large the wheel is across the tire bead. Common sizes range from 15 inches to more than 20 inches.

Wheel offset is the position of the mounting surface compared to the center of wheel depth. Offset is shown in Figure 5-28. Wheel offset can be zero, positive, or negative. Offset plays a large part in wheel clearance. OE wheels are designed to fit around the brake system, prevent the tire from making contact anywhere in the inner fender or suspension system, and place the weight and driving forces on the wheel bearings. Installing wheels with large amounts of offset can cause early wheel bearing failure and cause the wheel and tire to interfere with the suspension or inner fender.

■ ***Wheel Defects.*** New original equipment (OE) and aftermarket wheels are generally defect free, though there can be production problems which lead to a slightly misshaped wheel. Wheels should be perfectly round and have no side-to-side or lateral movement, called runout.

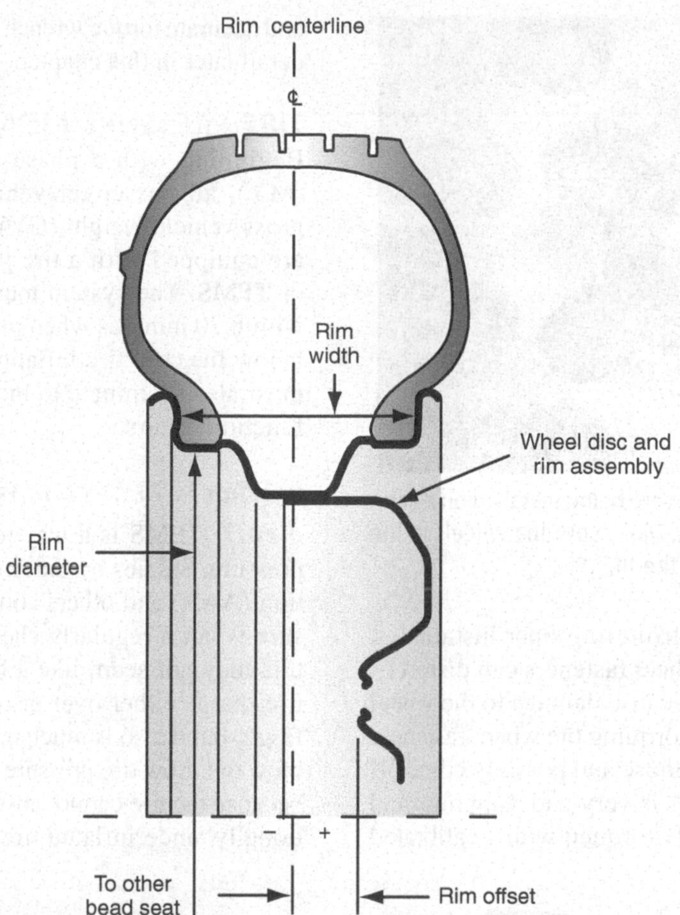

FIGURE 5-28 The parts and dimensions of a wheel.

FIGURE 5-29 The lug hole is tapered to accept the bevel of the lug nut. The contact between these surfaces helps to center the wheel on the lugs.

FIGURE 5-30 Aftermarket wheels may need adapters to fit properly to the hub. When removing aftermarket wheels, check to see if spacers are used and make sure they go back on.

Most problems that you will encounter will be caused by outside sources.

Wheels are susceptible to damage from potholes, curbs, low tire pressure, and other factors involved in driving the vehicle. **Figure 5-33** and **Figure 5-34** shows wheel damage from potholes and other road conditions. Cars with low-profile tires are more prone to wheel damage due to the small sidewall being able to absorb less road shock. In addition, drivers who get too close to curbs can easily damage the outside edge of the rim.

FIGURE 5-31 A set of wheels were being installed on a vehicle but did not quite fit correctly. Make sure the wheels fit the hub or install spacers to correct the fit.

Wheels can be damaged from improper installation as well. Over torquing the wheel fasteners can distort or warp the hub, which also can cause damage to the wheel hub and brake rotors. Under torquing the wheel fasteners will allow the wheel to work loose and possibly come off while driving. Obviously, this is very bad. Lug nuts and studs must always be properly torqued with a calibrated and accurate torque wrench. This will be covered in more detail later in this chapter.

TIRE PRESSURE MONITORING SYSTEMS

Beginning with a phase-in in the 2008 model year (MY), all passenger vehicles under 10,000 pounds gross vehicle weight (GVW) sold in the United States are equipped with a tire pressure monitoring system, or TPMS. The system must be able to alert the driver within 20 minutes when pressure is 25 percent or more below the cold tire inflation pressure. A warning light must also illuminate to indicate if a TPM system malfunction occurs.

■ ***What Is TPMS and Why Is It Standard Equipment?*** TPMS is a way to alert the driver of low tire pressure. Studies by the American Automobile Association (AAA) and others consistently show that tire pressure is not a regularly checked item by drivers. While this may not seem like a big problem, consider that a tire that is either over or underinflated not only wears faster, but it also is much more likely to be damaged and blow out. Low tire pressure can also cause wheel damage because the tire cannot absorb bumps as it should. Additionally, underinflated tires increase fuel consumption

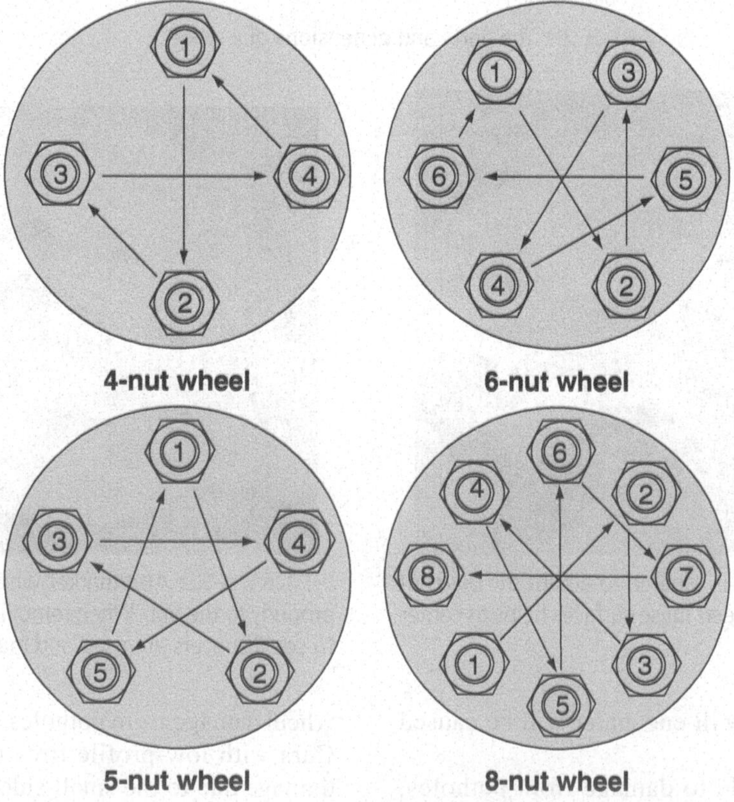

FIGURE 5-32 Common lug bolt patterns and tightening sequences.

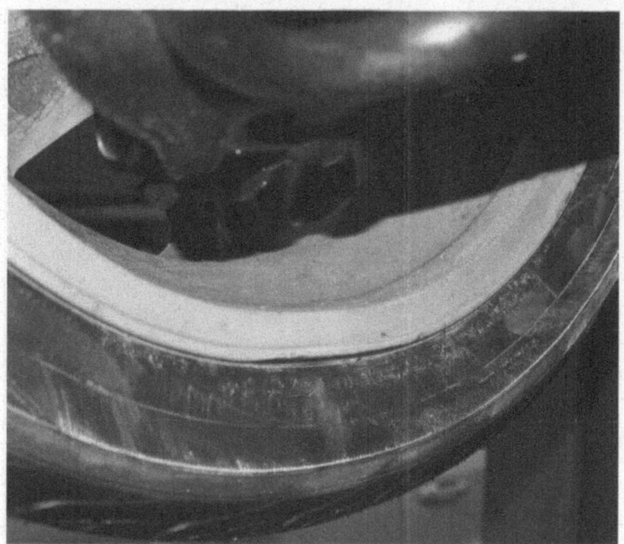

FIGURE 5-33 Potholes, curbs, and other hazards are common causes of wheel damage.

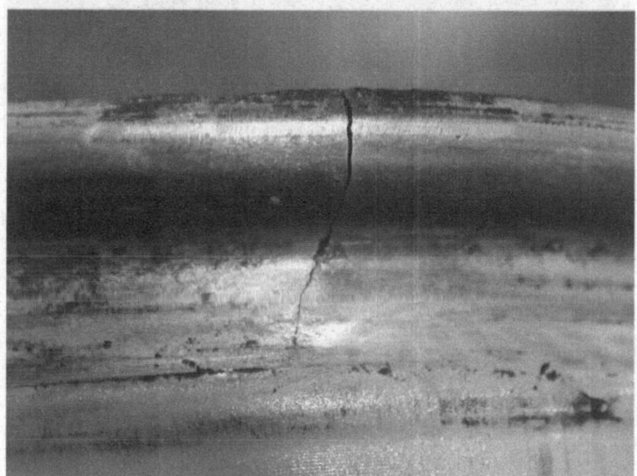

FIGURE 5-34 Aluminum wheels can crack. The vehicle this is from had two cracked wheels. This is more common on vehicles with low-profile tires.

and are more likely to fail due to overloading and damaging the sidewall of the tire. In 2000, Congress passed the TREAD Act, legislation that mandated TPMS on new passenger vehicles and light trucks starting with the 2008 models.

■ ***Types of TPMS.*** There are two types of TPMS currently in use, the indirect and the direct systems. Both can alert the driver of low-pressure conditions, but do so in different ways.

Indirect TPMS uses the vehicle's antilock brake system, or ABS, to monitor tire pressure. As tire inflation drops, the tire will ride lower and have a larger contact

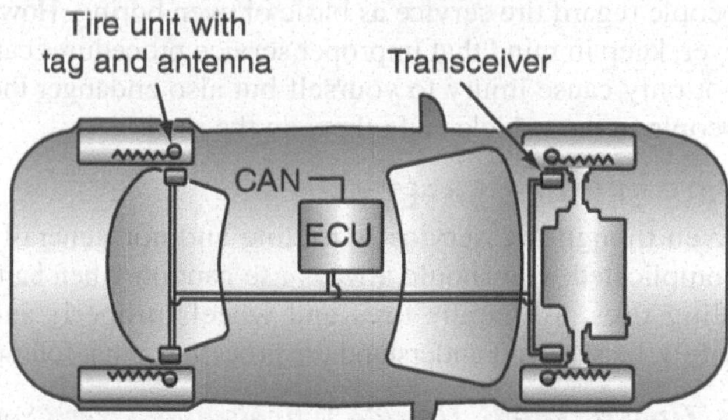

FIGURE 5-35 A direct TPM system has sensors in each wheel and a typically uses the keyless entry module as the radio receiver for tire data.

patch or footprint. This decreases the rolling diameter of the tire. Because the tire is now slightly smaller in diameter, it will rotate more times per mile than the properly inflated tires. If a wheel is turning faster than the other three, the system perceives this as low pressure and turns on a warning light on the dash.

Indirect TPMS provides a low-cost method of checking for low tire pressure, but it does have some disadvantages. An indirect system does not indicate which tire is low, just that a low-pressure condition exists. Also, if all four tires are low on pressure, the system may not recognize a problem as there will not be enough variation in the pressure and rotation speeds. In addition, because the system only works while the vehicle is moving, it also takes some time before the system can obtain readings and decide to turn on the light.

Direct TPMS eliminates the shortcomings of the indirect systems but at a cost. Direct systems use sensors placed in each wheel to directly measure and transmit the pressure to the vehicle's onboard computer system. **Figure 5-35** shows an illustration of a direct system. Each wheel has a battery-powered pressure sensor that wirelessly sends data back to an onboard computer. Because of this, direct systems are very accurate and can display real-time information to the driver. However, this accuracy adds cost to the vehicle and increased maintenance costs as the sensors fail and need to be replaced. Sensors are also prone to damage if the wheel and tire are not serviced properly.

Wheel and Tire Diagnosis and Service

Tires are some of the most commonly serviced components on the vehicle. As an entry-level technician, you will likely perform a lot of wheel and tire services. Most people regard tire service as basic or even boring. However, keep in mind that improper service procedures can not only cause injury to yourself but also endanger the people in the vehicle and others on the road.

TIRE SERVICE SAFETY

Even though tire service is routine and not generally complicated, you should always use caution when handling tires. To handle tires and wheels properly and safely, be sure you understand the procedures that follow.

■ ***How to Safely Handle Wheels and Tires.*** You may think that wheels and tires present no danger to you or others, but you can easily be hurt or damage the wheel and tire if you do not use caution when working with them.

First, tires operate in a very harsh environment and can pick up sharp objects that become embedded in the tread. When you are inspecting a tire, resist the urge to run your hands around the tread. Damaged treads and exposed belts can cause hand injuries. Small stones, pieces of metal and glass, and other road debris often lodge in tires. Running your hands around the tread is an easy way to get your hands cut from this debris. When you are inspecting a tire that is still mounted to the vehicle, wear mechanic's gloves and turn the tire by grasping the tire sidewalls. This reduces your risk of being injured. A severely worn tire has exposed steel belts (**Figure 5-36**). These pieces of steel are very sharp and can cause serious cuts if you happen to run your hand over them.

FIGURE 5-36 Use caution when handing tires. When inspecting tires, do not run your hands around the tread. Exposed belts and other hazards can cause painful injury.

Wheels and tires, especially on larger trucks and SUVs can be heavy. Use a back brace or get the help of another person to help you lift a wheel and tire off the vehicle if necessary. Even smaller and lighter wheels and tires can be dangerous to lift if they are wet, as the tire can slip through your hands and bounce off the floor, so use caution whenever you are handling one.

When you need to remove the wheel and tire from a vehicle, be sure you are wearing your mechanic's gloves. Because the brakes generate a lot of heat, and this heat transfers to the wheel and lugs, you can easily get your fingers burned if you handle these parts before they have a chance to cool. Also, many lug nuts have a thin metal shell over a solid core. Over time, this shell can rust and deteriorate, leaving jagged edges (**Figure 5-37**). These can be dangerous to your hands.

FIGURE 5-37 Capped lug nuts can be a hazard when removing wheels.

Use the proper tools, such as impact sockets if you are using an impact gun, when removing wheel fasteners. Many vehicles use a locking wheel fastener to deter theft. These require special sockets to remove and should only be used with hand tools (**Figure 5-38**). Using an impact can damage the key or the fastener, making removal very difficult.

Use caution when you are removing the center caps. These are often plastic and may use a snap ring to retain them in the centerbore. They are easily damaged if you handle or remove them improperly. Once the wheel and tire are off, do not lay the wheel face down on the floor. This can cause scratches or other damage to the wheel.

FIGURE 5-38 Some vehicles come equipped with locking lug nuts, as shown in the lower opening. A special key is required to remove these lugs. Do not use an impact on these lugs as they can be damaged by the impacting action.

TIRE SERVICE EQUIPMENT

There are several types of tire equipment you will learn to use, including dismounting and mounting machines, balancers, and repair equipment.

Tire dismounting and mounting machines usually use both electricity and compressed air to operate. An example of a tire dismounting machine is shown in **Figure 5-39**. Before using this equipment, you must be properly trained in its safe use. Check the electrical power cord, if applicable, before using the machine and make sure it is grounded properly. Make sure there are no air leaks and that all moving parts of the machine move smoothly and without binding. Do not use any piece of equipment that is not functioning properly.

FIGURE 5-39 An example of a tire dismount/mounting machine.

Tire balancing machines spin the tire on a shaft to determine how to correct for any static and/or dynamic balance issues (**Figure 5-40**). Make sure the hood of the machine fully covers the tire so that the hood stops any debris that flies off the tire.

FIGURE 5-40 An example of a tire balancing machine.

With both tire mounting machines and balancers, the tire must be lifted onto the machines for service. Use a back brace or get help from a coworker to lift the wheel and tire into position if necessary.

WHEEL AND TIRE SERVICE

Wheel and tire service means more than just changing tires. It also includes inspecting the tires, checking tire treadwear, checking and setting tire pressure, and balancing and rotating the tires.

■ ***Inspect Tire Condition.*** Visually inspect the tire for damage and for objects lodged in the tire. Look for signs of damage to the sidewall from either curbing or from driving with very low tire pressure (**Figure 5-41**). Look for evidence of belt separation or belt failure, as discussed earlier in this chapter.

■ ***Identify Tire Wear Patterns.*** A properly maintained tire should provide many thousands of miles of use before it needs to be replaced. Unfortunately, tires often are not properly maintained and will show evidence of premature wear (**Figure 5-42**). The most common causes of premature tire wear are inflation and alignment problems.

If the wear is caught early, it may be possible to correct the problem and reduce the effect of the wear on the tires. Maintaining the correct inflation, performing rotations, and setting the wheel alignment will help keep the tires from wearing out too quickly.

■ ***Check and Adjust Tire Pressure.*** Every new vehicle has a decal that provides information about the correct

FIGURE 5-41 Inspect sidewalls for signs of damage from driving on a flat or severely underinflated tire. This tire must be replaced due to the damage to the sidewall.

Conditions	Rapid wear at shoulders	Rapid wear at center	Cracked treads	Wear on one edge	Feathered edge	Diagonal wipe rear tire FWD vehicles	Scalloped wear
Effect							
Causes	Underinflation or lack of rotation	Overinflation or lack of rotation	Underinflation or excessive speed	Excessive camber	Incorrect toe	Incorrect wheel toe	Lack of rotation of tires or worn or out-of-alignment suspension
Corrections	Adjust pressure to specifications when tires are cool. Rotate tires.			Adjust camber to specs	Adjust toe to specs	Perform rear wheel alignment	Rotate tires and inspect suspension

FIGURE 5-42 Common tire wear patterns. Reading tire wear can provide a lot of information about the maintenance condition of the car.

tire, rim size, and proper inflation pressures. This decal is often located on a doorjamb (**Figure 5-43**) but can also be found in the glove box, console, or even the fuel door. This decal may show optional tire sizes available for that vehicle; so, pay close attention, so you know you are looking at the right information for the tires that are installed on the car or truck. The example in **Figure 5-44** shows the tire size options available for this particular vehicle.

These decals list cold tire pressure. This refers to what the pressure should be set to after the vehicle has been parked and the tires have cooled and pressure has returned to its lowest point. The amount of time necessary for the tires to cool and to provide an accurate reading will depend upon how long the vehicle was driven, how long it has sat, and what the ambient temperature is. A tire that has been driven on a hot summer day may need to sit overnight, while a tire may need to only sit outside on a cold winter day a few hours before it can be accurately checked. When you are in doubt, try to check

TIRE AND LOADING INFORMATION

SEATING CAPACITY | TOTAL 4 | FRONT 2 | REAR 2

The combined weight of occupants and cargo should never exceed 302 kg or 666 lbs.

TIRE	ORIGINAL SIZE		COLD TIRE PRESSURE	SEE OWNER'S MANUAL FOR ADDITIONAL INFORMATION
FRONT	305/30ZR19	Y	220 kPa, 32 PSI	
REAR	305/30ZR19	Y	220 kPa, 32 PSI	
SPARE	NONE		NONE	

2G1FS1EE5E9699047

FIGURE 5-43 The tire decal contains size and pressure information about the regular tires and spare tire. This decal can be located in several different places on the vehicle.

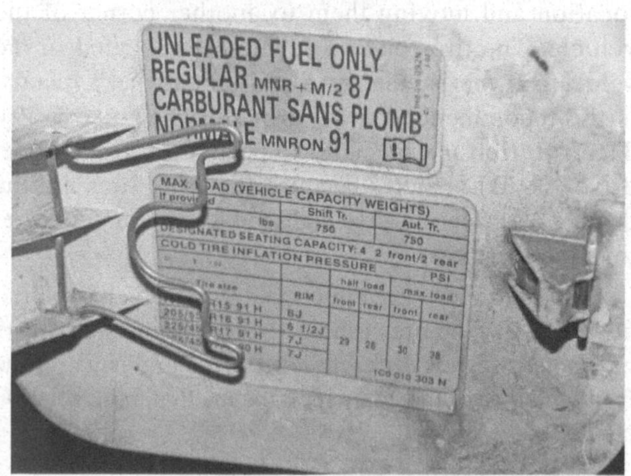

FIGURE 5-44 This fuel door-mounted tire decal gives optional tire size information. Verify the size of the tires installed on the vehicle when checking inflation pressure as different tire sizes may have different pressure specifications.

the tire pressure before the vehicle is first driven for the day. If you are checking the pressure after the vehicle has been driven, to get an accurate pressure reading you will also need to determine the tire's temperature and use a pressure/temperature chart. Refer to the shop's service information for pressure/temperature compensation charts. A temperature/pressure chart is used to determine the inflation pressure of the tire based on its current temperature and pressure.

There are several types of tire pressure gauges you can use (**Figure 5-45**). Regardless of the type, the gauge should be of good quality and accurate. To ensure you are accurately checking and setting tire pressure, obtain a tire pressure gauge certified to meet the American National Standards Institute (ANSI) for accuracy.

To test the air pressure in the tire, remove the valve stem cap if there is one, and firmly press the open end of the tire gauge to the opening of the valve stem. There will be a slight hiss when the two match up, but there should not be a continued hissing sound. If the hiss continues, remove and retry attaching the gauge until the hiss stops. **Figure 5-46** and **Figure 5-47** show two different gauges and their readings.

If the air pressure is low, connect a tire inflator to an air hose and inflate the tire (**Figure 5-48**). If the inflator has a built-in pressure gauge, monitor the tire pressure while you are inflating the tire. If you are using a separate inflator and gauge, stop frequently to recheck the pressure. If the pressure is excessive, deflate the tire until the correct pressure is reached.

Remember to check and adjust the pressure for the spare tire. This may require removing the contents of the trunk, so be extra careful when you are handling

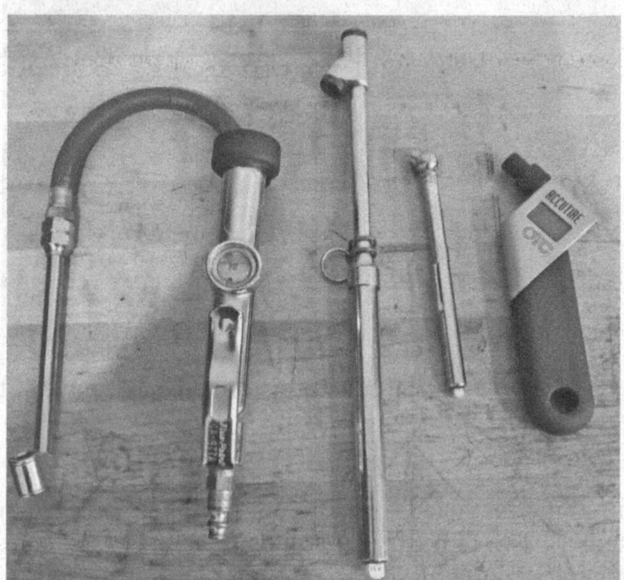

FIGURE 5-45 An assortment of tire pressure gauges.

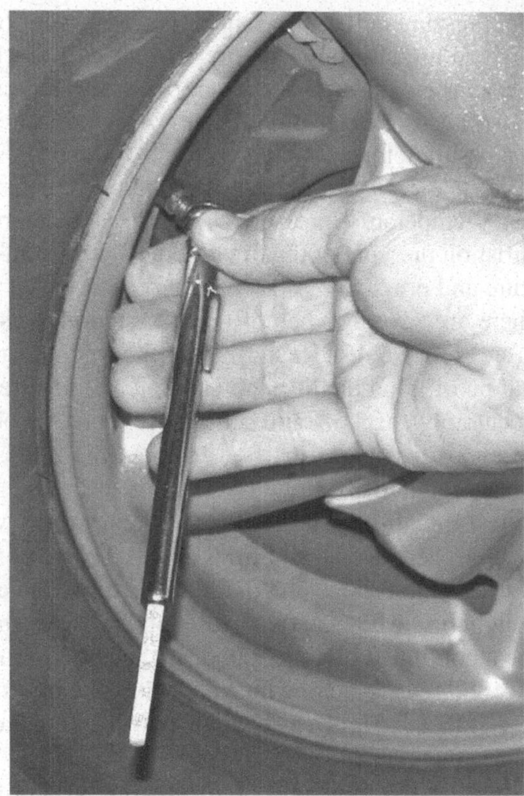

FIGURE 5-46 Checking tire pressure with a common type of gauge.

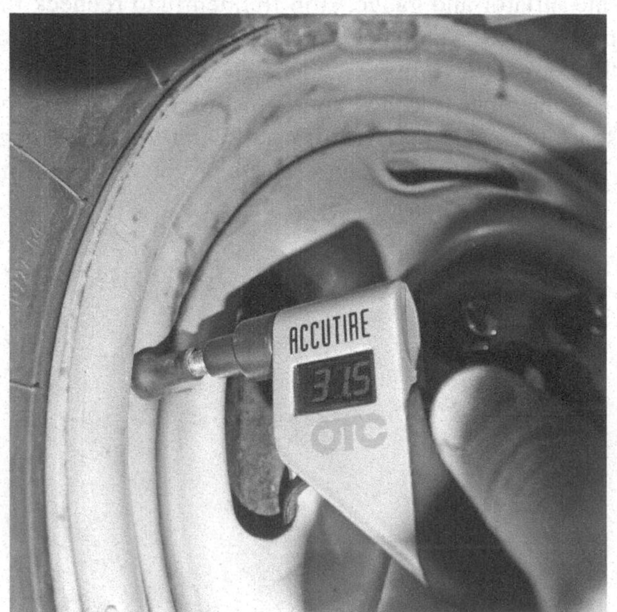

FIGURE 5-47 Digital pressure gauges are quick and easy to read.

anything that needs to be removed to get access to the spare. Also, remember that most temporary spare tires are inflated to 60 psi.

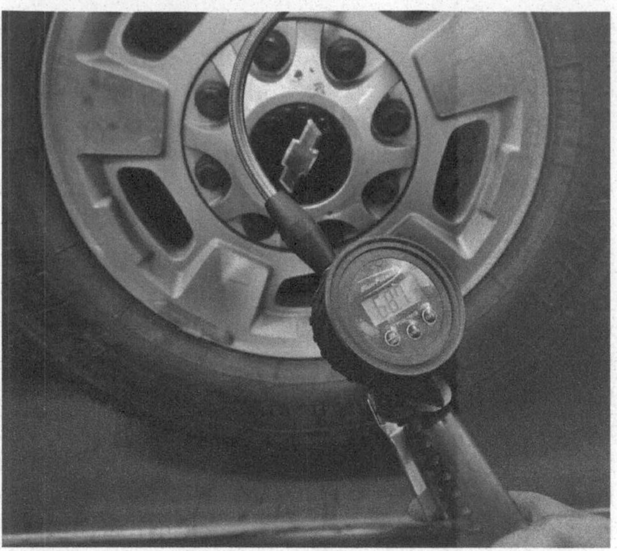

FIGURE 5-48 Inflators that lock onto the valve stem are popular and easy to read and use.

The California Air Resources Board mandated that beginning September 1, 2010, all auto service and repair businesses must check and document tire pressure on every vehicle serviced. This is because low tire pressures increase rolling resistance, decrease fuel economy, and ultimately increase exhaust emissions. Whether mandated in your state or not, checking tire pressure when a vehicle is serviced is important because of the negative effects of driving with low tire pressure.

■ ***Rotating Tires.*** To get the maximum life from the tires, they should be rotated periodically. This means removing the wheels and tires from their current location and moving them to another corner of the vehicle. The tire rotation schedule is located in the vehicle's owner's manual. Also in the owner's manual is the recommended rotation pattern (**Figure 5-49**). **Tire rotation** is especially important on front wheel drive (FWD) vehicles because the front tires are the driving and steering tires and tend to wear much faster than the rear tires.

Some vehicles have directional tires, which can limit how, if at all, the tires can be rotated. Some sports cars have staggered or different sized tires and rims on the front and rear of the car, which means that a tire rotation cannot be performed.

■ ***Dismount, Inspect, and Remount Tire on Wheel.*** **Photo Sequence 2** shows the use of a common tire machine to dismount and remount a tire.

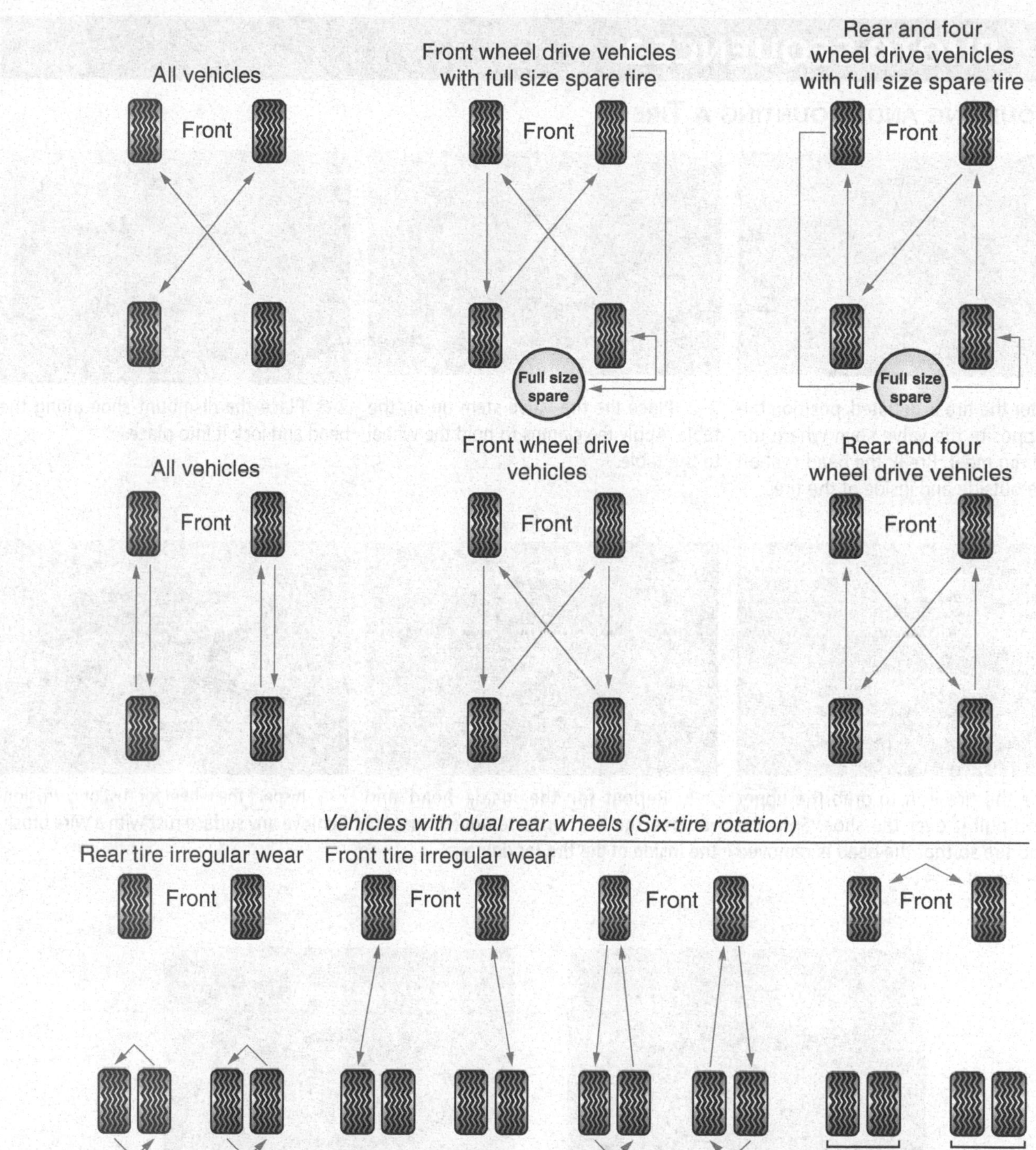

FIGURE 5-49 Common tire rotation patterns. Check the vehicle's owner's manual for the proper rotation pattern.

■ ***Special Service Precautions.*** Direct TPMS vehicles have the pressure sensor either strapped to the drop center section of the wheel or as part of the valve stem. Both types require proper service techniques so that the sensors are not damaged. Technicians must know the location of the sensor before they attempt to dismount the tire from the wheel.

Once the tire is dismounted from the rim, inspect the beads and inside of the tire for damage. A damaged bead can cause air loss and even prevent the tire from maintaining proper grip against the rim. Dust or debris inside the tire means that the tire has been driven on either flat or with very low pressure (**Figure 5-50**). If the bead or inner sidewalls are damaged, the tire must

PHOTO SEQUENCE 2

Dismounting and Mounting a Tire

2-1 After the tire is deflated, position the blade opposite the valve stem where the tire and rim meet. Break the bead seal on both the outside and inside of the tire.

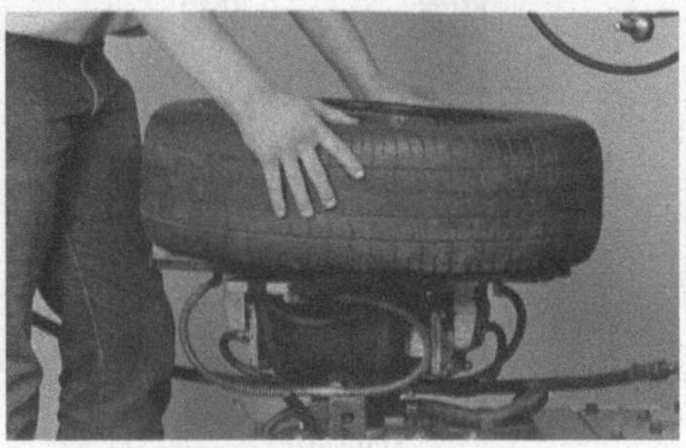

2-2 Place the tire valve stem up on the table. Apply the clamps to hold the wheel to the table.

2-3 Place the dismount shoe along the bead and lock it into place.

2-4 Use the tire iron to grab the upper bead and pull it over the shoe. Spin the table and tire so that the bead is removed from the wheel.

2-5 Repeat for the inside bead and remove the tire. Once removed, inspect the inside of the tire for damage.

2-6 Inspect the wheel for rust or corrosion. Remove any surface rust with a wire brush.

2-7 Lubricate the beads with tire lubricant to prevent damage during mounting. Do not over lubricate the tire. This can allow the tire to slip on the wheel. Place the inside bead over the mounting shoe and carefully rotate the tire onto the rim. Repeat for the upper bead. Keep the sidewall pressed down into the drop center of the wheel when installing the upper bead.

2-8 Locate the correct inflation pressure and inflate the tire to specifications.

be replaced. **Figure 5-51** shows how the inside of a tire wears from driving on a flat tire. This type of damage requires tire replacement.

When you are inspecting a dismounted tire, use caution because foreign objects can protrude either out from the tread or into the inside of the tire.

Before remounting the tire, determine if the tire is a directional or asymmetric tire so that it is mounted correctly on the rim and back on the vehicle.

When reinstalling the tire, avoid using too much tire lube on the bead or remove excess lubricant once the tire is on the rim. Excessive lubrication can allow the tire to slip and rotate around the wheel under high-torque acceleration or braking. This will cause the wheel and tire to get out-of-balance and cause a vibration.

Once the tire is mounted onto the wheel, inflate the tire to the proper pressure. Be careful not to overinflate the tire during mounting. Overinflation can cause the tire to rupture and explode, causing serious injuries.

FIGURE 5-50 This tire was driven on with very low air pressure. The result was that the sidewalls carried the vehicle weight, which destroyed the sidewalls.

FIGURE 5-51 This tire was driven on when flat. The weight disintegrated the sidewalls.

■ ***Balancing the Wheel and Tire Assembly.*** **Photo Sequence 3** shows how a typical wheel balancer is used to check and balance the wheel and tire.

Modern balancers check and correct both static and dynamic imbalance problems. Some balancers can detect runout problems. If excessive runout is detected, the balancer may show how to match mount the wheel and tire to reduce runout. Some wheel balancers detect other tire conditions, such as road force variations and conicity and even tell the user where to reinstall the tires on the vehicle to reduce vibration and pulling.

■ ***Reinstalling the Wheel.*** When the wheel and tire are ready to be reinstalled, perform the following checks and procedures:

1. Check the mounting area on the hub and the wheel for excessive rust or corrosion (**Figures 5-52** and **5-53**). This rust or corrosion needs to be removed to ensure proper fit and tightening. Clean the hub and mounting face of the wheel as needed. Make sure any hub spacers are properly installed and seated on the clean hub.
2. Inspect all wheel fasteners, lug nuts, and studs before installing. A damaged wheel fastener should be replaced if the threads cannot be cleaned up with a thread file or tap and die.
3. Carefully set the wheel into position on the hub and studs. Start the lug nuts by hand and make sure they start easily. If the vehicle uses lug bolts, place the wheel onto the hub and keep it in position while you start a lug bolt into the hub.
4. Continue to tighten and seat each lug until the wheel is fully seated against the hub. It is good practice to seat the lugs in the same pattern that is used to torque them to the manufacturer's specifications. Always tighten the lugs in a criss-cross pattern and not a circular pattern. Once the lugs have been tightened to spec, go over them again to ensure that the wheel is fully seated and the lugs are indeed tight.
5. Reinstall any center caps or wheel covers as necessary.

Service Note

With some aftermarket wheels, the wheel manufacturer states that the torque should be rechecked after 50 to 100 miles of driving. Be sure to inform the customer if this is necessary for the wheels being installed.

PHOTO SEQUENCE 3

Balancing a Wheel and Tire

3-1 If a wheel and tire is covered in mud or snow, clean it thoroughly before attempting to perform a wheel balance.

3-2 Using the correct mounting cones, mount the wheel onto the balancer shaft and tighten the wingnut.

3-3 Remove any old weights.

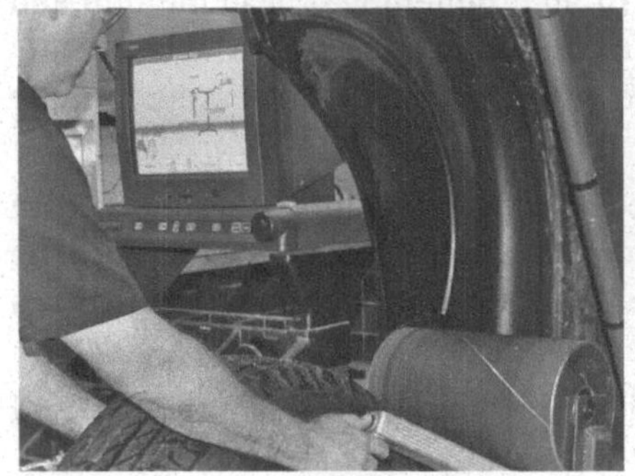

3-4 Input the wheel dimensions into the balancer.

3-5 Drop the hood and spin the tire. Watch the wheel and tire assembly as it spins and note any signs of runout.

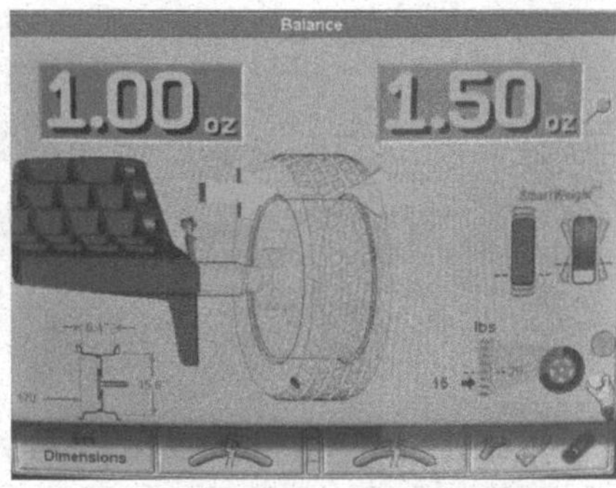

3-6 Once the check is complete, locate the correct type of weight for the wheel and the amount indicated by the balancer.

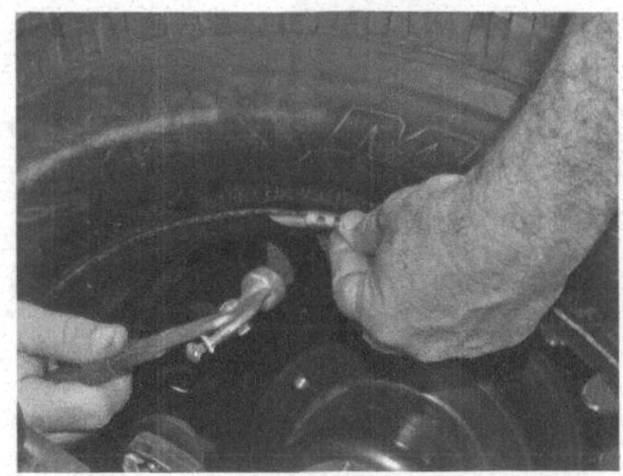

3-7 Install the weight onto the wheel at the locations specified by the balancer.

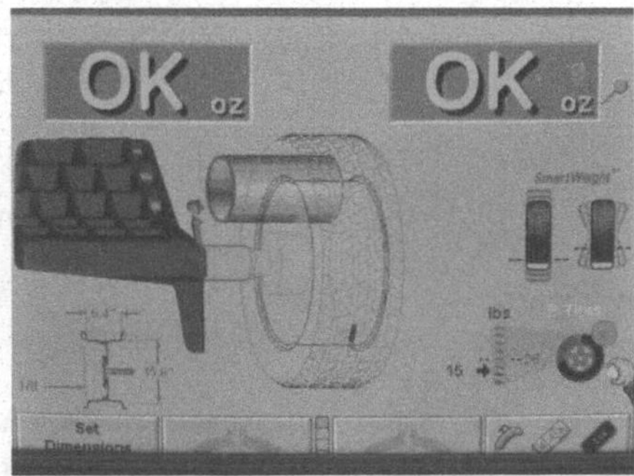

3-8 Perform a check spin to make sure the wheel and tire are balanced. Results of OK or 0.0 weight needed.

3-9 Remove the wheel and tire from the balancer and reinstall on the-vehicle.

FIGURE 5-52 Rust and corrosion can make getting the wheels and tires off the vehicle a lot of work. Always clean the hub area to make sure a good, clean, and flat contact area between the wheel and the hub on the vehicle.

FIGURE 5-53 Rust and corrosion forms inside on the hub area of the rim and needs to be cleaned before reinstalling the wheel.

■ ***Difficult to Remove Wheel Assemblies.*** It is not uncommon to find that a wheel is stuck to a vehicle, refusing to budge even after pulling and light rapping on the tire sidewall in an attempt to dislodge the stuck wheel. When you are faced with this situation, do not beat on the tire with your hands as this can lead to injury and pain and a wheel and tire that still are not off the vehicle. First, rethread a lug nut several threads onto a stud; this will prevent the tire from falling to the floor once it is free. Next, if it is possible, spray the hub and lug area with a penetrant, and let it soak for several minutes. Using a soft-faced dead blow mallet, hit the tire on the inside sidewall in several places. Do not hit the rim as you do not want to damage the rim. This may require several rounds of hitting the tire, but it will eventually come off.

Once the wheel is off the hub, inspect the wheel mounting face and the hub. An example of a corroded wheel and hub is shown in Figure 5-52. The corrosion is a result of an aluminum wheel reacting with the steel and iron of the hub and brake rotor. To make sure the wheel mounts and torques correctly, clean the wheel mounting face and hub. Cleaning both the back of the wheel and the hub is important to proper reseating of the wheel onto the car and for making sure the wheel gets properly torqued. Rust or corrosion between the wheel and hub can prevent the wheel from seating properly and can lead to wheels coming loose while driving.

■ ***Inspect for Air Loss.*** As discussed before, tires are subject to damage from all sorts of objects lying in the road and from rust or corrosion buildup around the bead area. There are two ways to check a tire for air loss: by spraying the tire with a soap and water solution and looking for bubbles or by placing the tire into a tire dip tank filled with water and looking for bubbles. Depending on the size of the leak, the escaping air may make a lot of bubbles that are easy to see. If the leak is slow, it may be difficult to find the exact cause of the problem. The following are common air loss problems:

- Bead area leaks. A common cause of leaks around the bead area is rust or corrosion buildup around the rim bead (**Figure 5-54**). Usually, this can be removed with a wire brush or scouring pad and the leak corrected.
- Over time, the valve stem and/or core can deteriorate and leak. If either is leaking, it is best to replace the entire stem with a new stem (**Figure 5-55**). Using soapy water is a good way to find a small leak because of the amount of bubbles generated from the air loss.
- Tires that scrub curbs often are subject to sidewall damage. Any damage to the sidewall that causes air loss or weakens the sidewall means the tire must be replaced.
- Tires, over time, will eventually start to dry rot. This problem accelerates if the tires are always exposed to direct sunlight day after day. There is no cure for dry rot other than replacement. Some car manufacturers now recommend that tires older than a certain age, often six years, be replaced, regardless of treadwear, due to the effects of dry rot.

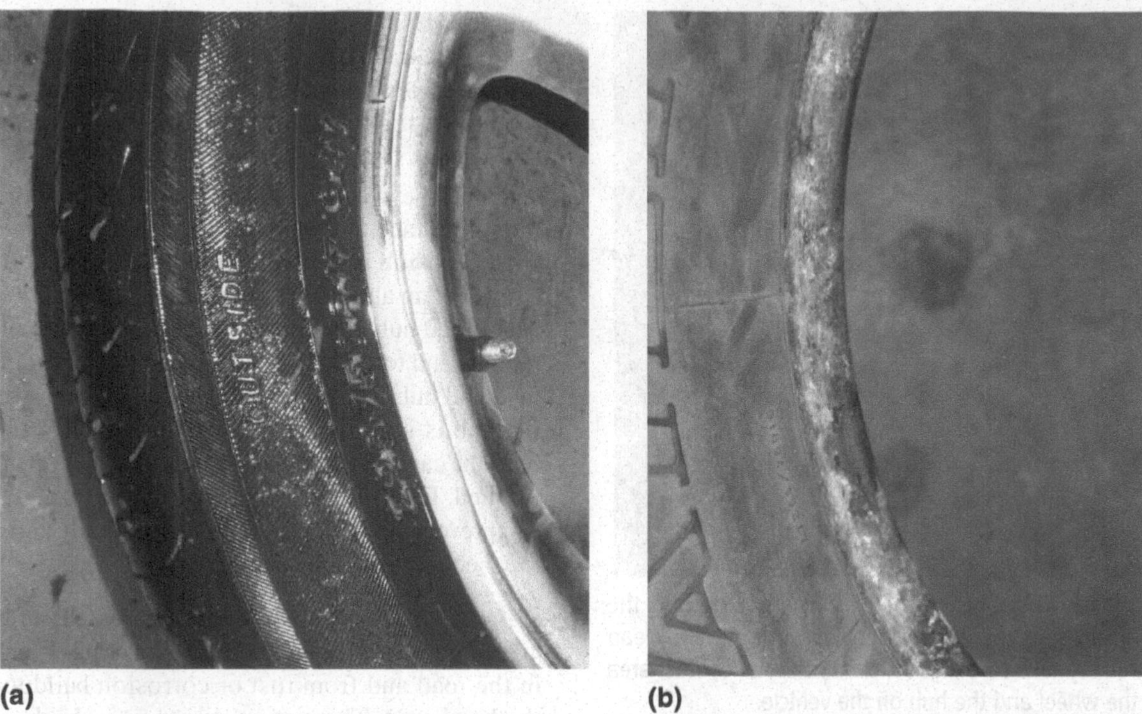

(a) (b)

FIGURE 5-54 (a) Evidence of air leaking from the bead. (b) Corrosion build up between the aluminum wheel and the bead of the tire.

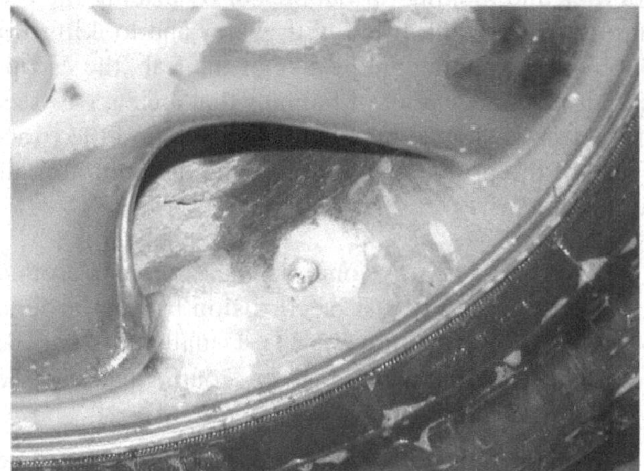

FIGURE 5-55 The bubbles show this valve stem is leaking where it passes through the wheel.

- Aluminum wheels are prone to cracking, especially those with low profile tires. Two wheels on this customer's Jeep (**Figure 5-56**) were found to have cracks in the bead area. Although uncommon, steel wheels can develop leaks where the center wheel disc attaches to the rim section. Depending on the type of wheel and why it is leaking, wheel replacement may be required.

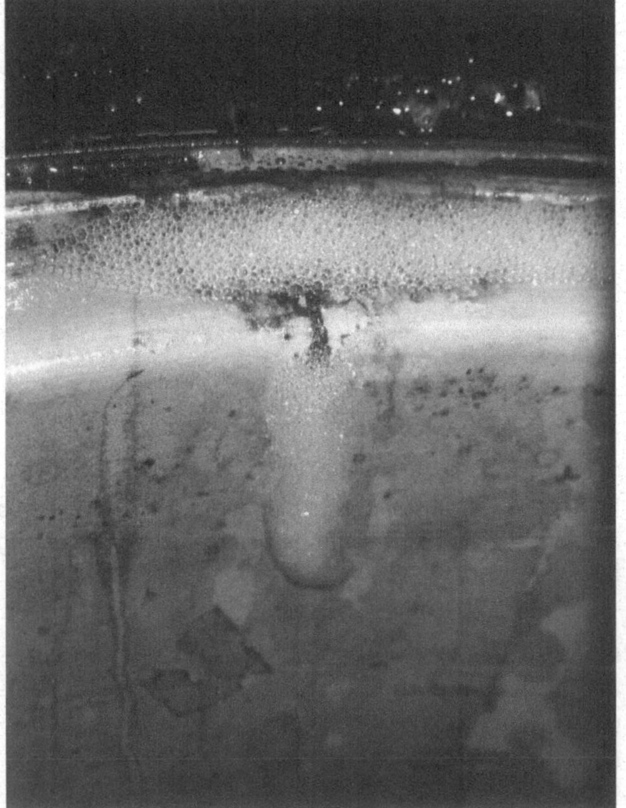

FIGURE 5-56 The wheel is leaking due to a crack in the aluminum.

■ ***Repairing a Tire with an Internal Patch.*** When a tire tread has been punctured, and if the hole is not too large, the tire may be able to be patched. **Photo Sequence 4** shows how to internally patch a tire. It should be noted with the customer and documented on the repair order that any type of puncture repair voids any speed rating the tire may have.

PHOTO SEQUENCE 4

Patching a Tire

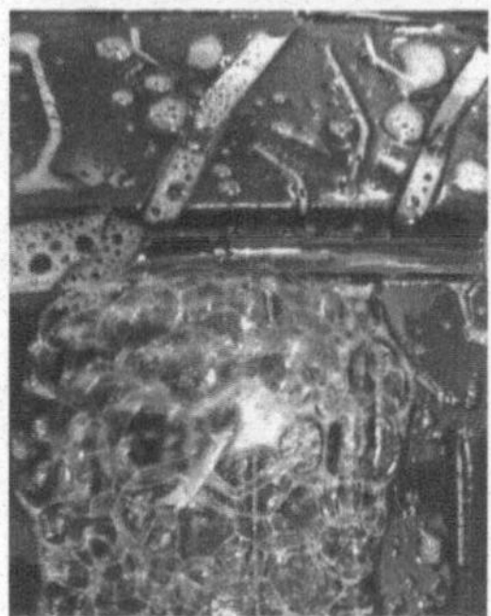

4-1 Once the leak has been found, in this case a puncture, you can determine if the tire can be repaired. Mark the puncture on the outside of the tire with a tire crayon or grease pencil. Note: Punctures in the sidewall or the last tread blocks at the sidewall cannot safely be repaired.

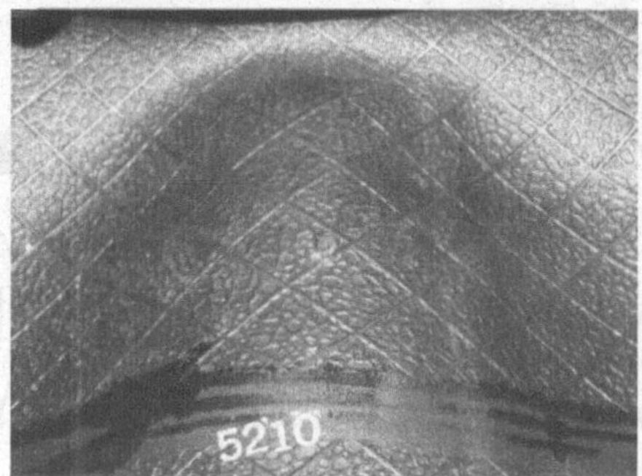

4-2 With the tire dismounted from the wheel, locate the puncture inside the tire. Mount the tire in a spreader to allow more room to work on the under tread lining.

4-3 Buff the lining to remove any texture from the surface. Be careful not to go past the lining into the tire's belts. When complete, blow the rubber dust from the tire using shop air.

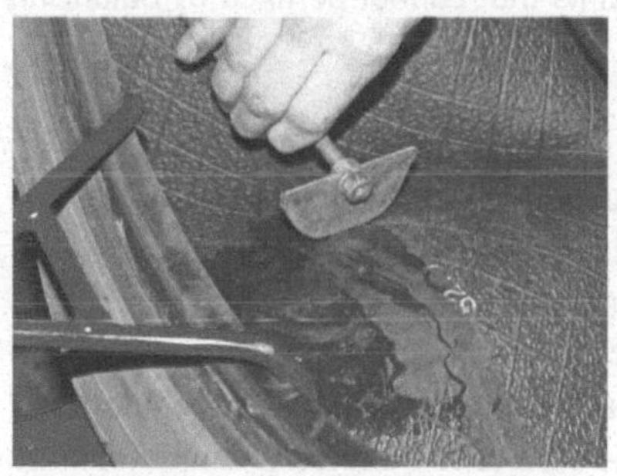

4-4 Apply a coat of rubber cement to the buffed area and scrape off the cement. This cleans the lining and removes any leftover debris or dust from the patch area.

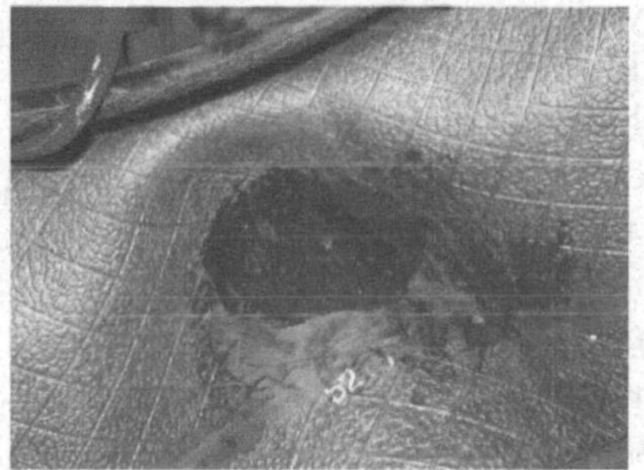

4-5 Apply a light coat of rubber cement and let dry. Place the patch-plug over the puncture and insert the plug stem into the hole. Pull the stem through the tire and seat the patch against the lining.

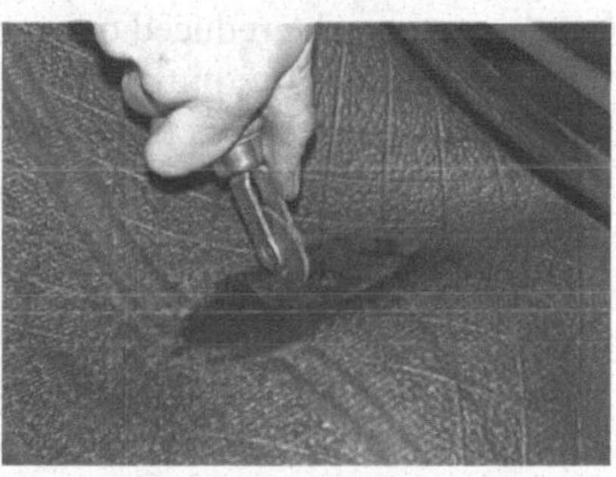

4-6 Using a stitching tool, work the patch against the lining to remove any air. Once this is done, reinstall the tire, inflate to specifications, and check for leaks.

VIBRATION AND PULLING

As discussed before, vibration concerns are often caused by tire balance problems, but tire balance is not the only possible cause of a vibration. To determine the cause, begin by talking with the customer and get as much information as possible about when the vibration occurs. It may be necessary to test drive the vehicle with the customer so that both of you are experiencing the same condition.

■ ***Diagnose Wheel and Tire Vibration or Shimmy.*** A wheel/tire that is statically out-of-balance will bounce or hop as it is driven. This is because the heavy or light spot will cause the tire to accelerate and decelerate unevenly during its rotation. As a result, the imbalance will cause a shake or vibration that can be felt throughout the vehicle. A wheel/tire that is dynamically out-of-balance will cause a side-to-side shaking of the steering wheel, called **shimmy**. Both are usually corrected easily by rebalancing the wheel and tire assembly.

Performing a wheel balance will not fix a bent wheel, an out-of-round wheel or tire, or a tire belt problem. Manufacturers often specify the maximum amount of wheel and tire runout that is acceptable for a particular vehicle. If the wheel or tire shows signs of runout, you should measure the runout and compare your readings to specifications. **Figure 5-57** shows how to set a dial indicator to measure wheel and tire runout. Total runout can sometimes be reduced or even nearly eliminated by remounting the tire on the wheel. This is called *match mounting* and relocates the points of high or low runout of the tire with the corresponding high or low spots on the wheel. The overall effect is to reduce the total runout of both together.

A bent wheel may be able to be straightened. Wheel specialist companies advertise wheel repair and straightening services. This may be a lower-cost option than replacing a bent wheel. Do not try to straighten a bent wheel in the shop. This can cause further damage to the wheel.

If the wheels and tires are in balance and the vibration persists, it could be caused by hub or axle **lateral runout**. If the hub or axle flange is bent, it will cause the wheel and tire to move laterally or side-to-side. Loose wheel bearings and over tightening wheel fasteners can cause hub flange distortion, which can cause vibration problems. The hub or axle flange can be checked for runout with a dial indicator. As with wheel and tire runout specifications, refer to the manufacturer's service information. If the hub has excessive runout it may need to be replaced. Be sure to check the wheel bearing adjustment, if applicable, before replacing the hub.

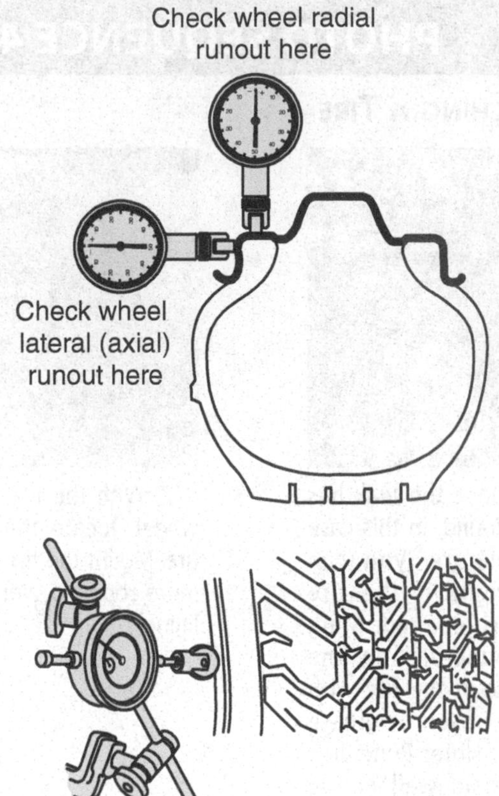

FIGURE 5-57 Wheel balancing cannot cure vibration problems caused by excessive runout. If the wheel/tire shows signs of runout, measure and compare the reading to the manufacturer's specifications. A slight amount of runout can cause vibration complaints that cannot be fixed by balancing the wheel and tire.

Radial runout exists if the wheel and/or tire are out-of-round. This forces the tire to roll unevenly and can cause severe vibrations throughout the vehicle.

Depending on the type of tire balancing machine in your shop, it may measure runout as part of the balancing procedure (**Figure 5-58**). Some balancers can determine the best way to remount the tire on the rim to reduce the total runout. The balancer will indicate where to reposition the tire relative to the valve stem.

Tire vibration problems can also be caused by excessive tire stiffness within certain sections of the tread. Also called *road force variation* (**Figure 5-59**), the differences in how the tire reacts to the force against the ground can cause vibrations to occur in the vehicle.

■ ***Diagnosing Tire Pull Concerns.*** Imperfections during tire construction can cause the tire to lead or pull while driving. If a customer states that the pull started after the tires were rotated, there is a good chance the problem is in the tires. To check for tire pull, switch the

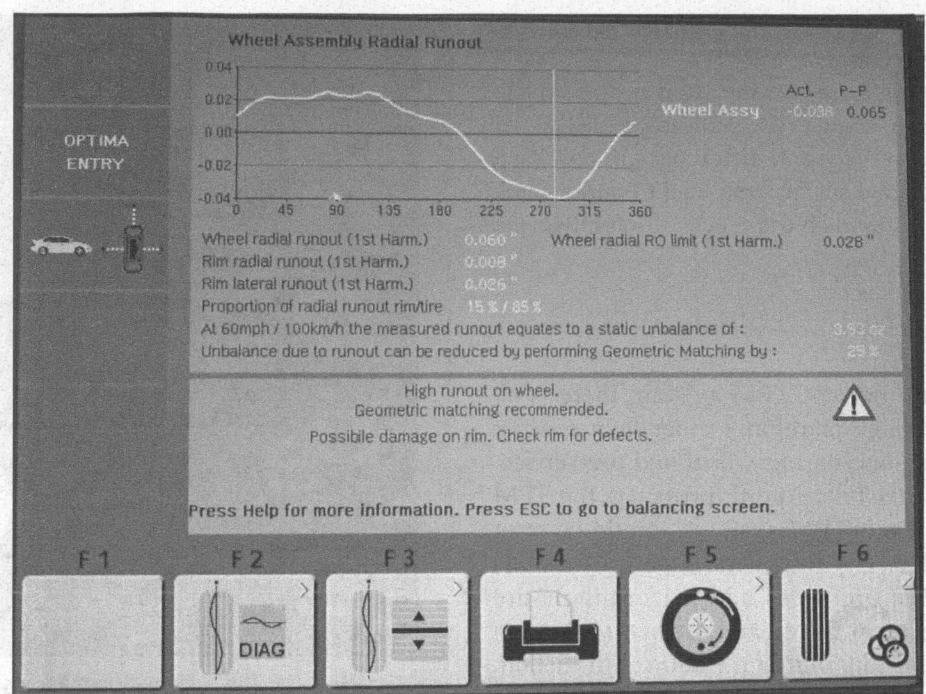

FIGURE 5-58 Some wheel balancers detect wheel and tire runout and measure it for you.

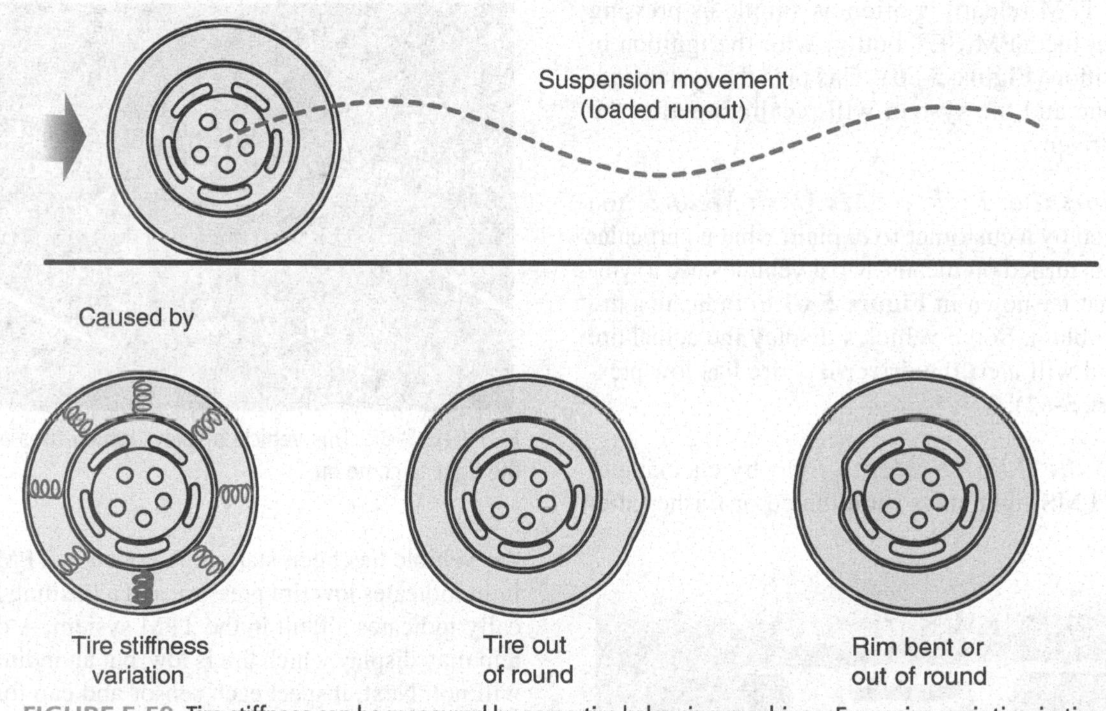

FIGURE 5-59 Tire stiffness can be measured by some tire balancing machines. Excessive variation in tire stiffness can cause vibration.

front tires from side-to-side, and see if the pulling stops. If so, the problem most likely is in the tire on the side in which the vehicle was pulling. If the pull switches sides, try rotating the front tires back to the rear of the vehicle. If the tire causing the pull is new, then the offending tire should be replaced under warranty. If the tire has been in service for some time and the pulling is a new situation, you will need to discuss with the customer his or her options, such as tire replacement, keeping the pulling tire on the rear, or living with the pull when the tire is installed on the front. These steps for diagnosing tire pull assume that tire pressure, wheel alignment, suspension, steering, and brake components have been checked and ruled out as possible causes.

When possible, both tires on the front of the vehicle should be the same size, type, and tread design. If dissimilar tires are on the front, differences in how each tire responds to the road surface can lead to pulling and/or vibration problems as each tire reacts differently to the road surface.

TIRE PRESSURE MONITORING SYSTEM (TPMS) SERVICE

If a vehicle has an indirect TPM system, there is no additional maintenance or repairs other than reset or relearn procedures done during wheel and tire service. If the ABS system is functioning properly, the TPM system should as well. TPM systems should be reset or relearned anytime any tire work is performed, this includes setting tire pressures and performing a tire rotation. A **TPMS reset** or relearn is when the system learns new tire position and/or inflation information, such as after a tire rotation is performed. Some vehicles, such as many Honda models, require no action and the system relearn takes place as the vehicle is driven.

Indirect TPM relearn is often as simple as pressing and holding the TPM SET button with the ignition in the ON position (**Figure 5-60**). This puts the system into a learn mode and the system will recalibrate after the vehicle is driven.

■ ***What Does that Light on My Dash Mean?*** You may be asked by a customer to explain what a particular light that has turned on means. Most vehicles use a symbol like what is shown in **Figure 5-61** to indicate a tire pressure problem. Some vehicles display the actual tire pressure and will alert the driver if a tire has low pressure (**Figure 5-62**).

■ ***Inspect the TPM System.*** Start by checking to see if the TPMS light stays illuminated or flashes after the vehicle has been started. A constant TPM warning light indicates low tire pressure and a flashing light typically indicates a fault in the TPM system. A direct system may display which tire is low, but an indirect system will not. Next, inspect each sensor and cap for damage. Note any signs of sensor or stem damage or missing or incorrect caps.

Locate the tire pressure specification on the tire decal, and check the pressure in each of the tires. If only one tire has low pressure, perform a thorough tire inspection to determine why the pressure has gone down. If all the tires are low, the driver may not be regularly checking and adjusting the pressure. An inspection of each tire should still be performed.

FIGURE 5-61 The exclamation mark inside the tire profile is the tire pressure monitor warning light.

FIGURE 5-62 This vehicle displays which tires are causing the light to come on.

FIGURE 5-60 A TPM reset button for a Toyota. This button is used to perform a reset after tire services are performed.

■ ***TPM Service.*** Before working on the tires or TPM system, verify system operation by observing the TPM warning light on the dash. Ensure the light comes on at Key-On bulb check and goes out once the engine starts. If the TPM light remains on or flashes, test the sensors and system to determine the fault before continuing.

TPM systems typically require a relearn after work is performed. They may include setting tire pressure, rotating the tires, or replacing a sensor. Before performing a tire rotation on a vehicle with a direct TPM system, refer to the proper service information for any special procedures that need to be followed to relearn the system. **Table 5-1** shows common TPM relearn procedures.

A malfunctioning TPM system will set fault codes in the vehicle's onboard computer system and cause the TPMS warning light to flash and illuminate as well. The most common cause of malfunctions are nonfunctioning pressure sensors. The sensors cannot be serviced and must be replaced if faulty or when the battery is too weak to broadcast the signal. Most manufacturers recommend replacing the sensors when the tires are replaced or every five to seven years.

To determine exactly why the TPM light is on, a scan tool or TPM tool is used to access stored fault codes. Always follow the diagnostic and repair procedures as outlined in the service information when you are working on the TPM system.

■ ***Dismounting and Mounting Tires with TPMS Sensors.*** Special tools and procedures are used when working on wheels and tires that have TPMS sensors installed (**Figure 5-63**). Failure to follow the service procedures can cause damage to the sensors.

Wheels that have the clamp-in valve stem type sensors usually require that you remove the valve stem retainer

TABLE 5-1

GM and some Asian Vehicles · Scan tool reset	Ignition ON Install scan tool and initiate TP sensor learn mode. Horn will sound twice and LF turn signal light will turn on. Start at LF sensor and activate with TPM tool and wait for horn to sound. Once horn sounds, continue and repeat for RF, RR, and LR wheels.
Some GM - Driver Information Center reset	Ignition ON Activate TPM relearn mode in DIC. Horn will sound twice and LF turn signal light will turn on. Start at LF sensor and activate with TPM tool and wait for horn to sound. Once horn sounds, continue and repeat for RF, RR, and LR wheels.
Some Ford	Ignition OFF Press and release brake pedal Turn ignition OFF to RUN three times Press and release brake pedal Turn ignition OFF Turn ignition OFF to RUN three times, horn should sound once. Activate LF tire with TPM tool, horn will sound when sensor is learned. Activate RF, RR, LR sensors.
Various	Press and hold the TPM reset button on the dash. The TPM light will blink. Once the vehicle is driven approximately 20 minutes, the TPM system will calibrate.
Various	Press Brake and release Cycle ignition on/off, on/off, on Press Brake and release Cycle ignition off/on, off/on, off/on Horn honks indicating now in learn mode and dash will indicate which wheel to start with (normally front left, then clockwise around the car) Use TPM tool to trigger each sensor Car horn will sound as each tire's sensor is learned and honk twice when completed successfully
Some GM Vehicles	Cycle ignition ON (on not start) Press "lock/unlock "on keyfob at the same time or use DIC menu Vehicle horn sounds twice indicating now in learn mode. Turns on lights in order of relearn (normally front left, then clockwise) Use TPM tool to trigger each sensor Car horn will sound as each tire's TPMS is relearned A double horn honk when completed.
Various Indirect Systems	Set pressure to specifications. Press and hold down button "RESET" button. Press and hold down TPM SET button until warning light flashes three times. Drive vehicle (speed and times vary by make, model, year).

FIGURE 5-63 Examples of TPMS sensor tools.

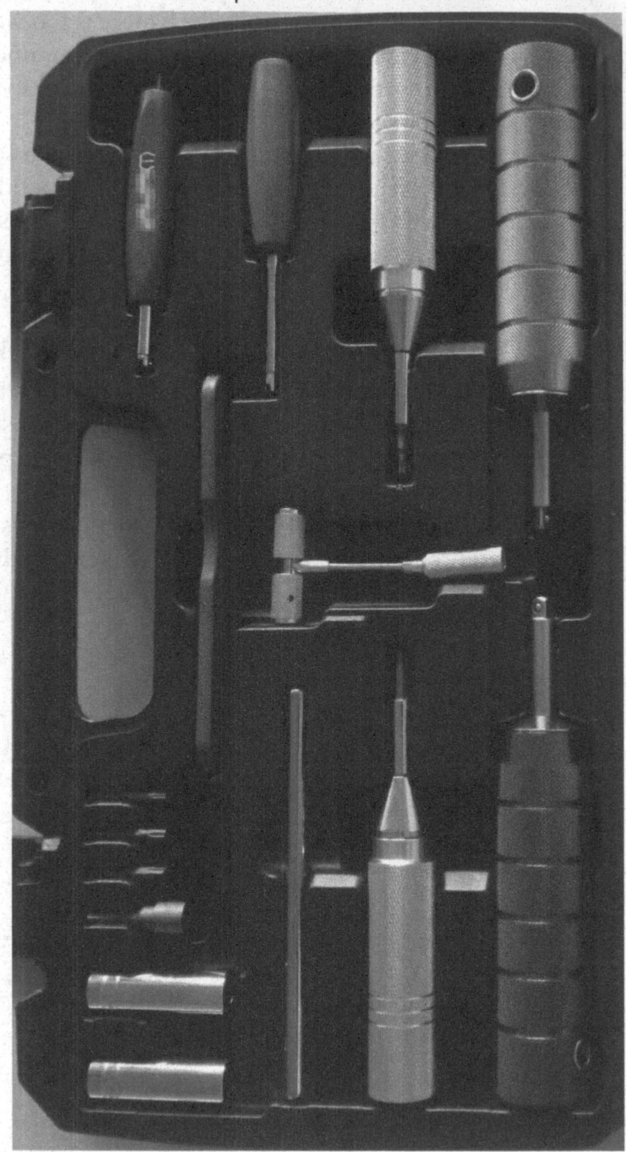

(a)

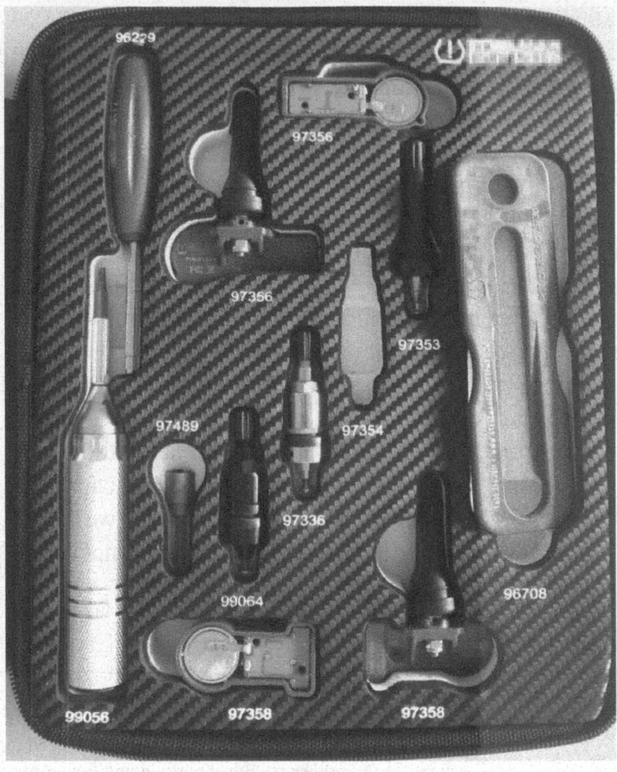

(b)

and push the sensor into the tire before attempting to dismount the tire (**Figure 5-64**). Once the sensor is out of the way, the beads can be unseated and the tire removed. Special tire deflators are available to deflate TPM sensors. The deflators thread onto the stem like the valve stem cap but protrude down against the core to deflate the tire. These are used because the TPM sensor can be damaged if you try to remove the valve core.

Many TPM sensors are easy to identify by their metal stem and cap (**Figure 5-65**). This is called *a clamp-in sensor*. Some vehicles use a rubber stem in the sensor (**Figure 5-66**). These are called *snap-in sensors* and the rubber stem and sensor are two separate parts that are fastened together. This type is easily mistaken for a standard rubber valve stem, which can be damaged when you are dismounting the tire. To avoid a costly mistake when dismounting a tire from the rim, place the blade opposite from the valve stem (**Figure 5-67**). This will allow the blade to miss the sensor during bead separation.

Whenever a clamp-in sensor is removed, a replacement service kit should be installed. A service kit contains a new retainer, seals, and cap. When you are reinstalling the sensor, torque the stem retainer to specifications. The valve core also requires proper torque. Use the proper valve core tool to torque the core into the stem.

Remind your customers that non-TPM caps, especially metal caps on clamp-in sensors, valve extensions and Fix-A-Flat type products are not recommended for use with direct TPM systems. **Figure 5-68** shows a TPM sensor that was ruined due to the owner installing a chrome metal valve stem cap. The cap reacted with the stem and seized in place, breaking the stem when the cap was removed to check air pressure.

■ ***TPM Sensor Programming.*** Each TPM sensor has its own unique ID code and replacement sensors may need to be programmed to the vehicle. An example of a TPM ID code is shown in **Figure 5-69**. Depending on the vehicle and the replacement sensor, it may be necessary to record the ID code of the failed sensor and program the code into the new sensor for it to be recognized by the onboard system. This is called *sensor cloning* and is common with aftermarket replacement sensors. Typically, installing an OE-type sensor does not require

(a)

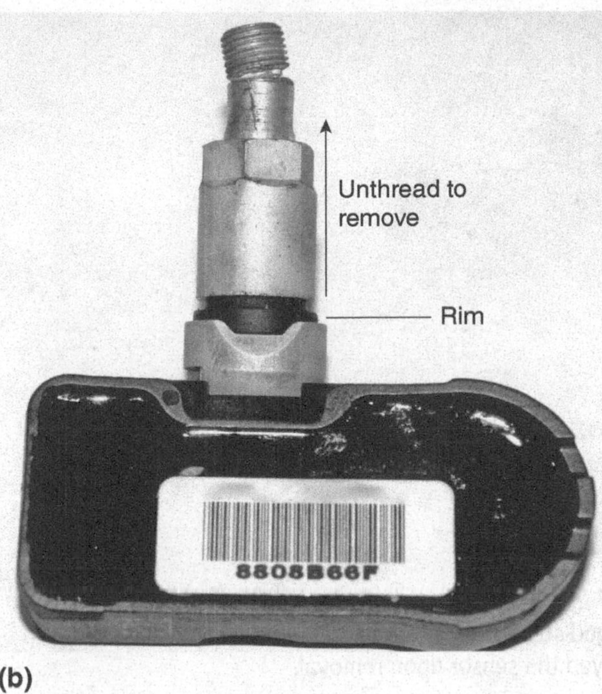

(b)

FIGURE 5-64 (a) A common type of TPM sensor. (b) The cap is unscrewed to the sensor and torqued down to seal and hold the sensor in place against the stem opening in the wheel.

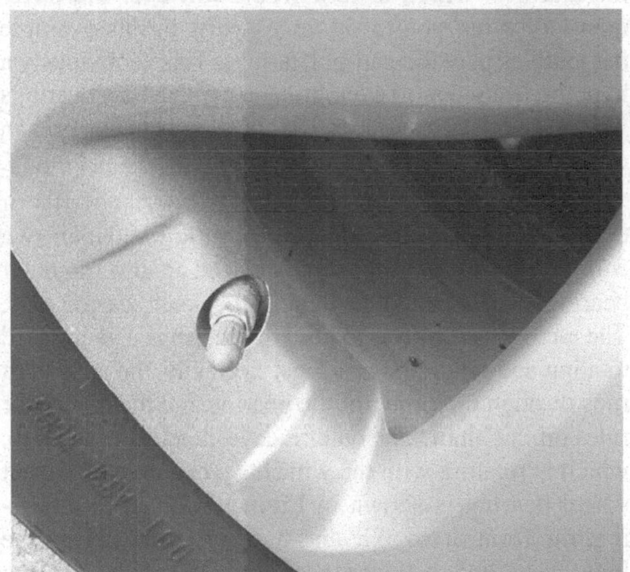

FIGURE 5-65 Clamp-in or metal stem sensors are very common.

FIGURE 5-66 Stealth or snap-in sensors have replaceable rubber stems that are secured in place with small screws.

FIGURE 5-67 Breaking the tire bead loose with a blade on a TPM sensor-equipped wheel.

programming; performing a relearn will allow the TPMS to identify and learn the new sensor IDs. However, refer to the vehicle and sensor manufacturer's service information for the specific procedures.

Wheel Bearing Principles and Operation

Bearings reduce the amount of friction between two objects. Without bearings, the friction produced by the rotating axles would destroy the axle and other components. Bearings, such as those used as wheel bearings,

FIGURE 5-68 Do not install aftermarket metal caps. Note the damaged stem. This sensor had the metal cap installed, which destroyed the sensor upon removal.

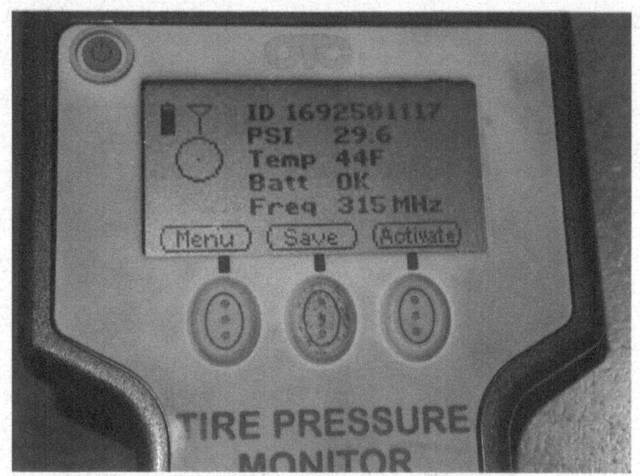

FIGURE 5-69 An example of TPM sensor data.

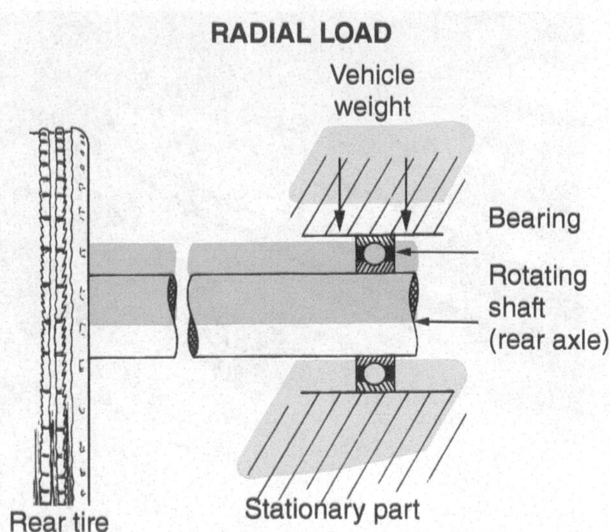

FIGURE 5-70 Radial loads are placed vertically through a wheel bearing. Thrust loads on a wheel bearing are horizontal.

use two races separated by ball or roller bearings. A lubricant, often oil or heavy grease, is used to reduce the friction between the moving parts and to carry heat away.

WHEEL BEARING FUNCTIONS

Bearings of all types are used to reduce friction and wear, and wheel bearings are no different. **Wheel bearings** generally use either roller bearings or tapered roller bearings to support the hub and wheel or axle shaft and wheel. An example of a bearing used to support an axle is shown in **Figure 5-70**.

■ ***How Bearings Reduce Friction.*** Bearings allow components to roll over or against each other instead of sliding. Imagine standing on a paved road with skis attached to your feet. It would be difficult to push yourself along smoothly with the skis sliding over the pavement. Now imagine attempting the same thing but with the pavement covered in marbles. The skis would roll over the marbles, which in turn would roll over the pavement, allowing you to ski very easily. In this example the marbles play the part of bearings. This is obviously a hypothetical scenario attempting to explain how bearings work between two surfaces. Do not attempt to perform an experiment such as this as it may result in personal injury.

Most automotive bearings use either steel balls or cylindrical rollers to reduce the friction between two components. One component is placed against the outer bearing race and the other against the inner bearing race. The races have grooves to hold the bearings and are used to align and retain the bearings. Typically, the axle shaft runs through the center of the bearing and the inner race rides on this shaft. The outer race is attached to the hub, which is rotating with the wheel. An example of a front wheel bearing is shown in **Figure 5-71**. In this bearing, the front drive axle and hub are connected to the inner races, and the outer races are bolted to the steering knuckle. The bearing allows the components to move easily in relation to each other. Since there is movement involved, the bearing needs some type of lubrication to operate. Depending on what the bearing is used for, it may be lubricated by oil, such as differential fluid or motor oil, or may be packed with a heavy grease. Many wheel bearings are sealed and lubricated for life, while others require periodic repacking with grease.

■ ***Forces Acting upon Wheel Bearings.*** A wheel bearing, when carrying the weight of a stopped vehicle, is subject only to a radial load. This type of load is carried perpendicular to the bearing on the axle (**Figure 5-72**).

FIGURE 5-71 An inside view of a double-row front wheel bearing assembly.

When the vehicle is moving, the bearing must be able to withstand thrust and axial loads. Thrust loads are applied sideways to the bearing as the vehicle turns. The combination of both the radial and thrust loads place a high demand on the wheel bearings, which are also subject to heat from the brakes and road shock.

WHEEL BEARING DESIGN AND OPERATION

The wheel bearings used in automotive applications are usually called *frictionless bearings*. They are called frictionless because the balls or rollers are placed between two races and allow for very easy movement. Most frictionless bearings have the same three parts, the inner and outer races and the bearings, which are either steel ball bearings or steel roller bearings.

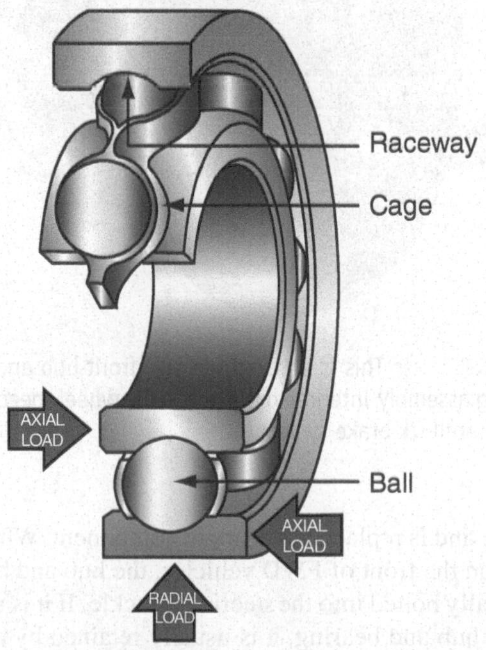

FIGURE 5-72 Ball bearing components.

■ ***Ball Bearings.*** Ball bearings provide good radial and some thrust load-carrying capacity and smooth, low-friction movement. But because the ball bearings have just a very small amount of contact with the races, they cannot take large amounts of thrust loading.

■ ***Roller Bearings.*** Roller bearings are often used as rear axle bearings in rear wheel drive (RWD) vehicles (**Figure 5-73**). Roller bearings, because of the much larger surface contact area, can take much larger radial loads than ball bearing designs. The cylindrical rollers and inner race support the axle shaft, while the outer race is pressed into the rear axle tube. When they are used as a rear axle bearing in RWD vehicles, the bearing is lubricated by the differential lubricant.

■ ***Tapered Roller Bearings.*** Tapered roller bearings were commonly used as front and rear wheel bearings due to their excellent radial and thrust load-carrying abilities. Many older RWD vehicles used a set of inner and outer tapered roller bearings for each front wheel (**Figure 5-74**). Some FWD vehicles used tapered bearings in the rear hubs. The angles of the rollers and bearing races allow this type of bearing to be able to withstand large amounts of both radial and thrust loads associated with driving. Tapered roller bearings are lubricated with heavy high-temperature grease and require periodic service.

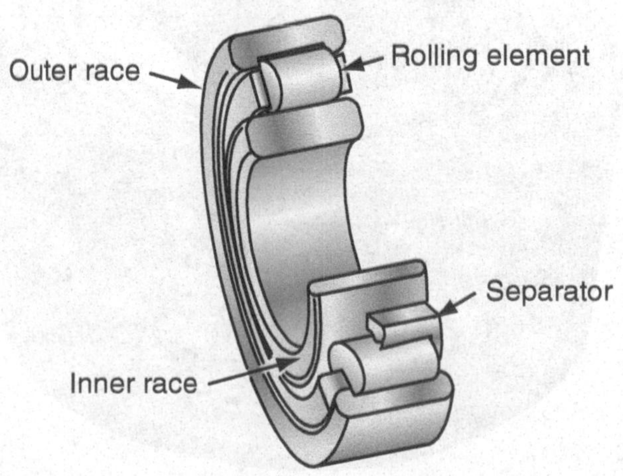

FIGURE 5-73 Roller bearings are common as rear wheel bearings on RWD vehicles.

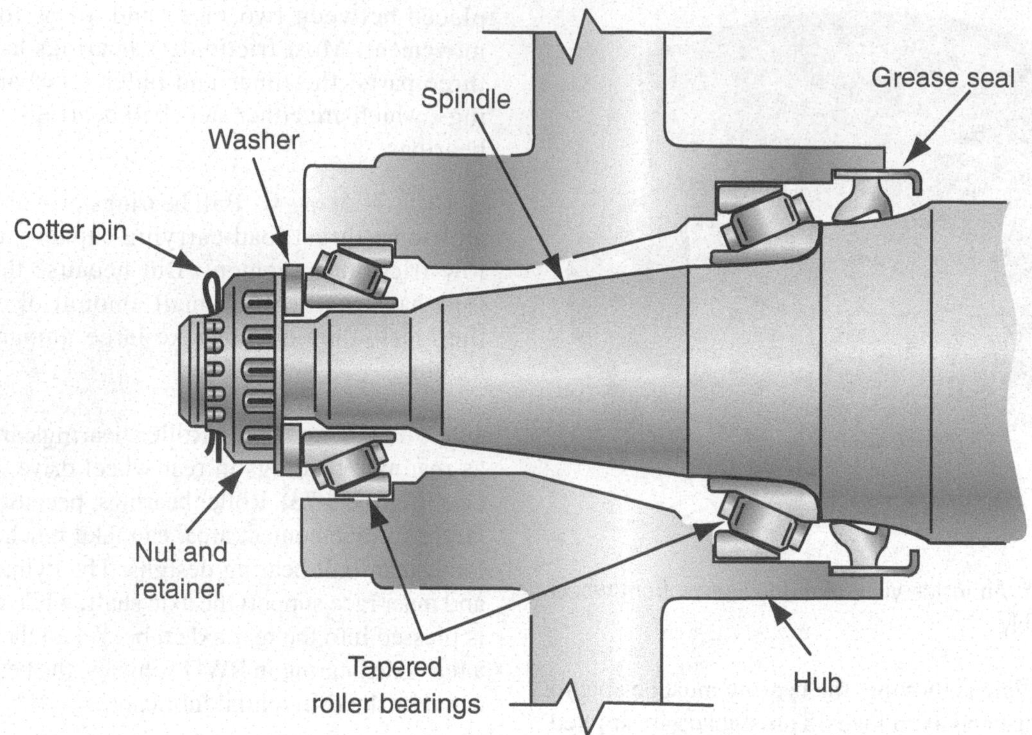

FIGURE 5-74 An illustration of how tapered roller bearings are used on some vehicles.

■ ***Double Row Bearings.*** Front wheel bearings on FWD vehicles are double-row bearings (**Figure 5-75**). This type of bearing is usually press fit, sealed in the hub, and is not serviceable other than for replacement.

■ ***Sealed Bearings.*** Another popular wheel bearing design is the hub and bearing unit. As shown in **Figure 5-76**, this combines the wheel bearing and hub flange and is replaced as a single component. When it is used on the front of FWD vehicles, the hub and bearing is usually bolted into the steering knuckle. If it is used as a rear hub and bearing, it is usually retained by an axle nut (**Figure 5-77**).

FIGURE 5-75 A double-row ball bearing that is pressed into the steering knuckle.

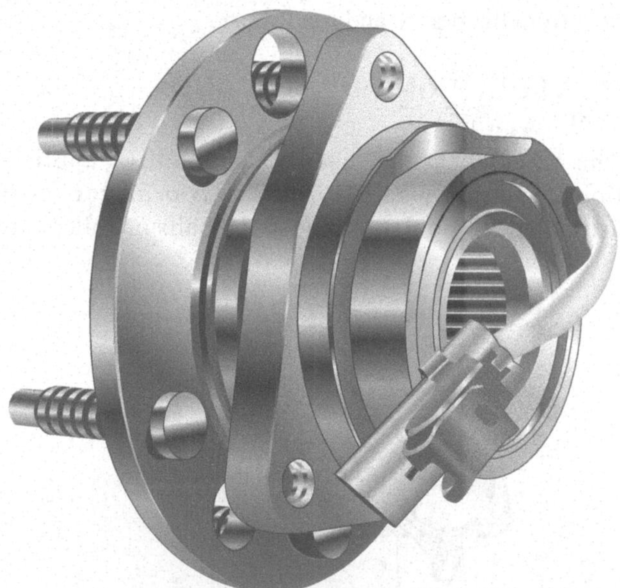

FIGURE 5-76 This is an example of a front hub and wheel bearing assembly. Integral to the hub is the wheel speed sensor for the antilock brake system.

FIGURE 5-77 A rear hub and bearing assembly on the front of a FWD car.

■ ***Bearing Preload and End Play.*** Wheel bearings require a specific amount of bearing preload. Preload places a thrust load against the bearings to prevent axial or side-to-side movement and to eliminate end play. Preload increases friction within the bearing and consumes power. Excessive end play allows the bearings room to move and wobble. A slight amount of end play is sometimes desired to allow for expansion as the bearings heat up during operation. Some bearings require preloading to prevent any movement. Setting bearing preload and end play is discussed in the wheel bearing service section of this chapter.

■ ***Bearing Grease Seals.*** Most wheel bearings have some type of grease seal to keep dirt out of the bearing (**Figure 5-78**). On some designs, such as those that use serviceable tapered roller bearings, the grease seal keeps the grease in the hub and helps prevent dirt and moisture from getting into grease and bearings. On other designs, the seal is only used to help keep dirt out.

FIGURE 5-78 Grease seals keep dirt and moisture out of the bearing.

Wheel Bearing Diagnosis and Service

As with most components on modern vehicles, wheel bearings have become more reliable and often require no maintenance. That does not mean that wheel bearings will always last the life of the car or truck, and you will need to be able to accurately diagnose wheel bearing failures.

WHEEL BEARING DIAGNOSIS

Wheel bearings, because of the function they perform, tend to wear and need replacement as part of normal vehicle operation. Being able to correctly diagnose wheel bearing concerns is important for both entry-level and more experienced technicians.

■ ***Noise Concerns.*** Faulty wheel bearings often are very noisy, and the customer may be complaining of a growling, grinding, or humming noise while driving. Wheel bearing noise will usually start with low-speed driving and become louder and change pitch as wheel speed increases. The bearing may not make much noise when the customer is driving straight ahead but it will get louder when turning a corner. If the noise increases when turning right, the left front wheel bearing is likely the cause as the right turn increases load on the left bearing. After a test drive to confirm the complaint, begin your inspection by raising the vehicle so the tires can spin freely. Spin each wheel by hand and feel for roughness or looseness while listening for any noise from the bearing. If no roughness or noise is apparent, it does not mean that the bearing is good, it just is not appearing faulty because it is not under sufficient load with the wheels off the ground.

To diagnose a front wheel bearing on a FWD vehicle, you may need to use the engine to drive the front wheels while you listen at the bearing for noise. This is much easier when using a stethoscope. Electronic stethoscopes are very useful in this situation, however, a long screwdriver or prybar will also work to help locate the noise.

SERVICE TIP

Try spinning the tire while grasping the coil spring to detect a worn wheel bearing. Many times, the bearing will vibrate and can be felt by holding the coil spring.

Service Warning

Use extreme caution when working near moving components, such as the wheels and tires.

Sometimes the faulty wheel bearing will feel loose. To determine if the bearing is loose, raise and support the vehicle so that weight is off the bearing, and the tire can rotate freely. Next grasp the tire at the three and nine o'clock positions and shake the tire back and forth (**Figure 5-79**). If any looseness is felt, closely examine the wheel bearing and other steering and suspension components. Side-to-side play at the tire can be a faulty wheel bearing, but it can also be loose steering linkage components, such as a worn tie rod.

■ ***Wheel Bearing—Caused Wheel Shimmy and Vibration.*** A loose front wheel bearing, in addition to making noise, can also cause vibrations in the steering wheel. If a sealed front or rear wheel bearing has any play, it should be replaced. A loose wheel bearing can allow the entire hub, wheel, and tire to wobble and change position while driving. The driver may feel this movement as a steering wheel shimmy or vibration.

If the loose bearing is an adjustable tapered roller bearing design, the hub and bearings should be removed, cleaned, and inspected for wear or damage. This type of service is covered in the next section of this chapter.

WHEEL BEARING SERVICE

Most vehicle manufacturers have phased out the use of the serviceable tapered roller bearing designs and now use sealed bearing units. This design change helps reduce the cost of maintenance and vehicle ownership over time. Even so, many vehicles still require periodic wheel bearing maintenance.

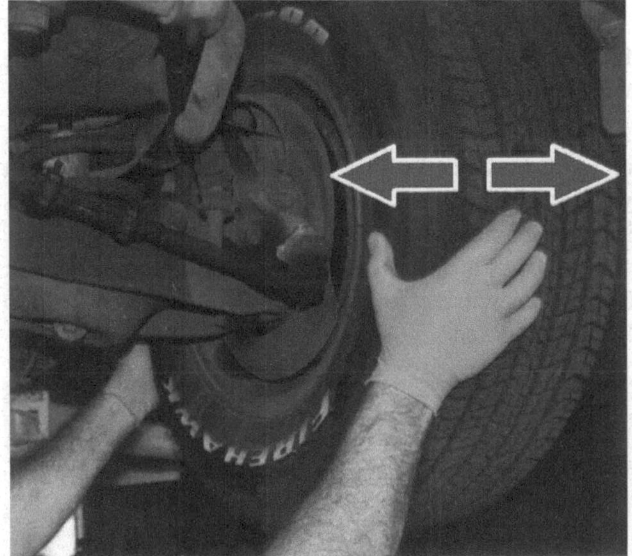

FIGURE 5-79 Checking for loose wheel bearings.

■ ***Tapered Roller Bearing Service.*** Tapered roller bearings should be serviced when the brakes are serviced or as outlined in the vehicle's maintenance schedule. Many manufacturers specify that bearings be cleaned, inspected, and repacked with new grease every 30,000 miles. The basic steps to service these bearings are as follows:

1. Remove the bearing dust cap, cotter pin, retainer, and outer bearing. Set aside for cleaning.
2. Pull the rotor or drum off the spindle, and place it on a workbench. Using a seal puller, remove the grease seal from the inside of the hub, and remove the inner bearing.
3. Remove as much grease as possible from the bearings, washer, nut, dust cap, and hub with a shop rag. Next, thoroughly clean all the parts in a solvent tank. Make sure all traces of grease are removed and dry the parts.
4. Determine the correct grease for the application and using a bearing packer, repack the bearings with new grease.
5. Apply grease to the inside of the hub. Place a small amount of grease inside the dust cap. Apply a light coating of new grease to the spindle. Install the inner bearing into the hub, and install a new grease seal using a seal installer.
6. Install the rotor or drum onto the spindle, and install the outer bearing and washers. Thread the nut onto the spindle. Locate the vehicle manufacturer's recommended tightening and torquing procedure. This normally requires tightening, loosening, and retightening the bearings to adjust end play. Follow the recommended procedures.
7. Once the bearings are seated and torqued, install the retainer and new cotter pin. Bend the ends of the cotter pin to prevent it from working loose. Install the dust cap. Always install a new cotter pin.

When you are replacing tapered roller bearings, the bearing race is replaced with the bearing. This is because the bearing and race have developed a shared wear pattern over time. Any damage to the bearing rollers can also damage the bearing races. The bearing races are driven out of the hub with a brass punch and hammer, and the new races are installed using a bearing driver kit. The bearing driver is usually aluminum so that it does not damage the race during installation.

■ ***Sealed Wheel Bearing Service.*** As discussed earlier, if a sealed wheel bearing is noisy or has roughness, it must be replaced as a unit. Do not attempt to remove and repack the grease in a sealed bearing unit.

Service Note

Always follow the service procedures as described by the manufacturer when you are servicing tapered roller bearings, especially for end play and preload adjustments. Always make sure that the proper grease is used when you are repacking tapered roller bearings. While many manufacturers allow the use of high-pressure wheel bearing grease, some vehicles require the use of a specific grease to provide the proper lubrication under all operating conditions.

Many front wheel bearings on FWD vehicles are pressed into the steering knuckle. There are a couple of methods for removing this type of bearing depending on the equipment you have available to you.

One method is to use an on-vehicle bearing removal and installation kit (**Figure 5-80**). The adapters and parts of the kit can be used to remove the hub flange, press the bearing from the knuckle, press the new bearing into place, and press the hub flange back into the bearing. If this tool is not available, you will probably have to remove the steering knuckle and use a hydraulic press to remove the bearing. With the knuckle removed and supported, use the press to push the hub flange from the bearing. With the hub flange removed, determine if the bearing is removed from the front or rear of the knuckle. Remove any seals and retaining rings as necessary. Using the appropriate adapters, press the bearing from the knuckle.

Once the old bearing is removed, compare the replacement bearing with the old bearing. Make sure the new bearing has the same inside and outside diameters. Some bearings are directional, meaning they must be installed a certain way. Make sure you orient the new bearing correctly before pressing it into place. Position the new bearing with the knuckle, and use the press and adapters to push the new bearing into place. Once the bearing is fully seated, support the inner bearing race, and press the hub back into the new bearing. Failure to properly support the bearing will cause the bearing to be damaged or destroyed when you press the hub flange back into place. Once the hub is pressed back into the bearing, ensure the bearing spins smoothly. Reinstall any seals and retaining rings and reassemble the steering knuckle to the vehicle.

Some FWD vehicles and trucks have bolt-in hub and bearing units (**Figure 5-81**). This type of bearing does not require the use of a press because the entire hub is also replaced, which often makes replacement

Service Warning

A hydraulic press develops thousands of pounds of pressure when it is in use. Follow all safety and service precautions provided by the manufacturer when you are using a hydraulic press.

FIGURE 5-80 This kit is used to remove and replace pressed-in wheel bearings with the steering knuckle still attached to the vehicle.

FIGURE 5-81 An example of a hub and bearing unit on a FWD car.

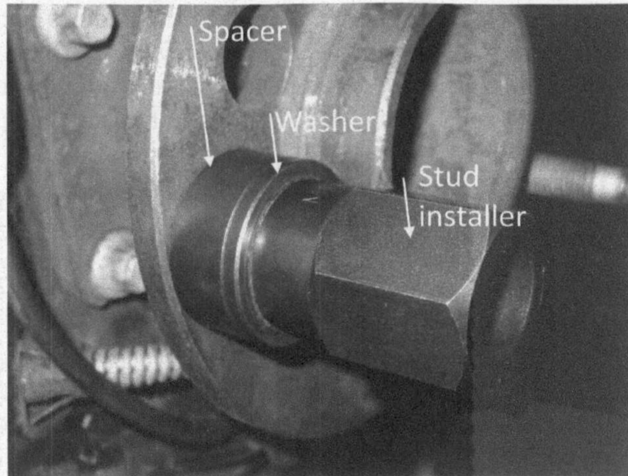

1. Place the new stud through the opening in the flange.
2. Install the spacer, washer, and stud installer over the new stud.
3. Tighten the installer with a wrench or socket until the new stud is fully seated.
4. Loosen and remove the stud installer, washer, and spacer.

FIGURE 5-82 Pressing a new stud into a hub flange using a stud press.

simpler and the labor less costly. To replace a bolt-in unit, follow all the manufacturer service procedures. In general, the steps to replace this type of bearing are as follows:

1. Lift and support the vehicle, and remove the wheel and tire assembly.
2. Remove the axle nut.
3. Remove either a ball joint to steering knuckle connection or the lower strut to steering knuckle bolts. Separate the knuckle from either the lower control arm or strut.
4. Remove the axle shaft from the wheel bearing. This may require using a puller to push the axle out of the bearing.
5. Remove the bolts securing the bearing to the steering knuckle, and remove the bearing. If the bearing also contains the ABS wheel speed sensor, disconnect the electrical connection from the bearing.
6. Verify the replacement bearing is correct.
7. Install the new bearing and reassemble the components in reverse order.
8. Torque all fasteners to specifications.
9. Check and adjust the wheel alignment.
10. Test drive the vehicle.

■ ***Inspect and Replace Wheel Studs.*** Other common problems are broken or damaged wheel studs. Broken studs are caused by overtightened wheel fasteners or by a cross-threaded lug nut that, when tightened, binds on the stud and breaks it off. Damaged threads on the wheel studs can be caused by someone attempting to install the incorrect lug nuts, by cross-threading the lug nuts on the studs, or by using an air impact gun to start the lug nuts instead of starting them by hand.

If the threads on a wheel stud are only slightly damaged, you may be able to correct them using a thread file. Match the threads of the stud with the correct side of the file and, using firm pressure, file the threads until they are restored to their normal shape. Be sure you keep the file exactly in line with the threads so that you do not cause further damage to the threads.

If the threads are severely damaged, you may need to use a tap and die set to fix the threads. Select the correct die for the stud, and place it into the die handle. Carefully start the die over the threads and work the die down the entire length of the threads, stopping every half turn to back the die up before resuming your downward motion.

A seriously damaged stud will need to be replaced. On some vehicles, the stud can be driven out with a hammer and punch or by using a wheel stud press. To install the new stud, use a stud press (**Figure 5-82**). You can also use a lug nut and flat washers to pull the new stud into place. Depending on how the hub is mounted, you may be able to easily replace the stud without removing the hub or flange. On some vehicles, you will have to remove the hub to remove and install the stud. This is because there is not enough room behind the hub flange to remove the stud without backing into the steering knuckle.

SUMMARY

The air inside the tire acts against the inner surface of the tire to support the weight of the vehicle.

Radial tires, having belts around the circumference of the tire, are the most common type of tire used.

Run-flat tires may contain an additional support ring or have extra stiff side-walls to support the vehicle's weight if air pressure is lost.

Information regarding tire size, quality, and loading is imprinted on the tire sidewall.

Misalignment of the belts in the tire can cause the tire to pull to one side.

A tire inspection should consist of a visual check of the treadwear, sidewall condition, and air pressure.

Periodic tire rotations help extend the life of the tire.

Wheel and tire assemblies that are statically out-of-balance can cause wheel hop.

Dynamic wheel and tire out-of-balance causes wheel shimmy.

Direct TPM systems use pressure sensors mounted in each wheel to measure tire pressure.

Indirect TPM systems use the antilock brake wheel speed sensors to determine if tire pressure is too low.

A TPM reset or relearn should be performed any time tire services are performed.

Bearings are used to reduce friction between components.

REVIEW QUESTIONS

1. The _______________ section of the wheel is where it matches to the hub of the vehicle.
2. When a vehicle with TPMS has the tires rotated and the tire pressure set, a TPMS_______________ should be performed.
3. A tire that is _______________ out-of-balance can cause a side-to-side shaking of the steering wheel.
4. The last two numbers in the tire size indicates the _______________ diameter.
5. A tire that is _______________ out-of-balance can cause the wheel/tire to tramp or hop.
6. Which of the following is not a part of a tire's size designation??
   - **a.** Rim diameter in inches
   - **b.** Tread width in inches
   - **c.** Aspect ratio
   - **d.** Tire design type
7. Which of the following are used with an indirect TPM system?
   - **a.** Pressure sensor
   - **b.** Speed sensor
   - **c.** Wireless receiver
   - **d.** Wireless transmitter
8. *Technician A* says that any high-pressure, high-temperature grease can be used to repack tapered wheel bearings. *Technician B* says that some vehicles require a specific grease for the wheel bearings. Who is correct?
   - **a.** Technician A
   - **b.** Technician B
   - **c.** Both A and B
   - **d.** Neither A nor B
9. Two technicians are discussing the cause of a steering wheel shimmy while driving on the highway: *Technician A* says a wheel may be bent. *Technician B* says the tires may be out-of-balance. Who is correct?
   - **a.** Technician A
   - **b.** Technician B
   - **c.** Both A and B
   - **d.** Neither A nor B
10. *Technician A* says a TPMS relearn should be performed as part of a tire rotation. *Technician B* says some vehicles do not require a TPMS relearn after a tire rotation. Who is correct?
    - **a.** Technician A
    - **b.** Technician B
    - **c.** Both A and B
    - **d.** Neither A nor B

CHAPTER 6

Suspension System Principles

Chapter Objectives

At the conclusion of this chapter, you should be able to:

- Describe the functions and operational principles of modern suspension systems.
- Identify the types of front and rear suspensions.
- Identify the components and their functions of the front and rear suspension systems.
- Explain how changes in wheels and tires affect the suspension.
- Explain the basic operation of electronically controlled suspension systems.

KEY TERMS

air springs
coil spring
dead axle
dependent suspension
frame
frequency
independent suspension
jounce
leaf spring
live axle
load-carrying ball joint
MacPherson strut
modified strut
multilink
oversteer
rebound
semi-independent
shock absorber
spring rate
sprung weight
stabilizer bar
torsion bar
track bar
understeer
unibody
unsprung weight

Functions and Basic Principles

Suspension systems have evolved significantly since the earliest adaptations from horse-drawn buggies to self-powered automobiles, but the basic requirements remain the same. The Duryea was a very early automobile, and not too different than a horse buggy (**Figure 6-1**). Note the very simple suspension and steering systems, capable of safe operation of only a few miles per hour. Just as in the horse and buggy days, today's suspension systems must provide for safe handling and maximum traction while being able to sustain passenger comfort. To accomplish these goals, modern suspension systems rely on various types of springs, shock absorbers (dampers), control arms, and other components.

All of the components of the suspension system must work together to provide the proper ride quality and handling characteristics expected by the driver and passengers. Each component is engineered to work as a part of the overall system. If one part of the system fails, it can lead to faster wear or damage to other components. Therefore, a complete understanding of each component and how it functions as part of the whole suspension system is critical.

FUNCTIONS OF THE SUSPENSION SYSTEM

All suspension systems have the same basic functions to perform, regardless of the type of suspension system used on the vehicle.

- The tires must be able to rise and fall, relative to the body, to allow the springs and dampers to reduce bump and road shock (**Figure 6-2**).
- The suspension allows the springs and dampers to absorb the energy of a bump for a smooth ride while not allowing uncontrolled movement of the tires.

FIGURE 6-1 For many years, suspensions were very basic, such as on this 1896 Duryea.

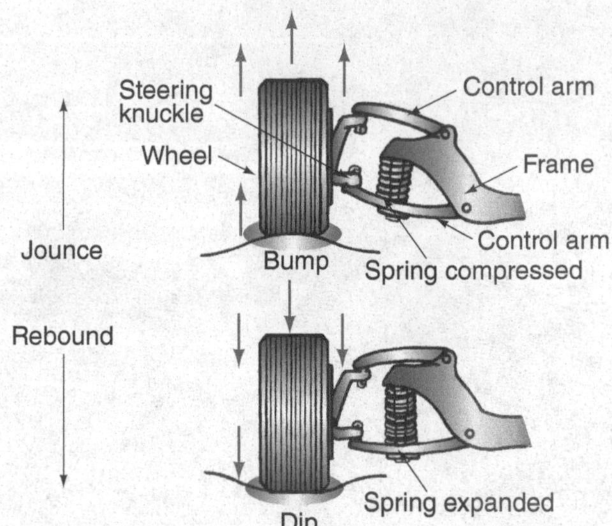

FIGURE 6-2 The suspension must allow for the upward and downward movements of the tires over irregular road surfaces.

- The suspension must handle movements caused by vehicle acceleration, braking, and cornering.
- The springs must be able to safely carry the weight of the vehicle. **Figure 6-3** shows an example of a failed spring and its obvious effect on ride height. The failure of the spring not only affects the ride height for that corner of the vehicle, the spring's ability to carry weight is now gone. This results in a very rough ride and increases the loads and stresses placed on other components.
- It is important for the suspension to keep the alignment of the tire as correct as possible so that maximum contact is maintained between the tire and the road.
- The rear suspension must carry the weight of the rear of the vehicle and any additional loads in the trunk, cargo area, or bed.

Regardless of type, the suspension carries the weight of the vehicle. Through the springs and other suspension components, the weight of the vehicle and its occupants is transferred to the wheels and tires. While this may sound simple, carrying around a two- or four-ton vehicle

FIGURE 6-3 A weak spring has a very noticeable effect on ride height and will affect ride quality.

is not an easy task. Not only is the vehicle's weight a load, but the additional forces of cornering, braking, accelerating, and negotiating every bump and dip in the road are applied to the suspension and tires. Engineers must balance weight carrying, ride control, comfort, and handling when they are designing a vehicle.

VEHICLE FRAMES

In use for decades, the body-over-frame design has the vehicle body as a separate component that, when assembled, is bolted to the frame. The **frame** is often ladder shaped, with two long frame rails that run the length of the vehicle and several crossmembers attached to the frame rails (**Figure 6-4**). The crossmembers carry the engine, transmission, and front suspension and tie the frame rails together. The front suspension is bolted to the frame and front crossmember. The rear suspension bolts to the rear of the frame and rear axle. The combined weight of the body, frame, passengers, and any other loads push down on the springs, which in turn pass the weight through parts of the suspension and finally to the tires. Rubber bushings are placed between the frame and the body to help isolate noise, vibration, and harshness (NVH) from the suspension system. This body-over-frame design continues to be

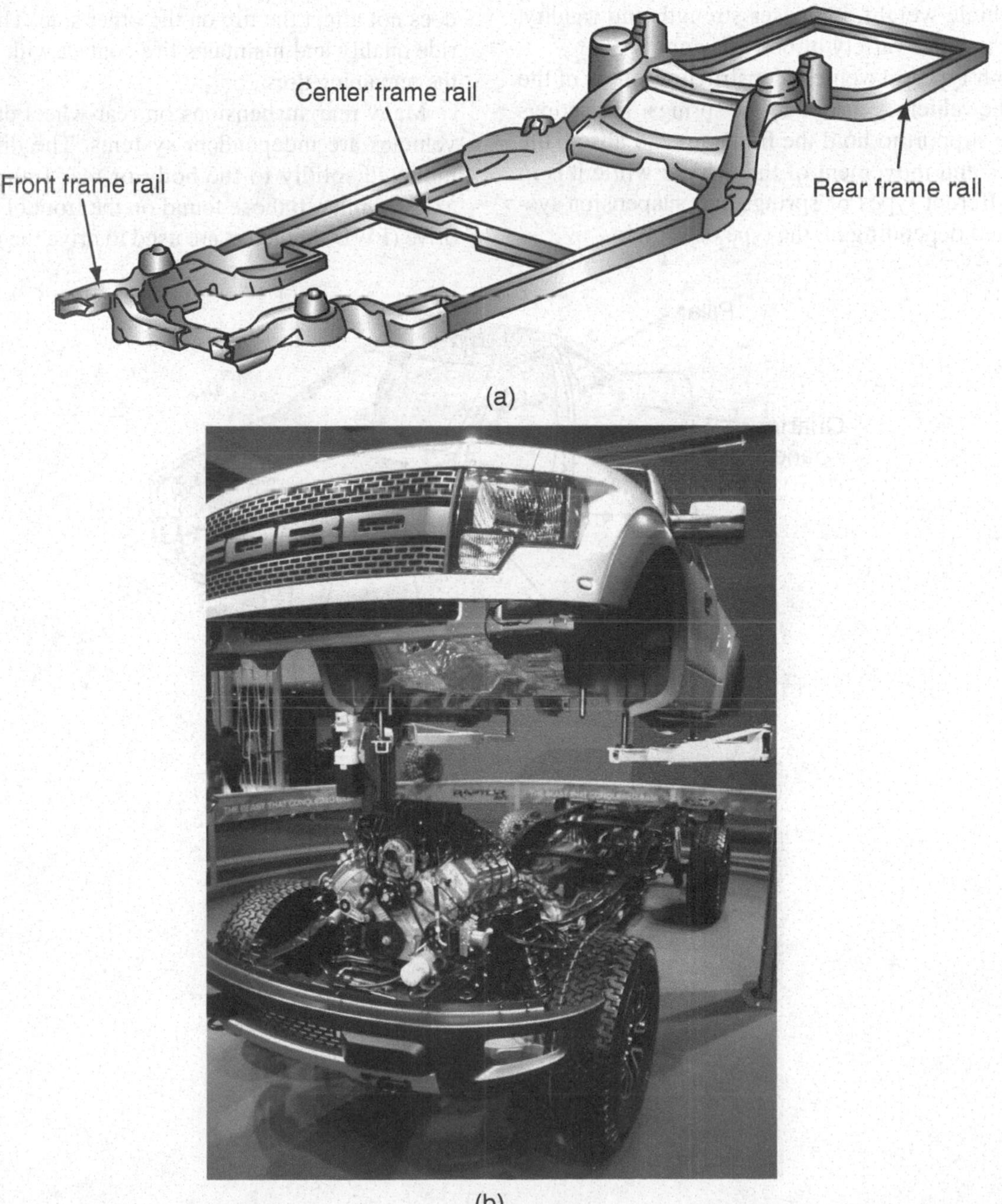

FIGURE 6-4 (a) A ladder frame has side rails and crossmembers to build a strong foundation for the vehicle. (b) Today, only full-size trucks and larger SUVs use ladder frames.

used today on trucks and some SUVs. Body-over-frame construction is strong but is also heavy and can allow unwanted flexing or twisting of the frame.

Most modern cars and small SUVs use a space frame or unitized body. This design does not have a separate frame. The body is constructed of many parts that are then assembled into a single unit, which is also the frame. This is called **unibody** construction (**Figure 6-5**). Once assembled, the outer body panels are attached to the unibody. The front and rear suspension systems are attached to the space frame with insulating rubber bushings, as in body-over-frame vehicles. This reduces the transmission of noise and vibration. The unibody frame reduces vehicle weight, increases strength and rigidity, and allows a wider variety of body designs.

With both types of vehicle construction, most of the weight of the vehicle is carried by the springs. The springs provide the support to hold the frame up and absorb the road shocks and movement of the vehicle while it is in motion. Different types of springs and suspension systems are used depending on the type of vehicle.

■ ***Independent Suspensions.*** To provide the best possible ride quality, many vehicles use fully independent front and rear suspension systems. This allows the vehicle to respond to varying road conditions much more effectively. Nearly all front suspensions found on modern cars and light trucks are independent. Even four-wheel drive (4WD) vehicles often have independent front suspensions to improve their ride and handling qualities. The top image in **Figure 6-6** shows how each wheel is able to move in an independent suspension while the bottom image illustrates the movements of a dependent or rigid axle. In an **independent suspension**, each wheel can move independently, so a bump on one side of the vehicle does not affect the tire on the other side. This improves ride quality and maintains tire contact with the road for the remaining tires.

Many rear suspensions on rear-wheel drive (RWD) vehicles are independent systems. The differential is mounted solidly to the body or rear frame and short axles, similar to those found on the front of front-wheel drive (FWD) vehicles, are used to drive the rear wheels.

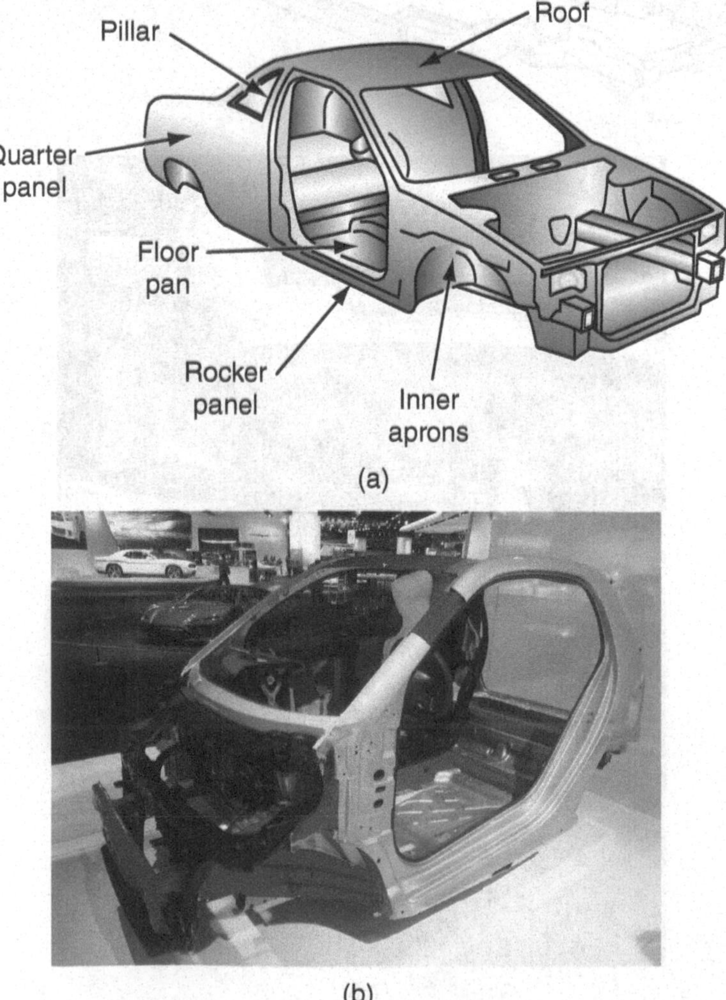

FIGURE 6-5 (a) Unibody frames provide the structure of the vehicle and are very rigid. (b) Modern cars, SUVs, crossovers, and even some light trucks are unibody construction.

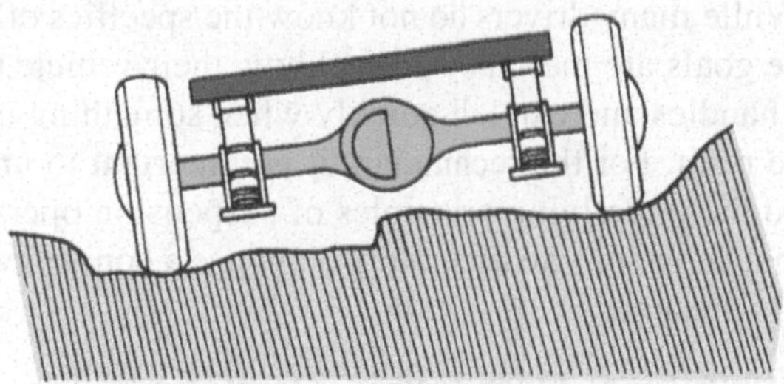

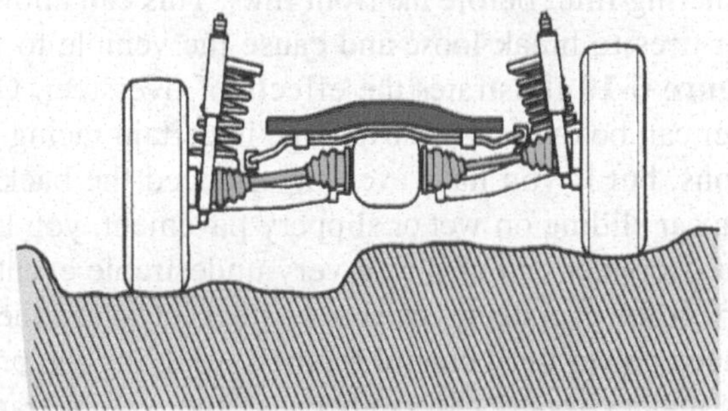

FIGURE 6-6 Rigid and independent suspensions and their effect on wheel movement. Independent suspensions allow much improved ride quality and handling compared to dependent suspensions.

This provides improved ride quality and handling. These suspension types are discussed later in this chapter. Many FWD cars have independent rear suspension systems as well. This improves ride quality and handling. Independent rear suspension arrangements also allow for lower overall vehicle ride height, which improves handling and aerodynamics.

■ ***Dependent Suspensions.*** Still found on the rear of many vehicles and on the front of most heavy-duty vehicles, **dependent suspensions** sacrifice ride quality for strength. Because the movement of one wheel affects the opposite wheel, ride quality, and handling suffer on these systems. A large, straight I-beam is often used on the front of heavy-duty vehicles, such as buses and semi trucks. An example of the front of a heavy-duty 4WD truck is shown in **Figure 6-7**. This design is used for its strength and durability but does not provide the best ride quality.

The rear axle on many RWD cars, light trucks, and SUVs is a dependent live axle (**Figure 6-8**). In this arrangement, the dependent **live axle** is the primary rear suspension member and is driving the rear wheels. Because a live axle is one large assembly it is a dependent system. Live rear axles are mounted on leaf springs, coil springs, or air springs.

A vehicle with a solid rear axle that does not drive the rear wheels has what is called a **dead axle** or a rigid or straight axle. An example of this type of rear suspension is shown in **Figure 6-9**. A dead axle supports the weight of the rear of the vehicle and can be fitted with coil, leaf, or air springs. A dead axle is a dependent form of suspension.

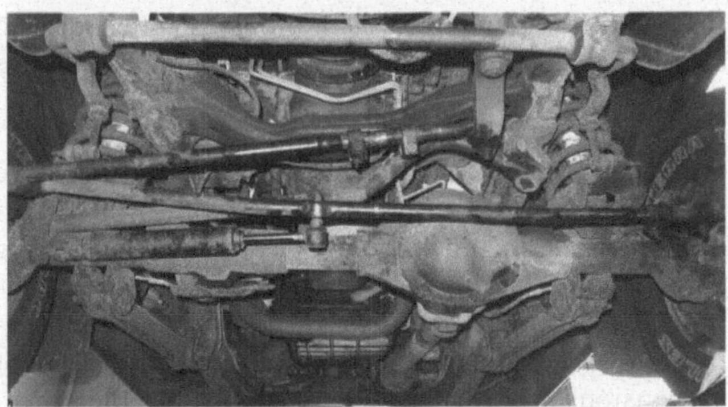

FIGURE 6-7 Some heavy-duty 4WD trucks have live axle front suspension systems.

FIGURE 6-8 A live axle is used to drive the rear wheels and act as part of the rear suspension.

■ ***Semi-Independent Suspensions.*** Found on the rear of many FWD vehicles, this type of system uses a fixed rear axle that twists slightly under loads. This allows for semi-independent movement of the rear wheels. This system typically uses coil springs or struts.

FIGURE 6-9 An example of a nondriving dependent rear suspension common on FWD vehicles.

The **semi-independent** system provides better ride and handling than a straight axle while not being as costly as a fully independent system.

■ ***Front Suspensions.*** The main purpose of the front suspension is to provide safe, comfortable handling while allowing wheel movement for the steering and enabling the driver to react to various road conditions. To accomplish this, several different front suspension styles are used in modern vehicles. The front suspensions on FWD vehicles also must be able to handle the additional torque of driving the front wheels. Additionally, during braking, as much as 70% of the vehicle weight is transferred to the front, adding additional loads to the front suspension.

Vehicle type and intended use are the main considerations when engineers begin to design the suspension systems. Many cars have suspensions that look very similar, but actually have many differences. The exact size and placement of components have a large effect on individual vehicle driving characteristics.

■ ***Rear Suspensions.*** The rear suspension must be able to carry any additional loads placed in the rear of the vehicle while still maintaining the correct ride height. The rear suspensions on many FWD and RWD vehicles are similar in that a solid type of axle is used. Though strong, a solid axle does not provide the level of handling and ride quality that an independent rear suspension does. The rear suspension on RWD vehicles must be able to handle the torque of the driveline. This can be difficult because torque tries to twist the vehicle and rear suspension.

BASIC PRINCIPLES

The components of the suspension system, while important separately, must operate as a whole for the system to provide all of the requirements during normal driving conditions. As stated before, the suspension is responsible for carrying vehicle weight, absorbing road shocks, providing a smooth ride, and allowing good handling qualities.

While many drivers do not know the specifics of how these goals are met, they do feel how their vehicle rides and handles and can tell quickly when something is not quite right. For the technician, it is important to understand the underlying principles of suspension operation so that he or she can accurately diagnose a concern when one is present.

■ ***Oversteer.*** **Oversteer** is a term used to describe a driving condition where the rear tires reach their cornering limit before the front tires. This can allow the rear tires to break loose and cause the vehicle to spin. **Figure 6-10** illustrates the effects of oversteer. Oversteer can be used as an advantage in certain racing situations, but if you have ever experienced the back end of a car sliding on wet or slippery pavement, you know that oversteer can also be a very undesirable event! To correct for oversteer, you should steer into the slide and reduce power until control returns. Applying the brakes can make oversteer worse because the weight transfer from the rear wheels can reduce rear tire traction.

■ ***Understeer.*** The opposite of oversteer is **understeer**. This condition occurs when the front of the vehicle cannot make a turn through the desired turn radius because the front tires have lost traction. **Figure 6-11** shows how a vehicle will continue in a somewhat straight line instead of making the intended turn. This causes the vehicle to overshoot the turn. If you have ever tried to make a turn in slippery or snowy conditions and the vehicle continued in a straight line instead of turning the corner, you have experienced understeer.

Understeer is measured by the difference between the angle the tires are pointing and the angle needed to make the turn. Most cars are designed to have understeer. This is because understeer can be reduced by reducing vehicle speed, which is safer for the average driver.

■ ***Neutral Steering.*** If a vehicle turns at the same rate that the steering wheel is turned, it is said to have neutral steering. This means that the vehicle does not exhibit a tendency to either over- or understeer.

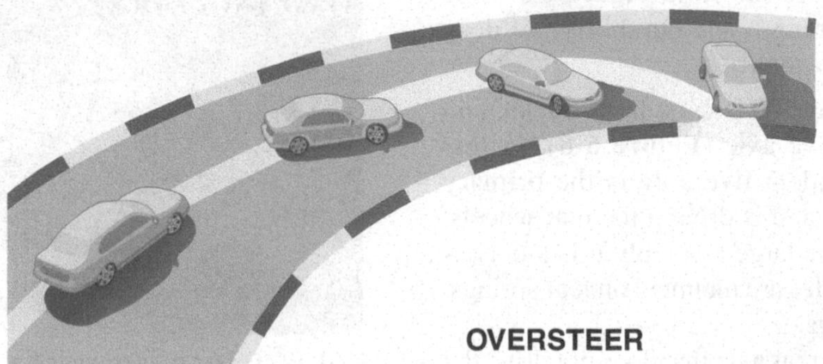

FIGURE 6-10 Oversteer occurs when the back of the vehicle slides out and puts the vehicle into a spin.

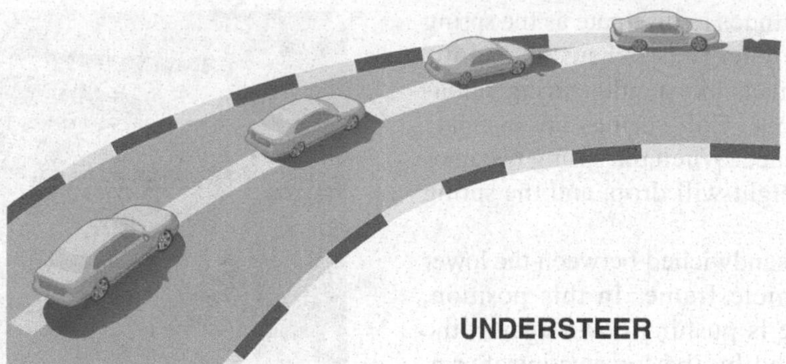

FIGURE 6-11 Understeer occurs when the front wheels cannot provide enough traction to move the vehicle through the desired turn.

■ ***Lateral Acceleration.*** Lateral acceleration is the measurement of the vehicle's ability to corner. What we feel during a corner is that a force pushes the vehicle and its occupants to the outside of a turn. In reality, as both the car and occupants turn, the people inside are still subject to Newton's First Law of Motion and continue to move in a straight line. The effect is that we feel pushed toward the outside of the corner. Centripetal force, meaning "toward the center," is the force that pulls an object toward the center of a circle as the object rotates. Imagine swinging a ball over your head and that the ball is attached to a string. The ball travels in a circle because the centripetal force is pulling the ball toward the center. Obviously, cars do not have strings pulling them in toward the center of a circle while turning but they do have tires. The tires are exerting the force toward the center. The lateral (sideways) force is perpendicular to the direction the car is traveling. This is where the term lateral acceleration comes from for vehicle test scores. The test is performed by driving the car on a large test-track circle at ever-increasing speeds. The faster the car can go around the circle, the greater the lateral acceleration. This means the better the vehicle will handle when cornering. **Figure 6-12** shows an illustration of how this test is performed. Low riding, wide wheelbase sports cars can achieve a much higher lateral acceleration than a vehicle that is higher off the ground, such as a minivan or SUV.

SPRINGS

The springs in the suspension have two important functions. Springs support the vehicle weight and absorb the bumps and movements that occur when driving. There are four types of springs used in suspension systems.

Coil springs—are a length of steel wound into a coil shape. Used on most front and many rear suspensions, coil springs are large pieces of round steel formed into a coil (**Figure 6-13**). The spring absorbs energy as the coils are forced closer together. This is called *compression*. The stored energy is released when the coil extends

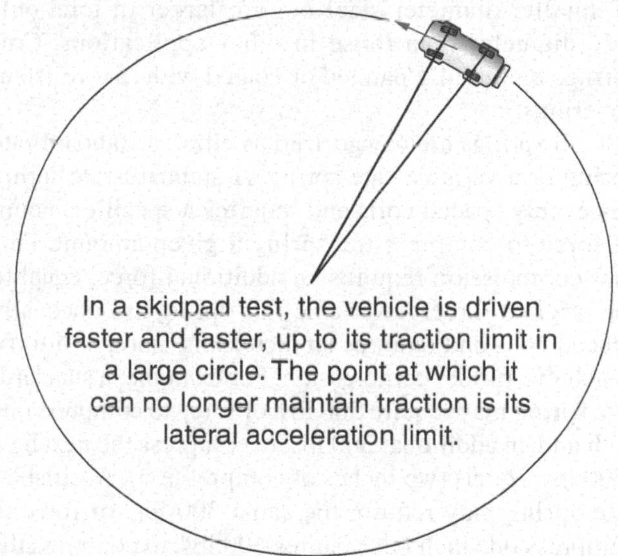

In a skidpad test, the vehicle is driven faster and faster, up to its traction limit in a large circle. The point at which it can no longer maintain traction is its lateral acceleration limit.

FIGURE 6-12 Lateral acceleration tests a vehicle's ability to grip the road and corner.

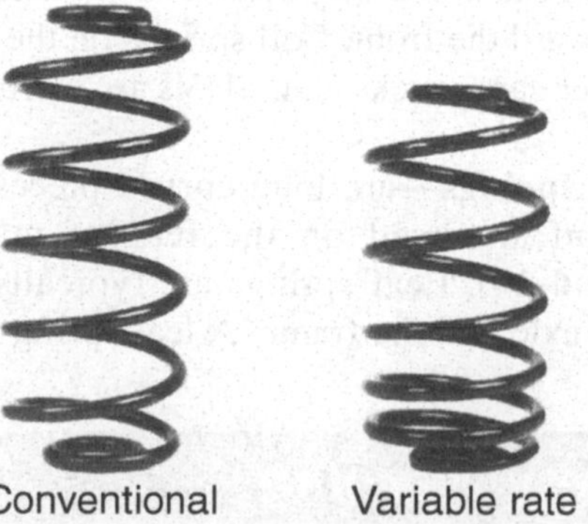

FIGURE 6-13 An example of two types of coil springs. The spring on the left is a standard- or linear-rate spring. The spring on the right is a variable-rate rear coil spring with unequally spaced coils.

back out. The energy continues to dissipate as the spring bounces. Eventually, the energy is exhausted and the spring stops bouncing. This is like a guitar string vibrating until it eventually stops. Coil springs are compact and do not need maintenance. When the spring becomes fatigued or weak, ride height will drop, and the spring will need to be replaced.

Coil springs are often sandwiched between the lower control arm and the vehicle frame. In this position, the weight of the vehicle is pushing down against the spring, which is supported by the lower control arm. This configuration allows movement of the suspension while the spring carries the weight and dampens out road shock. Coil springs often use rubber insulators between the spring and the frame to reduce noise.

The coil springs used in strut suspensions appear similar to those used in other applications, but are not interchangeable. Most strut coil springs are made of smaller diameter steel but are larger in total outside diameter than those in other applications. Coil springs are usually painted or coated with rust-resistant coverings.

Coil springs are categorized as either a standard-rate spring or a variable-rate spring. A standard-rate spring has evenly spaced coils and requires a specific amount of force to compress the spring a given amount. Further compression requires an additional force, equal to the original force. A variable-rate spring has unequally spaced coils and requires an increasing amount of force to achieve further compression. For example, a standard-rate spring may require 300 lbs. of force to compress one inch and an additional 300 lbs. to compress the next inch (600 lbs. equals two inches of compression). A variable-rate spring may require the same 300 lbs. of force to compress one inch but requires 500 lbs. to compress the next inch (800 lbs. equals two inches).

Coil springs used in passenger car rear suspensions are usually lighter duty than those found at the front. This is because the majority of the vehicle's weight is often toward the front. Coil springs on the rear of larger passenger cars, trucks, and SUVs are often variable-rate springs.

Leaf springs—are long curved pieces of flattened steel and are used on the rear of many vehicles (**Figure 6-14**). Leaf springs are typically mounted to the rear axle and the frame. A leaf spring is a long, flat piece of spring steel, shaped into a semicircle. The spring is attached to the frame through a shackle or bracket assembly that permits changes in the effective length of the spring as it is compressed. To carry heavier loads, additional leaves can be stacked below the master leaf. Increasing the number of leaves increases load carrying capacity but makes the ride stiffer. Some suspensions use transverse leaf springs that are mounted perpendicular to the frame. In a transverse arrangement, one leaf spring supports both sides of the suspension. This style was used for many years on the Corvette and on some FWD vehicles with independent rear suspensions.

FIGURE 6-14 Leaf springs are used in rear suspensions and can be placed either above or below the axle.

Safety Warning

Coil springs contain stored energy. Improper service of the suspension or coil springs can release the spring, causing serious injury.

Safety Warning

Vehicles with air springs often require special lifting and jacking procedures. Do not attempt to raise a vehicle with air springs until you have read and followed all applicable warnings and procedures.

Air springs—are thick, tough bags filled with air that act as springs. Air springs are used on some trucks, SUVs, sedans, and most large commercial semi trucks and trailers. Air springs may be used on just rear or on both the front and the rear (**Figure 6-15**). Like torsion bars, air springs are adjustable. On many vehicles, the on-board computer system uses a ride height sensor to determine suspension load. As additional weight is added to the trunk, the suspension will drop. When the computer senses this drop, it can turn on an on-board air compressor to supply more air to the air springs. The increased pressure in the springs will restore the ride height to the desired position. Some systems may use the adjustment of air pressure to the air springs to control ride height based on the vehicle's speed or driver input.

Torsion bars—are lengths of round steel bar fastened to a control arm on one end and the frame on the other end. Movement of the control arm causes the torsion bar

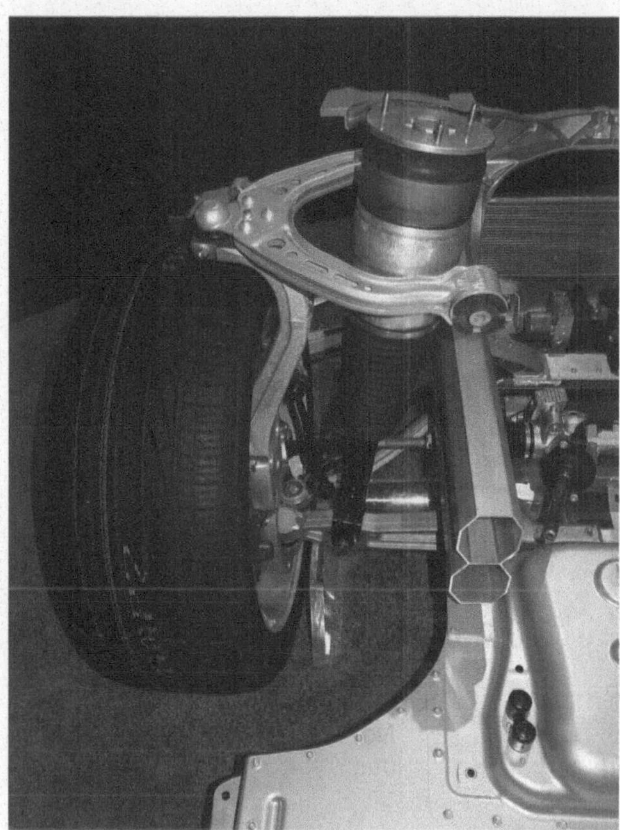

FIGURE 6-15 Air springs are often used in luxury cars, in some rear suspensions and are very common on heavy-duty trucks.

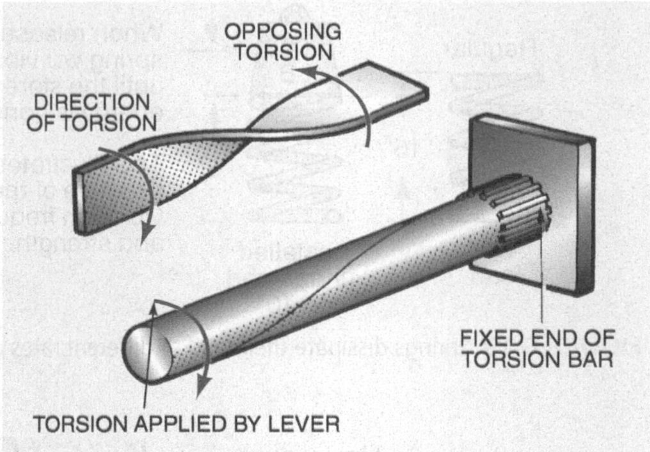

FIGURE 6-16 Torsion bars are long, straight steel bars that twist to absorb energy.

FIGURE 6-17 A torsion bar used on the front of a 4WD truck.

to twist (**Figure 6-16**). The absorption of the twist is similar to compression of a coil spring. As the torsion bar untwists, the control arm returns to its normal position. Torsion bars are used in many 4WD vehicles where a front drive axle occupies the space where the coil spring normally sits. The torsion bar is typically mounted to the lower control arm and the transmission crossmember (**Figure 6-17**). The control arm acts as a lever against the torsion bar, twisting the bar. The bar twists since it is rigidly mounted in a crossmember. As it releases energy and untwists, the torsion bar returns to its original shape, forcing the control arm back into position.

An advantage of torsion bars is that they are adjustable. At the rear torsion bar mount is an adjustment mechanism. If a torsion bar–equipped vehicle is sagging, the torsion bar may be able to be adjusted to bring the vehicle back into specification. When a torsion bar is replaced, it must be tightened to provide the necessary lift to support the vehicle.

■ ***Spring Ratings.*** Automotive springs are rated for their frequency and their load rate. When installed, springs are compressed and are under tension and they store energy. Hitting bumps in the road further compress the springs. When the tension is released, the springs will attempt to return to their original condition. Upward movement of the wheel that compresses the spring is called **jounce**. The stored energy is released when the spring rebounds. This downward movement of the tire, as the spring extends out, is called **rebound**.

As you probably know, a compressed spring will rebound many times before all of the energy is dissipated. The number of times a spring bounces (or oscillates) before returning to its rest point is called *the spring **frequency*** (**Figure 6-18**). The size of the spring and the spring material contribute to the spring frequency. Ideally, a spring should dampen out its oscillations quickly enough to provide a smooth ride but not so fast that it causes a harsh, jarring ride. If left to bounce or oscillate on its own, the spring will cause the vehicle to bounce excessively, probably to the discomfort of the passengers.

The amount of force it takes to compress or twist a spring a certain amount is called **spring rate**. Springs can have either linear or variable rates. **Figure 6-19** illustrates the differences in spring rate. When a vehicle is designed, the engineers will factor spring size, rate, and frequency based on the intended vehicle use, tire size and type, suspension style, and many other factors. The

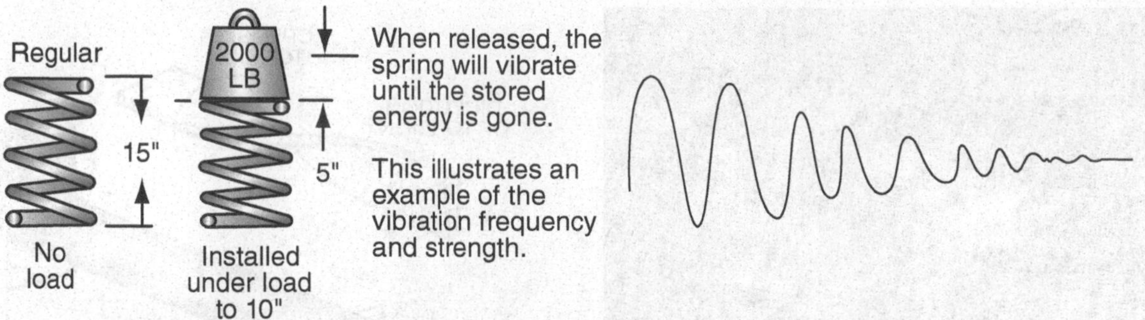

FIGURE 6-18 Springs dissipate their load at different rates depending on the size of the spring.

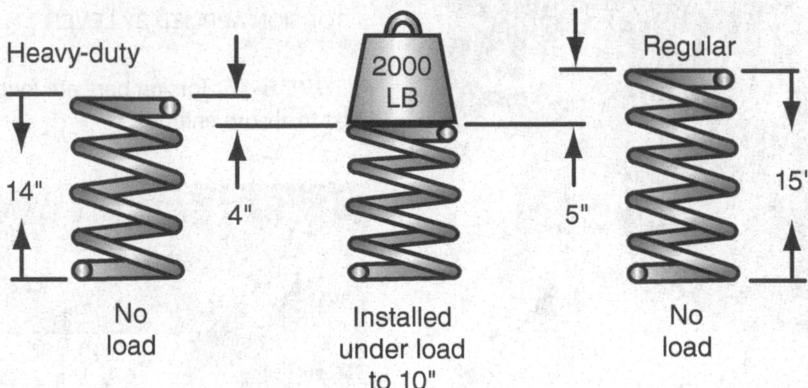

FIGURE 6-19 Standard-rate springs will compress at a linear rate under load. Variable-rate springs stiffen as the load increases.

goal is to have the best compromise between component weight, vehicle cost, and the ride and handling qualities desired for the vehicle.

■ *Sprung and Unsprung Weight.* Weight carried by the springs is called **sprung weight**. Weight not carried by the springs is called **unsprung weight**. The less unsprung weight a vehicle has, the better the handling and ride will be. Some examples of unsprung weight are the wheels and tires, brake components, control arms, steering knuckles, and rear axles. **Figure 6-20** shows examples of unsprung weight. Sprung weight includes the vehicle body, engine and transmission, the passengers, and in general, is an item above the axles (**Figure 6-21**). The amount of sprung weight should be high and unsprung weight should be low.

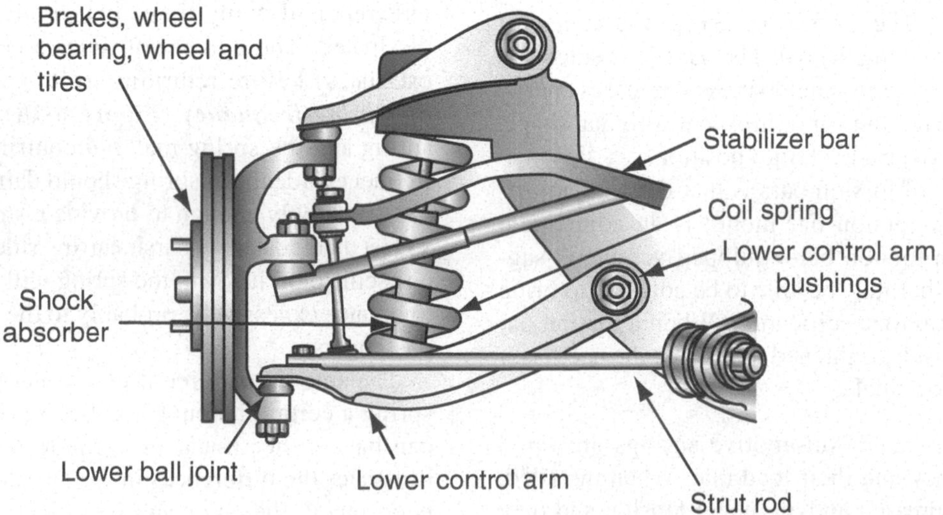

FIGURE 6-20 Unsprung weight is generally a component below the axles. The less unsprung weight, the better the vehicle will ride and handle.

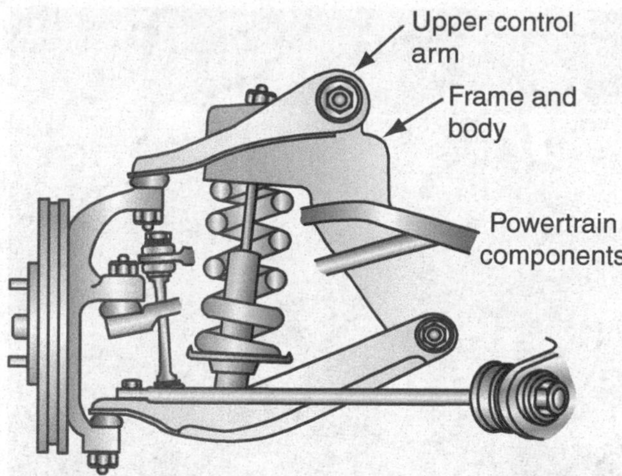

FIGURE 6-21 Sprung weight is the weight carried by the springs.

■ ***Shock Absorbers.*** **Shock absorbers** are actually dampers, meaning that they reduce or make something less intense. The springs do the shock absorbing while the shocks dampen the spring oscillations. Without the dampers, vehicles would continue to bounce for a long time after every bump, dip, and change in body movement.

The most common type of damper is the direct double-acting hydraulic unit. This means that they are used to directly act on motion; double acting means that they work in both compression and extension modes, and hydraulic means that a fluid is used to perform work. Compression is upward wheel travel, also called jounce. Extension is downward wheel motion and is also called rebound.

Dampers are typically mounted near the springs, with the lower end mounted on a lower control arm or axle. A front suspension and damper is shown in **Figure 6-22** and rear suspension in **Figure 6-23**. The top of the damper, which is connected to the piston, is mounted to

FIGURE 6-22 Dampers are mounted close to the springs. Exactly how and where varies extensively from vehicle to vehicle.

FIGURE 6-23 An example of a rear damper.

the vehicle body. Inside are two chambers, each partially filled with oil **(Figure 6-24)**. The piston moves up and down in the main chamber. This movement displaces the oil into a second chamber. A set of one-way valves control the flow of oil from the chambers. Moving the oil is difficult. This is where the damper's resistance to movement comes from. Allowing more oil to flow will provide less dampening and a smoother ride. By restricting oil flow, the damper will be more resistant to movement and provide a stiffer ride.

A damper may have an equal amount of resistance during both compression and extension, or it may have more resistance during extension. This is because the spring naturally resists compression, and the damper does not need to add much resistance to that of the spring. Because the spring will easily extend out, the damper's greater resistance on extension can help better control spring action.

■ ***Control Arms.*** Control arms are used to control wheel movement. Used on both front and rear suspensions, they are commonly referred to by their position, such as the upper and lower control arms. Common control arm configurations are shown in **Figure 6-25** and **Figure 6-26**. Control arms are also called A-arms or wishbones due to their similarity to being A or wishbone shaped.

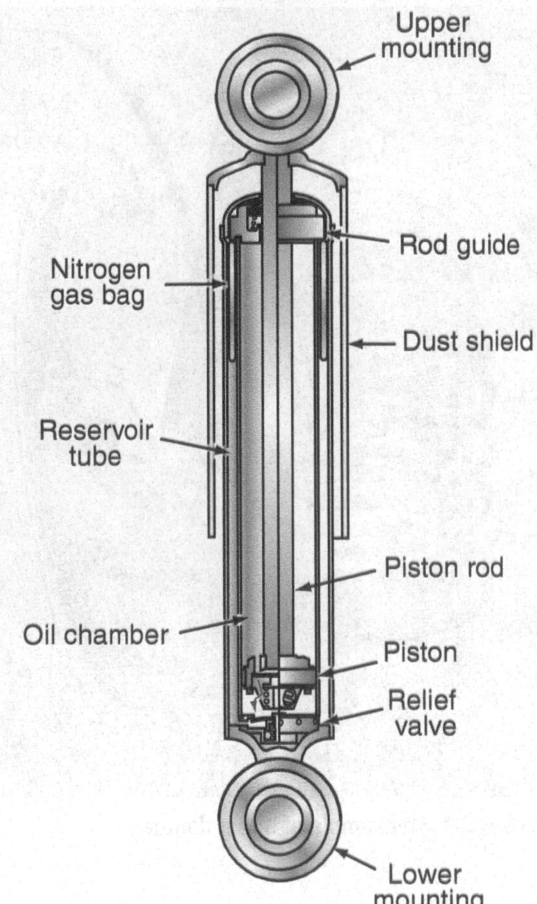

FIGURE 6-24 The damper contains the piston, valves, and oil. Oil flow through the valves controls the resistance of the movement of the piston.

FIGURE 6-25 Lower control arms connect the steering knuckle to the frame and come in a wide variety of sizes and designs. Many are open or wishbone shaped.

A-arms typically have two connections to the frame and a ball joint for connecting to the steering knuckle. The control arm mounts to the frame with bushings (**Figure 6-27**). These bushings allow for up and down movement of the arm while controlling back and forth motion. The bushings are generally rubber and steel and are pressed into the control arms. In addition to acting as pivots for the control arms, the bushings act as dampers, twisting and untwisting to return the control arm to its original position.

FIGURE 6-26 Another example of a front lower control arm.

Some suspensions use a lower control arm with a single frame mounting point. In this case, a strut rod will also be used as an additional mount and stabilizer for the control arm as shown in **Figure 6-28**.

Also connected to lower control arms are the stabilizer bar links. The stabilizer bar links join the lower control arms to the stabilizer bar. These links can be a set of bushings and washers or a solid link with ball-and-socket joints.

■ ***Ball Joints.*** Ball joints allow the steering knuckle to turn while providing a tight connection to the control arms and prevent any up and down or sideways movement. Ball joints use a ball-and-socket joint to allow a wide range of motion, similar to a shoulder or hip joint. An illustration of a ball joint is shown in **Figure 6-29**.

Ball joints can be one of two types, load carrying or nonload carrying. **Load-carrying ball joints** support the weight carried by the springs. Because of this, these joints tend to wear faster and need replacement more often than nonload-carrying joints. Nonload-carrying joints provide a steering pivot and component connection with a wide range of movement just like load-carrying joints, but without the sprung weight applied to them. **Figure 6-30** illustrates how weight is carried by a ball joint.

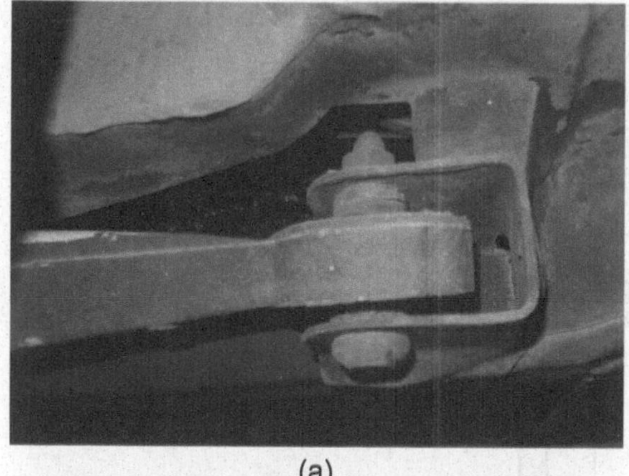

(a)

(b)

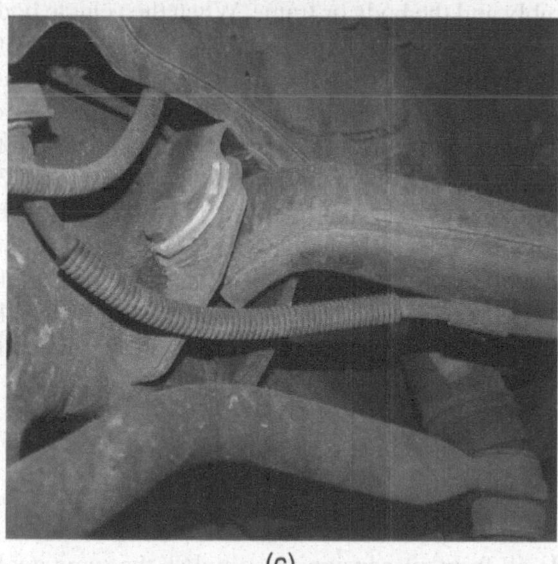

(c)

FIGURE 6-27 Control arm bushings also come in different shapes and styles. (a) A common lower control arm bushing on a FWD. This bushing helps reduce longitudinal movements of the control arm. (b) Another common FWD lower control arm bushing. This bushing limits lateral movements. (c) A common upper control arm busing arrangement.

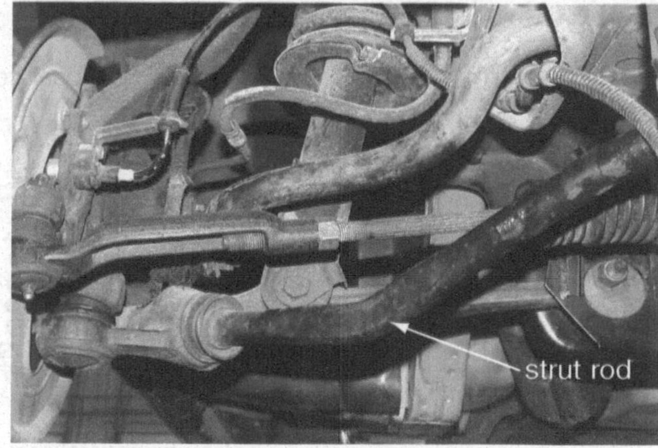

FIGURE 6-28 A strut rod is used when the lower control arm has only one mounting location. The strut rod forms part of the control arm and prevents forward and backward movement of the tire.

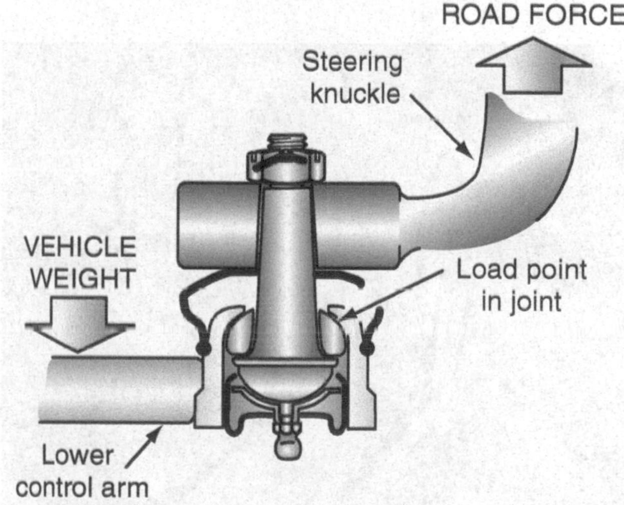

FIGURE 6-29 Ball joints allow for radial (turning) motion for steering and connect control arms to steering knuckles.

Ball joints are mounted to the control arms in a variety of ways. The most common ways are a press fit, bolt in, and rivets. Some older vehicles had threads on the ball joint itself, which was then threaded into the control arm. Joints that are riveted at the factory are replaced with joints that bolt into the control arm.

■ ***Steering Knuckles.*** Steering knuckles support the wheel and tire, brakes, and sprung weight of the vehicle. A steering knuckle can be mounted in a variety of ways for both front and rear suspensions. **Figure 6-31** shows an example of a common steering knuckle configuration. The steering knuckle also has an attachment point for the outer tie rod end. A wheel bearing or set of bearings mount to the steering knuckle to provide the mounting of the wheel hub.

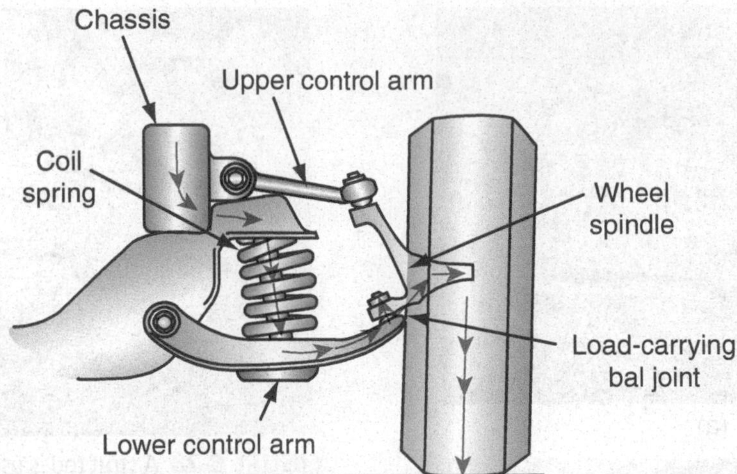

FIGURE 6-30 In this suspension, the sprung weight is carried through the spring to the lower control arm, through the lower ball joint and through the steering knuckle to the tire.

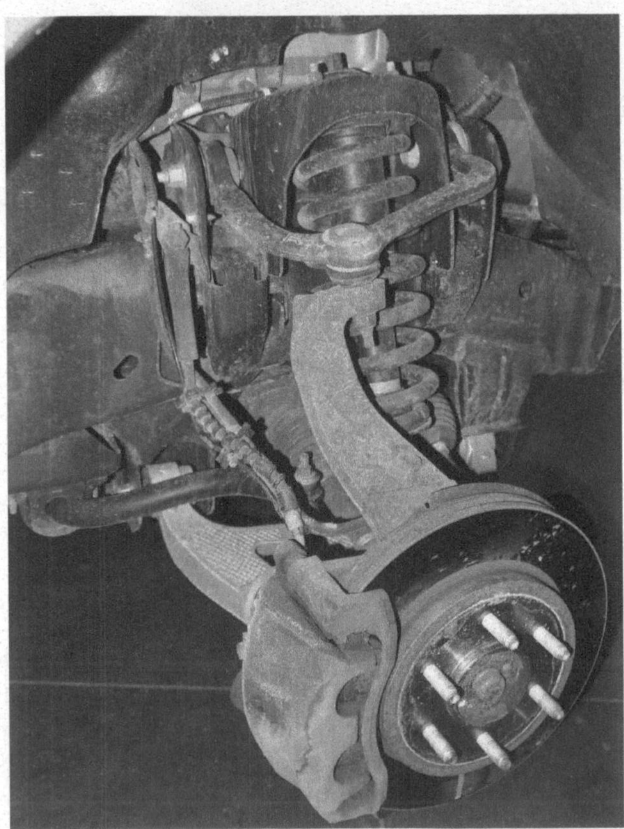

FIGURE 6-31 The steering knuckle on multilink suspensions are typically much taller than on other suspension types. Notice the aluminum lower control arm. Modern suspension systems often use aluminum parts to reduce weight and improve ride quality.

Steering knuckles are also sometimes called *hub carriers or spindles.* The spindle portion of the steering knuckle is where the wheel bearings and brake components are mounted. The spindle supports those components and allows the wheel to rotate on the wheel bearings.

To save weight, many vehicles use aluminum steering knuckles instead of cast iron. Using aluminum reduces unsprung weight and gross vehicle weight, which saves fuel. However, working on aluminum knuckles requires different service practices, which are discussed in Chapter 7.

■ ***Stabilizer Bars.*** **Stabilizer bars**, also called *sway bars* or *anti-roll bars*, reduce body roll (**Figure 6-32**). These steel bars attach to the lower control arms or axle assembly and the body or frame. When the vehicle body starts to lift while cornering, the bar tries to move with the body. Because the outer ends of the stabilizer bar are connected to the control arms or axle, and the control arms cannot move upward, it forces the stabilizer bar to pull the body back down, limiting body roll. **Figure 6-33** shows one way the stabilizer bar is connected to the control arm. Some vehicles have adjustable stabilizer bar links, while some modern vehicles use electronic anti-roll systems to reduce body movement (**Figure 6-34**). Regardless of the type, broken stabilizer bar links will cause excessive body roll while cornering.

Front Suspension System Design and Operation

While all front suspensions accomplish the same goals, different designs and arrangement are used on cars and light trucks to maximize qualities for specific vehicles. For example, a Chevrolet Corvette and a Chevrolet truck may both use types of short/long arm (SLA) suspensions; however, the size, location, and mountings of the components will differ because of the two very different purposes of each vehicle.

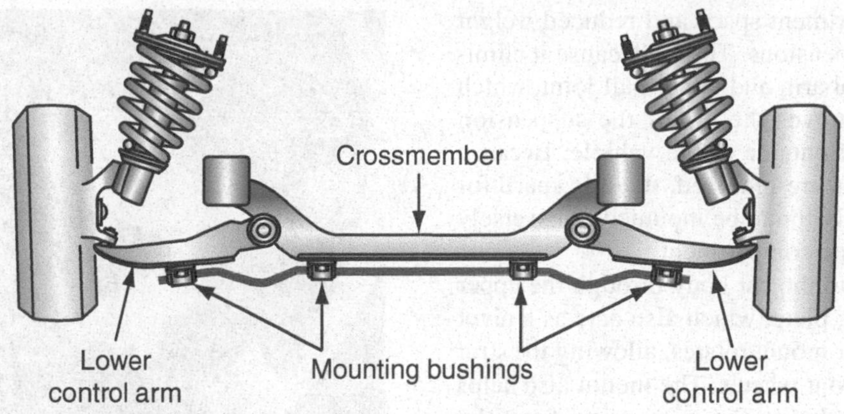

FIGURE 6-32 The stabilizer bar reduces body roll during turns and connects the lower control arms to the body with linkages and bushings.

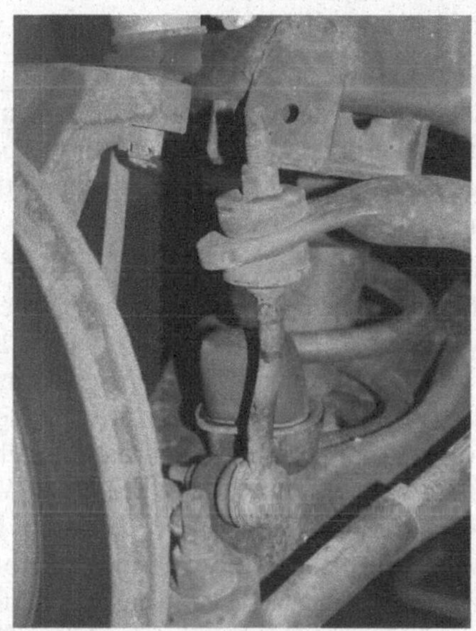

FIGURE 6-33 An example of a stabilizer bar link, connecting the bar to the control arm.

FIGURE 6-34 An active or adjustable stabilizer bar. Stiffness can be adjusted or the bar even unlocked to allow greater wheel travel up and down.

Even though there are many different suspension setups, most types can be categorized into one of the following types:

- Strut based; including MacPherson strut, modified strut, and multilink.
- Short/long arm.
- Solid and/or swing axles.

Regardless of the type, all suspensions try to accomplish the same goals of good ride quality and handling.

STRUT-BASED SUSPENSIONS

Most modern cars and light trucks use some form of strut-based suspension. A strut is a term used to describe either a damper and spring as a unit or a system where the damper is load carrying and works as part of the suspension and steering systems.

MACPHERSON STRUTS

The most common type of front suspension system used today is the **MacPherson strut** suspension. These combine a coil spring, damper, and bearing plate into a single unit (**Figure 6-35**). The MacPherson strut allows for

FIGURE 6-35 A MacPherson strut assembly.

greater engine compartment space and reduced weight compared to SLA suspensions. This is because it eliminates the upper control arm and upper ball joint, which reduces weight and moves the top of the suspension higher and toward the outside of the vehicle. Because the upper control arms are removed, there is space for the engine and transmission to be mounted transversely (sideways) in the engine compartment.

The strut connects to the car body through the upper strut mount or bearing plate, which also acts as a pivot and damper. The upper mount rotates, allowing the strut to turn and steer the front wheels. The mount also helps to reduce NVH by isolating the suspension from the body (**Figure 6-36**).

The damper in a MacPherson strut is larger than that of a standard damper. This is because the body of the strut is a load-carrying component, which must withstand the loads placed on it while it is driving and turning. Some newer vehicles are using a new form of MacPherson strut that uses a modified steering knuckle (**Figure 6-37**). This design improves handling performance and reduces torque steer by relocating the steering axis with a second ball joint. This takes the steering loads off the strut to improve handling.

■ ***MacPherson Strut Suspension System Design and Operation.*** The main advantages of strut suspensions are weight and space savings, simplicity, and low cost. The disadvantages include alignment change when moving vertically, reduced handling capabilities, and noise and harshness transmitted through the upper strut mount.

The strut mounts to the body, often called *the strut tower*, which is in the engine compartment. The connection between the strut and the body contains a mounting plate or a bearing plate. This plate contains either a bushing or a bearing that allows for steering movement. The top of the damper piston is secured to the upper bearing plate with a large nut (**Figure 6-38**). When the steering wheel is turned, the upper section of the bearing plate stays in place while the lower section rotates on the bushing or bearing. The lower section of the strut assembly is bolted to the steering knuckle. The steering knuckle is attached to the lower control arm with a ball joint. The line formed through the upper bearing plate and the lower ball joint is called *the steering axis* (**Figure 6-39**).

FIGURE 6-37 A strut suspension with a modified steering knuckle and two lower ball joints.

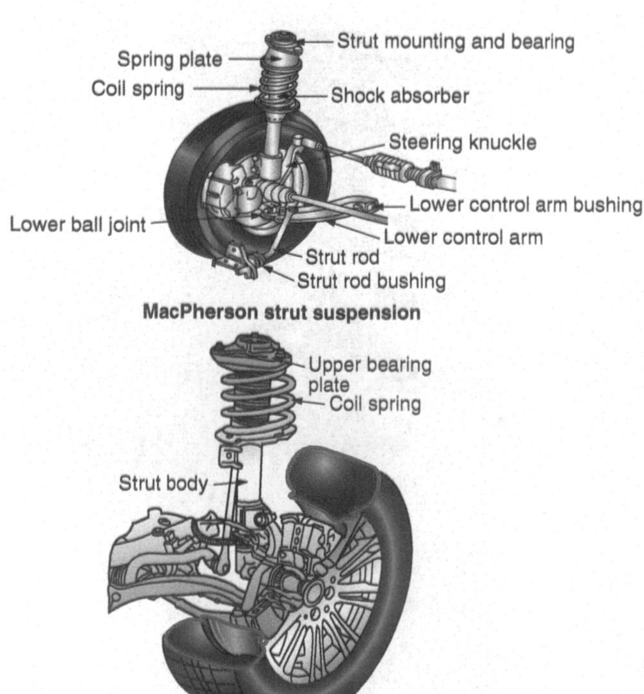

FIGURE 6-36 A view of the components of the MacPherson strut suspension.

FIGURE 6-38 The upper strut mount/bearing plate allows the strut to turn as part of the steering system.

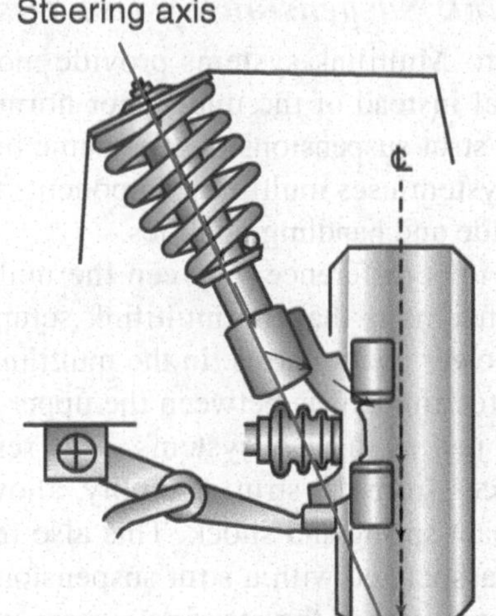

FIGURE 6-39 **The MacPherson strut combines the damper, spring, and upper steering pivot into one unit. The load is carried through the spring and strut to the steering knuckle to the wheel and tire. The ball joint is not a load-carrying ball joint.**

Safety Warning

Because struts retain the spring under tension, special service procedures must be followed to prevent injury. Strut service is covered in detail in Chapter 7.

The strut replaces the upper control arm and ball joint, reducing the number of parts and vehicle weight. Because the strut mounts almost vertically to the body, the strut tower can be moved further out toward the wheels, which increases room in the engine compartment compared to SLA suspensions.

The sprung weight of the vehicle is applied to the spring via the upper strut tower and strut mount. This weight is then applied to the lower spring seat of the strut body, down the strut to the steering knuckle. The load is then transferred to the wheel and tire to the ground (Figure 6-39). The lower ball joint in this system is not a load-carrying joint. It acts as a friction joint only.

MODIFIED STRUTS

Some vehicles use a strut-style damper but relocate the spring. These are not true MacPherson struts. Called a **modified strut**, this system has the spring mounted separate from the strut. The strut performs the function of the damper and is connected to an upper bearing plate at the top and to the steering knuckle at the lower end. The coil spring is located between the frame and the lower control arm. This design has the weight and space saving advantages of the MacPherson strut suspension but can contain larger springs. Relocating the spring also can allow for a wider distance between the wheel wells, increasing engine compartment room.

■ ***Modified Strut Suspension System Design and Operation.*** A modified strut suspension is similar to a MacPherson strut system, except that the coil spring is relocated, typically to where the coil spring normally sits in SLA suspensions. The coil spring is mounted between the lower control arm and the vehicle frame. Because the spring is now located on the lower control arm, the lower ball joint becomes a load-carrying ball joint. **Figure 6-40** shows a typical modified strut system and how it carries the weight of the vehicle. As with a MacPherson strut, the steering axis is through the upper bearing plate and the lower ball joint.

The strut assembly now consists of the damper and the bearing plate mounted to the strut tower. As in MacPherson systems, the steering knuckle connects to the lower end of the strut body. Used extensively on older GM Camaro and Firebird bodies as well as Ford Mustangs, this system offers the compactness of a strut suspension with improved ride and handling qualities.

■ ***Multilink.*** Many vehicles use a multilink system. With a **multilink** suspension, the steering knuckle is taller than a traditional strut or SLA suspension, often reaching the height of the top of the tire. The strut does

Safety Warning

Because the struts retain the spring under tension, special service procedures must be followed to prevent injury. Modified strut service is covered in detail in Chapter 7.

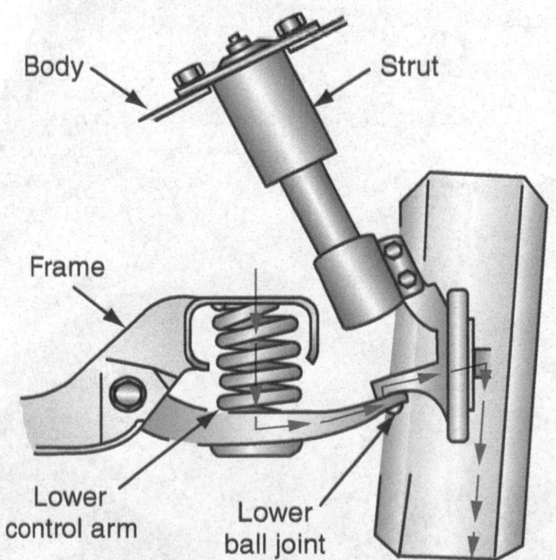

FIGURE 6-40 **The modified strut suspension uses a strut connected to the upper body and steering knuckle but moves the coil spring off the strut assembly.**

not turn with the steering axis; rather it is mounted rigidly to the body at the upper strut mount. This is because the steering knuckle pivots on the upper and lower ball joints for steering action. Multilink systems are designed to produce neutral steering on FWD vehicles, which tend to exhibit understeer with traditional MacPherson strut suspensions. This suspension is also commonly used on both FWD and RWD cars, light trucks, and SUVs. **Figure 6-41** shows a common multilink arrangement.

Multilink suspensions are also found on the rear of many vehicles, both FWD and RWD (**Figure 6-42**). Several control arms are used to reduce rear axle movements and provide better handling and ride qualities than a traditional rear strut system.

FIGURE 6-41 A multilink combines a nonrotating strut and two or more control arms to control wheel movements.

FIGURE 6-42 An example of a rear multilink suspension.

■ ***Multilink Suspension System Design and Operation.*** Multilink systems provide more neutral steering feel instead of the understeer normally associated with strut suspensions. As the name implies, the multilink system uses multiple components to achieve improved ride and handling qualities.

One obvious difference between the multilink and strut suspensions is that the multilink setup has both upper and lower control arms. In the multilink arrangement, the steering axis is between the upper and lower ball joints, just as in SLA systems. This removes the turning forces from the strut assembly, allowing for a more compact spring and shock. This also reduces the understeer associated with a strut suspension. Because the strut is not part of the steering system, there is no upper bushing or bearing in the upper strut mount. The load is carried by the lower ball joint because the strut is bolted to the lower control arm (**Figure 6-43**).

■ ***Short/Long Arm.*** Short/long arm suspensions, also called *SLA suspensions*, are typically used on RWD vehicles. This suspension consists of two unequal length control arms connected with a steering knuckle. The control arms are generally triangular and are often called *wishbones* or *A-arms*. A steering knuckle, control arm bushings, and ball joints comprise the rest of the suspension (**Figure 6-44**). Control arm design is matched with the spring for tire control and ride characteristics. The control arms are mounted to the frame with control arm bushings.

FIGURE 6-43 This illustrates how the weight is carried through a multilink suspension.

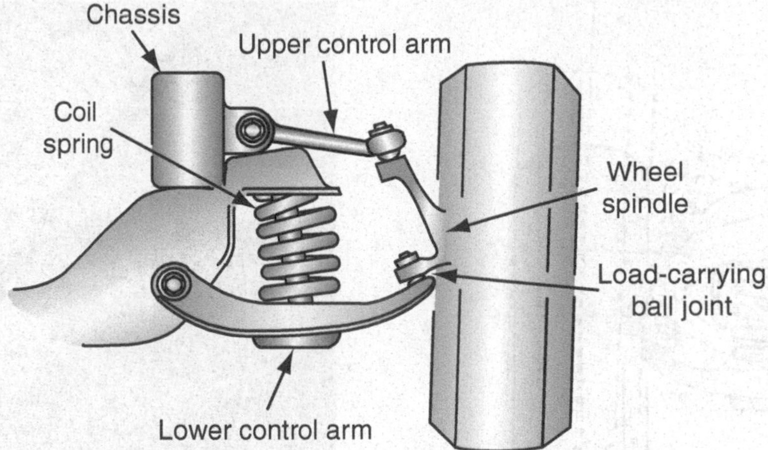

FIGURE 6-44 The SLA suspension uses two unequal-length control arms to control tire movements.

SLA systems use two ball joints, one of which carries the sprung weight of the vehicle. The other ball joint provides a friction and pivot point and does not carry weight. The load-carrying joint is located in the control arm in which the spring sits. The other ball joint is called the friction or following ball joint. **Figure 6-45** shows how the weight is carried by the load-carrying ball joint in an SLA suspension. SLA suspensions are not as common as they once were due to the popularity of FWD vehicles. These suspensions tend to intrude into the engine compartment, causing space problems with FWD drivetrains.

■ ***Short Arm Long Arm Suspension Design and Operation.*** Short/long arm suspensions have been in service for many years. The advantages of this type of suspension are that it is compact, provides good handling and ride quality, and provides for substantial adjustment to fine-tune wheel alignment and ride characteristics.

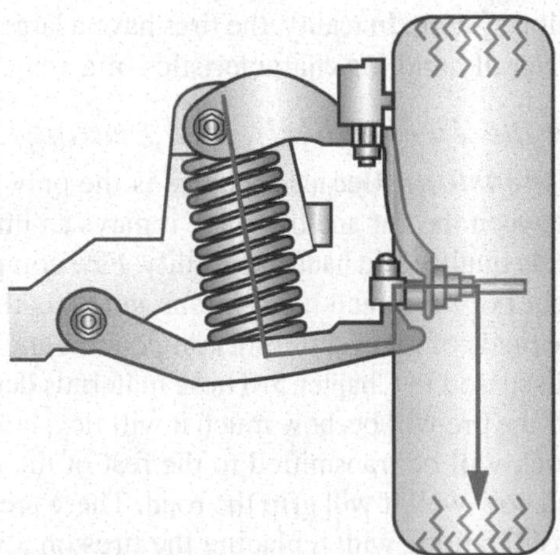

FIGURE 6-45 The sprung weight is carried from the spring to the lower control arm to the lower ball joint. The load then passes through the ball joint to the steering knuckle to the bearings, wheel, and tire.

Typically, a coil spring is sandwiched between the lower control arm and the frame, although many 4WD trucks and SUVs use torsion bars instead of coil springs. **Figure 6-46** shows a typical SLA configuration. The weight of the vehicle is pressing down on the coil spring where the spring meets the frame. The lower control arm, with the spring sitting on it, is then supporting the vehicle. The steering knuckle is connected to the upper and lower control arms by ball joints.

As the tire moves over a bump in the road, the lower control arm is forced up against spring tension (**Figure 6-47**). The size and position of the A-arms help to keep the tire vertical as it moves up and down over the road. The spring absorbs the bump and limits how much the bump is transferred to the body. The damper is typically mounted inside of the open center of the coil spring.

■ ***4WD Suspensions.*** For many years, the front suspensions on 4WD vehicles were nearly identical to the

FIGURE 6-46 An example of an SLA suspension.

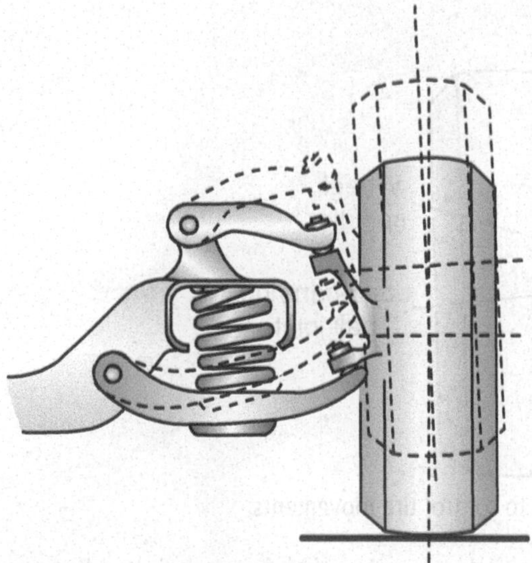

FIGURE 6-47 An illustration of the movements of the control arms and steering knuckles as the SLA moves over bumps.

FIGURE 6-48 Many 4WD trucks use the SLA suspension and a torsion bar.

rear suspensions. A large live axle supported with either leaf or coil springs was standard for most 4WD trucks, an example of which is shown previously in Figure 6-47. While strong, these systems did not have outstanding ride quality. To improve the ride and handling of 4WD trucks, manufacturers began to redesign the front suspensions to allow for independent wheel movement. One novel approach to this was Ford's Twin-Traction Beam or TTB. This system uses a live front axle that contains U-jointed axle shafts that allow for independent wheel movement for improved ride and handling while still retaining the durability and strength of traditional 4WD.

Manufacturers of 4WD vehicles today often mount the front differential directly to the chassis. Short FWD drive shafts then connect the differential to the wheels. This allows fully independent suspension movement. Full live front axles can still be found on heavy-duty light trucks, but most trucks now have independent front suspensions whether they are 2WD or 4WD.

■ ***4WD Suspension Design and Operation.*** Many 4WD trucks and SUVs use the SLA suspension, though typically with a torsion bar instead of coil springs (**Figure 6-48**). This is because the front drive axles pass through where the spring and damper are normally mounted in the SLA systems. Torsion bars are typically mounted to the rear of the lower control arm. This means that the lower ball joint is the load-carrying joint. The rear of the torsion bar is mounted in the rear crossmember that supports the transmission. An adjustment mechanism is installed at the rear torsion bar mount. The adjustment is used to remove or apply tension to the torsion bar during suspension service. As the torsion bar weakens over time and the ride height begins to drop, this adjustment can be used to apply more tension to the bar and restore ride height.

Many heavy-duty 4WD trucks use a live front axle. This is because of the strength and durability of the axle. The live axle does not permit independent front wheel movement, so ride quality and handling suffer compared to a vehicle with an independent suspension. Typically, only the heavy-duty versions of a 4WD truck will have a live front axle because of this. Front live axles can be mounted with either coil or leaf springs, depending on the vehicle. The ends of the axle connect to the steering knuckle with ball joints.

TIRES, WHEELS, AND BEARINGS AS PART OF THE SUSPENSION

Tires and wheels are often thought of as independent components, not necessarily as part of the steering and suspension system. In reality, the tires have a large impact on the overall handling characteristics of a vehicle.

■ ***How the Tires, Wheels, and Bearings Affect the Suspension.*** Because the tire is the only contact point between the car and the road, it plays an important part in ride quality and handling ability. Few components can affect both ride and handling as much as the tires. Tires are made of many different components and materials, as discussed in Chapter 5. These materials determine how stiff the tire will be, how much it will flex, how much road shock will be transmitted to the rest of the suspension, and how well it will grip the road. These are important considerations with replacing the tires on a vehicle. Different tread designs behave differently under normal operating conditions. A vehicle with two or more different types of tires installed can experience handling, noise, and pulling concerns due to the dissimilar tread patterns

(**Figure 6-49**). Vehicles can experience unpredictable and dangerous handling problems due to the differences in the size, type, and tread designs of the tires.

Tire pressure also plays a role in ride quality and handling. Every modern vehicle has a recommended tire inflation pressure, which is found on a tire information decal, typically located on a doorjamb or the glove box. Overinflating the tires can increase ride harshness and cause the center of the tread to wear more rapidly. Severe overinflation will bulge out the center of the tire and can be dangerous due to increased pressure and temperatures during driving. Underinflation of tires can also cause a rough ride, as well as increased rolling resistance and rapid wear and possibly wheel damage. An underinflated tire will have less cushioning than a properly inflated tire. Striking bumps and potholes can cause the tire to flex enough that the rim is damaged. Underinflation also places more load on the tire sidewall. This can lead to severe tire damage and tire failure in a very brief time.

Before attempting to diagnose any suspension or steering concern, a very careful inspection of the tires should always take place. Always check the following:

- For correct tire size, type, and construction
- Tire inflation pressures
- Tire tread patterns and treadwear patterns

Wheels can affect the operation of the suspension system when the original wheels are replaced with different-sized

FIGURE 6-49 Tire size, type, and tread design should be the same on each corner of the vehicle. Adverse ride and handling problems can result from mixing tires on the vehicle.

wheels and tires. When the vehicle is designed, the suspension, steering, and alignment geometry are designed around specific wheel and tire dimensions. When wider, taller, or different offset wheels are installed in place of the original wheels, clearance problems can arise, improper loading of wheel bearings can occur, and steering and suspension geometry can be altered with unexpected consequences. Changing wheel offset can affect scrub radius (**Figure 6-50**). Scrub radius is where the angle of the steering axis and wheel center cross. Most FWD cars use negative scrub radius and RWD cars and trucks typically have a positive setting. Changing scrub radius can affect steering effort, stability, and handling.

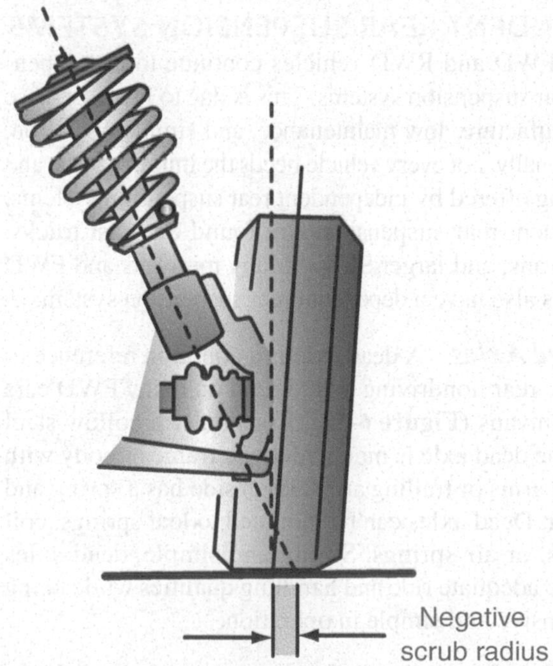

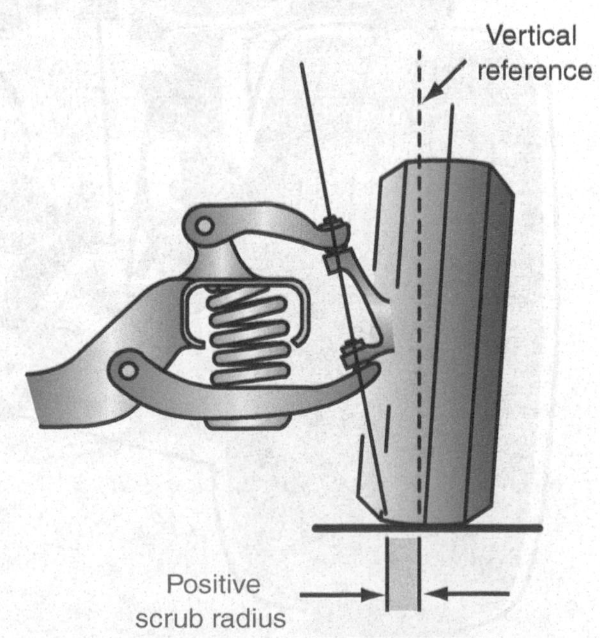

FIGURE 6-50 Changing wheel and tire size can affect scrub radius, which can change the driving and handling qualities of the vehicle.

Any time a vehicle's alignment is being checked and aftermarket wheels are installed, do a thorough check of all alignment angles, not just the normal caster, camber, and toe. Incorrect wheel and tire size can affect scrub radius, which can cause severe pulling.

Wheel bearings typically only cause suspension problems when they are excessively worn and become loose. Faulty wheel bearings can cause noise and allow wheel movement, which can result in noise, pulling, and vibrations. Aftermarket wheel and tire installations can lead to damaged wheel bearings by changing wheel offset, and consequently, bearing loading.

■ ***Sprung and Unsprung Weight.*** As discussed earlier, sprung weight is weight carried by the springs. This includes the vehicle's frame and body, passengers, the powertrain, and most major components. Unsprung weight is weight not carried by the springs. This usually includes the wheels, tires, wheel bearings, brakes, the springs themselves, lower control arms, ball joints, and other similar components. The amount of sprung weight should be high, and the amount of unsprung weight should be low.

FIGURE 6-51 Keeping unsprung weight low improves overall handling and vehicle response.

But how does this affect ride quality and handling? Every time a car hits a bump, the wheel, tire, brakes, and other unsprung parts move in response to the bump. This mass of parts accelerate upward, stop, and accelerate back down. The parts outlined in red in **Figure 6-51** illustrate many of the parts that move but are not carried by the spring. The more mass that moves, the greater the impact on ride quality. By decreasing the unsprung weight, the suspension will be able to respond more rapidly and effectively to bumps and road conditions. This means the tire can maintain contact with the road better after encountering a bump. In addition, because the springs and dampers still have to respond to movement by the unsprung weight, reducing this weight allows to better control the sprung weight of the vehicle, which is also responding to bumps, dips, potholes, and other road conditions.

Rear Suspension Systems

The rear suspension must meet the same requirements as the front, keeping the tires in contact with the ground, providing a smooth ride, and absorbing road shocks. However, the rear suspension does not have to accommodate steering except in a few cases.

Even though it is not usually part of the steering, the rear suspension often has a much harder job keeping the tires in contact with the road due to the decreased amount of weight in the rear of most vehicles. In addition, providing adequate traction to the rear driving wheels can also be a challenge for the rear suspension due to unequal weight balance and overall suspension design.

DEPENDENT REAR SUSPENSION SYSTEMS

Many FWD and RWD vehicles continue to use dependent rear suspension systems. This is due to low cost, ease of manufacture, low maintenance, and simple operation. Additionally, not every vehicle needs the improved ride and handling offered by independent rear suspension systems. Dependent rear suspensions are found on most trucks, RWD vans, and larger SUVs. Many minivans and FWD vehicles also have a dependent rear suspension system.

■ ***Dead Axles.*** A dead axle is a common reference to a single rear nondriving axle, found on many FWD cars and minivans (**Figure 6-52**). Essentially a hollow steel tube, the dead axle is mounted to the frame or body with control arms or trailing arms. Each side has a spring and damper. Dead axles can be mounted to leaf springs, coil springs, or air springs. Strong and simple, dead axles provide adequate ride and handling qualities while being inexpensive and simple in operation.

■ ***Live Axles.*** Rear live axles can be either dependently or independently mounted (**Figure 6-53**). Most trucks and SUVs use a dependent rear live axle with either leaf or coil

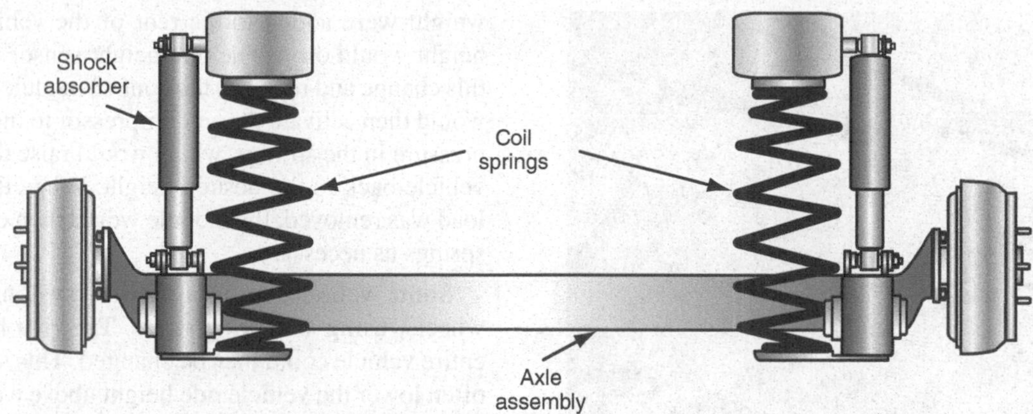

FIGURE 6-52 Dead axles are low cost, simple, and effective for some applications where ride and handling are secondary concerns.

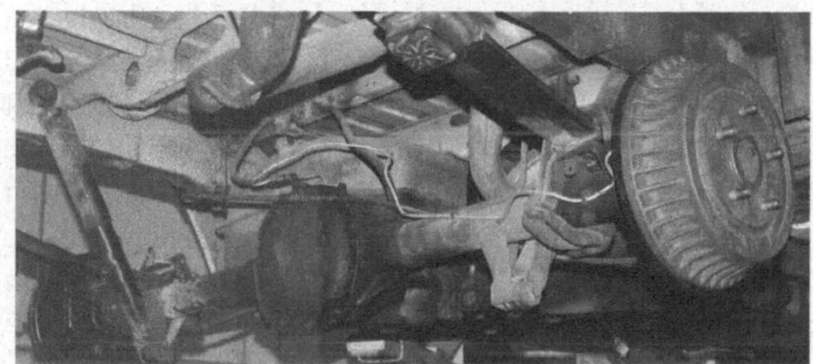

FIGURE 6-53 An example of a dependent RWD live axle suspension.

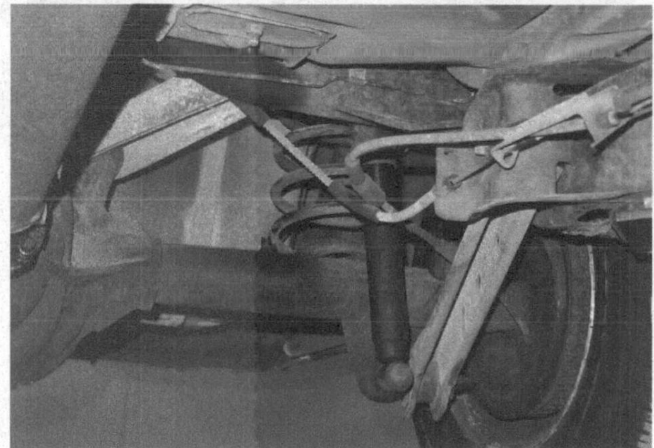

FIGURE 6-54 Rear suspensions that use coil springs also use control arms to limit rear axle movement.

springs. Some larger luxury cars use air springs with the rear axle. RWD axles with coil or air springs use some type of control arm to limit lateral movement and help control axle wind-up (**Figure 6-54**). Wind-up occurs when torque is applied to the rear axle, causing it to try to twist in response.

■ ***Independent Rear Suspension Systems.*** Many sports cars and some trucks and SUVs use an independent rear live axle (**Figure 6-55**). By mounting the differential directly to the body or frame, two short drive shafts can be used with an independent rear suspension to improve ride and handling. Independent RWD systems can use multiple control arms and trailing arms to limit unwanted lateral movements.

■ ***Track Bars and Watt's Links.*** Many rear suspensions use a track bar or Panhard bar (**Figure 6-56**). The **track bar** is used to limit side-to-side movement of the rear axle. Control arms or trailing arms are used with a track bar to limit front-to-rear axle movements.

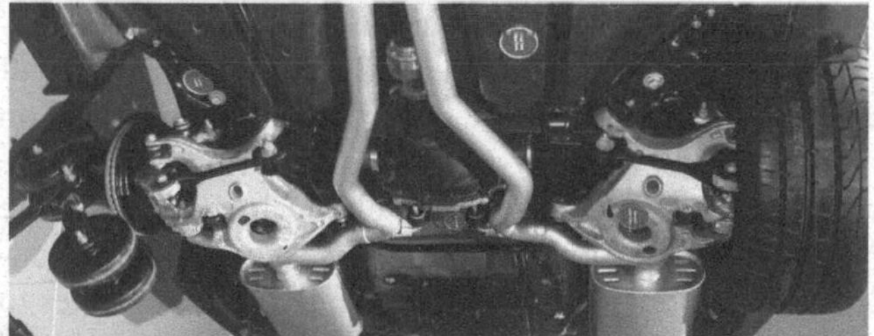

FIGURE 6-55 An example of an independent rear live axle system.

FIGURE 6-56 The track bar is used to control both axle and body movements caused by up and down movement of the axle.

FIGURE 6-57 **Watt's links are used to reduce body movements caused by the rear suspension moving.**

Some RWD vehicles use a Watt's link instead of a track bar. A Watt's link is also used to limit rear axle side-to-side movement while the axle moves up and down. An example of a Watt's link is shown in **Figure 6-57**.

Electronic Suspension Systems

Electronic suspension systems have been in use for many years. Basic systems use a ride height sensor and air-adjustable dampers or air springs to increase ride height as vehicle load increases. Newer vehicles go beyond this, using magnetorheological fluid in the dampers and active roll stabilization to greatly enhance ride quality and handling.

PURPOSE OF ELECTRONIC SUSPENSION SYSTEMS

Electronic suspension systems are now available on many vehicles, though they are most often found on luxury sport sedans and sport cars. The current generation of electronic suspensions are different than those found on older vehicles. Older systems used air springs to improve ride and to compensate for vehicle loading and ride height changes. Modern systems use advanced electronics and computer systems to make real-time changes to the steering and suspension systems to provide increased handling and safety.

■ ***Types of Electronic Suspension Systems.*** In the 1980s and 1990s, many large front-and rear-wheel drive cars had electronically operated air ride suspensions. Instead of traditional rear coil or leaf springs, air springs were used. By having an on-board air compressor and mounting a ride height sensor to the rear suspension, a module could detect changes in ride height and increase or decrease air pressure in the rear springs as needed. For example, if additional passengers and weight were added to the rear of the vehicle, the ride height would drop. The ride height sensor would detect this change and report to the control module. The module would then activate the air compressor to increase the air pressure in the springs, which would raise the rear of the vehicle back to the desired height. When the additional load was removed, the module would then deflate the air springs as necessary.

Some vehicles used a similar system at all four wheels, using air spring struts. The ride height of the entire vehicle could then be changed. This system would often lower the vehicle ride height above a certain speed to improve aerodynamics (**Figure 6-58**).

In the 1990s, Delphi introduced Magneride. Magneride uses a magnetorheological fluid in the dampers. Basically, a magnetorheological fluid has magnetic properties; it is like an oil with a very fine metallic dust added. Because magnetic fields can be controlled, the Magneride system controls the viscosity of the fluid in the damper. By allowing the fluid to be easier or more difficult to flow through the valves, the ride quality can be changed (**Figure 6-59**).

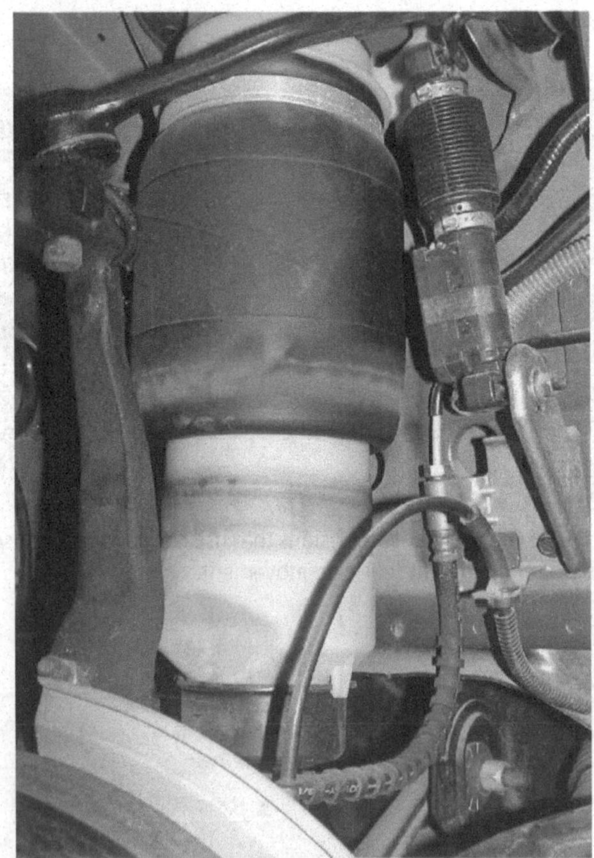

FIGURE 6-58 **Adaptive and active suspensions use sensors to transmit suspension movement information to the on-board computer system.**

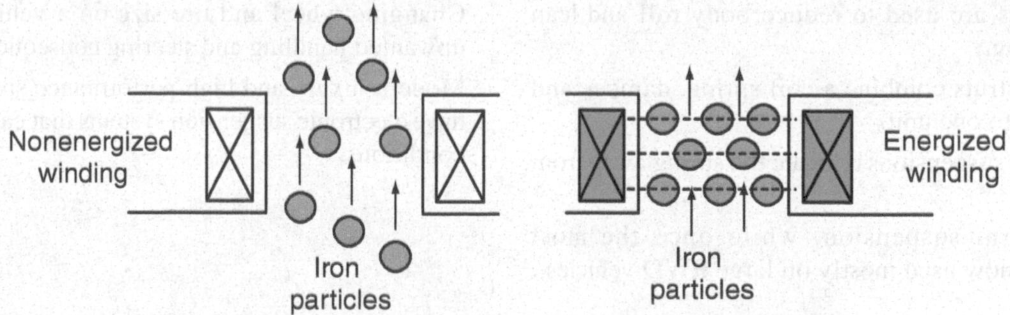

FIGURE 6-59 The fluid in the shock can be allowed to pass freely or be restricted. This changes the resistance of the shock and how the vehicle rides.

By employing a network of sensors, the ride control system monitors vehicle performance and operating conditions. **Figure 6-60** shows an example of a Magneride damper. Under normal driving conditions, the fluid allows for normal, gentle cushioning for a smooth ride. If the road becomes rough, the system detects this and can change how the damper will operate in about five milliseconds, fast enough to change how the fluid and damper will respond while the wheel and tire are still negotiating the bump or dip in the road surface. Another advantage of this system is that it allows for selectable suspension operation. The driver can select a soft, smooth ride, or by merely moving a switch, select a firm, sporty feel that improves vehicle handling.

Modern vehicles may use either adaptive or active suspension systems. Adaptive systems monitor the vehicle and road conditions and adapt to maintain ride quality. These systems are typically less complex and less expensive than active systems and do not perform functions such as reducing body roll when cornering. Active systems use more sensors and can not only adjust dampening rates but also reduce body roll and can work with brake, steering, and powertrain systems to correct for driver error and help keep the vehicle under control.

FIGURE 6-60 Active suspension systems can adapt and change how a vehicle is riding and cornering in real time. This type of damper has the ability to change its resistance and damping qualities based on driving conditions and driver settings.

SUMMARY

The suspension system carries vehicle weight, reduces road shocks, and provides safe handling and a comfortable ride.

Independent suspensions allow each wheel to move without affecting the other wheels.

When one wheel on a dependent suspension axle moves over a bump, the tire on the opposite side also moves.

Oversteer is when the rear of the vehicle loses traction and causes the rear of the vehicle to slide during a corner.

Understeer is when the front tires lose traction and the vehicle continues to travel in a line that does not follow the path of a turn.

Sprung weight is the weight that is carried by the springs.

Unsprung weight is the weight that is not carried by the springs.

Shock absorbers are used to dampen spring bounce or oscillations.

Ball joints located on the same control arm as a spring are load-carrying ball joints.

Stabilizer bars are used to reduce body roll and lean when cornering.

MacPherson struts combine a coil spring, damper, and upper pivot into one unit.

Modified strut suspensions relocate the spring away from the strut.

Short/long arm suspension, while once the most common, are now used mostly on large RWD vehicles.

Changing wheel and tire size on a vehicle can cause unwanted handling and steering consequences.

Modern luxury and high-performance sports cars often have electronic suspension systems that can adapt to road conditions.

REVIEW QUESTIONS

1. When springs compress, the movement is called ______________, and when the springs extend back out the movement is called ______________.
2. Most passenger cars and SUVs use ______________ construction, which is very rigid and strong.
3. ______________ weight is the weight carried by the springs.
4. The type of suspension that uses a strut without an upper control arm is the ______________ ______________ suspension.
5. The ______________ ______________ are used to reduce or dampen spring oscillations.
6. When discussing suspension components: *Technician A* says sway bars are used only on large cars and trucks to reduce how much the body moves when turning corners. *Technician B* says sway bars are used to reduce the bouncing of the springs after going over a bump. Who is correct?
   a. Technician A
   b. Technician B
   c. Both A and B
   d. Neither A nor B
7. Where is the load-carrying ball joint in a multilink system located?
   a. On the upper control arm
   b. On the lower control arm
   c. Either the upper or the lower control arm
   d. None of the above
8. A front suspension system that has a lower control arm with only one connection point to the frame will also use which of these other components?
   a. Sway bar
   b. Strut rod
   c. Watt's link
   d. Track bar
9. *Technician A* says the wheels and tires are unsprung weight. *Technician B* says the brake calipers, pads, and rotors are unsprung weight. Who is correct?
   a. Technician A
   b. Technician B
   c. Both A and B
   d. Neither A nor B
10. *Technician A* says control arms can be used to limit rear axle movements. *Technician B* says Watt's links are used in place of dampers on some live rear axle arrangements. Who is correct?
    a. Technician A
    b. Technician B
    c. Both A and B
    d. Neither A nor B

Suspension System Service

Chapter Objectives

At the conclusion of this chapter, you should be able to:

- Identify the tools and their correct usage for servicing the suspension system.
- Perform suspension system component inspections, including:
  - o Inspect upper and lower control arms, bushings, and shafts. (ASE Education Foundation MLR 4.B.9)
  - o Inspect track bar, strut rods/radius arms, and related mounts and bushings. (ASE Education Foundation MLR 4.B.11)
  - o Inspect upper and lower ball joints. (ASE Education Foundation MLR 4.B.12)
  - o Inspect suspension system coil springs and spring insulators. (ASE Education Foundation MLR 4.B.13)
  - o Inspect suspension system torsion bars and mounts. (ASE Education Foundation MLR 4.B.14)
  - o Inspect strut cartridge or assembly. (ASE Education Foundation MLR 4.B.16)
  - o Inspect front strut bearing and mount. (ASE Education Foundation MLR 4.B.17)
  - o Inspect rear suspension system lateral links/arms, control arms. (ASE Education Foundation MLR 4.B.18)
  - o Inspect rear suspension system leaf spring(s), spring insulators (silencers), shackles, brackets, bushings, center pins/bolts, and mounts. (ASE Education Foundation MLR 4.B.19)
- Diagnose suspension system concerns, including incorrect ride height, noises, excessive bounce and sway, and increased tire wear.
- Service components of the suspension system including:
  - o Inspect and replace rebound and jounce bumpers. (ASE Education Foundation MLR 4.B.10)
  - o Inspect and replace front stabilizer bar bushings, brackets, and links. (ASE Education Foundation MLR 4.B.15)
  - o Inspect, remove, and replace shock absorbers; inspect mounts and bushings. (ASE Education Foundation MLR 4.B.20)
  - o Replace upper and lower ball joints.
  - o Remove, inspect, and service strut.
- Perform prealignment inspection and measure vehicle ride height. (ASE Education Foundation MLR 4.C.1)
- Describe wheel alignment angles. (ASE Education Foundation MLR 4.C.2)

KEY TERMS

axle puller
camber
caster
geometric centerline
grease fitting
prealignment inspection
ride height
spring compressors
strut spring compressor
thrust line
toe
visual inspection

Repairs on the suspension system are some of the most common types of repairs technicians perform. The abuse of everyday driving conditions means that suspension components, such as springs and shocks, ball joints, and sway bar links, require frequent inspection and service.

Many technicians begin their careers servicing suspension systems before advancing into other aspects of vehicle service and repair. This is because suspension repairs utilize basic hands-on skills and help develop the critical thinking necessary for all technicians.

Tools and Safety

Every service and repair made to a vehicle must begin and end with safety in mind. A technician who does not perform his or her work safely is not an asset; he or she is a danger. Safe working practices include proper tool use and care, following the proper repair procedures, staying focused on the tasks at hand, and taking the time to perform your work properly.

Several special tools are used to service the suspension system, and using them properly will allow you to work safely and efficiently.

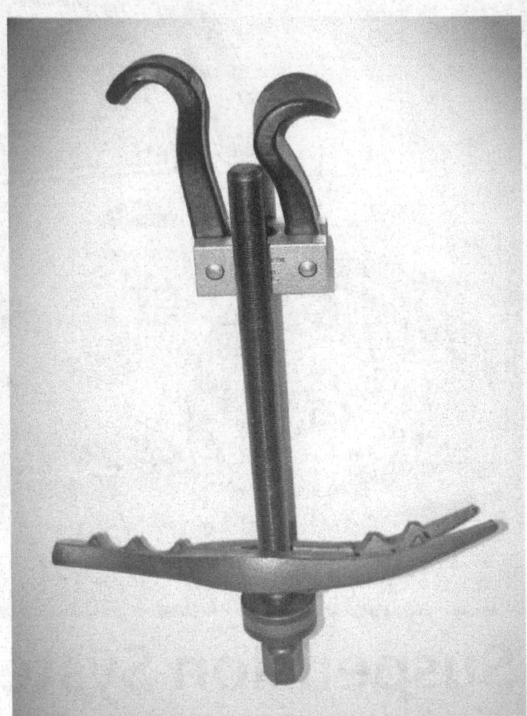

FIGURE 7-1 This type of spring compressor is placed inside the spring and compresses it between the hooks at the top and the fork at the bottom.

TOOLS

Without the correct tools, many of the common repairs made to the automobile would not be possible. When you are servicing the suspension system, common hand tools as well as some specialty tools, described below, are used.

■ ***Tools for Suspension Service.*** Figure 7-1 through Figure 7-7 show and explain the uses of many of the tools you will use when you are working on the suspension system.

Working with coil springs often requires using a **spring compressor**. **Figure 7-1** shows a spring compressor that attaches to the spring from the inside. This type of compressor is commonly used with front coil springs. **Figure 7-2** is of a coil spring compressor that attaches to the outside of the coils. This type of compressor can be used on struts and other coil springs.

When you are replacing MacPherson strut shock absorbers, often a spring compressor is used (**Figure 7-3**). The strut is removed from the vehicle and mounted in the compressor, where it can safely be disassembled for service.

Installing press-fit ball joints and certain types of suspension bushings requires a special press (**Figure 7-4**). The press is needed to install new pressed-in ball joints without damaging the joint or the control arm. Adapters for the press may be used to remove and install control arm bushings.

FIGURE 7-2 This is a type of external spring compressor. The two clamps attach opposite to each other on the spring and are tightened evenly to compress the spring.

Some control arm bushings will require a specific tool for service (**Figure 7-5**). This tool is used to remove and reinstall the bushings without damaging the bushing or the control arm.

Separating tie rods, ball joints, and other components is often easier using a separator tool (**Figure 7-6**). This tool is used so that the studs, threads, and grease boots of ball joints are not damaged when removing them without replacing them.

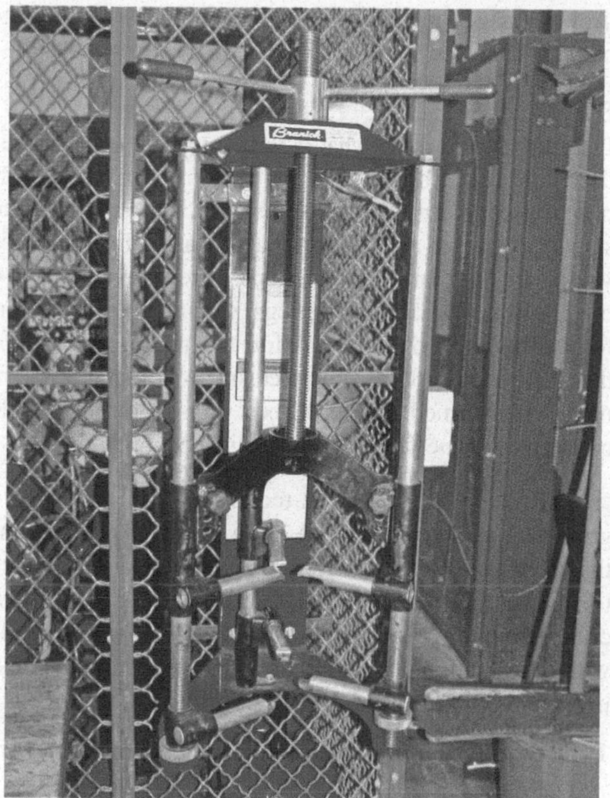

FIGURE 7-3 A strut compressor holds a strut assembly and allows the spring to be compressed and the strut disassembled for replacement.

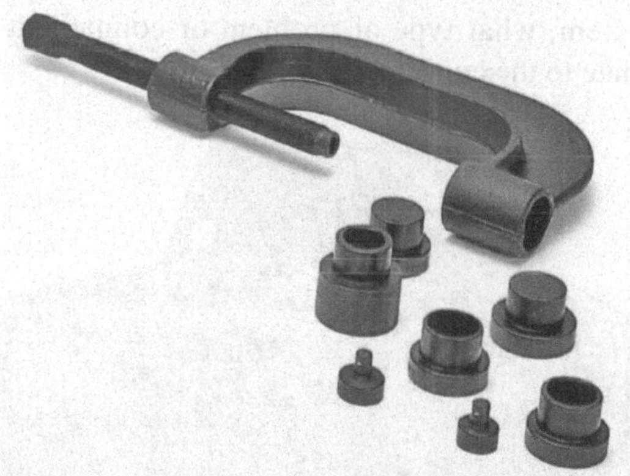

FIGURE 7-4 Ball joint presses are used to install press-fit ball joints and may have adapters to press in control arm bushings.

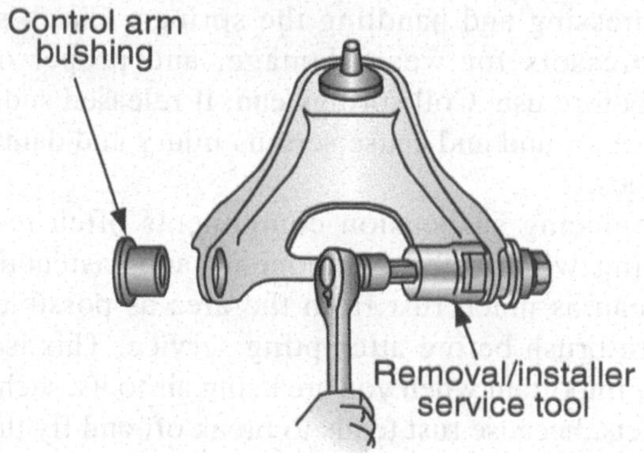

FIGURE 7-5 A special tool is used to remove and install some types of control arm bushings. Some bushings can be driven out with a hammer and the new bushings pressed back in.

FIGURE 7-6 This tool is used to separate ball joint studs from components without damaging the studs or grease boots.

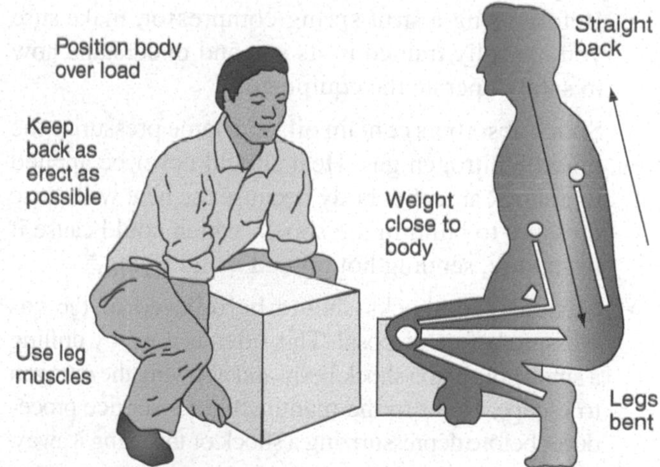

FIGURE 7-7 Proper lifting requires using your leg muscles and not those in your back.

SAFETY

Working on the suspension system often requires lifting heavy objects, working with springs that are compressed, using air-powered tools, and working with rusted components. These conditions require extra attention to safety to prevent injury and damage to the vehicle.

Wheel and tire assemblies can range from around 30 pounds to more than 50 pounds. Always use your leg muscles when lifting (**Figure 7-7**). Ask for help when needed to lift a wheel and tire. Back injuries are common among service technicians and can cause long-term pain and disability.

When you are working with suspension springs, always follow the proper service procedures for safely compressing and handling the springs. Check spring compressors for wear, damage, and proper operation before use. Coil springs can, if released suddenly, bounce around and cause serious injury and damage to vehicles.

Replacing suspension components often requires fighting with rusted components and fasteners. Try to clean as much rust from the area as possible with a wire brush before attempting service. This is especially important when you are using air tools, such as air impacts, because rust tends to break off and fly through the air. Apply a penetrant to rusted nuts and bolts before attempting to remove them.

■ ***Safe Work Practices.*** When you are performing work on the suspension system, you will often be working with springs and other components that require special handling procedures. The following are some general safe work practices for working on the suspension system.

- Make sure that the vehicle is properly raised and supported before beginning any work.
- Use the proper tool for the job and ensure that the tool is undamaged and in proper working condition before use.
- Use the appropriate spring compressors when servicing springs and struts. Suspension springs can contain a lot of stored energy that, if released accidentally, can cause serious injury to people and damage to the vehicle.
- Before using a strut spring compressor, make sure you are fully trained in its use and understand how to safely operate the equipment.
- Shock absorbers contain oil, and some pressurize the oil with nitrogen gas. Heat should never be applied to a shock absorber body because the heat will cause pressure to build in the shock, which could cause it to rupture, sending hot oil and metal flying.
- Gas-charged shocks should be relieved of the gas pressure before disposal. This often is done by drilling a small hole in the shock body and allowing the nitrogen to escape. Refer to the manufacturer's service procedures before depressurizing a shock or throwing it away.
- Do not use steel hammers, punches, or chisels on other steel components.

Inspection and Service

Inspection of the suspension system includes looking at the tires and the steering system. This is because problems with the suspension system may show up as tirewear issues and because there is some sharing of parts between the steering and suspension systems.

BASIC INSPECTION PROCEDURES

If trying to determine what is wrong with a vehicle, your first step should be to verify the customer's complaint. If necessary, perform a test drive with the customer to make sure that you both are talking about the same problem. Once you have done that, your next step is determining the cause of a concern. The best way to begin your diagnosis is to perform a thorough inspection. Before any repairs are made to the suspension system, a complete inspection of the system must be performed. One of the best ways to locate possible concerns is by performing a visual inspection.

■ ***Visual Inspection.*** A technician who performs a good **visual inspection** can often discover many items that need attention. Use a logical and systematic approach to visual inspections rather than randomly looking at components that may or may not have anything to do with the situation at hand. Begin your inspection by looking over the entire vehicle, and consider the following:

1. Does the vehicle appear to lean like the truck in **Figure 7-8**?
2. How do the tires look? Look at the tread for signs of abnormal wear. Take a close look at the tire in **Figure 7-9**. Do you see any wear pattern that may be caused by the suspension system?
3. Based on your knowledge of tires and the suspension system, what type of problem or component may relate to the concern?

FIGURE 7-8 This truck has a noticeable lean to the left, caused by a broken spring.

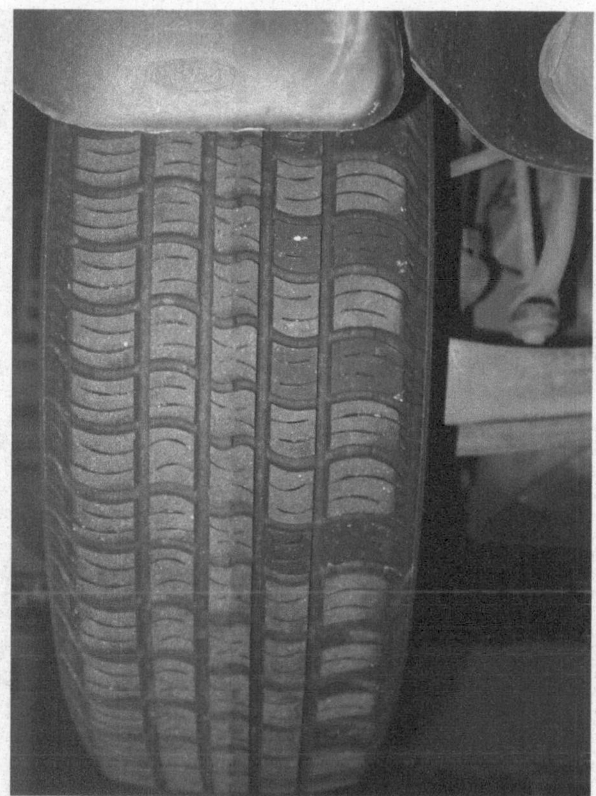

FIGURE 7-9 Reading tire wear can provide information about suspension components, such as dampers, that are worn and need to be replaced. Many times, tires provide information about the wheel alignment, tire pressure, and rotation maintenance of a vehicle.

FIGURE 7-10 Some damage is obvious, like bent strut shown here. Other times, components can have very slight amounts of damage but still cause problems.

Once you have completed this initial overall inspection, raise and support the vehicle, and look at the suspension parts. Look for any obvious damage, such as bent or broken parts, or leaking shocks (**Figure 7-10**). If nothing stands out as being wrong, proceed to more specific inspection items.

■ ***Ride Height.*** As springs age and weaken, vehicle **ride height**, the measurement of how high the vehicle sits off the ground, decreases. Broken springs (**Figure 7-11**) will cause a very quick and very noticeable drop in height, but sometimes broken springs do not cause as much of a difference (**Figure 7-12**). As springs weaken, the owner may not notice a gradual drop over years of service. Once the vehicle ride height falls below a certain spot, it may be impossible to bring the wheel alignment back into specification. Additionally, as the springs weaken and sag, their ability to handle the vehicle's weight and the constant jounce and rebound actions that occur while driving are reduced. A weak spring will be more likely to allow bottoming out of the suspension, resulting in a much harsher ride.

FIGURE 7-11 A failed coil spring in a MacPherson strut.

FIGURE 7-12 A broken coil spring but the break was not affecting the ride height.

Some trucks are prone to the rear leaf spring shackles rusting out and breaking (**Figure 7-13**). Without looking under the truck, it appears that the rear spring has failed. However, once inspected, it is obvious that the shackle has rusted apart.

Vehicle ride height can be checked in several different places depending on what the manufacturer specifies. One common way to check ride height is to measure the height from the top of the wheel opening to the ground (**Figure 7-14**). Another common method, especially on trucks and SUVs, is to measure at the lower control arm (**Figure 7-15**). Ride height measurements should usually be close to equal from front to rear and side-to-side. Always refer to the manufacturer's measurements, procedures, and specifications as some vehicles specify different ride heights at the front and rear.

As a rule, ride height should not vary more than about one-half of an inch (13 mm) from side-to-side or from front to rear. If the ride height is incorrect, carefully inspect the entire suspension, wheels, and tires before condemning the springs. Remember that replacement wheels and tires that have a different profile than the original equipment wheels and tires can affect ride height. Also, check the tire pressure; low tire pressure will also affect ride height measurements.

■ ***Shock Bounce Test.*** Severely worn dampers (shock absorbers) can be found by performing a bounce test. Bounce a corner of the vehicle up and down three times and count the number of bounces until the vehicle settles to a stop. Good dampers will usually settle out in two to three bounces. More than three bounces may mean the dampers are weak and not controlling the spring very well.

A test drive should also be performed to check damper condition over bumps, dips, and other body movements. How the dampers react to changing road conditions depends on how fast and how much they move. A simple bounce test in the shop does not check damper action under the different types of conditions experienced when driving so it is important to check the dampers over several different operating conditions.

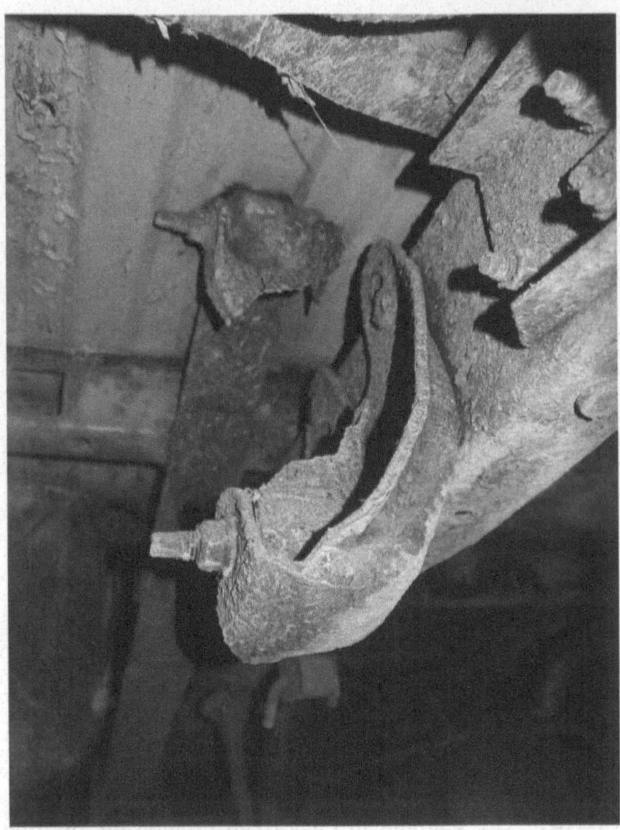

FIGURE 7-13 A rusted out and broken leaf spring shackle had a serious effect on ride height and ride quality.

FIGURE 7-14 Checking vehicle ride height to determine if the springs are good.

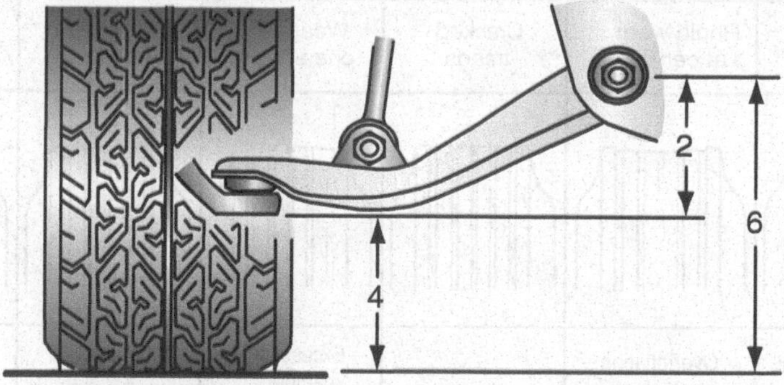

Item	Description
1	Distance between the ground and the lower knuckle surface near the ball joint
2	Distance between the ground and the center of the lower arm rearward mounting bolt
3	Ride height = 6–4

FIGURE 7-15 **Many vehicles require checking the ride height at specific locations in the front suspension.**

Visually inspect the dampers for signs of oil leaks. Any more than a very slight oil film around the upper piston seal means it is leaking excessively and needs to be replaced. You may have to move the dust boot covering the strut piston to check for a leaking strut (**Figure 7-16**).

Check the damper's upper and lower mounts and bushings. While it is not a common problem, the mounts can break, and the bushings can deteriorate, leaving the damper unsecured and rattling around.

■ ***Reading Tire Wear.*** A very important skill for anyone who is doing suspension, steering, and alignment repairs is the ability to read tire tread wear. The wear on a tire can provide a lot of information about the vehicle's condition and how well it is maintained (**Figure 7-17**). Being able to determine what factors are causing tire wear will help you to accurately diagnose suspension and steering problems.

It is not unusual for a vehicle to show more than one type of tire wear pattern. For example, an FWD car that does not have the tires rotated often enough and is only slightly out of alignment can show excessive wear on the front tires while the rear tires appear like new. This is because the front tires on FWD vehicles wear faster due to vehicle weight distribution and because of the turning and driving forces applied to them.

FIGURE 7-16 **Leaking dampers can no longer effectively control the spring and must be replaced.**

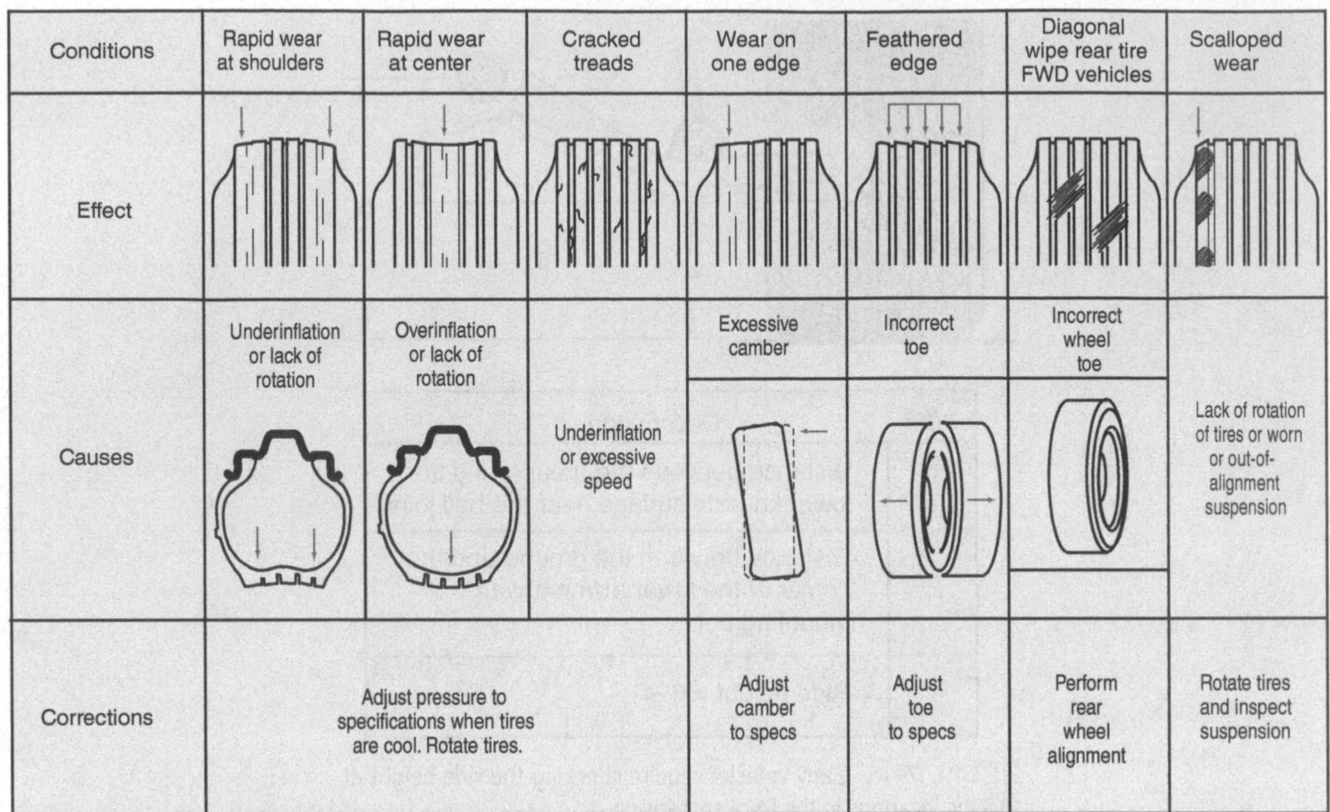

Conditions
Rapid wear at shoulders
Rapid wear at center
Cracked treads
Wear on one edge
Feathered edge
Diagonal wipe rear tire FWD vehicles
Scalloped wear
Effect
Causes
Underinflation or lack of rotation
Overinflation or lack of rotation
Underinflation or excessive speed
Excessive camber
Incorrect toe
Incorrect wheel toe
Lack of rotation of tires or worn or out-of-alignment suspension
Corrections
Adjust pressure to specifications when tires are cool. Rotate tires.
Adjust camber to specs
Adjust toe to specs
Perform rear wheel alignment
Rotate tires and inspect suspension

FIGURE 7-17 Examples of tire wear and the common causes.

■ ***Noises.*** Noises can be a challenge for even the most experienced technician, but if you have a thorough understanding of the vehicle, its systems, and its components, you will be much more likely to be able to quickly determine the causes of most noises.

Noises can be caused by worn suspension components, such as loose ball joints and strut bearing plates. A test drive is usually needed to pinpoint a noise. Try to determine under what conditions the noise occurs, such as over bumps, when turning, or when accelerating or decelerating. You may need to perform the test drive with the customer so that you know exactly what the customer's complaint is. If the customer is not available, perform a test drive with another technician so both of you can listen and discuss what could be the cause.

Sometimes, even more help is needed to pinpoint a noise. Use of an electronic stethoscope, especially the type with several individual microphones, can be extremely helpful. **Table 7-1** provides a list of common suspension noises and the typical causes.

■ ***Test Driving.*** A test drive is often necessary when a technician is trying to determine the cause of a noise or vibration. Even though as a student you are not likely to perform test drives, it is an expected and necessary part of being a professional technician. As a student, you may be able to drive the vehicle around the lab or parking lot enough to confirm the customer's complaint. Test drives can be extremely important to diagnose many concerns, especially noises and vehicle handling complaints. Once the problem is diagnosed and repaired, a final test drive confirms that the repair is complete, and the vehicle is ready to be returned to the customer.

■ ***Body Lean or Sway.*** While test driving the vehicle, pay attention to how the car body reacts while cornering. Excessive lean or sway is caused by broken sway bar

TABLE 7-1 Common suspension noises

Knocking noise over bumps	Loose ball joint, worn strut, worn strut bearing plate, worn tie rod, worn shock, or broken sway bar link
Clunking or knocking on acceleration/deceleration	Worn control arm bushings, worn ball joints, or worn FWD CV joint
Creaking or groaning	Worn control arm bushings, worn ball joints or tie rods, broken spring, or worn or damaged upper strut plate

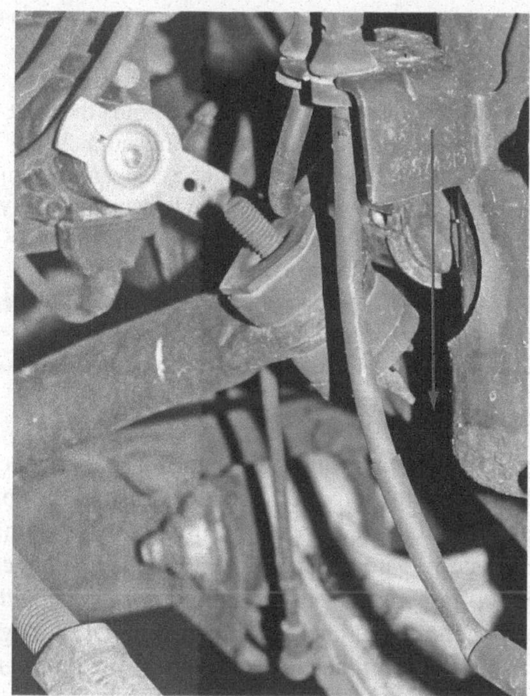

FIGURE 7-18 Broken stabilizer bar links are a common concern and usually cause noises and increased body roll when cornering.

links (**Figure 7-18**). If the sway bar links are broken, the body will pitch or roll much more than normal when turning a corner.

Many sway bar links are one piece and use a ball-and-socket connection, while other sway bar links are attached with a link kit, made up of bushings and washers and secured by tightening the nut and bolt. If one sway bar link is broken, both sides should be replaced.

■ ***Service History and TSBs.*** You should also perform a search of the vehicle's service history and for any relevant technical service bulletins (TSBs). If the vehicle is regularly serviced by your shop, there should be a service history available. Look through previous repair orders to check for prior repairs to the suspension or steering systems. This is called checking the vehicle's service history, which is often an important source of information about the vehicle (**Figure 7-19**). Also, check the service information for any TSBs related to the suspension or steering systems, especially if you are attempting to diagnose a noise complaint (**Figure 7-20**). The few minutes it takes to

RECOMMENDED SERVICES

OPERATION	OPERATION DESCRIPTION	MO/MI	TOTAL	OPERATION	OPERATION DESCRIPTION	MO/MI	TOTAL

SERVICE HISTORY

DATE	REPAIR ORDER	MILEAGE	ADVISOR	TECHNICIAN	TYPE	OPERATION	OPERATION DESCRIPTION
05/13/15	360871	39368	453	99	C	10JAZ01	DRIVEABILITY CONCERN
				99	C	01JAZZIS	MULTI-POINT INSPECT
03/03/15	355034	36272	453	258	W	10JAZ06	CHECK ENGINE LIGHT
				258	W	45JAZ	STEERING/SUSPENSION
02/10/15	353564	36094	434	258	W	01JAZ	MINOR OIL SERVICE
				258	C	01JAZZIS	MULTI-POINT INSPECT

SALESPERSON NO.

S E R V I C E

STATE REG# 034

FIGURE 7-19 An example of service history information on a repair order.

Technical Service Bulletins

Advanced Filters

System: Suspension

Symptom: ALL

Trouble Code: ALL

Quick Search

By TSB #:

Keyword:

Search

Show Recalls Only

2 matching items were found 1

TSB Description	TSB #	Issue Date	Type
Suspension Squeak/Squawk - Front Stabilizer Bar Bushings	02-002-04 REV. A	5/25/2004	Information, Labor, Training
Special Offset Ball Joint - Allows Adjustment To Caster And Cambe...	02-006-04	8/20/2004	Information, Labor, Technical, Warranty

FIGURE 7-20 Checking for TSBs can save you time and trouble when trying to solve a customer concern.

check these items could save you hours of work by identifying a known problem for you.

FRONT SUSPENSION SERVICE

Inspecting and repairing suspension concerns is often some of the first work an entry-level technician will perform. This is because much of the common suspension work, such as replacing shocks and struts, is generally simple and straightforward work.

Remember, the suspension system does not work alone. It works with the frame, steering system, wheels, and tires. Therefore, it can be affected by problems in these other areas.

■ ***Strut Service.*** The most common suspension type in use today is the strut suspension, whether a MacPherson strut, modified strut, or part of a multilink arrangement. Strut components suffer from the same types of problems as the components of any other type of suspension, leaking dampers, and worn springs. Struts also can develop problems with the upper bearing plates that require service or replacement. To conduct a thorough inspection of the strut to determine exactly what repairs are necessary, check the following:

- Inspect for leaks and conduct a bounce test.
- Measure vehicle ride height to check for spring sag.
- Test drive the vehicle to listen for noises over bumps and while turning.
- Inspect the strut assembly to see if it is bent from a collision or other impact.

Strut service, such as replacing the damper, usually requires removing the strut from the vehicle. Removing a typical MacPherson strut is shown in **Photo Sequence 5**.

Once the strut is removed, mount it in a strut spring compressor. Use a grease pencil or paint pen to index the mount and coil spring for proper alignment. When it is properly secured and the spring is compressed, remove the nut holding the strut piston rod to the upper bearing plate. With the nut removed, slowly remove the strut from the spring and bearing plate. Remove the upper mount and inspect it for wear and damage. Inspect the bearing or bushing closely. Worn or damaged strut bearings can cause noise, hard steering, and poor steering return problems. Turn the bearing by hand, and note any roughness. Replace the bearing or mount assembly if there are any signs of excessive wear or damage (**Figure 7-21**).

When you are replacing the damper, in many cases, you will need to remove the spring insulator, bumpers, and piston rod cover from the old strut and transfer them to the new unit. The spring insulator goes between the bottom of the coil spring and the spring seat attached to the strut. This insulator helps reduce noise as the spring moves on the strut. Inspect the insulator for damage, and replace it if necessary. The bumper prevents completely collapsing the spring, called *bottoming out*, on hard impacts. If the bumper shows signs of damage or rot, it should be replaced. Also, inspect the bumper for signs that the spring has been collapsing. The spring itself may show signs of wear as the coils come into contact with each other. This can happen as the spring ages and weakens, allowing the spring to collapse. The piston rod cover protects the piston from damage during operation. Make sure the new strut matches the old strut and reassemble the components.

Service Note

Do not remove the nut that holds the strut piston to the upper plate. This nut is keeping the strut spring compressed and the unit together. Removing this nut before the strut is safely compressed is dangerous and allows the spring to release. This situation can damage vehicle components, such as drive axle boots, brake lines, and wheel speed sensor wiring.

Install the new strut into the spring and upper mount, and guide the piston through the mounting hole. Place the washer (if used) over the piston and install a new retaining nut. Most struts require the retaining nut to be torqued after the strut is reinstalled on the vehicle and the weight is back on the suspension. It is usually easier to place the upper strut mount into position and install the bolts or nuts to hold the strut assembly when reinstalling. Once the strut is held in place, reattach the lower section of the strut to the steering knuckle. Torque all fasteners to specifications. Wheel alignment needs to be checked and set once the strut replacement is complete.

Some vehicles have a replaceable shock cartridge. In some vehicles, the strut cartridge can be replaced while the strut is still installed on the vehicle. To replace the cartridge, a special tool kit may be needed to keep the components aligned properly during disassembly and reassembly. Always follow the service information for procedures. In general, remove the upper retaining nut and then the cartridge (**Figure 7-22**). Ensure that the correct amount of oil remains in the strut body when you are replacing the cartridge. The oil is there to remove heat from the shock during operation.

Once the worn components have been replaced, reassemble the strut. Install the upper retaining nut, but do

PHOTO SEQUENCE 5

Removing, Replacing, and Reinstalling a MacPherson Strut

PS5-1 Before removing the strut, mark the position of the strut to the body. Some struts can only be installed in one way, but others allow for the mount to be in different positions. Mark the strut to realign the mount to the body.

PS5-2 Raise and secure the vehicle on a lift.

PS5-3 Remove the wheel and tire to access the strut.

PS5-4 Locate the strut-to-knuckle attachment and determine the tools needed to remove the strut.

PS5-5 Remove the strut-to-knuckle bolts.

PS5-6 Remove the knuckle from the strut. Be careful not to overextend the front axle when pulling the knuckle out.

PS5-7 Following the instructions on the strut compressor, mount the strut in the compressor so that it is secured properly.

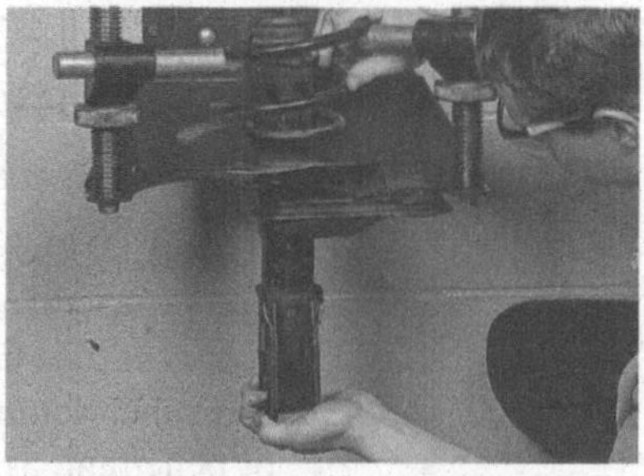

PS5-8 Compress the spring and remove the shock piston-to-mount nut. Remove the strut body and compare the new and old parts. Transfer any covers and insulators to the new strut and install it into the spring. Hand-tighten the shock piston nut and release the spring tension.

PS5-9 Reinstall the strut onto the vehicle. Torque all mounting bolts and nuts to specifications and torque the piston-to-mount nut with the vehicle on the floor and the weight on the suspension.

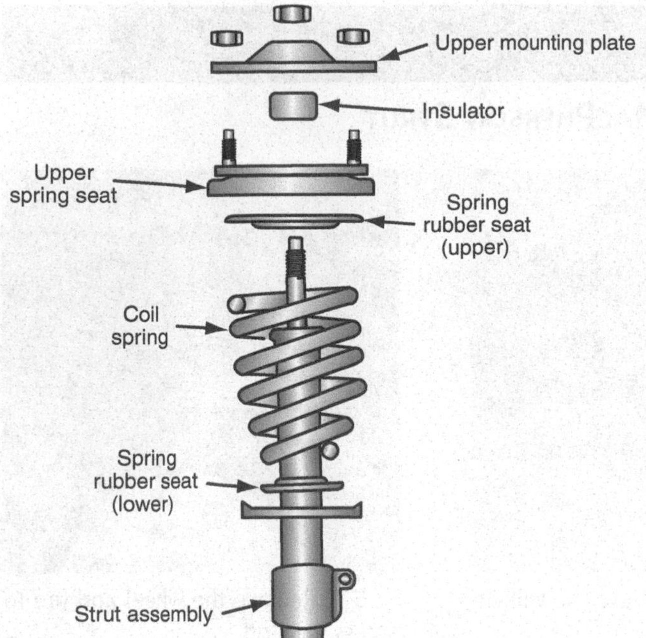

FIGURE 7-21 An exploded view of strut components. Each piece should be checked during strut service.

not torque it until the strut is reinstalled on the vehicle and weight is placed on the strut.

When you are replacing gas-charged struts or shocks, the gas charge needs to be released before throwing the strut or shock away. Refer to the shock manufacturer's service information before attempting to release the pressure in the shock as the location and size of hole varies. In general, to release the gas pressure, a small hole is drilled in the lower section of the strut or shock to allow the gas to escape. The location to drill the hole may be marked on the strut body for reference.

Depending on the age and condition of the vehicle, replacing the strut assembly may be more cost-effective than repairing it. This is an option to give to the customer. A replacement assembly contains the entire strut unit, which simplifies the removal and installation.

■ ***Damper Inspection and Replacement.*** Once a damper starts leaking oil or fails to adequately dampen the springs, it should be replaced. In most cases, replacement is very straightforward and simple, but always refer to the service information for procedures and tightening specs.

For front dampers on nonstrut vehicles, locate and remove the upper mount, then remove the lower mount and remove the damper. When you are installing the new unit, be sure to replace the bushings and washers in the correct order, and do not overtighten the bushings. Rear damper replacement is similar but may require the use of a jack to support the rear axle or control arm. Common mounting arrangements are shown in **Figure 7-23**.

Because dampers are exposed to the outside world, you should expect the mounting hardware to be rusted and possibly frozen into place. Apply a penetrant to the mounting bolts and nuts to ease removal. If the hardware is severely rusted, applying heat can often help with removal. If available, an induction heater, such as the Mini-Ductor, is an excellent tool to heat rusted and seized fasteners (**Figure 7-24**). This tool uses high-frequency magnetic fields flowing through a flexible conductor to heat metal objects. After a few seconds, the metal is heated and can be easily removed without any flames. If this type of tool is not available, you may need to use a traditional torch to apply heat to a fastener.

Install the new damper and torque all fasteners to specs. Lower the vehicle and bounce each corner of the vehicle. Make sure there are no noises as the vehicle is bounced.

■ ***Spring Inspection.*** As suspension springs age, they become less capable of supporting the weight of the vehicle; this will cause the vehicle to sag. Locating the vehicle's ride height specs and measuring the ride height is the best way to check for weak and sagging springs. Also, visually inspect the springs and the rebound bumpers (**Figure 7-25**). A bumper that appears shiny means that it has been contacting the frame, which can indicate

Service Note

Do not attempt to depressurize a gas shock or strut without reading and understanding the procedure described in the service information. Personal injury may result from improper service.

Safety Warning

Always follow all safety precautions when using any type of torch. Do not use the torch without proper instruction and supervision. Do not apply heat to any chemical or penetrant, as a fire may result. Using a torch to heat rusted shock hardware can result in the rubber bushings catching on fire. Be careful to only heat the hardware enough to allow it to break loose. Do not apply heat directly to the shock body as the increase in temperature will increase the pressure in the shock. This can cause the shock to burst, causing severe injury and burns.

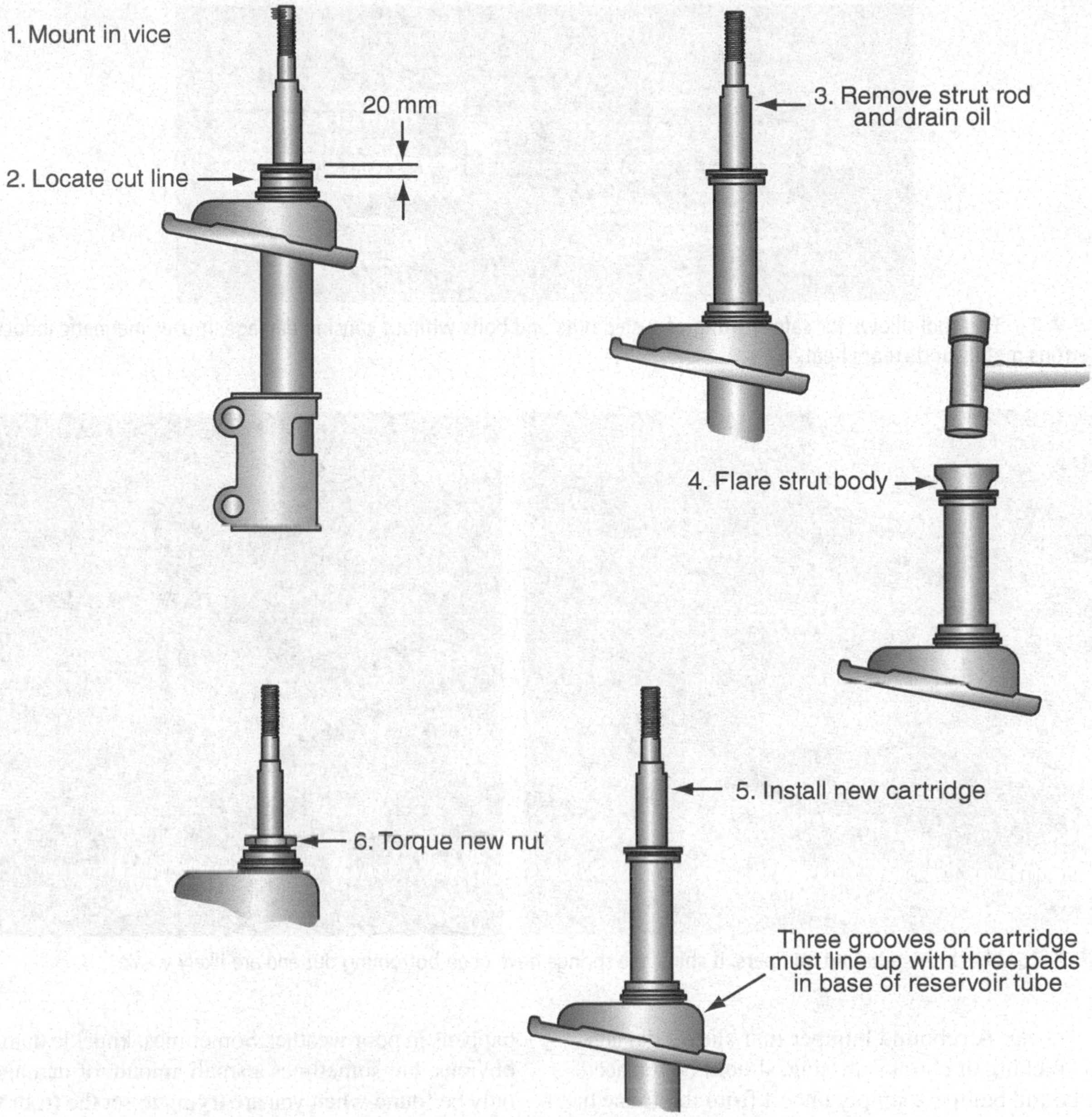

1. Mount in vice
20 mm
2. Locate cut line
3. Remove strut rod and drain oil
4. Flare strut body
5. Install new cartridge
6. Torque new nut
Three grooves on cartridge must line up with three pads in base of reservoir tube

FIGURE 7-22 Removing a strut cartridge may require cutting the shock housing. Always follow the manufacturer's service procedures when working on struts.

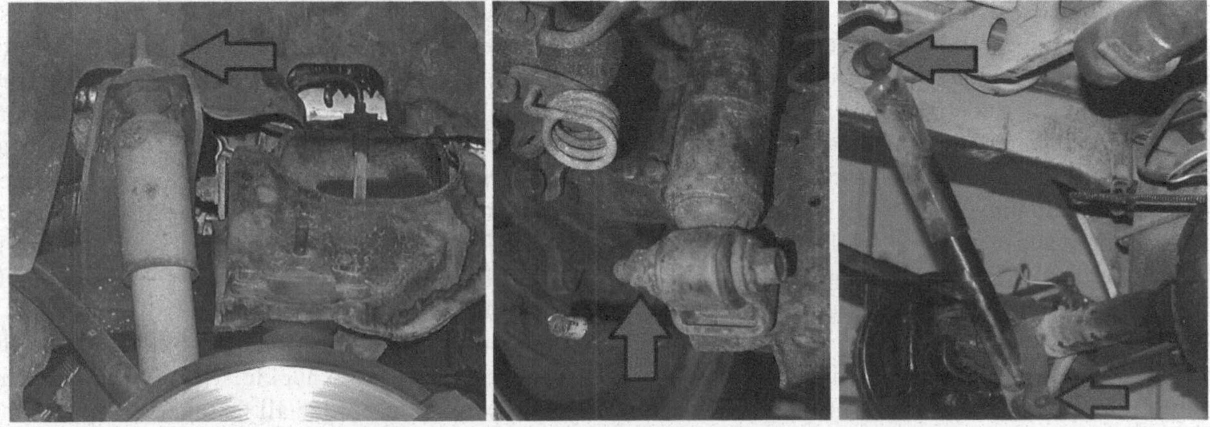

FIGURE 7-23 Examples of shock mounts. Inspect the mounting bushings for rot and cracking.

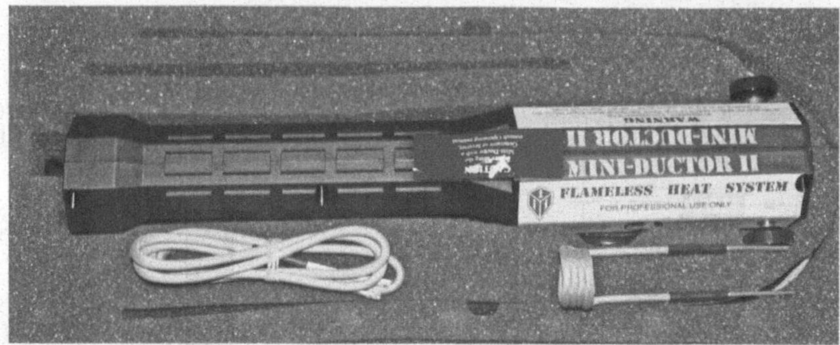

FIGURE 7-24 This tool allows for safe heating of rusted nuts and bolts without causing damage. It uses magnetic induction to excite ferrous metals and create heat.

FIGURE 7-25 Check the rebound bumpers. If shiny, the springs have been bottoming out and are likely weak.

a weak spring. A rebound bumper that shows damage, such as cracking or chunks missing, should be replaced. Most rebound bumpers simply unbolt from the frame or body for replacement.

Many vehicles have rubber spring insulators mounted between the spring and the body. These insulators are used to reduce noise when the spring twists slightly as the vehicle jounces and rebounds. Check the insulators to make sure they are intact and positioned correctly between the springs and the frame.

Inspect torsion bars and their mounts. The mounts can deteriorate over time and allow the torsion bar to move. This often causes a clunking noise when going over bumps. Check both the front and rear mounts for damage. Check also the crossmember for any possible damage. Impact damage to the crossmember can affect the rear mount and adjuster.

■ ***Steering Knuckles.*** Steering knuckles are subject to damage from collisions. This is common if the vehicle slides into a curb or other solid object, as sometimes happens in poor weather. Sometimes, knuckle damage is obvious, but sometimes a small amount of damage will only be found when you are trying to set the front wheel alignment.

Many vehicles have aluminum steering knuckles (**Figure 7-26**). Aluminum steering knuckles cannot take the same amount of impact force as older cast iron knuckles. Consequently, when you are working with aluminum knuckles, use extra caution when separating the ball joints, so you do not damage the softer aluminum.

Steering knuckle replacement is usually straightforward. On an RWD vehicle, remove the wheel and tire and unbolt and remove the brake caliper and rotor. Remove the tie rod end, and support the lower control arm with a jack. Remove the upper and lower ball joint nuts, and separate the ball joint studs from the steering knuckle. Once the ball joints are loose, simply remove the knuckle. When you are reinstalling the knuckle, torque all fasteners to specs, and use new cotter pins.

FIGURE 7-26 Many suspension components are made of aluminum to save weight. Servicing these components requires special care to not damage the soft aluminum.

On FWD vehicles, steering knuckle replacement is similar but requires removing the drive axle. Remove the wheel and tire and unbolt and remove the caliper and rotor. Remove the axle nut and the outer tie rod. Remove the lower ball joint connection to the knuckle, and remove the bolts that hold the strut to the knuckle. An **axle puller** may be needed to push the axle out of the bearing in the steering knuckle (**Figure 7-27**). Depending on the type of wheel bearing, you may be able to unbolt the bearing and install it into the new knuckle. If the bearing is press-fit, you may need to replace the bearing. Install the new knuckle, and torque all fasteners to specs. Use new cotter pins where applicable. Bearing replacement is covered in detail in Chapter 5.

■ ***Ball Joint Inspection and Replacement.*** How you check for worn ball joints depends on what type of suspension you are working on and where the load-carrying ball joint is located. Remember that the load-carrying ball joint is located in the control arm on which the spring sits.

Most MacPherson strut suspensions do not have a load-carrying ball joint, but that does not mean that those joints do not wear or need to be replaced, just that you have to check them for wear differently. **Figure 7-28** shows how to check the ball joint on a MacPherson strut suspension.

FIGURE 7-27 An axle puller is often necessary to remove an FWD hub and steering knuckle.

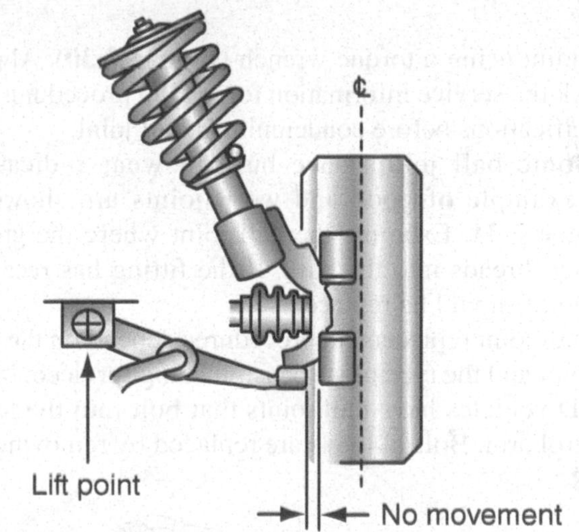

FIGURE 7-28 Check for play in the ball joint by supporting the front end under the frame, not the control arm.

Jacking the vehicle in the correct location unloads the ball joint so it can be checked for wear and movement. To check ball joints on SLA, modified strut, and many multilink systems, jack the vehicle and support the lower control arm (**Figure 7-29**). Place a pry bar under the tire, and pry up against the bottom of the tire. Most ball joints should have zero vertical movement when checked this way, though there are some applications that do allow a slight bit of movement. Check the service information for ball joint movement specifications before condemning a slightly loose joint as worn out and needing replaced. If necessary, attach a dial indicator to measure ball joint play to determine if the joint needs to be replaced.

This has been a typical method of checking ball joint condition; however, some manufacturers specify checking ball joint wear by measuring the rotational torque of

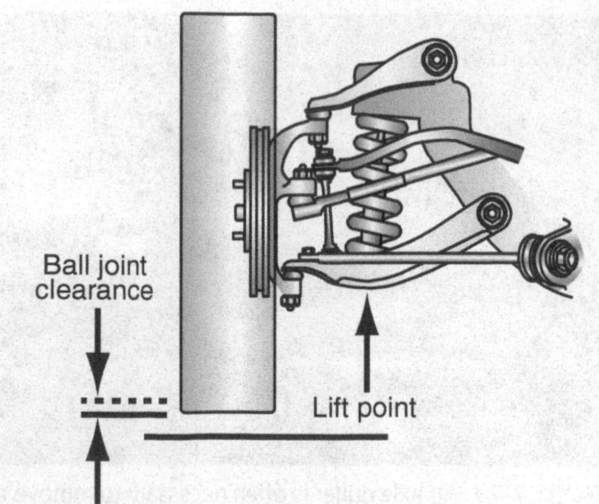

FIGURE 7-29 On SLA and multilink suspensions, support the lower control arm to check for ball joint wear.

the joint using a torque wrench (**Figure 7-30**). Always check the service information for testing procedures and specifications before condemning a ball joint.

Some ball joints have built-in wear indicators. An example of good and worn joints are shown in **Figure 7-31**. Examine the ball joint where the grease fitting threads into the base. If the fitting has receded, the joint should be replaced.

Ball joint replacement procedures depend on the type of joint and the type of suspension being serviced. Many FWD vehicles have ball joints that bolt into the lower control arm. Bolt-in joints are replaced by removing the ball joint to steering knuckle fastener and then unbolting the ball joint. Some vehicles have the ball joint permanently secured to the lower control arm, and the entire arm must be replaced when the joint is worn.

Many vehicles have press-fit ball joints. These require a ball joint press to install the new joint. The worn joints are either pressed from the control arms or hammered out to save time. An illustration of a ball joint press to install a ball joint is shown in **Figure 7-32**. To remove a press-fit joint, first select the correct size adaptor to allow the joint to pass out of the arm and into the adaptor. Install the press over the joint, and tighten it until the joint pushes out of the arm. To install the new joint, select an adaptor to fit the base of the joint and an open adaptor to allow the joint to pass through the arm into the adaptor. Position the joint and press so the joint is centered in the hole in the control arm. Tighten the press until the ball joint is fully seated. Loosen and remove the press and ensure the joint is fully seated in the arm.

Some ball joints are riveted into the control arm at the factory. These rivets must be removed and then the new ball joint is bolted back into the control arm. Rivets are usually drilled out and then driven out with a punch.

Before installing the new ball joint, check the installation instructions. Some GM and Ford vehicles require the joint be installed a certain way or indexed toward a specific location on the control arm due to the design of the opening in the joint. Failure to install the joint correctly can cause the joint to fail or snap off.

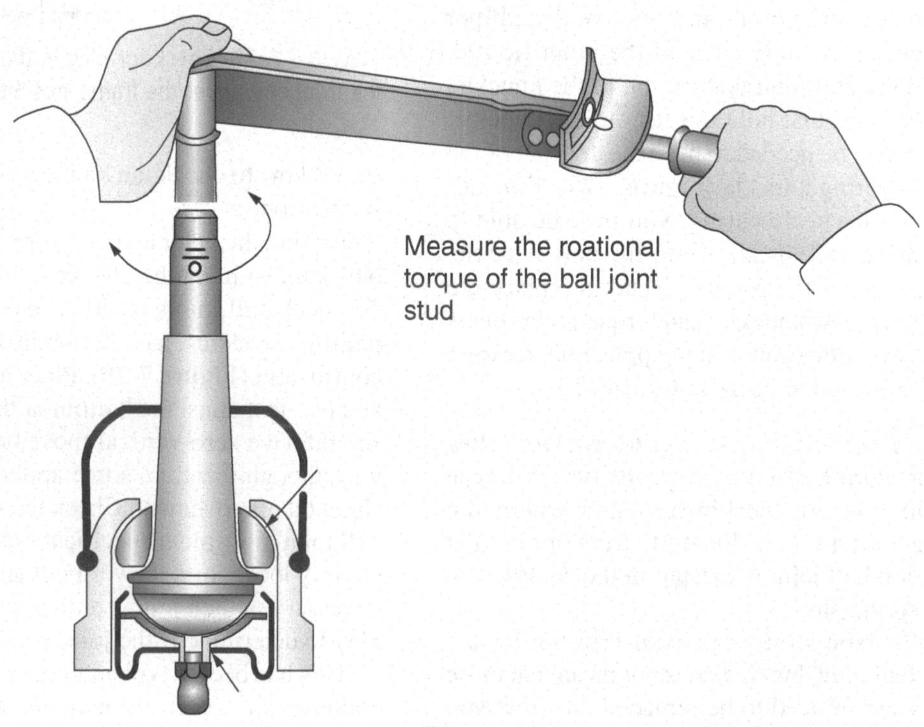

FIGURE 7-30 An excessively worn ball joint will be loose and will have a lower rotational torque than specified.

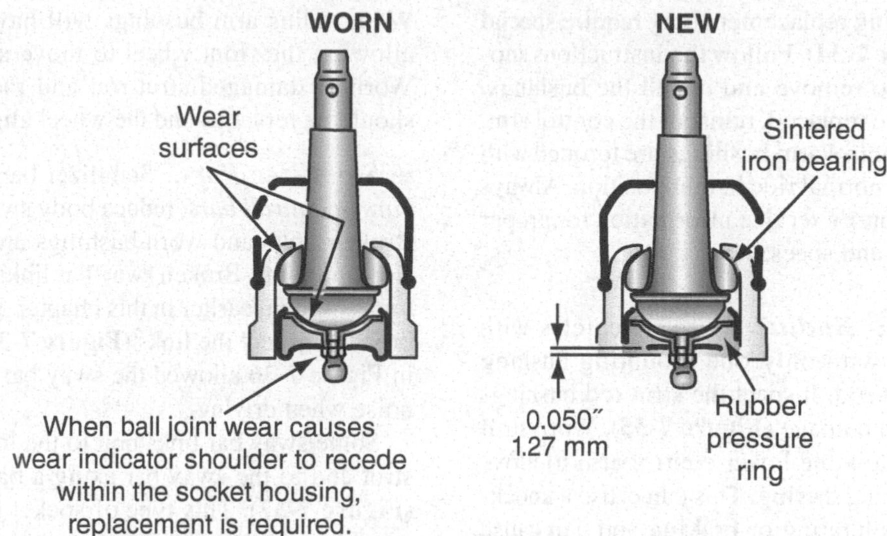

FIGURE 7-31 Wear-indicating ball joints use the grease fitting to show wear.

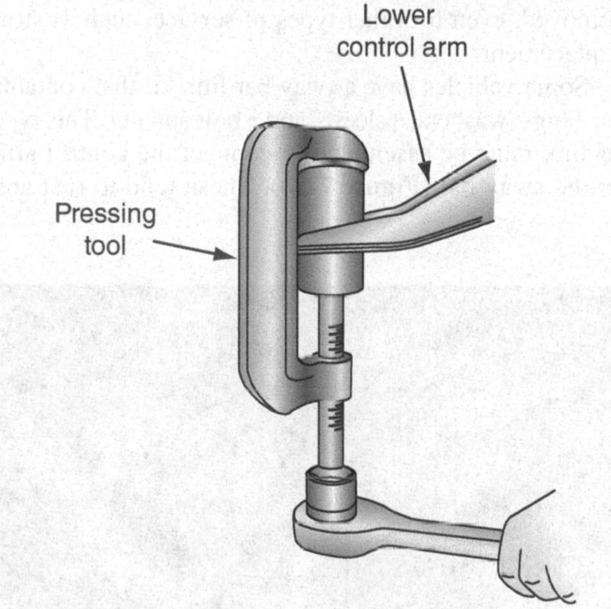

FIGURE 7-32 Using a ball joint press to install a press-fit ball joint into the control arm.

Once the new ball joints are installed, be sure to install the grease fitting and grease the joints. Tighten all fasteners to the correct torque spec, and replace all cotter pins with new pins. If the new cotter pin does not go into the stud because the holes do not align, tighten the nut until the cotter pin fits. Do not loosen the nut to install the cotter pin.

■ ***Control Arm Inspection and Service.*** Control arms usually do not require replacement unless damaged by a collision, though some must be replaced if the ball joint is part of the arm and not separately replaceable. Control arm bushings are common wear items and require replacement.

Inspect control arms for damage, such as cracking, rust-through, and bending, whenever the suspension is inspected. Cracks and bending can result from severe impacts or collisions. Control arm bushings deteriorate over time and can separate, causing noises and handling complaints (**Figure 7-33**). Bushing replacement generally requires that the control arms be removed from the vehicle.

Removing lower control arms on SLA and modified strut suspensions will require removing the coil spring and shock absorber. Once the coil spring and shock are removed, disconnect the lower ball joint at the steering knuckle and the sway bar link. Unbolt the control arm from the crossmember and remove the arm. Removing the upper control arm does not require spring removal, but the lower control arm will need to be supported. Disconnect the upper ball joint from the steering knuckle, and then unbolt the upper control arm from the frame.

FIGURE 7-33 A worn-out lower control arm bushing.

Control arm bushing replacement may require special bushing tools (**Figure 7-34**). Follow the instructions supplied with the tool to remove and install the bushings. Once the bushings are replaced, reinstall the control arm. In many cases, the control arm bushings are torqued with the control arm in its normal ride height position. Always follow the manufacturer's service information for proper torquing procedures and specs.

■ ***Strut Rods and Radius Arms.*** Vehicles with lower control arms with only one mounting bushing will also have a strut rod. Inspect the strut rod bushings for rot, cracking, and damage (**Figure 7-35**). Worn strut rod bushings will allow the lower control arm to move forward and back during driving. This can cause a knocking sound when accelerating or braking and can cause the vehicle to pull to one side as wheel position changes.

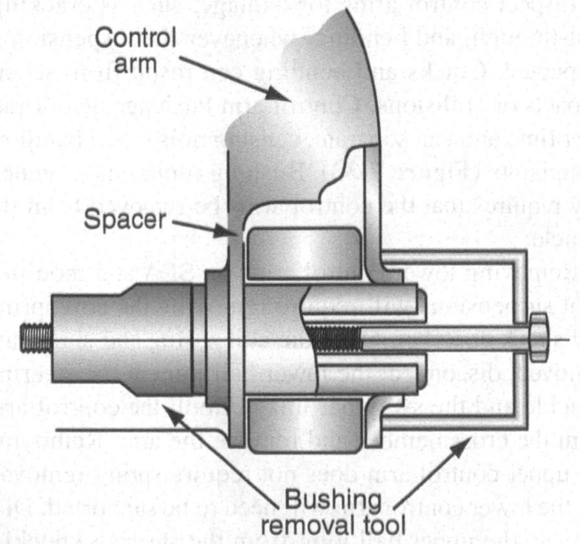

FIGURE 7-34 Using a bushing to tool to press out a control arm bushing.

FIGURE 7-35 A worn and falling apart strut rod bushing.

Worn radius arm bushings will have the same affect, allowing the front wheel to move and to cause noise. Worn or damaged strut rod and radius arm bushings should be replaced and the wheel alignment checked.

■ ***Stabilizer Bars.*** Stabilizer bars, also called *sway bars* or *antiroll bars*, reduce body sway when cornering. Broken links and worn bushings are common suspension problems. Broken sway bar links are usually easy to see, as shown earlier in this chapter. Even if they are not broken, inspect the links (**Figure 7-36**). The link shown in Figure 7-36 allowed the sway bar to move and make noise when driving.

Some sway bar links bolt to the lower control arm or strut and to the sway bar using a ball-and-socket joint (**Figure 7-37**). This type of socket often does not disassemble easily as the stud tends to spin in the socket when the nut is turned. Because of this, this type of sway bar link usually requires replacement if it needs to be removed, even for other types of service, such as strut replacement.

Some vehicles have a sway bar link kit that contains bushings, washers, a sleeve, and a bolt and nut. This type of link must be assembled to connect the control arm to the sway bar (**Figure 7-38**). These tend to rust and

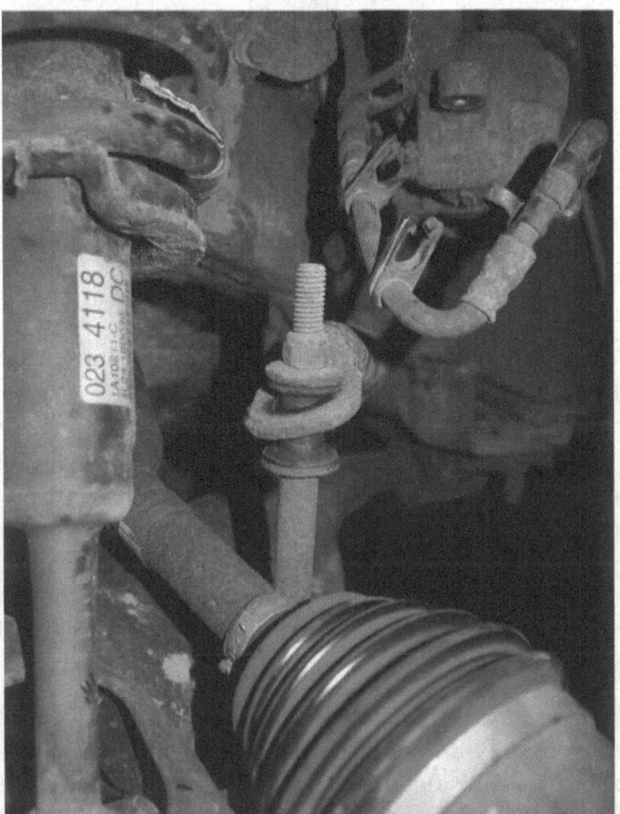

FIGURE 7-36 Even if not broken in half, worn-out bushings can cause concerns with the sway bar.

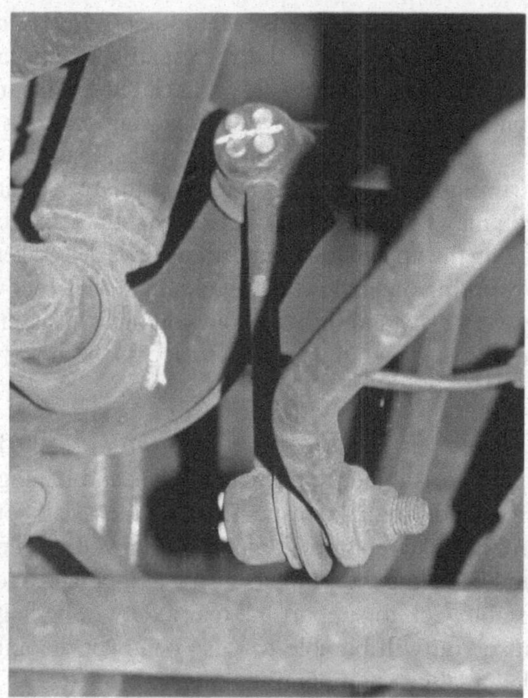

FIGURE 7-37 This type of sway bar link uses ball-and-socket joints, which often do not come apart easily and usually require replacement once removed.

FIGURE 7-38 This type of sway bar link is a kit of bushings, washers, bolts, and spacers.

seize, making disassembly difficult. It is not uncommon for the link bolt to break during an attempt to loosen and remove the link. Because both types of links often do not disassemble without damage, it is a good idea to include the cost of replacing these components into an estimate for services whenever the sway bar link needs to be removed.

To replace sway bar links, such as those shown in Figure 7-37, first try to remove the nut securing the ball studs. In many cases, the stud will spin, making removal difficult. Some of the studs have an internal hex built into the end of the stud. This allows you to install a hex driver or Allen wrench to hold the stud from turning while loosening the nut. If this does not work, you may have to cut the nut or stud. Once the old link is removed, install the new link and tighten the nuts to specifications.

To replace sway bar links like those shown in Figure 7-38, remove the nut, washers, bushings, sleeve and bolt. If it is seized, you may have to cut the link in half to remove it. When installing the new parts, make sure you place the washers and bushings in the correct order. Washers are placed against bushings to isolate metal components and prevent noise. Once assembled, tighten them to specifications or until the bushings compress to the diameter of the washers. Do not overtighten and flatten the bushings.

Inspect the sway bar bushings and mounts. The bushings isolate the sway bar from the body or frame. Replace the sway bar bushings if they are rotted, cracked, or falling apart. Make sure the bushing mounts or brackets hold the sway bar securely in place. A loose bracket will allow the sway bar to knock against the frame or body.

■ ***Suspension System Lubrication.*** Even though most modern vehicles have sealed and lubricated-for-life suspension and steering joints, there are some that still require periodic lubrication. **Grease fittings**, or Zerk fittings, are small, nipple-like fittings threaded into a component for the injection of grease. A small, spring-loaded check ball inside the fitting allows grease in but keeps it from being pushed back out. **Figure 7-39** shows a grease fitting on an outer tie rod end.

For most vehicles, high-pressure chassis grease is acceptable for joint lubrication, but always refer to the manufacturer's service information to ensure that the correct grease is used.

Figure 7-40 shows how to lubricate front suspension and steering components. When you lubricate suspension and steering joints, first wipe the old grease and dirt from the grease fitting. This prevents the dirt from being injected into the joint with the new grease. Then, attach the end of the grease gun to the fitting and slowly inject grease. Watch the grease boot as you put grease in.

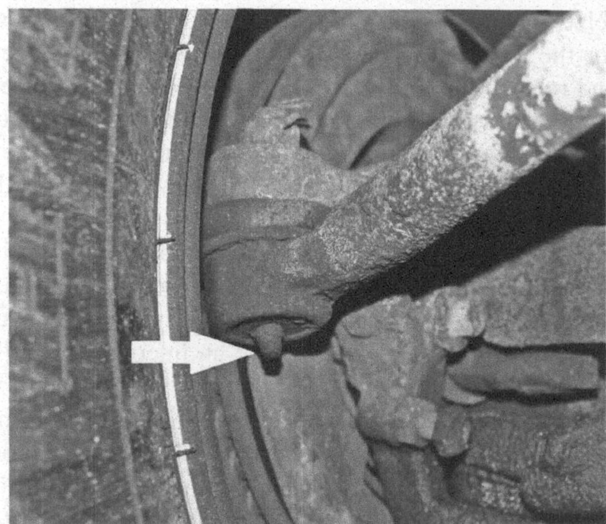

FIGURE 7-39 An example of a grease fitting in an outer tie rod end.

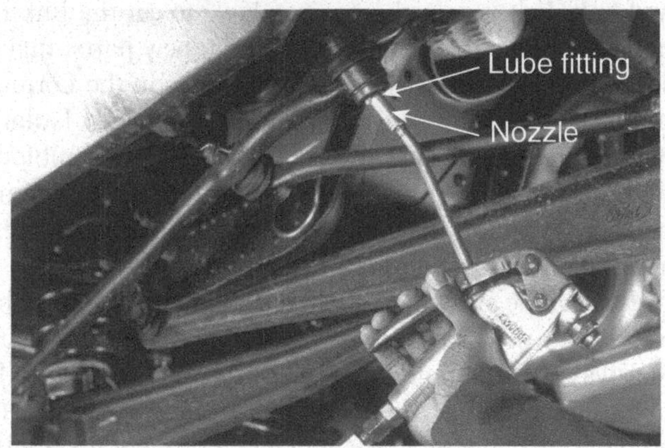

FIGURE 7-40 Using a grease gun to inject grease into a steering component.

Many boots will expand and release excess grease when full, but some will continue to expand until the pressure ruptures the boot. Wipe any excess grease from the grease fitting and around the boots so that it does not get onto other parts. Your customer will not be happy if the excess grease from a tie rod spills over onto the wheel and makes a greasy mess.

REAR SUSPENSION SERVICE

Most of the repairs to the rear suspension are similar to those performed on the front suspension.

■ ***Rear Strut Service.*** Rear strut replacement is nearly identical to that for front struts. Rear struts, unless they are on a four-wheel steering car, do not have bearings in the upper strut mount, so strut service is typically just replacement of the damper portion or replacing the entire unit if both the spring and damper need to be replaced.

Many rear struts bolt either into the trunk or into the rear firewall area where the rear seatback and parcel shelf are located (**Figure 7-41**). Check the service information about how to remove the rear dampers or struts before beginning to work. In some cases, the rear seats and trim must be removed to replace the rear struts. If so, you may want to start the replacement procedures by removing the upper strut connections first, before getting dirty working on the lower strut mount. You do not want to get very dirty and then find out you must disassemble the rear passenger compartment to access the upper strut mount.

■ ***Damper Inspection and Replacement.*** Rear dampers are subject to the same wear and damage as the front and should be inspected for oil loss and bounce control just like the front dampers. When you are removing and installing the dampers, using a floor jack or other adjustable jack to support the rear axle will be helpful, because you will be able to raise or lower the axle as needed to align the mounting bolts.

Place a jack or jack stands under the rear axle before removing the mounting hardware. Remove the mounting bolts, nuts, and any other related hardware. When you are installing the new dampers, do not overtighten and flatten the bushings. Tighten all fasteners to specs. If a tightening spec is not available, tighten until the rubber bushing is squeezed to the same diameter as the washer it is against.

■ ***Track Bars and Control Arms.*** Many rear suspension systems use track bars or control arms to limit rear axle movements. Both track bars and control arms are subject to damage from collisions and impacts and can be bent. This will cause the rear axle

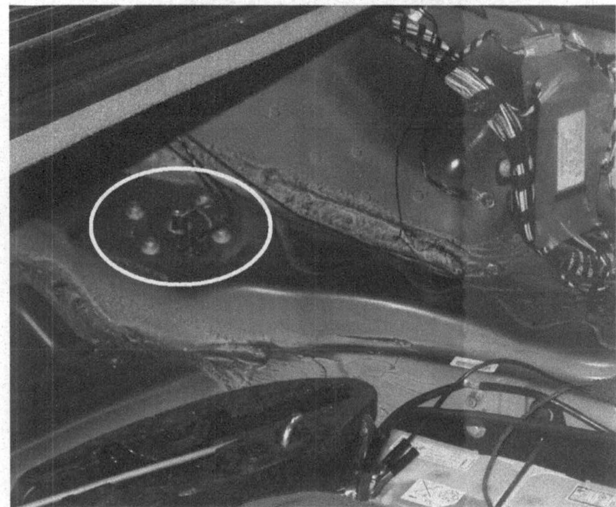

FIGURE 7-41 Some vehicles mount the rear struts behind the rear seats or in the trunk/cargo area.

FIGURE 7-42 Bent or broken rear control arms can allow the wheels to be far out of position.

to be misaligned to the body and will affect wheel alignment (**Figure 7-42**). Inspect for damage and replace as necessary.

As with other suspension parts, track bars and control arm are attached with bushings. Bounce the vehicle and listen for any noise from the bushings. Raise the vehicle and inspect the bushings for rot, cracking, and damage. Worn bushings can make noise and allow for changes in rear axle position, which will affect wheel alignment. Replace any worn or damage bushings.

■ ***Rear Spring Service.*** Inspect the rear coil springs for sagging by checking ride height. Check the spring insulators between the spring and body and the spring and axle. Worn insulators can cause noise as the vehicle goes over bumps. Check rear leaf springs for broken leaves, broken brackets that hold the springs together, damaged insulators between leaves, and worn or damaged shackles and bushings. Inspect the area of the frame where the leaf spring shackles are mounted as this is a common rust-through location.

On RWD vehicles, check where the leaf springs attach to the rear axle housing. A centering pin is used to keep the axle from moving but this pin can rust or break. This allows the rear axle to shift, which causes serious thrust angle and wheel alignment concerns.

Wheel Alignment

Even though performing wheel alignments is not within the scope of this text, you should be able to perform a **prealignment inspection**. This is done to determine whether the vehicle is able to be aligned. If any of the steering or suspension components that affect wheel alignment are worn or damaged to the point where the alignment will be affected or will not be able to be set, those components will need to be replaced before attempting to align the wheels.

PREALIGNMENT INSPECTION

Before a vehicle can have a wheel alignment performed, it must be inspected. Typically, a two-part inspection will take place. The initial check is the visual inspection. If the vehicle passes this, then a more detailed inspection is performed.

■ ***Visual Inspection.*** A good visual inspection should include looking at the tire wear patterns and making sure the wheels and tires are the correct size and are inflated properly. Check the tire placard for size and inflation information. A vehicle with different tire tread designs can experience ride and handling issues due to the different ways each tire responds to the road. Except for a few models of sports cars, front and rear tires should be the same size, type, and have the same tread pattern to prevent any tire-induced pulling or vibration issues. Also, check the tire sidewalls and wheels for damage. A bent wheel may not have a large impact on the overall wheel alignment, but it will probably cause a vibration the customer can feel, which could be interpreted as you not having correctly performed the wheel alignment.

A broken spring will usually be obvious because the vehicle will be sitting lower in one corner. However, spring sag that occurs over time is generally not as noticeable. Locate the ride height specs, and measure the ride height. If the height is incorrect, either too high or too low from modifications or weak springs, the wheel alignment will probably be affected. Check for signs of modifications, such as lowering springs, cranked up torsion bars, or lift kits (**Figure 7-43**). Be sure that the customer is aware that incorrect ride height may mean that the wheel alignment may not be able to be set to specs or even to a decent position.

■ ***Detailed Inspection.*** Perform a careful inspection of steering and suspension components. Be sure to check ball joint wear and play, wheel bearings, control arm bushings, and all steering linkage ball-and-socket joints.

One method of checking the steering linkage is the dry-park check. This involves placing the vehicle on a drive-on-style lift, such as an alignment lift. Have a helper sit in the driver's seat, and raise the vehicle until you can comfortably stand under the front suspension. With the engine off, have your helper rock the steering wheel back and forth just off steering wheel center as you watch and listen to the steering linkage. Loose components will often be easily visible and may even make popping noises. By moving the steering with the engine off, the weight of the front of the vehicle puts stress on the linkage as the tires are pushed back and forth, allowing you to identify loose components.

FIGURE 7-43 Check trucks with torsion bars for excessive ride height. Tightening the torsion bars will increase ride height but also decrease camber.

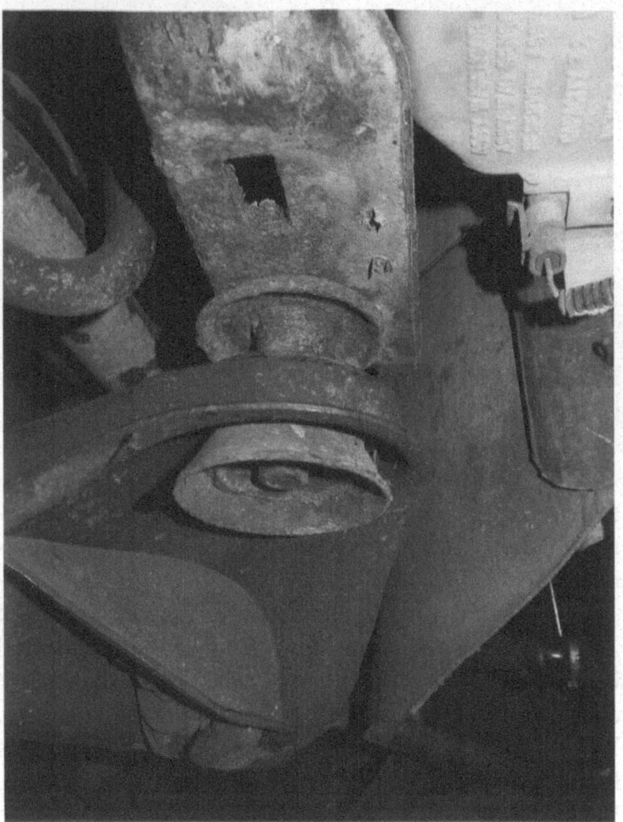

FIGURE 7-45 This vehicle had all four cradle bushings falling apart, allowing the entire front end to move when turning, braking, and accelerating.

Locate the vehicle ride height specs and measure the ride height as indicated (**Figure 7-44**). Some vehicles may not have specific measurements but will instead state that the ride height should not differ by more than a certain amount, often about one-half inch, from either side-to-side or front to rear.

Inspect frame and sub-frame bushings. Older vehicles can have worn or even missing frame bushings (**Figure 7-45**). This can cause noise, pulling, torque steer, memory steer, and numerous other concerns as the suspension shifts against the body.

Test driving is the best way to get a feel for how the vehicle rides and handles. Perform test drives for alignments over several types of road conditions to get an adequate understanding of how the vehicle performs. You should drive where you can take several corners or sharp turns and drive on both flat and crowned roads. Pay attention to how the vehicle responds to steering input, listen for any noises, and feel if there is any pulling or leading to the right or left. If the vehicle is FWD, accelerate quickly but safely to check for torque steer. Torque steer is the tendency of the vehicle to pull either right or left under acceleration due to having two different length front dive axles.

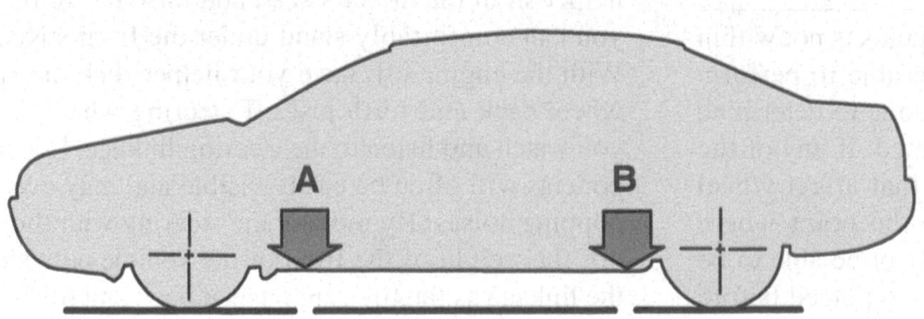

Measure at Point A and Point B. Reading should vary no more than 0.5″ (13 mm) from A to B or from side-to-side.

Vehicle		
Turbo X	A. 7.4″	B. 7.2″
Sedan	A. 8.75″	B. 9.0″
Truck	A. 9.8″	B. 10″

FIGURE 7-44 An example of ride height specifications.

Noises can be caused by loose steering and suspension components, defective dampers, broken sway bar links, loose wheel bearings, and even from the springs themselves. Determining the exact cause of a noise can be a challenge. You may need to enlist the help of another student or technician to help narrow down the possibilities. Using a stethoscope or Chassis Ear is another way to help identify the cause of a noise.

Broken sway bar links will allow the vehicle to sway or roll excessively when cornering. When test driving, pay attention to how the body responds when making a turn. A broken sway bar link can also cause a knocking sound from the front end as the sway bar moves up and down during turns and contacts the broken link.

WHEEL ALIGNMENT ANGLES

Wheel alignment is a general term referring to the relationship of the wheels and tires to the suspension and steering components and to various reference lines. One purpose of the suspension and steering systems is to maintain the best overall wheel alignment for the vehicle. This reduces tire wear, improves driving dynamics, and improves fuel economy. As a technician, you should understand the basic alignment angles and how they connect to vehicle performance.

■ ***Geometric Centerline and Thrust Line.*** An imaginary line that runs through the middle of the vehicle from front to rear is called the **geometric centerline**. From that, reference point, the thrust line, is measured. The **thrust line** is the line that extends from the direction of travel of the rear axle (**Figure 7-46**). When the thrust line and geometric centerline are the same, a vehicle has a zero-thrust angle. If the rear axle or rear tires do not track perfectly along the centerline, either a positive or negative thrust angle exists (**Figure 7-47**). If the thrust line and centerline do not overlap each other, the rear axle is pointing either to the left or to the right. This can cause the vehicle to track or pull to one side and cause the steering wheel to be off-center when driving.

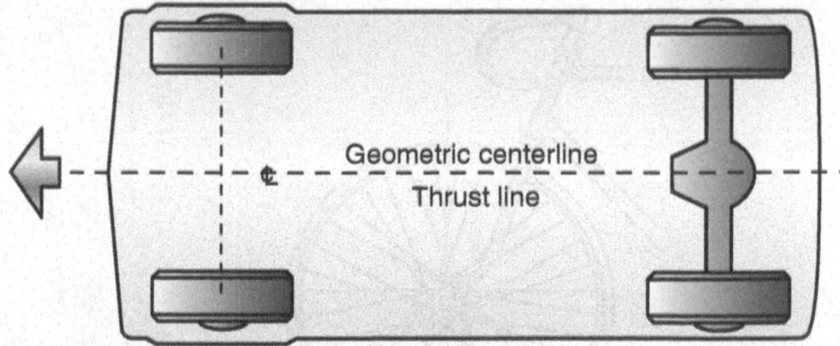

FIGURE 7-46 The geometric centerline is a line running through the center of the vehicle from front to rear. The thrust line should overlap the centerline.

To correct for negative or positive thrust angle, rear wheel toe is adjusted. If rear toe is not adjustable, something has caused the rear of the vehicle, either the axle or the frame, to move.

■ ***Caster.*** **Caster** is a steering and suspension angle and is responsible for straight ahead tracking and helping steering return. Caster is easily visualized in **Figure 7-48** with bicycle forks. The caster angle is formed between the steering pivot points, such as the upper strut mount bearing and the lower ball joint. A zero-caster angle places the two pivots directly in line vertically with each other (**Figure 7-49**). Caster is not a tire angle as

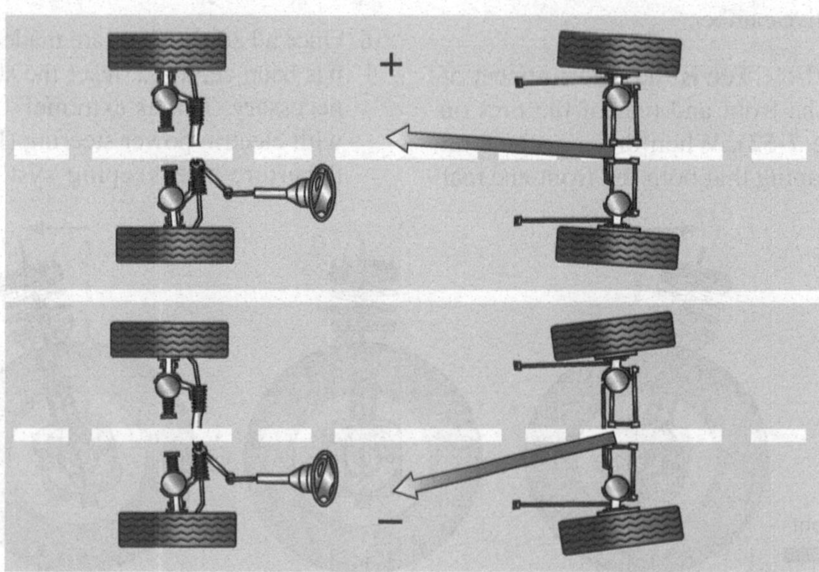

FIGURE 7-47 If the rear axle or rear tires are not aligned properly, the thrust line will not be along the centerline. This can cause the vehicle to pull in the direction of the thrust line.

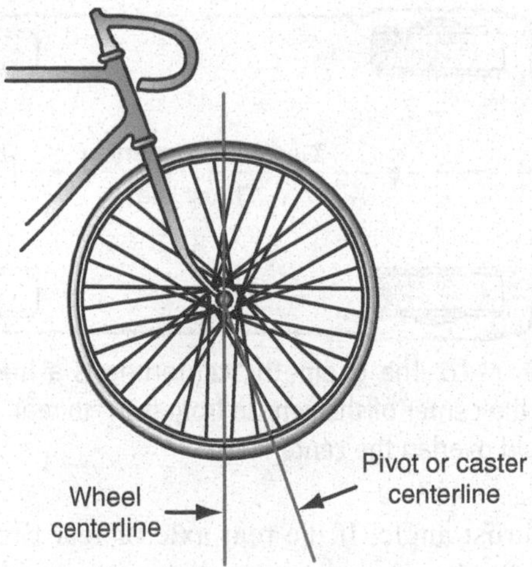

FIGURE 7-48 The caster line is drawn through the steering pivots. Bike have positive caster, which gives them straight-ahead tracking while riding.

camber and toe are and does not typically cause tire wear if incorrect. Nearly all modern vehicles have a positive caster setting, which places the upper pivot behind the lower. This increases stability but also increases steering effort. Negative caster reduces steering effort but at the expense of stability. Incorrect or uneven caster can cause a lead or pull to the side with the least amount of caster.

■ *Camber.* **Camber** is the tilt of the tire away from 0° vertical (**Figure 7-50**). Camber settings vary depending on the type of vehicle but are usually set either slightly (0.5°) positive or negative. Decreasing camber (negative) can improve handling and cornering ability but can increase the wear on the inside edges of the tires (**Figure 7-51**). Excessive camber can cause tire wear and pulling to the side with the most positive camber.

■ *Toe In and Toe Out.* **Toe** is the measurement of the distance between the front and rear of the tires on the same axle (**Figure 7-52**). When driving, the total toe should be zero, meaning that both the front and rear tires are perfectly in line and parallel to each other (the first image of **Figure 7-53**). Toe in or positive toe occurs when the tires are closer together at the front compared to the rear (the second image in Figure 7-53). Toe out or negative toe is when the tires point away from each other at the front. Incorrect toe can cause rapid tire wear as the entire contract patch of the tire is pushed or scrubbed sideways across the road surface. Incorrect rear toe affects the thrust angle (**Figure 7-54**).

■ *Basic Wheel Alignment Procedures.*

1. Perform a full inspection of the vehicle to check for any worn or damaged steering and suspension components.
2. Check the tire condition and set the pressure to specifications.
3. Check the vehicle ride height.
4. Attach the alignment targets or sensors to the vehicle and compensate the alignment computer and vehicle to obtain alignment readings (**Figure 7-55**).
5. If adjustments are necessary, adjust in this order:
   - **i** Rear camber
   - **ii** Rear toe
   - **iii** Front caster
   - **iv** Front camber
   - **v** Front toe

   This order is used because adjustments to caster and camber change the position of the steering knuckle, which will affect toe. If you set toe before caster or camber, you will most likely have to adjust the toe a second time.
6. Once all adjustments are made and the steering wheel has been centered, reset the steering angle sensor if necessary. This is extremely important on vehicles with electric power steering (EPS), any type of lane departure/lane keeping system, self-parking, and

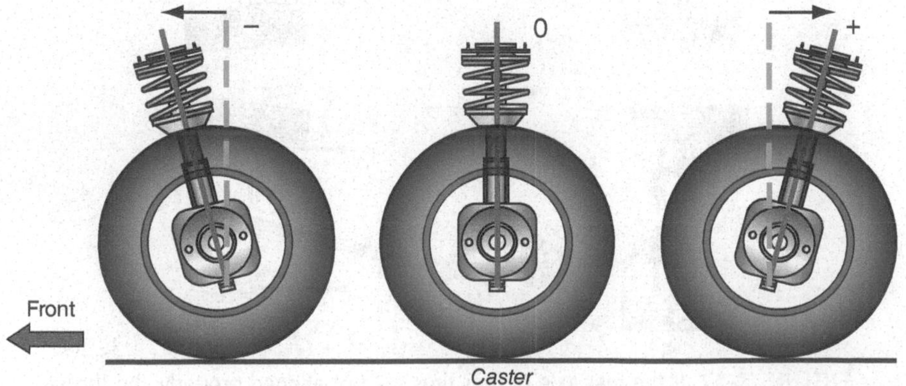

FIGURE 7-49 Caster as seen by the angle of the steering axis from the upper strut mount to the lower ball joint.

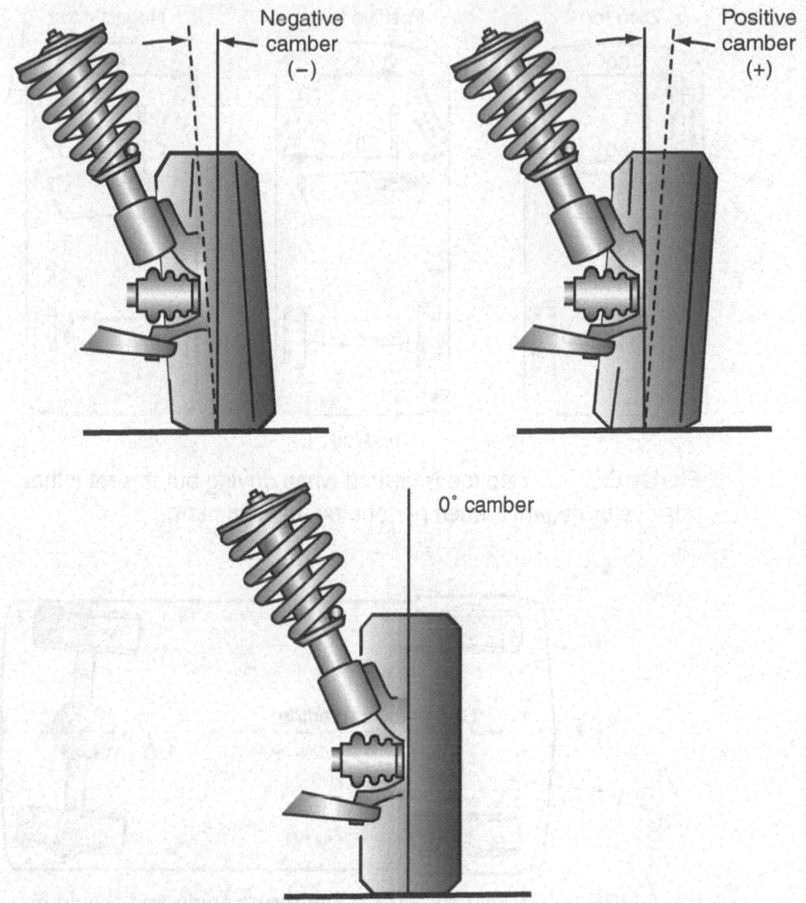

FIGURE 7-50 Camber is the inward or outward tilt of the tire at the top.

FIGURE 7-51 Excessive negative camber will wear the insides of the tire.

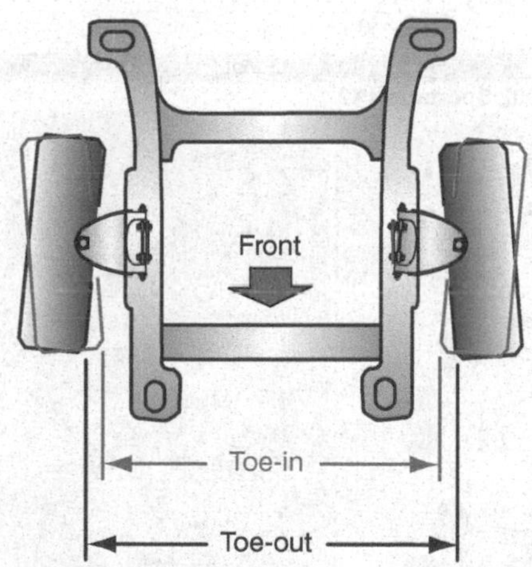

FIGURE 7-52 Toe as seen from over top of the front tires.

other active safety systems that rely on steering input and control.

7. Test drive the vehicle and make sure the steering wheel is straight and there is no pull or other handling concern.

These steps are general and are not meant to take the place of actual service information. Always refer to the manufacturer's service information and follow the procedures as specified.

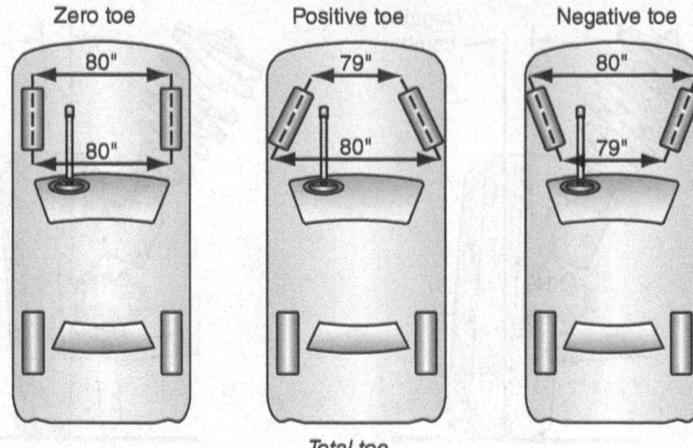

FIGURE 7-53 Zero toe is desired when driving but it is set either positive or negative when performing an alignment.

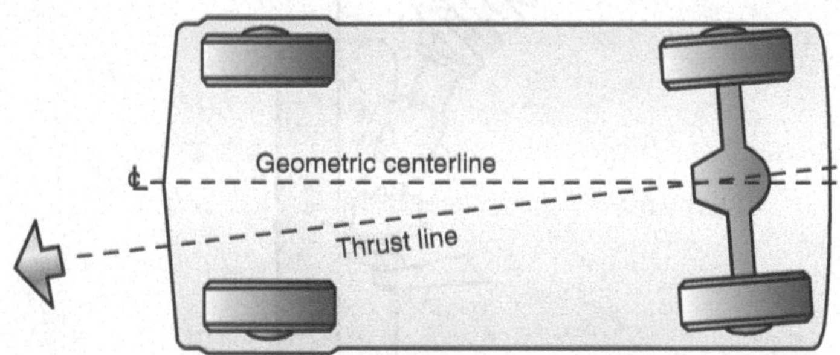

FIGURE 7-54 Rear toe affects the thrust angle and should be set (if possible) before setting the front wheel alignment.

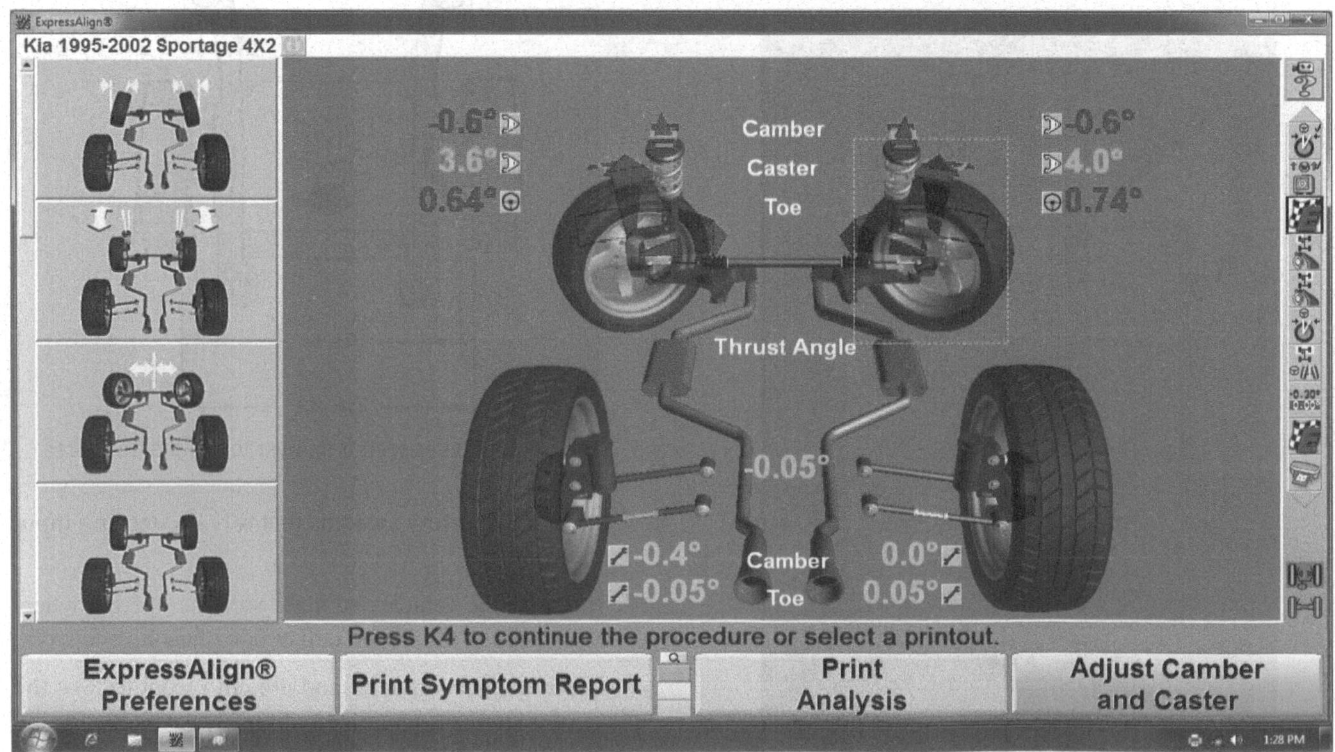

FIGURE 7-55 An example of alignment readings on an RWD vehicle with a nonadjustable rear suspension.

SUMMARY

Suspension system repairs are some of the most common types of repairs technicians perform.

Safe working practices include proper tool use and care, following the proper repair procedures, staying focused on the tasks at hand, and taking the time to perform your work properly.

Follow the proper service procedures for safely compressing and handling the springs.

A thorough inspection of the suspension system must be performed before beginning any repairs.

As the springs age and weaken, vehicle ride height will decrease.

Worn dampers will generally bounce more than two to three times when performing a bounce test.

A test drive is often necessary when trying to determine the cause of a noise or vibration.

Special tools may be required to remove and install a torsion bar.

When working with aluminum knuckles, use extra caution so you do not damage the softer aluminum.

Caster is the angle most responsible for straight tracking of the car.

Camber is the inward or outward tilt of the tire at the top.

Toe is the distance between the front of the tires compared to the distance at the back of the tire.

Make sure a steering angle sensor reset is performed after the wheel alignment is adjusted.

REVIEW QUESTIONS

1. If the vehicle ride height is lower than specification, it is likely due to weak or worn _______________.

2. When servicing MacPherson struts, a spring _______________ is needed to safely disassembly the strut assembly.

3. The _______________ angle or line should be parallel with the geometric centerline or the vehicle may pull to one side when driving.

4. _______________ loss, noise, or excessive bouncing indicates a damper needs to be replaced.

5. A _______________ or _______________ fitting is used to lubricate ball joints and other front-end components.

6. *Technician A* says wear on the inside half of a front tire indicates excessive positive camber. *Technician B* says wear on the inside half of a front tire indicates excessive negative caster. Who is correct?
   a. Technician A
   b. Technician B
   c. Both A and B
   d. Neither A nor B

7. A knocking sound from the front suspension is heard during turns: *Technician A* says a worn ball joint may be the cause. *Technician B* says a broken sway bay link may be the cause. Who is correct?
   a. Technician A
   b. Technician B
   c. Both A and B
   d. Neither A nor B

8. Two technicians are discussing wheel alignments: *Technician A* says toe is set first to get the steering wheel straight, then caster and camber are set. *Technician B* says that if rear toe is incorrect, the thrust angle will be affected. Who is correct?
   a. Technician A
   b. Technician B
   c. Both A and B
   d. Neither A nor B

9. *Technician A* says if a vehicle's ride height is below specifications, wheel alignment can be checked but may not be able to be set to specifications. *Technician B* says if ride height is too low, the dampers are weak and need to be replaced. Who is correct?
   a. Technician A
   b. Technician B
   c. Both A and B
   d. Neither A nor B

10. All of the following statements about wheel alignment are correct except:
    a. A vehicle will pull to the most positive camber.
    b. Toe may be adjustable on both the front and the rear.
    c. Caster is adjusted last because it does not affect tire wear.
    d. Excessive positive camber can wear the outside edges of a tire.

CHAPTER 8

Steering System Principles

Chapter Objectives

At the conclusion of this chapter, you should be able to:

- Identify the components and functions of the steering system.
- Describe the operation of the rack and pinion and recirculating ball gearboxes.
- Explain the various types of steering linkage arrangements and components.
- Explain the operation of power assist systems.
- Explain the operation of electric power assist systems.
- Describe the function of the power steering pressure switch. (ASE Education Foundation MLR 4.B.23)
- Describe the function of steering angle sensors. (ASE Education Foundation MLR 4.B.23)

KEY TERMS

collapsible steering column	**parallelogram linkage**	**steering angle sensor (SAS)**
drive belt	**power steering pump**	**steering ratio**
end-takeoff rack	**rack and pinion**	
flexible coupler	**recirculating ball gearbox**	

The steering system, along with the suspension system, allows the driver to safely and easily control the vehicle's direction while driving. To accomplish these goals, the steering system works with components of the suspension to provide for the turning movement of the wheels. In addition to connecting the driver to the wheels, the steering system also provides feedback to the driver from the front tires. This feedback, called *road feel*, is used by the driver to determine how the vehicle is handling.

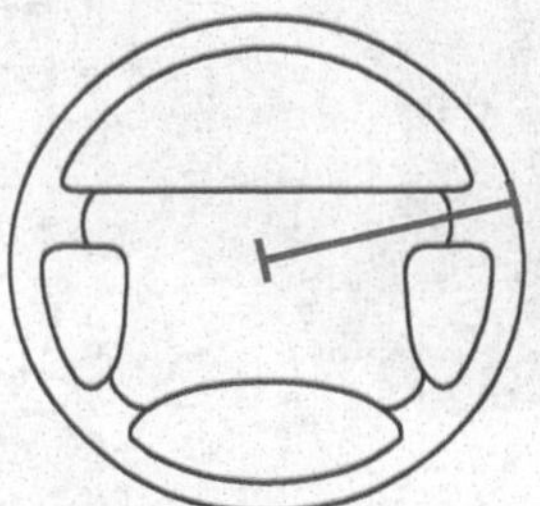

The distance from the center to the outside acts as a lever to increase the force applied to the steering shaft running through the center of the column.

FIGURE 8-1 The steering wheel is a lever, used to increase the force applied by the driver.

Functions and Basic Principles

The most basic function of the steering system is to allow the driver to safely and precisely steer the vehicle. Beyond this, the steering system also provides a way to reduce driver effort by making the act of steering the vehicle easier. The components of the steering system also absorb some of the road shock before it gets to the driver. In many ways, very little has changed in the operation of the steering system or in some of the system's components since the earliest automobiles. Things that have changed include the use of electric assist and the integration with various safety systems.

BASIC PRINCIPLES

Power steering systems use several basic principles to decrease driver effort. These include leverage, hydraulics, and electricity.

■ ***Mechanical Advantage of the Steering System.*** Leverage, or mechanical advantage, is used at the steering wheel and in the steering gearbox to increase the force supplied by the driver. The gears inside the gearbox act as levers, increasing mechanical advantage and reducing driver effort to turn the wheels.

> **Word Wall**
>
> *Steering Gearbox*—Steering gearbox is the general term for any type of mechanical system used to turn the rotary motion of the steering wheel and shaft into linear or side-to-side movement of the steering linkage.

Leverage is quite visible in the steering system. Think of the steering wheel as a lever (**Figure 8-1**). The force that you apply to the wheel while turning is applied over the radius of the steering wheel, which allows the steering wheel to act as a lever and to increase the force applied to the steering shaft.

The steering shaft in turn uses the force exerted on it by the steering wheel to act as the input for the steering gearbox. The gearbox uses two gears to further increase the mechanical advantage and decrease driver effort. The gearbox also converts the rotary motion of the steering wheel and shaft into a linear or back-and-forth motion that moves the wheels. **Figure 8-2** shows how a gearbox changes rotary motion into linear motion. The pinion gear rotates and drives the rack gear side-to-side. The number of the teeth on the pinion gear compared to the rack gear creates the gearbox **steering ratio**. This ratio creates leverage to make steering easier for the driver.

Why is the use of leverage important in the steering system? Try this experiment: Ask your instructor to provide you with a lab vehicle. Turn the engine off but leave the steering column unlocked and the vehicle sitting normally on the shop floor. Next, try to push the front wheels side-to-side by yourself. How easy was this to do? Most likely, you were not able to move the wheels much, and the little bit they did move required a lot of

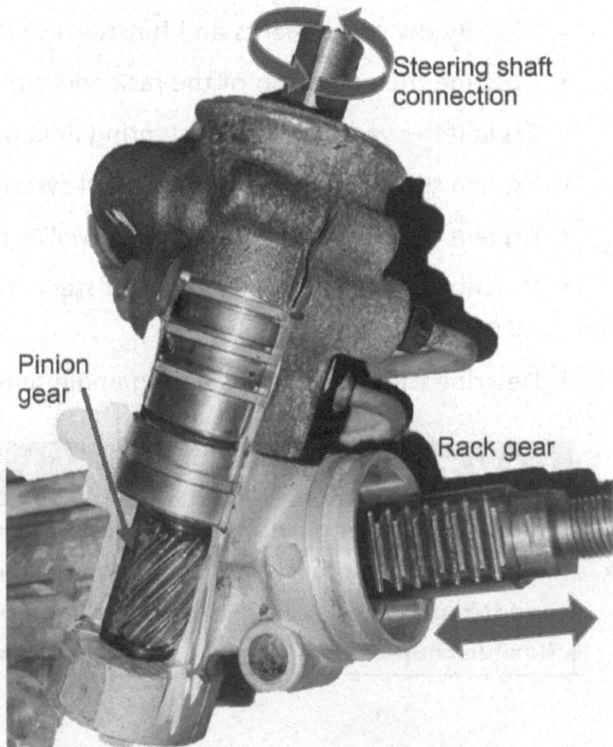

FIGURE 8-2 When the steering wheel is turned, the input shaft to the steering gearbox turns. This motion is then converted into side-to-side motion and the turning of the front wheels.

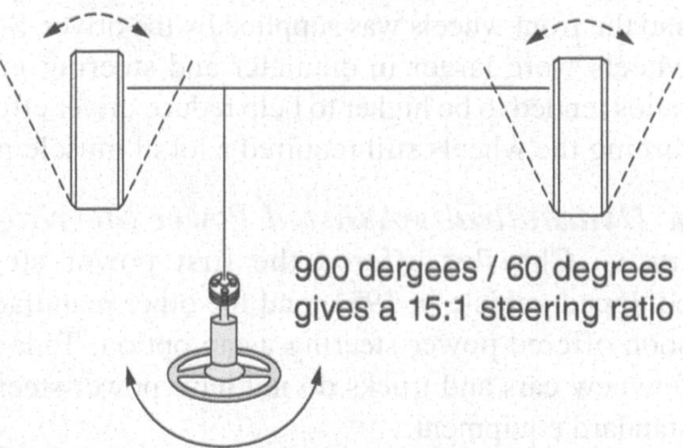

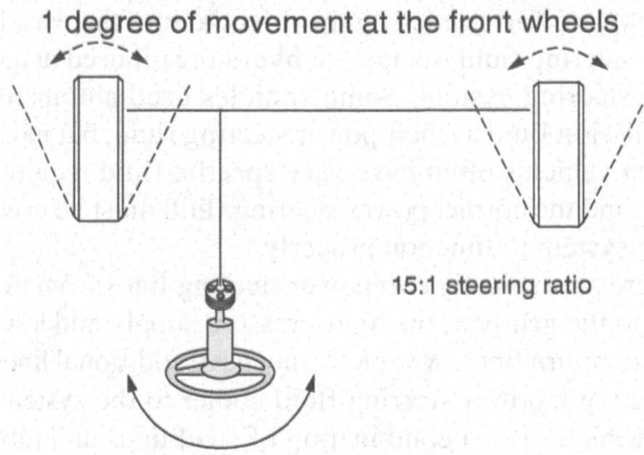

FIGURE 8-3 Gearbox ratios vary depending on the type of vehicle. Slow ratios provide easy but imprecise steering, while fast ratios provide feedback and precision but increase driver effort.

effort. Moving the wheels takes a lot of effort, which is why leverage is important to the steering system.

■ ***Steering Ratio.*** On the average vehicle, the steering wheel will turn about two and a half to three times completely around, from the right steering lock to the left steering lock, but the front wheels do not turn nearly as much as the steering wheel. This is because the steering gearbox is using gear reduction to gain mechanical advantage. When the gearbox transfers the several turns of the input gear into the smaller movement of the output gear, driver effort is reduced. The number of complete turns of the steering wheel compared to the total amount of wheel and tire movement is called *the steering ratio* (**Figure 8-3**). The steering ratio is found by dividing the total number of degrees the steering wheel turns by the total number of degrees of front wheel movement. The ratio determines how much advantage the gearbox will provide and how the steering will feel to the driver.

For example, you may have noticed that very large vehicles, such as school buses and semi-trucks, have large-diameter steering wheels and require a lot of turning of the steering wheel to go around a corner. This is because the steering gearbox has a very high numerical ratio. A high ratio provides easier steering but requires more turns of the steering wheel. A high ratio tends to have less feel or feedback to the driver and is not as responsive to driver input.

Sports cars usually have very responsive steering that also provides a lot of feedback to the driver. This is due to the gearbox having a lower gear ratio. The trade-off is that steering effort is increased as the gearbox ratio decreases.

■ ***Hydraulics.*** Power steering systems reduce driver effort and can be either hydraulic or electric. Electric power steering systems are discussed later in this chapter.

Traditional hydraulic power steering systems use a belt-driven power steering pump, power steering fluid, lines and hoses, and a power steering gearbox (in **Figure 8-4**).

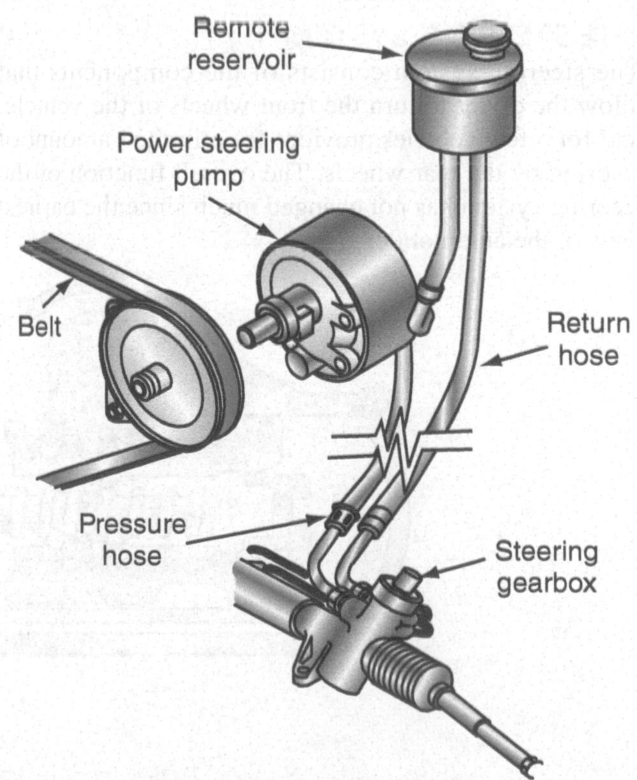

FIGURE 8-4 The hydraulic power steering system contains a pump, fluid lines, power steering gearbox, and hydraulic fluid.

The pump supplies high-pressure fluid to the steering gearbox. The fluid is used to drive a piston, which reduces the effort needed by the driver to turn the wheel. Power steering fluid is a type of hydraulic oil used in the power steering system. Some vehicles used automatic transmission fluid as their power steering fluid, but most modern vehicles often have very specific fluid requirements, and the correct power steering fluid must be used for the system to function properly.

There are typically two power steering lines from the pump to the gearbox; the high-pressure supply and low-pressure return lines. A vehicle may have additional lines connecting a power steering fluid cooler to the system. Most vehicles use a combination of steel lines and rubber hoses to allow for movement of the engine during operation.

The gearbox can be either a rack and pinion design or the recirculating ball type. With either type, when the engine is running, power steering fluid is pressurized by the power steering pump and directed to the gearbox by the high-pressure supply hose. In the gearbox, the fluid is directed against a piston. The pressurized fluid pushes on the piston, which decreases the amount of effort needed by the driver. The fluid leaves the gearbox through the power steering return or low-pressure hose return line, and returns to the power steering fluid reservoir. **Figure 8-5** shows a simplified power steering gearbox.

THE STEERING SYSTEM

The steering system consists of the components that allow the driver to turn the front wheels of the vehicle, and for a few vehicles provides for a limited amount of steering by the rear wheels. The overall function of the steering system has not changed much since the earliest days of the automobile.

■ ***Manual Steering Systems.*** Cars and trucks built before the 1950s had manual steering systems, meaning that most the effort needed to turn the steering wheel and the front wheels was supplied by the driver. Steering wheels were larger in diameter and steering gearbox ratios tended to be higher to help reduce driver effort, but turning the wheels still required a lot of muscle power.

■ ***Hydraulically Assisted Power Steering Systems.*** Chrysler offered the first power steering-equipped vehicle in 1951, and the other manufacturers soon offered power steering as an option. Today, very few new cars and trucks do not have power steering as standard equipment.

Hydraulic power-assisted steering uses a belt-driven hydraulic pump, called the **power steering pump**, to supply pressurized fluid to the steering gearbox (Figure 8-5). The pressurized fluid then applies force to a piston inside the steering gearbox. With the addition of the force applied by the fluid, the effort required by the driver to turn the wheels is reduced.

The power steering pump, which is belt-driven from the engine crankshaft, consumes a small amount of engine power to operate. This results in a slight loss in engine power and economy, but most people agree that the benefits are well worth it.

■ ***Electrically Assisted Power Steering Systems.*** A recent change in power steering is the replacement of the belt-driven hydraulic pump by electric motor assist. There are currently three types of electric assist in service: electrically powered hydraulic steering, column drive electric steering, and the electric motor-assisted rack and pinion steering gearbox.

Each of these systems offers the ability to provide variable amounts of assist based on driving conditions

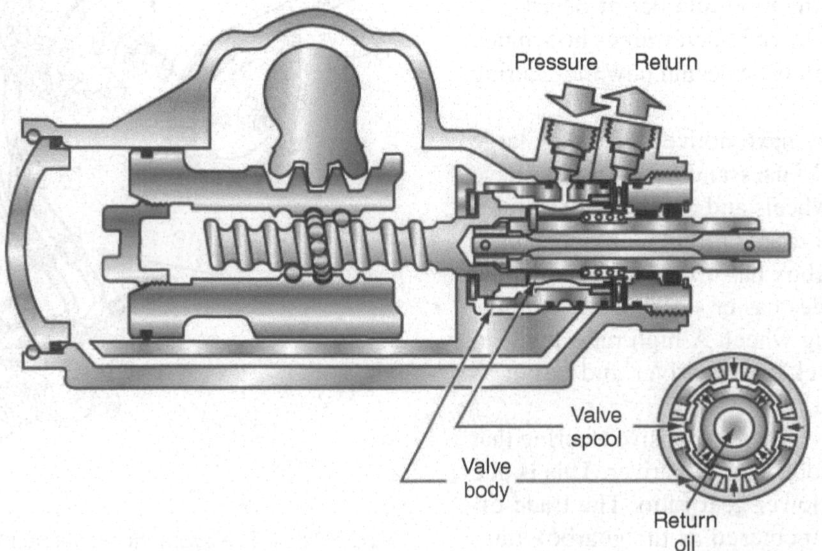

FIGURE 8-5 An example of the recirculating ball gearbox design. The use of the ball bearings greatly reduces friction in the gearbox and allows for easier steering.

and driver preference. The completely electric types do not use any type of fluid, and so they are more environmentally friendly because no fluid leaks can occur.

■ ***How Wheels, Tires, and Bearings Affect Steering.*** Because the suspension and steering systems rely on the tires as the only contact point with the ground, it is important to remember that the wheels, tires, and wheel bearings also affect steering performance.

Like many systems on the vehicle, the steering system does not work by itself. The steering and suspension systems work together and share some components that are necessary for both to operate efficiently.

One of the biggest factors in vehicle ride and handling is the tires. Because the tires are the contact point between the car and the road, it only makes sense that tires play a big part in ride and handling, as discussed in Chapters 5 and 6. Sports cars typically have a low-ratio rack and pinion gearbox for increased feel and responsiveness, performance springs and shocks, and low-profile sport tires that maximize grip and cornering. If the tires are replaced with a poorer performing tire, the overall performance of the suspension and steering will be affected, and the performance of the car may be drastically reduced by the choice of tire.

Wheel and tire choices can affect the steering if a wider-than-stock tire is installed. A wider tire has more contact area, which will increase steering effort.

Loose wheel bearings affect the steering because the wheel and tire position changes due to the play in the bearing. This also means that the wheel alignment will always be incorrect.

STEERING COLUMNS AND SHAFTS

The steering column in modern vehicles does more than allow the driver to steer the front wheels. Today's columns contain an airbag and controls for many other components, such as the cruise control, radio, exterior lighting, and windshield wipers (**Figure 8-6**). Most modern vehicles have steering columns that can tilt or adjust up and down. Many newer vesicles have electric tilt and telescoping functions as part of personality settings for different drivers. **Figure 8-7** shows an example of an electrically adjusted column. Vehicles with an automatic transmission may also have the shifter attached the column.

The steering shaft is mounted on bearings inside the column. The shaft transmits the motion of the steering wheel to the steering gearbox. There is usually at least one coupler or joint between the column and gearbox to allow for changes in angles and to reduce binding when turning the wheel. **Figure 8-8** shows how a steering shaft links the steering column and steering gearbox.

■ ***Basic Operation and Construction.*** Modern steering columns are made of plastic and metal with a steel steering shaft. The column is bolted to the dash panel on a reinforcement brace that supports the dash assembly. A coupler or joint is usually located where the column passes through the firewall. This joint allows the steering shaft to change angle to reach the steering gearbox.

Plastic covers cover the switches and other electrical components that are attached to the column (**Figure 8-9**). These covers are usually held together by several screws, which are removed to gain access to the column and any attached components.

The steering wheel is secured to the top of the steering shaft with a nut or bolt. The steering wheel and steering shaft are usually aligned by a locating notch or splined groove that can only be installed in one location. This prevents the steering wheel from being installed incorrectly. A few vehicles, mostly older model four-wheel drive (4WD) trucks, do not have an alignment notch for

FIGURE 8-6 The driver's side airbag is mounted to the steering wheel, along with other switches and controls for various systems.

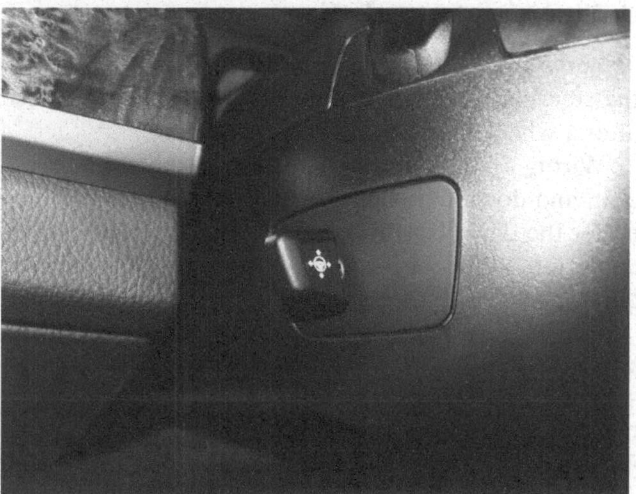

FIGURE 8-7 An example of a tilt/telescope switch.

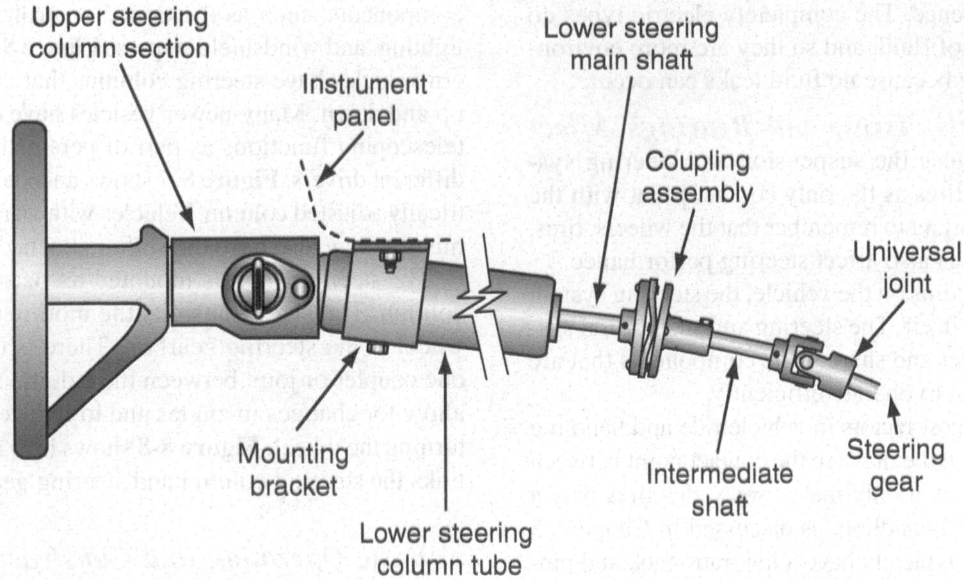

FIGURE 8-8 An illustration of the joints and couplers used in the steering system to connect the steering wheel to the gearbox.

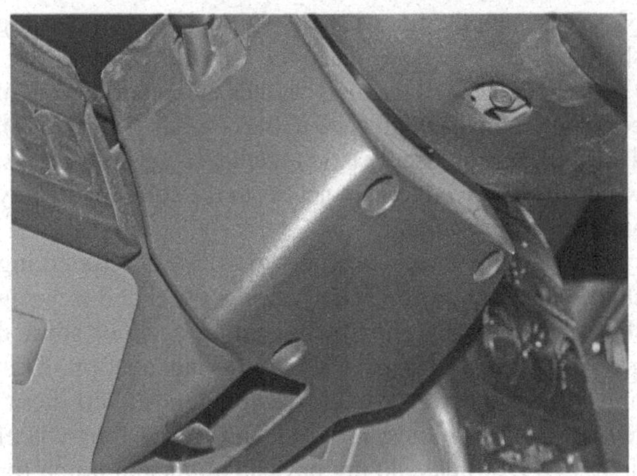

FIGURE 8-9 Covers around the steering column hide the electrical connections and wiring.

FIGURE 8-10 Collapsible columns compress in the event of a forward collision to help prevent injury to the driver.

the steering wheel. This is to center the steering wheel after a wheel alignment is performed.

Wiring for the electrical controls runs through the column and down to connecters at the base of the column under the dash. Narrow channels are designed into the column to allow wiring and components to be removed and reinstalled if necessary.

■ ***Collapsible Columns.*** In 1968, the National Highway Traffic Safety Administration (NHTSA) issued new safety requirements to help reduce driver injuries. Thousands of severe chest, neck, and head injuries were caused by impacting the steering wheel during a crash. This lead to the use of **collapsible steering columns** to help prevent injuries during front-end collisions. **Figure 8-10** shows an example of a collapsible column and the location of the shear points. Some designs use the flexible coupler as the collapsible section instead of the steering shaft inside the column.

The column is often made of two tubes with a plastic layer between the inner and outer tube. Figure 8-10 shows typical designs. In a frontal collision, the plastic that normally holds the two tubes rigidly together breaks, allowing the tubes to move in relation to each other. This prevents the column from being pushed into the driver's chest during impact. Once a column has collapsed, it must be replaced.

Another method used to prevent injury from the steering during a collision is using a folding steering shaft (**Figure 8-11**). In this system, a section of the steering shaft may break or deform. This prevents the steering

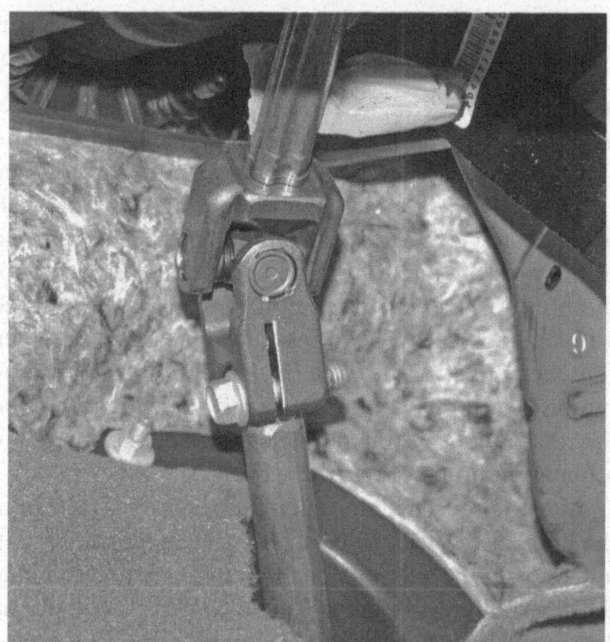

FIGURE 8-11 An example of a steering u-joint. Some of these designs are meant to collapse or break in a collision to prevent the column from injuring the driver.

shaft and column from being pushed rearward toward the driver in a collision.

■ ***Airbags.*** Airbags have been installed in steering wheels since the late 1980s. The purpose is to prolong or decrease the rate of deceleration of the driver during a collision and prevent him or her from being thrown through the windshield.

The airbag is mounted in the center of the steering wheel, beneath a cover that splits open when the bag deploys. The airbag is bolted into brackets built into the steering wheel center section. The electrical connection for the airbag is maintained by a component called *a clock spring*. **Figure 8-12** shows the clock spring and steering column.

If any service needs to be performed to the steering column, such as to the tilt mechanism, the airbag will have to be disabled and removed before the steering wheel can be removed. This is covered in detail in Chapter 9.

■ ***Steering Shafts and Couplers.*** Because the steering column is rarely perfectly aligned to the steering gearbox, an intermediate steering shaft and a coupler or joint is used to connect the column and the gearbox together. **Figure 8-13** shows a **flexible coupler** and a universal joint-equipped steering shaft. These are used to allow changes in the angle of the steering shaft and to absorb a small amount of vibration and road shock.

■ ***Tilt and Telescoping Functions.*** Most modern vehicles have a tilt steering wheel as standard equipment. The tilt wheel allows the driver to adjust steering wheel

FIGURE 8-12 A clock spring maintains the connection between the airbag system and the driver's side airbag.

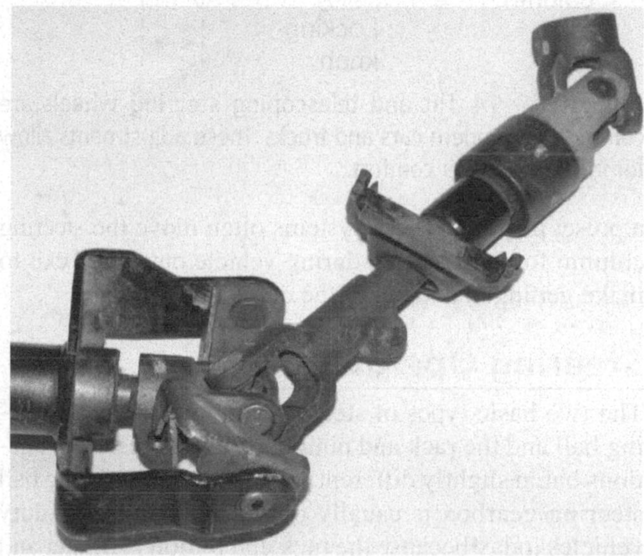

FIGURE 8-13 A steering shaft and u-joint.

position to increase comfort while driving. There are three types of tilt wheel, steering wheel tilt only, upper column and wheel tilt, and lower column and wheel tilt. All three change the position of the steering wheel by allowing a pivoting action inside the column. A release lever is mounted either on the side of the steering column or on the underside of the column. The release lever unlocks the tilt mechanism and allows the driver to adjust wheel position. Releasing the lever or pushing it back into its original position locks the column or wheel back into place.

Some vehicles are also equipped with a telescoping steering wheel. A telescoping wheel can be moved closer or farther from the driver's seat, again to increase driver comfort. **Figure 8-14** shows both tilt and telescoping column functions.

■ ***Power/Memory Columns.*** Some vehicles offer power tilt and telescoping columns. Many of these cars have memory functions that save driver preferences for seat, steering wheel, and mirror positions. Vehicles equipped with a memory steering column have motors that move the column up, down, forward, and rearward to

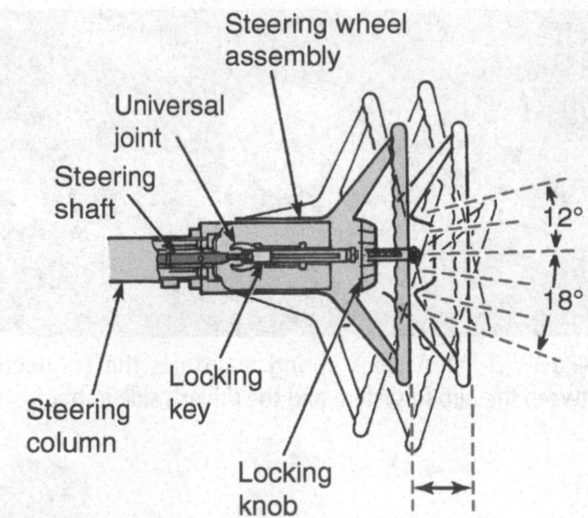

FIGURE 8-14 Tilt and telescoping steering wheels are common on modern cars and trucks. These adjustments allow for increased driver comfort.

a preset position. These systems often move the steering column forward and up during vehicle entry and exit to make getting in and out of the driver's seat easier.

Steering Operation

The two basic types of steering systems, the recirculating ball and the rack and pinion, perform the same functions but in slightly different ways. The recirculating ball steering gearbox is usually only found on heavy-duty vehicles today because the rack and pinion is lighter and uses fewer linkage components.

RACK AND PINION STEERING

Most cars and light trucks now use rack and pinion gearboxes. This is because the rack and pinion design is lighter and has fewer components than a recirculating ball gearbox. Even though both types of gearboxes perform the same functions, they do so differently.

■ ***Rack and Pinion Operation.*** A **rack and pinion** assembly has two main components, the rack gear and the pinion gear, enclosed in an aluminum housing (**Figure 8-15**). In a rack and pinion, the steering shaft is connected to the top of the pinion gear, which is held in the rack by a set of pinion bearings. The bottom of the pinion shaft has the pinion gear, which is meshed with the rack gear. The rack gear is a gear that has been flattened out with the teeth in a straight line. When the pinion gear turns, it moves the rack gear side-to-side so the rotary motion from the steering wheel is turned into a linear motion to move the wheels.

Inner tie rods are connected to the rack gear, either at the ends of the rack or bolted to the center of the unit, and transfer the side-to-side motion to outer tie rods and the steering arms or strut assembly. When the inner tie rods are attached to the ends of the rack gear, it is called an **end-takeoff rack**. When the inner tie rods are bolted at the center of the rack, it is called *a center-takeoff rack*.

Benefits of the rack and pinion gearbox include the reduced weight of the component and the elimination of several pieces of steering linkage. The rack gear attaches to inner tie rods, which in turn connect to outer tie rods. The rack and pinion assembly eliminates the Pitman arm, idler arm, and center link found on parallelogram linkage systems, discussed later in this chapter.

Power-assisted rack and pinion gearboxes use a spool valve that is integrated into the pinion gear, (**Figure 8-16**). A torsion bar runs through the pinion gear and is used inside the spool valve. The lower end of the torsion bar is connected to the end of the pinion gear, which meshes with the rack gear. As the driver turns the steering wheel, the top of the pinion gear begins to turn, but the lower

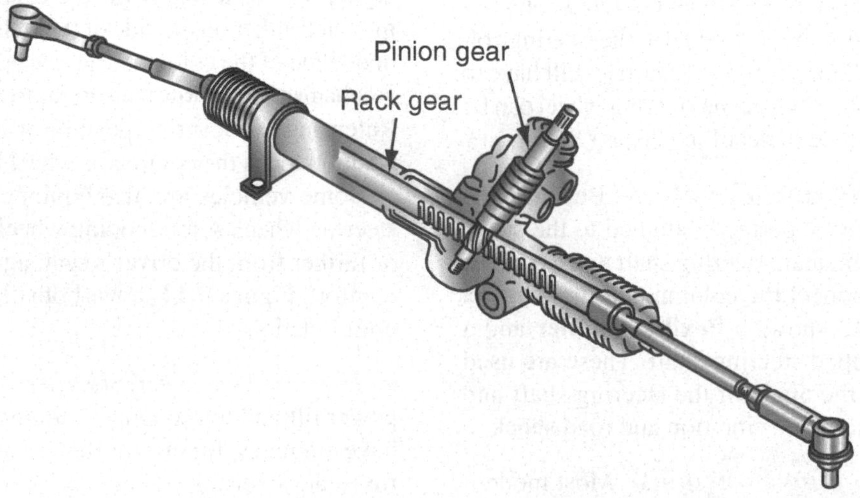

FIGURE 8-15 The rack and pinion gearbox is named for its two primary components, the rack gear and the pinion gear. Rack and pinion systems use only inner and outer tie rods, eliminating the center link, Pitman, and idler arms.

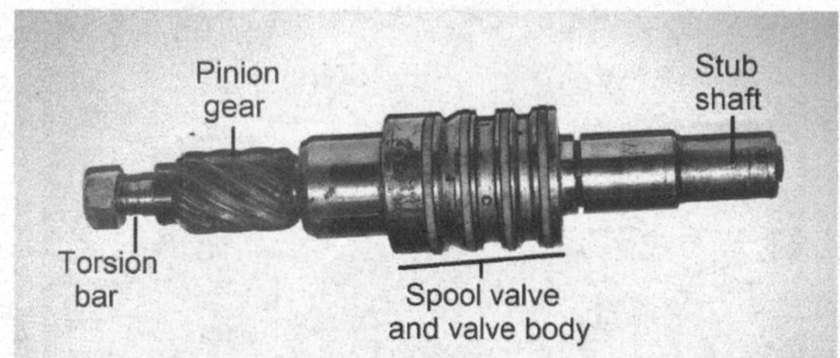

FIGURE 8-16 The spool valve is part of the pinion shaft inside the rack and pinion gearbox.

end at the rack gear does not. This is because the weight on the front suspension makes the rack gear resist moving. This difference in torque between the top and bottom of the pinion causes the torsion bar inside to twist. As the torsion bar twists, ports align in the spool valve that allow fluid to pass through the valve and to one side of the piston attached to the rack gear (**Figure 8-17**). When the driver turns the wheel back, the process repeats and allows fluid to travel to the other side of the rack piston. When the wheels are straight ahead, the ports are not aligned, and the fluid returns to the power steering pump reservoir.

■ ***Linkage Components.*** End-takeoff rack and pinion units have two inner and two outer tie rods (**Figure 8-18**). Both use ball-and-socket joints, such as in a ball joint. The inner tie rod that is threaded to the end of the rack (**Figure 8-19**). Covering the inner tie rod is the bellows (**Figure 8-20**). The bellows keeps dirt and moisture off the inner tie rod ball socket and the end of the rack assembly. A crossover vent tube typically connects the bellows boots together to allow air to flow between them when the wheels are turned. The

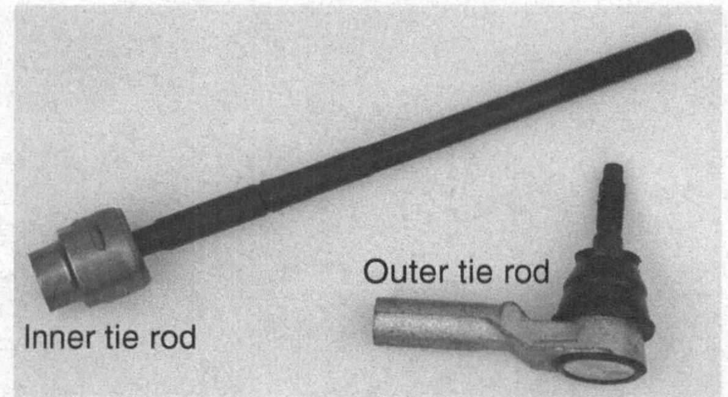

FIGURE 8-18 An example of an inner and outer tie rod for a rack and pinion gearbox.

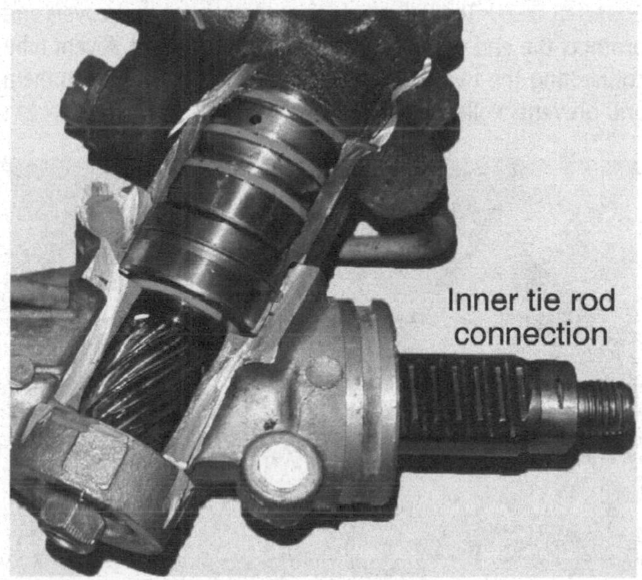

FIGURE 8-19 The inner tie rod end is threaded onto the end of the rack gear.

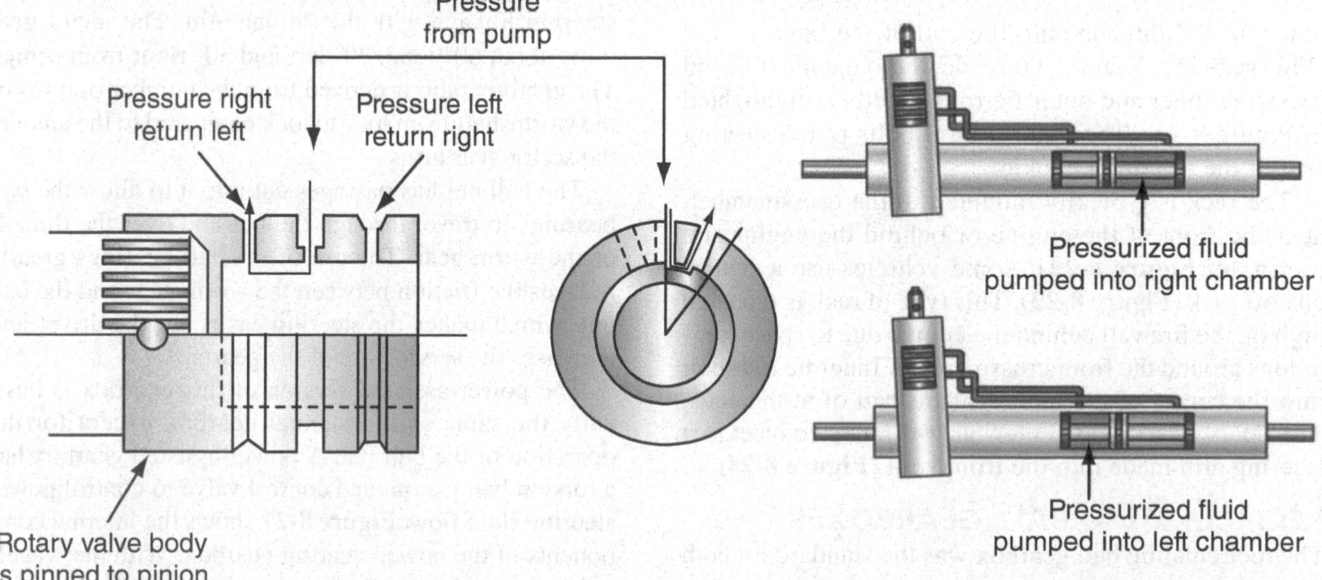

FIGURE 8-17 The spool valve directs fluid flow to either side of the rack piston based on the turning of the pinion gear and the torsion bar inside the valve.

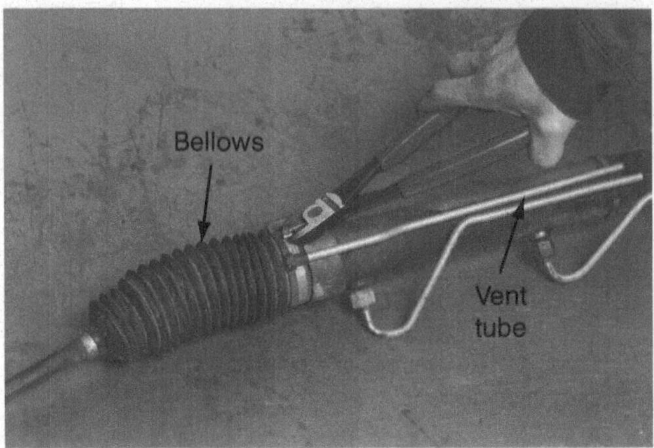

FIGURE 8-20 The bellows is the plastic boot that covers and protects the end of the rack and the inner tie rod. A vent tube connecting the two bellows allows air to flow between them and prevents collapse.

FIGURE 8-21 A jam nut secures the inner and outer tie rods together. When adjusting the wheel alignment, this nut is loosened to set the toe.

outer tie rod threads onto the end of the inner tie rod (**Figure 8-21**). A jam nut threaded onto the inner tie rod locks the inner and outer tie rods together, highlighted in Figure 8-21. The outer tie rod bolts to the steering arm of the steering knuckle.

The rack is typically mounted to the crossmember, near the front of the engine or behind the engine and transaxle (**Figure 8-22**). Some vehicles use a center-takeoff rack (**Figure 8-23**). This type of rack is mounted high on the firewall behind the engine due to space limitations around the front crossmember. Inner tie rods bolt into the center of the rack gear instead of at the ends. The outer tie rods on center-takeoff racks connect to a steering arm made into the front strut (**Figure 8-24**).

RECIRCULATING BALL GEARBOXES

The recirculating ball gearbox was the standard for both manual and power steering systems for many years. That started to change as vehicles got smaller, lighter,

FIGURE 8-22 An example of a common rack and pinion location.

and needed to obtain better fuel economy. Recirculating ball gearboxes are heavy, made from steel and cast iron, and require a more complex linkage arrangement than a rack and pinion system. Because of these issues, this type of gearbox is only used on heavy-duty light trucks and medium to heavy-duty trucks.

■ ***Recirculating Ball Steering Gearbox Operation.*** A **recirculating ball gearbox** has three major components: the worm gear, the sector gear, and the ball nut (**Figure 8-25**). The steering shaft connects to the worm gear, usually with a splined and indexed clamp. The worm gear is like a large bolt, with very large and smooth threads. The ball nut threads over the worm gear and moves back and forth along the wormshaft just like a nut on a bolt. The outside of the ball nut has deep teeth cut into it to mesh with the sector gear (**Figure 8-26**). As the ball nut moves back and forth on the wormshaft, the rotary motion of the wormshaft drives the ball nut and the sector gear.

The sector gear forms the top of the sector shaft. The sector shaft is the output shaft and is connected to the steering linkage with the Pitman arm. The sector gear turns about 60° total, 30° left and 30° right from center. The gearbox ratio is derived from the number of turns of the wormshaft from lock to lock compared to the amount the sector gear turns.

The ball nut has passages cut into it to allow the ball bearings to travel through the nut and over the threads of the worm gear. The use of the ball bearings greatly reduces the friction between the worm gear and the ball nut, which makes the steering easier for the driver and increases the service life of the gearbox.

The power-assisted version of this gearbox is basically the same as the manual gearbox except for the operation of the ball nut. A power-assisted gearbox has a torsion bar, piston, and control valve to control power steering fluid flow. **Figure 8-27** shows the internal components of the power steering gearbox. With the wheels pointed straight ahead, there is no torque on the torsion bar, and the ports inside the spool valve are all open and

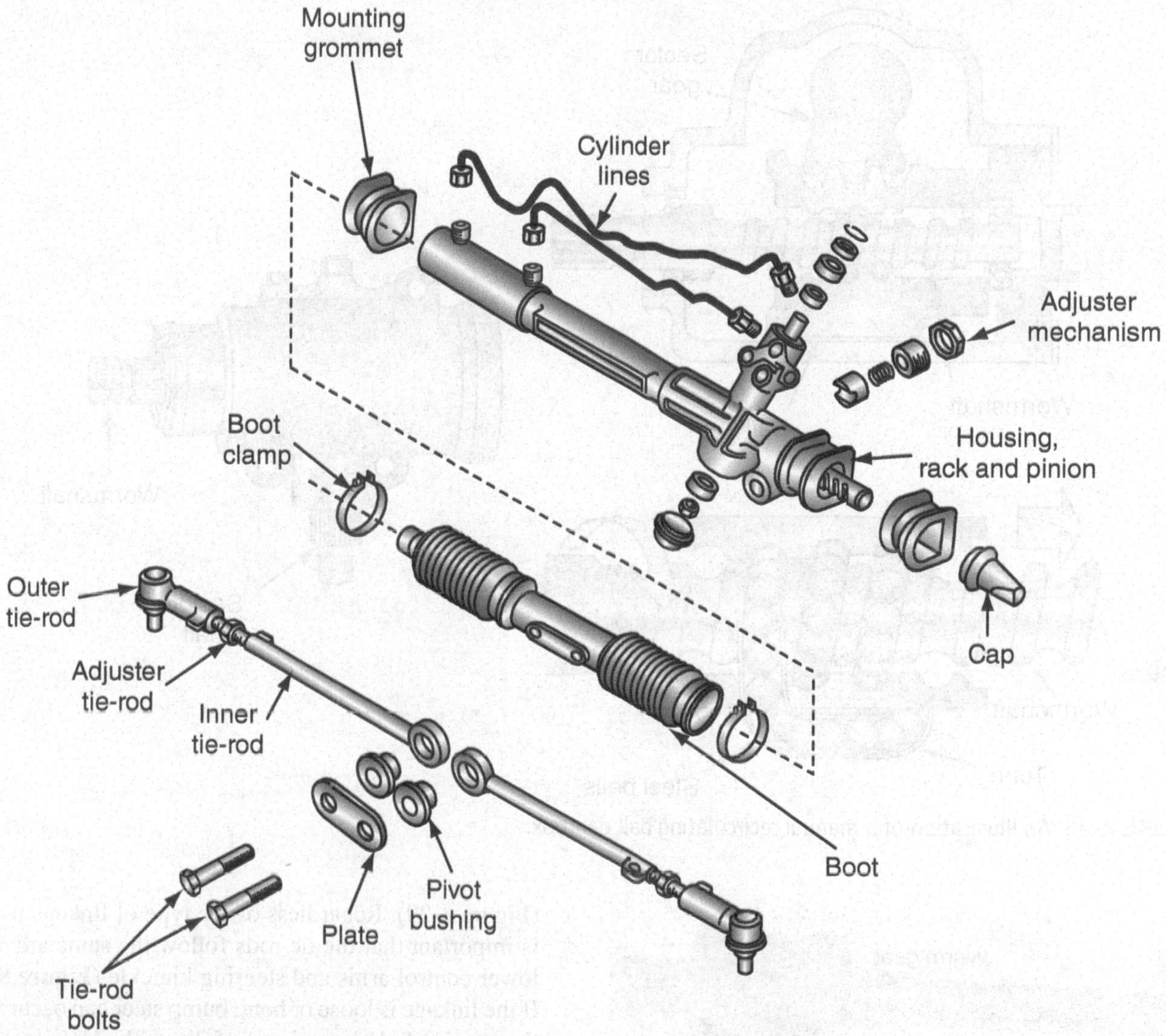

FIGURE 8-23 An illustration of a high-mounted center-takeoff rack and pinion.

FIGURE 8-24 This shows an example of a center-takeoff rack and how the tie rods connect to the strut.

fluid enters and returns to the power steering pump. When the steering wheel is turned enough to twist the torsion bar, the ports in the spool valve realign, and fluid is directed to the piston within the ball nut. The pressure applied by the fluid on the piston makes moving the ball nut easier, which reduces the effort needed by the driver to turn the wheels.

■ ***Linkage Components.*** There are two basic linkage arrangements used to connect the recirculating ball gearbox with the steering knuckle, the parallelogram and the crosslink types.

The gearbox is connected to the front wheels by the steering linkage. The opposite end of the sector shaft, outside the gearbox, connects to a Pitman arm. The Pitman arm connects to a center link or drag link. Tie rods complete the connection to the steering knuckles.

A common linkage arrangement, called the **parallelogram linkage (Figure 8-28)**. The parallelogram

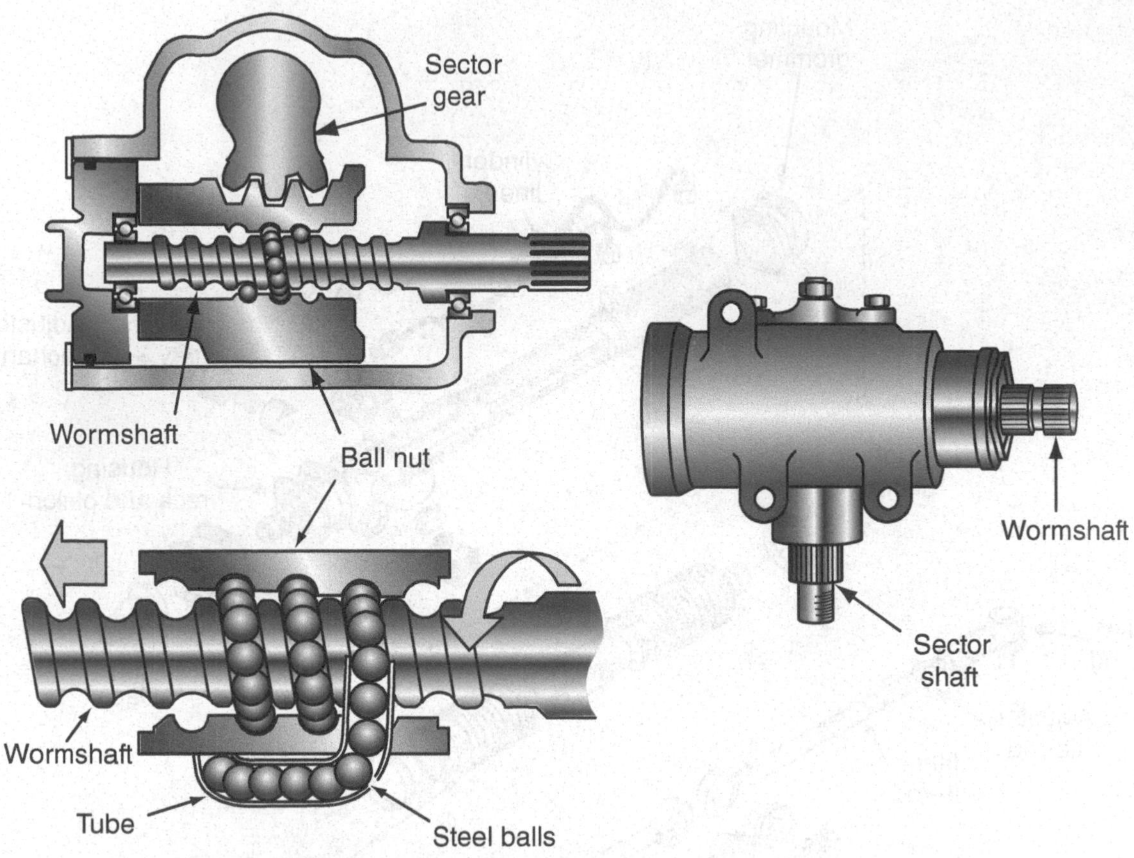

FIGURE 8-25 An illustration of a manual recirculating ball gearbox.

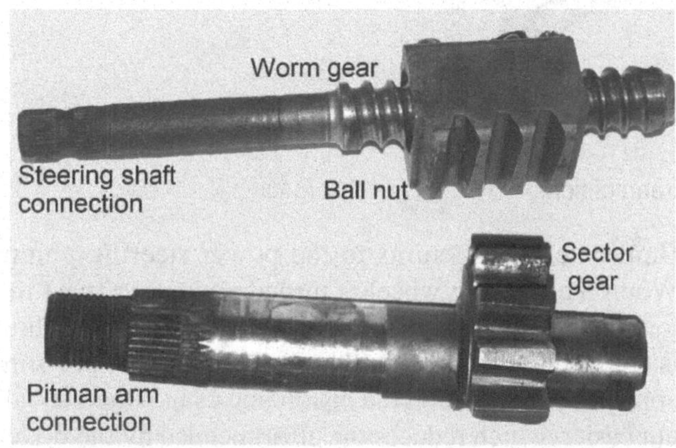

FIGURE 8-26 The main components of a manual recirculating ball gearbox.

linkage gets its name from the angles formed by the center link, Pitman, and idler arms as the wheels are turned. This type of linkage is used with the short/long arm type of suspension.

When the driver turns the steering wheel, the Pitman arm pushes the center link left or right (**Figure 8-29**). As the linkage turns, the center link and Pitman arm remain horizontal. This is necessary so that the toe angle does not change. The tie rods remain at the same angle as the lower control arms when the linkage moves side-to-side (Figure 8-29). Regardless of the type of linkage used, it is important that the tie rods follow the same arc as the lower control arms and steering knuckle (**Figure 8-30**). If the linkage is loose or bent, bump steer can occur when the steering linkage no longer follows the same arc as the lower control arms.

An example of two Pitman arms is shown in **Figure 8-31**. One end of the Pitman arm is splined to match the splines on the sector shaft and the other end connects to the center link. The Pitman arm may have an opening for a stud, as in the arm on the right, or it may have a ball socket stud like the arm on the left.

At the end of the center link opposite the Pitman arm is the idler arm. The center link can only move side-to-side because both the Pitman and idler arms prevent vertical movements. An example of an idler arm is shown in **Figure 8-32**. How the idler arm attaches to the linkage and the frame is shown in **Figure 8-33**.

Two tie rod assemblies connect the center link to the steering arm of the steering knuckle (**Figure 8-34**). These assemblies consist of an inner tie rod, a tie rod adjustment sleeve, and an outer tie rod. The inner tie rod is attached to the center link while the outer tie rod connects to the steering arm. The adjustment sleeve is used to connect the tie rods and allow for toe angle adjustment. The outer tie rod typically has right-handed

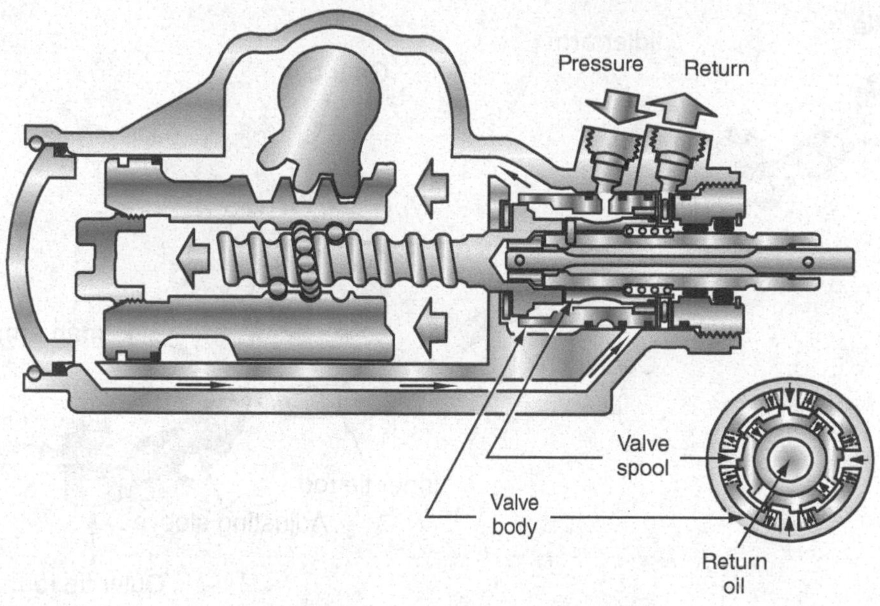

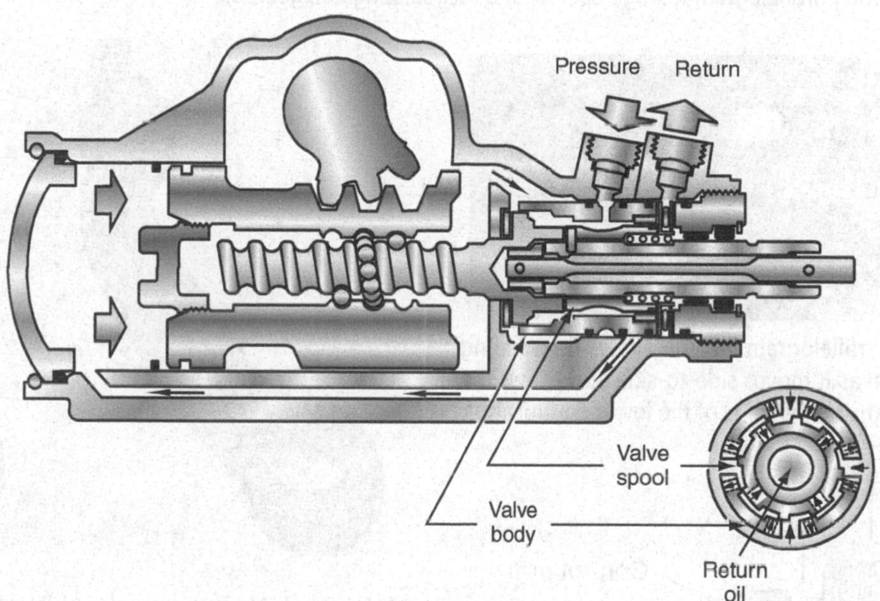

FIGURE 8-27 Inside the power steering gearbox, fluid is used to apply pressure against a piston, reducing the amount of effort required by the driver. End-takeoff racks have the inner tie rods threaded to the ends of the rack gear.

threads, and the inner tie rod has left-handed threads. The sleeve is placed between the tie rods, and when turned, the movement either shortens or lengthens the tie rod assembly.

To prevent bump steer, the tie rods are parallel with the lower control arms. Bump steer is the term used to describe the sudden darting of the vehicle to the left or right when a bump in the road is hit. The tie rods must travel in the same arc as the lower control arms while the arms respond to road conditions. This keeps the toe angle constant when the tire is moving up and down over bumps. If the tie rods do not travel exactly with the lower control arms, bump steer will occur.

The parallelogram linkage has been in use for many years and is a very effective linkage arrangement, but it does not work with all types of suspensions. Because it relies on the tie rods following the movement of the lower control arms, it is not the best choice for all front suspensions, particularly those used on 4WD vehicles.

■ ***Crosslink Linkage.*** Vehicles with a twin I-beam or a live front axle do not have a control arm arrangement like the SLA suspension. Because of this, the parallelogram linkage will not work. **Figure 8-35** shows an example of a crosslink arrangement as used on a 4WD truck with a live front axle.

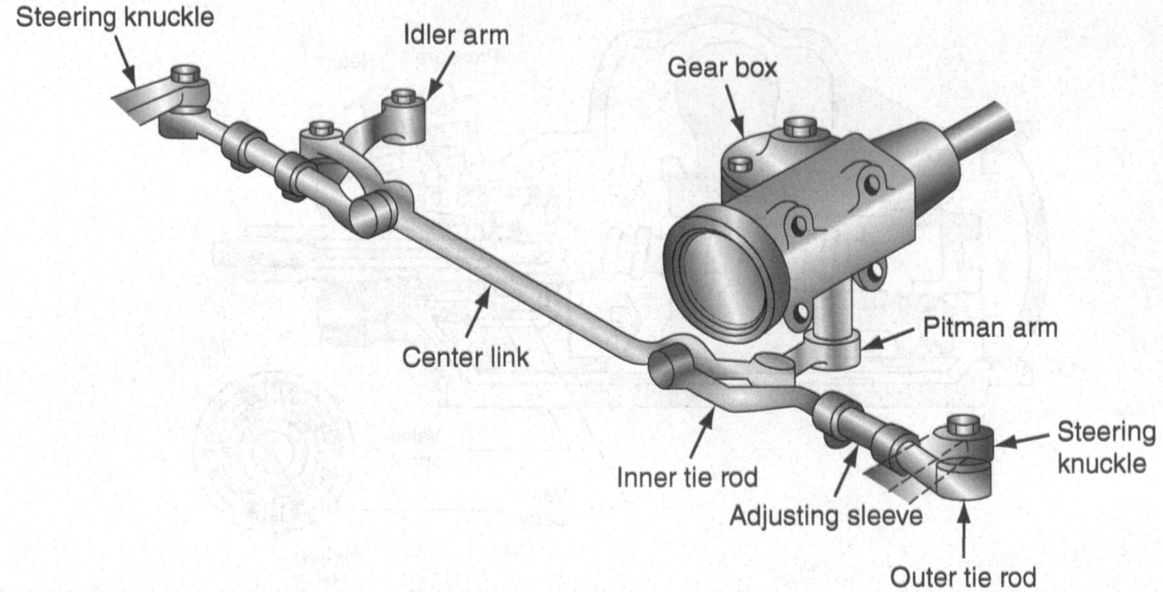

FIGURE 8-28 The parallelogram linkage used with a recirculating ball gearbox.

FIGURE 8-29 The parallelogram linkage allows the steering linkage to remain level as it moves side-to-side and allows the tie rods to travel with the movements of the lower control arms.

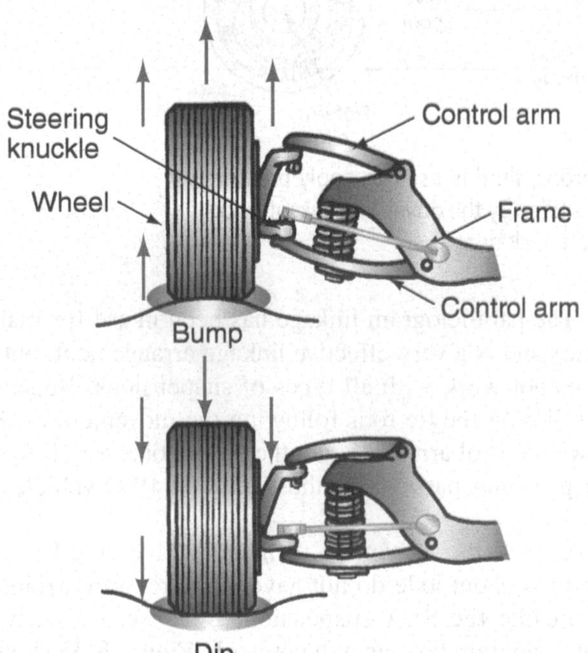

FIGURE 8-30 This illustrates the movements of the parallelogram linkage with the movements of the control arms.

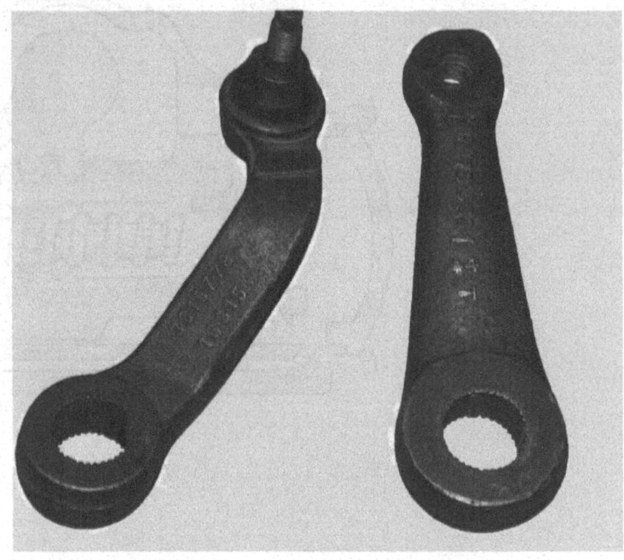

FIGURE 8-31 Two examples of Pitman arms.

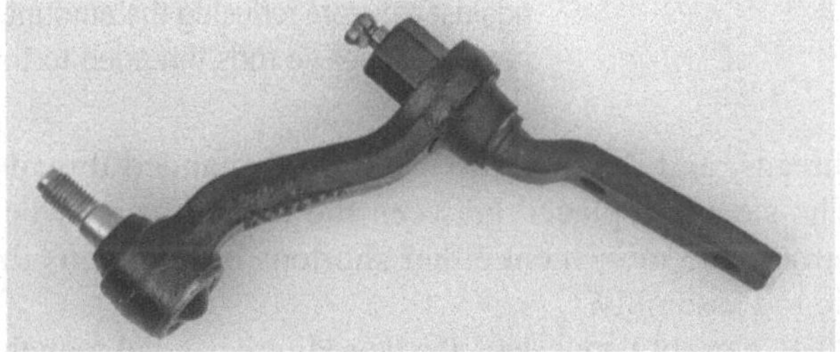

FIGURE 8-32 Idler arms keep the center link level as it moves side-to-side. This is to eliminate bump steer.

This design still uses a Pitman arm connected to the steering gearbox but does not use a center link or idler arm. Instead, the Pitman arm connects to a drag link. The drag link looks like a very long inner tie rod because it connects to a tie rod sleeve and outer tie rod. The tie rod

FIGURE 8-33 The idler arm keeps the linkage level and supported to prevent bump steer.

FIGURE 8-34 An example of an inner tie rod, tie rod sleeve, and outer tie rod in a parallelogram linkage.

assembly is also connected to the drag link. This arrangement allows the tie rods to follow the movement of the axle without causing bump steer.

Many dependent 4WD steering linkage arrangements use a steering damper (**Figure 8-36**). The damper attaches the frame and to a tie rod or drag link and is used to reduce vibration in the linkage. 4WD steering and suspension systems can experience a problem of excessive vibration called *death wobble*. Loose suspension bushings and weak steering dampers can contribute to the vibration. This is discussed in more detail in Chapter 9.

HYDRAULIC POWER ASSIST

Hydraulic power steering assist has been available since the 1950s and has not changed significantly since then. Most hydraulic assist systems use a belt-driven pump to supply pressure to a piston in the gearbox.

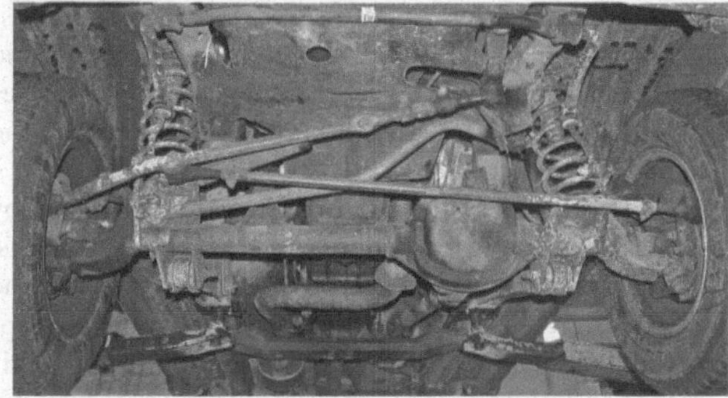

FIGURE 8-35 The crosslink system is used on 4WD trucks with live front axles.

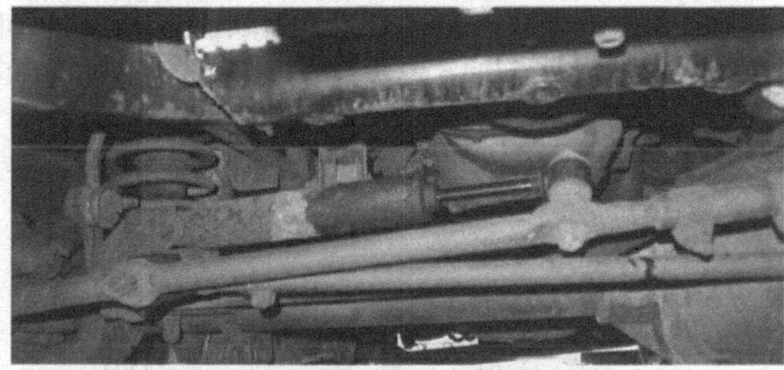

FIGURE 8-36 An example of a steering damper, used to reduce vibration in the linkage.

FIGURE 8-37 A belt-driven hydraulic power steering pump.

■ ***Power Steering Pumps.*** Most power steering systems use a belt-driven hydraulic pump (**Figure 8-37**). When the engine is running, the crankshaft drives the accessory drive belt, which drives the power steering pump.

Power steering pumps are positive displacement pumps, meaning that they pump a specific volume of fluid for each revolution of the pump. All the fluid that enters the pump is pressurized and exits the pump under pressure. The amount of fluid that can be pressurized at any time is based on the shape and size of chambers within the pump. The pump chamber's design and size depends on the type of pump used. The most common types of power steering pumps currently in use are the vane type and slipper type. **Figure 8-38** shows an illustration of a vane pump. Fluid is drawn into the pump and forced between the rotor and cam ring. This pressurizes the fluid, which then exits the pump and is sent to the steering gearbox.

In the both types of pumps, the power steering pulley drives a rotor that is located in a ring. As the rotor spins, fluid is drawn into the space between the rotor and the ring. As the rotor continues to rotate, the vanes or slippers reduce the area between the rotor and the ring. This decrease in volume places the fluid under pressure. An example of how this operates is shown in **Figure 8-39**.

A flow control valve controls fluid flow and maximum pressure. Because all the fluid entering and exiting the pump is not needed all the time, the flow control valve reroutes fluid back to the reservoir at higher pump rpm to reduce fluid flow and fluid temperature. An example of fluid flow in the system is shown in Figure 8-39.

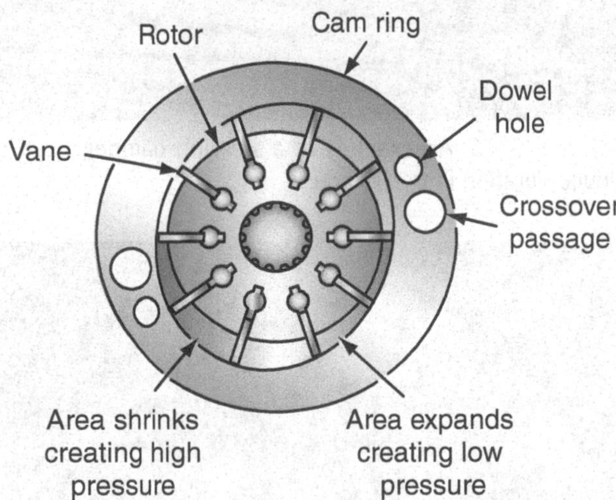

FIGURE 8-38 A common pump is the vane pump. Fluid is drawn in and pressurized as the rotor spins and the volume of the cavities decreases.

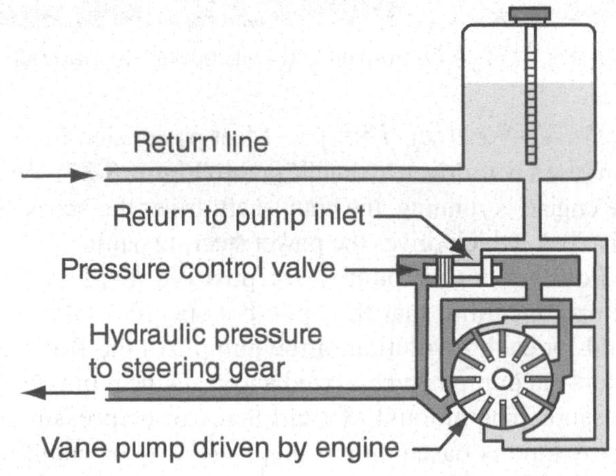

FIGURE 8-39 An illustration of fluid movement through the power steering pump.

Some vehicles control the flow of power steering fluid from the pump using an electrically operated flow control valve. This system opens or closes the fluid outlet based on power steering demand. When the vehicle is driving at low speed, such as when parking, fluid flow is increased for greater assist. At highway speeds, when less assist is needed, fluid flow is reduced. These systems are controlled by the engine control module (ECM) based on input from sensors in the steering column and various powertrain inputs.

■ ***Power Steering Pressure Sensors.*** Many power steering pumps have a pressure switch installed into them or the sensor may be installed into the power steering pressure hose (**Figure 8-40**). This sensor is used as an input for the ECM. Because the power steering pump can place a significant load on the engine during low-speed driving, especially during parking, the ECM uses the input from the sensor to increase engine speed. This prevents engine speed from dropping too low during parking and causing the engine to run roughly or stall.

■ ***Electric Pump Hydraulic Assist.*** Some vehicles use an electric pump instead of a belt-driven pump. This system takes the mechanical load of the pump off the engine, which increases fuel economy. This type of system can activate the pump only when needed and can provide a type of variable assist based on driving conditions and driver input.

■ ***Power Steering Fluid.*** Modern vehicles use power steering fluid in the power steering system but some older vehicles still use automatic transmission fluid in the power steering. Power steering fluid is low-viscosity hydraulic oil specially formulated for power steering systems. The low viscosity reduces the amount of power

FIGURE 8-40 A power steering pressure switch. These are typically located either in the high-pressure supply line or in the pump itself.

required to pump the fluid and reduces heat buildup in the system. Some vehicle manufacturers, such as Honda/Acura, require special power steering fluids, and the use of general-purpose power steering fluid can result in system damage and the loss of power assist. Always refer to the manufacturer's service information for specific fluid requirements before servicing the power steering system.

■ ***Power Steering Hoses.*** Power steering systems typically have two hoses to connect the pump to the gearbox: a high-pressure supply hose and a low-pressure return hose. These hoses often have steel tubing and reinforced rubber sections (**Figure 8-41**).

The high-pressure hose has threaded fittings for connection to the pump and gearbox and high-pressure crimp connections to join the steel and rubber sections. This is due to the high pressure the fluid is under during operation.

The low-pressure hose, while being a high-strength, reinforced design, does not carry high pressure like that of the supply hose, and as such does not usually have the same types of fittings. Many systems attach the return hose to the power steering pump return port with an ordinary worm or spring clamp.

A power steering cooler may be placed between the pump and the gearbox. The fluid becomes hot when the system is under load. The cooler removes heat to prolong the life of the fluid and other hydraulic components.

A filter may also be installed in the pressure line before the gearbox. This is used to trap any debris that may be in the system and prevent internal damage of the gearbox.

■ ***Power Steering Drive Belts.*** For the power steering pump to operate correctly, the power steering drive belt must be installed and tensioned correctly. Older vehicles often use a V-belt design, with one belt driving one or two accessories (**Figure 8-42**). Newer vehicles use serpentine or multi-rib belts. An example of how a modern accessory drive belt is used is shown in **Figure 8-43**. A **drive belt** uses the energy and motion of the engine's crankshaft to power accessories such as the power steering pump and air conditioning compressor.

Correct tension must be maintained between the belt and the drive pulley. If the belt is loose, it can slip around the pulley instead of actually driving the pulley. This will cause the belt to make noise and cause the power assist to be erratic and jumpy, especially during low-speed operation and when parking. Inspecting and servicing the drive belt is covered in detail in Chapter 9.

ELECTRIC POWER-ASSISTED STEERING

A recent trend is the use of electrically assisted power steering. The advantages of electric assist are the elimination of a belt-driven pump and power steering fluid. Turning the power steering pump takes horsepower away from the engine, which decreases fuel economy. Power steering fluid leaks are also a concern. Vehicle manufacturers are under pressure to decrease the amount of fluid loss due to leaks as these leaks are harmful to the environment. Electric power steering eliminates the power steering fluid and the possibility of leaks. Another

FIGURE 8-41 The power steering pressure and return lines at a rack and pinion. The high-pressure line has the compression fittings and the low-pressure line uses the spring clamp to secure the hose to the steel line.

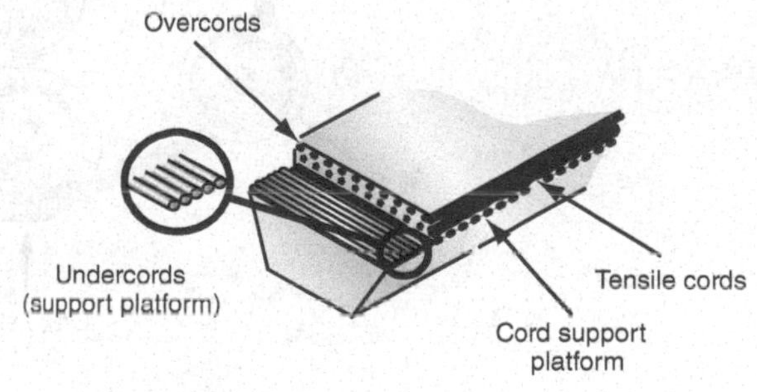

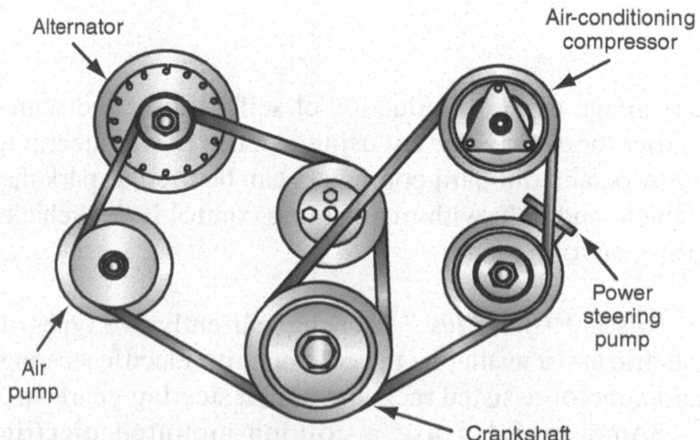

FIGURE 8-42 Many older vehicles are still in service that have V-belts. V-belts tend to be short, only driving one or two pulleys, compared to multi-rib belts.

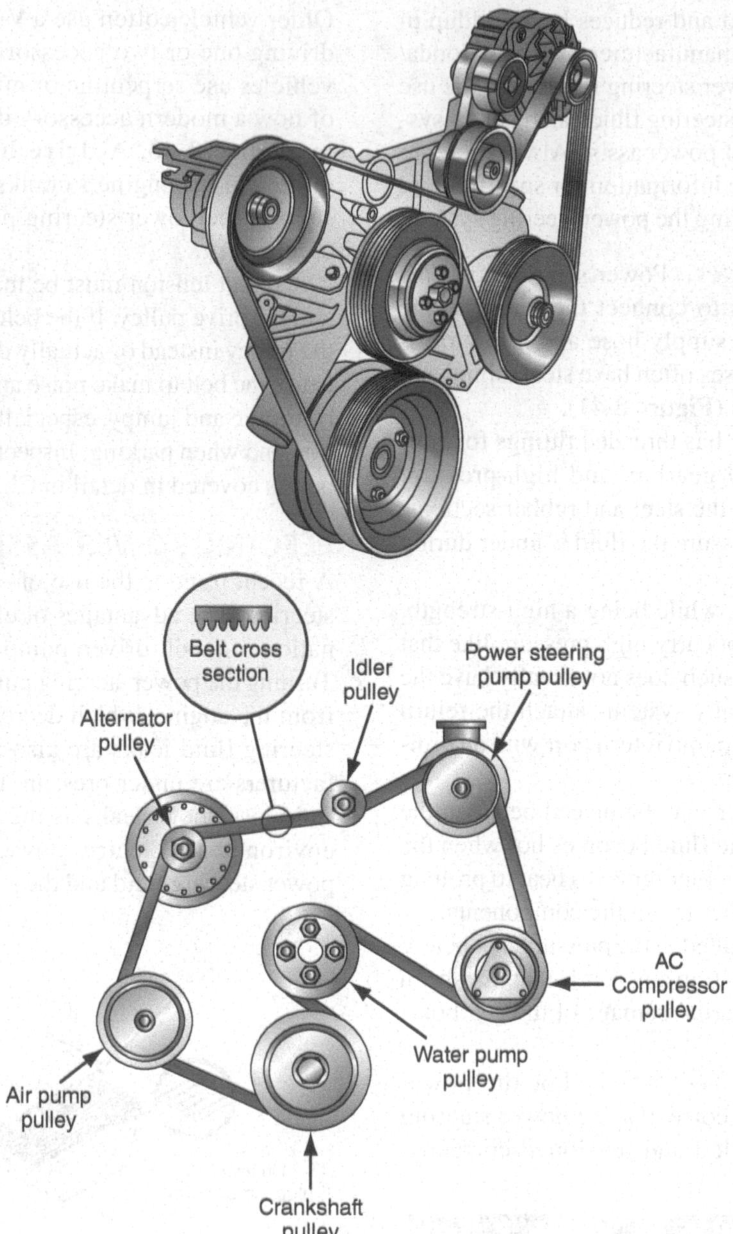

FIGURE 8-43 Serpentine or multi-rib drive belts are the most common.

advantage is the introduction of self-parking and semi-autonomous driving. By using electric power steering, sensors, and onboard computers can be used to park the vehicle and help with maintaining control if the vehicle drifts out of its lane.

■ ***Basic Principles.*** There are currently two types of electric assist available: the column drive electric steering and a motor-assisted rack and pinion steering gearbox.

Some vehicles use a column-mounted electric power steering (EPS) system (**Figure 8-44**). Sensors mounted in the steering column determine steering wheel position, rate of movement, and torque on the steering shaft. A power steering control module (PSCM) receives these and other inputs and determines how much assist is needed. Assist can be tailored to suit driving conditions as necessary. This type of assist tends to be used on smaller vehicles.

Many vehicles use an electric motor in the rack and pinion. One type of assist uses the electric motor to drive a belt and ball screw. The belt drives the ball screw, which then applies torque to the rack gear to provide assist. This type of assist tends to have a larger rack and pinion unit because of how the electric motor and belt are attached to the unit (**Figure 8-45**). Another rack drive system uses the electric motor to drive the pinion gear. An example

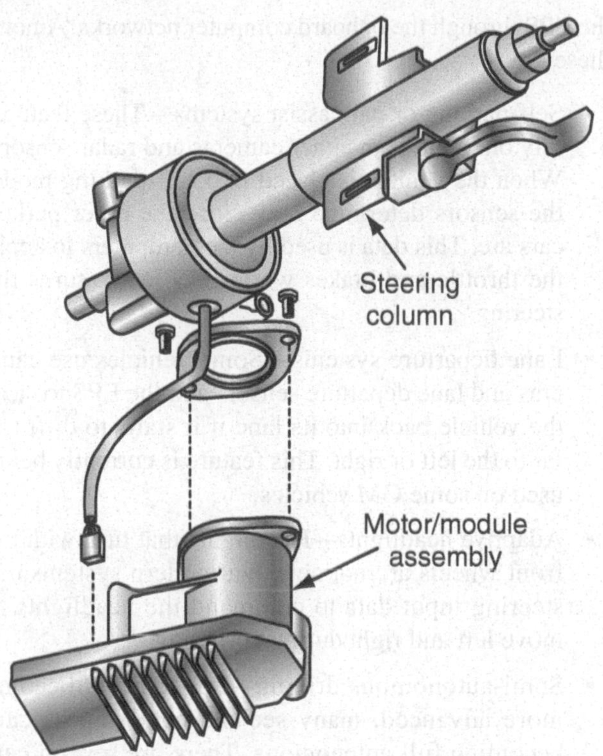

FIGURE 8-44 An illustration of an electrically assisted rack and pinion.

FIGURE 8-45 A belt-driven electric power assist rack and pinion gearbox,

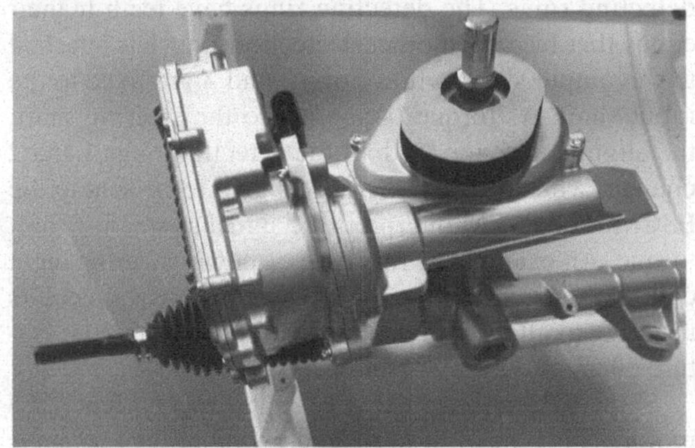

FIGURE 8-46 A rack with the electric motor driving the pinion gear.

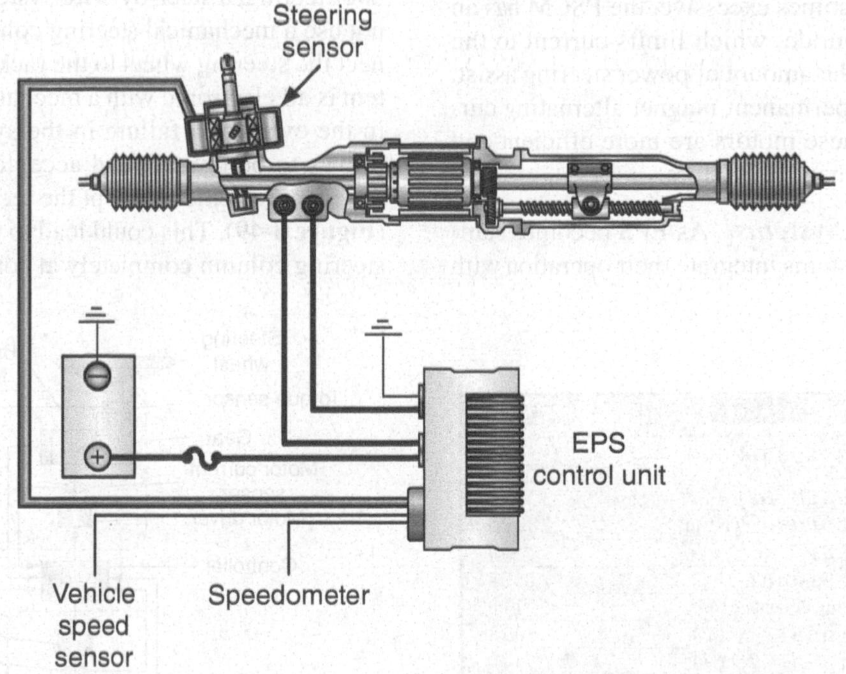

FIGURE 8-47 An illustration of a rack that uses an electrically driven ball nut to drive the rack gear.

of this type of EPS is shown in **Figure 8-46**. A third type drives the rack gear with a ball nut and drive screw parallel to the rack gear (**Figure 8-47**). All systems uses several sensors to determine how much assist is needed.

Regardless of which system is used, the PSCM only operates the assist motor when it is actually needed. This means that when the wheels are pointed straight ahead, no assist is necessary and the motor is not active.

This saves power because the electrical system is not loaded.

■ ***Operation.*** In general, the PSCM uses input from sensor(s), often called **steering angle sensors (SAS)**, to determine direction of travel, rate of turning, and how much assist is needed. In the EPS system used by GM, the input from the steering wheel, through the steering shaft, is transferred to the sensor. The output shaft from the sensor is attached to the steering coupler. The sensor uses a compensation coil, a detecting coil, and three detecting rings. The detecting rings have teeth in their edges that face each other. Detection ring 1 is attached to the output shaft, while rings 2 and 3 are fixed to the input shaft. As torque is applied to the steering shaft, the alignment of the teeth of the detection rings 1 and 2 change, which causes a voltage signal to be sent to the PSCM. The PSCM interprets the signal as steering shaft torque. An example of this type of power steering input data is shown in **Figure 8-48**. The compensation coil is used for to allow for changes that occur from temperature changes during operation.

Some EPS systems use a 12-volt direct current (DC) motor. This type of motor can draw more than 50 amps and can become quite hot in operation. Because of this high current demand, motor current draw is monitored by the PSCM. In the event the current flow overheats the motor or draw becomes excessive, the PSCM has an overload protection mode, which limits current to the motor and decreases the amount of power steering assist. Other EPS units use permanent magnet alternating current (AC) motors. These motors are more efficient and consume less power than DC motors.

■ ***EPS and Other Systems.*** As EPS becomes standard, other vehicle systems integrate their operation with the EPS through the onboard computer networks. Among these are:

- Self-parking or park assist systems—These features rely on input from either cameras and radar sensors. When the vehicle is placed into self-parking mode, the sensors determine how close the other parked cars are. This data is used by the computers to apply the throttle and brakes while EPS motor turns the steering.
- Lane departure systems—Some vehicles use cameras and lane departure sensors and the EPS to steer the vehicle back into its lane if it starts to drift too far to the left or right. This feature is currently being used on some GM vehicles.
- Adaptive headlights—Headlights that turn with the front wheels are not new, but modern systems use steering input data to command the headlights to move left and right during cornering.
- Semi-autonomous driving—As vehicles become more advanced, many see the next step as cars becoming full autonomous. There are several cars for sale now that offer semi-autonomous features, including models from Tesla and Mercedes-Benz.

Infiniti is currently selling vehicles with direct adaptive steering, a steer-by-wire system. These vehicles do not use a mechanical steering column and shaft to connect the steering wheel to the rack and pinion. The system is all electronic with a mechanical safety as backup in the event of a failure in the system. If this system proves to be reliable and accepted by drivers, expect to see more vehicles adopt the technology in the future (**Figure 8-49**). This could lead to the elimination of the steering column completely at some future point.

GM 1980-2009

PSCM Data

Steering Shaft Torque (N-m)	-8.75
Vehicle Speed(MPH)	0
Battery Voltage Signal	11.58
Calculated System Temp(°F)	118
Torque Sensor Signal 1 (V)	0.9
Torque Sensor Signal 2 (V)	4.0
Steering Position Sensor 1 (V)	3.0
Steering Position Sensor 2 (V)	0.5
Steering Wheel Position (°)	4
EPS Motor Command Amps	0

1 2 3 4

FIGURE 8-48 An example of data from the EPS module.

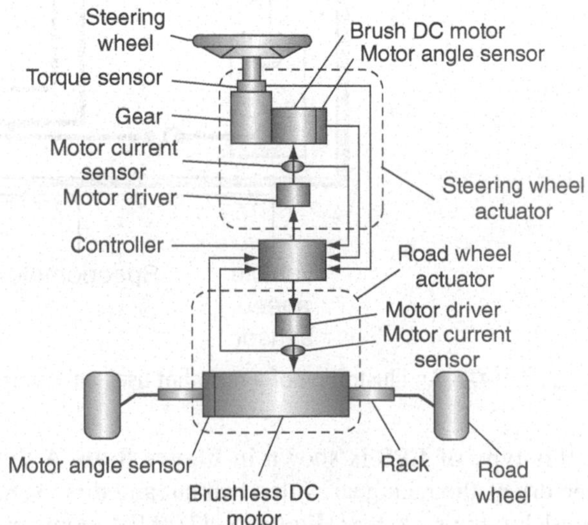

FIGURE 8-49 An illustration of a steer-by-wire system.

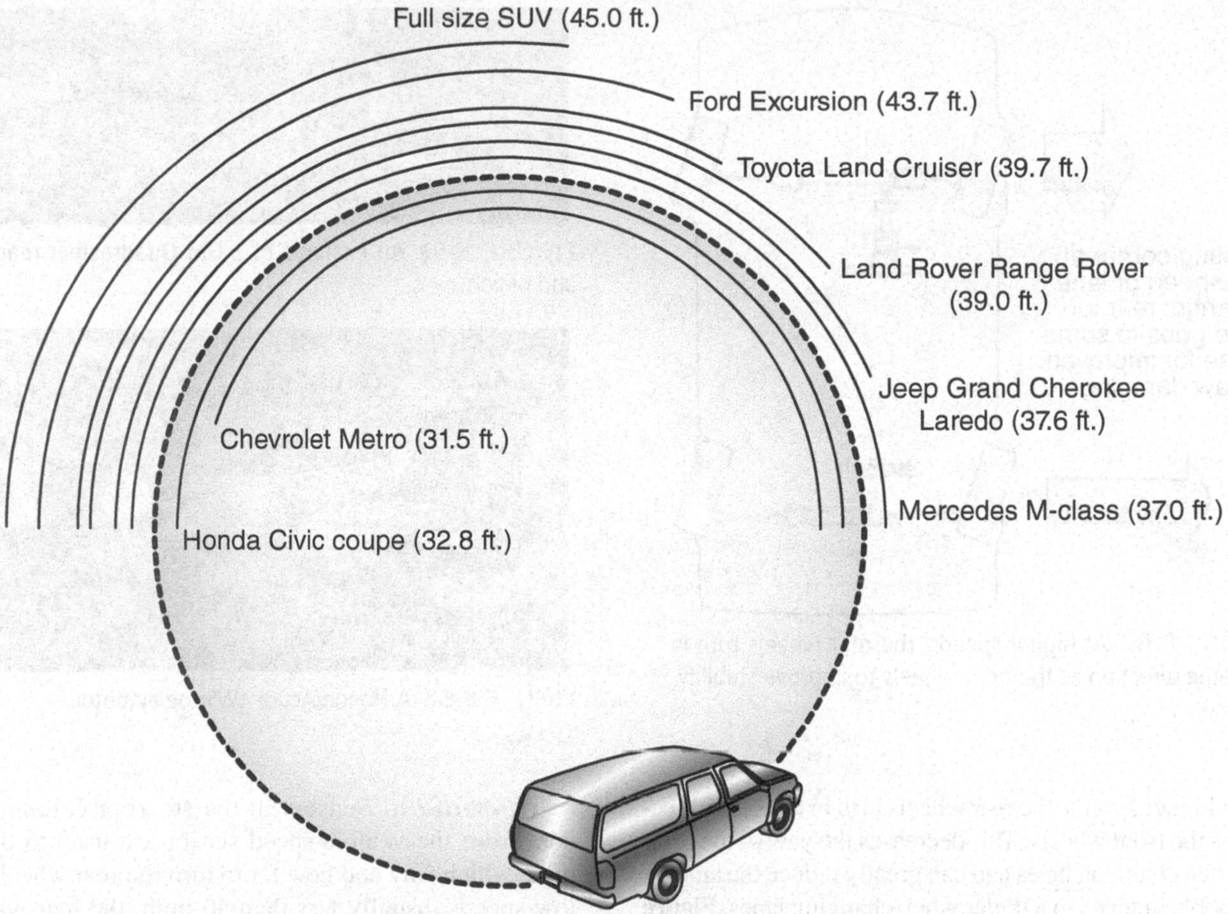

FIGURE 8-50 Four-wheel steering decreases turning radius and improves maneuverability.

Because of the increasing use of EPS for safety-related actions, the calibration of the steering angle sensor is critical after any steering system repairs or a wheel alignment have been performed. The onboard computer system must know the exact relationship between the steering shaft and the front tires to correctly act on inputs from the driver and for proper power steering operation.

FOUR-WHEEL STEERING SYSTEMS

Four-wheel steering (4WS), while not common, has been available on vehicles ranging from the Chevrolet Silverado to certain Acura, BMW, and Nissan models and is currently offered by Acura and Porsche. Four-wheel steering systems improve both low- and high-speed maneuvering by allowing the rear wheels to either countersteer the front wheels at low speed or steer in the same direction as the front at higher speeds. This improves handling and decreases turning radius (**Figure 8-50**).

■ ***Basic Principles.*** Allowing the rear wheels to turn in the opposite direction from the front wheels during low-speed turns and when parking enables the turning radius to be greatly reduced. That means a full-size truck can maneuver and park like a much smaller vehicle. This is especially helpful when pulling a trailer. Low-speed operation is illustrated in **Figure 8-51**.

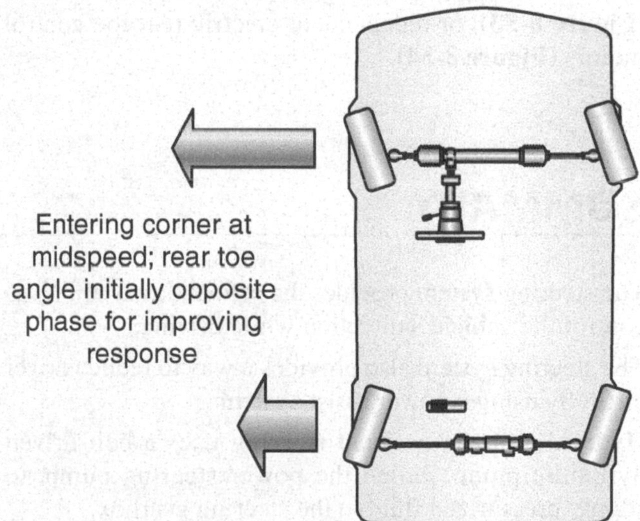

FIGURE 8-51 At low speed, the rear wheels turn opposite the front wheels to decrease turning radius.

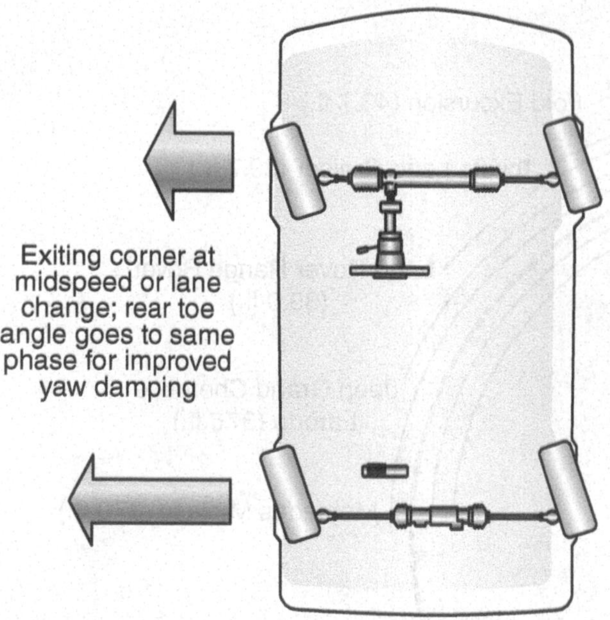
Exiting corner at midspeed or lane change; rear toe angle goes to same phase for improved yaw damping

FIGURE 8-52 At higher speeds, the rear wheels turn in the same direction as the front wheels to improve stability.

FIGURE 8-53 An example of a GM Quadrasteer rear rack and pinion.

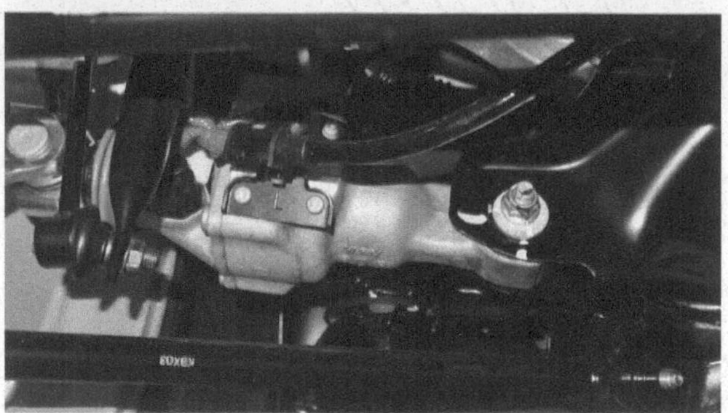

FIGURE 8-54 A Honda/Acura 4WS toe actuator.

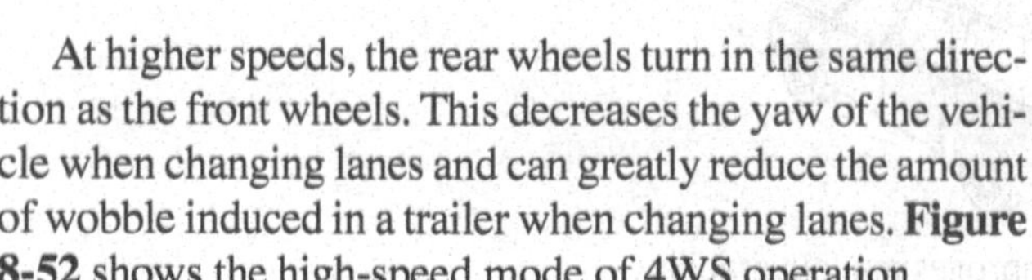

At higher speeds, the rear wheels turn in the same direction as the front wheels. This decreases the yaw of the vehicle when changing lanes and can greatly reduce the amount of wobble induced in a trailer when changing lanes. **Figure 8-52** shows the high-speed mode of 4WS operation.

Vehicles equipped with 4WS may use an electrically operated rack and pinion gearbox for the rear wheels (**Figure 8-53**), or independent electric rear toe control motors (**Figure 8-54**).

■ ***Operation.*** Sensors in the steering column and data from the vehicle speed sensor are used to determine which way and how far to turn the rear wheels. At low speeds, usually less than 40 mph, the rear wheels are counter or negative steered. At higher speeds the rear tires are turned in the same direction. Rear steering systems are also used to decrease stopping distance. This is done by moving both rear tires to point in, called *toe-in*, when braking.

SUMMARY

- The steering system provides the safe and easy ability to control the vehicle's direction while driving.
- The steering system also provides a way to reduce driver effort by using a power assist system.
- Hydraulic power-assisted steering uses a belt-driven hydraulic pump, called the power steering pump, to supply pressurized fluid to the steering gearbox.
- The number of degrees of the steering wheel from lock to lock compared to the total amount of wheel and tire movement is called the steering ratio.
- The gearbox is connected to the front wheels by the steering linkage.
- The rack and pinion system eliminates the Pitman arm, idler arm, and center link.
- In a frontal collision, the two tubes of the steering column break, allowing the tubes to move in relation to each other.
- The parallelogram is used with the short/long arm type of suspension.
- Most cars and light trucks now use rack and pinion gearboxes.
- Power steering systems have a high-pressure supply hose and a low-pressure return hose.
- The use of electrically assisted power steering is increasing and will likely become the standard in the near future.
- Four-wheel steering systems improve both low- and high-speed maneuvering.

REVIEW QUESTIONS

1. The amount of turning of the front wheels compared to the amount of at the steering wheel is called the ________________________________ ________________________________.

2. Two types of steering gearboxes in use are the reciprocating ball gearbox and the ____________ ________________________________ ________________________ gearbox.

3. A serpentine or ______________ belt is used to drive the power steering pump.

4. The power steering ______________ ______________ is an input to the computer system and is used to increase idle speed during parking situation.

5. Vehicles with memory steering columns may move both the ______________ and the ______________ functions as set by driver preference.

6. *Technician A* says most power steering fluids and automatic transmission fluids can both be used in any power steering system. *Technician B* says universal power steering fluid can be used in many makes and models of vehicles. Who is correct?

   a. Technician A **c.** Both A and B
   b. Technician B **d.** Neither A nor B

7. *Technician A* says power rack and pinion gearboxes may use hydraulic power assist. *Technician B* says some power steering gearboxes have electric motors to provide assist. Who is correct?

   a. Technician A **c.** Both A and B
   b. Technician B **d.** Neither A nor B

8. Which of the following are not used in rack and pinion steering linkages?

   a. Inner tie rods **c.** Outer tie rod ends
   b. Center link **d.** All of the above

9. A vehicle has suffered severe front-end damage in a collision: *Technician A* says the steering column may need to be replaced. *Technician B* says the steering linkage should be carefully inspected for damage. Who is correct?

   a. Technician A **c.** Both A and B
   b. Technician B **d.** Neither A nor B

10. *Technician A* says 4WS allows cars to more quickly go around corners. *Technician B* says 4WS can be used to improve braking and handling. Who is correct?

    a. Technician A **c.** Both A and B
    b. Technician B **d.** Neither A nor B

CHAPTER 9

Steering Service

Chapter Objectives

At the conclusion of this chapter, you should be able to:

- Identify and safely use tools for steering system service.
- Disable and enable the supplemental restraint system. (ASE Education Foundation MLR 4.A.2)
- Remove and replace a driver's side airbag.
- Inspect, remove, and replace steering linkage components, including:
  - o Inspect rack and pinion steering gear inner tie rod ends (sockets) and bellows boots. (ASE Education Foundation MLR 4.B.1)
  - o Inspect pitman arm, relay (centerlink/intermediate) rod, idler arm and mountings, and steering linkage damper. (ASE Education Foundation MLR 4.B.7)
  - o Inspect tie rod ends (sockets), tie rod sleeves, and clamps. (ASE Education Foundation MLR 4.B.8)
- Perform a prealignment inspection. (ASE Education Foundation MLR 4.C.1)
- Determine proper power steering fluid type; inspect fluid level and condition. (ASE Education Foundation MLR 4.B.2)
- Flush the power steering system. (ASE Education Foundation MLR 4.B.3)
- Inspect for power steering fluid leakage; determine necessary action. (ASE Education Foundation MLR 4.B.4)
- Remove, inspect, replace, and adjust power steering pump drive belt. (ASE Education Foundation MLR 4.B.5)
- Inspect and replace power steering hoses and fittings. (ASE Education Foundation MLR 4.B.6)
- Inspect electric power steering assist system. (ASE Education Foundation MLR 4.B.21)
- Identify hybrid vehicle power steering system electrical circuits and safety precautions. (ASE Education Foundation MLR 4.B.22)
- Explain the operation of nonhydraulic power steering assist systems.
- Identify hybrid vehicle power steering system electrical circuits.

KEY TERMS

belt wear gauge	**EPS warning light**	**power steering fluid flush**
drive belt tensioner	**inner tie rod socket**	**prealignment inspection**
dry-park check	**Pitman arm puller**	**steering damper**

Tools and Safety

Safe working practices include proper tool use and care, following the proper repair procedures, staying focused on the task, and taking the time to perform your work properly.

Several special tools are used to service the steering system, and using them properly will allow you to work safely and efficiently.

TOOLS

As with all aspects of automotive repair, the steering system has its own special tools used to diagnose and service its components. Although basic hand tools are used for most repairs, some jobs cannot be safely or correctly performed without the right tools.

■ ***Tools for Steering Service.*** The following figures show and explain the uses of many of the tools you will use when working on the suspension system.

Replacing inner tie rod ends on rack and pinion gearboxes requires special tie rod sockets (**Figure 9-1**). Some of these sockets fit specific tie rods, like the upper socket in Figure 9-1. The other tie rod socket is a universal tool, used on a variety of rack and pinions. Another type of tie rod tool clamps around the ball socket (**Figure 9-2**). Once the tool is attached to the tie rod, a ratchet is used to loosen and remove the tie rod.

Removing a Pitman arm requires a Pitman arm puller (**Figure 9-3**). This is because the Pitman shaft in the gearbox is tapered, making the fit between the arm and the shaft very tight.

Removing outer tie rod ends from the steering knuckle without damaging the tie rod or an aluminum knuckle is accomplished by using a tie rod separator (**Figure 9-4**).

Checking drive belt tension requires using a belt tension gauge (**Figure 9-5**).

SAFETY

Working on the steering system is not any more dangerous than working on other parts of the vehicle. However, when working around running engines, such as when checking power steering pumps, hoses, and lines, it is important to keep safe working practices in mind.

■ ***Safe Work Practices.*** When you are performing work on the steering system, you may have to service components that require special handling procedures. The following are some general safe work practices for working on the steering system.

- Use the proper tool for the job. Do not use pliers in place of sockets and wrenches.
- Ensure that the vehicle is properly raised and secure before attempting to work with the vehicle off the ground.

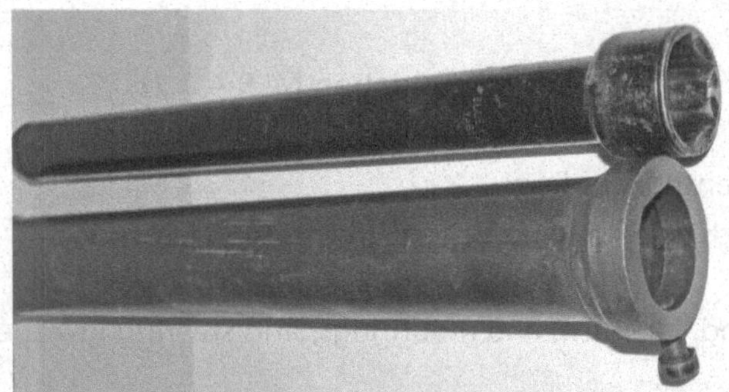

FIGURE 9-1 Examples of inner tie rod sockets.

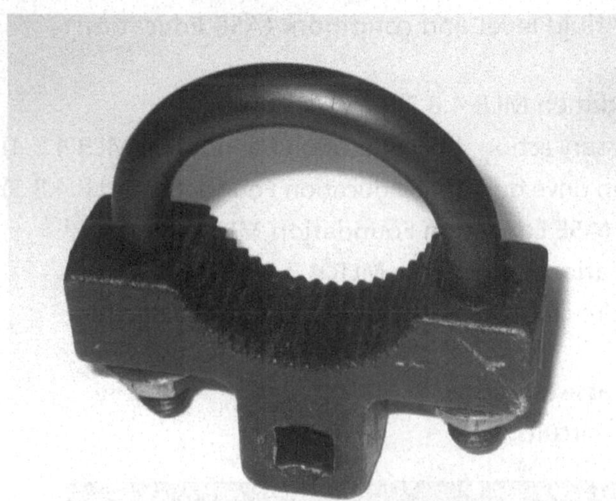

FIGURE 9-2 A type of universal inner tied rod tool.

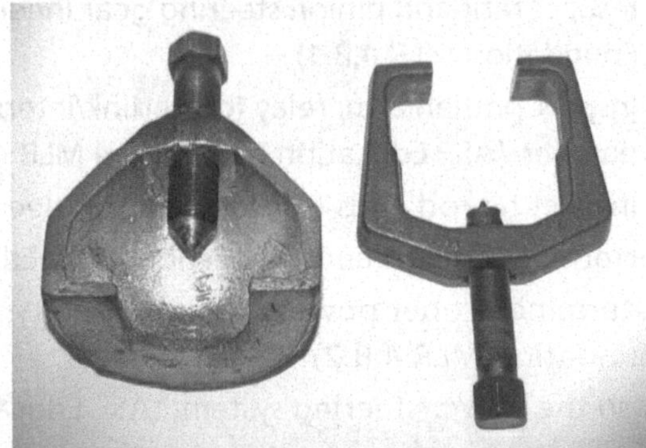

FIGURE 9-3 Pitman arm pullers. These are necessary due to the taper of the Pitman shaft making removal difficult.

FIGURE 9-4 A tie rod/ball joint separator tool.

- Do not use steel hammers, punches, or chisels on other steel components.
- Always identify all special procedures and warnings in the service information for the system on which you are working.
- Do not attempt to service a system or component if the correct tools are not available.
- Do not attempt to open or repair any airbag or airbag system component.
- Disable the airbag system before performing any repairs to the steering column.
- Follow the manufacturer's service procedures exactly when working on the airbag system.

Steering Columns

Cars and trucks use the steering column for more than holding the steering wheel in place. Even though the primary purpose of the column is for the driver to control the direction of the front wheels, there is often much more going on than just steering.

There are many switches and controls mounted on the column; the steering wheel contains the driver's airbag, and many cars and trucks now have electrically operated power steering assist that works through the steering column. Because of the combination of controls located around the column, it is common to perform repairs on either the column or some of its attached components.

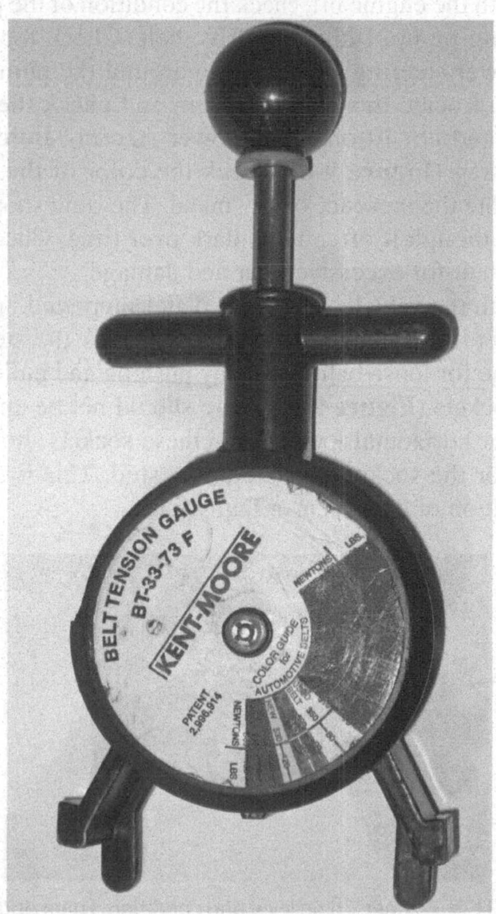

FIGURE 9-5 A drive belt tension gauge.

STEERING COLUMN SERVICE

There are situations where services and repairs have to be made to a steering column that is not related to the steering system. When you are working on the steering column, it is necessary to disable the driver's side airbag.

■ ***Disabling the Supplemental Restraint System.*** Always refer to the correct service information and follow the manufacturer's procedures to disable the supplemental restraint system (SRS). The steps listed below are general and do not apply to any specific vehicle but are common steps for many vehicles.

To disable the SRS temporarily:

- Remove the SRS fuse. Fuses may be labels SRS, SIR, or AIRBAG (**Figures 9-6** and **9-7**).
- Unplug the yellow SRS connector at the steering column.
- Wait approximately 5 minutes to let any residual voltage dissipate from the system.

Some manufacturers may have you disconnect the battery negative cable. This should only be done if the service information directs you to. **Disconnecting the battery on modern vehicles may require the reinitialization of numerous systems and can cause antitheft, airbag, hybrid, and other system problems.**

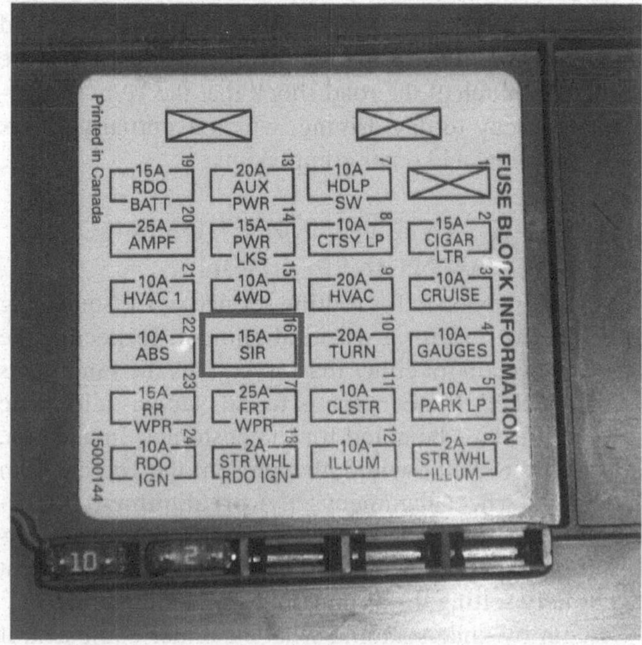

FIGURE 9-6 An example of an airbag system fuse.

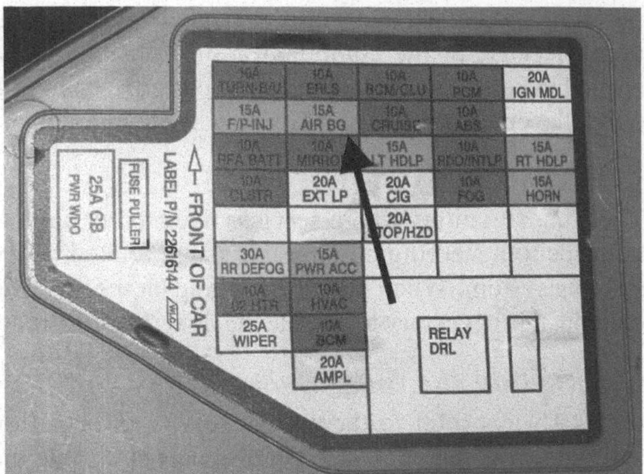

FIGURE 9-7 An example of an airbag system fuse.

Service Note

Disconnecting the battery negative cable will result in the loss of all memories retained by the vehicle's various systems and can cause antitheft systems to activate once the battery is reconnected. Loss of battery power can lock out the audio and passenger entertainment systems, reset engine and transmission adaptive memory, alter personal settings, and more. Make certain that disconnecting the battery is necessary as part of the airbag procedure before proceeding.

Steering System Inspection and Service

Because of the stresses placed on the steering system, many of the system's components are prone to wear and damage. Steering gearboxes and linkage components are subject to much of the road shock that the vehicle experiences in day-to-day driving, which eventually causes wear on joints and other components.

STEERING SYSTEM INSPECTION AND PREALIGNMENT INSPECTION

A vehicle may be in for routine service or for something specific, such as for a wheel alignment. Inspecting the steering system for either reason requires the same items to be checked. If a concern is found, such as a loose tie rod end, it must be noted for the customer. In addition, a loose tie rod end would also prevent the vehicle from having the wheel alignment set. A **prealignment inspection** is necessary to determine if any suspension or steering components are worn or damaged and would prevent accurately setting the alignment to specifications.

Begin by understanding what customer complaint, if any, needs to be addressed. If a vehicle is in for a possible steering and suspension system problem, you will need to determine the cause of the complaint and recommend the necessary repairs to correct the condition. Try to get as much information as possible from the customer. You may need to take a test drive with the customer to fully understand what he or she is concerned about. Spending time gathering information can often save time in the actual diagnosis.

■ ***Steering Inspection.*** Inspecting the steering system is also an inspection of the suspension system because the two share components and work together. You should also look at the tires and check tire pressure as part of your basic inspection. A customer complaint of hard steering may be caused simply by low tire pressure. Begin with a visual inspection of the steering system components, and then perform a closer, more detailed check of individual parts.

With the engine running, turn the steering wheel side-to-side. Pay close attention to the feel of the movement and listen for any noises. The steering wheel should move smoothly from lock-to-lock without binding, jerking, roughness, or looseness. If any of these conditions are present, this can indicate problems in the steering column, power steering system, or steering linkage, and a more thorough inspection will be necessary.

With the engine off, check the condition of the power steering pump, fluid, and drive belt. Check for signs of power steering fluid leaking around the pump and hoses. Locate the fluid reservoir, and check the fluid level and condition. Wipe power steering fluid on a clean rag (**Figure 9-8**). Check the color of the fluid, and note the presence of any metal. The fluid should be clear, though it often turns dark over time. Check the drive belt for excessive wear and damage.

With the vehicle safely raised and supported, inspect the steering linkage and gearbox. Check the steering linkage for loose ball sockets by pushing and pulling on the sockets (**Figure 9-9**). There should not be any vertical or horizontal looseness in these sockets. It is normal for the socket to rotate on its stud. This is not an indication of a problem or faulty joint.

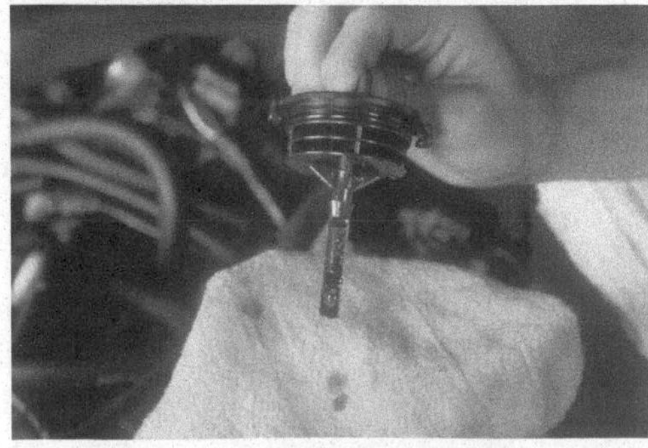
FIGURE 9-8 Check fluid level and condition. Foam and debris in the fluid are indications of problems in the system.

If the vehicle is on a drive-on lift, perform a **dry-park check** to locate loose ball sockets and worn rack and pinion bushings. This test places a high load on the steering and suspension components to check for loose joints and other parts. To perform a dry-park check, leave the wheels on the lift ramps and have a helper shake the steering wheel rapidly side-to-side while you listen and feel for loose linkage components. Performing this test with the engine off places a significant load on the front suspension and steering parts and is useful for finding worn and loose ball sockets.

If the vehicle is on a swing-arm lift, grasp the tires at the three and nine o'clock positions and move back and forth (**Figure 9-10**). This will help identify loose steering ball sockets and wheel bearings. Inspect the tie rods and other ball socket joints. Check grease boot condition, if the joint looks like it needs lubrication, and overall appearance. **Figure 9-11** shows an example of a neglected and damaged tie rod end. Even if the joint is not loose yet, with a damaged grease boot, it does not take long for water, dirt, and other contaminants to get into and damage the joint.

If the vehicle has been in an accident, hit a curb, or similar situation, check the linkage for bent components. **Figure 9-12** shows a vehicle that hit a curb while driving in snowy conditions. This damage is easy to spot but a slight bend or twist may not be visible.

Look for signs of fluid loss from the power steering hoses and gearbox. Depending on the results of your visual inspection, you may need to perform more specific tests of the steering system.

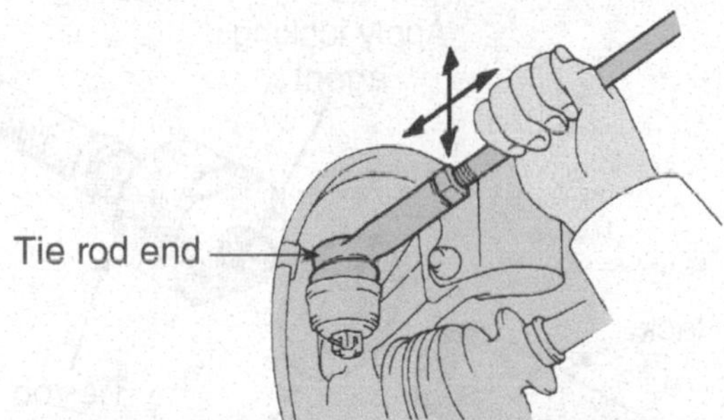

FIGURE 9-9 Checking for a loose tie rod socket by moving the tie rod up and down. Moving back and forth checks inner tie rod sockets.

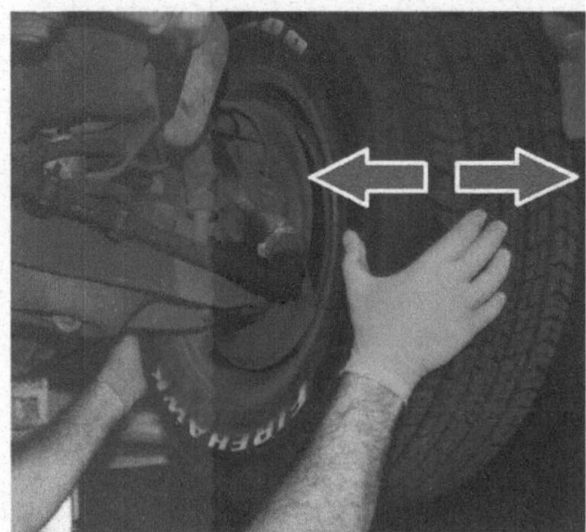

FIGURE 9-10 Moving the tire side-to-side can help locate loose tie rods, idler arms, centerlinks, and Pitman arms.

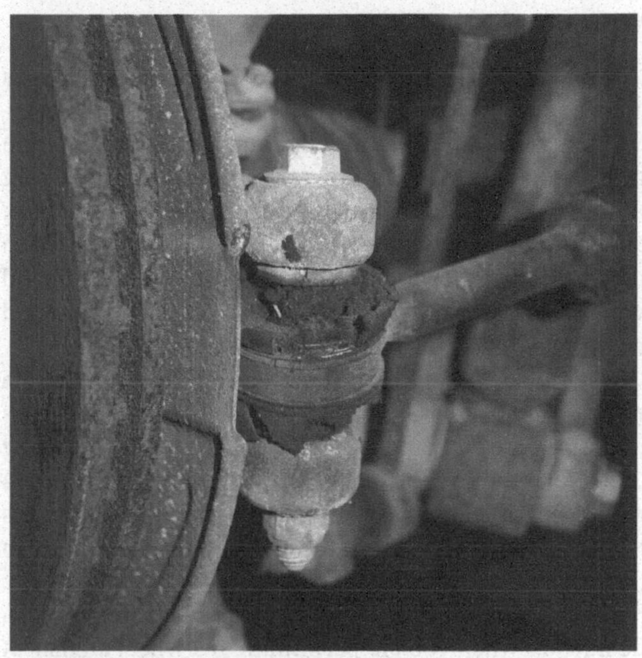

FIGURE 9-11 The damaged boot is going to eventually cause the joint to wear out.

FIGURE 9-12 An example of an inner tie rod damaged by hitting a curb a little too hard.

STEERING SYSTEM SERVICE

Once the cause of the complaint has been identified, you will likely need to remove and replace the defective component. The procedures discussed here are not specific to any particular vehicle but are common service practices. Always refer to the vehicle manufacturer's service information for the specific procedures.

■ ***Rack and Pinion Inner Tie Rods.*** Inner tie rods are a common wear item. The inner tie rods on end-take-off racks wear out just like the outer tie rods. Inner tie rods can be checked during a dry-park check, by trying to move the tie rod laterally, or by moving the tire back and forth quickly with the engine off. You may need to place your hand over the bellows to feel the inner tie rod while a helper moves the steering linkage. There should be no play in the inner tie rod socket.

To replace this type of inner tie rod, first remove the outer tie rod from the steering knuckle. **Figure 9-13** shows the removal of an outer tie rod using a tie rod separating tool. Next, remove the outer tie rod end from the inner tie rod. If possible, count or measure the number of threads the tie rod jam nut is from the end of the inner tie rod so that it can be reinstalled as close as possible to the same place (**Figure 9-14**). This will make setting the toe a little quicker when you are performing the wheel alignment after replacing the tie rod. Remove the jam nut and bellows clamps. These clamps are often factory crimped and cannot be reused. Remove the bellows, and examine the tie rod for any type of retention rivet or similar device (**Figure 9-15**). Remove the retainer from the tie rod before attempting to loosen the tie rod.

A special **inner tie rod socket** is usually needed to remove and install the tie rod (Figure 9-1). The rack gear should be held from twisting when you are removing the inner tie rod. This is often done by using a wrench on the rack gear and prevents the gear from twisting and possibly binding (**Figure 9-16**). Refer to the manufacturer's

FIGURE 9-13 Using a tie rod separator.

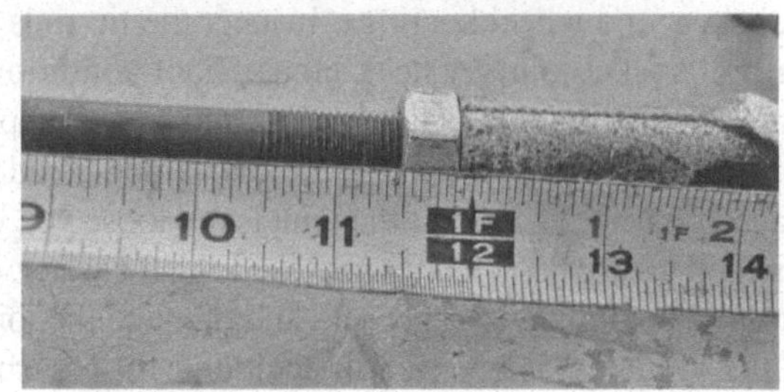

FIGURE 9-14 Measuring or marking the tie rod will help with setting the alignment after the repair.

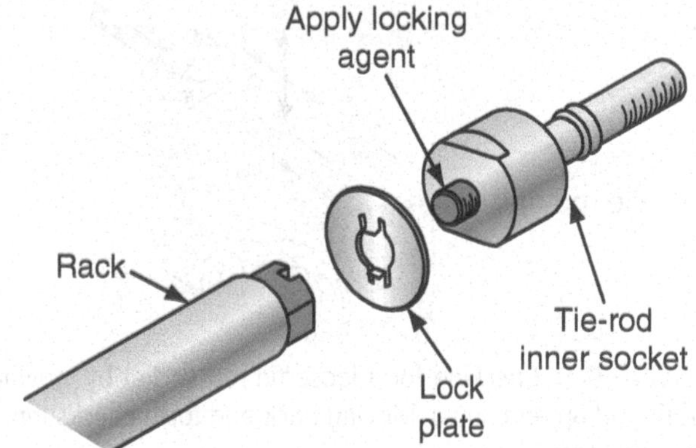

FIGURE 9-15 Many inner tie rods use a rivet, set screw, or lock plate to prevent the tie rod from loosening up.

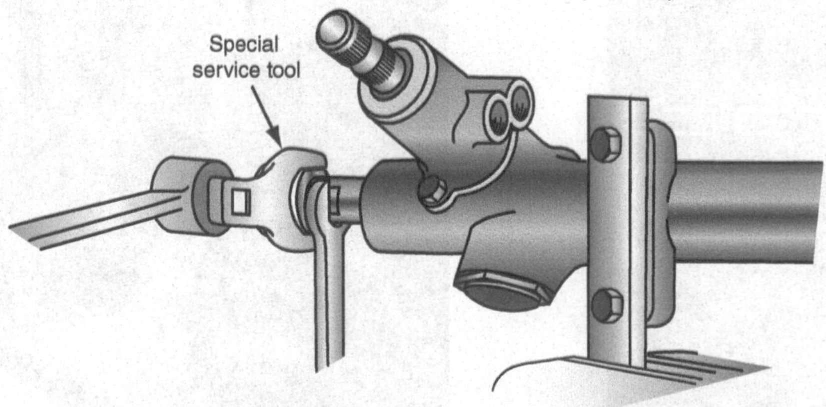

FIGURE 9-16 The service procedure may state to hold the rack gear steady with another tool when removing and installing the inner tie rod. This is to prevent twisting the rack gear and damaging the gearbox.

service information for the exact procedure. In general, select the correct inner tie rod socket tool and place it over the inner tie rod. Next, remove the inner tie rod.

Once the inner tie rod is removed, inspect the rack gear and the rack seal. There should not be any power steering fluid on the rack gear or leaking past the inner seal. The presence of fluid indicates that the seals are leaking, and the rack and pinion unit should be replaced (**Figure 9-17**).

Install the new tie rod, torque the tie rod to spec, and install any retention device as provided with the tie rod. Reinstall the bellows, jam nut, and outer tie rod end. Install the bellows clamps, and secure the bellows to the rack. This is important to prevent dirt and debris from getting into the inner tie rod socket. Many racks have a tube connecting the two bellows together. This tube allows air to pass back and forth between the bellows and prevents them from collapsing as the wheels are turned and air is forced out when the bellows shrinks. Make sure the tube is installed correctly and is secure in the bellows.

If the outer tie rod end is secured with a castle nut and cotter pin, torque the nut to specifications, and check the alignment of the nut to the hole in the tie rod stud. If the opening in the nut and hole do not align, tighten the nut until the cotter pin can be installed (**Figure 9-18**). Bend the ends of the cotter pin to prevent it from working out of the nut. This secures the pin in the tie rod and prevents the nut from loosening.

Many outer tie rod ends use a friction nut that contains a nylon ring (**Figure 9-19**). This type of nut is a one-time use nut and needs replaced if it is removed. If a new outer tie rod was installed, install a new grease fitting if necessary. Replacement outer tie rod may be sealed or may require periodic lubrication. If it is not sealed, use a grease gut to inject chassis grease into the new tie rod and wipe away any excess grease when complete.

Because the position of the tie rods will not be exactly as before replacement, toe will need to be checked and adjusted. Once the alignment has been set, tighten the jam nut to lock the two tie rods together.

■ ***Non Rack and Pinion Steering Linkage.*** Vehicles with recirculating ball gearboxes, such as those with SLA suspensions, usually use the parallelogram steering

Service Note

Jam nut torque is typically low, commonly around 40 to 50 ft.lbs. Do not overtighten the jam nut. Overtightening can lead to the nut seizing to the inner tie rod and making future alignments very difficult.

FIGURE 9-17 The fluid around the end of the rack gear indicates that the rack seals are leaking.

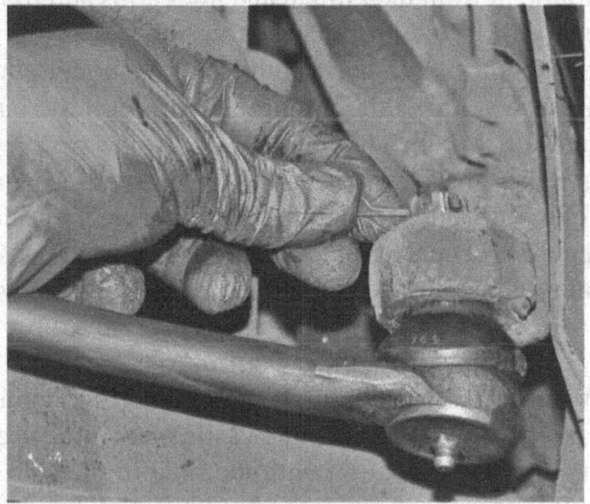

FIGURE 9-18 Torque the castle nut to specs and then align the hole in the stud by tightening the nut if needed. Install a new cotter pin and bend the open ends over to prevent the pin from working loose.

FIGURE 9-19 Some vehicles use friction nuts instead of castle nuts. These should not be reused.

linkage, which contains a centerlink and/or relay rod, idler and Pitman arms, inner and outer tie rods, and tie rod sleeves.

■ ***Centerlink.*** Centerlinks and Pitman arms can be classed as either wear or nonwear types. A wear-type centerlink or Pitman arm contains a ball-and-socket joint like those used on tie rods. Because this joint wears over time, looseness can develop in the steering from these components. Nonwear centerlinks and Pitman arms do not have ball-and-socket joints, just an opening for a stud to tighten into. Check for looseness by pulling on the linkage at the ball-and-socket joint (**Figure 9-20**).

Replacement of a worn centerlink or relay rod requires separating it from the Pitman and idler arms and from the inner tie rods. To do this, remove the cotter pin from the castle nut. Begin by straightening the cotter pin, and then use a pair of side-cutting pliers to pull the pin out of the tie rod stud. This may require some work with a wire brush and penetrating oil to remove the pin. On vehicles with a nylon friction nut, this type of nut should be replaced once it is removed and not reinstalled, as the nylon retainer is not designed to be reused. Remove the nuts, and using a tie rod separator, remove the other linkage components from the centerlink.

When you are installing the new centerlink, torque all nuts to specs. When tightening castle nuts, if the opening in the nut does not align with the hole in the stud once the nut is torqued, tighten the nut until it aligns. Insert a new cotter pin and bend the ends over to prevent it from falling out.

■ ***Pitman Arm.*** To replace the Pitman arm, first remove the nut securing it to the centerlink. Next, remove the nut holding the Pitman arm to the Pitman shaft of the gearbox. Removing the Pitman arm will require using a **Pitman arm puller** because the arm is splined to the tapered Pitman shaft. Install the puller and tighten (**Figure 9-21**). The Pitman arm should break free from the shaft and slide off the Pitman shaft. Note that Pitman arms may be able to be installed in four different directions. Make sure you install the Pitman arm indexed correctly so that it can be reattached to the centerlink. Install the Pitman arm and torque the retaining nut to specs.

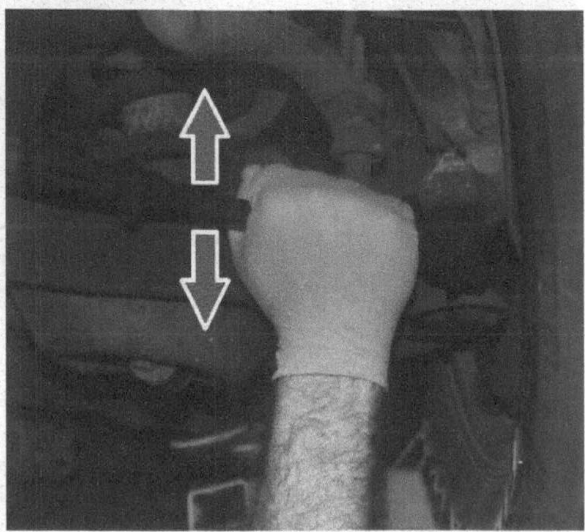

FIGURE 9-20 Check tie rods for play by moving them up and down.

■ ***Idler Arm.*** Idler arms are bolted to the frame to support the end of the centerlink opposite the Pitman arm. When you are checking for a worn idler arm, it is important to note that some manufacturers allow a slight amount of play in the idler arm socket. To determine if the socket is worn beyond specs, you will need to install a dial indicator to measure the amount of movement as a specific amount of force is applied (**Figure 9-22**). Do not replace the idler arm unless the total movement is beyond specs.

Service Note

Never reuse an old cotter pin. Always install a new cotter pin whenever one is removed.

FIGURE 9-21 Using a Pitman arm puller. The tapered shaft and high torque on the retaining nut means Pitman arms can be difficult to remove.

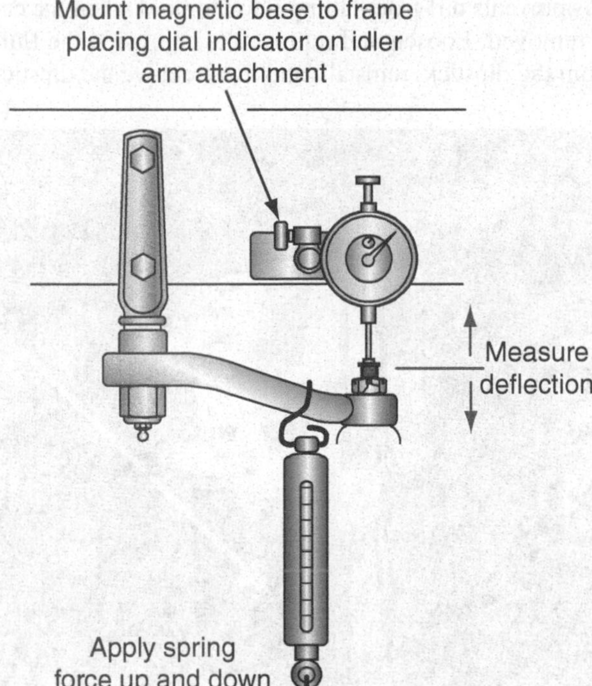
Mount magnetic base to frame placing dial indicator on idler arm attachment
Measure deflection
Apply spring force up and down

FIGURE 9-22 Many idler arms are allowed to have a slight mount of play. Checking the play requires applying a specific amount of force to the arm and measuring the movement with a dial indicator.

To replace an idler arm, first remove the nut holding the idler arm to the centerlink. Separate the idler arm from the centerlink, and then remove the bolts holding it to the frame. Install the new arm, and torque the bolts and nut to specs.

■ ***Steering Damper.*** Some heavy-duty 4WD trucks have a **steering damper** installed between the steering linkage and the frame. This damper, which looks like a shock absorber, helps reduce wheel shimmy on cross-link steering arrangements. Inspect the damper for leaks (**Figure 9-23**). Replacing a damper is like replacing a shock absorber: remove the mounting hardware from the damper, and then remove the damper itself. Install the new damper and torque the fasteners to specs.

■ ***Tie Rods.*** The tie rod assemblies, the inner, outer, and sleeve, often become rusted and seized over time, which makes setting toe difficult. Using too much force on a seized tie rod sleeve can cause damage to the sleeve. Inspect the sleeves, clamps, and bolts for damage. The tie rod sleeves can be damaged by trying to turn a stuck sleeve by using too much force or by using the wrong tool to try to turn the sleeve. The sleeve clamps can be damaged by overtightening the bolts or by rust. If the bolts and nuts of the clamps are damaged or rounded, replace them with new bolts and nuts.

FIGURE 9-23 Steering dampers, like suspension dampers, are oil filled and can leak over time.

Some technicians, when faced with an unmoving tie rod assembly, will remove the assembly from the vehicle and work on freeing the seized parts at a workbench. Clamp one of the tie rods in a vice and loosen the sleeve clamp bolts and nuts. Thoroughly lubricate the tie rod and sleeve threads, and start working on moving the sleeve and tie rods. Once the sleeve can move easily, reinstall the assembly on the vehicle, and torque the ball-and-socket studs and nuts to specs. While doing this makes it easier to work on the tie rods and set the toe during a wheel alignment, it should not be done without first having obtained authorization for additional work.

Power Assist Service

Power assist systems, whether hydraulic or electric, can suffer from insufficient assist, uneven assist, or no assist. Determining the cause of the complaint is important because problems with power assist can be caused by the power steering pump, electric assist system, or faults in the gearbox. Also, a problem that appears to be in the power assist system, such as insufficient assist or other concerns when turning the wheel, can be caused by problems in other areas, such as strut mounts, ball joints, and tie rods.

POWER STEERING SYSTEM SERVICE

The service of traditional hydraulic power assist systems has not changed significantly in many years, but service and repair still require proper tools and procedures to be used. Because many newer vehicles are using electric assist so you must first determine what type of power steering is used on the vehicle being inspected.

■ ***Determine Power Steering Assist Type.*** One easy method to determine which type of assist a vehicle has is to turn on the ignition and look for an **EPS warning light**, like that shown in the lower right corner of the dash in **Figure 9-24**. This light will come on to warn the driver if a fault is present in the electric power steering system. When you turn the ignition ON, the display of this warning light on the dash means that this vehicle has electric power steering assist. You can also look under the hood for a power steering pump and fluid reservoir. If there is no pump or reservoir, it means the assist is provided electrically.

Electric assist can be applied either at the steering column or to the rack and pinion. Because neither the entire column nor the rack is usually visible, you should check the service information systems for the description and operation of the type of system used in the vehicle being inspected. The description and operation will provide you with information about the components of the system, how the system operates, and should include any service warnings and precautions to be followed.

■ ***Power Steering Fluid.*** For many years, standard power steering and automatic transmission fluids were the only fluids used in the power steering system. Now, several manufacturers require specific power steering fluid blends in their systems. Use of the incorrect fluid can lead to damage to the power steering system and loss of assist. Because of these fluid requirements, you need to determine what the correct fluid is for any given vehicle.

To find the correct fluid type, locate the power steering fluid reservoir. The reservoir either is located on the power steering pump (**Figure 9-25**), or is mounted remotely for easier access (**Figure 9-26**). Inspect the cap for fluid specifications and instructions on how to correctly check the fluid level. If the cap does not state a specific fluid type, refer to the vehicle owner's manual or the manufacturer's service information.

Some reservoirs, usually those that are mounted remotely, are transparent enough that the fluid level can be seen through the plastic (**Figure 9-27**). The level should be above the minimum and below the maximum lines. Reservoirs that are mounted with the pump usually have a combination fluid cap and dipstick. Before removing the cap, use a shop rag to clean any dirt from the cap; this prevents dirt from falling into the fluid when the cap is removed. Loosen and remove the cap, wipe the fluid from the dipstick, reinstall, and then remove the dipstick

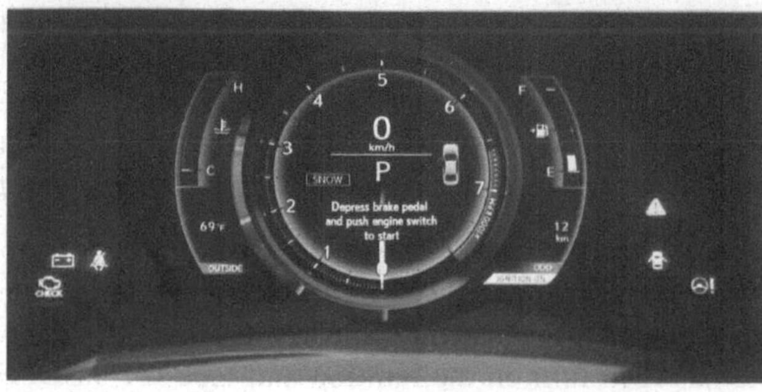

FIGURE 9-24 An example of an EPS warning light.

FIGURE 9-25 A power steering pump and reservoir. Many vehicles use remotely mounted fluid reservoirs.

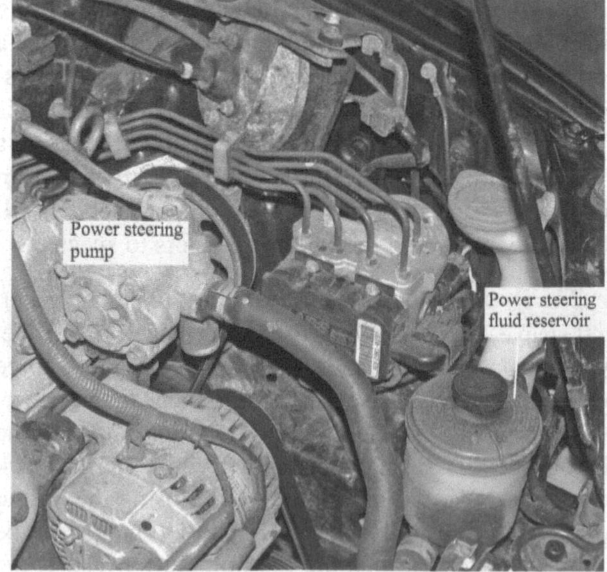

FIGURE 9-26 An example of a power steering pump and remotely mounted reservoir.

FIGURE 9-27 This remote reservoir allows for easy checking of the fluid level.

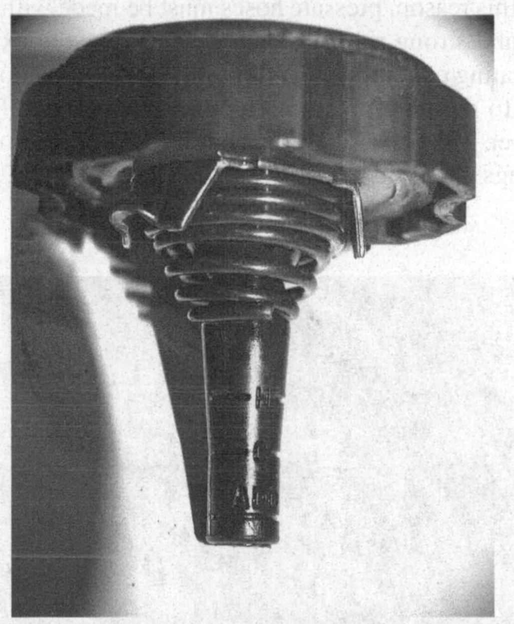

FIGURE 9-28 This dipstick is marked for hot and cold readings of the fluid level. After the steering has been turned lock-to-lock several times, the fluid should be checked at the hot level.

a second time. Note the fluid level and wipe the fluid onto a clean shop rag. Check for fluid color and for signs of metal shavings. It is normal for the fluid to darken with age. Metal shavings are an indication of pump or steering gear wear. With the engine running, have a helper turn the wheels side-to-side. Remove the reservoir cap and check for bubbles in the fluid. Bubbles in the fluid indicates that there is air in the system, which can be caused by a leak.

If the fluid level is low, top it off using the specified fluid. Many power steering dipsticks will have separate full level marks for hot versus cold fluid (**Figure 9-28**). The fluid level will show higher if the fluid is hot due to heat expansion. Do not overfill the system. Overfilling can cause the fluid to leak when it is hot, as there will not be room for expansion.

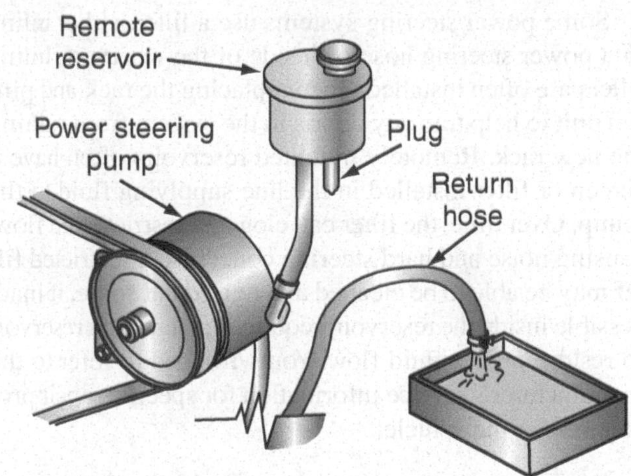

FIGURE 9-29 Flushing the power steering system by adding new fluid to the reservoir while pumping the old fluid out.

■ ***Flushing, Filling, and Bleeding the Power Steering System.*** A **power steering fluid flush** is needed whenever the power steering pump or gearbox has been replaced. This removes any debris from the system and prevents repeat failures. A failed pump or gearbox can allow metal shavings to contaminate the fluid and can remain in the system if not flushed out.

Flushing the system can be performed by disconnecting the fluid return line to the pump, placing the return hose into a container to catch the old fluid, and allowing the pump to draw new fluid from a gallon jug of new fluid (**Figure 9-29**). Start the engine, and turn the steering wheel so the pump will force fluid through the system. As the old fluid leaves the steering gear, new fluid is drawn in and pumped through the system. Once all the old fluid is removed, reattach the power steering hose, and top off the system.

After flushing, there is likely to be some air in the system. This will require the system to be bled. With the cap off the reservoir, start the engine and turn the wheels from lock-to-lock repeatedly while topping off the fluid level as necessary. When you are turning the wheels to the lock positions, do not leave them at full lock for more than a few seconds, as this generates a lot of heat in the pump and in the fluid. Holding the steering at a lock position for too long can damage the power steering pump. The system is bled when there is no longer any foam showing on the fluid, and the pump no longer makes a whining or humming sound when the wheels are turned.

Some power steering systems tend to keep air trapped in them and require that the vehicle be driven up to highway speeds to purge the air from the system. Be sure that the system is thoroughly bled and is full before returning the vehicle to the customer. Refer to the manufacturer's service information for the power steering bleeding procedures for the particular make and model on which you are working.

Some power steering systems use a filter, either inline in a power steering hose or inside of the reservoir. Inline filters are often installed when replacing the rack and pinion unit to help trap any debris in the system from ruining the new rack. Remotely mounted reservoirs often have a screen or filter installed in the line supplying fluid to the pump. Over time, the filter can clog and restrict fluid flow, causing noise and hard steering concerns. A restricted filter may be able to be cleaned and reused but some, if inaccessible inside the reservoir, require replacing the reservoir to restore proper fluid flow. You will need to refer to the manufacturer's service information for specific repair procedures for the vehicle.

■ ***Diagnose Power Steering Fluid Leaks.*** A very low power steering fluid level often causes the power steering pump to whine, which gets worse when the wheels are turned. This is caused by air in the system. The fluid may also appear milky or foamy; this too is because of the air in the system. Very low fluid level or an empty system can cause a lack of power assist or uneven assist.

Power steering fluid leaks can develop from any part of the system. Start by visually inspecting the power steering components for signs of a leak. **Figure 9-30** shows an example of fluid accumulation from a leaking high-pressure rubber hose and a leaking steering gearbox.

To diagnose where a leak is coming from, it may be necessary to clean the engine compartment of old fluid. Leaks may be more easily found if a helper is used to turn the wheels to the lock position with the engine running. This increases the pressure in the system and may increase the rate of the leak. Common leak locations are the power pump, steering hoses, and rack and pinion seals. The leak shown in **Figure 9-31** is from the power steering pump. **Figure 9-32** shows a leak that was just visible from the steel section of the pressure line only when the wheel was turned. Though hard to see in this image, a very thin stream of fluid was released from the line. This reinforces the necessity of wearing safety glasses at all times when you are working on cars and trucks. This type of leak could be dangerous if not noticed and fluid were to get sprayed into your eyes.

Power steering hoses, like cooling system hoses, wear from the inside out, though rust-through of the steel sections is also a common cause of failure. A leaking power steering hose must be replaced with an equivalent hose, both in size and construction. Power steering pressure hoses operate under high pressure and can get very hot. For this reason, pressure hoses must be made with durable and strong rubber compounds to take the extreme operating conditions. Fluid return hoses, while not subject to the high pressure, are also made of reinforced rubber. Most hoses have special fittings and seals on the fittings to provide a positive fit and prevent leaks.

FIGURE 9-30 A leaking power steering gearbox.

FIGURE 9-31 An example of a leaking power steering hose.

FIGURE 9-32 A rusted-through and leaking high-pressure power steering line.

When a leaking hose is being replaced, it is a good idea to replace both hoses. Because both have most likely been in service the same length of time, the failure of one hose can be followed shortly by failure of the other hose. To prevent this, you should recommend to the customer that both hoses be replaced at the same time. This can decrease repair costs because replacing both hoses at once can be less expensive than replacing one hose and then the other later.

Other possible sources of fluid leaks include the Pitman shaft seal on recirculating ball gearboxes and the seals in the pump itself. Leaking Pitman shaft seals are replaced by removing the Pitman arm, seal cover, snap ring, and then the seal. Install the new seal, snap ring, and cover. Reinstall the Pitman arm, and torque the nut to specifications. A fluid leak from a power steering pump generally requires pump replacement.

■ ***Replacing Power Steering Hoses.*** After you have determined which hose is leaking, you will need to replace either the entire hose or a section of hose. Some vehicles have power steering hoses in several sections, each being separately replaceable. Most hose connections use threaded fittings (**Figure 9-33**), which require using a line wrench to remove without stripping the fitting (**Figure 9-34**). Place a fluid pan under the area you are working on to catch the power steering fluid that will leak out of the line. Once you break the fittings loose, locate and remove any clamps that hold the line to the body or other components. Remove the old line, and compare it with the new line. Closely check the fittings to make sure the new line will thread properly into the existing components. If the new part is correct, begin to reinstall the new line by starting the fittings by hand. Start both fittings, and then secure the hose back into place using the clamps you removed earlier. Once the hose is in place, perform the final tightening of the fittings. Be careful not to overtighten the fittings as you may strip the threads and damage the aluminum rack and pinion.

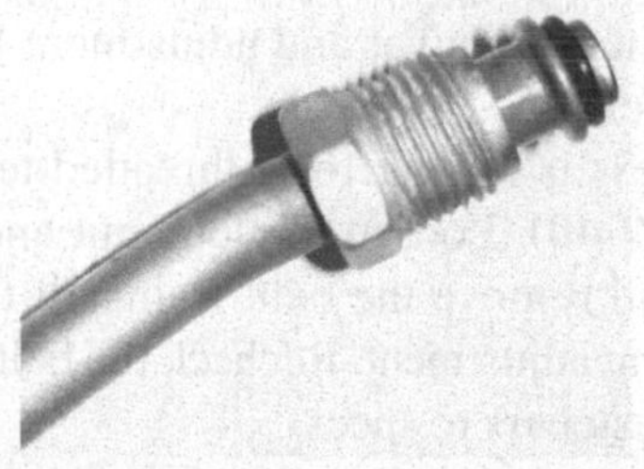

FIGURE 9-33 An example of a fitting on a power steering hose. Connections like this are necessary due to the in the system.

FIGURE 9-34 Use a line wrench when loosening and tightening power steering lines to avoid stripping the fittings.

Once installed, refill the reservoir with the correct fluid and bleed the air from the system. When you are finished, recheck your work and make sure the replacement line is not leaking.

■ ***Power Steering Pump Belt.*** The power steering belt, as with any accessory drive belt, requires periodic replacement. Belts should be inspected during routine service and often require replacement every 60,000 to 90,000 miles. A severely worn or damaged power steering belt can cause erratic assist, especially at low speeds, because the belt can slip around the drive pulley instead of actually turning the pulley. This can also cause a high-pitched screeching or squealing noise as the belt slips around the pulleys.

Inspect the belt for fraying, burning, cracks, and missing pieces of rubber. **Figure 9-35** shows examples of several types of belt wear problems. Older belts, made of neoprene rubber, tend to crack much more than newer belt designs (**Figure 9-36**).

Examine the belt closely. If the belt is shiny, has chunks missing, or has rubber deposited in the grooves, the belt tension is likely insufficient and you should examine the belt tensioner and other pulleys. Worn and weak belt tensioners can cause the same symptoms as a faulty belt. Many technicians replace the automatic tensioner when the drive belt is replaced to help eliminate problems with insufficient belt tension.

Use a **belt wear gauge** to inspect the grooves and sides of the ribs for wear. As the belt wear, the ribs will become narrower (**Figure 9-37**). This reduces the amount of contact between the belt and the pulley groove, which can allow the belt to slip. Place the tool along each groove in the belt and check the depth of the tool against the depth of the ribs. Using one type of belt wear tool is shown in **Figure 9-38**.

To replace the belt, first determine how tension is applied to the belt. Belts are either mechanically

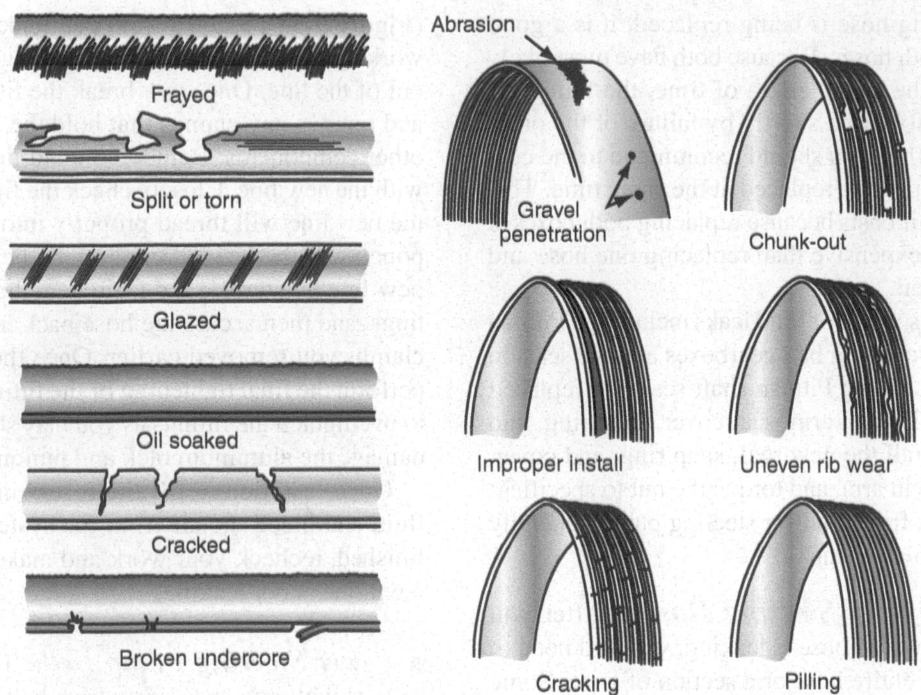

FIGURE 9-35 Common types of belt wear and damage. Check the entire belt when performing an inspection as part of the belt may look good but other parts may have problems.

FIGURE 9-36 Neoprene belts are prone to cracking with age. This belt has excessive cracking and needs to be replaced.

tightened by moving an accessory, like the power steering pump, against the belt, by a tensioning idler pulley, or by a spring-loaded automatic drive belt tensioner. The next few figures show how these three methods are applied to removing and installing a belt. For belts as shown in **Figure 9-39**, locate the bolts that hold the power steering pump in place. Select the correct size sockets or wrenches and loosen the pivot and adjusting bolts enough so that the pump can move. Move the pump away from the belt, and remove the belt. When you are reinstalling the belt, make sure it is sitting in each pulley, and then pry the pump to apply tension to the belt (Figure 9-39). Snug the pivot and adjustment bolt in place to hold the belt.

The process is similar for a threaded tensioner pulley (**Figure 9-40**). Loosen the lock nut and the adjustment bolt and remove the belt. Reinstall the new belt, and tighten the adjustment. Recheck the belt tension, and tighten the fasteners to specs.

Once the new belt is installed, it is important that the correct tension be applied. If too much tension is applied to the belt, it can damage the belt and the power steering pump shaft bearings. If the belt is too loose, it can slip, causing poor steering and noise. To check belt tension, locate the belt tension spec in the service information, and then use a belt tension gauge to check the belt tension (**Figure 9-41**). If necessary, loosen the pivot and adjustment bolts and retension the belt. Once the belt is properly tensioned, tighten the pivot and adjustment bolts to specs.

Most engines use an automatic or spring-loaded **drive belt tensioner** (**Figure 9-42**). This type of tensioner can maintain correct tension on the belt during changing engine rpm and loads on the belt. Many of these tensioners have a built-in slot for a square drive from a ratchet (**Figure 9-43**). This allows you to use a ⅜-inch or ½-inch drive ratchet to pry the tensioner away from the belt for removal. If the tensioner does not have a place for a ratchet, then a wrench or socket on the pulley bolt is

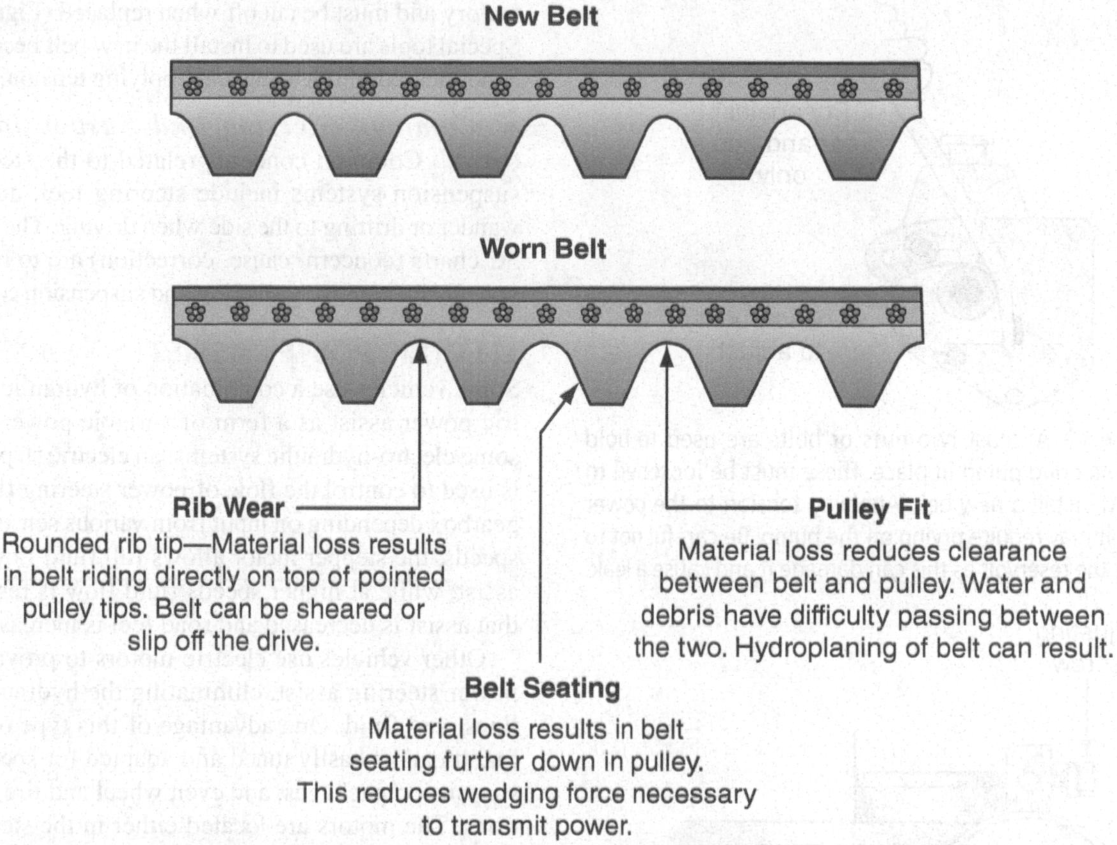

New Belt

Worn Belt

Rib Wear

Rounded rib tip—Material loss results in belt riding directly on top of pointed pulley tips. Belt can be sheared or slip off the drive.

Pulley Fit

Material loss reduces clearance between belt and pulley. Water and debris have difficulty passing between the two. Hydroplaning of belt can result.

Belt Seating

Material loss results in belt seating further down in pulley. This reduces wedging force necessary to transmit power.

FIGURE 9-37 This illustrates how multirib belts wear within the ribs.

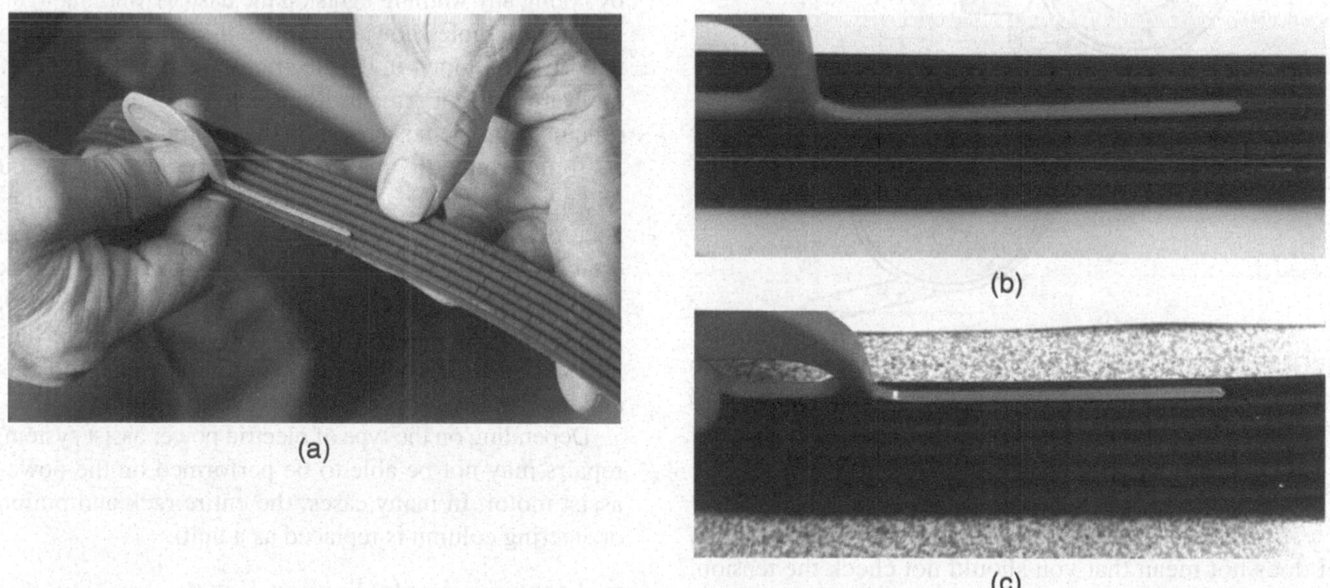

FIGURE 9-38 (a) Using a belt wear gauge to check rib wear. (b) The gauge rides above the ribs indicating this belt is still good. (c) When the gauge fits below the ribs, the belt is worn and due for replacement.

used to release the belt tension. Determine which way to move the tensioner and apply pressure with a wrench or ratchet until the belt can be removed. Once the belt is off, inspect the belt and pulleys. Each pulley, other than the crankshaft pulley, should spin easily and without noise. Noise or a roughness in the pulley often indicates worn bearings. Check that the pulleys are aligned, and if no other problems are present, install the new belt.

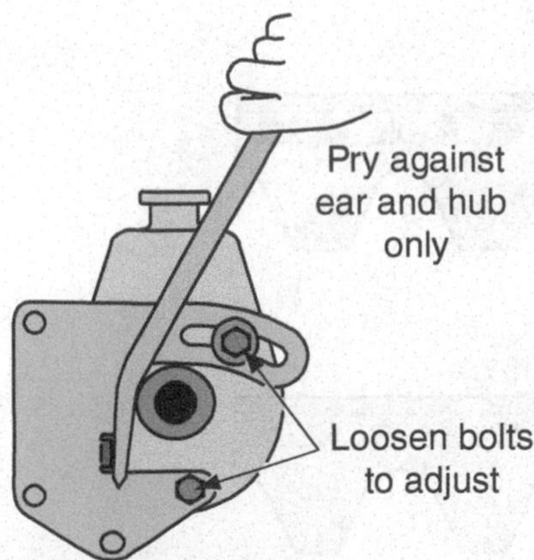

FIGURE 9-39 At least two nuts or bolts are used to hold the power steering pump in place. These must be loosened to remove and install a new belt. Applying tension to the power steering belt may require prying on the pump. Be careful not to pry against the reservoir as this can damage it and cause a leak.

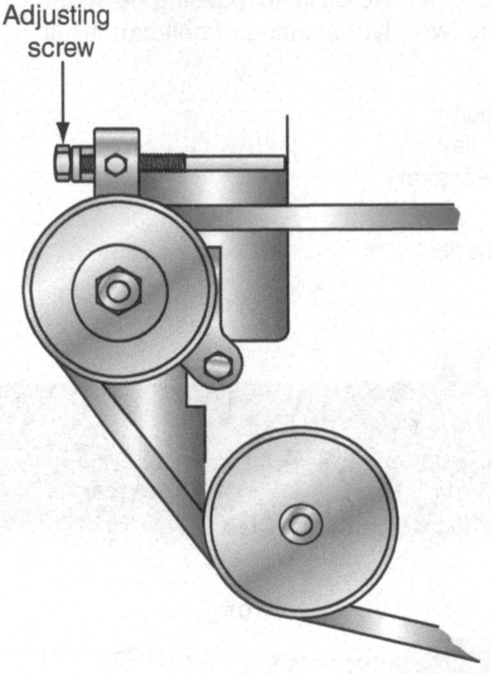

FIGURE 9-40 A threaded tensioner will have at least two nuts or bolts holding it in place that need to be loosened to replace or tighten the belt. concerns.

Just because a vehicle has an automatic belt tensioner, it does not mean that you should not check the tension. The spring in the tensioner weakens over time, decreasing the tension on the belt and requiring both the belt and tensioner to be replaced. Because of this, some technicians replace automatic belt tensioners when the drive belt is replaced.

Some late-model cars and trucks have stretch-fit belts installed on one of the accessories, commonly on the air conditioning compressor. These belts are installed at the factory and must be cut off when replaced (**Figure 9-44**). Special tools are used to install the new belt because there is no method of releasing and applying tension.

■ ***Common Steering and Suspension Concerns.*** Common concerns related to the steering and suspension systems include steering feel, noises, and wander or drifting to the side when driving. The following 3C charts (concern, cause, correction) are to help guide you when diagnosing steering and suspension complaints.

ELECTRIC POWER ASSIST

Some vehicles use a combination of hydraulic and electric power assist as a form of variable power assist. In some electro-hydraulic systems, an electric stepper motor is used to control the flow of power steering fluid to the gearbox depending on input from various sensors. At low speeds, the stepper motor allows full fluid flow and full assist, while at higher speeds fluid flow is restricted so that assist is decreased and road feel is increased.

Other vehicles use electric motors to provide all the power steering assist, eliminating the hydraulic pump, lines, and fluid. One advantage of this type of assist is that it can be easily tuned and adapted for specific cars, suspension packages, and even wheel and tire combinations. The motors are located either in the steering column or as part of the steering rack assembly.

■ ***Inspect the Electric Power Steering System.*** Begin by noting any warning lights on the dash. A warning light, like the example shown in **Figure 9-45**, is used to warn the driver of a fault in the electric power assist system. If the warning light remains illuminated once the engine is running, a problem is present in the electric power assist.

To determine the reason the warning light is on, you will need to connect a scan tool to the diagnostic link connector (DLC). Once connected and communications are established, select the power steering system. Next, check for stored diagnostic trouble codes (DTCs) (**Figure 9-46**). Before starting repairs, check for technical service bulletins (TSBs). These can help you find known problems and repairs, saving you time and energy (**Figure 9-47**).

Depending on the type of electric power assist system, repairs may not be able to be performed on the power assist motor. In many cases, the entire rack and pinion or steering column is replaced as a unit.

■ ***Steering Angle Sensor Reset.*** Any time that repairs are performed to the steering system and a wheel alignment is performed, it is critical that a steering angle sensor reset or relearn is performed. A reset or relearn is when the EPS control module relearns the position of the steering shaft in relation to the front wheels. Failure to perform this reset can cause uneven power assist on some vehicles. For vehicles with systems such as active lane

TABLE 9-1

Condition	Possible Cause	Correction
Loose steering	1. Steering linkage connections loose 2. Steering linkage ball stud rubber deteriorated 3. Flex coupling-to-steering-gear attaching bolt loose 4. Intermediate shaft-to-steering-column-shaft attaching bolt loose 5. Flex coupling damaged or worn 6. Front and/or rear suspension components loose, damaged, or worn 7. Steering rack adjustment set too low	1. Tighten as required. 2. Replace as required. 3. Tighten the bolt to specifications. 4. Tighten the bolt to specifications. 5. Replace as required. 6. Tighten or replace as required. 7. Check the steering rack adjustment and readjust as required.
Noise: Knock, clunk, rapping, squeaking noise when turning	1. Steering wheel-to-steering-column shroud interference 2. Lack of lubrication where speed control brush contacts steering wheel pad 3. Steering column mounting bolts loose 4. Intermediate shaft-to-steering-rack attaching bolt loose 5. Flex coupling-to-steering-rack attaching bolt loose 6. Steering rack mounting bolts loose 7. Tires rubbing or grounding out against body or chassis 8. Front suspension components loose, worn, or damaged 9. Steering linkage ball stud rubber deteriorated 10. Normal lash between pinion gear teeth and rack gear teeth when the steering gear is in the off-center position permits gear tooth chuckle noise while turning on rough road surfaces	1. Adjust or replace as required. 2. Lubricate as required. 3. Tighten the bolt to specifications. 4. Tighten the bolt to specifications. 5. Tighten the bolt to specifications. 6. Tighten the bolt to specifications. 7. Adjust/replace as required. 8. Tighten or replace as required. 9. Replace as required. 10. This is normal condition and cannot be eliminated.
Vehicle pulls/drifts to one side *Note*: This condition cannot be caused by the manual steering gear.	1. Vehicle overloaded or unevenly loaded 2. Improper line pressure 3. Mismatched tires and wheels 4. Unevenly worn tires 5. Loose, worn, or damaged steering linkage 6. Steering linkage stud not centered within the socket 7. Bent spindle or spindle arm 8. Broken and/or sagging front and/or rear suspension springs 9. Loose, bent, or damaged suspension components 10. Bent rear axle housing 11. Excessive camber/caster split (excessive side-to-side variance in the caster/camber settings) 12. Improper toe setting 13. Front wheel bearings out of adjustment 14. Investigate tire variance (conicity for radial tires and unequal circumference for bias-ply and belted bias-ply tires)	1. Correct as required. 2. Adjust air pressure as required. 3. Install correct tire and wheel combination. 4. Replace as required (check for cause). 5. Tighten or replace as required. 6. Repair or replace as required. 7. Repair or replace as required. 8. Replace as required. 9. Tighten or replace as required. 10. Replace as required. 11. Adjust camber/caster split as required. 12. Adjust as required. 13. Adjust as required. 14. Repair or replace as required.

TABLE 9-2

Condition	Possible Cause	Correction
Wander: The vehicle wanders side-to-side on the roadway when it is driven straight ahead, while the steering wheel is held in a firm position. Evaluation should be conducted on a level road (little road crown)	1. Loose/worn tie rod ends or ball socket 2. Inner ball housing loose or worn 3. Gear assembly loose crossmember 4. Excessive yoke clearance 5. Loose suspension struts or ball joints 6. Column intermediate shaft connecting bolts loose 7. Column intermediate shaft joints loose or worn 8. Improper wheel alignment 9. Tire size and pressure 10. Vehicle unevenly loaded or overloaded 11. Steering gear mounting insulators and/or attachment bolts loose or damaged 12. Steering gear adjustments 13. Front end misaligned 14. Worn front end parts or wheel bearings 15. Unbalanced or badly worn steering gear control valve	1. Replace tie rod end or tie rod assembly. 2. Replace tie and assemblies. 3. Tighten the two mounting nuts to specification. 4. Adjust yoke clearance. 5. Tighten or replace as required. 6. Tighten bolts to specification at gear and column. 7. Replace intermediate shaft assembly. 8. Set alignment to specification. 9. Check tire sizes and adjust tire pressure. 10. Adjust load. 11. Replace insulators and/or attachment nuts and bolts. Tighten to specification. 12. Refer to steering gear section of Shop Manual. 13. Check and align to specifications. 14. Inspect and replace affected parts. 15. Inspect and replace affected parts.

TABLE 9-3

Problem	Symptoms	Possible Causes
Excessive steering wheel free play	Steering wander when driving straight ahead	1. Loose worm shaft preload adjustment 2. Loose sector lash adjustment 3. Loose steering gear mounting bolts 4. Worn steering linkage components
Excessive steering effort	Excessive steering wheel turning effort when turning a corner or parking	1. Worn flexible coupling or universal joint in steering shaft.
Underhood noise	Rattling noise when driving over road irregularities Squawking noise while turning a corner	1. Lack of steering gear lubricant 2. Tight worm shaft bearing preload adjustment 3. Tight sector lash adjustment 4. Worn flexible coupling or universal joint in steering shaft 5. Loose steering gear mounting bolts 6. Cut or worn dampener 7. O-ring on spool valve
Erratic steering effort	Erratic steering effort when turning a corner	1. Low power steering fluid level 2. Air in power steering system 3. Worn, damaged worm shaft, ball nut, or sector teeth
Death wobble	Excessive or violent front-end shaking	1. Worn steering linkage sockets 2. Worn suspension bushings 3. Worn steering damper

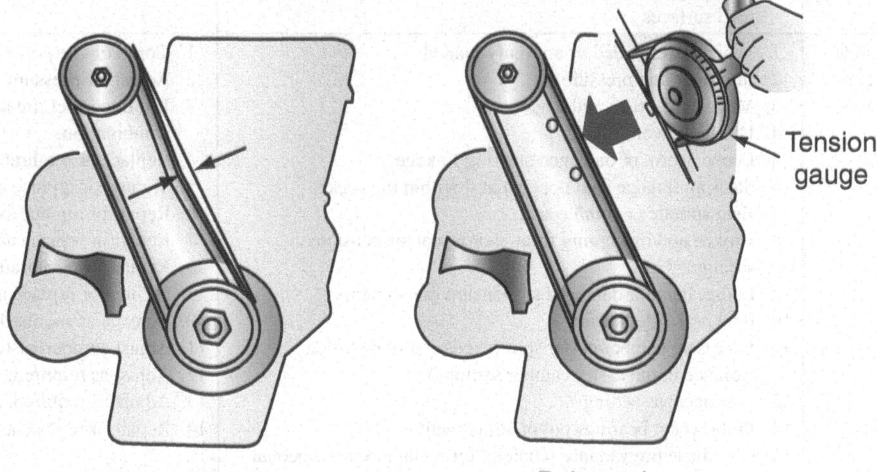

FIGURE 9-41 Belt tension should be checked whenever replacing or tightening a belt. Overtightening the belt can cause damage to bearings and rapid belt wear. A loose belt will slip, causing power steering issues.

FIGURE 9-42 An example of a spring-loaded automatic tensioner.

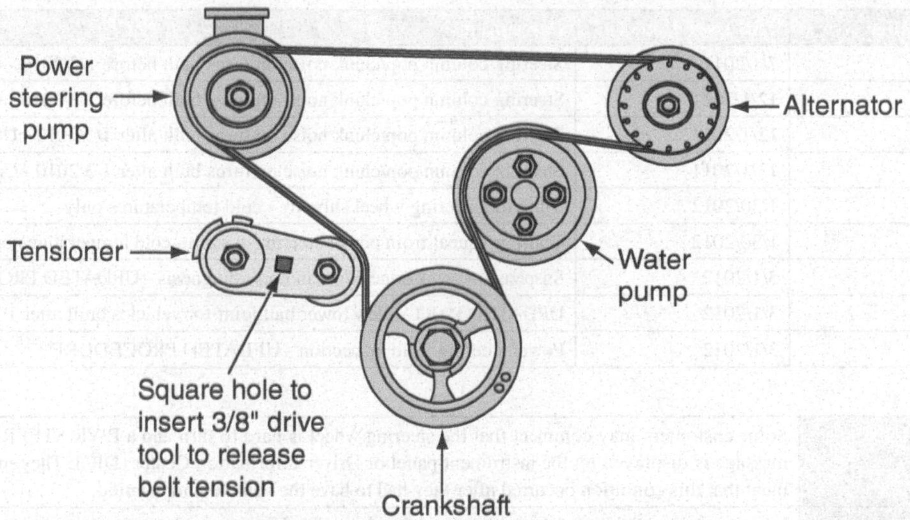

FIGURE 9-43 An illustration of how to remove the tension from the belt.

FIGURE 9-44 Many new vehicles have stretch-fit belts driving an accessory. The belts cut off and a new belt is stretched into place using special service tools.

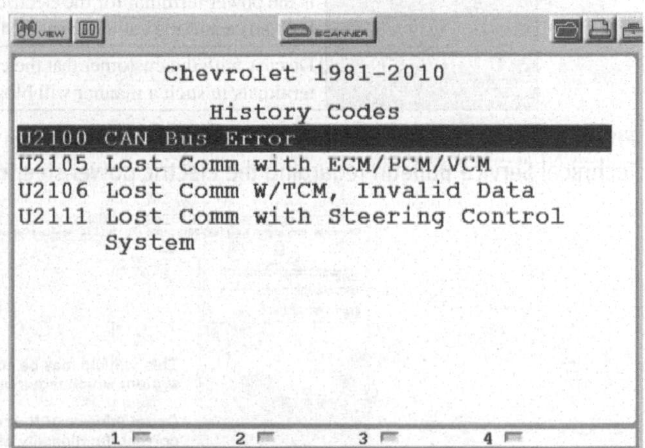

FIGURE 9-46 Steering system trouble codes. This vehicle had jump-started another vehicle and lost power steering.

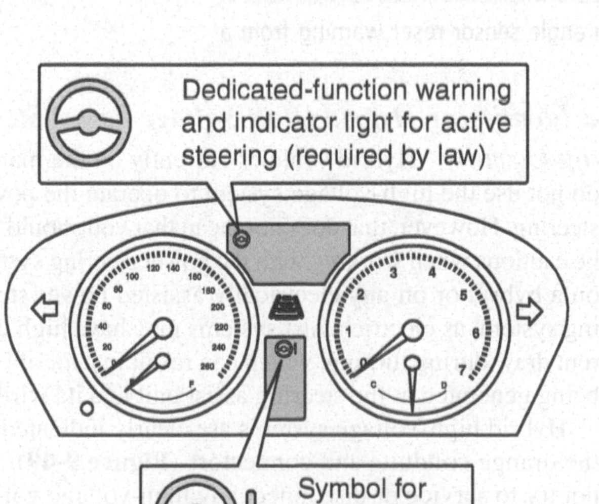

FIGURE 9-45 The steering wheel indicator light is used to warn the driver of a fault in the electric assist system.

keeping assist or self-parking, a system that is out of calibration can cause the vehicle to drive or steer incorrectly.

Once the vehicle has been repaired and the wheel alignment is being checked, the alignment computer should alert you if the vehicle requires a reset (**Figure 9-48**). Follow the steps on the aligner or in the manufacturer's service information. Three types of relearns are as follows:

- Self-calibrating—Meaning the system will relearn on its own after turning the steering wheel from lock-to-lock.
- Toyota/Lexus—Using a jumper wire to short specific terminals together at the data link connector.
- Scan tool relearn—Many vehicles using either a scan tool or the software with the alignment machine to perform a relearn.

These relearn types are not meant to take the place of the applicable service information or procedures. **Follow all the manufacturers' relearn procedures as outlined in the service information.**

TSB Number	TSB Issue Date	Description
10-15-11	7/1/2011	Steering column pop/clunk noise on turns built before 1/3/2010 - US
12-1-11	12/1/2011	Steering column pop/clunk noise on turns built before 1/3/2010 - Canada
12-1-11	12/1/2011	Steering column pop/clunk noise on turns built after 1/3/2010 - US
12-1-11	12/1/2011	Steering column pop/clunk noise on turns built after 1/3/2010 - Canada
1-30-12	1/30/2012	Vibration/steering wheel shimmy - cold temperatures only
1-30-12	1/30/2012	Chirp or squeal from power steering system - cold temperatures only
3-1-12	3/1/2012	Suspension creak/squeak/groan noise diagnosis - UPDATED PROCEDURE
4-15-12	3/1/2012	UPDATED PART - New lower ball joint for vehicles built after 10/15/10
5-1-12	3/1/2012	Power steering fluid inspection - UPDATED PROCEDURE

CONDITION	Some customers may comment that the steering wheel is hard to turn and a PWR STEERING warning message is displayed on the instrument panel or Driver Information Center (DIC). They may also comment that this condition occurred after they had to have the vehicle jump-started.
CAUSE	Improper jump-starting of the vehicle has been determined as a cause for the main power steering fuse to open (blow). Customers may mistakenly believe that the Underhood Junction Box connections are the battery connections and connect the jumper cables to the posts of the UHJB. The smaller post on the UHJB is the power terminal for the electric power steering assist motor. When the jumper cables are attached to both posts and the cables connected to another vehicle, the power steering fuse will blow.
CORRRECTION	Discuss with the customer that the connections at the UHJB are not for jump-starting and that using these terminals in such a manner will blow the power steering fuse and disable the electric assist.

FIGURE 9-47 (a) An example of steering and suspension Technical Service Bulletins (TSBs) for a vehicle. (b) An example of a Technical Service Bulletin regarding the electric power steering system.

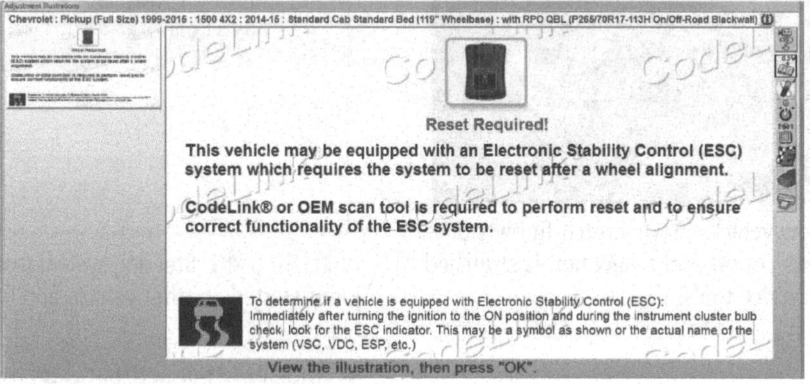

FIGURE 9-48 An example of a steering angle sensor reset warning from a Hunter alignment computer.

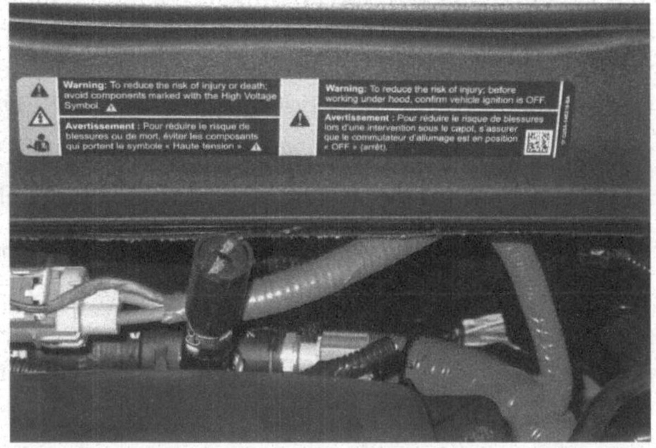

FIGURE 9-49 Orange wiring and connectors indicate high voltage. Never try to open any orange wiring or connections.

■ ***Identifying Hybrid High-Voltage Power Steering Circuits.*** Hybrid vehicles currently on the market do not use the high-voltage system to operate the power steering. However, that does not mean that you should not be cautious when working with the power steering system on a hybrid or on any electrically assisted power steering system, as electric assist systems may have high current draw during turning, which can result in a lot of heat being generated at the steering assist unit and its wiring.

Hybrid high-voltage systems are clearly indicated by the orange conduits and connectors (**Figure 9-49**). Do not try to service or disconnect any high-voltage wiring until all of the vehicle manufacturer's procedures to isolate the high-voltage system have been performed, and the system has been proven safe.

SUMMARY

Disable the airbag system before performing any repairs to the steering column.

Hard steering can occur if the power assist is not providing adequate fluid flow or pressure.

Worn strut bearing plates can cause many of the same problems as a worn rack and pinion gearbox.

Tie rods are a very common wear item for any type of steering system.

A special inner tie rod socket is usually needed to remove and install the tie rod.

Centerlinks and Pitman arms can be either wear or non-wear types.

Removing the Pitman arm will require using a Pitman arm puller.

Electric assist can be applied either at the steering column or to the rack and pinion.

The power steering fluid should be flushed when the power steering pump or gearbox is replaced.

Belts should be inspected during routine service and often require replacement every 60,000 to 90,000 miles.

REVIEW QUESTIONS

1. To remove _______________ tie rods on a rack and pinion, a special tool may be necessary.
2. Power steering assist can be either _______________ or _______________.
3. A steering _______________ is used on some larger trucks to reduce vibration in the steering linkage.
4. The _______________ - _______________ test is one method of inspecting the steering link for loose ball sockets.
5. When installing a castle nut onto a steering component, always replace the _______________ _______________ with a new one.
6. Which of the following is not part of the procedure to disable a driver's side airbag?
   a. Disconnect the battery
   b. Remove the airbag fuse
   c. Remove the airbag
   d. Disconnect the airbag
7. A vehicle has a hard steering complaint: *Technician A* says low tire pressure may be the cause. *Technician B* says a blown fuse may be the cause. Who is correct?
   a. Technician A
   b. Technician B
   c. Both A and B
   d. Neither A nor B
8. Which of the following is not a special tool for power steering system service?
   a. Pitman arm puller
   b. Tie rod separator
   c. Inner tie rod tool
   d. Cotter pin removal tool
9. *Technician A* says the electric power steering on most hybrid vehicles is part of the high-voltage system. *Technician B* says the power steering hoses on a hybrid are bright orange so that they are easily identified on a hybrid vehicle. Who is correct?
   a. Technician A
   b. Technician B
   c. Both A and B
   d. Neither A nor B
10. An older model vehicle has a rusted-through power steering pressure line: *Technician A* says both high- and low-pressure power steering lines should be replaced. *Technician B* says only the leaking line needs to be replaced. Who is correct?
    a. Technician A
    b. Technician B
    c. Both A and B
    d. Neither A nor B

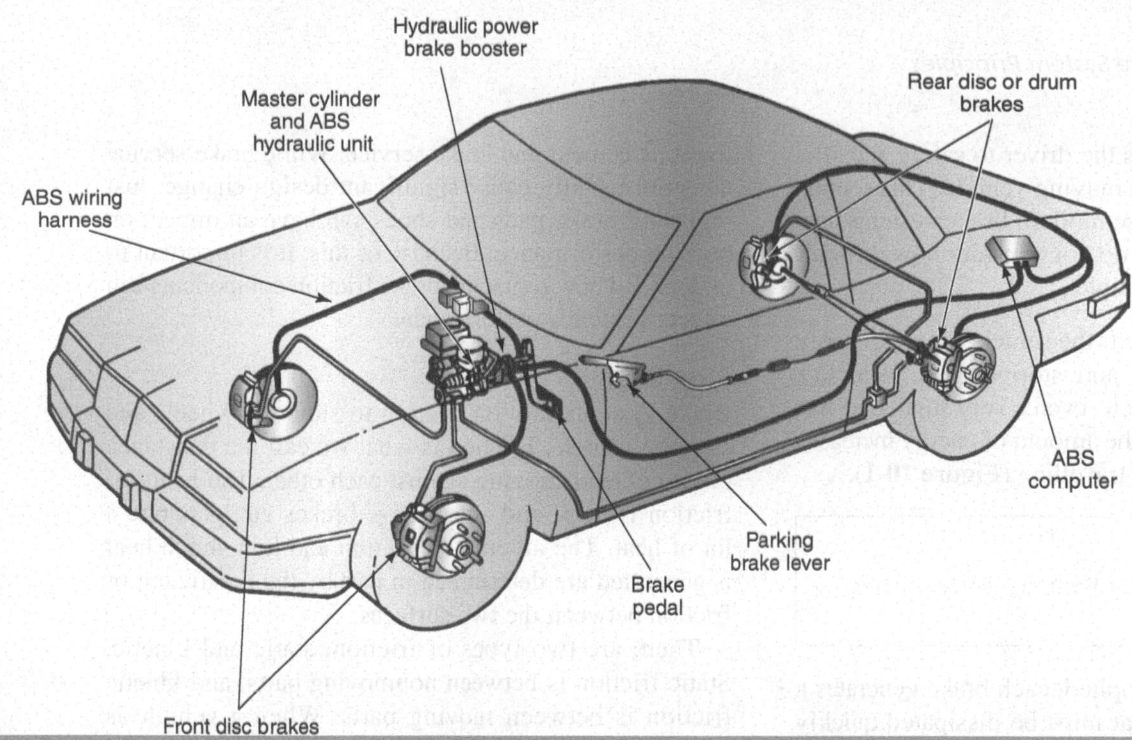

CHAPTER 10

Brake System Principles

Chapter Objectives

At the conclusion of this chapter, you should be able to:

- Describe how leverage and hydraulic principles are used in brake system operation.
- Explain how the master cylinder operates.
- Discuss the construction and purpose of brake lines and hoses.
- Describe how brake calipers and wheel cylinders function.
- Describe the operation of the regenerative braking system.

KEY TERMS

brake fade	**coefficient of friction (CoF)**	***P = F/A***
brake fluid	**hydraulics**	**wheel cylinder**
brake light switches	**hygroscopic**	
calipers	**master cylinder**	

The brake system allows the driver to safely and efficiently slow and stop a moving vehicle. This sounds simple but it is only because modern brake systems perform so well that we often do not consider how difficult the job actually is. For example:

- The brake system converts the vehicle's kinetic energy into heat by slowing and stopping the vehicle's wheels. A moving vehicle, even a very small one, has enormous energy, and the amount of energy increases the faster the vehicle is travelling (**Figure 10-1**).

> **Word Wall**
>
> *Kinetic energy*— Energy that something has by being in motion.

- When the brakes are applied, each brake generates a lot of heat, and the heat must be dissipated quickly to prevent damage and failure of brake parts. **Figure 10-2** shows in infrared image of a disc brake in operation.
- We expect the brakes to work every time, without noise or vibration. The friction must be consistent when the weather is cold or hot, when the brakes are wet, cold, or have been used for hours, or whatever the driving conditions are.

Modern brakes use mechanical, hydraulic, and electronic components to provide power-assist, and stability-control features such as collision avoidance. As vehicles become more complex, so do the demands placed on the brake system and the technicians who service them.

Factors Involved in Brake System Design and Operation

For the service technician, very little work performed on the car makes any change to the vehicle's design or operation, with a few notable exceptions; these include tire replacement and brake service. While brake service does not usually cause significant design change, just replacing brake pads and shoes can have an impact on braking performance. Because of this, it is important to understand how replacing brake friction components can affecta vehicle's performance.

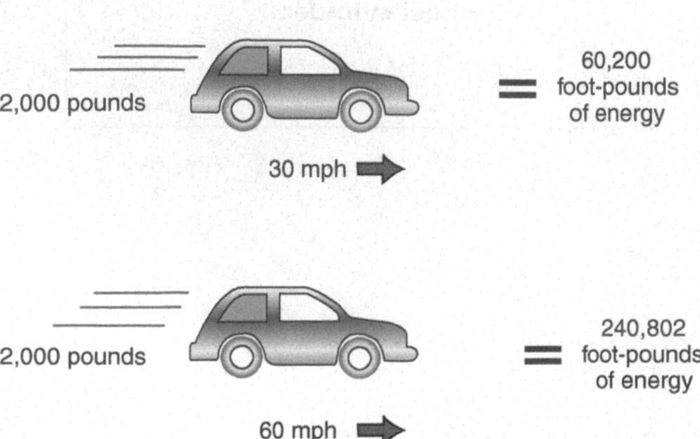

FIGURE 10-1 A vehicle in motion has a lot of energy. The larger the vehicle is or the faster it moves, the more kinetic energy is created. The brake system must be able to absorb and dissipate that energy.

FRICTION

Brake systems rely on friction to slow the wheels and stop the vehicle. Friction is what we call the resistance of two objects moving against each other. The result of friction is heat, and automotive brakes can generate a lot of heat. The amount of friction and how much heat is generated are determined in part by the coefficient of friction between the two surfaces.

There are two types of friction, static and kinetic. Static friction is between nonmoving parts, and kinetic friction is between moving parts. When a vehicle is parked, static friction between the tires and the ground keeps the car or truck from moving. While driving, kinetic friction between the brake components slows the vehicle (**Figure 10-3**).

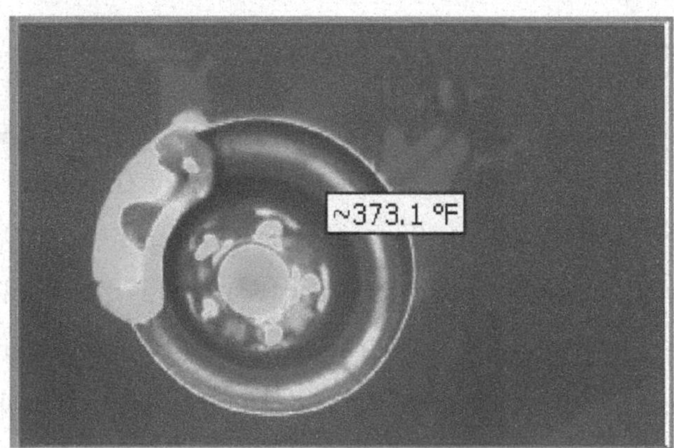

FIGURE 10-2 An infrared image of brake heat generation. In operation, the brakes may reach 500°F or more under normal driving conditions.

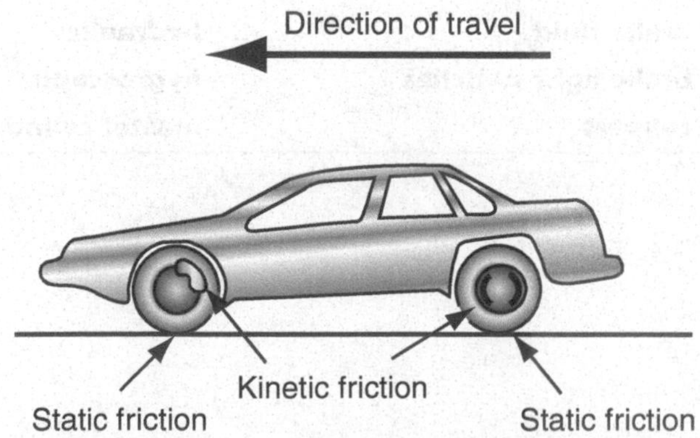

FIGURE 10-3 Kinetic friction is between the brake parts when the vehicle is in motion. Static friction between the tires and the ground prevents the vehicle from sliding down the street when parked.

■ ***Coefficient of Friction.*** The **coefficient of friction (CoF)** is a number that expresses the ratio of force required to move an object divided by the mass of the object. **Figure 10-4** shows an example of how the CoF is different depending on the type of surface the movement is across. As you can see, the CoF of the block of ice is significantly less than that of rubber.

For the brake system to operate safely and effectively, the CoF cannot be too high or too low, as in the previous example. If the CoF is too high, the brakes will be very touchy and may grab with the slightest application. If the CoF is too low, the friction between the brake pads will be insufficient to quickly slow the wheel, causing extended stopping distances and increased heat generation due to longer brake application times. In addition, car makers must also balance brake feel, noise, dust generation, federal guidelines for stopping distances, customer expectations for braking performance, and service life. These factors lead to some compromising in brake friction materials.

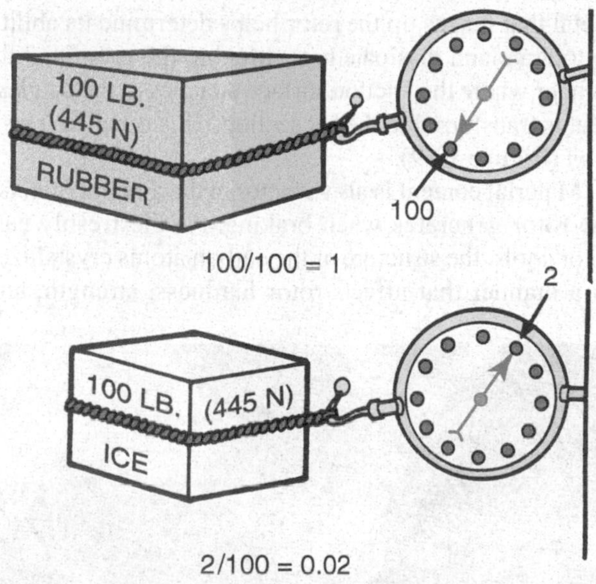

FIGURE 10-4 An illustration of the coefficient of friction. The greater the coefficient, the harder it is to move one object across another.

FIGURE 10-5 The coefficient of friction of the brake linings has a large impact on how a vehicle stops and how quickly the parts wear. New brake pads and shoes are stamped with friction codes for both cold and hot operations. The code on this pad is FF, meaning its friction falls into category F for both cold and hot operations.

New brake pads and shoes may have the CoF codes stamped onto them (**Figure 10-5**). The chart in **Figure 10-6** shows the relationship between the CoF and the codes. The pad and shoe shown in Figure 10-5 have CoF ratings of FF. This means that for both cold and hot operations, both have a CoF between 0.45 and 0.55. These codes are standardized based on testing performed to SAE test standards. Cold application is tested with the brakes between 200°F and 400°F. Hot application is with the brakes between 400°F and 600°F.

The coefficient increases with temperature until a certain point, at which point it drops significantly. This is due to brake fade. **Brake fade** occurs when the friction surfaces become hot enough that the brakes lose effectiveness or even fail. **Figure 10-7** shows the general relationship

Contacting surfaces	Static CoF
Teflon on steel	0.04
C	Not over 0.15
D	Over 0.15 not over 0.25
E	Over 0.25 not over 0.35
F	Over .35 not over 0.45
G	Over 0.45 not over 0.55
H	Over 0.55
Steel on steel	0.8
Rubber tires on dry asphalt	1
Aluminum on aluminum	Over 1.0

FIGURE 10-6 Examples of lining ratings and their coefficient of friction.

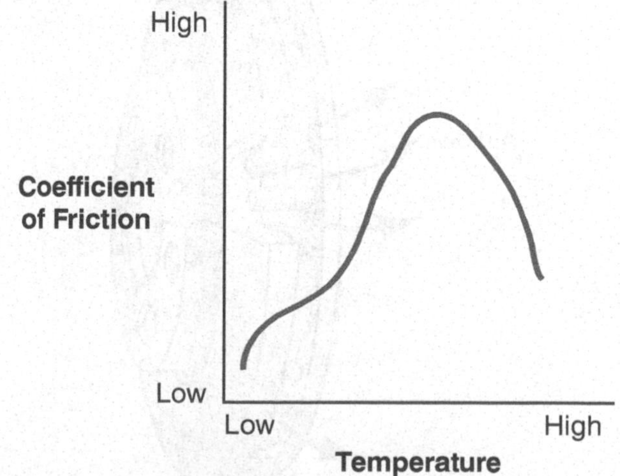

FIGURE 10-7 As lining temperature increases, the CoF increases until the linings overheat and the CoF drops. Overheating the brakes can cause poor stopping as the CoF decreases from high temperatures.

between the CoF and temperature. There are three types of brake fade, mechanical, lining, and gas fade. Mechanical fade occurs when the brake drum overheats and expands away from the brake shoes, causing increased pedal travel. Lining fade is when the pad or shoe lining material overheats and the CoF decreases. Gas fade is rare but occurs when a thin, hot layer of gas forms between the pads and rotor, acting as a lubricant and decreasing friction.

This is important because as a technician, you have the responsibility to install the correct brake parts to maintain safe vehicle operation. Just because a set of replacement pads or shoes fit the vehicle does not mean that they necessarily are the best choice. Vehicles are designed and sold with specific brake qualities that can, if parts with different CoF are used, change significantly.

■ ***Heat Dissipation and Noise.*** Because heat is the byproduct of friction, it is critical for the brake system to dissipate heat quickly and efficiently. This is accomplished by how and where the brake parts are located, how much air flows across the parts, and the design and construction of the rotors and fins on the drums.

Heat dissipation is critical for the continued safe operation of the brakes. Excessive heat buildup can cause the brake fluid to boil, resulting in brake fade and loss of braking. In addition, excessive heat can damage the pad and shoe friction material, and distort or even crack the rotors and drums. Excessive heat from the brakes can also have negative secondary effects on wheel bearings, wheels, and even hubcaps.

All modern cars and light trucks use vented front brake rotors, which pull air from the center of the rotor called the hat, through the vents, and out of the rotor (**Figure 10-8**). This helps to remove heat from the rotor and pads, prolonging their service life. To improve airflow over the brakes, some vehicles are designed with air ducts or passages that route air from the front of the vehicle to the brakes.

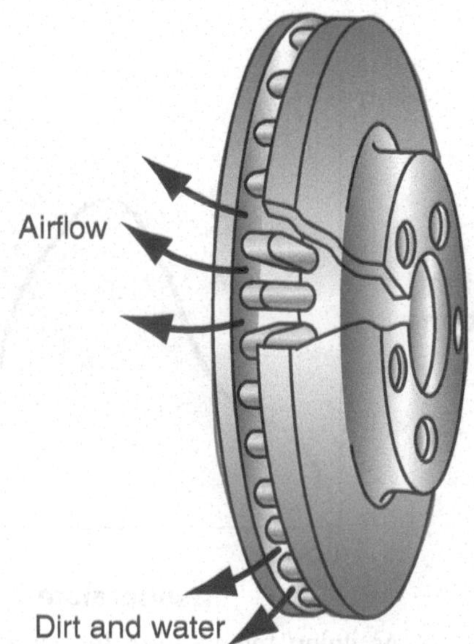

FIGURE 10-8 Airflow, heat dissipation, and cooling capacity of the brakes are important factors in brake design.

Some cars use directional rotors that have curved vents to further increase airflow. In addition to curved vents, some rotors use segmented vents. Many of these types of rotors are different for the left and right sides due to the shape of the cooling vents in the rotors.

Other factors in heat dissipation include how the rotor is cast and whether there is a heat dam between the friction surface and the hat. Many automotive parts, including brake rotors, are cast parts. Casting means molten metal is poured into a form to create the basic shape of the rotor. The part is then machined into its final shape and size. The amount of iron and other materials in the metal that makes up the rotor helps determine its ability to tolerate and dissipate heat. Making the rotor slightly thinner where the friction surface attaches to the hat, less heat is transferred to the hat section. This is called *a heat dam* (**Figure 10-9**).

Material content is also a factor in the amount of noise the rotor generates when braking. As the freshly cast rotor cools, the structure of the carbon atoms crystallizes in a manner that affects rotor hardness, strength, and

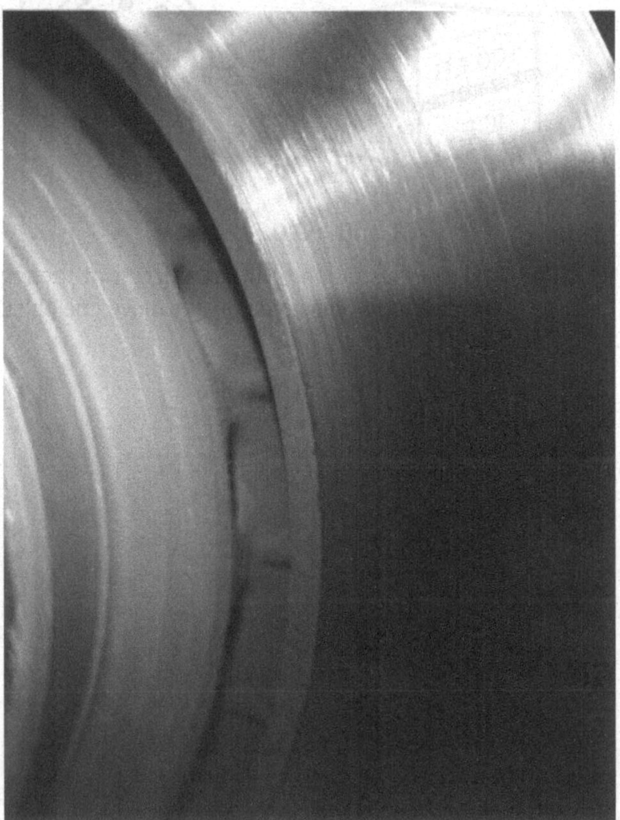

FIGURE 10-9 An example of a heat dam built into the brake rotor. This gap helps reduce heat transfer from the friction surfaces to the hub.

how the rotor vibrates during braking. Vibrations during braking occur because of the rotational movement of the rotor and the force of the pads against the rotor. These vibrations, even if they are not felt by the driver, are still present and can cause brake noise. To reduce both vibration and the noise, manufacturers use different grades of iron in the rotors and different compounds in the pad friction material. Adding shims and other vibration dampening hardware to the brake assemblies help to further reduce noise.

■ ***Vehicle Type and Use.*** Even though all vehicles must meet the federal requirements for brake system performance, it does not mean that all brake systems perform the same. Brake system performance is often based on the type of vehicle and its general use (**Figures 10-10** and **10-11**). The brakes in Figure 10-10 are carbon ceramic brakes on an Audi R8, a car capable of reaching 200 mph. The brakes shown in Figure 10-11 are from a mid-sized SUV. The difference in brake size is obvious.

In addition, less expensive cars often have combination front disc/rear drum brake systems. This is because the overall performance of the vehicle does not require the use of four-wheel disc brakes. The brakes on these smaller vehicles tend to be small and inexpensive to replace. In contrast, the brakes on sports cars, such as a Corvette, are much larger and can generate significantly greater stopping power and allow greater heat dissipation due to the driving conditions that can be achieved by this type of car.

Something that is sometimes overlooked by those who modify their cars for increased performance is that the standard brake equipment may not capable of meeting the needs of high-performance operation. This oversight can have disastrous results for people who try to operate their modified vehicles on the track as the brakes are incapable of repeated high-speed, high-pressure stops. This type of driving can easily overheat the brakes and boil the brake fluid, resulting in brake fade. The increased heat can cause the brake pads and rotors to also overheat, crack, and possibly even fall apart.

FIGURE 10-10 An example of high-performance brakes on a car that is built to go fast.

FIGURE 10-11 Disc brakes from a mid-sized SUV. The brakes on this vehicle are much smaller compared to those in the previous image based to the capabilities of the vehicles.

BRAKE PEDALS AND LEVERAGE

Modern brake systems combine the principles of leverage, hydraulics, and electronics. Leverage, also called *mechanical advantage*, is used by the brake pedal assembly and by drum brake systems. Fluid, when under pressure, transmits both force and motion. Electronics provide antilock braking and traction control functions. The combination of these means that today's vehicles stop better than ever before.

Leverage, in the mechanical sense, is defined as using a lever to gain mechanical advantage. The amount of advantage gained depends on what kind of lever is used and how it is used. If necessary, review the section in Chapter 4 about levers and how they are used.

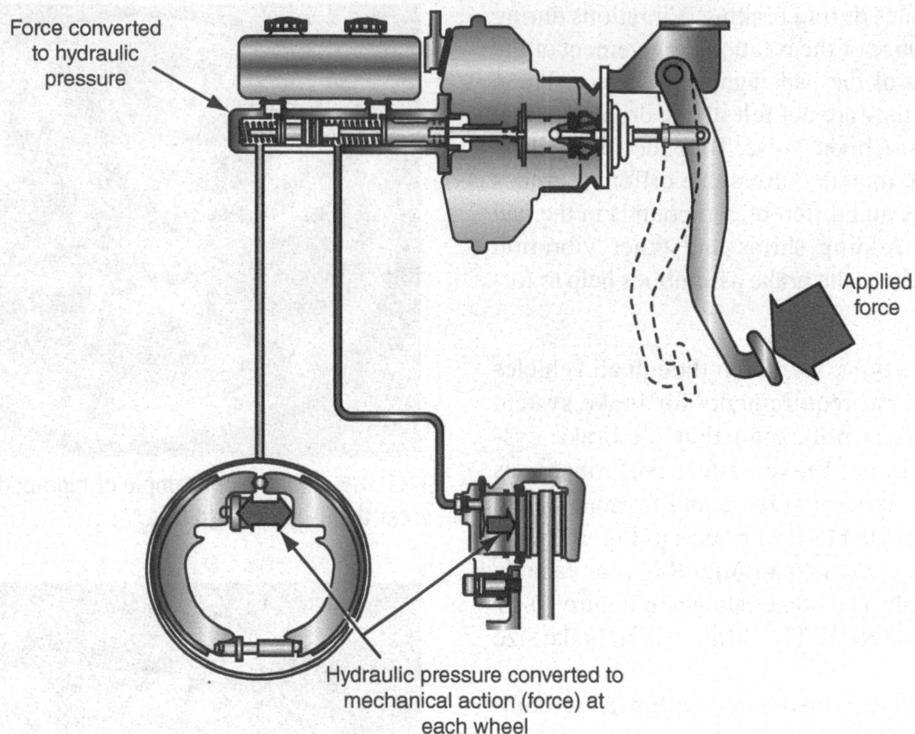

FIGURE 10-12 An illustration of the hydraulic brake system. The input force from the driver and brake pedal is applied to the master cylinder.

■ ***Brake Pedal Design and Operation.*** The brake pedal is a lever used to decrease the amount of driver effort needed to apply the brakes. The size of the pedal and the amount of leverage obtained are based on the overall design of the brake system. The brake pedal is also used to operate electrical switches for the stop lights, cruise control, transmission torque converter, antilock brakes, and traction control systems.

Most people do not think too much about the brake pedal in their car or truck, only that when it is pressed, the vehicle should slow and come to a stop. Brake pedals, by way of how they are mounted, allow for an in-crease in brake force in addition to the force provided by the driver pushing on the pedal. This is to reduce driver effort, and ultimately, driver fatigue. **Figure 10-12** shows how a typical brake pedal is mounted. The pivot is located at the top of the pedal, and the pushrod connects just below the pivot. When the driver pushes on the pedal, several things take place:

1. The force applied to the pushrod is increased. How much the force increases depends on the ratio of the distance between the pivot and the pushrod and the distance between the pushrod and the pad where the driver's foot pushes the pedal. This ratio is shown in **Figure 10-13**. In this example, the distance at the top is 2 inches, and the distance below is 10 inches. This gives a ratio of 10:2 or 5:1. This means that the force applied by the driver will be increased five times.
2. Because leverage involves a trade-off, increasing force requires moving a greater distance; the distance the brake pedal moves is reduced at the pushrod by the same ratio in which the force is increased. In this case, because the force is multiplied by 5, the pushrod's movement decreases by a factor of 5. Therefore, when the driver presses the pedal and it moves 2½ inches at the lower end, the pushrod only moves ½ inch. This is important because the actual movement of the pistons in the master cylinder is quite small compared to the movement of the brake pedal.

> **Word Wall**
> *Piston*—A moving part that fits closely in a cylinder to either pressurize a liquid or compress a gas.

The pedal is mounted to a bracket assembly that often has the accelerator pedal and clutch pedal (for a manual transmission) attached as well. The brake pedal pivot rides on a bushing to reduce friction, wear, and noise. The brake pedal position sensor or brake light switch is also mounted on this bracket.

■ ***Pedal Travel.*** Brake pedal travel, how far the pedal moves when the brakes are applied under normal pressure, depends on the condition of the hydraulic system and the condition of the brake friction components. Manufacturers often specify that the brake pedal should travel from

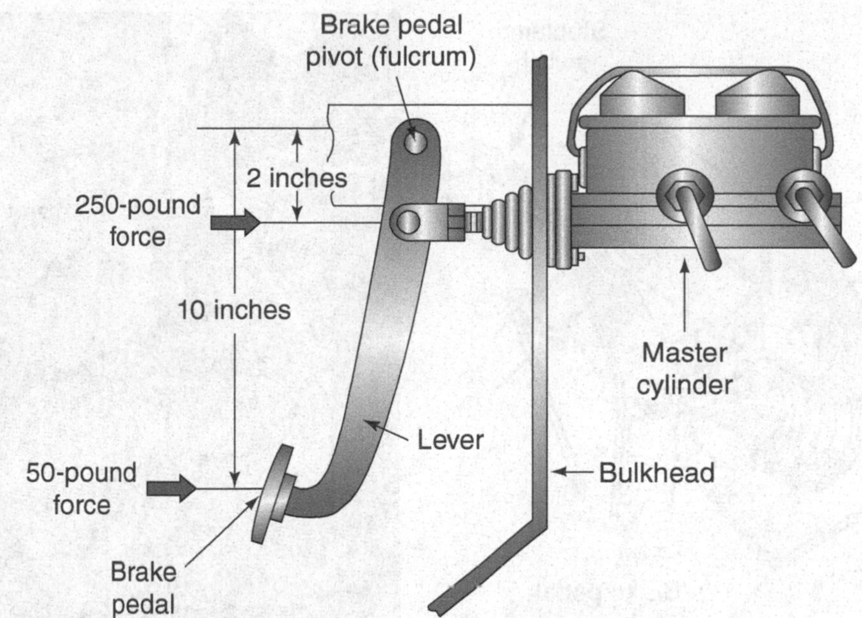

FIGURE 10-13 This shows how the size of the brake pedal and how it is mounted affect the ratio of leverage provided by the pedal. The greater the ratio, the greater the leverage generated.

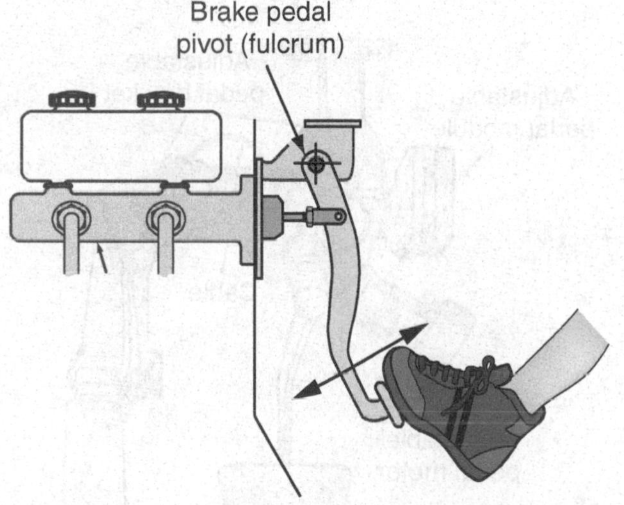

FIGURE 10-14 An illustration of normal brake pedal travel. In general, pedal travel is between 2 and 3 inches (51–76 mm).

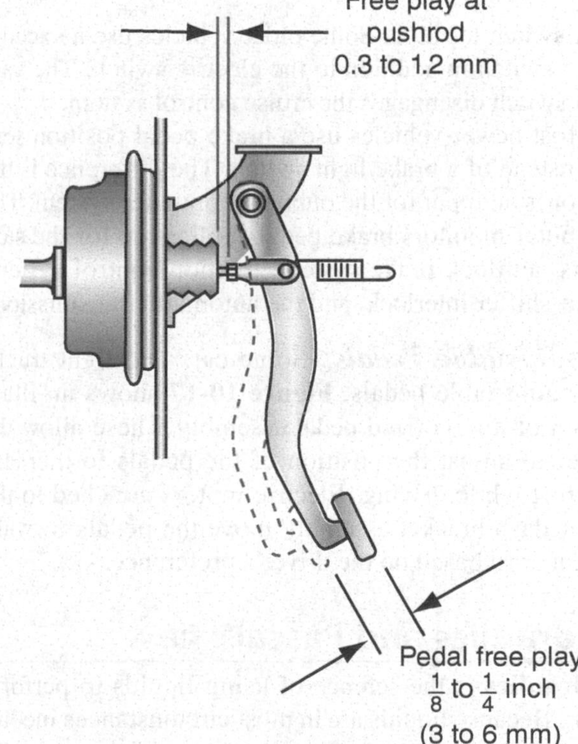

FIGURE 10-15 Free play is the slight amount of movement of the pedal before it pushes on the pushrod.

2 to 3 inches. **Figure 10-14** illustrates typical pedal travel. To accurately measure pedal travel, a specified amount of force is applied and the distance the pedal moves is measured. This is discussed in Chapter 11. If the pedal moves more than the specified distance, the hydraulic system and brake components should be inspected.

■ ***Brake Pedal Free Play.*** **Free play** is the slight amount of pedal movement at the released position before the pushrod begins to move into the booster and master cylinder (**Figure 10-15**). A slight amount of free play is usually built into the brake pedal and pushrod.

■ ***Brake Light Switches.*** **Brake light switches** control the operation of the brake lights at the rear of the vehicle and may be an input for the computer system. They are mounted to the pedal bracket and are activated when the pedal is pressed. **Figure 10-16** illustrates common brake

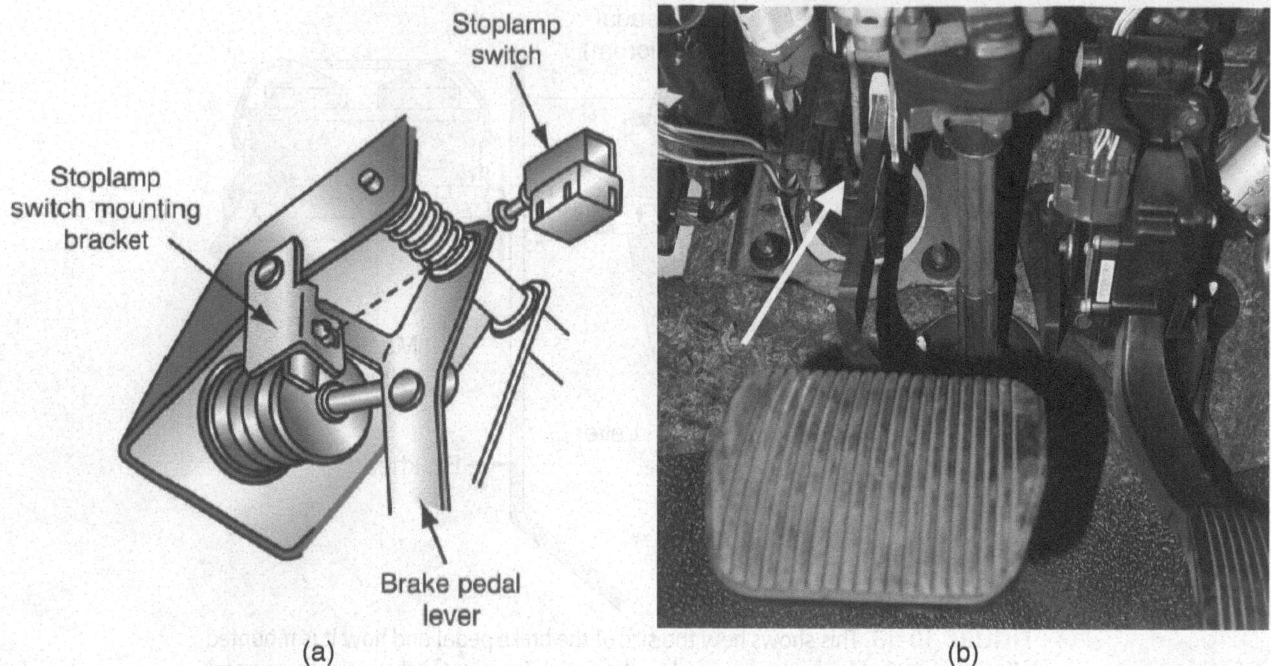

FIGURE 10-16 (a) An illustration of common brake light switch/pedal position sensor location. (b) A brake pedal position sensor from a late-model vehicle.

light switch location. Some older vehicles use a vacuum brake switch in addition to the electric switch. The vacuum switch disengages the cruise control system.

Most newer vehicles use a brake pedal position sensor instead of a brake light switch. The difference is the sensor is an input for the onboard computer system. The computer monitors brake pedal application for the stop lights, antilock brake system, traction control system, brake-shifter interlock, and the automatic transmission.

■ ***Adjustable Pedals.*** Some cars and light trucks offer adjustable pedals. **Figure 10-17** shows an illustration of a motorized pedal assembly. These allow the driver to adjust the position of the pedals to increase comfort while driving. Electric motors attached to the under dash bracket assembly move the pedals forward or rearward based on the driver's preference.

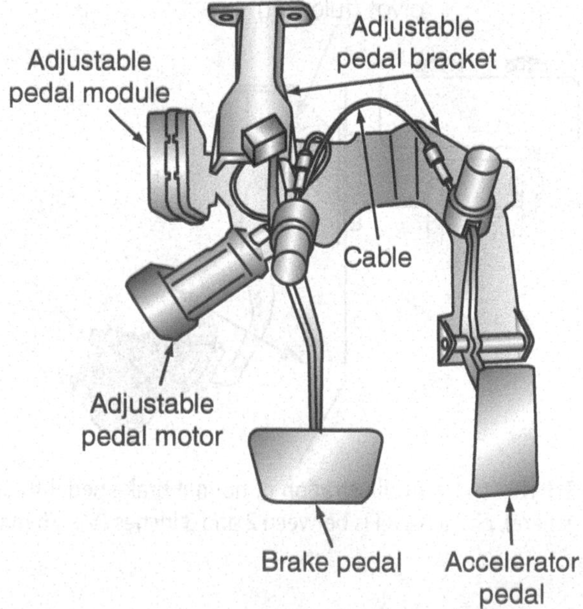

FIGURE 10-17 An illustration of adjustable pedals, used to allow the driver to better customize his or her driving position.

Hydraulics and Pascal's Law

Hydraulics is the science of using liquids to perform work. Because liquids are in most circumstances incompressible, we can use them to perform work by transmitting force and pressure.

BRAKE SYSTEM HYDRAULICS

In the 1600s, French scientist Blaise Pascal determined that pressure exerted on a confined liquid caused an increase in pressure at all points. This meant that in a closed container or system, any pressure against the liquid would be transmitted without any pressure loss through the system. This led to what is called *Pascal's Law*. Pascal's Law states that pressure = force/area or $P = F/A$. This is the basis for hydraulics.

The reason why pressure is not lost is because liquids, for practical purposes, can be put under pressure but not compressed. Compared to air, the water molecules are packed together as nearly as tightly as possible, leaving no room to squeeze them more closely together (**Figure 10-18**). When a fluid is pressurized in a closed system, meaning there both the fluid and the pressure remain trapped or contained within the system, the fluid can transmit both motion and force (**Figure 10-19**).

A simple hydraulic system has two cylinders of equal size (**Figure 10-20**). The two cylinders are connected with a tube. If one piston is pushed downward with 100 lbs. (45 kg) of force and moves down 5 inches (25.4 cm), the piston in the second cylinder will move upward 5 inches (25.4 cm) with the same 100 lbs. (45 kg) of force. Because the pistons are the same size, any force and movement applied on one piston will cause the same reaction to the second piston.

Where the use of hydraulics really provides an advantage is when the sizes of the pistons are different. This results in force and movement that can be increased or decreased as needed. The pressure generated by the piston is dependent on piston size. Input pressure, like what is created when the driver steps on the brake pedal, is found by using the formula $P = F/A$, or pressure is equal to force divided by piston area. The smaller the input piston surface area, the larger the force is from that piston. The larger the input piston surface area, the less the force is from that piston (**Figure 10-21**). This is because the force must act on the area of the piston. If the area is small, the force produces a lot of pressure over the small area. If the piston is large, the force is spread out over the larger surface, which reduces the overall amount of pressure produced.

In this way, a hydraulic system is like the lever. If you have used a hydraulic floor jack, you have applied

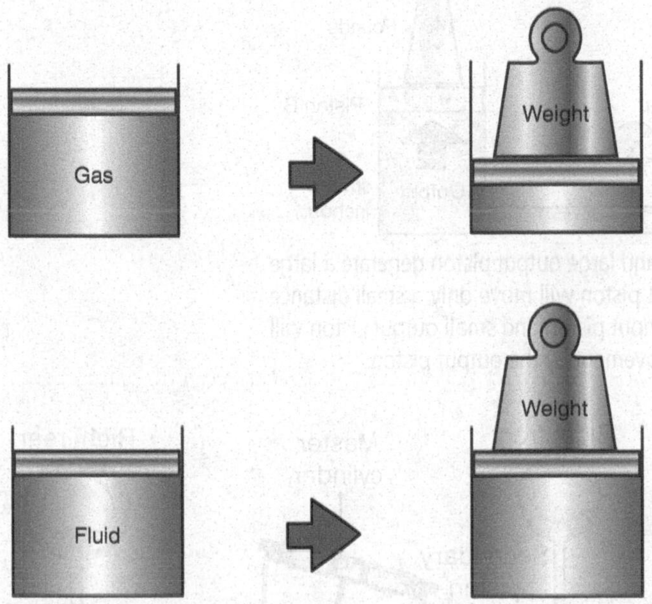

FIGURE 10-18 Gases, when placed under pressure, can compress or reduce their volume as shown in the upper image. Liquids, under most conditions, are not compressible but can be placed under pressure.

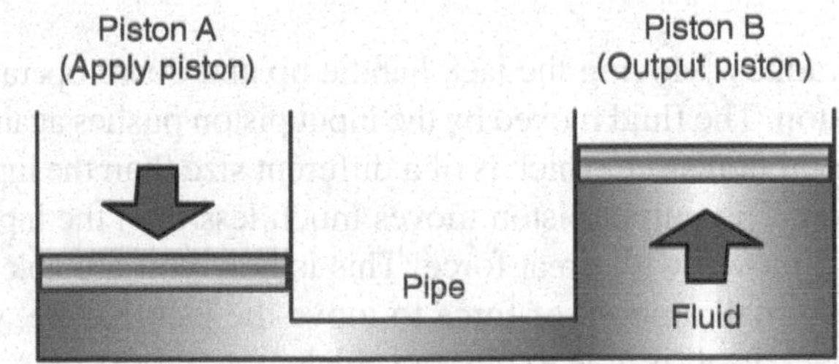

FIGURE 10-19 Using a liquid to transfer motion and force is the basis of hydraulic systems. In this example, the input and output are the same size, so movement and force will be equal for both.

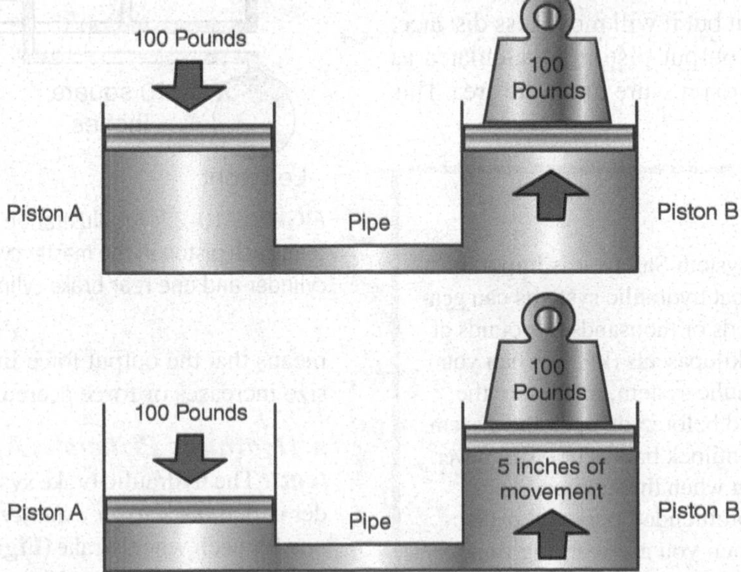

FIGURE 10-20 Pistons and cylinders of the same size will transmit force and motion without loss or gain.

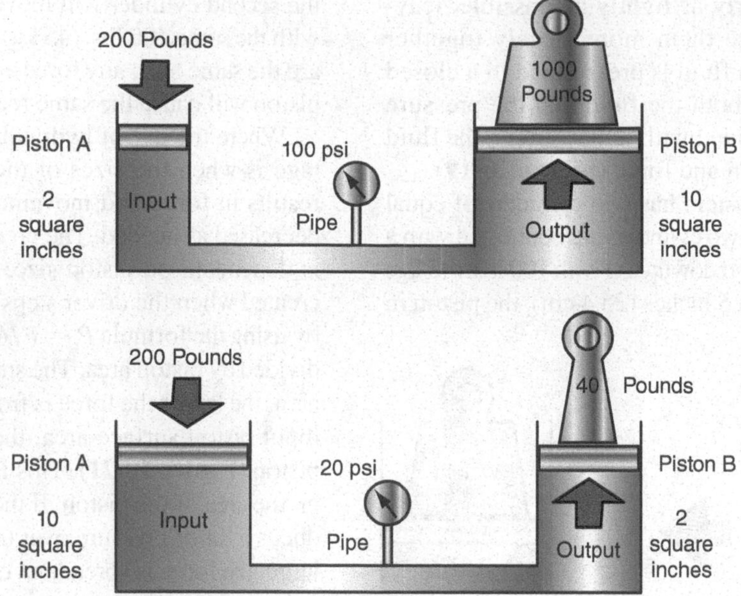

FIGURE 10-21 A small input piston and large output piston generate a large increase in output force, but the output piston will move only a small distance compared to the input piston. A large input piston and small output piston will generate less output force but more movement by the output piston.

hydraulics. Moving the jack handle up and down operates a piston. The fluid moved by the input piston pushes against the output piston, which is of a different size than the input piston. The output piston moves much less than the input, but it moves with great force. This is how you are able to use a small amount of force to move the jack handle, yet the jack itself is able to raise a vehicle.

Just as system pressure is based on piston size, so is output force. The output piston force is proportional to the pressure against the surface area of the piston. An output piston that is larger than the input piston will move with greater force than the input but it will move less distance. The force created by an output piston is calculated as $F = PA$, or force is equal to pressure times the area. This means that the output force increases if the output piston size increases or force decreases if piston size decreases.

Service Note

Hydraulic System Safety. It is important to remember that hydraulic systems can generate hundreds or thousands of pounds of pressure or kilopascals (kPa). When you are working on any hydraulic system, make sure the system pressure is relieved before you open the system. On cars and trucks with antilock brakes or ABS, never open the hydraulic system when the ignition is on.

Always wear the recommended personal protective equipment (PPE) when you are working on a vehicle. It is especially important to wear eye protection when you are working with liquids.

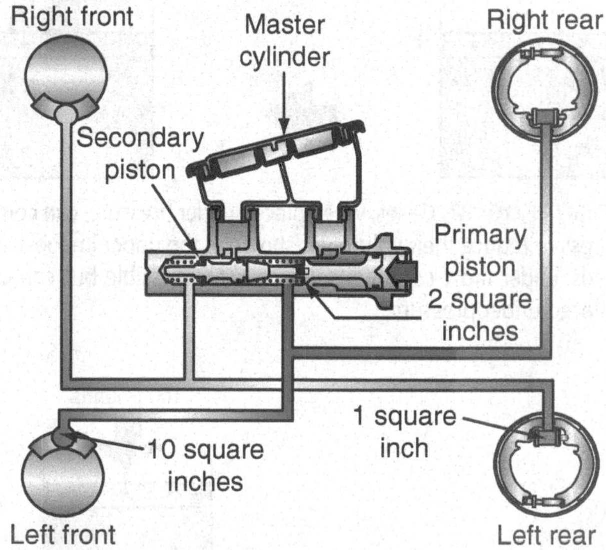

FIGURE 10-22 An illustration of a split-hydraulic brake system. Each piston in the master cylinder supplies one front brake cylinder and one rear brake cylinder.

■ ***Hydraulic Principles Applied to the Brake System.*** The hydraulic brake system contains an input cylinder called *the master cylinder*, and four output cylinders, one for each wheel brake (**Figure 10-22**). When the driver presses on the brake pedal, force is applied to the pushrod and to the rear piston in the master cylinder. The pistons inside the master cylinder move forward, pushing on the fluid. Because the fluid cannot be compressed, the pressure on the fluid increases. Threaded into the master cylinder are brake fluid lines. The fluid pressurized by the pistons transmits that pressure through these brake lines, which connect to the pistons at the wheel brakes.

The pressure on the brake fluid can be easily calculated. In Figure 10-22, the area of the master cylinder pistons is exactly 2 square inches. If the force applied to the piston by the pushrod is 100 lbs. (45 kg), then using $P = F/A$, we know that the fluid is at 50 lbs. of pressure per square inch, or 50 psi (345 kPa). Because the fluid is not compressible and the pistons move in the master cylinder bore, the fluid also transmits this motion along with the pressure. In this example, the piston moved forward ½ (12.7 mm) inch.

It is mostly just the pressure that is transmitted through the lines, though some of the fluid will leave the master cylinder through the brake lines. These lines are very small in diameter compared to the size of the master cylinder pistons. This is to maintain the pressure on the fluid.

At the front, the hydraulic output is a disc brake caliper (**Figure 10-23**). The caliper has one or more pistons that are much larger than the master cylinder pistons. This is to increase the force the output piston can apply. Because the output force is increased, the total movement of each piston is decreased. In our example, the caliper piston has a surface area of 10 square inches. The force of the output piston is $F = PA$, or 500 lbs. (227 kg) of force. Because the force generated by the piston is increased by a factor of 5, the distance the piston travels is decreased by 5 also. Because the movement of the master cylinder piston is ½ inch (12.7 mm), the movement of the caliper piston is one-fifth of that, or about 3/32 of an inch (2.5 mm).

Why do we use a larger piston for the front brakes? Disc brake calipers squeeze the brake pads against the spinning brake disc, and the contact between the brake pads and the brake rotor is small (**Figure 10-24**). This requires a large amount of clamping force, and consequently, a large piston to apply that force. Because the brake pads are positioned very close to or just against the brake disc, very little movement is required. By using a larger output piston, brake force is increased and movement is decreased.

Some vehicles use drum brakes on the rear of the vehicle. The output of the hydraulic system in the drum brake is the wheel cylinder (**Figure 10-25**). The pistons in the wheel cylinder are typically small, smaller than the pistons in the master cylinder. In this example, the wheel cylinder pistons have a surface area of 1 square inch (25.4 mm). The brake fluid pressure of 50 psi (345 kPa),

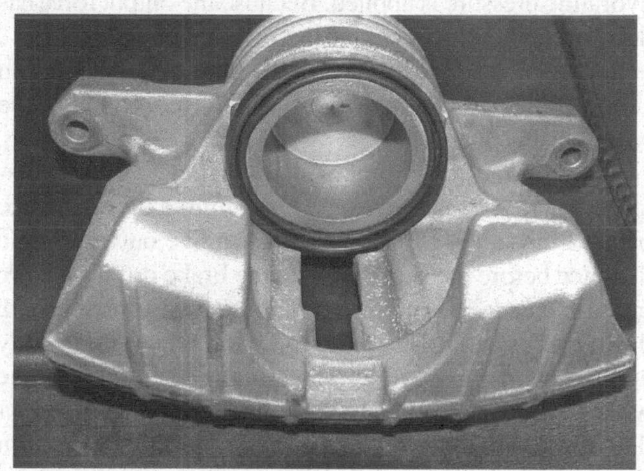

FIGURE 10-23 An example of a disc brake caliper. The caliper is the hydraulic output for the disc brake system.

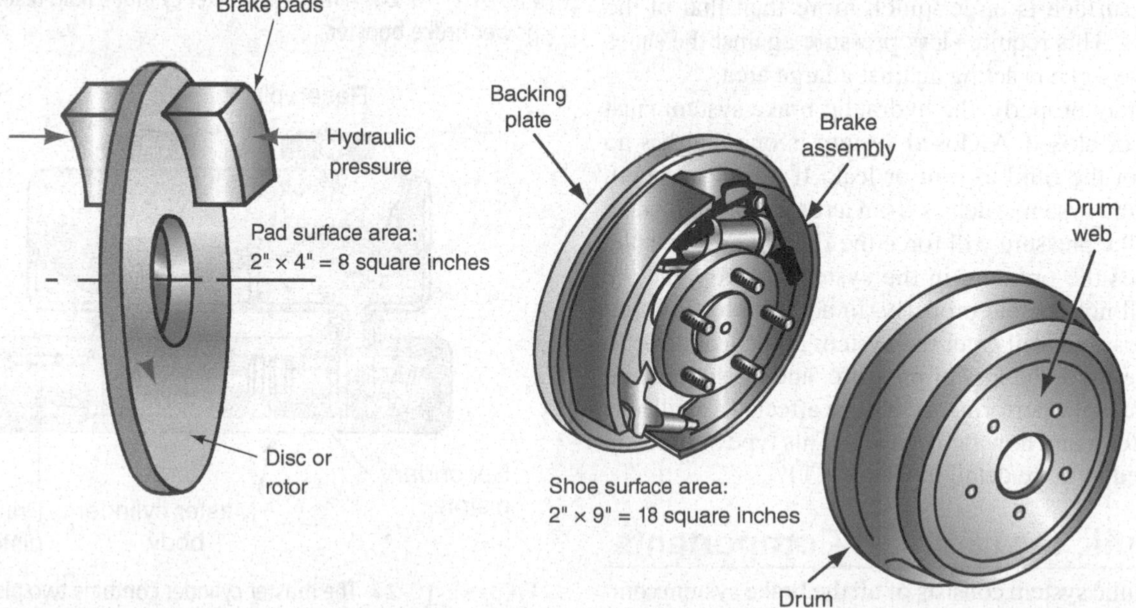

FIGURE 10-24 This illustrates the surface contact area differences between disc and drum brake designs and why disc brakes require more output force.

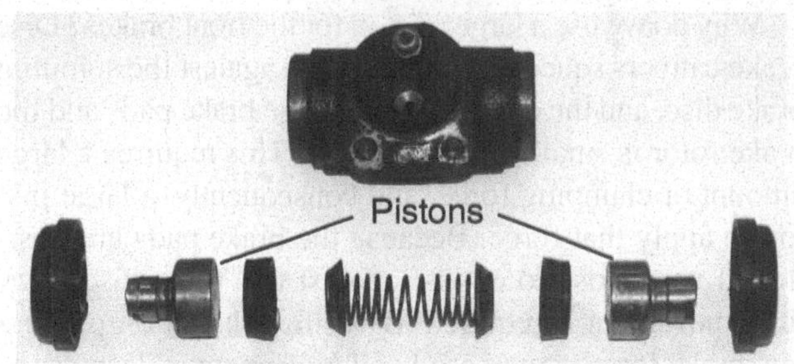

FIGURE 10-25 An example of a wheel cylinder. The wheel cylinder is the hydraulic output for the drum brake system. Most wheel cylinders have two pistons, which move outward to press the shoes against the inner drum surface.

when it is acting on the wheel cylinder pistons, it produces an output force of 50 lbs. (23 kg). The decrease in output force is due to the smaller size of the piston on which the hydraulic pressure is applied. Because the output force has decreased, the movement of the pistons increases. The ½ inch (12.7 mm) of movement of the master cylinder piston is now increased to 1 inch (25.4 mm) of movement by the wheel cylinder piston.

Why do the drum brakes use smaller hydraulic pistons? There are three reasons. First, unlike the disc brake pads, drum brake shoes often must move outward some distance before the shoes contact the brake drum, so more travel is needed for the wheel cylinder pistons. Second, drum brakes can increase the force the brake shoes apply against the drum beyond what force is supplied by the wheel cylinder piston. This is called a self-energizing or duo-servo brake design and is covered in more detail in Chapter 12. Because of the duo-servo action, less force is required by the wheel cylinder, so smaller pistons can be used. Finally, the contact area between the shoes and the drum surface is large, much more than that of the disc brakes. This requires less pressure against the shoes because the force is acting against a large area.

To operate properly, the hydraulic brake system must be sealed or closed. A closed system is one that has no opening for the fluid to vent or leak. If a leak occurs in the hydraulic system, such as from a rusted through brake line, then the pressure will force the fluid out of the hole. This causes the pressure in the system to drop and the brakes will not operate properly. In addition to fluid and pressure loss, air will enter the system at the leak. Due to the leaking fluid, the loss of pressure, and air getting into the system, brake performance will be affected as will how the pedal feels and how far it travels. This type of problem is discussed in more detail in Chapter 11.

Hydraulic System and Components

The hydraulic system consists of all the brake system components that operate using brake fluid to transfer force. This generally means the master cylinder, hydraulic valves, lines and hoses, calipers, and wheel cylinders. Hydraulic antilock brake system components are discussed in Chapter 16.

MASTER CYLINDERS

The **master cylinder** is the hydraulic system input and is responsible for generating the pressure for the system (**Figure 10-26**). Usually the brake fluid reservoir is attached to the top of the master cylinder. In some vehicles, the reservoir is mounted remotely due to space limitations.

■ ***Types.*** Modern master cylinders have two pistons and two chambers (**Figure 10-27**). Until the 1960s, the master cylinder was a single-chamber design. In the late 1960s, the U.S. government mandated the use of dual-piston master cylinders. This is so that if a leak develops in one of the hydraulic circuits, the other circuit will still have pressure to slow and stop the vehicle.

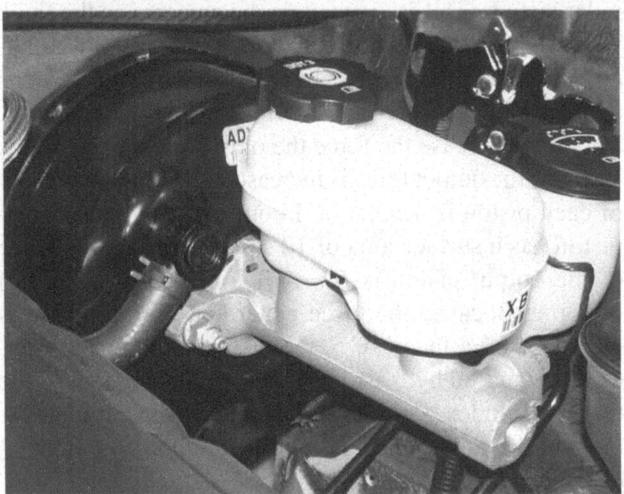

FIGURE 10-26 A modern master cylinder, fluid reservoir, and power brake booster.

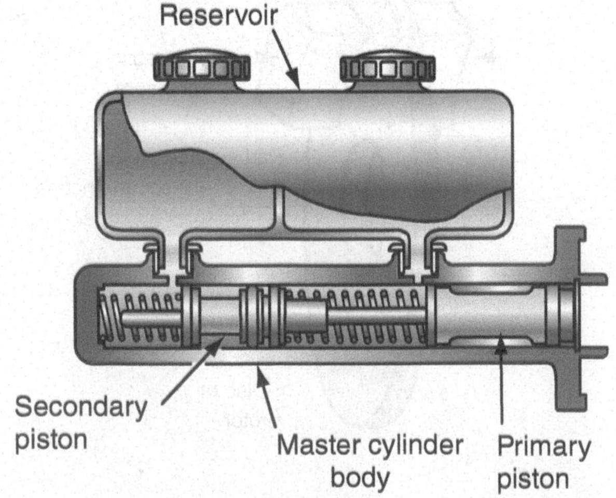

FIGURE 10-27 The master cylinder contains two pistons, the primary and the secondary. The pushrod actuates the primary piston at the rear of the cylinder.

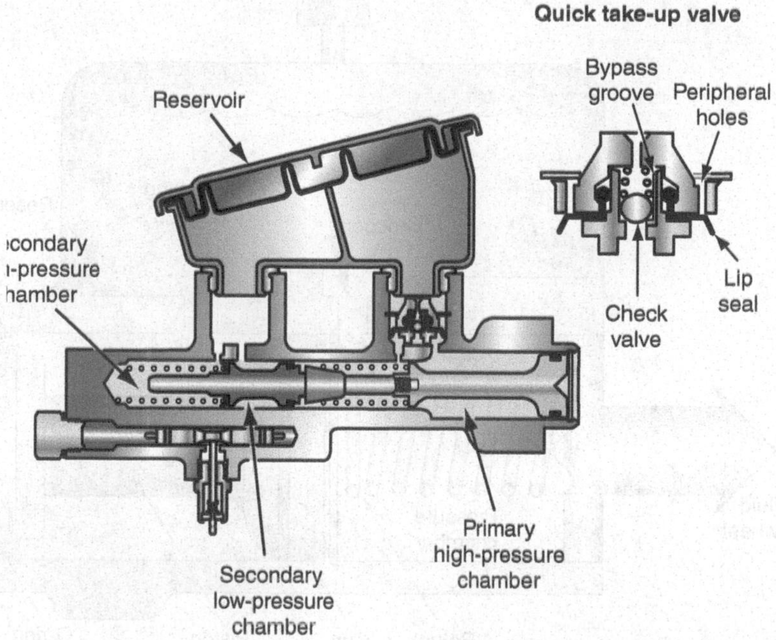

Quick take-up valve
Bypass groove
Peripheral holes
Lip seal
Check valve
Reservoir
›condary
ı-pressure
hamber
Primary high-pressure chamber
Secondary low-pressure chamber

FIGURE 10-28 A step-bore master cylinder has pistons of different size.

Standard master cylinders have two pistons of the same size, each producing the same amount of pressure when the brake pedal is pressed. Some master cylinders, called *step-bore cylinders*, have different-sized pistons and chambers. Step-bore master cylinders are also called *quick take-up master cylinders* and are used with low-drag calipers (**Figure 10-28**). In this design, the caliper piston is retracted back into the caliper bore slightly so that the pads do not stay in contact with the brake disc under normal driving conditions. The master cylinder design pushes more fluid to the caliper to take up the extra space between the pads and the rotor when the brakes are applied. This system reduces pad and disc wear and provides a slight fuel economy increase.

Some master cylinders are integral with the ABS, meaning that the master cylinder, ABS, and power assist are one unit (**Figure 10-29**). Special service procedures are required for integral systems, even for checking the brake fluid level. This is discussed in more detail in Chapter 16.

FIGURE 10-29 A modern integral master cylinder. This unit functions as the master cylinder, power assist, and antilock brake hydraulic unit.

■ ***Construction and Operation.*** Master cylinders are constructed of aluminum alloys, which are lightweight and resist corrosion. On older vehicles, those built in the 1980s and earlier, they are often made out of cast iron. The fluid reservoirs can be either mounted directly to the cylinder or mounted remotely, and are typically plastic so that the fluid level is easily checked without removing the cap.

The master cylinder reservoir cap or caps have seals to keep dirt and moisture out of the fluid. The master cylinder caps are vented to allow pressure to be released. Pressure can develop from the heat transfer at the wheel brakes to the brake fluid. The increase in temperature will cause an increase in pressure in a closed system. The reservoir cap seals are often accordion seals, meaning that they expand as the fluid level in the reservoir drops. The seals extend downward, taking up the space as fluid level drops. This reduces the volume of air that can be trapped between the seals and the fluid, which limits the amount of moisture the brake fluid absorbs from the air in the reservoir.

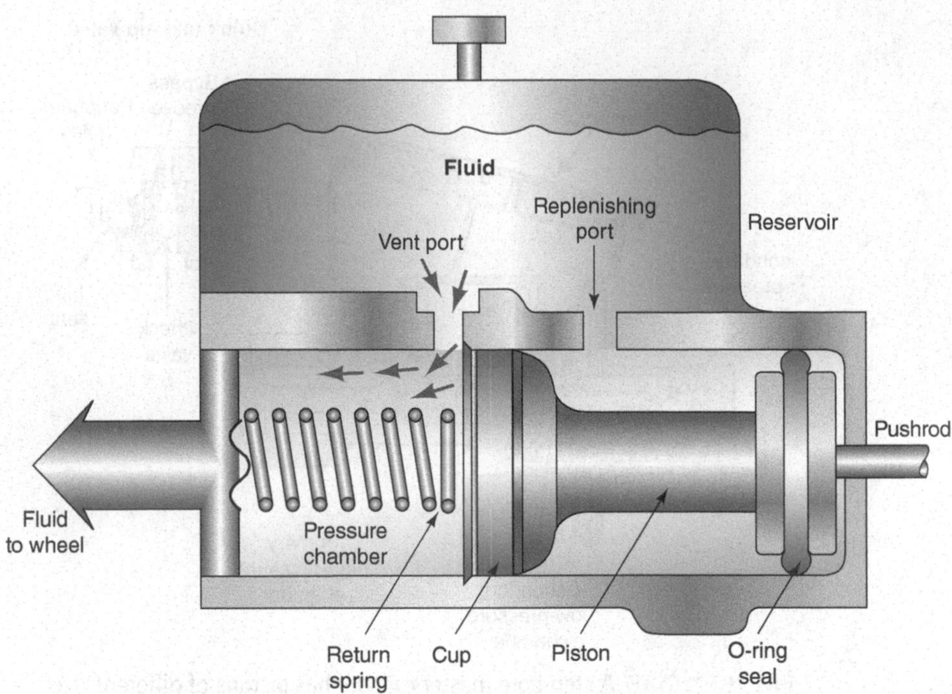

FIGURE 10-30 The ports inside the master cylinder allow fluid to pass between the cylinder and the reservoir during braking.

Inside the master cylinder are two chambers and two pistons, the primary and secondary. Rubber seals around the pistons keep the fluid trapped in the primary and secondary chambers during brake application. Two ports, the vent and replenishing port, connect the reservoir to each chamber (**Figure 10-30**). The vent port allows fluid to pass from the master cylinder into the reservoir when the brakes are released. This allows the fluid to expand at the wheel brakes from heat buildup. As the pistons move forward, the piston and seal move past the port openings. This traps the fluid in each chamber so that the pressure exerted on the fluid can only go out the ports for the brake lines, which go to the wheel brakes. When the brake pedal is released, the springs in front of each piston provide for piston return.

When the brakes are applied, the pistons move forward and block the vent port. Fluid from the reservoir then fills in behind the piston via the replenishing port, eliminating a low-pressure area from forming behind the piston. When the brakes are released, the pressure in the brake system and springs in the master cylinder bore push the pistons backward to their unapplied positions. Fluid behind the pistons either travels over the piston seals or back through ports into the reservoir.

HYDRAULIC LINES, HOSES, AND VALVES

To get the fluid from the master cylinder to the wheel brakes, a network of steel lines and reinforced rubber hoses is used. Older systems (non-ABS) used valves to control fluid pressure to the front and rear brakes. Modern vehicles use the antilock brake system to control pressure to the individual wheel brakes.

■ ***Brake Line.*** Brake line is made from steel—jacketed copper tubing or a copper–nickel alloy. This combination allows for flexibility, strength, and durability. Steel line is used to supply the brake fluid from the master cylinder or ABS hydraulic unit across the vehicle and to a connection to a brake hose. The ends of brake lines have specially formed ends, called *flares*, to provide a positive, leak-proof connection to other hydraulic components. Two types of flares are used in modern vehicles, the double or English flare and the ISO or metric flare (**Figure 10-31**). Fittings are specific for each type of flare and are not interchangeable.

■ ***Brake Hoses.*** Rubber brake hoses are used where there is movement, such as between the vehicle body and a suspension strut or axle. Brake hose is made of layers of reinforced rubber and is designed to handle the high pressure of the brake fluid, high temperature

Word Wall

Fitting—A fitting is a part used to connect different items together. In automotive brake applications, fittings are threaded parts used to join sections of brake line.

generated during braking, and the extreme conditions of being exposed to the weather and outside environment. At the ends of the hose are connections. The connections may be male or female threads or a banjo fitting. **Figure 10-32** shows examples of brake hoses.

The steel line threads into a connection on the hose (**Figure 10-33**). The hose may have a threaded fitting or an opening, called *a banjo fitting*. **Figure 10-34** shows a banjo hose connection. The lower end of the banjo hose has an opening for a hollow bolt to pass through. The bolt attaches the hose to the brake caliper.

■ ***Junctions.*** RWD vehicles may have a junction at the rear axle. One brake line runs from the master cylinder or ABS to the rear for both rear brakes. The junction splits the single line so that each rear wheel brake receives brake fluid. This is illustrated in **Figure 10-35**.

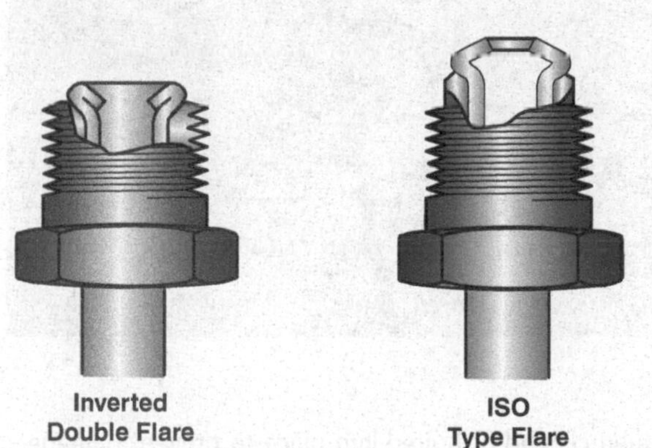

FIGURE 10-31 This illustrates the differences between SAE double flares and ISO bubble flares. In addition to the flares being different, the threads on the fittings are not interchangeable. SAE fittings use English thread pitch and ISO use metric thread pitch.

■ ***Brake System Valves.*** For many years, different types of brake system valves were used to control brake application, mostly to prevent excessive application pressure and wheel lockup. On most vehicles, these valves have been eliminated by the antilock system.

Older vehicles with combination brake systems, front disc brakes and rear drum brakes, two hydraulic valves are commonly used to help control brake application. The proportioning valve is used to limit hydraulic pressure to the rear drum brakes under heavy braking. Too much pressure on the rear brakes can cause the rear wheels to lockup, and the vehicle control is then compromised. To reduce the possibility of rear wheel lockup, the proportioning valve is used to limit pressure to the rear brakes once the pressure reaches a certain point (**Figure 10-36**). When hydraulic pressure reaches a specified amount, fluid flow is reduced, which limits the pressure to the rear wheel cylinders. Under light to medium braking, the fluid passes through the valve and the rear brakes operate normally.

Some vehicles use a height - or load-sensing proportioning valve. Used on minivans, larger passenger cars, and light trucks, the load-sensing proportioning valve can vary the amount of pressure that can pass to the rear brakes based on the position of the lever attached to the valve. As the increased weight pushes the rear of the vehicle down, the lever moves the valve. This allows more pressure to the rear brakes to offset the increased vehicle load, which increases the demands on the brakes and increases the stopping distance.

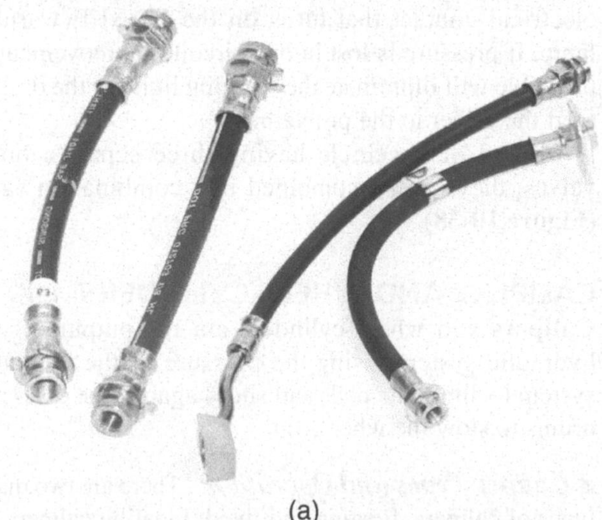

(a)

(b)

FIGURE 10-32 (a) Brake hoses are reinforced high-pressure rubber hoses used to connect the steel brake lines to the wheel brakes. The rubber hoses allow for up, down, and side-to-side movements. (b) A cutaway of a brake hose shows the layers of reinforcing materials.

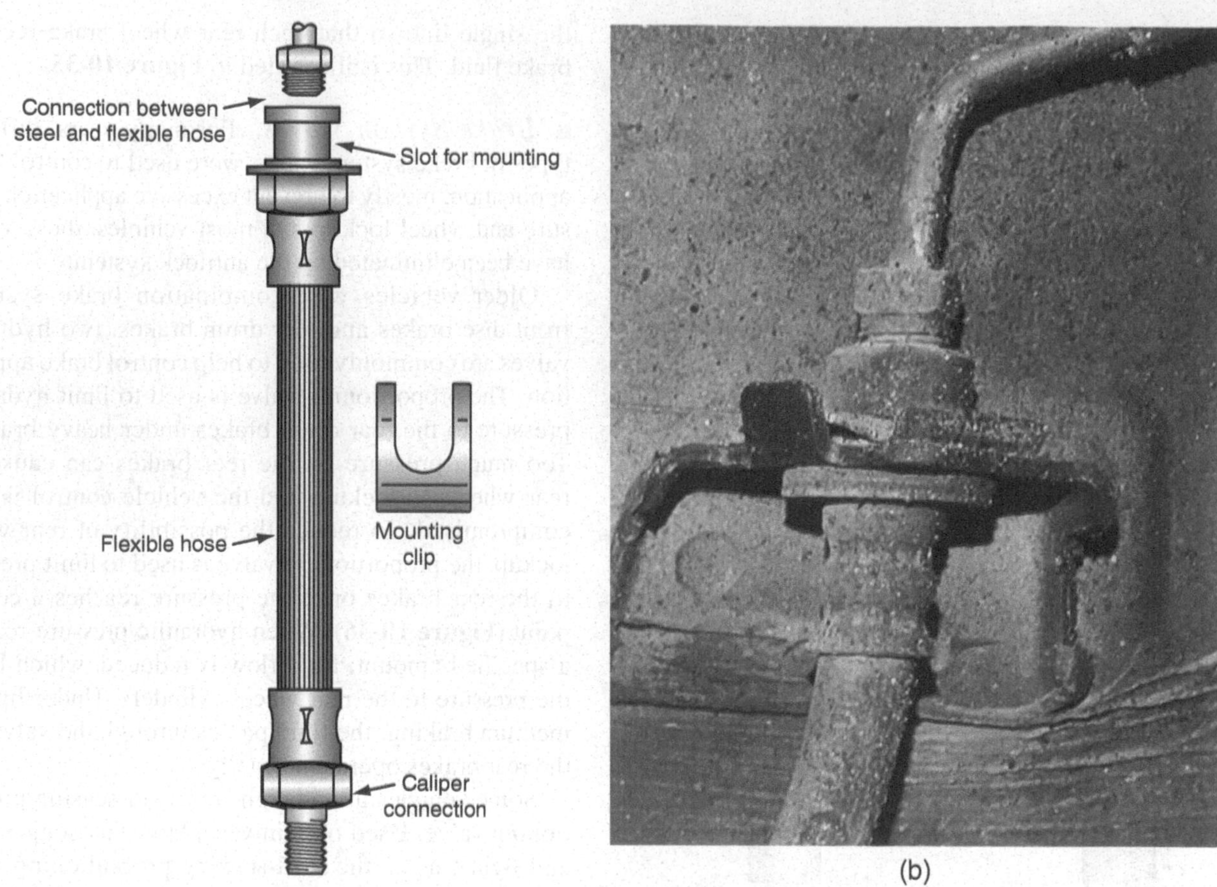

FIGURE 10-33 (a) Brake hoses are connected to lines and then clipped or bolted into place to prevent damage. (b) An example of a brake line to hose connection.

Another valve used in combination brake systems is the metering valve. The metering valve is used to delay slightly the application of the front disc brakes. Because the drum brake shoes must overcome the tension by the return springs and travel farther to contact the brake drum, there is a lag between when the front and rear brakes actually apply enough to slow the wheels. The metering valve holds pressure to the disc brakes so that the drum brakes can overcome return spring pressure, move out, and start to apply.

In the event of a leak and loss of hydraulic pressure, the pressure differential valve closes off circuit with low pressure. This is to allow the other circuit to provide enough pressure to slow and stop the vehicle. The valve operates by using the principles of hydraulic pressure (**Figure 10-37**). When the system is closed and operating normally, pressure is equal between the two circuits and the valve is centered. When a leak occurs, the circuit with the leak has essentially no pressure while the other circuit has normal pressure. This pressure difference forces the check valve to move from high pressure to low pressure, sealing off the ruptured circuit. This way one circuit is still available to provide brake pressure, although only for two of the four wheels, so the stopping distance increases greatly.

The pressure differential valve usually contains an electrical contact that turns on the BRAKE warning lamp. If pressure is lost in one circuit, the movement of the valve will illuminate the warning light on the dash to alert the driver to the problem.

Instead of a vehicle having three separate brake valves, they can be combined in a combination valve (**Figure 10-38**).

CALIPERS AND WHEEL CYLINDERS

Calipers and wheel cylinders are the outputs of the hydraulic system, using the pressure of the hydraulic system to apply the pads and shoes against the discs and drums to slow the wheel.

■ ***Caliper Types and Operation.*** There are two major types of calipers, floating and fixed. Floating calipers are the most common type of brake caliper used in modern cars and light trucks. This caliper gets its name because it floats on its mounting hardware and moves back and forth

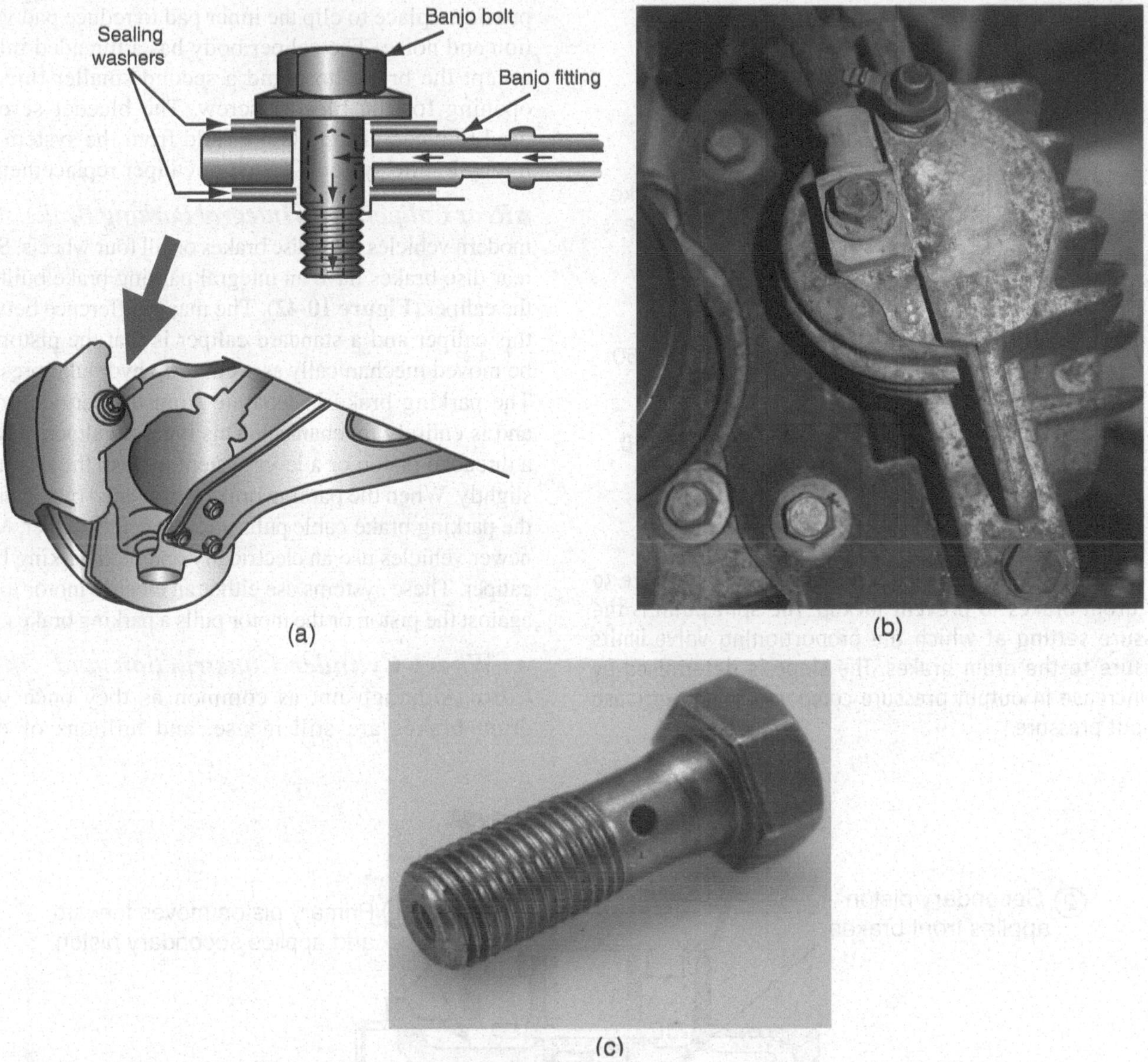

FIGURE 10-34 (a) An illustration of a bolt and banjo fitting. (b) These are commonly used to connect a brake hose to a caliper. (c) The hollow bolt allows fluid to flow from the hose to the caliper.

as the brakes are applied (**Figure 10-39**). A floating caliper has to be able to move so that both brake pads are applied.

Fixed calipers have at least two pistons and are mounted directly to the steering knuckle (**Figure 10-40**). Each piston receives equal fluid pressure and pushes on a brake pad. This type of caliper is used mainly in sports cars and high-performance applications.

■ ***Caliper Construction.*** A floating caliper consists of the caliper body, fluid passage, piston, square seal, dust seal, and bleeder screw (**Figure 10-41**). Most calipers are a one-piece design, meaning the entire caliper is formed out of one piece of aluminum or cast iron. The caliper shown in Figure 10-41 is a two-piece design. Caliper pistons can be made of steel, a special plastic, or aluminum. The pistons are hollow to save weight and

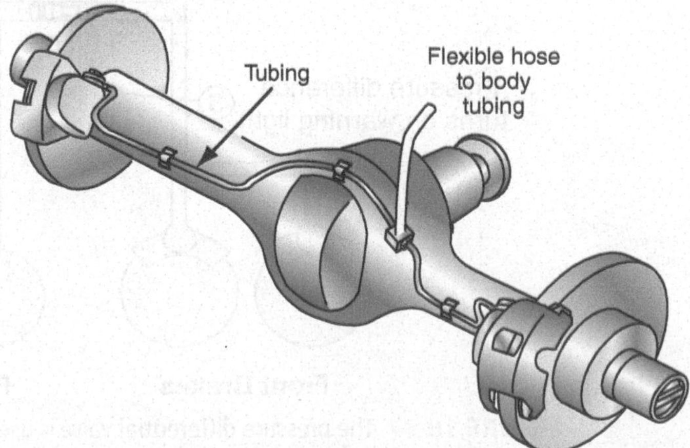

FIGURE 10-35 A junction often connects a rubber brake hose to the steel lines at the rear suspension.

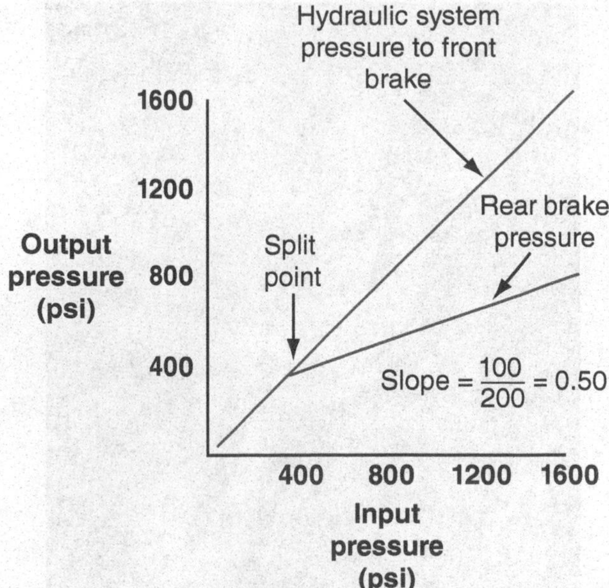

FIGURE 10-36 A proportioning valve limits pressure to rear drum brakes to prevent lockup. The split point is the pressure setting at which the proportioning valve limits pressure to the drum brakes. The slope is determined by the increase in output pressure compared to the increase in input pressure.

provide a place to clip the inner pad to reduce pad vibration and noise. The caliper body has a threaded inlet to accept the brake hose and a second smaller threaded opening for the bleeder screw. The bleeder screw is used to bleed air and brake fluid from the system during hydraulic system service or caliper replacement.

■ ***Rear Calipers with Integral Parking Brake.*** Most modern vehicles have disc brakes on all four wheels. Some rear disc brakes have an integral parking brake built into the caliper (**Figure 10-42**). The major difference between this caliper and a standard caliper is that the piston can be moved mechanically as well as by hydraulic pressure. The parking brake is separate from the service brakes and is entirely mechanical. This type of caliper may use a threaded piston or a lever system to push the piston out slightly. When the parking brake is activated by the driver, the parking brake cable pulls a lever on the caliper. Many newer vehicles use an electrically operated parking brake caliper. These systems use either an electric motor to turn against the piston or the motor pulls a parking brake cable.

■ ***Wheel Cylinder Construction and Operation.*** Although not as common as they once were, drum brakes are still in use, and millions of drum

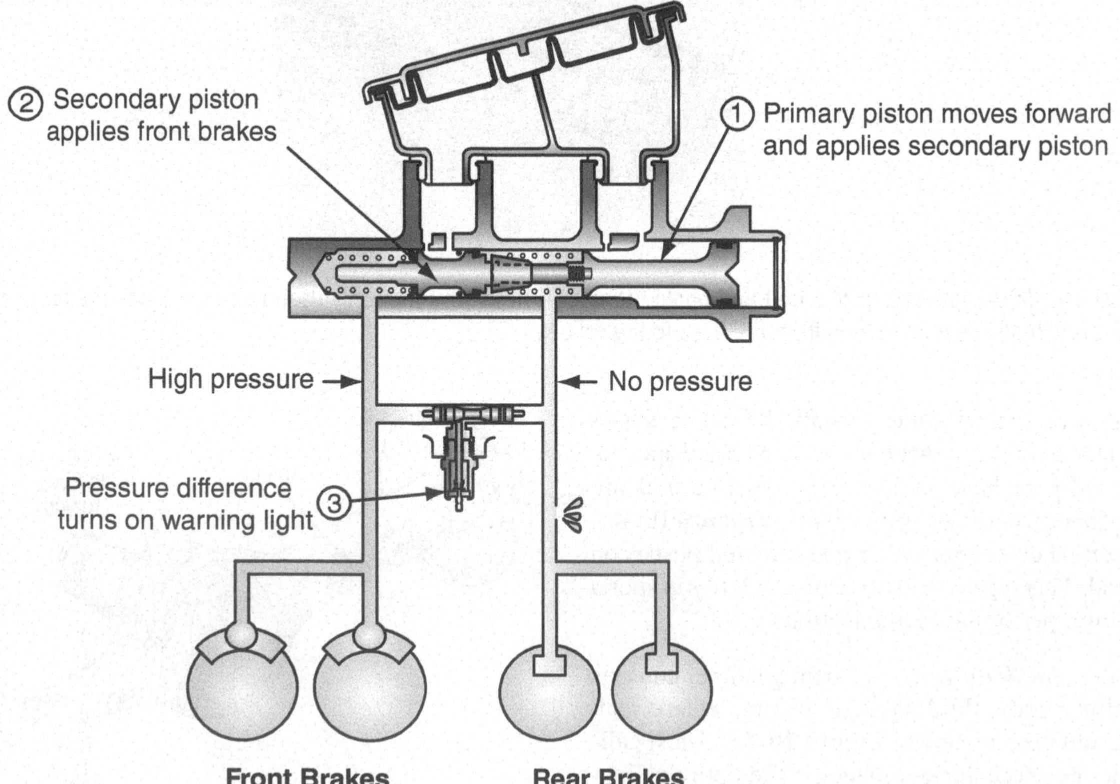

FIGURE 10-37 The pressure differential valve is used to turn on the BRAKE warning light and shut off the part of the hydraulic system that has a leak.

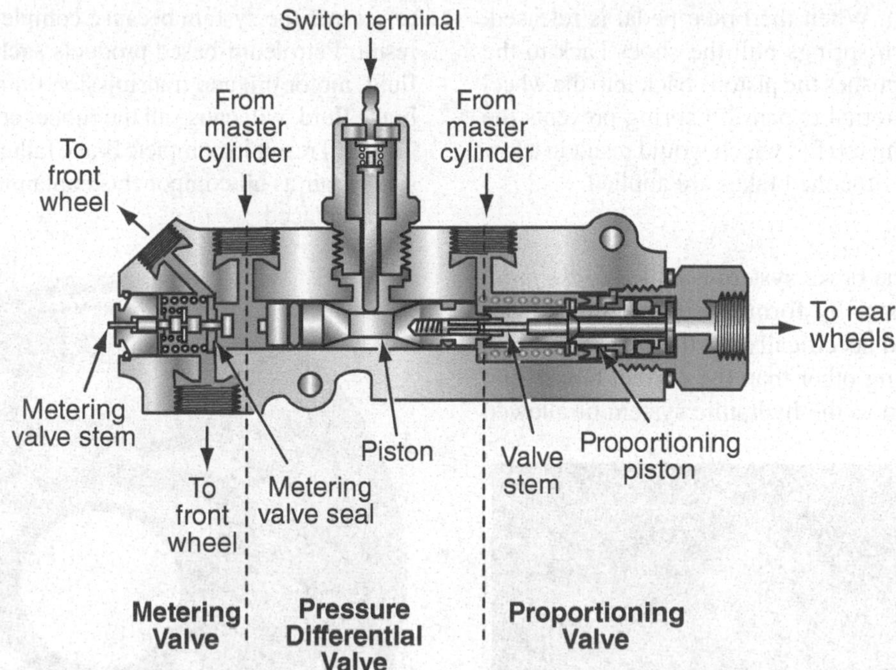

FIGURE 10-38 A combination valve contains the metering, proportioning, and pressure differential valves. These have been replaced by the ABS controlling brake pressure.

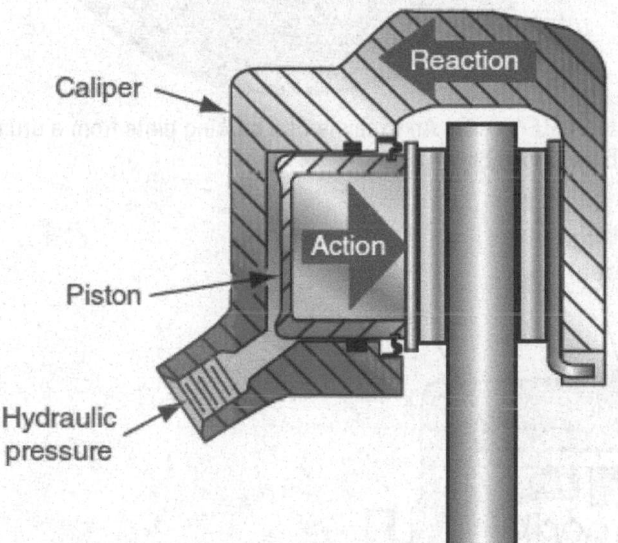

FIGURE 10-39 When the floating caliper is applied, brake fluid pressure pushes the piston outward from the caliper piston bore. This causes the caliper body to move backward, which results in both brake pads being pressed against the rotor. The most common caliper type is the floating caliper, which floats or moves on bolts or pins.

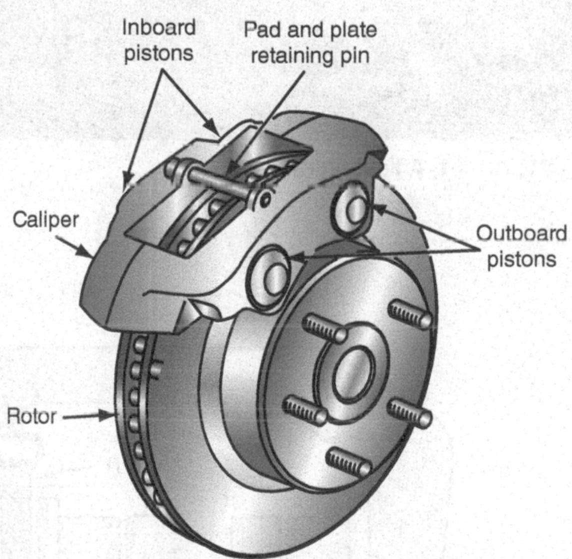

FIGURE 10-40 Fixed calipers are mounted rigidly to the steering knuckle and contain at least two pistons, one for each brake pad.

brake-equipped vehicles are still on the road. Modern drum brake systems use a dual-piston **wheel cylinder** to apply the brake shoes. A typical wheel cylinder is shown in **Figure 10-43**.

When the brakes are applied, the fluid pushes the pistons outward. On some drum brakes, the pistons apply directly against the brake shoes, while on other types a wheel cylinder link is placed between the piston and the shoe (**Figure 10-44**). When the brake pedal is released, the brake shoe return springs pull the shoes back to the rest position, which pushes the pistons back into the wheel cylinder bore. An internal expansion spring prevents the pistons from retracting too far, which would result in a low brake pedal the next time the brakes are applied.

BRAKE FLUID

The liquid used in the brake system is called *brake fluid*. **Brake fluid** is a specially formulated, nonmineral oil-based fluid, designed specifically for the demands of the brake system. Nothing other than the correct brake fluid should ever be added to the hydraulic system or allowed to get into the system because complete brake failure can result. Petroleum-based products such as power steering fluid, motor oil, and transmission fluid, if mixed with the brake fluid, will cause all the rubber components to swell. This will result in complete brake failure and a very expensive repair as all components containing rubber will have to be replaced.

FIGURE 10-41 A floating caliper.

FIGURE 10-43 An example of a backing plate from a drum brake assembly.

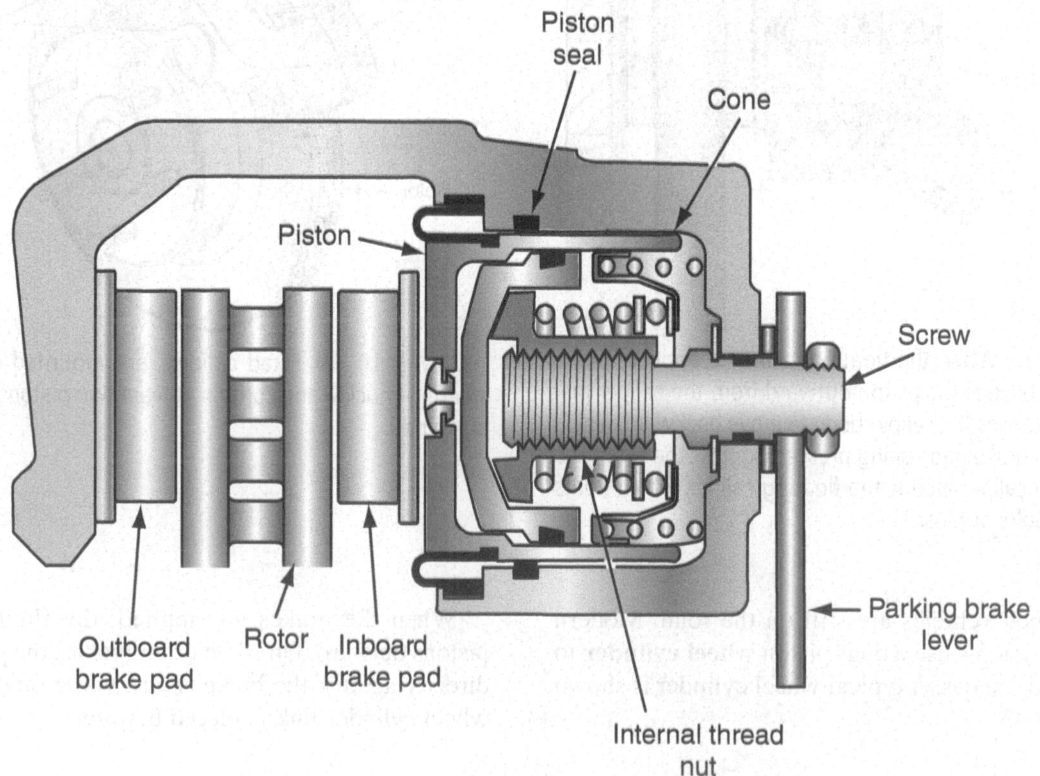

FIGURE 10-42 An integral rear parking brake caliper. The parking brake is mechanically operated for safety.

FIGURE 10-44 An example a drum brake assembly.

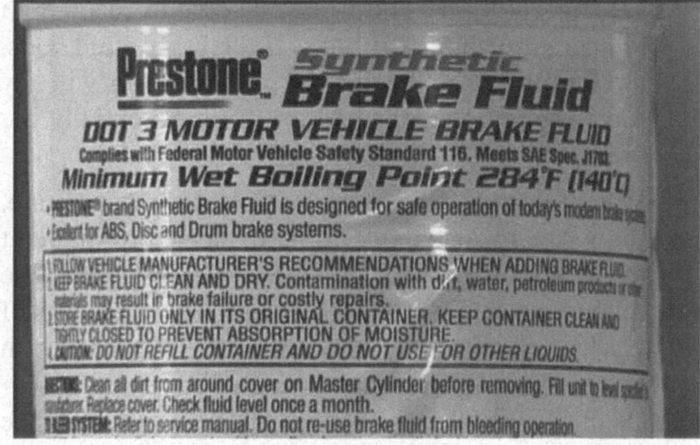

FIGURE 10-45 A brake fluid bottle label with boiling point specification.

■ ***Types.*** The brake fluid type most commonly used today is the glycol ester-based fluid, used in DOT 3, DOT 4, and DOT 5.1 fluids, with DOT 3 and 4 being the most common. DOT 3 is a rating, in which DOT means Department of Transportation. Brake fluids must meet performance criteria per DOT standards. The current specs are DOT 3, 4, and 5.1.

DOT 3 and DOT 4 fluids differ in their respective boiling points and are interchangeable, but DOT 3 should not be used in place of DOT 4 due to its slightly lower boiling point.

DOT 5 brake fluid is silicone based, is not compatible with any other brake fluid, and is not commonly used except in older, non-ABS vehicles and some limited-use or show cars.

■ ***Ratings.*** DOT 3 brake fluid is certified for use with both disc and drum brake systems and has a dry boiling point of at least 401°F (205°C) and a wet boiling point of 284°F (140°C) (**Figure 10-45**). DOT 4 fluid has a dry boiling point of at least 446°F (230°C) and a wet boiling point of 310°F (155°C). Dry boiling point means that the fluid has not absorbed more than about 3% of its volume in moisture. Once the moisture content is 3% or greater, the wet boiling point specification applies. The absorption of moisture, which has a much lower boiling temperature, dilutes the brake fluid and reduces its effectiveness. Water boils at 212°F (100°C) but brake fluid has a minimum boiling point of 401°F. As the percentage of water increases, the boiling point of the brake fluid decreases. Fluid with more than about 2% or 3% water should be flushed out and new fluid installed.

DOT 5.1 is a relatively new fluid specification and is also a glycol ester-based fluid, but it has a dry boiling point of 518°F (260°C) and a wet boiling point of 356°F (180°C). DOT 5.1 is used in some high-performance cars because of its high boiling point.

DOT 5 is a silicone-based fluid and is not compatible with any of the other fluids nor with antilock brake systems. It was used in some high-performance vehicles because the silicone did not absorb moisture, and therefore it did not lower the boiling point the way traditional glycol ester-based fluids do. It is slightly compressible compared to DOT 3, 4, or 5.1, which causes the brake pedal to travel farther and to be spongier. While DOT 5 is not used by vehicle manufacturers as standard equipment in modern vehicles, it is often used by owners of collector or show cars to prevent moisture from damaging the brake systems.

There are many brands of brake fluid available, some of which have much higher boiling points than listed here. Many of these fluids are for high-performance or racing applications. When selecting a brake fluid, always refer to the vehicle manufacturer's recommendation for the correct fluid.

■ ***Brake Fluid Properties.*** Brake fluid, by nature of being part of the brake system, must meet many requirements and operate under extreme conditions. To accomplish this, brake fluid has the following properties:

- Noncorrosive. Brake fluid must be able to remain stable and not react with many different types of materials, such as cast iron, steel, aluminum, plastic, rubber, brass, copper, and others.
- Maintain the correct viscosity. Brake fluid cannot thicken or freeze at low temperatures but also cannot get thin at high temperatures. The brake fluid may start out at below-freezing temperatures on a cold winter day, and then within minutes be several hundred degrees as heat transfers from the brake components to the fluid.
- High boiling point. Because the brakes create heat as a byproduct of the friction necessary to slow the

vehicle, the brake fluid must be able to reach high temperatures before it boils. If the brake fluid boils, the fluid vapor will compress and no longer will any force be applied to the pads or shoes. Essentially, the brakes fail.

- Water tolerant. Because brake fluid absorbs moisture, the water must be able to be dispersed throughout the volume of the fluid instead of separating, like oil and water. This is so there will not be pockets of water in the brake system. Because water has much higher freezing and much lower boiling points, pockets of water will cause serious problems for the brake system.
- Lubricant. Brake fluid acts as a lubricant to the moving parts, such as pistons and seals in the brake system.

An unwanted property of glycol-based brake fluids is that it is **hygroscopic**, meaning it absorbs moisture. Over time, brake fluid, even in the sealed brake system of the vehicle, will absorb moisture. The moisture will then allow rust and corrosion to form, causing damage to the hydraulic system components. In addition, the additive package in the brake fluid contains rust and corrosion inhibitors. These inhibitors, over time, break down and become depleted. Because of this, some vehicle manufacturers recommend that the brake fluid be flushed and new fluid installed periodically, usually every two to three years. This is especially important for ABS-equipped vehicles because the moisture and resulting rust and corrosion can foul the small passages and components found in the ABS hydraulic system.

Another negative quality of brake fluid is its ability to damage paint and some plastics. Brake fluid that gets spilled on a painted surface, if not promptly cleaned, can cause the paint to bubble and peel (**Figure 10-46**). Some plastics when exposed to brake fluid can dissolve. Always use fender covers, fluid pans, chemical gloves, and shop towels when you are working with brake fluid. This will prevent both skin irritation and damage to the vehicle.

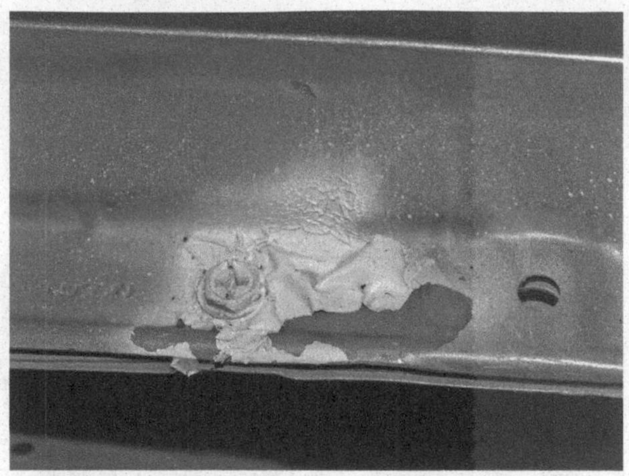

FIGURE 10-46 Brake fluid can damage paint, plastics, and other materials. Always use a fender cover when handling brake fluid, and clean any spills immediately.

■ ***Handling.*** Brake fluid should be kept in its original, sealed container, until it is needed. Brake fluid is sealed to prevent the absorption of moisture. Once the bottle is opened, the fluid should be used immediately. Leftover fluid should not be stored and used at a later date. An example of a brake fluid label and handling instructions is shown in **Figure 10-47.**

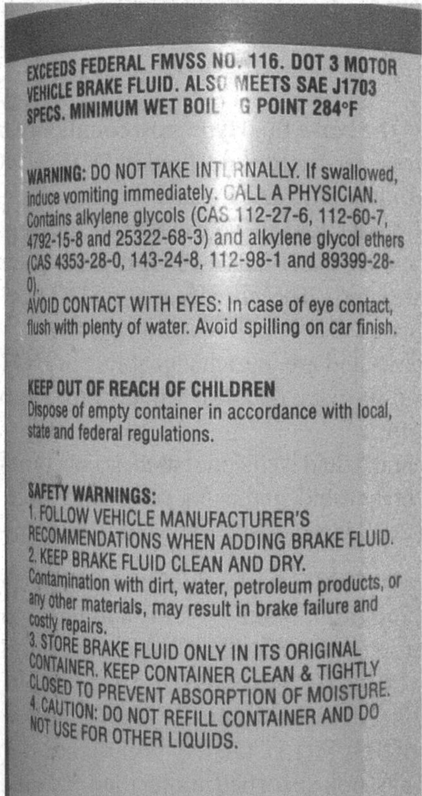

FIGURE 10-47 The brake fluid bottle provides handling instructions and precautions.

Because brake fluid absorbs moisture, you should wear chemical gloves when working with the fluid. Exposing your skin to brake fluid will cause your skin to dry out and can cause severe skin irritation.

When inspecting brake fluid, first, install a fender cover on the vehicle. Before checking and/or adjusting brake fluid levels in the master cylinder, clean the reservoir cap and area around the cap to prevent dirt from falling into the fluid when the cap is removed. Check

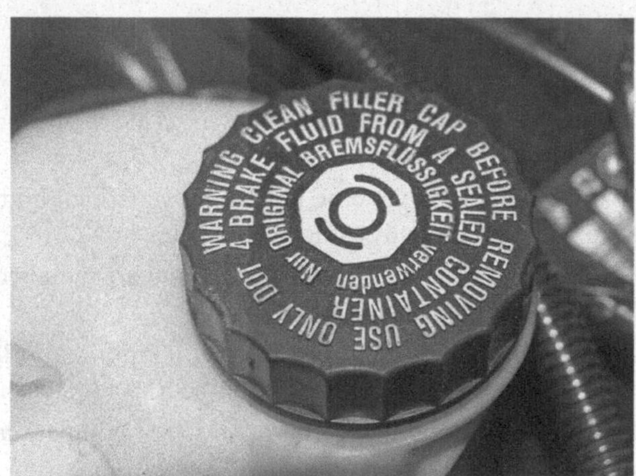

FIGURE 10-48 The brake fluid reservoir cap will indicate what type of brake fluid to use.

FIGURE 10-49 B Mode or regenerative braking is used to recapture energy from the brakes that is normally lost. The electric motors become generators to recharge the battery when using regenerative braking.

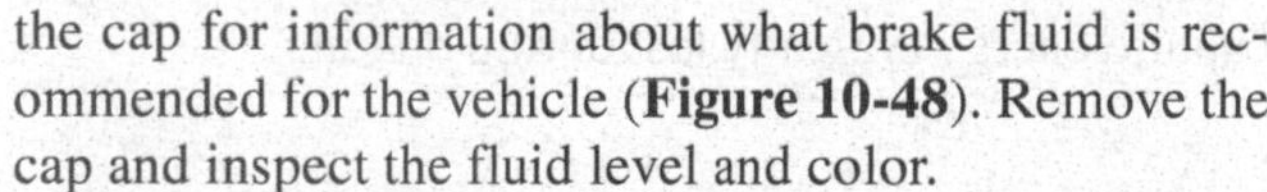

the cap for information about what brake fluid is recommended for the vehicle (**Figure 10-48**). Remove the cap and inspect the fluid level and color.

If the fluid needs to be topped off, use fluid from a sealed container, and add until the level reaches the maximum level. Do not overfill the reservoir. Brake fluid testing and service are covered in more detail in Chapter 11.

ELECTRONIC BRAKE SYSTEMS

As all systems on modern cars have been changed because of the addition of electronics and onboard computer systems, so has the brake system. The adoption of antilock braking and traction control systems as standard equipment means that steering and braking inputs are used by the computer system to determine the best application of the brakes and, if necessary, to alter torque to the driving wheels.

■ ***Antilock Brake and Electronic Stability Control Systems.*** All new cars and light trucks are equipped with ABS and some form of electronic stability control (ESC). These systems use sensor input to monitor wheel speeds, steering input, and brake input. Depending on the driving conditions, the ABS and ESC may control one or more wheel brakes to help prevent wheel lockup and to help the driver maintain control of the vehicle. ABS and ESC are discussed more in Chapter 16.

■ ***Regenerative Braking.*** Hybrid and electric vehicles use the braking system to recover what in other vehicles is wasted energy, and use it to recharge the high-voltage batteries.

In most conventional vehicles, when the brakes are applied, the kinetic energy of the vehicle is converted into heat energy at the brakes, which is then dissipated into the air as the brakes cool. Hybrid and electric vehicles use electric motors as a method of propulsion, and the electric motors can also act as electrical generators, called motor/generators. The electronics that control the high-voltage system and the electric motor can, when the driver presses the brake pedal, turn the electric motor into a generator. In addition, the driver can select a driving mode that will more aggressively use the electric motors for braking to recover energy at an increased rate. This is shown as the B mode on the gear selector in **Figure 10-49**. As an electric motor, the motor/generator uses high voltage to create strong magnetic fields that interact to drive the wheels. As a generator, those strong magnetic fields within the motor/generator are used to slow the vehicle. By operating this way, the motor/generator also acts as an electric brake. This greatly decreases the demands on the standard service brakes, which only operate at speeds below about 10–12 mph (16–19 kph). Because of this very limited operation, the service brakes tend to last much longer on hybrid vehicles compared to conventional vehicles.

Regenerative braking is also being used on non-hybrid vehicles. Mazda uses the I-ELOOP system to recover energy when braking, stores the energy in a large capacitor, then uses it to power the electrical system during acceleration. This helps increase fuel economy and performance as some of the load to turn the generator is taken off of the engine. It is likely that the use regenerative braking will increase as manufacturers adopt new ways to improve fuel economy.

SUMMARY

Brake pedals provide an increase in brake application force, in addition to that provided by the driver.

Free play is the slight amount of pedal movement at the released position before the pushrod begins to move into the booster and master cylinder.

Hydraulics is the science of using liquids to perform work.

Hydraulic brake systems contain an input cylinder called the master cylinder.

The output of the hydraulic system in the drum brake is the wheel cylinder.

Modern master cylinders have two pistons and two chambers.

Step-bore master cylinders have different-sized pistons and chambers.

The proportioning valve is used to limit or proportion hydraulic pressure to the rear drum brakes.

The metering valve is used to delay slightly the application of the front disc brakes.

The pressure differential valve is used to close off one circuit in the event of pressure loss.

Floating calipers have to be able to move so that both brake pads are applied against the rotor.

Fixed calipers have at least two pistons and are mounted directly to the steering knuckle.

Brake fluid is a specially formulated, nonmineral oil-based fluid designed specifically for the brake system.

DOT 5 is a silicone-based fluid and is not compatible with any of the other brake fluids.

Brake fluid is hygroscopic, meaning it readily absorbs moisture.

Brake fluid should be kept in its original, sealed container until needed.

Hybrid and electric vehicles use regenerative braking to capture energy typically lost during braking.

REVIEW QUESTIONS

1. The term ______________ refers to the science of using fluids to perform work.
2. A ______________ caliper has one or two pistons, and is mounted so that it can move when the brakes are applied.
3. Mechanical advantage or ______________ is used to apply greater force to the master cylinder pistons than that applied solely by the driver.
4. The brake shoes are pushed out against the drum by the wheel cylinder ______________.
5. A ______________ valve is often used on older vehicles to prevent rear wheel lockup during hard braking.
6. All of the following are components of the brake hydraulic system except:
   a. Wheel cylinder
   b. Brake lines
   c. Caliper
   d. Brake pedal pushrod
7. *Technician A* says the brake pedal is used to increase force applied by a driver. *Technician B* says the movement of the brake pedal turns on the brake lights. Who is correct?
   a. Technician A
   b. Technician B
   c. Both A and B
   d. Neither A nor B
8. *Technician A* says DOT 4 brake fluid is silicone based and cannot be mixed with DOT 3 brake fluid. *Technician B* says that DOT 4 brake fluid is petroleum based and can be mixed with DOT 3 brake fluid. Who is correct?
   a. Technician A
   b. Technician B
   c. Both A and B
   d. Neither A nor B
9. *Technician A* says a leak in the hydraulic system will reduce the pressure applied to the caliper or wheel cylinder pistons. *Technician B* says a leak in the hydraulic can affect pedal feel. Who is correct?
   a. Technician A
   b. Technician B
   c. Both A and B
   d. Neither A nor B
10. All of the following are advantages of regenerative brakes except:
   a. Longer brake pad life
   b. Decreased brake heat generation
   c. Decreased fuel economy
   d. None of the above

Brake System Service

Chapter Objectives

At the conclusion of this chapter, you should be able to:

- Describe procedure for performing a road test to check brake system operation, including an antilock brake system (ABS). (ASE Education Foundation MLR 5.A.2)
- Inspect and adjust brake pedal height, pedal travel, and free play. (ASE Education Foundation MLR 5.B.1)
- Check master cylinder for external leaks and proper operation. (ASE Education Foundation MLR 5.B.2)
- Inspect and service brake lines, hoses, and fittings. (ASE Education Foundation MLR 5.B.3)
- Select, handle, store, and fill brake fluids to proper level. (ASE Education Foundation MLR 5.B.4)
- Identify and inspect components of the brake warning light system. (ASE Education Foundation MLR 5.B.5)
- Flush and bleed the brake hydraulic system. (ASE Education Foundation MLR 5.B.6)
- Inspect brake fluid condition and test for contamination. (ASE Education Foundation MLR 5.B.7)

KEY TERMS

bench bleeding	**bypassing**	**pressure bleeding**
brake fluid flush	**flare tool**	**vacuum bleeding**
brake pedal position sensor	**free play**	
brake pedal height	**manual bleeding**	

Eventually, a vehicle's brake system will need service. Brake lines rust through, brake hoses crack and rupture, and other hydraulic components leak and need to be replaced. This chapter focuses on servicing and repairing the hydraulic system.

Tools and Safety

Following proper service procedures include using the correct tools to perform the job and using the tools correctly. This allows for safe and productive work without injuries or time lost due to poor work practices.

TOOLS FOR HYDRAULIC BRAKE SYSTEM SERVICE

Servicing the hydraulic system requires basic hand tools, such as wrenches, and sockets as well as specialized tools such as brake bleeding equipment. The most common tools you will need are shown in Figure 11-1 through Figure 11-6.

- Line or flare nut wrenches are used on brake line fittings (**Figure 11-1**). Because of the shape of the wrench, fittings are less likely to strip when loosening and tightening them with a line wrench.
- A flaring tool is used to make new flares on a brake line. Flaring tools are available for both SAE and metric lines and flares (**Figure 11-2**).
- Bleeding the brakes can be easier with tools such as the vacuum bleeder (**Figure 11-3**). Shop air is connected to the tool and a vacuum draws air and fluid out of the brake system.
- A pressure bleeder applies pressure to the brake fluid at the master cylinder and forces air and fluid out of the bleeder valves (**Figure 11-4**).

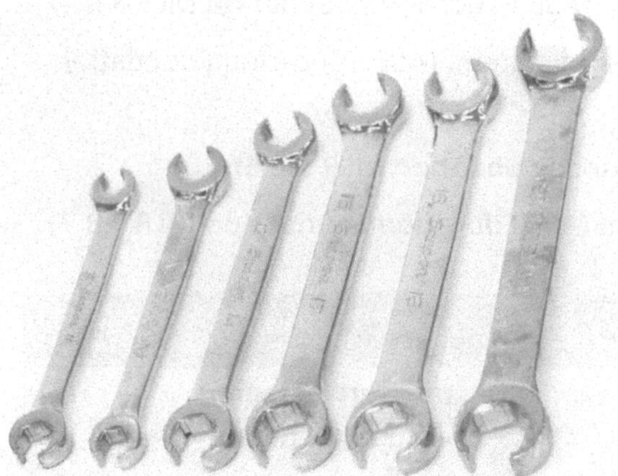

FIGURE 11-1 Line wrenches, also called tubing or flare nut wrenches, are used on brake line and similar fittings. The extra grip reduces the chance of stripping the hex on the fitting.

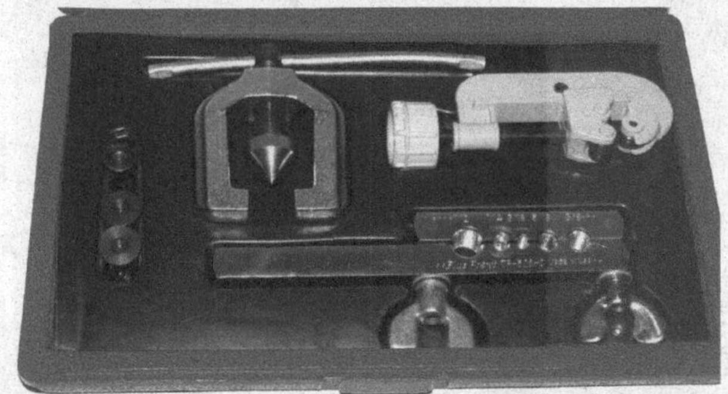

FIGURE 11-2 A flaring tool is used to make a flare on the end of a brake line. Flare tools are either SAE or ISO.

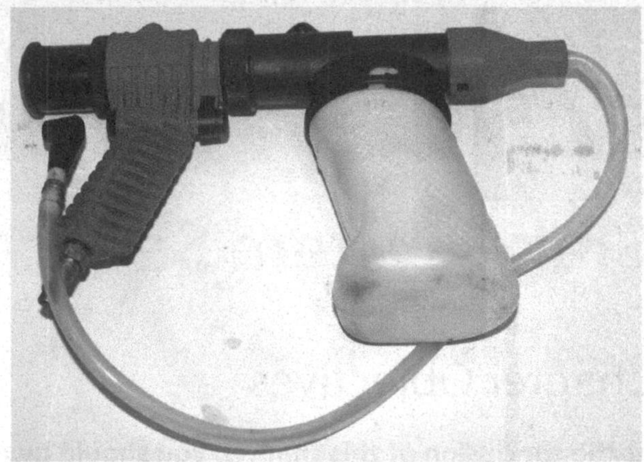

FIGURE 11-3 This vacuum bleeder is used when flushing or bleeding the hydraulic system.

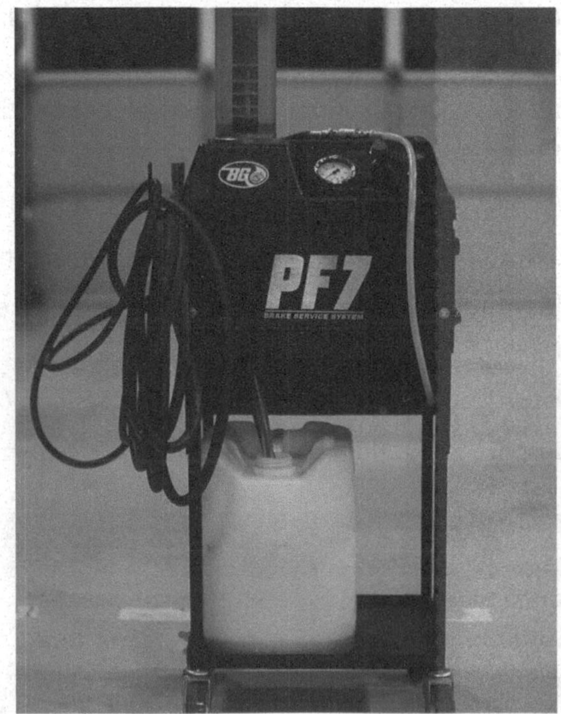

FIGURE 11-4 A pressure bleeder forces brake fluid through the system under pressure to remove any trapped air.

- When you are making new brake line, a tubing bender is used to bend the brake line without kinking it (**Figure 11-5**).
- A special tool is used to test the brake fluid for moisture content (**Figure 11-6**).

TOOL SAFETY AND SAFE WORK PRACTICES

Tools are designed to be used in specific ways. When a tool is used incorrectly or for something other than its intended purpose, accidents can occur. Although the service tools for the hydraulic system do not generally present any significant danger in their typical usage, improper use can lead to personal injury and damage to equipment and vehicles. To reduce the chances of having an accident when you are working on the brake system, follow these tool safety precautions:

- Use the correct wrenches and other hand tools for the job, and use the tools correctly. Do not use open-end wrenches to break loose tight fasteners, and pull wrenches and ratchets toward your body when you are loosening fasteners.

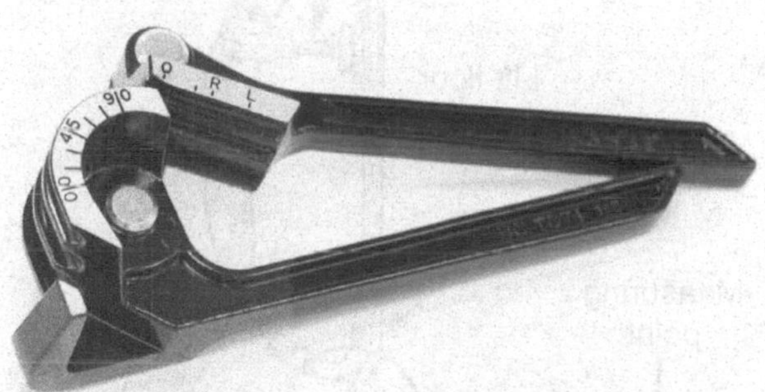

FIGURE 11-5 A tubing bender is used when fabricating new brake lines. Using a bender reduces the chance of kinking the line.

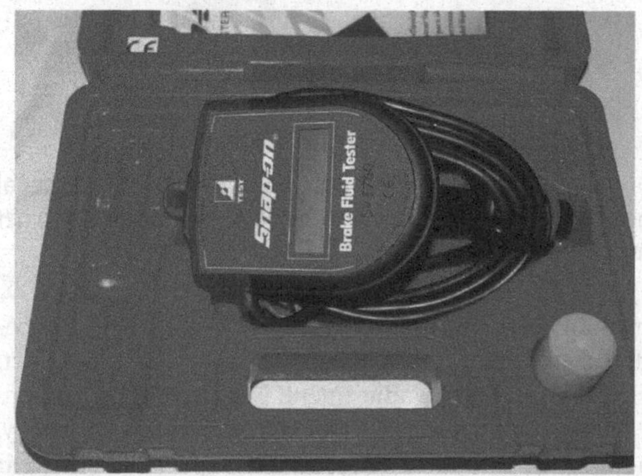

FIGURE 11-6 A moisture tester heats a sample of brake fluid to determine its moisture content.

- Ensure the brake system is depressurized before you open the hydraulic system on vehicles with integral antilock brake systems or systems (ABS) with high-pressure accumulators.
- Read and follow all the instructions for the setup and use of a pressure bleeder. Do not overpressurize the bleeder or the brake system.
- Use caution when using a brake fluid moisture tester. Both the tester and the fluid become very hot.
- Allow the brakes to cool before you attempt to remove or service any components.
- Use fender covers to protect the vehicle from brake fluid.
- Clean up brake fluid immediately; do not allow fluid to remain in contact with painted surfaces, wiring, plastics, or any other parts of the vehicle.
- Do not reuse brake fluid or use brake fluid from an open container.
- Use the correct tools for the job.
- Do not reuse damaged parts.
- Ensure the brakes are working properly before moving the vehicle.
- Read and follow all service procedures related to the brake system service.

Determining Brake System Concerns

As a technician, you will be required to test drive vehicles to verify customer concerns and to verify repairs have been completed properly. When it is possible, a test drive should be performed before and after all repairs. However, common sense needs to be used; if a vehicle has no brakes, such as from a hydraulic system leak, a test drive is not needed to confirm the complaint. Once repairs are made and the brake system is functioning, a test drive should be performed to determine if other problems, such as brake noise or a pulsation, are present.

Many technicians develop a test drive route that covers different types of driving conditions. This is to check and test the brakes, handling, and performance of the engine, and the drivetrain. When you are test driving for brake system concerns, you will want to be able to check for the following:

- Noise during driving and braking. Some technicians find that driving where noise can be reflected, such as behind stores, makes diagnosing some noise complaints easier.

- Pulsation during stopping. This may require driving on the highway or other roads with higher speed limits to feel a pulsation.
- Pulling during stopping. This should be checked on a section of smooth, flat road where there is little traffic.
- False ABS activation. This concern should also be checked on a section of smooth, flat road or in a parking lot where there is little traffic.

The main requirements for a test drive are that it is done safely and that all driving laws are observed. At no time should a vehicle be operated in any unsafe manner or in a way that violates speed limits or any other motor vehicle laws. It is important to remember that even though accidents do happen, an accident involving a customer's car is not going to be tolerated if it is your fault. If it is necessary, have a coworker ride along to help with locating noises so that you are not dividing your attention between driving and diagnosing.

As a student, you are probably not going to be able to take extended test drives off the school's campus. However, you may be able to test drive within the school's parking lots. If so, make sure you observe the speed limit within the parking lot, and pay close attention for vehicles pulling in and out of parking spaces while you are test driving.

BRAKE PEDALS AND HYDRAULIC SYSTEM

Most modern vehicles do not require adjustment to the brake pedal or pushrod unless components such as the vacuum power assist unit are replaced. However, some vehicles permit adjustments of the brake pedal pushrod and pedal free play. Some brake switches will require adjustment if replaced.

Unless a vehicle has power-adjustable pedals, typically the only inspection for the brake pedal assembly is for pedal height, travel, and free play. It is possible that a customer may experience a noise when the brake pedal is applied that is caused by the bushing located on the pedal pivot. The pedal should move freely but without play within the pedal assembly. If noise or play is present, inspect the pedal mounting components (**Figure 11-7**).

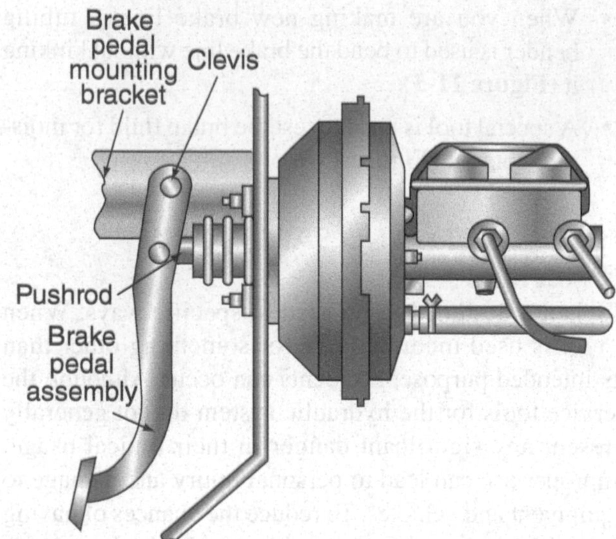

FIGURE 11-7 An illustration of the brake pedal assembly. Inspect the pedal and bracket for wear and noise as part of a brake inspection.

■ ***Pedal Height and Free Play.*** **Brake pedal height** is measured as the distance from the pedal to the floor with the pedal depressed (**Figure 11-8**). Some manufacturers may provide pedal height specs from the steering wheel down to the pedal (**Figure 11-9a**). In addition, a pedal effort gauge may be required (**Figure 11-9b**). The gauge is used to apply a specific amount of pressure, such as 100 pounds, to the brake pedal when testing pedal height. Always refer to the manufacturer's service information for the correct procedure for checking pedal height. If the height is incorrect, check the rear brake adjustment. Excessive shoe-to-drum clearance will cause a low brake pedal. If it is necessary, adjust the rear brakes and recheck the pedal height. Rear brake adjustment is discussed in Chapters 13 and 15.

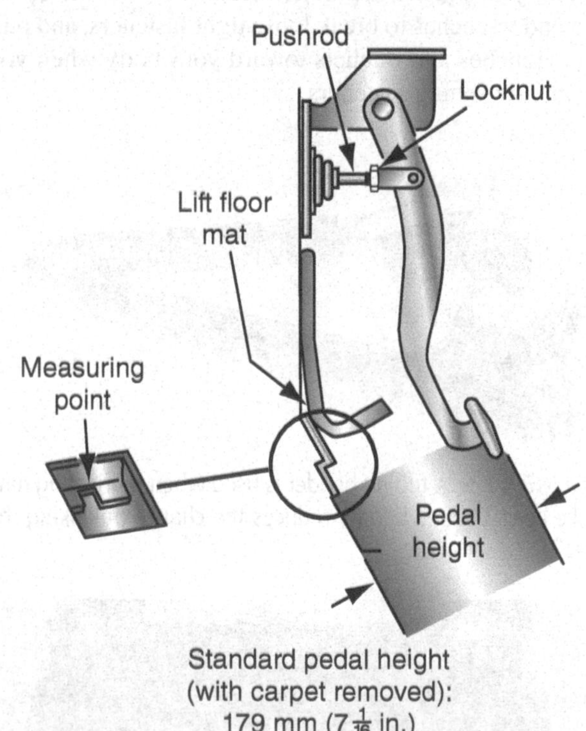

FIGURE 11-8 This illustrates how to check brake pedal height. It is important to move the floor mat and carpet to get an accurate measurement.

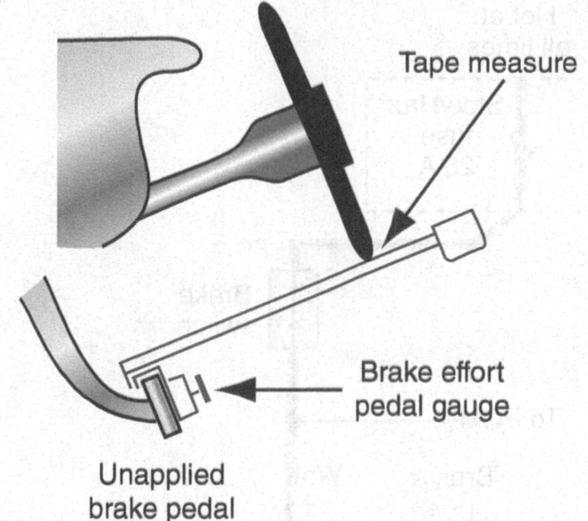

FIGURE 11-9a Some manufacturers specify checking pedal height from the steering wheel to the pedal.

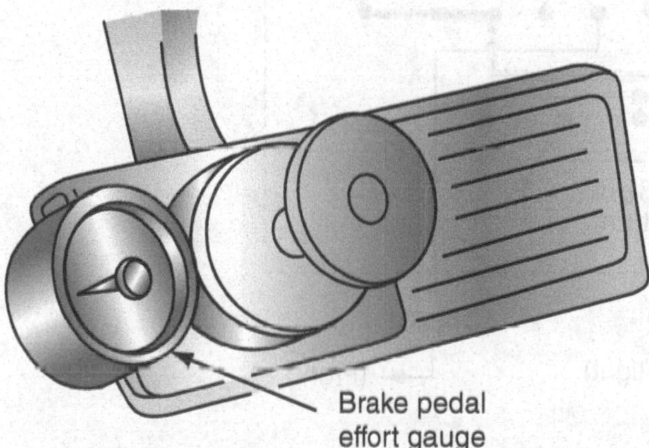

FIGURE 11-9b An example of a brake pedal effort gauge. When testing pedal height, press on the plunger until the specified effort is shown on the gauge and check the pedal height.

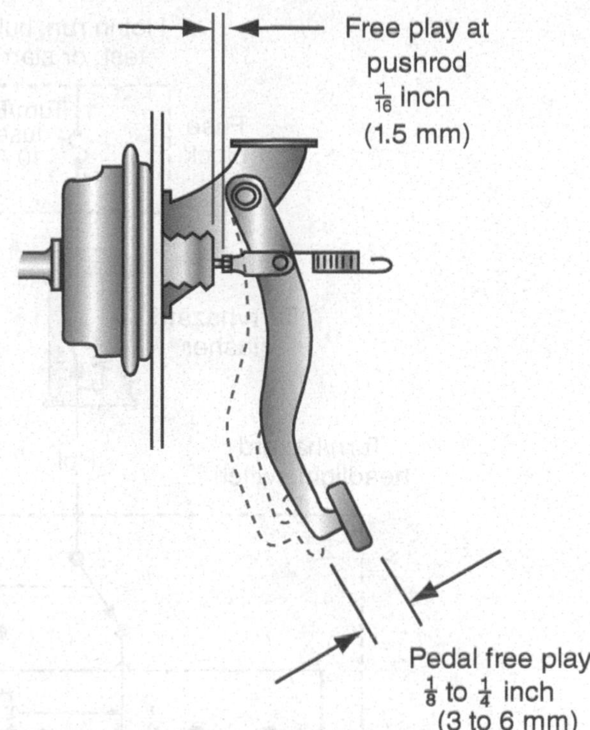

FIGURE 11-10 Free play is measured as the slight amount of movement of the pedal when released.

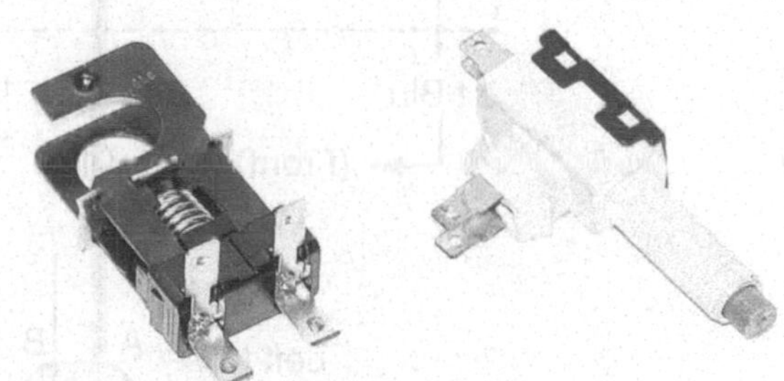

FIGURE 11-11 Examples of brake light switches.

Brake pedal **free play** is the very slight movement of the pedal before the brake pushrod begins to move (**Figure 11-10**). Free play on some vehicles is adjusted by loosening a jam nut on the brake pedal pushrod and adjusting rod length.

■ ***Brake Light System.*** The brake or stop lights alert drivers that the vehicle is slowing. As part of a brake system inspection, you should check that all brake lights work properly. Mounted on the brake pedal bracket is the brake pedal position sensor or brake light switch (**Figure 11-11**). The switch is either open or closed. When it is open, the circuit is open, and the brake lights are off. When it is closed, as when the driver presses the brake pedal, the circuit closes, and the brake lights at the rear of the vehicle should illuminate. Have an assistant press the brake pedal while you watch brake light operation.

The **brake pedal position sensor** is used for more than activating the rear brake lights. It is an input for the powertrain control module (PCM), body control module (BCM), the ABS, and the traction control (TC) system. Brake switch activity is also monitored as an input for the lockup torque converter clutch used in automatic transmissions. A nonfunctioning brake light switch, aside from being a potential driving hazard for those driving behind the vehicle, will affect more than just the brake lights.

In older vehicles, power from the switch flows to the brake lights and other circuits. An example of a brake switch circuit is shown in **Figure 11-12**. If all the vehicle's brake lights fail to operate, locate and check the brake light circuit's fuse first. If the brake light fuse is good, then check for power to and from the brake switch

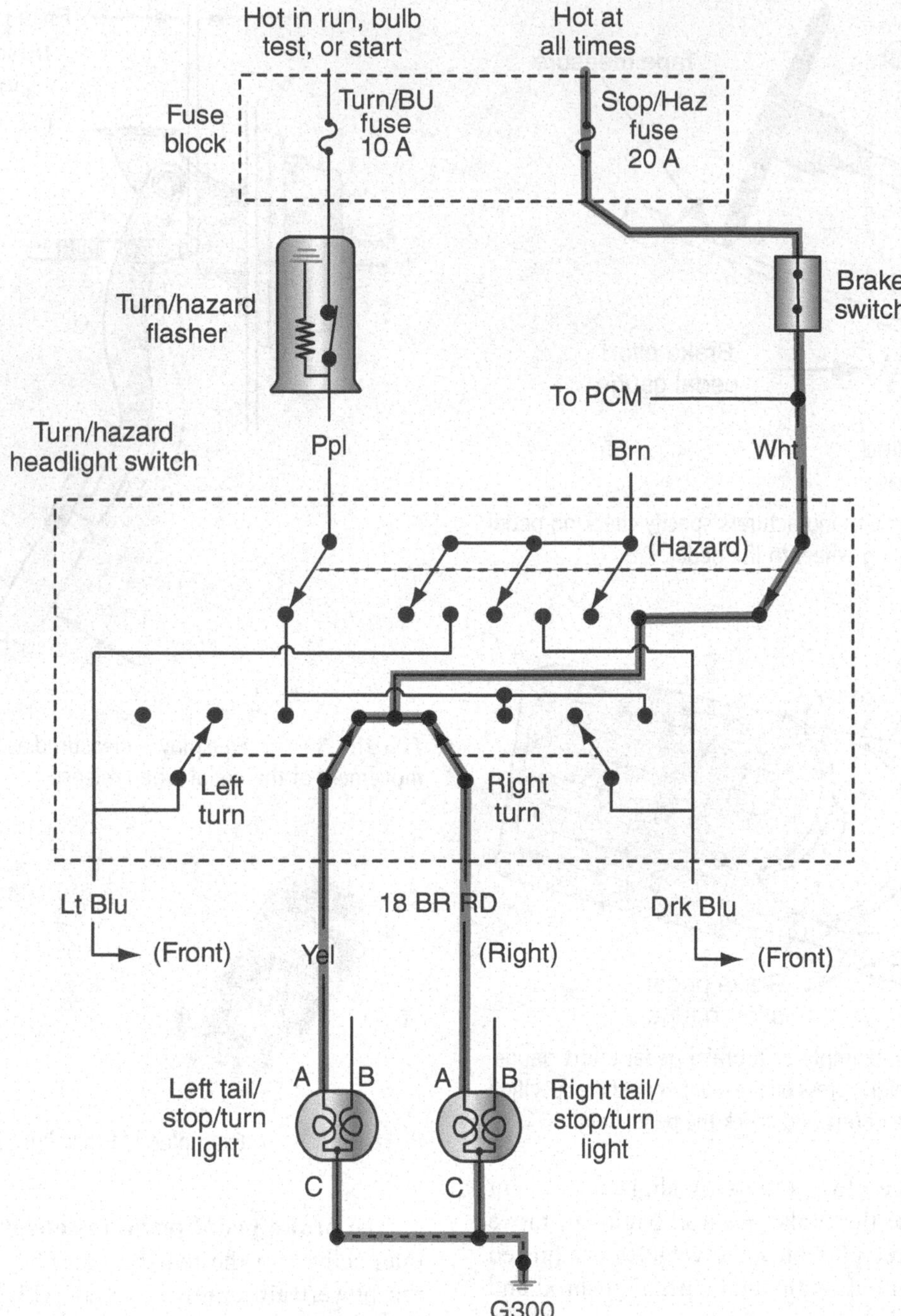

FIGURE 11-12 In some vehicles, the brake light circuit is completed through the turn signal switch. Other vehicles do not use combination brake/turn signal bulbs and the brake light circuit does not connect through the turn signal switch.

(**Figure 11-13**). If power is present to the switch but no power is present at the output wire with the brake applied, the switch is open. If there is no power to the switch and the fuse is good, you will need to look for an open in the wiring between the fuse and the switch.

On most newer vehicles, the brake pedal position (BPP) sensor is an input into the BCM. The BCM uses the BPP sensor signal to turn on the stop lamps. Sensor data is also sent out on the network for other modules to use as needed. For these systems, you will need to use a scan tool to look at the brake switch data when diagnosing a problem with the brake light circuit. **Figures 11-14** and **11-15** show BPP data from a scan tool. As you can see in the data, the BCM looks at more than just whether the switch is open or closed. Diagnosing sensor and network concerns is discussed in Chapter 21.

On older vehicles with a brake light switch, the circuit is usually a hot-at-all-times circuit. This means that battery voltage is present to the switch at all times, even when the key is off and the vehicle is parked. Power to the switch is typically from a dedicated fuse. This means the fuse supplies power only to the brake light

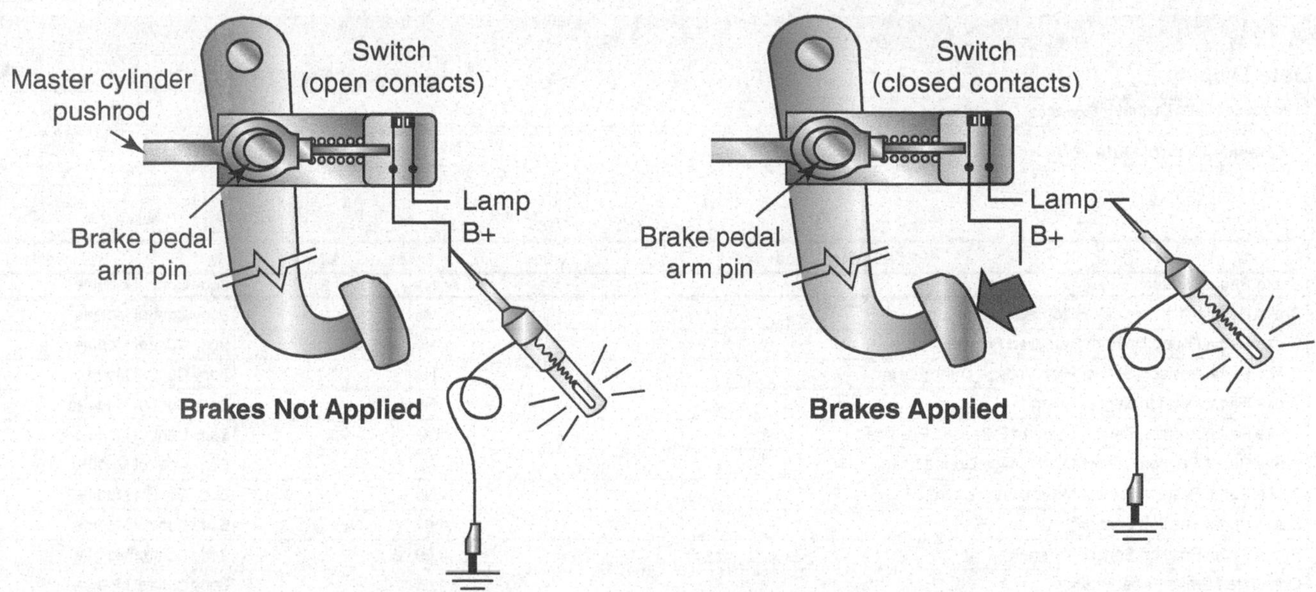

FIGURE 11-13 This illustrates how to test for power to and through a brake light switch with a test light.

GDS 2

Data Display

Create Report

Diagnostic Data Display | Graphical Data Display | Line Graph

Chassis Control Data

Parameter Name	Value	Unit	Control Mc
Brake Pedal Applied	Inactive		Body Control Module
Brake Pedal Initial Travel Position Achieved	No		Body Control Module
Brake Pedal Pulled Up from Released Position	No		Body Control Module
Brake Pedal Position Sensor High Voltage During Learn	No		Body Control Module
Brake Pedal Position Sensor Learn	No		Body Control Module
Brake Pedal Position Sensor Learned Released Position	Yes		Body Control Module
Brake Pedal Position Sensor Low Voltage During Learn	No		Body Control Module
Brake Pedal Position Sensor Move During Learn	No		Body Control Module
Brake Pedal Position Sensor	0.96	V	Body Control Module
Brake Pedal Position Sensor Reference	4.89	V	Body Control Module
Calculated Brake Pedal Position	0.06	V	Body Control Module
Calculated Brake Pedal Position	1	%	Body Control Module
Brake Transmission Shift Interlock Solenoid Actuator Command	Inactive		Body Control Module
Park Brake Status	Released		Body Control Module

FIGURE 11-14 Using a scan tool to monitor brake pedal position sensor operation. In this image, the brake is not applied.

circuit. This is illustrated in the brake light circuit shown in Figure 11-12. Power passes through the switch when the pedal is pressed a slight amount. On vehicles with a BPP sensor, these circuits usually require the ignition to be ON before the stop lights will illuminate.

On most vehicles, the stop lights should turn on if the brake pedal is pressed ¼ inch (6 mm). If the pedal has to travel excessively before the brake lights illuminate, the switch may need to be adjusted. Some switches thread into the brake pedal bracket and are adjusted by loosening the jam nut and turning the switch until the lights come on within about ¼ inch (6 mm) of pedal travel. An example of adjusting a threaded switch is illustrated in **Figure 11-16**. Some switches and BPP sensors are not adjustable and require replacement if not working properly.

■ ***Adjustable Pedals.*** Adjustable pedals are an option on some vehicles, often in combination with power-adjustable steering columns and memory seat functions. The pedal assembly is mounted on a motorized carriage that allows for about 3 inches of travel forward and back ward (**Figure 11-17**). This adjustment is helpful for

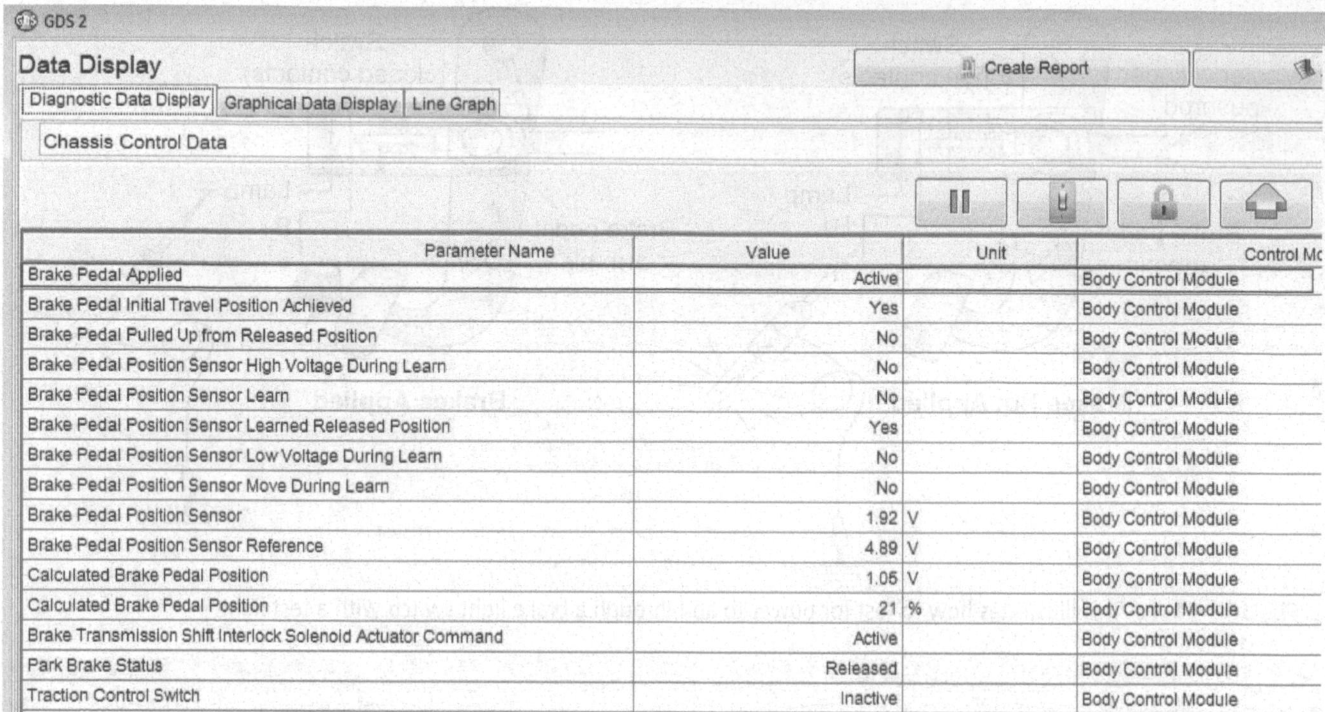
GDS 2

Data Display

Create Report

Diagnostic Data Display | Graphical Data Display | Line Graph

Chassis Control Data

Parameter Name	Value	Unit	Control Mc
Brake Pedal Applied	Active		Body Control Module
Brake Pedal Initial Travel Position Achieved	Yes		Body Control Module
Brake Pedal Pulled Up from Released Position	No		Body Control Module
Brake Pedal Position Sensor High Voltage During Learn	No		Body Control Module
Brake Pedal Position Sensor Learn	No		Body Control Module
Brake Pedal Position Sensor Learned Released Position	Yes		Body Control Module
Brake Pedal Position Sensor Low Voltage During Learn	No		Body Control Module
Brake Pedal Position Sensor Move During Learn	No		Body Control Module
Brake Pedal Position Sensor	1.92	V	Body Control Module
Brake Pedal Position Sensor Reference	4.89	V	Body Control Module
Calculated Brake Pedal Position	1.05	V	Body Control Module
Calculated Brake Pedal Position	21	%	Body Control Module
Brake Transmission Shift Interlock Solenoid Actuator Command	Active		Body Control Module
Park Brake Status	Released		Body Control Module
Traction Control Switch	Inactive		Body Control Module

FIGURE 11-15 Using a scan tool to monitor brake pedal position sensor operation. In this image, the brake is applied.

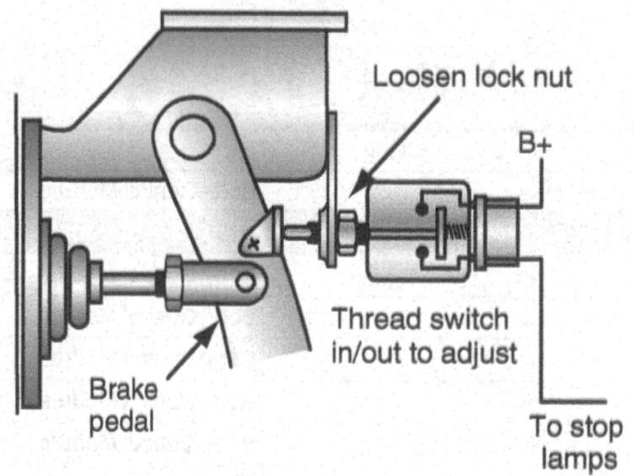

FIGURE 11-16 An example of how to adjust a brake light switch.

shorter drivers who would need to sit too close to the steering wheel—and the air bag—to reach the pedals.

If a vehicle has power-adjustable pedals, you may see concerns regarding improper pedal position or a failure of the pedals to adjust position. For adjustable pedal concerns, a scan tool is necessary to access data and stored diagnostic trouble codes (DTCs). Begin your inspection by activating the system and checking that the assembly is able to move forward and backward along the entire range of motion. The pedals should move without binding. If the pedals do not operate, first check the system fuse. If the fuse is intact, connect a scan tool to the data link connector (DLC), and access the menu for the pedal system. First check for any current and history DTCs related to the pedal system. If no DTCs are present, activate the pedal control switch, and note the appropriate data on the scan tool. If the switch input does not change on the data display, inspect the switch and wiring. If the input from the switch is correct, try to control the movement of the pedal assembly through the scan tool. If the pedals do not respond to active commands, the drive motor or wiring may be at fault.

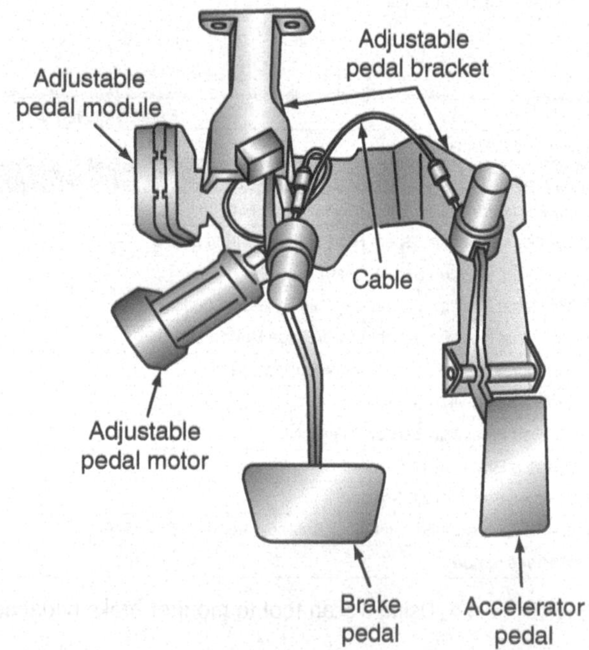

FIGURE 11-17 An illustration of power-adjustable pedals.

BRAKE FLUID INSPECTING AND TESTING

Until the widespread adoption of ABS, brake fluid sometimes remained in the vehicle over the entire life of the vehicle. Now, vehicle manufacturers recommend periodic flushing of the brake fluid to remove moisture and other contaminants. This prolongs the life of the hydraulic system components and ultimately can reduce repair costs.

■ ***Checking and Adjusting Brake Fluid Level.*** As discussed in Chapter 10, the most common brake fluids are DOT 3 and DOT 4. Both are chemically similar and can be mixed in vehicles that specify the use of DOT 3. If a vehicle requires DOT 4, use only DOT 4 because it has a higher boiling temperature than DOT 3. The reservoir cap usually indicates the recommended fluid (**Figure 11-18**).

When it is necessary to add brake fluid to the reservoir, always use new brake fluid from a sealed container. Also, check the label to make sure you are using the correct fluid. Examples of brake fluid bottles are shown in **Figure 11-19**. Before adding brake fluid to the master cylinder reservoir, install a fender cover to protect the vehicle. Next, clean around the cap to prevent any dirt from getting into the brake fluid when the cap is removed. Remove the cap and inspect the seal. If the fluid is low, the reservoir seal will likely be extended. This is normal. Check the seal for damage or swelling as these can indicate fluid contamination. If the cap and seal are good, fill until the fluid reaches the MAXIMUM line on the reservoir. (**Figure 11-20**). If the reservoir caps use accordion seals that have extended out, carefully push the seals back together before you reinstall the cap.

Most vehicles use a brake fluid level sensor in the master cylinder reservoir. The sensor turns on the red BRAKE warning light on the instrument panel when the fluid level drops to a certain point (**Figure 11-21**). If the brake fluid level is low, you should recommend a brake system inspection to the customer. Worn friction components, specifically the brake pads and rotors, can cause a low brake fluid level. As the pads and rotors wear, the caliper

FIGURE 11-19 Examples of common brake fluids.

FIGURE 11-20 Many reservoirs allow you to check fluid level without opening the cap. Do not open the reservoir unless adding or checking fluid condition as this can allow dirt to get into the reservoir.

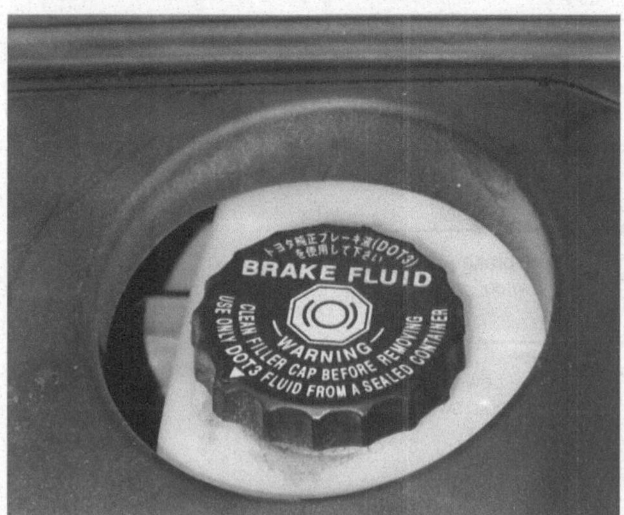

FIGURE 11-18 Check the brake fluid reservoir cap for the recommended type of brake fluid. Clean the cap and area before removing to prevent dirt from getting into the fluid.

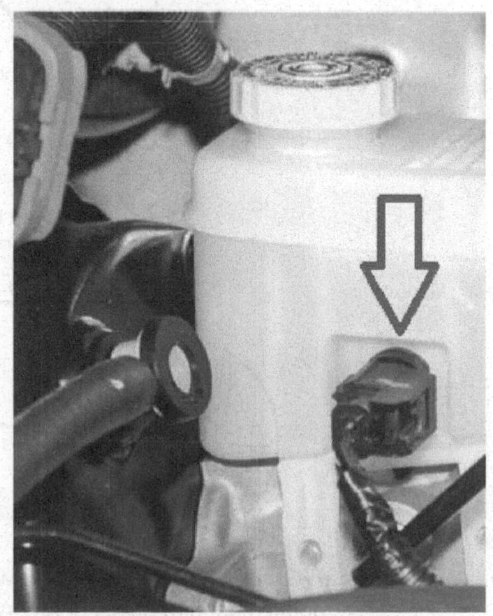
FIGURE 11-21 A brake fluid level sensor.

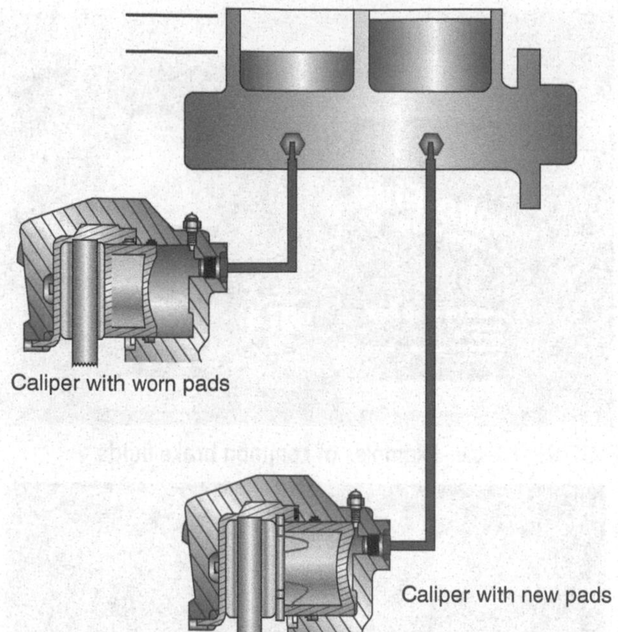

FIGURE 11-22 As the pads and rotors wear, the pistons extend further out of the bores. This reduces the fluid level in the master cylinder.

pistons move farther out of their bores to compensate for the wear. The increased volume of fluid in the calipers reduces the amount of fluid in the master cylinder reservoir (**Figure 11-22**).

Low fluid level can also be caused by a leak in the system. However, a leak in the hydraulic system is usually very noticeable due to the very low brake pedal and poor stopping power. If a leak is suspected, perform a thorough inspection of the hydraulic system.

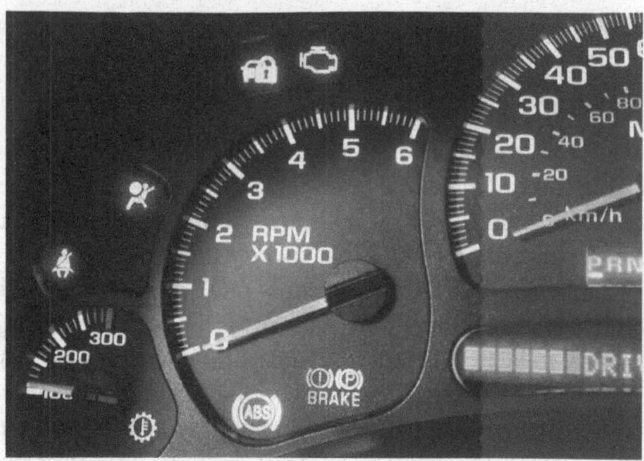

FIGURE 11-23 The red BRAKE warning light. The light should also come on during bulb check when the ignition is turned to the ON or RUN position.

■ ***Brake Warning Light.*** The red BRAKE warning light on the dash should illuminate during key-on bulb check (**Figure 11-23**). Once the engine starts, the light should go out. If the light remains on, there are three possible reasons: low brake fluid level in the reservoir, low pressure in the hydraulic system, or the parking brake is applied. An illustration of a BRAKE warning light circuit is shown in **Figure 11-24**. As discussed previously, many vehicles use a fluid level sensor in the master cylinder reservoir. If the fluid level drops to the minimum, the float inside the reservoir closes the circuit and turns on the

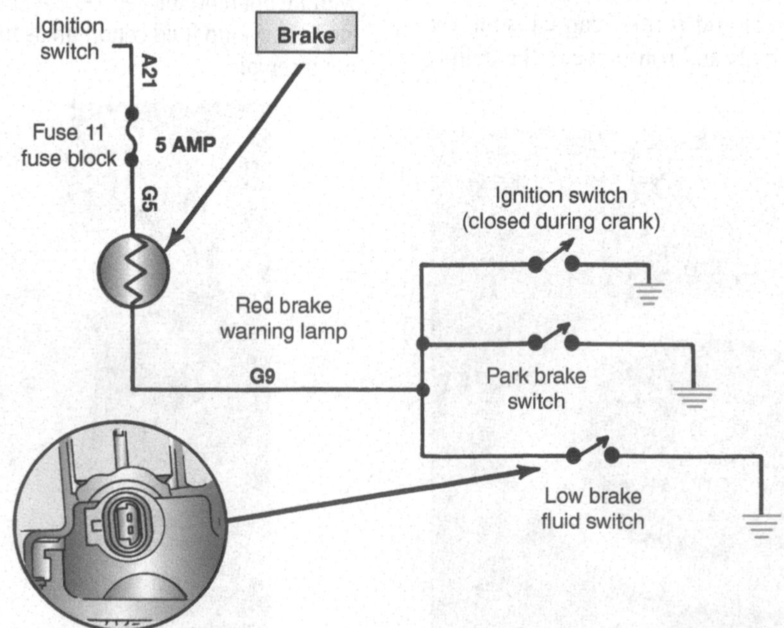

FIGURE 11-24 The BRAKE warning light can illuminate due to low fluid level, an applied parking brake, or low system pressure.

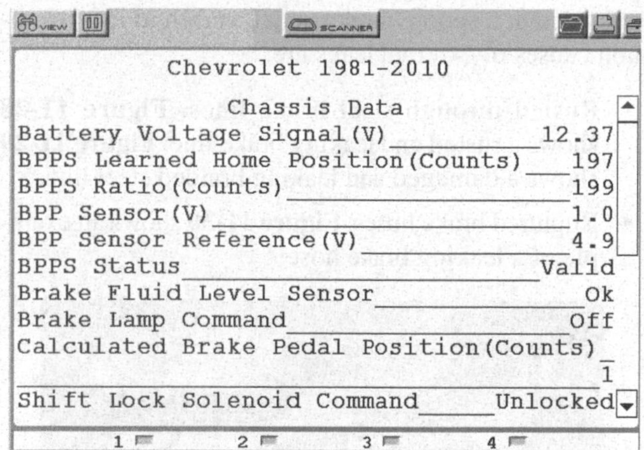

FIGURE 11-25 Scan tool data showing the brake fluid level sensor input and BRAKE light command.

FIGURE 11-26 The rubber master cylinder cap seal in the middle is swollen and distorted from exposure to petroleum-based products in the brake fluid. A normal cap and seal is shown on the right.

warning light. Topping off the brake fluid should turn off the light. If the light remains on after the fluid is refilled, the float may be sticking. Disconnect the sensor connector, and note if the light remains on or goes out. If the light remains on, there may be a problem with the sensor wiring, or another part of the circuit may be turning on the light. On late-model vehicles, you can check the operation of the warning light using a scan tool (**Figure 11-25**).

On older vehicles, a leak in the hydraulic system will unbalance the pressure differential valve and the warning light will illuminate. Once the leak is repaired and the system is bled, the valve should reset and the light will go out. Note that vehicles that use a brake fluid level sensor do not use a brake warning light switch as part of the pressure differential valve.

A switch located at the parking brake handle or pedal is also used to turn on the BRAKE warning light. This is to alert the driver so he or she does not try to drive the vehicle with the parking brake applied. Vehicles with electronic parking brakes also turn on the BRAKE light when the parking brake is applied. Make sure the parking brake is releasing completely before diagnosing a BRAKE light staying on.

If the BRAKE warning light remains on and the fluid level is correct, there are no leaks, and the parking brake is fully releasing, then there may be a short to ground in the light circuit. You will need to locate a wiring diagram and test the circuit to locate the fault. This usually involves disconnecting each part of the circuit until the light goes out.

■ ***Test Brake Fluid Contamination.*** It is critical that only brake fluid is added to the brake hydraulic system. Petroleum-based fluids, such as engine oil and power steering fluid, if allowed to enter the brake system, will cause rapid and irreversible damage to the rubber seals and hoses. These parts will begin to swell and deteriorate, becoming very spongy, distorted, and enlarged, and the brakes will not operate. An example of how this affects the rubber parts is shown in **Figure 11-26**. If this occurs, every component that contains rubber will have to be replaced and all the steel lines flushed. As you can imagine, this is a very expensive repair.

Brake fluid also becomes contaminated with moisture. Because glycol-based brake fluid is hygroscopic, it will absorb moisture over time, lowering the boiling point of the fluid. The amount of moisture the fluid has absorbed can be tested with a brake fluid refractometer, special test strips, or a brake fluid moisture tester (Figure 11-6). Once the moisture level reaches about 3%, the boiling point of the fluid is reduced more than 100°F. This amount of moisture accumulation can occur in just a couple of years, so periodic flushing is important.

As the corrosion inhibitors in the brake fluid diminish, corrosion within the system can begin to form. The internal components, such as in the ABS solenoid valves, require very close tolerances to operate properly. If the space between these components is clogged from corrosion or rust, the ABS will not be able to function properly or may not work at all.

To test the brake fluid, remove the reservoir cap and install the moisture tester into the fluid. If the reservoir has a screen preventing the tester from going down into the fluid, remove a small amount of fluid, and place it into the container that comes with the tester. Refer to the tester's operating manual for the specific steps for using the tester. These steps usually require connecting the tester to the battery and pressing and holding the TEST button. Once the fluid has been tested, note the results (**Figure 11-27**).

If you are using fluid test strips, place the strip with the test pads down into the fluid for 1 second. Remove the strip and check the color of the test pads against the chart with the test strips. If corrosion is indicated or if the fluid has a low boiling point, it should be flushed out and new fluid installed.

Hydraulic System Inspection and Service

When a vehicle needs inspected or brake work performed, perform a complete inspection of all brake system components and their operation. It is common for a vehicle to have more than one brake problem. Performing only a partial inspection, such as just removing the front wheels to check the front brakes, is unprofessional and dangerous as other problems may remain undiscovered.

HYDRAULIC SYSTEM INSPECTION

Problems in the hydraulic system will typically be in one of these areas: external fluid loss, blocked hoses, internal seal bypassing in the master cylinder, or a faulty valve. When you suspect a hydraulic system problem, as indicated by concerns such as a very low, spongy brake pedal, begin your inspection for external leaks from the brake system components, brake lines, and hoses.

■ ***Inspect Brake Lines.*** Problems with the hydraulic system range from the obvious, such as leaks, to the less obvious, such as restricted brake hoses and seized caliper or wheel cylinder pistons. Leaks result in a loss of hydraulic pressure and braking ability, a very low, soft brake pedal, a spongy brake pedal, and fluid loss. Common causes of external leaks are:

- Rusted through steel brake lines. **Figure 11-28** shows a rusted and leaking brake line. **Figure 11-29** shows a damaged and leaking braided steel line.
- Ruptured brake hoses. **Figure 11-30** shows an example of a leaking brake hose.

FIGURE 11-27 A moisture tester in use. This type of tester heats the fluid. Use caution to not burn your fingers when using this tool.

FIGURE 11-28 A rusted and leaking rear brake line.

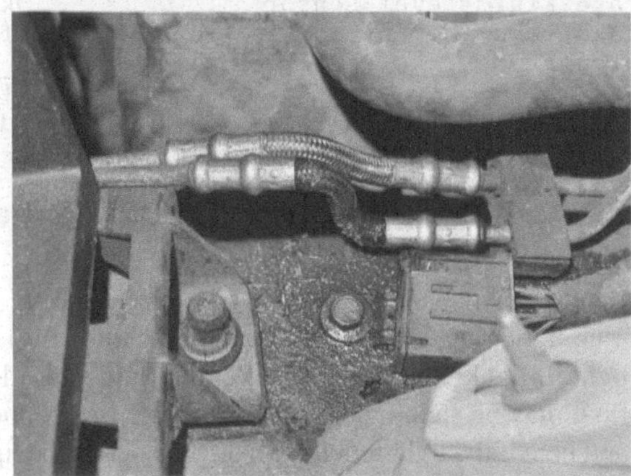

FIGURE 11-29 A leaking steel brake line caused by a faulty battery ground connection.

FIGURE 11-30 An example of a leaking front brake hose.

FIGURE 11-31 A leaking master cylinder primary seal allows brake fluid to run out onto the power booster. The vacuum booster shows evidence of the leak due to the brake fluid damaging the paint.

- Leaking master cylinder. **Figure 11-31** shows where fluid has leaked from the rear of the master cylinder onto the power brake booster.
- Calipers. **Figure 11-32** shows a leaking rear caliper.
- Wheel cylinders. **Figure 11-33** shows a leaking wheel cylinder.

Another cause for inadequate hydraulic pressure is restricted or clogged rubber brake hoses. Over time, the interior of the brake hose can deteriorate and begin to come apart (**Figure 11-34**). Pieces of rubber can dislodge and clog the hose, restrict or even block the hose completely. This can keep a wheel brake or both rear brakes from applying with enough pressure, or prevent them from engaging at all if the hose is completely blocked. If a front caliper hose is blocked, only one caliper will apply. This can cause a serious pull in the direction of the operating caliper when the brakes are applied. If a rear hose is blocked that feeds both rear brakes, neither rear brake receives pressure, forcing the front brakes to carry the full load of stopping the vehicle. This condition will cause the vehicle to nose dive when stopping and will wear the front brake components much faster than normal.

Brake hoses can also fail internally and keep pressure applied to a wheel brake after the brakes are released. Also, rust can form between the hose and mounting brackets built into the hose, putting pressure on the hose and trapping fluid in the hose. If the pressure is not released from a brake caliper, the brake stays applied. This causes the vehicle to pull in the direction of the stuck brake and also causes overheating and very rapid brake pad wear.

Over time, enough moisture can be absorbed by the fluid to rust pistons into their bores. This can happen to either caliper or wheel cylinder pistons. To check for seized pistons, try to push the pistons back into the caliper

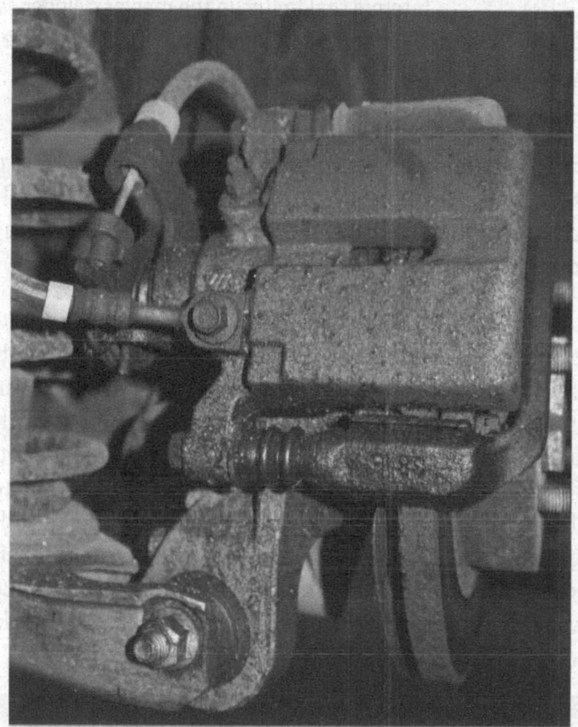

FIGURE 11-32 A leaking rear brake caliper.

FIGURE 11-33 A leaking wheel cylinder .

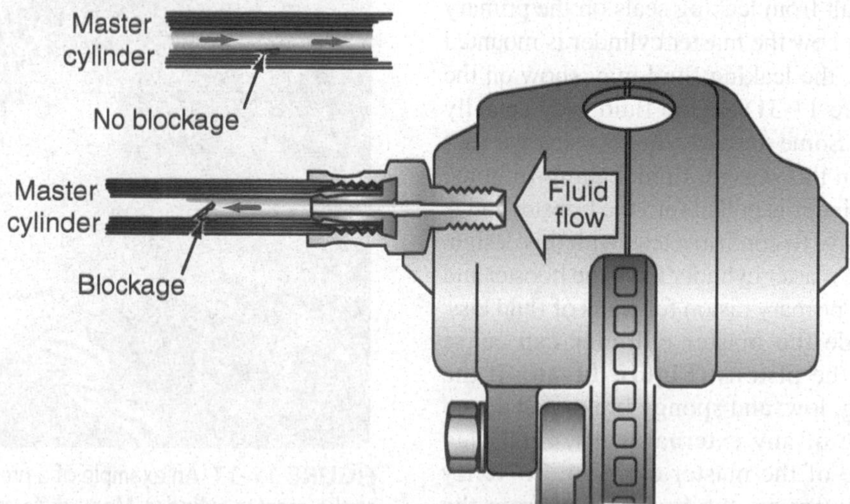

FIGURE 11-34 Brake hoses can become restricted and block fluid from passing to the wheel brake or keep the fluid from releasing pressure in the wheel brake.

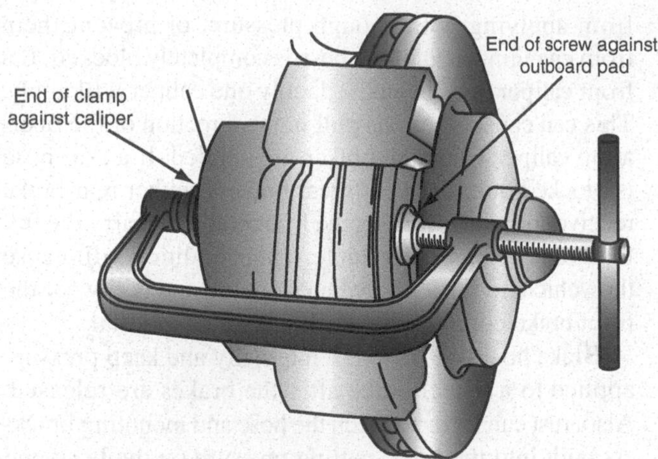

FIGURE 11-35 Checking for a seized caliper piston by attempting to retract it with a C-clamp. Do not try this on a rear integral brake caliper, like that shown in Figure 11-32.

with a C-clamp (**Figure 11-35**). To check wheel cylinder pistons, remove the drum and try to push the pistons back into the cylinder when the brakes are being inspected. If the pistons will not move back into the cylinder, the wheel cylinders need to be either rebuilt or replaced.

A quick way to check the hydraulic system is to raise the vehicle on a hoist and to have a helper press the brake pedal, and try to rotate each tire. If a tire or tires rotate with the brakes applied, you should perform a thorough hydraulic and service brake inspection and determine the cause for the brake's not holding.

■ ***Check Master Cylinder for Internal and External Leaks.*** Even though master cylinders are more reliable than ever, the seals can harden over time or can be damaged from contaminants in the brake fluid, and begin to leak. A master cylinder can leak either externally or internally, causing conditions ranging from a low, sinking brake pedal to a complete loss of brake pressure.

External leaks result from leaking seals on the primary piston. Depending on how the master cylinder is mounted to the power booster, the leaking fluid may show on the power booster (Figure 11-31), or the fluid may actually leak into the booster. Some master cylinders seal the vacuum in the booster. In this design, fluid leaking from the rear of the master cylinder is pulled into the booster and is not be visible externally. To confirm a leak with this design, you need to unbolt the master cylinder from the booster and inspect the rear of the primary piston for signs of fluid loss.

Seal failure inside the master cylinder can cause the fluid to bypass the pistons (**Figure 11-36**). If the vehicle has a sinking, low, and spongy brake pedal, and there is no evidence of any external leaks, fluid may be **bypassing** inside of the master cylinder. To verify if this is occurring, remove the brake lines from the master cylinder and plug the outlet ports with threaded

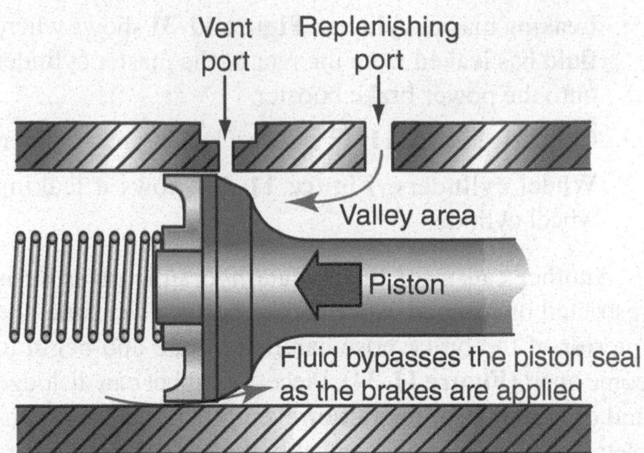

FIGURE 11-36 Internal seal failure in the master cylinder can allow the fluid to bypass the seal and piston. This will cause a low or sinking pedal and poor stopping performance.

plugs. This isolates the master cylinder from the rest of the hydraulic system. Next, press the brake pedal. If the pedal continues to drop, the internal seals are leaking, and the master cylinder needs to be replaced.

■ ***Master Cylinder Replacement.*** To replace the master cylinder, first install fender covers on the vehicle and place a shop pan under the master cylinder. Remove the reservoir cap and remove most of the brake fluid and discard it. Using the correct sized line wrenches, loosen the brake line fittings at the master cylinder ports. Next, loosen and remove the mounting nuts that secure the master cylinder to the brake booster. Disconnect the fluid level sensor wiring if necessary. Remove the fittings from the master cylinder and remove the cylinder from the vehicle.

FIGURE 11-37 An example of a pressure bleeder connection at the master cylinder. Many different types of adapters are available to fit different types of caps.

When you are replacing a master cylinder, the replacement should be bench bled before being installed on the vehicle. **Bench bleeding** removes the air, prefills the master cylinder, and reduces the amount of bleeding that needs to be done on the vehicle (**Figure 11-37**)

Once the master cylinder has been bench bled, reinstall it on the vehicle. Be sure you have fender covers installed to protect the vehicle. If possible, place a shop pan under the master cylinder to catch any fluid that leaks out during installation. Thread the brake lines into the outlet ports first. This allows for easier alignment and less chance of cross-threading the fittings if the master cylinder can be moved as needed to align with the fittings. Once the fittings are started by hand, install the master cylinder over the mounting studs on the power assist unit. Tighten the fittings and torque the mounting nuts to specs. The hydraulic system needs to be bled to remove any air trapped in the system. Bleeding the hydraulic system is discussed later in this chapter.

■ ***Replacing Brake Lines and Hoses.*** As discussed in Chapter 10, brake lines and hoses are manufactured to withstand the high pressure of the hydraulic system. When you are replacing brake lines and hoses, the correct size line, types of fittings, and replacement parts must be used.

A rusting and leaking brake line must be replaced. Do not attempt to repair a leaking steel line with a compression fitting. The pressure in the brake system can rupture any type of patch job or compression fitting, resulting in brake failure.

For some vehicles, replacement brake lines are available. These premade lines are bent, flared, and ready to install. However, in many cases, you will have to make your own custom line out bulk line or from sections of line. If you are replacing a section of brake line, like a rear brake line, start by determining the size and type of line that is needed. Brake line is sold in pieces that can range from as small as 6 inches long to bulk line 25 feet long (**Figure 11-38**). Brake line is also sized by its diameter and by the type of flare used. Brake line can be either SAE or ISO, which means it is measured in fractions of an inch or in millimeters. SAE line has a double flare, and ISO line uses a bubble flare. The two types of lines and fittings are not interchangeable (**Figure 11-39**).

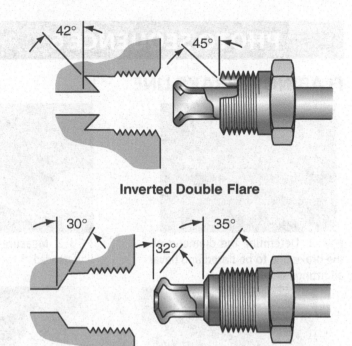

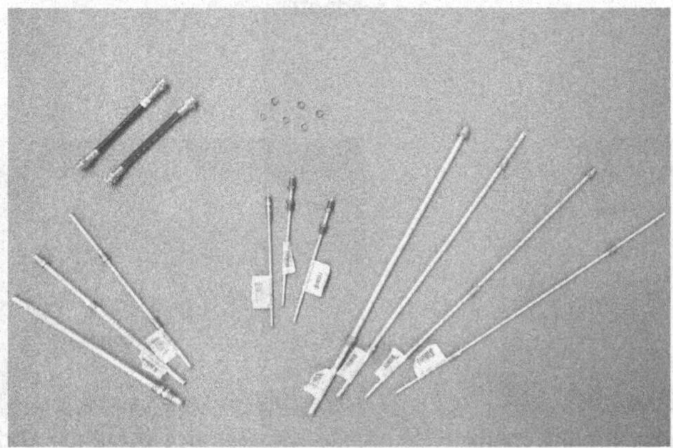

FIGURE 11-38 A selection of replacement brake line.

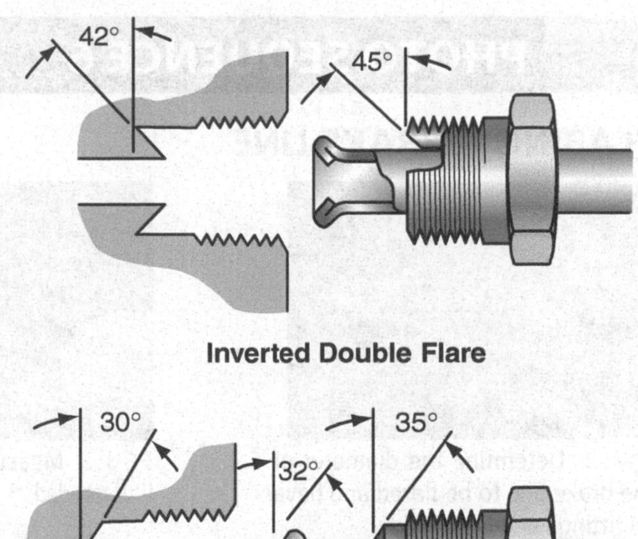

FIGURE 11-39 Examples of the SAE double flare and ISO bubble flare. The two types of flares and fittings are not interchangeable.

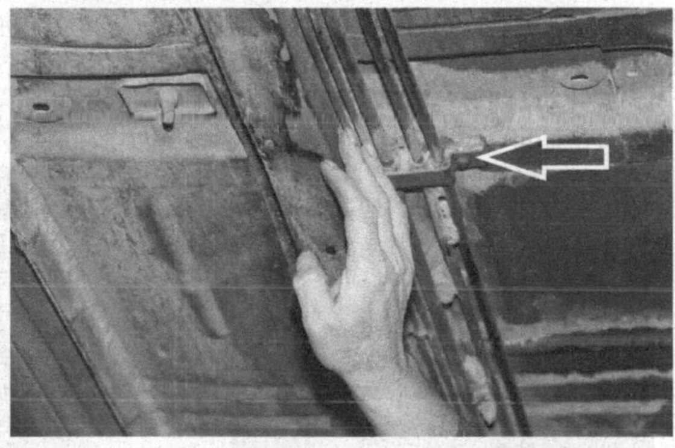

FIGURE 11-40 When replacing a brake line, remove the clamps that hold the line to the frame or body. Reinstall these clamps with the new line.

First remove the damaged section of brake line. You will need to remove it from clamps that secure the brake (and possibly fuel lines) to the body or frame (**Figure 11-40**). If you are replacing an entire section of line, from connection point to connection point, remove the line and compare the fittings of the new line to the old line. Ensure the replacement fittings have the correct type of flare and threads. To make removal of severely

PHOTO SEQUENCE 6

FLARING A BRAKE LINE

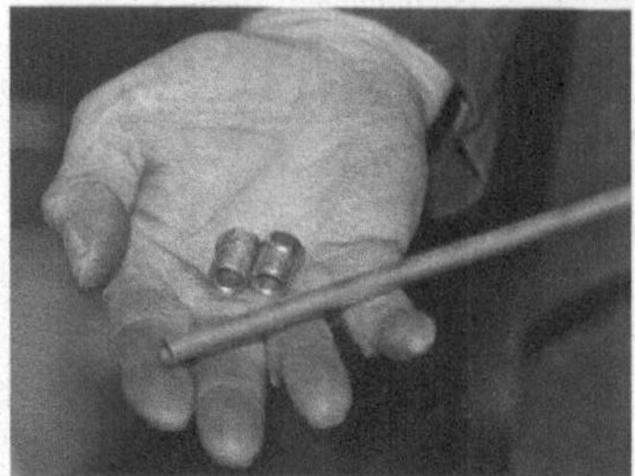
PS6-1 Determine the diameter of the brake line to be flared and have all fittings and lines ready.

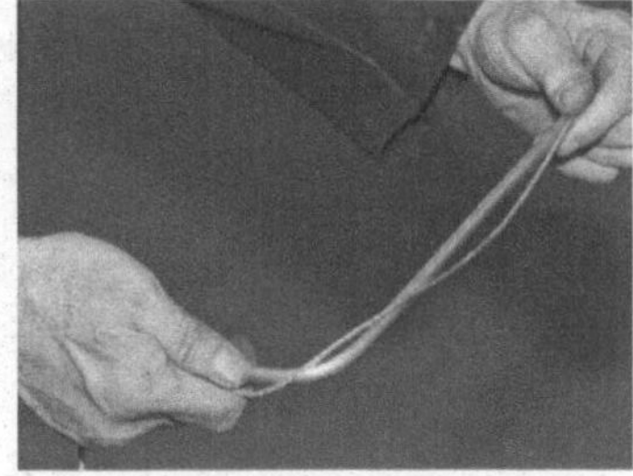
PS6-2 Measure the length of line needed.

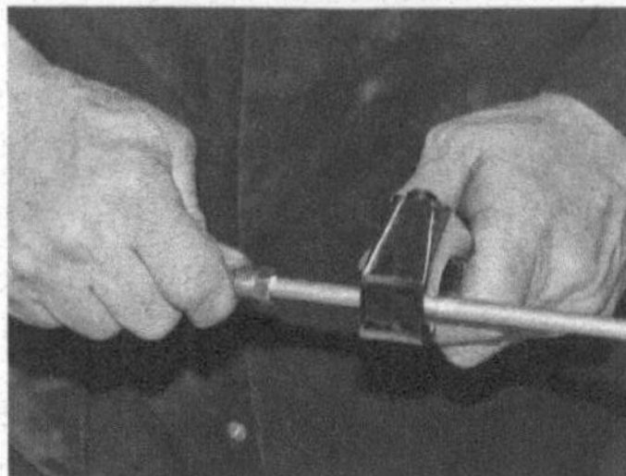
PS6-3 Using a tubing cutter, cut the line to length. Do not use a hacksaw or cutting pliers as these will damage the line.

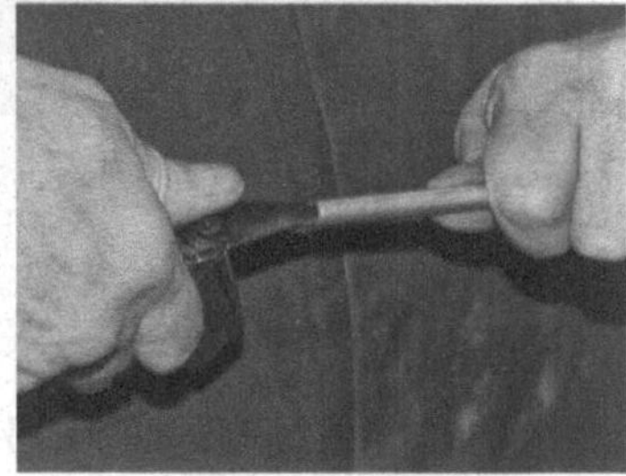
PS6-4 Deburr the end of the line and ensure that the opening is round and not damaged by the tubing cutter. You may need to ream the opening slightly to remove any metal that may obstruct the opening. Most tubing cutters have a reaming tool built into the cutter.

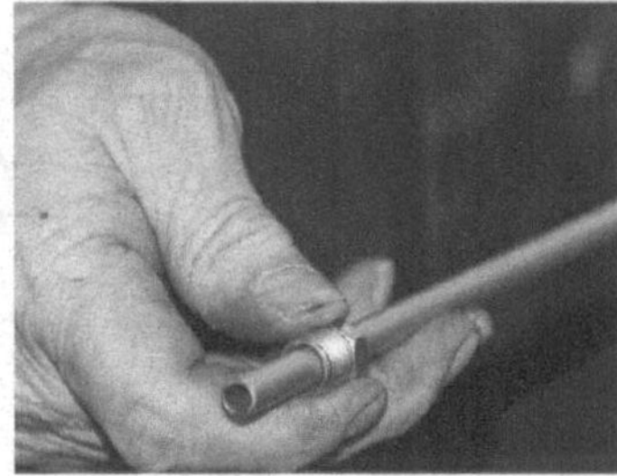
PS6-5 Install the fitting onto the line and ensure that it is going in the right direction.

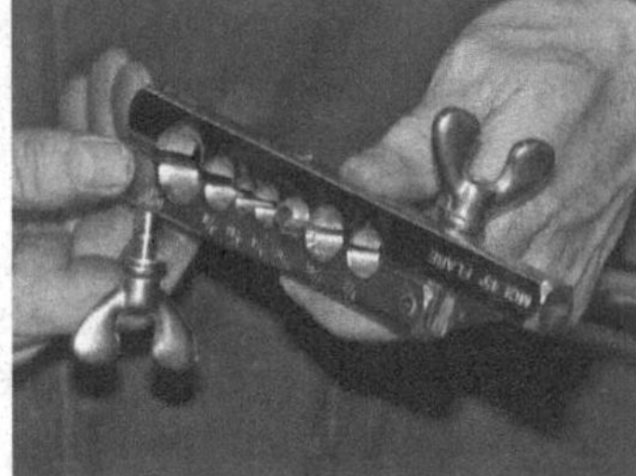
PS6-6 Place the line into the clamping fixture. Check the directions for the flaring tool to determine how much line to extend out of the fixture.

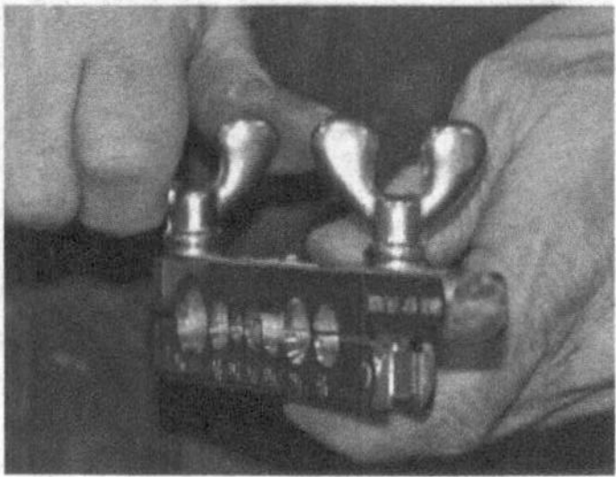
PS6-7 Once properly positioned, tighten the clamps to secure the line in the fixture. Do not overtighten the clamps as this will break the pins in the clamp. However, failure to tighten the clamps enough will allow the line to push through the clamp while flaring the end.

PS6-8 Install the adapter for the size of line being flared.

PS6-9 Install the anvil over the clamping fixture and against the adapter.

PS6-10 Tighten the anvil until the adapter is flush with the clamping fixture. Remove the anvil and adapter.

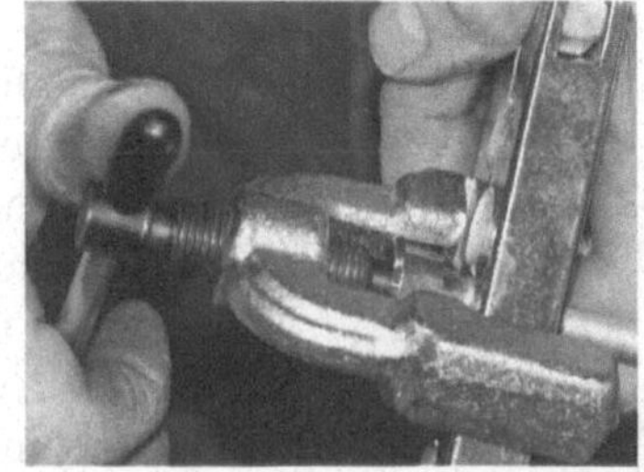
PS6-11 Place the anvil into the opening of the brake line and tighten to roll the flared metal into the opening.

PS6-12 Remove the anvil and inspect the flare.

rusted lines easier, cut the line just behind the fittings where the line is to be disconnected. Use a socket to remove the fitting while holding the other section of line in place with a wrench.

If you are replacing a section of line between connection points, you may need to cut the new line to length and flare the ends of the lines to install a union. A **flare tool** is used to shape the end of the lines so that the flared end matches exactly with the fittings to provide a leak-free connection. **Photo Sequence 6** shows how to create a double flare.

Once the new line is ready to install, carefully begin to thread the fitting into the connection. Start the fitting by hand and continue to tighten it until the fitting is seated. You should not have to use a wrench to start the fitting. If it is difficult to thread into the connection, remove the fitting and inspect the threads for dirt or damage. Be careful not to overtighten the fitting; this can crack the flare or strip the threads, resulting in a leak. Once the line is installed, refill the master cylinder reservoir and bleed the brakes. Recheck for leaks once the system is bled.

When replacing brake hoses, you need to remove the hose from its bracket where it connects to the steel line. Hoses are usually held in place with a clip (**Figure 11-41**). Use a line wrench when you remove the fittings, and hold the hose with a second wrench (**Figure 11-42**). This helps prevent twisting and damaging the lines and hoses and reduces the chances of stripping the fittings.

Compare the new and old hoses before you install the new ones. Make sure the new hose matches the old hose at the fittings and brackets. Most front brake hoses are specific to the left and right sides of the vehicle, so make sure you have the correct parts before you try to install the hose. Install the hose by threading the fitting together by hand. Once seated, tighten the fitting using a line wrench. Most front brake hoses use a hollow bolt to

(a)

(b)

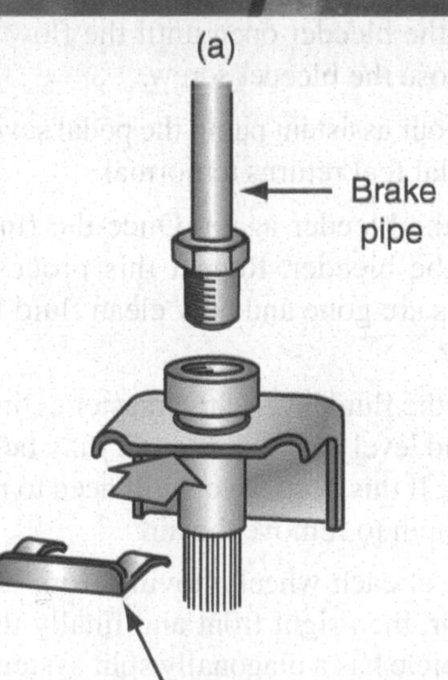

(c)

FIGURE 11-41 (a) Brake hoses are often bolted and clipped to the body, (b) strut, or other components to prevent contact with moving parts, such as the wheels and tires. (c) Be sure to reinstall all mounting bolts and clips when installing a new hose.

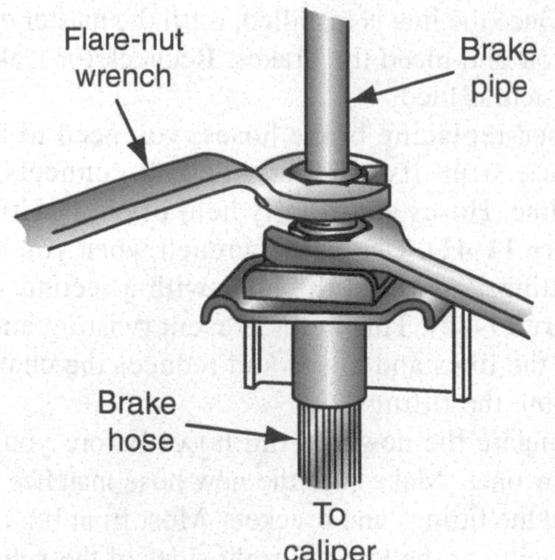

FIGURE 11-42 Use line wrenches to steady the line or hose when removing. This helps prevent damage to the fittings and line.

secure the hose to the caliper. Install new washers, one between the bolt and hose and one between the hose and the caliper, and thread the bolt into the caliper. Make sure to tighten the hose bolt to specs. If a little loose, the connection will leak. Overtightening a little can snap the bolt in two.

Be sure you reconnect all brackets and clips that hold the hose in place during reinstallation. These brackets and clips are necessary to keep the hose from being damaged by contacting moving parts and tires. Once complete, fill the reservoir with new brake fluid and bleed the air from the system.

HYDRAULIC SYSTEM FLUSHING AND BLEEDING

A **brake fluid flush** removes all the old fluid and replaces it with new. This is done as a maintenance service to prolong the life of the hydraulic brake components. Bleeding the brakes removes air trapped in the system, whether from a leak or from repairs. If the system is opened for bleeding, many technicians perform a flush at the same time.

■ ***Reasons to Flush the Brake System.*** Because brake fluid is hygroscopic, it should be flushed periodically to prevent rust and corrosion buildup within the system. Car and truck manufacturers often recommend the brake system be flushed every two years and whenever brake services are performed. Rust or corrosion of the hydraulic components can cause the ABS to malfunction, cause the pistons to bind or seize in calipers and wheel cylinders, and increase the rate of deterioration of metal components. Additionally, because water has a much lower boiling point than brake fluid, moisture trapped in the system can cause brake fade or a loss of braking if the fluid boils.

■ ***Bleeding the Brake System.*** Bleeding the brakes removes air from the system and removes old fluid. Bleeding and flushing are really two operations but are often performed as one since both remove old fluid and any air trapped in the system. There are several methods for bleeding the brake system, including manual and pressurized procedures.

Manual bleeding usually requires two people, one to pump the brakes and the other to open and close the bleeder screws. Many vehicles have specific bleeding procedures that need to be followed, so refer to the correct service information before bleeding the brake system. Some vehicles can be bled by using the following general procedure:

1. Remove as much of the old fluid from the master cylinder reservoir as possible using a syringe or suction tool. Fill the master cylinder with new brake fluid. Reinstall the reservoir cap(s).
2. Start at the right rear wheel. Install the correct size box-end wrench on the bleeder screw. Place a hose over the bleeder screw and put the other end into a bottle or other container to catch the old fluid (**Figure 11-43**).
3. Have your assistant pump the brake pedal several times and hold the pedal down. Open the bleeder using the box-end wrench. When the bleeder is opened, the pedal will drop. Do not allow the pedal to go completely to the floor.
4. Leave the bleeder open until the flow of fluid stops, then close the bleeder screw.
5. Have your assistant pump the pedal several times until the pedal feel returns to normal.
6. Open the bleeder again. Once the fluid flow stops, close the bleeder. Repeat this process until all air bubbles are gone and only clean fluid flows from the bleeder.
7. Check the fluid level at the master cylinder. Do not let the fluid level get too low or air may be pulled into the system. If this occurs, you will need to repeat bleeding the system to remove the air.
8. Repeat at each wheel, moving from the right rear to left rear, then right front and finally the left front. If the vehicle has a diagonally split system, the bleeding sequence is usually RR, LF, LR, and RF. When complete, all air should be removed from the system and only new, clean fluid present.

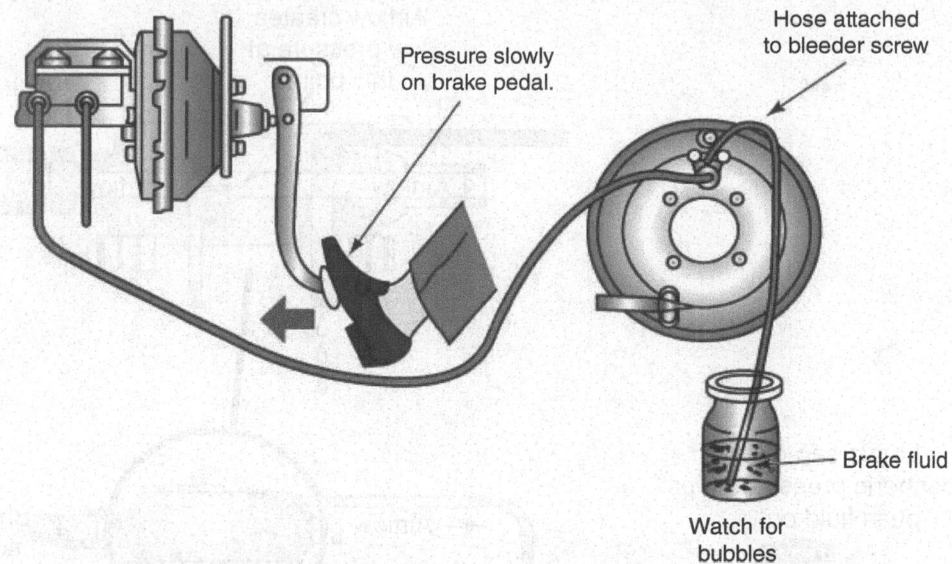

FIGURE 11-43 This illustrates manual bleeding to remove air from the hydraulic system. Using a container of clean brake fluid allows one person to bleed the brakes.

Another way to bleed and flush the brakes is like the previous procedure, but instead of opening and closing the bleeder only to let fluid out, the bottle used to catch the fluid is used to also hold a reservoir of fluid.

1. Remove as much of the old fluid from the master cylinder reservoir as possible using a syringe or suction tool. Fill the master cylinder reservoir with new fluid. Reinstall the reservoir cap(s).
2. Install a hose over the bleeder, and place the other end of the hose into the bottle. Fill the bottle half full with clean new fluid. Open the bleeder screw at the right rear wheel.
3. Pump the brake pedal repeatedly until the new fluid entering the bottle no longer has any air bubbles.
4. Close the bleeder and refill the master cylinder as needed. Then proceed to the left rear wheel and repeat. Bleed all four-wheel brakes. When bleeding is complete, only new fluid should remain in the system and all air will be removed.

Another method of bleeding and flushing uses a vacuum evacuation tool (Figure 11-3). This tool is used to pull the fluid through the brake system to remove air and to flush out the old fluid. An illustration of how **vacuum bleeding** works is shown in **Figure 11-44**.

1. Remove as much of the old fluid from the master cylinder reservoir as possible using a syringe or suction tool. Fill the master cylinder with new fluid. Leave the reservoir cap loose.
2. Open the right rear bleeder screw, and install the hose of the vacuum bleeder over the bleeder screw.
3. Attach a shop air hose to the bleeder air fitting and pull the trigger. As the air flows through the vacuum bleeder, brake fluid is pulled from the system and is collected in the tool's reservoir.
4. Continue to bleed until all air is removed and clean brake fluid is flowing from the bleeder. Remember to check the fluid level in the master cylinder often to prevent air from being pulled into the system.
5. When complete, close the bleeder screw and move to the left rear wheel.

When using a vacuum bleeder, you will likely see small air bubbles or foam being pulled into the bleeder even after all of the air has been removed from the hydraulic system. This is because the vacuum pulled through the bleeder valve can pull air from around the threads of the valve, causing the fluid to foam (**Figure 11-45**). Once you are finished with the vacuum bleeder, manually bleed each wheel once to ensure that no air remains trapped in the system.

Another way to flush and bleed the brake system is by using a pressure bleeding tool. **Pressure bleeding**, as the name implies, uses pressure to force new fluid through the brake system. A special adapter is installed over the master cylinder reservoir (**Figure 11-46**). Many different adapters are available to fit various master cylinder arrangements. The pressure tank is then filled with brake fluid, and an air hose is connected to the tank. Pressure is usually limited to less than 30 psi. When a bleeder screw is opened, the pressure on the system forces air and fluid out. This method requires only one person to bleed the brake system, so

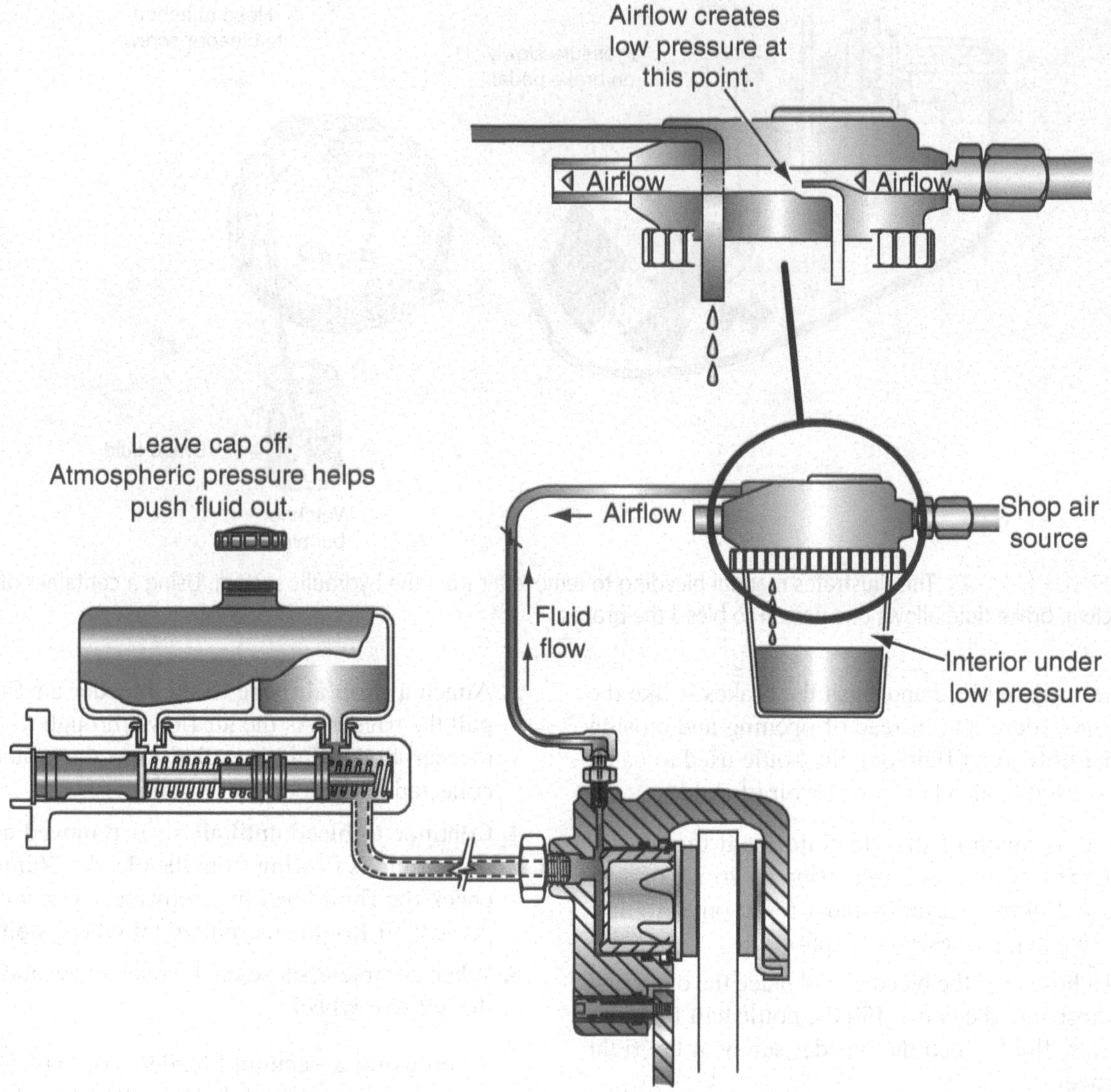

Airflow creates
low pressure at
this point.
Airflow
Airflow
Leave cap off.
Atmospheric pressure helps
push fluid out.
Airflow
Shop air
source
Fluid
flow
Interior under
low pressure

FIGURE 11-44 A vacuum bleeder uses shop air to pull a vacuum at the bleeder valve. This tool allows the brakes to be bled by one person.

Vacuum draws
foamy fluid
into bleed
canister.
Bleed screw
Air drawn
past threads
Fluid mixed
with air
Fluid from
wheel circuit

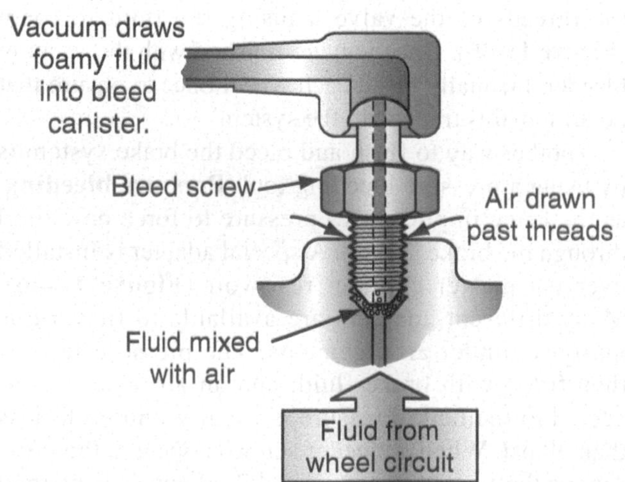

FIGURE 11-45 Very small bubbles often develop around the bleeder screw when using a vacuum bleeder.

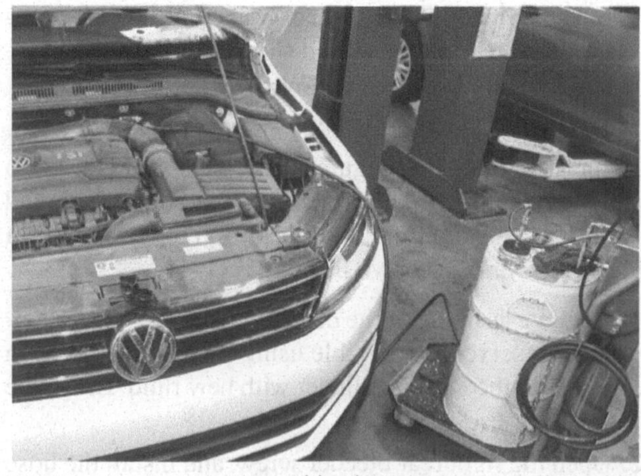

FIGURE 11-46 A pressure bleeder connection to the master cylinder.

FIGURE 11-47 The technician hangs the fluid catch bottle from the car with the bleeder open, leaving him or her to perform other tasks while the brakes are being flushed.

it is convenient in situations where a helper is not available (**Figure 11-47**).

■ ***Bleeding ABS.*** The procedures listed previously apply to most non-ABS vehicles and may also be used on many ABS-equipped cars and trucks. Before attempting to bleed the brake system on an ABS-equipped vehicle, refer to the manufacturer's service procedures. Some ABS, especially those that have integral ABS, require special bleeding procedures that must be followed to prevent vehicle damage and even personal injury. An example of an ABS service warning decal is shown in **Figure 11-48**.

Some vehicles require the use of a scan tool connected to the on-board computer system to activate a specific hydraulic bleeding mode before the system can be bled. **Figure 11-49** shows one example of this scan tool procedure. Once the system bleed is started, follow the onscreen commands to bleed the air from the system.

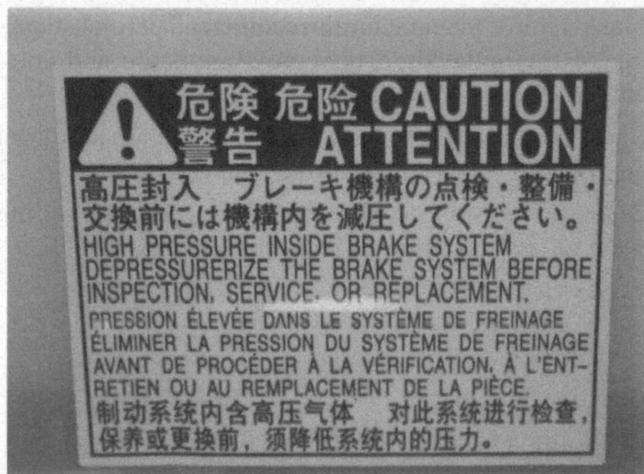

FIGURE 11-48 An example of an ABS high-pressure warning decal. Always follow the manufacturer's service procedures when working on the ABS to prevent injury and damage to the system.

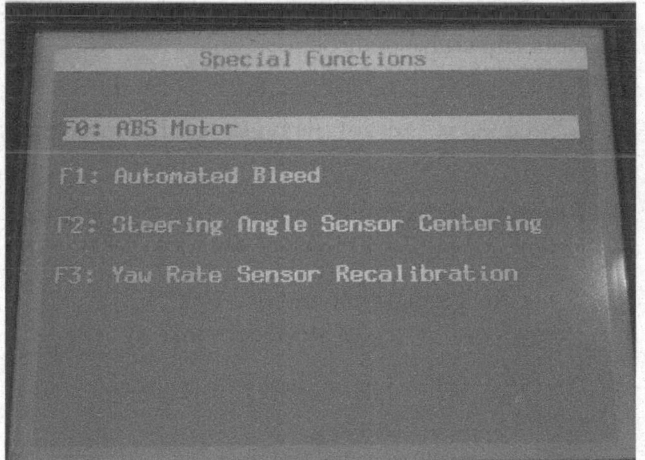

FIGURE 11-49 Some vehicles require using a scan tool to activate the ABS to bleed the hydraulic system.

SUMMARY

If the brake pedal height is incorrect, check the rear brake adjustment.

Inspect adjustable pedals by activating the system, and check that the assembly can move forward and backward along the entire range of motion.

Many vehicle manufacturers recommend periodic flushing of the brake fluid to remove moisture and other contaminants.

Clean around the reservoir cap to prevent any dirt from getting into the brake fluid when the cap is removed.

Once the moisture level reaches about 3%, the boiling point of the brake fluid is reduced more than 100° F.

Leaks will result in a loss of hydraulic pressure and braking ability, a very low, soft brake pedal, a spongy brake pedal, and fluid loss.

A master cylinder can leak either externally or internally.

Bench bleeding removes air, prefills the master cylinder, and reduces the amount of bleeding that needs to be done on the vehicle.

Bleeding the brakes removes air from the system and removes the old fluid.

Some vehicles will require the use of a scan tool to activate a specific hydraulic bleeding mode.

REVIEW QUESTIONS

1. Because brake fluid is ______________, it will absorb moisture.

2. The brake ______________________________
____
____
sensor is also used as an input for engine, transmission, and cruise control operation.

3. Many manufacturers recommend periodic ______________ of the brake fluid to remove moisture and contaminants.

4. Bleeding the hydraulic system removes trapped ______________.

5. Some ABS require the use of a ______________ ______________ to bleed the brake system.

6. All of the following are brake bleeding methods except:
   - **a.** Pressure
   - **b.** Vacuum
   - **c.** Reverse
   - **d.** Manual

7. *Technician A* says low brake pedal height can be caused by worn brake pads and rotors. *Technician B* says a low, sinking brake pedal can be caused by worn brake shoes and pads. Who is correct?
   - **a.** Technician A
   - **b.** Technician B
   - **c.** Both A and B
   - **d.** Neither A nor B

8. The red BRAKE warning light is illuminated on a vehicle with four-wheel disc brakes. *Technician A* says the parking brake lever may be the cause. *Technician B* says this means there is a problem with the stop light system. Who is correct?
   - **a.** Technician A
   - **b.** Technician B
   - **c.** Both A and B
   - **d.** Neither A nor B

9. All of the following are common hydraulic system concerns except:
   - **a.** Restricted brake hose
   - **b.** Brake line rust through
   - **c.** Master cylinder bypassing
   - **d.** Seized master cylinder pistons

10. *Technician A* says bleeding an ABS may require a scan tool. *Technician B* says vacuum bleeding requires using a special tool connected to the shop compressed air system. Who is correct?
    - **a.** Technician A
    - **b.** Technician B
    - **c.** Both A and B
    - **d.** Neither A nor B

CHAPTER 12

Drum Brake System Principles

Chapter Objectives

At the conclusion of this chapter, you should be able to:

- Identify servo and nonservo drum brake designs.
- Describe the operation of servo and nonservo brake systems.
- Describe the types and operation of drum parking brake systems.

KEY TERMS

anchor	**duo-servo**	**return springs**
backing plate	**holddown springs and pins**	**self-adjuster assembly**
brake drum	**leading–trailing brakes**	**wheel cylinder**
brake shoes	**parking brakes**	

Drum brakes have been in use since the earliest days of the automobile. Even though most new vehicles come with four-wheel disc brakes, drum brakes are still in production and millions of drum-equipped vehicles are on the road. Drum brakes are simple in operation and can, by using leverage, apply more braking force than what is supplied by the driver.

Drum Brake Design and Operation

Drum brakes have had to adapt from purely mechanical applications to modern hydraulic and electronically controlled systems. Modern systems use longer lasting friction linings, but the overall design and operation has not changed.

DRUM BRAKE DESIGN

Drum brakes systems, such as disc brakes, all share similarities in design and operation. On cars and light-duty trucks, the hydraulic drum brake system has not changed very much since hydraulics replaced mechanical brake linkages in the 1930s. The same basic components are still used, and they perform the same functions regardless of the vehicle.

■ ***Advantages and Disadvantages.*** Drum brakes (**Figure 12-1**), use a set of brake shoes that expand outward against the inside of a rotating brake drum. Hydraulic pressure pushes against pistons in the wheel cylinder. The pistons then move, pressing the shoes outward against the drum. When the brake pedal is released, return springs pull the shoes back to their rest position.

One advantage of drum brakes is that when applied, the energy of the rotating drum can be used so that the forward shoe acts as a lever against the rear shoe, increasing the amount of brake application force. This allows the drum brake to achieve greater stopping force than what is supplied by the driver pushing on the brake pedal. Because of this advantage, vehicles with four-wheel drum brakes do not require a power-assisted brake system like those using disc brakes.

FIGURE 12-1 An example of a common drum brake assembly.

Disadvantages of drum brakes include mechanical brake fade as the drum expands from heat generation, poor heat dissipation compared to disc brakes, dust buildup within the drum that often causes brake noise, and the need for an adjusting system to maintain shoe-to-drum clearance and correct brake pedal height.

■ ***Components.*** Most modern drum brake systems use the following components.

- **Backing plate**—The plate, attached to the knuckle or axle housing, holds the components of the drum brake assembly. A backing plate is stamped steel and has various holes for springs, the parking brake cable, and wheel cylinder attachment, and support pads for the shoes (**Figure 12-2**). The channel or labyrinth seal around the outside of the backing plate may be used to keep water from entering the brake assembly.
- **Brake shoes**—The shoes are the metal backing on which the lining material is attached (**Figure 12-3**). The brake lining is either riveted or glued to the shoe (**Figure 12-4**). Depending on the brake design, all four shoes may be the same, allowing them to be installed on either side in either the leading or the trailing position. Some applications require a specific shoe be used in one location only. The brake shoe lining material varies but often contains abrasives, such as aluminum, iron, and silica; friction modifiers including, graphite and ceramic compounds, fillers, and binders. For many years, linings contained asbestos to improve friction at very high temperatures. Research showed that asbestos dust could lodge in the lungs and lead to lung cancer. Because of the health concerns, asbestos was phased out of use in brake linings. However, it is possible that linings produced outside of the United States contain asbestos, and it is impossible to know whether asbestos dust is present in the brake system. For this reason, special brake dust collection and containment procedures are used when working on the brake system. These procedures are discussed in Chapter 13.
- **Brake drum**—The drum has an internal friction surface for the shoes to rub against. Compared to a brake rotor, a drum has significantly less contact area (**Figure 12-5**). However, the contact area between the shoes and drum is large, much larger than the contact area between brake pads and the

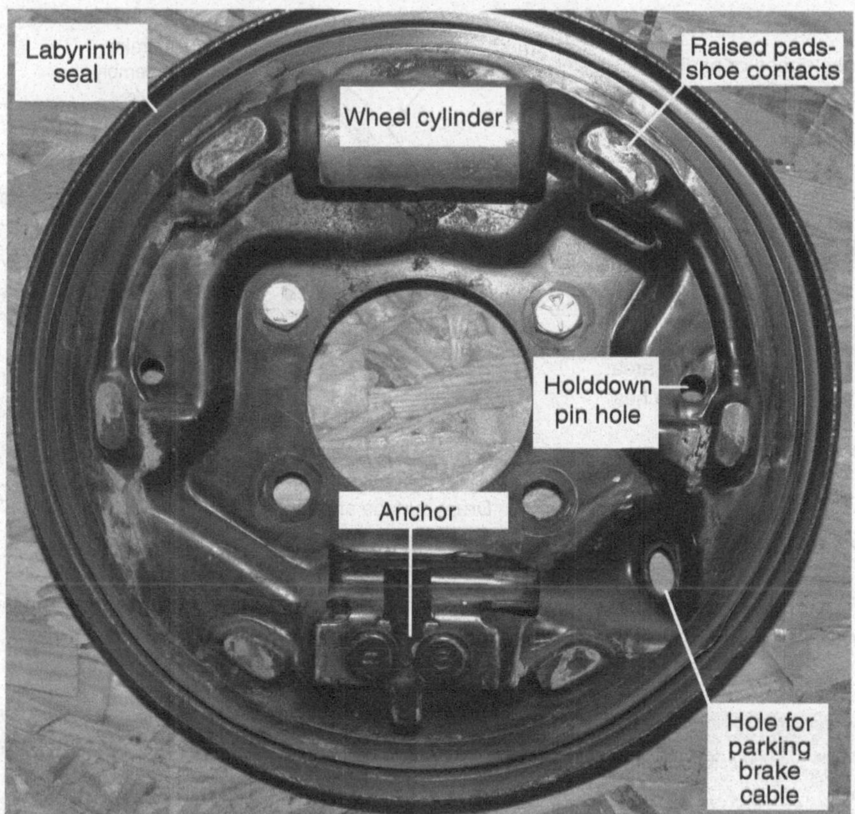

FIGURE 12-2 The backing plate holds the drum brake components and bolts to an axle or knuckle.

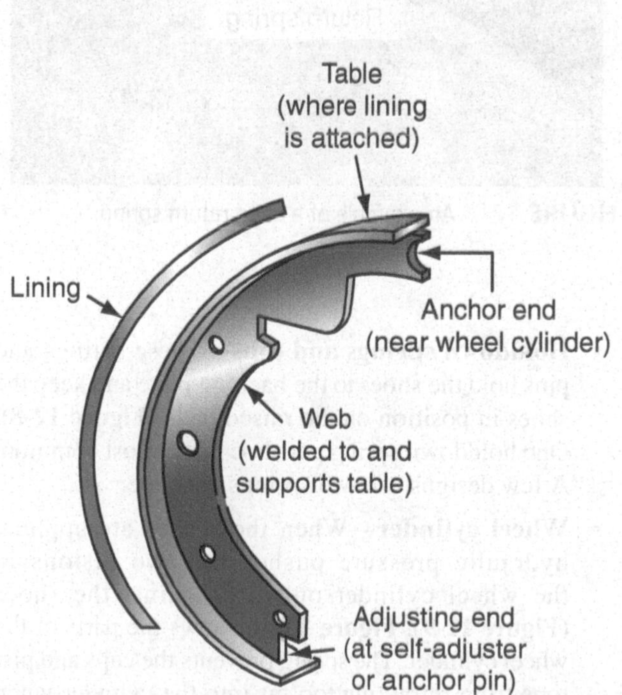

FIGURE 12-3 The brake shoe is the metal piece that supports the lining. The shoe is specifically shaped and configured for each application.

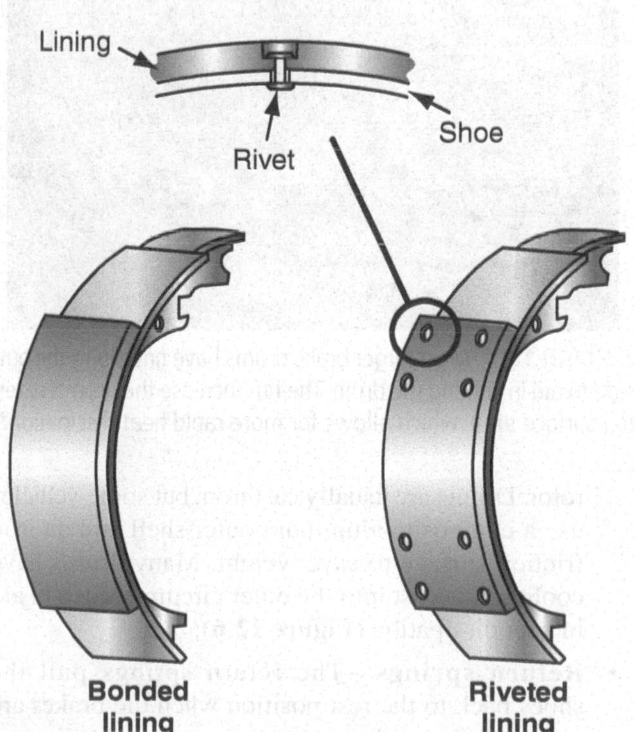

FIGURE 12-4 Brake linings are either bonded (glued) or riveted to the shoes.

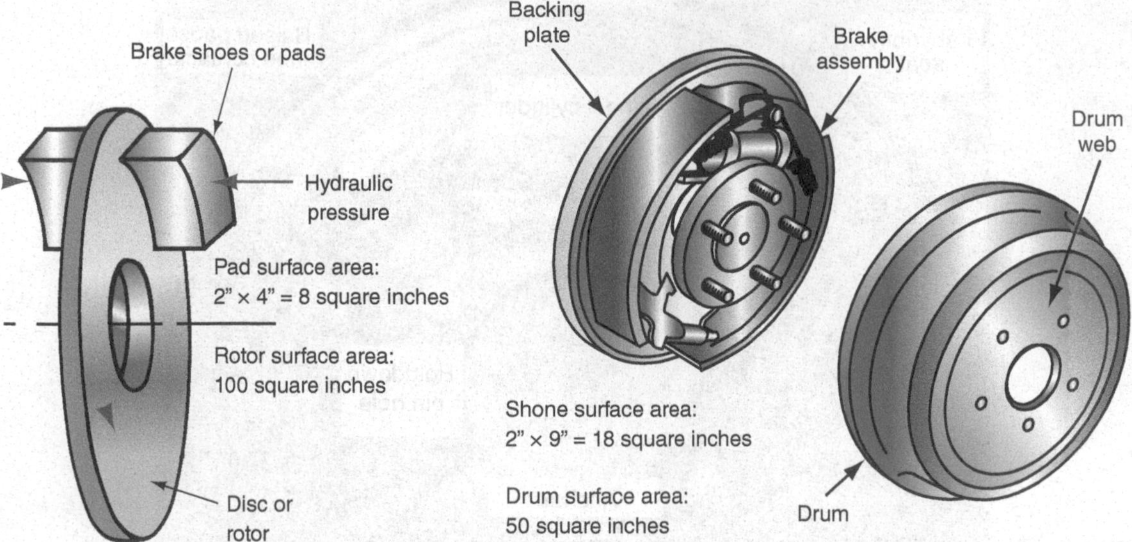

FIGURE 12-5 A brake drum has less surface area than a brake rotor, but there is much more contact area between the brake shoe linings and the drum than there is between the brake pads and rotor.

FIGURE 12-6 Most larger brake drums have fins along the outside to aid in cooling the drum. The fins increase the drum's external surface area, which allows for more rapid heat dissipation.

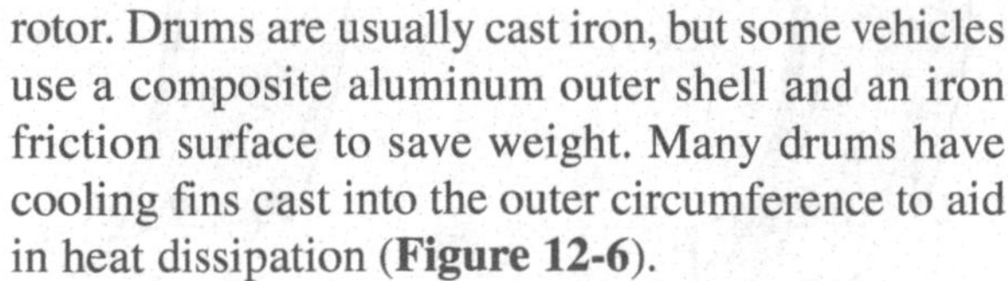

rotor. Drums are usually cast iron, but some vehicles use a composite aluminum outer shell and an iron friction surface to save weight. Many drums have cooling fins cast into the outer circumference to aid in heat dissipation (**Figure 12-6**).

- **Return springs**—The return springs pull the shoes back to the rest position when the brakes are released. Some designs use one return spring per shoe, while others use a spring bridging both of the shoes (**Figure 12-7**).

FIGURE 12-7 An example of a shoe return spring.

- **Holddown springs and pins**—These springs and pins hold the shoes to the backing plate and keep the shoes in position on the raised pads (**Figure 12-8**). One holddown spring per shoe is the most common. A few designs use two springs per shoe.
- **Wheel cylinder**—When the brakes are applied, hydraulic pressure pushes the two pistons in the wheel cylinder outward against the shoes (**Figure 12-9**). **Figure 12-10** shows the parts of the wheel cylinder. The spring prevents the cups and pistons from retracting too far into the cylinder when the brakes are released. The cups seal the fluid in the cylinder. External dust boots prevent brake dust and debris from entering the cylinder.

FIGURE 12-8 Examples of two different types of holddown springs.

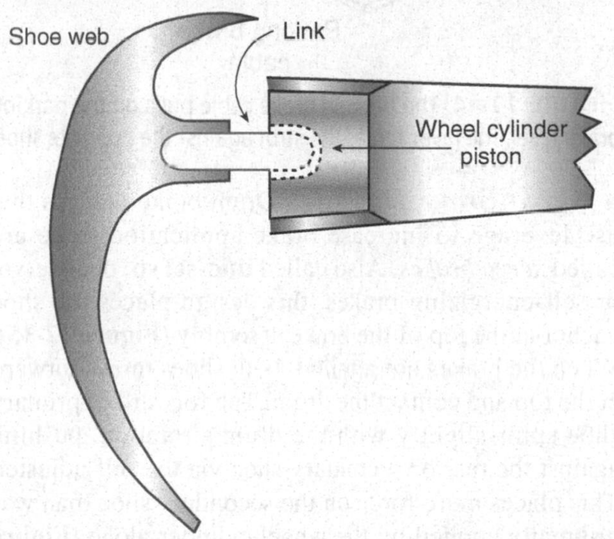

FIGURE 12-9 The pistons in the wheel cylinder push out against the brake shoes, which apply against the drum to slow the vehicle.

- **Self-adjuster assembly**—Drum brakes require adjustment to maintain the shoe-to-drum distance. A self-adjuster is used to expand the shoes slightly when the vehicle is driven in reverse and the brakes are applied or when the parking brake is applied and released. Several methods of self-adjusting linkages are used, and they vary by manufacturer, though the most common type is the threaded self-adjuster (**Figure 12-11**).
- **Anchor**—The anchor is used to prevent shoe movement or twisting as the brakes are applied. The torque of the brake drum will try to twist the brake shoes

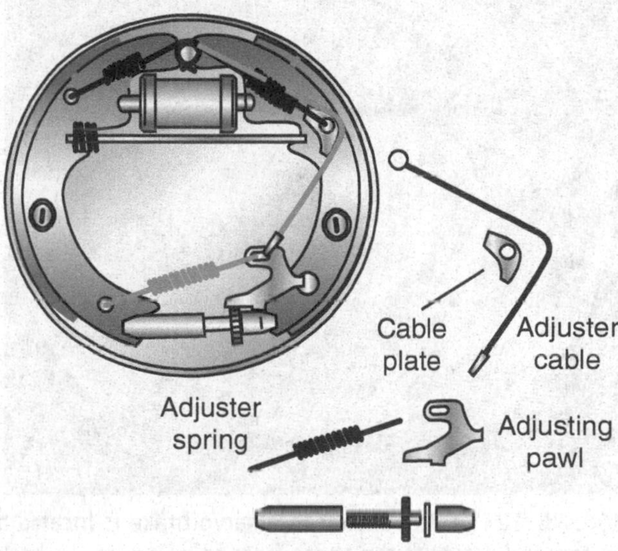

FIGURE 12-11 Self-adjusters are typically located at the bottom of the servo brake assembly. Rods, cables, and levers are used to actuate the self-adjuster.

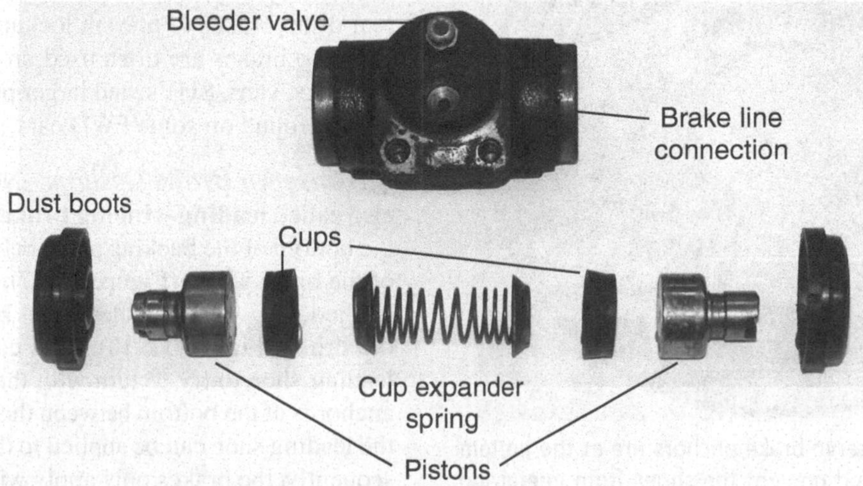

FIGURE 12-10 The insides of a wheel cylinder.

during application. The anchor is used to either limit or eliminate this movement. On servo brake designs, the anchor is located at the top (**Figure 12-12**). Nonservo brake designs place the anchor at the bottom of the backing plate (**Figure 12-13**).

- **Parking brake**—A parking brake lever and strut are used to force the shoes apart and against the drum with the parking brake applied (**Figure 12-14**). Many nonservo brake designs use the self-adjuster mechanism as the parking brake strut.

FIGURE 12-12 The anchor in a servo brake is located at the top and prevents the shoes from spinning on the backing plate.

FIGURE 12-13 Nonservo brake anchors are at the bottom of the backing plate and prevent the shoes from energizing each other.

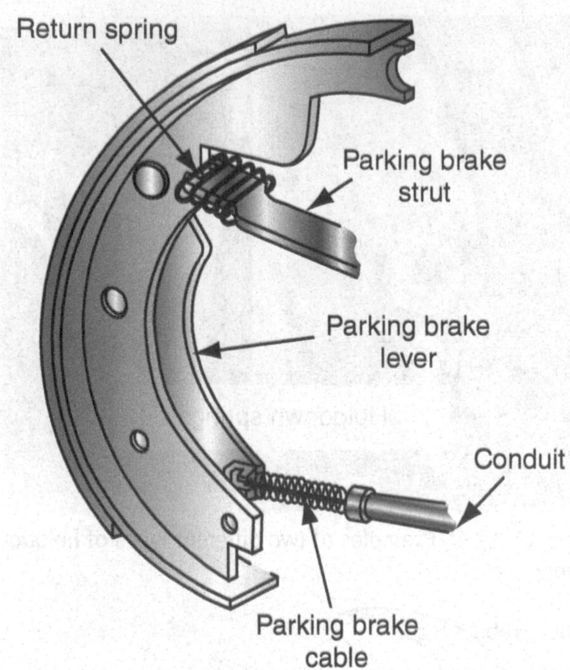

FIGURE 12-14 The parking brake cable pulls on the parking brake lever. The lever forces the strut against the opposite shoe.

■ ***Servo Brake Designs.*** Drum brake designs that use leverage to increase brake application force are called *servo brakes*. Also called **duo-servo**; dual-servo, or self-energizing brakes, this design places the shoe anchor at the top of the brake assembly (**Figure 12-15**). When the brakes are applied, both shoes move outward at the top and contact the drum. The forward or primary shoe spins slightly with the drum's rotation, pushing against the rear or secondary shoe via the self-adjuster. This places more force on the secondary shoe than was originally applied by the wheel cylinder alone (**Figure 12-16**). Because of this servo action, the secondary shoe has a longer brake lining than the primary shoe. Because of the ability to increase braking pressure, proportioning valves are typically used to limit fluid pressure to the rear drum brakes to prevent lockup under hard braking.

Servo brakes are often used on larger vehicles, such as trucks, vans, SUVs, and larger passenger cars but can also be found on some FWD cars.

■ ***Nonservo Brake Designs.*** Nonservo drum brakes, also called **leading–trailing brakes**, place the anchor at the bottom of the backing plate, between the lower edges of the brake shoes (**Figure 12-17**). When the brakes are applied, the wheel cylinder forces both shoes out against the drum (**Figure 12-18**). This causes the forward or leading shoe to try to turn with the drum, but since the anchor is at the bottom between the shoes, no force from the leading shoe can be applied to the trailing shoe. Consequently, the brakes only apply with the force provided by the wheel cylinder.

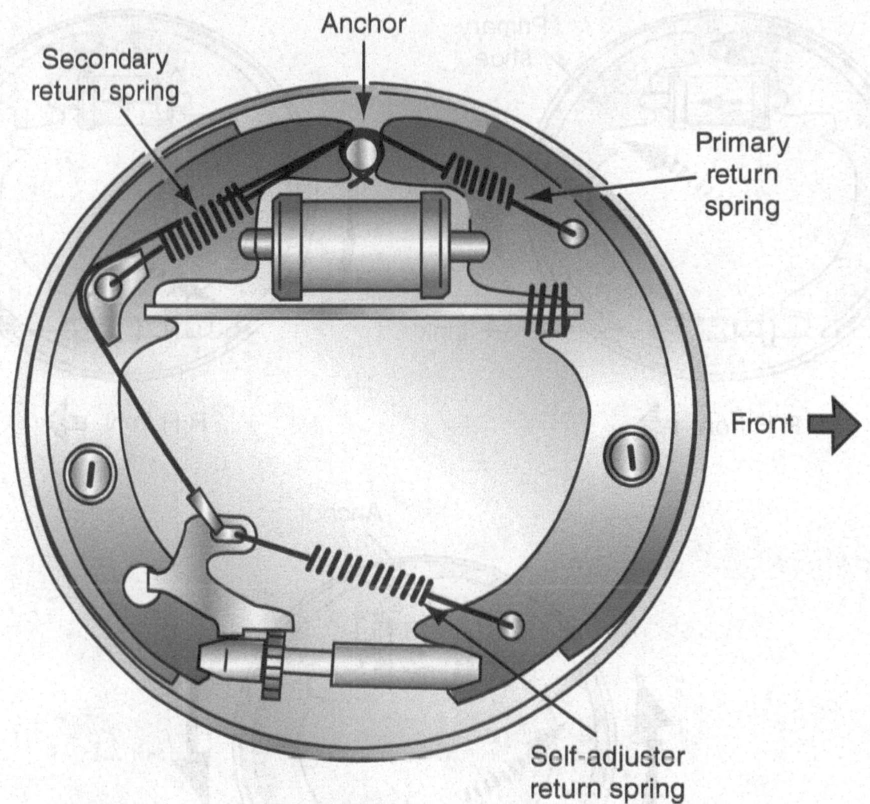

FIGURE 12-15 An example of a servo brake assembly. Note the differences in the size of the linings on the shoes. The primary shoe is mounted toward the front of the vehicle and has a shorter lining than the secondary shoe.

■ ***Self-Adjustment Mechanisms.*** Unlike disc brakes, drum brakes do not automatically compensate for wear. This is because the return springs pull the shoes back to their rest position after each application. This also brings the wheel cylinder pistons back into the cylinder bore. To maintain the correct shoe-to-drum clearance, various types of self-adjusting mechanisms are used.

The most common self-adjusters (**Figure 12-19**) are threaded on one end and can rotate freely on the other. Often called *star wheel adjusters*, these are usually mounted between the lower sections of the shoes on servo-type brakes (**Figure 12-20**). A spring holds the pieces tightly together. A lever, usually held in place against a shoe by a return or holddown spring, and a link or cable attached to the anchor are used to turn the adjuster. When the vehicle moves in reverse and the brakes are applied, the shoes twist opposite their normal rotation. This pulls the shoe away from the link, which pulls on the adjuster lever. Because the lever is in contact with the star wheel, it moves the wheel a small fraction of a turn. Over time, these very small movements thread the adjuster outward, expanding the brake shoes, maintaining the shoe-to-drum clearance.

Nonservo adjuster mechanisms are similar (**Figure 12-21**). In many designs, the adjuster also acts as the parking brake strut. Ratcheting self-adjusters are used on some nonservo brake vehicles. This type of adjuster is activated by using the parking brake.

PARKING BRAKE SYSTEM

The parking brake, also called the *emergency* or *e-brake*, is used primarily to lock the brakes when the vehicle is parked. In the event of hydraulic brake failure, the brake can be used to slow and stop the vehicle, though over a much longer distance. To help maintain its operation, the parking brake should be used on all vehicles, though it is often only used on vehicles with manual transmissions. Customers who have vehicles with automatic transmissions should also be using the parking brake, at least periodically, to make sure that it remains functional. Additionally, whenever a vehicle with an automatic transmission is parked on a grade, the parking brake should be set before the vehicle is placed in Park. If the vehicle is placed into Park and then rolls into place against the parking pawl in the transmission, it can be difficult to shift the transmission out of Park. When the parking brake is used first, it holds the vehicle in place without adding stress to the parking pawl in the transmission.

■ ***Components.*** In vehicles with rear drum brakes, either a hand-operated lever or a foot-operated pedal sets the parking brake.

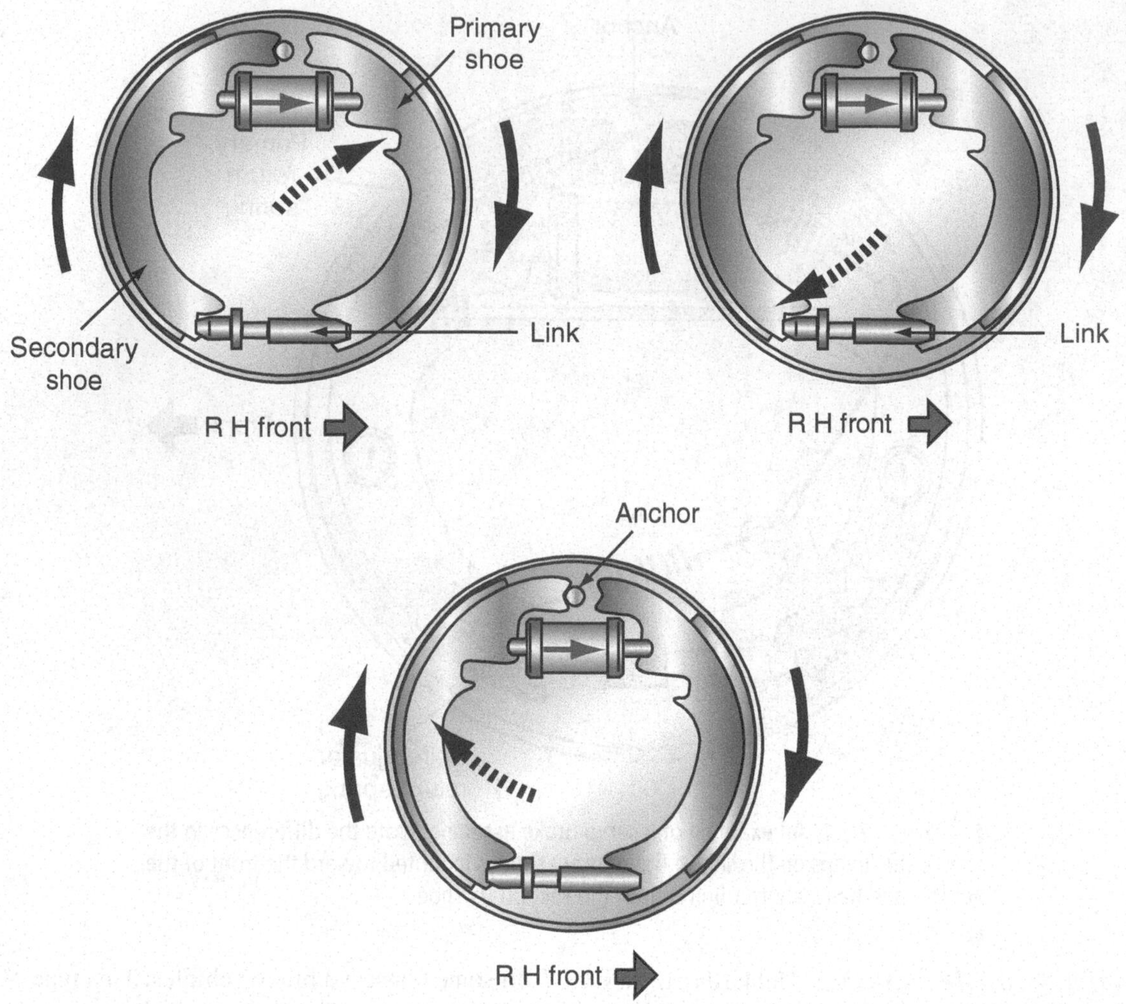

FIGURE 12-16 Servo brakes use leverage from the primary shoe to increase the force on the secondary shoe, which increases overall braking power.

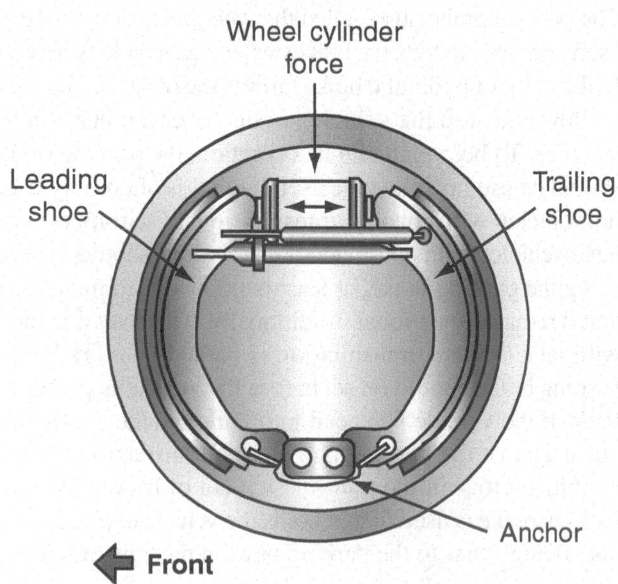

FIGURE 12-17 Nonservo brakes have leading and trailing shoes. The leading shoe is mounted to the front and the trailing shoe toward the rear.

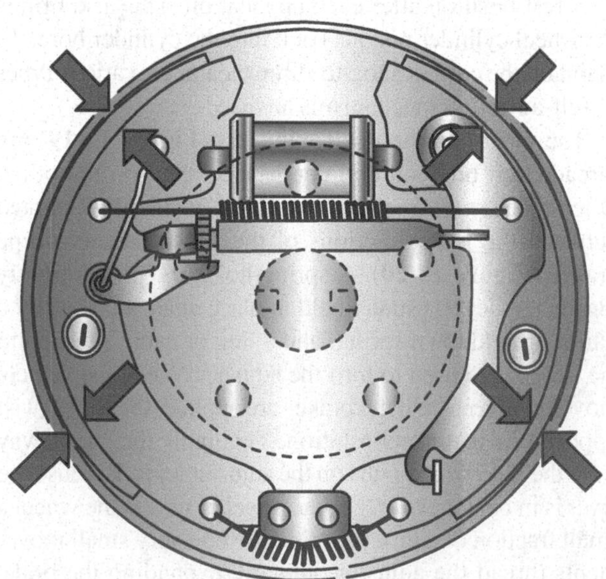

FIGURE 12-18 A nonservo brake does not allow one shoe to act on the other shoe to increase brake force. The anchor prevents the servo action. The anchor acts as a pivot for the shoe but prevents the servo action between the shoes.

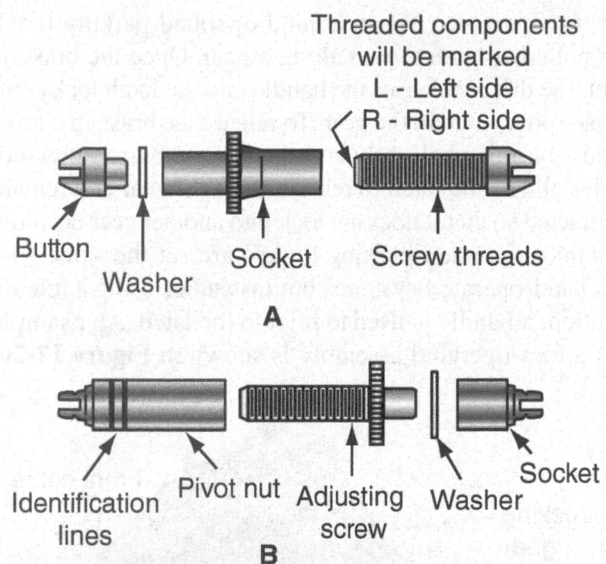

FIGURE 12-19 An illustration of a typical servo brake self-adjuster. Note that adjusters are either right or left and cannot be swapped side-to-side. Threaded adjusters are common on both servo and nonservo designs.

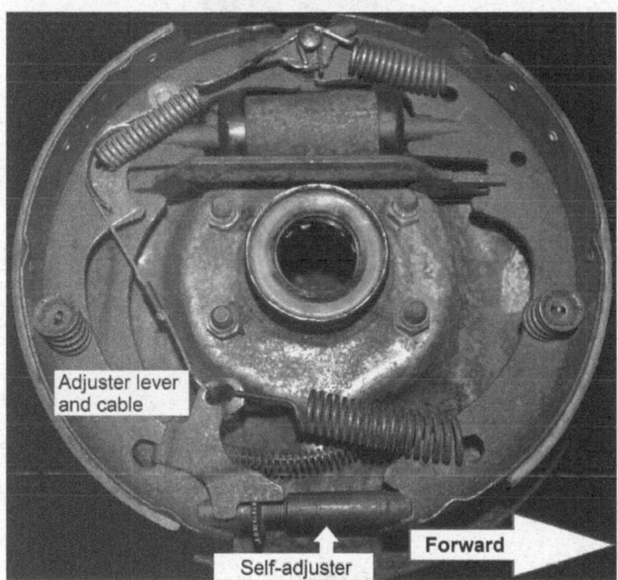

FIGURE 12-20 Self-adjusters are used to maintain the shoe-to-drum clearance. Either rods or cables are used to actuate the adjuster, typically when applying the brakes when reversing the vehicle.

Most hand-operated levers are between the front seats in the center console, although a few vehicles place the handle on the dash (**Figure 12-22**). Regardless of location, when the handle is pulled, it also pulls on the parking brake cables. Most vehicles use a front cable that connects to the parking brake handle and in the rear, to an equalizer or Y-shaped connector that connects the front and rear cables (**Figure 12-23**). The rear brake cables are routed into the backing plate and to the parking brake lever in the drum brake assembly.

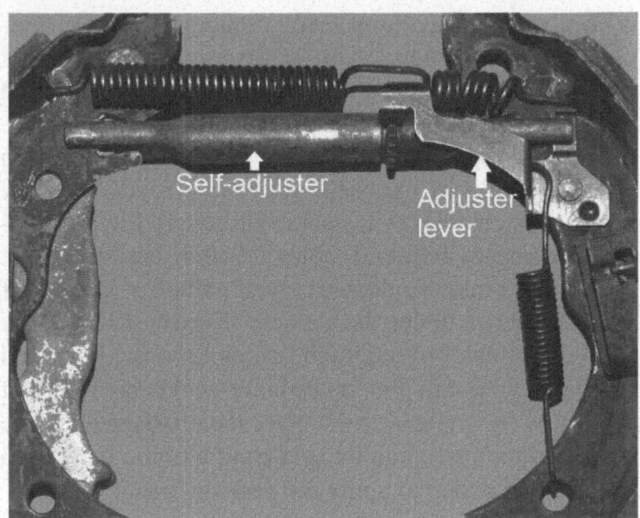

FIGURE 12-21 A common self-adjuster arrangement on a nonservo design. The self-adjuster also acts as the parking brake strut.

FIGURE 12-22 A hand-operated parking brake.

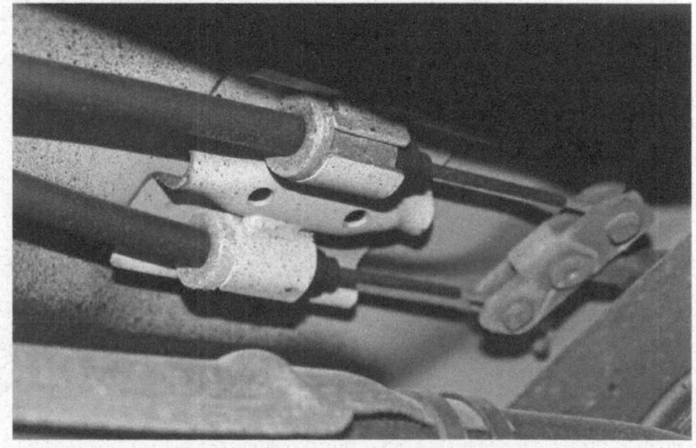

FIGURE 12-23 An example of a connection between the front parking brake cable and the two rear cables.

Foot-operated parking brakes operate like the hand-operated type. When the pedal is depressed, the front cable is pulled forward. This type may use two or three cables, as in the hand-operated system.

The parking brake cables are generally one of two types: a length of exposed steel cable made of many strands of steel wire or a steel cable covered in a protective outer sheath. Exposed cables are often used to connect the parking brake handle to the rear cables or to connect cables together under the vehicle (**Figure 12-24**). The outer sheath is used to attach the cable to the body and backing plate and to protect the inner steel cable from rust and corrosion. By bolting the outer sheath to the body, the outside of the cable can be held rigid and the inner steel cable can be pulled to apply the parking brake.

The parking brake cable may have an adjustment bolt to allow for loosening or tightening the cables as needed. Some parking brakes are self-adjusting and any slack is taken up in the pedal or handle assembly.

In the drum brake assembly, the cable attaches to the parking brake lever. The lever is attached to one of the brake shoes with a pin or a hook. The parking brake strut or self-adjuster screw is placed over the lever (**Figure 12-25**).

FIGURE 12-24 Parking brake cables and cable connections.

■ ***Operation.*** When a hand-operated parking brake is pulled, a latch moves along a gear. Once the brake is set, the driver releases the handle and the latch locks into place on a notch in the gear. To release the brake, the lever must be raised slightly and the release button pressed. This allows the latch to release from the gear and remain retracted so that it does not lock into another gear position.

Foot-operated parking brakes are set the same way as hand-operated systems, but instead of using a release button, a handle is used to release the latch. An example of a foot-operated assembly is shown in **Figure 12-26**.

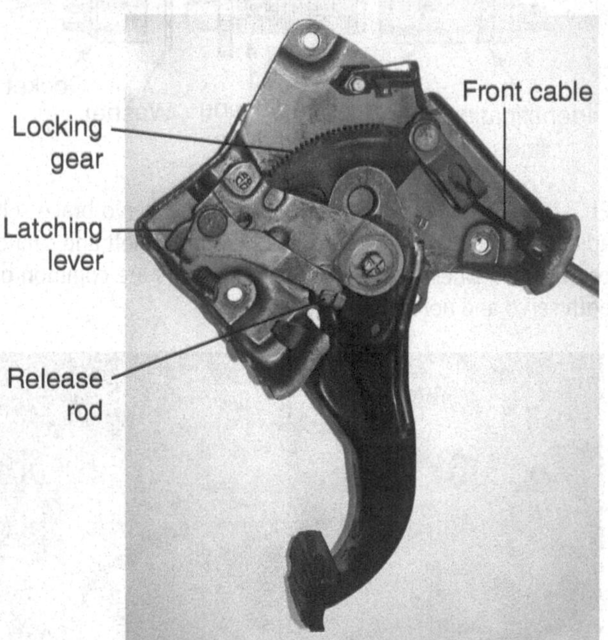
Front cable
Locking gear
Latching lever
Release rod

FIGURE 12-26 The components of a foot-operated parking brake pedal.

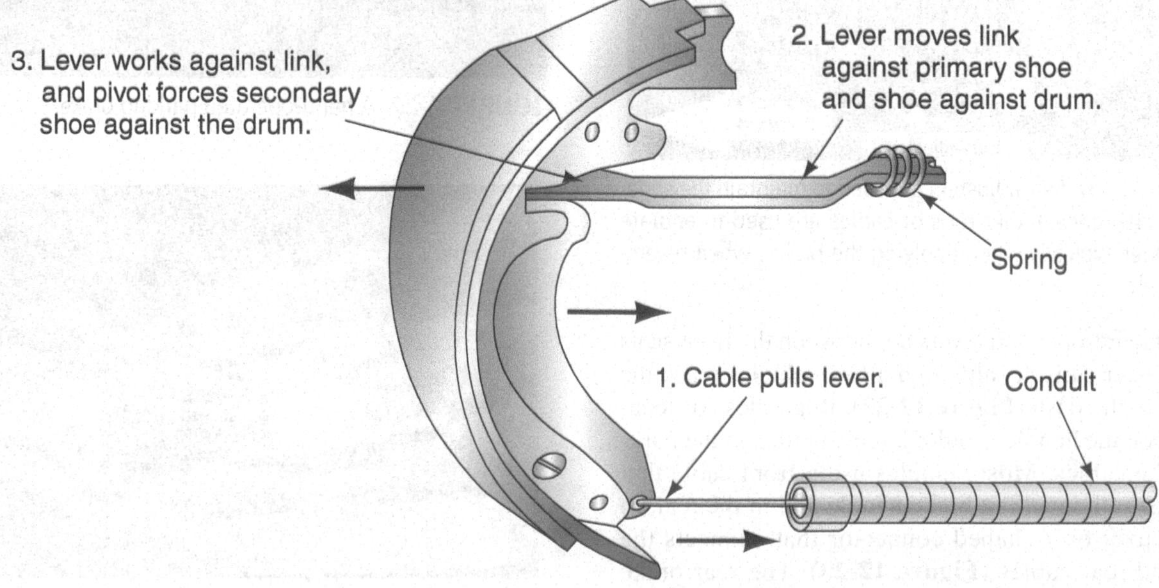
3. Lever works against link, and pivot forces secondary shoe against the drum.
2. Lever moves link against primary shoe and shoe against drum.
Spring
1. Cable pulls lever.
Conduit

FIGURE 12-25 The parking brake cable connects to the parking brake lever. When applied, the lever pushes against the parking brake strut, which applies the forward shoe against the drum.

FIGURE 12-27 An example of a drum-in-hat parking brake assembly.

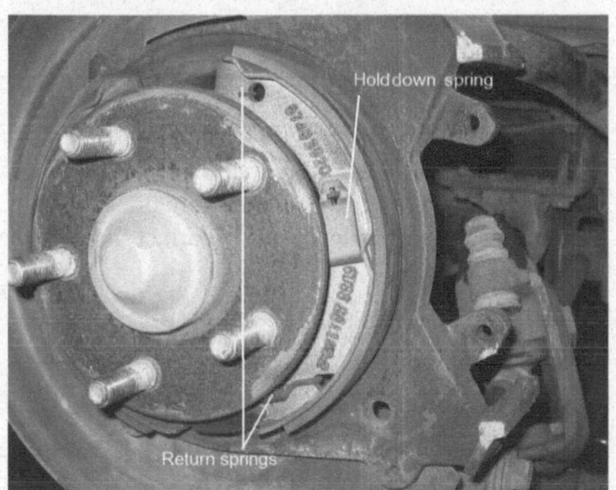

FIGURE 12-28 The springs and shoes of a drum-in-hat parking brake.

FIGURE 12-29 Another type of drum-in-hat parking brake.

On some vehicles, to release the parking brake you press the parking brake pedal down and it disengages. Many foot-operated systems now use automatic brake releases. These systems release when the vehicle is shifted out of Park.

■ ***Drum-in-Hat Parking Brake.*** Many cars and trucks use a drum-in-hat parking brake system (**Figure 12-27**). This system gets its name from its location—in the hat section of the disc brake rotor. The operation is like a standard drum brake: holddown springs and return springs keep the shoes in place (**Figure 12-28**). Some styles combine the actuator with the adjuster. The brake shown in **Figure 12-29** has the actuator on the bottom and the adjuster at the top. Regardless of arrangement, the drum-in-hat brake provides a mechanical parking brakes system tucked neatly away in space otherwise unused. A benefit of this design is that it eliminates a more costly rear brake caliper with integral parking brake function.

SUMMARY

Drum brakes use a set of brake shoes that expand outward against the inside of the rotating brake drum.

When the brake pedal is released, return springs pull the shoes back to their rest position.

The backing plate is attached to the axle assembly and holds the components of the drum brake assembly.

Shoes are the metal backing on which the lining material is attached.

The drum has an internal friction surface for the shoes to rub against.

The return springs pull the shoes back when the brakes are released.

Holddown springs and pins hold the shoes to the backing plate.

Hydraulic pressure forces the wheel cylinder pistons outward against the shoes.

The self-adjuster assembly is required to maintain the correct shoe-to-drum clearance as the shoes and drum wear.

Duo-servo brakes place the anchor at the top of the brake assembly and allow the shoes to rotate slightly during braking.

Nonservo or leading–trailing brakes place the anchor at the bottom of the backing plate and prevent the shoes from rotating during braking.

REVIEW QUESTIONS

1. The type of drum brakes that use leverage between the primary and secondary shoe to increase brake application force are called duo-_______________ brakes.
2. Linings are either riveted or ___________ to the brake shoes.
3. The _______________ _______________ is the foundation of the drum brake assembly to which all the other parts are attached
4. The _______________ _______________ is the hydraulic output for the drum brakes.
5. The _______________ springs pull the shoes back to the anchor when the driver releases the brake pedal.
6. All of the following are components of the drum brake assembly except:
   - **a.** Wheel cylinder
   - **b.** Self-adjuster
   - **c.** Holddown pin
   - **d.** Servo actuator
7. Which of the following seals the fluid in the wheel cylinder?
   - **a.** Piston
   - **b.** Dust boot
   - **c.** Spring
   - **d.** Cup
8. *Technician A* says the drum parking brake may be either electrically or hydraulically operated. *Technician B* says the drum parking brake is mechanically operated. Who is correct?
   - **a.** Technician A
   - **b.** Technician B
   - **c.** Both A and B
   - **d.** Neither A nor B
9. *Technician A* says the location of the anchor determines if a brake design is servo or nonservo. *Technician B* says modern vehicles may have leading–trailing or servo drum brakes. Who is correct?
   - **a.** Technician A
   - **b.** Technician B
   - **c.** Both A and B
   - **d.** Neither A nor B
10. When the brakes are applied, which component is responsible for pushing the shoes outward toward the drum?
    - **a.** Holddown springs
    - **b.** Cup expander
    - **c.** Parking brake strut
    - **d.** Wheel cylinder

Drum Brake System Inspection and Service

Chapter Objectives

At the conclusion of this chapter you should be able to:

- Determine concerns associated with drum brake systems, such as pulling, dragging, and noise.
- Inspect and measure brake drums. (ASE Education Foundation MLR 5.C.1)
- Set up and machine a drum on a brake lathe. (ASE Education Foundation MLR 5.C.2)
- Remove, inspect, and install brake shoes and hardware. (ASE Education Foundation MLR 5.C.2)
- Inspect and replace wheel cylinders. (ASE Education Foundation MLR 5.C.4)
- Inspect and adjust drum parking brake components and operation. (ASE Education Foundation MLR 5.C.5, 5.F.2 & 5.F.3)

KEY TERMS

brake hardware	**floating drum**	**pulsation**
clearance gauge	**grabbing**	**return spring tool**
drum brake micrometer	**holddown spring tool**	**wet sink**
drum in hat	**maximum diameter**	

Being able to inspect and service the brake system is often a requirement for entry-level technicians. Even though brake system work is often thought of as easy because of the relative ease with which many brake system repairs are accomplished, it is critical that all repairs are performed safely and properly.

Service Tools and Safety

As with disc brakes and other systems on modern vehicles, there are special tools for drum brakes that make their service safer and easier for the technician. Using the correct tool for the job and using the tool safely are important aspects of doing any job well. Incorrect tool usage can damage the tool and the components being worked on and can cause personal injury.

Brake service tools are used to make the process of working with the brake system safer and easier. While some technicians prefer to use basic hand tools when working on brake springs and other components, it is best to use the tools specifically designed to perform certain tasks.

DRUM BRAKE SERVICE TOOLS

A few tools are used to make drum brake spring removal and installation easier and safer. These tools can often be used on most types of servo and nonservo brake designs (Figures 13-1 through 13-5).

- A vacuum enclosure or a **wet sink** is used to clean brake dust from the brake assembly (**Figure 13-1**). This is to trap airborne dust and asbestos fibers that may be present in the linings.
- The **return spring tool** is used to safely remove and install the high-tension return springs (**Figure 13-2**).
- The **holddown spring tool** is used to remove coil-type holddown springs (**Figure 13-3**).
- A **drum brake micrometer** is used to measure drum diameter and to check for out-of-round (**Figure 13-4**).

FIGURE 13-1 A wet sink is used to clean the drum brakes and trap dust and to prevent the dust from becoming airborne.

■ ***Tool Safety.*** Drum brake service requires removing and replacing springs that can have a lot of tension on them. Because of this tension, improper tool use can result in damage to components and personal injury. Do not use pliers and screwdrivers in place of the proper brake tools. Pliers can damage or break the springs, and screwdrivers can slip, causing you to injure yourself. Proper tool use is demonstrated later in the service section of this chapter.

■ ***Brake Service Precautions.*** Drum brake service involves working with many small parts and springs, sometimes connected in ways that are not obvious. Take time to determine how the parts fit together before disassembling the shoes. Locate a diagram of the brake assembly, or take pictures before you start to work; this will help you to correctly reassemble the components. In addition to wearing standard personal protective equipment (PPE), such as safety glasses and work boots, gloves are also recommended when you are working on drum brake assemblies. Nitrile gloves will not protect your hands against cuts but they will keep the brake dust from working into your skin.

Before you begin working on the brakes:

- Always allow the drum brakes to cool before beginning to service them.
- Familiarize yourself with the brake system before attempting service. If necessary, research system operation before performing any type of work on the system.
- Locate and follow all the manufacturer's service procedures for working on the drum brake system.
- Do not reuse damaged parts.
- Do not reuse drums that are damaged or worn beyond their service limit.

■ ***Asbestos.*** Even though asbestos has been phased out of brake lining materials, there is no way to know if a set of brakes contains asbestos, so treat all drum brakes with equal caution. When servicing drum brakes,

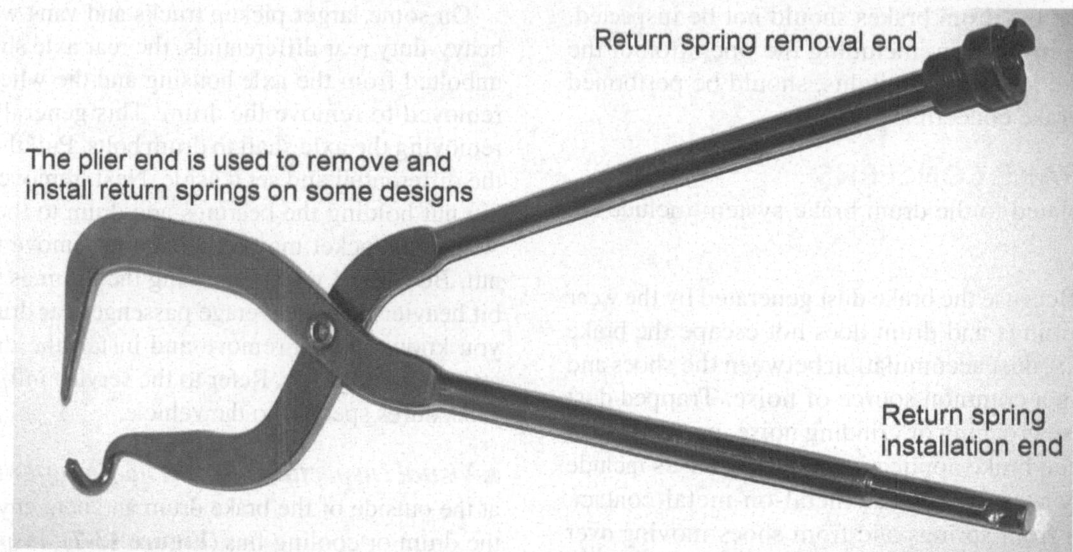

FIGURE 13-2 An example of a return spring tool.

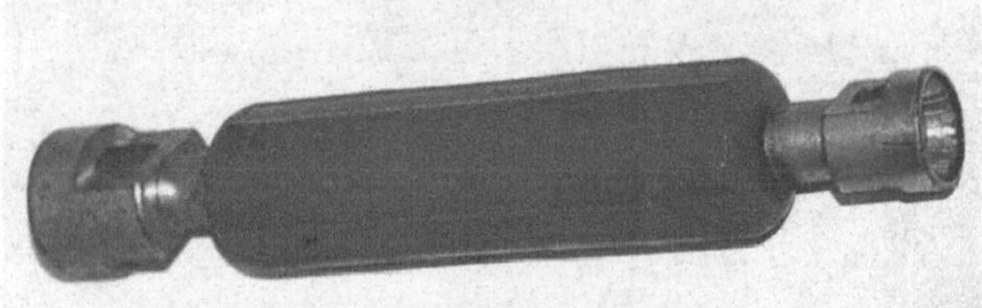

FIGURE 13-3 This tool is used to remove coil-type hold-down springs.

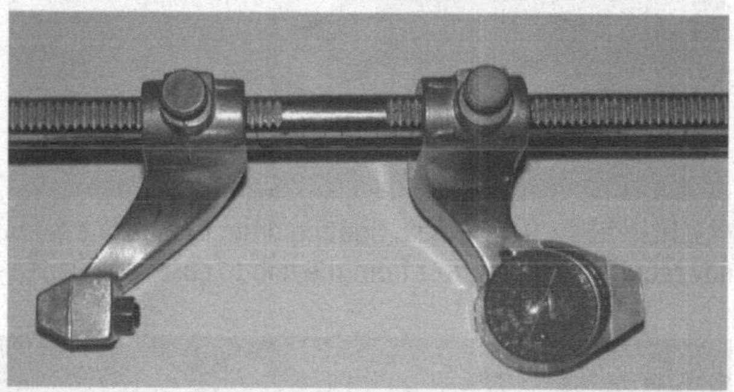

FIGURE 13-4 A drum micrometer is used to measure drum diameter and out-of-round.

use an approved method of brake dust collection. These methods include the following:

- A shop may use a vacuum system (**Figure 13-5**). These use a high-efficiency particulate air (HEPA) filters to trap brake dust so that it does not become airborne.
- Wet sinks (Figure 13-1) commonly used to trap the dust and clean the brake components.
- Spray solvents/spray cans can be used to wet the brakes and trap the dust.

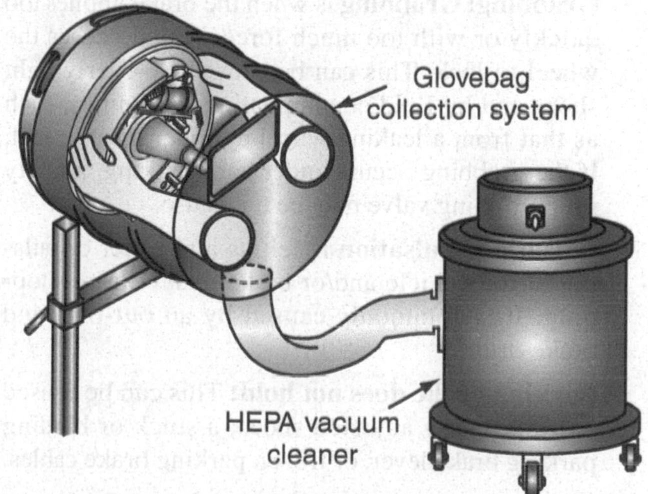

FIGURE 13-5 A special brake vacuum enclosure is used to remove and trap brake dust.

Regardless of the type of system you use, follow the equipment's directions for proper use. **Never use compressed air to clean off brake dust.** Even if the linings do not contain asbestos, they still contain other forms of lining fibers, dust, and debris, and should not be inhaled.

Drum Brake Inspection

Any time a vehicle is checked for a brake system complaint, all wheel brake assemblies should be inspected. This will require removing the brake drums to fully inspect the brake linings, wheel cylinders, springs, and other components.

If the customer's concern is not specific to the drum brakes, such as noise or other issues that indicate problems with the disc brakes or hydraulic system, that does not mean that the drum brakes should not be inspected. A full brake inspection, including the operation of the parking brake and warning lights, should be performed any time a brake concern is present.

DRUM BRAKE CONCERNS

Problems related to the drum brake system include the following:

- **Noise:** Because the brake dust generated by the wear on the linings and drum does not escape the brake assembly, dust accumulation between the shoes and drums is a common source of noise. Trapped dust can cause groaning or grinding noise, particularly at low-speed brake applications. Other causes include severely worn linings as metal-on-metal contact, squeaks from springs and from shoes moving over dry backing plate pads, and pieces of broken springs and clips rattling around inside the drum.
- **Grabbing: Grabbing** is when the brake applies too quickly or with too much force, which causes the wheel to lock. This can be caused by overly tight shoes and by fluid contaminating the linings, such as that from a leaking wheel cylinder or axle seal. If the grabbing occurs under hard braking, a faulty proportioning valve may be the cause.
- **Pulsation: Pulsation** is felt as a shudder or pulsing of the vehicle and/or brake pedal during stopping. It is commonly caused by an out-of-round brake drum.
- **Parking brake does not hold:** This can be caused by improperly adjusted shoes, a stuck or binding parking brake lever, or frozen parking brake cables.
- **Low, soft, or spongy brake pedal:** A low brake pedal is usually caused by excessive shoe-to-drum clearance or worn shoes and drums. If the pedal is spongy or drops very low, there is likely a leak in the hydraulic system.

Table 13-1 shows common brake concerns and possible causes.

DRUM BRAKE INSPECTION

Once you have an idea about what the concerns are and what their possible causes may be, it is time to remove the wheels and begin to inspect the brake assemblies. Before you attempt to remove the brake drum, check the service information for the procedure to remove the drum. Most vehicles use floating drums that slip over the hub and are held in place by the wheel fasteners (**Figure 13-6**). Some HD 4WD vehicles hold the drums in place with the rear axle and wheel bearings, which must be removed to remove the drum.

On some larger pickup trucks and vans with one-ton heavy-duty rear differentials, the rear axle shaft must be unbolted from the axle housing and the wheel bearings removed to remove the drum. This generally involves removing the axle shaft to drum bolts. Pull the axle from the differential and set it aside. Next, remove the retaining nut holding the bearings and drum to the axle tube. A special socket may be needed to remove the bearing nut. Be careful when removing the drum as it is quite a bit heavier than an average passenger car drum. Be sure you know how to remove and install the drum before attempting to do so. Refer to the service information for procedures specific to the vehicle.

■ ***Visual Inspection of External Components.*** Look at the outside of the brake drum and note any damage to the drum or cooling fins (**Figure 13-7**). Inspect the rear

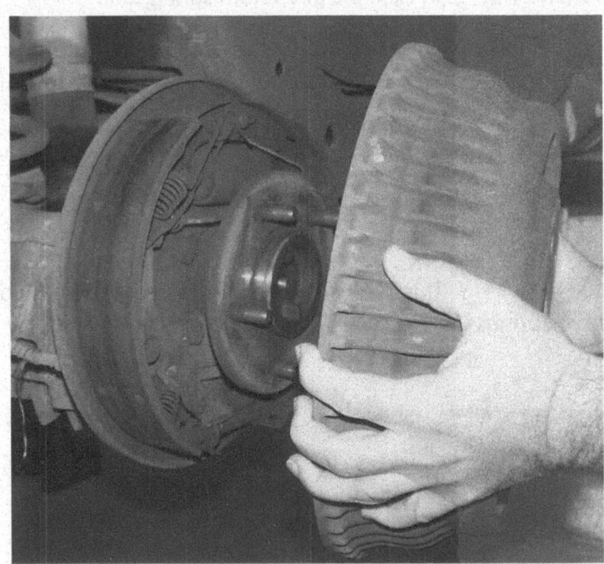

FIGURE 13-6 Removing a floating drum is usually easy but may require removing rust from the hub to ease removal.

FIGURE 13-7 A drum with external damage.

TABLE 13-1

Symptom	Possible Cause	Correction
Brake warning light on	Low fluid level in master cylinder, check for leaks	Refill master cylinder. Repair leak.
Excessive pedal travel Pedal goes to the floor	Worn shoe linings, worn drums, excessive shoe-to-drum clearance Leak in hydraulic system	Inspect/adjust/replace brake shoes. Repair leak.
Brakes drag Brakes grab Spongy pedal	Misadjusted shoes, weak return springs, stuck parking brake, contaminated linings, incorrect linings, air in hydraulic system	Adjust shoes. Replace brake hardware. Repair parking brake. Inspect linings, repair/replace. Check for leaks. Repair system.
Premature rear wheel lockup	Faulty proportioning valve, misadjusted shoes	Test/replace proportioning valve. Adjust shoes.
Excessive pedal effort Brakes chatter (rough)	Glazed linings, contaminated linings, power vacuum assist fault, vacuum leak to power assist, weak holddown springs, scored linings	Inspect/replace linings. Adjust shoes. Inpsect/replace hardware.
Brake noise (rear) Brake pulsation (surge)	Worn linings, loose or broken shoe hardware, excessive dust in drum Brake assembly Out-of-round drums	Inspect/clean/repair shoes and hardware. Measure/machine/replace drums.

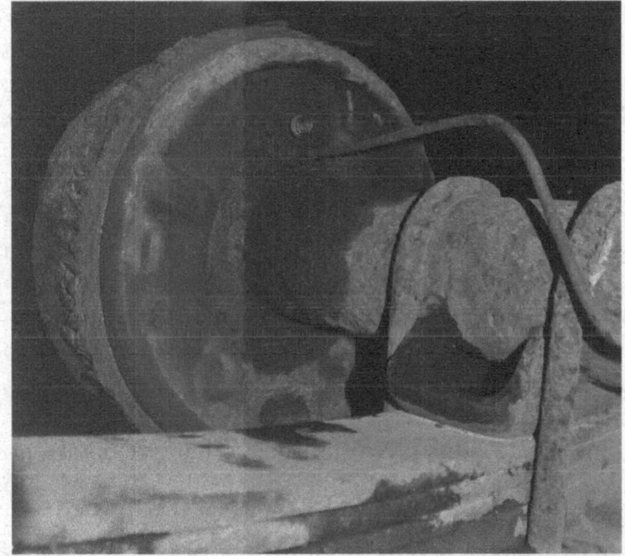

FIGURE 13-8 External evidence of a leak in the drum brake assembly.

FIGURE 13-9 An example of a leaking rear axle seal.

of the backing plate for signs of damage, such as bending or rust-through. If the backing plate is wet (**Figure 13-8**), it can mean the wheel cylinder is leaking. If it is a RWD vehicle, it could also mean that an axle seal is leaking (**Figure 13-9**). Even though technically not a drum brake problem, leaking rear axle seals can contaminate and ruin rear brake shoes and cause several types of brake concerns.

Check the brake lines and hoses for signs of leaks, kinks, or damage. Locate and check the parking brake cables. In some cases, you can tell if the parking brake is used by looking at the cable where it enters or leaves the outer sheath. If the cable is clean and not rusted, this usually indicates that the cable is moving in and out of the sheath and is operating normally. If the cable appears rusted and uniform in color, the parking brake is likely not being used and should be inspected more closely.

DRUM REMOVAL

If a brake drum has never been removed for inspection, you may find stamped steel retainers on the wheel studs (**Figure 13-10**). These clips are used to hold brake drums and rotors in place during vehicle assembly and can be removed and discarded. If the drum has not been removed before, use a grease pencil or paint pen to mark the drum's location to a wheel stud so that it can be reinstalled in the same position.

FIGURE 13-10 The Tinnerman clips used to hold the drums in place during vehicle assembly can be discarded when removing the drum.

Depending on the wear and how well the shoes are adjusted, you may need to loosen or back off the self-adjuster to remove the drum. As the shoes wear into the drum, a ridge or lip is formed along the extreme inside and outside edges of the drum's friction surface (**Figure 13-11**). To unadjust the shoes, use a brake spoon or small screwdriver to back off the self-adjuster (**Figure 13-12**).

■ ***Floating Drums.*** **Floating drums** are designed to slide off the hub; however, years of exposure to the weather tends to rust the drums to the hub, making their removal slightly more difficult. If the drum is rusted in place, apply a penetrant to the hub and around the lug studs. Use a wire brush to remove as much rust around the hub as possible to ease the removal. Use a mallet to tap on the drum to help break the rust bond. Do not use a steel hammer. A steel hammer can damage the wheel studs and brake drum. Do not pry the drum from between the drum and the backing plate as this can damage the brake shoes and bend the backing plate. If the shoes are well adjusted to the drum, you may need to retract the self-adjuster to allow the drum to slide over the shoes.

If the drum is really stuck on the hub, a drum puller may be required (**Figure 13-13**). First, place the puller fingers around the outer lip of the drum where it rides along the backing plate. Next, secure the fingers into the bar and thread the puller bolt against the hub (Figure 13-13). Tighten the puller with a wrench or socket to remove the drum.

On some drums, you may have to remove one or two screws that hold the drum tight to the hub (**Figure 13-14**). These screws usually need to be removed with an impact driving screwdriver (**Figure 13-15**). Often drums also have threaded holes in which bolts can be threaded to push the drum from the hub (**Figure 13-16**). In most applications, thread a metric 8 mm × 1.25 mm bolt into each of the holes and tighten slowly and evenly. This will push the drum from the hub.

Once the drum is removed, wash or vacuum any brake dust from the drum and brake assembly. Do not use compressed air to blow dust from the drum brake assembly. Cleaning the drum brake is discussed in more detail in the "Drum Brake Disassembly" section of this chapter.

■ ***Nonfloating Drum Removal.*** Some vehicles secure the rear drums along with the rear wheel hub and bearings, which must be removed to remove the drum and inspect the brakes. The three most common arrangements are the sealed bearing, tapered wheel bearing, and full-floating rear axle types.

Sealed rear bearings are secured with a large axle nut that must be removed before the drum can be removed. **Figure 13-17** shows one type of this rear bearing setup. The axle nut may be staked to the axle; if this is so, use a punch to unstake the nut before removing it. Staking means that a thin portion of the nut is bent down into a groove cut into the spindle. This prevents the nut from working loose and threading off the spindle. Many manufacturers recommend replacing this type of axle nut whenever it is removed, so refer to the service information to determine if the nut is reusable. Once the nut is removed, the drum can be removed and the brakes inspected.

When the drum is reinstalled, axle nuts on this assembly often have a high torque spec as the nut is used to properly load the wheel bearing. Always refer to the manufacturer's service information for proper installation procedures and torque specs.

For many years, tapered roller bearings were used on the rear hubs of FWD cars, which also held the drums in place. **Figure 13-18** shows how these bearings are assembled. To disassemble:

1. Remove the dust cap and cotter pin. Discard the cotter pin; it will not be reused.
2. Loosen the retaining nut and remove the washer and outer bearing. Place the parts in the bearing cap or in a shop rag.
3. Slide the drum off the spindle, and remove the brake dust from the drum and brake assembly using the appropriate procedures.

If the bearings, grease, grease seal, and hub are free of dirt and debris and the bearings do not need to be

FIGURE 13-11 (a) Drums will often have two ridges, one on the inside and one on the outside of the friction surface. (b) The outside ridge can make drum removal difficult.

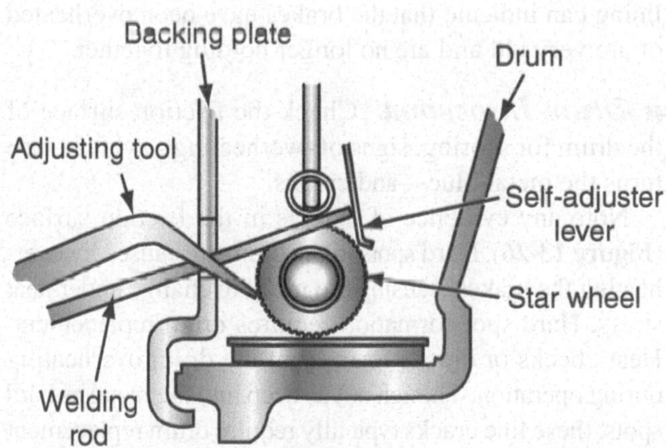

FIGURE 13-12 If the drum will not slide over the shoes, you may have to retract the self-adjuster to gain additional clearance.

FIGURE 13-13 The drums that are really stuck may require a drum puller to remove.

cleaned and repacked with new grease, you can reinstall the drum and the bearings. If the bearings need to be cleaned and repacked, refer to Chapter 5 for the steps to service these bearings. When you are reinstalling the drum and bearings, follow the manufacturer's adjustment procedures to correctly load the bearings, and always use a new cotter pin.

Full-floating rear axle assemblies found on many one-ton series trucks and vans require you to remove the axle and axle bearings to remove the drum (**Figure 13-19**). Always follow the manufacturer's service procedures for the proper removal and installation steps. A common method involves unbolting the axle flange from the hub and removing the axle. Next, remove the bearing

FIGURE 13-14 Some drums have screws that hold the drum in place. Remove the screws before trying to remove the drum.

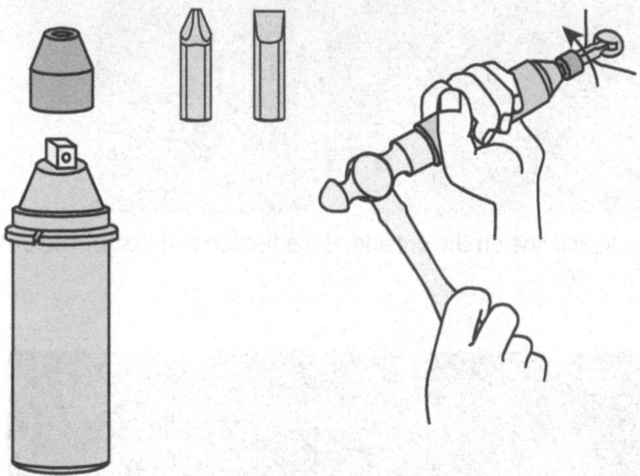

FIGURE 13-15 (a) To remove the screws holding the drum in place, an impact driver is often necessary. The impact driver hammers and twists the screw when hit with a hammer. This dual action usually will break tight screws loose. (b) Place the correct bit in the impact driver and hold the tool firmly against the screw. Hit the back of the driver with a hammer to break the screw loose. Be careful to not hit your hand.

retaining nut and the axle bearings. Once the bearings have been removed, the drum will slide off the brake shoes. Caution should be used as these drums tend to be significantly heavier than those on passenger cars.

PRELIMINARY INSPECTION

Once the drums are removed, begin your inspection by looking at the dust present in the drum and on the linings. **Figure 13-20** shows what brake dust typically looks like, meaning it is dry and contains only lining and drum residue. **Figure 13-21** shows brake dust that is wet from a leaking wheel cylinder. As you can see, the brake dust in Figure 13-21 is much darker due to the presence of the brake fluid. On RWD vehicles, look closely at the axle seal behind the hub, as this often leaks rear differential lube onto the brakes (**Figure 13-22**). Axle lubricant is much darker and thicker than brake fluid, making identifying the source of the leak easier. Any leak from either the wheel cylinder or the axle seal must be repaired before new brakes are installed.

■ ***Lining Inspection.*** Inspect the linings for wear. The linings should show even wear along the length of the shoes (**Figure 13-23**). It is not uncommon for non-servo shoes to wear in a taper from top to bottom. Also, look for uneven wear across the inside to outside edges of the lining. Uneven wear across the shoe can indicate a bent backing plate or a problem with the drum.

To accurately check lining wear, measure the thickness of the linings above the shoe or rivets (**Figures 13-24** and **13-25**). Brake lining thickness gauges or a tire tread depth gauge can provide an accurate measurement of lining thickness. This will provide you with an actual measurement of the amount of lining remaining for comparison to the manufacturer's service information. Typically, brake linings should be replaced when they are worn to the point that about 1⁄16 inch (1.5 mm) remains.

Check for cracks in the lining material, especially around the rivets. Severe cracking or crumbling of the lining can indicate that the brakes have been overheated or are very old and are no longer holding together.

■ ***Drum Inspection.*** Check the friction surface of the drum for scoring, signs of overheating—which often turns the metal blue—and cracks.

Note any evidence of defects in the friction surface (**Figure 13-26**). Hard spots in the drum are caused by overheating the brakes, causing the metal to change under heat stress. Hard spot formation requires drum replacement. Heat checks or cracks form from the drum overheating during operation. Though not as deep into the metal as hard spots, these fine cracks typically require drum replacement to fix. Because both conditions are caused by excessive heat, a thorough inspection of all other brake components is necessary. In addition, talking with the driver of the vehicle may help you determine how these problems occurred and help to prevent their happening again.

Cracks in the drum facing, such as around lug holes, can occur from extreme stress or from a collision. Cracking in the friction surface is a result of overheating and extreme stress when braking. A cracked drum should be replaced.

FIGURE 13-16 (a) Many drums have threaded holes in the face, which can be used to push the drum off the hub. (b) Thread the correct size screws into the holes. (c) Tighten the screws evenly to remove the drum.

FIGURE 13-17 Some vehicles require removing the axle nut to remove the drum. Some drums do not separate easily from the hub. Removing the hub bearing nut allows for easier drum removal.

Other types of drum problems include scored, bell-mouthed, concaved, and convexed friction surfaces (**Figure 13-27**). Light scoring, shallow lines and grooves worn into the drum's surface, results from normal lining wear and from dust and debris being trapped between the drum and the brake linings. Heavy scoring usually results from metal-on-metal contact between the shoe and the drum. A bell-mouthed drum is one in which the inside diameter is less than the diameter around the outside of the drum. This is caused when mechanical fade occurs and the drum expands. Increasing brake pressure expands the drum, which does not contract back into its original shape when it cools. Reusing drums with friction surfaces worn beyond service limits is a common cause of bell-mouthing.

Concave wear results from extreme braking pressure against the shoe distorting the shoe and lining so that more pressure is exerted in the center of the lining than at the inner and outer edges. Convex wear can occur when the drum friction surface is too thin and/or too hot, and the pressure of the shoe during braking widens the open end of the drum.

Determining if a drum is bell-mouthed, concave, or convex requires using a drum micrometer to measure the drum diameter. This is discussed later in this chapter.

■ ***Inspect Brake Hardware.*** Drum **brake hardware** consists of the springs and related parts of the drum brake assembly. It is often difficult to determine

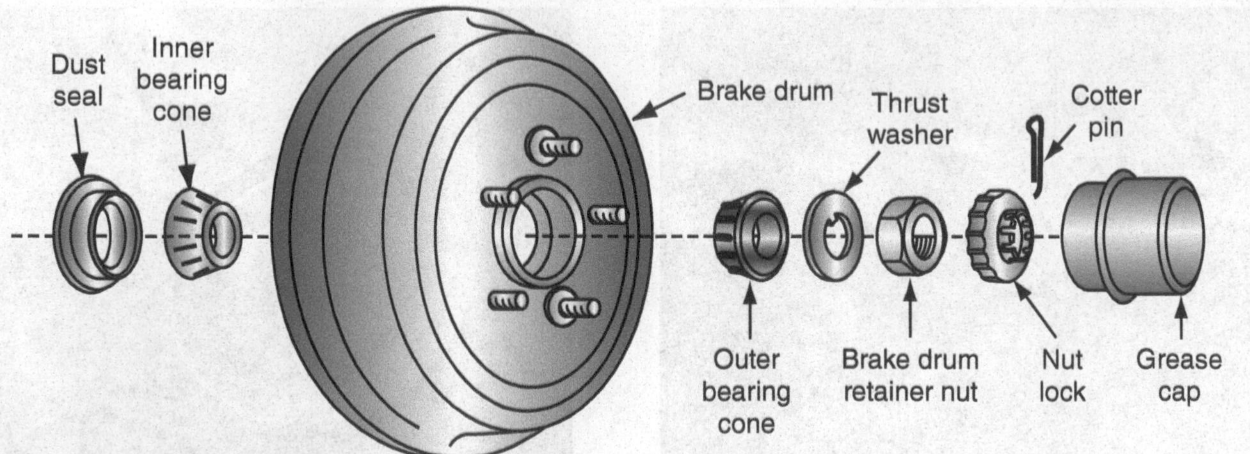

FIGURE 13-18 Some vehicles require removing the wheel bearings to remove the brake drums. Be careful when handling bearings as dropping them can damage the outer race, requiring bearing replacement.

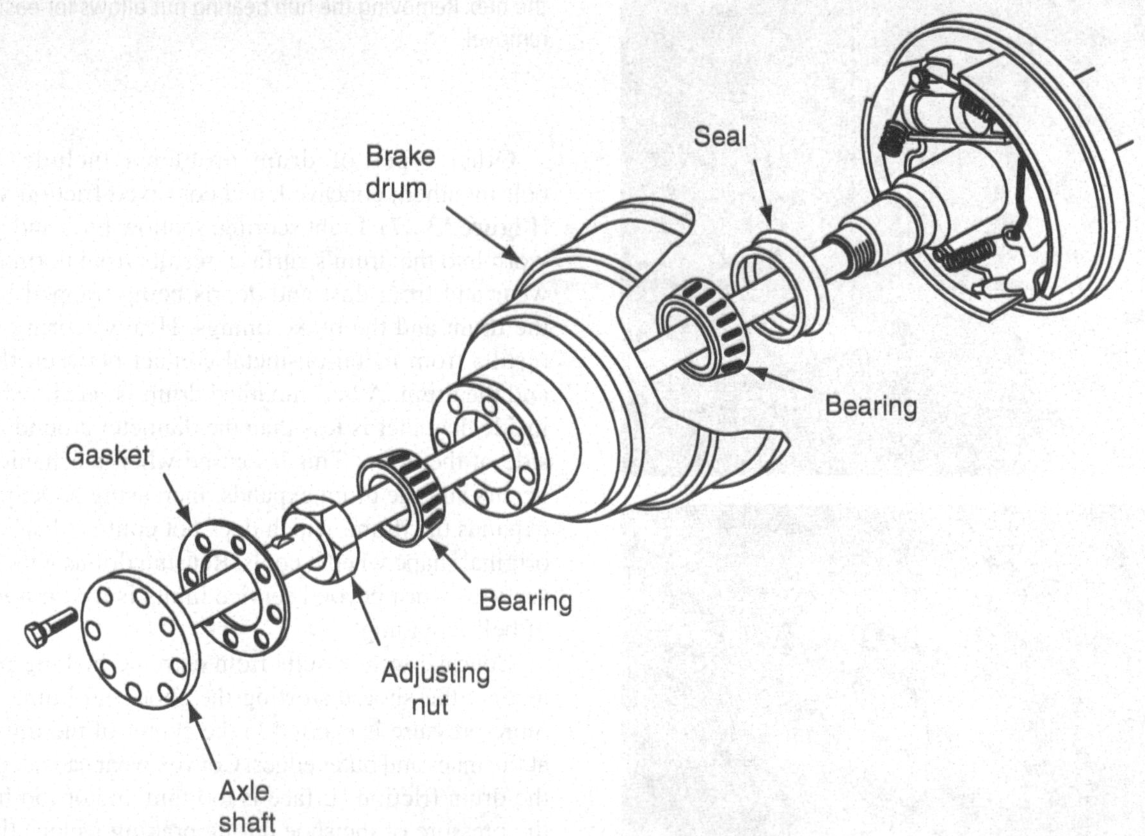

FIGURE 13-19 Removing the drum from a full-floating rear axle requires removing the axle shaft, axle nut, and wheel bearings.

the condition of these parts unless the brakes are disassembled, which will be covered in more detail in the Servicing section later in this chapter. Look for obvious problems like broken springs or adjuster cables. Try to rotate the star wheel adjuster in both directions to determine if it is operational. It is common for the star wheel adjuster to seize. This prevents the shoes from adjusting properly as the linings and drum wear.

■ ***Wheel Cylinder Inspection.*** Even if the wheel cylinder does not show any obvious signs of leakage, this does not mean it is not faulty. Begin by carefully pulling the dust boots away from the cylinder to check for fluid trapped under the boots (**Figure 13-28**). If fluid drips from the dust boot, this means the cups are leaking, and the cylinder needs to be rebuilt or replaced. Next, carefully try to push the pistons back

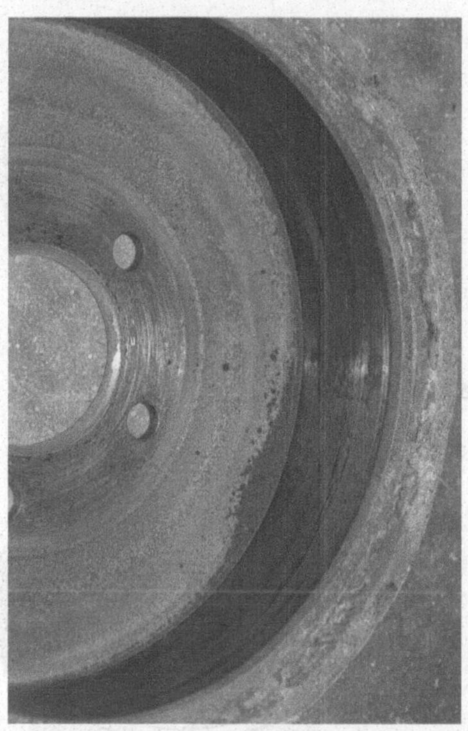

FIGURE 13-20 Dust accumulates in the drum as the linings and drum wear. Use a brake vacuum or wet sink to collect the dust.

FIGURE 13-22 Rear-wheel-drive vehicles can leak differential lubricant past the axle seals and onto the brake assembly.

FIGURE 13-21 Wet brake dust and components indicate a leak, in this case, from the wheel cylinder. Pull the wheel cylinder dust boot back to check for leaking cups.

into the bore. Moisture in the brake fluid can, over time, allow the wheel cylinder pistons to rust into place in the bore (**Figure 13-29**). This prevents the rear brakes from applying, which greatly increases lining life, but is not good for overall braking performance as it causes the front brakes to perform 100 percent of the braking.

■ ***Checking Parking Brake Operation.*** If you have not checked the parking brake operation yet, inspect the cables and note signs of use as discussed earlier in this chapter. With the drum removed, have an assistant slowly and carefully start to apply the parking brake. If the parking brake is operating correctly, the shoes will start to expand slightly. If they do, stop and release the parking

FIGURE 13-23 With the drum removed, inspect the lining wear and note any damaged or broken springs.

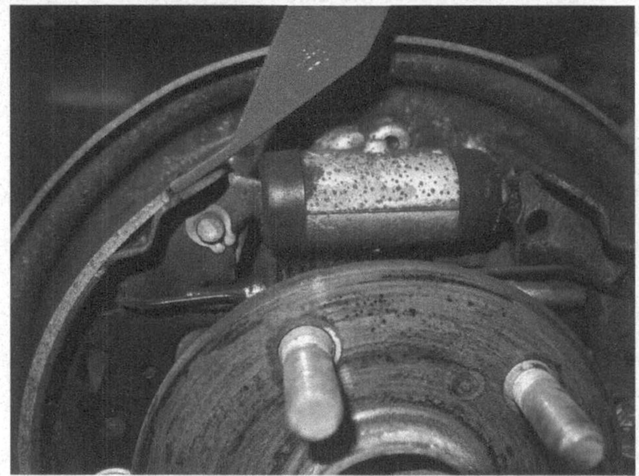

FIGURE 13-24 Using a lining thickness gauge to measure lining wear.

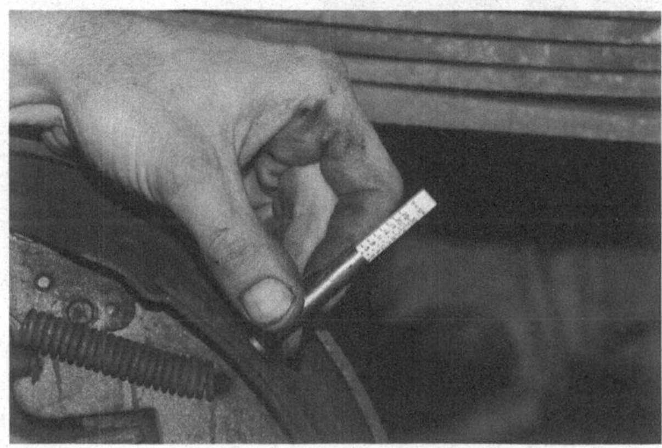

FIGURE 13-25 Measuring lining thickness using a tire tread depth gauge.

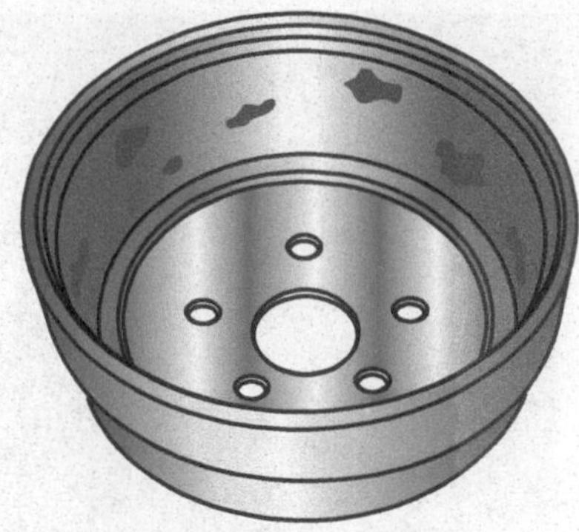

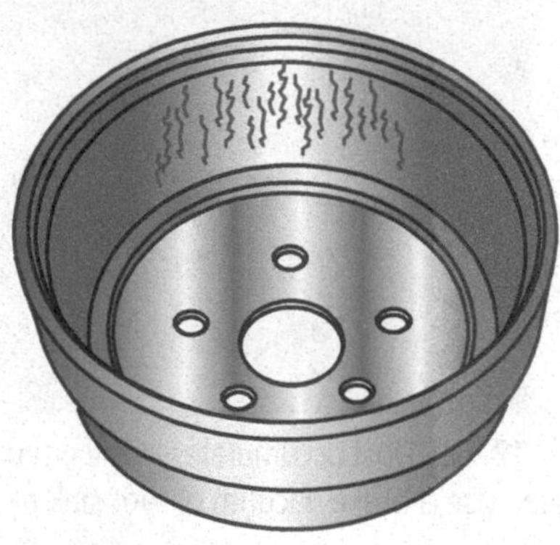

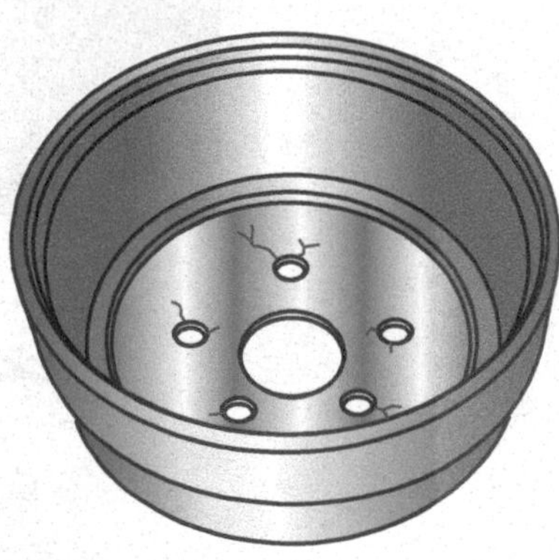

FIGURE 13-26 Examples of drum defects that typically require replacement.

brake, and make sure the shoes retract fully. A parking brake that is stuck and applied, even slightly, can prevent you from reinstalling the drum after the brake shoes have been replaced. Always check the parking brake operation as part of the drum brake inspection to determine if there are problems with the parking brake before you begin to estimate repairs and begin service. Both you and the customer will be unhappy if after you replace the shoes and hardware you find out that the drums cannot be reinstalled because the parking brake is stuck partially applied.

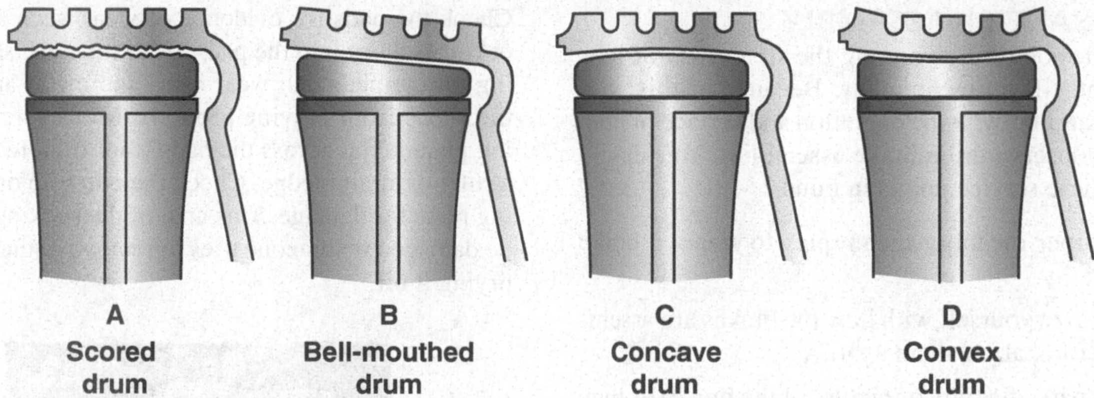

FIGURE 13-27 Examples of drum wear patterns.

FIGURE 13-28 Inspecting a wheel cylinder by pulling back the dust boot.

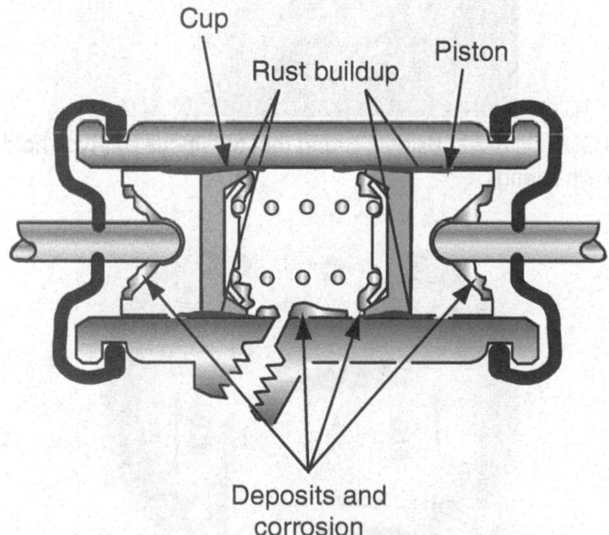

FIGURE 13-29 Rust can build up in the wheel cylinder, seizing the pistons. The rust can also cause the cylinder to leak after shoe replacement as the cups are pushed back into the cylinder bore and over the rust.

FIGURE 13-30 An example of using a wet sink to clean and trap the brake dust.

Drum Brake Service

As stated at the beginning of the chapter, servicing either servo or nonservo brakes often requires special brake spring tools, which should be used to help prevent damage to components and personal injury.

Begin by cleaning as much brake dust from the linings and hardware as possible with either a wet sink, aerosol brake cleaner, or brake dust vacuum (**Figure 13-30**). Take a close look at the brake assembly and note how the parts are arranged, in which direction the springs are placed, and how the parts fit together. If you are working on brakes that you are not familiar with, you may want to locate a parts view diagram of the brake assembly from your service information. You can also use your phone or a digital camera to take pictures before you begin disassembly. If a diagram is unavailable, you can draw a rough sketch of how the brake is assembled. The reason for this is that drum brakes can have many pieces that, although they may go back together in many ways, really only have one way that they are correctly assembled.

DRUM BRAKE DISASSEMBLY

Brake services are often among the first types of jobs given to entry-level technicians. Because of this, you need to be familiar with the operation and service of both servo and nonservo drum brake assemblies. You should also keep these service points in mind:

- Use proper cleaning techniques to remove brake dust.
- Familiarize yourself with how the brakes are assembled before taking them apart.
- Have a parts diagram or picture of the brakes as they are before disassembly.
- Take one side apart at a time; this allows you to keep the other side intact for reference.
- Use the appropriate brake tools for disassembly and reassembly.
- Carefully inspect all components as you take the brakes apart.

As each part is removed, lay it aside so that it is arranged as the brakes are assembled.

■ ***Remove and Inspect Servo Brake Components.*** There are many types and styles of servo brakes and many ways in which one can service them. The steps shown here represent one type of servo brake and common service steps. You may find that this procedure works for you, or you may develop your own. As you remove the components, place them on a work space as they are arranged on the vehicle.

- Remove the return springs from the anchor (**Figure 13-31**). If the return springs hold the self-adjuster cable guide, remove the guide also.
- Remove the holddown spring on the shoe that secures the self-adjuster lever, and remove the self-adjuster links or guides.
- Remove the shoe, self-adjuster, and parking brake strut.
- Remove the remaining holddown spring (**Figure 13-32**).
- Disconnect the parking brake lever from the shoe and remove the shoe.
- Arrange the components as they are installed (**Figure 13-33**).

■ ***Inspect the Backing Plate.*** Once the shoes and hardware are removed, you will need to inspect the backing plate (**Figure 13-34**). Use a wire brush to remove rust from the raised pads and around the anchor. Check the pads for evidence of wear, such as grooves or ridges worn into the pads. Over time, the shoes moving on the pads can wear away the metal and restrict the shoes from moving properly. Make sure the backing plate is flat across the pads from front to rear using a ruler or straightedge. Check the outer lip of the backing plate for damage. This area of the backing plate can be damaged if someone tries to remove a stuck drum by prying it off.

FIGURE 13-31 Using a spring tool to remove the return springs from the anchor.

FIGURE 13-32 Using a spring tool to remove coil-type holddown springs.

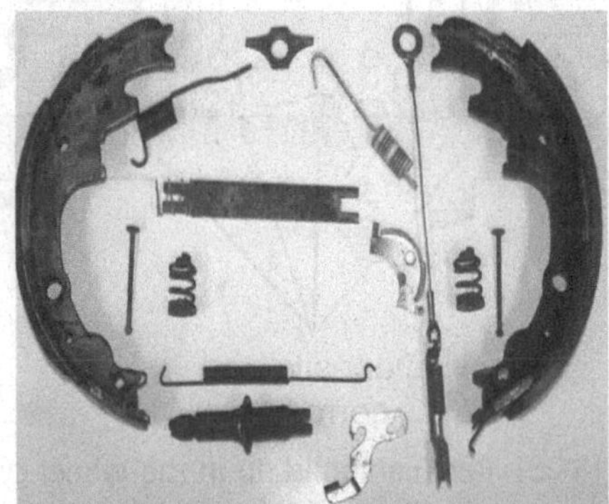

FIGURE 13-33 Once the parts are removed, lay them out to make sure you know how everything fits back together and to inspect the parts for wear and damage.

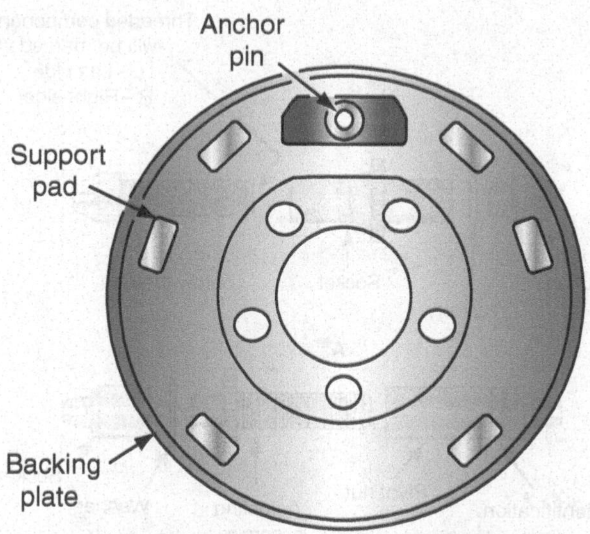

FIGURE 13-34 An illustration of the backing plate. Inspect and clean the support pads before reassembling the shoes. Lubricate the pads per the manufacturer's service information.

FIGURE 13-35 A suggested method of disassembling a nonservo brake.

■ ***Nonservo Brake Service.*** Just like servo brakes, there are many styles of nonservo brakes, each with their own service procedures. In general, servicing nonservo brakes is nearly identical to servicing servo brakes. The main differences are that nonservo brakes typically combine the self-adjuster and parking brake strut functions. This means that nonservo brakes usually do not have a separate strut between the shoes for the parking brake. The general steps to remove the shoes include the following:

- Remove the upper return spring first. This is usually the spring near the self-adjuster, labeled 1 in **Figure 13-35**.
- Remove the holddown springs, labeled 2 in Figure 13-34.
- Remove the self-adjuster, labeled 3 in Figure 13-34.
- Disconnect the parking brake lever from one of the shoes. The shoes and lower spring may now be removed while connected together.

COMPONENT INSPECTION

Once the shoes, springs, and other hardware are removed, you will need to make a close inspection of the parts to see if they can be reused or should be replaced.

■ ***Brake Linings.*** If you are removing the brake shoes, it is most likely because they are being replaced. However, brake shoes may be removed so that another component can be replaced, such as a parking brake cable, and then reinstalled if there is sufficient lining remaining.

Inspect the linings for wear. Bonded linings should be replaced when the lining reaches $\frac{1}{16}$ inch or about 1.5 mm, and riveted linings should be replaced when the lining is $\frac{1}{16}$ inch or 1.5 mm above the rivet. Refer to the manufacturer's service information for the exact wear specification. Check for cracks in the lining, especially around the rivets. If the linings are shiny or blue, this indicates glazing. Glazed linings decrease stopping performance and can make noise. Glazing can be caused by poor-quality brake lining materials or by overheating of the brakes.

Check for signs of uneven wear (**Figure 13-36**). Uneven wear, either from top to bottom or across the surface usually indicates a problem. Linings worn unevenly from top to bottom can indicate shoes that are sticking on the backing plate or are not returning properly. Check for a bent backing plate if there is tapered wear across the lining.

■ ***Spring and Hardware Inspection.*** Springs and hardware should be cleaned and inspected once the brake is disassembled. Examples of spring damage are shown in **Figure 13-37**. Many shops recommend replacing all the springs and spring hardware every time the shoes are replaced, and there is merit to this practice. Because the rear brakes tend to last longer than the front brakes, the springs and hardware are in service for a longer time and are subject to many more heat cycles. This can weaken the springs. In addition, rust eventually eats away at the springs, further weakening them. Rear spring and hardware kits are inexpensive and provide insurance against future brake problems that can result from weak springs.

Some parts are not usually replaced during routine brake service, such as the self-adjuster, levers, links, and parking brake components. That does not mean that these parts cannot sustain wear or damage or need

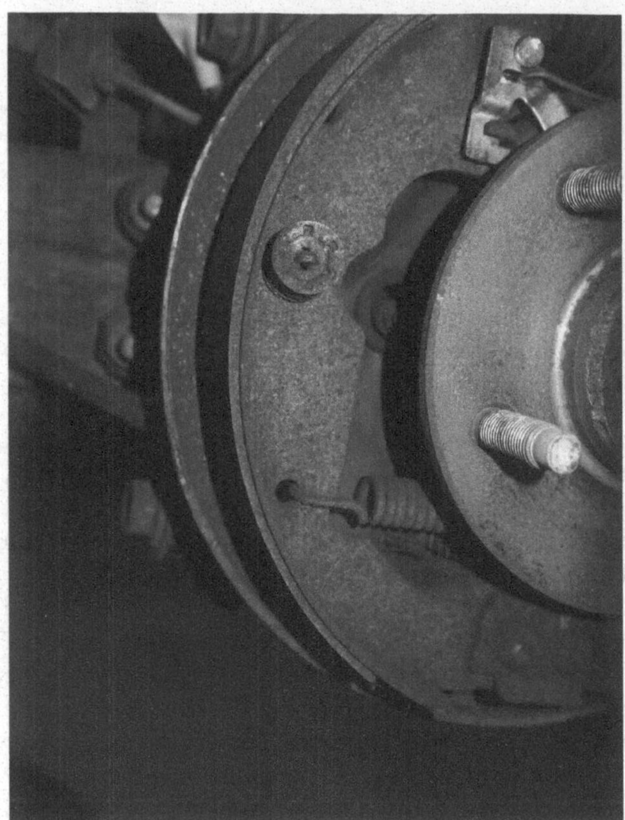

FIGURE 13-36 An example of uneven lining wear.

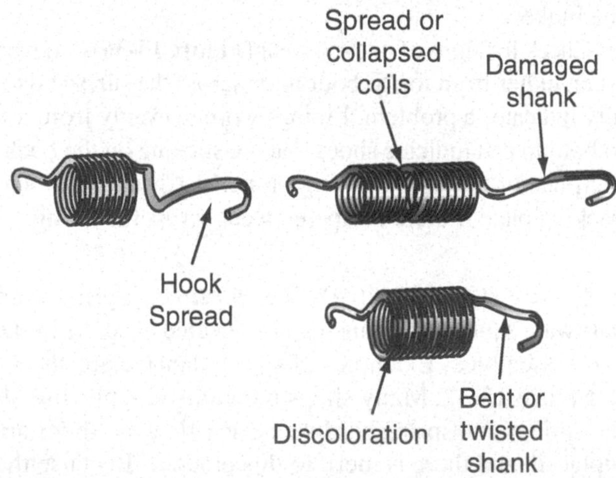

FIGURE 13-37 Return springs rust, bend, and break. Many technicians replace the springs each time the shoes are replaced.

replacement. All the components need to be carefully inspected, and any part that is damaged or worn beyond usefulness should be replaced.

■ ***Self-Adjuster Inspection.*** Many vehicles use threaded star wheel self-adjusters (**Figure 13-38**). Check the star wheel for wear on the contact points. If the contacts are worn down, the adjuster lever may

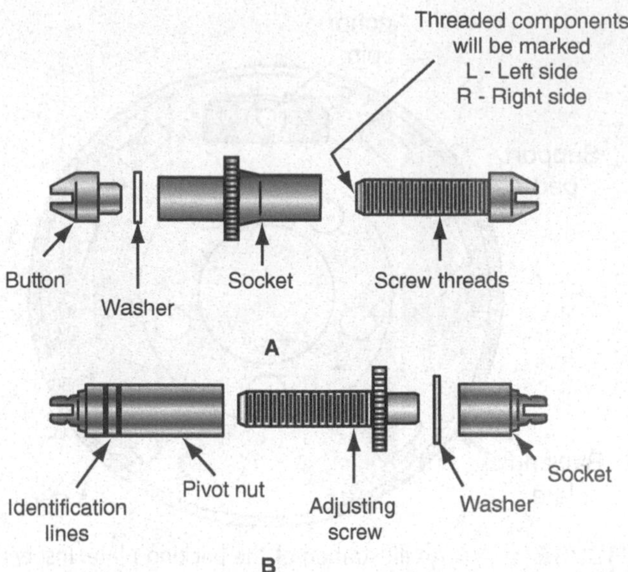

FIGURE 13-38 Inspect the self-adjuster for damage to the star wheel, and clean and lubricate the adjuster threads and posts.

not be able to actuate the self-adjuster. Damage to the star wheel itself is common when pliers are used to turn the adjuster instead of a brake spoon or screwdriver. Unscrew the threaded section from the body and remove the unthreaded end. If there is a buildup of rust on the internal parts, use a wire brush to clean off the rust. Before reassembly, apply a very light amount of brake lubricant to the threads and the unthreaded end of the adjuster. This will make the adjuster easier to operate and will slow rust formation.

On vehicles with ratcheting self-adjusters, move the adjustment lever to ensure that it operates freely. If the adjuster is severely rusted, you may need to disassemble the pieces to completely remove the rust and lubricate the parts. Inspect the teeth or splines on the adjuster. Movement between the parts can wear away the teeth, preventing the adjuster working properly.

BRAKE DRUM INSPECTION AND MEASUREMENT

Brake drums are inspected for problems just as brake rotors are, but the types of problems are different. As the drums wear, their inside diameter increases, and the amount of metal that makes up the friction surface decreases. This reduces the amount of heat the drum can withstand before deforming or cracking.

■ ***Clean and Inspect Drum.*** Examine the outside of the brake drum for damage, which is often caused by a hammer being used to remove the drum for inspection. Look at the friction surface inside the drum and check for scoring, signs of excessive heat, and cracks. Normal wear and scoring is shown in **Figure 13-39**. Deep

FIGURE 13-39 Normal drum wear and scoring.

scoring occurs when the linings wear down into the shoe, allowing metal-on-metal contact. This usually requires drum replacement. If the surface is blue, it means that the brakes have been overheated. Cracked drums can also be caused by overheating or by the friction surface wearing beyond its service limit. If any of these conditions exist, the drum should be replaced.

■ ***Drum Measurement.*** Brake drums become larger as they wear. As the friction surface of the drum wears away, the space between the shoes and the drum increases. Drums often have two wear specifications, machine-to and maximum diameter. The machine-to limit is typically about 0.030 inch (0.8 mm) less than maximum diameter. This is the wear buffer used to allow the drum to remain in service before it must be replaced. If a drum is between the machine-to and maximum diameter during your inspection, it should be replaced. All drums have a **maximum diameter** spec which is stamped into the drum, which is the absolute maximum size the drum can reach before replacement is necessary (**Figure 13-40**). Drum specifications can also be found in brake specification guides and in the vehicle's service information.

To use a metric drum micrometer, begin by determining the maximum diameter of the drum. In this example, the drum shown in Figure 13-40 will be used. The maximum diameter is 226.3 mm or 22.63 cm. A metric drum micrometer has two sets of graduated markings on the shaft, one set in even-numbered centimeters (**Figure 13-41**), and the other in odd-numbered centimeters on the opposite side of the shaft. Each small mark between the longer, numbered marks is equal to 2 mm or 0.2 cm. To measure this drum, the mic will be set initially at 224 mm.

The dial-indicator part of the mic (**Figure 13-42**), is used to fine-tune the diameter, down to 0.1 mm. As the plunger is pushed in, the needle swings counterclockwise.

FIGURE 13-40 An example of a brake drum wear specification.

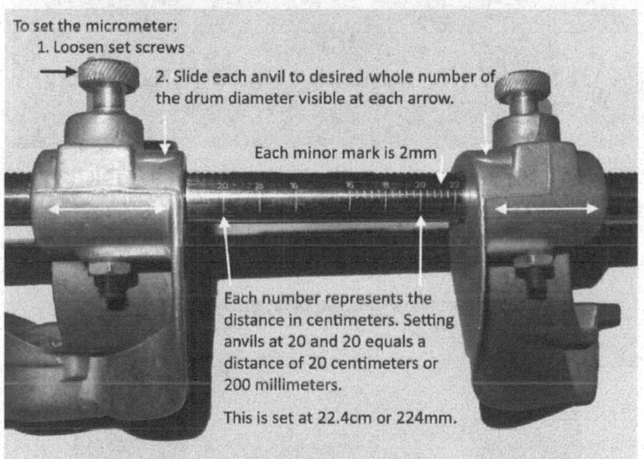

FIGURE 13-41 The metric micrometer has two scales on the shaft, even- and odd-numbered centimeters. Each mark between centimeter marks is 2 mm.

The number indicated by the gauge, from 0 to 3 mm in 0.1 mm increments, is then added to the initial size setup from the anvil positions on the main shaft. For example, the micrometer is set at 22.4 cm (224 mm), and the gauge needle is pointing at the second minor mark, which equals 0.2 mm (**Figure 13-43**).

Place the micrometer into the drum (**Figure 13-44**). Hold the dial indicator in place, and rotate the opposite end of the micrometer slightly back and forth in the drum until the maximum diameter is shown on the gauge. Add the gauge reading (**Figure 13-45**) to the original setting to find the maximum diameter. In this example, the drum is measuring 225.7 mm. This is obtained by adding the 1.7 mm on the dial to the 224 mm setting on

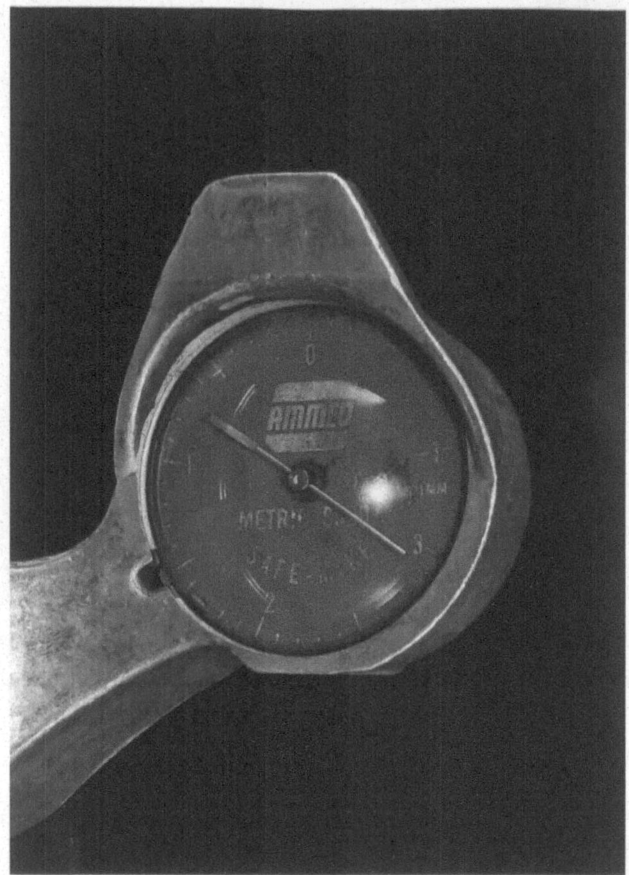

FIGURE 13-42 The dial indicator on the metric micrometer measures to 0.1 mm.

FIGURE 13-43 The dial indicator plunger can move inward, decreasing the diameter reading by up to 1 mm. Moving the plunger outward adds to the measurement.

FIGURE 13-44 Measuring the brake drum. Make sure you move the mic around slightly inside the drum to obtain the largest diameter reading.

FIGURE 13-45 The final reading of the drum's diameter is 225.7 mm.

the mic between anvils. Note this reading, remove the micrometer, and reinstall it 90° from the first position and remeasure. This is because drums are checked for out-of-round. Out-of-round drums cause brake pulsations when the brakes are applied (**Figure 13-46**). This second measurement is compared to the first to determine if the drum is out-of-round.

If the drum diameter is larger than machine-to or maximum, it will need to be replaced. If the drum is out-of-round, has excessive scoring or other surface defects, but is within the wear specification, it will need to be machined before returning to service.

■ ***Machining a Brake Drum.*** Because there are different types of brake lathes in use, the steps shown here are general and are not meant to be specific to any lathe or drum type. Before you attempt to machine a drum, read the lathe manufacturer's safety and operating manual. Brake lathes present a danger from the spinning components and from the metal shavings coming off the drum during the

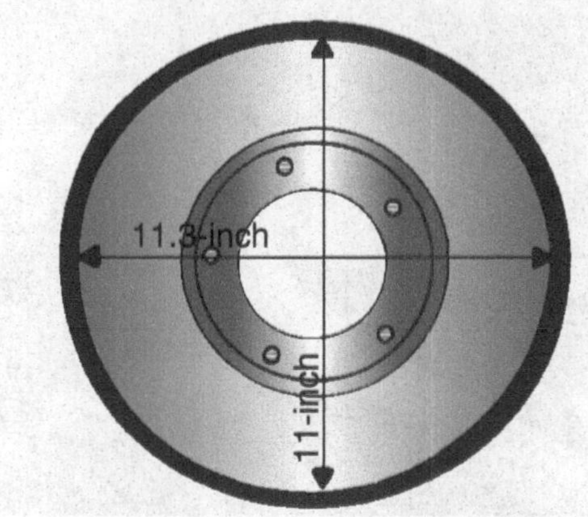

FIGURE 13-46 Check the drum across two diameters to check for out-of-round.

FIGURE 13-47 Examples of adaptors to fit different-sized drums and rotors.

FIGURE 13-48 Install a cone and spring before the hub adaptor. The cone supports the drum and the hub adaptor keeps it centered on the lathe.

FIGURE 13-49 Once the drum is mounted, secure the vibration band around the drum, covering the entire width of the friction surface.

machining process. Always wear safety glasses or a face shield while you are operating a brake lathe.

To machine a brake drum, first determine the correct mounting method. **Figure 13-47** shows examples of different types of lathe mounting adapters. Typically, a floating drum will require using a hub adapter, spring, and two open mounting adapters. Install an open adapter, spring, and hub adapter onto the main arbor (**Figure 13-48**). Next, place the drum on the hub adaptor, install the outer open adapter, spacers, and arbor nut. Wrap the drum vibration damper belt around the outside of the drum and secure it in place (**Figure 13-49**). Start the lathe and ensure that the drum is not wobbling or vibrating as it spins.

Next, position the drum as close to the lathe as possible by moving the arbor feed, then move the drum out slightly. Placing the drum close to the lathe helps reduce vibration in the drum during machining. Place the cutting arm near the inside edge of the drum where the friction surface and face come together. Turn on the lathe, and slowly feed the cutting arm and bit until the bit just contacts the drum surface. Now back the bit off the drum, and turn the lathe off. Loosen the drum on the arbor, and reposition the drum 180° on the arbor, then retighten the arbor nut. Turn the lathe on, and turn the cutting arm until the bit just contacts the drum a second time. Turn the lathe off, and compare the scratch cuts. Both cuts should be parallel on the drum surface (**Figure 13-50**). If the cuts are parallel, the drum is mounted correctly and is ready to machine. If the cuts are opposite each other, 180° apart on the drum's surface, dismount the drum and inspect the mounting components for damage. Scratch cuts opposite each other indicate there is runout in the drum setup. Remount the drum, and perform the two scratch cuts again.

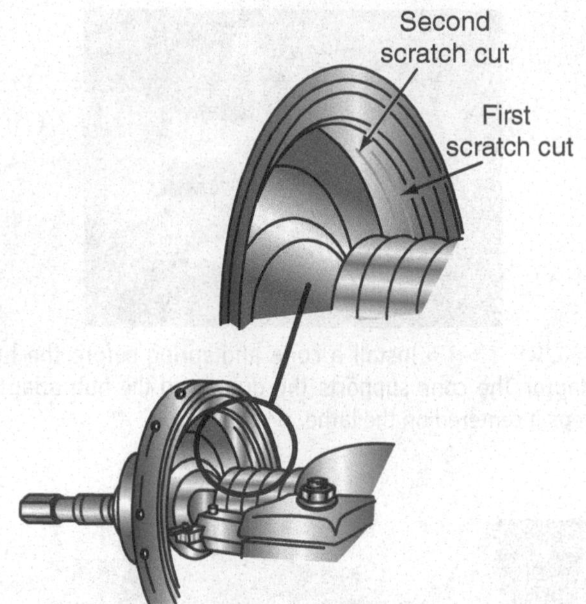

FIGURE 13-50 Making a scratch cut to check lathe setup. If the marks align, the setup is acceptable.

To machine the drum, position the cutting arm near the center of the drum surface, and move the bit until it just touches the drum. Next, move the drum so that the cutting bit is positioned at the inside edge of the friction surface. Turn the cutting arm feed so that the bit will cut into the drum about 0.004 to 0.006 inch (0.1 to 0.15 mm) Engage the lathe feed speed for a rough or fast cut. Once the cut is complete, turn the lathe off and inspect the drum. If the surface defects and/or low spots are removed, the drum is ready for the final slow cut. If the drum shows surface defects or low spots, repeat the rough cut until the surface is free of defects. Performing the final slow cut is the same as a rough cut except the arbor feed speed is much slower.

Once the final cut is done, turn the lathe off and inspect the drum. The surface finish should be smooth and free of all defects. Dismount the drum from the lathe, and clean the friction surface with soapy water to remove all metal dust. Dry the drum and remeasure its final diameter. If the drum diameter is beyond the machine-to specification, the drum must be replaced.

■ ***Wheel Cylinder Replacement.*** If during your inspection you find a leaking wheel cylinder, it must be either rebuilt or replaced. Most technicians replace leaking wheel cylinders, as the cost is often less for a new cylinder than it is for a rebuild kit and the labor to overhaul the cylinder.

To remove the wheel cylinder, first remove the brake line using a flare or line wrench. Next, remove the two bolts or the clip that secures the cylinder to the backing plate (**Figure 13-51**). Remove the cylinder and compare

FIGURE 13-51 Wheel cylinders are typically bolted to the backing plate.

the new and old parts. If correct, install the new one into the backing plate. Thread the brake line fitting into the cylinder by hand, make sure it threads in straight and sets properly. Installing the fitting first allows you to make small adjustments to the cylinder's position to make getting the fitting started easier. Next, install the bolts or clip that secure the cylinder in place and tighten to specifications, then tighten the brake line fitting. Do not overtighten the fitting as this can crack the flare or strip the fitting's threads. Once the brakes are reassembled, you will need to bleed the hydraulic system of air. Bleeding the brake system is discussed in Chapter 11.

■ ***Brake Reassembly.*** Before you start to reassemble the brakes, lay out all of the parts as they appear when installed. This can help you determine in what order the parts should be installed and ensures that you have all the parts.

Apply a very light coating of brake lubricant to the raised pads on the backing plate (**Figure 13-52**). This allows the shoes to move back and forth without making noise. Apply a very light coat of lube to any points where there is movement between components, such as at pivot points for adjusters or parking brake levers.

There are several types of synthetic and silicone-based brake lubricants available, designed for high-temperature brake applications. For backing plates, high-temperature synthetic grease is often used, but always refer to the manufacturer's service recommendations before applying any type of lubricant to brake parts. Do not use a low-temperature grease or any grease not suited for brake applications. Using the incorrect lubricant may cause the lube to migrate or move when hot, contaminating the brake linings.

Before beginning to install the shoes, you may cover the linings with masking tape (**Figure 13-53**). This prevents lube and any leftover dirt from contaminating the linings during installation.

FIGURE 13-52 Apply a light coat of brake lubricant to the support pads before reassembling the shoes.

FIGURE 13-53 Many technicians tape the linings to keep dirt from contaminating the new shoes.

Begin to reassemble the pieces in the reverse order from which you took it apart. For servo brakes, install the shoes and holddown springs first, then attach a return spring. Install the parking brake strut, self-adjuster levers or guides, and then the second return spring. Once the shoes are in place, install the self-adjuster assembly. Check that all parts are installed correctly and that the shoes are seated on the pads and against the anchor. Nonservo brake reassembly is similar. Install the parts in reverse order of disassembly.

Once the entire assembly is together, you will need to adjust the shoe-to-drum clearance. There are two ways this is done: using a clearance gauge or by repeating installing and testing the drag of the drum.

A **clearance gauge** is used to measure the drum diameter and then transfer that dimension to the shoes (**Figure 13-54**). If the distance across the shoes is smaller than the drum's diameter, turn the self-adjuster until the two are about the same. Then install the drum and check the amount of drag between the drum and the shoes.

Some technicians perform this adjustment by installing the drum and feeling the drag between the drum and the shoes. If the drum rotates freely without any drag from the shoes, then the adjuster is moved out until the shoes just begin to touch the drum. Many brake designs have an access hole in the lower section of the backing plate (**Figure 13-55**). This allows access to adjust the shoes with the drum installed. The self-adjuster on nonservo brakes is usually located up near the top of the brake shoes and an access hole is sometimes provided (**Figure 13-56**).

Regardless of how you perform this adjustment, it is important that the manufacturer's procedures are followed. Typically, the drum will have a slight amount of drag from contacting the shoes. However, the vehicle manufacturer may specify adjusting the shoes to contact the drum and then backing the adjustment in slightly. It is important not to adjust the shoes out too far and create excessive drag between the shoes and the drum. The proper adjustment provides a high and firm brake pedal and allows the parking brake to be applied properly. If the shoe-to-drum clearance is excessive, the brake pedal will be low and parking brake travel will be excessive. Make sure that the shoes are not adjusted too far, making the drum tight. This can make the brakes grab, overheated, and lead to glazing and damage.

Once the brakes are properly adjusted, remove the drum and perform a final inspection of your work. If everything is correct, reinstall the drum. If the wheel cylinder was serviced, you will need to bleed the system. This should be done before the wheel and tire are reinstalled. Once it is complete, install the wheel and tire, and torque the wheel fasteners to specs.

Be sure to test-drive (if able) the vehicle before you return it to the customer. Ensure proper parking brake operation as specified by the manufacturer.

■ ***Drum Parking Brake Service.*** The effectiveness of a parking brake on a vehicle with rear drum brakes depends on how well the shoes are adjusted. When the parking brake is applied, the rear parking brake cables pull a lever attached to one of the brake shoes

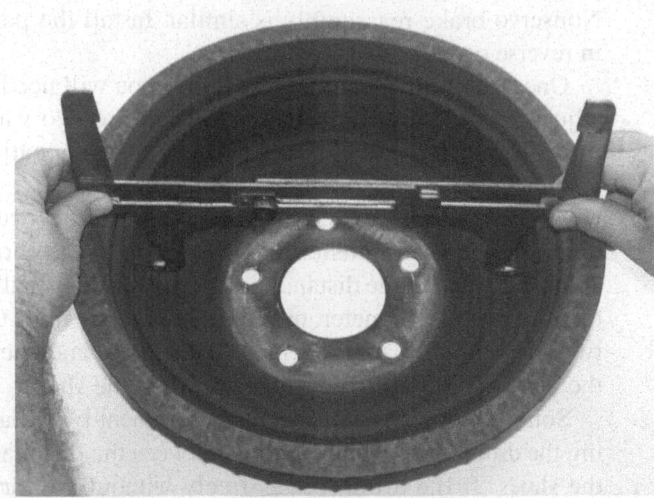

FIGURE 13-54 (a) Setting the adjustment tool to the drum's diameter. (b) Transferring the drum's diameter to the shoes. Preadjusting the shoes saves time once the drum is installed.

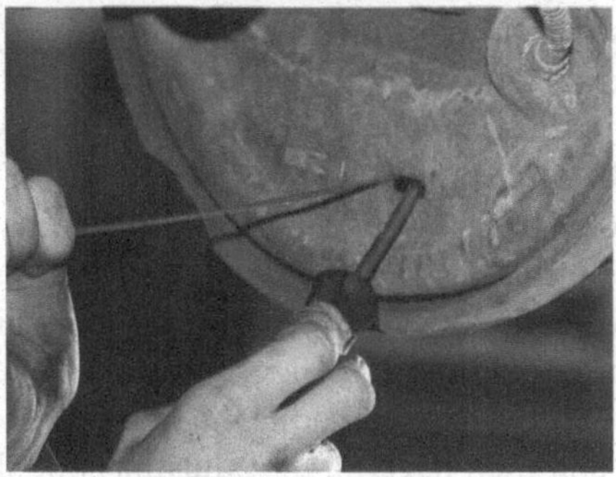

FIGURE 13-55 Adjusting the brake shoes from the access hole in the backing plate.

FIGURE 13-56 An example of an adjustment hole and rubber plug.

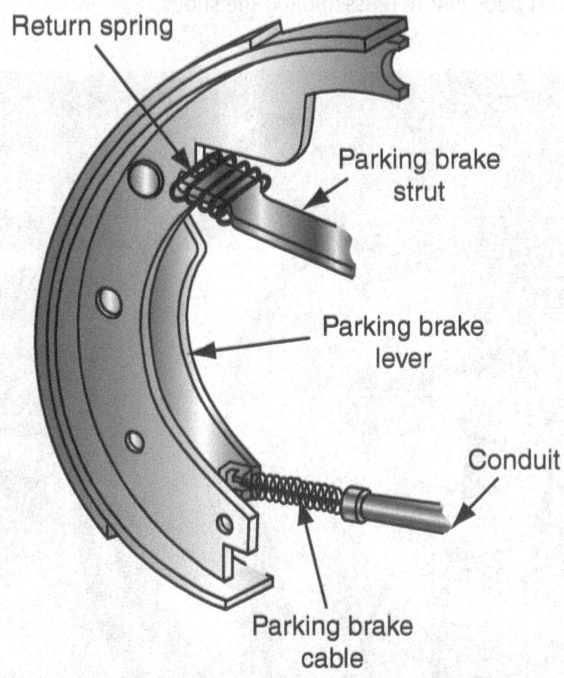

FIGURE 13-57 When the parking brake cable pulls on the lever, the shoe pushes against the strut, which applies the forward shoe against the drum. The rear shoe is pushed against the drum by the action of the parking brake lever.

(**Figure 13-57**). As the parking brake lever moves, it pushes against either a parking brake strut or the self-adjuster, depending on the particular style of brakes on the vehicle. If the shoe-to-drum clearance is small, as with properly adjusted brakes, then the parking brake will be able to force the shoes tightly against the drum, locking it in place. However, if the shoe-to-drum clearance is excessive, the parking brake travel may not be sufficient to force the shoes tightly against the drum, and the parking brake will not hold the vehicle in place.

Because the parking brake cables are located on the underside of the vehicle, they tend to rust and seize over time, making the parking brake inoperative. For those who live in areas of the country where rust is an issue, stuck parking brakes are a common problem. For technicians, it is important to check parking brake operation before servicing the rear drum brakes. If the parking brake is stuck in the applied position, from seized cables, and you replace the brake shoes, you may not be able to reinstall the brake drum over the new shoes. This is because the parking brake being applied forces the shoes apart, increasing the shoe diameter.

■ ***Inspect Parking Brake Operation.*** Checking parking brake operation can be tricky. If you apply the parking brake on a vehicle in which the customer does not regularly use the parking brake, you may set the brake and find that it will not release.

Begin by asking the customer if he or she uses the parking brake. If it is not regularly used, you should warn them that as part of a complete brake inspection, the parking brake must be checked, and that it is possible that when applied it may not release.

To check parking brake operation and adjustment, the vehicle manufacturer will have a procedure to follow. A common example is to raise the vehicle off the ground, and apply the parking brake a certain number of clicks, then check to see if the brake has applied. If you are confident that the parking brake is not seized, you should follow the service provided by the manufacturer.

If, however, you are uncertain about the brake releasing, you will need to advise the customer that testing the parking brake may result in it getting stuck and that additional repair work will be required to correct the problem.

■ ***Inspect Cables.*** Parking brake cables are strands of steel cable encased in a weather-resistant outer shell. The cables and adjusters are prone to rusting and seizing, and breaking (**Figure 13-58**). You can check cable operation, if you are careful, by pulling the cable while you are under the vehicle. **Figure 13-59** shows an exposed brake cable where you can attempt to pull to check cable operation. The cable should be tight and should move slightly when pulled. If the cable does not move, do not continue to pull. It does not take much force to actually pull the cable to start to set the parking brake. If it feels tight and you find you will have to strain to move the cable, the cable is probably seized.

It is sometimes possible to service a sticking cable and restore it to its proper working condition, although it is usually more cost and time effective to replace the cable.

■ ***Replacing Parking Brake Cables.*** Because the type and location of parking brake cables vary from vehicle to vehicle, the steps described here are general and are not specific to any particular make or model.

FIGURE 13-58 A broken cable, like this example, can keep the parking brake from setting firmly enough even on the working side of the car.

FIGURE 13-59 Connection points, like that shown here, can be used to check if the cables are seized by gently pulling on the connection and watching cable movement.

Replacing a rear cable means removing the cable from the drum brake assembly. This may require at least partial disassembly of the brake shoes to allow access to the parking brake lever. To remove the cable from the lever, hold the crimped end of the cable with pliers and pull the spring to allow the cable to slide out of the lever (**Figure 13-60**). Once the lever is off the cable, you will need to remove the cable from the backing plate. Most cables use a three-pronged retainer (**Figure 13-61**). Squeeze the prongs flat with pliers or a hose clamp, and slide the cable from the backing plate.

Some cables connect to an equalizer or adjuster between the passenger compartment and rear brakes, whereas others connect to other cables using a union (**Figure 13-62**). To separate the cable from the connector, hold each in a pair of pliers and pull the cable end

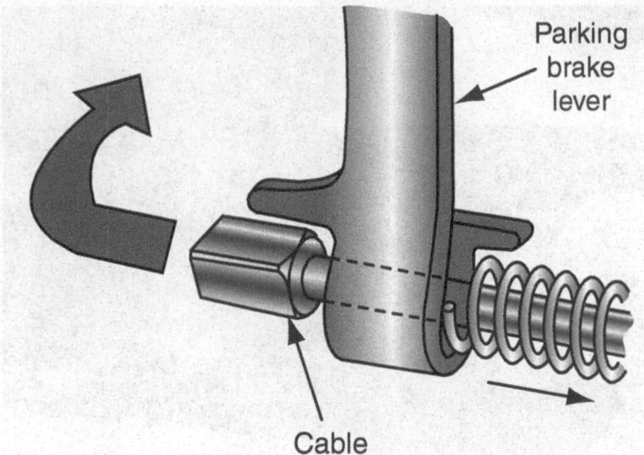

FIGURE 13-60 To remove the cable, pull the anchor out of the hook in the parking brake lever.

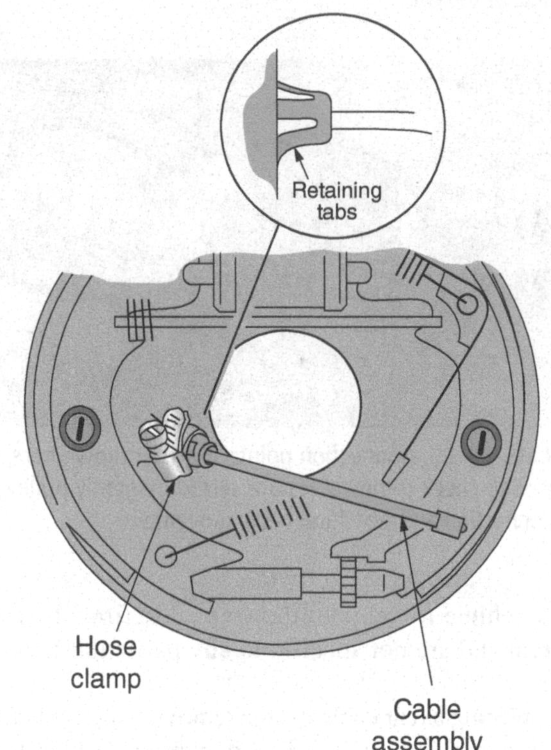

FIGURE 13-61 An example of how a parking brake cable attaches to a backing plate.

FIGURE 13-62 An example of a connection between the two rear cables.

from the union. The cable may require tapping out of the union with a punch and hammer if the two do not easily separate.

Replacing the front cable will require removing the cable from the hand or foot brake. Begin by removing the rear connection of the front cable, either at the equalizer or where it connects to another cable. Once loose, pull the cable out of the parking brake actuator lever. Remove the cable from its mounts and brackets, and remove the cable from the vehicle.

Before installing the new cable, match it to the old cable to ensure it is the correct length and has the same types of ends. Route the new cable into place, and secure it by reattaching any clamps or brackets. Connect the cable ends, and make sure the end is secure in its connection.

Once it is installed, you need to adjust the parking brake. This is usually done by tightening an adjustment nut at the equalizer (**Figure 13-63**). To set the final tension, refer to the manufacturer's service information for the procedure to test parking brake operation. This usually requires setting the brake a certain number of clicks, and then checking to see if the wheels spin. Proper adjustment of the parking brake is important because if it is left too loose, the brake will not hold. If the adjustment is too tight, this can cause the parking brake to remain applied all the time.

■ ***Drum in Hat Parking Brakes.*** Some rear disc brake systems use a set of miniature shoes housed within the hat of the rotor for the parking brake. This arrangement is called **drum in hat**.

Because it is used as a parking brake, the shoes are applied mechanically, through the parking brake lever or pedal and parking brake cables. The cable pulls on a cam, which pushes the shoes apart. The assembly uses two shoes, holddown springs, return springs, and

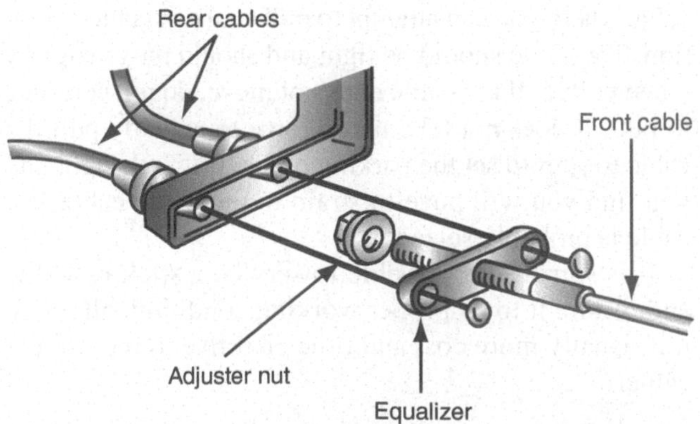

FIGURE 13-63 Adjustments are typically made at the equalization point where the cables come together.

an adjuster mechanism nearly identical to a regular rear drum brake system.

Though it is used only as the parking brake, the shoes and hardware will eventually wear or rust and need to be replaced. An example of a high-mileage drum in hat system is shown in **Figure 13-64**. Note that the linings are worn and are falling apart. Another concern with some vehicles is that the backing plate that holds the drum in hat assembly tends to rust apart (**Figure 13-65**).

Servicing these brake shoes is similar to regular rear drum brake assemblies. Remove the return springs and holddown springs to remove the shoes. Clean and lubricate the backing plate and adjuster. Reinstall the shoes and rotor, and adjust the shoe clearance. Shoe clearance may be adjusted through the backing plate, as in standard drum brakes and as shown earlier in Figure 13-55.

■ ***Parking Brake Warning Indicator*** When the parking brake is set, the red BRAKE warning light should illuminate on the dash (**Figure 13-66**). This alerts the driver that the parking brake is set. Some vehicles also use an audible warning, such as beeping chime, to alert the driver if the brake is set and the vehicle is in motion.

If the warning light does not illuminate with the parking brake set, determine if the same light is used by the brake fluid level sensor or pressure differential valve. If it is the same light, try to trigger the light by unplugging or grounding the level sensor or pressure differential switch. If the light still does not illuminate, suspect a burned-out bulb, which requires at least partial instrument panel disassembly to correct.

If the light illuminates from the fluid level or pressure differential switch, locate the parking brake indicator switch, which is located with the parking brake lever or pedal. Using a wiring diagram, you can test the switch and wiring. Testing switches is covered in detail in Chapter 18.

FIGURE 13-64 A drum in hat parking brake that has seen a few miles and shows some wear.

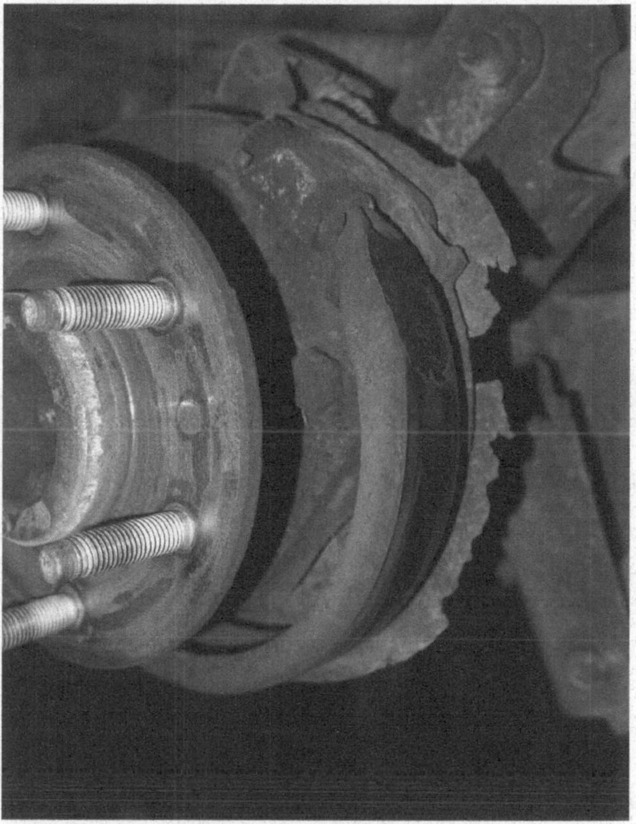

FIGURE 13-65 Some vehicles have problems with the backing plate for the drum in hat parking brake rusting away.

FIGURE 13-66 The BRAKE light should illuminate when the parking brake is set. If it does not, check the electrical switch mounted to the parking brake actuator for power.

SUMMARY

A special brake drum micrometer is used to measure the wear of the drum's friction surface.

Drum brake pulsation occurs if the drum becomes out-of-round.

Dust accumulation in the drum brake assemblies can cause noise complaints.

When you are replacing brake shoes, take a picture of the brakes before disassembly.

Lay out the old parts, and compare them with the new parts during lining replacement.

Measure and adjust the rear shoe clearance before adjusting the parking brake.

Refer to the manufacturer's service information for the correct type of lubricant to use on drum brake components.

Measure the brake drum with a micrometer to determine if it is within wear limits to remain in service.

REVIEW QUESTIONS

1. An out-of-round brake drum can cause a brake pedal ______________ during braking.

2. Before servicing the drum brakes, clean the ______________ from the brake assemblies.

3. Dust accumulation between the shoes and drum can be a source of brake ______________.

4. Before adjusting the parking brake, you should first check the ______________ to ______________ clearance.

5. Full-floating rear differentials may require removing the ______________ ______________ to remove the drums to inspect the brakes.

6. A vehicle with rear drum brakes has a grinding noise when the brakes are applied. *Technician A* says worn rear brake linings can cause this. *Technician B* says a leaking wheel cylinder can cause this. Who is correct?

- **a.** Technician A
- **b.** Technician B
- **c.** Both A and B
- **d.** Neither A nor B

7. The service brake pedal is low but firm and the parking brake does not hold the vehicle in place. Which is the most likely cause?

- **a.** Broken parking brake cable
- **b.** Misadjusted parking brake
- **c.** Parking brake stuck on
- **d.** Excessive shoe-to-drum clearance

8. The parking brake does not lock the rear wheels when fully applied. *Technician A* says excessive shoe-to-drum clearance may be the cause. *Technician B* says the parking brake may need to be adjusted. Who is correct?

- **a.** Technician A
- **b.** Technician B
- **c.** Both A and B
- **d.** Neither A nor B

9. When the rear brake shoes are replaced, a return spring breaks from age and rust. *Technician A* says only the broken spring should be replaced. *Technician B* says all of the brake springs should be replaced. Who is correct?

- **a.** Technician A
- **b.** Technician B
- **c.** Both A and B
- **d.** Neither A nor B

10. *Technician A* says to correct a brake pulsation concern, the rear drums may need to be machined. *Technician B* says an on-car brake lathe can be used to correct out-of-round drums. Who is correct?

- **a.** Technician A
- **b.** Technician B
- **c.** Both A and B
- **d.** Neither A nor B

CHAPTER 14

Disc Brake System Principles

Chapter Objectives

At the conclusion of this chapter, you should be able to:

- Identify types of disc brakes.
- Identify disc brake components.
- Describe disc brake operation.
- Describe rear disc brake and parking brake operation.

KEY TERMS

brake pads	**dust boot**	**shims**
brake rotors	**fade**	**square seal**
caliper hardware	**fixed calipers**	**squealer**
caliper pistons	**floating calipers**	
drum-in-hat	**caliper hardware**	

Even though disc brakes are not a new design, they were not widely adopted for use in passenger vehicles until the 1960s. Drum brakes were used on the front and rear of many vehicles, and disc brakes did not become standard equipment on many domestic vehicles until the 1970s. Since then, disc brakes have become standard on the front and rear of most passenger cars and light trucks sold today.

Disc Brake Systems and Components

Front disc brakes are standard on all modern cars and light trucks. The majority of modern vehicles use disc brakes for the rear brakes as well. The main advantages of disc brakes compared to drum brakes are:

- Increased resistance to brake fade
- Self-cleaning of dust and debris
- Self-adjusting

A disadvantage of disc brakes is that it requires a lot of force to clamp the pads against the brake rotor. This increases the effort by the driver to slow and stop the vehicle. Because of this, disc brake-equipped cars use a power assist system to increase brake force and decrease driver effort and fatigue.

Disc Brake Components and Operation

All disc brakes are similar in design and operation: pressurized brake fluid forces the caliper piston outward from the caliper bore, which applies pressure against the brake pads (**Figure 14-1**). This squeezes the two brake pads against the rotor, also called *a disc*. The pressure and friction slow the rotor and create heat. The heat is dissipated into the air.

DISC BRAKE CALIPERS

The caliper is the hydraulic output for the disc brake system. Each caliper contains one or more pistons. The number of pistons depends on caliper design and its application on the vehicle. The caliper houses the piston(s) and pads and attaches to the steering knuckle, partially covering the rotor (**Figure 14-2**).

When the brakes are applied, hydraulic pressure pushes on the back of each piston. A square-cut seal surrounds each piston and seals fluid within the piston bore (**Figure 14-3**). As the piston moves outward in its bore, the square-cut seal deforms slightly, twisting to follow the piston. When the brakes are released, the seal returns to its original shape, pulling the piston back into the bore. By doing this, the square-cut seal acts as a return spring for the disc brakes (**Figure 14-4**).

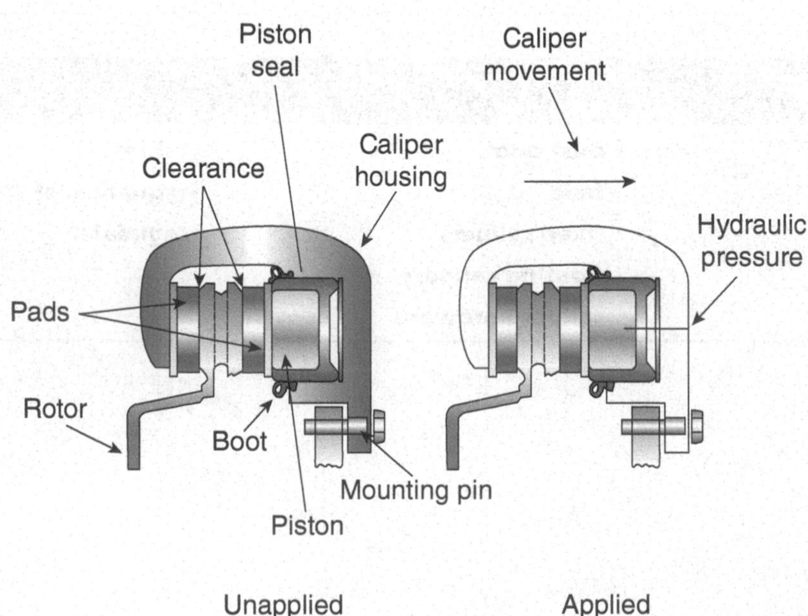

FIGURE 14-1 Basic operation of the disc brake system; hydraulic pressure pushes the pistons against the pads, which rub against the rotor to slow its rotation speed.

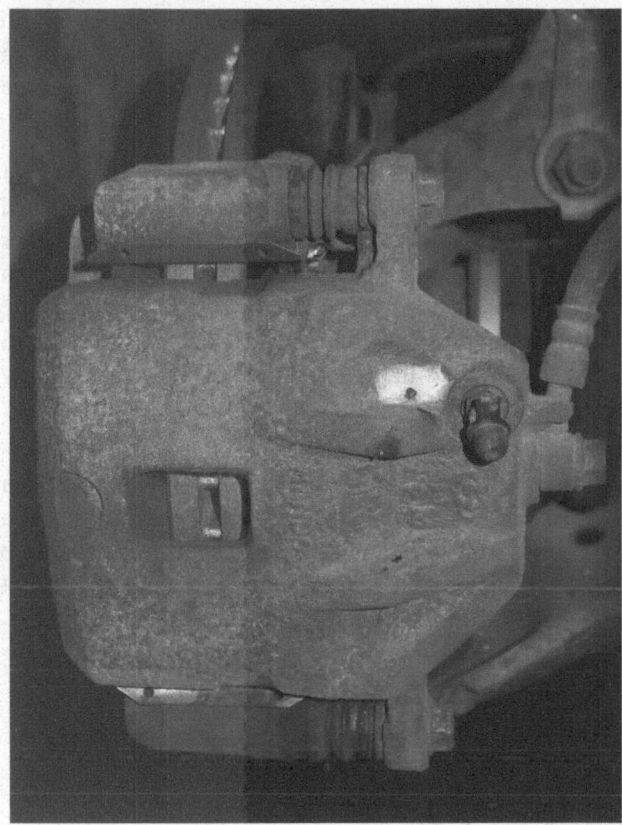

FIGURE 14-2 An example of a common type of a front disc brake assembly. The caliper covers the pads and part of the rotor.

As the brake pads and rotors wear, the pistons will move further out of their bores to compensate for the space left by the worn pads and rotors. Over time, as the pistons move further out, brake fluid fills the void in the piston bore. When the brakes are released, the piston seal returning to its normal shape cannot force the piston back more than a very slight amount. So, each time the brakes are applied and the pads wear slightly, the piston also moves a very slight amount further out of the bore (**Figure 14-5**).

FIGURE 14-3 An inside view of the caliper piston and square seal.

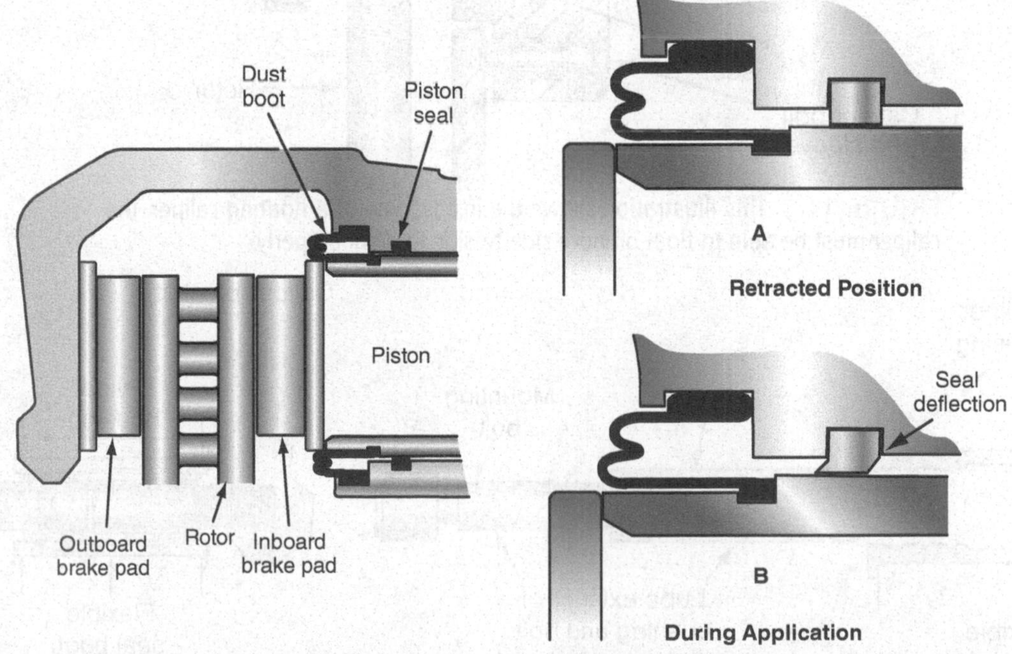

FIGURE 14-4 During operation, the square seal deforms slightly. When pressure drops, the seal returns to its original shape, which pulls the piston back into its bore.

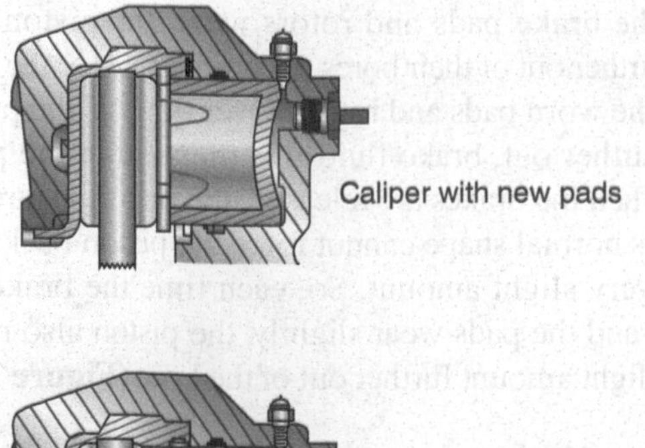

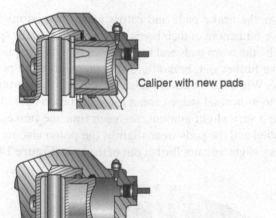

FIGURE 14-5 As the pads and rotors wear, the caliper pistons move further out of their bores. This reduces the fluid level in the master cylinder as the volume of fluid behind each piston increases.

A **dust boot** protects the outside of the piston and the piston bore from dirt and debris. The dust boot is an accordion-type seal, meaning it will expand to cover the piston as the piston moves outward in the bore.

■ ***Floating Calipers.*** There are two major types of calipers: floating and fixed. The most common type used in modern cars and light trucks are **floating calipers** (**Figure 14-6**). This caliper gets its name because it floats on its mounting hardware and moves side-to-side as the brakes are applied. A floating caliper must be able to move so that both the inner and outer brake pads apply against the rotor.

Floating calipers are smaller, lighter, and less expensive than fixed calipers and less prone to pulsation from rotor runout. Lateral runout is the side-to-side wobble of a brake rotor as it spins on the hub and is discussed in more detail in Chapter 15. Floating calipers are mounted so that it can move side-to-side on bolts or pins (**Figure 14-7**).

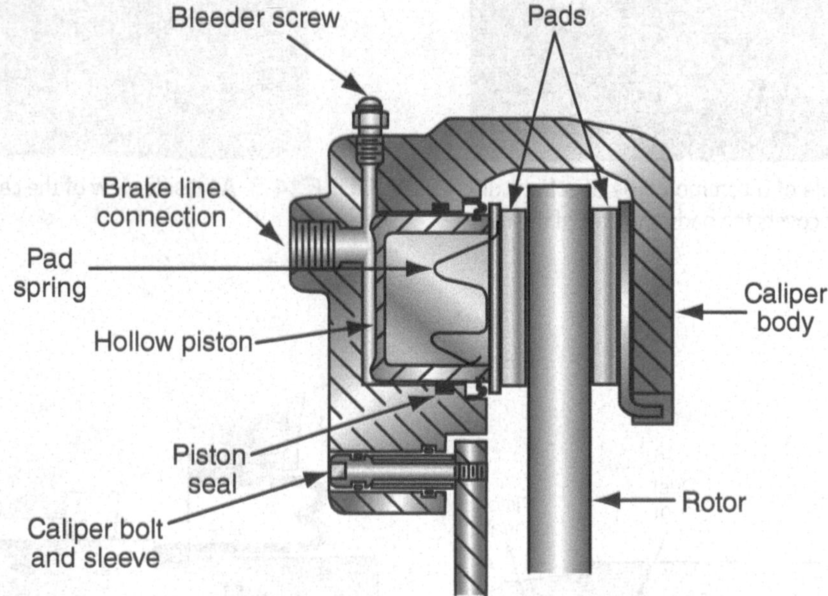

FIGURE 14-6 This illustration shows the inside view of a floating caliper. The caliper must be able to float or move side-to-side to work properly.

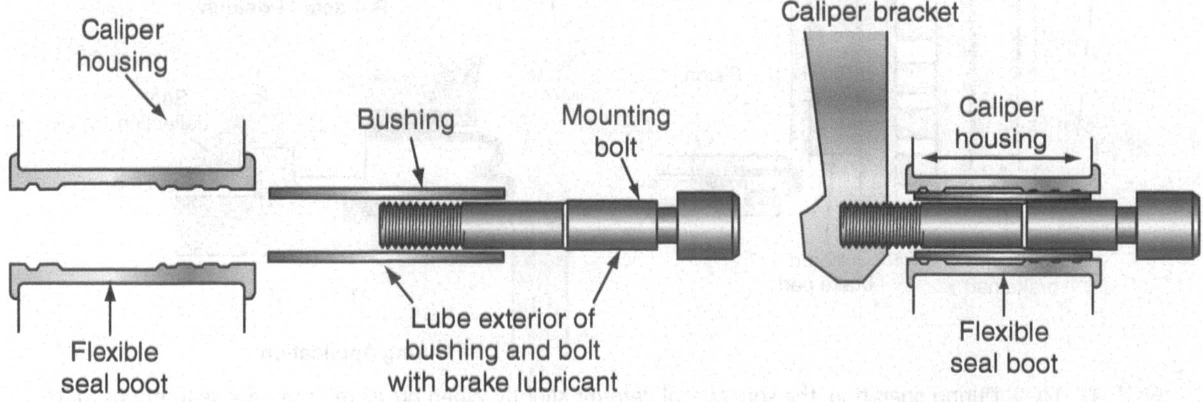

FIGURE 14-7 An example of mounting bolts on a floating caliper. The bushings and bolts or pins allow the caliper to move during application and when the brakes are released.

To operate, floating calipers take advantage of Newton's Third Law of Motion: for every action there is an equal and opposite reaction (**Figure 14-8**). When the fluid pressure pushes on the back of the caliper piston, the piston pushes on the inner pad and the inner pad against the brake disc. As the fluid pushes against the piston, it also pushes against the rear of the caliper piston bore. This pushes the caliper rearward or opposite the movement of the piston. The backward movement of the caliper pushes against the outer brake pad, applying equal force to both pads. For this caliper to work correctly, it must be able to move easily on the mountings and the pads must be able to move within the caliper or on the caliper bracket.

A variation of the floating caliper is the sliding caliper (**Figure 14-9**). The operation of the sliding caliper is the same as that of the floating caliper, except instead of being mounted with bolts and bushings, and it is mounted on sliding keyways.

■ ***Fixed Calipers.*** **Fixed calipers** are mounted directly to the steering knuckle and have at least two pistons (**Figure 14-10**). Because hydraulic pressure is exerted equally, each piston receives equal pressure and pushes on the brake pad with the same force (**Figure 14-11**). This type of caliper is used mainly in sports cars and high-performance applications.

Fixed calipers have at least one bleeder screw, and some designs have more than one to provide complete bleeding of air from the system. A brake hose supplies brake fluid to the caliper. Transfer tubes or internal passages in the caliper supply fluid to the outer pistons.

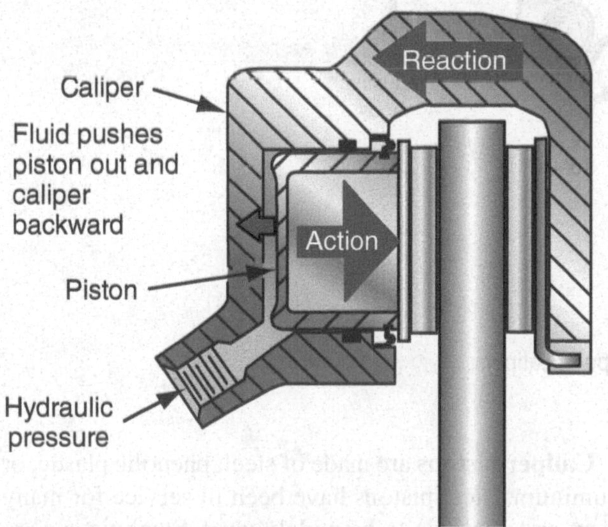

FIGURE 14-8 Floating calipers use Newton's Third Law of Motion—for every action there is an equal and opposite reaction. If the caliper is floating properly, the hydraulic force on the piston also pushes the caliper backward to apply force against the outboard pad.

■ ***Caliper Construction.*** A basic caliper consists of the caliper housing body, internal fluid passage, piston, square seal, dust boot, and bleeder screw (**Figure 14-12**). The piston bore is machined to fit very closely to the piston, usually within a few thousandths of an inch

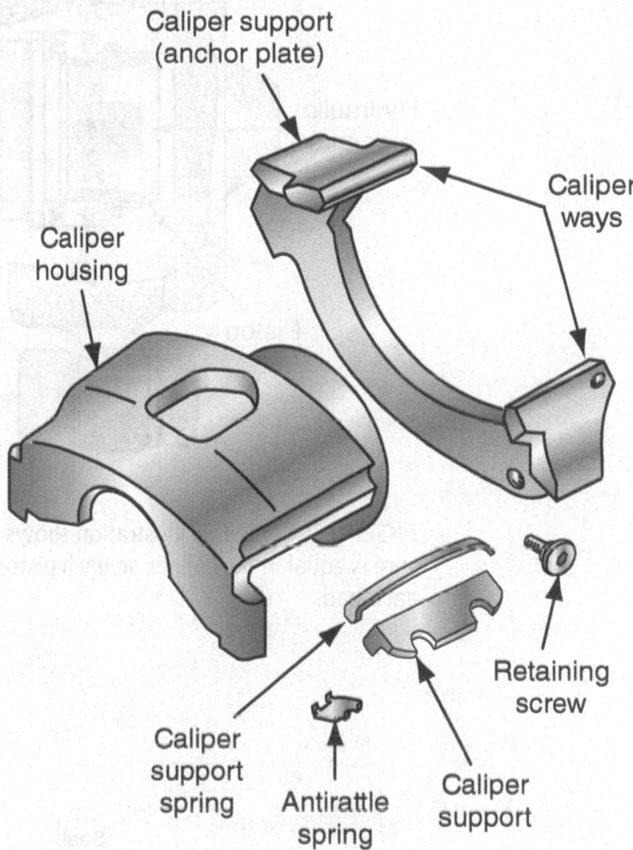

FIGURE 14-9 An illustration of a type of sliding caliper.

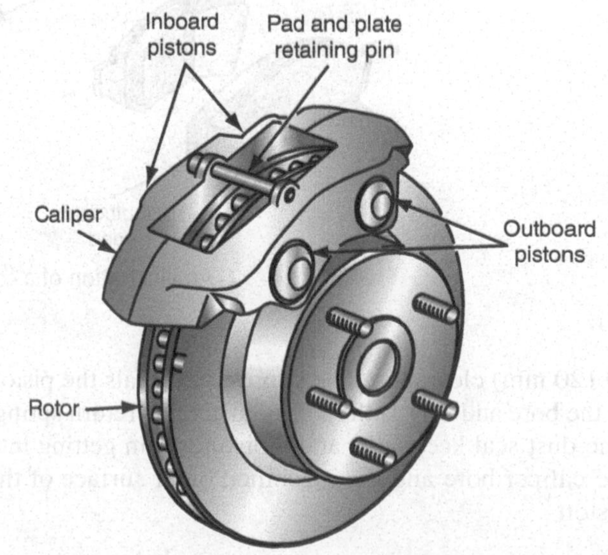

FIGURE 14-10 An illustration of a fixed caliper. This design has pistons on both sides of the rotor.

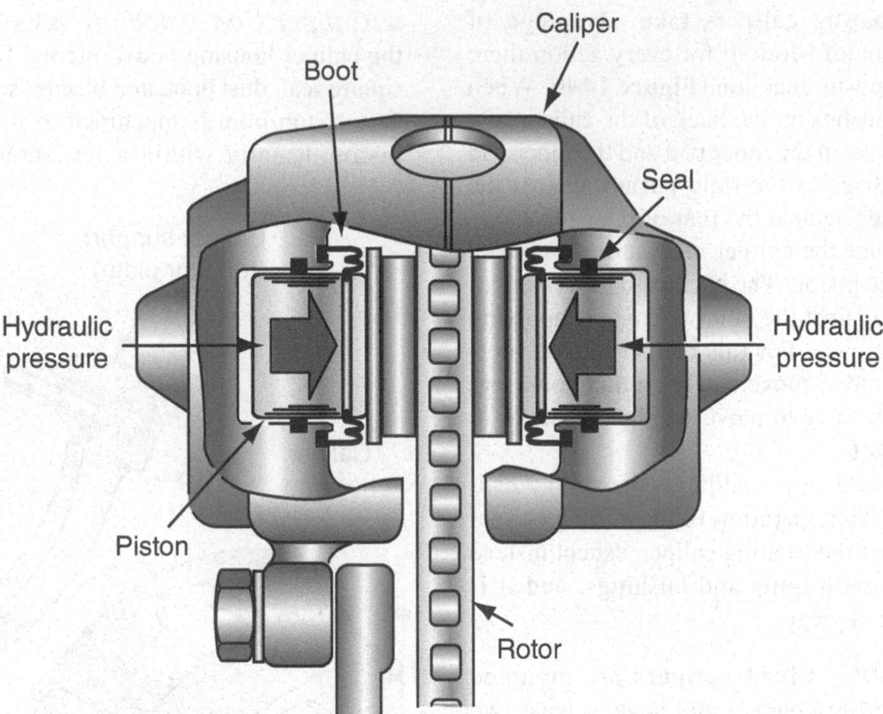

FIGURE 14-11 This illustration shows the operation of a fixed caliper. Hydraulic pressure is equal in the caliper, so each piston moves with the same amount of force against each pad.

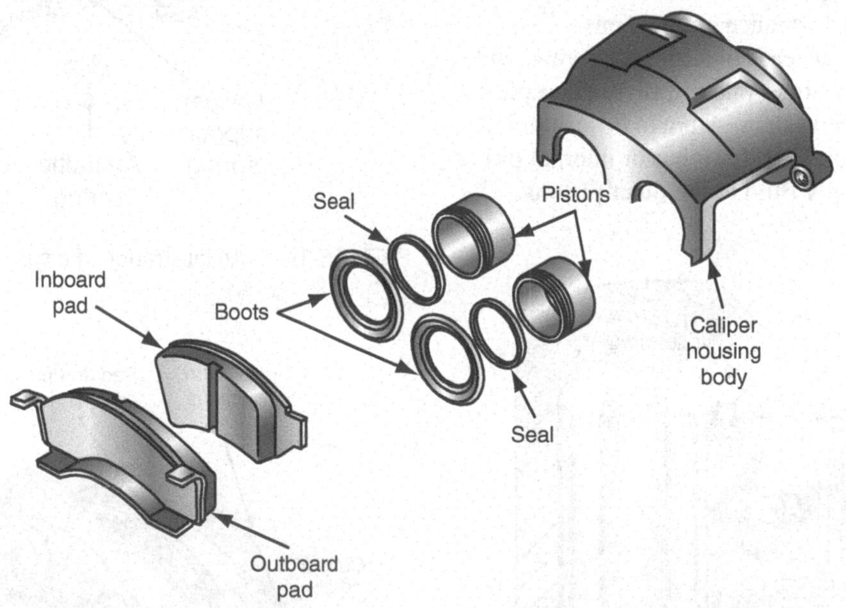

FIGURE 14-12 An illustration of a common type of caliper.

(0.120 mm) clearance. The **square seal** seals the piston in the bore and acts as the self-adjuster and return spring. The dust seal keeps dirt and moisture from getting into the caliper bore and the machined outer surface of the piston.

Caliper pistons are made of steel, phenolic plastic, or aluminum. Steel pistons have been in service for many years and continue to be widely used. Steel pistons are strong and can be fitted very close in the caliper bore, usually within 0.005 inches. Phenolic plastic pistons have also been in use for many years and are used to reduce weight. Plastic pistons, while lighter, have much thicker walls than comparable steel pistons to increase strength. Plastic pistons tend to expand more than steel pistons and require a larger clearance to the piston bore, typically 0.008 to 0.010 inches. Many calipers use aluminum pistons. Aluminum pistons are lightweight and strong, having the advantages of both steel and plastic pistons.

The caliper piston seal is a square-cut seal, housed in a recess in the piston bore (**Figure 14-13**). The inner surface of the seal encircles the piston, prevents fluid loss, and acts as a piston return spring. The dust boot is attached to the outside edge of the piston bore, and its inner surface attaches to the outer lip of the piston (**Figure 14-14**). As the piston moves outward, the dust boot extends to protect the outside surface of the piston. Dust boots may be secured to the caliper with a snap ring, by a press-fit between the seal and the caliper, or with the lip of seal placed into a groove cut into the caliper body.

FIGURE 14-13 The square seal keeps fluid in the caliper and acts as the return spring for the piston.

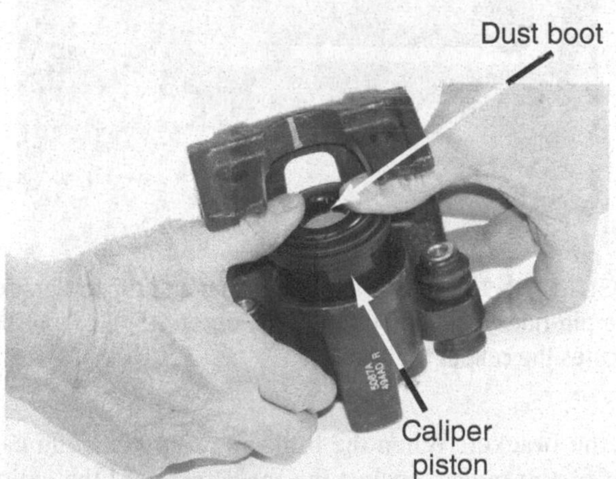

FIGURE 14-14 The square seal keeps the fluid inside the caliper (the seal is removed from the caliper for clarity). The dust boot keeps dirt and debris from damaging the piston and the caliper bore. The dust boot extends further out of the bore as the pads and rotor wear.

The caliper may have bores for the mounting hardware. This type often uses some type of bushing or rubber O-ring on which the mounting bolts can move. There are many different types of mounting bolts, pins, and bushing configurations, but all of them perform the same function, which is allowing the caliper to move back and forth during application so that both pads apply pressure to the disc. When these mounting components seize from rust and corrosion, the caliper cannot move properly, and uneven brake pad wear will occur.

Floating calipers have mounting bolts, pins, or bushing bores (**Figure 14-15**). Some calipers bolt directly to the steering knuckle (**Figure 14-16**). Most vehicles use a caliper support bracket to hold the pads and caliper to the knuckle (**Figure 14-17**).

Calipers that are bolted to the knuckle often use a combination mounting bolt and sleeve that rides in the caliper body (**Figure 14-18**). The bolt and sleeve allow the caliper to move backward as the piston is pushed against the inner pad.

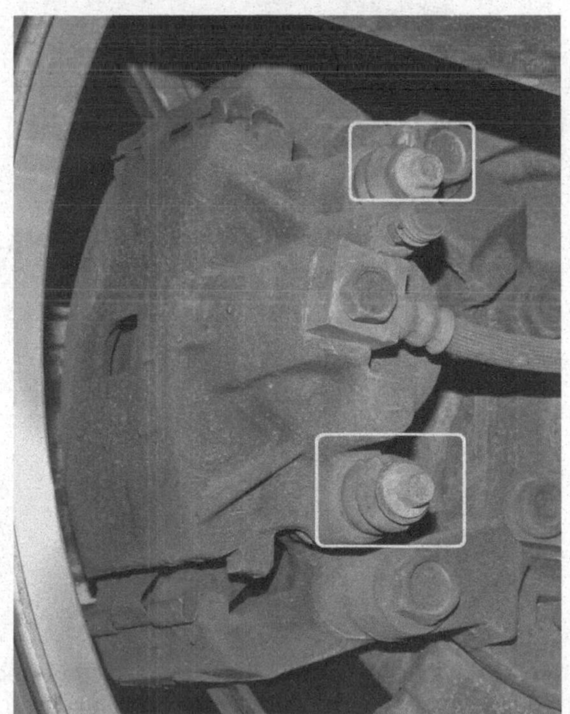

FIGURE 14-15 The bolts and sleeves allow the caliper to float in operation, meaning the caliper can move side-to-side as the piston moves out and back into the bore.

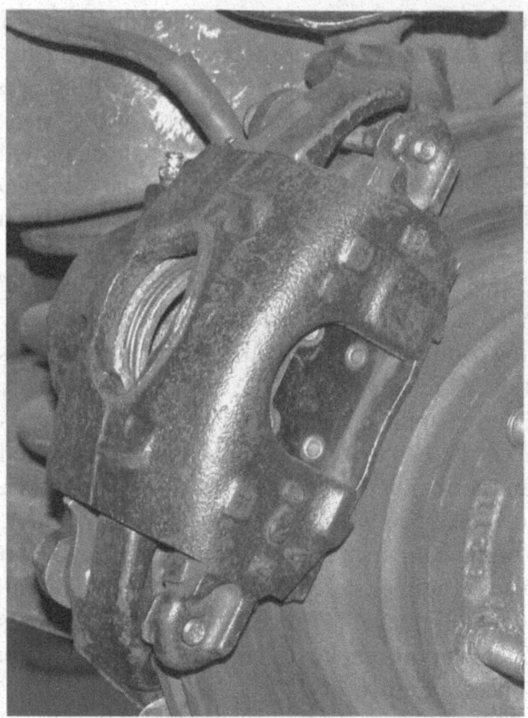

FIGURE 14-16 Some vehicles mount the caliper directly to the steering knuckle, as in this example.

This bushing
and bolt secure
the caliper to
the bracket.
These bolts hold
the bracket to the
steering knuckle.

FIGURE 14-17 Many cars and trucks use a support bracket that mounts to the knuckle, and the caliper then bolts to the bracket. The bracket often houses the pin bushings that allow the caliper to float.

Sleeve
Mounting bolt

FIGURE 14-18 An example of caliper mounting bolts with the sleeves inside the caliper housing.

Bracket
Caliper pin
Caliper
mounting
bolt

FIGURE 14-19 An example of a caliper and support bracket. The pin rides inside a busing in the bracket. The caliper bolt secures the caliper to the pin.

Vehicles that use a mounting bracket often use a type of floating pin that connects the caliper to the bracket (**Figure 14-19**). The caliper bolts to the pins, which float in the bracket. When the brakes are applied, the caliper piston moves against the inner pad, and the caliper moves backward as the pins extend out of their bores in the bracket.

Most brake arrangements use some type of **caliper hardware** to reduce pad noise. Pad clips or guides are used to keep the pads secure in the bracket and to reduce vibration and noise (**Figure 14-20**). The pads must be loose enough to move side-to-side in the clips, but held snug enough to prevent noise. Many pads also use **shims** attached to the pad backing to alter the frequency of pad vibrations so that it is inaudible to humans (**Figure 14-21**). Many replacement pads have the shims attached to the backing plate, while other pad sets have shims with adhesive backs that need to be applied to the pads when they are installed. Regardless of the type of shims and hardware, it is important that the pieces be installed correctly.

FIGURE 14-20 Caliper and pad hardware are used to hold the pads in place and reduce noise.

■ ***Disc Brake Rotors.*** The brake pads are half of the friction components in disc brake systems; the rotors are the other half. Rotors, also called *brake discs*, are mounted to the hub and rotate with the wheel and tire. As the caliper clamps the brake pads against the rotor, a substantial amount of friction is created between the pads and the two friction surfaces on the rotor (**Figure 14-22**). This friction is what slows the wheel and

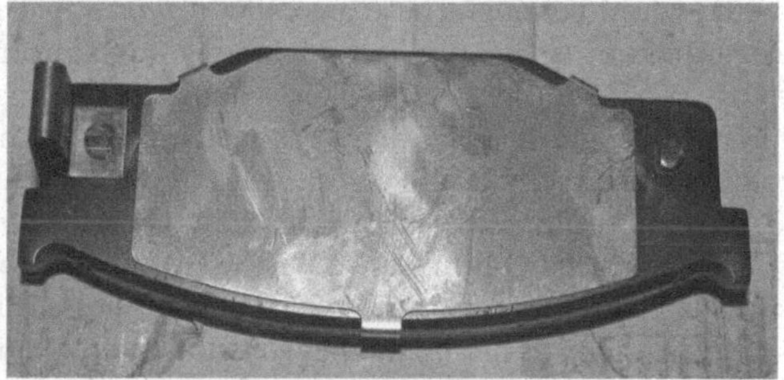

FIGURE 14-21 Shims are used to alter the frequency of pad vibrations so that it is inaudible to humans.

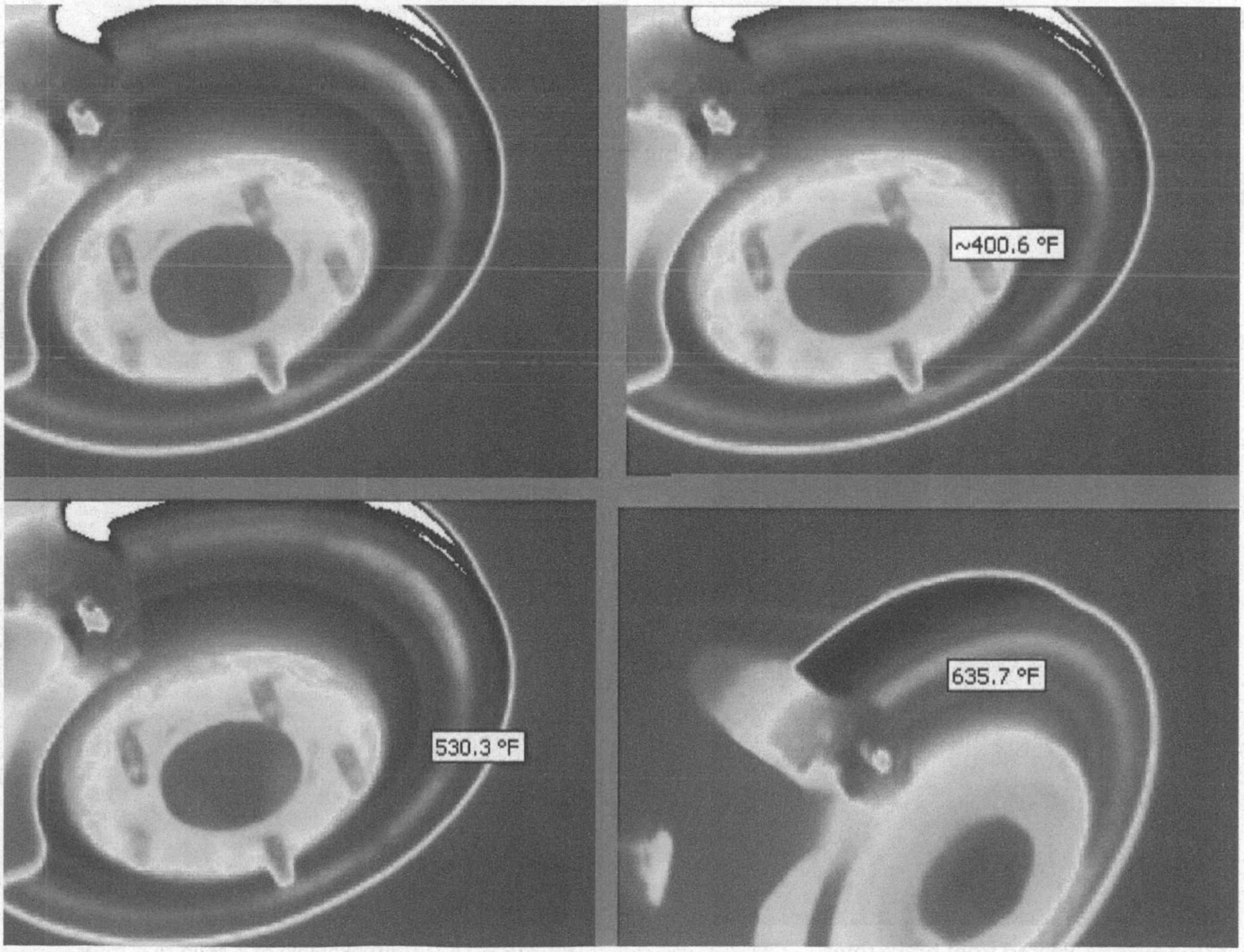

FIGURE 14-22 A thermal image of disc brakes in operation.

creates the intense heat generated by the brakes. Because of the stresses of braking and the heat that is generated, brake rotors must be strong and able to withstand high operating temperatures.

The most common **brake rotors** are made of cast iron and are vented in the center between the friction surfaces. The central part of the rotor, called the hat, provides the mounting point for the rotor on the hub. The center hub hole is designed to fit precisely to the hub. Rotors that slide onto the hub and are held in place by the wheel and lugs are called *floating rotors*. Some rotors, often on 4WD and some older FWD vehicles, are held in place by the wheel bearing. These are called *trapped rotors*. Regardless of which type a vehicle uses, replacement rotors should always be checked against the original rotors to ensure that the hat areas match and that the rotors will correctly fit the vehicle.

A vented rotor has vents located between the two friction surfaces (**Figure 14-23**). The overall friction surface area is large, but the contact area of the pads is small. As the pads are pressed against the rotor, heat is generated all around the rotor's surfaces. As the rotor spins, air is pulled through the center of the rotor and out through the vents to remove heat from the friction surfaces and cool the rotor (**Figure 14-24**). The front brake rotors on all modern cars and trucks are vented. Nonvented or solid rotors (**Figure 14-25**) are used on the rear of some vehicles, and they can be found on the front of some older, smaller vehicles. Nonvented rotors can be used in the rear because the rear brakes are doing less work than the front brakes, and the additional cooling is not necessary.

For many vehicles, the vents in the rotors are simply straight passages from the outside to the inside of the hub or hat section of the rotor. Some vehicles use rotors that

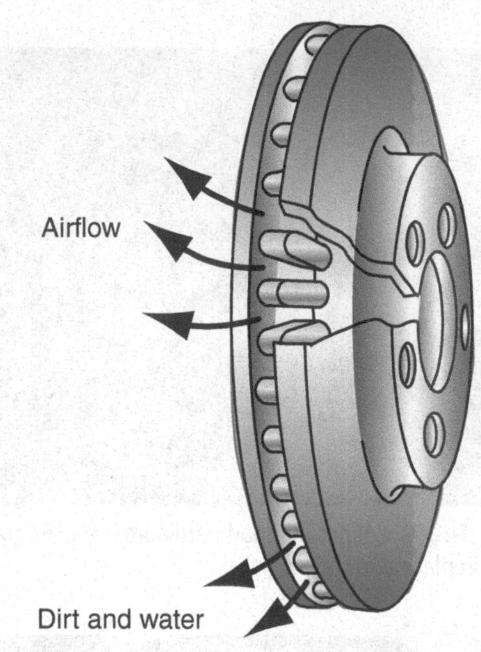

FIGURE 14-24 An illustration of how airflow through the rotor draws heat away from the brakes. As the rotor spins, it acts like a fan, drawing air in from the center and out through the vents.

FIGURE 14-23 A vented front brake rotor.

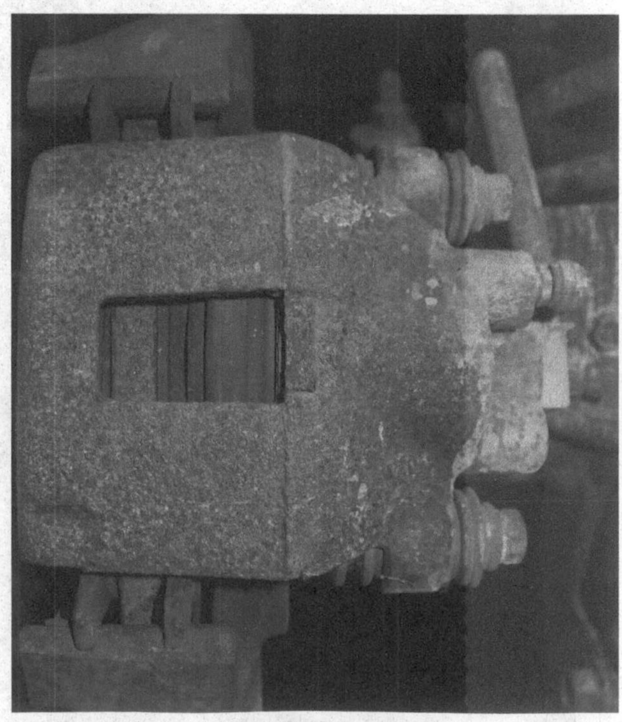

FIGURE 14-25 An example of a solid rear rotor.

have curved or directional fins in the vents (**Figure 14-26**). Rotors with directional vents have improved airflow for better heat dissipation, but they must be installed on the correct side of the vehicle to work correctly.

Many GM vehicles are using ferritic nitro-carbonized (FNC) rotors. FNC rotors are specially heat-treated during manufacture to last longer and resist rust and corrosion better than standard rotors. The heat-treating process creates a very thin layer on the rotor that is removed if the rotor is machined. Once removed, the rotor loses its ability to resist wear and corrosion.

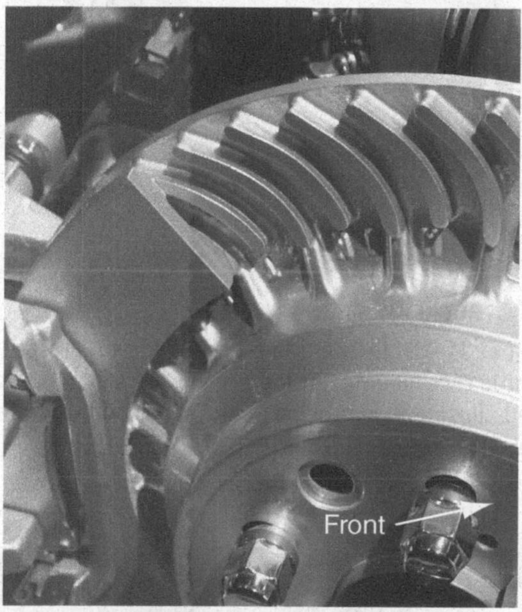

FIGURE 14-26 Many rotors use directional fins in the vents to improve cooling of the rotor. These rotors are left and right specific and must be installed in the correct location to cool properly.

Brake rotors on some high-performance vehicles are drilled and slotted. Drilled and slotted rotors are common on motorcycles, but are not generally found on average passenger cars or trucks. Drilled rotors have a series of holes drilled through the friction surface (**Figure 14-27**). Slotted rotors are similar except that the slots do not go through the entire depth of the friction surface (Figure 14-27). When the brakes are applied, the point of application between the pads and rotors can create a gas barrier that reduces braking ability. The holes and slots in the rotors allow the gas to escape, which improves braking.

Some high-performance vehicles offer ceramic-composite brake rotors (**Figure 14-28**). These rotors and pads offer extreme heat dissipation, reduced weight, and the ability to withstand high-speed braking with reduced fade or distortion. **Fade** is the term used to describe the loss of braking power or performance, usually as a result of heat. Ceramic-composite brakes are optional equipment on cars such as the Corvette Z06 and on many other vehicles, but they can add several thousand dollars to the price of the car.

■ ***Disc Brake Pads.*** **Brake pads**, which are made up of the friction material or pad lining and the backing or support plate for the lining, clamp down on the brake rotor to slow the wheel (**Figure 14-29**). The materials that make up the pad friction material determine its coefficient of friction and ultimately, how well the

(a)

(b)

FIGURE 14-27 An example of high-performance rotors, cross-drilled and slotted to vent gases and dust.

FIGURE 14-28 Examples of high-performance rotors available from GM.

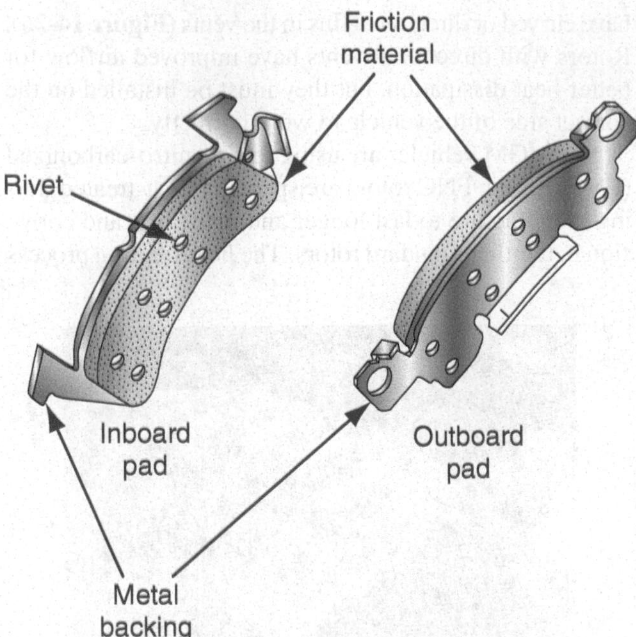

FIGURE 14-29 The parts of a brake pad. Not all pads have antirattle clips mounted on them; many use shims and spring-loaded clips between the pad and caliper bracket to reduce noise.

FIGURE 14-30 Pads are curved to fit the contour of the rotor.

vehicle stops. As with brake shoe linings, brake pad friction materials vary depending on the manufacturer and the application. Common ingredients include iron, steel, copper, synthetic fibers, and ceramic compounds.

Brake pads are shaped to follow the curve of the brake rotor (**Figure 14-30**). When the caliper piston(s) move outward, the pads are pressed against the rotor with great force, which results in friction and a lot of heat. The pads must be made from materials that can not only withstand the friction and heat of braking but also operate effectively when the brakes are cold. Brake pad linings are a compromise between several factors, such as pad life, noise generation, and cold and hot coefficients of friction. Brake pads that use soft friction compounds offer quiet operation and will not wear the brake rotors very fast, but will themselves wear quickly and may not have the stopping power of semi-metallic or metallic pads that contain metal shavings and other materials that increase stopping performance. Semi-metallic pads, however, will cause more rapid wear of the rotor and tend to generate more noise and rust-colored brake dust.

Pads are often categorized as asbestos, organic, semi-metallic, or ceramic. Asbestos pads were made with asbestos fibers, which tolerated the high temperatures generated by the brake system but also produced dust that, if inhaled, could cause lung damage and cancer. Asbestos has been phased out of brake linings since the 1990s. Organic pads are made of natural compounds such as glass and Kevlar. These pads are softer and create less noise but also wear faster and make a lot of brake dust due to the softer compounds. Semi-metallic pads contain iron, steel, copper, and graphite, which are mixed together into the friction compound. Semi-metallic pads offer longer service life but also wear the brake rotors faster and can generate more brake noise and dust. Many brake pads use ceramic fibers—and other materials, such as copper—and are often considered superior to other pads due to long service life and their ability to dissipate heat, provide quiet operation, and generate low dust.

However, due to environmental concerns, brake pads are also being made with much less or zero copper content. The U.S. states Washington and California have passed laws requiring much lower amounts of heavy metals, specifically copper, in brake lining materials. Because of this, brake pads now have copper content labels, rated Level A, Level B, and Level N. The major provisions of the legislation include:

- Level A: Brake pads and shoes manufactured after January 1, 2015, must not contain asbestos, hexavalent chromium, mercury, cadmium, or lead.
- Level B: Brake pads and shoes comply with Level A and reduce copper content to less than 5% of material content by weight by 2021.
- Level N: Brake pads and shoes will have zero copper (less than 0.5% total content by weight) by 2025.

Brake pads often have grooves and chamfers cut into the lining (**Figure 14-31**). Grooves are used to help remove the dust generated between the pad and the rotor during braking. Chamfering the leading edges of the pads helps decrease noise.

Many pads have a built-in wear indicating device, sometimes called a **squealer** due to the noise it makes when it contacts the rotor (**Figure 14-32**). **Figure 14-33** shows an example of a wear indicator tab as it has just

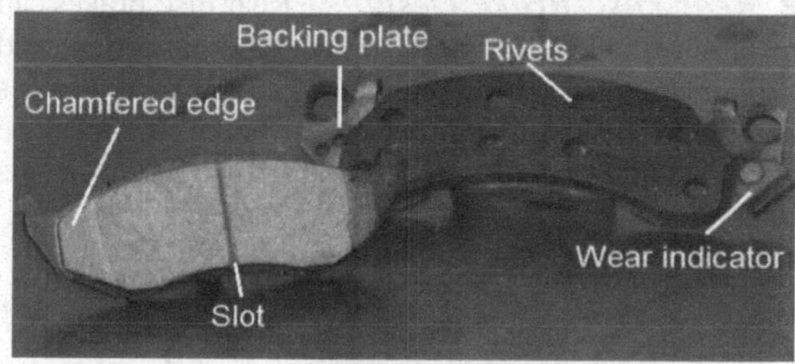

FIGURE 14-31 An example of pads with grooves for dust dispersion and chamfered edges to reduce noise. The groove also provides an indication of wear. When the pad is worn to the point where the groove disappears, the pad needs to be replaced.

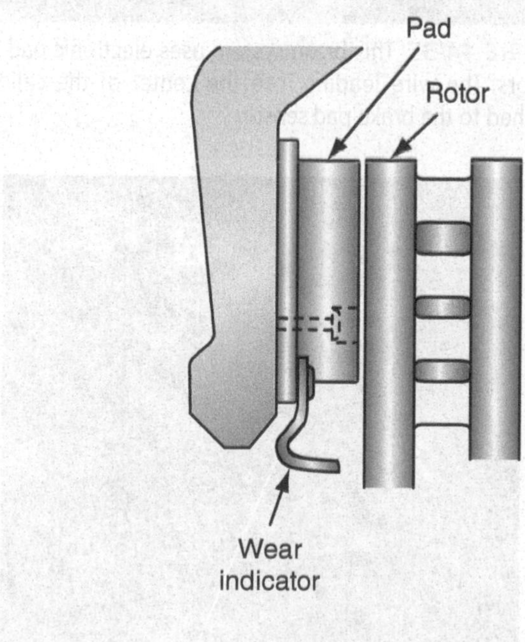

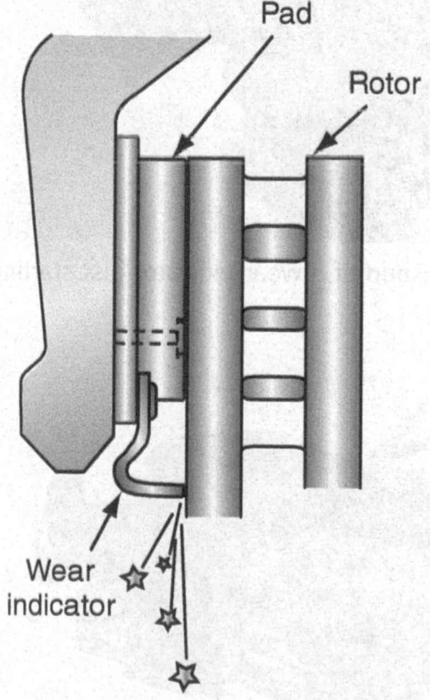

FIGURE 14-32 The squealer-type wear indicator makes a high-pitched noise when it contacts the spinning rotor.

started contacting the rotor. Some pads use an electrical pad wear indicator that is embedded into the pad's friction material. The pad has the electrical connector and sensor used to alert the driver when the pads are worn out (**Figure 14-34**). An example of how this looks installed on the vehicle is shown in **Figure 14-35**. This sensor activates a warning light on the dash to alert the driver that the brake pads need to be inspected or replaced (**Figure 14-36**).

REAR DISC BRAKE SYSTEMS

Rear disc brakes are replacing rear drum brakes on more and more cars and trucks. Rear disc brakes operate exactly like those on the front, but rear disc brakes are smaller in both rotor and pad size.

■ ***Rear Disc Brake Designs.*** Rear disc brakes can be either fixed caliper or floating caliper designs. Fixed calipers tend to be used on higher-performance vehicles compared to floating calipers. Rear brake rotors can be either vented or solid, depending on the requirements of the vehicle. **Figure 14-37** shows a front and rear disc brake assembly for comparison. The front rotor is vented, and the pads are larger than those used on the rear brake. The rear rotor is also a solid rotor.

■ ***Rear Disc Brakes and Parking Brakes.*** All vehicles sold in the United States are required to have a parking or emergency brake. The parking brake is a mechanical brake that often uses components of the hydraulic service brakes but does not rely on the hydraulic system to operate. This is so that in the event of a loss of hydraulic pressure, the parking brake can be used to slow and stop the vehicle.

When the parking brake is part of the rear disc brake caliper, it is called an integral parking brake caliper or **integral caliper**. (**Figure 14-38**). The piston, under normal braking, is applied by hydraulic pressure, but when the parking brake is applied, a lever pushes the caliper

FIGURE 14-33 A closeup of a wear indicator just starting to do its job.

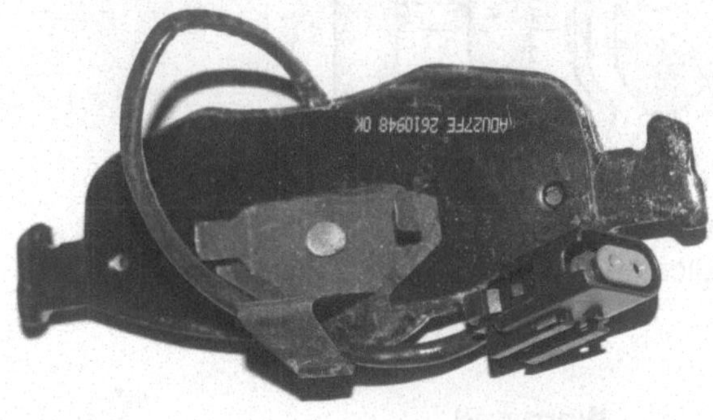

FIGURE 14-34 An example of a pad with an electronic wear indicator.

FIGURE 14-35 This brake system uses electronic pad wear sensors. The wire leading into the center of the caliper is attached to the brake pad sensor.

FIGURE 14-36 When the pad wears to the point when the sensor touches the rotor, the warning light on the dash is illuminated.

FIGURE 14-37 An example of a rear (a) and a front (b) brake assembly for comparison.

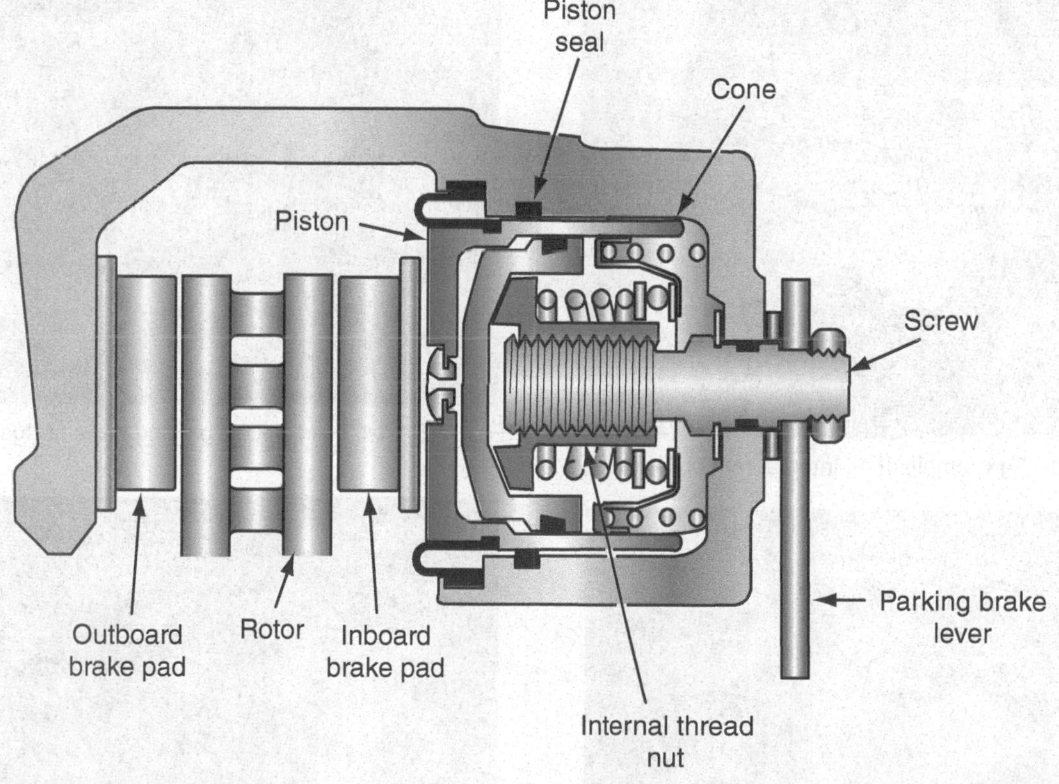

FIGURE 14-38 An illustration of an integral parking brake caliper. The piston is threaded to the parking brake lever to provide a mechanical connection, not dependent upon the hydraulic system.

piston out slightly, which locks the pads against the rotor. There are two methods for applying the caliper piston with the parking brake, a threaded piston and a ball-and-ramp design.

A threaded caliper uses a threaded piston and a screw that passes through the caliper body (Figure 14-38). When the parking brake is set, the brake cable pulls on the lever at the back of the caliper (**Figure 14-39**).

The lever is attached to the rear of the screw, and the front of the screw is seated against the rear of the caliper piston. When the lever rotates the screw, the screw turns in the caliper and pushes against the piston. The piston moves out slightly, setting the pads against the rotor. A spring, often located at the rear of the caliper at the lever, retracts the lever and screw when the parking brake is released.

The ball-and-ramp design uses an actuator with reliefs that are cut into its surface. The reliefs are tear drop shaped and are deeper at the large end and shallow at the point. Ball bearings are placed between the piston and the actuator and sit in the reliefs. When the parking brake is applied, the actuator turns and the ball bearings move from the deep to the shallow part of the relief. This action pushes the bearings against the piston to set the brake. When it is released, the actuator rotates back and the bearings recess back into the reliefs.

■ ***Drum in Hat Designs.*** Another type of rear parking brake system is called the **drum in hat** design (**Figure 14-40**). In this arrangement, the rear disc brakes operate exactly as the front disc brakes. The only difference is that the inside of the hat of the rear rotors contains a machined surface. This surface is the brake drum for a set of small brake shoes, which are used only as the parking brake. Servicing this type of brake is covered in detail in Chapter 13.

■ ***Electrically Operated Parking Brake.*** Many new vehicles use electronic parking brakes. To set or release the brake, the driver simply presses the parking brake button (**Figure 14-41**). A module commands the electric motor(s) to apply or release the parking brake. Application can be direct, with the motor attached to the back of the rear caliper or the motor may pull a parking brake cable attached to a traditional parking brake caliper (**Figure 14-42**). These systems typically have an

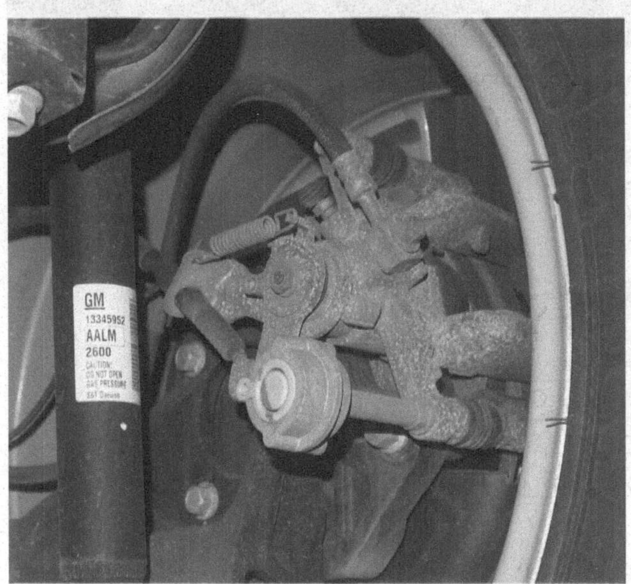

FIGURE 14-39 An example of an integral rear brake caliper.

FIGURE 14-41 An electric parking brake button.

FIGURE 14-40 An example of a drum in hat rear brake system.

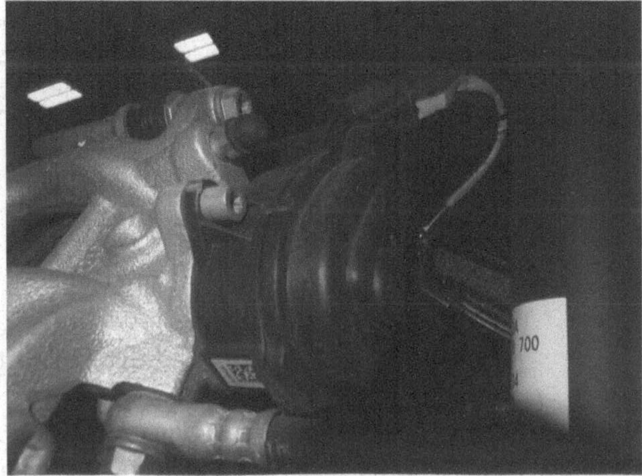
FIGURE 14-42 A rear brake caliper containing an electrically operated parking brake motor.

automatic release function. If the driver puts the vehicle into gear and starts to drive without pressing the button to release the parking brake, the onboard computer will automatically release the brake to prevent damage. In both designs, the motors are not designed to be used to stop the vehicle in the event the service brakes fail; instead, the system is used only for holding the vehicle in place when it is parked.

Servicing these systems, even installing rear brake pads, often requires the use of a scan tool. Refer to the vehicle's service information for specific service procedures.

■ ***Hybrid Vehicle Disc Brakes.*** As discussed in Chapter 10, hybrid and electric vehicles can recover braking energy to recharge the high-voltage (HV) batteries. Because of this, the service brakes tend to last much longer on these vehicles than on nonhybrid cars. On most vehicles, the driver can select the Regen or B mode while driving, which allows even more brake energy to be recaptured. The B mode is designed to utilize more regenerative braking than the mode for normal driving conditions.

The disc brakes on hybrid and electric vehicles operate exactly like those on nonhybrid vehicles. The disc brakes on the Toyota Prius, Honda Insight, and Civic hybrids, and others are single-piston floating caliper designs. What is different about how the brakes operate on these vehicles is that the on-board computer system monitors brake pedal input and then controls braking through the regenerative process and the service brakes. This means that even though the driver is pressing the brake pedal, the computer system is determining how much braking will be done by the hydraulic brake system. The split between hydraulic braking and regenerative braking depends on many factors but is mostly dependent upon the HV battery state of charge. The idea is to recapture as much energy as possible to recharge the HV battery. However, when the battery is charged, the system will greatly reduce or stop the amount of energy recaptured by the brakes. In the event of a malfunction in the braking circuits, the regenerative system is disabled, and normal brake operation remains.

One special brake service note: some hybrid vehicles perform periodic brake system integrity tests when the vehicle is parked and off. During this test, the brake system is pressurized for short periods of time. To perform brake services on these vehicles, it is necessary to place the vehicle into pad service mode. This will prevent the self-test from operating. To perform brake service and place the vehicle into service mode, refer to the manufacturer's service procedures.

SUMMARY

Disc brakes are less prone to fade than drum brakes, but are more prone to making noise.

There are two major types of disc brake calipers: the fixed and floating caliper designs.

Fixed calipers are bolted directly to the steering knuckle and have multiple pistons.

Floating calipers move laterally on bolts or pins.

Pad hardware is used to keep the pads secure in the bracket and reduce vibration and noise.

Some pads use an electrical pad wear indicator that is embedded into the pad's friction material.

When the parking brake is part of the rear disc brake caliper, it is called an integral parking brake caliper design.

REVIEW QUESTIONS

1. The ______________ seal around the caliper piston prevents fluid loss and acts as the piston return spring.
2. Floating calipers often mount to the bracket using bolts or ______________ and bushings that allow movement.
3. A rotor that is held in place by the wheel bearing is called a ______________ rotor.
4. A caliper that is mounted directly to the steering knuckle and has a piston on each side of the rotor is called a ______________ caliper.

5. ______________ mounted on the back side of the pads are commonly used to limit the noise generated by disc brakes.

6. Rotors can be which of the following types?
   - **a.** Floating
   - **b.** Vented
   - **c.** Nonvented
   - **d.** All of the above

7. *Technician A* says the rear disc brakes on trucks and SUVs are typically larger than the front brakes. *Technician B* says some rear disc brakes have auxiliary drum brakes inside the rotor to provide additional braking when required. Who is correct?
   - **a.** Technician A
   - **b.** Technician B
   - **c.** Both A and B
   - **d.** Neither A nor B

8. All of the following statements about disc brakes are true except:
   - **a.** Disc brakes require more application force than drum brakes.
   - **b.** Disc brakes are commonly used on the rear wheels.
   - **c.** Disc brakes resist fade better than drum brakes.
   - **d.** Disc brakes are not self-adjusting.

9. Which of the following statement about disc brake operation is not correct?
   - **a.** Hydraulic pressure forces the piston(s) out against the inner pad(s).
   - **b.** On a floating disc brake system, the outer pad is applied by the caliper moving inward.
   - **c.** Air enters the rotor's vents and leaves through the hat to remove heat.
   - **d.** Fixed calipers require at least two pistons.

10. *Technician A* says pad wear indicators always make a loud squealing noise when the pad is worn too thin. *Technician B* says that pad wear indicators may turn on a warning light on the dash. Who is correct?
    - **a.** Technician A
    - **b.** Technician B
    - **c.** Both A and B
    - **d.** Neither A nor B

CHAPTER 15

Disc Brake System Inspection and Service

Chapter Objectives

At the conclusion of this chapter, you should be able to:

- Identify tools used to service the disc brake system.
- Diagnose common disc brake concerns.
- Remove, clean, and inspect disc brake assemblies for wear and determine needed repairs. (ASE Education Foundation MLR 5.D.1, 5.D.2, & 5.D.3)
- Lubricate and reinstall caliper, pads, and related hardware. Seat pads and inspect for leaks. (ASE Education Foundation MLR 5.D.4)
- Remove, inspect, and replace a disc brake caliper, rotor, and pads.
- Inspect, measure, and refinish brake rotors. (ASE Education Foundation MLR 5.D.5, 5.D.6, 5.D.7, & 5.D.8)
- Retract a rear disc brake caliper piston on a vehicle with an integrated rear disc parking brake caliper. (ASE Education Foundation MLR 5.D.9)
- Check brake pad wear indicator systems. (ASE Education Foundation MLR 5.D.10)
- Describe importance of operating vehicle to burnish/break-in replacement brake pads according to manufacturer's recommendations. (ASE Education Foundation MLR 5.D.11)

KEY TERMS

Dial indicator	**piston retraction tool**	**taper**
disc brake micrometer	**pulsation**	**thickness variation**
on-car lathe	**rough cut**	**wear indicator**
parallelism	**lateral runout**	

Disc brake service, while usually not very difficult, does require that you take your time to thoroughly inspect the brake system and understand all the details necessary for proper service and repair of the system. Performing a brake job means more than just installing a set of brake pads, especially if diagnosing a noise or pulsation complaint is required.

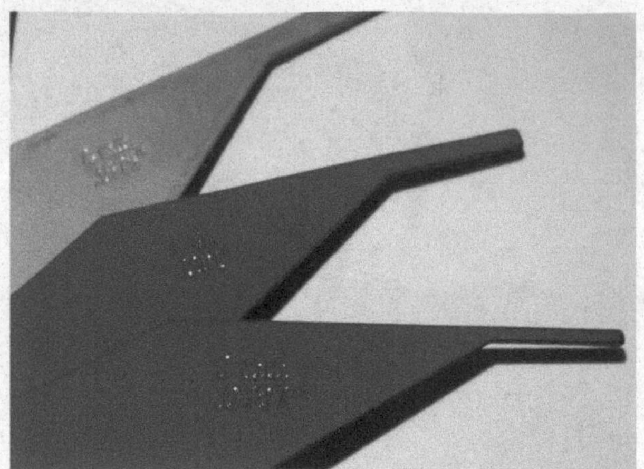

FIGURE 15-1 Brake wear gauges are used to determine the amount of lining left on pads and shoes.

Service Tools and Safety

Performing brake system services, such as brake pad replacement, is among the most common work technicians perform. Most employers expect that entry-level technicians can correctly diagnose and service disc brakes. While pad replacement is in itself an easy task, there is more to disc brake service to ensure that the brakes perform properly and provide the noise- and vibration-free operation that the customer expects.

TOOLS

The proper use of tools and procedures is an important aspect of performing your work safely and efficiently. Improper use of tools and taking shortcuts can lead to vehicle damage and can also cause personal injury. Always use the proper tools for a particular service or repair, and follow all manufacturer service procedures when you are working on the brake system. While there are no unimportant parts on a vehicle, the components of the brake system are, arguably, among the most important parts of any car or truck.

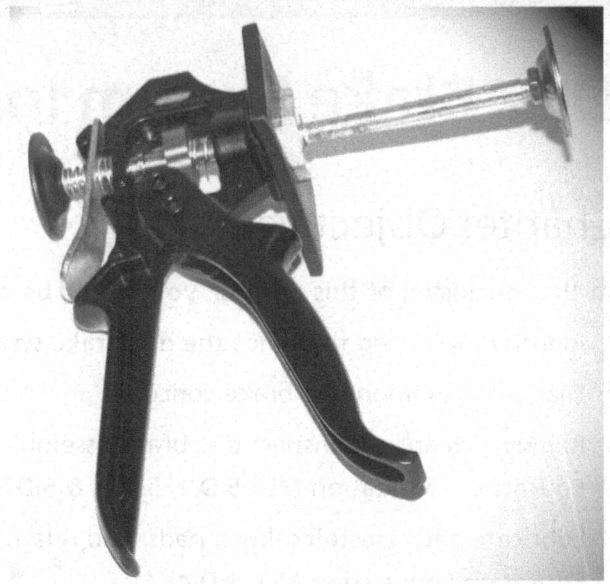

FIGURE 15-2 This tool retracts the caliper piston back into the bore.

■ ***Disc Brake Service Tools.*** The most commonly used tools for brake service are your basic hand tools. Specialty tools for brake service include the following:

- Lining wear gauges are used to measure the amount of lining material left on the brake pads (**Figure 15-1**). These gauges allow you to measure lining thickness with the pads still installed on the vehicle. In place of these, a tire tread depth gauge or machinist ruler can be used once the pads are removed.
- A piston retraction tool is used to push the brake caliper piston back into its bore (**Figure 15-2**).
- The piston retraction tool is used on rear disc brake systems that use a parking brake integral with the rear caliper (**Figure 15-3**).
- **Disc brake micrometers** are used to measure rotor thickness (**Figure 15-4**). Rotor thickness and thickness variation should be checked anytime brake work is performed.
- **Dial indicators** are precision tools used to measure small amounts of movement. In brakes, it is used to measure rotor runout (**Figure 15-5**).

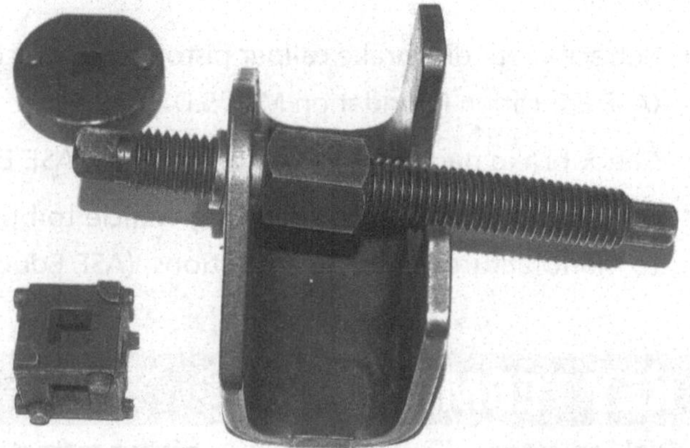

FIGURE 15-3 This tool retracts the caliper piston on integral parking brake calipers.

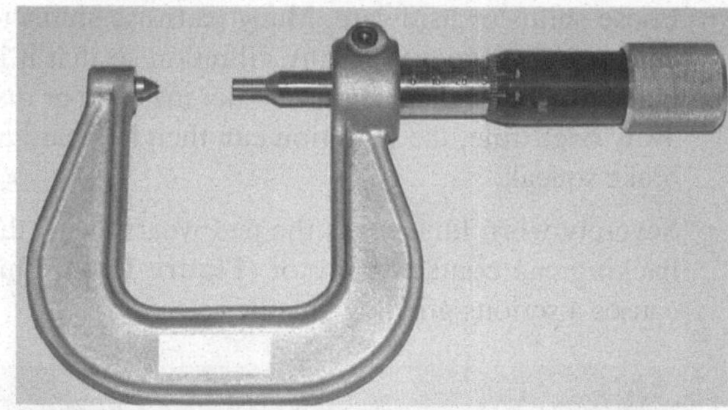

FIGURE 15-4 An example of a metric brake rotor micrometer.

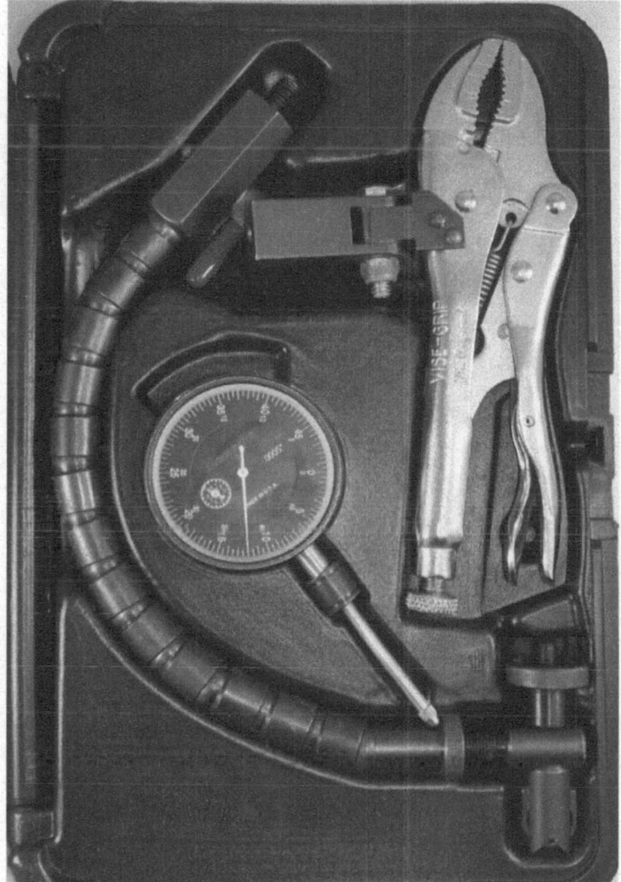

FIGURE 15-5 A dial indicator is used to measure rotor and hub runout.

In addition to basic hand tools and those listed above, you also use both on-car and off-car brake lathes, which are discussed later in the chapter.

■ ***Brake Service Precautions.*** Aside from using the correct tools properly for brake service, you should also observe the following disc brake system service precautions:

- Let the brakes cool before you service the components.
- Use calibrated torque wrenches when you tighten the fasteners.
- Do not let the brake calipers hang by their hoses.
- Do not reuse worn-out or damaged parts.
- Double check your work. Test the brakes before moving or driving the vehicle.

SAFETY

Even though most brake services are routine, do not take for granted the work that you perform. When you are working on the vehicle, remember to do the following:

- Double check the lift or jack contacts before you begin your work.
- Do not use your back muscles to lift when you are removing and installing the wheels and tires. Use your legs or ask for help when lifting.
- Wear mechanic's gloves to avoid damaging your hands. Brake service is often dirty, and wearing gloves helps protect your hands and helps prevent you from getting new components dirty during installation.
- Check all the replacement parts against the old parts.
- Make sure the replacement pads, rotors, calipers, hoses, and other parts are correct before installing them.
- Once work is complete, pump the brake pedal and make sure the brakes work before moving the vehicle.

Double check all of your work by doing the following:

- Be sure of the torque specifications, and recheck to make sure that all fasteners are properly torqued.
- Ensure that all air is bled from the system if the hydraulic system has been opened.
- Check the operation of the brake warning light on the dash and the stop lamps at the rear of the vehicle.

Remember, the customer is counting on you to make sure that their vehicle is properly serviced and in 100% operating condition when you are finished. It is important to inform the customer—and the customer understands—that the brakes may feel different and that stopping performance may have changed after brake service. Pads with different lining compounds can have a different feel when the brakes are applied. In some cases, even different brake fluids can change the feel of the brakes and the feedback from the brake pedal.

Disc Brake Diagnosis

Disc brake system problems can be grouped into a couple of specific types of concerns, as listed next. Often a problem can present itself as several of these concerns at once. A thorough understanding of the brake system is required to be able to make accurate diagnosis of brake system complaints. If necessary, review the disc brake system components and operation in Chapter 14.

COMMON BRAKE SYSTEM CONCERNS

Disc brakes are prone to noise problems because of the way in which they operates. As the rotor spins and the pads apply pressure against the rotor surface, the pads can vibrate and cause various types of squeaks and squeals. Also, disc brakes commonly experience pulsation concerns from rotor distortion.

■ ***Brake Noise.*** One of the most common complaints is of noise from the brakes. Disc brakes can make noise when applied and when not applied. A common customer concern is of a high-pitched squeal or squeak when driving. Because disc brake noise can have several causes, it is important to understand how noises occur and what to look for during an inspection. Common concerns include:

- Pad contact against the rotor surface. Even though a rotor may appear and feel smooth to the touch, it actually has very small peaks and valleys (**Figure 15-6**). As the rotor moves across the pads, these imperfections make the pad vibrate, which causes brake noise. *Chamfering* the edges of the pad can help reduce this noise (**Figure 15-7**). Shims and improved pad designs are used to alter these vibrations so that any noise becomes inaudible to the passengers.
- Pad **wear indicator**. Many pads have this metal tab secured to the pad backing (**Figure 15-8**). As the pad wears to within approximately 1/16 inch (1.5 mm), the tab will begin to contact the rotor, causing a high-pitched squeal to alert the driver to have the brakes inspected.
- Friction material composition. Pads with high amounts of metal often cause more noise than softer organic pads.

> **Word Wall**
>
> *Chamfering*—Chamfering means to angle or bevel the edges of a hole or surface to prevent sharp edge contact between two surfaces. Chamfering reduces noise between the pads and rotor.

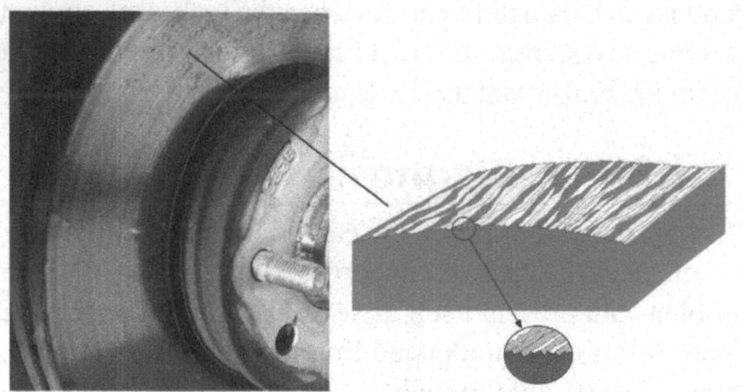

FIGURE 15-6 A rotor's surface, even when it feels smooth, has very small peaks and valleys.

- Loose shims or hardware. Many pads use shims to change the frequency of any vibration so that it is inaudible, but if these shims are not installed or rust away over time, the vibration can then be heard as brake squeak.
- Severely worn linings. As the pad wears away, the backing can contact the rotor (**Figure 15-9**). This causes a serious grinding sound.

FIGURE 15-7 Examples of chamfered edges on brake pads.

FIGURE 15-8 A pad wear indicator or squealer.

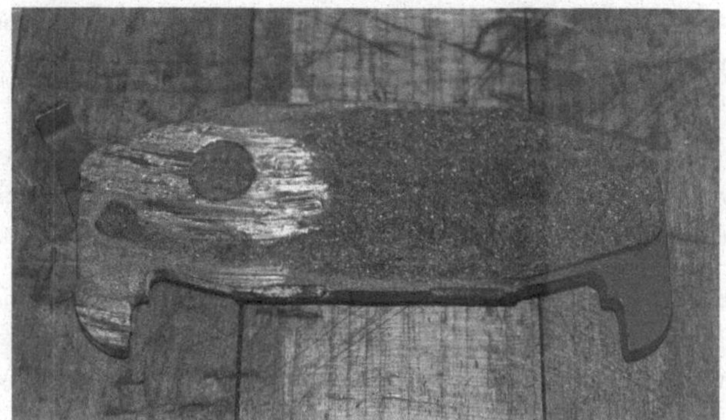

FIGURE 15-9 This pad has worn past the lining and into the backing plate. Note the uneven wear of the lining across the pad.

- Disc brakes also make grinding noises if the brakes form a layer of rust on the pads and rotor. This occurs when a vehicle is driven in wet conditions and is then parked. The rust causes a grinding noise when the brakes are applied the first few times, but then stops as the rust wears off.
- A rhythmic sound like a wire brush against metal may be heard as the wheels spin and the brakes are not applied. This noise can be caused by rotor runout or thickness variation. As the rotor spins and rubs against the pads, a slight amount of noise may be generated. The noise will increase in frequency as rotor speed increases.
- Pad and rotor knock occurs when there is excessive runout. With the brakes applied, rotor runout can knock the pads and caliper side-to-side, causing vibration and noise.
- Other causes of brake noise include contact between the rotor and pad hardware (**Figure 15-10**), or against the splash shield or other (**Figure 15-11**). A thorough inspection of the brake assembly indicates where there is contact between components.

■ ***Vibration.*** Disc brake vibration or **pulsation** is a common problem that is often described by the customer as a pulsating or shaking brake pedal and steering wheel shake when the brakes are applied. The vibration is often worse at higher speeds, such as on the highway. There are two main causes of this problem, brake rotor parallelism and runout.

Rotor **parallelism**, also called *rotor thickness variation* or taper, refers to the thickness of the rotor at points around the friction surface (**Figure 15-12**). Over time, runout in the rotor, hub, or bearing creates a slight wobble in the rotor as it spins. The wobble causes the rotor to rub against the pads, which wears away the rotor. As the rotor wears, the friction surfaces wear unevenly. Where the rotor is thinner, the caliper pistons and pads move further out. Where the rotor is thicker, the pads and piston are pushed back slightly. The movement of the piston displaces brake fluid in the caliper and back up to the master cylinder, causing the brake pedal to pulsate. To correct excessive thickness variation, the rotor must either be machined or replaced depending on how thick the friction surfaces are.

Lateral runout, either in the rotor or in the hub, can also cause brake pulsation concerns. **Lateral runout** is the side-to-side movement of the rotor, and it can be caused by a warped rotor, a bent or damaged hub flange, or loose wheel bearings. **Figure 15-13** shows an example of runout and how runout can cause excessive rotor thickness variation. Runout causes the rotor to wobble as it rotates, forcing the pads to follow the side-to-side movements. If the caliper and pads are capable of floating properly on the mounting hardware, a slight amount of runout should not cause a pulsation. Over time, however, this slight bit of contact between the pads and rotor wears away at the rotor. This leads to runout-induced thickness variation, which causes a pulsation concern.

As the pads and hardware are exposed to the outside elements and rust develops, the caliper and pads may begin to stick in place. This can keep the pads and the caliper from being able to slide or float. When this happens, the rotor wears from the pads remaining against the rotor surface. Eventually, this runout leads to rotor thickness variations and a brake pulsation. **Figure 15-14** shows and example of a brake pad that was stuck in its bracket. Note the **taper** of the lining, the difference in wear from the top to the bottom of the lining.

FIGURE 15-10 Check the fit of replacement parts. These new guides were slightly different than the original parts and made contact with the rotor.

FIGURE 15-11 The wheel speed sensor wiring bracket on this vehicle was contacting the rotor, creating a noise similar to a pad wear indicator.

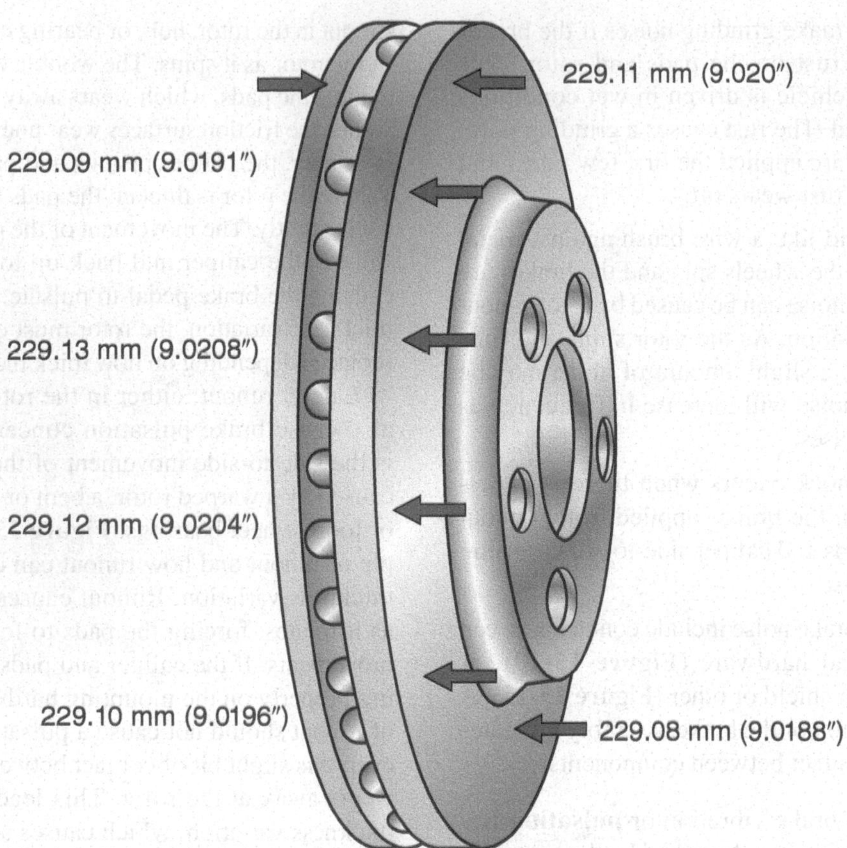

FIGURE 15-12 An example of excessive rotor parallelism.

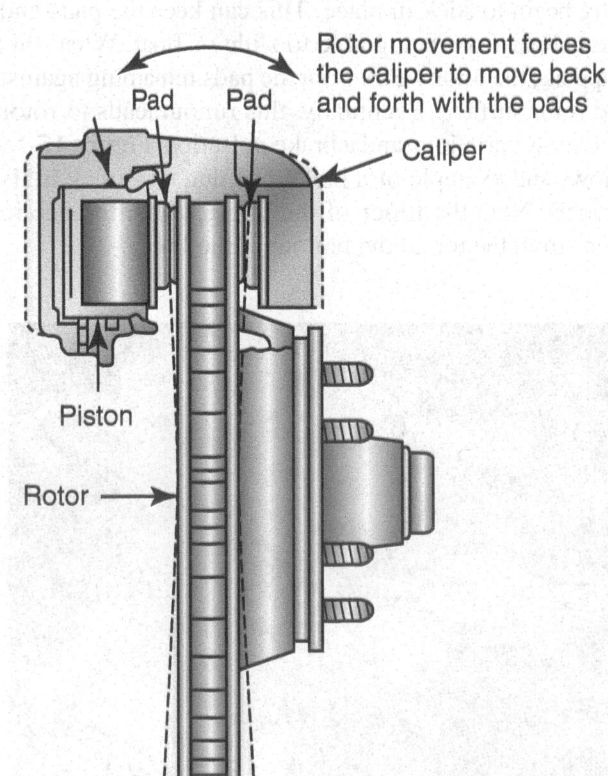

FIGURE 15-13 Runout is side-to-side variation from vertical in the rotor as it spins. As the rotor spins, the runout causes contact between the rotor and pads, which wears on the rotor. This then causes parallelism problems as the rotor wears from the contact.

Rotor runout can be caused by overtorquing wheel fasteners, especially by those who use air impact tools to install wheels. The excessive torque pulls on the hub flange and through the rotor mounting face (**Figure 15-15**). This distorts the hub and hat section of

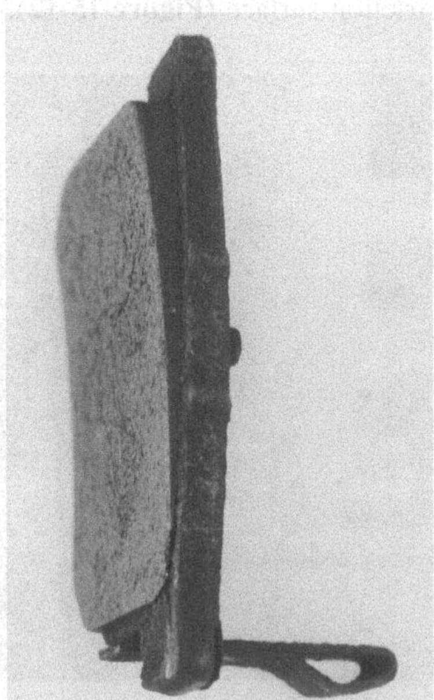

FIGURE 15-14 Uneven or tapered wear on the pads indicates either the caliper is not floating properly or the pads are stuck in the bracket.

the rotor. Runout is checked by using a dial indicator and is covered in detail later in the chapter.

■ ***Vehicle Pulling When Braking.*** The most common causes of disc brake pull are restricted hydraulic hoses and seized caliper pistons. Begin your diagnosis by verifying that the pull is related to the brake system. Perform an inspection of the tires and front suspension and steering components. Uneven or low front tire pressure and worn-out control arm bushing can cause the vehicle to pull when braking (**Figure 15-16**). Ensure that all suspension and steering components are in good condition.

Raise the vehicle so that the tires can rotate freely, and then spin each wheel. Note the effort needed to turn the wheel and how well it rotates after you let go. Any wheel that requires significantly more effort to turn—or does not continue to rotate on its own—requires a closer inspection.

If all wheels turn freely, have an assistant apply the service brakes while you try to rotate each wheel again. A restricted brake hose or seized caliper piston can prevent the brake from applying, and it will continue to spin even with the brakes applied. If one wheel continues to rotate, you need to inspect the caliper and hose to determine the cause of the concern. With your assistant keeping the brakes applied, open the bleeder screw of the problem caliper. If no fluid or very little fluid escapes, suspect a collapsed brake hose. If fluid flows out of the bleeder, suspect a seized piston. If the hose is the cause, replace all of the rubber brake hoses. If the caliper is

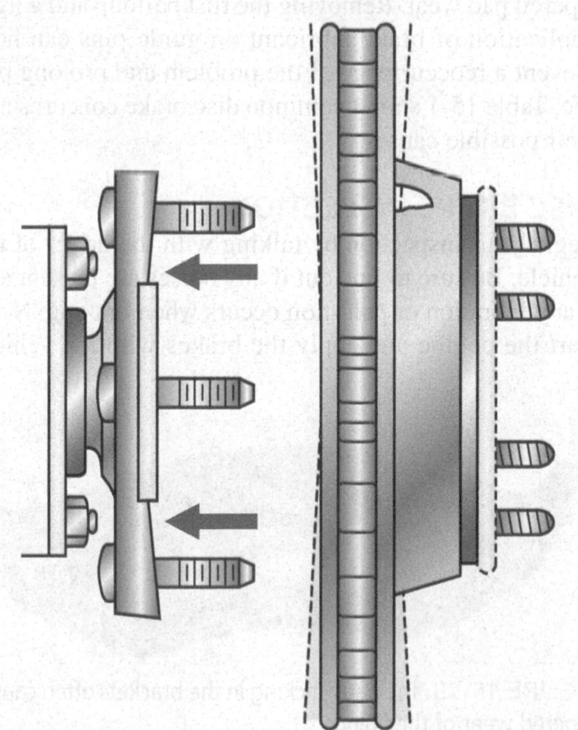

FIGURE 15-15 Overtorquing lug nuts can distort the hub and hat of the rotor, causing runout.

the cause, rebuild or replace both calipers on the axle. Depending on the age and mileage of the vehicle, it is usually a good idea to replace the brake hoses when you are replacing the calipers.

A brake staying applied after the brakes are released can also cause a brake pull. This is typically caused by a restricted or partially collapsed brake hose. Fluid passes through the hose when under pressure but then blocks the fluid release. To check for it, raise the vehicle and rotate each tire. A tire that does not turn or is much harder to turn has the problem brake. To determine if the hose is keeping pressure on the caliper, open the bleeder screw. If fluid shoots out under pressure, the hose is the problem. If fluid does not come out under pressure, the caliper piston may not be retracting into the caliper when the brakes are released. If the brake is still tight with the bleeder valve open, use a C-clamp to try to press the piston back into its bore (**Figure 15-17**). Even if the piston does return, the caliper should be rebuilt or replaced.

■ ***Brake Grab.*** Brake grab, also called *grabbing*, refers to a brake that applies with too much braking force, causing the wheel to lock up easily. This condition can be caused by fluid contaminating the linings, using pads with too high a coefficient of friction, and if the hydraulic system is applying excessive pressure.

■ ***Brake Drag.*** Brake drag, or dragging, occurs when a brake remains applied after the brake pedal is released. Dragging brakes can be caused by an improperly adjusted brake pedal pushrod. However, dragging is typically caused by a problem with the brake hose or by the caliper piston not retracting. Dragging brakes can cause the pads and rotors to severely overheat, *glaze* the surfaces, and lead to brake fade and failure. Linings and

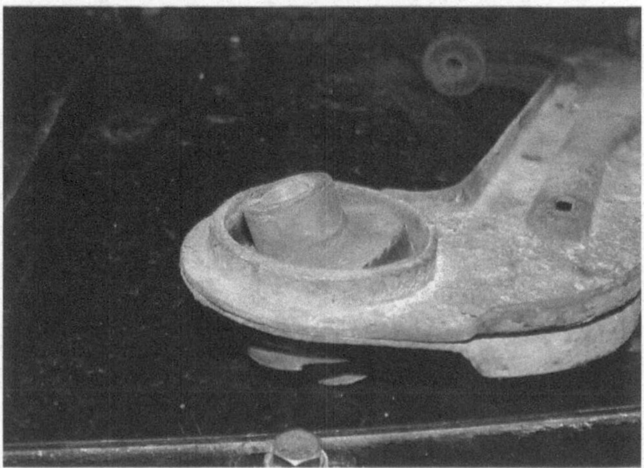

FIGURE 15-16 Nonbrake problems, like this worn-out control arm bushing, can cause a pull when braking. Perform a thorough inspection of the vehicle when diagnosing a concern.

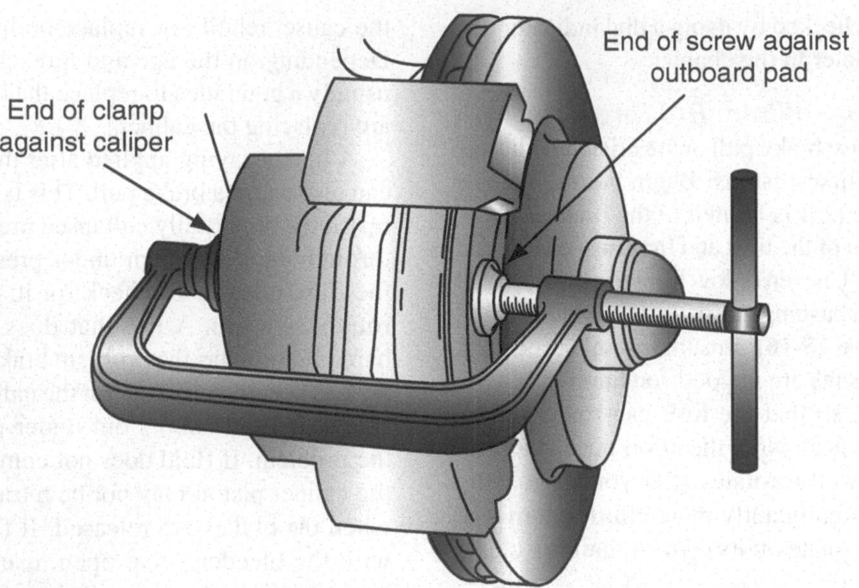

FIGURE 15-17 Use a C-clamp to retract the caliper piston when checking for dragging brakes.

> **Word Wall**
>
> *Glaze*—Glaze is a surface condition, often caused by overheating the pads and rotors, that discolors the friction surfaces, reduces brake efficiency, and can generate noise and vibration concerns.

rotors that are discolored are an indication that the brakes are dragging (**Figure 15-18**).

■ ***Abnormal Pad Wear.*** Brake pads, both the inner and outer, should wear equally and from side-to-side on the vehicle though a slight bit of tapered wear is not unusual. Uneven pad wear can be caused by problems in the hydraulic system, such as restricted brake hoses and seized caliper pistons. Abnormal wear is often caused by the pads not being able to move within the mounting hardware in which they are secured (**Figure 15-19**). Sticking calipers, stuck caliper pins, and rusted pad hardware prevent the pads from sliding properly. This can cause the pads to stay in contact with the rotor after the brakes are released, causing rapid and tapered pad wear. Removing the rust buildup and a light application of brake lubricant on guide pins can help prevent a reoccurrence of the problem and prolong pad life. Table 15-1 shows common disc brake concerns and their possible causes.

FIGURE 15-18 This vehicle had a caliper that would not release. This overheated the pads and rotor.

DISC BRAKE INSPECTION

Begin your inspection by talking with the driver of the vehicle. Be sure to find out if any noises are present and if any vibration or pulsation occurs when braking. Next, start the engine and apply the brakes with the vehicle

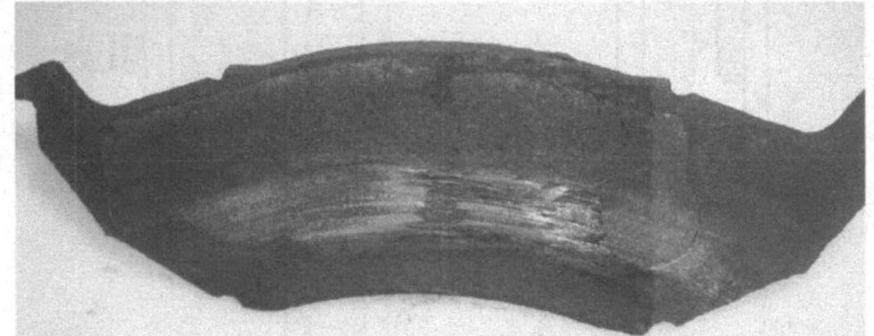

FIGURE 15-19 The pads sticking in the brackets often causes tapered wear of the linings.

Symptom	Possible Cause	Correction
Brake pull or drift	Incorrect tire pressure Incorrect wheel alignment Contaminated brake linings Damaged brake linings Binding or sticking caliper/pads Worn suspension components; ball joints and control arm bushings	Inspect and set tire pressure Inspect and set wheel alignment Clean or install new linings Replace linings Inspect and clean caliper/pads Inspect and replace worn components
Vibration when braking	Excessive rotor thickness variation Excessive rotor runout Excessive hub runout Loose wheel bearings	Machine/replace rotor Machine/replace rotor Replace hub Tighten/replace wheel bearings
Brake grab	Binding or sticking caliper Loose caliper	Clean or replace caliper Inspect and repair/replace caliper
Brake drag	Binding or sticking caliper Binding or sticking parking brake	Clean or replace caliper Clean, repair, or replace parking brake components
Abnormal pad wear	Binding or sticking caliper/pads Binding or sticking parking brake	Clean or replace caliper/pads Clean, repair, or replace parking brake components
Brake noise	Glazed or worn linings Glazed rotors Damaged splash shield Foreign object in contact with rotor Loose/damaged caliper/mount	Replace linings Replace rotors Repair/replace shield Inspect/remove foreign object Inspect and repair/replace caliper/mount
Brake warning light on	Low fluid level in master cylinder Parking brake applied	Check fluid level and fill if needed, and check for leaks
Excessive pedal travel	Binding pads or stuck caliper guides Air in the system	Inspect and service pads and hardware Bleed hydraulic system

TABLE 15-1

stopped. Ensure that the pedal is firm and has plenty of reserve. Check if the BRAKE or ABS warning lamps are operational. Both should illuminate during engine start and bulb check and then go out after a few seconds. Some vehicles have electronic brake pad wear indicators that turn on a dash warning light when the pads need to be replaced. This light remains on after the engine is started. When it is possible, perform a test drive, listening for any noises during driving and braking, and note any pull or vibration.

■ ***Calipers.*** When you are performing a visual inspection of the brake system, always look for signs of fluid loss (**Figure 15-20**). Brake calipers can leak fluid from a faulty square seal inside the caliper. This allows the fluid to leak around the piston and out of the dust boot onto the pads and rotor. The loss of fluid reduces brake application pressure, pulls air into the hydraulic system, and severely reduces the stopping ability of the brakes. A leaking caliper must be either rebuilt or replaced.

> **Word Wall**
>
> *Pedal reserve*—Pedal reserve is the measurement from the floorboard to the brake pedal when applied. Pedal travel is typically about 2½ to 3 inches and reserve is often about 2 to 3 inches.

Calipers can also fail and leak due to excessive wear on the pads and rotor. The rear brakes shown in **Figures 15-21** and **15-22** had enough wear that the caliper piston came out of the piston bore.

■ ***Hoses.*** Like any component, brake hoses also fail and leak. Inspect the hoses for cracks in the external covering layers and leaks around the fittings and clamps (**Figure 15-23**). When a hose is leaking, all brake hoses should be replaced since all of them are likely the same age and may fail in the near future. Carefully inspect the hoses for signs of rubbing or contact with other components, especially if steering or suspension work has recently been performed on the vehicle. A missing hose retainer or bracket bolt can allow a hose to contact the tire, which will quickly wear through the hose.

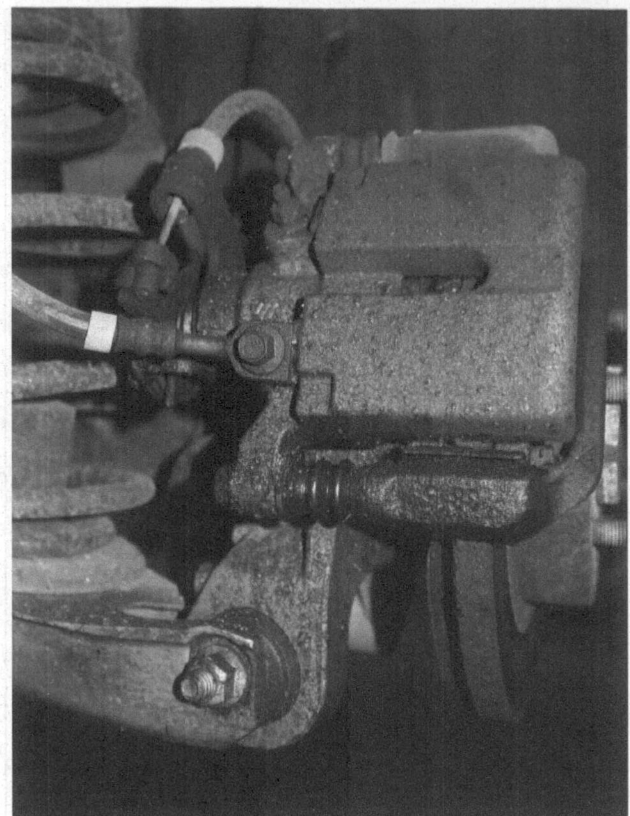

FIGURE 15-20 A leaking rear brake caliper.

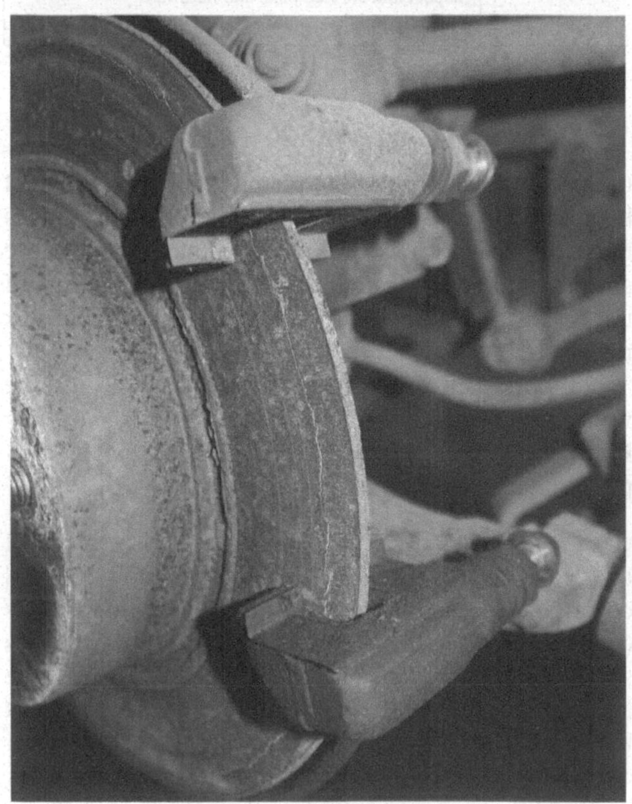

FIGURE 15-21 A severely worn brake rotor.

FIGURE 15-22 The pads and rotor (from **Figure 15-21**) were worn so much, the caliper piston extended out the piston bore.

FIGURE 15-23 A leaking brake hose. Always check the entire brake system during a brake inspection.

Check hoses for evidence of twisting. Many hoses have a stripe that is used to help prevent it from being twisted during brake service (**Figure 15-24**).

■ ***Inspect Pad Wear.*** Pad lining thickness can often be checked with the caliper in place (**Figure 15-25**). However, a thorough inspection requires the caliper be at least partially removed so that the pads can be removed and examined. **Figure 15-26** shows a pad that requires replacement, but it could not be thoroughly checked with the caliper in place. The pad in Figure 15-26 is severely cracked. Just checking the lining thickness with the pads installed does not provide the complete story of the pads' condition.

Lining thickness can be measured with special tools, such as those shown in **Figure 15-27**. This measurement can be used to determine approximately how much life remains in the pads. In most cases, the brake pads should

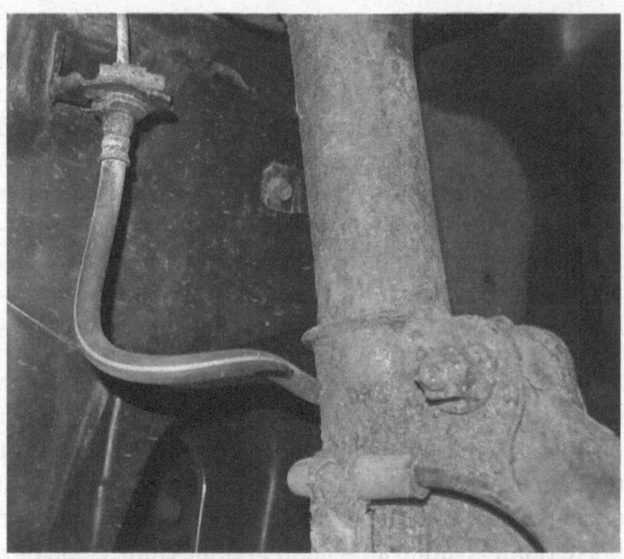

FIGURE 15-24 Check the hoses for twisting. Improper installation of the caliper can twist the hose and block fluid flow. Look for the line on the hose to tell if it is twisted.

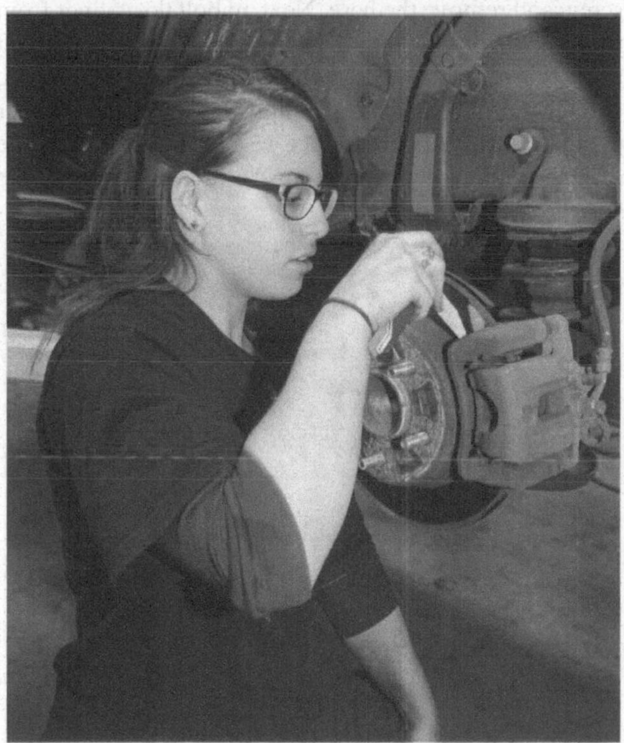

FIGURE 15-25 Checking pad lining thickness.

FIGURE 15-26 This pad is cracking and crumbling. This will not be noticeable by just looking at pad thickness with the pads installed on the rotor. Remove the caliper to get a look at the pads during an inspection.

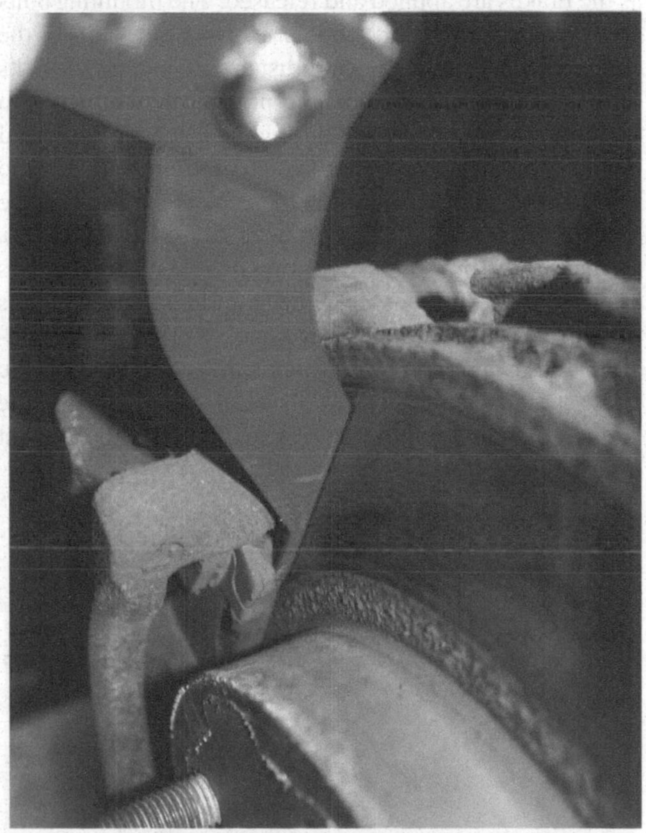

FIGURE 15-27 Using a lining thickness tool.

be replaced when the lining material reaches about 0.060 inch (1.5 mm). Many manufacturers have specifications for minimum lining thickness. If the lining wears below this minimum amount, the stress and heat of braking can cause the lining to crack and crumble. You should make sure the customer understands that the thinner the pad lining is, the more rapid the wear will take place. For example, a pad with half of its lining worn away may have been in service for 30,000 miles. As the pad gets thinner, the remaining half of the lining will likely not last an additional 30,000 miles. This is due to the acceleration of wear as the lining becomes thinner and is less able to dissipate heat.

Inspect the pads for uneven wear. This means that both the inner and outer pads should have the same amount of lining remaining and that the lining is wearing evenly along the length of the backing material. Uneven wear indicates that the pads are not sliding properly and are not fully releasing from the applied position against the rotor (Figure 15-19). Uneven inner and outer pad wear can indicate a problem with the caliper piston not retracting or the caliper itself not releasing on the mounting pins. Careful inspection of the pads allows you to accurately diagnose pad wear and correct the causes when you service the brake system.

When you find uneven wear, check all hardware carefully. Many floating calipers attach to brackets with bolts that float in the brackets. This allows the caliper to move as the brakes are applied and released. The mounting bolts are protected with rubber boots. However, in parts of the country where snow, salt, and rust are a concern, the bolts often seize in the brackets, and prevent the caliper from floating. If it is not checked during the inspection, a seized bolt may not be apparent until you try to remove it or try to reinstall the caliper. Too often these bolts break off in the bracket when you attempt to remove them (**Figure 15-28**).

Inspect the bracket or steering knuckle where the pads attach. Some vehicles do not use guides, and the pads ride directly on the bracket or knuckle. This can eventually wear away the metal, which can lead to noise as the pad gets loose and rattles. This wear can also cause the pads to bind and not to move properly along the bracket or knuckle.

FIGURE 15-28 Caliper pins often rust and seize into the brackets.

DISC BRAKE DISASSEMBLY

Before you disassemble the brakes, first make sure that you understand how the brakes are assembled. Note the locations and appearance of the bolts, clips, pins, and other parts that secure the caliper and the pads in place. You may want to take a picture or obtain a diagram or exploded view of the disc brakes from the service information in your shop before you begin to remove any parts. If provided, follow the service instructions for brake service supplied by the vehicle manufacturer.

■ ***Caliper and Pad Service.*** On vehicles with floating or sliding calipers, begin by pressing the caliper piston back into the caliper bore using a C-clamp over the caliper (Figure 15-17). Loosen the bleeder screw and attach a hose from the bleeder into a bottle to catch the brake fluid. Position a C-clamp over the caliper, and turn the clamp so that the piston is forced into the caliper. Slowly tighten the clamp to retract the piston. Going slowly reduces the stress on the square seal inside the caliper as the piston is forced back into its bore.

Once the piston is back in the bore, close the bleeder screw and remove the hose. Next, unbolt the caliper from the knuckle or bracket. An example of how a caliper and bracket are attached is shown in **Figure 15-29**. Once the caliper is unbolted, use a bungee cord or wire hanger to support the caliper (**Figure 15-30**). Do not allow the caliper to hang on the brake hose as this can damage the

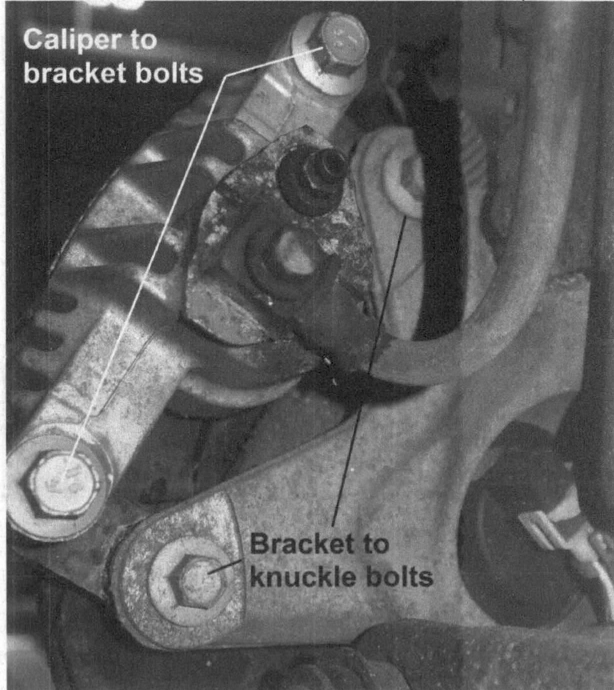

FIGURE 15-29 A common caliper and bracket assembly. Remove the two caliper to bracket bolts to access the pads.

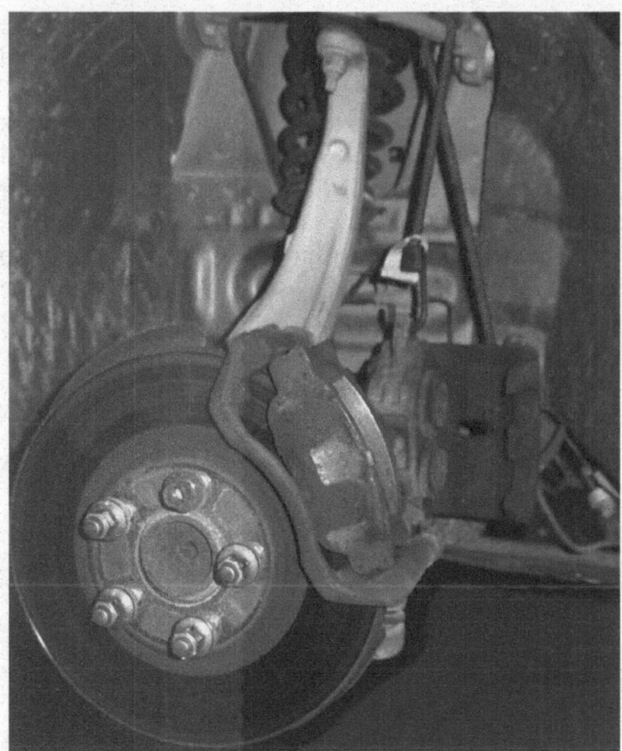

FIGURE 15-30 Do not let calipers hang by their hoses. This can stress and damage the hose. Always support the caliper with a wire or bungee cord.

hose. Hang the caliper from the strut or other suspension component. Do not attach the bungee cord or wire to the fender or other piece of body work.

If the caliper mounts directly to the knuckle, remove and inspect the mounting bolts, sleeves, or other components that hold the caliper in place. Check that all mounting components are free to move. Many shops include replacing all of the pad and caliper mounting parts in the estimate when replacing brake pads. These parts deteriorate and wear out over time. Inspect where the pads mount to the caliper body. These locations should be cleaned of rust before the new pads are installed. If the pad attaches to the caliper, clean these contact points thoroughly (**Figure 15-31**). If the inner pad rests in the steering knuckle, clean this area and make sure the new pad moves properly when installed.

Calipers that mount to a bracket often use a sliding sleeve or pin between the caliper and the bracket (**Figure 15-32**). These must be free to move or the caliper cannot float properly. Inspect the pins, pin boots, and the bore in the bracket for rust and corrosion that can prevent movement. An illustration of the pins and sleeves is shown in **Figure 15-33**. The bracket has two bores, one for each pin. The pins fit closely into the bore in the bracket. A rubber bushing is often used on one of the pins (Figure 15-33). The caliper mounting bolt threads into

FIGURE 15-31 If the pads attach to the caliper as these do, clean the contact areas before installing the pads.

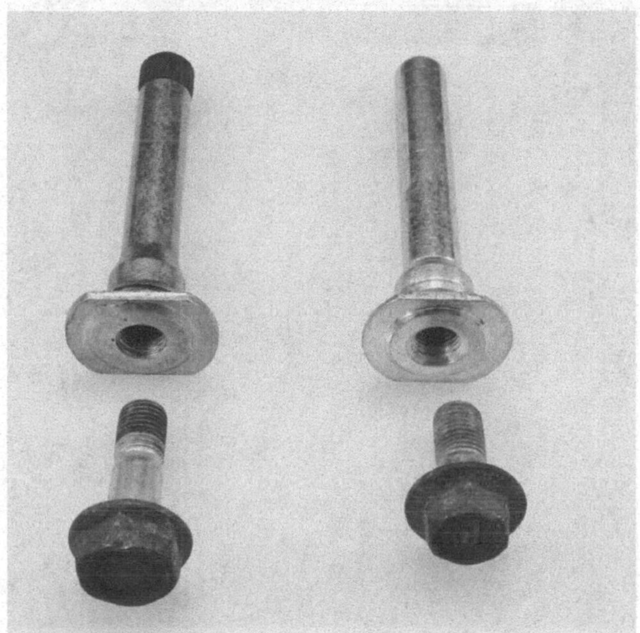

FIGURE 15-32 Examples of caliper pins and bolts. The pins must be free to float or slide in the bracket for the caliper to work properly.

the pin and secures the caliper to the bracket. Inspect the pins, pin bores, and the rubber components. Clean and lubricate the pins and the bracket so that the pins can move freely when installed. Use the lubricant specified by the vehicle manufacturer or a high-quality synthetic brake lubricant made for use with rubber parts.

Many of the brackets use thin pieces of steel hardware between the bracket and the pad (**Figure 15-34**). The guides are often spring-loaded to fit tightly to the bracket to prevent vibration and noise. If in good condition, the guides can be cleaned and reused. However, it is a good idea to replace these hardware pieces as part of the brake service. If damaged, loose, or rusted out, the pieces must be replaced. The guides hold the pads securely to the bracket to prevent vibration and noise, but also allow the pads to move as the brakes are applied and released.

Many technicians replace all the caliper hardware during brake pad service and many brands of replacement brake pads come with replacement guides. An example of a hardware kit is shown in **Figure 15-35**. This is a low-cost method of ensuring that the brakes will be in like-new condition, and it eliminates the problems of worn, rusted, or damaged parts compromising the quality of the repair.

The pads may just ride within the bracket (**Figure 15-36**). When servicing the pads, thoroughly clean the surfaces where the pads sit in the bracket. When reinstalling the pads, verify what, if any, lubricant is specified for the contact areas between the pads and the bracket. Do not apply a lubricant if none is required.

If you are removing but not replacing the rotor, mark the location of rotor to a wheel stud with grease pencil (**Figure 15-37**). This will allow you to reinstall the rotor back onto the hub in the same place it was installed, which can help reduce runout. Remove the rotor to inspect it. Floating rotors may slide off the hub with little

FIGURE 15-33 An illustration of a mounting bolt and sleeve for a floating caliper. Caliper mounting bolts and sleeves should be cleaned, inspected, and lubricated before installation.

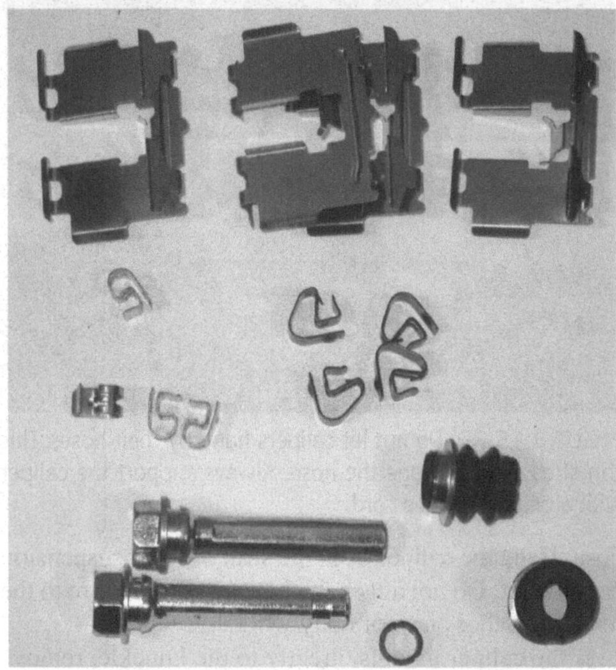

FIGURE 15-35 Examples of new caliper hardware and pins.

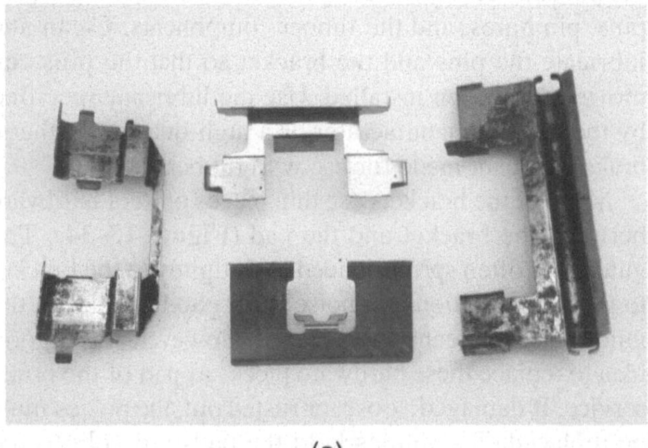

(a)

(b)

FIGURE 15-34 (a) Many pad sets come with new guides. Here, both new and used are shown for comparison. (b) If reusing the old guides, thoroughly clean and inspect them before reinstallation. If the guides are rusted through, are not firmly attached to the bracket, or damaged in any way, discard them and install new guides.

trouble. However, rotors can rust and stick to the hub, making removal difficult. First, make sure the rotor is a floating rotor, meaning it is not held on by the wheel bearing. Also, check for retaining screws holding the rotor in place (**Figure 15-37**). If screws are installed, remove them first. You may need to clean around the hub to ease rotor removal. Rust tends to build up in this area and can make rotor removal difficult. If necessary, apply a penetrant to the hub area and lightly tap the rotor with a plastic hammer or rubber mallet.

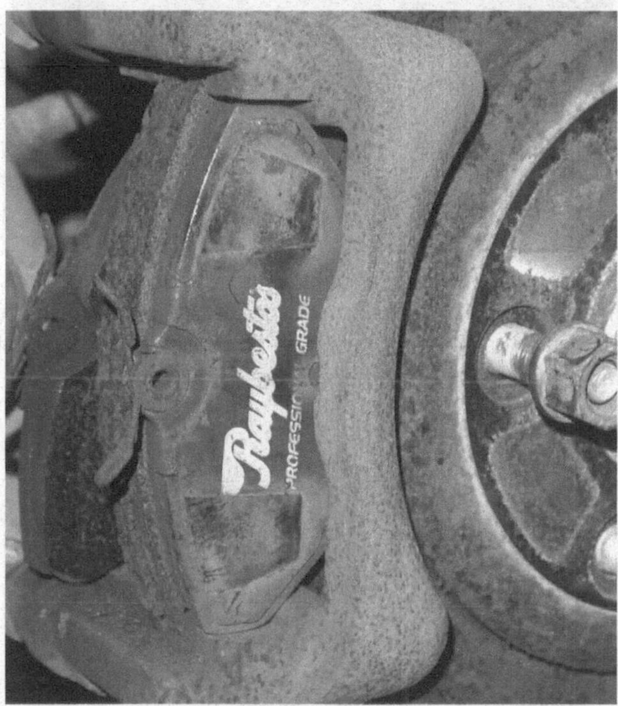

FIGURE 15-36 Some pads, like these, ride just inside bracket without additional hardware. Thoroughly clean the surfaces and ensure the pads can move freely when installing new pads.

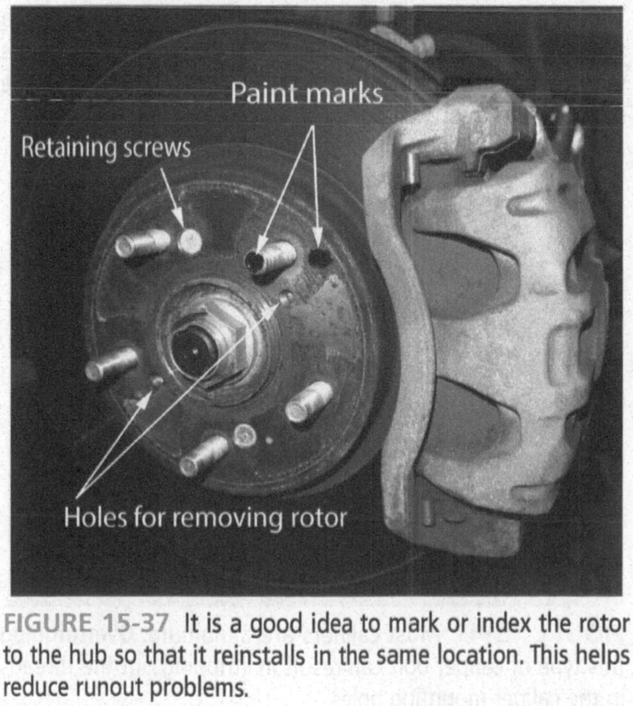

FIGURE 15-37 It is a good idea to mark or index the rotor to the hub so that it reinstalls in the same location. This helps reduce runout problems.

Some rotors have two threaded holes in the hat in which bolts, typically 8 mm × 1.25 bolts, can be threaded into these holes and tightened to push the rotor off the hub (Figure 15-37). Do not hit the friction surface of the rotor with a steel hammer unless you are replacing the rotor. Once off, clean the inner and outer surfaces of the hat and the hub mounting surface before you reinstall the rotor (**Figure 15-38**).

If the rotor is trapped, meaning held in place by wheel bearings, the bearings must be disassembled to remove the rotor. This is discussed in Chapter 5. If the rotor is trapped with a sealed wheel bearing, the bearing and rotor are removed as an assembly. You need to follow the manufacturer's service procedures for the specific vehicle on which you are working. Removing this type of bearing often requires using a slide hammer to separate the bearing from the steering knuckle and usually requires replacing the bearing. Removing this type of rotor is covered later in the chapter.

When you are ready to reassemble the brakes, you may need to use a small amount of brake lubricant on places where the pads or calipers must move. **It is important that only a lubricant approved for brakes is used, that the lubricant is used correctly and in the correct locations, and that excessive amounts are not used.**

FIGURE 15-38 Once removed, clean both sides of the rotor's hub area and clean the hub flange on the vehicle. The hub and rotor need to have as good a mating surface as possible to prevent runout and vibration concerns.

Not just any type of lubricant can be used on brake systems. Ordinary chassis grease and similar petroleum-based lubricants should not be used in any brake system application. Disc brake lubricants are special, high-temperature greases that are either synthetic or silicone based and often contain a form of moly or graphite. The advantages of moly and graphite lubricants are that they do not migrate, meaning they stay where they are applied. The high brake temperatures do not cause the moly or graphite lubricants to get thin and become mobile. Brake lubricants should only be used as directed by the manufacturer's service information. Do not apply lubricants to the linings or backing plates between the caliper and the plate. Lastly, do not apply large amounts of lubricant on every surface. The excess lubricant will be forced out and will end up on the linings or rotor surface, which is the last place you want any type of lubricant.

If the brake caliper attaches to a support bracket, ensure that the contact points between the bracket and pads are clean. Apply lubricant if it is indicated in the service information. Next, install the pads on the bracket and against the rotor surface. Make sure the pads are aligned properly and that the linings are against the rotor's friction surfaces. Typically, the pad with the wear indicator is the inside pad. Be sure the pads are installed correctly or the brake may not reassemble correctly. Carefully install the caliper over the pads, align the caliper mounting bolts with the bracket, and install the bolts. Check that the caliper and mounting bolts can slide properly once they are installed. Torque the caliper bolts to specifications.

Here is a good place to mention common sense. Common sense means to use good judgment in practical matters and is extremely valuable when working on cars and trucks. In many modern brake systems, the caliper is held in place by two small bolts, typically 8 mm in diameter (**Figure 15-39**). These bolts thread into the caliper pin and often have a torque spec between 12 and 20 ft.lbs. The caliper bracket uses larger bolts, often 12 mm or larger in diameter, and can have a torque spec between 60 and 140 ft.lbs. An example of these specs is shown later in Figure 15-51. Before torquing either set of bolts, check the torque specifications and think about which spec is applied to which bolt. Mixing up the torque spec will result in broken caliper pin bolts (**Figure 15-40**). Additionally, sometimes the wrong spec gets printed, which may show the small caliper bolts with a torque spec of 80 ft.lbs. Common sense should kick in here because an 8-mm caliper bolt cannot be torqued to 80 ft.lbs., which is the same as the lug nuts on many Honda vehicles. Caliper bolts that thread into the aluminum caliper housing also have low torque specs (**Figure 15-41**). Overtorquing these bolts will strip the threads in the soft aluminum of the caliper mounting ear.

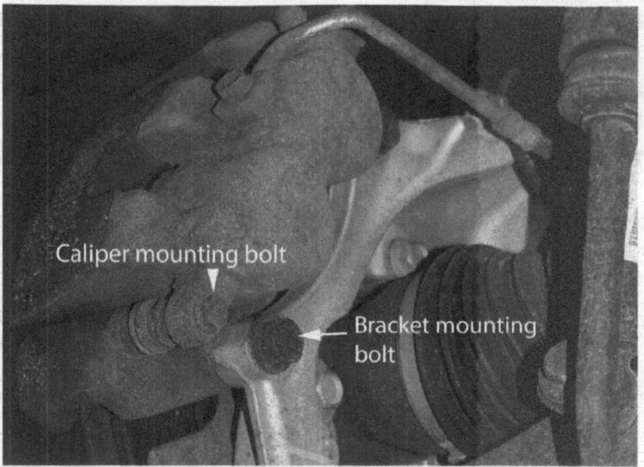

FIGURE 15-39 When looking up torque specifications, make sure you have the correct spec for the correct fastener. Tightening caliper bolts to bracket bolt specs will often result in bolts that look like those in Figure 15-40.

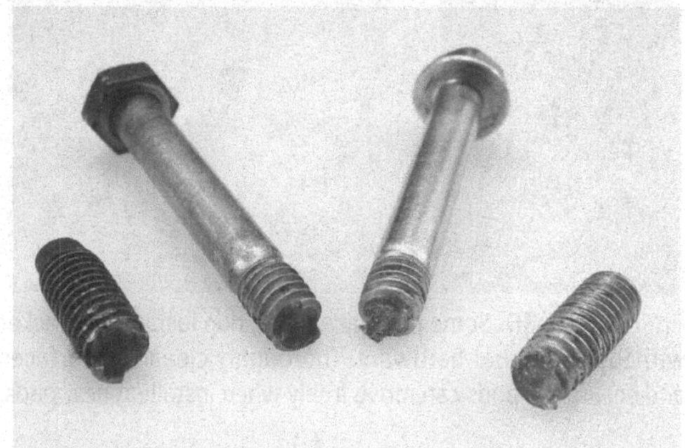

FIGURE 15-40 Use common sense when torquing fasteners. These caliper bolts are 8 mm in diameter and usually have a torque spec between 12 and 20 ft.lbs. Confusing these bolts with caliper bracket or other bolts, with much higher torque specs, results in broken bolts.

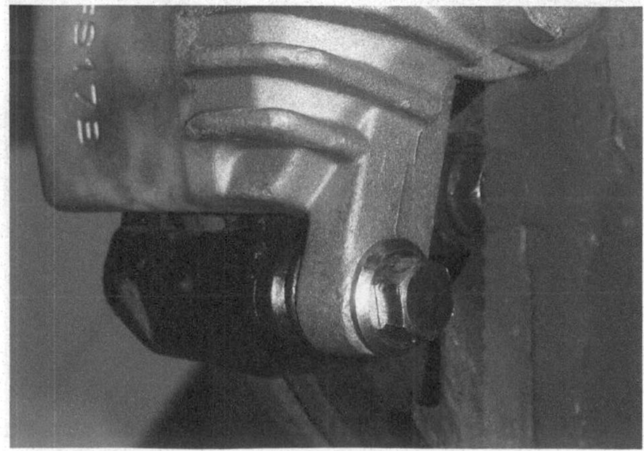

FIGURE 15-41 Most calipers are aluminum. Overtorquing this type of caliper bolt can result in stripping out the threads in the caliper mounting holes.

Once the brakes are together and torqued, it is time for check their operation. Because the caliper pistons are retracted completely into the caliper bore, the brake pedal will be soft and will likely go to the floor for the first several brake applications. **Before you start and move the vehicle, pump the brake pedal several times until the pedal pumps up and is firm. Failure to pump up the brakes before moving the vehicle can result in an accident, serious injury or death, and damage to the vehicle and equipment. Never let the customer be the first person to drive the vehicle after the brakes have been serviced.** Always make sure the brake pedal is firm and the brakes are working before moving the vehicle. Perform a test drive and bed-in the pads before you return the vehicle to the customer.

■ ***Pad Break-In or Bed-In.*** New pad break-in, bed-in, burnishing, and conditioning are terms describing the process of effectively breaking in the new pads to the brake rotors. Break-in should be performed by the technician before the vehicle is returned to the customer. However, it may require several hundred miles of normal driving to fully bed in the pads. Performed properly, this will extend the life of the pads and reduce noise and vibration problems as an even layer of pad material is transferred to the brake rotors. If done properly, the rotors will suffer less wear, and the possibility of rotor thickness variation is reduced.

How brake pads are bedded in depends on the pad manufacturer's recommendations. In general, 20 to 30 stops are made from low speed at 20 to 30 mph, with a cooling time between stops. Read the documentation provided with the new pads to perform the correct procedure. Improper bed-in does not create the transfer layer between the pads and rotors, or it may overheat and damage the pads. Explain to the customer that the brakes may feel slightly different after pad replacement. New or different pads can change the feel and stopping ability of the vehicle, especially just after replacement.

Brake Rotor Inspection

During the life of the brake pads, the material of the brake rotors wears away. Eventually, the brake rotors wear out and require replacement. When to leave the rotor alone, machine it, or replace it depends on how much wear the rotor has and the internal and external conditions of the rotor.

While a visual inspection of the brake rotors can provide you with information about scoring, rust buildup, or even overheating, it should not be the only method of inspection. Rotors must be carefully inspected and measured to determine if they are fit to remain in service.

VISUAL INSPECTION

Begin your inspection by looking over the rotor friction surfaces and hat. Do not just look at the outer surface; often the inner friction surface is in bad condition while the outer surface appears normal. Obvious defects, such as scoring, rust pits, and heat discoloration, should be noted.

Light scoring, also called *ribbing,* is normal and generally is not a concern (**Figure 15-42**). The materials in the brake pads and normal wear will cause shallow score marks around the rotor surface. The outlines of the pads shown on this rotor indicate a transfer of pad material to the rotor took place, likely after keeping the brakes applied after a very hard and hot stop. This can also indicate that proper new pad break-in was not performed after pad replacement.

Heavier scoring results from normal driving over a long period of time (**Figure 15-43**). Deep scoring occurs when the pad lining wears down to the rivets or backing, and results in metal-on-metal contact between the pad

FIGURE 15-42 Light scoring of the rotor is normal and is not considered a problem.

FIGURE 15-43 An example of a more heavily scored rotor.

and rotor (**Figure 15-44**). Rotors that are heavily scored should be machined or replaced. Refer to the vehicle manufacturer's service information to determine at what point scoring is excessive and requires service. General Motors service information states that scoring less than 0.060 inch (1.5 mm) is acceptable and the rotor, if free of other defects, can be reused without machining.

Rust pits can come from inside the metal of the rotor and from exposure to the salt used on roads in the winter (**Figure 15-45**). Rust deposits form inside the cast iron, and as the friction surface wears and flakes away, pits begin to show in the surface. Machining the rotor will not fix this problem as it is a defect in the iron.

Severe rusting of the rotor surfaces is another common problem (**Figure 15-46**). This rear brake was not applying, which kept the pads and rotor creating enough friction to keep the rotor clean and free of water and other elements. In locations that use salt on the roads, failing to keep the vehicle clean during the winter, can cause the rotors to rust away the inside out. From the outside, the rotor looks normal, but upon inspecting the vents and the inboard side, the amount of the rust is obvious (**Figure 15-47**). In severe cases, the rotor can rust in half (**Figure 15-48**). These rotors rusted through the vents and separated the inboard and outboard halves of the rotors.

Rotors do not rust only on the outside; serious rust can also occur inside vents (**Figure 15-49**). To inspect the entire vents, use a bore scope or video bore scope. As discussed above, rotors can completely rust away from the inside. This leads to the friction plates actually sepa-

FIGURE 15-44 Heavy scoring from metal-on-metal contact between the pad backing plate and the rotor surface.

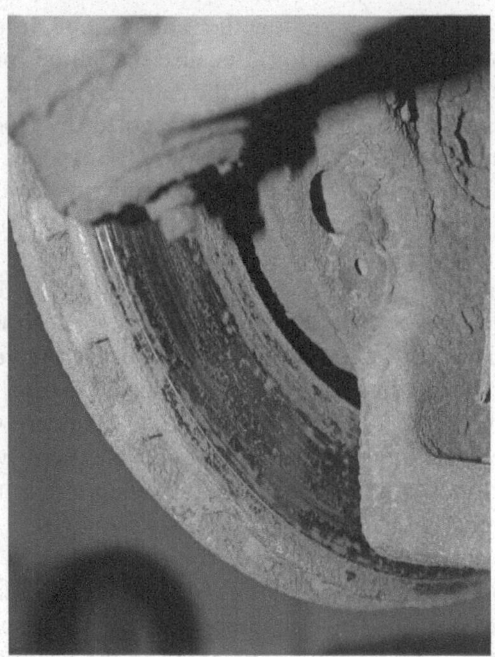

FIGURE 15-45 Always check the inboard side of the rotor for defects. The outboard surface may look good, but the inboard can be in very bad condition.

FIGURE 15-46 Severe rusting of the rotor typically requires rotor replacement.

FIGURE 15-47 Long exposure to road salt can cause severe rusting in the rotor.

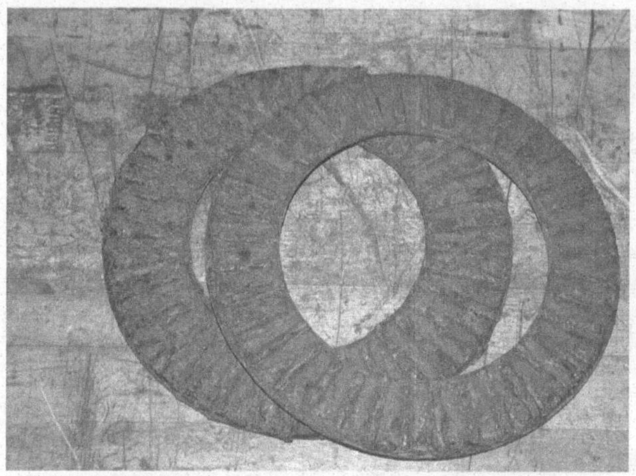

FIGURE 15-48 Rotors can rust in half through the vents. This usually locks the brake as the two halves separate and then wedge into the caliper and pads.

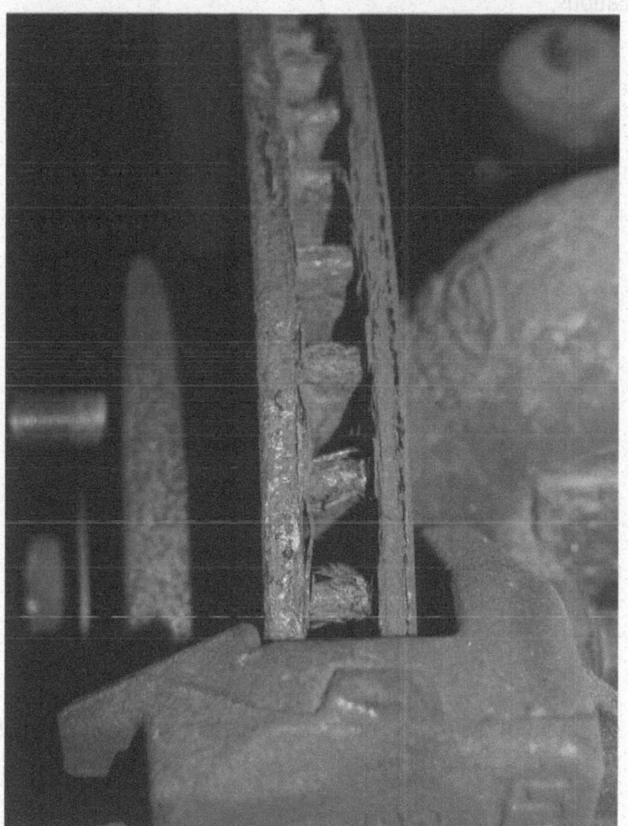

FIGURE 15-49 An example a rotor rusting apart from the inside. Take a look inside the vents to check for rust-through.

rating in two (Figure 15-48). Once the rotor separates, it will be easy to diagnose due to the noise and pulsation. However, your customer will be much happier if you determine that the rotor is severely rusted internally before it actually fails.

Overheated brakes can cause the rotor to glaze and turn blue (Figure 15-18). This vehicle had a sticking brake caliper that kept pressure on the pads, resulting in overheating of both the pads and rotor. This allowed excessive transfer of material between the pads and rotors.

FIGURE 15-50 Performance rotors often show small cracks around the drilled holes. If the cracks do not extend between holes or to the edges of the rotor, the rotor may still be serviceable.

If working on a performance vehicle with cast iron drilled or slotted rotors, inspect the surfaces around the slots and holes. Small, shallow cracks around the holes are common and are not considered a problem unless the crack extends between two holes or to either the inside or outside edges (**Figure 15-50**).

Inspecting the outside surface of the rotor is easy, but inspecting the inner friction plate requires removing the caliper to get access to the rotor. There can be a significant difference between the inner and outer friction surfaces, so always check both sides of the rotor closely. To thoroughly inspect the rotor, it should be removed. With the rotor removed, inspect the hub for damage and rust buildup that can cause the rotor to sit improperly and cause runout. Check the inner hat of the rotor for rust, and clean before machining or reinstalling it.

ROTOR SPECIFICATIONS

All brake rotors have a minimum thickness or discard specification, meaning how thin the friction plates can be or how much the rotor can wear down to before it must be replaced. Some manufacturers will also provide a machine-to or minimum refinish spec (**Figure 15-51**). Depending on the condition of the rotor, you may be able to find the minimum thickness spec stamped into the hat section or around the outside edge of the friction surface.

■ ***Minimum Thickness.*** Rotors have a minimum thickness spec because as the rotor friction plates wear away, the rotor less able to absorb and dissipate heat and stress. As the friction plates wear, more heat is left to be absorbed by the brake pads. This increases pad wear and can lead to cracks in the lining material. When a rotor wears down to or past the minimum or discard spec, it must be replaced.

Year	Model		Brake Disc			Brake Drum Diameter			Minimum Lining Thickness	Brake Caliper	
			Original Thickness	Minimum Thickness	Maximum Runout	Original Inside Diameter	Max. Wear Limit	Maximum Machine Diameter		Bracket Bolts (ft. lbs.)	Mounting Bolts (ft. lbs.)
2006	Accent	F	0.870	0.790	0.001				0.079	62-69	16-23
		R				8.000	①	①	0.039		
	Azera	F	1.100	1.040	0.002				0.079	58-72	16-23
		R	0.390	0.310	0.002				0.080	58-72	16-23
	Elantra	F	1.020	0.940	0.002				0.079	58-72	16-23
		R	0.390	0.330	0.002				0.079	36-43	16-23
	Elantra	F	1.020	0.940	0.002				0.079	58-72	16-23
		R				8.000	①	①	0.039		
	Sonata (2.4L)	F	1.024	0.961	0.002				0.120-0.160	59-74	18-22
		R	0.390	0.330	0.002				0.120	59-74	18-22
	Sonata (3.3L)	F	1.100	1.040	0.002				0.120-0.160	59-74	18-22
		R	0.390	0.330	0.002				0.120	59-74	18-22
	Tiburon	F	1.024	0.961	0.003				0.079	48-55	16-24
		R	0.400	0.330	0.002				0.080	48-55	16-24

① Drum roundness Service Limit: 0.00236 inch

FIGURE 15-51 An example of brake pad and rotor specifications.

The machine-to spec is often 0.020 to 0.030 inch (0.5 to 0.76 mm) larger than the discard spec and is used to provide a wear buffer for the rotor to remain in service. If a rotor is larger than the machine-to spec and is in otherwise good condition, it can remain in service. If, however, the rotor is above machine-to but requires resurfacing, the rotor's thickness after refinishing must be above the machine-to spec. This 0.020 to 0.030 inch (0.5 to 0.76 mm) of friction surface is what is worn away during normal service. Once the vehicle needs the brake pads replaced again, the rotor will likely be below the machine-to spec and will need to be replaced.

■ ***Thickness Variation.*** Rotors also have a spec for thickness variation, parallelism, or taper, referring to the varying thickness of the rotor around the friction surfaces. This spec is often quite small, around 0.0005 inch or one-half of a thousandth of an inch (0.012 mm). For perspective, an average piece of printer paper is 0.003 to 0.004 inch (0.0762 to 0.1016 mm). This means the average allowance for thickness variation is about 1/6 to 1/8 of the thickness of a sheet of paper. Variations in rotor thickness result in brake pedal pulsation as the pads move in and out to compensate for the rotor.

Thickness variation and taper can be caused by sticking pads or calipers that do not retract. This leaves the pads in contact with the rotor as it spins. Thickness variation can also be caused by runout in the rotor, hub, or wheel bearing (**Figure 15-52**).

■ ***Lateral Runout.*** Rotor runout specs, the amount of side-to-side or lateral wobble of the rotor, is often 0.002 to 0.004 inch (0.5 to 0.1 mm). An illustration of lateral runout is shown in **Figure 15-53**. Overtightening lug

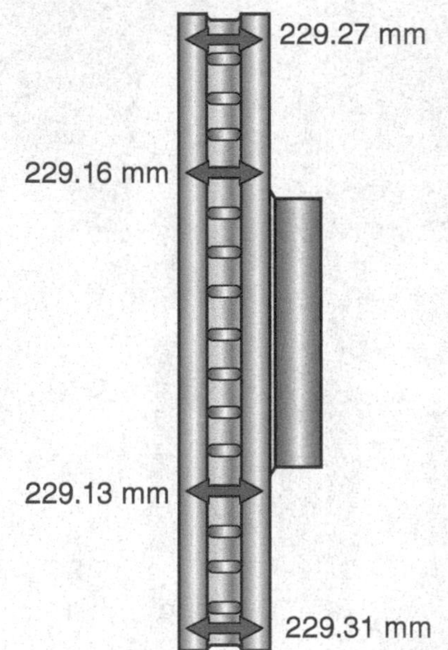

FIGURE 15-52 An illustration of rotor taper. Careful measurement of the rotor is necessary to determine how much it is worn and if it can be reused.

fasteners, rotor overheating, worn wheel bearings, and damage or rust buildup in the hub can cause rotor runout, which can also cause brake pedal pulsation.

Runout can also cause thickness variation issues. As the rotor spins and wobbles due to runout, the friction surfaces can contact the pads. This wears away at the rotor at those contact points, which results in thickness variations around the rotor.

ROTOR MEASUREMENT

Rotors must be measured any time disc brake service is performed. Rotor thickness is measured with a disc brake

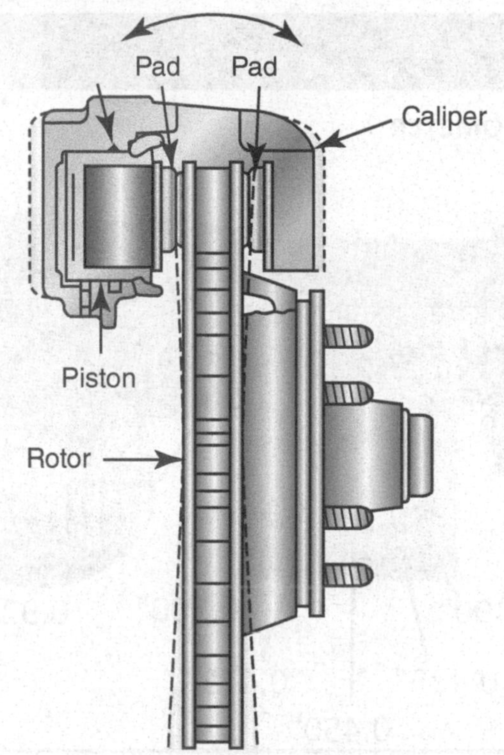

FIGURE 15-53 An illustration of lateral runout.

micrometer (**Figure 15-4**). Rotors that are smaller than machine-to or below the discard limit must be replaced. Rotors above machine-to may be able to be placed back in service depending on their parallelism, runout, and surface condition. A rotor that is above machine-to and is within parallelism and runout specs and has no surface defects can often be reused without machining. Many manufacturers state that rotors that meet all specs and have acceptable surface conditions should not be machined and should be reinstalled as they are. This is because machining removes material, and makes the rotor less able to dissipate heat. Also, improper machining methods can cause pulsation, increased braking noise, and customer dissatisfaction.

Measuring rotor parallelism requires a micrometer that can measure at least to 0.0005 inch or about 0.01 mm. Many brake micrometers use a graduation that indicates half-thousandths (**Figure 15-54**). The graduation allows you to determine if the rotor is parallel to within 0.0005 inch, which is a common spec, although some manufacturers specify parallelism to be less than 0.0005 inch. An example of the graduations on a metric mic is shown in **Figure 15-55**. Measure the rotor at 8 to 12 places around the friction surface as shown in **Figure 15-56**. The reason so many measurements are made is that you are measuring approximately the size of the lining contact area between each point. A rotor with excessive parallelism will need to be either machined or replaced, depending on its overall thickness.

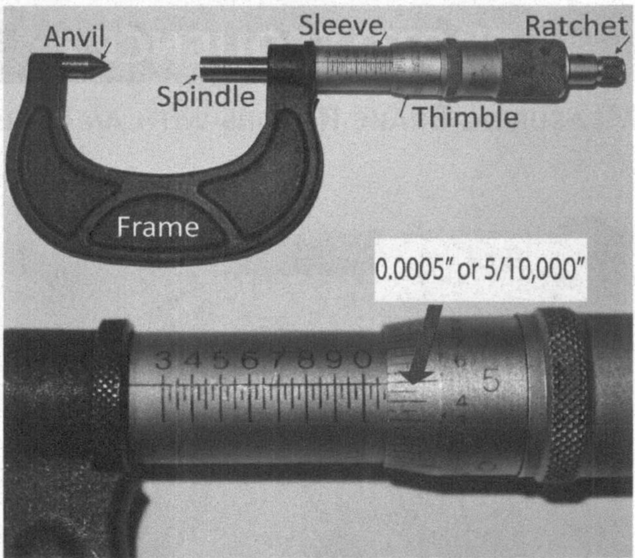

FIGURE 15-54 An English rotor micrometer. This mic can read to 0.0005 inch. This mic starts at 0.300 inch and measures up to 1.300 inches.

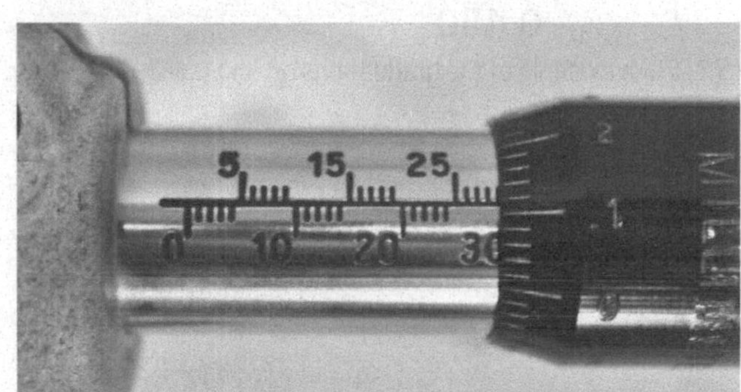

FIGURE 15-55 An example of a metric rotor micrometer. This mic reads to 0.02 mm.

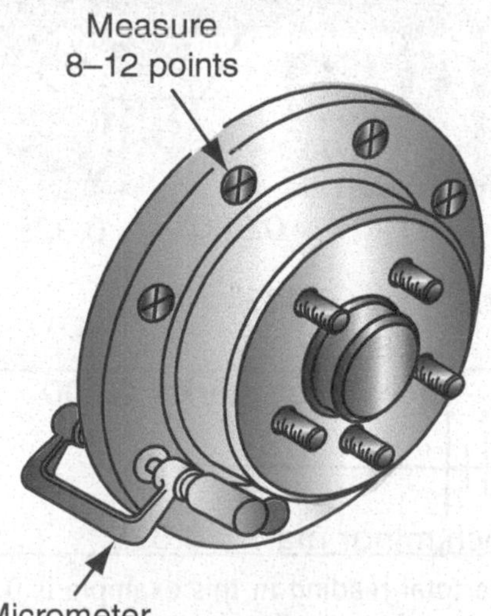

FIGURE 15-56 Measure the rotor at 8 to 12 points with the micrometer. Subtract the smallest measurement from the largest to get total thickness variation.

PHOTO SEQUENCE 7

Measuring Brake Rotors with an English Micrometer

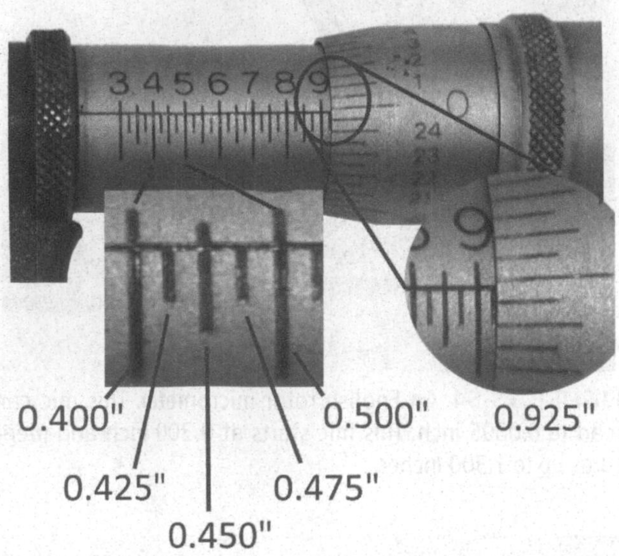

PS7-1 An example of the graduations on an English micrometer.

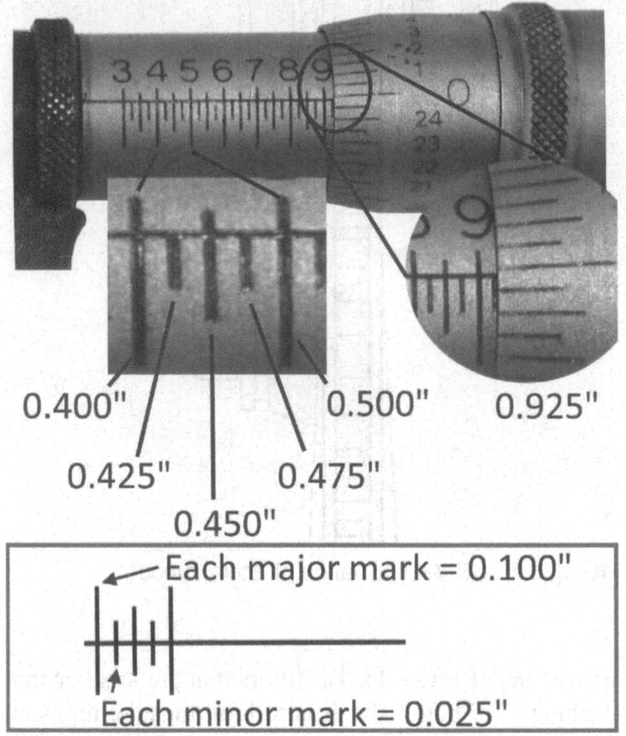

PS7-2 Note the major graduations on the sleeve. Each number represents 0.100 inch and each minor mark is equal to 0.025 inch. As the thimble moves to the right, the readings become larger as more gradations are exposed.

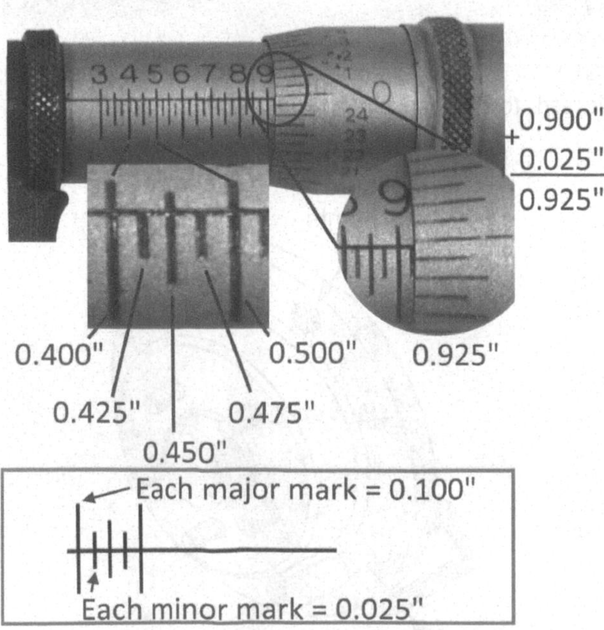

PS7-3 The total reading in this example is 0.925 inch or nine-hundred and twenty-five thousandths of an inch. Where the thimble aligns with the horizontal scale line on the sleeve determines what number on the thimble is added to the sleeve.

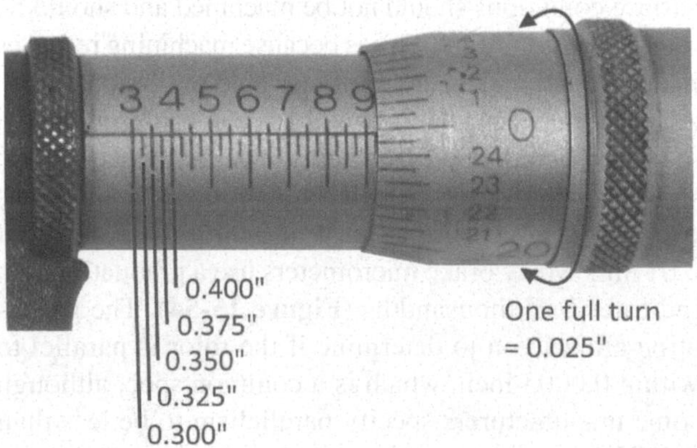

PS7-4 Another view of reading the mic. As the thimble makes one full turn, the measurement either increases or decreases by 0.025 inch. Each graduation on the thimble is equal to 0.001 inch or one-thousandth of an inch. Graduations between each 0.001 inch mark are 0.0005 inch. These marks represent one-half of one-thousandth of an inch or 5/10,000 inches.

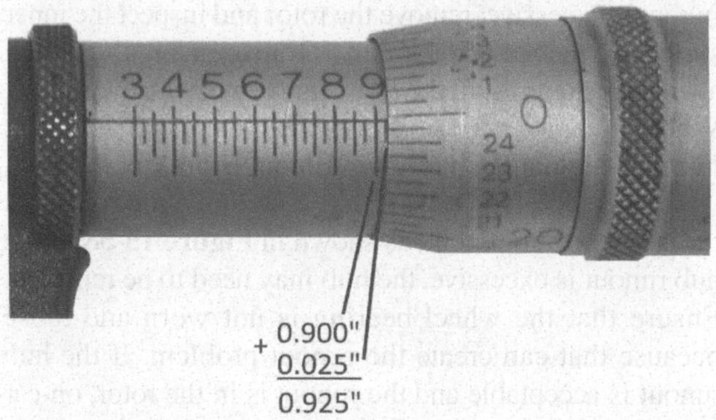

PS7-5 Because the thimble is one full turn past the 0.900-inch line, the next 0.025-inch mark is showing. This number is added to the 0.900-inch reading for a total of 0.925 inch.

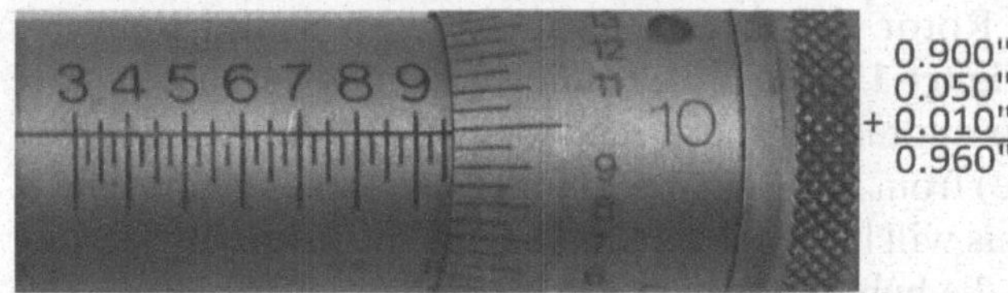

PS7-6 This example shows the thimble two marks past the 0.900-inch line. In addition, the line on the thimble closest to the horizontal scale line is 0.010 inch. The three numbers added together give a reading of 0.960 inch.

PHOTO SEQUENCE 8

Measuring Brake Rotors with a Metric Micrometer

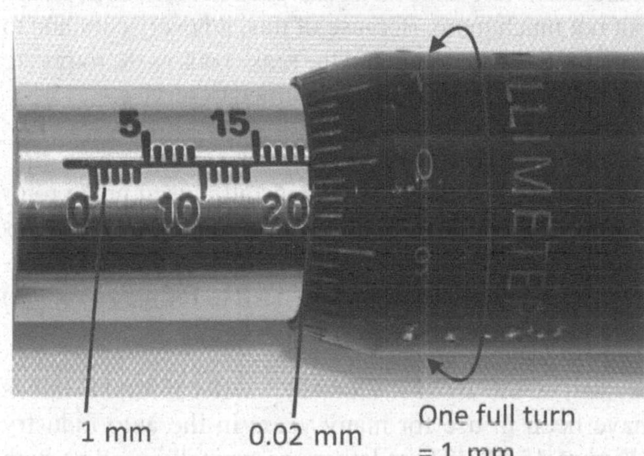

PS8-1 This example of a metric micrometer has whole millimeters on the horizontal scale on the sleeve. Each turn of the thimble adds 1 mm.

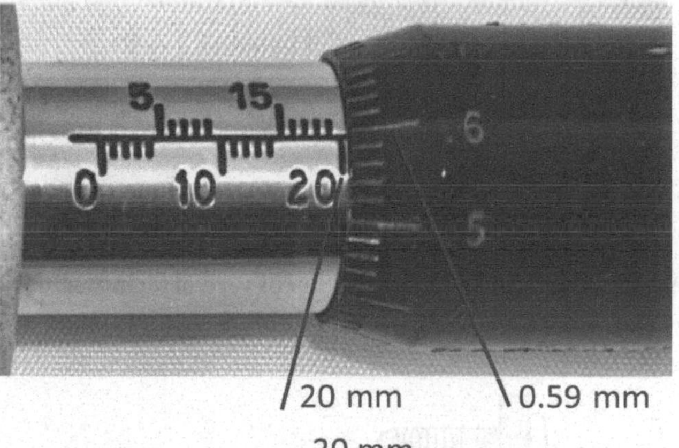

PS8-2 Major graduations on the thimble are 0.1 mm and each minor graduation is 0.02 mm. The numbers on the spindle in this image between the 0.5 mm and 0.6 mm are 0.52 mm, 0.54 mm, 0.56 mm, and 0.58 mm. Based on these graduations, readings down to 0.01 mm can be made for checking thickness variation.

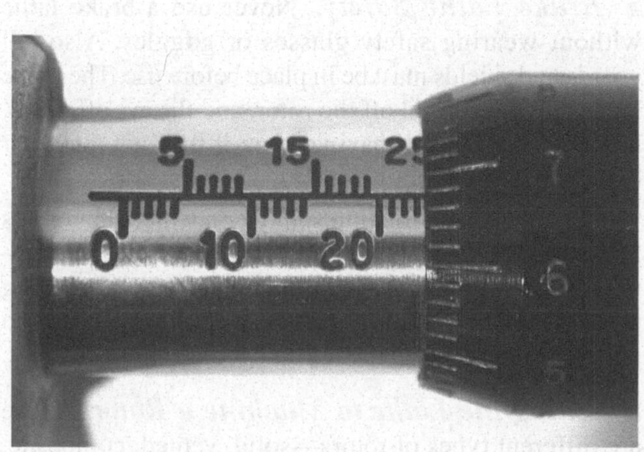

PS8-3 This mic reading is 23.68 mm.

Rotor runout is measured with a dial indicator (**Figure 15-57**). If the reading is beyond the spec, remove the rotor and reindex it on the hub 180 degrees (if possible) from its original position, then remeasure the runout. This will help you determine if the runout is with the rotor or the hub. Rotor runout can also be caused by excessive rust buildup between the inside of the hat and the hub. If runout is excessive, remove the rotor and inspect the inner surface of the hat and the hub. Remove rust buildup with a wire brush, reinstall the rotor, and remeasure the runout. When the brake service is complete, the rotor should be installed so that the minimum runout is present.

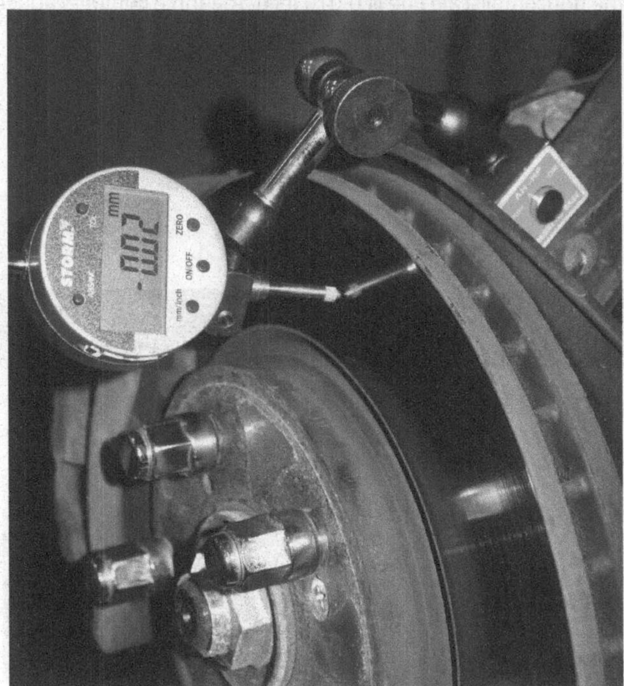

FIGURE 15-57 Measure rotor runout with a dial indicator. Total runout can be a combination of several factors, so do not assume the rotor is the cause without further testing.

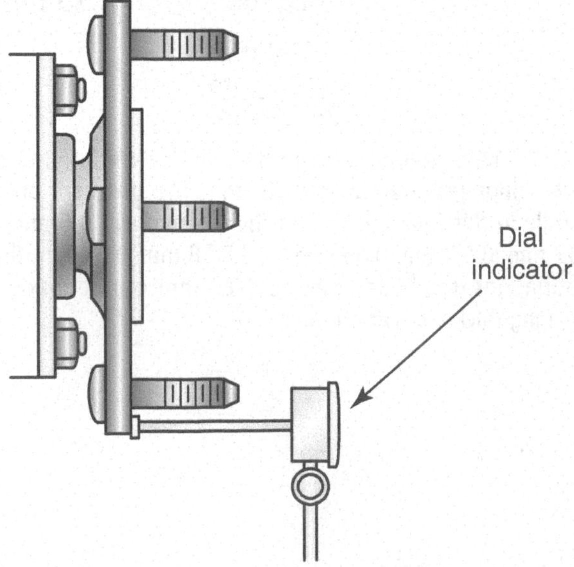

FIGURE 15-58 Excessive rotor runout can be caused by runout in the hub or a loose wheel bearing. Do not replace the rotor until the true cause of the runout is found.

If the reading is still excessive, remove the rotor and measure the hub runout, as shown in **Figure 15-58**. If the hub runout is excessive, the hub may need to be replaced. Ensure that the wheel bearing is not worn and loose because that can create the runout problem. If the hub runout is acceptable and the runout is in the rotor, on-car machining of the rotor may solve the runout. If the rotor is below thickness specs, you need to replace the rotor and remeasure the runout. If it is necessary, reindex the new rotor until the smallest amount of runout is obtained.

Rotor Refinishing and Brake Lathes

Once a common practice, performed every time the brake pads were replaced, brake rotor machining is now done less frequently due to the smaller and lighter rotors in service today. Many rotors on today's cars and trucks have enough material to last until the brake pads need to be replaced but not much more. Because of this, it is very common to replace the brake rotors and the brake pads as the rotors are too thin to be machined and returned to service.

REFINISHING A ROTOR

Some manufacturers say to reinstall a used brake rotor without refinishing it if it is within wear specs and does not have any major surface condition problems. However, there are times when you need to refinish a rotor to correct a problem.

■ ***Off-Car Brake Lathes.*** Off-car brake lathes have been in use for many years in the auto industry (**Figure 15-59**). These lathes are typically used on both brake rotors and brake drums. Before you attempt to use a brake lathe, some basic safety must be observed.

■ ***Brake Lathe Safety.*** Never use a brake lathe without wearing safety glasses or goggles. Also, all guards and shields must be in place before use. The brake lathe shaves material off the rotor, and throws off small, hot, and sharp metal shavings that, if they get into your eyes, can cause serious eye damage and loss of sight.

The lathe uses a rotating spindle on which the rotor is mounted. Be sure that you do not have any loose clothing, jewelry, or hair that can be pulled into the rotating parts. The motor on the lathe has a lot of torque and can pull you into the machine if you become entangled in it.

■ ***Setting the Lathe to Machine a Rotor.*** There are different types of rotors—solid, vented, composite, hub, and hubless—and each requires specific mounting

FIGURE 15-59 An example of a common bench brake lathe.

and machining procedures. Refer to the documentation that comes with the lathe you are using before you attempt to mount a rotor and make any cuts. The steps shown in **Photo Sequence 9** are typical for mounting a cast iron hubless or floating rotor. Once it is mounted, follow these steps:

- Center the cutting head over the rotor. Place the cutting bits about halfway across the rotor surface, and turn the lathe power on.
- Adjust the cutting bits until each just touches the rotor and makes a scratch cut. Back the bits off the rotor and turn the lathe off.
- Loosen the rotor mounting nut, turn the rotor 180 degrees on the arbor, and retighten.
- Perform a second scratch cut. The two scratch cuts should be parallel; if not, remove and inspect the mounting components, and clean and remount as needed.
- Set the cutting bits halfway across the rotor and turn them until they just touch the surface. Set the depth-of-cut collars to zero.
- Turn the lathe on and move the cutting head in toward the hat. Set the depth-of-cut between 0.006 inch and 0.010 inch (0.15–0.25 mm). Engage the feed for a fast or rough cut.
- If the rough cut cleans up the surface and removes the low sections, reset the lathe to perform the slow or finish cut. This cut is usually a shallower cut, between 0.004 inch and 0.006 inch (0.10–0.015 mm). Repeat the cutting process for the final cut.
- Once complete, apply a nondirectional finish to the rotor. Remeasure rotor thickness and clean thoroughly.

FIGURE 15-60 Once the rotor has been machined, a non-directional surface finish is applied to help reduce pad nose.

Rotors are usually rough cut, and then a finish cut is performed. The **rough cut** removes the surface defects and restores the surfaces to parallel. If the surface is heavily scored, a second rough cut may be necessary. Once the surface is cleaned of defects, a finish cut is performed. This cut is much slower and cuts less material than the rough cut. Once the finish cut is complete, a nondirectional surface finish is applied using a light abrasive. This breaks up the lines that appear in the rotor surface from the cutting bit. **Figure 15-60** shows a rotor after the nondirectional finish is applied. Breaking up the lines in the rotor surface helps reduce noise produced by the pads and rotor during braking.

Once the rotor is finished, recheck its finished thickness with a micrometer to ensure that it can be put back into service. If the rotor is usable, it must be cleaned before being reinstalled on the vehicle. Even though the rotor appears clean and like new, the surface has a very fine layer of metal dust embedded in the valleys of the metal. Wash the rotor with a warm soapy water solution to remove the dust, and dry completely.

ON-CAR BRAKE LATHES

On-car brake lathes perform the same function as off-car lathes except that the **on-car lathe** attaches to the vehicle in order to machine the rotor.

■ ***Why Use On-Car Lathes?*** Many vehicle manufacturers require the use of an on-car lathe to machine the brake rotors. The main reason is to eliminate rotor runout. **Figure 15-61** shows an example of what

PHOTO SEQUENCE 9

Setting Up a Rotor on a Brake Lathe

PS9-1 Before mounting the rotor on the lathe, clean both the inner and outer mounting surfaces. Excessive rust buildup on these surfaces can cause improper mounting and runout in the cut.

PS9-2 Determine the correct adapters required to mount the rotor to the brake lathe.

PS9-3 Install the open cone onto the spindle followed by the spring and the hub adapter.

PS9-4 Position the rotor onto the spindle with the hat outward, just as the rotor is installed on the vehicle.

PS9-5 Install a second open cone against the outside of the hat.

PS9-6 Install a spacer as needed to take up space on the spindle out to the threads.

PS9-7 Install the spindle nut. Note that the threads are left-handed.

PS9-8 Install the vibration dampening strap around the rotor vents and secure it in place.

PS9-9 Turn the lathe on and check for excessive rotor movement. If the rotor moves excessively from side-to-side or up and down, turn off the lathe and remove the rotor. Check to see if the rotor is sitting properly on the hub adapter.

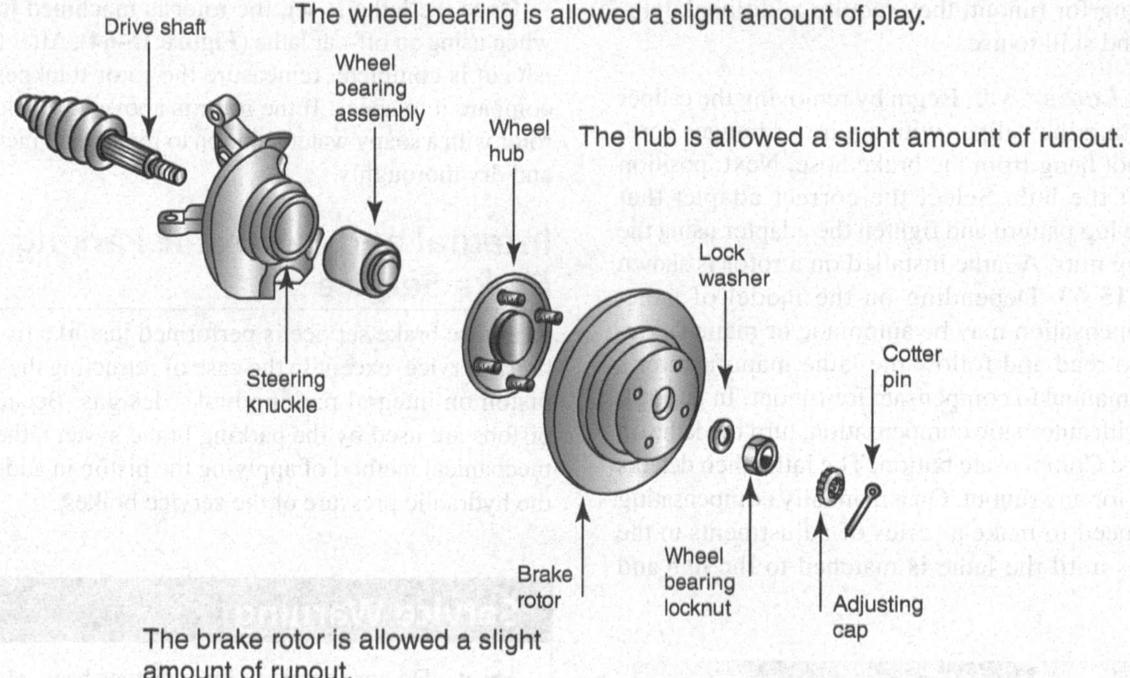
The wheel bearing is allowed a slight amount of play.
Drive shaft
Wheel bearing assembly
Wheel hub
The hub is allowed a slight amount of runout.
Lock washer
Cotter pin
Steering knuckle
Brake rotor
Wheel bearing locknut
Adjusting cap
The brake rotor is allowed a slight amount of runout.
All of these can add up to an excessive amount of rotor runout even though the rotor itself does not have significant runout.

FIGURE 15-61 Excessive rotor runout can be caused by stacked tolerances, where each component is allowed a slight amount of runout, but the sum of the tolerances results in excessive rotor runout.

is called *stacked tolerances*, meaning that several components, all with a spec for allowable runout, are mounted together. This means the runout of each component is added together until the total runout is excessive. Removing the rotor and machining it with an off-car lathe does not remove the runout present at the rotor. For example, if rotor runout measures 0.010 inch (0.254 mm), it is possible that the runout is caused only by the wheel bearing and hub. Removing and machining the rotor will not fix the problem because the runout is not in the rotor itself. An on-car lathe compensates for the runout present at the rotor, making the finished rotor true to the vehicle. This eliminates runout problems.

A second reason for using an on-car lathe is that improper setup of a rotor on an off-car lathe can actually make rotor runout worse. Using the incorrect adapters and/or not having clean mounting surfaces on the rotor can cause runout on the lathe arbor. Machining will not correct any technician-induced runout, and the rotor, when it is reinstalled, will have as much or even more runout than when it was removed. Also, machining a rotor with runout in the hub with an off-the-car lathe does not remove the runout in the rotor.

On-car lathes are also used because on some vehicles, the rotor is trapped or captured, meaning that it mounts inboard of the hub and bearing, as shown in

FIGURE 15-62 Some rotors are trapped or held on by the wheel bearings. Machining these rotors with an on-car lathe is much easier than removing them for machining.

Figure 15-62. To remove the rotor, the bearing must be removed. This is time consuming and often causes damage to the bearing. Instead of removing the rotor, it is machined on the car. This eliminates runout issues and saves the technician time.

The disadvantages to on-car lathes are that they are more expensive than off-car lathes, and if not self-compensating for runout, they require additional training, time, and skill to use.

■ ***On-Car Lathe Use.*** Begin by removing the caliper and supporting the caliper with a wire or bungee cord, so it does not hang from the brake hose. Next, position the lathe at the hub. Select the correct adapter that matches the lug pattern and tighten the adapter using the vehicle's lug nuts. A lathe installed on a rotor is shown in **Figure 15-63**. Depending on the model of lathe, runout compensation may be automatic or manual. You will need to read and follow the lathe manufacturer's instruction manual to compensate for runout. In general, on a lathe with automatic compensation, turn the lathe on and press the Compensate button. The lathe then detects and adjusts for any runout. On a manually compensating lathe, you need to make a series of adjustments to the runout dials until the lathe is matched to the hub and rotor.

FIGURE 15-63 Setting up an on-car lathe requires installing the correct adaptor to the hub and then installing the lathe onto the adaptor.

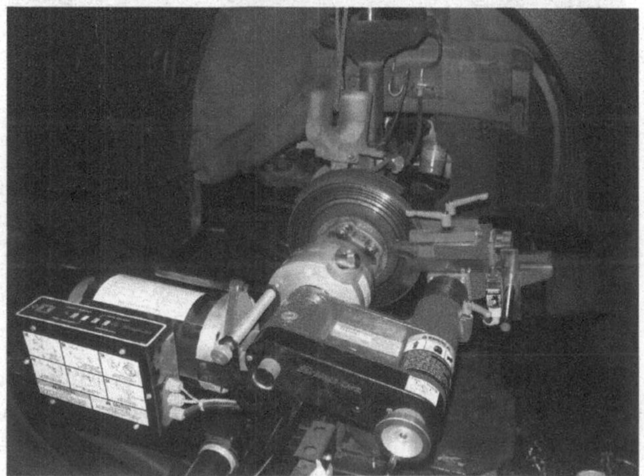

FIGURE 15-64 Once the on-car lathe is set up, using it to machine a rotor is very similar to using a bench lathe.

Once the lathe is set, the rotor is machined just like when using an off-car lathe (**Figure 15-64**). After the finish cut is complete, remeasure the rotor thickness, and compare it to specs. If the rotor is above spec, clean the rotor with a soapy water solution to remove all metal dust and dry thoroughly.

Integral and Electronic Parking Brake Service

Rear disc brake service is performed just like front disc brake service, except in the case of retracting the caliper piston on integral parking brake designs. Because the pistons are used by the parking brake system, there is a mechanical method of applying the piston in addition to the hydraulic pressure of the service brakes.

Service Warning

Do not attempt to force pistons back into electronic parking brake calipers. This can damage the caliper, requiring its replacement.

■ ***Integral Piston Caliper Service.*** There are several types of tools available to retract the caliper piston on a mechanical system. Many technicians use a square drive tool (**Figure 15-65**). Align the correct side of the tool with the notches in the piston and rotate the piston back into the caliper bore. The **piston retraction tool** pushes the piston back into the bore as it rotates, so it is important to keep the tool secure in the caliper body during use. Once the piston is fully retracted, remove the tool from the caliper. Make sure that the piston is properly aligned. If the piston is not aligned correctly, the notch does not fit against the tab on the back of the pad (**Figure 15-66**). It would cause the brake to drag slightly when released.

Install the caliper and pads over the brake rotor, and test-fit the pads. There may be a slight gap between the pads and the rotor, especially if the rotor is not replaced. If the gap is excessive, use the piston tool to rotate the piston one-quarter to one-half of a turn, depending on the number of pad-locating notches the piston has. Your goal is to have the pads as close to the rotor as possible without causing drag on the rotor. If turning the piston causes the pads to be too far out, to the point where the pads and rotor will not fit over the rotor, return the caliper piston to its fully retracted position.

Once the pads are installed, install the caliper and torque all fasteners to specs. Apply the brakes several

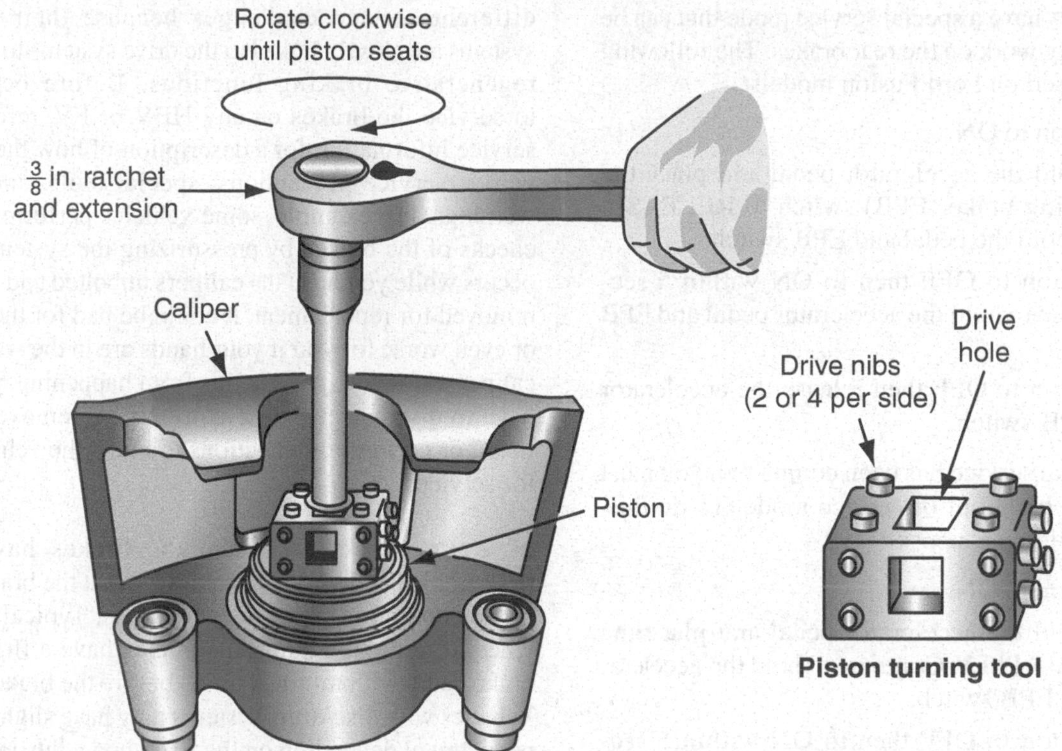

FIGURE 15-65 When servicing integral rear calipers, retracting the piston requires threading the piston back into its bore. *Do not try to seat the piston using just a C-clamp or similar tool as this will damage the caliper.*

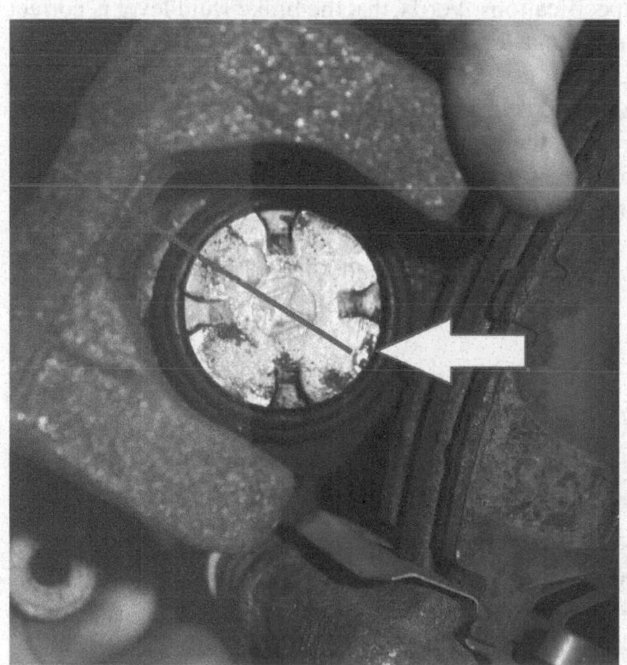

FIGURE 15-66 An example from a vehicle brought in with a rear brake concern after the customer replaced the rear brake pads. Note the piston position. For the inner pad to seat correctly, the notches in the piston must align properly to the caliper housing, as shown by the red line.

times and apply the parking brake. Make sure the parking brake applies fully and locks the wheel. In some cases, the gap between the pads and rotor will need to be adjusted before the brake applies fully. This may require applying the parking brake many times to adjust the piston out to take up the gap.

■ ***Electronic Parking Brake Caliper Service.*** Many newer vehicles have electronic parking brake systems or hill-assist braking systems that require special service. Before attempting to service the brakes on a vehicle equipped with either of these features, check the service information for how the system(s) operate and how to properly service the brakes. Failure to follow the service procedures can cause serious damage to the brake calipers.

If the rear calipers have traditional parking brake cables and an electric motor that pulls the cables to apply the parking brake, retracting the piston is performed using a piston retraction tool as shown in Figure 15-65.

To retract the piston on rear calipers with electrically driven pistons, you may need to connect a scan tool to the onboard computer. Depending on the vehicle being serviced, a scan tool may be required to place the electronic parking into service mode. This retracts the parking brake motors and allows the piston to be retracted back into the caliper bore.

Some vehicles have a special service mode that can be accessed to safely work on the rear brakes. The following is an example used on Ford Fusion models:

1. Set the ignition to ON.
2. Press and hold the accelerator pedal and place the electric parking brake (EPB) switch to RELEASE. Continue to hold the pedal and EPB switch.
3. Set the ignition to OFF then to ON within 5 seconds. Continue to hold the accelerator pedal and EPB switch.
4. Set the ignition to OFF then release the accelerator pedal and EPB switch.

Once the brake service has been completed, you must then take the vehicle out of service mode. To do this, complete the following steps:

1. Set the ignition to ON.
2. Press and hold the accelerator pedal and place the EPB switch to APPLY. Continue to hold the accelerator pedal and EPB switch.
3. Set the ignition to OFF then to ON within 5 seconds. Continue to hold the accelerator pedal and EPB switch.
4. Release the accelerator pedal and EPB switch.

This will fully apply the parking brake and set the air gap between the pads and rotors.

Always refer to the manufacturer's service information when working on any electronic parking brake or hill-assist systems for specific procedures.

■ ***Hybrid Vehicle Brake Service.*** Hybrid electric vehicles (HEVs) and electric vehicles (EVs) present different service challenges because their braking systems are closely linked to the drive systems to perform regenerative braking functions. Before beginning to service the brakes on any HEV or EV, refer to the service information for a description of how the system works, service precautions, special tools, and safety warnings. For example, some vehicles perform random checks of the brakes by pressurizing the system. If this occurs while you have the calipers unbolted and the pads removed for replacement, it could be bad for the caliper or even worse for you if your hands are in the way of the caliper pistons. To prevent this from happening, you may have to disconnect the 12-volt battery, remove certain fuses, or perform other actions to make the vehicle safe for service.

■ ***Final Checks.*** Once the brakes have been reassembled and torqued, make sure that the brake pedal pumps up and is firm under pressure. Typically, four-wheel disc brake equipped vehicles have a firm, high, brake pedal with minimal travel before the brakes apply. Vehicles with disc/drum systems may have slightly more pedal travel depending on the rear shoe adjustment.

Recheck for leaks around brake hoses and bleeder valves, and clean any brake fluid that may have escaped during bleeding. Once you have verified that everything is correct, reinstall the wheels and torque the lug nuts to specifications. Verify that the brake fluid level is correct in the master cylinder, and then test-drive the vehicle. During the test-drive, note the feel of the brake pedal, and listen closely for any noise during brake application. Follow the brake pad manufacturer's recommendations for breaking in or bedding-in the pads before returning the vehicle to the customer.

SUMMARY

Noisy brakes are a common complaint about disc brake systems. Rotor parallelism, also called rotor thickness variation or taper, refers to the thickness of the rotor at all points around the friction surface.

Lateral runout is the side-to-side movement of the rotor. It can be caused by the rotor being distorted, hub flange distortion, or worn wheel bearings.

Rotors, in addition to scoring of the friction surface, can also rust away from the inside out.

Rotor thickness is measured with a disc brake micrometer.

Measuring rotor parallelism requires a micrometer that can measure at least to 0.0005 inch or 0.01 mm.

Rotor runout is measured with a dial indicator.

Some manufacturers recommend reinstalling a used brake rotor without refinishing if it is within wear specs and does not have any major surface condition problems.

Special tools are necessary to retract the piston on an integral parking brake caliper.

REVIEW QUESTIONS

1. _______________ runout is the side-to-side wobble or distortion of a rotor.
2. A _______________ in the brake pedal and steering wheel is often caused by excessive thickness variation.
3. A _______________ _______________ is used to measure rotor runout.
4. _______________ refers to the distortion of the rotor surface cause by pad wear.
5. A collapsed brake hose or sticking caliper can cause the brakes to _______________ or feel like they are staying applied.
6. The measurement shown in **Figure 15-67** is being discussed: *Technician A* says thickness variation is being checked. *Technician B* says lining thickness is being checked. Who is correct?
   - **a.** Technician A
   - **b.** Technician B
   - **c.** Both A and B
   - **d.** Neither A nor B
7. A vehicle has a grinding sound when applying the brakes: *Technician A* says this may indicate the brake pads need to be replaced. *Technician B says* this may be caused by warped brake rotors. Who is correct?
   - **a.** Technician A
   - **b.** Technician B
   - **c.** Both A and B
   - **d.** Neither A nor B
8. A rotor shows twice the specified amount of runout with a dial indicator: *Technician A* says the hub runout should be checked. *Technician B* says the rotor should be machined or replaced to correct the runout. Who is correct?
   - **a.** Technician A
   - **b.** Technician B
   - **c.** Both A and B
   - **d.** Neither A nor B

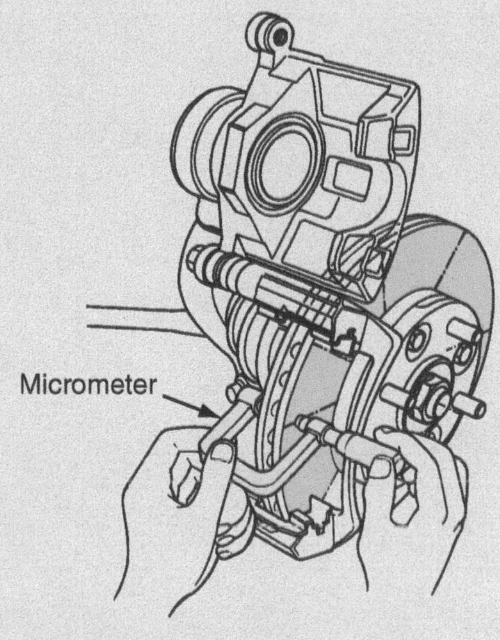

FIGURE 15-67

9. Which of the following is approved by all manufacturers as a disc brake lubricant?
   - **a.** Antiseize
   - **b.** Synthetic brake grease
   - **c.** Lithium grease
   - **d.** None of the above
10. Brake pad conditioning refers to which of the following?
    - **a.** Retracting the piston into the caliper bore
    - **b.** Resurfacing the brake rotors
    - **c.** Installing lubricant on the pads to prevent noise
    - **d.** Breaking-in or bedding the pads after installation

CHAPTER 16

Antilock Brakes, Stability Control, and Power Assist Systems

Chapter Objectives

At the conclusion of this chapter, you should be able to:

- Explain the purpose and operation of antilock brake and traction control systems.
- Identify components of the antilock brake and traction control systems. (ASE Education Foundation MLR 5.G.1)
- Describe procedure for performing a road test to check brake system operation, including an antilock brake system (ABS). (ASE Education Foundation MLR 5.A.2)
- Depressurize the antilock brake system.
- Identify the types of power brake assist.
- Check brake pedal travel with, and without, engine running to verify proper power booster operation. (ASE Education Foundation MLR 5.E.1)
- Check the power brake assist operation. (ASE Education Foundation MLR 5.E.2)

KEY TERMS

accumulator	**hydraulic assist**	**vacuum assist**
EBCM	**pressure increase mode**	**vacuum check valve**
electrohydraulic unit	**release mode**	**wheel speed sensors (WSS)**
hold mode	**tire slip**	

Antilock brakes began as an option on vehicles in the 1980s and have become standard equipment on passenger cars and light trucks. Antilock brake systems (ABS) operate with the regular service brakes, using a computer and sensors to monitor wheel speeds, and if necessary, to take control of brake the application. The system prevents the tires from staying locked up and causing a loss of vehicle control. Vehicle stability control systems work with the antilock brake system to help the driver maintain vehicle control when things go wrong.

ABS/ESC Principles and Operation

Antilock brake systems (ABS) and electronic stability control (ESC) systems are standard equipment on modern cars and light trucks sold in the United States. This is because ABS and ESC are mandated for all passenger cars and light trucks built since the 2012 model year. Because of the regulation requiring ABS and ESC, you will need to understand how these systems operate.

ABS PRINCIPLES

The purpose of ABS is to allow the driver to maintain steering control during braking in the event of wheel lockup. This is done by the ABS pumping the brakes for the driver. An average driver may be able to pump the brakes two to three times per second, while the ABS system can pump the brakes over a dozen times per second. This allows the ABS to pump and release the brakes to keep the tires from staying locked up.

The ability of the ABS is subject to the limits of physics. If you are driving on a wet road in a non-ABS-equipped car and an obstacle, such as a fallen tree, blocks the road ahead, overbraking can lock the wheels. This can cause the car to continue in a straight line regardless of the position of the steering wheel. The ABS allows the wheels to continue turning so the driver can slow down and steer around the obstacle. However, if the same scenario is repeated on an ice-covered road, the ABS cannot provide sufficient braking if there is no traction between the tire and the road. In this case, the car may continue in a straight line, and into the tree, regardless of whether the ABS controls wheel speed.

The amount of traction between the tire and the ground is called **tire slip**. A freely rolling wheel has zero tire slip, while a locked wheel moving over the pavement has 100% slip (**Figures 16-1** and **16-2**). Maximum braking occurs with the wheel rotating, but at the edge of locking up. The antilock system, by monitoring wheel speed and controlling hydraulic pressure to the wheel brakes, can maintain a low tire slip rate, typically below 20%, which keeps the wheel rotating and not skidding over the pavement.

Antilock brake systems use sensors, called **wheel speed sensors (WSS)**, to monitor the rotational speed of each wheel. These sensors provide this information to an electronic brake control module or (EBCM). The EBCM controls the operation of an electrohydraulic unit, which contains electric motors, solenoids, and valves that are used to control the flow of brake fluid to each wheel brake.

■ ***ABS Components.*** There are many types of ABS systems found on modern cars and trucks. Many share the same types of components and operate in similar ways. A typical system contains the following components:

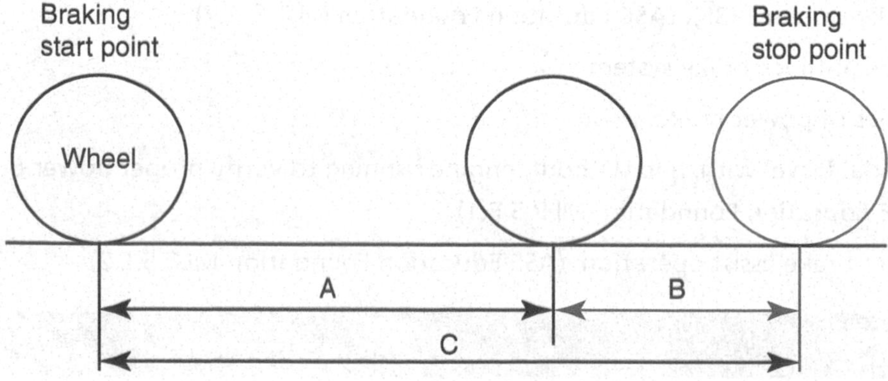
Braking
start point
Braking
stop point
Wheel
A
B
C
A is the distance covered without slip
B is the distance slipped
C is the total stopping distance

FIGURE 16-1 An illustration of wheel slip. A tire with zero slip is rolling freely, and a tire with 100% slip is locked. A wheel slowing faster than the other wheels has a slip rate between 0% and 100%.

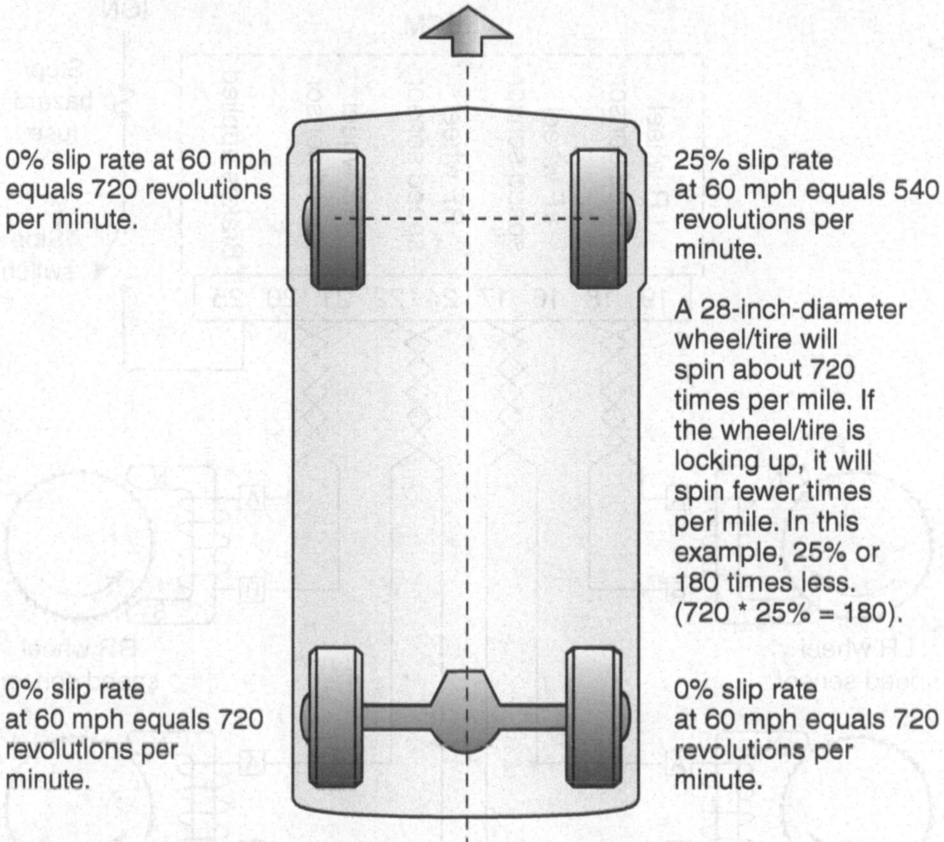
0% slip rate at 60 mph equals 720 revolutions per minute.
25% slip rate at 60 mph equals 540 revolutions per minute.
A 28-inch-diameter wheel/tire will spin about 720 times per mile. If the wheel/tire is locking up, it will spin fewer times per mile. In this example, 25% or 180 times less. (720 * 25% = 180).
0% slip rate at 60 mph equals 720 revolutions per minute.
0% slip rate at 60 mph equals 720 revolutions per minute.

FIGURE 16-2 The ABS monitors wheel slip rate through the wheel speed sensors.

- **EBCM**—This is the electronic control unit for the ABS system. Data from the wheel speed sensors is used by the EBCM to determine if one or more wheels are rotating at a lower speed than other wheels. An illustration of the EBCM and speed sensors is shown in **Figure 16-3**. Based on wheel speed input and its software, the EBCM commands the electrohydraulic unit to hold, open, or close the wheel brake hydraulic circuits. The EBCM has self-diagnostic capability, meaning it can determine faults in system operation, provide diagnostic data, and provide diagnostic trouble codes (DTCs).
- **Electrohydraulic unit**—This usually contains electric motors, solenoids, and valves that control the flow of brake fluid to the wheel brakes based on input from the EBCM. The electro hydraulic unit may be mounted near the master cylinder (**Figure 16-4**) or attached to the master cylinder (**Figure 16-5**).
- **Analog wheel speed sensors**—Two types of wheel speed sensors are currently in use, analog and digital sensors. Analog sensors are made of a permanent magnet and a winding of wire. This type of sensor produces an AC voltage signal as the teeth on a tone ring pass by the sensor (**Figure 16-6**). As the tone ring passes the sensor, it produces positive and negative voltages within the sensor (**Figure 16-7**). These pulses are AC voltage. The faster the wheel rotational speed is, the more pulses are generated each second. The EBCM interprets these pulses as wheel speed.
- **Active wheel speed sensors**—Active sensors produce a digital DC voltage signal. These sensors are more accurate, can produce signals at very low speeds, and tell which way the wheel is rotating (**Figure 16-8**). Regardless of which sensor is used, both provide a signal to the EBCM each time a segment of the tone ring or reluctor moves past the sensor. The EBCM measures the frequency of each sensor's output and determines the wheel speed.

■ ***Analog and Digital.*** An analog signal, like the AC voltage produced by a WSS, can have varying voltage or frequency. This means that the voltage can be negative or positive, or the voltage can vary, such as between 0 and 1 volt. A digital signal can change in frequency but the voltage does not change. Most digital sensors produce a fixed voltage, such as 5 volts or 12 volts. Frequency is

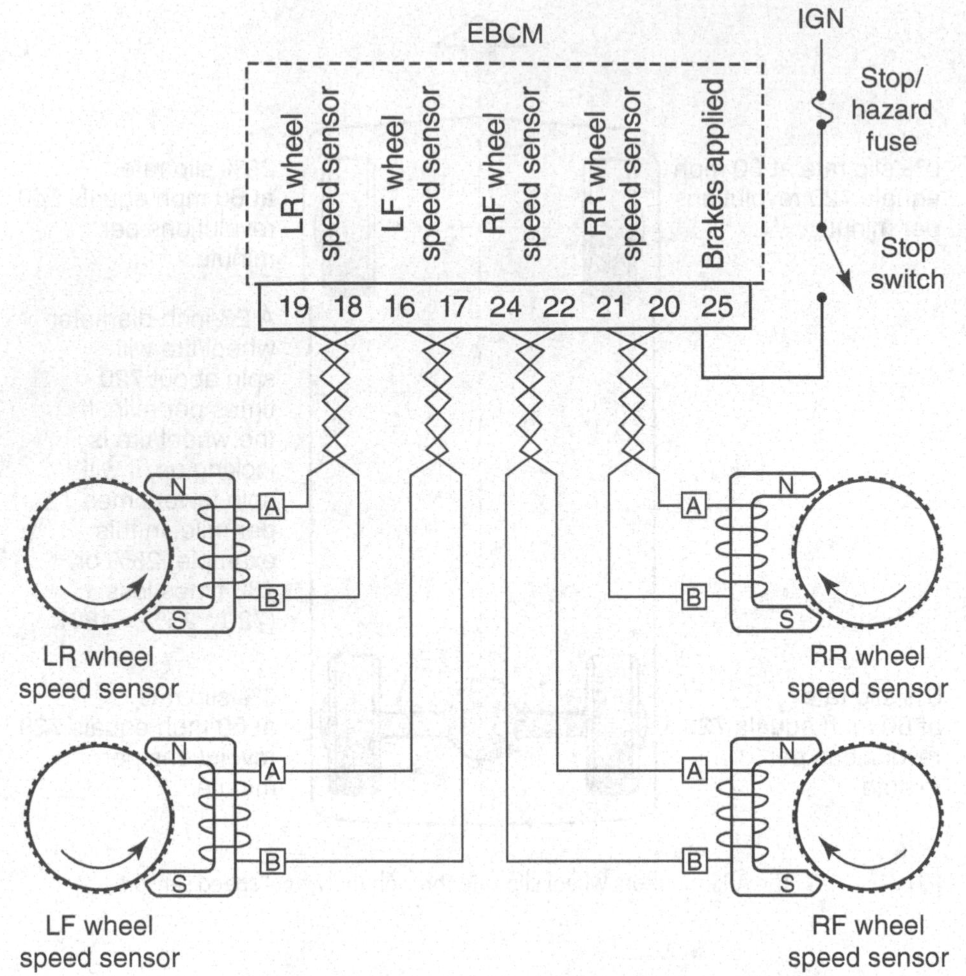

FIGURE 16-3 An illustration of an ABS system.

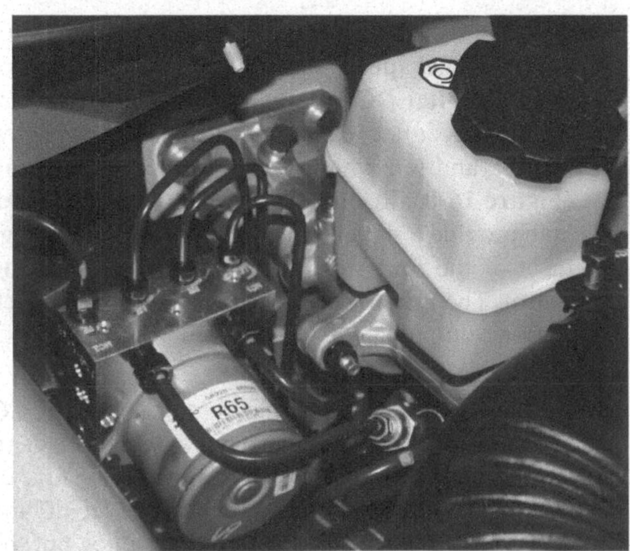

FIGURE 16-4 This ABS electrohydraulic unit and master cylinder.

FIGURE 16-5 This older ABS system has the ABS hydraulic control unit bolted to the master cylinder.

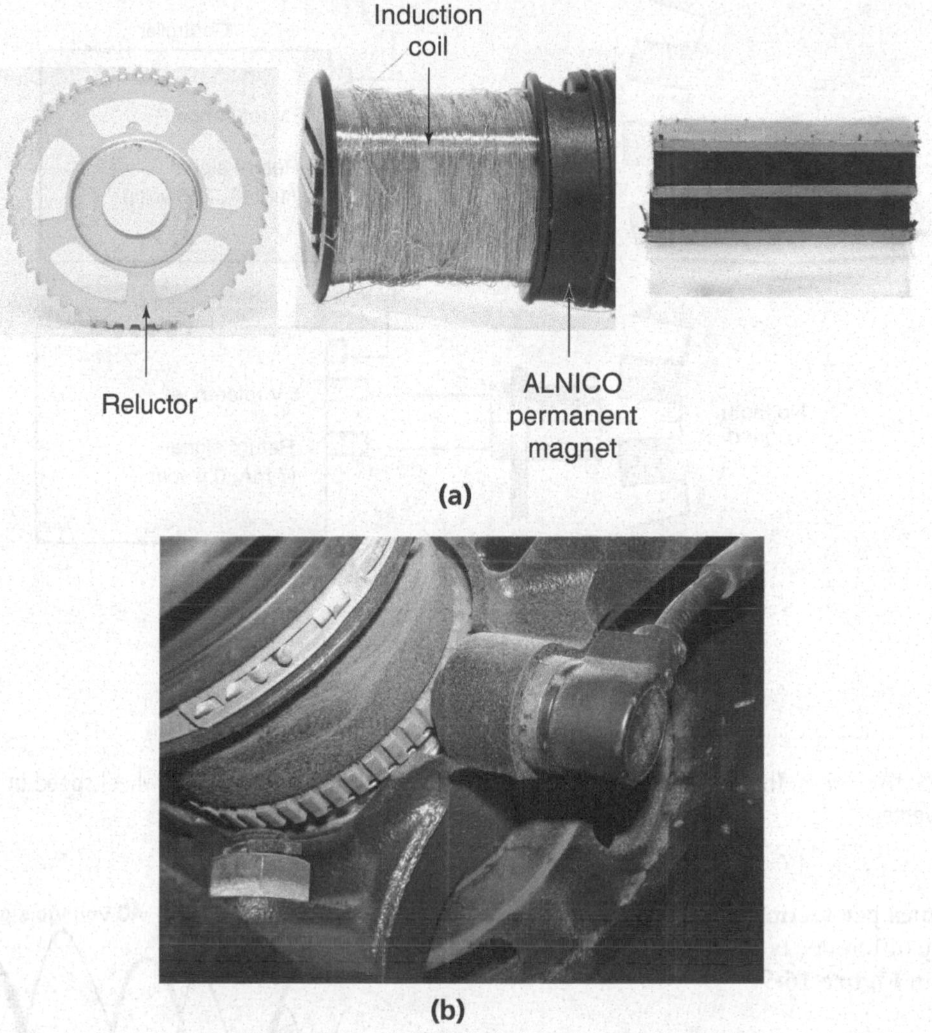

FIGURE 16-6 (a) An inside view of a wheel speed sensor. The tone ring induces a magnetic field into the coil as it passes the sensor. (b) A wheel speed sensor and reluctor or tone ring.

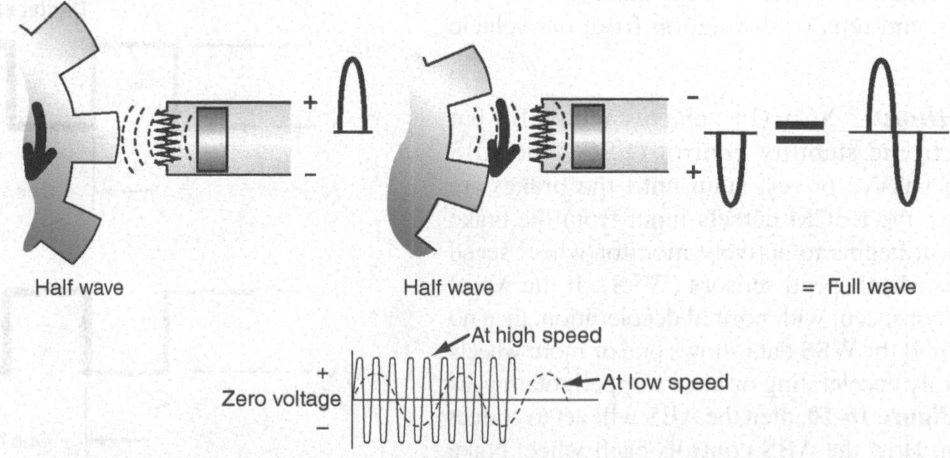

FIGURE 16-7 The magnetic field in the PM sensor creates an AC voltage, which produces positive and negative pulses. These pulses are read by the computer and converted into wheel speed data.

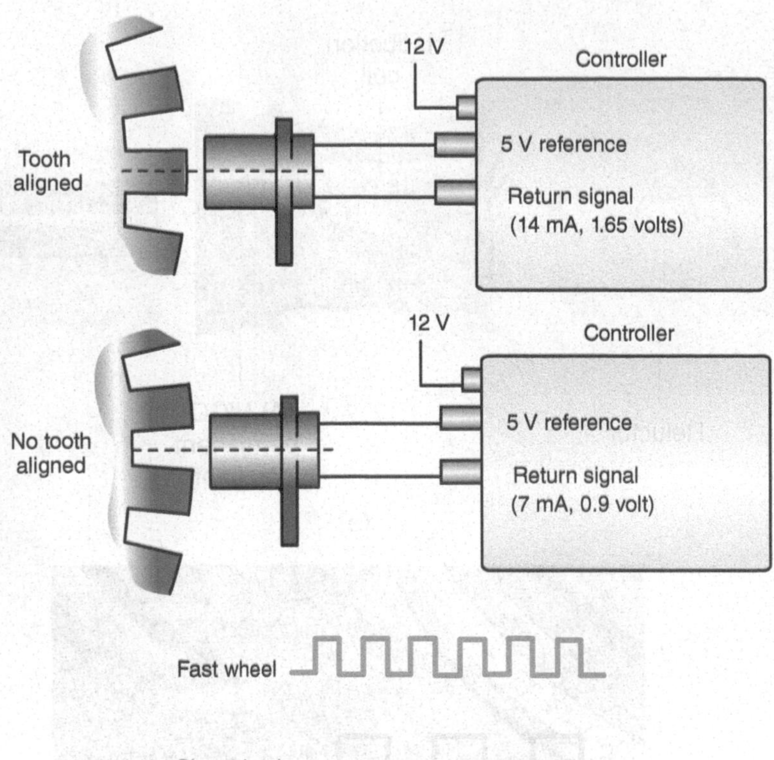

FIGURE 16-8 The active WSS produces a digital signal and can determine wheel speed in forward or reverse.

the number of times per second that the signal occurs. An example of the difference between analog and digital signals is shown in **Figure 16-9**.

OPERATION OF THE ABS

There are many different antilock brake systems in use on modern cars and trucks from several different manufacturers. Each system has specific operating, diagnosing, and servicing procedures. The following is a general explanation of ABS operation. For specific systems, refer to the service and repair information from the vehicle manufacturer.

■ ***ABS without ESC.*** On vehicles with ABS but without electronic stability control (ESC), the ABS remains in a passive or rest state until the brakes are applied. When the EBCM detects input from the brake pedal switch, it begins to actively monitor wheel speed data from the wheel speed sensors (WSS). If the wheel speed data is consistent with normal deceleration, then no action is taken. If the WSS data shows one or more wheels are more rapidly decelerating or have stopped rotating, as indicated in **Figure 16-10**, then the ABS will act to correct the condition. How the ABS controls each wheel brake depends on the type of system. In general, there are three modes of operation: hold, release, and pressure increase.

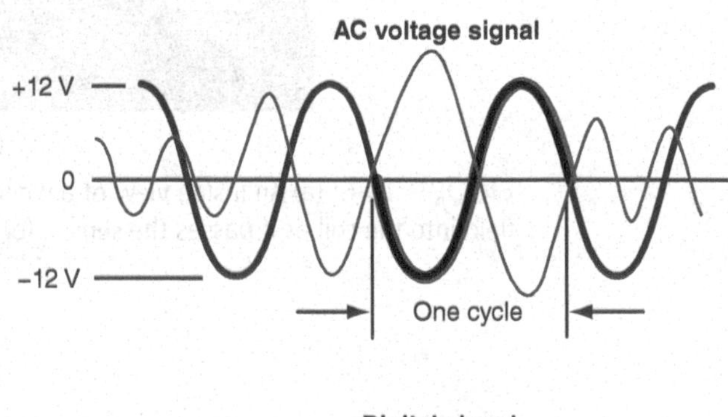

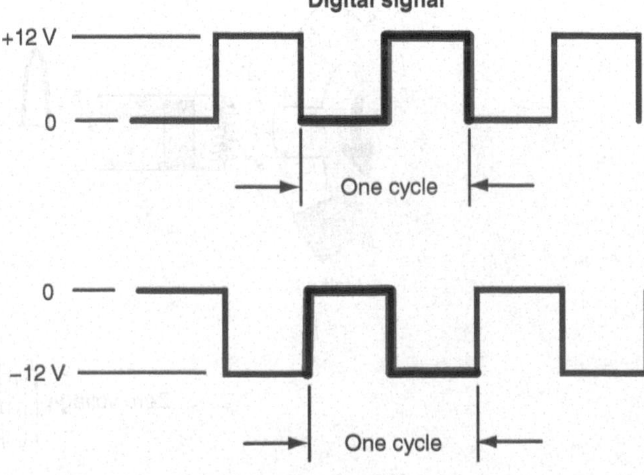

FIGURE 16-9 This illustrates the difference between analog and digital sensor signals.

- **Hold mode**, also called isolate mode, limits any further pressure increase in a wheel brake circuit. **Figure 16-11** shows how pressure is held.
- **Release mode**, also called pressure decay or dump mode, is used when the EBCM senses that even after holding pressure to a wheel brake, the wheel is still slowing too rapidly (**Figure 16-12**). The EBCM commands the pressure in the circuit be released so that the wheel can begin to rotate again.
- **Pressure increase mode** is used to reapply pressure to the brake circuit to slow the wheel again (**Figure 16-13**).

This cycle of hold, release, and apply allows a wheel to unlock and resume its rotation and allows the system to reapply brake pressure to slow the wheel again. This cycle takes place very rapidly, a dozen or more times per second during an ABS event.

■ ***ABS with ESC.*** Vehicles with ESC use the ABS to monitor wheel speeds to determine if a wheel or wheels are rotating faster than others. The function of the ABS remains the same. However, the system works with the ESC to provide increased capability, such as oversteer and understeer correction and torque vectoring. A detailed explanation of the ESC is covered later in the chapter.

■ ***Safety Precautions.*** Because the ABS system uses fluid under high pressure to rapidly pump the brakes, caution must be used when servicing the brake system. Never open a hydraulic line or bleeder screw with the ignition on, as this can cause injury from the release of high-pressure fluid. Always follow the manufacturer's service precautions and procedures when you are servicing the brake system and ABS system.

Some ABS systems require that the high-pressure accumulators be discharged before any service is performed (**Figure 16-14**). An **accumulator** holds fluid under high pressure. To discharge the pressure, turn the ignition off and pump the brake pedal at about 40 times. Failure to fully discharge the pressure can result in damage to the ABS system and injury if the system is opened.

ABS INSPECTION AND SERVICE

The ABS itself does not require normal service. When problems occur, it is often the result of a component that is either not functioning correctly or is not working at all.

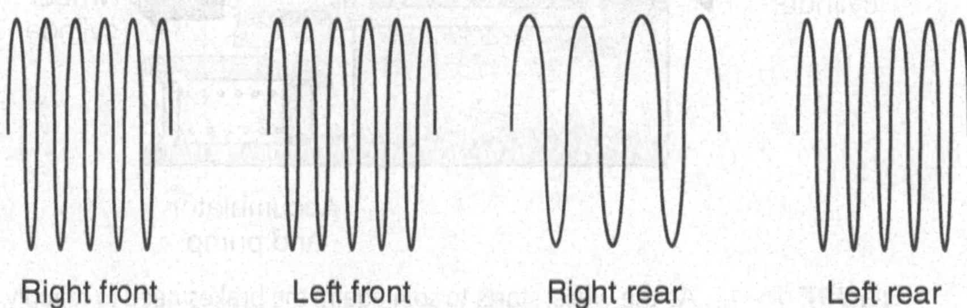

FIGURE 16-10 This illustrates the signals from four-wheel speed sensors. One wheel is spinning much slower than the other three.

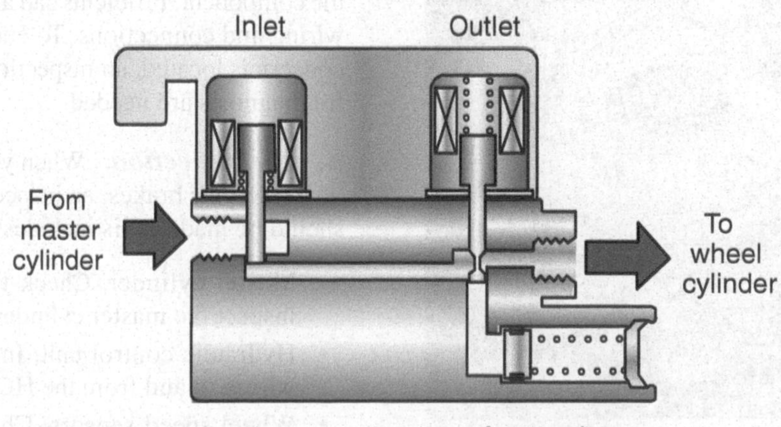

FIGURE 16-11 In hold mode, pressure from the master cylinder is blocked to prevent the wheel from locking.

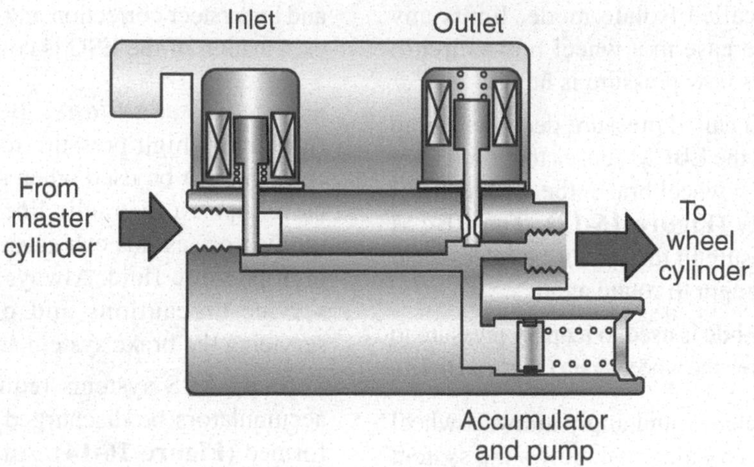

FIGURE 16-12 In dump mode, pressure from the wheel brake is released by the ABS to allow the wheel to spin.

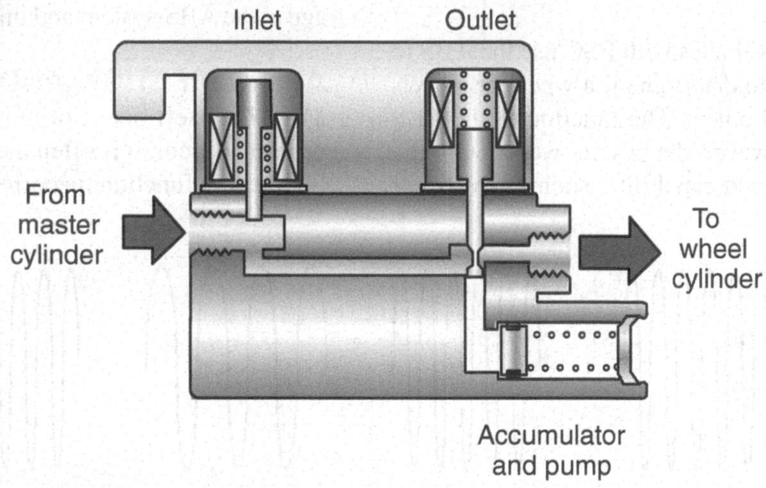

FIGURE 16-13 As the wheel starts to spin again, the brakes need to reapply. This is called pressure increase mode.

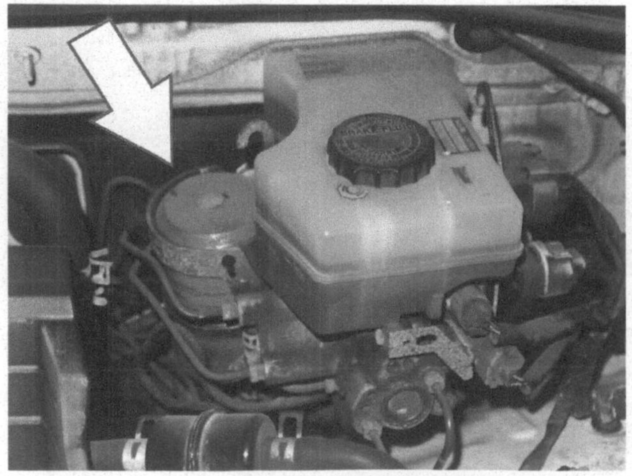

FIGURE 16-14 A high-pressure accumulator in an integral ABS unit. **Never open or attempt to service an integral system until the reserve pressure has been relieved.** Refer to the manufacturer's service information to determine how to relieve the pressure.

When this happens, you diagnose the fault and replace the component. Problems can also be caused by damaged wiring and connections. To ensure the actual cause of a concern is located, an inspection of the system and careful diagnosis are needed.

■ ***ABS Inspection.*** When you are performing service on the service brakes, an inspection of ABS components should be made. This includes:

- Master cylinder. Check the brake fluid level and inspect the master cylinder for leaks.
- Hydraulic control unit. Inspect the brake lines and wiring to and from the HCU.
- Wheel speed sensors. Check the connections and wiring of the wheel speed sensors and inspect the reluctor rings if visible.

In addition, brake fluid condition can affect the ABS components and operation. Because of this, many vehicle manufacturers recommend periodic brake fluid replacement as part of an inspection and maintenance program.

■ ***Flushing and Bleeding the ABS System.*** On many cars and trucks, the brake system can be flushed and bled using the same methods as used on non-ABS-equipped vehicles.

Because the brake fluid absorbs moisture, it should be periodically flushed and replaced. This is especially important on ABS-equipped vehicles because the solenoids and check valves operate within very small *tolerances*. Moisture in the fluid leads to rust and corrosion formation, which can cause the close-fitting parts of the electrohydraulic unit to stick or seize. Many vehicle manufacturers recommend brake fluid replacement every two to three years to prevent damage to the hydraulic system components. In most cases, flushing the hydraulic system on an ABS-equipped vehicle can be done using a power bleeder, vacuum bleeder, or manual bleeding methods, just as on non-ABS vehicles. It is important to refer to the manufacturer's service information for specific procedures before you work on the ABS system.

On ABS systems that use high-pressure accumulators, the pressure must be discharged before any service or bleeding is performed. Follow the manufacturer's service guidelines to relieve the system pressure, and follow the flushing and bleeding procedures exactly to prevent damage to the ABS system.

Some vehicles will require the use of a scan tool to bleed the ABS system properly. **Figure 16-15** shows some of the steps in bleeding a General Motors vehicle while using a scan tool. Flushing and bleeding the brake system is discussed in detail in Chapter 11. Once the system is flushed and bled, ensure that the fluid is full and test-drive the vehicle. The brake pedal should be firm and has adequate reserve. If the pedal is spongy, air may be trapped in the system, requiring you to rebleed the brake system.

> **Word Wall**
> *Tolerances*—The parts fit together very closely and even the smallest amount of dirt or foreign matter can interfere with their operation.

FIGURE 16-15 Bleeding some antilock systems requires using a scan tool. Failure to follow the procedures correctly can result in air staying trapped in the system.

Traction Control and ESC Systems

Traction control (TC) systems, in various forms, began to appear on vehicles in the 1990s. The range of functions of the types of TC systems varied widely based on the type of vehicle. Traction control has since evolved into ESC systems. ESC systems monitor and manage certain vehicle functions to help the driver maintain control.

TC SYSTEMS

Many early TC systems were essentially software add-ons to the ABS system. Since the ABS was already used to monitor wheel speeds when braking, by revising the software, the ABS began to monitor wheel speeds during regular driving. By doing this, the ABS could detect if a drive wheel had a faster rotation speed than other wheels (**Figure 16-16**). If this occurred, the ABS could then apply the brake on the wheel with reduced traction.

Other TC systems use a more active approach to reducing unwanted wheel spin. On some vehicles, if excessive wheel spin is detected, the powertrain control module reduces engine power by cutting fuel delivery to reduce torque to the drive wheels. When a vehicle with this ability is on ice or a similar low-traction surface, even with the throttle wide open, the engine speed drops to idle and power to the drive wheels drops to zero, and this stops the wheel spin and allows the tires to regain traction. Another benefit of TC systems is that, by reducing wheel spin, performance cars can accelerate even faster if the drive wheels are not subject to excessive spinning at takeoff.

ESC

TC has now become electronic stability control or ESC. ESC systems are similar to traction control but add monitoring of steering input and other vehicle motions. In 2006, the National Highway Traffic Safety Administration (NHTSA) issued a proposal to phase in ESC on

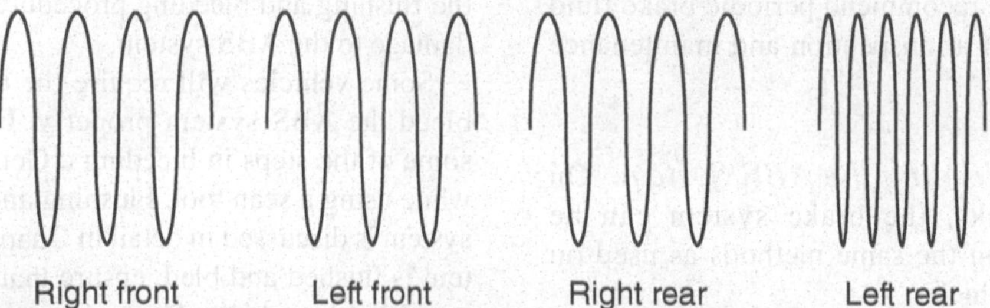

FIGURE 16-16 Traction and stability control systems use the wheel speed sensors to monitor wheel speed at all times. This illustration shows one wheel spinning faster than the others, indicating a low-traction condition.

all new vehicles under 10,000 lbs. gross vehicle weight (GVW). Starting in 2008, with full compliance by 2012, all new vehicles sold in the United States weighing less than 10,000 lbs. GVW have ESC as standard equipment. The idea is that ESC can save approximately 10,000 lives per year. This number represents fatal injuries from single-vehicle and rollover accidents where the ESC system could help the driver maintain control and prevent the accident from occurring.

■ ***ESC Components.*** To comply with the ESC standard, vehicles covered under the rule must meet the following requirements:

- Must be able to apply and adjust individual wheel brake torque. This means the need of an ABS system to monitor wheel speed and to control wheel brake operation.
- Able to modify engine torque. This allows the powertrain control module (PCM) to modify fuel delivery, spark timing, transmission gear selection, and other operating conditions to increase or decrease torque to the drive wheels as necessary.
- Monitor driver steering input. A steering angle sensor and a steering torque sensor are used to determine where the driver is trying to point the vehicle. Steering angle sensors may be rack-mounted or mounted on the steering column (**Figure 16-17**).
- Monitor yaw rate and sideslip. Yaw is the rate of change in rotation from around the vehicle's center of gravity (**Figure 16-18**). Yaw sensors are installed at the center of gravity, often in the center console of the vehicle. Slip rate is similar to yaw, but is measured as the difference between the lateral (sideways) speed and the longitudinal (along the centerline of the vehicle) speed. Slip rate is shown in **Figure 16-19**. Vehicles use a lateral acceleration sensor to measure slip rate.
- Limit oversteer and understeer. This is done by monitoring steering input and wheel speeds. For example, when oversteer occurs, the rear of the vehicle starts to break loose, causing a spinout. By monitoring steering angle, the ESC system can determine where the driver is trying to point the vehicle. Inside of the vehicle, yaw sensor data shows where the vehicle is actually headed. As the vehicle turns, the inside and outside wheels rotate at different speeds. If the speed difference is excessive, the computer can apply the brakes and reduce engine torque as necessary to reduce the oversteer, allowing the driver to regain control.

■ ***What Does that Light on My Dash Mean?*** As part of the ESC mandate, the system must be able to warn the driver if a fault is present and the system is not working. A fault in the ABS will turn on both the ABS and ESC warning lights on the dash (**Figure 16-20**).

GM 1980-2009

PSCM Data

Steering Shaft Torque (N-m)	-8.75
Vehicle Speed(MPH)	0
Battery Voltage Signal	11.58
Calculated System Temp(°F)	118
Torque Sensor Signal 1 (V)	0.9
Torque Sensor Signal 2 (V)	4.0
Steering Position Sensor 1 (V)	3.0
Steering Position Sensor 2 (V)	0.5
Steering Wheel Position (°)	4
EPS Motor Command Amps	0

1 2 3 4

FIGURE 16-17 Steering input data is used by the ESC system to determine driver intent. This information, combined with that from other sensors, is used to determine how to correct for unwanted vehicle motion.

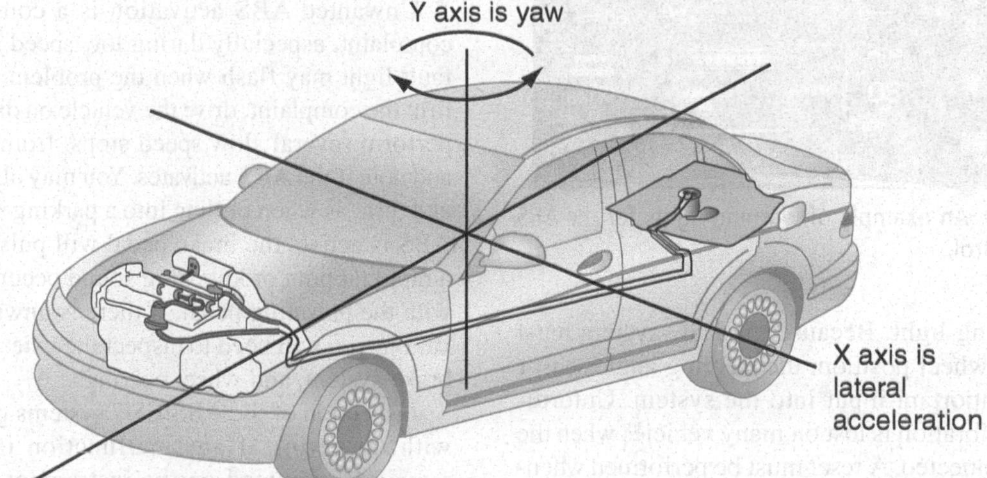
The Z axis is along the thrust angle and represents linear acceleration and decleration.
The X axis is the sideways lateral acceleration when cornering.
The Y axis is represented by yaw, and is the rotational force around the X axis.
Y axis is yaw
X axis is lateral acceleration
Z axis is along the thrust angle

FIGURE 16-18 Yaw is motion around the vehicle's center.

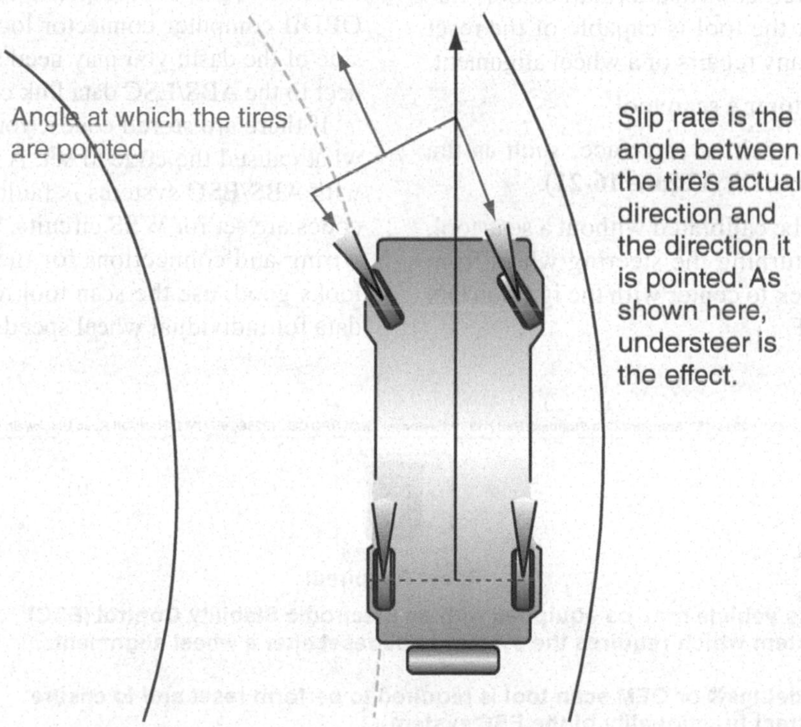
Angle at which the tires are pointed
Slip rate is the angle between the tire's actual direction and the direction it is pointed. As shown here, understeer is the effect.

FIGURE 16-19 Slip angle is the difference between where the tires are pointed and the actual direction of travel.

Because the ESC systems relies on the ABS for wheel speed input, a failure of a wheel speed sensor will disable the ABS and TC/ESC systems. The service brakes still work but without ABS function, and the TC/ESC system may be unable to provide assistance to the driver if the situation should arise.

A problem in the ESC system, such as a faulty sensor or an uncalibrated steering angle sensor can also turn on

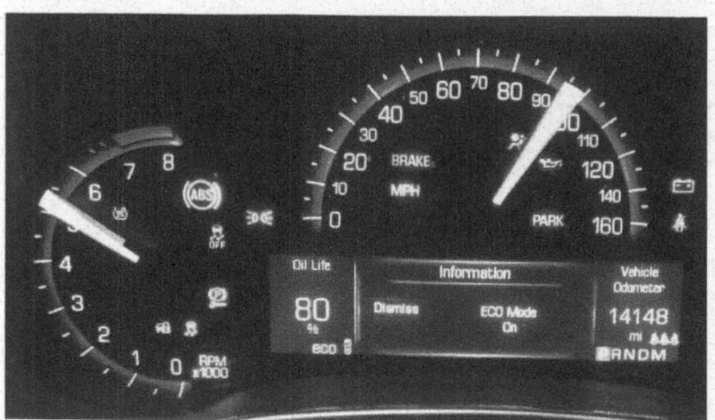

FIGURE 16-20 An example of warning lights for the ABS and traction control.

the dash warning light. Because the ESC system must know steering wheel position, the steering angle sensor (SAS) is an important input into the system. Unfortunately, SAS calibration is lost on many vehicles when the battery is disconnected. A reset must be performed whenever a wheel alignment is performed. There are several ways in that a SAS reset can be performed, these include:

- Using an aftermarket or generic scan tool, such as from Snap-On. Coverage with aftermarket tools varies so determine if the tool is capable of the reset before performing any repairs or a wheel alignment.
- Using the manufacturer's scan tool.
- Wheel alignment machine interface, such as the Hunter CodeLink system (**Figure 16-21**).
- Some vehicles can be calibrated without a scan tool. This may involve turning the steering wheel from lock to lock and back to center with the ignition ON and the engine OFF.

Regardless of how the reset is performed, what is important is that the process be successfully completed. Failure to reset the SAS and ESC systems can cause problems with the ESC, electric power steering assist, lane departure warning, active cruise control, park assist, and other systems.

Unwanted ABS activation is a common customer complaint, especially during low speed stops. The ABS fault light may flash when the problem occurs. To confirm the complaint, drive the vehicle on dry pavement and perform several slow speed stops, from about 15 mph, and note if the ABS activates. You may also need to brake and turn, as when pulling into a parking space. When the ABS is active, the brake pedal will pulsate quickly and a rapid tapping or knocking sound occurs that coincides with the pulsating pedal. If there is unwanted ABS activation, you will need to inspect the wheel speed sensors, brake system, and wheel bearings.

Inspection of the ABS/ESC systems generally begins with observing if any malfunction indicator lamps remain illuminated on the instrument panel with the engine running. If the ABS and TC/ESC lights are illuminated, connect a scan tool to the data link connector (DLC) to check for diagnostic trouble codes (DTCs). Not all vehicles provide ABS/ESC data through the standard OBDII computer connector located under the driver's side of the dash; you may need special adaptors to connect to the ABS/ESC data link connector.

If there are stored codes, you will need to diagnose what caused the code to set. A very common problem with ABS/ESC systems is faulty WSS. If one or more codes are set for WSS circuits, begin by inspecting the wiring and connections for the sensors. If the wiring looks good, use the scan tool to look at the ABS/ESC data for individual wheel speeds (**Figure 16-22**). Raise

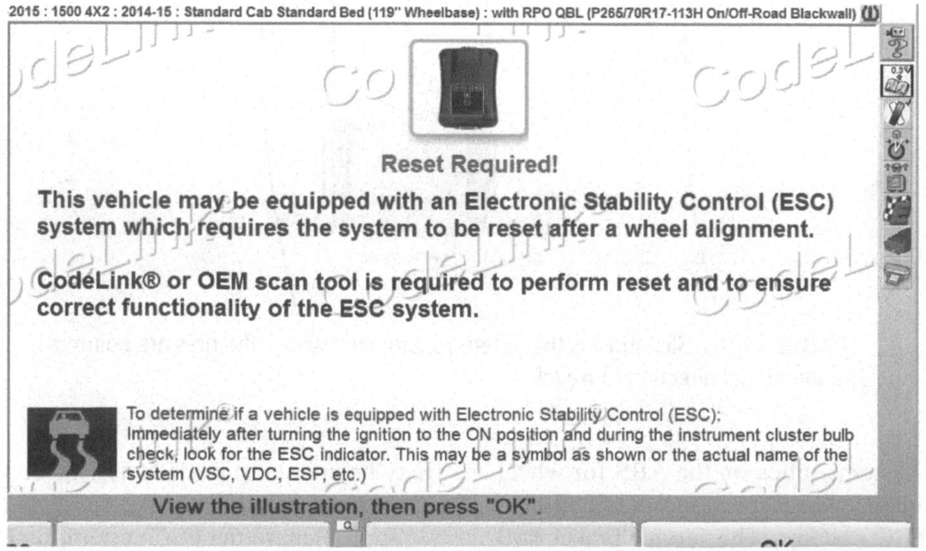

FIGURE 16-21 When performing a wheel alignment, steering angle sensors need recalibrated so the ESC system knows the true location of the steering shaft.

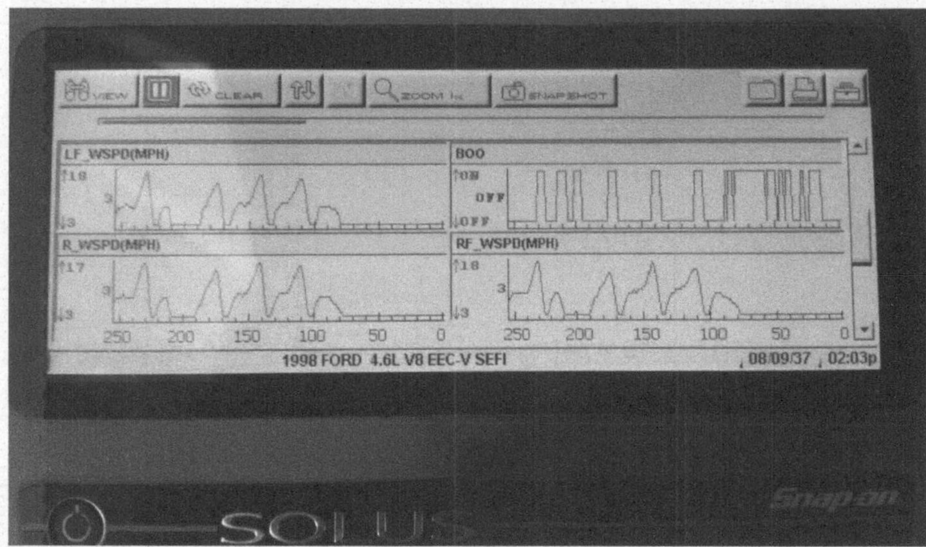

FIGURE 16-22 Using a scan tool to graph wheel speed sensor data.

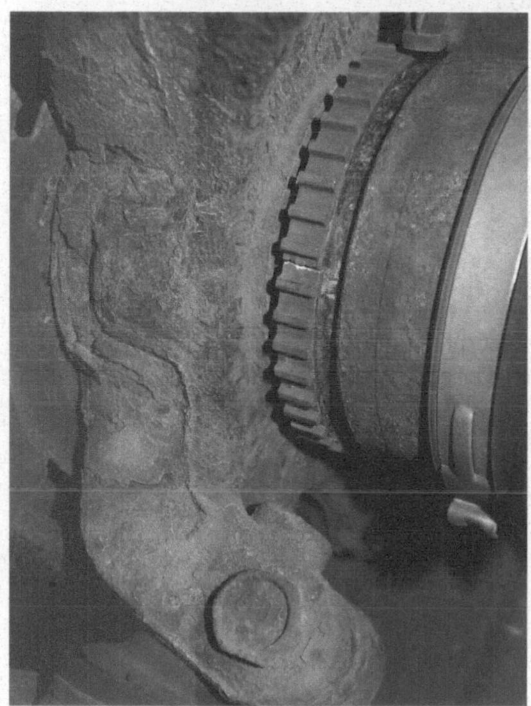

FIGURE 16-23 An example of a broken ABS tone ring.

the vehicle so the tires are off the ground, and spin each tire while you watch the WSS data. If no wheel speed is indicated, the sensor or its wiring are likely the cause.

Vehicles with unwanted ABS activation may not have any stored trouble codes. Loose wheel bearings can often cause this concern. Another common cause of this complaint is metal and rust buildup around the WSS in the wheel hubs. The rust buildup creates false speed data, which is interpreted by the EBCM as a difference in wheel speed. This causes the EBCM to activate and attempt to control a slipping wheel. The sensor and tone ring may be able to be cleaned and reused. Severe rusting requires replacement of the sensor or hub assembly.

Though not as common as problems with the sensor itself, the tone rings themselves can fail and cause problems with the ABS. An example of a broken ABS tone ring is shown in **Figure 16-23**.

Detailed testing requires locating the appropriate service information for the vehicle. Follow the vehicle manufacturer's diagnosis and testing procedures to determine the cause of the fault.

Power Assist Types and Components

There are two types of external power brake assist systems, vacuum assist and hydraulic assist. **Vacuum assist** uses vacuum supplied by the engine or a vacuum pump. **Hydraulic assist** is supplied by the power steering system to a hydraulic power brake booster. Some vehicles use an integral brake master cylinder, ABS, and power assist unit. The most common type of power assist is the vacuum assist since vacuum is readily available from gasoline-powered engines. Hydraulic assist is often used on diesel-powered vehicles. However, there are exceptions to this as some gasoline-powered vehicles also use hydraulic-assisted brakes and some diesel-powered vehicles have a vacuum pump to operate a vacuum booster.

VACUUM ASSIST

As discussed in Chapter 4, vacuum is air pressure that is less than atmospheric pressure. A pressure/vacuum chart is shown in **Figure 16-24**. Pressure moves from high to low; the difference in pressure can be used to generate force, which is applied to the brake pushrod to reduce the amount of force supplied by the driver.

■ ***Components.*** Vacuum for the assist unit is supplied by the engine or by a vacuum pump. Gasoline-powered engines produce vacuum in the intake manifold

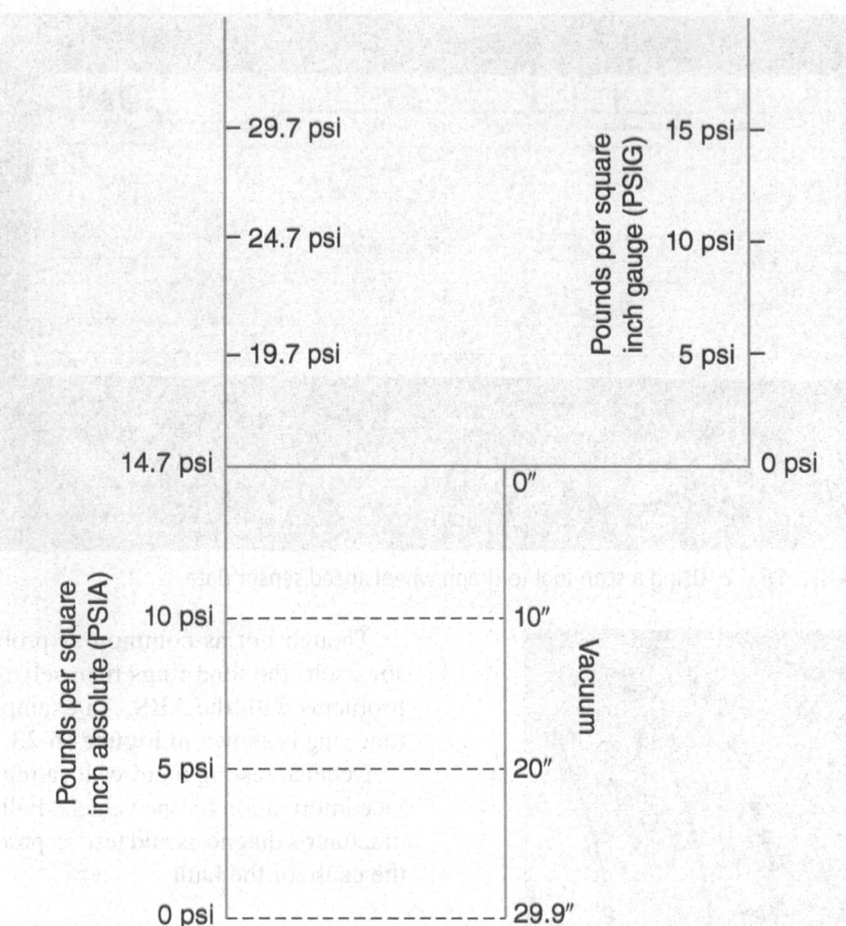

FIGURE 16-24 Vacuum is pressure that is less than atmospheric pressure. Vacuum, along with atmospheric pressure, is used to increase the force applied to the master cylinder.

(**Figure 16-25**). Typically, a large-diameter vacuum hose connects the assist unit to the intake manifold (**Figure 16-26**). A **vacuum check valve** is located either in the vacuum hose or at the connection to the booster assembly. **Figure 16-27** shows a check valve installed at the booster. The check valve allows vacuum to the booster but stops the release of vacuum from the booster back to the engine. This allows the booster to store a vacuum reserve. The reserve is to provide assist for a couple of pedal applications if the engine stalls. An illustration of the check valve is shown in **Figure 16-28**.

Some vehicles use a vacuum pump to supply vacuum to the assist unit. Vacuum pumps may be belt driven, gear driven, cam driven, or electric (**Figure 16-29**).

The vacuum power booster is bolted to the firewall in the engine compartment. The brake master cylinder is attached to the front of the booster. The brake pushrod connects to the booster under the dash (**Figure 16-30**).

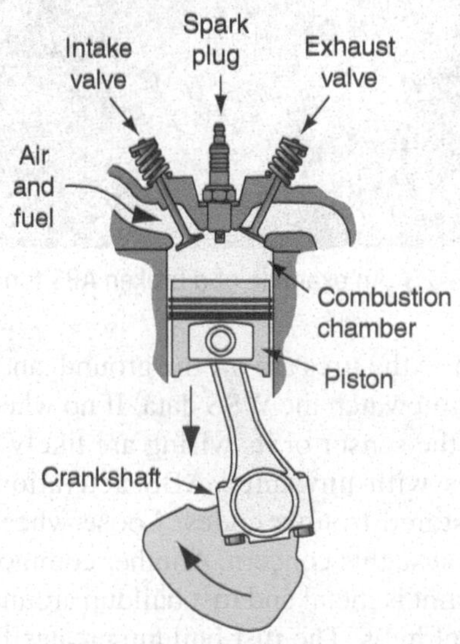

FIGURE 16-25 Vacuum is formed as the piston moves down in the cylinder. This reduces the pressure in the intake manifold.

FIGURE 16-26 A vacuum supply hose from the engine to the power assist unit.

FIGURE 16-27 The vacuum check valve traps vacuum in the booster. This allows for a couple of assisted stops in the event the engine stalls.

FIGURE 16-29 An example of a belt-driven vacuum pump.

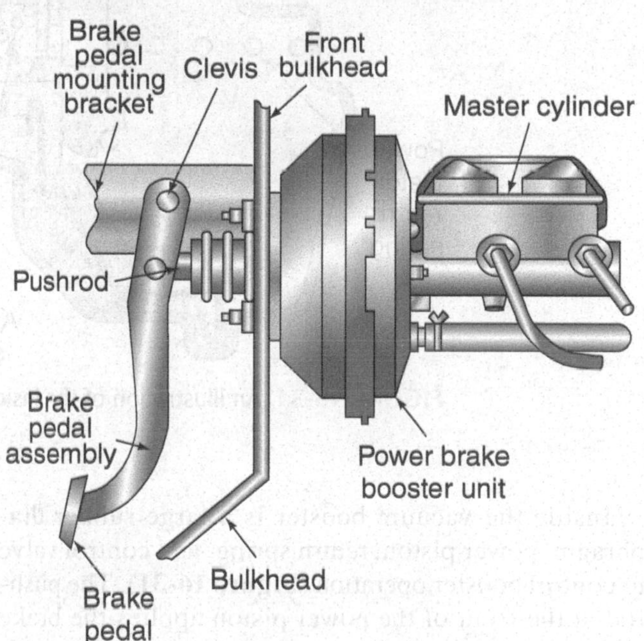

FIGURE 16-30 This illustrates the connection of the brake pedal to the vacuum booster.

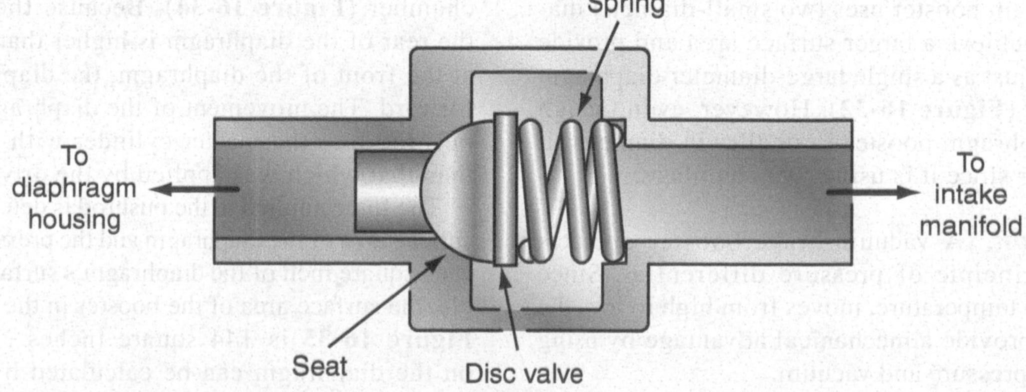

FIGURE 16-28 An inside view of a vacuum check valve.

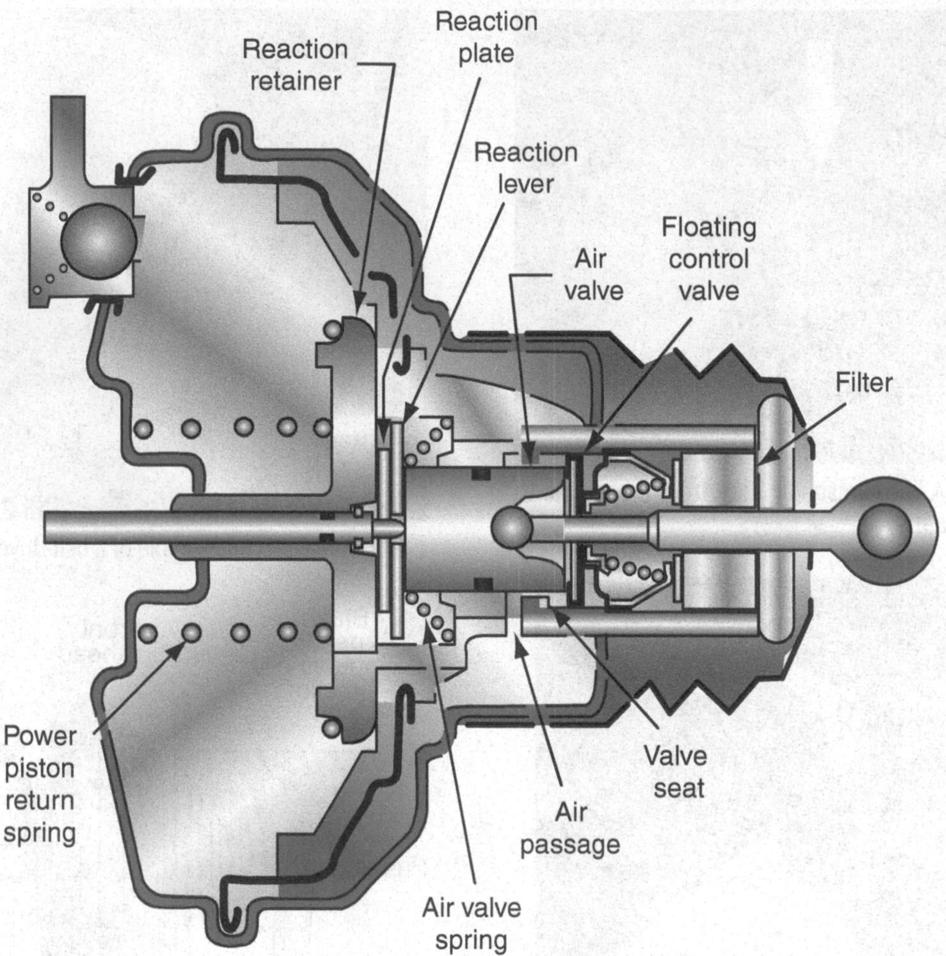

FIGURE 16-31 An illustration of the inside of a vacuum booster.

Inside the vacuum booster is a large rubber diaphragm, power piston, return spring, and control valve to control booster operation (**Figure 16-31**). The pushrod at the front of the power piston applies the brake force to the primary master cylinder piston. In vehicles where space limitations prevent the vehicle manufacturer from using a large-diameter vacuum booster, a tandem or dual-diaphragm booster may be used. A dual-diaphragm booster uses two small-diameter diaphragms to achieve a larger surface area and provide power assist just as a single large-diameter diaphragm booster does (**Figure 16-32**). However, even though the dual-diaphragm booster is smaller in diameter, it is also longer since it is using two chambers.

■ ***Operation.*** A vacuum brake booster operates using the principle of pressure differential. Since pressure, like temperature, moves from high to low, the booster can provide a mechanical advantage by using atmospheric pressure and vacuum.

Figure 16-33 shows an illustration of a typical vacuum booster with the brakes released. During normal operation with the engine running, vacuum is present on both sides of the booster diaphragm. Because the pressure is equal on both sides, the diaphragm remains in place. When the brake pedal is pressed, the pushrod moves a valve off its seat, allowing atmospheric pressure to enter the rear of the diaphragm chamber (**Figure 16-34**). Because the pressure at the rear of the diaphragm is higher than the vacuum at the front of the diaphragm, the diaphragm flexes forward. The movement of the diaphragm forces the pushrod into the master cylinder with greater force than that which was applied by the driver only.

The force applied to the pushrod is determined by the surface area of the diaphragm and the pressure applied to each square inch of the diaphragm's surface. For example, the surface area of the booster in the diaphragm in **Figure 16-35** is 144 square inches. The pressure on the diaphragm can be calculated by subtracting

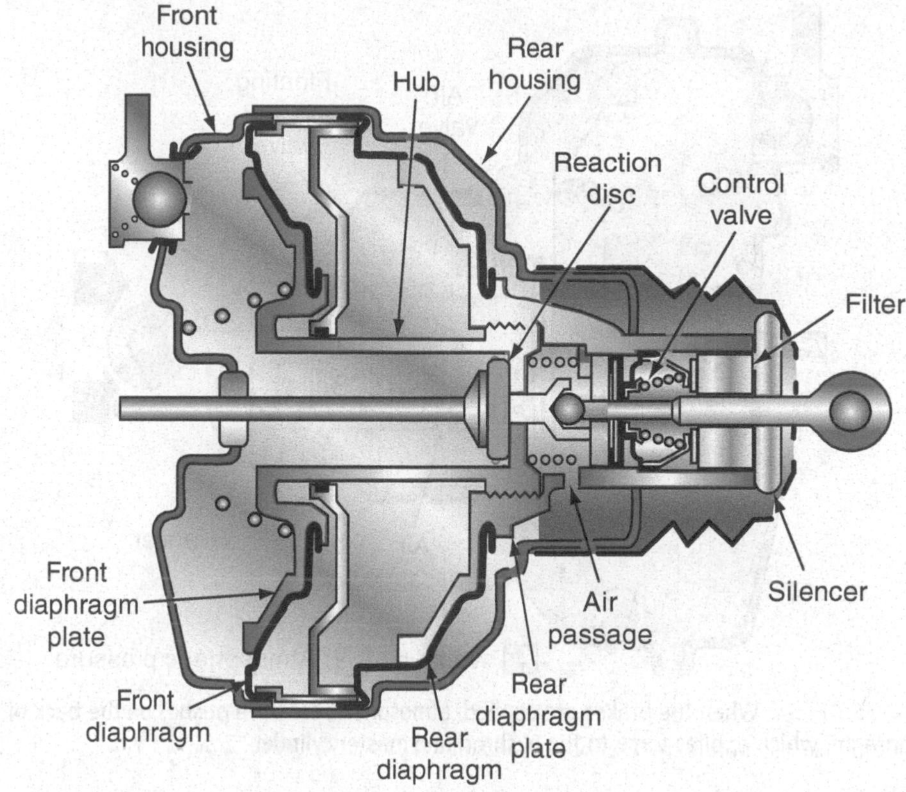

FIGURE 16-32 A tandem booster uses two diaphragms of smaller diameter instead of one large diaphragm.

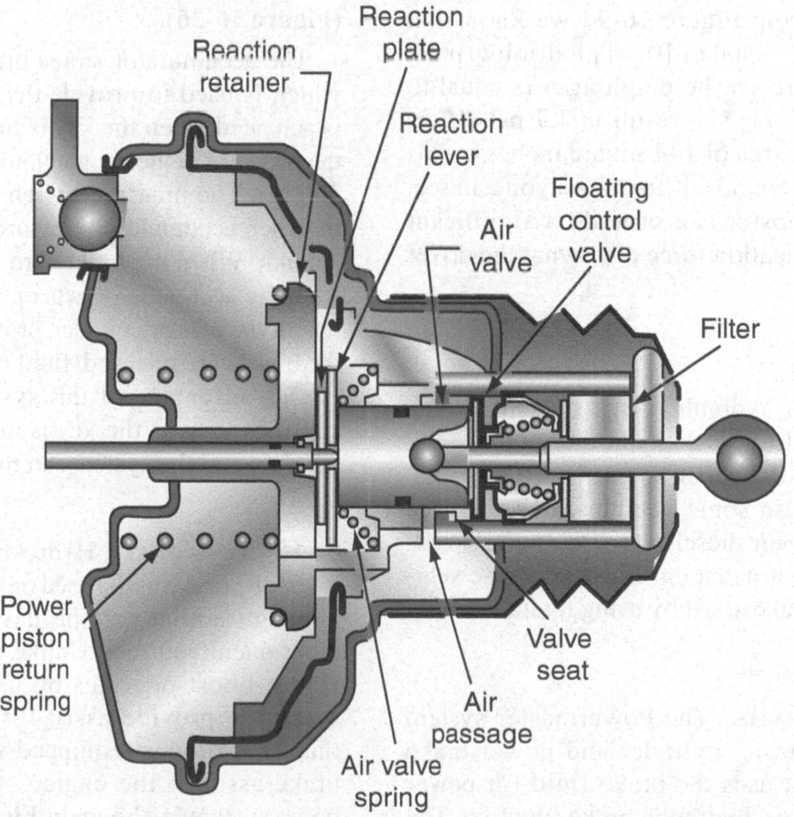

FIGURE 16-33 With the brakes unapplied, vacuum is present on both sides of the diaphragm. No movement or assist is provided because of the equal pressure.

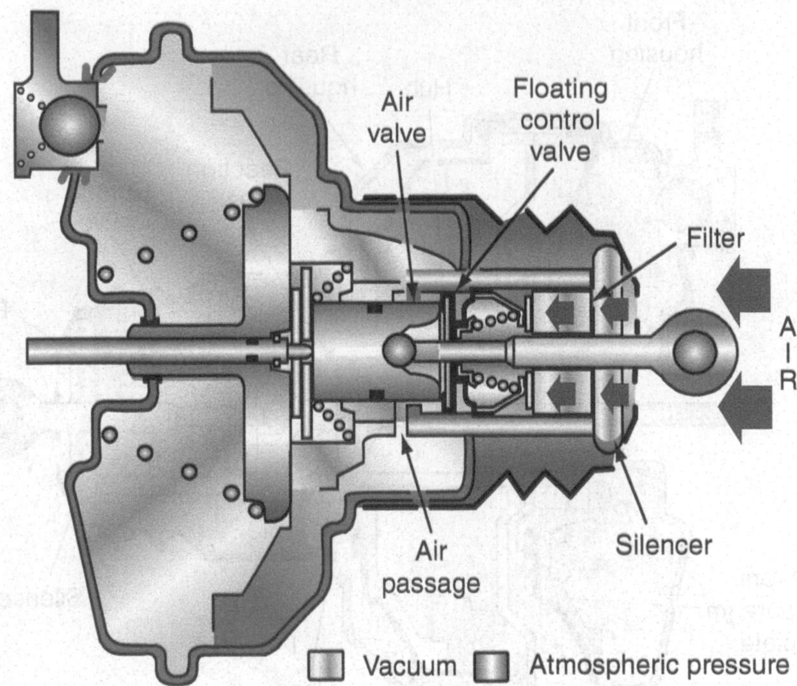

FIGURE 16-34 When the brakes are applied, atmospheric pressure pushes on the back of the diaphragm, which applies force to the pushrod and master cylinder.

the pressure on the vacuum side from atmospheric pressure. If 10 inches of manifold vacuum is supplied to the booster, and from Figure 16-23 we know that 10 inches of vacuum is equal to 10 psi of absolute pressure, then the pressure on the diaphragm is equal to 4.7 psi (14.7 - 10 = 4.7). The result is 4.7 psi difference times the surface area of 144 square inches equals slightly less than 677 pounds of force. As you can see, the vacuum brake booster can provide a significant increase in brake application force over what the driver alone can produce.

HYDRAULIC ASSIST

There are two types of hydraulic assist systems, which are generally referred to by their product names, the Powermaster and Hydro-boost power brake systems. These are used because some vehicles, due to space limitations or due to their diesel-powered engines, cannot use a vacuum power assist unit. Instead, these vehicles generate power brake assist by using hydraulic fluid pressure.

■ ***Powermaster Assist.*** The Powermaster system is a self-contained master cylinder and power brake booster assembly that uses the brake fluid for power assist as well as standard hydraulic brake function. The Powermaster unit has an electrically operated pump, an accumulator, pressure switch, and master cylinder (**Figure 16-36**).

The accumulator stores brake fluid under pressure, which is used to provide the power assist. The pump is activated when the key is turned on, and it maintains the pressure in the accumulator, which is approximately 500 psi. The pressure switch is used to turn the pump on when accumulator pressure drops below a specified amount. When the brakes are applied, pressurized fluid from the accumulator acts upon the power piston at the rear of the master cylinder, providing power assist. When the brakes are released, fluid returns to the reservoir.

One advantage of this system is that if the engine stalls, as long as the key is on, the pump can generate pressure and the system can maintain power assist.

■ ***Hydro-Boost.*** Hydro-boost is the name of the hydraulic brake assist used on General Motors products, although similar systems have been used by Ford and other manufacturers. Unlike the Powermaster system, Hydro-boost operates by using the power steering system to provide assist for the brakes. Because of this, Hydro-boost-equipped vehicles cannot maintain brake assist if the engine stops running. A Hydro-boost system is shown in **Figure 16-37**.

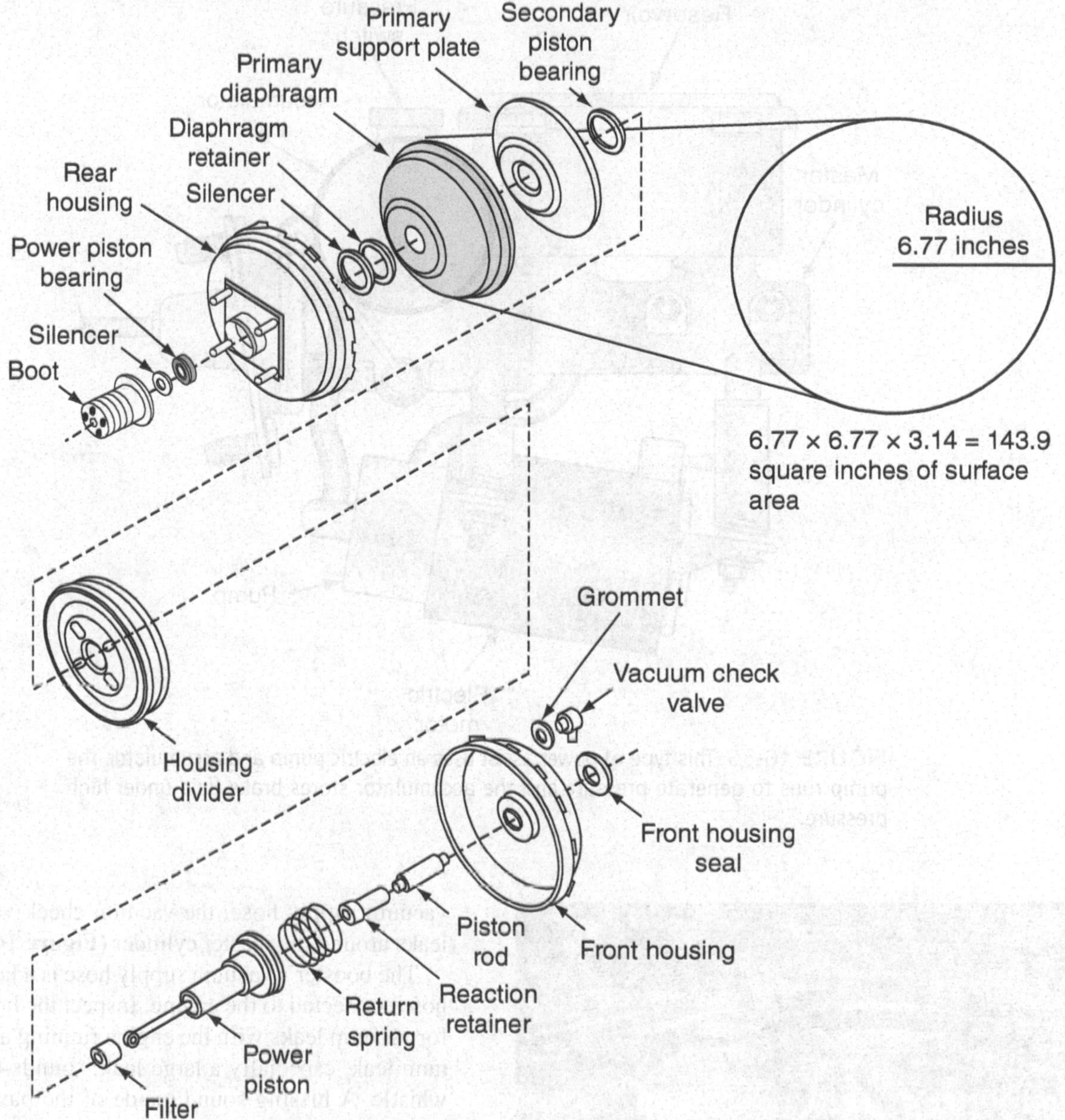

FIGURE 16-35 The amount of assist provided by the booster depends on the size of the diaphragm and the difference in pressure.

■ ***Components.*** The power steering pump supplies the hydraulic pressure for the Hydro-boost system. The Hydro-boost unit is mounted between the firewall and the master cylinder and is connected to the power steering pump and power steering gearbox via the power steering hoses (**Figure 16-38**).

■ ***Operation.*** When the brakes are not applied, fluid enters and exits the Hydro-boost unit (**Figure 16-39**). When the brakes are applied, the spool valve closes the fluid return port, building pressure in the power chamber. This exerts force on the power piston, providing brake assist (**Figure 16-40**).

POWER ASSIST INSPECTION

Customer complaints about the power assist system are often related to how the brake pedal feels. The customer may not know that the problem is in the power assist, just that the pedal does not feel correct. The most common complaint is that the brake pedal is very hard to

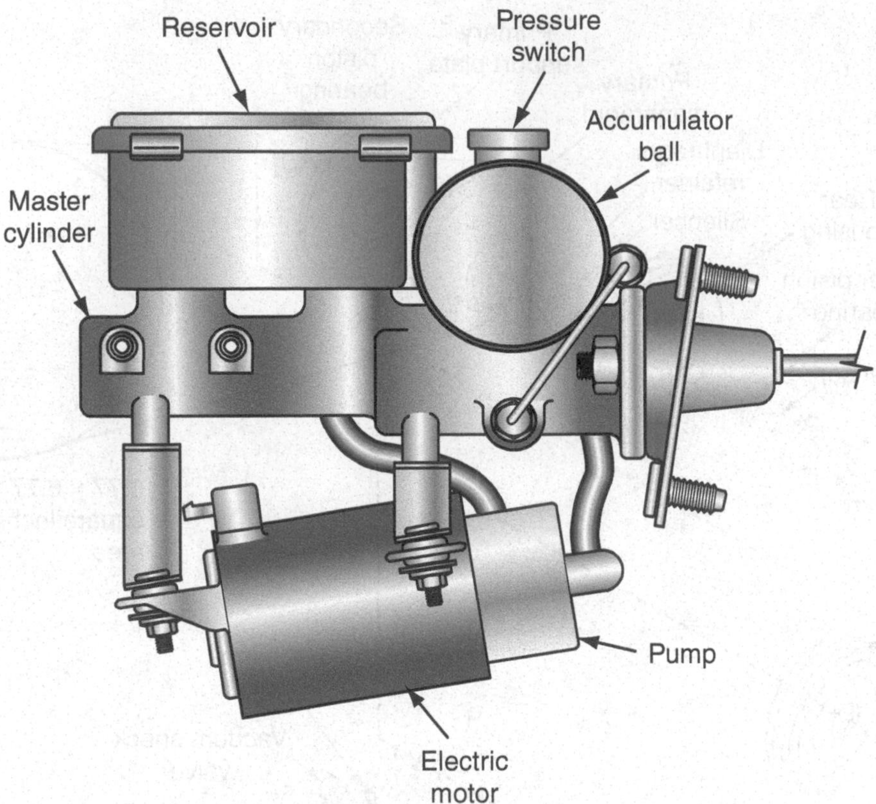

FIGURE 16-36 This type of power assist uses an electric pump and accumulator. The pump runs to generate pressure and the accumulator stores brake fluid under high pressure.

FIGURE 16-37 This system uses the power steering pump to supply high-pressure fluid for power brake assist.

press and that there is little or no power brake assist. After confirming the complaint, begin with a visual inspection of the system.

■ ***Vacuum Booster Inspection.*** A visual inspection of the vacuum brake booster should include checking the vacuum supply hose, the vacuum check valve, and for leaks around the master cylinder (**Figure 16-41**).

The booster's vacuum supply hose is a large-diameter hose connected to the engine. Inspect the hose and valve for vacuum leaks with the engine running at idle. A vacuum leak, especially a large leak, sounds like a hiss or whistle. A hissing sound inside of the passenger compartment when the brakes are not applied is often due to a leaking power assist unit. Listen for a vacuum leak around where the brake pushrod goes into the booster (**Figure 16-42**).

The vacuum valve keeps vacuum trapped in the booster. This is so that if the engine stalls, there will be some vacuum stored in the booster to allow power assist to stop the vehicle. To check the vacuum valve, run the engine for a minute, shut the engine off, and carefully remove the valve from the booster. If there is a rush of air when the valve is removed, then it is working properly. Another way to test the valve is to run and then shut off the engine. Pump the brake pedal a couple of times. If the effort to press the pedal is normal for two or three applications, the check valve is working.

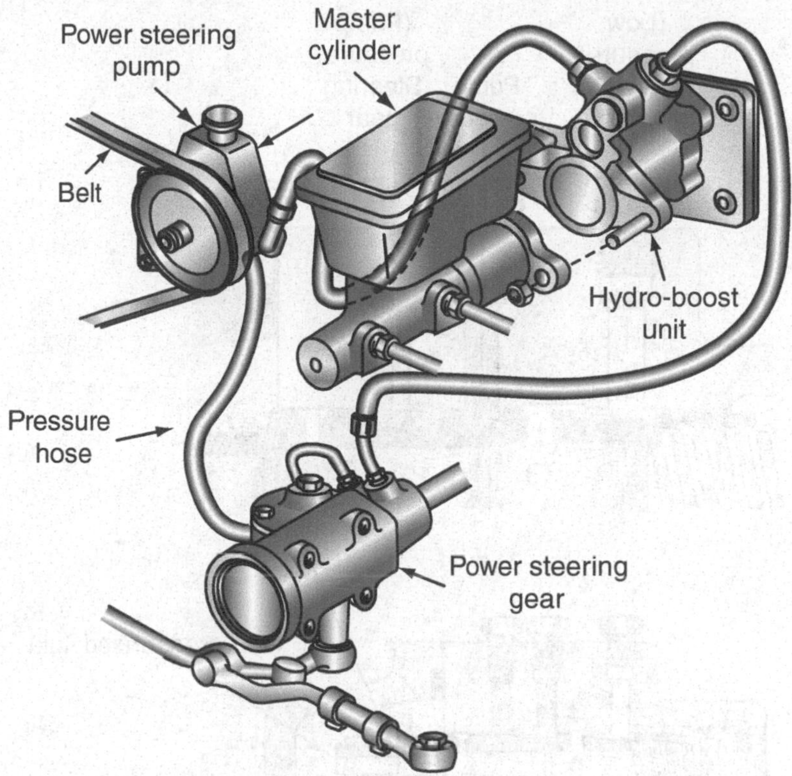

FIGURE 16-38 How the Hydro-boost is connected to the power steering system.

■ ***Hard Brake Pedal and No Power Assist.*** When a vacuum booster develops an internal leak, or if the vacuum supply to the booster is insufficient, the customer complaint is typically of a very hard to press brake pedal. Stopping requires significantly more effort to apply the brakes and longer stopping distances. There may also be a hissing noise from under the dash or from the engine compartment if there is a vacuum leak to the booster. When the diaphragm inside the booster ruptures, vacuum will occur on both sides of the diaphragm. When the brakes are applied, atmospheric pressure cannot build in the rear of the chamber due to the ruptured diaphragm. This means the pressure on both sides of the diaphragm is equal, and consequently, no power assist is provided.

Sometimes, a leaking booster creates enough of a vacuum leak to cause the engine to run rough, stumble, hesitate, or even stall. This is because the vacuum leak affects the amount of air entering the engine. Too much air for the fuel supplied by the fuel system causes a rough idle or running condition.

In areas of the country where rust is a problem, the booster shell can rust through or even come apart (**Figure 16-43**). This booster had developed a small leak, which over time got worse until the unit completely separated and fell apart.

■ ***Check Vacuum Booster Operation.*** With the engine off, pump the brake pedal several times until the pedal is very hard to press. Keep your foot on the brake pedal and start the engine. The pedal should drop a couple of inches once the engine starts. This indicates vacuum is supplied to the booster, and the booster is working normally. If the pedal remains very hard, check the vacuum to the booster. If the vacuum is sufficient, then the booster is defective.

■ ***Hydraulic Boost Inspection.*** Start the engine to test the power steering and brake assist. If neither have assist, the problem is likely with the power steering system. Visually inspect for leaks around the power steering pump, lines, and brake assist unit. Check the power steering belt and fluid because the brake assist depends on the power steering system for pressure. If there are no leaks and the drive belt is intact and tight, pressure testing the system will be necessary to determine if the power steering pump is working properly.

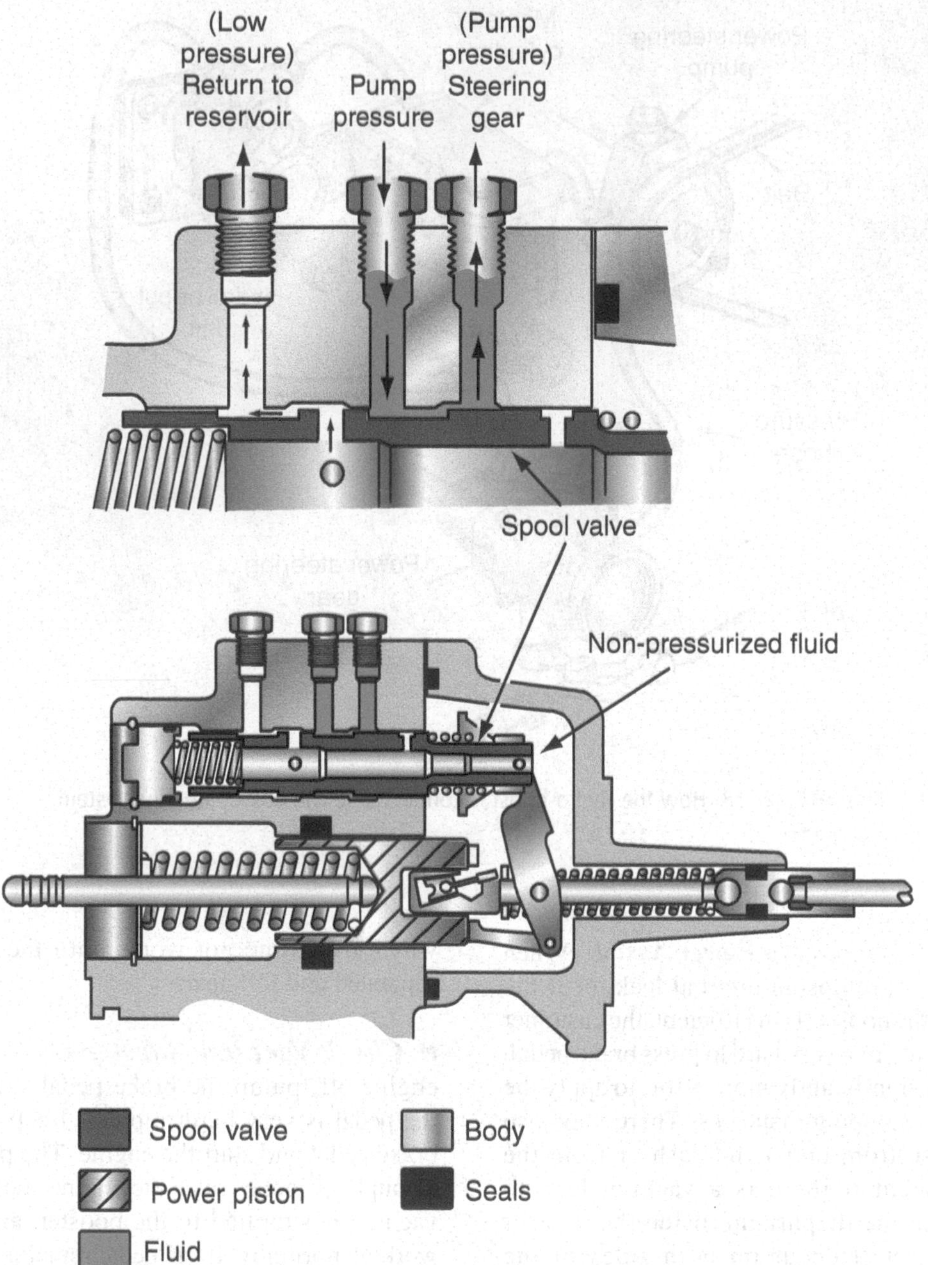

FIGURE 16-39 When the brakes are not applied, fluid to the booster simply returns to the power steering system.

■ ***Hybrid Vehicles and Integral ABS Units.*** Many vehicles use a combination master cylinder, brake booster, and ABS unit (**Figure 16-44**). These units, like the Powermaster system, use electric pumps to supply pressure for the system. One significant difference is that integral systems store brake fluid under very high pressure in accumulators.

Because integral units store fluid under high pressure, the system pressure must be discharged before work on the system is performed. With the ignition off, pump the brake pedal at least 40 times to discharge the accumulators. The pedal should become very firm once the system is depressurized. Do not turn the ignition on while you are working on the hydraulic system as this can cause the pump to operate and build pressure in the system.

On Toyota Prius models, any time the battery is connected, even if the power switch is off, the brake control

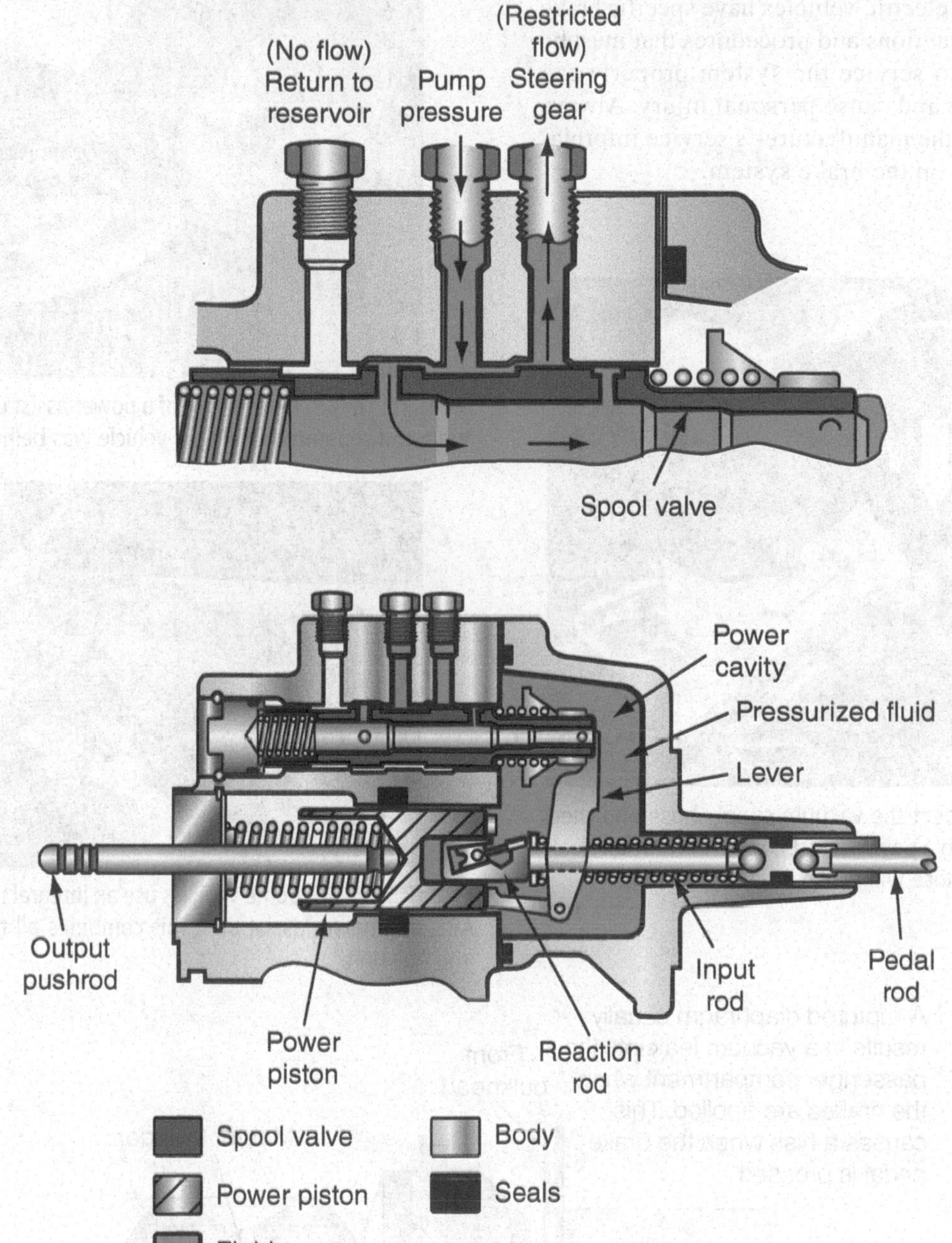

FIGURE 16-40 With the brakes applied, fluid is routed to the rear of the power piston to provide power assist.

system activates when the brake pedal is depressed. The system also activates when any door switch turns on. Do not operate the brake pedal or the doors while the battery is connected and you are servicing brake system components.

Some vehicles use a backup power source for the brake system. This often consists of capacitors that store a 12-volt charge in the event the regenerative braking becomes inoperative. This allows the brake system to function long enough to slow and stop the vehicle if a malfunction occurs with the 12-volt battery supply. If the backup power source voltage is depleted, the brake system will continue to function as a conventional hydraulic brake system. Because some models use capacitors as power backup, there may be an electrical charge left in the brake control system. Do not try to service the brake control system or the backup power supply until the capacitors have discharged.

Most hybrid and electric vehicles have specific brake system service precautions and procedures that must be followed. Failure to service the system properly can damage the vehicle and cause personal injury. Always refer to and follow the manufacturer's service information when working on the brake system.

FIGURE 16-43 An example of a power assist unit that rusted apart and separated while the vehicle was being driven.

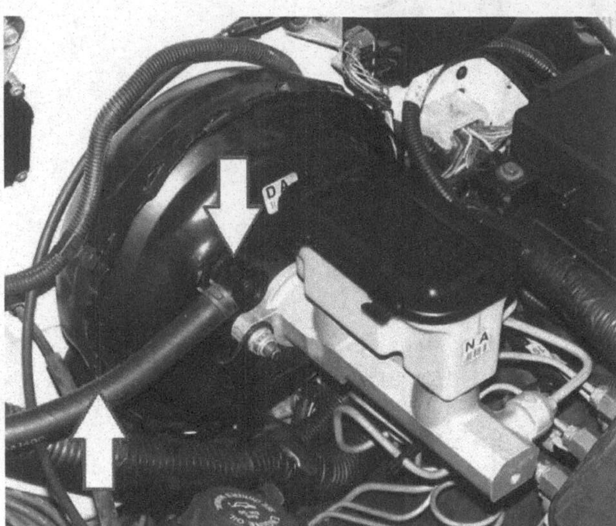

FIGURE 16-41 Inspect the vacuum supply hose and check valve when diagnosing a power assist concern. A blocked, leaking, or disconnected hose will cause little or no assist.

FIGURE 16-44 Some vehicles use an integral master cylinder, ABS, and power assist unit. This combines all three functions into one unit.

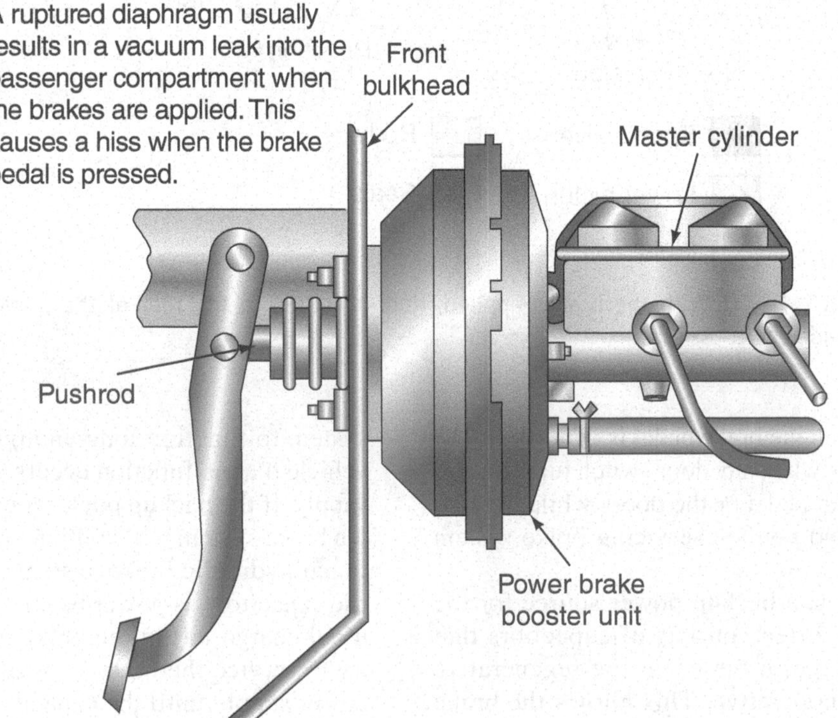
A ruptured diaphragm usually results in a vacuum leak into the passenger compartment when the brakes are applied. This causes a hiss when the brake pedal is pressed.
Front bulkhead
Master cylinder
Pushrod
Power brake booster unit

FIGURE 16-42 A leaking vacuum booster will have a hard brake pedal. Sometimes, the leaking vacuum can be heard inside the passenger compartment as a hiss with the brakes released.

SUMMARY

The ABS system prevents the wheels from staying locked by rapidly pumping the brakes for the driver.

The amount of traction between the tire and the ground is called tire slip.

Wheel speed sensors monitor the rotational speed of each wheel.

The EBCM commands the electrohydraulic unit to hold, open, or close wheel brake hydraulic circuits depending on wheel slip.

The electrohydraulic unit contains electric motors, solenoids, and valves that control the flow of brake fluid to the wheel brakes based on input from the EBCM.

Hold mode is used to limit any further pressure increase in a wheel brake circuit.

Release mode is used when the wheel is still slowing too rapidly.

Pressure increase mode is used to reapply pressure to the brake circuit to slow the wheel.

Never open a hydraulic line or bleeder screw with the ignition on as this can cause injury from the release of high-pressure fluid.

Integral ABS systems require that the high-pressure accumulators be discharged before any service is performed.

Some vehicles will require the use of a scan tool to properly bleed the ABS system.

REVIEW QUESTIONS

1. A _______________ check valve is used to keep a reservoir of _______________ in the power brake booster in the event the engine stops running.
2. ABS units that combine functions of the master cylinder, ABS control, and power assist are called _______________ units.
3. If a wheel locks during braking, wheel speed sensor data will show a _______________ in the number of rotations for the wheel.
4. An integral ABS unit may have high-pressure _______________ that store brake fluid under high pressure.
5. Vehicle movement around its center is measured by the _______________ sensor.
6. Two technicians are discussing bleeding the hydraulic brake system on an ABS-equipped vehicle. *Technician A* says some antilock systems require using a scan tool to bleed the system. *Technician B* says special bleeding procedures may be required to remove all the air from the system. Who is correct?
   a. Technician A
   b. Technician B
   c. Both A and B
   d. Neither A nor B
7. All of the following are modes of ABS operation except:
   a. Hold mode
   b. Pressure increase mode
   c. Recycle mode
   d. Release mode
8. Which of the following is not a common ABS configuration?
   a. Four wheel antilock
   b. Rear wheel antilock
   c. Three wheel antilock
   d. Front wheel antilock
9. A vehicle has illuminated red BRAKE and yellow ABS lights. *Technician A* says a thorough inspection of the brake system should be performed. *Technician B* says to check for stored DTCs. Who is correct?
   a. Technician A
   b. Technician B
   c. Both A and B
   d. Neither A nor B
10. Which of the following tools can be used to check wheel speed sensor operation?
    a. Digital multimeter
    b. Oscilloscope
    c. Scan tool
    d. All of the above

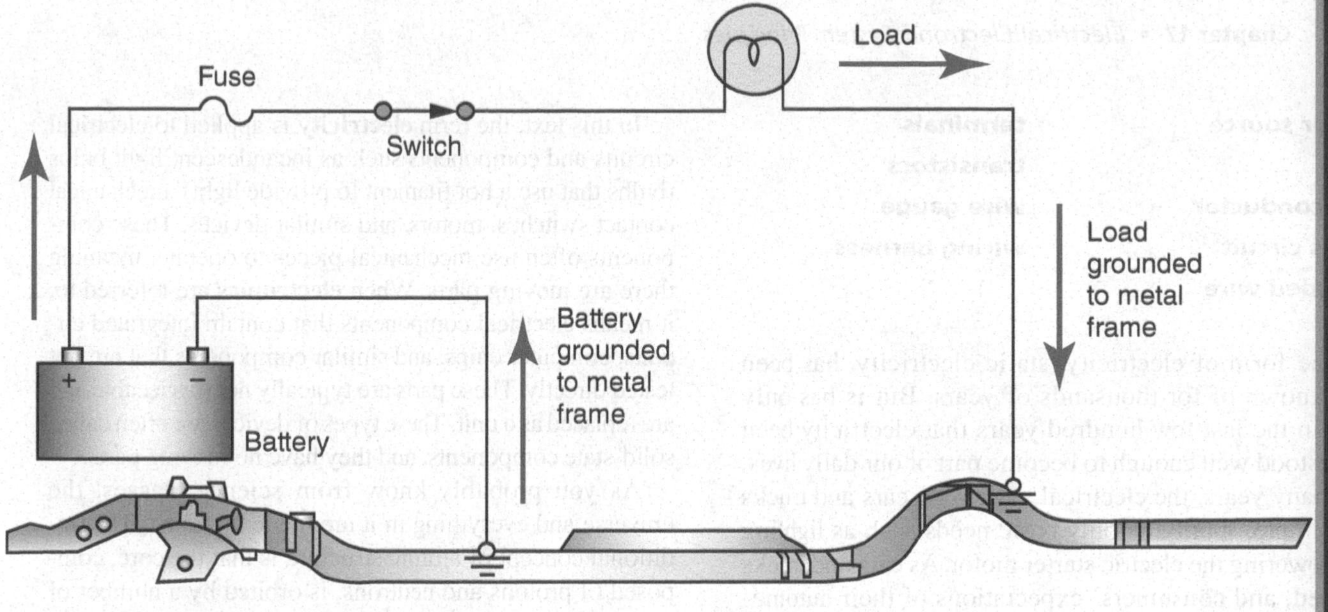

CHAPTER 17

Electrical/Electronic System Principles

Chapter Objectives

At the conclusion of this chapter, you should be able to:

- Explain the differences between AC and DC current flow.
- Describe how AC and DC are used in the modern automobile.
- Explain circuit types and their operation. (ASE Education Foundation MLR 6.A.2)
- Use Ohm's law to explain circuit operation. (ASE Education Foundation MLR 6.A.2)
- Describe circuit protection and types of protection devices.
- Explain types of circuit faults and their causes.
- Explain the fundamentals of electromagnetism.
- Explain the purpose and operation of relays.
- Describe automotive wiring, connections, and terminals.
- Describe basic electronic devices and their operation.

KEY TERMS

alternating current (AC)	**connectors**	**electronics**
amperage	**digital multimeter**	**fuse**
chassis ground	**diode**	**fuse links**
circuit	**direct current (DC)**	**ground**
circuit breakers	**electrical noise**	**ohm**
circuit protection	**electricity**	**Ohm's law**
conductor	**electromagnetism**	**parallel circuit**

power source
relay
semiconductor
series circuit
stranded wire
terminals
transistors
wire gauge
wiring harness

One form of electricity, static electricity, has been known of for thousands of years. But is has only been in the last few hundred years that electricity been understood well enough to become part of our daily lives. For many years, the electrical system on cars and trucks was simple, supplying only basic needs such as lighting and powering the electric starter motor. As cars and trucks evolved, and consumers' expectations of their automobiles increased, the electrical systems also evolved.

The modern vehicle now contains a complex system of computers, wiring, and components designed to make driving safer and more enjoyable. While this is good for everyone who enjoys GPS navigation, Bluetooth music device connections, and other common accessories, it does require occasional repair when the gadgets stop working. Over the last two decades, the trend has been to either enhance or replace mechanical components with electrical components. This further adds to the complexity of the modern automotive electrical system.

Principles of Electricity

For many people, electricity is a mystery, an unseen force that makes things work. Some people fear electricity, sometimes with good cause, but most just accept that it is there and that it works without a real understanding of how or why. As a technician, you need to have a good understanding of what electricity is, why it works, how it works, and what happens when things go wrong. Modern vehicles are dependent upon electronics to operate and the cars and trucks of tomorrow will likely continue to become even more complex.

WHAT IS ELECTRICITY?

Simply stated, electricity is the flow of electrons from a source of higher electrical potential to a source of lower electrical potential through a conductor. If that made sense, skip to the next heading. For everyone else, we will examine the first sentence in more detail.

■ ***Electricity Explained.*** You do not need to have an electrical engineering degree to work on today's electrical systems, but you do need to understand the basics of electrical systems and components.

> **Word Wall**
> *Potential*—Capable of being or becoming.

In this text, the term **electricity** is applied to electrical circuits and components such as incandescent light bulbs (bulbs that use a hot filament to provide light), mechanical contact switches, motors, and similar devices. These components often use mechanical pieces to operate, meaning there are moving parts. When **electronics** are referred to, it means electrical components that contain integrated circuits, computer chips, and similar components that are not tested directly. These parts are typically not serviceable and are replaced as a unit. These types of devices are often called solid-state components, and they have no moving parts.

As you probably know from science classes, the universe and everything in it are made of atoms. The traditional concept of atomic structure is that the core, composed of protons and neutrons, is orbited by a number of electrons (**Figure 17-1**). The number of protons, neutrons,

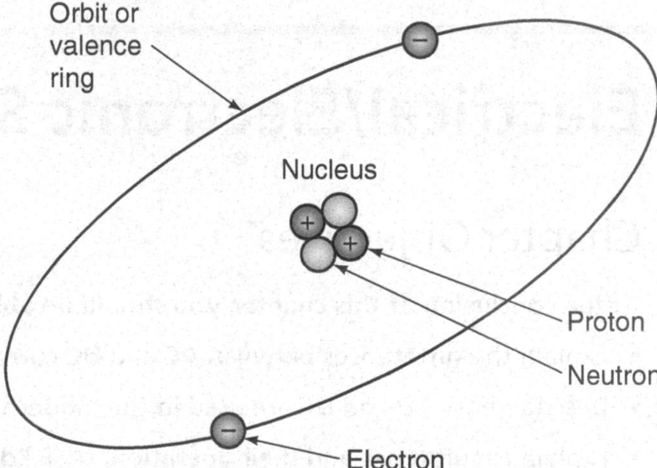

FIGURE 17-1 In this simplified illustration of an atom, the negatively charged electrons orbit the nucleus, which contains positively charged protons and uncharged neutrons.

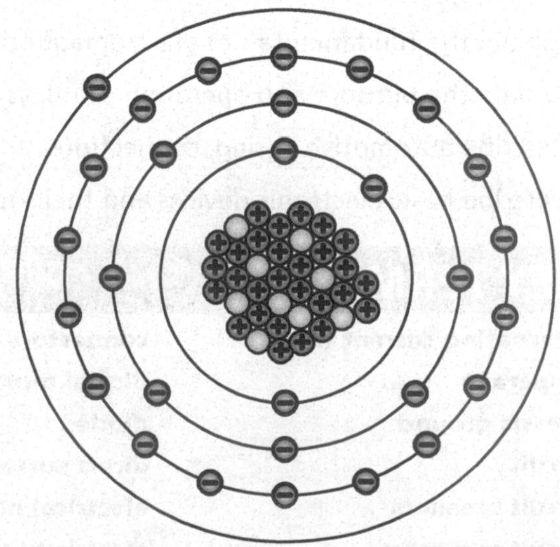

FIGURE 17-2 An illustration of a copper atom, which contains 29 electrons, of which only one is in the outer shell.

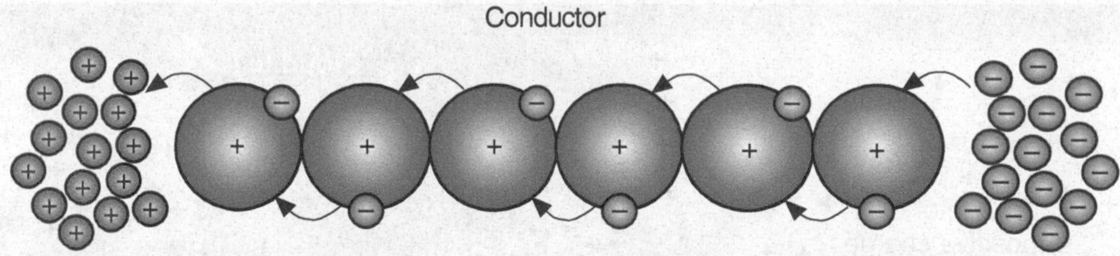

FIGURE 17-3 The movement of electrons along a conductor is the basis of electricity.

and electrons that make up the atom depends on what the atom is. Hydrogen, the simplest atom, has one proton and one electron and is the most abundant element in the universe. Copper atoms have 29 protons and 29 electrons, with several rings of electrons orbiting at different distances from the core (**Figure 17-2**). One of the reasons why copper is used so extensively in electrical wiring is that it only has one electron in its outer ring. The lone electron in the outer ring, given the right conditions, can easily move to another copper atom. The movement of this electron from atom to atom is the basis of electricity (**Figure 17-3**).

A copper atom, in its normal state with 29 protons and 29 electrons, is electrically balanced, having neither a positive nor a negative charge. If, however, the outer electron from one copper atom moves to another copper atom, a state of unbalance occurs. The atom losing the electron becomes positive, and the atom gaining the electron becomes negative. Because the copper atoms do not want to be unbalanced, the positively charged copper atom attracts another electron to its outer ring while the negatively charged atom loses one of its electrons to another nearby copper atom. This flow of electrons is called *current flow* and is measured in amperage.

■ ***Examples of Electricity in Action.*** Talking about electricity might make you think of static electricity. Much to the annoyance of friends and siblings, static electricity has been used for mischief for many years. Static electricity is the term used to describe the accumulation of an electrical charge by a something not normally conductive to electricity. For example, hair, sweaters, and balloons are normally neutral, having no electrical charge. However, under dry weather conditions, the lack of moisture in the air, which is usually responsible for the dissipation of built-up electrical charges, can cause hair and sweaters to become charged (**Figure 17-4**). Walking across a carpeted floor during the winter can cause enough of a static charge to build up to give someone quite a shock when you touch him or her. This is caused by the buildup of dissimilar charges, positive and negative, discharging back to a state of zero charge.

FIGURE 17-4 Static electric charges can cause attraction and repulsion between objects that normally do not conduct very well.

The same process, on a much larger scale, causes lightning. A buildup of electrical charges within the clouds, when released, is a spectacular and enormous display of the same static electrical shock you pass on to your friends and siblings. **Figure 17-5** shows how lightning occurs.

ELECTRICAL TERMS AND MEASUREMENT

Just as with any system on an automobile, there are terms that describe parts or functions, and electricity is no different. The electrical terms we use today take their names from the scientists who made significant contributions to the study of electricity.

■ ***Voltage.*** Voltage, volt, and volts are used to define an amount of electrical potential and to describe how much electrical energy something has, such as a 9-volt battery. Electrical potential is the capacity to do work. Just as a spring under tension can perform work, so does stored electrical energy. To illustrate electrical potential, think about yourself being well rested and fed. You can perform a lot of work, use energy, and get things done. When the work is complete and you are tired and hungry, your potential or ability to do a lot of work is decreased. This is like electrical circuits. The power source has

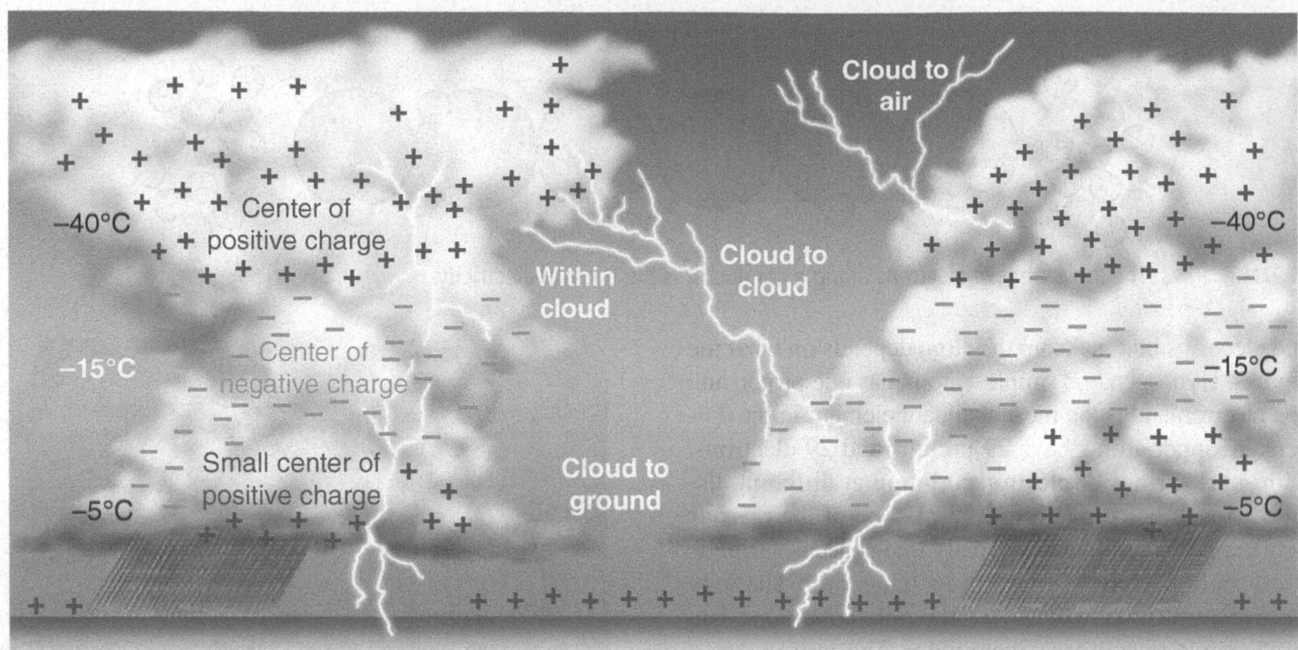

FIGURE 17-5 Large amounts of static electricity generate lightning, which can be millions of volts and thousands of amps in strength.

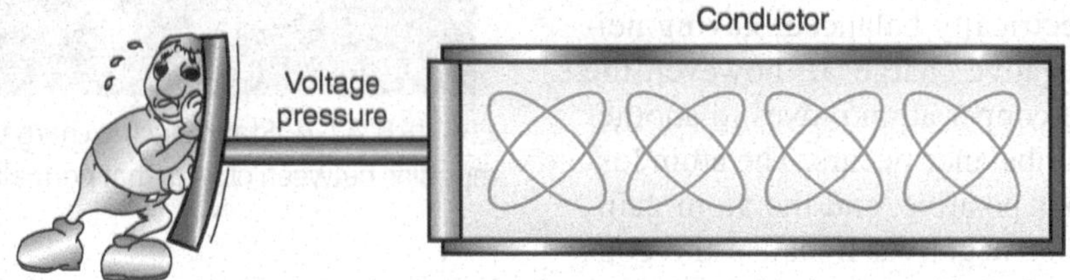

FIGURE 17-6 Voltage is the pressure or force that is exerted on electrons. The electrons move from a point of higher pressure to a point of lower pressure.

energy, and the load uses the energy to perform work. Once the work is done, there is little leftover energy.

The term *voltage* is taken from the Italian scientist Alessandro Volta, who discovered the electrical potential between dissimilar metals in an electrolyte, while he was constructing one of the first types of batteries. You will often see voltage represented by *E*, which is short for electromotive force or EMF.

Voltage is the force that causes the copper electron to jump from atom to atom. Voltage can be thought of as electrical pressure (**Figure 17-6**). Many people think of voltage as being similar to water pressure. Water under higher pressure, such as that from a pressure washer, can do more work, like clean a sidewalk, than water supplied from an ordinary garden hose.

Voltage is measured with a voltmeter and is shown as a capital V. When thousands of volts are discussed, the term kilovolts (1000 volts), abbreviated kV, is used. When very small amounts of voltage are measured, down to one thousandth of a volt, the term millivolts, abbreviated mV, is used.

■ ***Voltage Drop.*** Voltage drop refers to the amount of voltage that is being used by a component, or by a wire, or a connection. Voltage drop is also the measurement of voltage lost through wiring, components, and connections. Because all parts of the electrical system have some resistance, a little bit of voltage is wasted going through the wires, components, and connections. In automotive circuits, a slight bit of voltage drop is considered normal. This is discussed in greater detail later in the chapter.

■ ***Amperage.*** **Amperage**, *ampere, amp*, and *amps* are used to define the amount of electrical current flow that is occurring at a particular point. Amperage is the quantity of electrons actually flowing through a wire, and using the water analogy, it is represented by the amount of water flowing through a pipe or hose. The more electrons (or water) that can flow, the more work can be done. The term *amperage* is taken from André-Marie Ampère, a French scientist who made significant discoveries regarding electromagnetism.

You may see amperage represented as *I*, which means intensity. On electrical test equipment, amps are identified by the capital letter A. When a small amount of current flow is being measured, down to thousandths of an amp, the term *milliamps*, abbreviated mA, is used.

■ ***Resistance.*** Electrical resistance is anything that restricts the flow of electrons and is measured in ohms (pronounced Ōm) of resistance. Ohms are represented by the Greek symbol Ω (omega) on electrical test equipment and as *R* within text. Resistance is found in all electrical circuits (excluding superconducting circuits) and, along with voltage, is what determines how much current flows through the circuit. Resistance can be good or bad. An example of good resistance is the filament in an incandescent light bulb. The filament resists the flow of electrons in the circuit. The resistance to electron flow causes the filament to become very hot and glow, providing light. Unwanted resistance, such as corrosion, can cause circuits and components to operate incorrectly or not all.

The term **ohm** is taken from the German scientist Georg Simon Ohm, who, while working with Volta's newly invented electrochemical battery, determined the relationship among voltage, amperage, and resistance.

■ ***Ohm's Law.*** Volts, amps, and ohms are interrelated in electrical circuits. If any one of the three changes, there will be a change in circuit operation. The way the three components interact is defined by **Ohm's law.** **Figure 17-7** shows the Ohm's law circle. The circle visually shows how the three components, voltage, amperage, and resistance relate with each other. At its most basic, Ohm's law shows that one volt can push one amp through one ohm of resistance.

In practical use, Ohm's law is used as a formula (**Figure 17-8**). You can see from the circle of Ohm's law that $V = A \bullet R$ $(E = I \bullet R)$, which means that $A = V/R$ $(I = E/R)$ and $R = V/A$ $(R = E/I)$. This also means that amps and resistance (*I* and *R*) have an inverse relationship—if resistance increases then amperage decreases, and if resistance decreases amperage will increase, provided the voltage stays the same (**Figure 17-9**).

You will also see Ohm's law represented with *E*, *I*, and *R* for volts, amps, and resistance. In this text, we will use **V**olts, **A**mps, and **R**esistance instead. It makes no difference which letter you use to represent voltage, *E* or *V*, if you understand what *E* or *V* mean and what they represent.

It is important to understand that when working with electricity and Ohm's law, what operations can be performed between units. Voltage can be added to another voltage, as amperage can be added to amperage, and resistances can be added to resistances, as shown in **Figure 17-10**. This is because each can be added to like terms. Because voltage and amperage are not like terms, they cannot be added together. The use of Ohm's law requires that you understand and follow the mathematical relationships between the units and what functions can be performed between them.

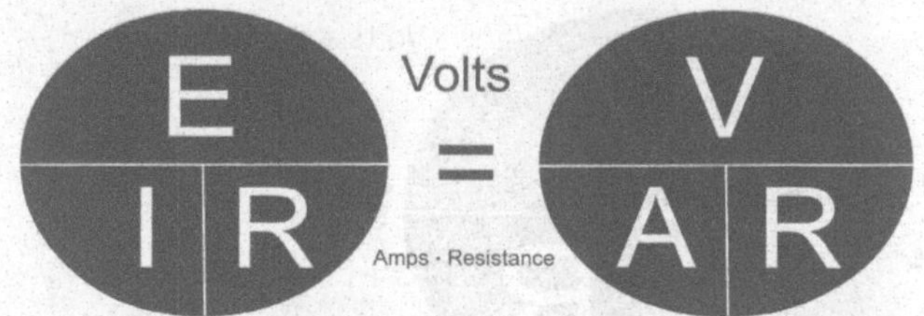

FIGURE 17-7 Ohm's law describes the basic properties of how voltage, amperage, and resistance respond in a circuit.

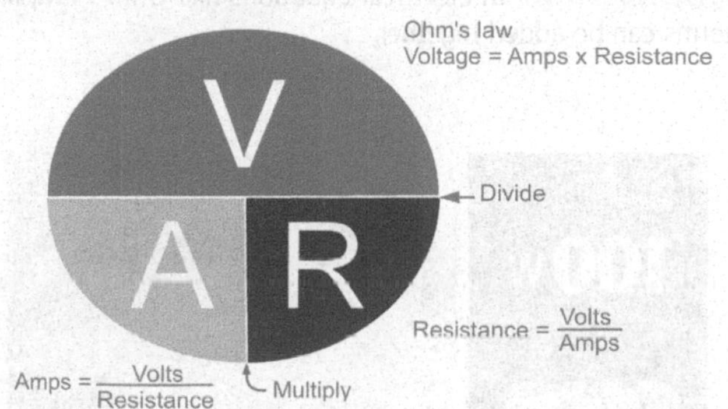

FIGURE 17-8 The relationship of Ohm's law among volts, amps, and resistance.

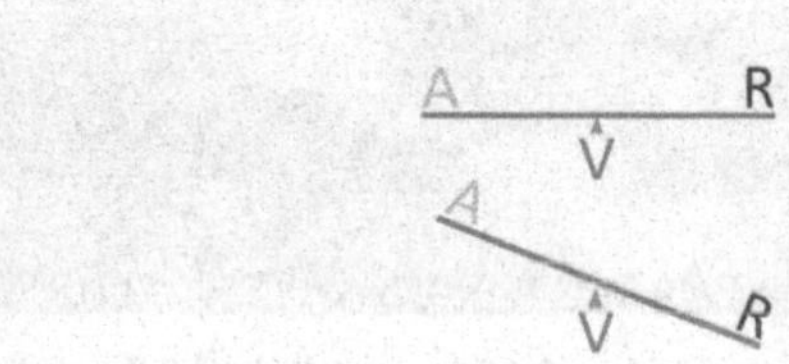

FIGURE 17-9 If voltage remains constant, the actions of resistance and current flow are inversely proportional. If resistance decreases by half, current flow will double.

■ ***Watts.*** If you have ever changed a light bulb, you probably know that bulbs come in different wattages, which determine how bright the bulb is (**Figure 17-11**). The consumption and production of electrical power, represented as P, is rated in watts. The amount of wattage used by a bulb or other component is calculated by the circuit voltage multiplied by the amperage. In automotive applications, most headlight bulbs use about 55 watts

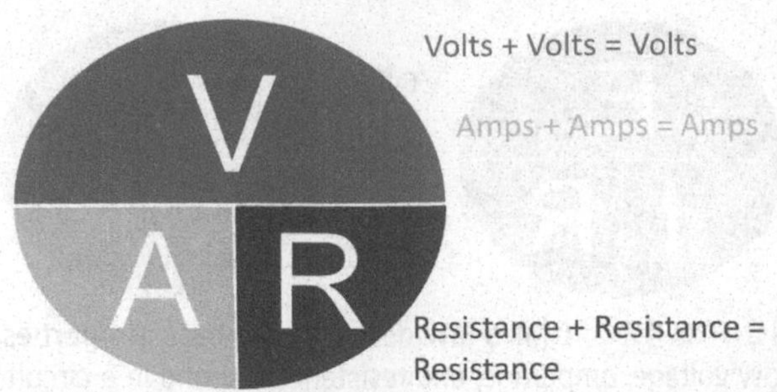

FIGURE 17-10 In electrical equations like Ohm's law, like terms can be added together.

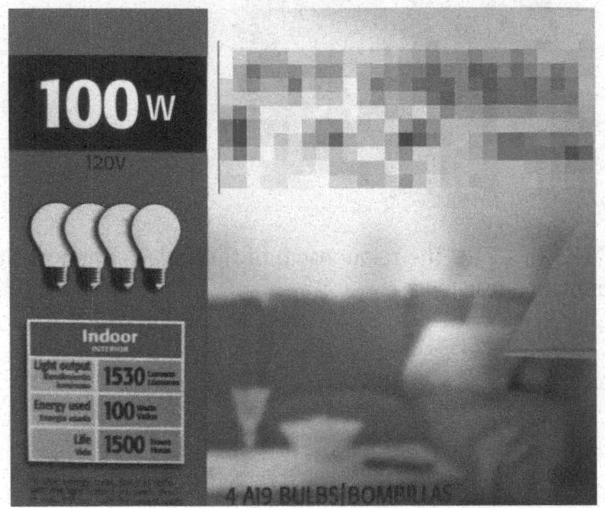

FIGURE 17-11 An example of 100-watt light bulbs, which are consumers of electrical power. 100 watts of power, in residential electrical systems, means the bulbs use about 0.8 amps at 120 volts.

of power. This means that a typical headlight bulb uses about 4.5 amps at 12 volts. The term *watt* is taken from James Watt, a Scottish inventor who made improvements to steam engines and who quantified the rate at which work is performed; what we call horsepower.

AC AND DC VOLTAGES

You may know that electricity comes in two forms, alternating current (AC) and direct current (DC). Devices that use batteries to operate run on DC, while AC powers your home, your school, and most of the electrical items found in homes and businesses.

■ ***AC.*** AC is generated at a power station, transmitted along the power lines, and supplied to your home and school. **Alternating current** gets its name from the way that the electrons flow, which is in an alternating positive and negative manner (**Figure 17-12**). In your home and

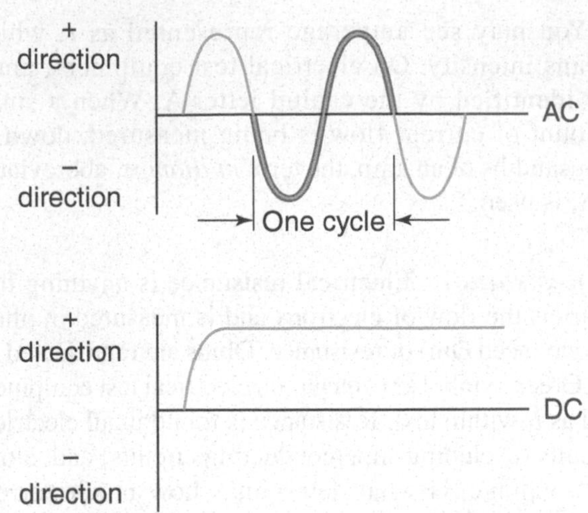

FIGURE 17-12 Alternating current has both positive and negative voltages. Direct current has only positive voltage.

school, AC is powering nearly every electrical device by switching the electron flow from positive to negative and then from negative to positive 60 times per second.

Why use AC? AC can be transmitted over very long distances with very little loss to heat generation, making it the most efficient method of electrical power transmission. Also, large industrial electric motors operate much more efficiently using AC.

In the automobile, the AC generator, which most people call the alternator, produces AC just like a power generation plant, only on a much smaller scale. The AC is then turned into DC to recharge the battery and operate the accessories. Some sensors, mostly for the ABS and engine, produce AC and are used to monitor wheel and engine speed.

■ ***DC.*** DC is used by anything that uses a battery as a power source. **Direct current** flows only from zero to positive and does not alternate between negative and positive as AC does (Figure 17-12).

Under many circumstances, whether AC or DC is used is not important, as both can perform the same work. For example, incandescent light bulbs do not operate any differently if AC or DC current powers them. DC, however, cannot be transmitted over long distances without significant loss from heat, making it ineffective for modern power transmission.

Many devices, such as computers and phone chargers, actually use DC but are plugged into the AC wall outlet. These devices use a transformer to convert the AC into DC (**Figure 17-13**).

■ ***Why Use AC and DC?*** Both AC and DC are commonly used in our day-to-day life. The advantage of using AC in automobiles is that it is easily generated and converted into DC and is then used to charge the

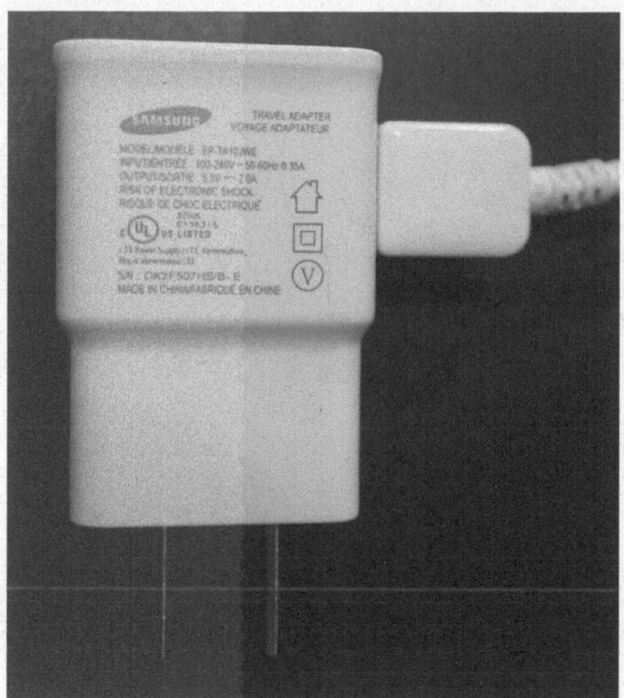

FIGURE 17-13 A transformer is used to convert AC into DC for many common electrical devices.

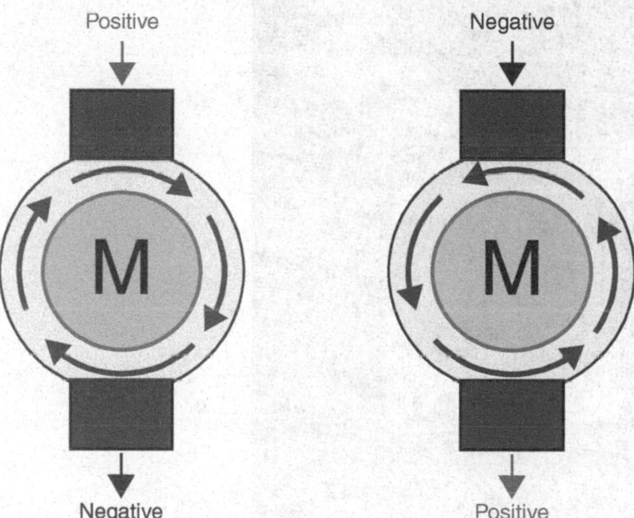

FIGURE 17-14 DC motors will spin in either direction depending on the polarity of the connection.

battery and power the electrical system. AC is also used in hybrid electric vehicles (HEVs) to power the traction motors that propel the car. This use of AC requires the use of expensive and complex electronic systems. The major disadvantage of AC is that it cannot be stored.

DC is used to power the vehicle's electrical system. An advantage of DC include that it can be stored for later use. The storage battery, whether used by a car or a cell phone, allows for electricity to be stored and used when and where as needed. Another advantage of DC is that by reversing the polarity of the connections to a DC motor, the motor can rotate in either direction. DC motors are small and inexpensive and used throughout modern cars and trucks. Reversing polarity means switching how the positive and negative are connected. If you are familiar with power windows and seats, then you have experienced DC motors being used to rotate one direction and then the other, based only on how they are connected to power and ground (**Figure 17-14**).

Circuits, Components, and Operation

Electrical components, such as light bulbs, speakers, and motors, are arranged in circuits. A circuit is a complete path for current flow. Much like a race track, which is also called *a circuit*, the path from beginning to end must be complete for the circuit to work correctly.

WHAT IS A CIRCUIT?

Just as the word circuit can be applied to mean a journey or a course that begins and ends in the same place, circuit refers to a complete electrical path from positive to negative, usually referred to as power and ground. In the vehicle, all electrical circuits begin and end back at the battery, which supplies the power and the ground paths through which electrons flow. There are several types of circuits, both good and bad. A good circuit has a path from the power source, through a load to ground. An example of a bad but complete circuit is one in which an unintended path from power to ground occurs, resulting in damage to the components.

> **Word Wall**
>
> *Load*—Any device or part of a circuit that consumes electrical power.

Think of a circuit as an energy conversion system. The circuit provides a way for one form of energy, such as electricity stored in a charged battery, to convert into another form of energy, such as a light bulb radiating light (**Figure 17-15**). An electrical **circuit** is a path for electrons to flow from a source of higher electrical potential to a source of lower potential. If there is not a complete path for the electrons to travel, then there is an incomplete circuit, and no current can flow. An electrical circuit can be very simple, such as when you get shocked by static electricity, or very complex, having many different components, connections, and modes of operation.

COMPONENTS OF A CIRCUIT

Electrical components, when hooked together to perform some function, become parts of a circuit. An individual component, such as a light bulb, cannot be a circuit alone; there must be other parts hooked together to make

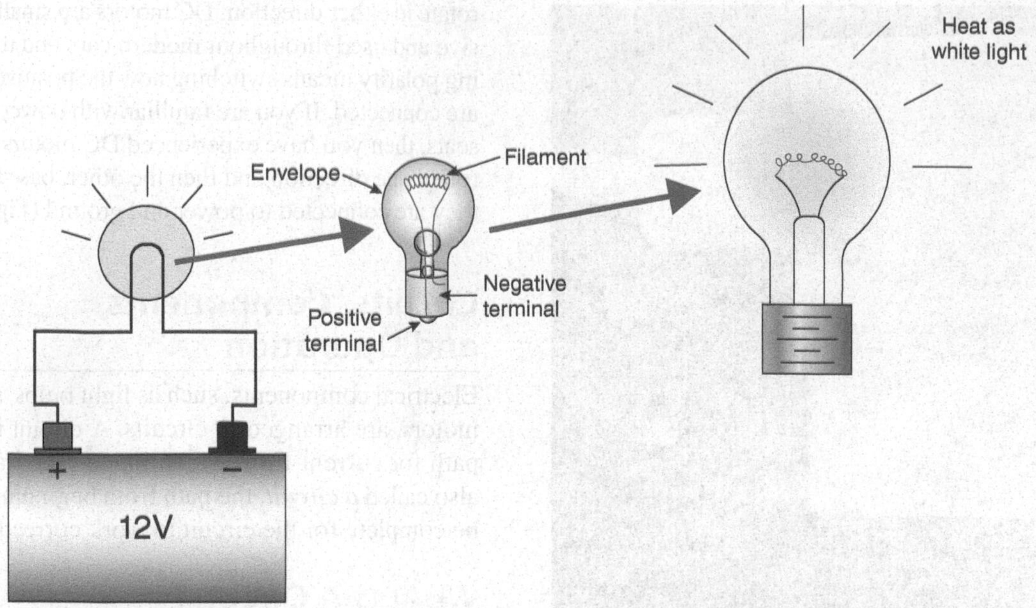

FIGURE 17-15 An electrical circuit converts electrical energy into something else. In this example, heat and light are produced.

FIGURE 17-16 An example of a 12-V battery used to power the electrical system in modern cars and trucks.

the light bulb operate. All circuits need some basic components to operate properly and safely.

■ ***Power Source.*** All electrical circuits need a power source. In the car, the battery is the power source when the engine is off (**Figure 17-16**). With the engine running, the AC generator is the power source.

The **power source**, which we will just call the battery, is the source of high electrical potential. The battery stores electrical energy to be used as needed but does not create energy. Instead, batteries store and release electricity in a chemical reaction between the positive and negative parts of the battery. Just as your cell phone, iPad, and other battery-powered devices need to be charged, so does the car battery. If the battery is allowed to run down, the source of high potential is gone, and the electrical system's components will stop functioning.

■ ***Ground or Negative.*** In DC systems, negative or the more commonly used term, **ground**, refers to the return path back to the battery or negative part of a circuit. In the automobile, the frame and body are part of the ground system (**Figure 17-17**). Many components connect to the frame or body for ground. Because the battery negative is also connected to the frame or body, both are equal. This means that the negative side of the circuit, the vehicle frame and/or body, and the negative side of the battery all should have nearly 0 volts or potential. For a circuit to work, there must be a difference in potential between the positive and negative parts of the circuit, an unbroken path from the battery's positive connection, through the load, and back to the battery's negative connection.

■ ***Electrical Load.*** Anything that consumes electrical power is a load. Light bulbs, heated seats, stereos, and anything else powered by the electrical system is a load. The load uses the flow of the electrons to perform work. In an incandescent light bulb, the electrons moving through the filament cause the filament to become hot and glow, consuming power. The wiring and connections that make up the circuit are also loads, though usually very slight loads. Voltage is used up or dropped by the wiring and connections because there is a slight amount of resistance in all the wiring and connections.

■ ***Conductors.*** **Conductors** are the wires, terminals, and connectors that connect all of the parts together. Most automotive wiring is stranded copper wire. **Stranded wire** is used to provide the flexibility needed in automotive applications (**Figure 17-18**). The wire provides a path for the electrons to move through the circuit. Conductors also include connectors, which attach the wiring to the component and pass through parts of the vehicle's body. The wiring is covered in a rubber/plastic coating called *insulation*. The insulation prevents the copper wire from touching ground or other parts of the electrical system.

■ ***Circuit Control.*** Every time you turn a light off or on, turn the key to start your car, or turn on a computer, you are using a type of circuit control. Controls, such as switches, are used to turn a circuit off or on as needed. Some, like the ignition switch, can be complex switches controlling several different circuits at the same time. Others, like a dome light switch, can be basic switches that open and close the path of electricity to the bulb from its power or ground source.

■ ***Circuit Protection.*** Fuses, fuse links, and circuit breakers are types of circuit protection. These help prevent a faulty circuit from causing damage to the vehicle's wiring and other components. Circuit protection is discussed in more detail later in the chapter.

BASIC CIRCUITS

A basic circuit has the following components: a power source, a control like a switch, a fuse, a load, and a ground, all connected with wiring (**Figure 17-19**). When the switch is closed, the circuit path is complete, and voltage travels from the battery through the wire to the load, in this case a light bulb. The ground provides the path for the electrons to return to the battery.

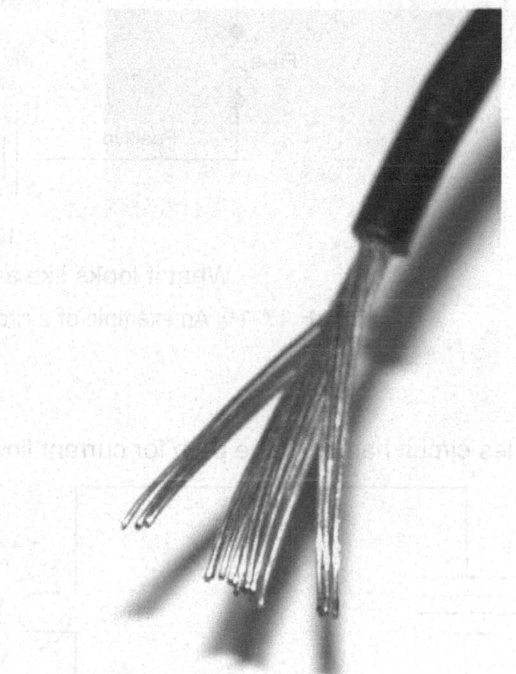

FIGURE 17-18 An example of stranded copper wire.

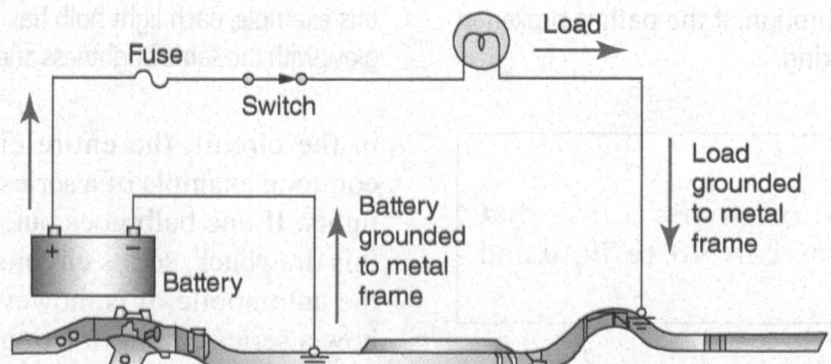

FIGURE 17-17 The electrical system must have a complete path for current flow. The negative side of the battery is connected to the vehicle's body or frame to allow electrons to complete the circuit.

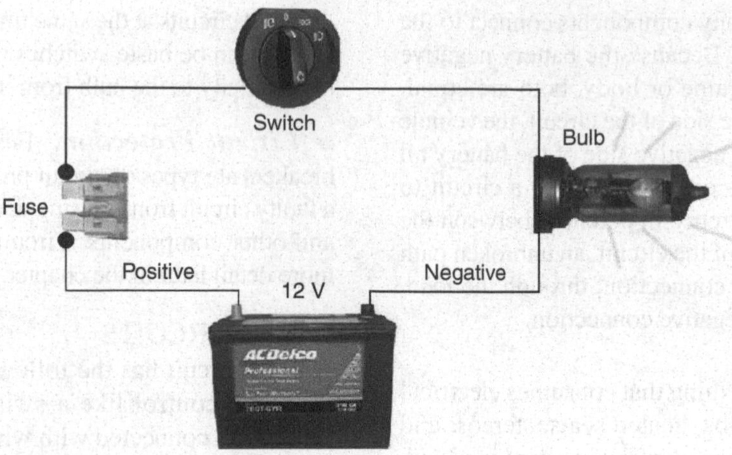

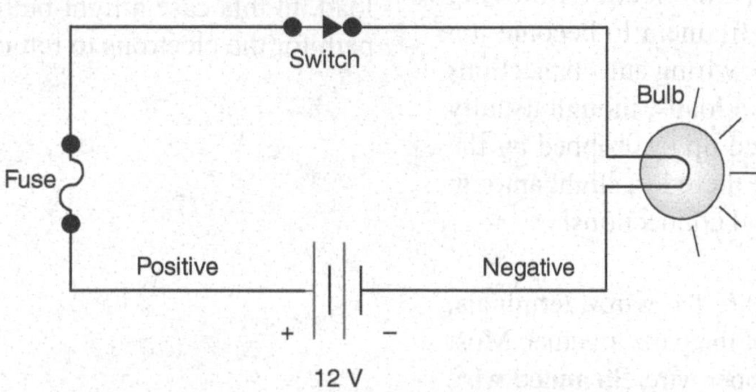

FIGURE 17-19 An example of a circuit containing all the components for safe operation.

FIGURE 17-20 A series circuit is the most basic and has only one path for electrons to flow through. If the path is broken in any place, the circuit stops working.

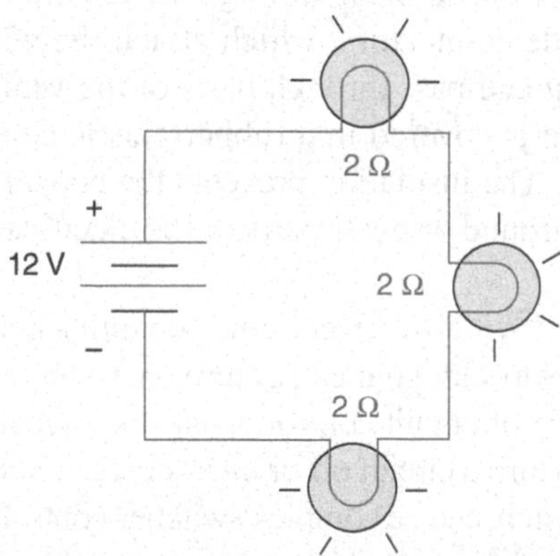

FIGURE 17-21 A series circuit may have more than one load. In this example, each light bulb has the same resistance, so each will glow with the same brightness and use the same amount of voltage.

Word Wall

Closed—A closed switch or circuit is one that is complete and can allow current to flow and work to be done.

■ ***Series Circuits.*** **Series circuits** have all the parts of the circuit in a single path (**Figure 17-20**). The problem with this type of circuit is that if there is a break anywhere in the circuit, the entire circuit stops functioning. A common example of a series circuit is a set of Christmas lights. If one bulb goes out, they all go out. Because of this drawback, series circuits are not used extensively in the automobile. It is, however, important to understand how a series circuit functions.

Figure 17-21 shows a series circuit with three light bulbs of all the same type as the loads. When the circuit is operating, the voltage supplied by the battery is shared by all three bulbs. Because of this sharing, each bulb only uses a portion of the total voltage from the battery, making all three bulbs dim compared to how the same bulbs would appear if they are hooked to the battery individually. To understand why the circuit operates in this way, you need to understand the basic principles of series circuits.

- There is only one path in which current can flow.
- The total resistance of the circuit is equal to the sum of all the resistances.
- The amperage of the circuit is based on the available voltage and the total resistance.
- The amperage is the same at all points in the circuit.
- The total available voltage is used in part by each load based on the resistance of the load—this use of the voltage is called *voltage drop*.

In **Figure 17-22**, we will provide some electrical values and apply the principles stated above.

1. The first principle from above applies to this circuit; there is only one path from power to ground, through the bulbs.
2. The second principle states that in a series circuit, total resistance is equal to the sum of all resistances, or $Rt = R1 + R2 + R3\ldots$, so the total circuit resistance is equal to 6 ohms. However, you should note that the bulbs in the circuit shown in Figure 17-22 do not have the same resistances. Because of this, the bulbs will glow with varying amounts of brightness.
3. The third principle says the amperage is based on voltage and resistance, so by using Ohm's law, we know that we can divide our resistance total into the voltage to determine the amperage, which in this case would equal 2 amps ($12\,\text{volts} / 6\,\text{ohms} = 2\,\text{amps}$). To double-check our numbers, we can also multiply the resistance, 6 ohms, by the amps, 2, to get voltage, which equals 12.

Because there is only one path for current flow, the current flow is the same at all points in the circuit. As you will see later in the chapter, it is possible to test this principle and see it in action.

At this point, we know how many volts we have at the power source, how much resistance each bulb and the circuit have, and what the current flow would be. What we do not yet have is an explanation of why the bulbs vary in brightness. To determine that cause, we need to know how much voltage each bulb is using. Determining the voltage used by a component is called the voltage drop.

4. The fourth principle states that the total voltage available is used by the loads in the circuit based on their resistance. The voltage drop of an individual load can be calculated by multiplying the current flow through the load by the load's resistance, or $Vdrop = A \cdot R$. In **Figure 17-23**, for bulb 1, 2 amps of current flow multiplied by 1 ohm equals 2 volts ($2\,\text{A} \times 1\,\Omega = 2\,\text{V}$). The second bulb will use 4 volts and the third bulb 6 volts. Because volts are like terms, we can add them together to double-check our work. Because the sum of the voltage drops also equals 12 volts, we have determined the voltage drops correctly. The third bulb is using the most volts, and it is the brightest of the three. No matter in what order we hook these three bulbs up to the battery, if they are in series, the results will be the same: the bulb with the highest resistance will be the brightest, regardless of its numerical order in the circuit.

But why was the bulb with the most resistance the brightest? On the surface that does not seem to make sense, but when we use Watt's law, we will see why. Wattage is abbreviated P for power. Power (in Watts) is equal to the voltage used times the amperage flowing through the load, $P = V \cdot A$. The Watt's law circle is like that of Ohm's law (**Figure 17-24**). In the case of bulb 3, the 6 volts used and 2 amps of current flow make the watts equal to 12 W. The other two bulbs consumed 8 W and 4 W. When all are added together, it equals the total wattage for the circuit, 24 watts. As a double-check,

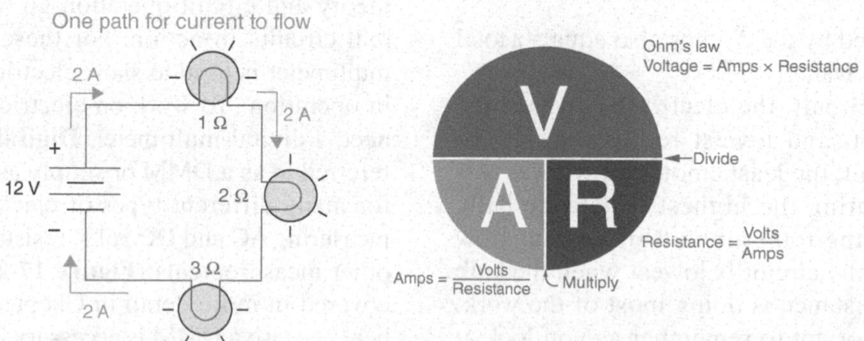

FIGURE 17-22 Using Ohm's law to understand how a series circuit operates. Since each bulb has a different resistance, each will shine with a different brightness.

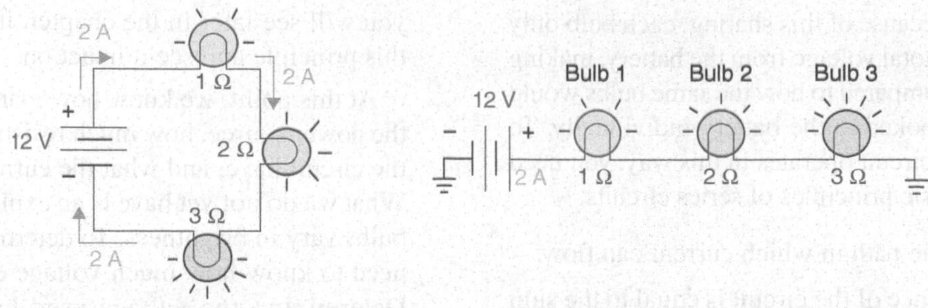

12 V

2 A 1 Ω

2 Ω

3 Ω

Voltage drop = Amps × Resistance

Bulb 1: 2 A × 1 Ω = 2 Volts
Bulb 2: 2 A × 2 Ω = 4 Volts
Bulb 3: 2 A × 3 Ω = 6 Volts

Total voltage drop = 12 Volts

FIGURE 17-23 The voltage drop for each bulb depends on current flow and resistance of the load.

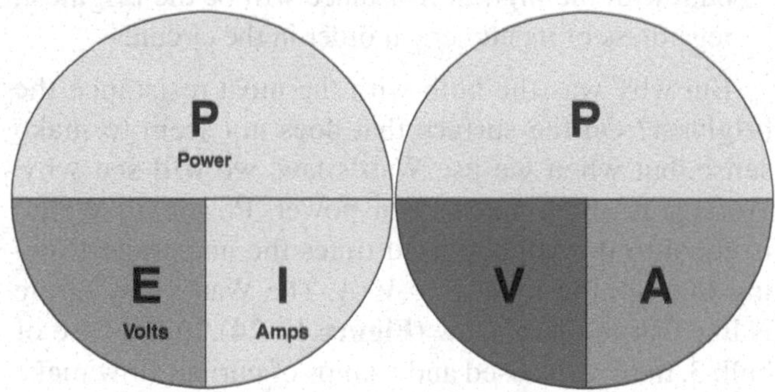

FIGURE 17-24 Watt's law states that power is equal to voltage multiplied by amperage. Notice the similarities with Ohm's law.

the 12 volts multiplied by the 2 amps also equals a total circuit wattage of 24 watts.

In an electrical circuit, the electrons will take the easiest, most direct, and lowest resistance path to ground. In this circuit, the least amount of work is performed by illuminating the highest resistance bulb. Stated differently, the total amount of current flow (amps) required by the circuit is lowest when the bulb with the highest resistance is doing most of the work. This concept is important to remember as you look at different types of circuit faults, which are discussed later in the chapter.

Refer to Chapter 17 in the Lab Manual for additional exercises with series circuits.

ELECTRICAL CIRCUITS IN OPERATION

Ohm's law, in addition to helping you understand how circuits work on paper, also defines how circuits function on the automobile. Once you have developed a good understanding of how electrical circuits operate, you need to put that knowledge into practice by testing circuits. The following section is an example of using Ohm's law and a multimeter to understand circuit operation.

■ ***Applying Ohm's Law to Series Circuits.*** This section will introduce you to how electrical circuit theory and circuit operation go together and apply to real circuits in action. For these examples, a digital multimeter is used to show electrical values for circuits in operation. To work on electrical systems, you will need a digital multimeter. **Digital multimeters**, often referred to as a DMM or simply as a meter, can be used for many different types of electrical tests, including measuring AC and DC volts, resistance, amps, and many other measurements (**Figure 17-25**). Using a meter is covered in more detail in Chapter 18. It is introduced here because a DMM is necessary to show how electrical circuits operate.

Figure 17-26 shows a simple series circuit constructed of automotive light bulbs and wiring, connected to battery positive and negative. The circuit is the same as that shown previously in Figure 17-21. Three bulbs of the same resistance are connected in series. As you can see, the bulbs do not glow very brightly. This is because in a series circuit, total resistance is the sum of all the resistances. In this real-world example, the total circuit resistance is 17 ohms (**Figure 17-27**).

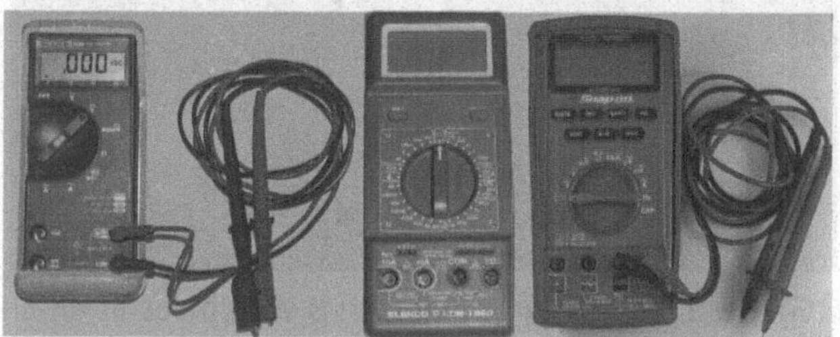

FIGURE 17-25 Examples of digital multimeters or DMMs.

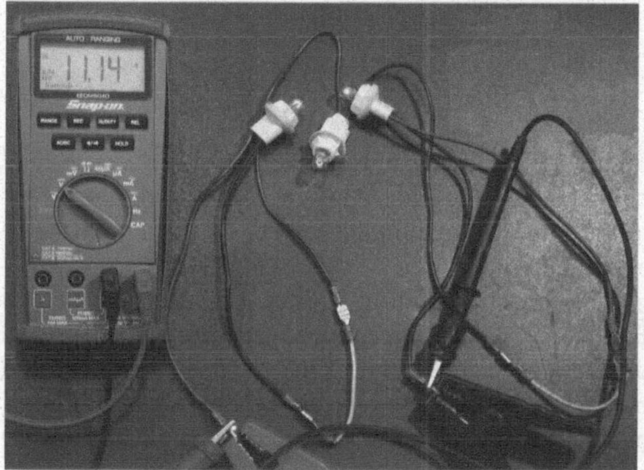

FIGURE 17-26 An example of a simple series circuit. The voltage used by the circuit is 11.14 volts. This is a real-world example of the circuit shown in Figure 17-21.

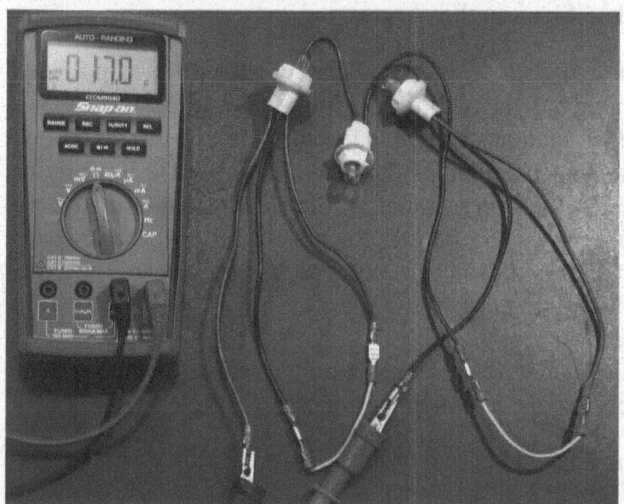

FIGURE 17-27 The total circuit resistance is equal to the sum of the three resistances.

Because this is a series circuit, there is only path for current flow. In **Figure 17-28**, the current flow is 0.125 amps. This is being measured at the power connection to the circuit. **Figure 17-29** shows the same circuit but with the amperage measured between bulbs 1 and 2. This proves the circuit principle that amperage is the same at all points in a series circuit.

In **Figures 17-30**, **17-31**, and **17-32**, the voltage drop across each bulb is shown. Because each bulb is connected in series with each other, the voltage is split up

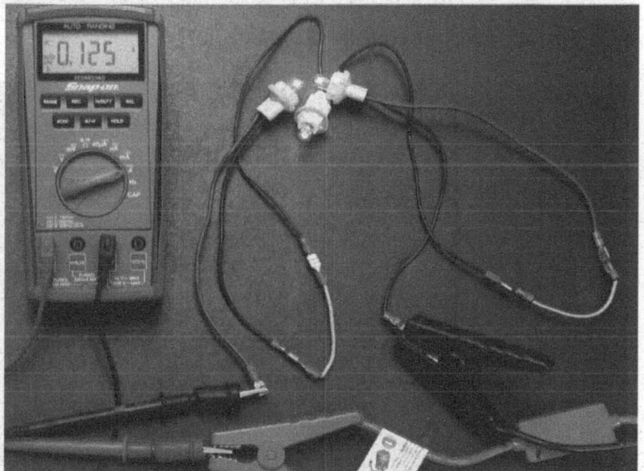

FIGURE 17-28 The amperage through this circuit is 0.125 amps. You may notice that this amperage reading does not equate with the resistance and voltage readings shown in the previous figures. That is because the resistance reading is taken on cold, nonoperating bulbs. Once the bulbs are connected and current flows through them, they get hot, which increases their resistance. Actual operational resistance is more than 89 ohms based on the voltage and amperage.

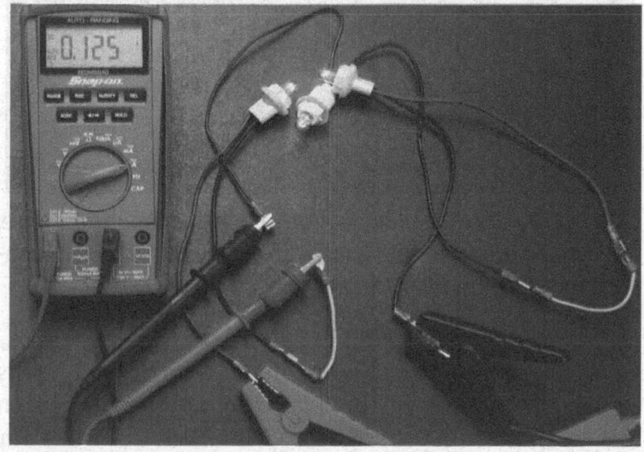

FIGURE 17-29 This shows that the amperage is the same at all points in the circuit.

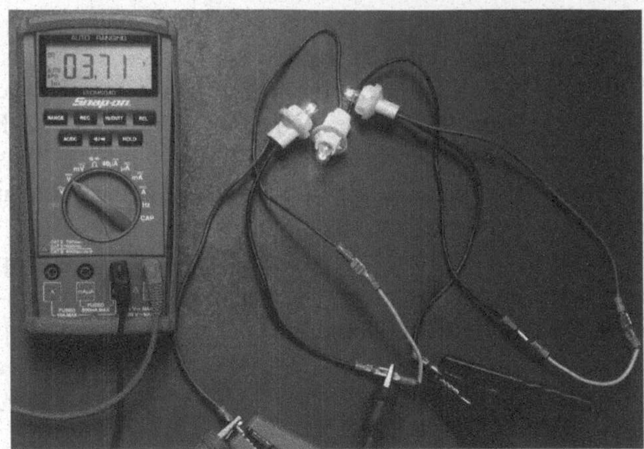

FIGURE 17-30 The voltage drop of bulb 1.

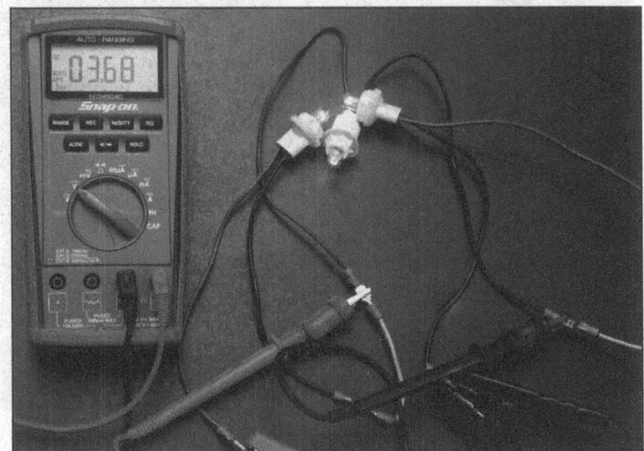

FIGURE 17-31 The voltage drop of bulb 2.

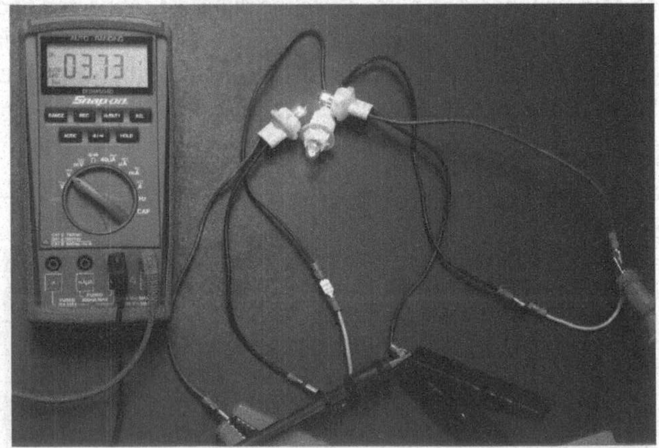

FIGURE 17-32 The voltage drop of bulb 3. Though very close, the voltage drops are not exactly the same due to small variations in the bulbs, bulb sockets, and wiring.

between each bulb. This means that each bulb drops or uses about 1/3 of the available battery voltage. In this example, the voltage drop is about 3.7 volts for each bulb. If you notice the battery voltage reading in

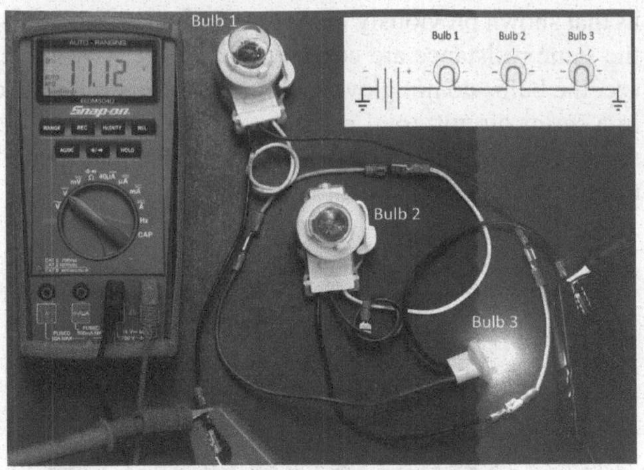

FIGURE 17-33 A three-bulb series with bulbs of different resistances. The two large (1157) bulbs have two filaments, one low (brighter light), and one high (dimmer light). You can see that circuit is complete but not functioning properly.

Figure 17-26, the meter reads 11.14 volts. Because each bulb is nearly the same resistance, each will drop close to the same voltage, in this case, 1/3 of 11.14 volts or about 3.7 volts.

The circuit shown in **Figure 17-33** is like that in Figure 17-23. The three bulbs in Figure 17-33 all have different resistances. Bulb 1 has the least resistance and bulb 3 has the most. As you can see, only bulb 3 is illuminated. There is sufficient voltage as shown on the meter, so why is the bulb with the most resistance glowing while the other bulbs are not? As discussed earlier, the high resistance bulb requires less work (electrical power) to light up. This is important to understand when trouble-shooting circuit faults and will be discussed in more detail later in the chapter.

■ ***Parallel Circuits.*** **Parallel circuits** have individual branches for each load (**Figure 17-34**). As you can see, each load has its own connection to power and ground, so that if either of the loads, in this case light bulbs, fails, the other branch of the circuit is unaffected and continues to operate. Because each bulb has its own power and ground supply, each bulb receives and uses the full battery voltage, and all bulbs can operate at full intensity.

Like series circuits, parallel circuits also have some principles on which they operate.

- There are two or more paths of current flow.
- Each branch has a voltage drop equal to the source voltage.

- The total resistance in a parallel circuit is less than the lowest branch resistance.
- The total amperage in the circuit is equal to the sum of the amperage used in each branch.
- The amperages in each branch do not have to be equal.

Unlike series circuits, the resistances of the loads in the parallel circuit are not added together. The resistance of the load determines the current flow for each branch. The sum of the current flow for all the branches is present at the common power and ground connection points (**Figure 17-35**). Because we do not add the resistances together to get total resistance, we have to use another method. For parallel circuits with two branches, the formula

$$R_t = \frac{R_1 \cdot R_2}{R_1 + R_2}$$

is used to find the resistance total. So, $\frac{2x2}{2+2} = \frac{4}{4} = 1$ or 1 ohm of resistance.

This conforms to our third principle of parallel circuits in that the resistance total is less than the lowest branch resistance. You will find that no matter how many loads are in a parallel circuit, and no matter what each load's resistance value is, the total resistance in a parallel circuit will always be less than the lowest resistance value of any of the loads. This is because as loads are added to the circuit, each requires its own power, ground, and amperage to operate. Based on Ohm's law, if amperage increases then resistance must decrease. So, even if the loads added to the circuit have high resistance, as long as the circuit is complete and the load operates, amperage will increase and resistance will decrease. If you have ever had a situation where plugging something in, such as a hair dryer or other appliance, and the circuit breaker or fuse blew for the whole circuit, you have experienced this principle in action.

To find the circuit amperage, 12 V/1 ohm gives us 12 amps. The voltage drop for each bulb is 12 V. Because the bulbs have the same resistance and are each using 12 V, both are equally bright.

Another way to solve this circuit is by going back to Ohm's law (**Figure 17-36**). 12 V / 2 ohms = 6 amps for the first branch, and 12 V / 2 ohms = 6 amps for the second branch. We can add the two amperages because they are like terms, so the total current flow is 12 amps. By dividing our 12 V by the 12 amps, we find the resistance is 1 ohm, just as when we figured the circuit out using the formula. The benefit to using the resistance formula is that it provides the resistance total even if the circuit voltage is unknown.

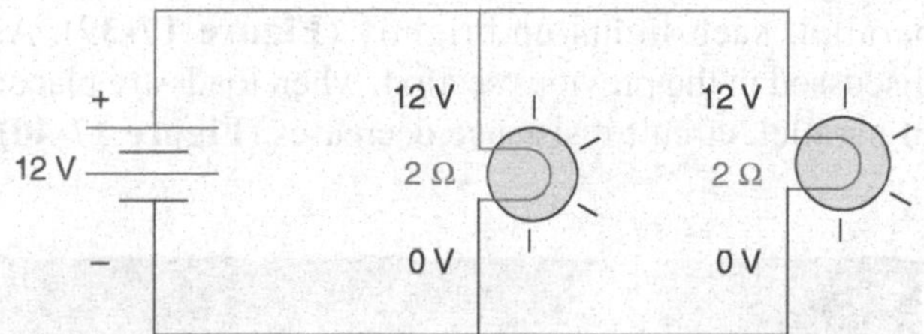

FIGURE 17-34 An example of a simple parallel circuit. Each load has its own connection to positive and negative.

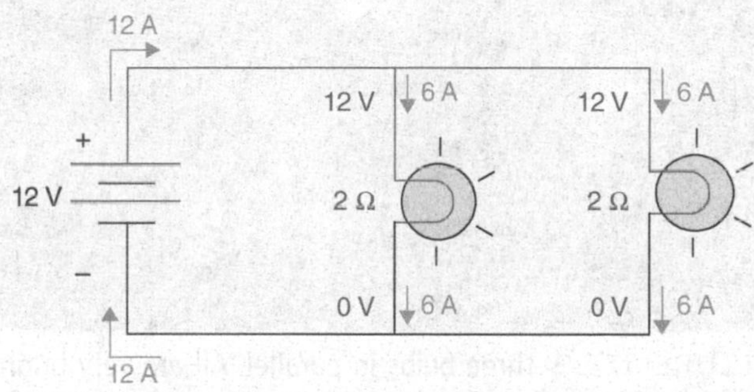

FIGURE 17-35 Parallel circuits operate differently than series circuits. Each branch receives battery voltage and total current flow is the sum of the current flows through each branch.

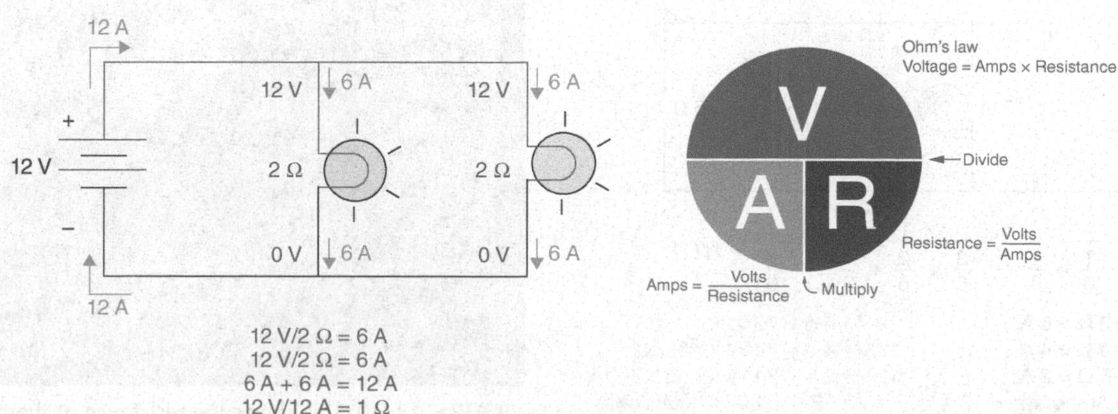

FIGURE 17-36 Using Ohm's law to understand how a parallel circuit operates. Another way in which Ohm's law can be used to determine the resistance and amperage in a parallel circuit.

For parallel circuits, which have more than two branches, we need to use a different formula (**Figure 17-37**). The total resistance can be found by using

$$R_{\text{Total}} = \frac{1}{\frac{1}{R_1} + \frac{1}{R_2} + \frac{1}{R_3}}$$

This is called *the reciprocal formula.* For the circuit in Figure 17-37, we have $\frac{1}{2} + \frac{1}{2} + \frac{1}{2} = \frac{3}{2}$ or 1.5 ohms. 1/1.5 equals 0.6667. Because the total resistance is 0.667 ohms, then the total amperage is 18 amps since 12 V/0.667 ohm equals 18 amps.

This circuit can also be solved by dividing the circuit resistances into the voltage and adding the amps for each branch together. When you are working the circuit out that way, 12 V/2 ohms equals 6 amps, 12 V/2 ohms equals 6 amps, and 12 V/2 ohms equals 6 amps. When the amperages for each branch are added together, 18 amps is the total. Doing the circuit both ways can help you find and correct any errors in your math. Because each bulb has its own path to power and ground, each receives the full 12 V.

If the bulbs are different resistances, you can use Watt's law to find which of the bulbs is the brightest (**Figure 17-38**). The 2-ohm bulb will be significantly brighter than the other two due to its much higher wattage and increased current flow.

Refer to Chapter 17 in the Lab Manual for additional exercises with parallel circuits.

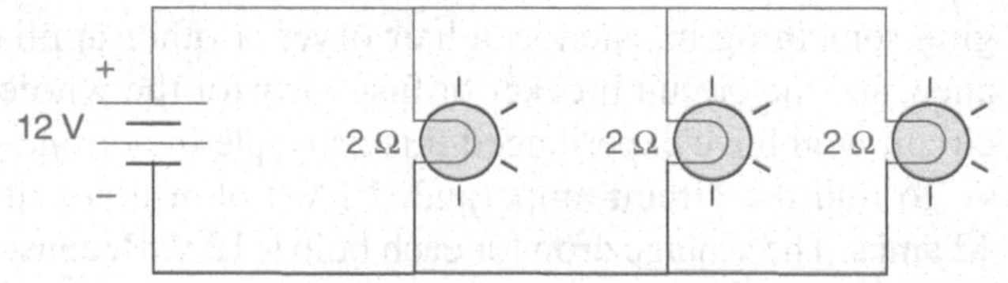

+
12 V
2 Ω 2 Ω 2 Ω
–

$$\frac{1}{\frac{1}{2} + \frac{1}{2} + \frac{1}{2}} = \frac{1}{\frac{3}{2}} \quad \frac{1}{3/2 = 1.5} = 0.667\ \Omega$$

12 V/2 Ω = 6 A 6 A + 6 A + 6 A = 18 A
12 V/2 Ω = 6 A 12 V/18 A = 0.667 Ω
12 V/2 Ω = 6 A

FIGURE 17-37 When calculating a parallel circuit with three or more branches, the reciprocal formula is used to determine total circuit resistance.

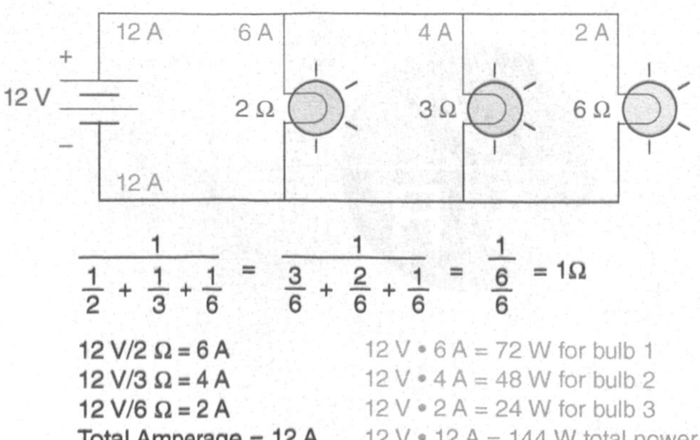

FIGURE 17-38 Using Ohm's law or the reciprocal formula to solve for resistance total. If the voltage is known, either method will work.

■ ***Applying Ohm's Law to Parallel Circuits.*** When three bulbs used in the series circuit are connected in parallel, each lights up brightly (**Figure 17-39**). As discussed in the previous section, when loads are placed in parallel, circuit resistance decreases (**Figure 17-40**).

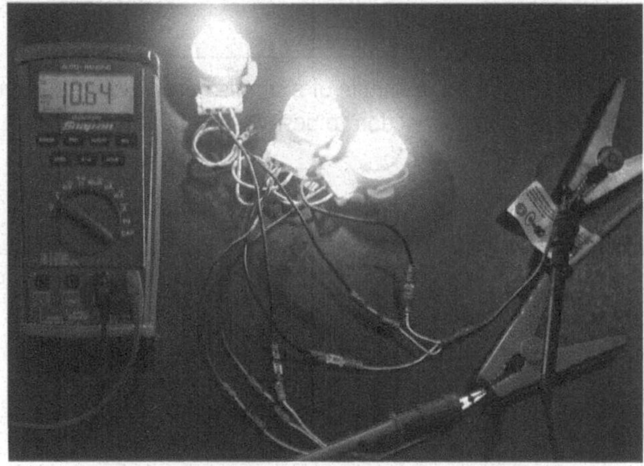

FIGURE 17-39 Three bulbs in parallel. All are very bright since each has its own power and ground.

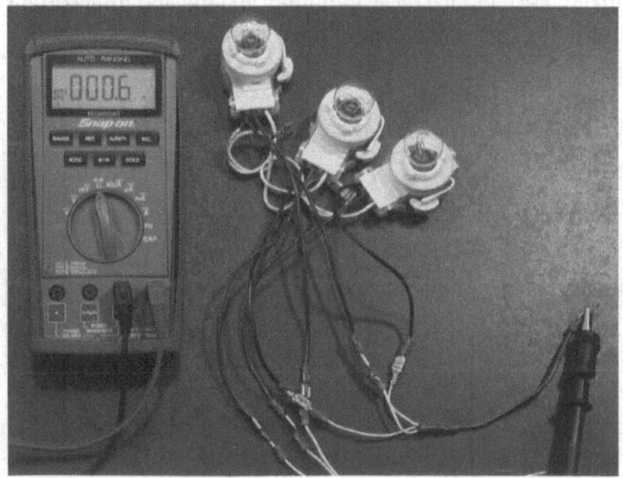

FIGURE 17-40 This shows the resistance of the three bulbs in parallel. Since each bulb has its own path for power and ground, the amperage for each bulb adds to the total amperage. As the amperage increases, the resistance decreases.

Because each bulb has its own power and ground connection, the resistance decreases and amperage increases, causing the bulbs to glow very bright.

■ ***Series–Parallel Circuits.*** You will find that automotive circuits are series–parallel circuits. This means that they contain both series and parallel circuit structure (**Figure 17-41**). Note that while the loads are not series–parallel, the components of the circuit are connected in series–parallel. In this example, the control section of the circuit, highlighted in red, is in series, and the load section is highlighted in blue. The horns are in series–parallel with the switch in the relay and the power from the battery and fuse.

Even though the loads are not in series, each section of wire, each switch, and each connection has a slight bit of resistance, so there are voltage drops occurring between the battery and the load. And while slight, all the small voltage drops add together and reduce the voltage available at the load by a small amount. This is shown in **Figure 17-42**.

There are two ways you may experience problems in series–parallel circuits; the first is shown in **Figure 17-43**. In Figure 17-43, there is an unwanted resistance in series with one of the light bulbs. This extra resistance will cause a voltage drop and decrease the current flow for that portion of the circuit, resulting in the bulb being dimmer than it should be. To solve this problem on paper, add the resistance of the bulb and the extra resistance because they are in series (**Figure 17-44**). This will give you a resistance for each branch, and the circuit can now be solved like those earlier in the chapter. **Figure 17-45** shows how the unwanted resistance affects the circuit by using voltage that should have been available for bulb 2. You could imagine this circuit representing a simplified headlight or brake light circuit. One bulb will be its normal brightness while the other bulb is dimmer than it should be due to the resistance in the circuit.

The second type of series–parallel circuit has an unwanted resistance in the ground side of the circuit after the loads (**Figure 17-46**). This type of problem will affect both bulbs because it is in the common or shared ground circuit. With this circuit, solve the parallel section for total resistance first; then it can be treated as a simple series circuit and solved accordingly. In this example, the unwanted resistance is causing a voltage drop of 6 volts. The means that the bulbs will only use 6 volts, making them quite dim.

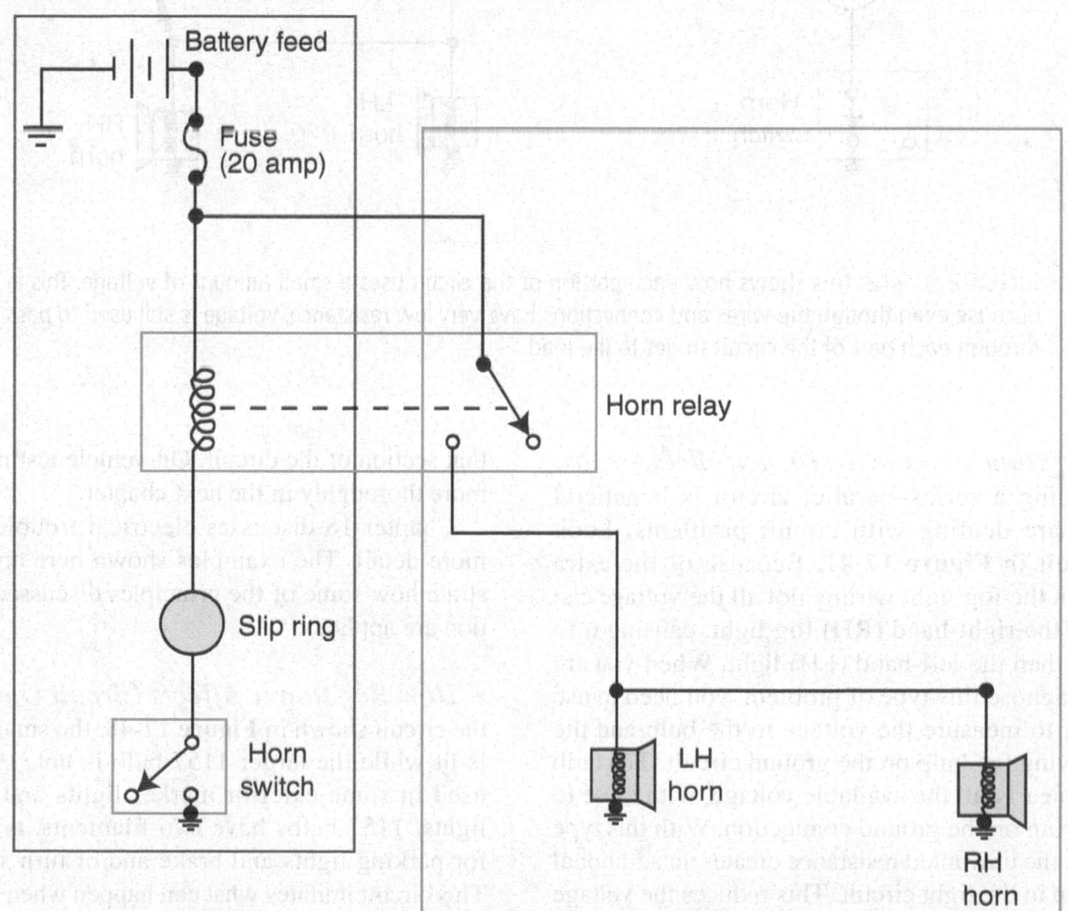

FIGURE 17-41 This illustrates how circuits contain components in series, parallel, and series–parallel.

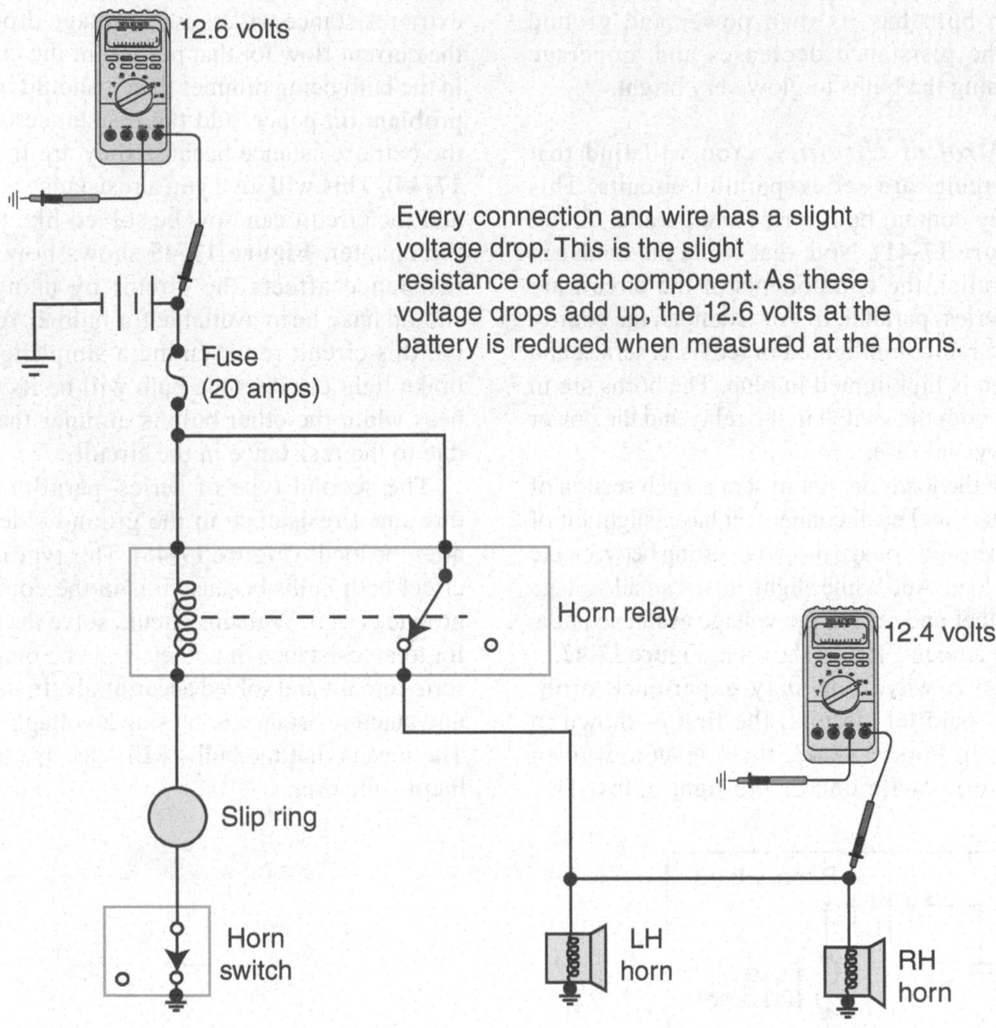
12.6 volts
Every connection and wire has a slight voltage drop. This is the slight resistance of each component. As these voltage drops add up, the 12.6 volts at the battery is reduced when measured at the horns.
Fuse (20 amps)
Horn relay
12.4 volts
Slip ring
Horn switch
LH horn
RH horn

FIGURE 17-42 This shows how each portion of the circuit uses a small amount of voltage. This is because even though the wires and connections have very low resistance, voltage is still used to pass through each part of the circuit to get to the load.

■ ***Applying Ohm's Law to Series–Parallel Circuits.*** Understanding a series–parallel circuit is beneficial when you are dealing with circuit problems. Look at the circuit in **Figure 17-47**. Because of the extra resistance in the fog light wiring, not all the voltage can be used by the right-hand (RH) fog light, causing it to be dimmer than the left-hand (LH) light. When you are trying to diagnose this type of problem, you need to use a voltmeter to measure the voltage to the bulb and the voltage leaving the bulb on the ground circuit. The bulb should use nearly all the available voltage, with close to 0 volt showing on the ground connection. With this type of problem, the unwanted resistance creates an additional voltage drop in the light circuit. This reduces the voltage available for the light and decreases the current flow in that section of the circuit. On-vehicle testing is covered more thoroughly in the next chapter.

Chapter 18 discusses electrical troubleshooting in more detail. The examples shown here are to demonstrate how some of the principles discussed in this section are applied.

■ ***How Resistance Affects Circuit Operation.*** In the circuit shown in **Figure 17-48**, the smaller 194 bulb is lit while the larger 1157 bulb is not. 194 bulbs are used in some exterior marker lights and for interior lights. 1157 bulbs have two filaments, typically used for parking lights and brake and/or turn signal lights. This circuit imitates what can happen when an unwanted resistance affects a circuit. For this example, the 1157

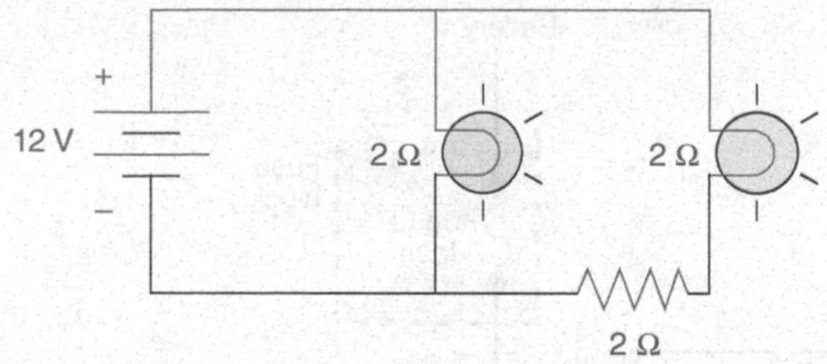

FIGURE 17-43 An example of unwanted resistance in a parallel circuit. In this situation, only one bulb is affected by the resistance.

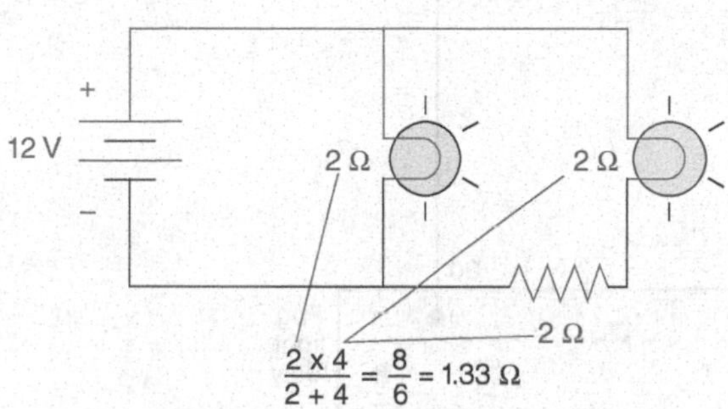

FIGURE 17-44 Calculating how the unwanted resistance affects the bulb in the circuit. Because the unwanted resistance creates a voltage drop, the bulb cannot operate correctly.

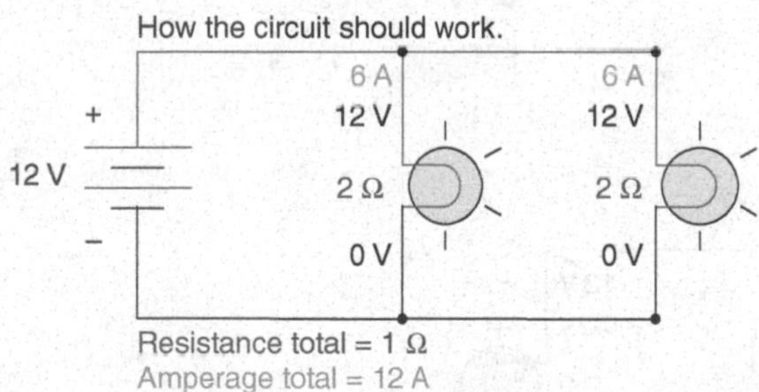

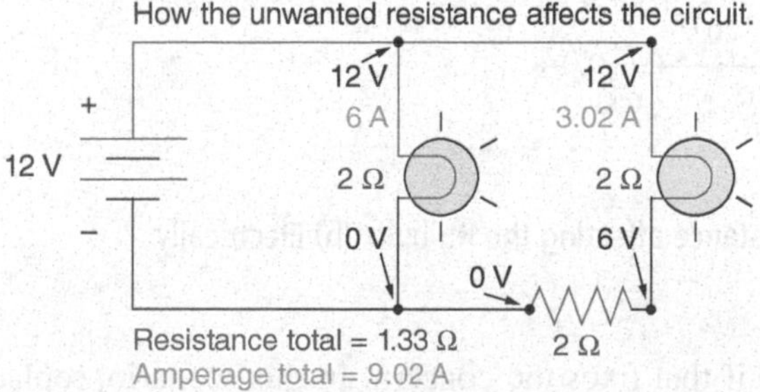

FIGURE 17-45 This shows how an unwanted resistance affects the operation of the bulb and overall circuit resistance and amperage.

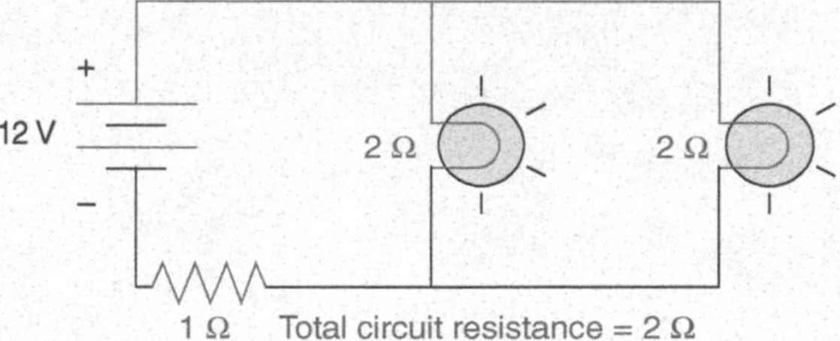

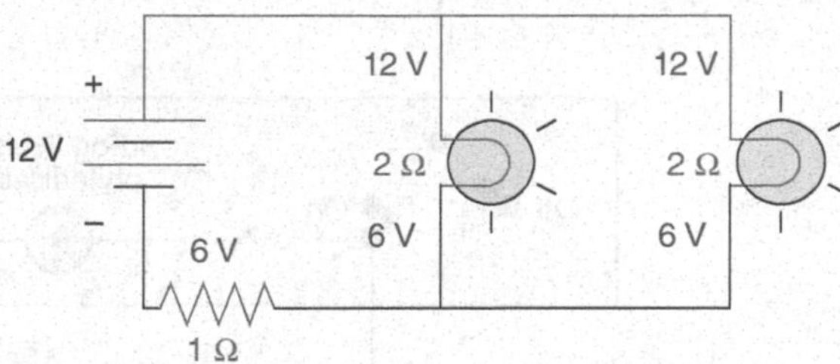

FIGURE 17-46 An example of unwanted resistance affecting an entire circuit. In this example, both bulbs are affected and will not shine as brightly as they are designed to because of the extra resistance.

bulb represents a turn signal or brake light bulb, and the 194 bulb is the unwanted resistance in the circuit, preventing the 1157 from illuminating. When faced with an inoperative bulb, most people just replace it. If this bulb is replaced, the new bulb will not work either, leaving a problem that needs to be diagnosed.

Using this circuit as an example of a real-world circuit problem, let us begin by examining the circuit and thinking about what you already know.

- Is this a complete circuit? Yes, it has to be complete for the 194 bulb to light up. Because this is a series circuit, if the circuit is incomplete, neither bulb would light up.
- Is there sufficient battery voltage? Yes, the small bulb is brightly lit, and the DMM confirms the voltage of the circuit.
- Is the 1157 bulb bad? No, if the bulb is bad (open filament), the 194 bulb would not light up. Because the bulbs are in series, an open in the 1157 opens the circuit.

Next, think about this as a problem with a vehicle. An example is shown in **Figure 17-49**. Because this is simulating a brake or turn signal circuit, you can assume for the moment that the other brake or turn signal bulb operates normally, as in Figure 17-49. The customer's concern is one brake or turn signal bulb does not light up: Your job is to understand what can cause this problem and figure out what keeps the bulb from lighting up. Possible causes include:

- A defective bulb
- Insufficient voltage and current to the bulb
- An open or incomplete ground path

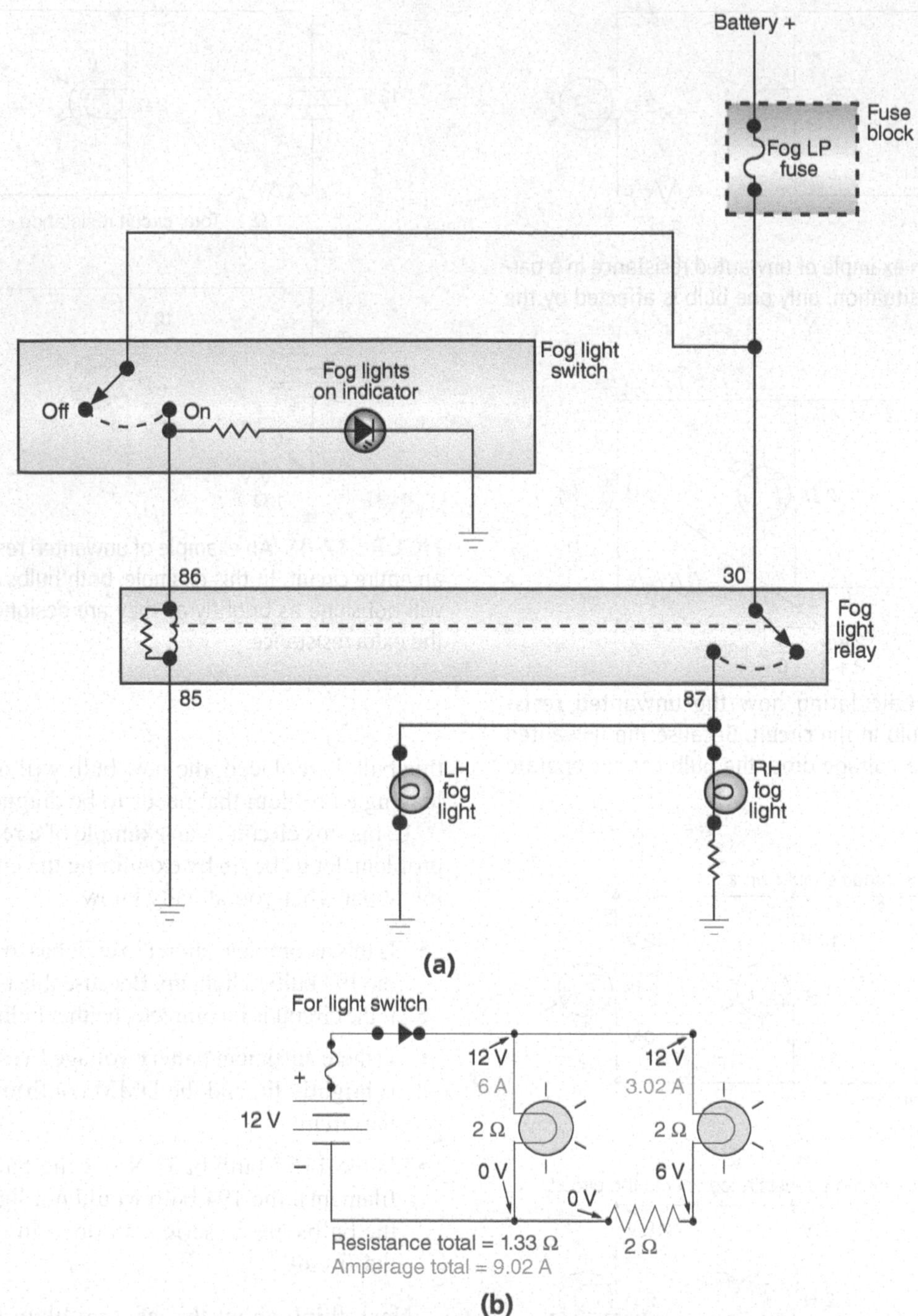

FIGURE 17-47 (a) A fog light circuit with unwanted resistance affecting the RH bulb. (b) Electrically, this is how the unwanted resistance affects the circuit.

To determine the cause of the problem, begin with testing the simplest explanation first: is the bulb defective? Can the bulb condition be tested with a meter? Yes, testing the bulb's resistance shows if the bulb is open or not. Most automotive bulbs have a resistance between 0.2 and 5 ohms. For this example, the bulb passes the resistance test.

Are voltage and ground easily tested? Yes, connecting a test light or meter provides information about the power and ground circuits.

When faced with a bulb that does not light up on a vehicle, most people will replace the bulb and check to see if that fixes the concern. In this scenario, replacing the bulb does not fix the problem. Forget for a moment that the cause of the unwanted circuit resistance (the 194 bulb) is quite obvious (and glowing brightly) in the image, and think about what you would do once you replace the nonfunctioning 1157 bulb only to have the replacement bulb also fail to work.

Figure 17-50 shows this circuit as it is electrically. Begin by checking the voltage to the 1157 bulb. Place the meter on DC volts, and connect the negative lead to a known good ground. Connect the positive lead to the

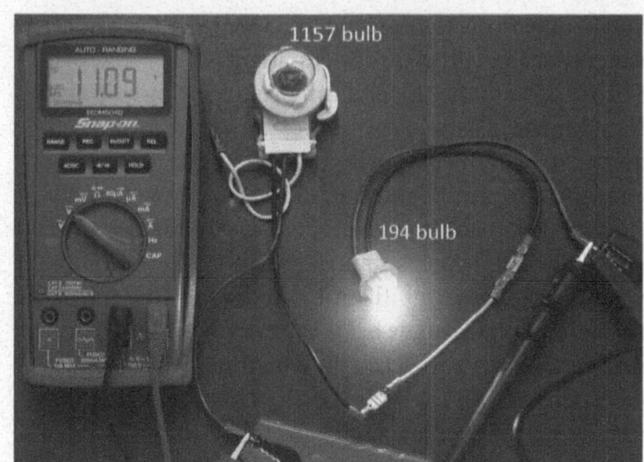

FIGURE 17-48 How an unwanted resistance can affect circuit operation. Instead of the large (1157) bulb lighting up, the small (194) bulb is using 11.09 volts, keeping the 1157 bulb from working.

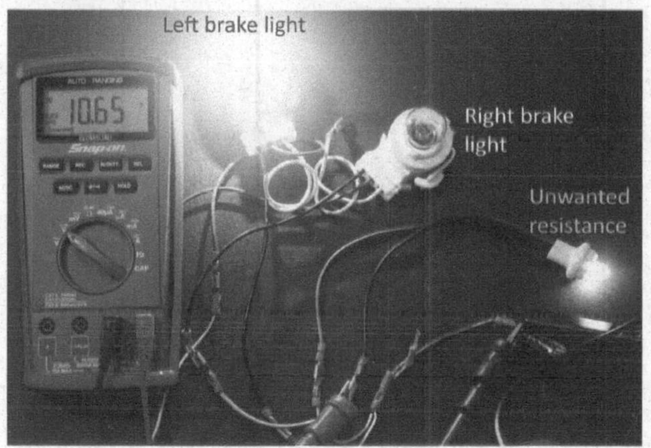

FIGURE 17-49 An example of two 1157 bulbs, acting as brake lights. The 194 is in series with one 1157 bulb and causing it not to work. If faced with a non working brake light bulb, most people will replace it. In this example, replacing the bulb will not fix the problem.

power supply wire of the 1157, and the negative lead to the bulb's ground wire (**Figure 17-51**). The meter should read very close to battery voltage, meaning the bulb is getting power and ground as it should. However, this bulb is reading 137.4 millivolts (mV) or 0.137 volt. This indicates that the circuit is complete but the voltage is very low. Because the bulb is only using 137 mV, it cannot glow brightly as it should. The meter reading shows that there is a problem with the circuit that is consuming voltage that should be used by the 1157 bulb.

An unwanted resistance can affect the whole circuit if it is in either the power supply or common ground circuit (**Figure 17-52**). The circuit in Figure 17-52 could be two headlights or brake lights, neither of which light up because of excessive resistance in the circuit. The meter in this image shows that the two 1157 bulbs are only using 0.133 volt.

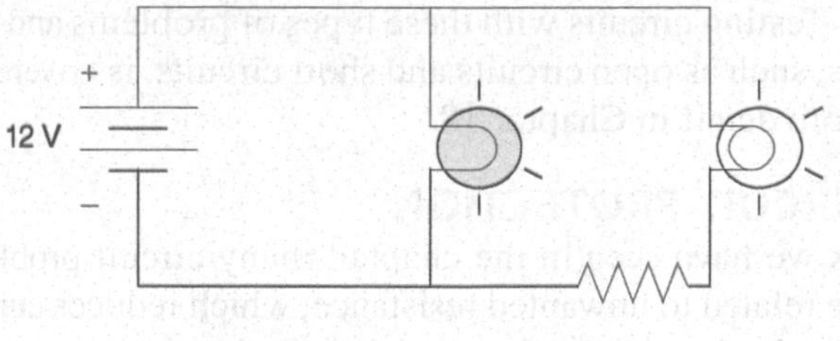

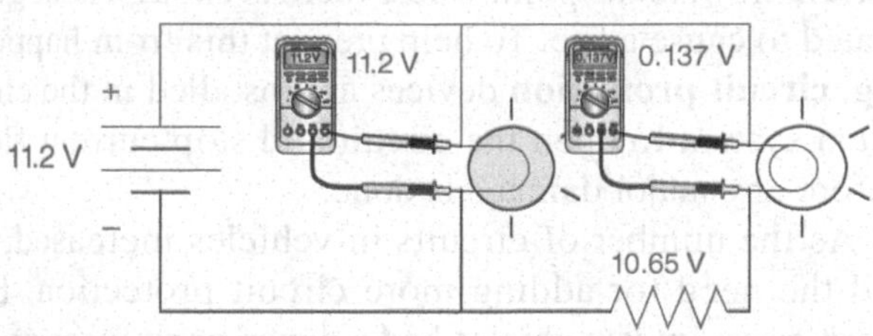

FIGURE 17-50 This shows the previous circuit as it is electrically.

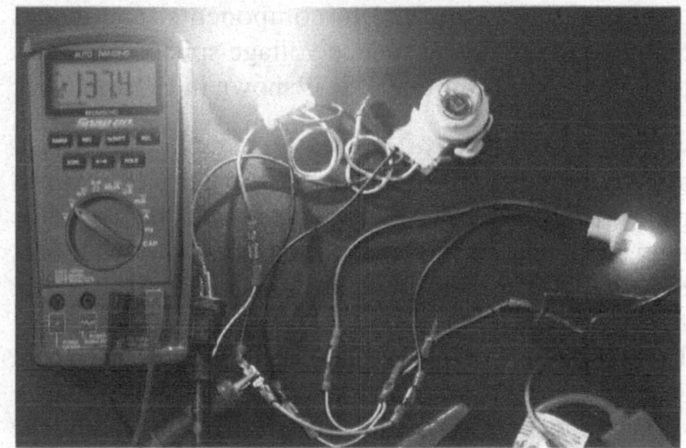
FIGURE 17-51 The 1157 bulb is only receiving 137 mV due to the resistance in the circuit.

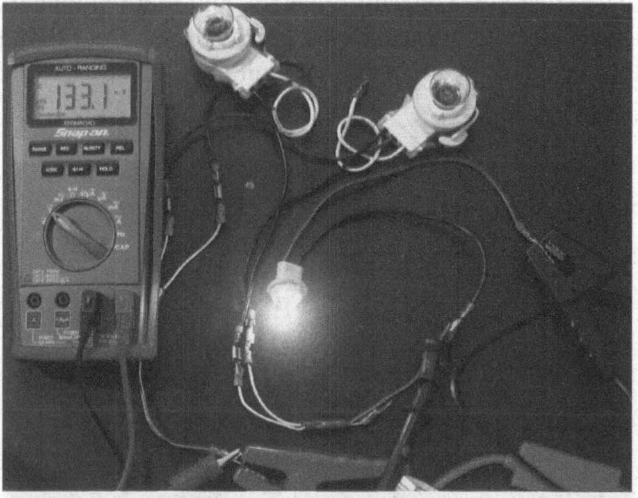
FIGURE 17-52 An unwanted resistance in the power or ground circuits can affect the entire circuit. Here, neither bulb lights up because of the unwanted resistance of the 194 bulb.

Testing circuits with these types of problems and others, such as open circuits and short circuits, is covered in more detail in Chapter 18.

CIRCUIT PROTECTION

As we have seen in the chapter, many circuit problems are related to unwanted resistance, which reduces current flow in the circuit. Some circuit faults cause increased current flow, to the point where there is enough heat generated to cause a fire. To help prevent this from happening, **circuit protection** devices are installed in the electrical system to open the circuit and stop current flow before any major damage is done.

As the number of circuits in vehicles increased, so did the need for adding more circuit protection. For many years, most vehicles had a dozen or so fuses and a couple of circuit breakers, depending on the options that were installed. Modern vehicles commonly have several fuse boxes and 30-plus fuses, as well as inline fuses, circuit breakers internal to components, and diodes installed in circuits to reduce voltage spikes. An example of a typical fuse/relay box is shown in **Figure 17-53**.

■ ***Why Do Circuits Need Protection?*** An average-size house may have supplied to it 150 amps of electrical power at 220 volts. The average car or truck battery, while only supplying 12 volts, has the capacity to provide 600, 800, or 1,000 or more amps. In terms of electrical power, our example house has about 33,000 watts of power, and a 1,000-amp car battery has 12,000 watts of power, or about 36 percent of the power found in a typical house.

If a power circuit accidently grounds, bypassing the load in the circuit, the current flows through the wiring to the point of ground can cause so much heat that the wire can turn red hot, igniting the insulation and anything near the insulation. The circuit protection components, fuses, fuse links, circuit breakers, and thermal limiters are sensitive to the heat developed by the current flowing through the circuit. When enough heat is generated by the current flow, the fuse melts, opening the circuit and protecting the vehicle from damage.

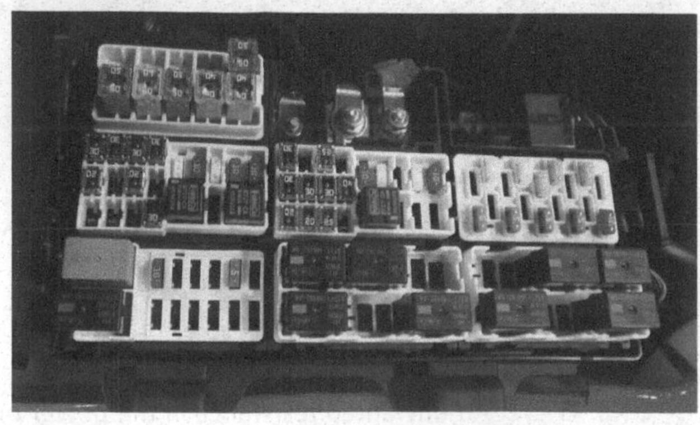

FIGURE 17-53 An example of a fuse/relay or junction box. Many vehicles have more than one fuse/relay box.

To illustrate the use and importance of circuit protection, **Figure 17-54** shows an example of a brake light circuit in which the positive wire to one of the lights shorts to ground. **Figure 17-55** shows how this direct short of power to ground affects the circuit. Because the short has such low resistance, the amperage through the circuit increases significantly. With the fuse in place, the increased amperage heats the circuit until the thermal limit of the fuse is reached, at which point the fuse melts and opens the circuit. Current flow stops and the circuit goes dead, preventing damage to the wiring and vehicle. If the fuse is bypassed or is not present in the circuit, the result can be catastrophic because the wiring will overheat

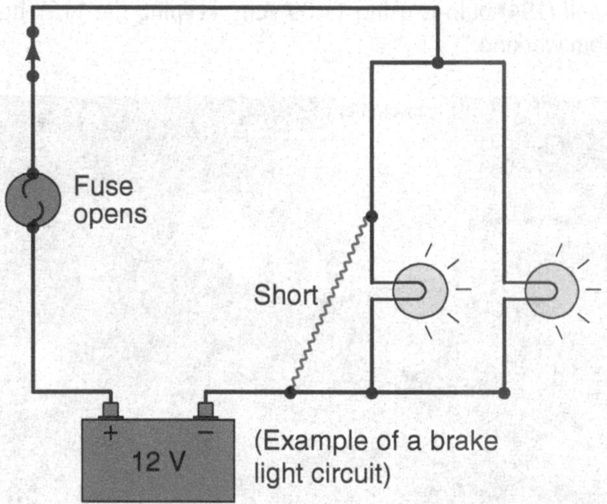

FIGURE 17-54 Without a fuse or other form of circuit protection, the short to ground will overheat the circuit. Current will continue to flow until the weakest part of the circuit melts. Hopefully, the weakest part is the fuse.

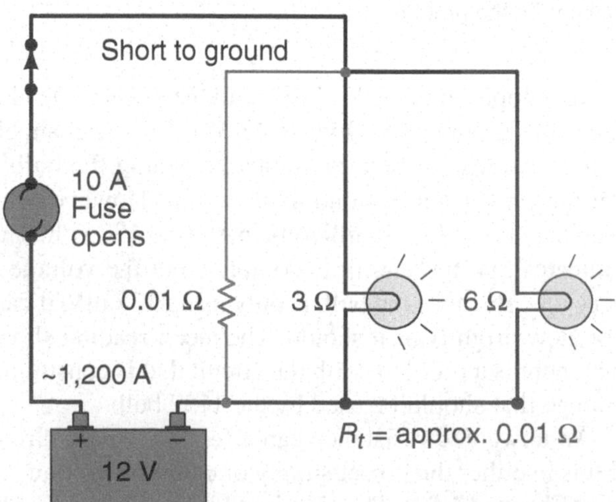

FIGURE 17-55 Because the short to ground has very low resistance, current flow is greatly increased. This overheats the fuse and causes it to burn open.

from the current flow (**Figure 17-56**). The circuit will not open until the weakest point melts, by which time it may be too late and the vehicle may be on fire.

TYPES OF CIRCUIT PROTECTION

As vehicles become more electronic and complex, the methods of protecting the systems must also evolve. Modern cars and trucks typically have several fuse/junction boxes and dozens of fuses. In addition to fuses, there are other methods of protecting electrical circuits.

■ ***Fuses.*** The most common automotive circuit protection device is the **fuse**. There are several types of fuses (**Figure 17-57**). Most vehicles use plastic blade fuses that plug into a fuse box, like those shown in Figure 17-56. The element in the fuse melts when the current flow in the circuit causes excessive heat. In a circuit protected by a 20-amp fuse, the expected current flow is about 15 to 18 amps. The fuse, being rated at 20 amps, provides a slight buffer for small surges in current flow that can occur when some components are turned on or off.

Fuses are rated for their amperage capacity (**Figure 17-58**). The ratings, such as 15 A, 10 A, and others, indicate the amperage rating for the fuse and its circuit. For blade fuses, certain colors are used to distinguish fuses by their amperage ratings. This fuse box cover correlates the fuse amperage rating with the color. This helps ensure that the correct fuse is installed for each circuit. As with many components, the blade fuse has downsized and is called *a mini fuse*. There are several types of blade fuses in use: standard size, mini, low-profile mini, micro, multi-circuit, and maxi fuses that are used on high-current circuits.

■ ***Circuit Breakers.*** Like fuses, **circuit breakers** are used to open a circuit with excessive current flow (**Figure 17-59**). Unlike fuses, circuit breakers reset after

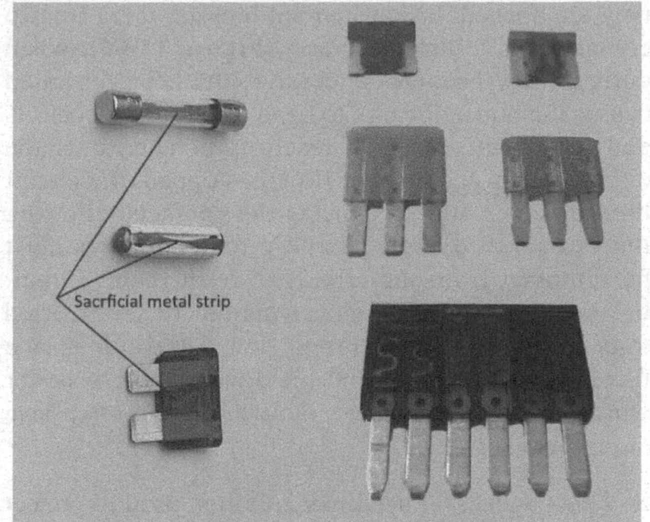

FIGURE 17-57 Examples of different types of automotive fuses. Multi-circuit fuses are becoming more common.

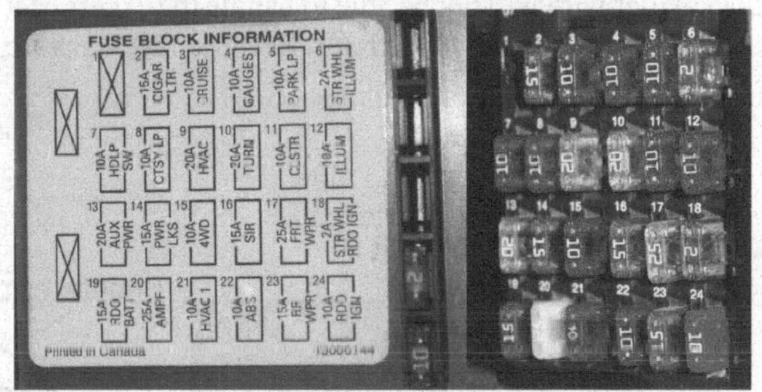

FIGURE 17-58 An example of a fuse box and the fuse legend. Both the fuses and the legend show the amperage rating of the fuse. Colors of fuses are standardized to their amperage rating. For example, red fuses are rated for 10 amps and blue fuses are rated for 15.

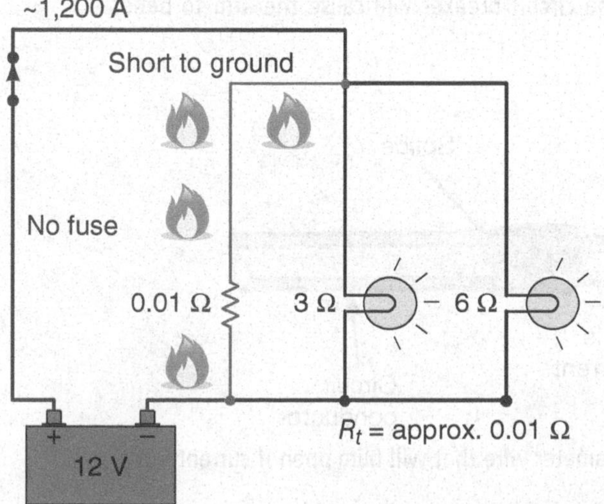

FIGURE 17-56 Without a fuse or other circuit protection, a short can overheat the wiring and cause a fire.

FIGURE 17-59 An automotive circuit breaker. Like fuses, circuit breakers are rated for their amperage capacity.

they are opened. Inside a circuit breaker, there are two contacts and a bimetallic arm (**Figure 17-60**). When current flow becomes excessive, the heat generated causes the bimetallic arm to bend, opening the contacts and the circuit. In a self-resetting breaker, once the circuit opens and current flow is stopped, the circuit breaker cools and reconnects the contacts, allowing the circuit to operate. Some circuit breakers must be removed from the circuit to reset them. Circuit breakers are used on circuits where there are expected to be sudden surges in current flow that would open a fuse. Circuits such as windshield wiper motors, power window motors, and other motorized circuits often have circuit breakers.

■ ***Fuse Links.*** Fuse links are often used to protect main power supply wiring, such as that from a power distribution point on the starter motor. A fuse link is a smaller gauge wire than the wiring used in the rest of the circuit (**Figure 17-61**). If excessive current flow occurs, the smaller wire will not be able to handle the excess heat and will melt, opening the circuit. Because fuse links are built into the wiring of the circuit, when one fails, it must be cut out and a new link soldered into the wiring. For this reason, many newer vehicles use fuse links that are essentially fuses (**Figure 17-62**).

■ ***PTC Circuit Breakers.*** Some vehicles have automatically resetting circuit breakers (**Figure 17-63**). There are solid-state devices, which means that they have no moving parts. These devices, in their normal state, allow current to flow through by having a conductive path through the carbon particles. When excessive current flows through the device, the particles heat and expand, opening the circuit. Once the carbon particles cool, they contract and the circuit will operate. This type of circuit breaker is often used on power window and power door lock circuits.

■ ***Diodes.*** A **diode** is also a type of solid-state electronic device. Diodes allow current to flow in one direction and are used to limit voltage spikes and reduce electronic "noise." Diodes are discussed in more detail in the Electronics section later in the chapter.

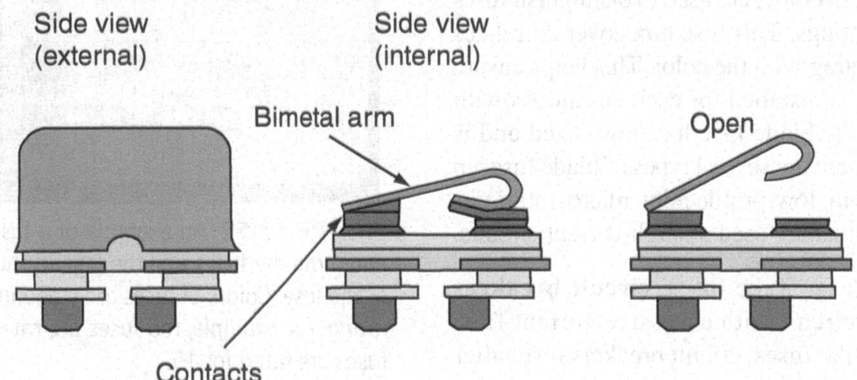

FIGURE 17-60 Excessive current flow through the circuit breaker will cause the arm to bend and open the contacts and the circuit.

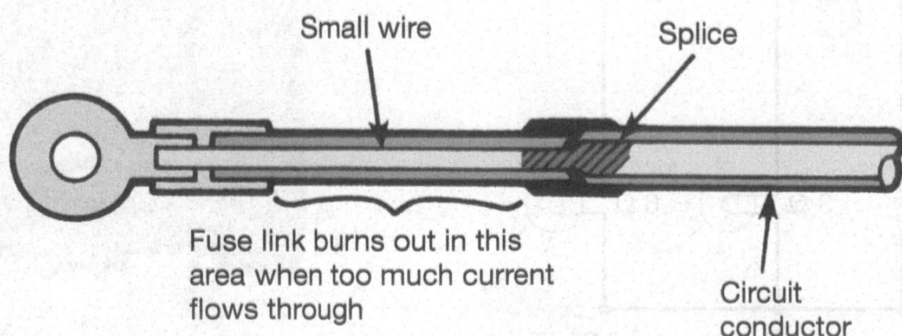

FIGURE 17-61 A fuse link is a section of smaller diameter wire that will burn open if current flow becomes excessive.

RESULTS OF IMPROPER CIRCUIT PROTECTION

As stated above, an unprotected circuit can, if grounded, allow enough current to flow that the wiring becomes hot enough to catch fire. Circuits and wiring are designed to be able to safely carry a certain amount of amperage. The smaller the size of the wire is, the less amperage it can carry safely. If a circuit is protected by a 10-amp fuse, the wiring will be just large enough to handle the 10-amp load. If the 10-amp fuse is replaced with a larger fuse, such as a 20-amp fuse, and a faulty component draws more than the 10 amps the wiring can handle, the wiring can overheat and catch fire. Never replace a fuse with one of a higher rating. If a fuse keeps blowing, it indicates there is a problem in the circuit causing excessive current flow. The fault must be isolated and corrected.

A common mistake made by some is to not use any type of circuit protection when adding an electrical accessory. Running power and ground wiring straight to battery power without installing any type of circuit protection is an invitation to problems. Without a fuse or other circuit protection device installed, an accidental grounding of the circuit before the load will cause the wiring to overheat and can lead to a fire. **Figure 17-64** shows an example of a stereo amplifier ruined by an unprotected circuit.

Additionally, when a fuse with too high amp rating is installed in a circuit, and a fault develops, the increased current flow can damage sensitive electronics before the fuse opens. This is because the fuse cannot open until current flows through the circuit. For current to flow through the circuit, there must be a load of some sort, drawing power. If the load is an electronic module of some type and is installed in a circuit rated for 10 amps and a 20-amp fuse is installed, much more current can flow before the fuse opens.

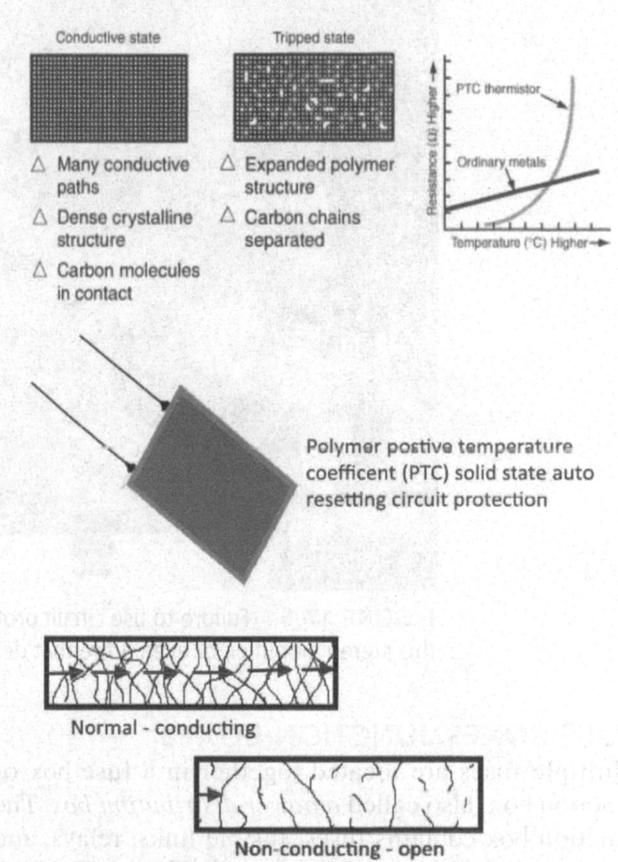

FIGURE 17-63 An illustration of a PTC circuit breaker and how it works.

FIGURE 17-62 Wire fuse links have largely been replaced by fuse links shown here.

FIGURE 17-64 Failure to use circuit protection can cause damage to components, such as this stereo amplifier, or even a fire that destroys the vehicle.

FUSE BOXES/JUNCTION BOXES

Multiple fuses are located together in a fuse box or junction box, also called *a power distribution box*. The junction box contains fuses, fusible links, relays, and may contain circuit breakers as well (**Figure 17-65**). A vehicle can have several fuse/junction boxes located in various places throughout the vehicle. Common locations are under the hood, in or under the driver's side of the dash, in or under the passenger side of the dash, along the center console, or in or behind the glove box. If the vehicle's battery is under the back seat, in the trunk, or other location not under the hood, there is often a fuse or junction box close to the battery, requiring you to figure out where all the junction boxes are located.

AUTOMOTIVE WIRING

The wiring in today's cars and trucks is more than just a lot of copper. In addition to the wire itself, there are hundreds of connections, thousands of terminals, and the plastic insulation used to protect the wiring from contact with the metal components of the vehicle. An example of just the visible underhood wiring and connectors on a late-model vehicle is shown in **Figure 17-66**. The average automobile contains several thousand feet of wire, adding up to 50 pounds or more of the vehicle's weight. Manufacturers are always looking for ways to reduce manufacturing costs and vehicle weight, and one way they have done both is by reducing the amount of wiring needed by each vehicle. However, the number of electrical devices continues to increase as more systems are networked and mechanical systems are replaced by electronic systems.

Compared to the wiring found in homes and businesses, automotive wiring has a tougher job. This is because the vehicle is subject to constant vibration, twisting and flexing, and extreme temperature changes.

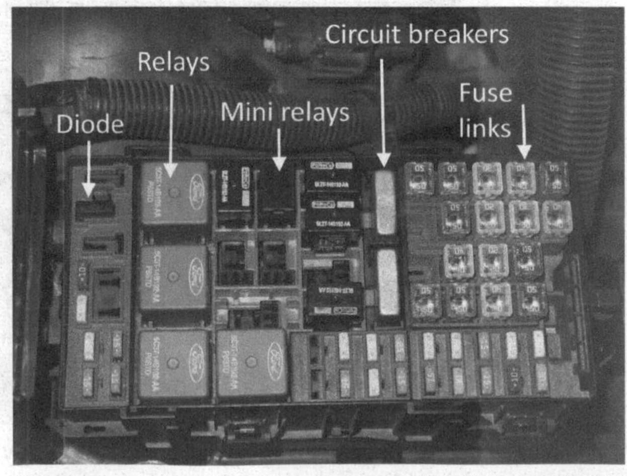

FIGURE 17-65 The many components inside a fuse/relay box.

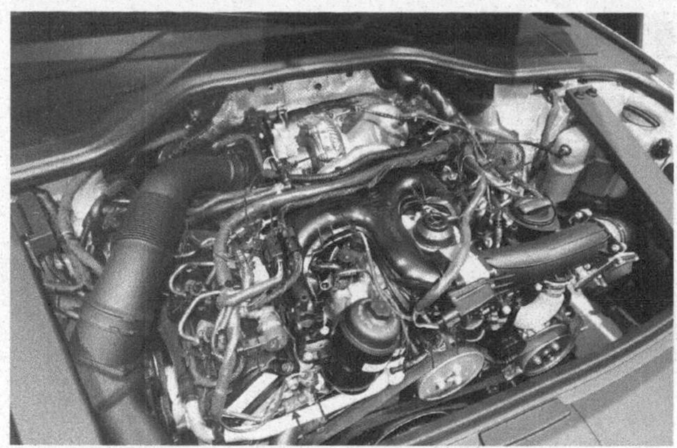

FIGURE 17-66 There are many connectors under the hood. These are just some of those that are visible.

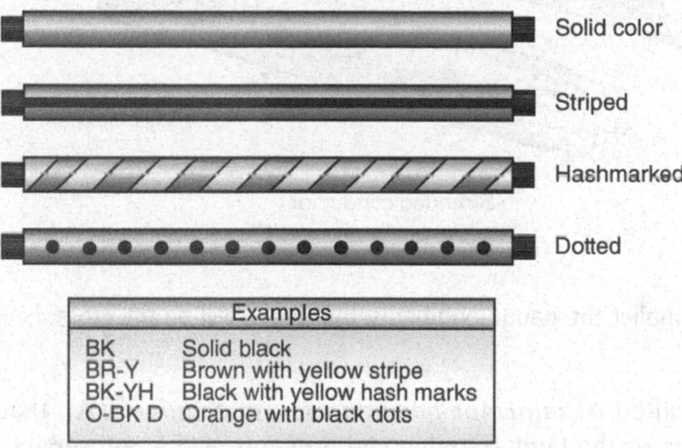

FIGURE 17-67 Wire colors, stripes, dots, and lines are used to identify one wire from another.

To be able to survive the harsh conditions found in car and truck electrical systems, the wiring has to be flexible and protected.

■ ***Automotive Wire.*** Automotive wire is made of copper strands. Stranded wire provides greater flexibility than solid wire, as used in residential and commercial constructions. The wire is covered in a synthetic rubber and plastic coating, called insulation. The insulation prevents the wire from touching other wires or metal parts. The insulation is also painted in various colors, with differently colored stripes or dots, so that individual wires in a bundle, called a harness, can be identified (**Figure 17-67**). Some newer vehicles use wiring that is all the same color, typically white, and only the ends have identification marks. This is likely to become more common as manufacturers look for ways to decrease manufacturing costs.

The size or diameter of the wire used in a circuit depends on the amount of current that flows through the circuit. To keep cost and weight down, the wire will only be large enough in diameter to handle the current load of the circuit. As wire diameter decreases, the resistance increases. As wire length increases, so does resistance. For this reason, only enough wire needed to connect the circuit without excessive voltage drop and heat loss is used.

Wire size, or **wire gauge**, refers to the cross-sectional size of a wire. There are two gauge systems commonly used, the American Wire Gauge (AWG) and metric sizing. AWG gauges go from Number 36 to 0000 (**Figure 17-68**). In automotive applications, you will often see wiring from 12 to 24 gauge. Battery cables are typically 00 to 04 gauge. Many manufacturers use metric wiring sizing, which refers to the diameter of the wire in millimeters. Regardless of which sizing method is used on a vehicle, it is important that if a wire is repaired or replaced, the correct size replacement is used. Installing a wire that is smaller in diameter than the original increases the circuit resistance and decreases the current-carrying capacity.

■ ***Harnesses.*** A **wiring harness** is bundle of wires and connectors (**Figure 17-69**). A harness typically has many connections to components and is used for an entire system or section of the vehicle. For example, an engine wiring harness includes all the wiring and connectors to connect the engine control module to the sensors, fuel injectors, and other components of the engine management system. Harnesses are covered in protective plastic conduit to help prevent damage from sharp pieces of metal and moving parts.

Some harnesses, like those used on wheel speed sensors, crankshaft position sensors, and computer communication data lines, use a twisted-pair wire harness (**Figure 17-70**). This type of wiring helps prevent electrical interference from other wires and radio waves.

■ ***Connectors.*** **Connectors** are used to connect or attach the wiring to components and to other sections of wire. Connectors come in many different types and styles and are either male or female (**Figure 17-71**). Connectors used inside the vehicle are different than those used outside the vehicle. This is because connectors outside the vehicle are exposed to more extreme weather conditions, water, snow, road debris, and other hazards.

All connectors use a latch mechanism to hold the two parts of the connector together (**Figure 17-72**). Some connectors, like those on fuel pumps, air bags, and other critical circuits, use a second retaining tab,

American Wire Gauge Sizes	
Gauge Size	**Conductor Diameter (inches)**
20	0.032
18	0.040
16	0.051
14	0.064
12	0.081
10	0.102
8	0.128
6	0.162
4	0.204
2	0.258
1	0.289
0	0.325
2/0	0.365
4/0	0.460

Metric Size (mm^2)	AWG (Gauge) Size	Ampere Capacity
0.5	20	4
0.8	18	6
1.0	16	8
2.0	14	15
3.0	12	20
5.0	10	30
8.0	8	40
13.0	6	50
19.0	4	60

PCV insulation

Solid conductor

Cross-section size (gauge)

PVC insulation

Stranded conductor

FIGURE 17-68 Wire gauge indicates the size of wire. The smaller the gauge number is, the larger will be the cross section of the wire.

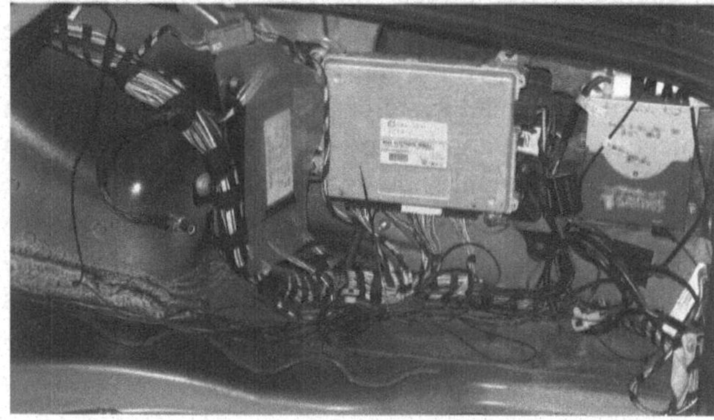

FIGURE 17-69 An example of a rear wiring harness.

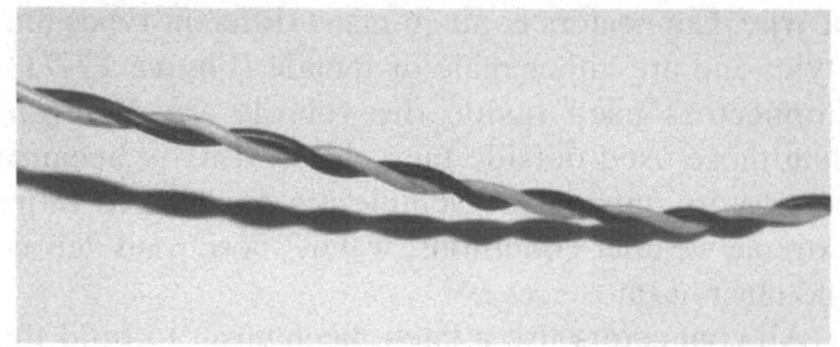

FIGURE 17-70 Twisted-pair wire is common for permanent magnet sensors and computer data bus wiring. Twisting the pair helps prevent stray electromagnetic interference from altering the signal.

called *a connector position assurance tab (CPA)*, that holds the latch together and prevents accidental separation (**Figure 17-73**).

To disconnect a connector, first find the latch. Next, carefully pull the latch away from the tang slightly, and pull the two halves of the connector apart. It should not require very much effort to pull the latch and separate the connector; if it does, check the connector for signs of damage or melting. When you are reconnecting the connector, first align the two halves correctly. Most connectors only have one way in which they will fit together. Next, carefully push the two pieces together until they fully seat. There should be a small click as the parts seat and the latch grabs onto the tang. Once they are together, give the two halves a slight tug to make sure they are fully latched and secure.

While most connections on the vehicle use a connector as discussed here, a few components, such as the starter motor and generator, often have bolt-on connections (**Figure 17-74**). These typically are high-current-draw connections that require more force to maintain the connection than what is provided by a plastic connector. System grounds are also bolted in place, to the body and engine, to make sure ground circuits are complete and held tight.

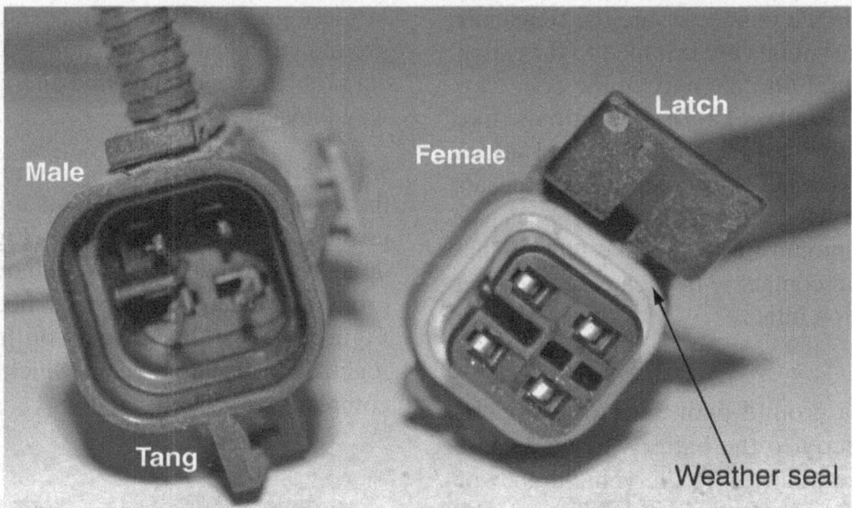

FIGURE 17-71 A male and female connector and terminals. All connectors use some type of latch to keep the connector secured together.

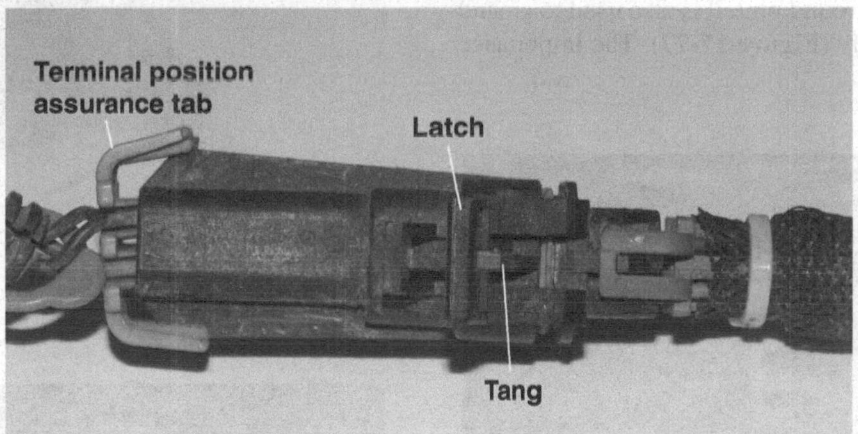

FIGURE 17-72 The outside of a connector. To disconnect, gently pull the latch until it clears the tang.

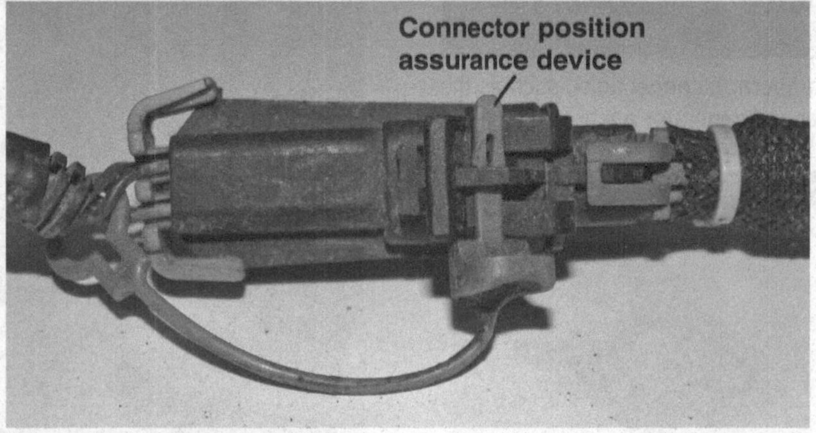

FIGURE 17-73 Some connections use assurance locks to prevent accidental disconnections.

■ ***Terminals.*** Inside the connector are the **terminals** (**Figure 17-75**). The terminals are crimped to the end of the wire and are secured into a cavity in the connector by a terminal latch (**Figure 17-76**). Terminals, like connectors, come in a wide variety of sizes and shapes and are either male or female. Like connectors, terminals have a specific way in which they fit together. Each pair of terminals has an amount of spring-like tension that keeps the two pieces in contact with each other, keeping the electrical connection intact.

GROUNDS

Every circuit needs a ground path, but not every circuit is grounded directly to the battery. Instead, there are several grounds in addition to the vehicle body or frame, which act as a ground path for the entire vehicle. This is often called **chassis ground**. The negative battery cable usually grounds to the engine block. The same cable may also connect to the vehicle body or frame or a second cable may be used to link the engine to the body. Often a small ground pigtail is also used to ground the battery to the body (**Figure 17-77**). The importance of these grounds cannot be overstressed. The loss of a ground strap can lead to strange and confusing electrical problems as circuits find alternate ground paths any way they can.

■ ***Common Grounds.*** Common grounds combine the grounds for several circuits together; these are then grounded to a common spot (**Figure 17-78**). This is done to reduce wiring and save money. Another common grounding method is using a splice pack, also shown in Figure 17-78. This connection has multiple grounds together in one connector, which bolts to the body to complete the ground.

FIGURE 17-74 High-amperage connections, such as the battery cables at the starter and generator, are bolted together.

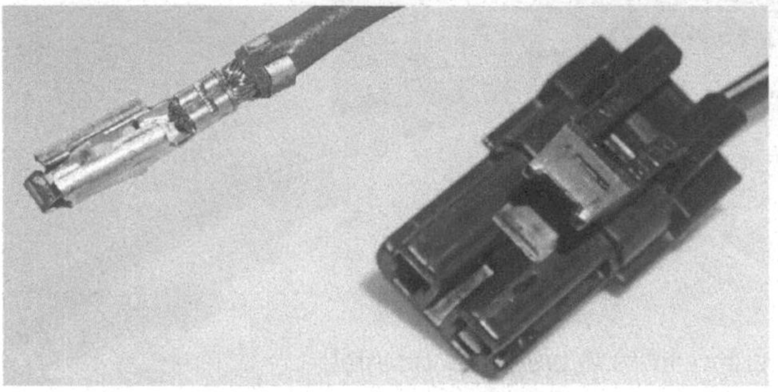

FIGURE 17-75 A female wire terminal.

FIGURE 17-76 The latch that keeps the terminal inside the connector.

FIGURE 17-77 A battery-to-body ground.

Common grounds, if removed or damaged, can cause problems for several circuits at once. If you are faced with multiple, seemingly unrelated circuits showing operating problems, refer to a wiring diagram to see if the circuits share a common ground location.

■ ***Floating Grounds.*** Some circuits, particularly computer sensor circuits, often use floating or isolated grounds. These ground circuits are typically shown as reference or sensor low in wiring diagrams (**Figure 17-79**). In these circuits, the ground is inside of the computer. The reference low circuit floats above chassis ground by passing through a fixed resistance in the computer. The computer uses the fixed ground reference voltage so it is not affected by noise generated by other, often high-current circuits. For example, inside of the computer, the reference low circuit "ground" voltage may be 0.75 V (750 mV) above chassis ground voltage. The computer can then compare the sensor signal high, which may be a 5-V to 12-V signal, to the 0.75-V low reference.

■ ***Electrical Noise and Electromagnetic Interference.*** Because the number of electrical components and systems continues to increase, so does the problem of **electrical noise** and interference. Electrical noise is the unwanted fluctuation or changing of the voltage in a circuit. Electrical interference is typically the result of unwanted electrical signals, voltages, or current flowing from one circuit to another. For example, whenever a relay or motor turns on or off, noise and/or interference is generated in the circuits connected to the devices. Most of the time, the amount of noise or interference is so small that it is not a problem. **Figure 17-80** shows a lab scope image of a power window motor circuit in operation. The lab scope is like a DMM except that it can measure and display voltages from very small fractions of a second. The arrows in the image point to areas where small spikes in voltage and current occur when the window turns off and on. Left on their own, these voltage spikes can feed back into the electrical system and cause problems with computer communication signals. Much of the noise created is by electromagnetic components, such as relays and motors.

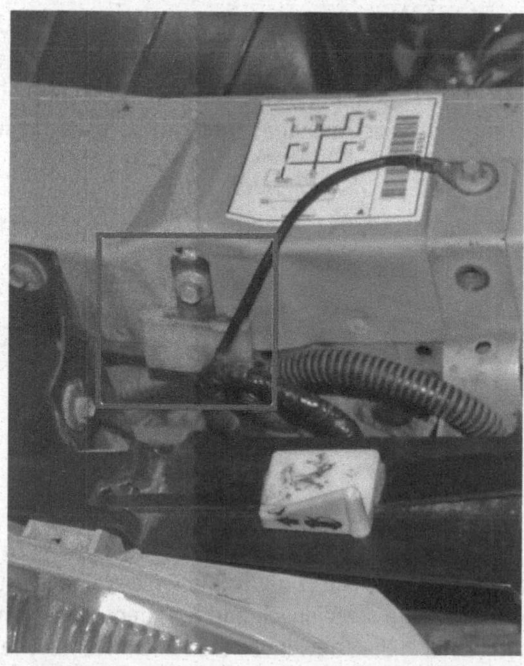

FIGURE 17-78 A common ground through a splice pack.

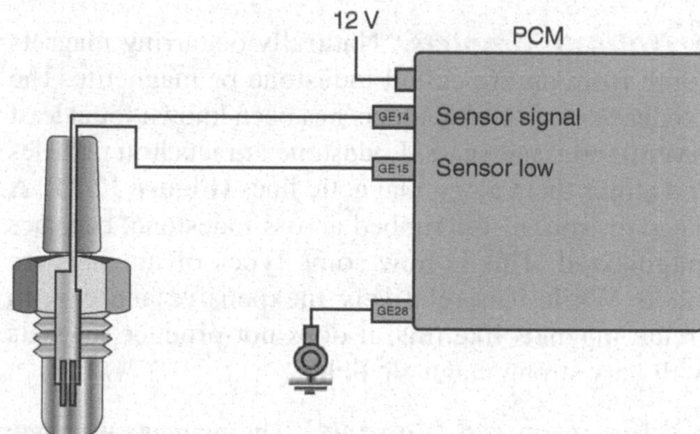

FIGURE 17-79 Some systems use floating grounds. These grounds are not connected to the chassis or body ground and are used as a low reference for the computer.

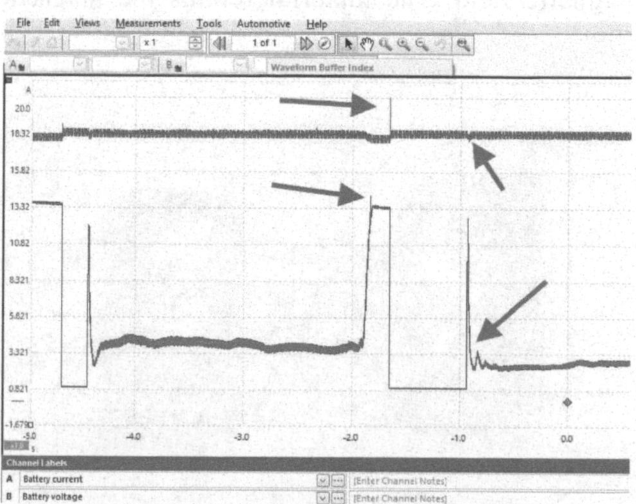

FIGURE 17-80 This scope pattern shows voltage spikes as the power window motor turns off and on.

Electromagnetism and the Vehicle

Electromagnetism is the term used to discuss the whole of electricity and magnetism. This is because electricity and magnetism are interconnected and inseparable; magnetism can be used to generate electricity, and the flow of current through a conductor generates a magnetic field around the conductor. Our modern world would not exist without having harnessed electromagnetism.

MAGNETS

There are two types of magnets, natural and manufactured. Natural magnets are not commonly used in automotive applications. Manufactured magnets, and the ability to make them very powerful yet very small, have revolutionized the modern world. Without these small, powerful magnets, many of the devices we consider common, such as personal computers, smartphones, and ear buds, would not exist.

■ ***Natural Magnets.*** Naturally occurring magnets come from an ore called lodestone or magnetite. The existence of natural magnets has been known for at least several thousand years. Lodestone attracts iron particles and aligns them along magnetic lines (**Figure 17-81**). A piece of iron, if it is rubbed across lodestone, becomes magnetized. This is how some types of magnets are made. While it is relatively inexpensive and easy to create magnets like this, it does not produce magnets with very strong magnetic fields.

■ ***Manufactured Magnets.*** The magnets we come into contact with are manufactured magnets. Magnets are made from various materials, based on the requirements of the strength of the magnetic fields. Once the basic structure and shape of the magnet are formed, it is subjected to the sudden application of a high-strength magnetic field. The material retains the magnetic properties and then can be used in motors, speakers, sensors, computer hard drives, and many other devices.

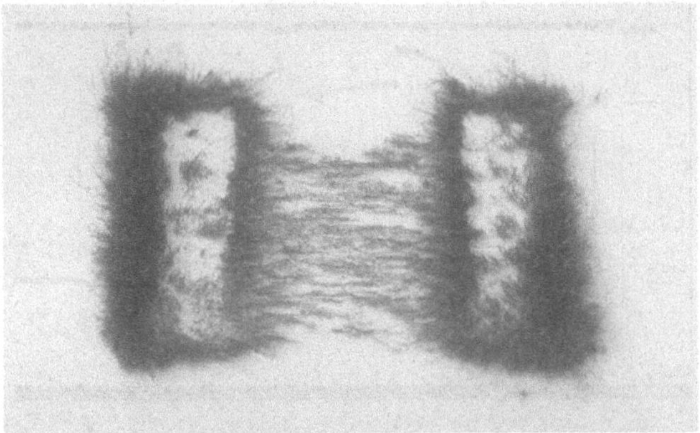

FIGURE 17-81 A magnet has a north and south pole, from which magnetic lines of force, called flux lines, originate.

■ ***Electromagnets.*** Electromagnets are magnets that can be turned on and turned off. Electromagnets consist of a coil of wire wrapped around an iron core that is easily magnetized but does not retain the magnetic field (**Figure 17-82**).

When current is supplied to the wire, the magnetic field around the wire is amplified by the iron core, making the magnet more powerful. When the current is turned off, the magnetic field dissipates, and the magnet turns off. Electromagnets are used in relays, DC motors, and AC generators. A simple electromagnet can be made using a screwdriver and a coil of wire (**Figure 17-83**). Depending on the metal used in the screwdriver, this method can be used to magnetize the screwdriver.

USE OF MAGNETISM IN THE AUTOMOBILE

Magnets, both permanent and electromagnets, are used bumper to bumper in modern vehicles. Magnets are used in the horns, starter motors, generators, relays, fan motors, speakers, power window motors, electric fuel

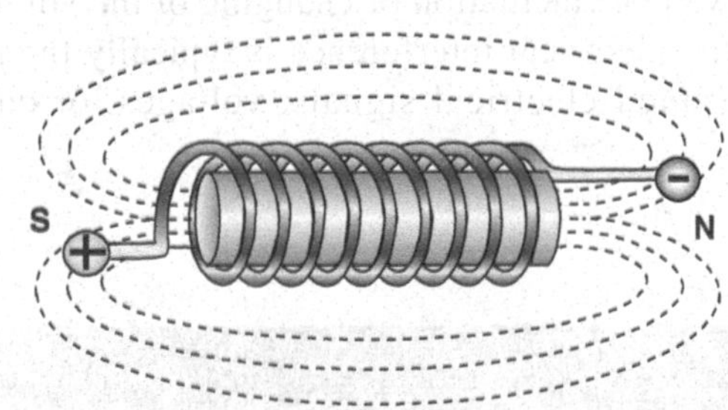

FIGURE 17-82 Current flowing through a coil with an iron core creates an electromagnet that can be used for many purposes.

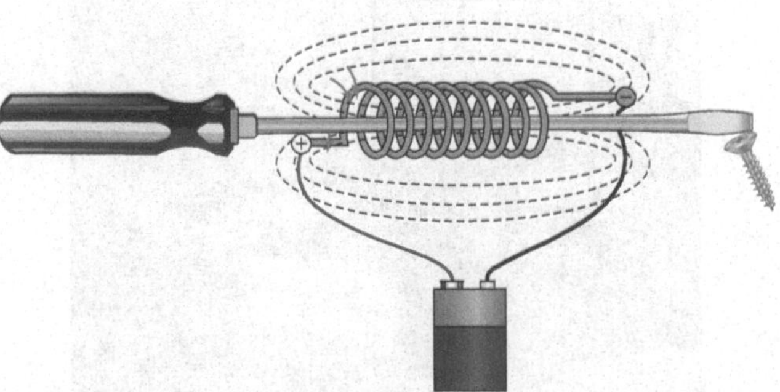

FIGURE 17-83 Constructing a simple electromagnet using a length of wire, a screwdriver, and a battery. Depending on the metal used in the screwdriver, the screwdriver may remain magnetic.

pumps, and many other devices. One of the most common devices used in electrical circuits is the relay.

■ *Relays.* **Relays** use electromagnetism and a switch to allow a small amount of current to control a large amount of current. Relays, like those shown in **Figure 17-84**, are used in headlight, fan motor, horn, and many other circuits. A relay has a coil of very fine wire that is wound hundreds to thousands of times around an iron core (**Figure 17-85**).

You may wonder why the wire, which is very small in diameter, when connected to battery voltage does not burn up. If the wire is unwound, straightened out, and then connected to the battery, it definitely burns up. However, since the wire is formed as a coil, the interaction of the magnetic field around each winding forms electrical resistance. This resistance is dependent upon the size of the wire and number of windings of the wire. The resistance of the coil of wire in a relay is usually between 50 and 150 ohms. High resistance in the coil means that current flow through the wire is low.

When current flows through the coil, the coil becomes magnetized and acts as an electromagnet. The magnetic field produced by the coil is used to pull a contact switch closed (**Figure 17-86**). The contacts and terminals for the load circuit are often large enough to handle 15 to 20 amps.

FIGURE 17-84 A typical automotive relay.

FIGURE 17-85 The inside of a relay. The coil of wire can be several hundred feet long and wrapped thousands of times around the core.

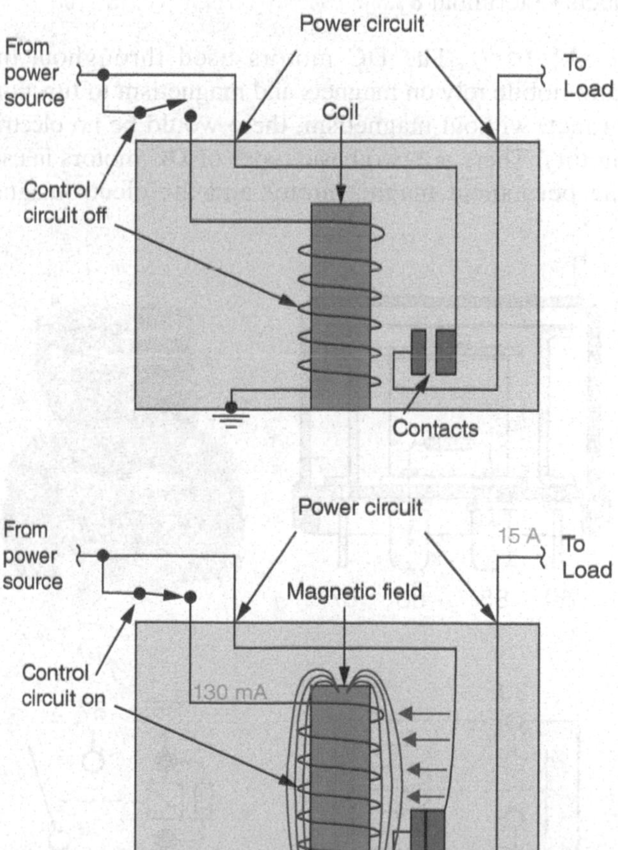

FIGURE 17-86 How a relay operates. The magnetic field pulls the contact switch closed to complete a circuit. A spring pulls the contacts open once the current flow through the relay stops.

Because of the low current flow through the coil, the relay allows a small amount of current to control a much larger current.

A typical four-terminal relay is shown in **Figure 17-87**. The circuit numbers shown in the figure represent common circuit designations developed by Bosch. In a circuit like that in **Figure 17-88**, power is supplied to terminals 86 and 30. Depending on the circuit, these may be from a single fuse, or they may be from separate fuses. Ground for the coil is supplied through the horn switch to terminal 85. When grounded, the current flows from the fuse, through the coil to ground. The coil generates a magnetic field and is a load for that part of the circuit. The magnetic field attracts the armature in the relay and pulls the contacts closed, allowing power to flow from terminal 30, through the contact, and out terminal 87 to the horns.

Some circuits use five-pin relays, like that illustrated in **Figure 17-89**. A five-pin relay has two possible output circuits, from either terminal 87 or terminal 87a. When the relay is unenergized, terminal 87a is typically connected to terminal 30. When energized, terminal 30 connects to terminal 87.

■ ***Motors.*** The DC motors used throughout the automobile rely on magnets and magnetism to function. In fact, without magnetism, there would be no electric motors. There are two basic types of DC motors in use, the permanent magnet motor and the electromagnet motor. Both types of electric motors use the interaction of magnetic fields to turn a rotating assembly, called *the armature* (**Figure 17-90**).

Motors operate because of the interaction of the magnetic fields. As you probably know, when two north or two south magnetic poles get close together, they repel. When a north and a south pole get close to each other, they attract. An electric motor uses these interactions to create movement. When current flows through a conductor, a magnetic field is generated around it (**Figure 17-91**). The armature, which is made of several conductors, is placed between stationary magnetic fields. The interaction of the fields causes the armature to move.

Until recently, only hybrid and electric vehicles had AC motors. That is starting to change as manufacturers are starting to use AC motors in other applications, such as for interior fan motors.

■ ***AC Generator.*** To power the electrical system and recharge the battery, modern vehicles use an AC generator, commonly called *an alternator*. Like motors, the AC generator uses electromagnetism to operate.

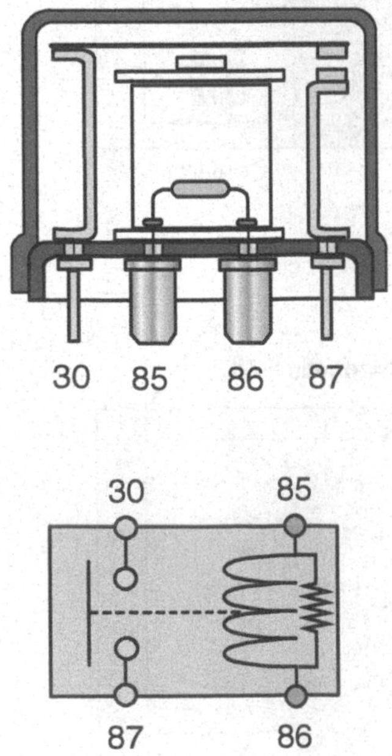

FIGURE 17-87 An inside and outside look at a common type of four-pin relay.

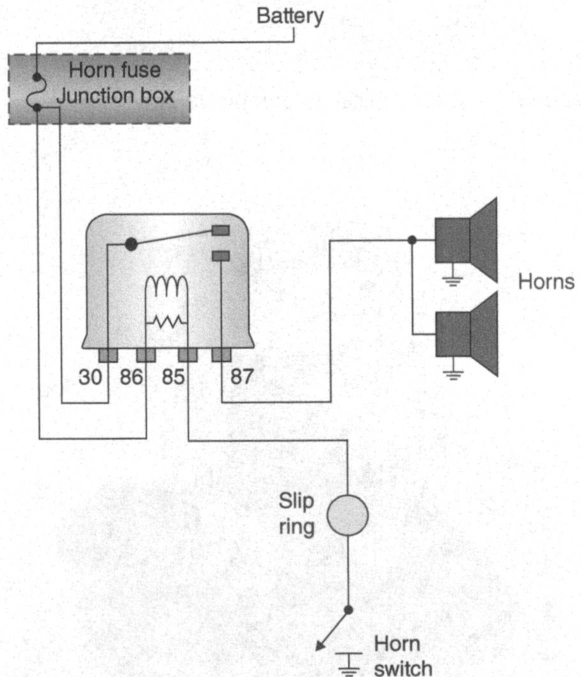

FIGURE 17-88 This shows how a relay can be used in a circuit. The magnetic field pulls the contacts closed, allowing current to flow to the load in the circuit. The amount of current needed to control the relay is very small compared to the current flows through the switch to the load.

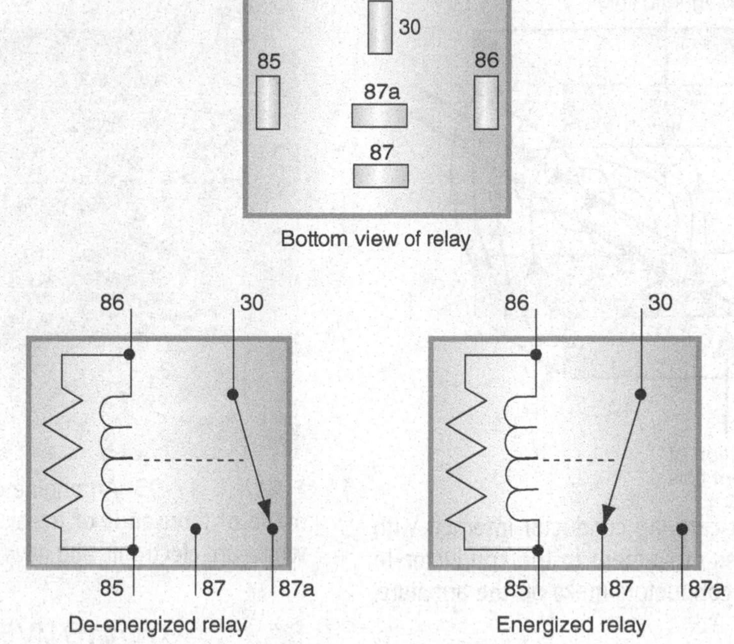

FIGURE 17-89 Typical relay configurations for four- and five-pin relays.

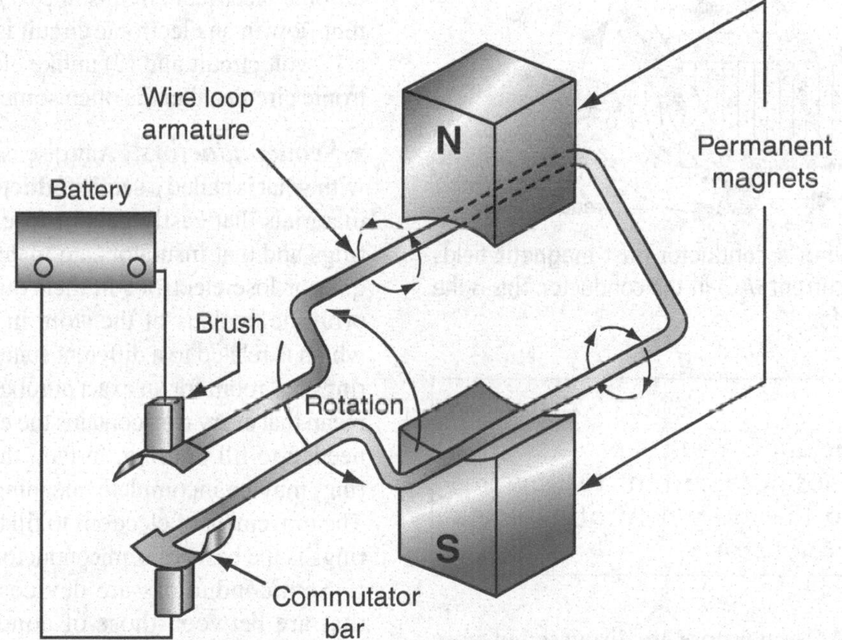

FIGURE 17-90 Electric motors use the interactions of magnetic fields to create motion.

Where an electric motor uses electricity to perform a mechanical function, generators use mechanical motion to generate electricity.

The AC generator operates by using induction. When a conductor moves through a magnetic field (**Figure 17-92**), AC voltage is induced in the conductor. In the AC generator, spinning magnetic fields induce current flow within stationary windings of wire. The AC voltage created in the windings is then turned into DC for use by the rest of the vehicle's electrical system.

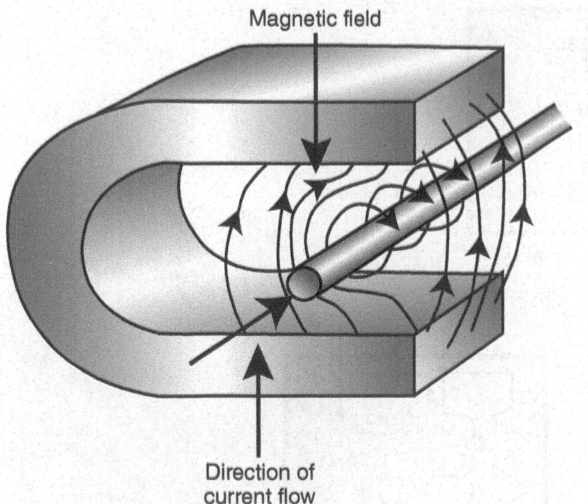

FIGURE 17-91 A current-carrying conductor interacts with magnetic fields. This causes movement in the conductor. In an electric motor, multiple conductors make up the armature, rotating part of the motor.

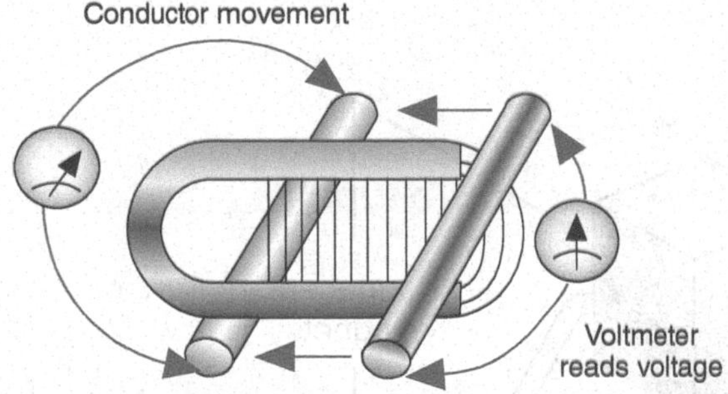

FIGURE 17-92 Moving a conductor past magnetic fields generates alternating current (AC) in the conductor. This is the basis of the AC generator.

> **Word Wall**
>
> *Inducing or Induction*—the process of a magnetic field causing a current to flow in a conductor due to the movement of magnetic fields across the conductor.

Both motors and AC generators are discussed in more detail in Chapter 19.

Electronics

As stated at the beginning of the chapter, the term *electronics* refers to integrated circuits, solid-state electronics, and components that are generally not repairable. **Figure 17-93** shows an example of an electronic circuit board from a modern electronic control module. As you can see, the circuit board and the components attached to it are not made to be serviced or repaired.

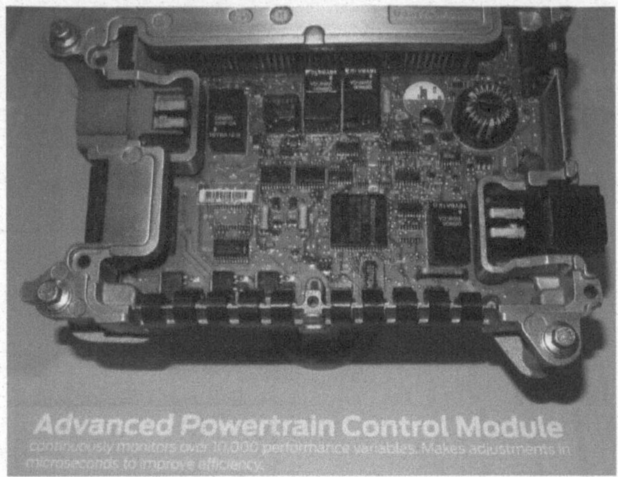

FIGURE 17-93 An engine control module. This module is made of thousands of transistors, resistors, and chips; all of which are electronic and have no moving parts.

BASIC AUTOMOTIVE ELECTRONICS

While it is rare to ever repair electronic circuits in the field, you should have a basic understanding of simple electronic circuits and functions. Electronic circuits differ from traditional electrical circuits in two major ways: (1) the current flow in an electronic circuit is much less compared to a 12-volt circuit and (2) unlike electrical circuits, in electronic circuits, there is often some type of microprocessor.

■ ***Semiconductors.*** Any discussion of electronics begins with what is called a **semiconductor**. Conductors are those materials that easily gain or lose electrons in their outer rings and that insulators are materials that do not readily gain or lose electrons in their outer rings. The electrons orbit the nucleus of the atom in rings or shells, each of which is related to a different state of energy. Each of these rings has room for an exact number of electrons. It does not mean that every ring contains the exact number of electrons needed to fill the ring. In fact, the outer ring, or valence ring, may be incomplete, meaning it is missing electrons. The movement of electrons to fill the gaps in some of these rings is the basis of semiconductor technology.

Semiconductors are devices made from materials that are between those of conductors and insulators. This means they are neither true conductors nor insulators. Metals tend to make good conductors due to their ability to gain and lose electrons from their outer rings. Most semiconductors are made from silicon and germanium. Both silicon and germanium atoms have four electrons in the outer ring, allowing the atoms to connect in a lattice structure. Because the outer ring has four electrons, each can bond with another germanium atom (**Figure 17-94**). This is called *a covalent bond*, and in this state, it is difficult to remove or add another electron. Because the electrons are not free to move,

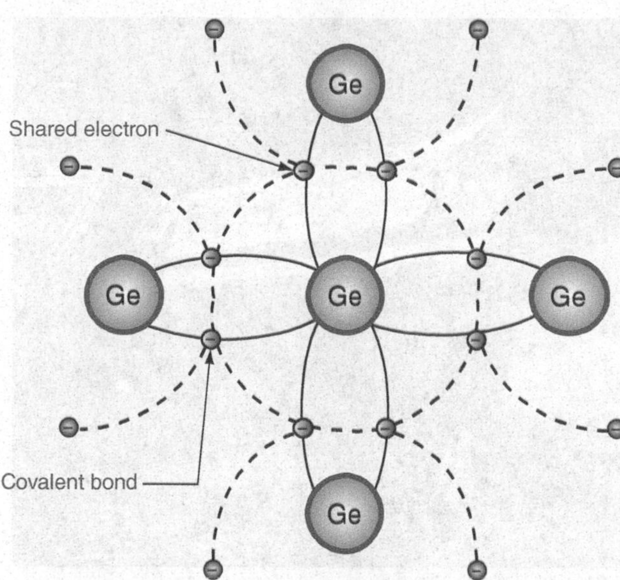

FIGURE 17-94 An example of a semiconductor, having four electrons in the outer shell. A covalent bond sharing electrons between germanium atoms.

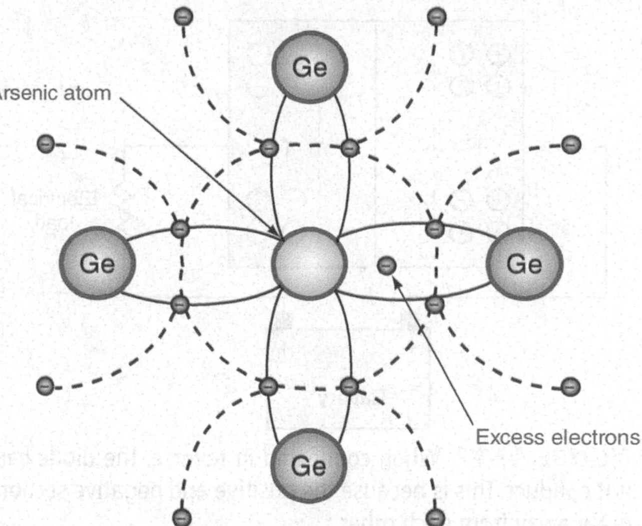

FIGURE 17-95 N-type semiconductors have an extra electron, causing a net negative charge.

and there is no opening for a free electron to enter, this makes the arrangement between that of an insulator and that of a conductor.

To turn the nonconducting germanium into a conductor, an impurity is added. This process is called *doping*, and it is used to create N-type (negative) and P-type (positive) semiconductors (**Figures 17-95** and **17-96**). N-type semiconductors have small quantities of impurities with five electrons in the valance ring. Because only four of the electrons can bond in the covalent lattice, the fifth electron is a free electron, able to move about. Because there is an extra electron, the germanium has a negative charge, which is why it is called an N-type semiconductor. If the impurities used to dope the germanium have three electrons in the valence ring, this leaves a gap where there is nothing for one of the electrons to bond with, as shown in Figure 17-96. This hole can accept a free electron. Because there is a missing electron, the P-type semi-conductor has a positive charge. Adding the impurities to the germanium allows it to conduct electricity, but not as well as a metal, such as copper, hence the name semiconductor is used.

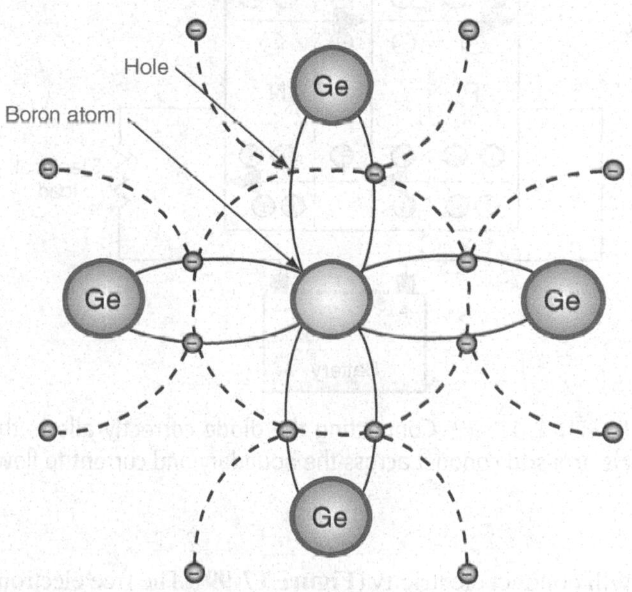

FIGURE 17-96 P-type semiconductors have a missing electron, causing a net positive charge.

■ ***Diode.*** Diodes are the simplest form of semiconductors (**Figure 17-97**). Formed from a P-type semiconductor and an N-type semiconductor (**Figure 17-98**), diodes are used as one-way electrical check valves. When they are connected, as shown in Figure 17-98, no current flows across the center connection of the diode. This is because the electrons on the N-type side are drawn to the positive side of the battery while the holes in the P-type

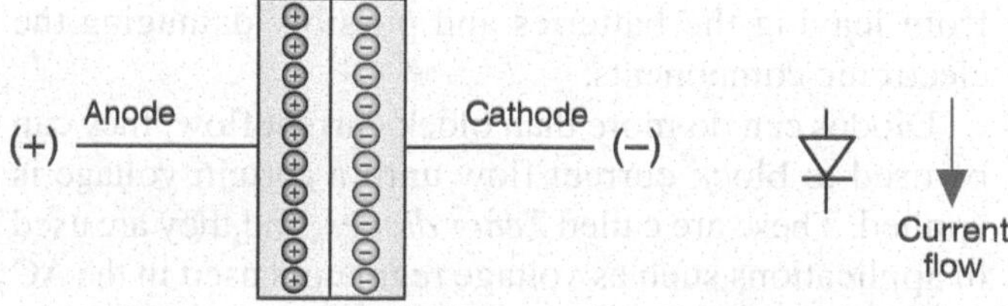

FIGURE 17-97 Diodes are formed with separated positive and negative areas.

side are drawn to the negative side of the battery. Since the electrons are moving away from the center of the diode, no current flows through the diode or the circuit. However, if you reverse the battery connection, the diode

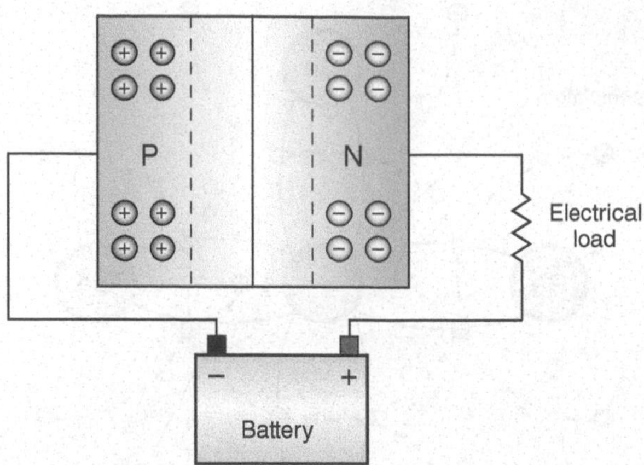

FIGURE 17-98 When connected in reverse, the diode cannot conduct. This is because the positive and negative sections draw away from each other.

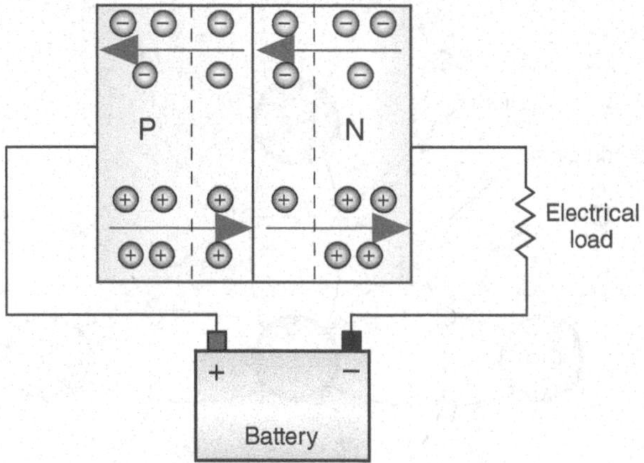

FIGURE 17-99 Connecting the diode correctly allows the electrons to conduct across the boundary and current to flow.

will conduct electricity (**Figure 17-99**). The free electrons will fill the holes, and current will be able to flow. If you have ever inserted batteries backward into something like a clock, diodes were responsible for blocking the current from leaving the batteries and possibly damaging the electronic components.

Diodes can do more than block current flow; they can be used to block current flow until a certain voltage is applied. These are called *Zener diodes*, and they are used in applications such as voltage regulators used in the AC generator and in some instrument panels.

Another common application for diodes is in components that use coils as electromagnets, such as air conditioner clutches and solenoids. Because there is a voltage spike whenever a coil is turned off, diodes are used to block the voltage spike from traveling back up the electrical system and causing damage. Diodes in this application are often called *clamping diodes*.

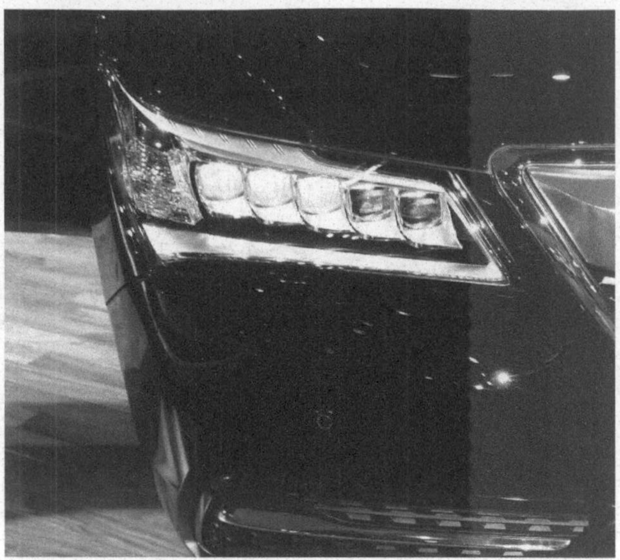

FIGURE 17-100 Light-emitting diodes are used in both interior and exterior lighting applications.

Discussed in more detail in Chapter 19, diodes are used in the AC generator to turn AC into DC for use by the rest of the electrical system.

Light-emitting diodes (LEDs) have seen an enormous increase in use in the last several years. LEDs are diodes that, when the electrons fill the holes in the P-type material, energy is released in the form of photons, or particles of light. LEDs are becoming widely used in interior and exterior lighting due to their low energy consumption, low heat generation, durability, and long life spans (**Figure 17-100**). A photodiode is a semiconductor that produces a voltage when exposed to light. This type of device is used with automatic headlamp circuits.

■ ***Transistors.*** **Transistors** are solid-state switches and are responsible for the electronic revolution of the 20th century. A transistor has three layers of semiconducting material, formed as either a PNP or NPN sandwich. There are three parts of the transistor, the base, the collector, and the emitter, and three electrical connections (**Figure 17-101**). When voltage is applied to the base, current can flow from the collector to the emitter and the load in the circuit (**Figure 17-102**). Transistors can also act as amplifiers, allowing a small current to control a large current.

Transistors are the basis of all computers. Millions of transistors are packed into smaller and smaller silicon chips, which make up the core of modern microprocessors. Silicon chips contain thousands to millions to hundreds of millions of transistors on a single chip and are

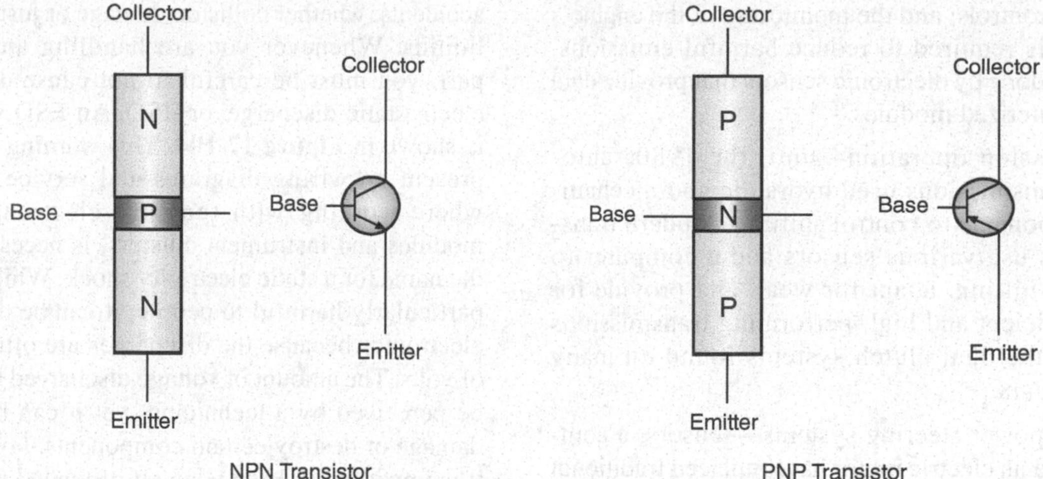

FIGURE 17-101 Transistors can switch circuits on and off using very small amounts of current.

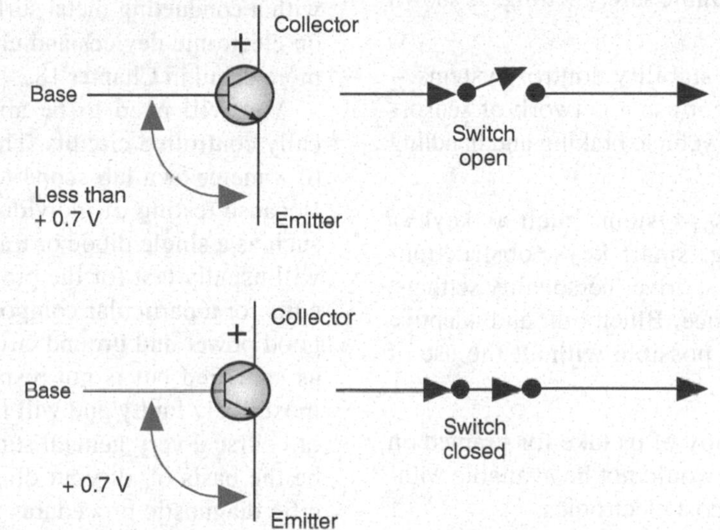

FIGURE 17-102 A transistor can be used to control the operation of a relay, a bulb, or many other devices.

capable of performing logic functions. These chips have enabled the development of personal computers, smart phones, and countless other electronic devices that many of us now take for granted.

The transistors, acting as switches, can create logic gates. Logic gates operate by taking information bits, 0s and 1s, and process the bits through the circuits; in this manner, input data are examined and output decisions are made. Some gates can be arranged to be able to remember input values. These are the basis for random access memory (RAM).

The electronic devices described here do not include every type of electronic component used in modern vehicles. This brief discussion just provides you with an overview of some basic electronic components.

■ ***Electronics Used in the Automobile.*** Modern cars and trucks increasingly rely on electronics to provide the comfort, safety, and performance expected by the consumer. In fact, the modern automobile is entirely dependent upon the use of electronics for almost all aspects of its operation. The following examples are of systems that once used mechanical and basic electrical components to operate but now use electronic monitoring and/or control.

- Engine management—mechanical fuel system components such as carburetors have been replaced by electronic fuel injection; contact points (a mechanical switch) once used to trigger the firing of the spark plugs were phased out in the mid-1970s by electronic ignition controls; and the monitoring of the engine's exhaust is required to reduce harmful emissions, which is done by electronic sensors that provide data to computerized modules.

- Transmission operation—until the 1980s automatic transmissions used hydraulic and mechanical components to control shifting. Modern transmissions use various sensors and a computer to control shifting, adapt for wear, and provide for more efficient and high-performing transmissions such as the dual clutch systems found on many modern cars.
- Electric power steering systems—sensors, a computer, and an electric motor have replaced traditional hydraulic power steering assist and now can provide parking assist and lane departure/correction ability. An example of configurable safety settings is shown in **Figure 17-103**.
- Antilock braking and stability control systems—now standard equipment use a network of sensors and modules to control vehicle braking and handling functions.
- Electrical accessories—systems such as keyless entry, remote starting, smart keys, obstruction-sensing power windows, driver personality settings, remote vehicle assistance, Bluetooth, and adaptive cruise control are not possible without the use of electronics.

The accessories that many of us take for granted on our cars and trucks simply would not be available without the availability of low-cost electronics.

■ ***Testing Electronic Components.*** Most electronic devices are not serviceable and are replaced if they are faulty. However, it is very important to properly diagnose the cause of the failure so that the new replacement part is not damaged. Electronic components do not wear out as mechanical parts do, but they are susceptible to damage from voltage spikes, current overloads, and from accidents, whether collision damage or just from spilled liquids. Whenever you are handling any electronic part, you must be careful to not cause damage from electrostatic discharge, or ESD. An ESD warning sign is shown in **Figure 17-104**. This warning sign is often present on wiring diagrams and service information where working with sensitive electronics, such as modules and instrument clusters, is necessary. ESD is the name for a static electricity shock. While ESD is not particularly harmful to people, it can be destructive to electronics because the discharges are often thousands of volts. The amount of voltage discharged may not even be perceived by a technician, yet it can be enough to damage or destroy certain components. For this reason, it is important to keep yourself grounded when you are handling any electronic components. This can be done by either wearing a grounding strap or keeping in contact with a conducting metal surface when you are working on electronic devices and circuits. This is explained in more detail in Chapter 18.

You will need to be able to diagnose electronically controlled circuits. This usually requires the use of a meter or a lab scope to monitor voltage signals. Because testing of individual electronic components, such as a single diode or transistor, is impractical, you will usually test for the proper power and ground circuits for a particular component. If a component has a good power and ground circuit and is receiving inputs as specified but is not responding, the component is most likely faulty and will need to be replaced. This is of course a very general statement and is not meant to be the basis of all your diagnostic testing. More specific diagnostic procedures are included in the following chapters as they relate to testing electronic systems and components.

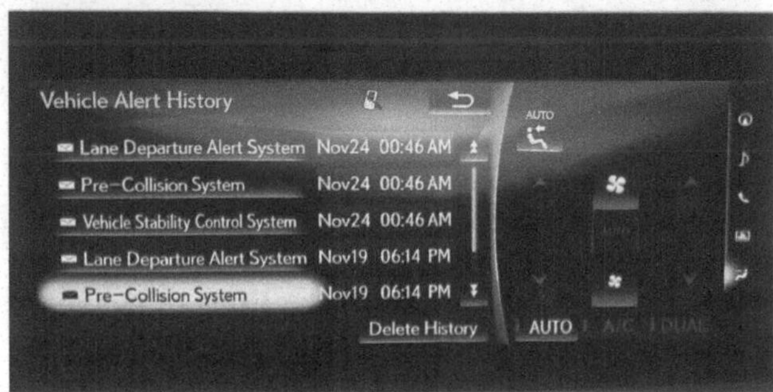

FIGURE 17-103 An example of electronics in a modern vehicle.

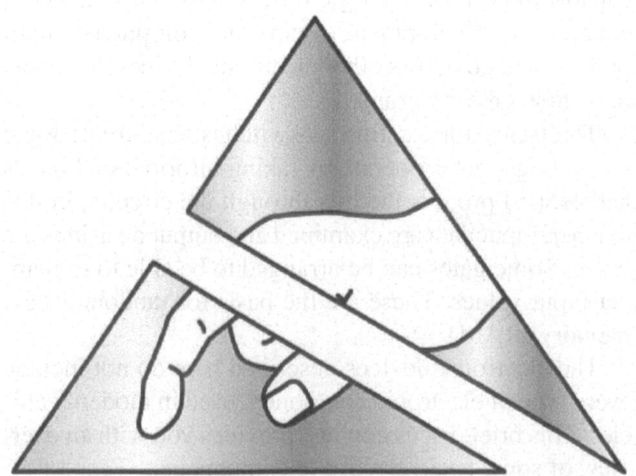

FIGURE 17-104 The symbol for electrostatic discharge (ESD) warning.

SUMMARY

Simply stated, electricity is the flow of electrons, from a source of higher electrical potential to a source of lower electrical potential.

Static electricity is the term used to describe the accumulation of an electrical charge by a substance not normally conductive to electricity.

Voltage and volts are used to define an amount of electrical potential.

Voltage drop is the amount of voltage that is being used by a component, a wire, or a connection.

Amperage is the amount of electrical current flow.

Electrical resistance is anything that restricts the flow of electrons.

Alternating current has electron flow that alternates from positive to negative and negative back to positive.

Direct current flows only from zero to positive.

A circuit is a path for electrons to flow from a source of higher electrical potential to a source of lower electrical potential.

Series circuits have all the parts of the circuit in a single path.

Parallel circuits have individual branches for each load.

Series–parallel circuits contain series and parallel circuit structures.

The most common automotive circuit protection device is the fuse.

Connectors join sections of wire and components. Terminals join wires together or to a component.

Digital multimeters are used to make several different types of electrical tests.

Unpowered test lights can be used to perform basic power, ground, and current flow tests.

Electromagnets are magnets that can be turned on and turned off.

Diodes are solid-state devices that allow current to flow in one direction.

Transistors are solid-state switches that contain no moving parts.

REVIEW QUESTIONS

1. The basis of electricity is the movement of ______________.

2. A ______________ is used to open a circuit if current flow becomes excessive.

3. Voltage ______________ is the term used to describe the voltage used by each part in a circuit.

4. The five components of an electrical circuit are:
   ____
   ____
   ____
   ____
   ____

5. Automotive circuits are wired in ______________ ______________.

6. *Technician A* says that if resistance increases in a circuit, current flow will also increase. *Technician B* says that if resistance in a circuit increases, current flow will decrease. Who is correct?
   a. Technician A **c.** Both A and B
   b. Technician B **d.** Neither A nor B

7. All of the following are types of circuit protection except:
   a. Fuse link **c.** Relay
   b. Multi-circuit fuse **d.** Circuit breaker

8. Which of the following statements about parallel circuits is incorrect?
   a. The total circuit resistance is less than that of the lowest resistor.
   b. The resistance total is the sum of the resistances.
   c. Full battery voltage is available for each load.
   d. The voltage drop of each load is equal to source voltage.

9. *Technician A* says a wiring harness contains connectors, terminals, and the wires that connect various components. *Technician B* says that terminals and connectors have both male and female ends. Who is correct?
   a. Technician A **c.** Both A and B
   b. Technician B **d.** Neither A nor B

10. Two technicians are discussing electromagnetism: *Technician A* says that a wire carrying current generates a magnetic field. *Technician B* says relays use electromagnetism to open and close a switch. Who is correct?
    a. Technician A **c.** Both A and B
    b. Technician B **d.** Neither A nor B

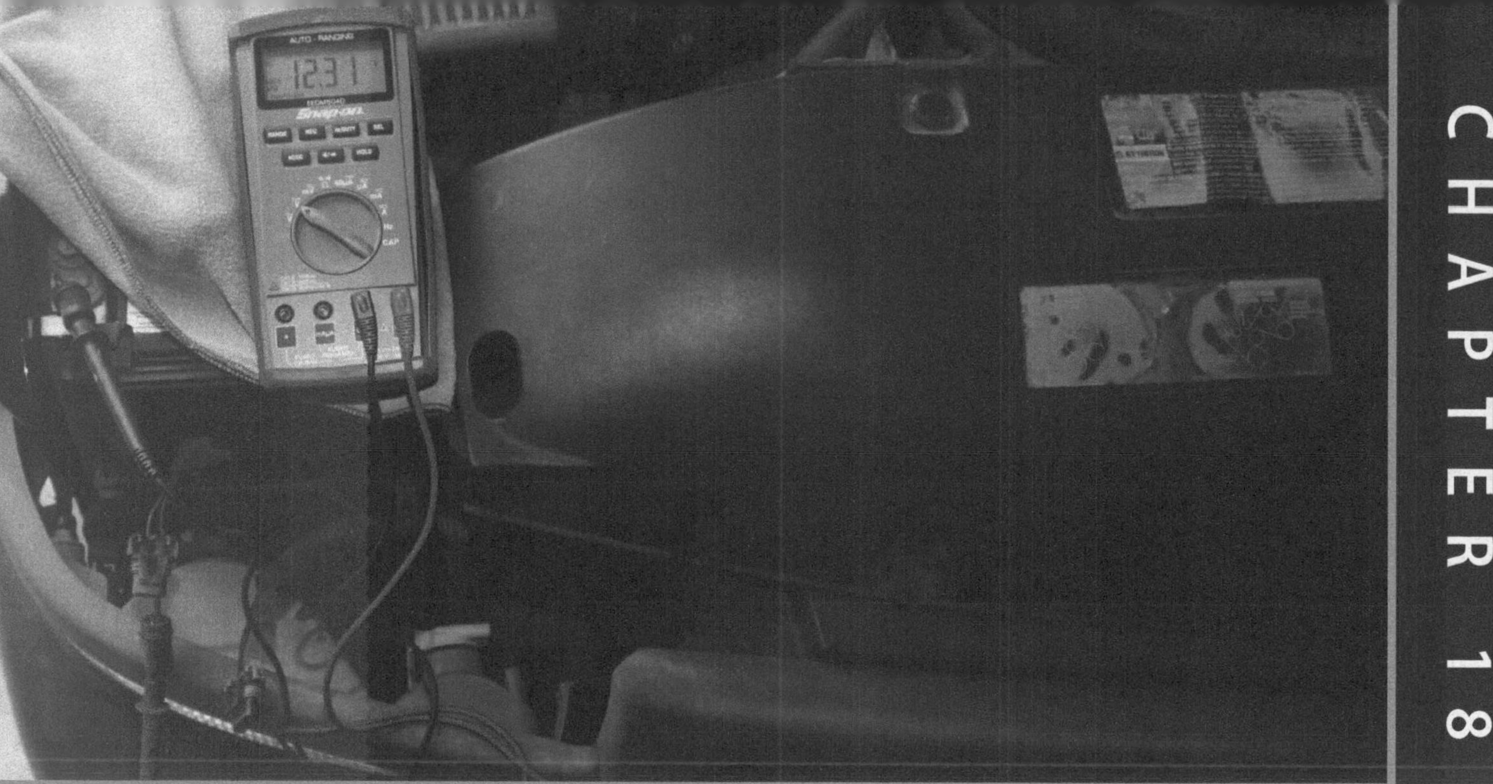

CHAPTER 18

Basic Electrical/Electronic System Service

Chapter Objectives

At the conclusion of this chapter, you should be able to:

- Describe and perform voltage, amperage, and resistance tests. (ASE Education Foundation MLR 6.A.4)
- Interpret and use wiring diagrams. (ASE Education Foundation MLR 6.A.3)
- Describe and diagnose circuit faults. (ASE Education Foundation MLR 6.A.5 & 6.A.7)
- Identify and test circuit protection devices. (ASE Education Foundation MLR 6.A.9)
- Check operation of circuits with a test light. (ASE Education Foundation MLR 6.A.6)
- Perform wiring repairs. (ASE Education Foundation MLR 6.A.10)
- Replace electrical connectors and terminal ends. (ASE Education Foundation MLR 6.A.11)
- Test relay operation.

KEY TERMS

circuit faults
continuity
electrical shock
fretting
inductive current clamp
open circuit
resistance test
schematic
short circuits
thermal cycling
unwanted resistance
voltage drop

Being able to diagnose and repair electrical system problems is probably the most important technical skill needed by current and future technicians. As automobiles continue to integrate electronic systems into all types of vehicle functions, the importance of being able to understand and service these systems is critical.

Electrical Safety, Tools, and Equipment

When you are working on the electrical system, safety should be your first priority. Electrical system service requires the use of specialized tools and equipment, and an extra awareness by the technician. This is because of the additional risks when working with electricity and electronic components.

SAFETY

Most people have been shocked by static electricity. You may have even been shocked by AC current or by an ignition system without serious injury. That does not mean that when you are working around electricity, you can ignore that there are dangers. Even small electrical shocks can be dangerous, if not to you then to the electronic components you are working on.

When you experience an **electrical shock**, more happens than just the tingle or pain. Because resistance to current flow causes heat, you can experience not just external burns but also electrical burns that go deep beneath the skin. Electric shock can also overcome the normal nerve pulses that control all our bodily functions. An electric shock can disrupt the operation of the heart, causing fibrillation. This means that instead of a normal heartbeat, the heart flutters and does not adequately pump blood. This can, if not quickly corrected, lead to death. Though death from electric shock is not likely from working on most cars and trucks, the risk of injury is always present. Getting shocked usually results in trying to pull away from the source very quickly. This movement can itself cause injury if you pull away but collide with something else. Before working on any vehicle, especially one you are not familiar with, locate any service precautions in the service information, and check for warning labels (**Figure 18-1**). This is a warning label for a high intensity discharge or HID headlight system.

In addition, electrical burns are possible if you contact an overloaded circuit. This can occur from extended engine cranking, which can overheat the starter and battery cables. Circuit breakers, though designed to open in the event of excessive current flow, sometimes remain closed and can become very hot. A circuit that is shorted to ground can very quickly overheat and catch fire. If this occurs, quickly disconnect the battery negative cable.

Even though the risk of death is low for automotive technicians from electric shock, the risk is increased if one is working on high-voltage electrical systems. Hybrid electric vehicles (HEVs) and electric vehicles (EVs) are increasingly common on the road and in repair shops. These vehicles have very visible orange wiring and warning stickers to alert you to the presence of high-voltage components (**Figure 18-2**). In addition, read the warning precautions and labels before you attempt any service on a hybrid vehicle. An example of a hybrid warning label is shown in **Figure 18-3**. *Never attempt to service the high-voltage system without proper training, tools, and the appropriate service information.*

FIGURE 18-1 Always read and follow all warnings and service guidelines to prevent damage to the vehicle and personal injury. The HID lighting system uses high voltage, which can be hazardous.

FIGURE 18-2 Hybrid and electric vehicles use bright orange to indicate high-voltage components. Do not attempt to service or repair any component of the high-voltage system without proper training and equipment. Serious injury may result from attempting to work on the high-voltage system.

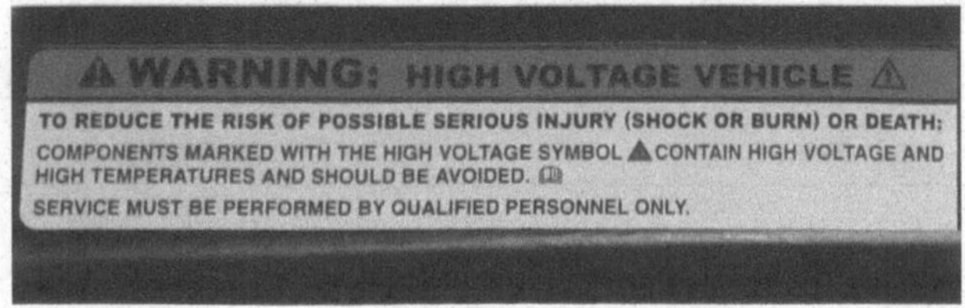

FIGURE 18-3 An example of a warning decal on a hybrid vehicle.

■ ***Working Around Batteries.*** Automotive batteries present several dangers from both electrical and physical factors (**Figure 18-4**). Follow these precautions when you are working near the battery:

- Never lay tools or parts on the battery or near where anything can contact the positive battery terminal and ground.
- Never smoke or have an open flame near the battery, as this can ignite any hydrogen gas present from charging.
- Do not overcharge the battery.
- Do not try to jump-start or charge a frozen battery.
- Charge batteries only in well-ventilated areas.
- Wash off any battery acid or corrosion that contacts your skin immediately.

TOOLS AND EQUIPMENT

Some of the tools used in electrical system service are discussed in Chapter 17. The figures here show examples of many of the tools used when a technician works on the electrical system.

- Digital multimeter. A digital multimeter (DMM) is the primary tool to make voltage, amperage, and resistance measurements (**Figure 18-5**). Most meters can perform other types of functions also, such as temperature measurement and diode testing, and more. A good-quality meter should have at least 10,000,000 (10 megohms or 10M ohms) of internal resistance. This is to ensure that when you are using the meter to test electronic components, the meter does not cause an increase in current flow, which can destroy sensitive electronics.
- Unpowered test light. A test light is used to check for power, ground, and current flow or continuity (**Figure 18-6**). Test lights are also called *continuity testers* or *circuit testers.* When the clip is attached to negative (or a good ground) and the probe end is touched to power, the light bulb lights. When the clip is attached to a power source and the probe end is touched to ground, the bulb also lights. Test lights are very useful for checking fuses quickly and checking for either power or ground to a component. Because test light bulbs are a relatively low resistance as a load, they should not be used when working with computers and other solid-state electronic circuits. The current flow through the test light is sufficient to damage solid-state electronics like computers and control modules.

FIGURE 18-4 Batteries pose several dangers to technicians; always take all necessary precautions when working on or near a battery. Never smoke or cause sparks around a battery as this can cause the battery to explode.

FIGURE 18-5 An example of a DMM. This type of meter can perform many different types of electrical measurements and is one of the most important tools you can learn to use.

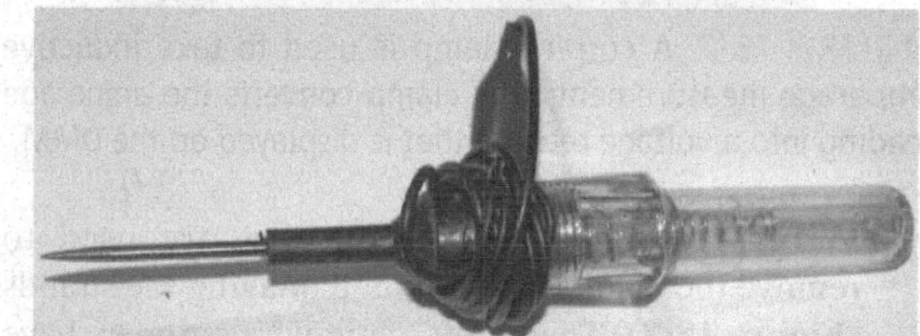

FIGURE 18-6 An unpowered test light is used to check for power, ground, and current flow.

- Current clamps. A current clamp is also called *a current probe* (**Figure 18-7**). These tools are used to make inductive amperage measurements by measuring the strength of the magnetic field generated around a current-carrying wire. Low-current probes are used to measure up to 60 amps. Larger versions can measure 500 amps or more (Figure 18-7).

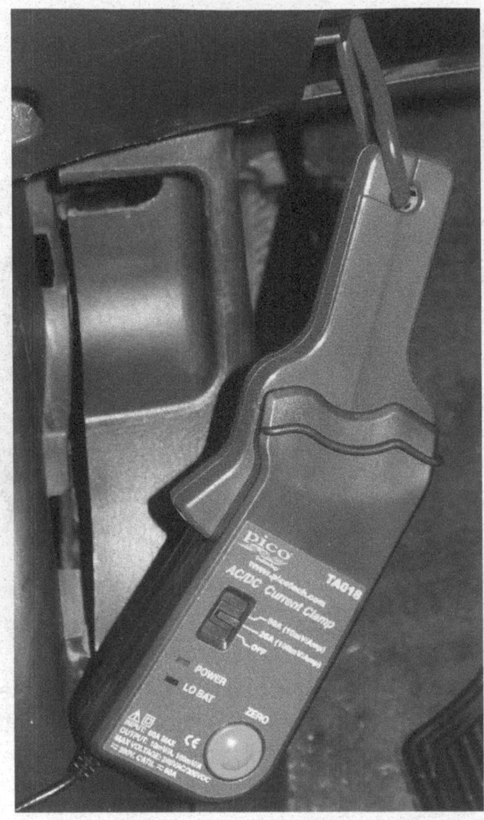

FIGURE 18-7 A current clamp is used to take inductive amperage measurements. The clamp converts the amperage reading into a voltage reading that is displayed on the DMM.

- **Wire strippers. Wire strippers** are used to remove the insulation without damaging the strands (**Figure 18-8**). These automatic wire strippers have several standard-sized openings for cutting specific wire gauges.
- Wire terminal pliers. These multifunction tools are used to strip and cut wire and crimp terminals (**Figure 18-9**).
- Soldering iron. Soldering irons or soldering guns are used to heat a terminal or connection to melt and flow solder into the wiring to form a permanent bond (**Figure 18-10**).
- Jumper wires. Jumper wires are used when you are diagnosing circuit problems, such as an open connection. When working on power circuits, a fused jumper should be used (**Figure 18-11**). The fuse protects the circuit for excessive current flow and possible damage.
- Fuse Buddy. When measuring amperage with a current clamp, a Fuse Buddy can be useful (**Figure 18-12**). The Fuse Buddy plugs into the fuse cavity and allows for easy connection of the current clamp around the loop of wire. This arrangement keeps the circuit operating and protected by a fuse during testing.

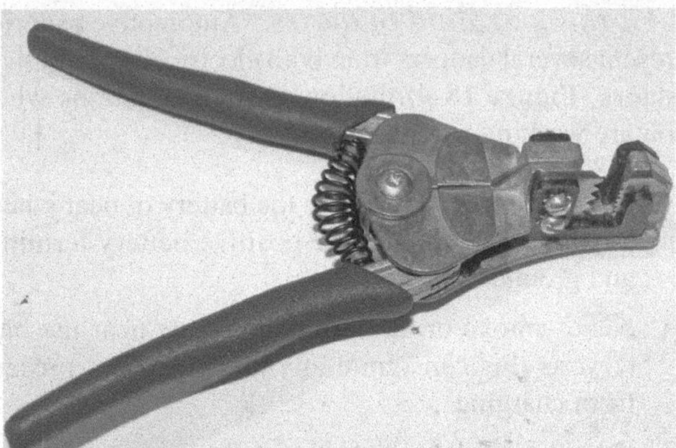

FIGURE 18-8 Wire strippers are used to safely and correctly remove insulation from wire.

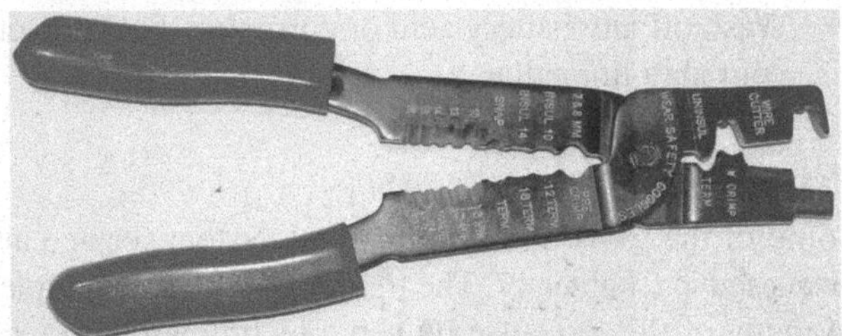

FIGURE 18-9 Wire terminal tools such as this one are used to strip wire and crimp connections.

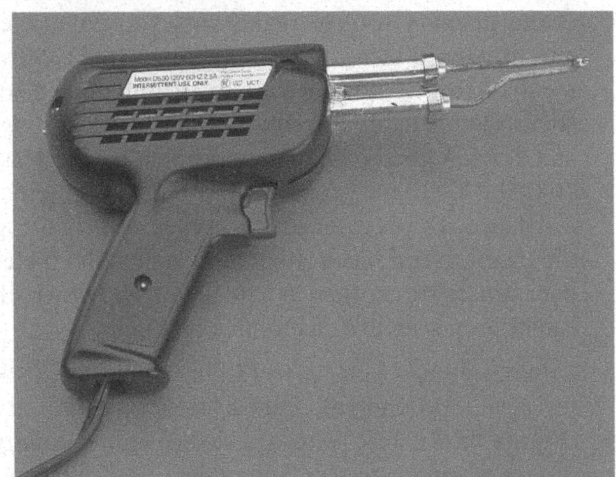

FIGURE 18-10 Soldering irons heat wire and connectors for solder to melt into to form permanent connections.

■ ***Electrostatic Discharge.*** Electrostatic discharge (ESD) is the accidental discharge of static electricity. Electronic components are very sensitive to voltage spikes and can easily be damaged or destroyed by an accidental static shock. Even a small amount of static electricity, not enough for you to feel the discharge, can be enough to damage transistors, diodes, and other electronic parts. Always ground yourself whenever you are handling electronic components. This can be done by frequently touching a bare metal surface to discharge any static charge you may accumulate.

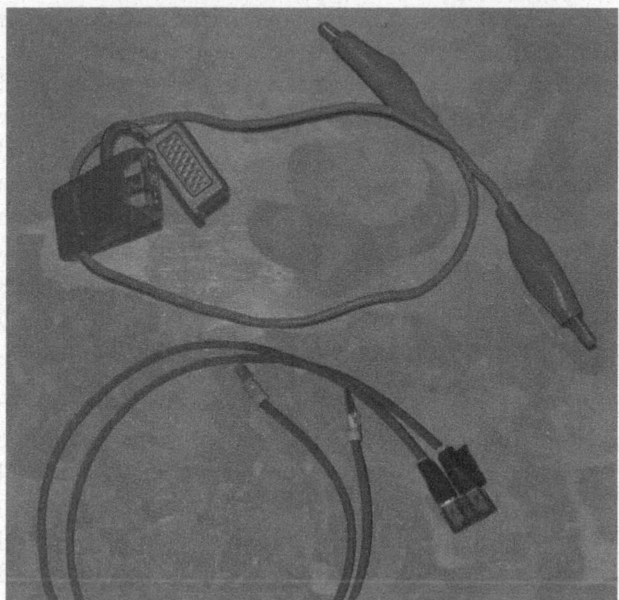

FIGURE 18-11 Fused jumper wires are used during diagnosis to bypass around suspected opens in the circuit and supply power and ground connections.

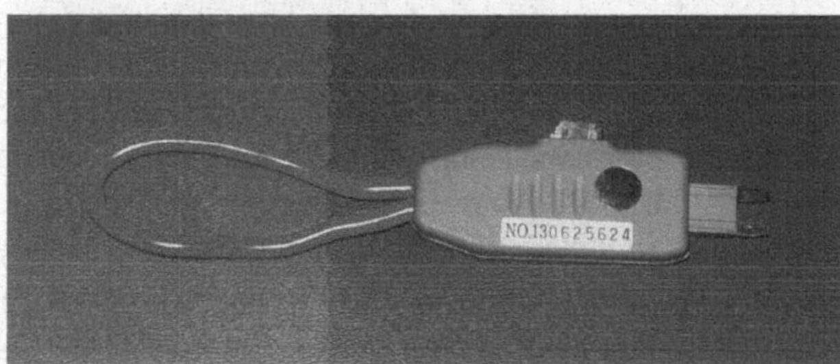

FIGURE 18-12 A Fuse Buddy allows you to insert a fused jumper into a fuse cavity and take amperage measurements.

Electrical Measurements

The core of working on electrical systems is the ability to take meaningful readings of voltage, resistance, and amperage. Being able to safely and properly make these measurements is essential for technicians because so much of the vehicle is electrically operated.

VOLTAGE

Voltage is measured as the difference in potential between two points. To measure voltage, a DMM is used. The meter can only measure voltage between two points, those being the ends of the positive and negative leads. Knowing where to place the leads and what the reading means is one of the most important diagnostic skills you need to have.

■ ***Safety Precautions for Voltage Measurements.*** As a technician, you will measure voltage countless times. To prevent damage to your equipment and possible personal injury, remember the following:

- Make sure the meter is set to correctly measure what you are testing, such as AC voltages (VAC) or DC voltages (VDC). Failure to use the proper setting can cause electric shock and damage to the meter.
- If you are testing AC voltage, be absolutely sure that the meter leads and measurement selection are correct. An incorrect meter setup while you are measuring AC voltage can cause damage to the meter, personal injury, or death.
- Make sure the meter is set to the correct range. Do not try to measure battery voltage if the meter is set on a low-voltage scale, such as on a 200-millivolt scale.
- Be sure that the meter leads are inserted into the correct locations in the meter. A common mistake is not placing the meter leads into the correct jacks. All meters should indicate the rating for each type of measurement (**Figure 18-13**). Pay careful attention to the different ratings for VAC, VDC, amps, and mA because an incorrect meter setup can cause damage to the meter or injury.
- **If you are working on a high-voltage system, be sure the meter and leads you are using are Category III rated for high-voltage systems.** Follow

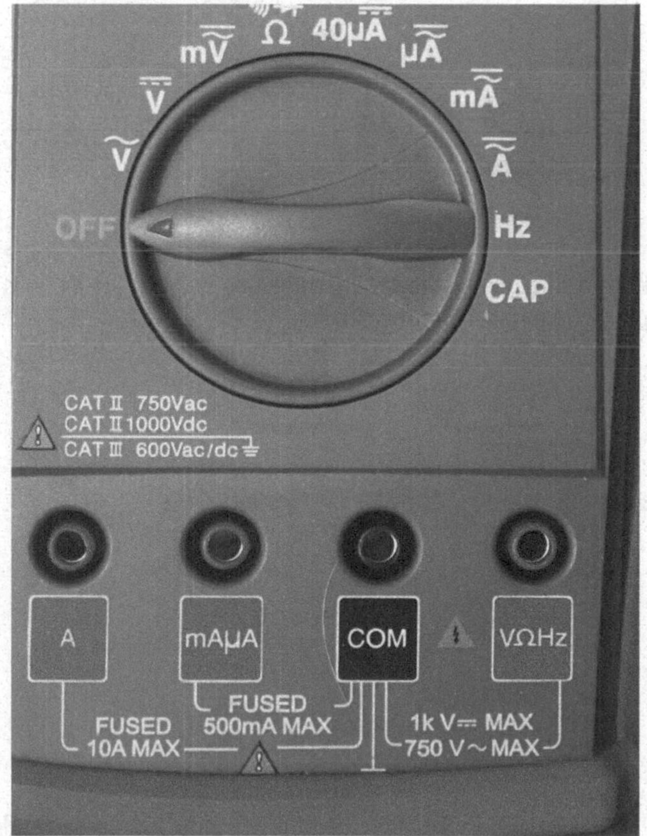

FIGURE 18-13 The input jacks on a DMM are labeled according to what type of measurement each is for and how much voltage and amperage the meter can safely handle.

all the vehicle manufacturer's service guidelines for working around the high-voltage system.

■ ***What Is a Voltage Measurement?*** It is important to understand what a DC voltage measurement means so that you correctly interpret what is shown on the DMM. When you use a meter to check voltage, the meter shows the voltage potential that exists at the two meter leads (**Figure 18-14**). The difference in potential between the meter's positive and negative leads is determined by the electronics in the meter.

Connecting the meter to a circuit places the DMM in parallel to the load in the circuit, which causes current to flow through the meter (Figure 18-14). To limit the current flow, DMMs must have very high internal resistance. Internal meter resistance should be 10 megohms (10,000,000 ohms) or more. Shown in **Figure 18-15**, the meter on the left is displaying the internal resistance of the meter on the right, 10.93 MΩ. The

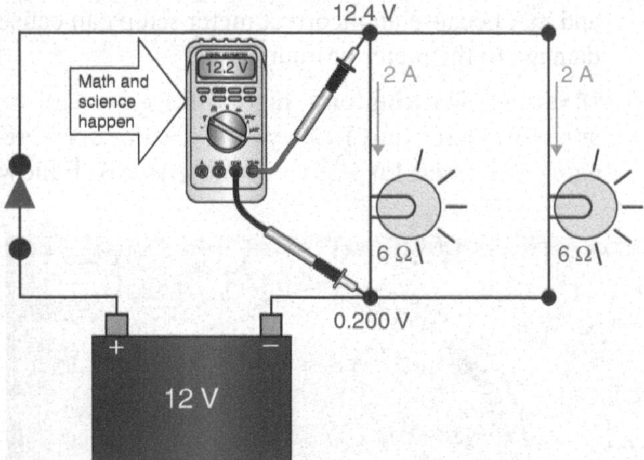

FIGURE 18-14 The meter measures what is present at the ends of the two test leads. When testing voltage, the meter is placed in parallel into the circuit.

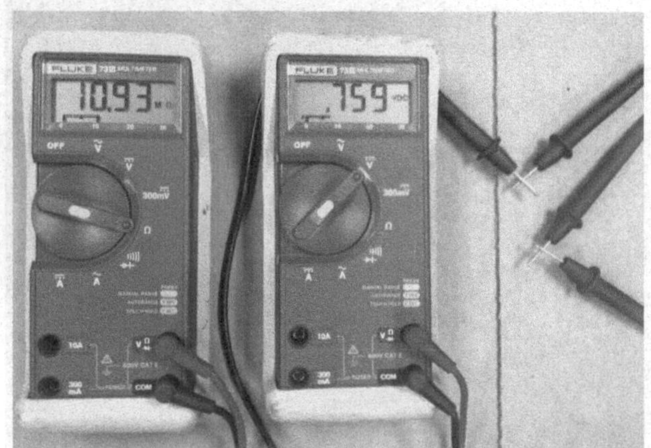

FIGURE 18-15 The meter on the left displays the internal resistance of the meter on the right, which is about 10,930,000 ohms. The meter on the right displays the voltage put out by the meter reading resistance.

meter on the right is displaying the voltage produced by the meter on the left that is used to measure resistance.

If the meter had low internal resistance, the operation of the circuit would change when placed into the circuit (**Figure 18-16**). This will have a very serious effect on the circuit and on the meter. Current flow will increase significantly due to the low resistance of the meter being placed in parallel with the load. To prevent this, the meter has very high internal resistance, which has virtually no effect on circuit operation (**Figure 18-17**). This internal resistance greatly reduces how much current can flow when the meter is connected to the circuit.

■ ***Setting Up the Meter to Measure Voltage.*** To measure voltage correctly, you must first make sure the meter is set correctly. Any meter you are likely to use will have settings for both VAC (volts AC) and VDC (volts DC) (**Figure 18-18**). It is important that you set the meter to read the correct type of voltage. Setting the meter to read VAC and then trying to measure battery voltage will cause an incorrect reading, such as 0.0 V, to be displayed. Setting the meter to read VDC and then checking AC voltage will also result in incorrect readings. This may cause damage to the meter, and worse, cause you to sustain a shock and possible injury.

In addition, be aware of how different meters switch between measuring AC and DC volts (Figure 18-18). The meter in the left image requires moving the selector switch between AC and DC. However, the meter in the right image can easily be switched between AC and DC on accident by bumping the button. It is important to set the meter correctly and confirm it remains correct when you are measuring voltage to prevent damage to the meter and personal injury.

Equally important to setting the correct type of voltage is to set the correct voltage range. Some meters, like that shown on the right in Figure 18-18, do not automatically detect and select the correct voltage range. To measure battery voltage, this type of meter needs to be set to the correct voltage scale. The voltmeter measures up to the voltage in each scale. If you try to measure a fully charged battery voltage on the 4-VDC scale, the meter will not register a reading because the battery voltage is above the scale. If using a manual-range DMM, always use the lowest scale possible to obtain the most accurate reading.

■ ***Measuring Voltage.*** As a technician, you will be measuring a lot—and a lot means over and over many times. This is to make sure your testing setup is correct, to make sure that your readings are correct, and that you are getting the readings you expect. Failure to take accurate voltage readings can cause hours of lost time, lost income, and frustration due to following the wrong diagnostic path.

Begin by making sure your meter is working by testing battery voltage on a known good battery. Once you have verified your meter works, you can move on to

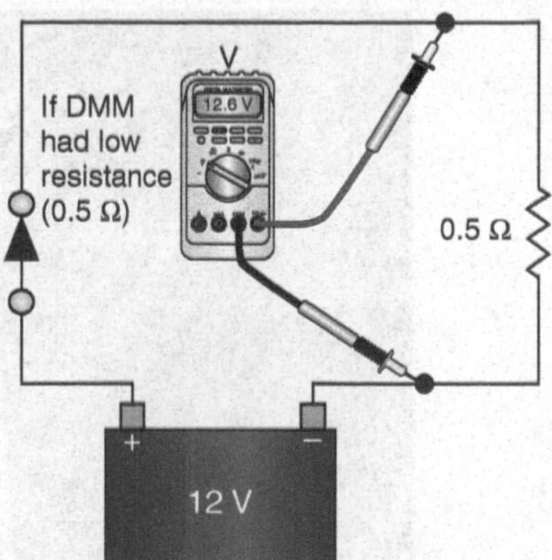
V
12.6 V
If DMM
had low
resistance
(0.5 Ω)
0.5 Ω
12 V
Why does a DMM need to have low resistance?
Using Ohm's law, the 0.5-Ω resistor draws 25.2 amps of current.
If the DMM, connected in parallel to measure voltage, had low resistance, the circuit resistance drops to 0.25-Ω amperage increases to 50.4 amps.
$\frac{0.5 \times 0.5}{0.5 + 0.5} = \frac{0.25}{1}$ or 0.25 Ω
This will cause very high current flow, which is dangerous for the circuit, the DMM, and you.

FIGURE 18-16 Low internal resistance in the meter will cause the placement of the meter into the circuit to change circuit operation. It would also allow for a lot of current to flow through the circuit and through the meter.

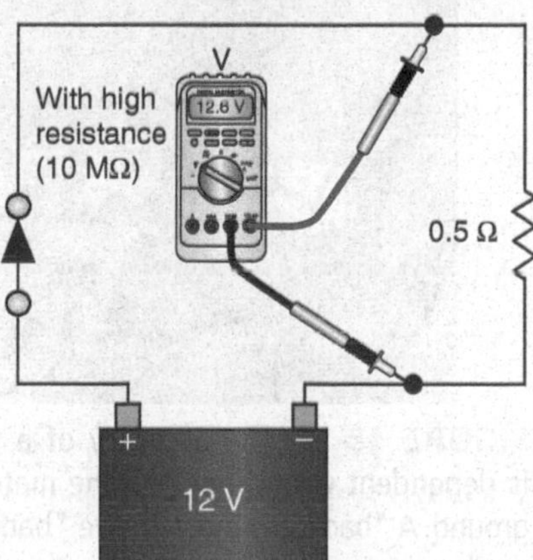
V
12.6 V
With high
resistance
(10 MΩ)
0.5 Ω
12 V
To safely measure the voltage in the circuit, the DMM has very high internal resistance, often 10,00,000 ohms (10 MΩ) or more.
When the leads are placed in parallel with the circuit, the DMM's resistance is so high that it has alomost no effect on the circuit.
$\frac{10\text{ M} \times 0.5 =}{10\text{ M} + 0.5 =} \frac{5{,}000{,}000}{10{,}000{,}000.5}$ or 0.499999975 Ω
Since the resistance remains unchanged, the current flow is unchanged.

FIGURE 18-17 The high internal resistance has, for all practical purposes, no effect on a 12-volt circuit.

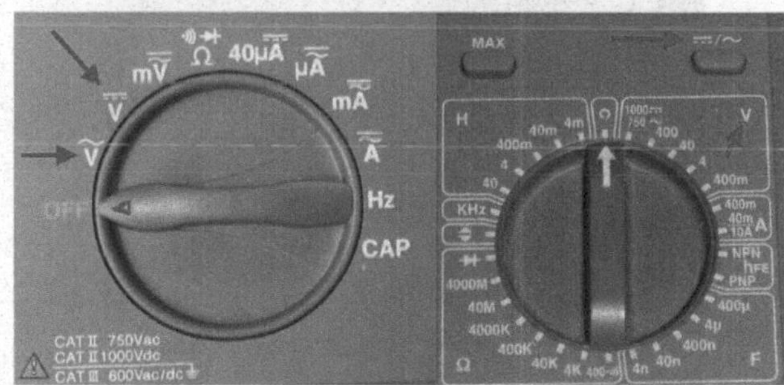
mV
Ω 40μA
μA
mA
V
A
Hz
CAP
OFF
CAT II 750Vac
CAT II 1000Vdc
CAT III 600Vac/dc
MAX

FIGURE 18-18 Two different methods of selecting VAC or VDC. It is important that the correct setting is used to prevent incorrect readings, damage to the meter, and personal injury.

performing some tests. To take a voltage reading, you should first determine three things:

1. **Where to take the reading**. This is probably the easiest of the three. If you diagnosing a light bulb that is not working and you have already eliminated the bulb itself as the problem, you will want to measure voltage to the bulb at its socket. However, you can check voltage with a component unplugged (**Figure 18-19**), or with the component installed and using a backprobe tool (**Figure 18-20**). If checking for power to a component, such as a light bulb, removing the part and checking at the connector is likely the best method. If you have to check for voltage at a component in operation, you will need to use a backprobe pin to push into the back of the connector along the wire to take a reading.
2. **How will you connect the meter to negative or ground?** This can be easy or difficult depending on the vehicle and where testing needs to take place. Failure to use a "good" ground can cause you to waste a lot of time looking for a problem that does not exist because you did not check the meter's ground connection. Examine the meter reading in **Figure 18-21**. The positive meter lead is connected to a taillight socket to check the voltage to the taillight bulb. The negative lead is attached to the body where the taillight housing attaches to the bed of the truck. The meter is reading 4.26 volts, which should seem like

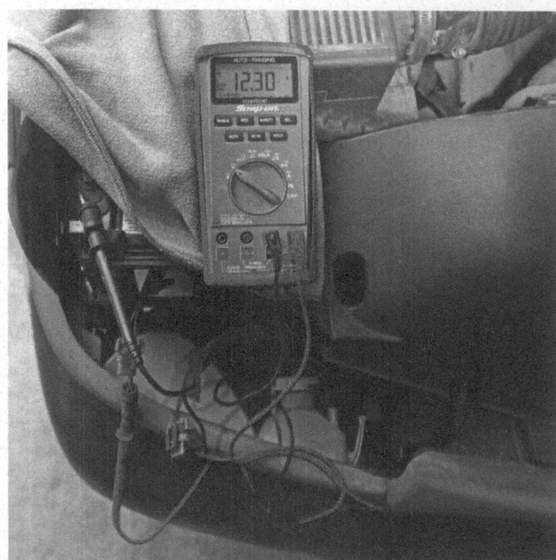

FIGURE 18-19 Measuring voltage at a connector. The setup shown here shows how to check for voltage at a specific location with the load in the circuit removed. This type of measurement is useful for checking for power to something such as a bulb.

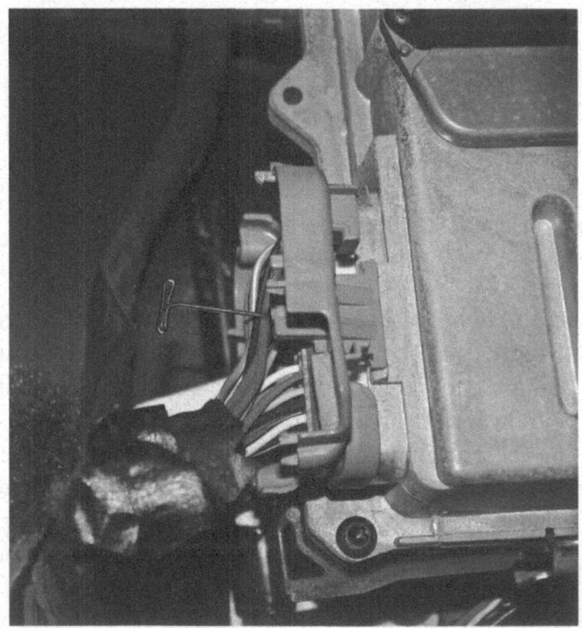

FIGURE 18-20 To check for power at a component with the connection still attached, a backprobe tool or T-pin can be used to touch the metal terminal inside the connector to take a reading.

a problem. However, the same test is also shown in **Figure 18-22**, but with the negative lead attached to a "good" ground. The voltage reading in Figure 18-22 is acceptable based on the conditions in which the test was performed. What is important to understand is that by attaching the meter's negative lead to just

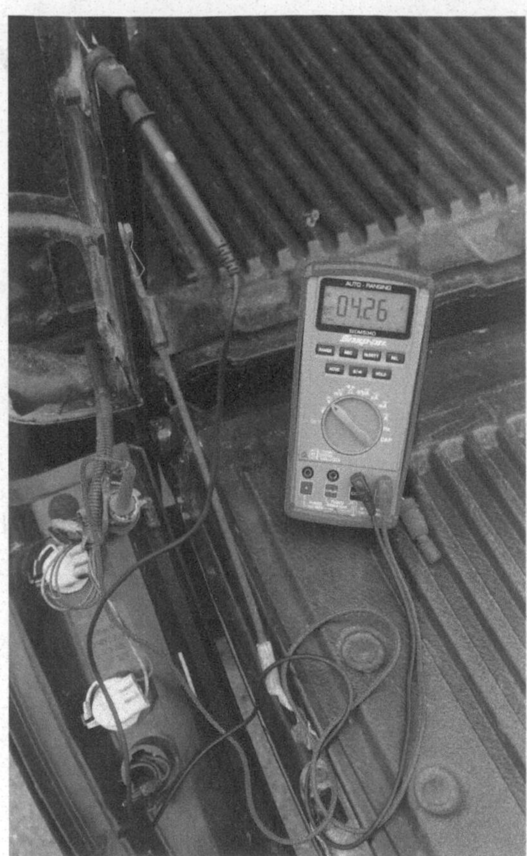

FIGURE 18-21 The accuracy of a voltage measurement is dependent upon how well the meter is able to reference ground. A "bad" ground will give "bad" readings.

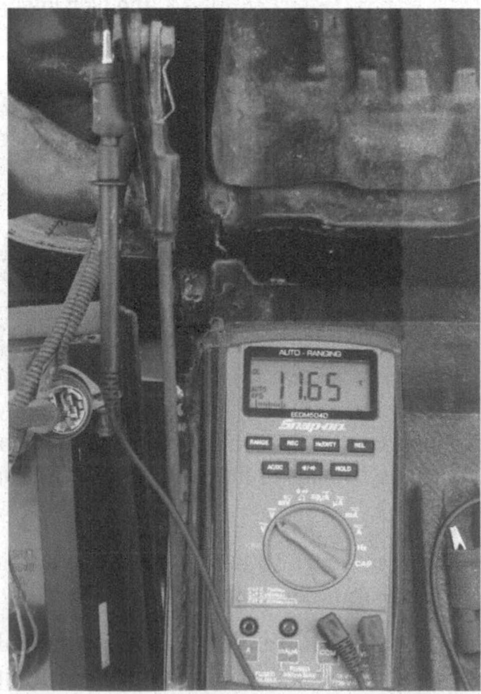

FIGURE 18-22 The same bulb is being tested from Figure 18-21. This time, the meter's negative lead is attached to a better ground location and the reading is acceptable.

any piece of metal is not the best way to ground the DMM. If a good location to ground the meter is not available where you are testing, connect a long jumper wire back to the battery and use it to provide the connection for the negative meter lead.

3. **What reading are you expecting to see?** This is where things can get interesting. The meter may or may not display what you think it should; this is most likely your fault (see Figure 18-21). The DMM, no matter what brand, how good, or how expensive, can only display readings based on what is present at the ends of the two leads. It is up to you to connect the meter correctly and determine if the readings are correct or not. In **Figure 18-23**, a DMM is connected to a light bulb and is displaying the voltage used by the bulb. Based on the image, would this be considered acceptable or is there a problem?

   Figure 18-24a is similar except that it shows what the meter may read at different locations in the circuit. With the DMM's negative lead connected to the battery negative, all reading will be referenced to battery negative or 0.0-volt. In **Figure 18-24b**, with the meter's negative lead connected to the bulb ground connection, the readings are different than in Figure 18-24a. The readings displayed by the DMM depend on how you connect the leads to the circuit. None of these readings are incorrect, but because of where the leads are placed, different readings can mean different things. This is where it is important for you to understand how to connect and read the meter.

   Once you have taken the measurement, you will also need to be able to interpret the readings displayed on the meter. For example, a voltage reading of 246 mV is the same as a reading of 0.246 volts, or slightly less than a quarter of a volt. How the meter displays low and high values will depend on the brand and model you are using. **Table 18-1** shows examples of how readings may be displayed and how to interpret them.

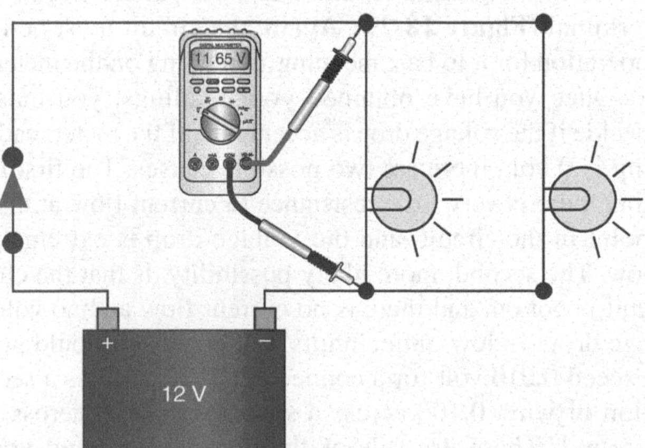

FIGURE 18-23 The meter displays what is present between the two test probes. In this example, it shows the voltage drop of the first light bulb.

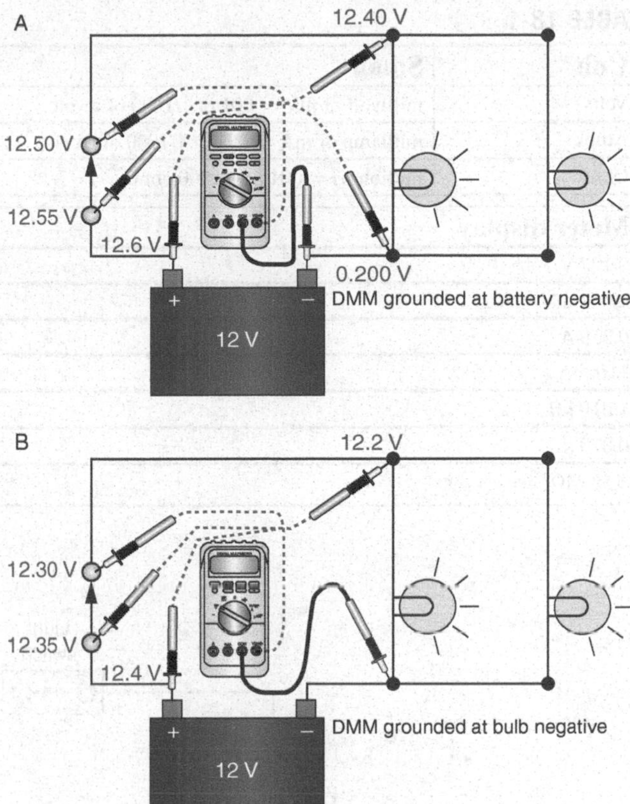

FIGURE 18-24 (a) Voltage readings at various points in the circuit based on the meter's negative lead connected to the battery negative. All voltages shown are referenced to 0.0 volt. (b) Voltage readings at the same points but with the meter's negative lead connected to the ground side of the bulb. Voltage readings are different because the voltage at the ground side of the bulb is greater than the voltage at the battery negative terminal.

■ ***Voltage Drop Testing.*** The term **voltage drop** defines how much voltage is being used by a component, connection, or conductor. In a perfect world, the load would consume all the voltage in a circuit and nothing would be lost by any other parts in the circuit. In reality, all components of an electrical circuit have resistance, and each part will consume some of the available voltage. This is normal and expected, as shown previously in Figure 18-24.

Voltage drop testing is important when there is a problem, such as a corroded connection, that affects the operation of the load (**Figure 18-25**). This means that the intended load will not receive all the voltage it should. When this occurs, light bulbs may appear dim or motors

TABLE 18-1

Unit	Small	Big
Volts	millivolt or mV = .001 or 1/1000 of a volt	kilovolt or kV = 1000 volts
Amps	milliamp or mA = .001 or 1/1000 of an amp	Kiloamps and megaamps not common in automotive
Ohms	milliohms = .001 or 1/1000 ohms	kilohms or kΩ = 1000 ohms, megaohms or MΩ = 1,000,000 ohms

Meter display	Actual reading
0.127 V	127 mV (millivolts)
231.1 mV	0.231 volts
0.270 A	270 mA (milliamps)
110 mA	0.110 A
1.010 kΩ	1,010 ohms
0.879 kΩ	879 ohms
2.56 MΩ	2,560,000 ohms

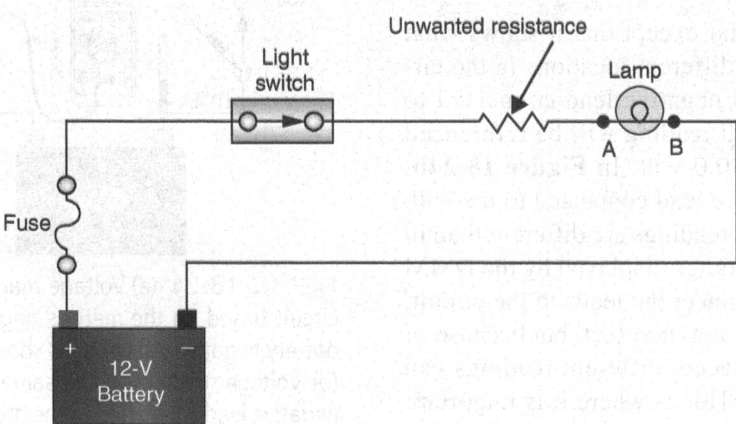

FIGURE 18-25 Unwanted resistance in the circuit will affect circuit operation. The resistance will consume voltage and decrease current flow to the bulb.

may move slowly. If the resistance is too great, the in tended load will not work at all.

Voltage drop testing is a dynamic test, meaning that the circuit must be in operation. If the circuit is off, there is neither current flow nor voltage being used, so voltage drop cannot be tested. To test voltage drop:

1. Make sure the meter is set to VDC and the leads are in the correct jacks before you connect the meter to the circuit. Test the meter on a battery.
2. Turn on the circuit to be tested.
3. Next, connect the meter leads in parallel to the part of the circuit being tested (**Figure 18-26**). The positive meter lead should be connected to the most positive side of the circuit. If the leads are reversed, you will get a negative sign on the meter display. This does not affect the accuracy of the reading, it only indicates that the leads are reversed to the polarity of the circuit.

To check the ground circuit voltage drop, connect the meter to the ground location and the battery negative terminal (**Figure 18-27**). Again, the circuit must be in operation for it to be a meaningful reading on the meter.

Once you have obtained your readings, you must decide if the voltage drop is acceptable. If the meter reading is 0 volt, there are two possible causes. The first is that there is very little resistance to current flow at that point in the circuit, and the voltage drop is extremely low. The second, more likely possibility, is that the circuit is not on, and there is no current flow and no voltage drop. In low-amp circuits, voltage drop should not exceed 0.010 volt for a connection, 0.100 across a section of wire, 0.100 across a switch, or 0.100 across a ground. These are rule-of-thumb numbers, and you should always refer to the service information for the vehicle being serviced for the manufacturer's specs on voltage drop limits.

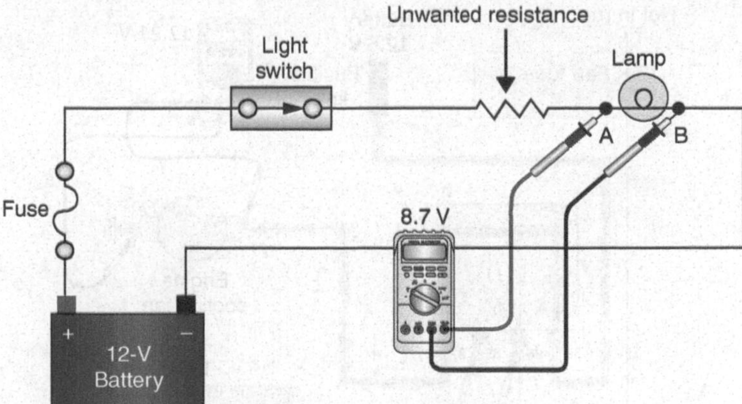

FIGURE 18-26 A voltage drop measurement is taken to determine how much voltage is being used at a particular location in a circuit. It is important to remember that the circuit must be in operation to perform a voltage drop test. The voltage shown on the meter indicates that there is something else in the circuit using voltage.

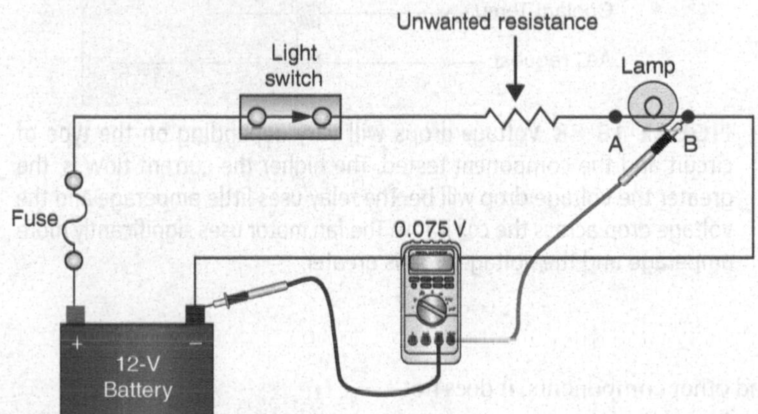

FIGURE 18-27 Checking the voltage drop of the ground circuit. In this example, the voltage drop on the ground is acceptable. This indicates that for this example, the problem is in the power side of the circuit.

On high-amp circuits, such as the starter motor and generator, up to 0.500 volt is considered acceptable during high-amp operation. It is important to remember that voltage drop will increase as the current flow through the circuit increases. This means that the voltage drop of a low-amp circuit, such as a relay coil, will often be lower than the voltage drops found in the circuit that is controlled by the relay (**Figure 18-28**). The relay control circuit, highlighted in yellow, is a low-current circuit and should have a very low voltage drop. The DMM reading at the relay coil in Figure 18-28 shows that of the available 12.6 volts at the battery, 12.55 volts are being used by the relay coil. This means that only 0.05 volts is lost in that circuit. The DMM connected to the fan motor reads 12.21 volts. This means that the voltage drop in the fan motor circuit, highlighted in red, is 0.39 volts. Because the fan draws more current than the relay, the total voltage drop in the motor circuit will likely be greater than that in the fan control circuit.

When you are working on computer circuits, the current flow is usually in the milliamp range. This means that increased voltage drop has a much larger effect on circuit operation than on other noncomputerized circuits. It also means that you need to use a good-quality meter with at least 10 megohms internal resistance. This ensures that the meter, when it is connected to the circuit, will not draw enough current to alter the operation of the circuit and the reading you receive.

■ ***Why Am I Testing Voltage Drop?*** As discussed earlier, all the components of a circuit have some resistance to current flow. You also know that a by-product of this resistance to current flow is heat. While you and I may not be able to feel any noticeable change in the temperature

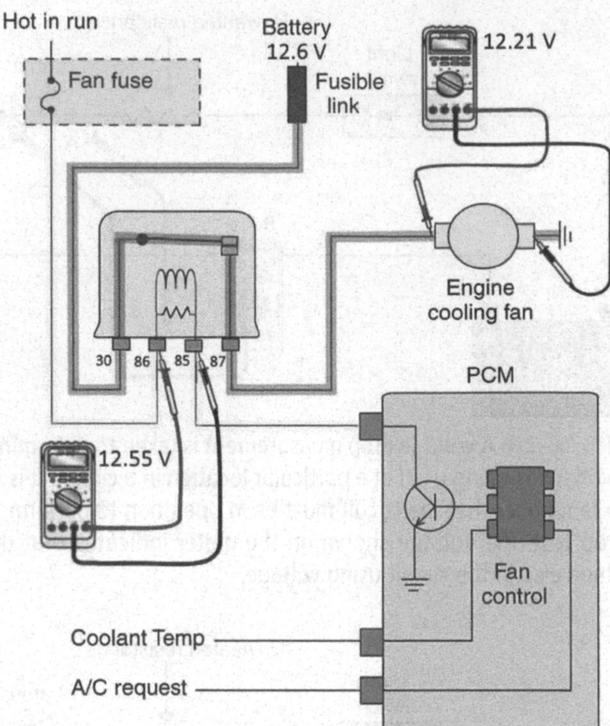

FIGURE 18-28 Voltage drops will vary depending on the type of circuit and the component tested. The higher the current flow is, the greater the voltage drop will be. The relay uses little amperage and the voltage drop across the coil is low. The fan motor uses significantly more amperage and the voltage drop is greater.

of the wire, connectors, and other components, it does not mean that it does not take place.

The heat generated during circuit operation and the cooling when the circuit is off, in addition to the normal temperature variations of the weather, cause thermal cycling. **Thermal cycling** is the warming and cooling of components during operation. This cycling, along with vibration, can create electrical connection problems that increase the resistance of the circuit. This increases the voltage drop at various points throughout the circuit. This means that over time, and even without any other factors involved, circuits can develop problems just from being used.

Terminals remain in contact with each other due to the tight fit of the metal-to-metal contact. An example of a female terminal and the wear on the tension plate is shown in **Figure 18-29**. Over time, thermal cycling and vibration can lead to a condition called fretting (**Figure 18-30**). **Fretting** is the very small movement of the terminals against each other that wears away the outer layer of the metal. This can cause an intermittent connection or even a complete loss of connection. Because a combination of fretting and vibration can cause the connection problem, the intermittent connection can appear and disappear randomly, making it very difficult to find. If the terminals are part of a computer circuit, an intermittent connection can cause problems such as misfires, hesitation and unwanted ABS activation. In a high-amperage circuit, the poor connection can lead to overheating of the terminal and connector, causing the connector to melt.

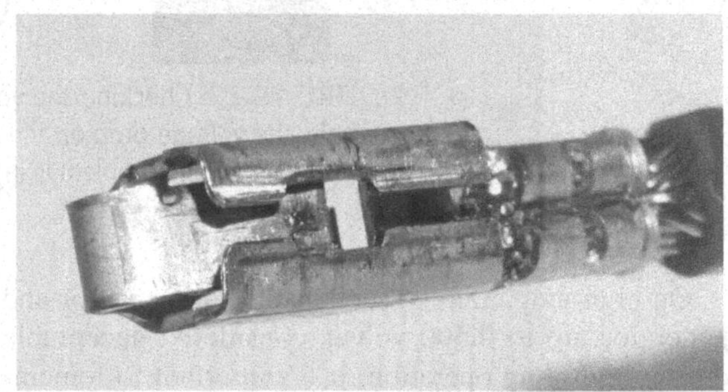

FIGURE 18-29 A closeup of a female terminal and the wear on the metal from disconnecting and reconnecting it several times.

Refer to Chapter 18 in the Lab Manual for exercises about performing voltage drop tests.

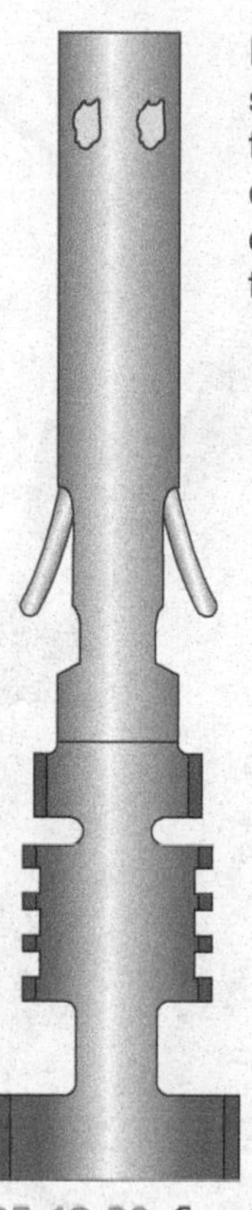

Fretting appears as small scuff marks or smudges on the terminal. These are caused by the abrasive contact between moving terminals.

FIGURE 18-30 Constant movement of terminals against each other causes fretting, which over time, can cause a resistive connection.

RESISTANCE

Resistance in an electrical circuit has many sources, but it originates from the empty space between the atoms that make up the components of the circuit. Even though atoms are extremely small and seemingly packed together so closely that matter appears solid, there is a lot of empty space between electrons. In practical application, resistance depends on many factors, such as conductor material, length, diameter, and temperature.

■ ***Safety Precautions for Measuring Resistance.*** To measure the resistance of a component or circuit, the power must first be removed. If you attempt to measure resistance on a live circuit, the readings will not be correct and damage to the meter can result. While in most automotive applications the risk of injury from performing a resistance test is minimal, if you are working on the high-voltage system of a hybrid vehicle, failure to remove the power from the circuit before testing can result in serious injury or death.

■ ***What Is a Resistance Measurement?*** The DMM you use has a battery to power the meter and the display. The battery also provides the power to measure resistance. When the meter is used to test resistance, voltage from the battery flows from the meter, through the leads, and back to the meter. The two meters shown in Figure 18-15 show the voltage generated by the meter, in this case 0.759 volt, when a technician is testing resistance. The meter on the left is sending out a small voltage, as measured by the meter on the right. The meter performs a voltage drop test when it is measuring resistance. This is why the power from the circuit must be removed before attempting to measure resistance. The voltage produced by the meter is very small and would be completely lost in a circuit operating at 12 volts and no meaningful readings would be obtained.

■ ***Measuring Resistance.*** Begin by making sure the power is removed from the component or circuit you are going to test. Next, select the resistance setting on the meter. This is shown as Ω, the symbol for Ohms of resistance. Place one meter lead on one connection to the component and the other meter lead on the other connection (**Figure 18-31**). When you are measuring resistance, the polarity of the meter leads does not matter as there is no negative resistance.

■ ***Why Am I Going to Measure Resistance?*** As discussed earlier the best way to find unwanted resistance in a circuit is by performing a voltage drop

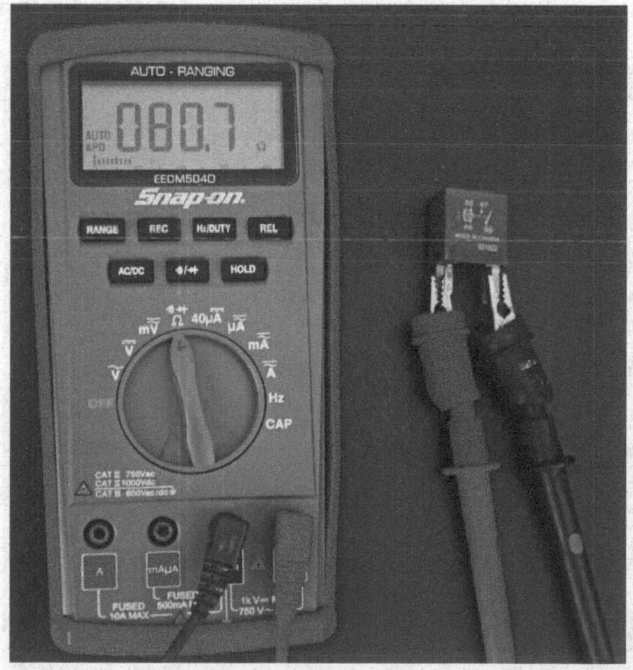

FIGURE 18-31 Testing the resistance: Make sure power is disconnected or the component is removed, as shown here. Place the meter selector on the Ohm's (Ω) setting and connect the leads.

test. This does not mean that a resistance test has no value; in fact, measuring resistance is often necessary in diagnosing electrical concerns. A resistance test can also be a continuity test. **Continuity** means that there is a complete circuit path, as in a continuous or unbroken path from point A to point B. Many meters have a continuity test function with an audible warning when continuity is present. This setting is indicated on the DMM shown in **Figure 18-32**. Depending on the DMM, there may be a button press (Figure 18-32), or a separate setting you select with the selector switch. The continuity setting can be helpful and timesaving when you are checking items such as fuses or connections because a beep from the meter indicates continuity, meaning you do not have to constantly look back at the meter screen to see what is displayed. Common resistance measurements include:

- Checking fuses for open circuits.
- Checking across connections and switches (**Figure 18-33**).
- Checking bulb resistances and for open filaments (**Figure 18-34**).
- Testing resistances of sensors, such as temperature and position sensors.

It is important to remember that taking a resistance measurement does not test the component or circuit under load. This means that a resistance test cannot always find the source of a problem. A **resistance test** is a passive or unloaded test of a circuit. This means that it can be used to confirm a problem, such as an open wire or connection, but it cannot prove that a component is "good" or is not faulty during operation. To

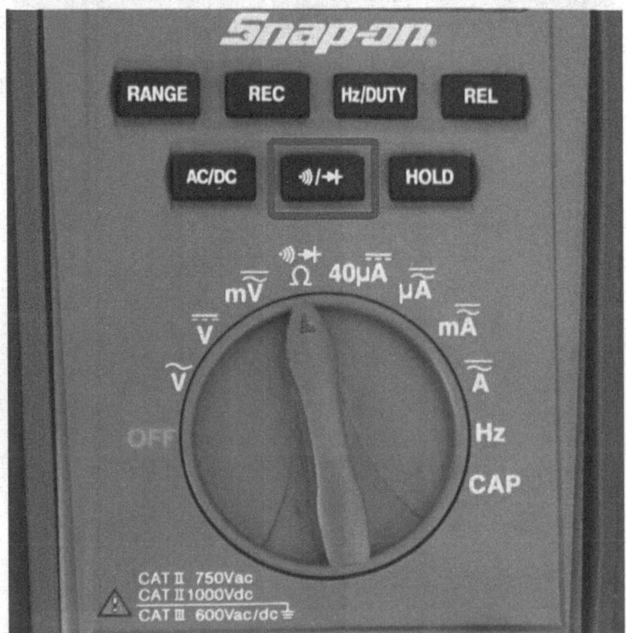

FIGURE 18-32 The continuity setting uses a tone to indicate that a circuit is complete and has low resistance, often less than 30 ohms. This is handy when checking switches and other components as you do not have to keep an eye on the display.

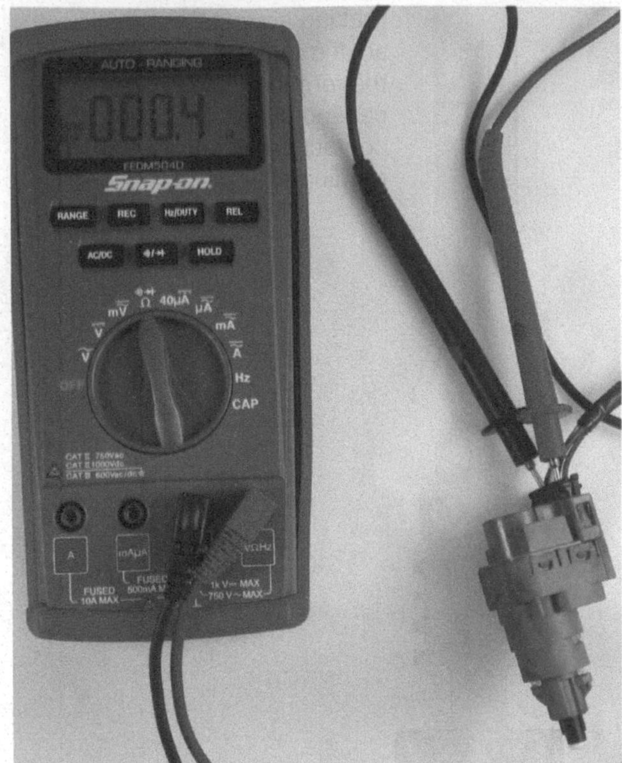

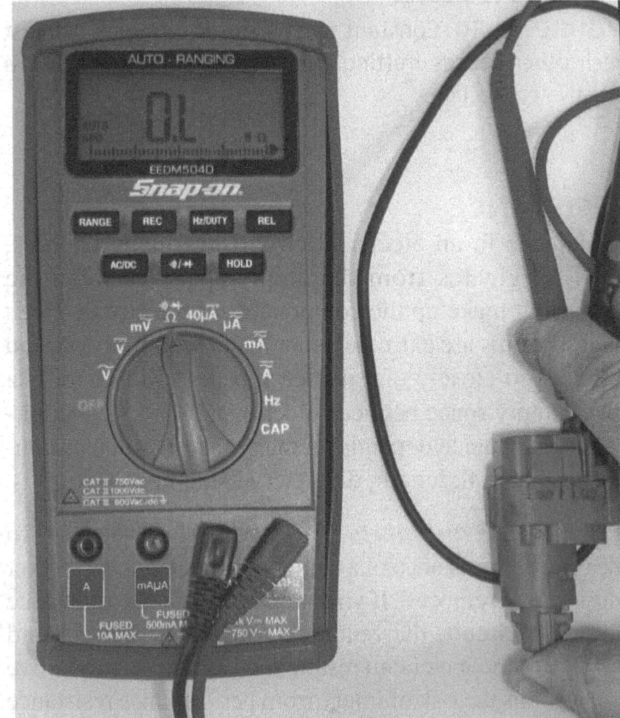

FIGURE 18-33 Testing the resistance of a switch. With the switch closed, there should be very low resistance (a), and with the switch open, very high or infinite resistance (b), as shown here.

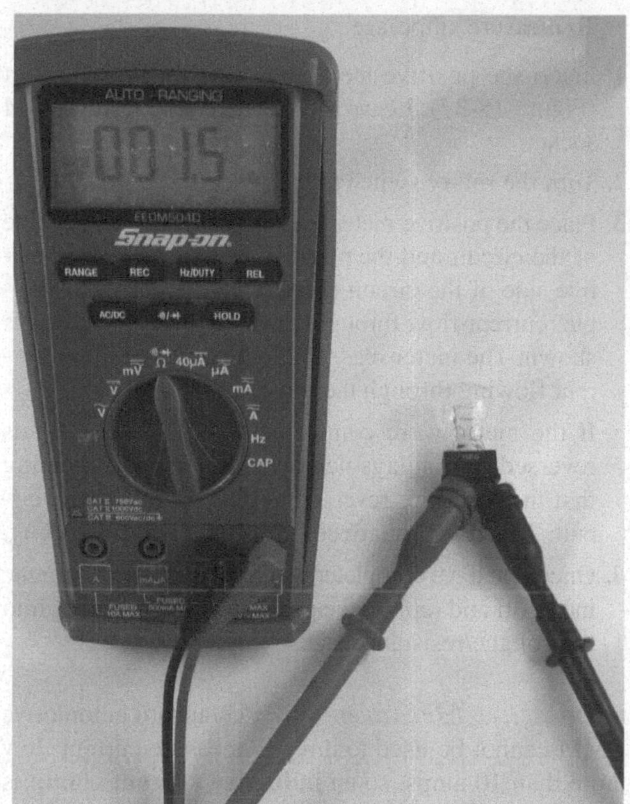

FIGURE 18-34 Testing a bulb filament. The resistance reading indicates the filament is intact.

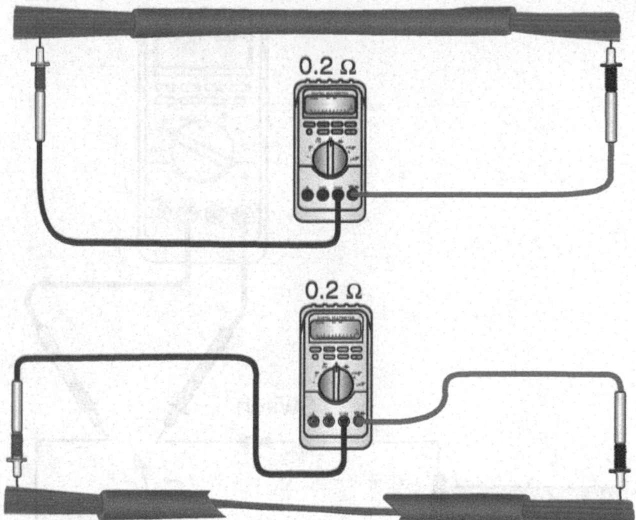

FIGURE 18-35 A resistance measurement can confirm something is open but cannot confirm something is good. A mostly broken wire will show the same resistance as an unbroken wire. In operation, the broken wire will have an impact on circuit operation.

illustrate this, think of a section of automotive wire as shown in **Figure 18-35**. The wire is made of strands of smaller wires bundled together and covered in insulation. Suppose this wire is made of 12 strands of wire and that it runs from the front of the vehicle to the back of the vehicle and powers the electric fuel pump. Damage to the wire could break 11 of the 12 strands. Measuring the resistance of the wire will not show a problem because one strand is sufficient to allow the meter to check resistance. The resistance of the wire will not show high because the one strand is able to carry the very low voltage and current from the meter for the resistance test. However, when the circuit is active and current needs to flow through the wire to power the fuel pump, the single strand will not be sufficient to carry the necessary current flow to drive the pump.

In this example, two outcomes are possible. The first possibility is that the single strand of wire will overheat and melt, which will open the circuit. This will then be able to be detected by a resistance test as an open circuit. The second possibility is that the single strand will be able to carry some current but at an increased resistance. The will result in an increased voltage drop, which will cause a decrease in power available for the pump. This will likely cause the pump to spin too slowly, reducing fuel pressure and volume.

When faced with a poorly performing fuel pump, many people replace the pump and expect the problem to be solved. In this situation, however, a new pump will not fix the problem. Performing a voltage drop test at the pump would show a problem and allow correct diagnosis. For this reason, it is important to understand the limits of a resistance test and its value in diagnosing electrical concerns.

See Chapter 18 in the Lab Manual for exercises on testing resistance.

AMPERAGE

Amperage refers to the amount of electrons actually flowing in a circuit and it is directly proportional to the voltage and resistance in the circuit. In modern vehicles, electronic circuits operate with milliamps, while the starter motor may use several hundred amps.

■ ***Safety Precautions When Measuring Amperage.*** The biggest hazard when measuring amperage in automotive service is to the meter. Incorrect meter setup will blow the fuse in the meter. In some cases, it may permanently damage the meter or can destroy the meter's test leads.

■ ***Direct Measurement.*** Low-amperage circuits, usually 10 amps or less, can be tested directly with the DMM. To measure amperage with the DMM, you must move the positive test lead to the amperage jack. This bypasses the high internal resistance of the meter and places the positive and negative leads in series within the meter. **Figure 18-36** shows a typical meter connection

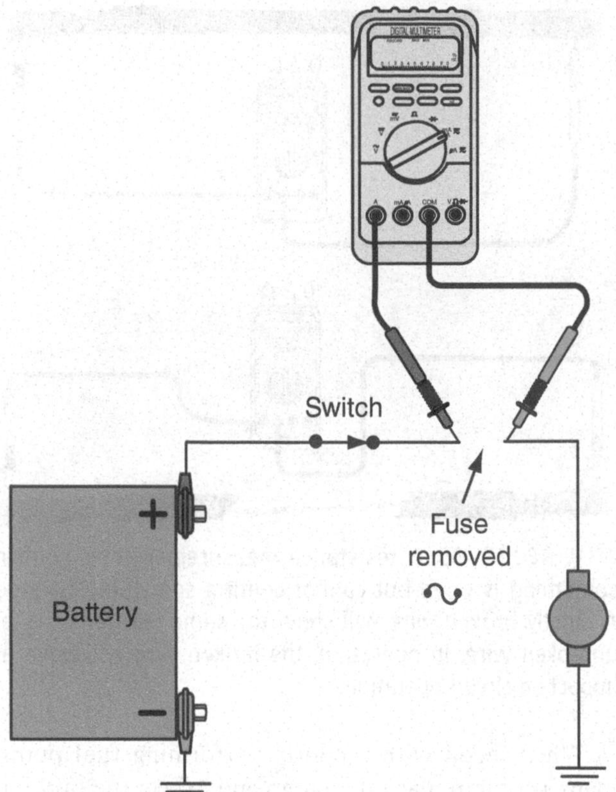

FIGURE 18-36 To measure current flow in a circuit, the meter must be placed in series into the circuit. Most meters have a maximum rating of 10 amps. This is to prevent damage to the meter and leads from excessive current flow. To measure amperage: Turn the circuit off, open it a convenient point, configure and connect the meter, and turn the circuit on.

for amp testing. If your meter is capable of both AC and DC amperage measurements, be sure to set the meter on DC amps when you are testing on the automobile.

The amperage test setup places internal amperage fuse in series with the positive and negative meter leads. **Figure 18-37** shows how the leads connect to the DMM and how the internal fuse is wired into the meter's leads connections. The fuse is there to protect the meter from damage if the measured amperage is greater than the meter's amp rating. **Do not replace this fuse with one of a higher rating. This can cause damage to the meter if the current flow exceeds the meter's rating but is below the rating of the replacement fuse.**

For the DMM to measure amperage, it must be placed in series with the circuit (Figure 18-36). This means that whatever current flows through the circuit also flows through the meter. The resistance of the meter must necessarily be very low for testing amperage, otherwise the load of the meter in the circuit will affect the results of the reading.

To measure amperage:

1. Insert the positive meter lead into the Amp socket (Figure 18-37). Leave the negative lead in the COM socket.
2. Turn the rotary switch to the A setting.
3. Place the positive meter lead on the most positive side of the circuit and the negative lead on the least positive side of the circuit (**Figure 18-38**). In this example, current flow through a basic light bulb circuit is shown. The meter reads 0.226 amp (226 mA) of current flowing through the three bulbs in series.

   If the meter were connected with the meter leads reversed, it displays negative amperage, indicating that the leads are reversed to the polarity of the circuit. This does not affect the accuracy of the reading.
4. Once you have completed your measurement, turn the meter off and replace the positive meter lead back into the voltage/resistance jack.

■ ***Inductive Measurement.*** A standard automotive DMM cannot be used to directly measure current flow more than 10 amps, so an **inductive current clamp** is

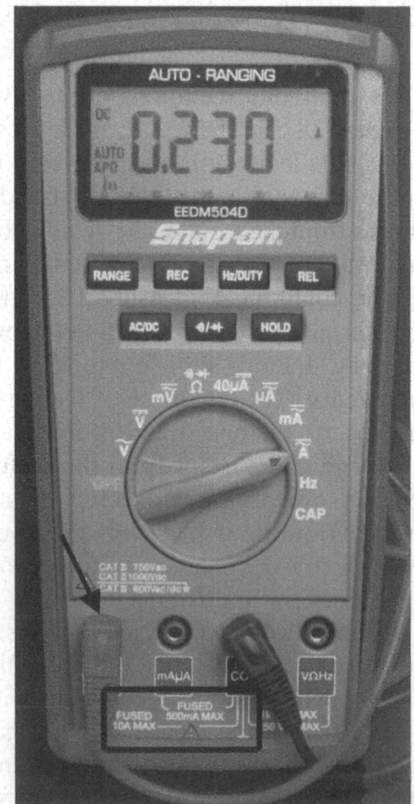

FIGURE 18-37 The positive meter lead plugs into the amperage jack. Note this meter has a maximum amperage test rating of 10 amps. The amp fuse is located inside the meter and is in series between the Amperage and COM jacks.

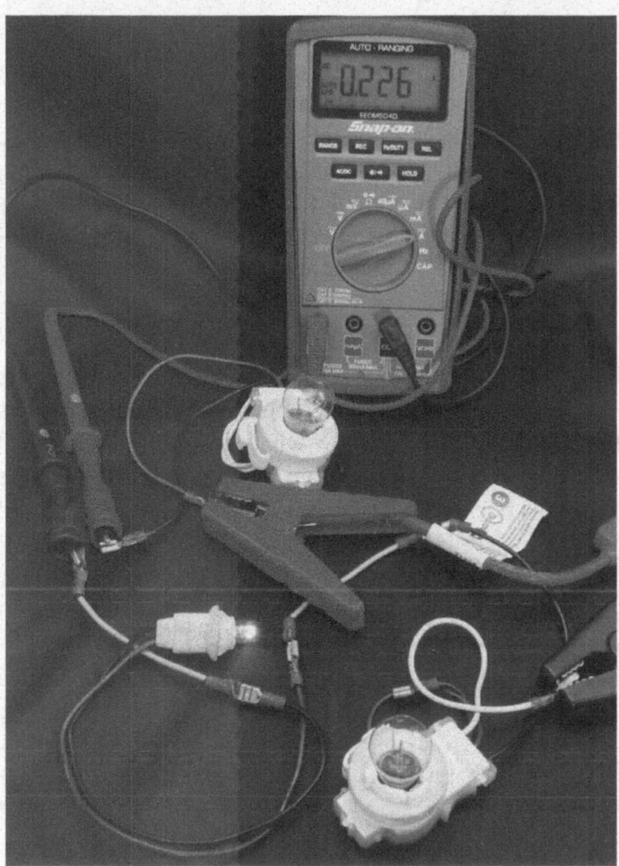

FIGURE 18-38 A meter connected to a series circuit and measuring current flow.

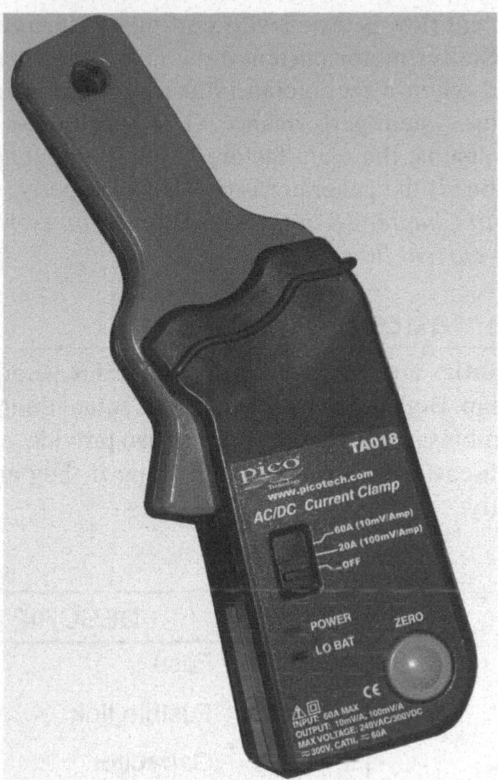

FIGURE 18-39 To measure current flow without opening the circuit or to measure more than 10 amps, an inductive current clamp is used. This tool converts the amperage measurement into a voltage reading. Connected to a scope, current clamps can provide invaluable information about circuit operation.

used (**Figure 18-39**). Also known as current probes, these devices measure current flow by detecting the strength of the magnetic field created by current flowing through a conductor. This type of ammeter is also used on battery/starter/generator testing equipment such as the VAT-40/45 and Snap-On AVR tester. By using a current probe, large amounts of amperage can safely be measured.

To use a current probe, remove the standard positive and negative leads from the voltmeter and insert the leads from the probe. Set the voltmeter to display VDC and turn on the clamp. This clamp has two settings, 100 mV equals 1 amp or 10 mV equals 1 amp. The setting you choose depends on the amount of current to be measured. In **Figure 18-40**, the clamp is set to 100 mV equals 1 amp, and the clamp is placed around the battery cable to measure key-off battery draw. The reading of 0.146 volt equals 1.46 amps because the clamp is set on 100 mV equals 1 amp.

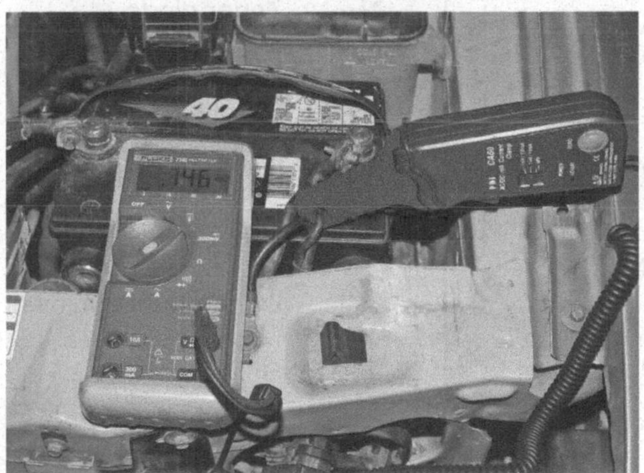

FIGURE 18-40 An example of using a current clamp and a DMM to measure amperage draw on a battery.

■ ***Why Am I Going to Measure Amperage?*** One reason to measure amperage is to diagnose key-off battery draw. If a vehicle's battery is going dead after sitting a day or two, there is probably a circuit that is not shutting down that is draining the battery. To determine the amount of the draw, key-off battery drain is measured with an ammeter. Another example of measuring amperage is for fan motor circuits. As the fan motor ages and spins more slowly, current flow increases in the fan circuit. This can damage the fan relay, blower resistor, or fan controller even though the fan may seem to operate normally.

Current flow is tested with starting and charging systems. Starter motor current draw is a basic check performed when a technician is testing the starting and charging system performance. Generator output is also tested against the manufacturer's rated output specs to determine if the generator is operating properly.

Go to Chapter 18 in the Lab Manual for exercises in testing current flow.

Schematics

Schematics are electrical wiring diagrams, similar to a road map. Both are representations of actual things, both use symbols and a legend, and both can provide you with a much easier way to get things done if they are used correctly.

ELECTRICAL COMPONENT SYMBOLS

To reduce confusion and to standardize electrical repair, symbols are used to represent components. Using drawings of electrical components as they actually look would be confusing and a waste of time. Instead, you will learn what the electrical symbol used for the component looks like. Most of the time, the symbol is similar for all manufacturers and their schematics.

■ ***Why Symbols Instead of Showing the Actual Component?*** Symbols used in wiring diagrams are representations of electrical components (**Figure 18-41**). If the diagrams tried to show what each part looked like, it would be very difficult to gain much familiarity with the diagrams because every manufacturer's diagrams would

SYMBOL	DESCRIPTION	SYMBOL	DESCRIPTION
	Fuse		Circuit breaker
	Fusible link		Relay
	Capacitor		Switch, push button
	Resistor		Switch, mechanical
	Variable resistor		Switch, pressure
	Coil		Switch with light
	Diode		Sender, oil/water/fuel
	Zener diode		Component case is directly attached to metal part of car
	Ground		5-cavity connector (5 out of 5 are used)
	Junction point	C103, 2	Female terminal
J1	Junction point ID		Connection reference number for component location list
	Light, single filament		Cavity number
	Light, double filament		Male terminal
	Splice		
	Wires crossed, not connected		
	Switch, N.O.		
	Switch, N.C.		
	Male/Female connection		
	Female connection		
	Male connection		

FIGURE 18-41 Examples of common wiring symbols. Most manufacturers use common types of symbols in their wiring diagrams.

have different-looking symbols. Also, components such as relays and DC motors perform the same functions no matter what type of vehicle in which they are installed. They may have some slight physical differences but electrically they operate the same.

READING A SCHEMATIC

Until the 1980s, most vehicles had four to five pages of schematics for the entire bumper-to-bumper electrical system. Now, just the engine operating systems often are four or five pages long. Being able to read a schematic can save you hours of diagnostic time. Once you understand what the vehicle concern is, you can start by looking at the wiring diagram. This can help you narrow down the possibilities of what and where the problem may be before touching the car.

Modern cars and trucks have a lot of wiring. This also means they need more wiring diagrams. Fortunately, each system is usually broken down into smaller pieces. This is so that you do not have to try to isolate a particular system from within pages of diagrams for the entire vehicle. This allows you to focus on just the system and components necessary without having to try to separate what you want to look at from what you do not need.

■ ***Parts of the Schematic.*** All circuits contain the same basic components: power, ground, circuit protection, circuit control, and the load. Each of these parts is shown in the wiring diagram along with the wiring connecting them altogether. **Figure 18-42** shows a wiring diagram for a horn circuit with its different sections labeled.

- It is a common practice that the power for the circuit is shown at the top of the page (A). Notice that the power is shown coming from a fuse box, which is drawn in a broken line. This indicates that the entire component is not shown in this diagram. This

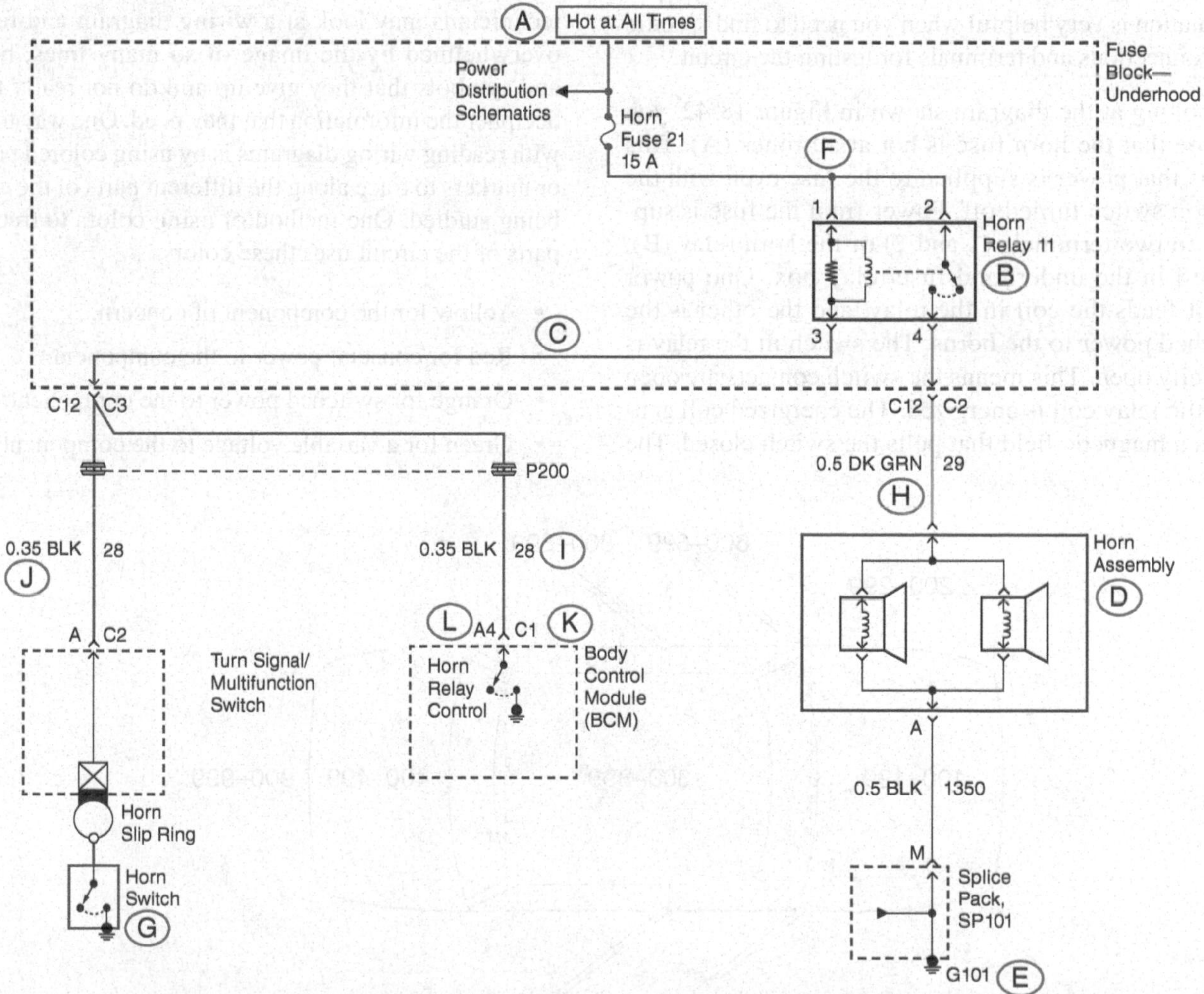

FIGURE 18-42 Learning to understand a wiring diagram is one of the most important skills for electrical system service. Correctly reading and understanding a diagram can save you hours of diagnostic time. Most diagrams share common traits, such as power at the top of the image and grounds at the bottom.

is because the entire fuse box is not relevant to this circuit and would just clutter up the diagram and possibly cause confusion.

- Below the power section is the horn relay (B). Connected to the relay are the control circuit (C) and horns (D).
- The grounds are shown toward the bottom of the diagram (E) and at the horn switch.
- Lines that cross each other are wires that make up the circuit but are not connected to each other. When an intersection of two wires has a dot, it means the wires are connected at this point (F).
- Switches and relays are shown in their normal operating state, meaning a normally open switch, like a horn switch, is drawn as an open circuit (G).
- Also provided are the wire colors (H), circuit numbers (I), wire size (J), connector ID (K), and terminal numbers within the connector (L). This example indicates connector C1, terminal A4 at the BCM. This information is very helpful when you need to find specific connections and terminals for testing the circuit.

Looking at the diagram shown in Figure 18-42, you can see that the horn fuse is hot at all times (A). This means that power is supplied to the fuse even with the ignition switch turned off. Power from the fuse is supplied to two terminals (1 and 2) of the horn relay (B), located in the underhood fuse/relay box. One power circuit feeds the coil in the relay, and the other is the switched power to the horns. The switch in the relay is normally open. This means the switch contacts are open until the relay coil is energized. The energized coil generates a magnetic field that pulls the switch closed. The relay coil ground circuit, at terminal 3 of the relay, is supplied by the horn switch in the steering wheel pad (G) or the BCM. When the driver presses the horn pad, contacts under the pad connect the wire to ground. This ground allows the current to flow through the relay coil, forming the magnetic field necessary to close the switch. When the relay closes, power flows through the switch contacts, out terminal 4 and to the horns (D) to ground G101 (E). When the driver releases the horn pad, the contacts open, removing the ground from the relay. With the ground removed, current no longer flows through the coil and the magnetic field collapses. This causes the switch contacts to open, turning off the horns.

Items such as connector, splice, and ground numbers are also shown in a wiring diagram. These numbers are based on a coding system for the different sections of the vehicle (**Figure 18-43**). Section or zone coding is used to group circuits, connections, grounds, and other electrical components into specific areas of the vehicle.

■ ***Using Color to Trace a Circuit.*** Students and technicians may look at a wiring diagram and feel so overwhelmed by the image of so many lines, boxes, and symbols that they give up and do not really try to decipher the information that they need. One way to help with reading wiring diagrams is by using colored pencils or markers to trace along the different parts of the circuit being studied. One method of using colors to trace the parts of the circuit uses these colors:

- Yellow for the component of concern.
- Red for constant power to the component.
- Orange for switched power to the component.
- Green for a variable voltage to the component.

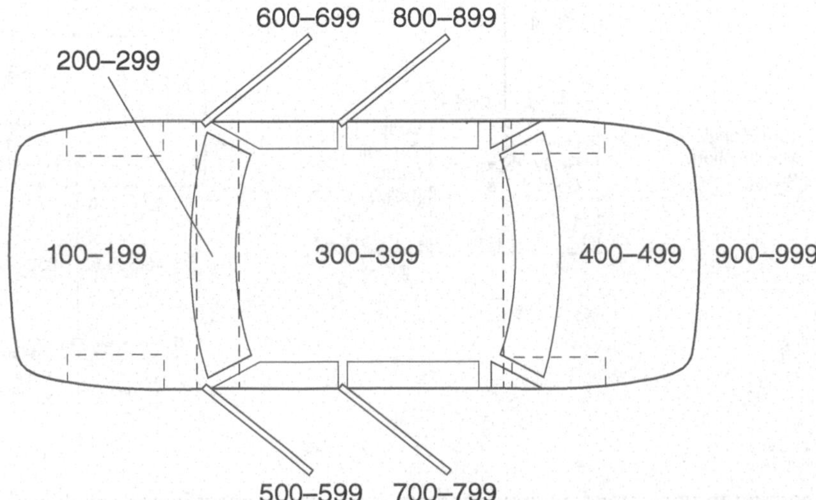

FIGURE 18-43 Wiring zones are used to simplify component, connection, and circuit numbers.

- The direct path to ground is colored black.
- Blue is for ground controlled by a switch or other control device.

Refer to Chapter 18 of the Lab Manual for practice in reading and tracing wiring diagrams.

By using colors to highlight the parts of the circuit, you can quickly determine what parts of circuit perform certain functions. This can be very useful when you are trying to isolate a certain part of a circuit that is shown in a larger diagram. Using the colors allows you to focus on just what components and wiring you need while being able to safely ignore other sections of the diagram.

The diagram shown in **Figure 18-44** uses colors to separate the sections of the circuit as described. Color coding the parts of the circuit allows you to break the circuit into small, easily identified sections. Because each section has its own purpose and function, this method often helps you to determine how the circuit functions piece by piece. Once you have identified the major parts of the circuit, trace each component and wire in the circuit with the color that corresponds to its function (Figure 18-44). After you have traced the circuit, you should have a good understanding of the path of current flow through the circuit. This can be invaluable as a diagnostic tool because when you understand how the circuit operates, it is easier to determine what is wrong when the circuit does not function.

For example, imagine you are presented with a vehicle with the circuit shown in Figure 18-44. The complaint is that the horns does not sound, and after confirming the complaint, you:

1. Check the horn fuse, and it is intact and has power on both sides.
2. Check for power to the horn with the horn button pressed and find there is no power to the horn.

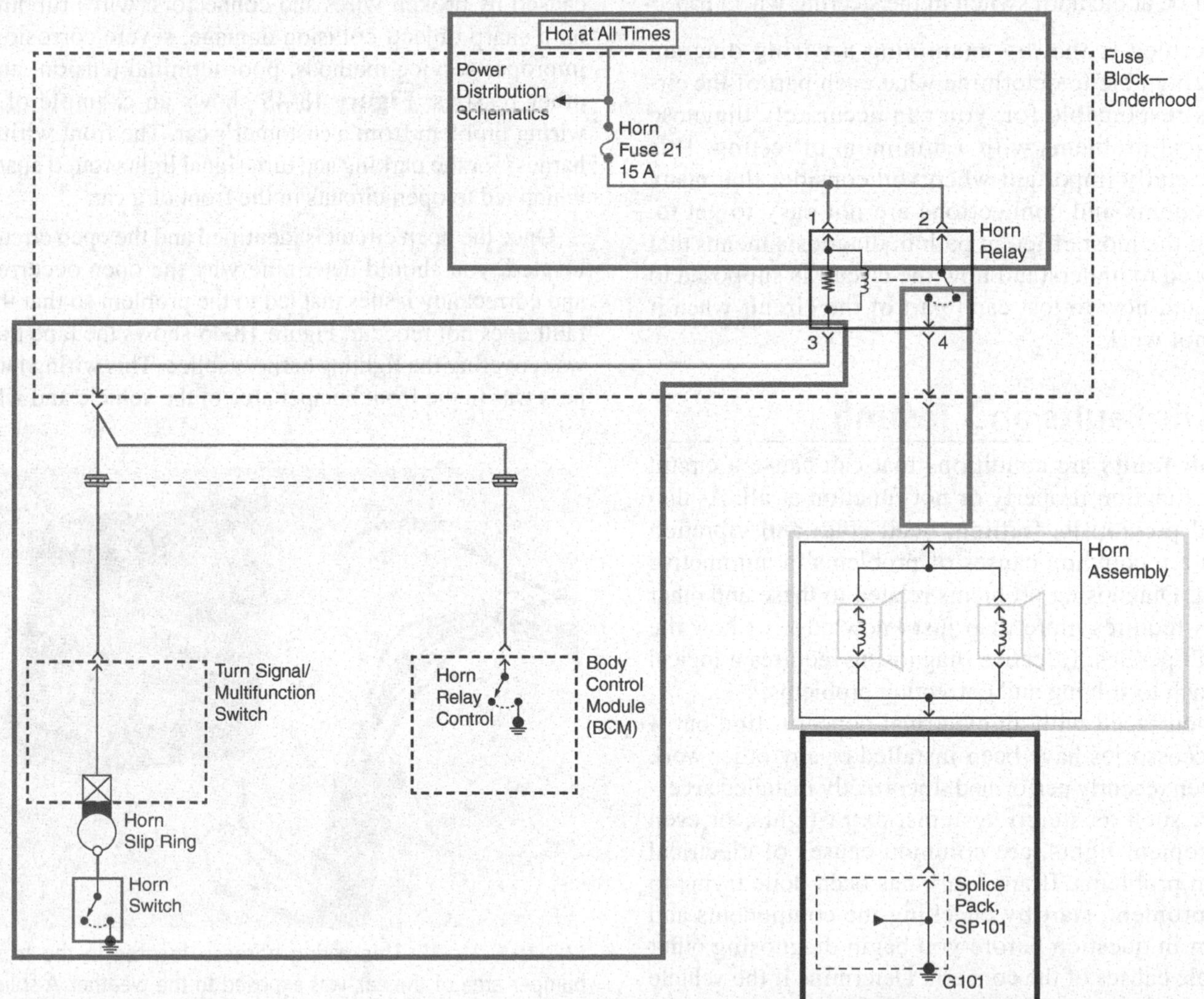

FIGURE 18-44 Using color to define parts of a circuit can be a helpful aid in learning how a circuit works and in understanding the diagram. Color helps to quickly identify the power and ground parts of this circuit. Once you have traced out the circuit, you will understand what to expect when you actually test the circuit on the vehicle.

If you are unfamiliar with this vehicle, your next step will likely be to get a wiring diagram of the horn circuit.

Using the wiring diagram, you can quickly separate the circuit into sections and begin testing. Because the relay is a central point in the circuit, it is a logical place to continue your diagnosis. If power is present at terminals 1 and 2 at the relay, then you can determine if the problem is in the relay, the relay control circuit, or in the path that supplies power to the horns.

3. With a meter or test light, check for power at relay terminal 4 with the horn button pressed. If power is present, there is an open circuit between the relay and the horn.
4. If there is no power, check for ground at relay terminal 3 with the horn pressed. If there is ground at terminal 3 but no power at terminal 4, the relay is defective. If there is no ground at terminal 3, the circuit between the relay and horn switch ground is open. Your next test will be at the horn switch in the steering wheel pad.

The idea is that by examining a wiring diagram and being able to determine what each part of the circuit is responsible for, you can accurately diagnose electrical problems with a minimum of testing. This is especially important when you consider that many components and connections are not easy to get to. Taking the most efficient path in diagnosis means that you need to understand how the circuit is supposed to work and how to test each part of the circuit when it does not work.

Circuit Faults and Testing

Circuit faults are conditions that can cause a circuit to not function properly or not function at all. As discussed previously, fretting, heat, cold, and vibration issues are common causes of problems in automotive wiring. Diagnosing problems related to these and other causes requires more than just knowledge of how the circuit operates. Effective diagnosing requires a logical approach to solving and correcting problems.

When faced with an electrical concern, find out if any accessories have been installed or any other work has been recently performed. Incorrectly installed accessories, such as stereo systems, extra lights, or even replacement lights, are common causes of electrical system problems. If any work has been done trying to fix a problem, start by checking the components and system in question before you begin diagnosing other possible causes of the concern. Determine if the vehicle has been in an accident or other situation where repairs have been made. Problems that develop after a wreck or other major repairs have taken place are often related to the prior work.

CIRCUIT FAULTS

A circuit fault is something that causes a circuit to not operate as it was designed. The results of circuit faults range from very basic, such as a burned-out light bulb, to complex. Regardless of the type of fault or the circuit affected, diagnosis requires a logical and systematic approach. Begin by determining what type of fault you are dealing with.

There are three types of circuit faults you may encounter: open circuits, short circuits, and unwanted resistance. Each has its own symptoms and diagnostic procedures.

■ ***Open Circuits.*** As the name implies, an **open circuit** is one in which the path for current flow is broken, or open, causing the circuit not to function. The obvious symptom of this problem is a nonfunctioning component or system.

■ ***How Open Circuits Occur?*** Open circuits can be caused by broken wires and connectors, wires rubbing on a sharp object, collision damage, severe corrosion, improper service methods, poor terminal tension, and other reasons. **Figure 18-45** shows an example of a wiring problem from a customer's car. The front wiring harness for the parking and turn signal lights rotted apart, which led to open circuits in the front of a car.

Once the open circuit is identified and the open circuit located, you should determine why the open occurred and correct any issues that led to the problem so that the fault does not reoccur. Figure 18-45 shows the tape that was covering the lighting harness splice. This wiring harness was in the front bumper area of the vehicle and was

FIGURE 18-45 This wiring harness, located in the front bumper area of the car, was exposed to the weather. A splice was supposed to connect three of the loose wires together, but it had rusted away. This tape was all that was protecting the splice that rusted away. Inadequate protection from the weather caused the circuit to fail.

exposed to the environment. The tape covering the splice was insufficient protection from the weather, which allowed water to enter the harness and corrode away the splice connecting the wires. Once the wires separated, a problem with the front parking and turn signal lights developed.

■ ***Diagnosing Open Circuits.*** The first step in diagnosing an open circuit is to confirm the complaint. If the customer complaint is that the power windows are not operating correctly, sometimes working and other times not working, first duplicate the problem. In some cases, the complaint is an intermittent problem, requiring you to try to re-create the conditions in which the problem presents itself. For this example, we will diagnose an inoperative power window complaint. Note that this example is for a vehicle that does *not* use a body control module (BCM) or other module to control power window operation.

When you are verifying the complaint, also make sure that the problem the customer describes is actually a problem and not a normal function of the vehicle. If the vehicle has a power window lockout switch, be sure that the customer is aware that the passenger windows do not operate, even from the driver's master switch, if the lock-out is engaged.

1. Start by verifying that with the key on, the power windows are inoperative. Try to open each window from the master switch and from each door switch. If none of the windows work, locate the fuse panel and look for a power window circuit fuse or circuit breaker. It is a good idea to refer to a wiring diagram to determine how power is supplied to the circuit, either with a circuit breaker or by how many fuses are used and their locations in the circuit.
2. Next, check for power on both sides of the fuse. This can be done by using a test light connected to a good ground. If power is present on both sides of the fuse, carefully remove the master power window switch from the door panel. Locate the power supply wire to the switch and test for voltage. This is at terminal 2 in the wiring diagram shown in **Figure 18-46**. If no voltage is present, refer to the wiring diagram for any connection locations between the fuse panel and the master switch. Locate the nearest connection to the switch and check again for voltage. If none is present, repeat at each connector until voltage is present. Once you have found a point where voltage is present in the circuit and where the voltage is not present, you have determined where the open circuit is located.

In this example, voltage is present on the interior side of the driver's door jamb but not on the door side of the jamb. This means that somewhere in between the body side of the door and the inside of the door is the open. The wiring for the driver's door is routed through the door jamb and is encased in a flexible conduit (**Figure 18-47**). Because the driver's door is the most used door, and the wiring is subject to movement at the jamb thousands of times over the life of the vehicle, it is likely that the open circuit is located in the harness between the body and the door. Try to operate the power windows while moving the harness. If the windows start to work or start and stop when moving the harness, you have located the problem area. **Figure 18-48** shows an example of damaged wiring in the door jamb conduit. The wiring on this vehicle had been previously repaired for a similar complaint, as shown by the red wire and blue connector. Once the fault is identified, follow the manufacturer's service information on how to remove the door harness for further inspection and repair.

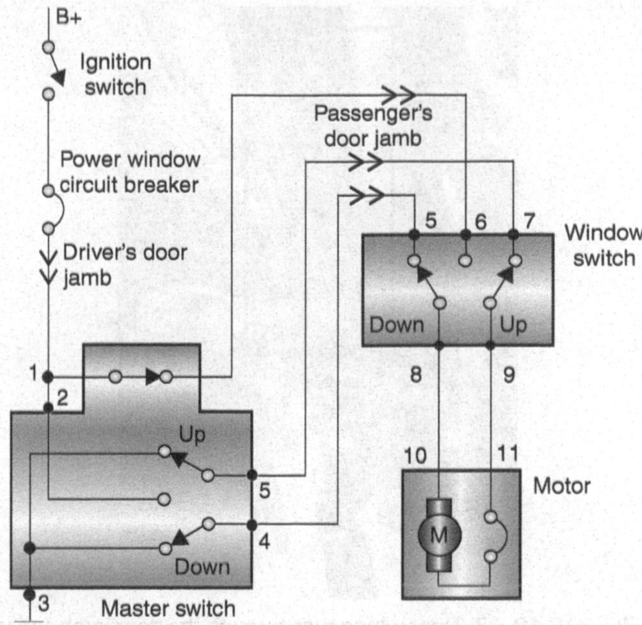

FIGURE 18-46 An example of a power window circuit. Using a diagram, you can identify the locations where to test the circuit.

■ ***Using Jumper Wires.*** Sometimes, when diagnosing an open circuit, a fused jumper wire is used to bypass the open part of the circuit to confirm the fault. Jumper wires are used to reroute current flow around the open in a circuit (**Figure 18-49**). This allows you to make sure that the circuit does in fact work properly when the open is corrected before you actually make the repair. The jumper wire should be fused to prevent circuit damage in the event of accidently shorting power to ground.

■ ***Repairing Open Circuits.*** Once the open has been located, you need to reconnect the wiring to restore proper circuit operation. This can be done in

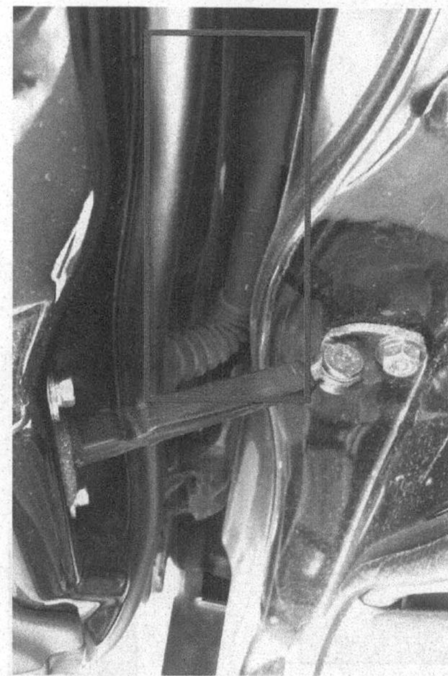

FIGURE 18-47 Door wiring runs through the door jamb and is subject to damage over time as the door is opened and closed.

FIGURE 18-48 An example of damaged door wiring. This vehicle had already had one broken wire repaired.

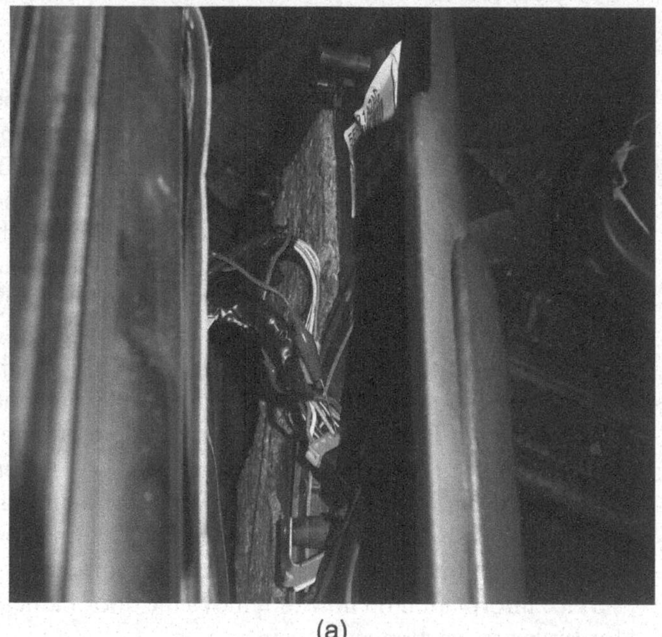

(a)

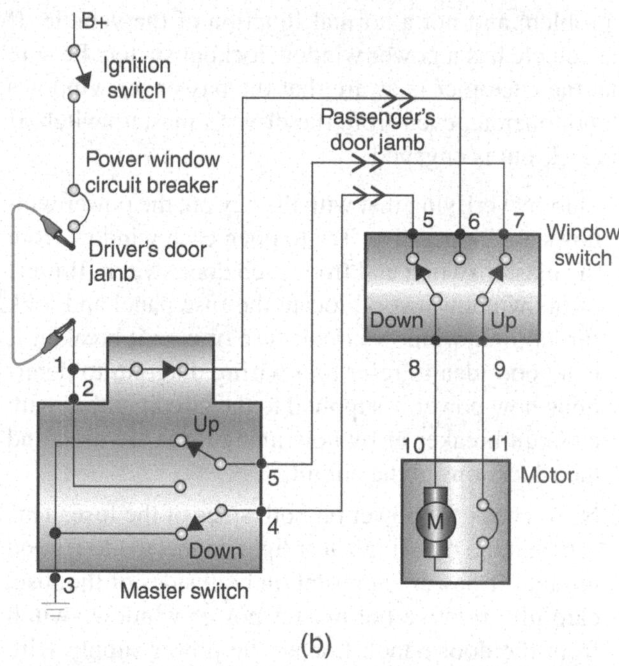

(b)

FIGURE 18-49 (a) A jumper wire can be used to bypass a switch or other suspected open in a circuit. (b) This shows how using a jump wire can be used to bypass the damaged wiring in the door jamb.

several ways, but it is usually done by using some type of wire connector. Many different types of connectors are available, each with specific applications. Wire repair is covered in more detail later in the chapter.

SHORT CIRCUITS

Short circuits can be either a short to power or a short to ground. A short to power often causes an unwanted connection between two different circuits. This causes a circuit to operate when another unrelated circuit is used. A short to ground occurs when the power supply of a circuit touches or connects to ground. While both types of shorts are problematic, a short to ground can lead to overheated wiring and a fire.

■ ***Short to Power.*** A short to power causes unwanted circuit interaction, such as pressing on the brakes causes the horn to sound (**Figure 18-50**). This can occur if a power or ground accidently bridges to another circuit, causing the second circuit to operate unexpectedly.

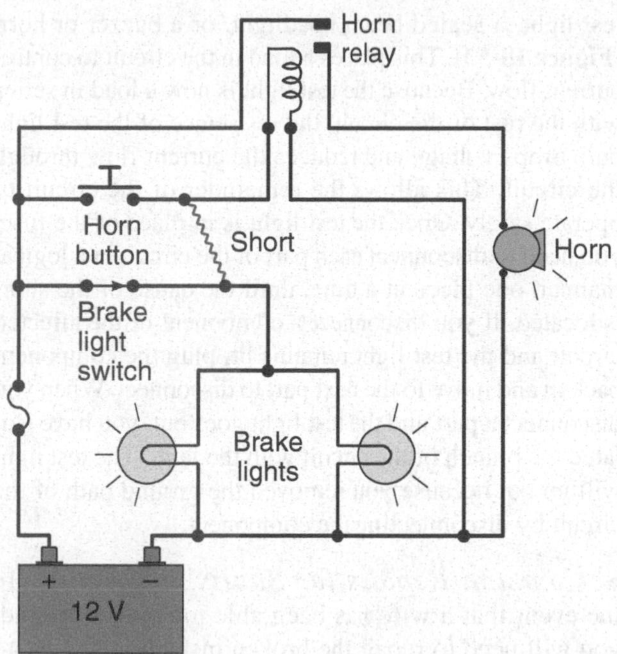

FIGURE 18-50 A short to power can cause a circuit to suddenly start operating when another circuit is active.

■ ***How a Short to Power Occurs?*** Most often, this type of problem occurs inside multifunction switches and in damaged or melted connectors. Sometimes, a short to power occurs when a wiring harness chafes against something sharp and a connection is made as exposed wiring comes into contact.

Sometimes, a short is caused by the customer or someone trying to fix a problem. A vehicle was in the shop because the rear lights were not working properly. A bulb was incorrectly installed and taped into a light socket (**Figure 18-51**). The single filament bulb was forced into a dual filament socket (Figure 18-51b). The result was that the bulb bridged the terminals in the socket and connected the parking and brake/turn signal circuits together.

■ ***Diagnosing a Short to Power.*** Because this problem creates an unwanted connection between two different circuits, this actually helps in the diagnosis because you know which circuits are directly affected. Start by locating a wiring diagram for the circuits and look for possible cross-connection points. Even if the two circuits do not share any connection, such as the brake lights and horn circuits in this example, there may be wiring located close together or even in the same multiterminal connector. Check wiring and connectors that are in common closely for damage, such as melted plastic, and for proper connection of the male and female halves of the connector. Open the connector and look for melted cavities, damaged and/or misaligned terminals.

■ ***Short to Ground.*** A short to ground that occurs before the intended load in the circuit usually causes the fuse for the circuit to open. This is because the short decreases the resistance of the circuit, which increases the current flow beyond the rating of the fuse. An example of a short to ground is illustrated in **Figure 18-52**. Circuits that short to ground are dangerous because of the potential for the circuit to overheat and start a fire. This is especially true if the circuit protection is bypassed in an attempt to get the circuit working.

■ ***Diagnosing a Short to Ground.*** Before you begin to diagnose a short to ground, check to see if any aftermarket accessories have been installed in the vehicle. Improper installation of aftermarket stereos and other

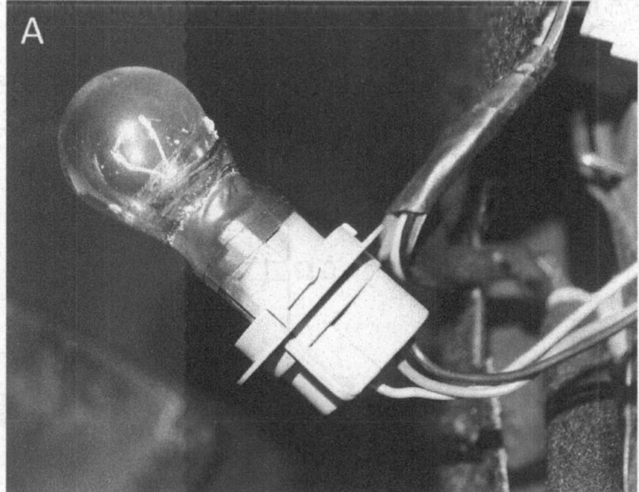

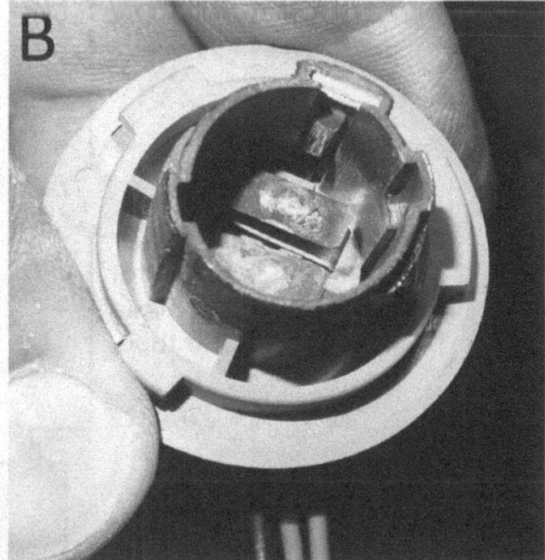

FIGURE 18-51 (a) Improperly secured or routed wiring is often the cause of short circuits. Another cause is from parts being incorrectly installed. Bulbs typically do not require tape to hold them in their sockets. (b) Use of the wrong bulb, a single filament bulb in a two-filament socket, created a short between the parking light and turn signal circuits.

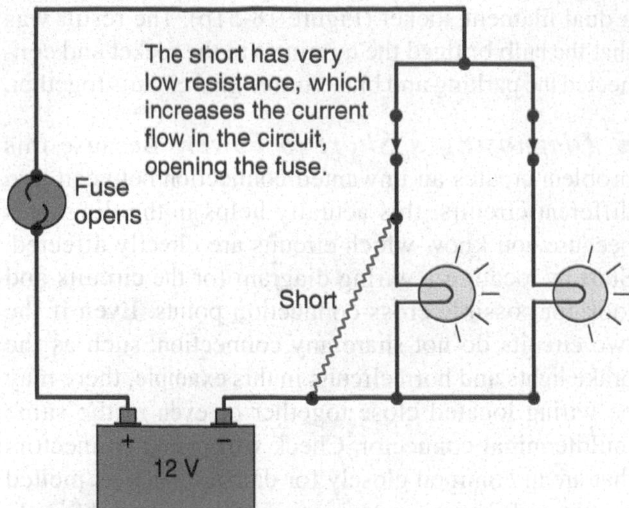

FIGURE 18-52 A short to ground usually will cause the fuse to blow and the circuit to stop working.

equipment can cause an overload on an existing circuit, causing a fuse to blow. Find out if the vehicle has been a collision or suffered other damage. Collision damage can crush wiring or chaff a harness, causing a short. Wiring can also be pinched in between parts, causing the insulation to rub off the wiring and create a short circuit. Always try to determine from the customer a detailed vehicle history and get as much information as possible about when the fault started, what other circuits seem to be affected, and if any recent repairs have been performed.

Because a short to ground usually causes a fuse to blow, you need to find a way to power the circuit without overloading it but without blowing fuse after fuse. One method that many technicians use is to replace the fuse with a test light, a sealed beam headlight, or a buzzer or horn (**Figure 18-53**). This places a load in the circuit to control current flow. Because the test light is now a load in series with the rest of the circuit, the resistance of the test light bulb drops voltage and reduces the current flow through the circuit. This allows the remainder of the circuit to operate safely. Once the test light is in place of the fuse, you need to disconnect each part of the circuit in a logical manner, one piece at a time, until the cause of the short is located. If you disconnect a component of the affected circuit and the test light remains lit, plug the component back in and move to the next part to disconnect. When you disconnect a part and the test light goes out, you have isolated the branch of the circuit with the fault. The test light will go out because you removed the ground path of the circuit by disconnecting the component.

■ ***Common Repairs for Shorts to Ground.*** In the event that a wire has been able to contact ground, you will need to repair the broken insulation and try to reroute the wiring so that it does not happen again. An example of damaged and shorted wiring is shown in **Figure 18-54**. This harness was able to rub against the body, near the trunk support. If the short is caused by a failed component, usually the component needs to be replaced. Once the repairs are made, reinstall the fuse and check circuit operation.

HIGH RESISTANCE/UNWANTED RESISTANCE

When excessive or **unwanted resistance** in the circuit occurs, an additional voltage drop is created in the circuit. This also reduces the current flow in that portion

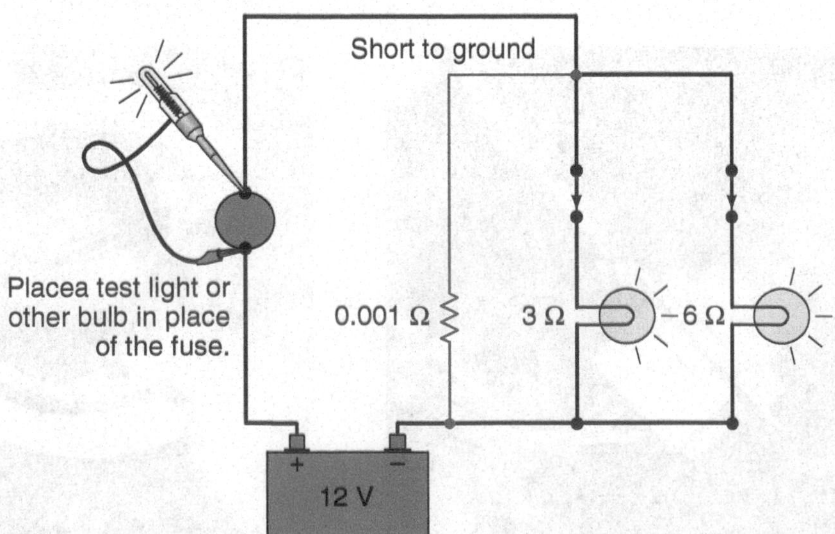

FIGURE 18-53 Diagnosing a short to ground using a test light in place of the fuse. This places a resistance in the circuit and reduces the current flow to safe levels.

FIGURE 18-54 A short to ground in the brake light circuit due to the harness pinching under the trunk hinge and lift support.

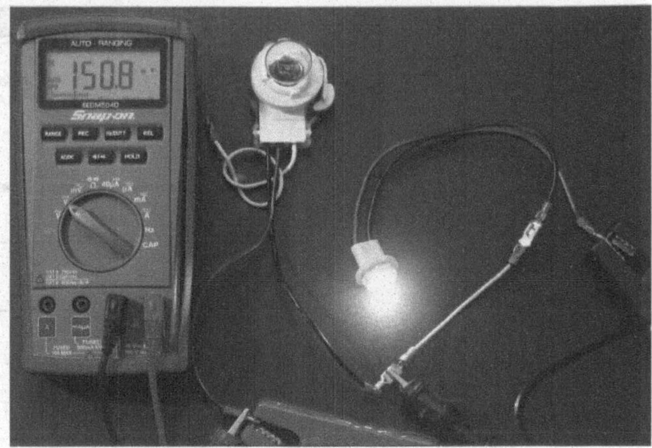

FIGURE 18-55 This illustrates how an unwanted resistance can affect a circuit or component. In this example, the large (1157) bulb represents a brake/turn and tail light bulb that should be working. However, the unwanted resistance the small (194) bulb creates a voltage drop that prevents the 1157 bulb from lighting up.

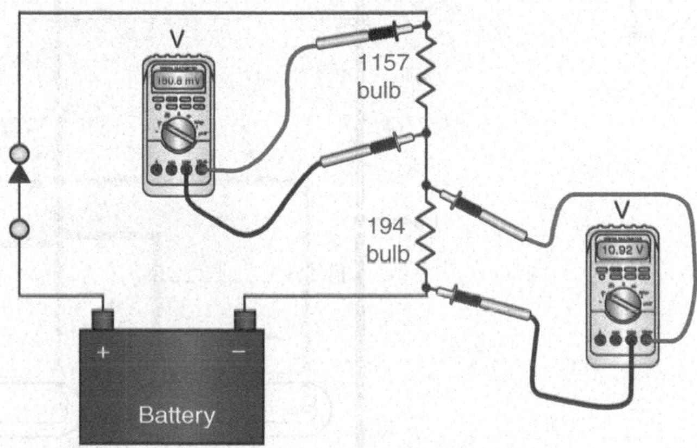

FIGURE 18-56 This illustrates the circuit in Figure 18-55 as it is electrically. Nearly all the power is being used by the unwanted resistance of the 194 bulb.

of the circuit. The resistance may cause the intended load in the circuit to operate improperly, or it may not work at all (**Figure 18-55**). The small (194) bulb in series with the larger (1157) bulb represents an unwanted resistance in the circuit. The 194 bulb prevents the 1157 bulb from working and is using nearly all the available voltage in the circuit (**Figure 18-56**).

■ ***How High Resistance Affects a Circuit?*** The location of the unwanted resistance in the circuit determines what effect the resistance has and on how many circuits. In **Figure 18-57**, unwanted resistance at point A affects both low-beam bulbs. If the resistance is at point B, only the left headlamp bulb is affected. Extra resistance causes the bulbs to be dim because the unwanted resistance is a voltage drop and is reducing current flow in the circuit.

In addition to affecting component operation, high resistance can also cause connectors to overheat and melt. At first, this may not seem to make much sense: increased resistance reduces current flow and makes bulbs dim, motors slow, and similar problems. However, until a circuit opens, a complete circuit will allow current flow. Even if the circuit is not operating correctly, such as with a dim bulb, work is taking place (**Figure 18-58**). The fog light circuit shown has an unwanted resistance affecting the left fog light. Depending on the amount of the resistance, this may cause the left light to be dim or to not illuminate at all.

In **Figure 18-59**, some electrical values are inserted into the circuit and it shows what can happen because of the resistance. The left bulb still has a path to power and ground and will still operate, but it will not be as bright as the right bulb. The unwanted resistance, because it is in series with the left bulb, creates a voltage drop, which reduces the voltage available for the bulb. In this example, the unwanted resistance uses the same voltage and consumes the same power, 18-watts, as the bulb. Bulbs consume power and turn it into heat and light. The unwanted resistance, because it is not a bulb, only generates heat. This is how damage to wiring and connectors can occur from unwanted resistance.

The following are examples of damaged wiring due to resistance problems. **Figure 18-60** shows a headlight connector and switch that was damaged due to a poor connection. **Figure 18-61** is a damaged interior fan motor connector. Any time you find a connector that

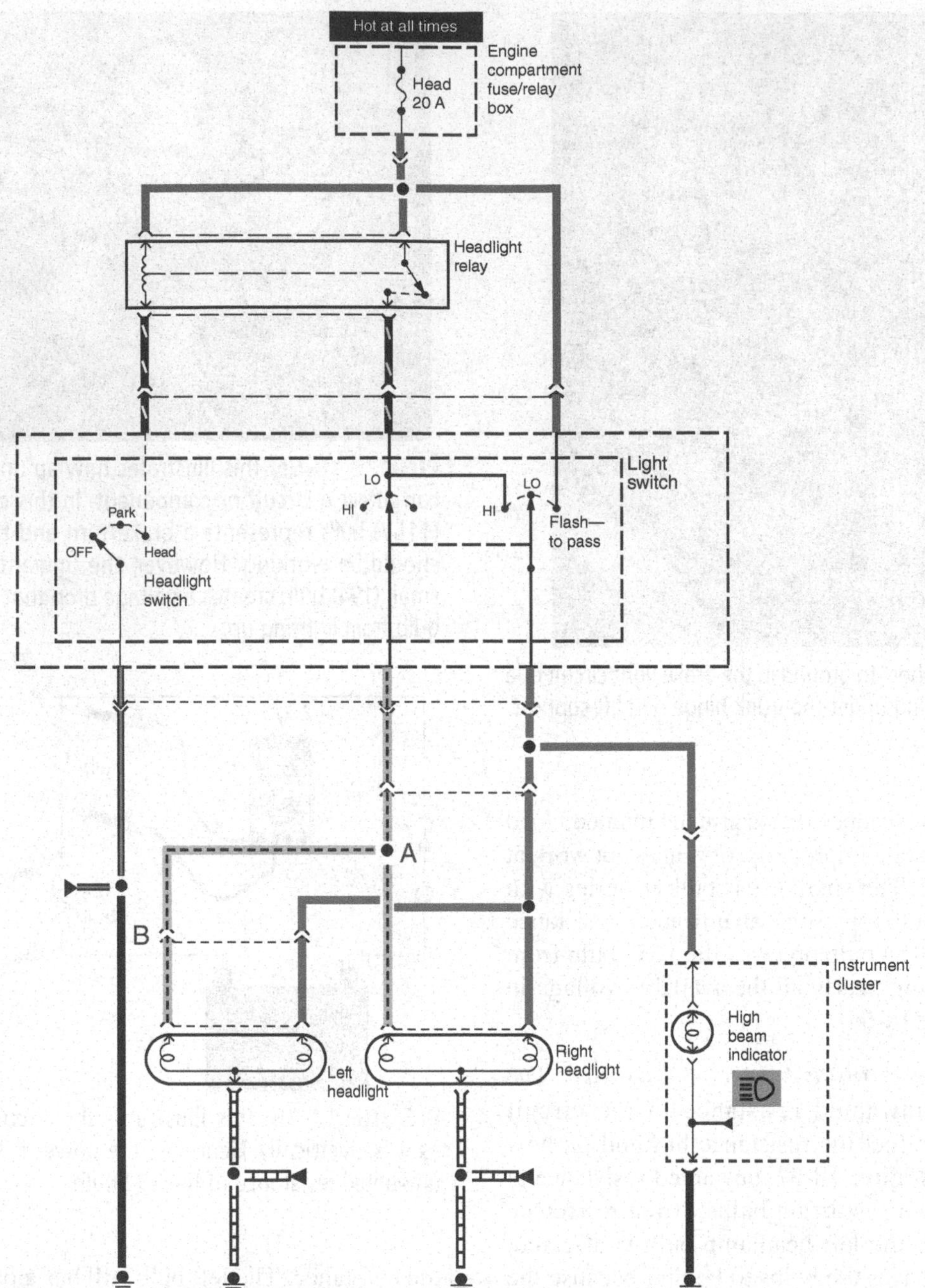

FIGURE 18-57 Unwanted resistance in a circuit can have different effects depending on the location of the resistance. At point A, both bulbs are affected, but at point B, only the left bulb is affected.

shows heat damage, be sure to inspect all of the wiring and components to find the source of the problem.

■ ***Common Causes of Unwanted Resistance.*** Corrosion is a common cause of unwanted circuit resistance (**Figure 18-62**). Corrosion around the battery cables can affect the starter circuit, charging circuit, and potentially the operation of every circuit on the vehicle (**Figure 18-63**). Corrosion can also form in connectors and on terminals that are exposed to the outside environment.

■ ***Diagnosing High Resistance Problems.*** Any time you have a complaint of dim bulbs or slow operation by a motor, you should suspect there is a resistance problem. Start by confirming the complaint. Next, perform a voltage drop on the affected circuit. For example, if one fog light is significantly dimmer than the other, check the voltage drop of the dim bulb (**Figure 18-64**). The voltage drop should be close to battery voltage. Unwanted resistance in the circuit causes a lower-than-normal voltage drop on the light. Determine if the resistance is on the power side of the circuit by measuring the voltage supplied to the bulb by placing the positive voltmeter lead on the power to the

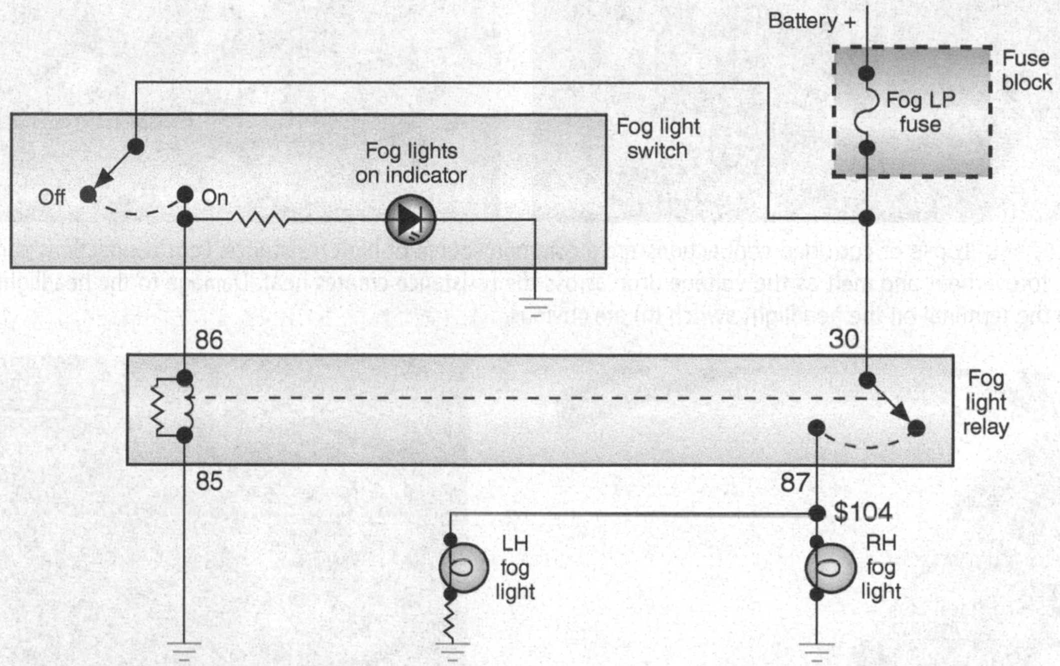

FIGURE 18-58 Resistance in the bulb's ground circuit creates a voltage drop that affects the left fog light and also consumes power.

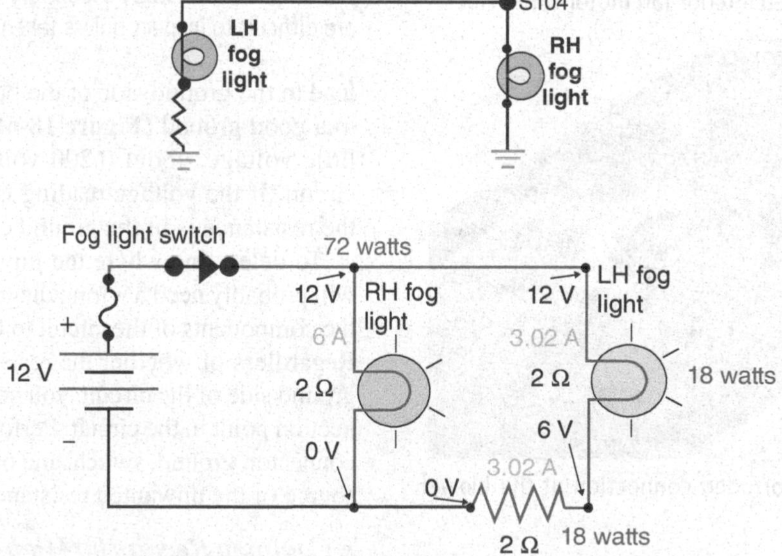

FIGURE 18-59 This shows how a resistive connection or corrosion can cause damage to wiring and connectors. The resistance consumes power, which is dissipated as heat.

FIGURE 18-60 Loose or corroded connections are a common source of high resistance. Poor connections can cause the connector to overheat and melt as the voltage drop across the resistance creates heat. Damage to the headlight connector (a) and to the terminal on the headlight switch (b) are obvious.

FIGURE 18-61 A damaged interior fan motor connector.

FIGURE 18-63 Side post battery terminals often corrode and are difficult to inspect unless taken off.

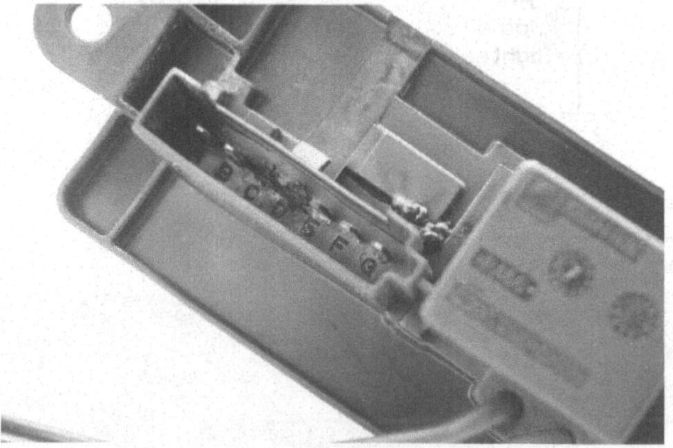

FIGURE 18-62 A very corroded connection at the blower motor resistor.

bulb and the meter's negative lead on the battery negative terminal (**Figure 18-65**). Again, the voltage should be close to battery voltage. If the voltage is less than battery voltage or less than is supplied to the other headlight, the resistance is in the power side of the circuit. If the voltage supply is normal, connect the DMM positive lead to the ground side of the bulb and the negative lead to a good ground (**Figure 18-66**). There should be very little voltage, about 0.200 volt or less, on the ground circuit. If the voltage reading is higher than 0.200 volt, the resistance is in the ground circuit.

To determine where the unwanted resistance is, you will probably need a wiring diagram so that you can locate the components of the circuit to further test voltage drops. Regardless of whether the resistance is on the power or ground side of the circuit, you need to test at the next connection point in the circuit. Perform voltage drops on each connector, ground, switch, and other components until the source of the unwanted resistance is located.

■ ***Common Repairs for High Resistance.*** Depending on where the problem is located, corrosion or rust can sometimes be removed using penetrants or battery cleaner on the terminals and connectors. Many times, individual terminals need to be removed from connectors and cleaned and then reinstalled. Loose grounds should be cleaned with a wire brush then reinstalled and the

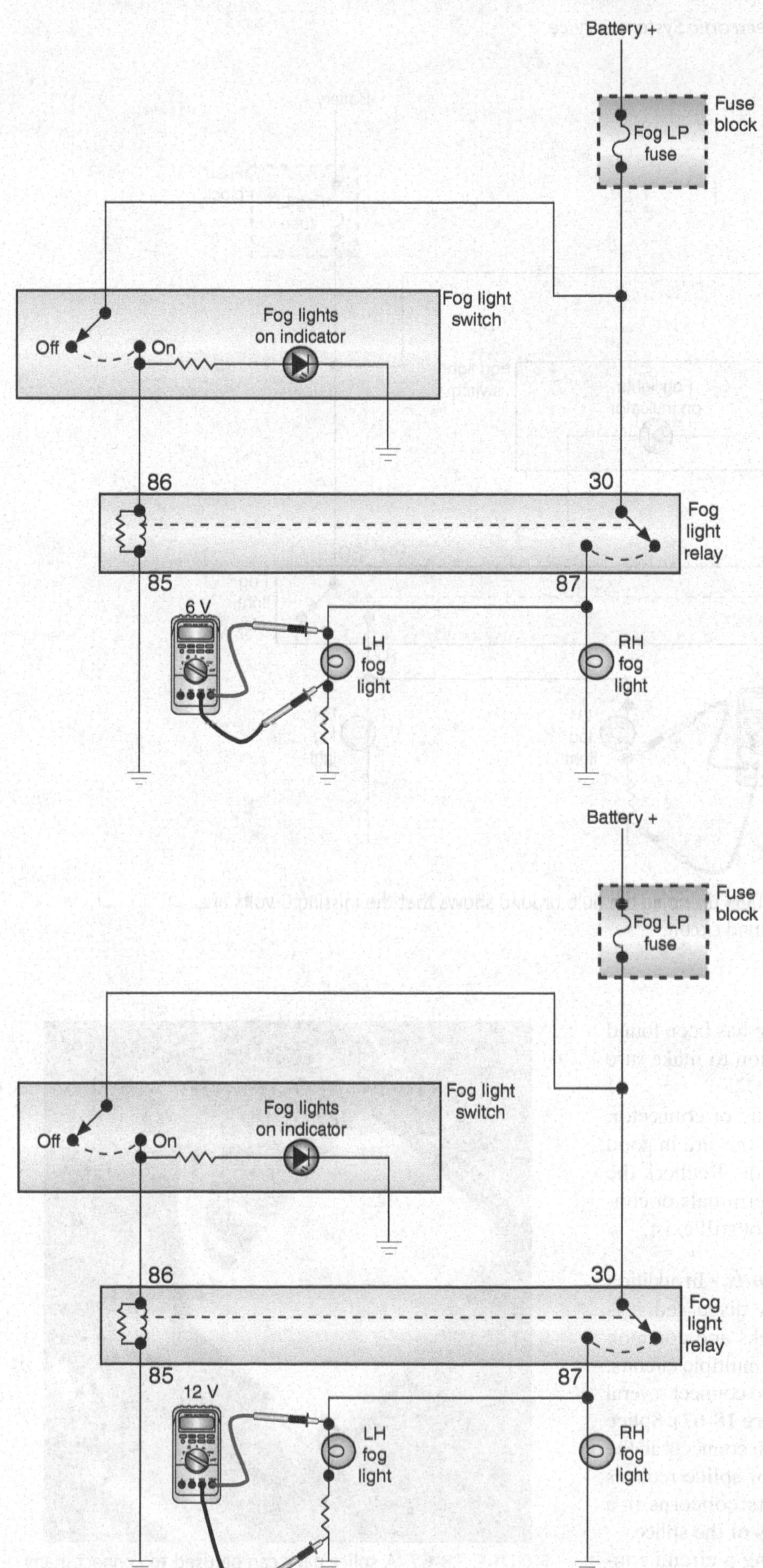

FIGURE 18-64 When diagnosing unwanted resistance problems, voltage drop test the component affected. The reading of 6 volts indicates that only about half of the normal amount of voltage is being used by the bulb. The next step is to figure out where the other 6 volts are going.

FIGURE 18-65 Since the meter shows 12 volts at the bulb when the negative test lead is placed on a good ground, the missing voltage must be on the ground side of the bulb.

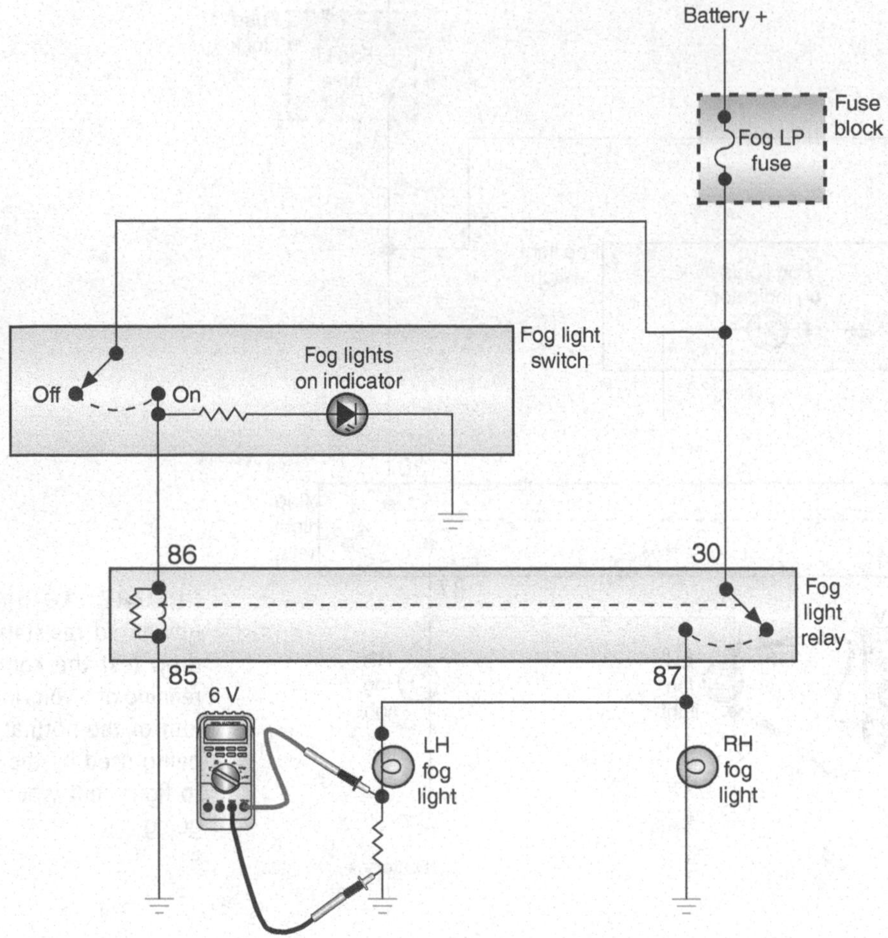

FIGURE 18-66 Voltage dropping the bulb ground shows that the missing 6-volts are being used in the ground circuit.

fastener torqued to specs. Once the cause has been found and repaired, recheck the circuit operation to make sure everything is working as designed.

If the problem has damaged the wiring or connector, replace both and make sure the connections are in good condition and tight once repairs are made. Recheck the voltage drop through the replacement terminals or connections to make sure a problem does not still exist.

■ ***Splice Packs and Common Grounds.*** In addition to the types of wiring problems already discussed, it is important to understand how splice packs and common grounds are used and how they can affect multiple circuits. A splice pack or splice connector is used to connect several circuits together, often as a ground (**Figure 18-67**). Splice packs take many different shapes, but each connects all the wires in the splice together. This type of splice reduces wiring in the vehicle but also presents concerns if a problem develops within the connections of the splice.

When diagnosing a concern involving a circuit connected with a splice pack, make a thorough inspection of

FIGURE 18-67 A splice pack can be used to connect many different grounds together.

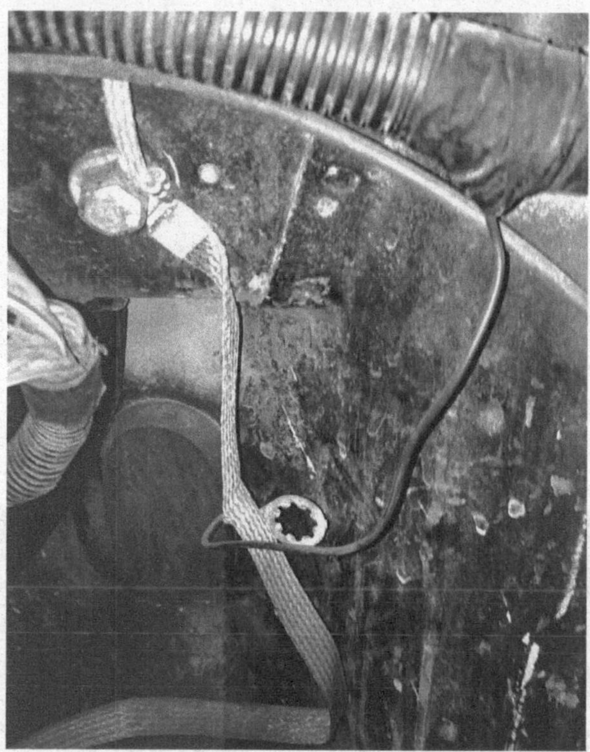

FIGURE 18-68 An example of common grounds and one that was left off from a previous repair. Missing grounds can cause several circuits to have problems and force them to find another path back to ground.

the terminals and connectors. You may need to check the continuity of the circuit while you move and wiggle the wires and splice pack connectors to check for intermittent connections.

Common grounds can also cause problems for many different circuits if the ground is loose, corroded, or disconnected (**Figure 18-68**). Common grounds are also used to reduce the amount of wiring in a harness, and like splice packs, they can affect several circuits if they are faulty.

Check common grounds for a tight, corrosion-free connection. If necessary, remove the fastener and clean the mounting surfaces and wiring eyelets. Common ground connections should be checked for voltage drop between the ground location and the battery negative terminal. Remember that the circuits need to be in operation to check the voltage drop.

CIRCUIT PROTECTION

Circuit protection components do not usually fail without a reason. Because fuses, circuit breakers, and fusible elements all operate by opening the circuit when excessive current flow generates more heat than the device can carry, excessive current flow is the cause of open circuit protection.

■ ***Testing Fuses.*** The first step when you are testing a fuse is finding the fuse. Most vehicles have multiple fuse/relay boxes, with some fuses under the hood, some in dash fuse boxes, and some located under seats, in kick panels, and other locations. When you are dealing with an inoperative circuit, you need to determine how many fuses are used and where they are so that you can test them. It is not uncommon for a system to have power from two or more fuses located in different locations on the vehicle.

> **Word Wall**
>
> *Kick panels*—Kick panels are the interior trim panels near the door jambs where the floor joins with the front fenders.

Fuses can be tested while they are still in the fuse box by touching a test light connected to ground to the back of the fuse (**Figure 18-69**). Plastic blade fuses have small windows in the plastic that allows access to the metal of the fuse for testing. Begin by attaching the test light clip to a good ground. Next, check the ground and test light operation by checking a known power source, such as the battery or fuses that are hot at all times. Always check test light operation every time you use one and every time you move from one location to the next. Too many times this basic step is forgotten, and a technician wastes time following false leads because he or she did not verify that the test light was connected properly.

Place the probe end of the test light against the metal of the fuse, testing at each side of the fuse. If the test light only lights when it is placed on one side of the fuse, the fuse is open. If the test light does not light on either side, refer to the wiring diagram for the circuit because the fuse may only be powered when the ignition is on or when another component is turned on. If the fuse should have power but does not, determine where the power to the fuse comes from. Many times fuses receive power from other larger fuses.

FIGURE 18-69 A test light is used to quickly check power on both sides of fuses.

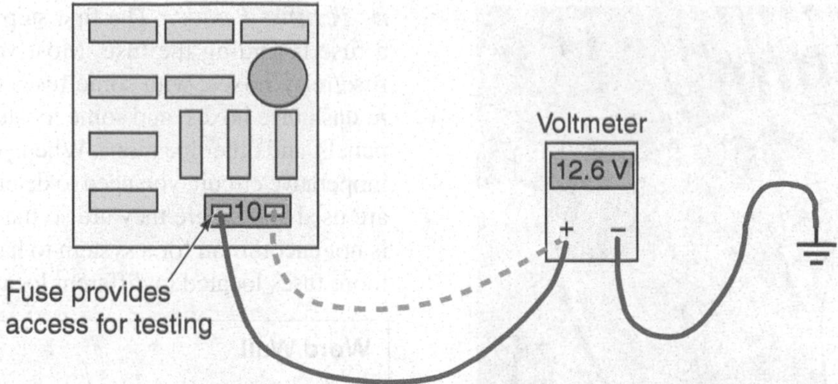

FIGURE 18-70 A DMM can be used to check power to a fuse. Make sure the meter has a good ground before testing at the fuses.

Using a DMM to test a fuse is similar. Connect the meter's negative lead to a known good ground and touch the positive lead to battery voltage to check your connections. Next, use the positive lead to check at the fuse (**Figure 18-70**).

■ ***Diagnosing Open Fuses.*** Fuses are rated for the amperage they can carry before they open. An open or blown fuse indicates that more current was flowing through the circuit than the fuse could safely carry. To diagnose repeated fuse failures, you need to determine what circuits are powered by the fuse. Then, try to find out what caused the fuses to blow

First, refer to a wiring diagram and examine the circuits connected to the open fuse. Trace the circuits from the fuse and make a list of possible causes. This can help you to narrow down the possibilities that can open the fuse. You may also need to inspect the wiring and the components of the circuit to look for signs of wire damage (**Figure 18-71**).

Replace the fuse and operate the circuits on the fuse to see if the fuse opens again. If all the circuits operate properly and the fuse does not open, you may have an intermittent problem or you could have experienced a random issue, such as a paper clip falling into a power outlet, which caused the fuse to blow.

■ ***What if the Fuse Opens Again?*** In this case, you need to perform more diagnostic work to determine the cause of the problem. If the fuse blows as soon as it is installed, you have a component or wire that is directly shorted to ground. If the fuse blows only after parts of the circuit are turned on, you need to examine the parts of the circuit after the switch for faults. This is because the short only occurs after the circuit is turned on, indicating the short is after the switch. For example, a power window motor that is binding draws more current than normal, which can cause the fuse to blow but only when that window is used.

If the fuse blows immediately upon installation, you have a short to ground in a power supply circuit. To diagnose this problem, you can install a test light, buzzer, or short finder in place of the fuse. This allows a small amount of current to flow through the circuit, which allows you to unplug each component until the short is found. If after all the components are unplugged the short is still present, there is likely a grounded power wire. Inspect the wiring harnesses for signs of damage and chafing against brackets and other parts that can rub through the harness and wire insulation.

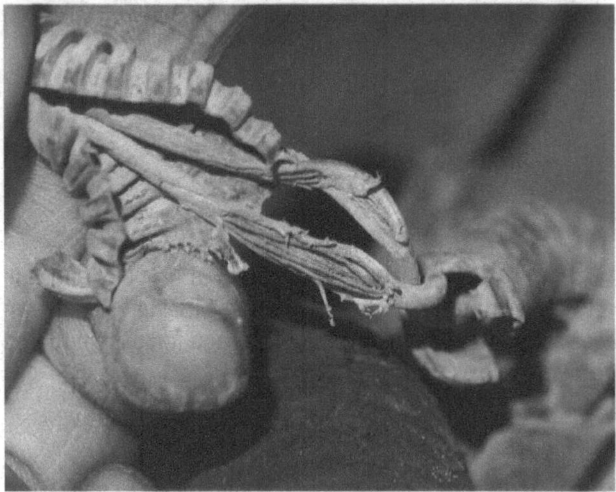

FIGURE 18-71 If diagnosing a short that is blowing a fuse, inspect the wiring of affected circuits. Damaged wiring can cause a short to ground.

■ ***Fusible Elements.*** Fusible elements can be tested by removing the fuse and checking for continuity or resistance across the terminals (**Figure 18-72**). On many of these fuses, the clear plastic cover can be removed to allow a close inspection of the fusible element. You can test them in place by checking for power on both sides of the element with a test light or DMM.

CIRCUIT BREAKERS

A circuit breaker, like a fuse, is used to open a circuit in the event of excessive current flow. Unlike a fuse, however, a circuit breaker can reset and close the circuit again, and thus restore its operation.

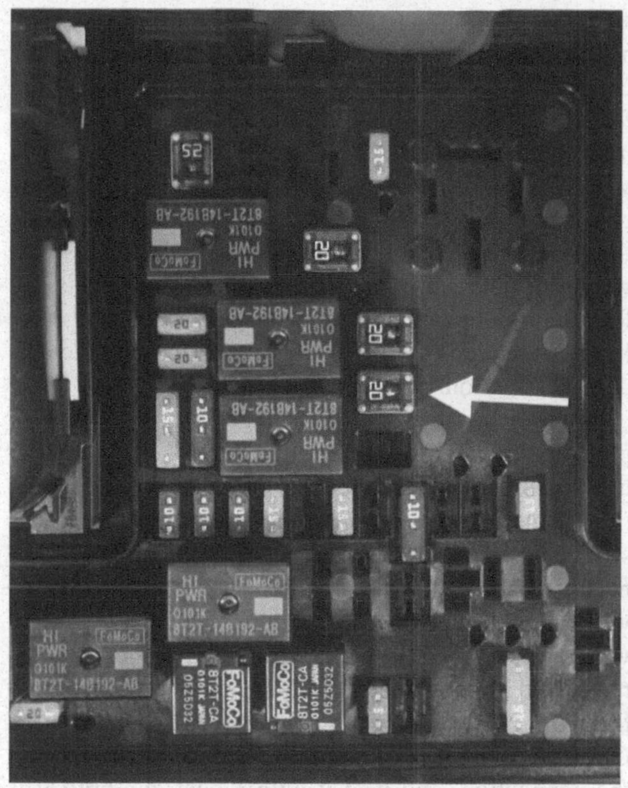

FIGURE 18-72 Fuses, fusible elements, circuit breakers, and relays are all often located in a junction box.

■ ***Why Use Circuit Breakers?*** Circuit breakers are used in circuits that can have a sudden high amperage draw that could blow a fuse but instead, the circuit breaker can open and then reset, and restore the circuit's operation. This is common in windshield wiper circuits, where the wiper motor can have a current spike, as when the wiper blades are frozen to the windshield. Circuit breakers are also found in power window and headlight circuits.

■ ***Testing Circuit Breakers.*** Circuit breakers can be checked with an ohmmeter for resistance or continuity (**Figure 18-73**). A circuit breaker should have very low resistance when it is closed. If the circuit breaker shows infinite resistance, the internal contacts are open. Replace a breaker that shows either high resistance or an open circuit.

FUSE LINKS

Fuse links, though not as common as they once were, are used to protect power distribution circuits, such as that from the battery connection at the starter solenoid. The fuse links are wired into the circuits from the starter to the fuse or junction blocks.

■ ***Testing Fuse Links.*** An open fuse link can be found by pulling on the fuse link. Once the wire of the fuse link burns open, the insulation is weakened, stretches, and pulls apart. You can also use a DMM to check the fuse link for continuity (**Figure 18-74**), or check for voltage on both sides of the link.

■ ***Replacing a Fuse Link.*** Fuse link replacement requires cutting out the burned fuse link and then soldering in a new fuse link. In general, the replacement fuse link should be four wire gauges smaller than the wiring for the circuit it is protecting. Because this takes more work than just replacing a fuse, you need to be sure that the problem that caused the fuse link to burn is repaired; otherwise, the new fuse link will also burn when overloaded.

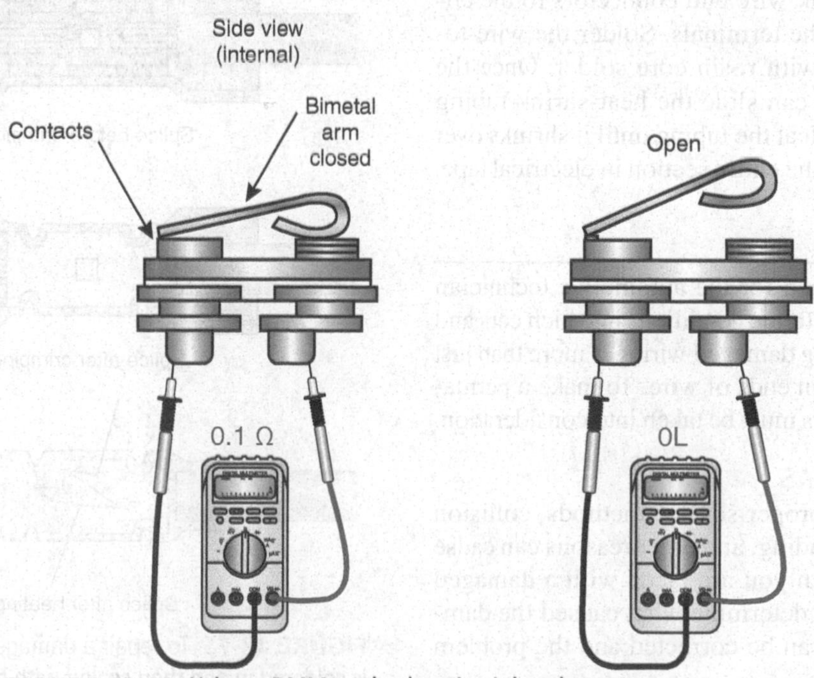

FIGURE 18-73 Using a DMM to check a circuit breaker.

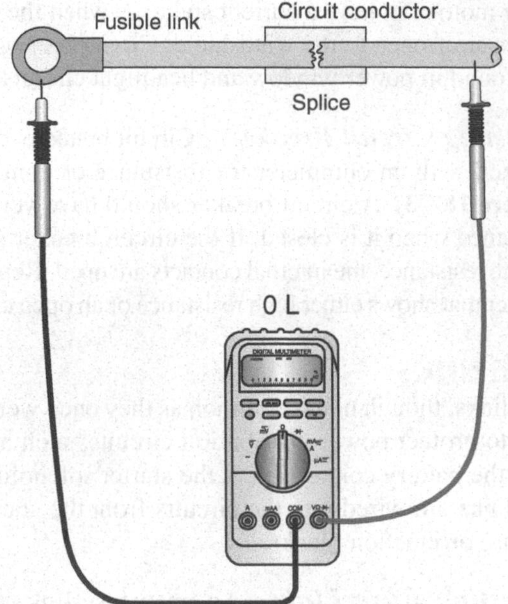

FIGURE 18-74 Fusible links, though not as commonly used, are usually tested by pulling on the link. If it pulls apart, it is open. A meter can also be used, but this may require opening the link or penetrating the insulation to test. Replacing a fusible link requires cutting out the old link and soldering in a new section. This should only be done after the cause of the fault that opened the link has been found and repaired.

Once the cause of the circuit problem has been located and corrected, cut out the burned fuse link and prepare the ends of the wire to which it connects. Cut a section of the appropriate gauge fuse link wire for the circuit and prepare the ends of the fuse link wire. Install and crimp two terminal connectors to the fuse link wire. Slide a section of heat-shrink tube over the wire, and then install the fuse link wire and connectors to the circuit wires and crimp the terminals. Solder the wire-to-terminal connections with rosin core solder. Once the connections cool, you can slide the heat-shrink tubing over the connections. Heat the tubing until it shrinks over the connections. Wrap the entire section in electrical tape.

Wiring Repair

Wiring repair is common for the automotive technician due to the sometimes extreme conditions in which cars and trucks operate. Repairing damaged wiring is more than just reconnecting two broken ends of wire. To make a permanent repair, many factors must be taken into consideration.

WIRING PROBLEMS

Misrouted wiring, improper service methods, collision damage, circuit overloading, and other reasons can cause wiring problems. When you are faced with a damaged wire, it is important to determine what caused the damage so that the cause can be corrected and the problem does not reoccur.

■ ***Damaged Wiring.*** Any time you are repairing a wire that is exposed to the outside elements, the connection should be soldered to prevent corrosion from forming at the repair point. **Photo Sequence 10** shows how to prepare and solder a wire connection. When you are repairing a damaged wire, make sure that the repair will not pull on the wire or harness in any way that will keep tension on the wiring or will cause the wiring to pull tightly against another object. This can lead to further damage to the wiring. If necessary, splice in a new section of wire of the same size to replace the damaged section.

Begin by cutting out the damaged section and stripping the insulation from the wire back about one-quarter of an inch. Next, install heat-shrink tubing over one section of the wire being repaired. Select the correct size splice connector for the wire gauge, and crimp the connector to one end of the exposed wire. Ensure that the wiring lies correctly before crimping the other section of wire to the connector. If the wiring and harness lie correctly and are not pulled tight, crimp the terminal onto the wire. Next, solder the connection, and let it cool. Once the solder is cool, slide the heat-shrink tubing over the connection, and heat it with a heat gun until it is completely sealed around the connection. An illustration of how the repair will look is shown in **Figure 18-75**.

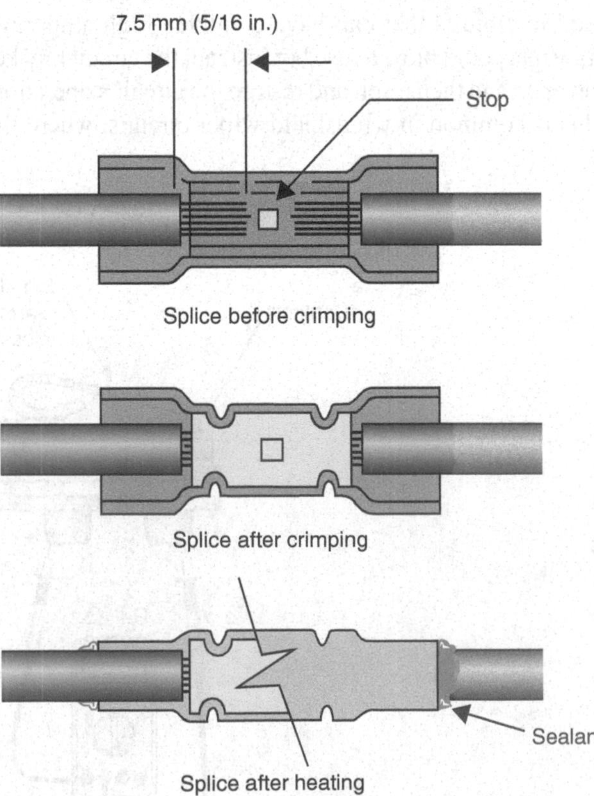

FIGURE 18-75 To repair a damaged or broken wire, a splice is soldered in and then sealed with heat-shrink tubing.

Finally, wrap the repaired section in electrical tape and place the wire back into the harness. Be sure to reinstall the harness into any brackets or other clamps as necessary to keep the harness from moving and chafing against other objects.

■ ***Replacing a Terminal.*** Over time, the terminal connection can weaken due to fretting and from being repeatedly disconnected and reconnected. Fretting is the term that applies to the movement of the connectors against each other due to thermal cycling. As current flows through the circuit, heat is generated in the wire, connectors, and terminals. In addition, the wiring is subject to changes in ambient temperature and to increased underhood temperatures. This heating causes a slight expansion when it gets hot and then a contraction when it cools. Over time, this slight movement can wear away the metal of the terminal, creating a poor connection. Also, because there is a slight bit of spring tension between the two terminals, repeatedly disconnecting and reconnecting a connector can decrease this tension, leading to a poor connection. It is important to note that terminals are also damaged by incorrect testing methods with test lights and DMM probes. Do not force the probes into the terminals while you are performing tests as this can spread the terminal apart or cause other damage, resulting in open circuits or intermittent connections.

Sometimes, you need to replace a damaged terminal. This requires removing the terminal from the connector. There are many different ways in which terminals are secured into the connector (**Figure 18-76**). You need to determine how the terminal is retained in the connector, and then locate the correct terminal tool to release the safety latch and allow the terminal to come out of the connector (**Figure 18-77**). Special tools terminal are used to release the terminal without causing damage to the terminal or connector, though if used carefully, items such as T-pins and other similar tools can be used to release the terminal locks.

Once the damaged terminal is removed from the connector, remove it from the wire. If possible, open the crimp and remove the wire from the terminal. It may not be possible to remove the old terminal, which will require cutting the wire to install the new terminal. Insert the wire into the new terminal and, using the appropriate tool, crimp the new terminal to the wire. Replace the terminal into the connector and ensure that the terminal latch holds the terminal securely.

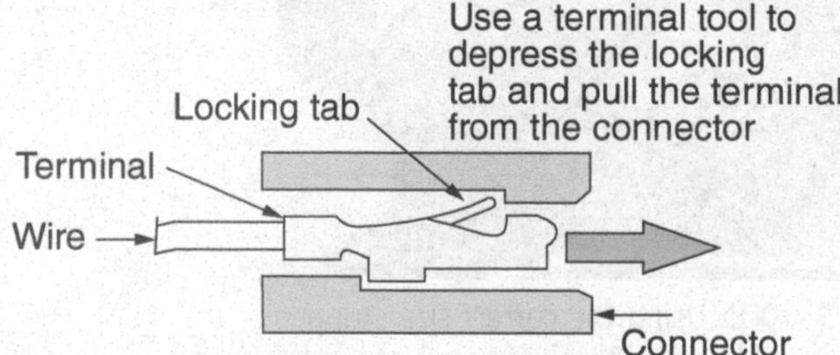

FIGURE 18-76 Terminals are secured in the connector with a locking tab or tang. These prevent the terminal from backing out of the connector.

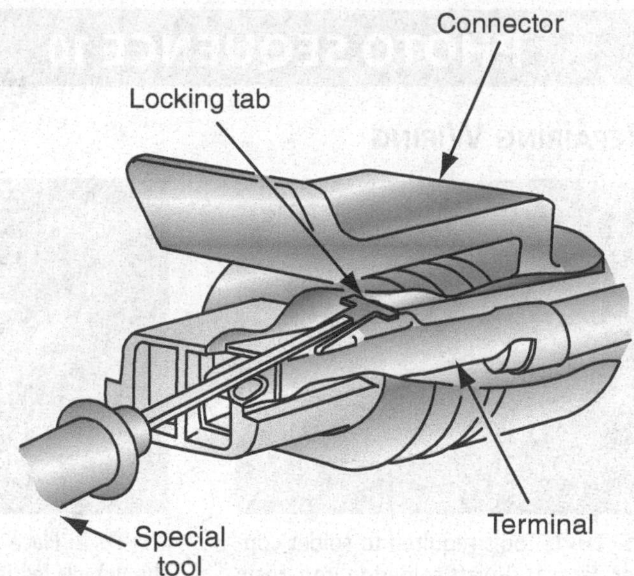

FIGURE 18-77 Removing a terminal requires figuring out how to release the locking tab and then using the correct tool to release the tab.

■ ***Replacing a Connector.*** Sometimes, a connector is damaged and requires replacement. Replacement connectors are available for many applications, making replacing a damaged connector easy and economical. When you are replacing a damaged connector with a new connector, first ensure that the circuit is off and the fuse removed. Next, cut the wires to the connector as close to the connector as possible. Match the replacement connector's wires to the original wires and strip all wire ends so that the wires can be reconnected. Place heat- shrink tubing over the wiring and connect the new and old wiring. Test the operation of the component with the new connector installed. If it is correct, heat the shrink tube to cover the repair, and then wrap the harness with electrical tape.

RELAY TESTING

Because relays are used extensively throughout the vehicle, it is important to understand how to test their operation. Normally quite dependable, relays can be damaged by excessive current flow, and the contacts in the switch can pit and burn over time.

■ ***Relay Construction.*** Relays, as discussed in Chapter 17, use an electromagnet to open or close a switch contact. Current flow through a relay coil is low

PHOTO SEQUENCE 10

Repairing Wiring

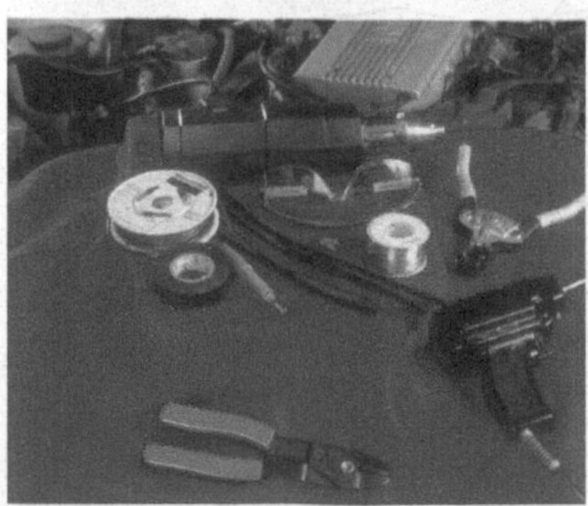

PS 10-1 Tools required to solder copper wire: 100-watt soldering iron, rosin core solder, crimping tool, splice clip, heat-shrink tube, heating gun, safety glasses, sewing seam ripper, electrical tape, and fender covers.

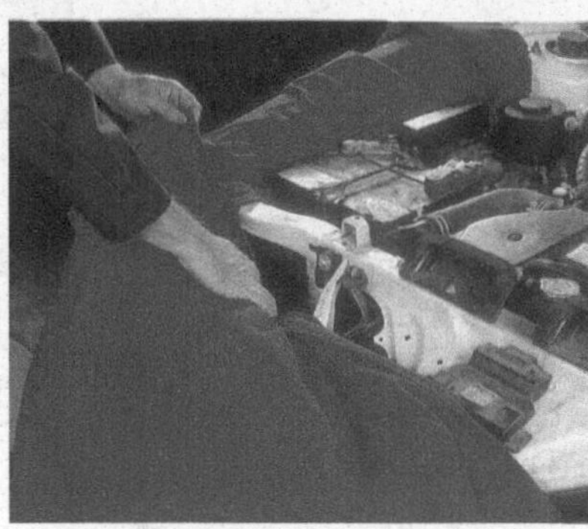

PS 10-2 Place the fender covers over the vehicle fenders.

PS 10-3 Disconnect the fuse that powers the circuit being repaired. Note: If the circuit is not protected by a fuse, then disconnect the battery.

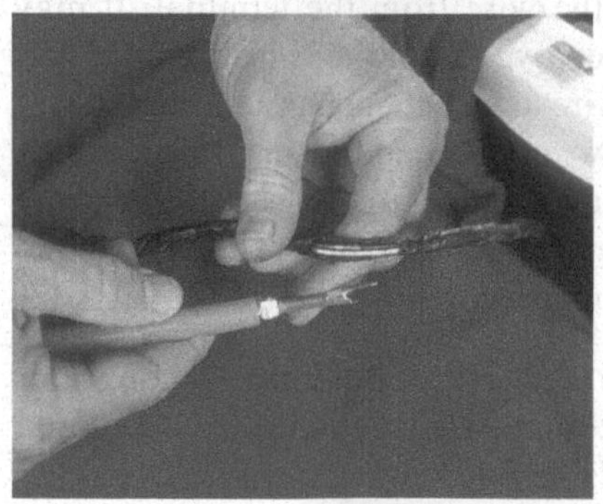

PS 10-4 If the wiring harness is taped, use a seam ripper to open the wiring harness.

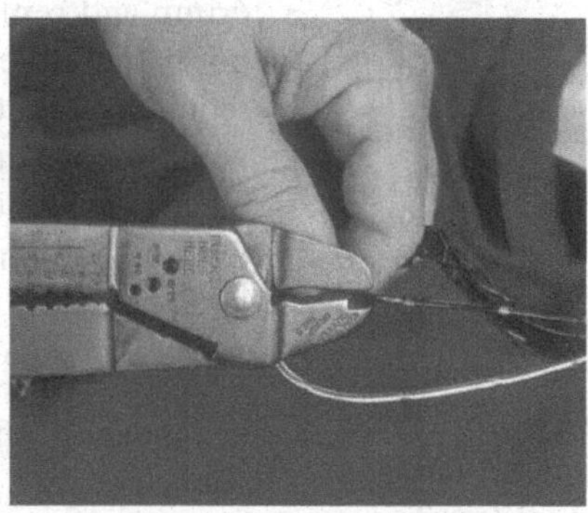

PS 10-5 Cut out the damaged wire using the wire cutters on the crimping tool.

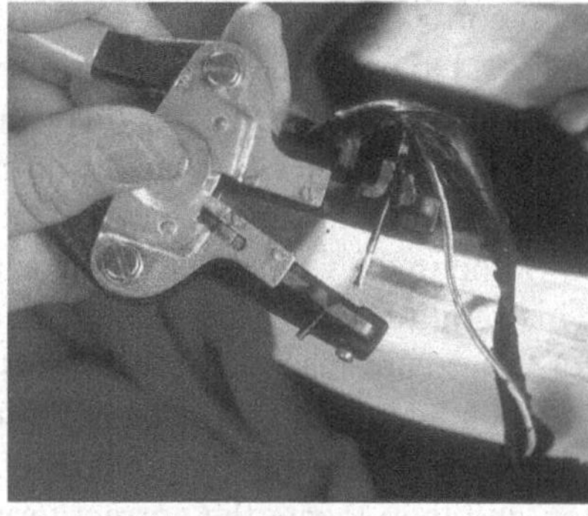

PS 10-6 Using the correct size stripper, remove about 1/2 inch (12 mm) of the insulation from both wires. Be careful not to nick or cut any of the wires.

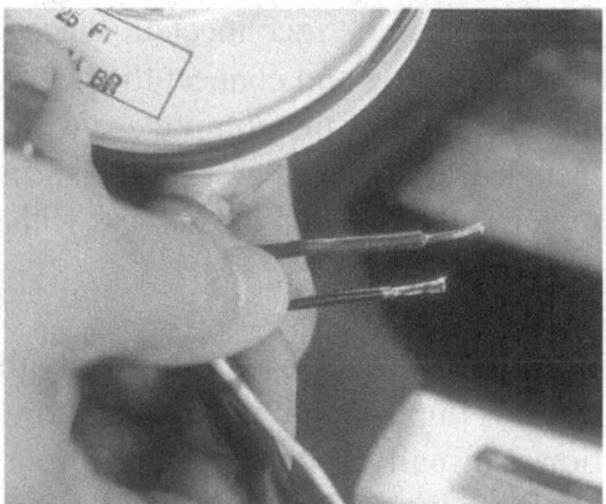

PS 10-7 Determine the correct gauge and length of replacement wire.

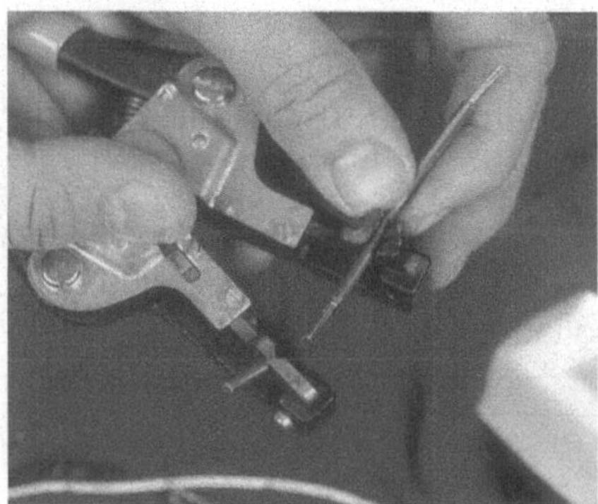

PS 10-8 Using the correct size stripper, remove about 1/2 inch (12 mm) of insulation from each end of the replacement wire.

PHOTO SEQUENCE 10 (CONTINUED)

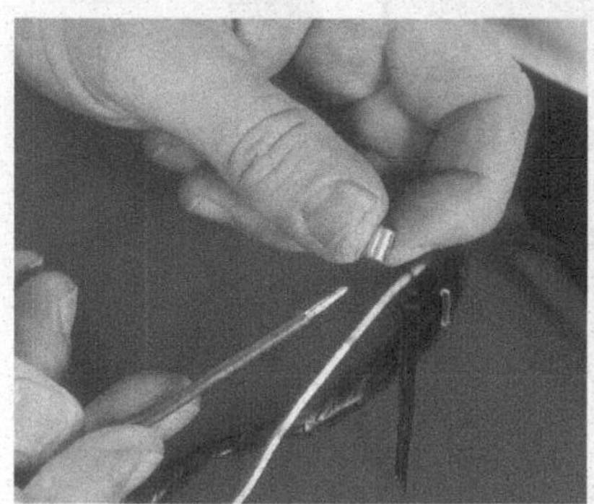

PS 10-9 Select the proper size splice clip to hold the splice.

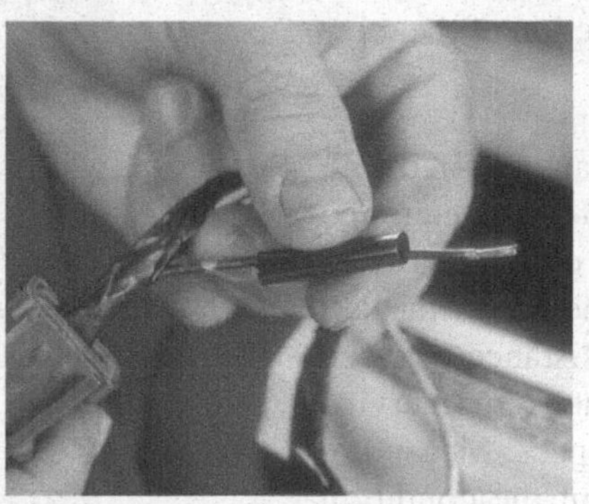

PS 10-10 Place the correct length and size of heat-shrink tube over the two ends of the wire. Slide them far enough away so they are not exposed to the heat of the soldering iron.

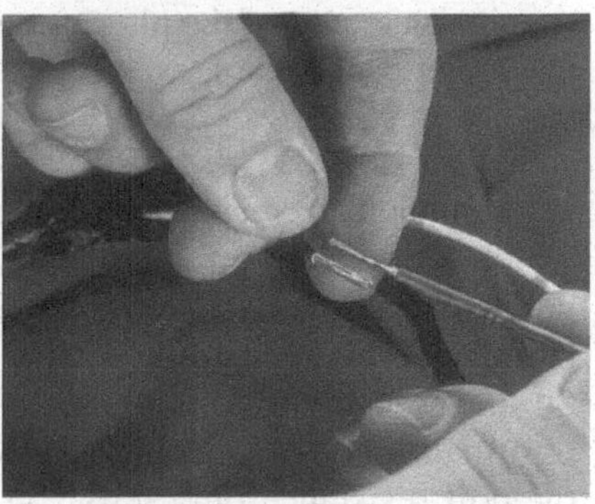

PS 10-11 Overlap the two splice ends and hold in place with thumb and forefinger.

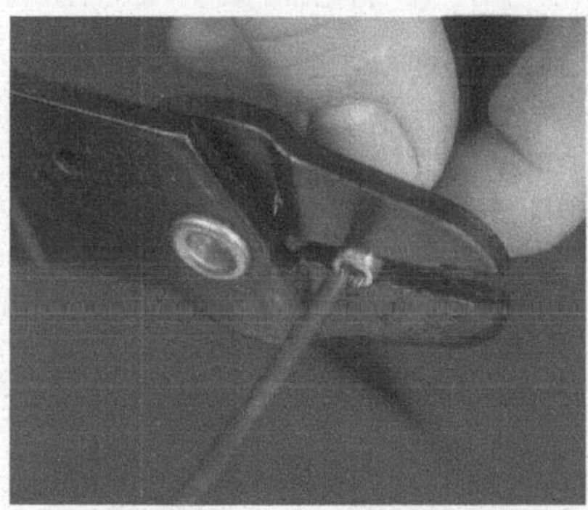

PS 10-12 Center the splice clip around the wires and crimp in place. Make sure that wires extend beyond the splice clip in both directions. Crimp the clip on both ends.

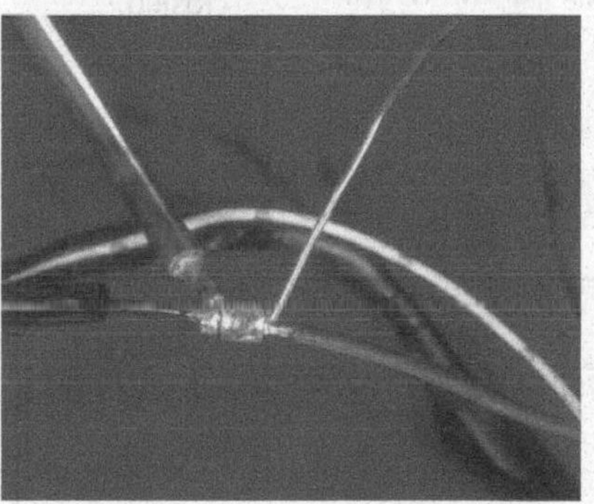

PS 10-13 Heat the splice clip with the soldering iron while applying solder to the opening in the back of the clip. Do not apply solder to the iron; the iron should be 180 degrees away from the opening of the clip.

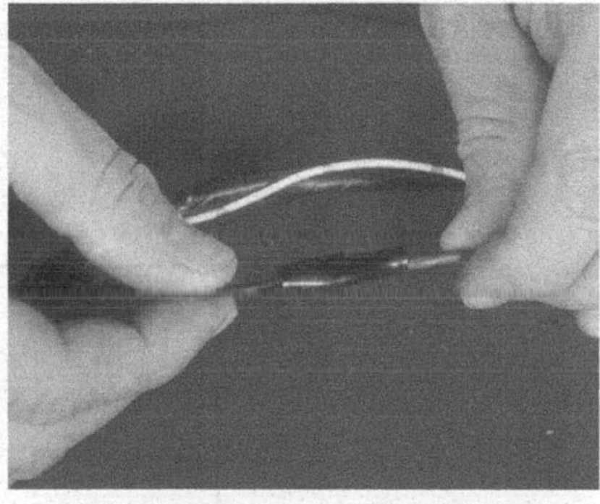

PS 10-14 After the solder cools, slide the heat-shrink tube over the splice.

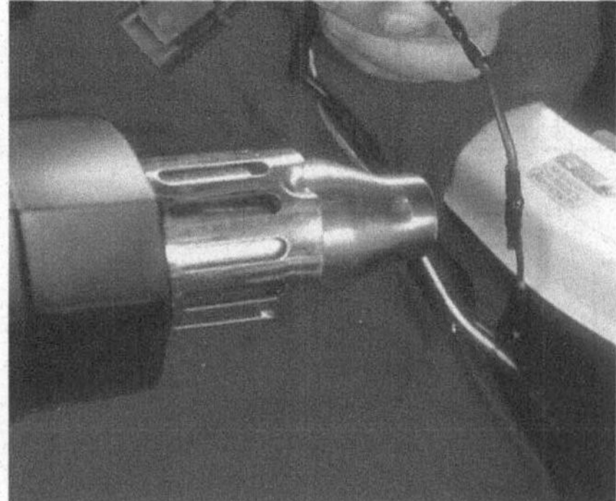

PS 10-15 Heat the tube with the hot air gun until it shrinks around the splice. Do not overheat the tube.

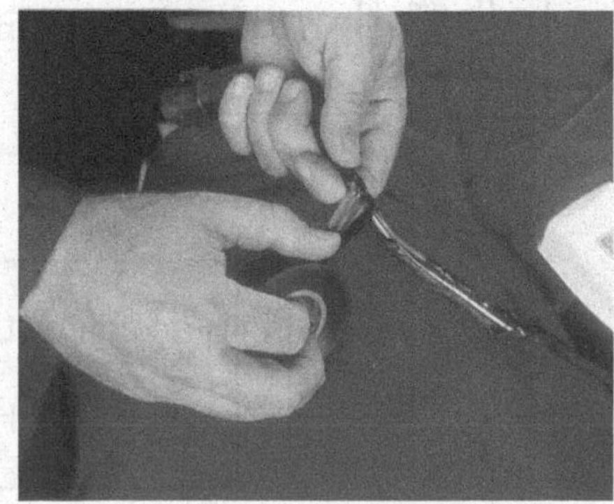

PS 10-16 Retape the wiring harness.

because it is often controlled by an electronic module. An example of a computer-controlled relay circuit is shown in **Figure 18-78**. When the coil is energized, the magnetic field closes the switch contacts, allowing current flow through a circuit. The coil, contacts, and terminals are part of the plastic construction of the relay and are not serviceable.

■ ***Relay Coil Testing.*** Relay coils can be tested with an ohmmeter for resistance (**Figure 18-79**). Determine which terminals are for the coil. The common Bosch numbering system uses terminals 85 and 86 for the coil (**Figure 18-80**). Most relays have a resistance between 50 and 150 ohms, though always check the service information for the specific vehicle's specs. Measuring the coil resistance only tells you about the condition of the coil with the relay unpowered, so this is not a comprehensive test.

■ ***Load Testing Relays.*** To determine if a relay is functioning properly, it should be tested under load, meaning the coil should be activated and a load placed on the contacts so that voltage drop can be tested. **Figure 18-81** shows one way to connect a relay to a battery

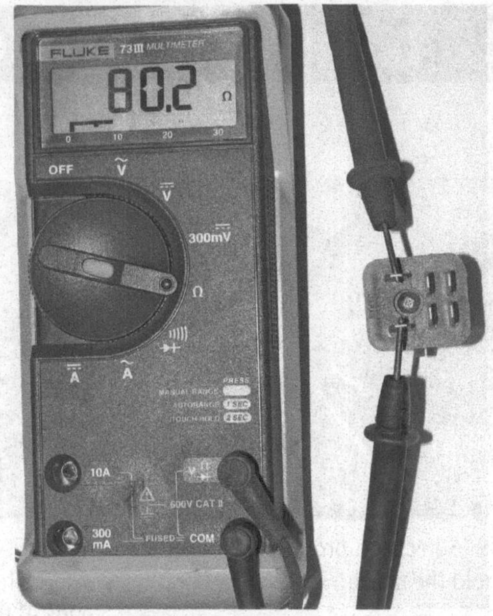

FIGURE 18-79 Testing the relay coil resistance. If the resistance is too low, current flow will increase, which can burn the wire or damage the relay control circuit. High resistance will weaken the magnetic field and may prevent the contacts from closing.

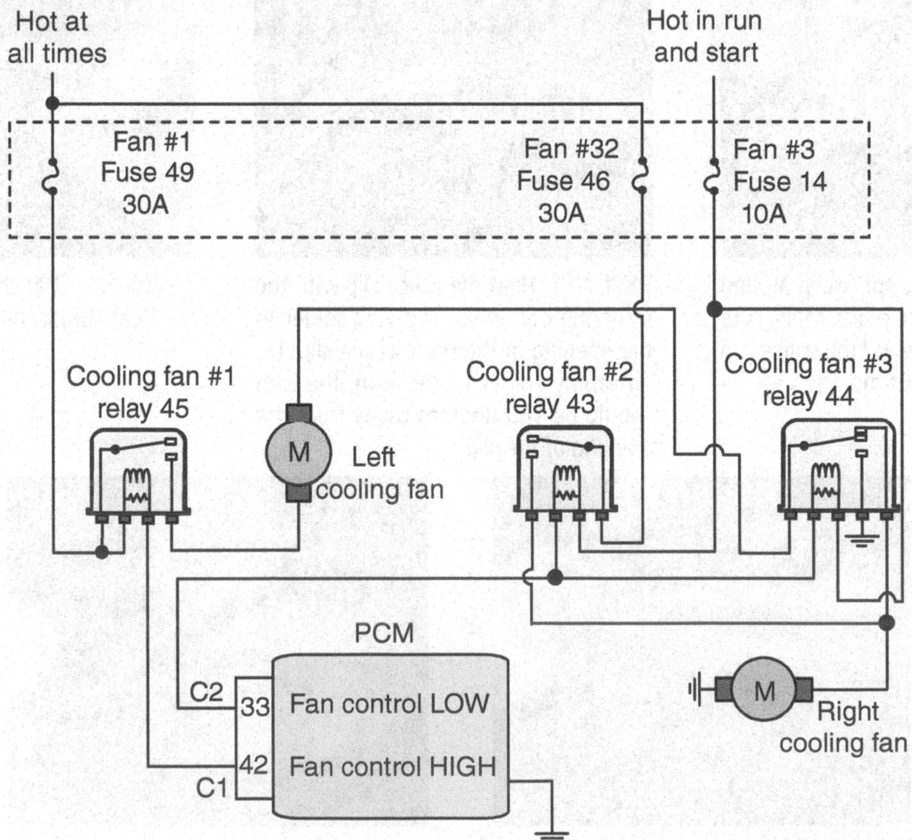

FIGURE 18-78 Relays are used extensively in modern vehicles. This arrangement allows for both high- and low-speed fan operations by switching the fans from being wired in series to parallel.

FIGURE 18-80 Bosch-type relays use standard terminal numbers. This relay provides a diagram of the terminals for testing.

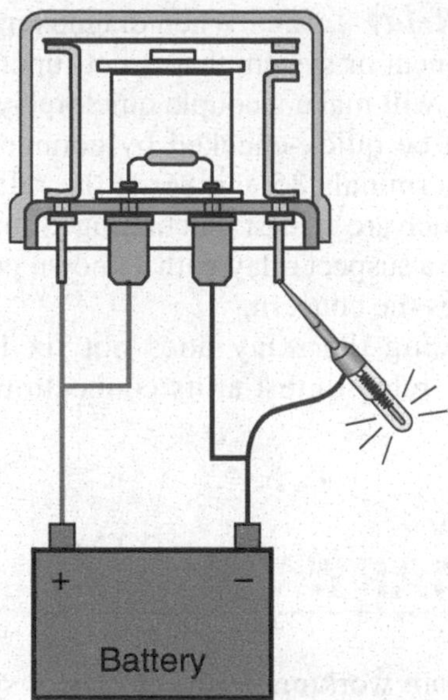

FIGURE 18-81 Load-testing a relay with a test light ensures that the relay can conduct current flow.

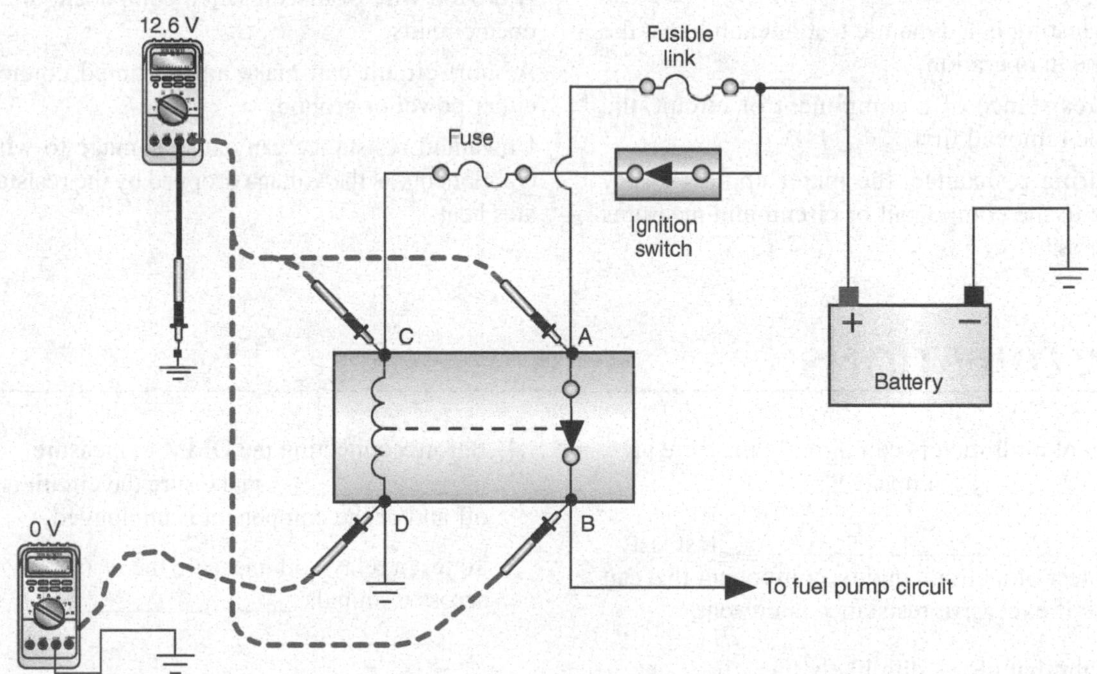

FIGURE 18-82 Since the relay is central to the circuit, each part of the circuit can be tested from the relay's connection. Both the power and control circuits can be easily tested by removing the relay.

with a test light and jumper wires. Some technicians make a small relay testing harness with four or five short wires with male and female terminals. This harness can be connected to a relay to test the relay, or it can be used to test at the relay box. The short wires allow you to remotely connect the relay and test at the terminals, to test the relay's circuits. This simple set of connection wires can be used to test four- and five-pin relays.

■ ***Other Relay Tests.*** When diagnosing a complaint of a component or system that is not functioning, many technicians will make a couple quick relay checks first. Relays can be quick-checked by connecting a 9-volt battery to terminals 85 and 86. If the relay clicks, the coil and switch are at least functioning. Some technicians will replace a suspect relay with a known good relay and see if it fixes the concern.

If replacing the relay does not fix the problem, remove the relay to test at its connection point to the circuit. This allows you to test nearly all parts of the circuit at one location (**Figure 18-82**). With the relay removed, there should be power on two terminals, as shown as points A and C in the figure. Control of the coil, point B, is typically provided by grounding the coil, though you need to refer to a wiring diagram to be certain. If both power circuits and the ground for the coil are present, the problem is either in the relay or the circuit to the load

SUMMARY

When you are working on the electrical system, safety should be your first priority.

ESD is the accidental discharge of static electricity that can damage electronic components.

A voltmeter shows the voltage potential that exists between the two meter leads.

Voltage drop defines how much voltage is being used by a component, connection, or conductor.

Voltage drop testing is a dynamic test, meaning that the circuit must be in operation.

To measure resistance of a component or circuit, the power must be removed first.

When measuring resistance, the meter applies a very small voltage to the component or circuit and measures the voltage drop.

To measure amperage, the meter must be placed in series with the circuit.

Most meters can directly measure up to 10 amps.

Schematics are electrical wiring diagrams, similar to road maps.

The three types of circuit faults are open circuits, short circuits, and high or unwanted resistance.

A broken wire of disconnected component are types of open circuits.

A short circuit can make an unwanted connection to either power or ground.

Unwanted resistance can cause damage to wiring and components as the voltage dropped by the resistance creates heat.

REVIEW QUESTIONS

1. Most digital multimeters can directly measure up to ______________ amps.
2. A ______________ ______________ test is a dynamic test of a circuit and/or component that can determine if excessive resistance is present.
3. A headlight that glows dim likely has ____________________________ in the circuit.
4. Before connecting the DMM to measure ______________, make sure the circuit is turned off and/or the component is unplugged.
5. To test a relay coil, measure the resistance across terminals ______________ and ______________.

6. A vehicle's interior accessory fuse blows repeatedly. *Technician A* says there is a short to power in the circuit. *Technician B* says unwanted resistance in the circuit can increase the current flow, causing the fuse to blow. Who is correct?
   a. Technician A
   b. Technician B
   c. Both A and B
   d. Neither A nor B
7. A vehicle has an inoperative interior fan and the fan motor connector shows signs of overheating and melting. *Technician A* says that a poor connection at the fan motor could cause the damage. *Technician B* says a short to ground in the fan motor connector could cause the damage. Who is correct?
   a. Technician A
   b. Technician B
   c. Both A and B
   d. Neither A nor B
8. What measurement is being taken in **Figure 18-83**?
   a. Current flow
   b. Resistance
   c. Voltage drop
   d. Source voltage
9. What is being measured in **Figure 18-84**?
   a. Resistance
   b. Current flow
   c. Voltage drop
   d. Source voltage
10. Referring to **Figure 18-85**, *Technician A* says that to test the relay coil, place the ohmmeter leads on terminals 30 and 87 of the relay. *Technician B* says testing terminals 85 to 86 should show less than 2 ohms of resistance. Who is correct?
    a. Technician A
    b. Technician B
    c. Both A and B
    d. Neither A nor B

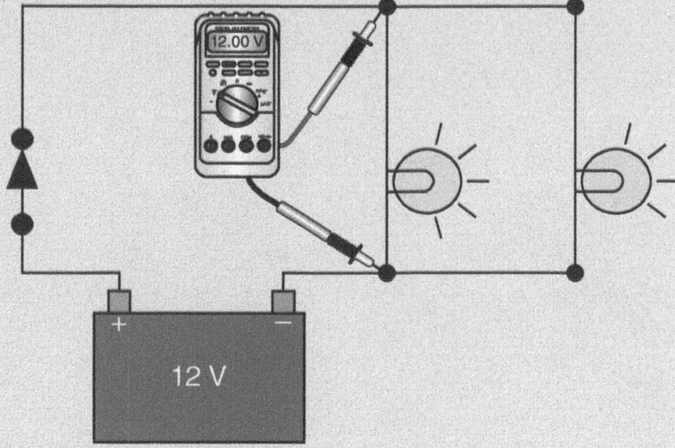

FIGURE 18-83 Review question image.

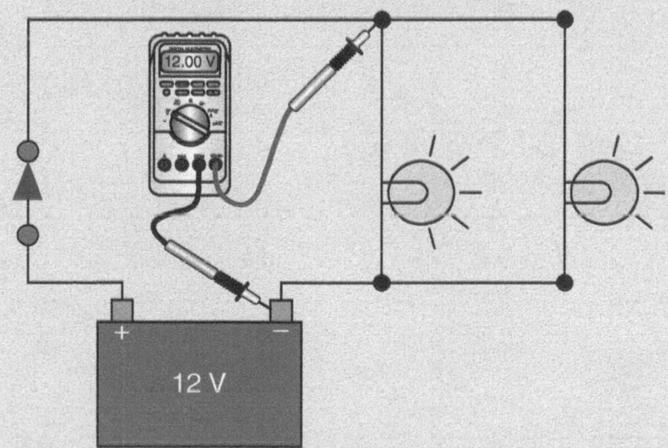

FIGURE 18-84 Review question image.

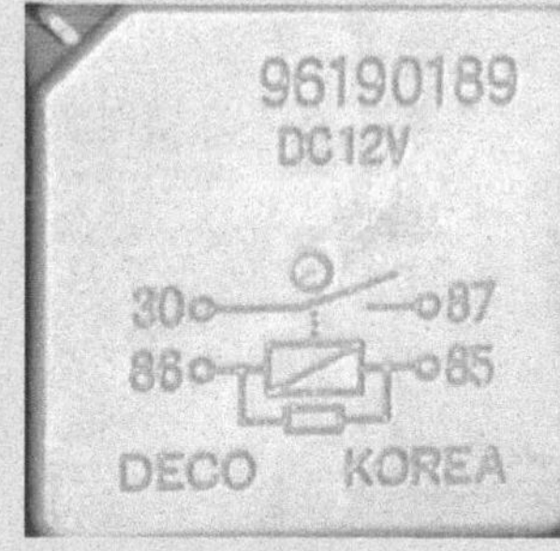

FIGURE 18-85 Review question image.

CHAPTER 19

Battery, Starting, and Charging System Principles

Chapter Objectives

At the conclusion of this chapter, you should be able to:

- Identify battery ratings and battery types.
- Identify battery service safety precautions.
- Identify starter types and starter components, and describe starter operation.
- Identify components of the starting system.
- Identify high-voltage starting system components in hybrid vehicles.
- Explain how AC generators operate.
- Identify components of the AC generator.
- Determine how generator output is regulated.
- Identify components of the high-voltage charging system on hybrid vehicles. (ASE Education Foundation MLR 6.B.7)

KEY TERMS

Absorbed glass mat (AGM) battery
amp-hour rating
anode
battery cycle
battery reserve capacity
BCI group
cathode
cold cranking amps (CCA)
cranking amps (CA)
diodes
electrolyte
induction
insulated circuit
lead-acid battery
low-maintenance battery
maintenance-free battery
permanent magnet motor
state of charge
starter control circuit
starter motor
stator
voltage regulator

Lead-acid batteries have been in use in automobiles since the very earliest days of self-propelled vehicles. In the early 1900s, all-electric, battery-powered vehicles made up a significant percentage of all vehicles sold.

Battery Principles

Batteries, regardless of type or construction, store and release energy by the chemical reactions inside the battery. Batteries can be made from a variety of materials and even from some items you may not think of, such as fruits and vegetables.

Refer to Chapter 19 of the Lab Manual for experiments with batteries.

The three components that make up a battery are the anode, cathode, and electrolyte (**Figure 19-1**). An **anode** is an electrode or electrical conductor *into* which current flows. A **cathode** is an electrode or conductor from which current flows *out*. **Electrolyte** contains free ions that make the substance electrically conductive. In automotive batteries, the negative electrode is the anode, the positive electrode is the cathode, and the electrolyte is the mixture of sulfuric acid and water.

AUTOMOTIVE LEAD-ACID BATTERIES

The basic design and operation of lead-acid batteries has not changed very much. However, improvements in construction and materials have made them much more powerful and dependable than ever before.

■ ***Battery Construction.*** The **lead-acid batteries** used in automobiles each with several plates submerged in a mixture of sulfuric acid and water (**Figure 19-2**). Each cell provides 2.1 volts when it is fully charged. For all battery types, cell voltage is determined by the materials that make up the plates. The current-producing ability of the battery is based on the surface area of the plates. The greater the surface area, the more amperage that can be produced. The positive plate contains lead peroxide and small amounts of other materials to strengthen the plates. The negative plate is sponge lead. Both plates are porous to allow the electrolyte to easily penetrate. An illustration of a simple battery cell is shown in **Figure 19-3**.

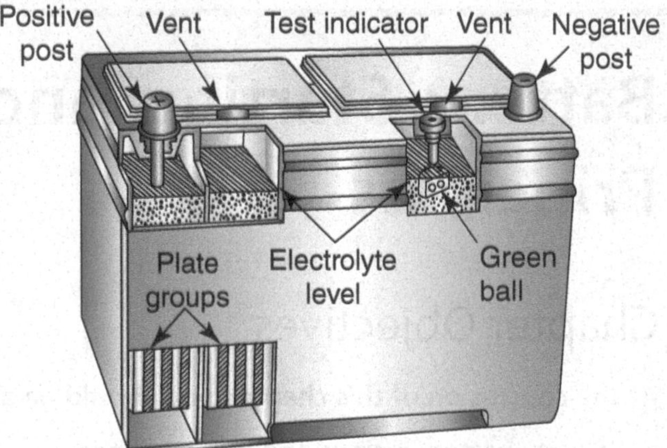

FIGURE 19-2 An illustration of a 12-volt automotive battery.

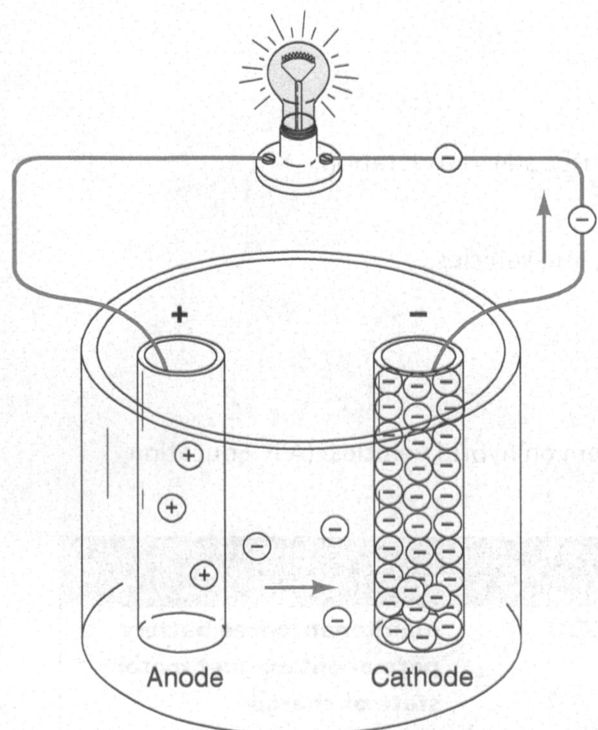

FIGURE 19-1 A battery consists of the anode, cathode, and electrolyte. The reaction between the dissimilar anode, cathode, and electrolyte causes electrons to be released.

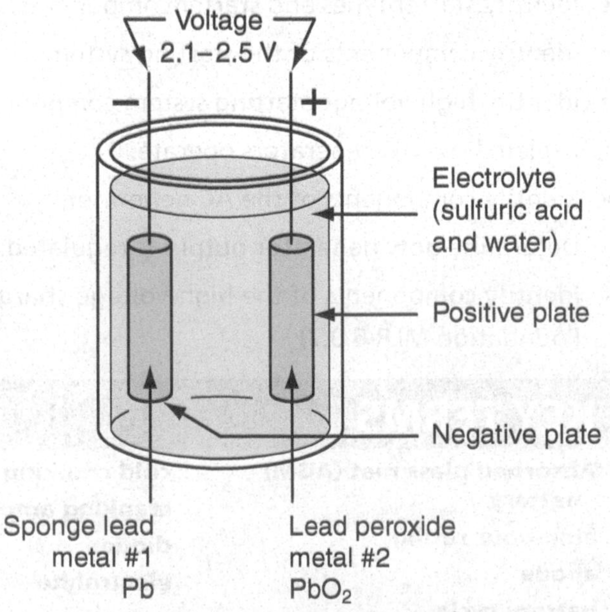

FIGURE 19-3 Lead-acid batteries produce about 2.1 volts per cell regardless of the number of plates that make up the cell.

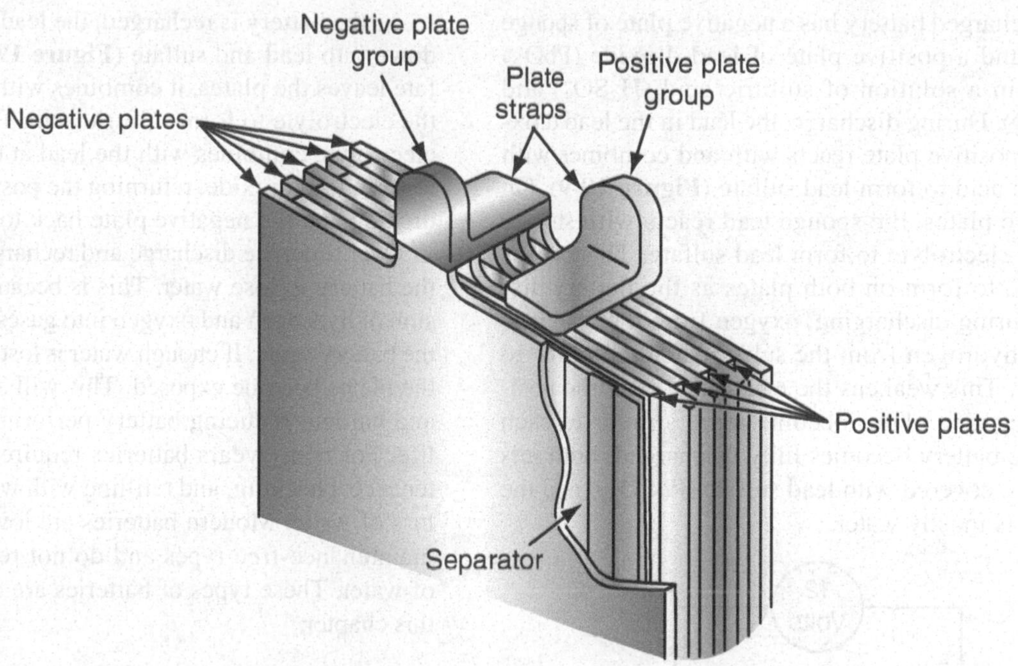

FIGURE 19-4 Plates are connected together in groups in each cell. The positive and negative plates are insulated from each other by separator plates.

The plates are arranged into elements. Elements contain both positive and negative plates (**Figure 19-4**). The positive plates are connected together in series. The negative plates in the element are also connected together in series. The voltage of each cell adds together to become the total battery voltage (**Figure 19-5**). By placing multiple plates together in parallel, the amperage capacity of the battery increases (**Figure 19-6**). To supply the voltage and current to crank the starter motor, an automotive battery uses several plate groups connected to each other (**Figure 19-7**).

Battery acid or electrolyte is a mixture of sulfuric acid and water. At full strength, the electrolyte is 64% water and 36% sulfuric acid (**Figure 19-8**). As the battery discharges, the percentage of water in the electrolyte increases as the sulfuric acid combines with the plates and hydrogen and oxygen are released. Recharging the battery pulls the sulfur off the plates and back into the acid.

■ ***Basic Operation.*** Lead-acid batteries provide large amounts of current for short periods of time to power the starter motor. This is the main function of the automotive battery—to power the starter motor. Cranking the engine partially discharges the battery. Once the engine is started, the charging system recharges the battery and powers the electrical system. The ability to partially discharge and recharge many times makes the lead-acid battery ideal for automotive use.

There are four stages of battery operation, called the **battery cycle**. These stages are charged, discharging, discharged, and recharging. However, in automotive use, the battery should not become completely discharged. If the battery is allowed to become completely discharged and recharged repeatedly, it will shorten battery life.

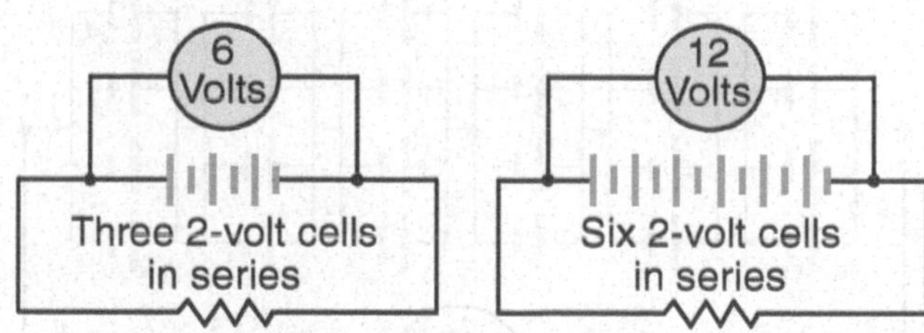

FIGURE 19-5 Adding cells in series increases the battery voltage. Older vehicles used 6-volt batteries, which used three cells in series. Modern 12-volt batteries use six cells in series.

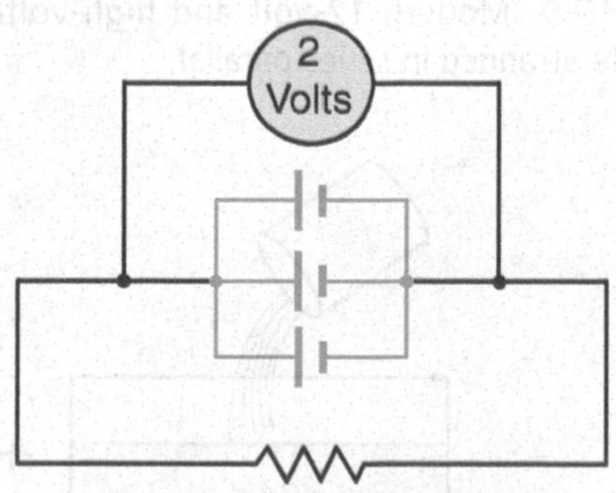

FIGURE 19-6 Connecting cells together in parallel increases the amperage capacity of the cell but does not affect cell voltage.

A fully charged battery has a negative plate of sponge lead (Pb) and a positive plate of lead dioxide (PbO_2) immersed in a solution of sulfuric acid (H_2SO_4) and water (H_2O). During discharge, the lead in the lead dioxide on the positive plate reacts with and combines with the sulfuric acid to form lead sulfate (**Figure 19-9**). On the negative plates, the sponge lead reacts with sulfate ions in the electrolyte to form lead sulfate. This causes lead sulfate to form on both plates as the battery discharges. During discharging, oxygen from the lead peroxide and hydrogen from the sulfuric acid combine to form water. This weakens the electrolyte, and the positive and negative plates become more similar to each other. If the battery becomes fully discharged, both sets of plates are covered with lead sulfate ($PbSO_4$), and the electrolyte is mostly water.

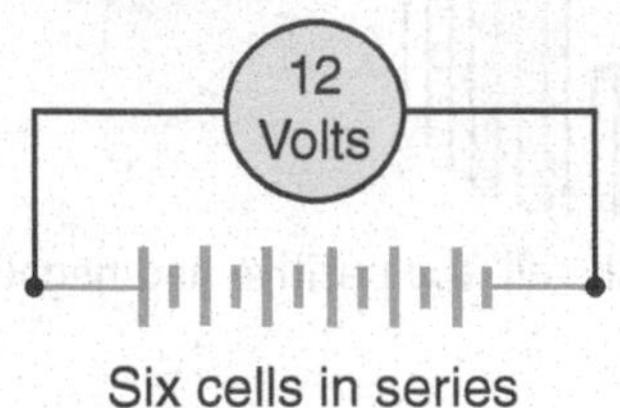

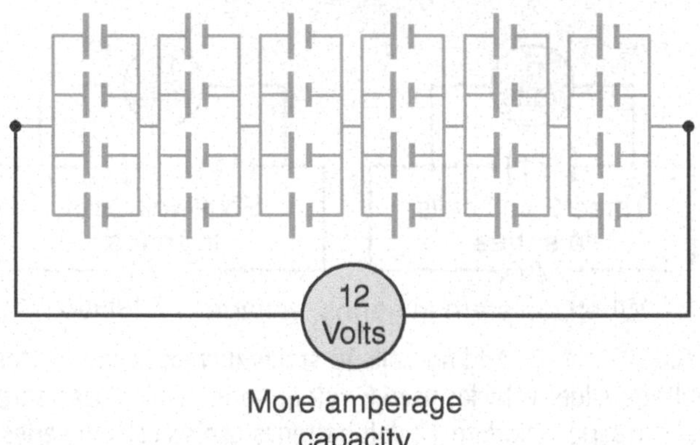

FIGURE 19-7 Modern 12-volt and high-voltage hybrid batteries are arranged in series-parallel.

As the battery is recharged, the lead sulfate is broken down into lead and sulfate (**Figure 19-10**). As the sulfate leaves the plates, it combines with the hydrogen in the electrolyte to form sulfuric acid. The oxygen in the electrolyte combines with the lead at the positive plate to form lead dioxide, returning the positive plate to lead dioxide and the negative plate back to lead.

Over time, the discharge and recharge process causes the battery to lose water. This is because of the conversion of hydrogen and oxygen into gases that escape from the battery vents. If enough water is lost and not replaced, the plates become exposed. This will allow them to dry and harden, reducing battery performance and service life. For many years batteries required periodic maintenance, checking, and refilling with water, to offset this loss of water. Modern batteries are low-maintenance or maintenance-free types and do not require the adding of water. These types of batteries are discussed later in this chapter.

■ ***Factors Affecting Battery Operation.*** How well a battery can deliver power and accept a charge depends on several factors, including battery temperature, battery state of charge, plate area, and cell construction.

Temperature affects battery condition because as the battery becomes colder, the chemical reaction between the cells and the electrolyte slows. This decreases output capacity and increases the amount of time it takes to charge the battery. **Figure 19-11** shows battery cranking power related to temperature. As the battery warms, it can deliver power more readily and accept a charge more quickly.

On the opposite end of the temperature scale, higher temperatures reduce battery life. **Figure 19-12** shows how operating temperature affects battery life. To help maintain acceptable battery temperatures, a vehicle may have fresh air routed to cool the battery or relocate the battery from under the hood to help reduce ambient temperatures. A battery cooling system is

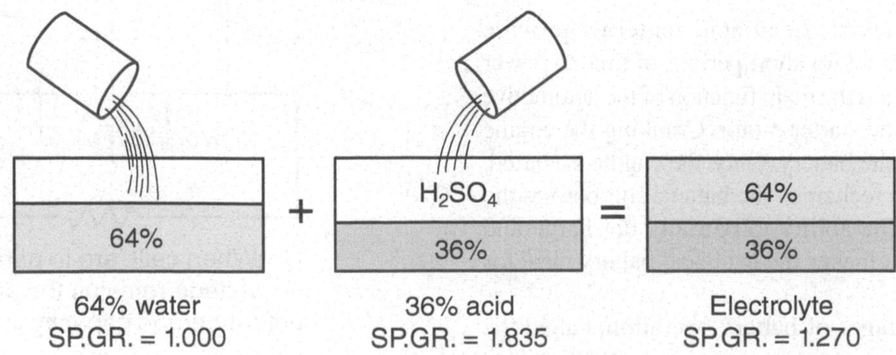

FIGURE 19-8 The electrolyte in a lead-acid battery is a mixture of sulfuric acid and water.

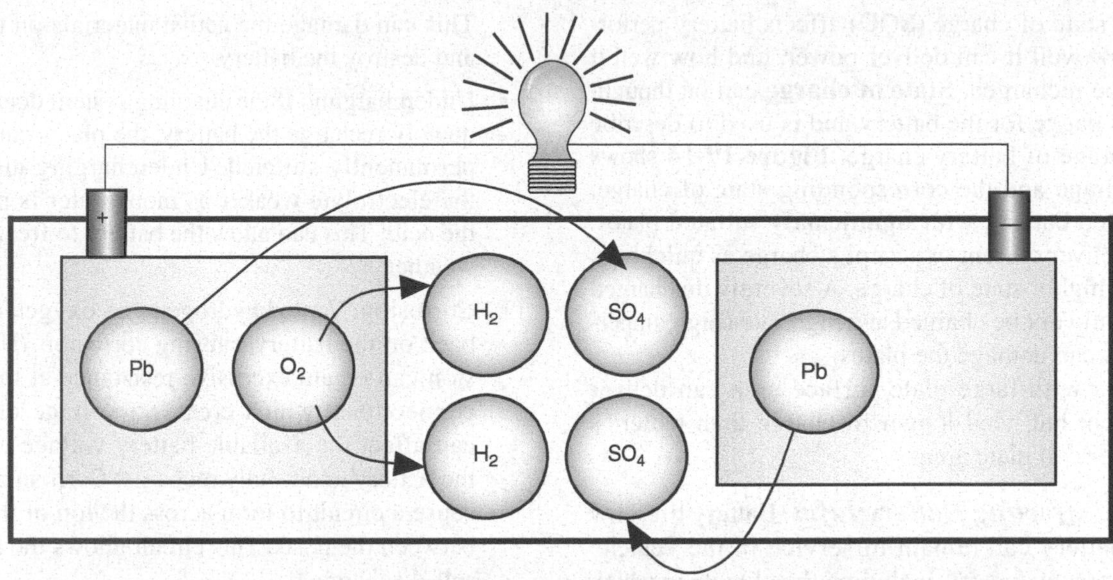

FIGURE 19-9 An illustration of a discharging battery. During discharging, the positive plate reacts with sulfuric acid to form lead sulfate. The oxygen and hydrogen combine to form water, diluting the acid.

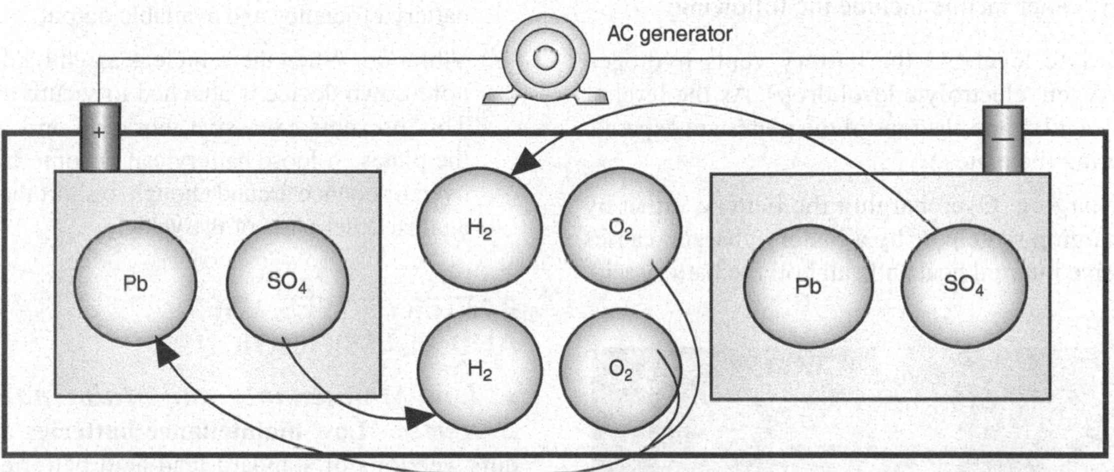

FIGURE 19-10 During charging, the lead sulfate breaks apart and the sulfate combines with hydrogen to form sulfuric acid.

Temperature	% of Cranking Power
80°F(26.7°C)	100
32°F (0°C)	65
0°F (−17.8°C)	40

FIGURE 19-11 An example of how battery power is reduced as the temperate decreases.

Temperature	Battery Life
77°F (25°C)	5 Years
92°F (33°C)	2.5 Years
107°F (42°C)	~ 1 Year

An increase of 15°F reduces battery life about 50%.

FIGURE 19-12 As battery temperature increases, its service life decreases.

shown in **Figure 19-13**. Another way to help maintain battery temperature is to directly monitor battery temperature or accurately calculate it based on other data. Older vehicles with noncomputer-controlled charging systems had to infer battery temperature to determine recharging rates.

Battery state of charge (SOC) affects battery performance, how well it can deliver power, and how well it is able to be recharged. **State of charge** can be thought of as a gas gauge for the battery and is used to describe the percentage of battery charge. **Figure 19-14** shows battery voltage and the corresponding state of charge. A discharged battery, with significantly sulfated plates, does not deliver current or accept a charge as quickly as one with a higher state of charge. A severely discharged battery should not be charged at a high rate (high amperage) as this can damage the plates.

Batteries with large plate surface area can deliver more current but need longer to charge than batteries with smaller cell plate areas.

■ ***Factors Affecting Battery Life.*** Battery life, how long the battery can remain in service in the vehicle, depends on many factors, including the climate in which the vehicle operates. For example, batteries used in hot climates, such as the American Southwest, tend to have shorter service lives than batteries used in other parts of the country. Other factors include the following:

- Electrolyte level. As the battery vents hydrogen and oxygen, electrolyte level drops. As the level of battery acid drops, the tops of the plates are exposed, hardening the plates.
- Overcharging. Overcharging the battery, either by the charging system or by a battery charger, causes excessive internal heat and can boil the battery acid. This can damage the active materials on the plates and destroy the battery.
- Undercharging. If the charging system does not adequately recharge the battery, the plates can become permanently sulfated. Undercharging also leaves the electrolyte weaker as more water is present in the acid. This can allow the battery to freeze in cold weather.
- Corrosion. Vented hydrogen and oxygen condense back on the battery, causing corrosion. This corrosion can create excessive resistance at the battery connections, which creates a voltage drop. This can affect the available battery voltage and cause the battery to not fully recharge. Corrosion can also cause a circuit to form across the top of the battery between the posts. This circuit allows the battery to self-discharge.
- Temperature. High temperatures, either from overcharging or high ambient and underhood temperatures, shorten battery life. Cold temperatures reduce battery efficiency and available output.
- Vibration. When the vehicle is assembled, a battery hold-down device is attached to secure the battery. This prevents excessive vibration and damage to the plates. A loose battery can become cracked, tip over, or bounce around enough to short the terminals against other parts of the vehicle.

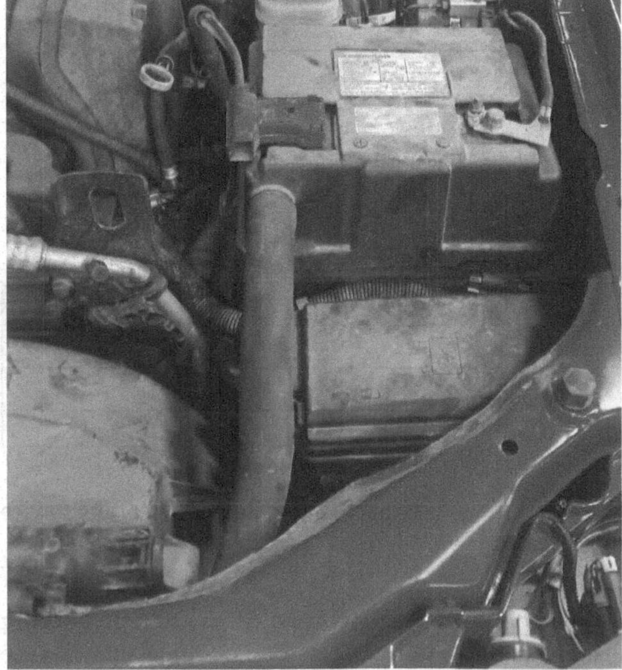

FIGURE 19-13 This battery is contained within a battery box, which has its own fresh air inlet to help keep the battery cool.

BATTERY TYPES, USES, AND CLASSIFICATIONS

■ ***Low-Maintenance and Maintenance-Free Batteries.*** **Low-maintenance batteries** are heavy-duty versions of standard lead-acid batteries. A low-maintenance battery (**Figure 19-15**) has removable vent caps so that the water level can be checked, and if necessary, water can be added. Because of the use of stronger construction materials for the plates, low-maintenance batteries require water much less frequently than standard lead-acid batteries.

Open Circuit Voltage (V)	State of Charge (%)
12.6 or greater	100
12.4–12.6	75–100
12.2–12.4	50–75
12.0–12.2	25–50
11.7–12.0	0–25
11.7 or less	0

FIGURE 19-14 Using battery voltage as a way to determine its state of charge.

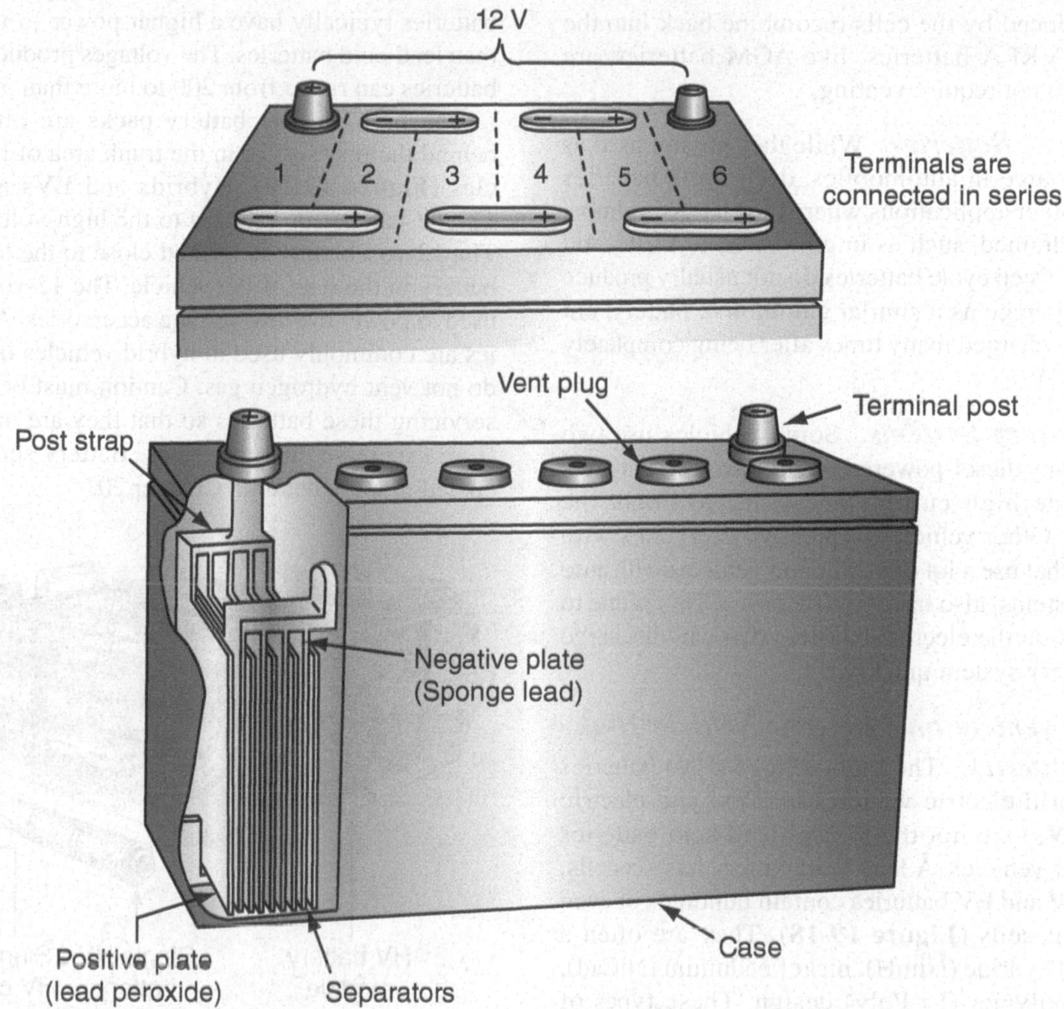

FIGURE 19-15 An illustration of a low-maintenance battery that has removable vent caps.

Maintenance-free batteries use slightly different plate materials and release almost no gas. This type of battery is sealed, as shown in **Figure 19-16**, and water cannot be added. If you attempt to open a sealed, maintenance-free battery, the case will be damaged and the battery will need to be replaced. These batteries do have gas relief vents to prevent excessive pressure from building within the case.

■ ***AGM Batteries.*** **Absorbed glass mat (AGM) batteries**, as shown in **Figure 19-17**, hold the acid in the absorbent mat, eliminating acid leaks. Another benefit is that the hydrogen that is normally given off during the charging of lead-acid batteries remains inside the battery. Because of this feature, AGM batteries are referred to as recombination batteries, meaning the hydrogen and oxygen recombine back into the battery instead of being released. Other benefits include quick recharge times and lower internal resistance, allowing for increased power output compared to traditional lead-acid batteries. AGM batteries require special charging procedures, as discussed in Chapter 20.

■ ***Valve-Regulated Lead-Acid Batteries.*** This type of battery, also referred to as a VRLA battery, is also a type of recombination battery. The hydrogen and

FIGURE 19-16 Maintenance-free batteries are sealed, and the electrolyte cannot be checked.

oxygen produced by the cells recombine back into the electrolyte. VRLA batteries, like AGM batteries, are sealed and do not require venting.

■ ***Deep Cycle Batteries.*** While they are not used as the power source in automobiles, deep cycle batteries are used in other applications where the charge is almost completely drained, such as in golf carts, forklifts, and boat motors. Deep cycle batteries do not usually produce as much amperage as a similar automotive battery, but they can be recharged many times after being completely discharged.

■ ***Two Battery Systems.*** Some vehicles use two batteries. Many diesel-powered trucks have two batteries to supply the high current necessary to crank the engine over. Other vehicles, typically luxury cars with accessories that use a lot of power and vehicles with auto stop/start systems, also have two batteries. This is due to the demands on the electrical system that can discharge a single battery system quickly.

■ ***Hybrid Vehicle and Electric Vehicle High-Voltage Batteries.*** The high-voltage (HV) batteries used in hybrid-electric vehicles (HEVs) and electric vehicles (EVs) are not the typical lead-acid batteries used in other vehicles. A lead-acid battery has six cells, modern HEV and EV batteries contain hundreds or even thousands of cells (**Figure 19-18**). They are often a nickel-metal hydride (NimH), nickel-cadmium (NiCad), or lithium-polymer (Li-Poly) design. These types of batteries typically have a higher power-to-weight ratio than lead-acid batteries. The voltages produced by these batteries can range from 200 to more than 500 volts.

The high-voltage battery packs are often located behind the rear seat or in the trunk area of hybrid vehicles (**Figure 19-19**). Hybrids and EVs also have a 12-volt battery, in addition to the high-voltage battery. The 12-volt battery is located close to the high-voltage battery in the rear of the vehicle. The 12-volt battery is used to power the low-voltage accessories. AGM batteries are commonly used in hybrid vehicles because they do not vent hydrogen gas. Caution must be used when servicing these batteries so that they are not damaged from excessive charging rates. Battery service is discussed in more detail in Chapter 20.

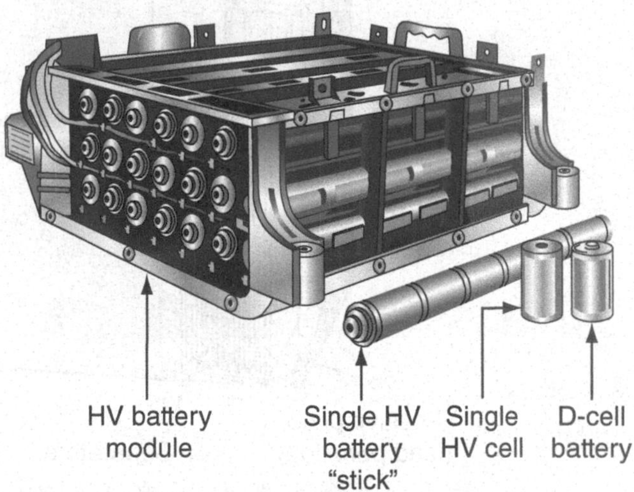

FIGURE 19-18 An illustration of a high-voltage battery for a hybrid vehicle.

FIGURE 19-17 Absorbed glass mat (AGM) batteries use an electrolyte in paste form and do not require any maintenance.

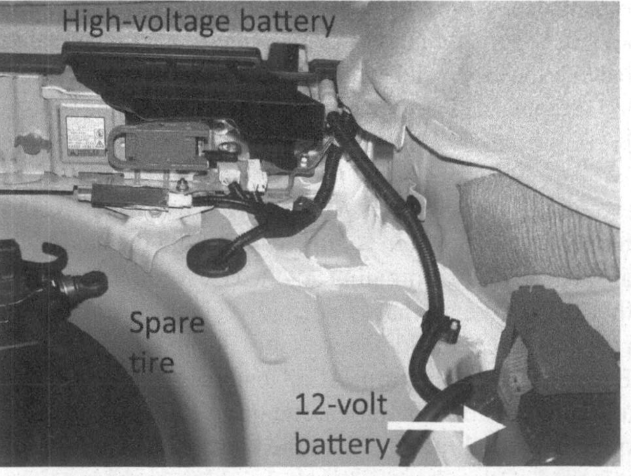

FIGURE 19-19 The rear compartment of a Toyota Prius houses both the high- and low-voltage batteries.

■ ***High-Voltage Battery Safety.*** The high-voltage (HV) wiring found on HEVs and EVs is covered in bright orange conduit, clearly identifying it as high voltage (**Figure 19-20**). Even though the risk of accidental shock from the HV system is very small, it is important to identify and be aware of the dangers of working around the HV system. With the vehicle ON or READY, the HV system is in operation and all HV components and wiring will be powered up. **Never attempt to open, disconnect, or work on any high-voltage wiring, connectors, or components while the system is ON.**

All hybrids and EVs have a method of disconnecting the high-voltage system. This is necessary when performing certain services and repairs on the high-voltage components. **Figure 19-21** shows the high-voltage service disconnect from a Chevrolet Volt. In this vehicle, the disconnect is hidden in the center console. Many hybrid vehicles use similar high-voltage disconnects to the one shown here. **Never disconnect the high-voltage system without following the proper service procedures.** Disconnecting a high-voltage battery is discussed in Chapter 20.

■ ***Battery Ratings.*** Automotive batteries are rated by cold cranking amps (CCA), cranking amps (CA), reserve capacity (RC), watts, and amp-hour (AH) ratings. Battery ratings are found on the battery sticker (**Figures 19-22** and **19-23**). Not all batteries have all of the ratings on the sticker. Most batteries have at least the battery type, called the BCI number, CCA, and CA.

The **BCI group** or Battery Council International Group number defines the physical qualities of a battery, such as terminal location and the height, width, and depth of the battery. This is a standardized system that

FIGURE 19-20 This shows some of the high-voltage wiring and components located under the hood of a Honda Accord Hybrid.

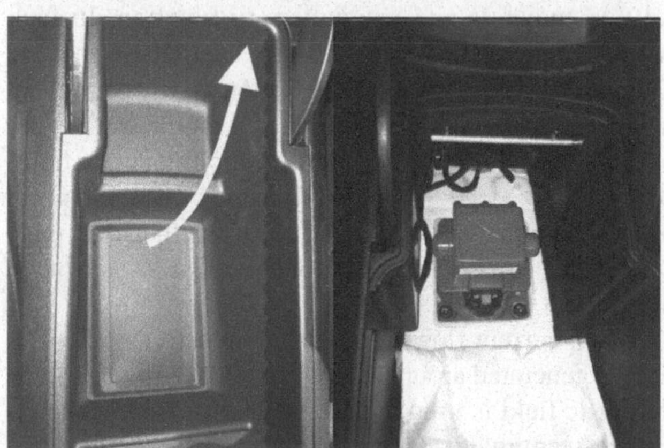

FIGURE 19-21 A high-voltage battery service disconnect. Do not attempt to remove this without following the manufacturer's service procedures.

FIGURE 19-22 This battery decal has the CCA, amp-hour rating, and the reserve capacity in addition to the BCI group number.

FIGURE 19-23 This battery is a group 86-7YR.

ensures that batteries within a group have similar physical characteristics no matter who manufactures the battery or where it is purchased (**Figure 19-24**).

The **cold cranking amp (CCA)** rating is based on the battery's ability to deliver current for 30 seconds at 0°F (−17.7°C) before the total voltage drops to 7.2 volts. The CCA rating is important for anyone who lives in a cold climate because battery output decreases as temperature decreases. Typical CCA ratings for modern vehicles range from 600 to 800 CCA.

The ability of the battery to produce amperage is determined by the amount of surface area of the positive and negative plates. A battery with an insufficient CCA capacity may perform well during warm weather but will likely keep the vehicle from cranking and starting when the weather turns cold.

The **cranking amps (CA)** rating is similar to CCA except that it is measured at 32°F (0°C) instead of 0°F. The CA rating is higher than the CCA rating.

Battery reserve capacity (RC) is a rating, in minutes, of how long the battery can supply 25 amps to the electrical system before voltage drops below 10.5 volts if the charging system fails. The RC rating is based on a battery temperature of 80°F. Typical RC ratings range from 50 to 120 minutes.

The **amp-hour (AH) rating** of a battery describes how many amp-hours of current the battery can supply for 20 hours before voltage falls below 10.5 volts. If a battery can supply 5 amps for 20 hours, the amp-hour rating equals 100 AH.

A battery's wattage or power rating is determined by multiplying the current the battery can supply times the voltage at 0°F. A typical range is 2,000 to 5,000 watts.

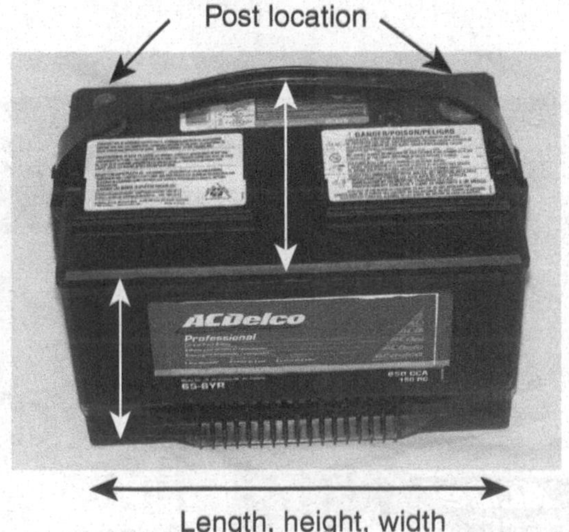

FIGURE 19-24 The BCI group number represents the physical characteristics of the battery.

In the automotive industry, the internal combustion engine, AC generator, and electric motors are now often described by wattage output or consumption. Expect to see watts used more in the future when you are dealing with power ratings.

Starting System

The starting system's only function is to spin the engine fast enough that the air-fuel mixture can be compressed and ignited within the engine. Once the engine can run on its own, the starter motor is no longer needed. To crank the engine over, the **starter motor** drives a large gear, called a ring gear or flywheel, which is bolted to the rear of the engine's crankshaft. The starting system is largely unchanged from early vehicles except that in modern cars and trucks, the starting system operates with the on-board computer and antitheft systems.

In the early days of internal combustion-powered vehicles, the driver had to use a special wrench to crank the engine over using muscle power alone to start the engine. Needless to say, this was not an easy task and it often led to serious injury if the engine kicked back against the bar. Electric starter motors became common in the 1920s, though many vehicles retained the ability to be hand cranked into the 1930s.

DC MOTORS

The electric starter motors used in most of today's cars and trucks are very much like those used in the 1920s and 1930s. The principles of DC electric motor operation remain the same.

■ ***Motor Components and Operation.*** The direct current (DC) motors used throughout the automobile rely on magnets and magnetism to function. In fact, without magnetism there would be no electric motors. There are two basic types of DC motors, the permanent magnet motor and the electromagnet motor. Both types of electric motors use the interaction of magnetic fields (**Figure 19-25**). As you probably know, when two north or two south magnetic poles get close together, they repel and when a north and a south pole get close to each other, they attract (**Figure 19-26**).

When current flows through a conductor, a magnetic field is generated around it (**Figure 19-27**). In a relay, the magnetic field is used to pull the armature and contacts closed to complete a circuit. An electric motor also uses magnetic interactions to create movement. The difference is that in a motor, the magnetic fields keep changing so that movement turns into rotation and rotation is turned into work.

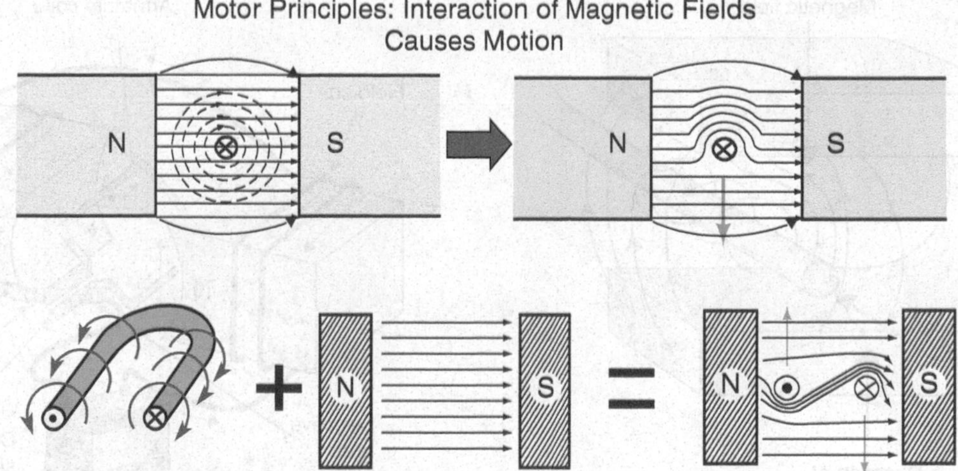

FIGURE 19-25 An electric motor uses the interaction of magnetic fields to produce motion.

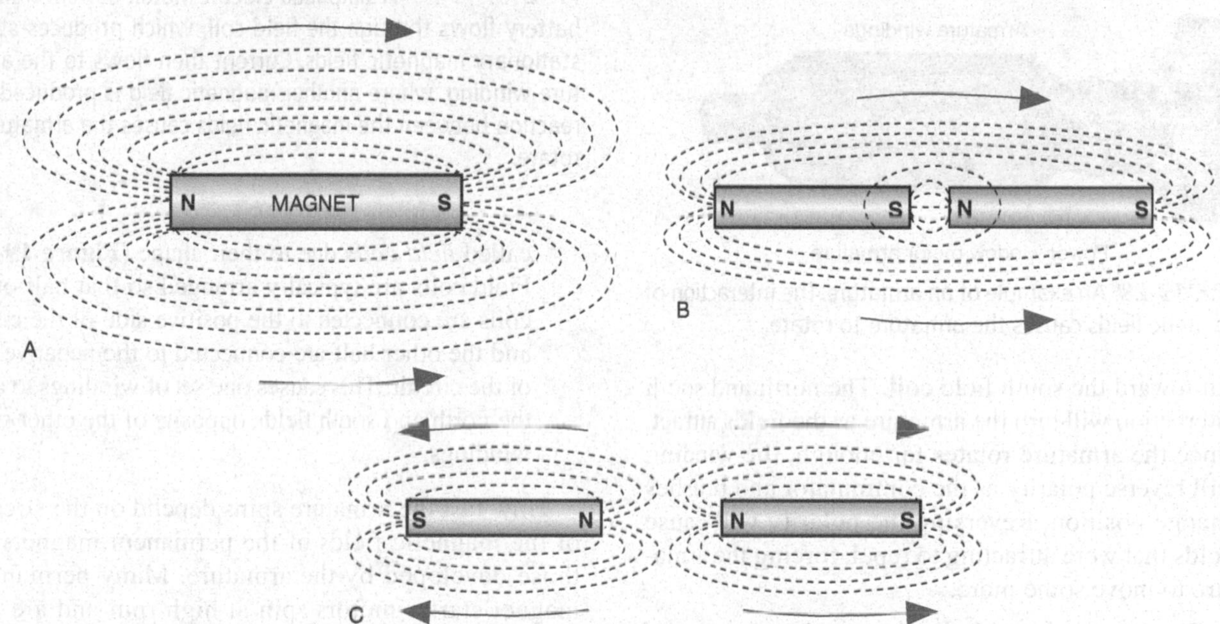

FIGURE 19-26 Opposite magnetic poles attract and similar poles repel each other.

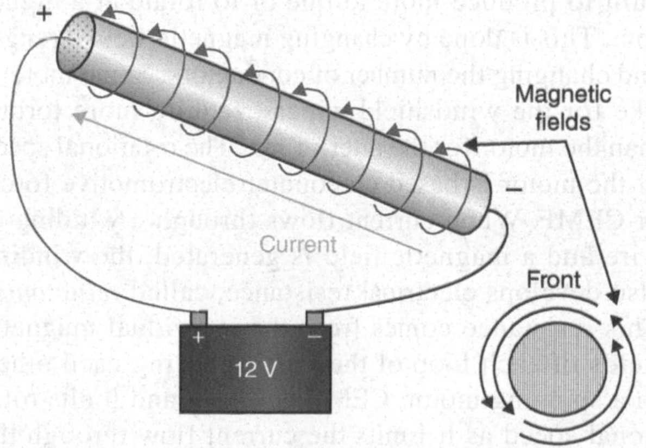

FIGURE 19-27 Current flow through a conductor creates a magnetic field around the wire.

Components of DC Motors

- **Armature**—In a DC motor, there are stationary magnetic fields and magnetic fields that form around conductors. The conductors form a rotating part called an *armature*. The armature is placed inside the stationary magnetic fields (**Figure 19-28**). When current flows through the armature, the interaction of the fields causes the armature to move.
- **Brushes and Commutator**—To keep the armature moving requires changing the polarity of the magnetic fields. Polarity means the orientation of positive and negative or north and south poles. As shown in **Figure 19-29**, the section of armature winding against the positive brush has north magnetic fields

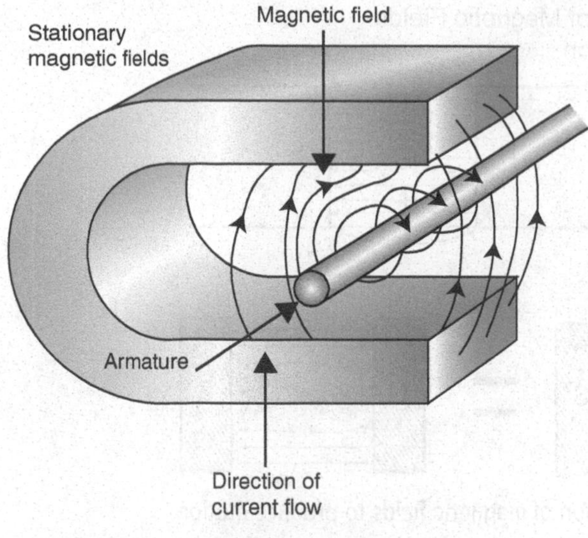

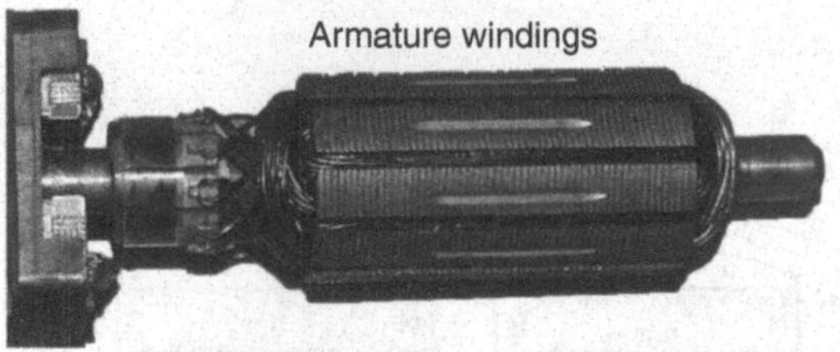

Power window motor armature

FIGURE 19-28 An example of an armature. The interaction of the magnetic fields causes the armature to rotate.

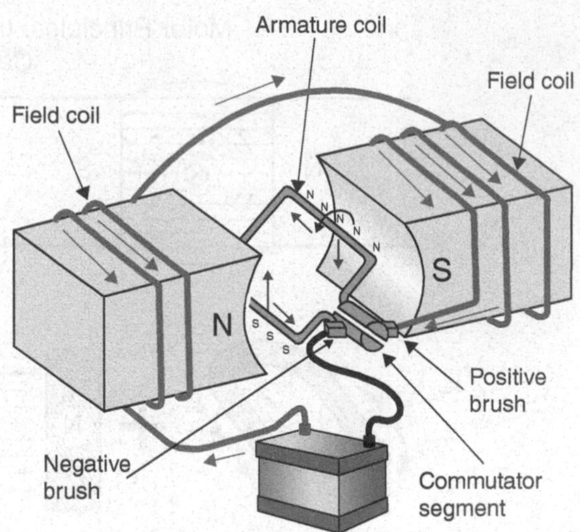

FIGURE 19-29 A simplified electric motor. Current from the battery flows through the field coil, which produces strong stationary magnetic fields. Current then flows to the armature winding, where another magnetic field is produced. The reaction between the magnetic fields causes the armature to rotate.

out toward the south field coil. The north and south interaction will turn the armature as the fields attract. Once the armature rotates far enough, the winding will reverse polarity as the commutator and brushes change position. Reversing the polarity will cause fields that were attracting to repel, forcing the armature to move some more.

The brushes provide the connection for current to flow to the armature, and the commutator is the connection point on the armature itself. As the armature rotates, a different commutator segment and different armature windings connect to power and ground (**Figure 19-30**). This means that as the armature moves, a different wire becomes magnetized so that the attraction and repulsion between the armature and the stationary fields keep changing. While the fields created around the armature change each time a different commutator segment contacts a brush. The stationary magnetic fields remain unchanged.

- **Stationary Fields**—The stationary magnetic fields are formed by either strong permanent magnets or electromagnets. Permanent magnets are secured to the motor housing and are aligned so the north and south poles alternate polarity. The electromagnets are often called *field coils* due to their shape (**Figure 19-31**). Field coils are typically arranged so that half of the coils are connected to the positive side of the circuit and the other half are connected to the negative side of the circuit. This causes one set of windings to align the north and south fields opposite of the other set of windings.

How fast the armature spins depend on the strength of the magnetic fields of the permanent magnets and those developed by the armature. Many permanevnt magnet starter motors spin at high rpm and are then geared down to decrease rotation speed and increase torque. Depending on its application, a motor can be built to produce more torque or to rotate at a higher rpm. This is done by changing magnetic field strength and changing the number of conductors. Some motors, like for the windshield wipers, require more torque than the motor for the interior fan. The rotational speed of the motor is based on counter electromotive force, or CEMF. When current flows through a winding of wire and a magnetic field is generated, the winding also develops electrical resistance, called *reluctance*. This resistance comes from the individual magnetic fields of each loop of the wire opposing each other. In a spinning motor, CEMF develops and limits rotational speed as it limits the current flow through the armature windings.

Refer to Chapter 19 of the Lab Manual for an exercise about building a small electrical motor.

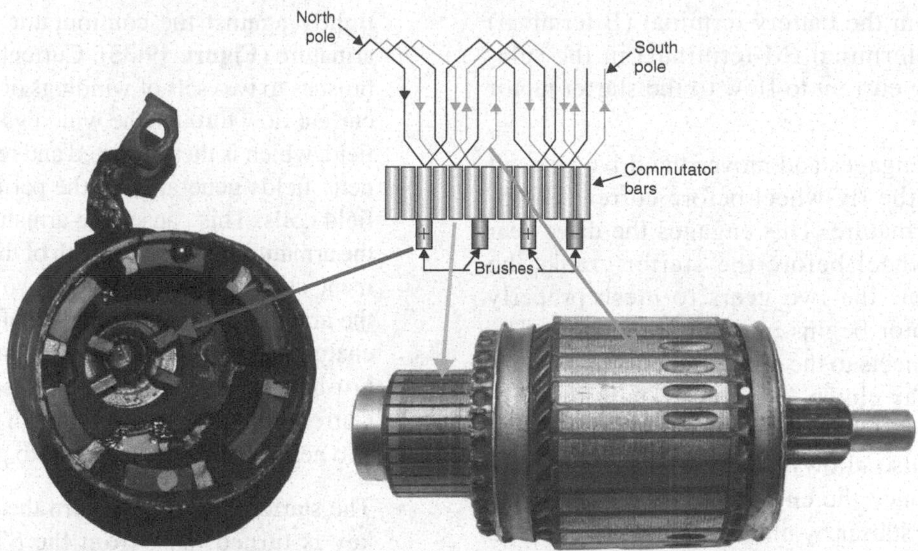

FIGURE 19-30 An example of the commutator and brushes from a starter.

FIGURE 19-31 A starter motor with electromagnet field coils. These powerful electromagnets are used to create very strong stationary magnetic fields.

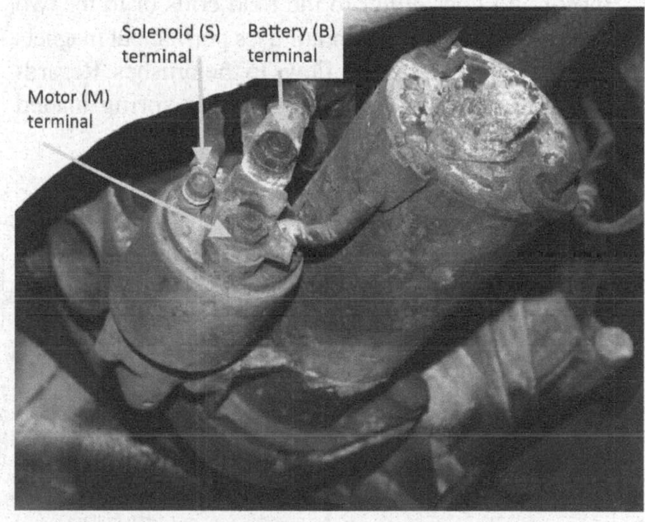

FIGURE 19-32 An example of the connections on the starter motor.

■ ***Starter Motor Operation.*** There are two basic types of starters, permanent magnet and electromagnet motors. Both operate the same, the difference is in the field coils that create the stationary magnetic fields. Basic motor operation is as follows:

- Power is supplied to the battery connection, usually referred to as the B (battery) terminal at the solenoid. Depending on the vehicle, this may be from where power is distributed to various systems by fuse links connected at the B terminal. The rear of a solenoid and its connections are shown in **Figure 19-32**.
- Voltage is applied to the S terminal when the control circuit is activated. The control circuit has the ignition and other switches that control when the starter operates. When in START, the voltage to the S terminal causes the pull-in winding of the solenoid to energize. This creates a magnetic field that attracts the plunger. Once the plunger is pulled into the solenoid, the shift lever pushes the drive gear out into the nose of the starter, and the pull-in winding circuit opens. A hold-in winding keeps the plunger retracted, as long as the ignition is in the START position. This is done because less current is needed to hold the plunger in place once it has been retracted, so a lighter winding and less current can be used. Once the plunger is retracted, the main contact at the rear of the solenoid completes

the circuit from the battery terminal (B terminal) to the motor terminal (M terminal) in the solenoid, allowing current to flow to the starter motor (**Figure 19-33**).

- The solenoid engages and moves the drive gear out to mesh with the fly-wheel before current is supplied to the armature. This engages the drive gear with the flywheel before the starter cranks the engine to allow the two gears to mesh properly before the motor begins to crank the engine. The drive gear connects to the armature with an overrunning or one-way clutch. The clutch keeps the drive gear locked to the armature shaft while cranking the engine but it also allows the drive gear to counter-rotate freely once the engine starts. This prevents damage to the starter by unlocking the drive gear to spin opposite its normal drive rotation. Starter drive gear operation is shown in **Figure 19-34**.
- Once the solenoid contacts close, current enters the motor and goes either to the field coils or to the two positive brushes. If the motor uses permanent magnets as the field coils, current flows to the brushes. Regardless of motor type, the brushes are spring loaded tightly against the commutator at the rear of the armature (**Figure 19-35**). Current flows through the brushes to two sets of windings of the armature. This current flow through the windings creates a magnetic field, which is then attracted and repelled by the magnetic fields generated by the permanent magnets or field coils. This causes the armature to rotate. Once the armature moves the width of the commutator segment, current flow through the two windings stops. As the armature spins, another set of windings become energized as the commutator segments align with the brushes and the attraction and repulsion continues. Current continues to flow through the windings to the two negative brushes and then to ground.
- The starter continues to turn the flywheel until the key is turned back from the START position or the powertrain control module (PCM) senses that the engine has started. Once this occurs, power is removed from the S terminal at the solenoid. This causes the magnetic field in the solenoid's hold-in winding to collapse. Without the magnetic field to hold the plunger back in the solenoid, the return spring pushes the shift lever out. This disengages

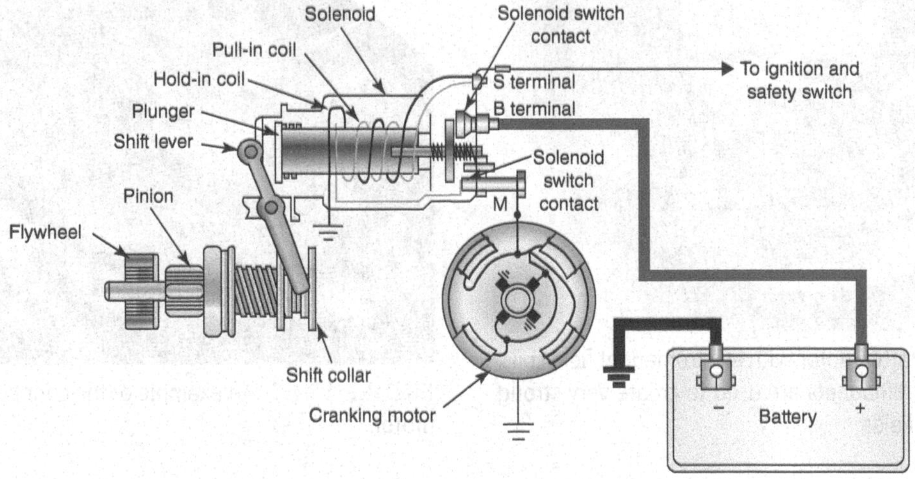

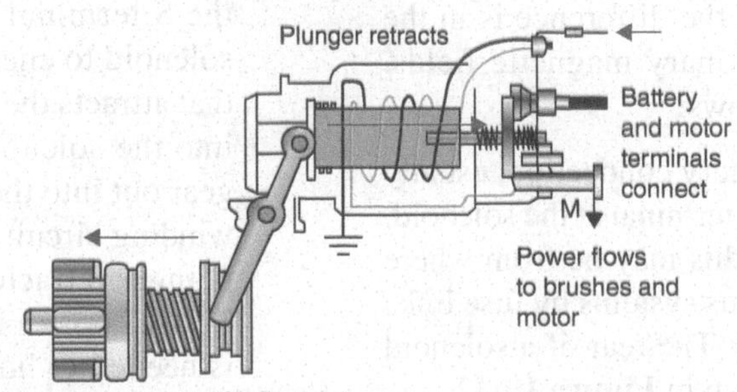

FIGURE 19-33 The solenoid contains two windings and the contacts that supply battery power to the starter motor. The solenoid pulls back on the shift lever, which forces the drive gear out against the flywheel.

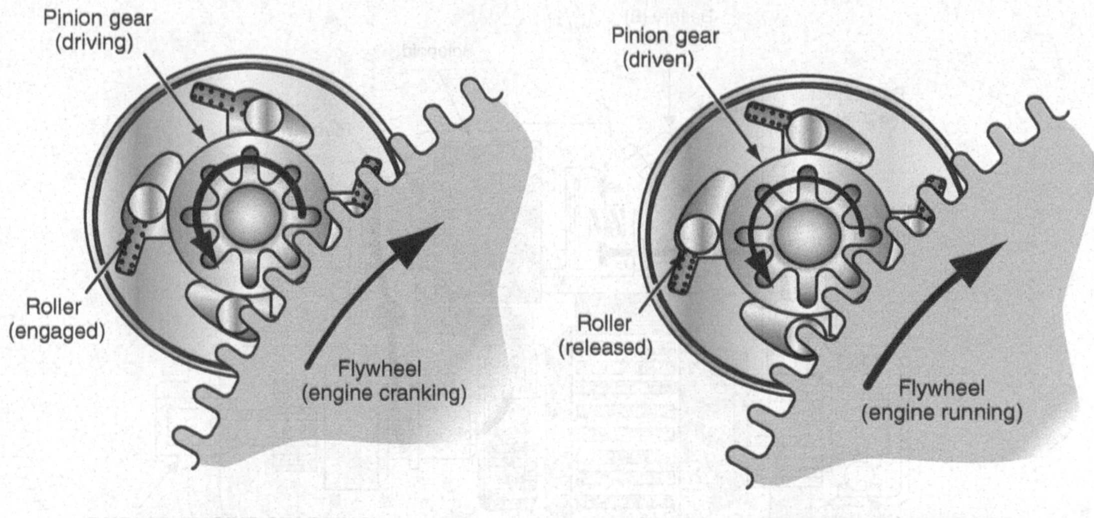

FIGURE 19-34 The drive gear is connected to an overrunning clutch. This allows the drive gear to be locked to the armature when driven in one direction and to spin freely in the other direction as the engine starts.

FIGURE 19-35 The brushes supply current to the armature. Two brushes and field coils are connected to positive and two are connected to ground.

the contacts in the solenoid and opens the circuit between the battery and motor terminals. The armature slows, and the drive gear is pulled back into the motor housing.

Permanent magnet motors use strong manufactured permanent magnets (PM) as the field coils (**Figure 19-36**). Permanent magnet motors draw less power than motors with electromagnetic field coils, reducing the load on the battery. PM motors also tend to be smaller overall than electromagnet motors, and this reduces weight.

Permanent magnet starters often use a gear reduction system, converting fast armature rotation speeds into slower, high-torque drive gear speeds. Two common gear reduction methods are the planetary gearset and the gear-to-gear system (**Figure 19-37**). The planetary gearset consists of three gears, the sun gear in the center, the planetary pinion gears, and the outer ring gear. The sun gear is attached to the armature and is the input gear. When the armature turns and the ring gear is held, the pinion gears rotate more slowly than the armature but with increased torque. This torque is applied to the starter drive gear, which is used to crank the engine. Gear-to-gear starters use a small gear on the armature as the drive gear, which is meshed with a larger gear that is attached to the starter drive gear. The armature spins at high rpm, transferring that speed into the lower speed but higher torque of the starter drive gear.

Some motors use electromagnets instead of permanent magnets. In the motor illustrated in **Figure 19-38**, current flows through two of the four field coils before passing to the brushes and armature. Two field coils are connected to the positive side of the circuit while the other two are connected to the negative side. When current flows to the motor, strong stationary magnetic fields are formed by the field coils. These fields attract and repel the fields formed by the armature. Aside from the field coils, an electromagnet starter operates exactly the same as a permanent magnet motor.

Starting System Components and Operation

Starting systems can be either basic, noncomputerized systems or computer controlled. Older vehicles typically have basic systems that contain the battery, battery cables, starter, starter solenoid, ignition switch, trans-

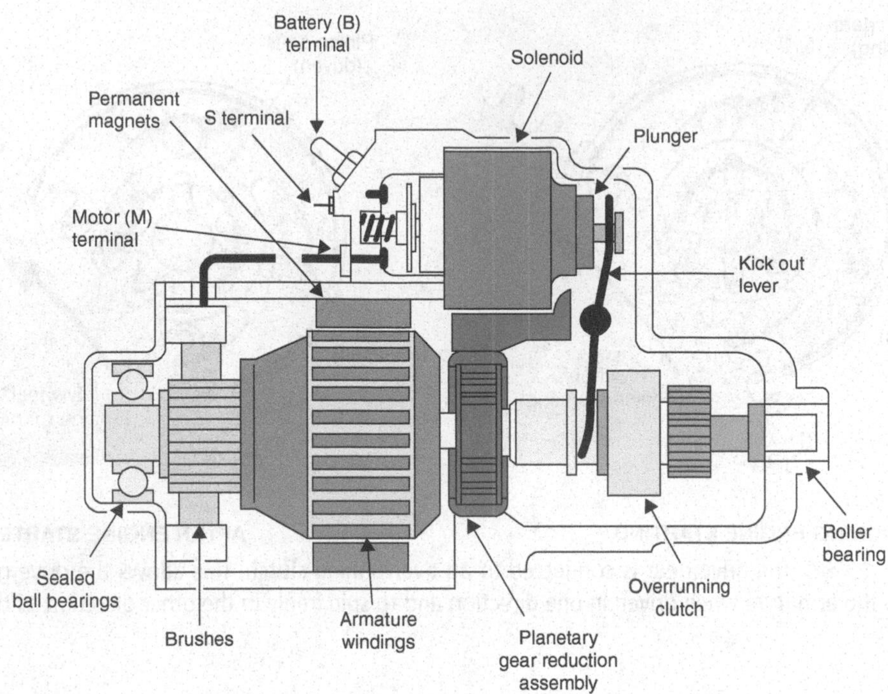

FIGURE 19-36 This illustrates the components of a gear reduction starter.

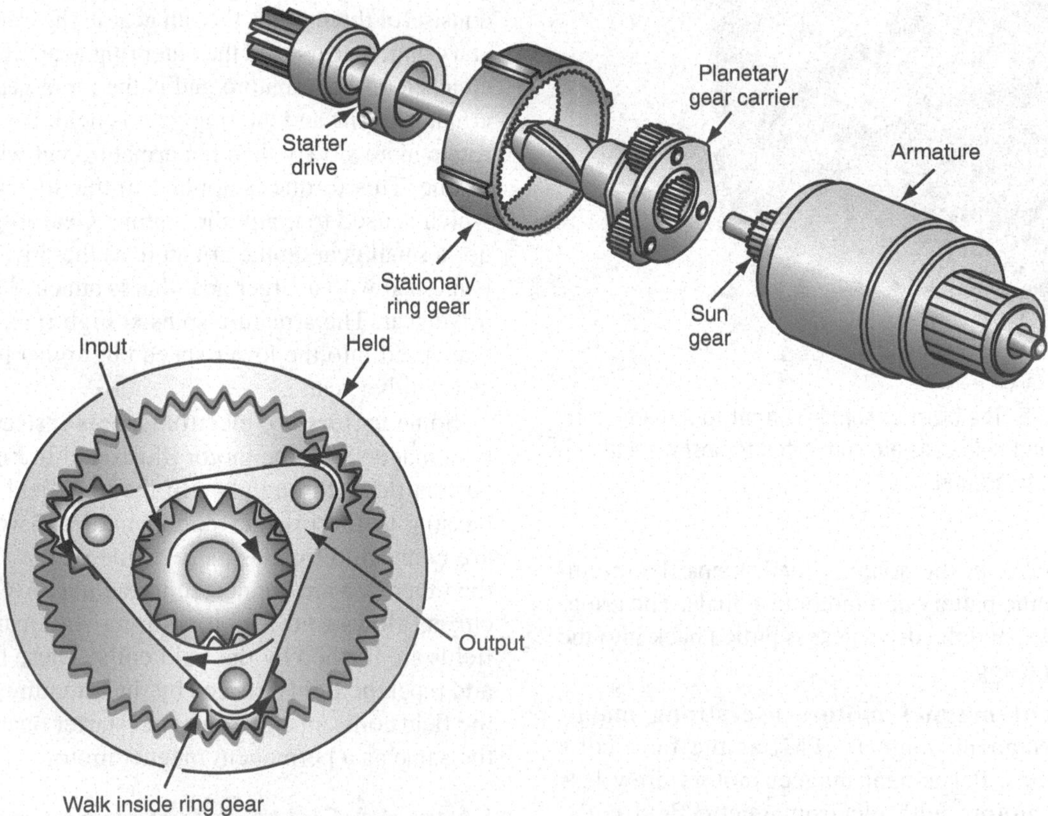

FIGURE 19-37 Gear reduction starters often use a planetary gearset to increase torque.

mission safety switch, and the related wiring. Most vehicles built in the 2000s and later have some type of antitheft and/or have a push-to-start system as standard equipment.

STARTER MOTOR CIRCUIT

There are two starter circuits, the starter motor circuit and the starter control circuit. The starter motor circuit includes the battery, battery cables, starter solenoid,

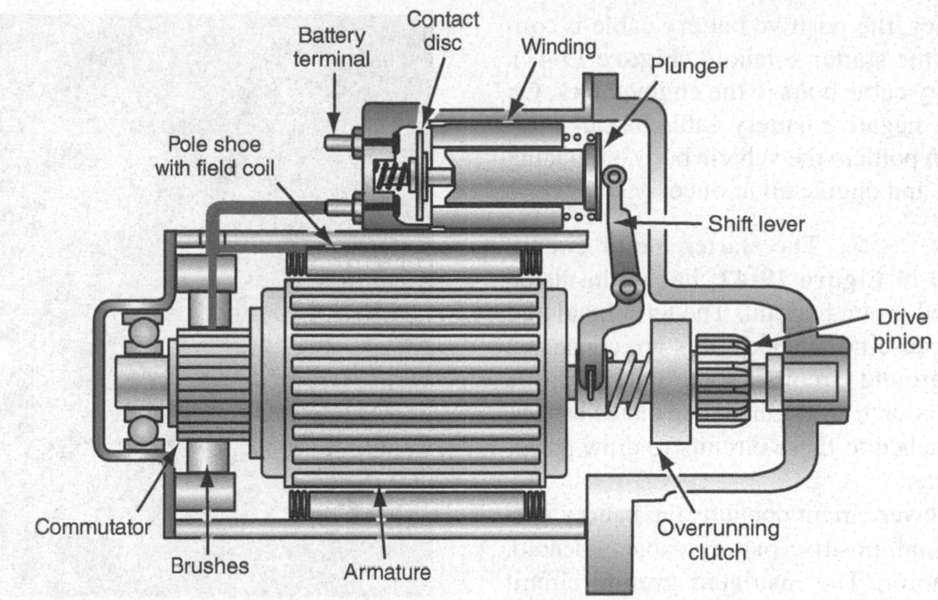

FIGURE 19-38 An illustration of an electromagnet starter design.

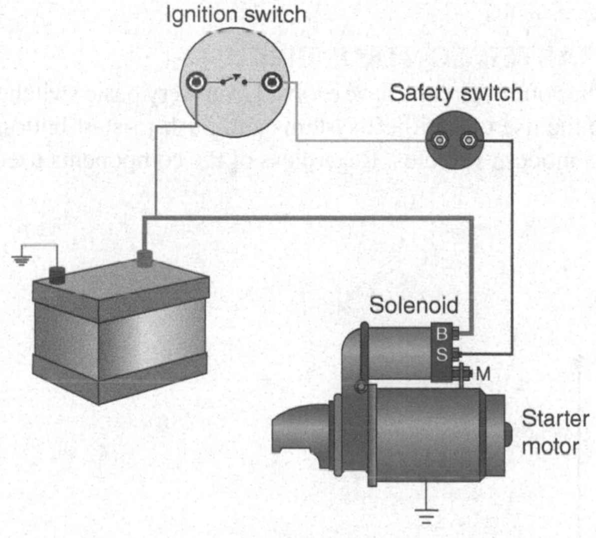

FIGURE 19-39 An illustration of a simple starter circuit. Starter operation is controlled by the ignition switch and either a clutch or a neutral safety switch.

and starter motor (**Figure 19-39**). The motor circuit is responsible for supplying the high current from the battery to the starter motor to crank the engine. The motor circuit consists of the battery, both the positive and negative battery cables, the starter solenoid, and the motor.

■ ***Battery and Cables.*** The battery supplies power for the starter, which is the largest single load on the electrical system. During cranking, the starter may draw 150 to 300 amps of current or more. The amount of current needed depends on engine size, compression ratio, and other factors. Because of the high amount of current used by the starter, the cables that connect the battery to the starter are large-diameter copper cables. The positive battery cable must provide a solid connection at the battery and the starter motor to prevent excessive voltage drop. If a poor connection occurs, the starter motor may not receive full battery voltage when cranking, which results in slow cranking speed. For this reason, battery cables are bolted in place, both on the battery and on the starter motor.

Battery cables come in a variety of types and sizes (**Figure 19-40**). The end of the cable that connects to the starter typically has an eyelet connection. This connection uses a metal terminal with a hole punched into it. The hole is for the battery stud on the solenoid. The eyelet is placed onto the solenoid, and the stud sticks through the hole. A nut is used to tighten the connection against the stud.

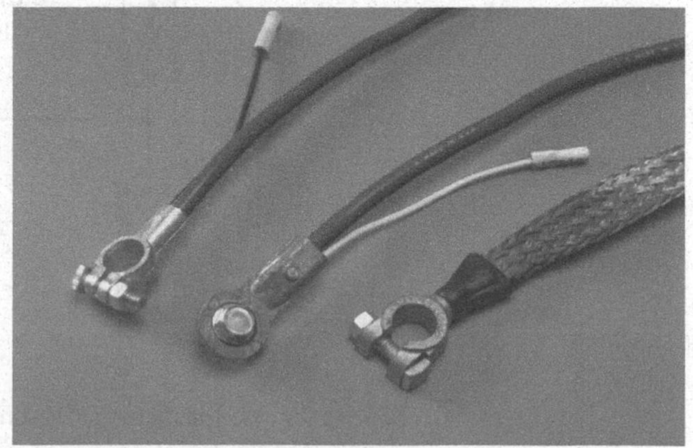

FIGURE 19-40 Battery cables are large diameter to handle the high current required to crank the engine.

On most vehicles, the positive battery cable is connected directly to the starter solenoid (**Figure 19-41**). The negative battery cable bolts to the engine block. On some vehicles, the negative battery cable has an intermediate connection point to the vehicle body, grounding the battery, frame, and engine all at once.

■ ***Insulated Circuits.*** The starter motor circuit, highlighted in red in **Figure 19-42**, has an insulated power and insulated ground circuit. The term **insulated circuit** is used because no other circuits are connected to the power and ground circuits for the starter. This is because the starter is such an extreme load on the battery, nothing else is attached to these circuits to draw power away from the starter.

The insulated power circuit contains the battery, battery positive terminal, positive battery cable, solenoid, and the starter motor. The insulated ground circuit includes the battery, negative battery terminal, negative battery cable, negative battery cable connection—usually on the engine block—and the starter motor.

During cranking, current flows from the battery, through the battery cable to the starter, and back to battery ground. If at any of the connections in the circuit there is a poor connection, a voltage drop will occur. This will reduce the voltage available for the starter motor, which will affect motor operation.

FIGURE 19-41 The positive battery connection at the starter solenoid. Because the starter connection carries high amperage, the battery cable is bolted to the starter to ensure a tight connection.

STARTER CONTROL CIRCUIT

The control circuits have evolved from very basic switches to the use of antitheft systems and push-to-start buttons on modern vehicles. Regardless of the components used,

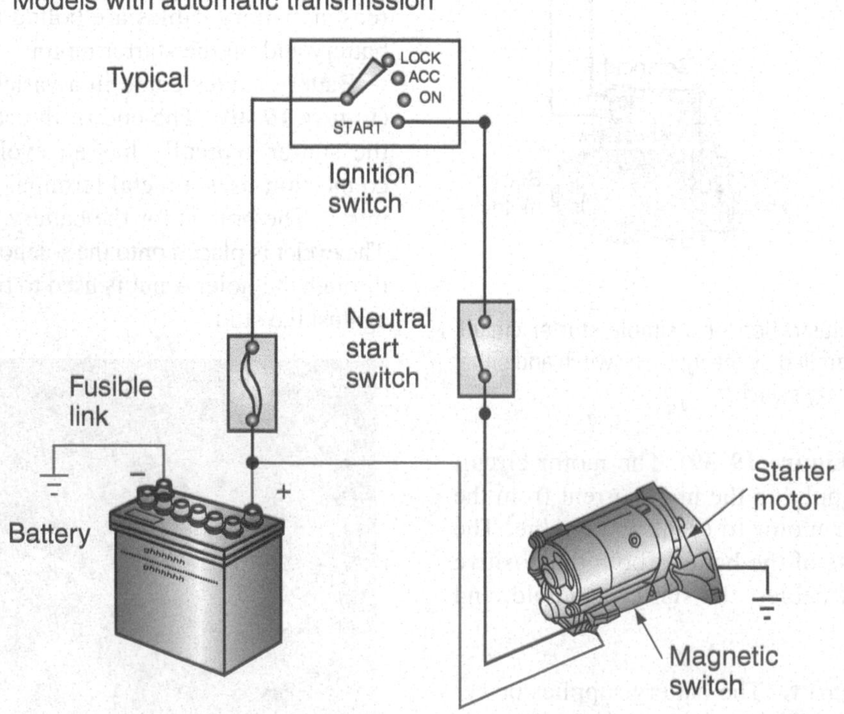

FIGURE 19-42 The motor circuit contains the battery, positive and negative cables, solenoid, and the motor.

the starter control circuit's only function is to control the operation of the starter based on the actions of the driver.

■ ***Non-Antitheft Control Systems.*** In older vehicles, without antitheft systems, the **starter control circuit** is a simple circuit containing the battery, ignition switch, transmission safety switch, starter relay or solenoid, and the related wiring (**Figure 19-43**). When the driver turns the ignition key, the contacts in the ignition switch close, which allows current to flow from the battery, through the ignition switch, and finally to the transmission safety switch.

In vehicles with automatic transmissions, the transmission gear selector must be in either Park or Neutral for there to be a closed connection through the switch. For this reason, many people call this switch the Park-Neutral switch. In vehicles with manual transmissions, a clutch safety switch is installed, usually on the clutch pedal under the dash. When the clutch is fully pressed, the switch contacts close to complete the circuit from the ignition switch to the solenoid. When the transmission is in Park or Neutral or the clutch is fully pressed and the ignition key turned, current then flows to the S terminal at the solenoid. The pull-in winding in the solenoid energizes and creates a magnetic field, pulling the plunger rearward. Once the plunger moves fully rearward, the main contacts in the solenoid close, which allows battery voltage and current to flow through the solenoid and into the starter motor.

■ ***Antitheft Starter Circuits.*** Since the 1980s, vehicles have been equipped with integrated antitheft systems, some of which disable the starter motor circuit. **Figure 19-44** shows a wiring diagram of a vehicle with a starter interrupt relay as part of the starter control circuit. In addition to the basic starter control components, an additional circuit is used to either allow or inhibit starter operation. In this example, the starter inhibit relay is used to control the starter solenoid circuit. For the starter to operate, the ignition switch, transmission safety switch, and antitheft system must close to supply power to the inhibit relay.

■ ***Ignition Switches.*** The ignition switch, like most other components, has evolved to meet various changes over the years. In many vehicles, the ignition switch is still key-operated by the driver. Locations of the ignition switch vary, from being mounted directly to the ignition lock cylinder to being remotely mounted on the steering column or in the dash (**Figure 19-45**). Ignition switches are multiple-pole multiple-throw switches, meaning that when the switch moves from position to position, several contacts are used to connect several different circuits (**Figure 19-46**). In the crank position,

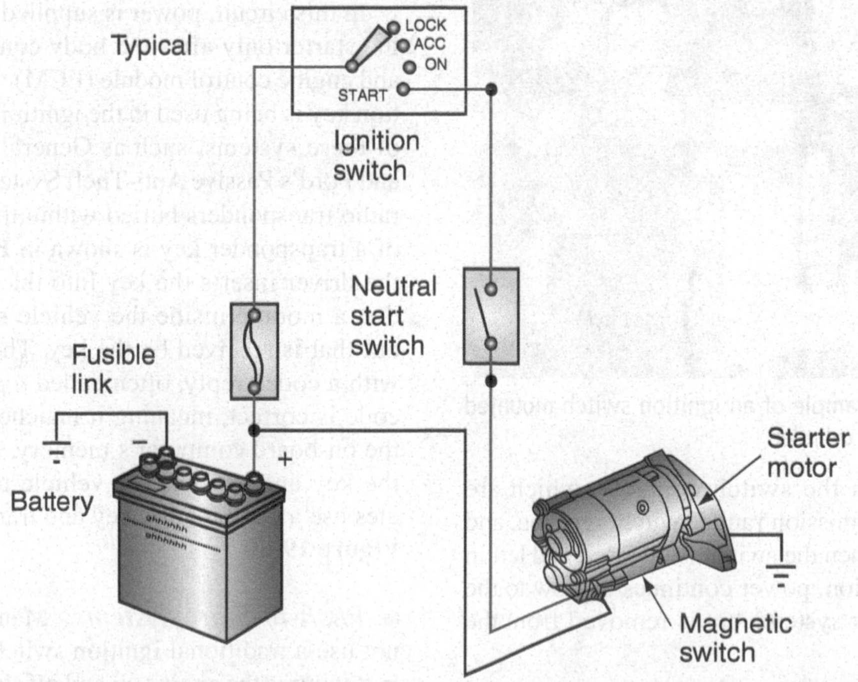

FIGURE 19-43 An illustration of the starter control circuit without an antitheft system.

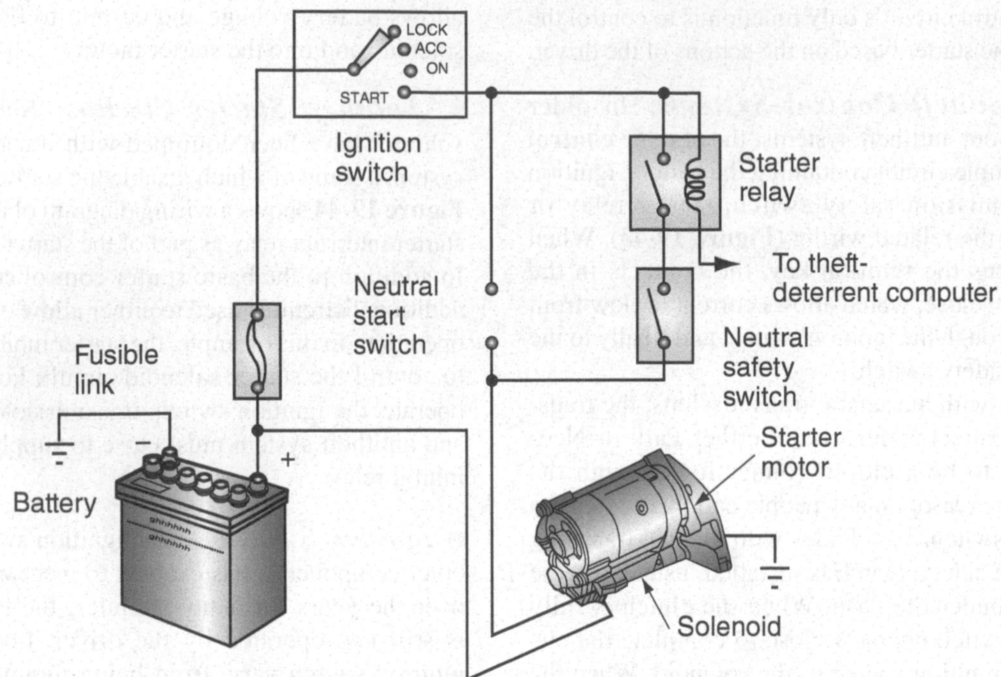

FIGURE 19-44 Vehicles with an antitheft system often include starter enable relays, which can prevent starter operation if the antitheft is active.

FIGURE 19-45 An example of an ignition switch mounted behind the ignition lock cylinder.

power flows through the switch contacts, which are connected to the transmission range switch, ignition, and computer systems. When the switch is released and left in the ON or RUN position, power continues to flow to the ignition and computer systems but is removed from the starter control circuit.

■ ***Computer-Controlled Starter Control Circuits.*** In newer vehicles, the ignition switch often does not supply power directly to the starter circuit; instead, the switch is used as part of the control circuit that includes starter relays and control modules. These switches are inputs to the BCM (**Figure 19-47**). Current does not flow through the switch to the starter circuit. **Figure 19-48** shows an example of a starter control circuit with this type of switch.

In this circuit, power is supplied to the S terminal on the starter only after the body control module (BCM) and engine control module (ECM) validate that the ignition key is being used in the ignition lock cylinder. Most of these systems, such as General Motors's Pass-Key and Ford's Passive Anti-Theft System (PATS), use small radio transponders buried within the keys. An example of a transponder key is shown in **Figure 19-49**. When the driver inserts the key into the ignition lock cylinder, a module inside the vehicle sends a radio signal out that is received by the key. The key then responds with a coded reply, often called *a password*. If the key code is correct, meaning it matches the code stored in the on-board computer's memory, the system validates the key and enables the vehicle to start. Other vehicles use a combination key and transmitter as shown in **Figure 19-50**.

■ ***Push-to-Start Systems.*** Many newer vehicles do not use a traditional ignition switch; instead, a START button turns the engine on and off (**Figure 19-51**). In this system, an on-board computer monitors the position of the START button to place the vehicle into RUN, START, and STOP modes.

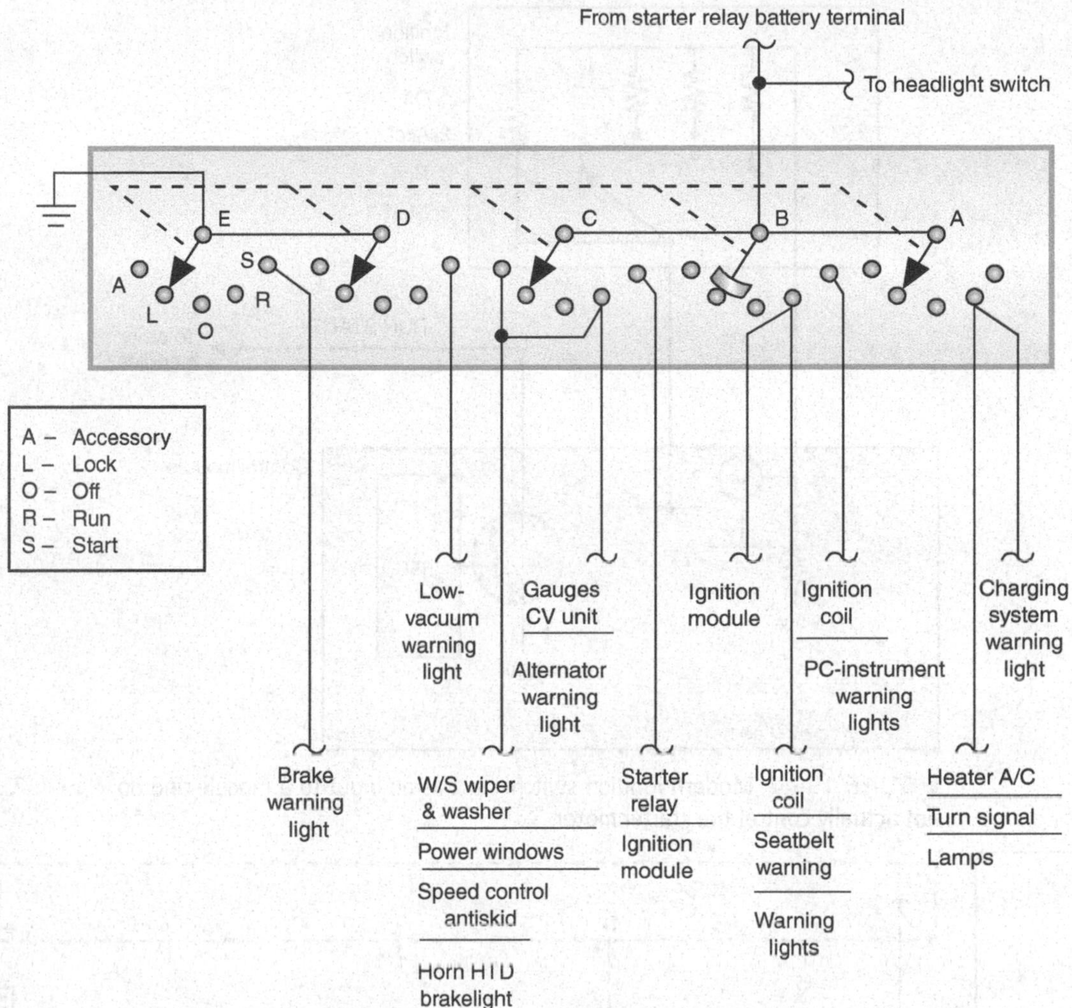

FIGURE 19-46 An example of a multiple-pole multiple-throw ignition switch.

In this type of system, the START button is simply a switch that is used to input to a computer the request to turn on and start the vehicle's operation. Vehicles that use a START button also use a type of smart key system to validate the vehicle startup process.

Keyless systems require a key fob or remote, also called *a smart key* (**Figure 19-52**). These fobs are used by the antitheft system to recognize that the vehicle should be allowed to start. Without the correct key fob being present, the push-to-start feature does not operate. The vehicle may feature remote starting as a function on the smart key, allowing the user to start the engine remotely. The button at the top of the remote in Figure 19-52 is to activate the remote start function.

Keyless systems do have a potential drawback; what if the battery is weak or goes dead in the key fob? Smart keys have a mechanical key that can be used to open a door and enter the vehicle (**Figure 19-53**). Once inside, the key fob will need to be inserted or held in a certain location so that the in-car key detectors can locate and validate the key (**Figure 19-54**).

■ ***Auto Stop/Start Systems.*** Several manufacturers have vehicles with auto stop/start or idle stop systems. These systems shut the engine off during idle periods and restart the engine very quickly when the driver lifts off the brake and presses the accelerator pedal. Designed to improve fuel economy, these systems are equipped with heavier-duty batteries and starter motors than vehicles without the feature. Due to increased fuel economy standards, expect auto stop/start systems to become standard on most cars and light trucks in the next few years.

An example of a starter motor for auto stop/stop is shown in **Figure 19-55**. This starter uses a tandem solenoid, meaning it can control the drive gear and starter motor operation as two separate operations. This separation allows the starter drive gear to reengage the flywheel on the engine even while the engine is spinning down. The starter can then reengage and restart the engine without waiting for the engine to stop moving. The reduces restart times and provides a quicker response when the driver wants to accelerate and drive off.

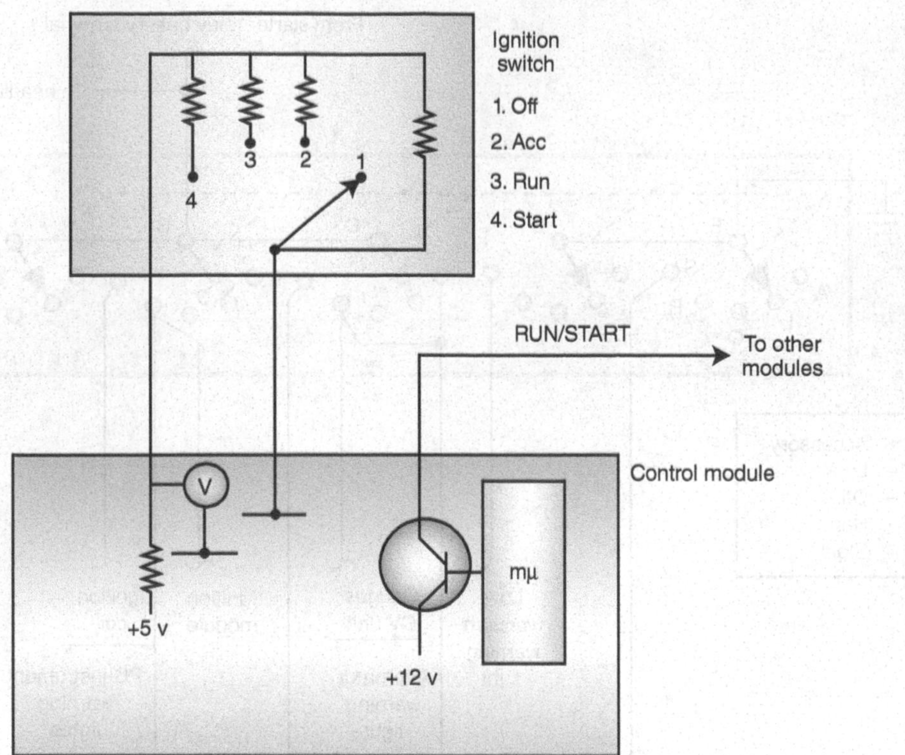

FIGURE 19-47 Modern ignition switches act as an input to a module and do not actually control the starter motor.

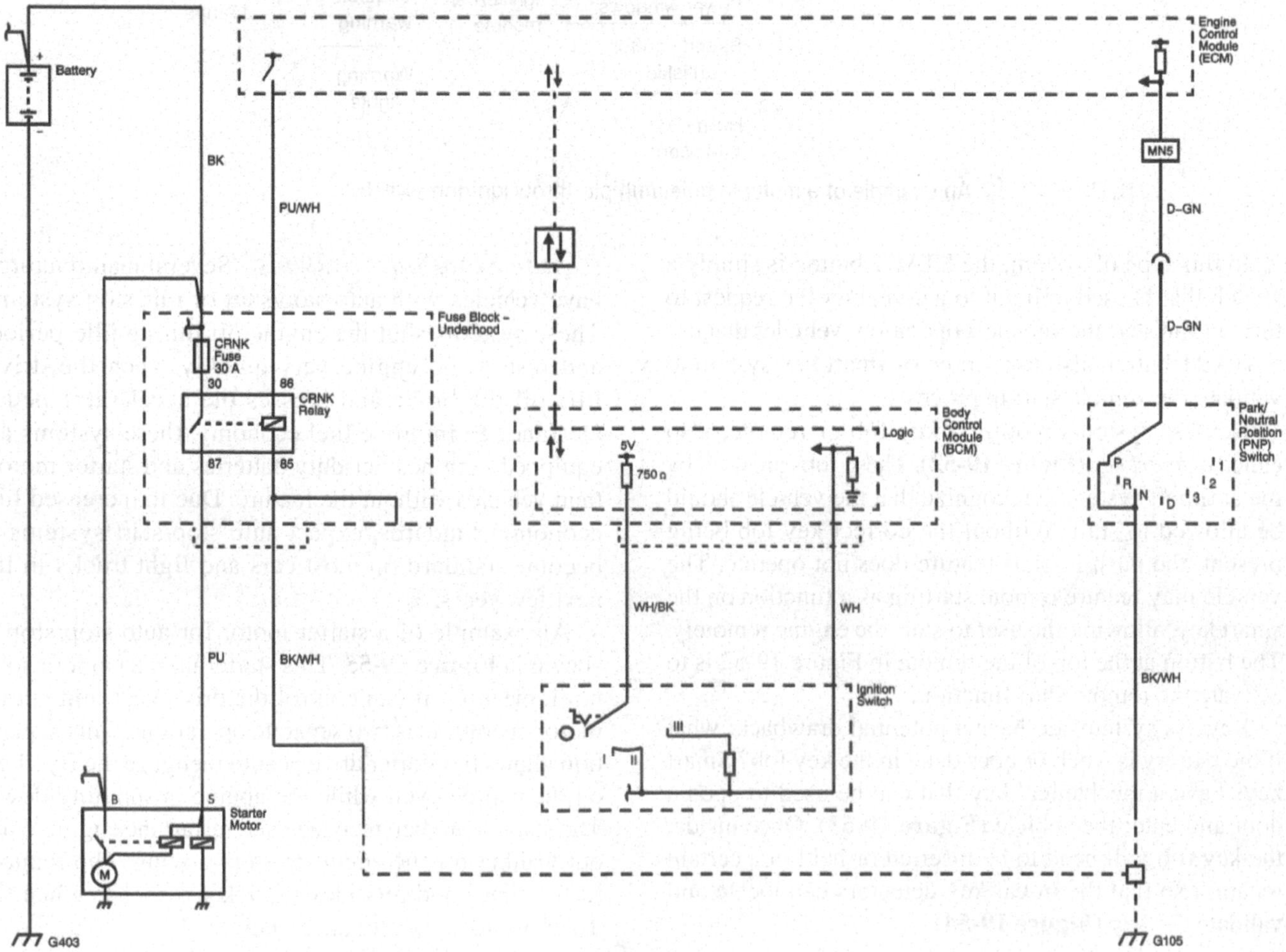

FIGURE 19-48 An example of a starting system controlled by a BCM.

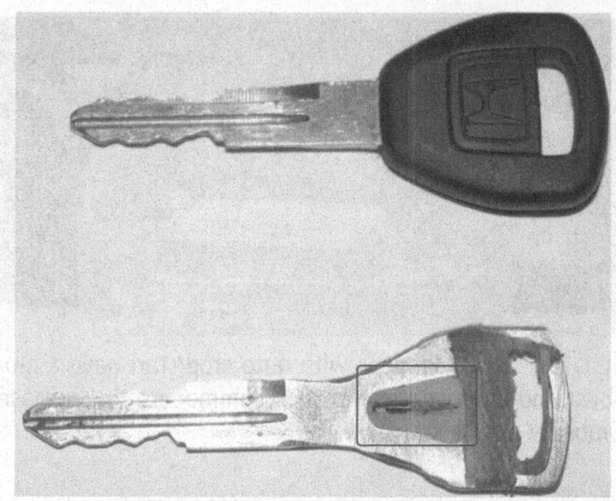

FIGURE 19-49 Ignition keys with radio transponders. Many keys use radio transponders that communicate with the antitheft systems. These chips must be close to the transmitter/receiver in the vehicle to operate.

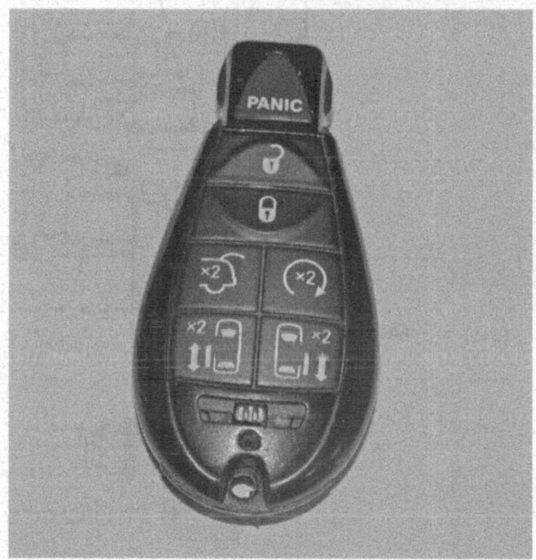

FIGURE 19-50 An example of a smart key which can be used to remote-start the engine, unlock the doors, and deactivate the antitheft system.

FIGURE 19-51 An example of a push-to-start ignition. The Start button both starts and shuts down the engine.

FIGURE 19-52 A smart key fob. The vehicle detects the presence of the fob to unlock the doors and to allow the engine to start.

Hybrid Vehicle Starting Systems

There are two major types of hybrids currently available, the full hybrid and the assist hybrid. Both have unique starting systems that differ from traditional systems. These vehicles use one or more motor/generators to start the engine, propel the vehicle in electric mode, and recharge the high-voltage batteries.

MOTOR/GENERATORS

Full hybrids use motor/generators to start the engine, propel the vehicle, and recharge the high-voltage battery packs. These vehicles, such as the Toyota Prius and Ford Fusion Hybrid, can be driven solely by the electric motors, depending on the state of charge of the high-voltage batteries. The motor/generators can then be used to recharge the high-voltage batteries during other types of vehicle operation. For the motor/generators to be able to propel the vehicle, they require high operating

FIGURE 19-53 This type of smart key unlocks the doors, disarms the antitheft, and allows the engine to start. No regular key is used in this vehicle. The key shown is used to unlock the driver's door in the event the battery in the remote fails.

(a)

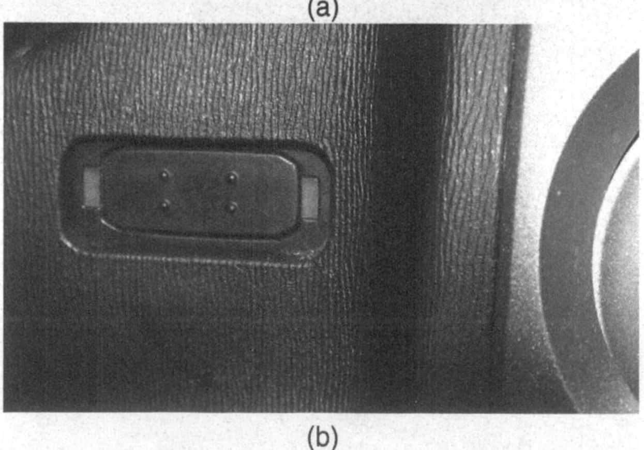

(b)

FIGURE 19-54 (a) If the remote battery is weak, it can be held close to the area indicated on the steering column to get the vehicle to start. (b) A slot for the key fob to be inserted if necessary.

voltages. In some systems, the voltage supplied to the motors can be as high as 600 VDC. These systems should only be inspected and serviced by trained and qualified technicians.

Full hybrids do not use a 12-volt starter motor to crank the engine. The integrated high-voltage motor/generator acts as the engine starter and can crank the engine much

FIGURE 19-55 Engines with auto stop/start have a more robust and improved starter to accommodate the increased number of engine starts.

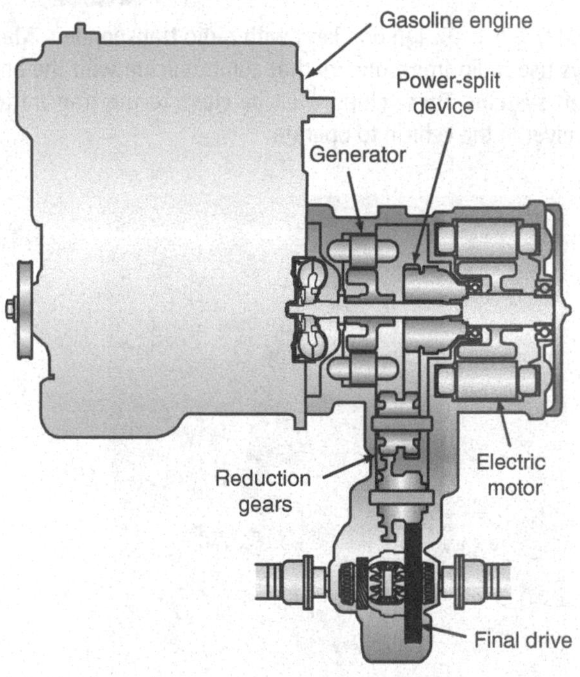

FIGURE 19-56 Many hybrids use electric motors for propulsion, energy generation, and to start the ICE.

faster than traditional starters for faster engine restart. The Prius, which has two motor/generators, called MG1 and MG2, uses the smaller MG1 as the engine starting motor. **Figure 19-56** shows the configuration of the transaxle and motor/generator components similar to that used by Toyota. The generator is also used as the starter motor for the gasoline engine. The motor/generator system is housed in the hybrid drive system, which in other vehicles is the transaxle assembly.

Vehicles that are assist hybrids, such as Honda IMA hybrids and GM eAssist vehicles, do not use the motor/generators for electric-only propulsion. Assist hybrids, as used on Honda models such as the Civic Hybrid, CR-Z, and Insight, have an integrated motor assist (IMA) unit. The IMA is a motor/generator and is located between the engine and transmission (**Figure 19-57**). In this system,

FIGURE 19-57 Honda uses an integrated motor assist (IMA) unit located in the rear of the engine to start the engine and provide additional torque.

FIGURE 19-58 This General Motors belt/alternator/starter is an older assist system used for engine starting and idle stop.

the electric motor acts as the engine starter and it provides additional torque to supplement the power produced by the gasoline engine.

A third type of hybrid system, also an assist system, uses what is called a belt/alternator/starter or BAS. Used by General Motors, these eAssist systems used a heavy-duty starter/alternator (**Figure 19-58**). The combination starter/alternator uses a large heavy-duty drive belt, which both drives the alternator and then cranks the engine depending on system operation. eAssist systems use a high-voltage battery, about 115 volts, which is less than that used on full hybrids. Early BAS systems used a 42-volt system and were available on select GM vehicles from 2006 through 2010.

FIGURE 19-59 The orange wiring indicates high voltage. Never attempt to work on the HV wiring or components unless properly trained and following the manufacturer's service procedures.

HIGH-VOLTAGE SAFETY

Because of the very high voltages, 115 to 600 volts, used by many of the hybrid drive systems, caution must be used when you are working on hybrid vehicles.

■ ***Identify High-Voltage System Components.*** The industry standard for identifying the high-voltage circuits in hybrid vehicles is the use of orange conduit, wiring, connectors, and covers (**Figure 19-59**). Do not attempt to open or service any wiring or component of the high-voltage system.

Charging Systems

Once the engine is started, the AC generator is used to recharge the battery and power the electrical system. Like electric motors, the AC generator uses electromagnetism to operate. Where an electric motor uses electricity to perform a mechanical function, generators use mechanical motion to generate electricity.

Over the last couple of decades, the increasing number of electrical accessories has required upgrading the vehicle's electrical system. In the 1980s, charging systems typically produced 40–60 amps. As computer systems and more electronic accessories and equipment became standard, generator output had to increase to meet the new requirements. In modern vehicles, typical AC generator output is 140–180 amps or more. And again, electrical power demand is pushing the limits of the 12-volt AC generator systems, causing manufacturers to look at 42-volt systems to keep up with electrical system demands.

COMPONENTS AND OPERATION

A small amount of current supplied to the generator is used to generate magnetic fields. The magnetic fields move past stationary conductors, wrapped in loops. It

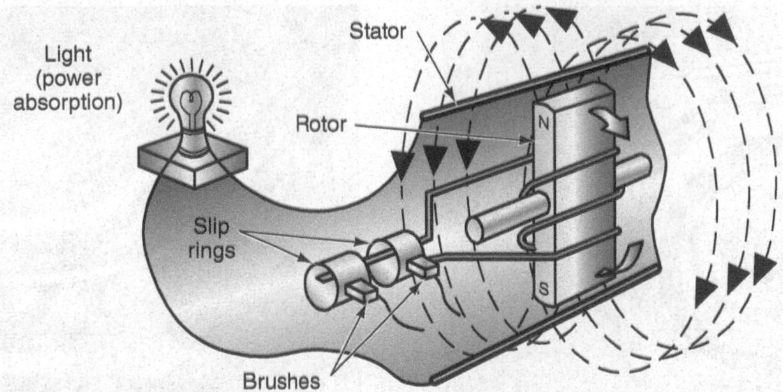

FIGURE 19-60 AC voltage is generated when a conductor passes across magnetic fields.

is the interaction of magnetic fields moving past the conductors that generate electricity. An illustration of a voltage generation is shown in **Figure 19-60**. As the magnetic fields pass the windings, AC voltage is produced. The AC is then converted into DC voltage for the battery and electrical system.

Generator Construction and Operation

The majority of generators in use in today's cars and trucks are air-cooled, though some manufacturers, such as Mercedes-Benz and BMW, have liquid-cooled generators on some models. Regardless of the construction and cooling type, all AC generators operate in the same way and use the same basic components.

■ ***Driveshaft.*** As you may have noticed, the drive pulley on the generator is much smaller in diameter than the drive pulleys on other accessories (**Figure 19-61**). This is because the generator is overdriven, meaning it turns at a higher rpm than the crankshaft pulley. This is because the generator does not produce much output at low rpm. Also, as generator speed increases, less current is required to produce the magnetic fields, so operation becomes more efficient at higher rpm.

The drive pulley is designed to accommodate multi-ribbed belts (**Figure 19-62**). On older vehicles, a V-belt may be used instead of a multi-ribbed belt. The drive belt must be in good condition to maintain adequate tension on the pulley during operation. As the demand on the generator increases, so does the force required to turn the pulley. A loose belt can slip, reducing the generator output.

Many vehicles use a clutch drive in the generator pulley (**Figure 19-63**). These are also called alternator overdrive (AOD) pulleys and alternator decoupler pulleys.

(a)

(b)

FIGURE 19-61 (a) The diameter of a typical generator pulley. (b) A typical crankshaft pulley. Because the generator pulley is about one-third the diameter of the crankshaft pulley, the generator will spin around three times faster than the crankshaft.

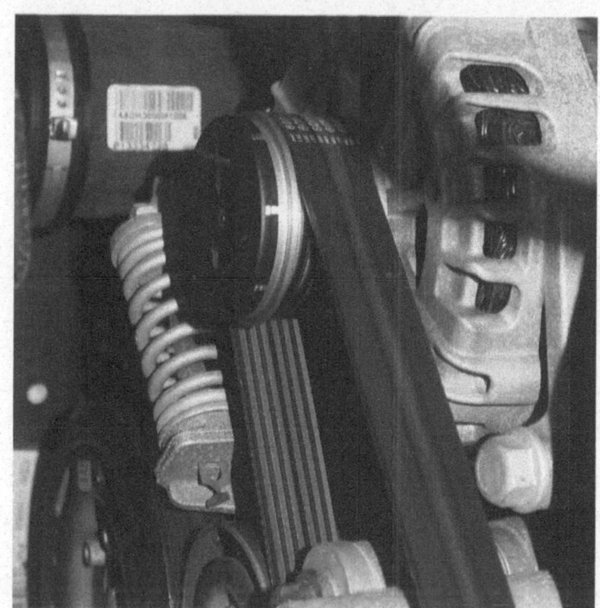

FIGURE 19-62 A multi-ribbed generator belt.

FIGURE 19-63 A generator with an overdrive pulley. Overdrive pulleys typically have a cap over the front to prevent dirt from entering the bearing and clutch assembly.

Regardless of the name used, an overrunning clutch is used within the pulley. The clutch helps even out belt tension fluctuations caused by the constant speed changes of the drive belt. This helps to increase belt life, reduce noise, and vibration.

The drive pulley is mounted to the front of the generator driveshaft. At the front and rear of the driveshaft are bearings. Air-cooled generators use a fan or set of fans to pull air through the generator housing. Some are located on the outside, behind the drive pulley while others are internal (**Figure 19-64**). Without the airflow from the fan, the generator can easily overheat under load, causing failure of the internal components. Some vehicles use air ducts to route incoming air from the front of the vehicle to the generator. This increases airflow to cool the generator.

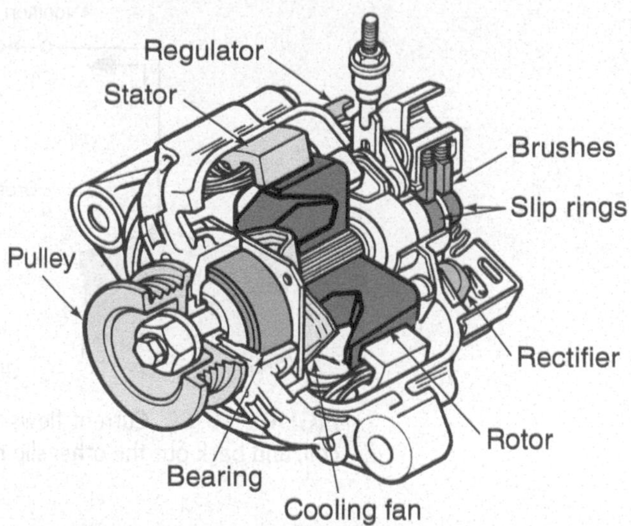

FIGURE 19-64 An illustration of the inside of a generator. Fans are used to pull air through and keep the internal components from overheating.

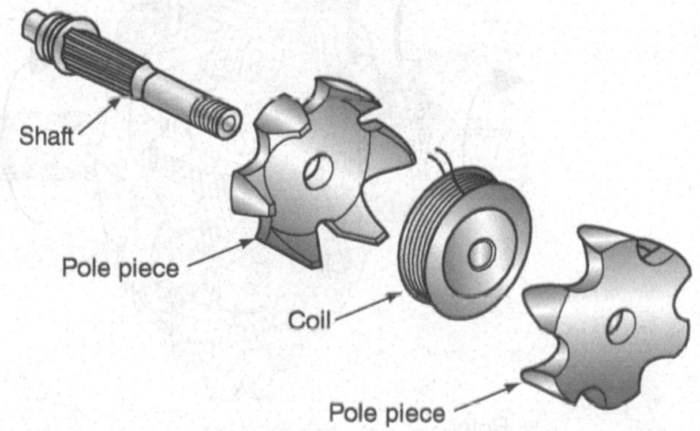

FIGURE 19-65 The rotor is made of the driveshaft, coil, and pole pieces.

Mounted in the center of the generator and press-fit to the driveshaft are two pole pieces (**Figure 19-65**). Sandwiched in between the pole pieces is the solid copper field wire is a coil. Current is supplied to the field wire by the positive brush and slip ring and a negative brush and slip ring (**Figure 19-66**). When current is supplied to the field wire, a magnetic field forms around the coil. The amount of current supplied to the field wire determines how strong a magnetic field is produced by the field and the pole pieces. The field extends to the fingers of the pole pieces, creating alternating north and south poles (**Figure 19-67**). Typically, the amount of current flowing through the coil is low, between 3 and 6 amps.

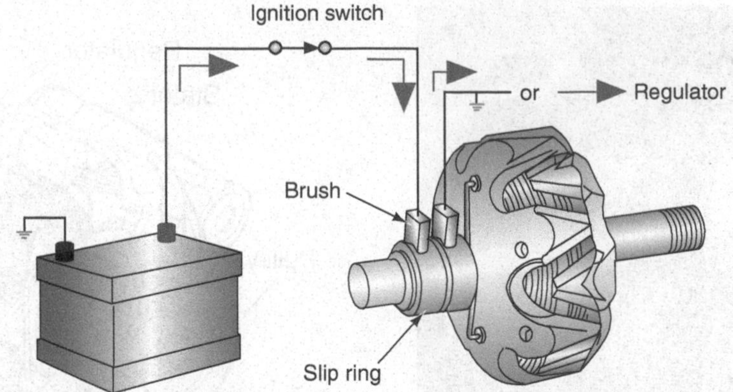

FIGURE 19-66 Current flows through the positive slip ring, through the coil, and back out the other slip ring.

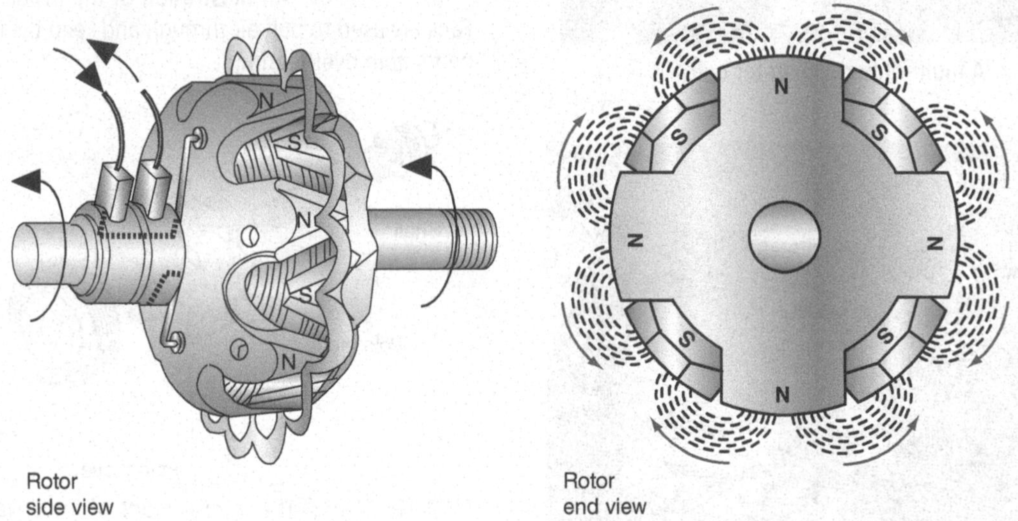

FIGURE 19-67 The current flow through the coil generates a magnetic field in the pole pieces.

The magnetic fields produced by the pole pieces move as the driveshaft spins. These fields move across the windings in the stator, where AC voltage is induced into the windings.

■ *Stator.* The stationary windings of wire in the generator form an assembly called the stator (**Figure 19-68**). The **stator** is typically composed of three windings of wire, looped in an offset pattern so that the windings are 120 degrees apart from each other. Some vehicles use six sets of stator windings. As the magnetized rotor spins, the magnetic field lines cut across the windings of the stator. This induces voltage into the windings. Since both a north and a south field pass each winding, a positive and a negative voltage are produced. This is AC voltage. The AC voltage induced into the stator flows out of the stator windings to the rectifier assembly. The rectifier contains diodes to turn the AC voltage into DC voltage.

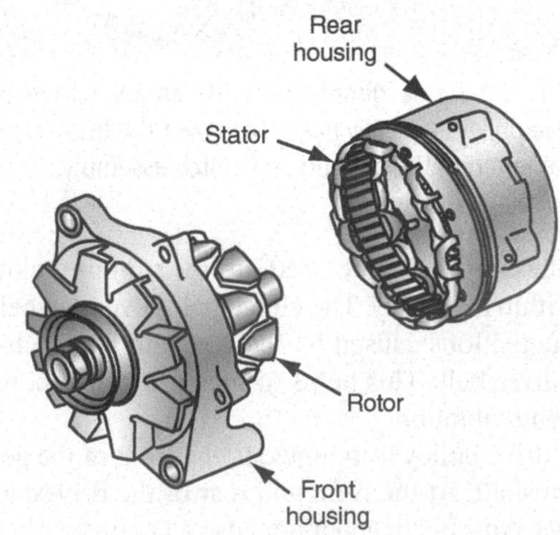

FIGURE 19-68 The stator has three windings offset 120 degrees from each other. Some newer designs use six stator phases.

Stator windings are formed into two different configurations: the wye winding (**Figure 19-69**) and the delta winding (**Figure 19-70**). The wye winding has a central neutral junction from which AC can be tapped if desired. Some manufacturers use this junction to use the available AC to power certain accessories. The delta winding can supply more current at higher speeds than the wye winding.

■ ***Diodes.*** Diodes are used in the generator to convert the AC produced in the stator into usable DC for the vehicle's electrical system. This is done by using the diodes to block unwanted current flow. **Diodes** allow current to flow in one direction (**Figure 19-71**). Because diodes can block the flow of electricity, sets of diodes are used to block the flow of the unwanted portions of the AC current (**Figure 19-72**). How diodes are used to convert the AC into DC is discussed later in this chapter.

■ ***Voltage Regulation.*** Left unregulated, an AC generator can produce 20 volts or more. This is bad for the generator and for the electrical system. Excessive output will overheat and damage the generator. Excessive voltage can blow bulbs and destroy electronic components. Regardless of the age or type of AC generator, all have similar methods of controlling output.

Many generators use internal voltage regulators (**Figure 19-73**). These systems rely on the voltage regulator sensing the demand on the charging system. As demand increases, voltage regulator resistance decreases. This allows more current to flow to the field. This type of regulation typically controls the ground circuit resistance of the field circuit. As the regulator decreases ground circuit resistance, charging output increases. How the voltage regulator is wired in the charging circuit is shown in **Figure 19-74**.

Most modern vehicles use the PCM to regulate generator output. This is because the PCM can tailor charging output to meet the needs of the battery and the electrical demands placed on the system. This improves fuel economy, reduces idle roughness, and prolongs battery life.

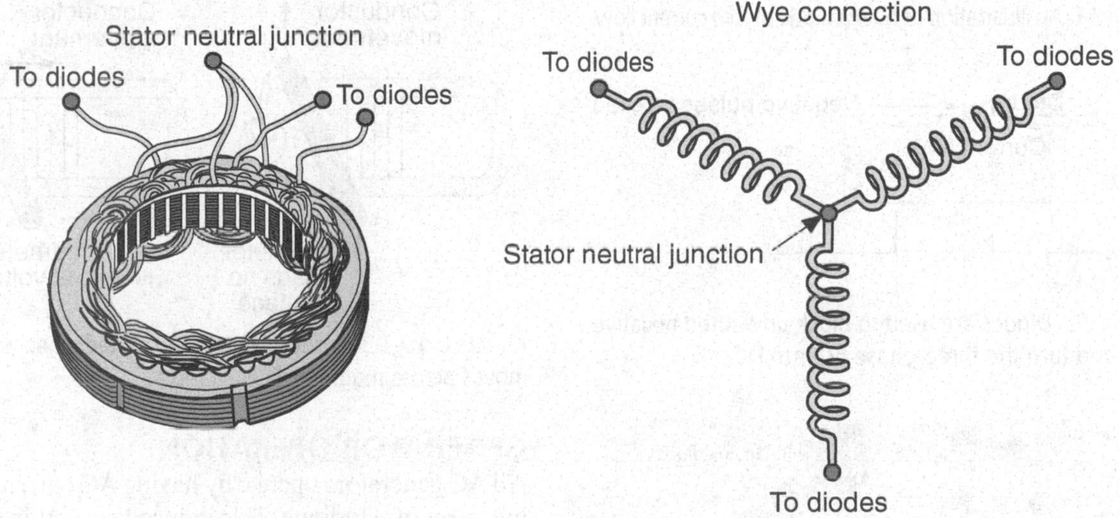

FIGURE 19-69 Wye windings have a center neutral junction.

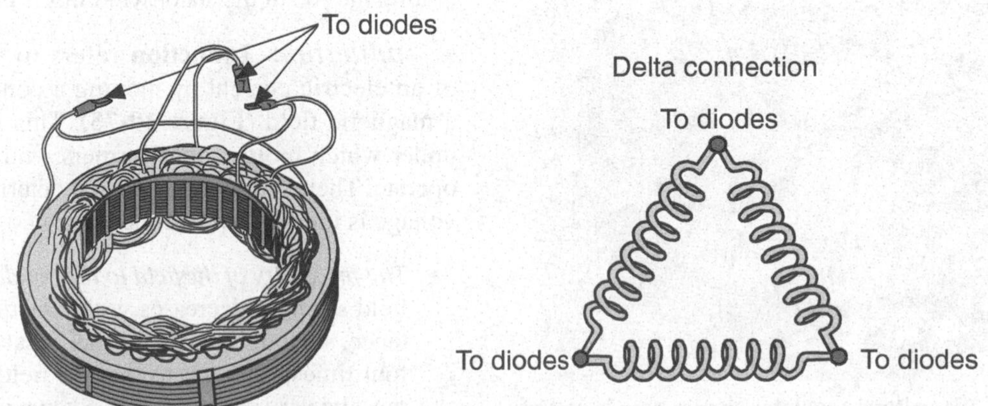

FIGURE 19-70 Delta windings are common as they produce more output than wye windings.

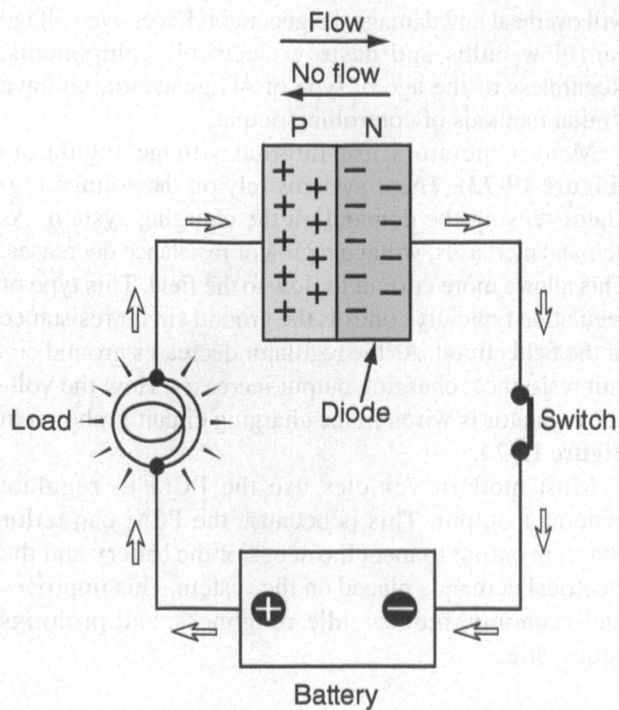

FIGURE 19-71 An illustration of how a diode blocks current flow.

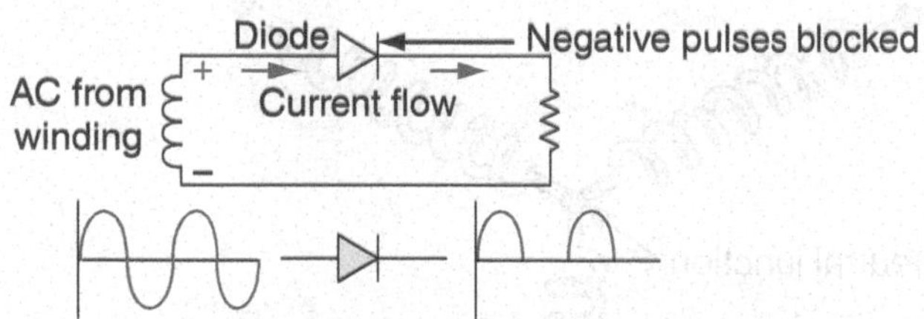

FIGURE 19-72 Diodes are used to block unwanted negative AC voltage and turn the three-phase AC into DC.

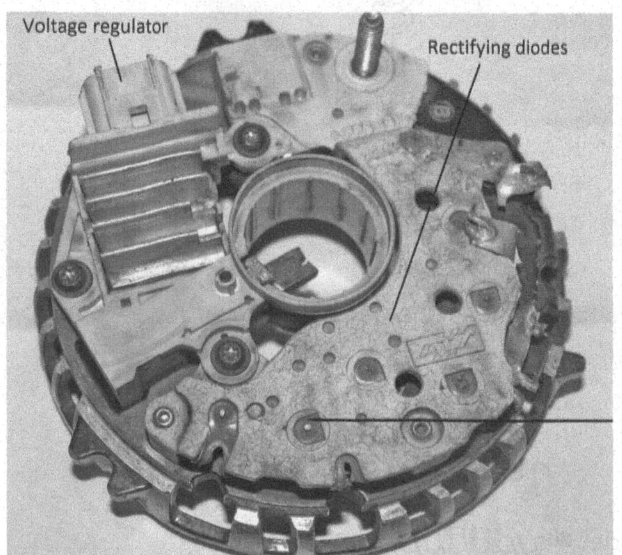

FIGURE 19-73 The voltage regulator shown here is part of the brush assembly. The regulator controls current flow to the field coil.

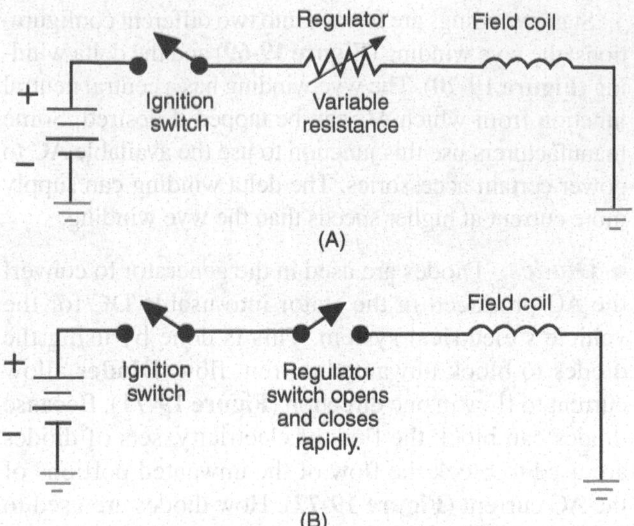

FIGURE 19-74 This illustrates two types of voltage regulators. (a) A variable resistor changes the amount of resistance and current flow to the field. (b) The field is turned on and off very rapidly to control current flow to the field.

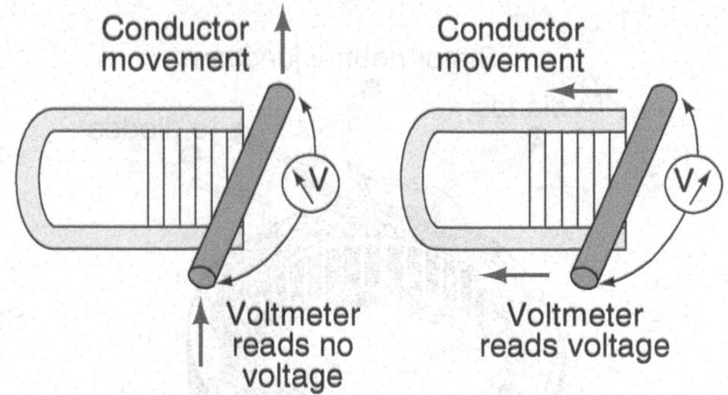

FIGURE 19-75 Induced voltage is created when a conductor moves across magnetic lines of force.

GENERATOR OPERATION

All AC generators operate by having AC current induced into a set of windings. This induced current is then converted into DC current by the diodes. The process of generating the AC in the stator windings is called *induction*.

■ ***Induction.*** **Induction** refers to the production of an electric current by moving a conductor through a magnetic field (**Figure 19-75**). This is the principle under which motors, transformers, and AC generators operate. There are four factors that determine how much voltage is induced into a conductor:

- *The proximity of the field to the conductor*. Magnetic field strength decreases with the square of the distance, so doubling the distance weakens the field by four times. This means that the field and conductor must be very close together. Shown in **Figure 19-76**, the fit between the rotor and stator is very close, typically less than 0.030 in. (0.76 mm).

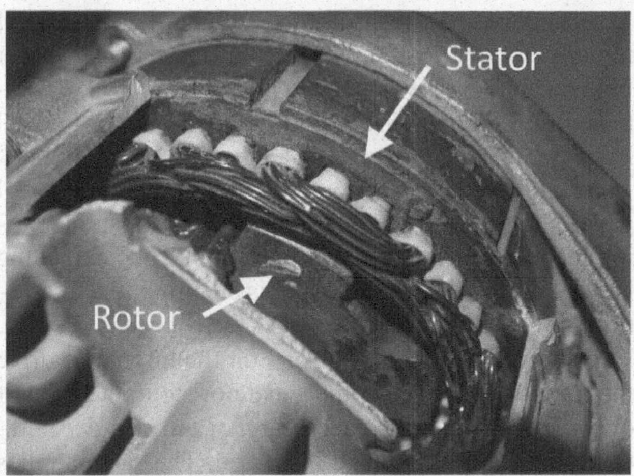

FIGURE 19-76 The rotor fits very close to the stator windings. This is because magnetic field strength decreases with the square of the distance.

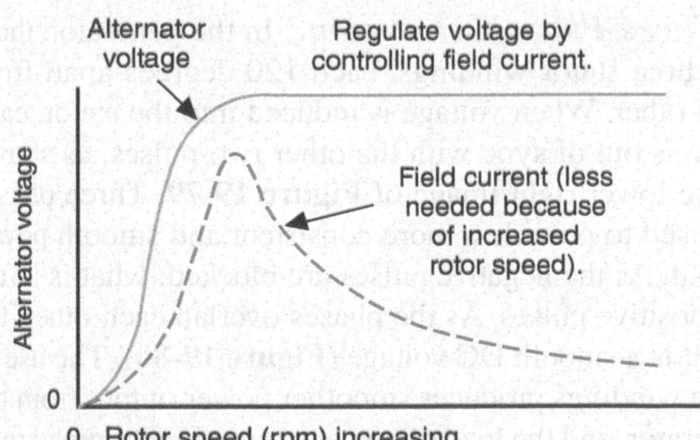

FIGURE 19-77 The field circuit does not require large amounts of current for generator output. This is because the faster the generator spins, the more voltage is induced into the stator windings.

- *The speed at which the fields cut across the conductor*. For voltage to be induced, the fields must cut across the conductor at 90-degree angles. The faster the fields cross the conductor, the more voltage can be induced.
- *The strength of the magnetic field*. The stronger the magnetic fields, the more voltage can be induced into the conductor.
- *The number of conductors*. The larger the number of conductors or loops of the conductor, the more voltage can be induced.

When the field coil in the rotor is supplied current, magnetic fields are generated that move past the stator windings. This movement of magnetic lines of force across the conductors induces AC current into the stator windings. The rotor is mounted so that it spins very close to the stator windings at very high rpm, often three or more times faster than crankshaft rpm. At low rpm, more current is supplied to the field to create stronger magnetic fields. This is necessary because of the low rpm. This creates more induced voltage in the stator windings and increases generator output. As engine speed increases, the magnetic fields move across the stator more frequently so more current is induced, and increases generator output (**Figure 19-77**). Because output increases with rpm, less current is needed by the rotor.

■ ***Single-Phase Rectification.*** When the magnetic field passes across one of the stator windings, positive and negative voltages are generated. To change the AC into DC, the negative pulses must be removed. To accomplish this, each stator winding is connected to two diodes wired in series. The diodes block the flow of the negative voltage but allow the positive voltage to pass (**Figure 19-78**). This is called *single-phase half-wave rectification* because only the negative pulses are blocked.

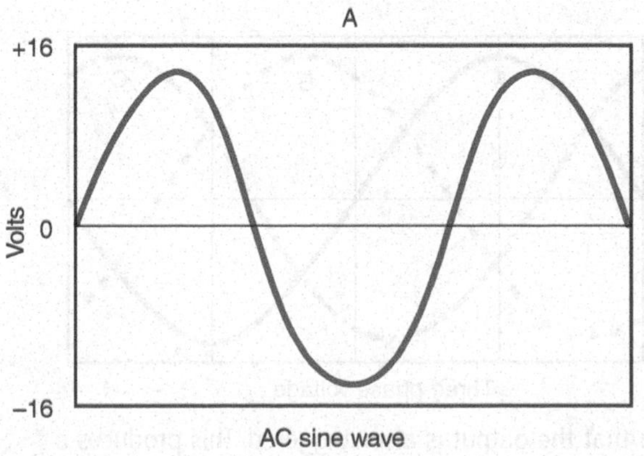

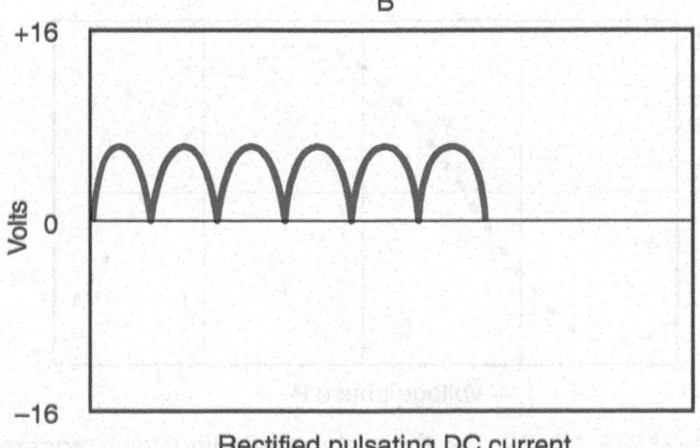

FIGURE 19-78 This illustrates how the AC output is changed into DC. When the diodes block the unwanted parts of the AC output, DC is what remains.

■ ***Three-Phase Generation.*** In the generator, there are three stator windings, each 120 degrees apart from each other. When voltage is induced into the stator, each pulse is out of sync with the other two pulses, as shown in the lower right image of **Figure 19-79**. Three phases are used to provide a more consistent and smooth power output. As the negative pulses are blocked, what is left is the positive pulses. As the phases overlap each other, the result is a smooth DC voltage (**Figure 19-80**). The use of three windings produces smoother power output from the generator, and the load of power generation is spread over the three windings instead of relying on just one phase to produce the necessary power. To rectify each winding's output from AC into DC, each stator winding has a pair of diodes (**Figure 19-81**).

■ ***Field Control.*** Field control is important for generator operation and for the rest of the electrical system. If too little current is supplied to the field, charging output may be insufficient, causing battery drain. Too much current to the field increases the generator output but also creates more heat inside the generator, which can damage the battery, light bulbs, and electronics on the vehicle. Eventually, the generator would overheat and burn out internally.

Generator control has become a function of the PCM module (**Figure 19-82**). The PCM can turn the generator field on and off very quickly to precisely control generator output. **Figure 19-83** shows the field control signal from a PCM to a generator in a Honda.

Manufacturers use the engine computer to control generator output for several reasons, including increasing fuel economy, improving idle quality, and extending battery life. When the generator is charging, the interaction of the magnetic fields makes the rotor hard to spin. This creates a mechanical load on the engine to drive the rotor. By controlling the generator's output so that it only charges when it is necessary, or only on deceleration, fuel economy can be increased. The same applies to improving idle quality; by turning the generator off at idle, the engine is under less of a load. This can improve idle smoothness and reduce fuel consumption. Generator output is also tailored to the needs of the battery. A cold battery does not accept a charge as easily as a warm battery, so generator output can be increased at a cold-start to quickly warm the battery (**Figure 19-84**).

■ ***Voltage Regulators.*** The **voltage regulator** controls the amount of current that is supplied to the field winding on either the power supply or the ground circuit.

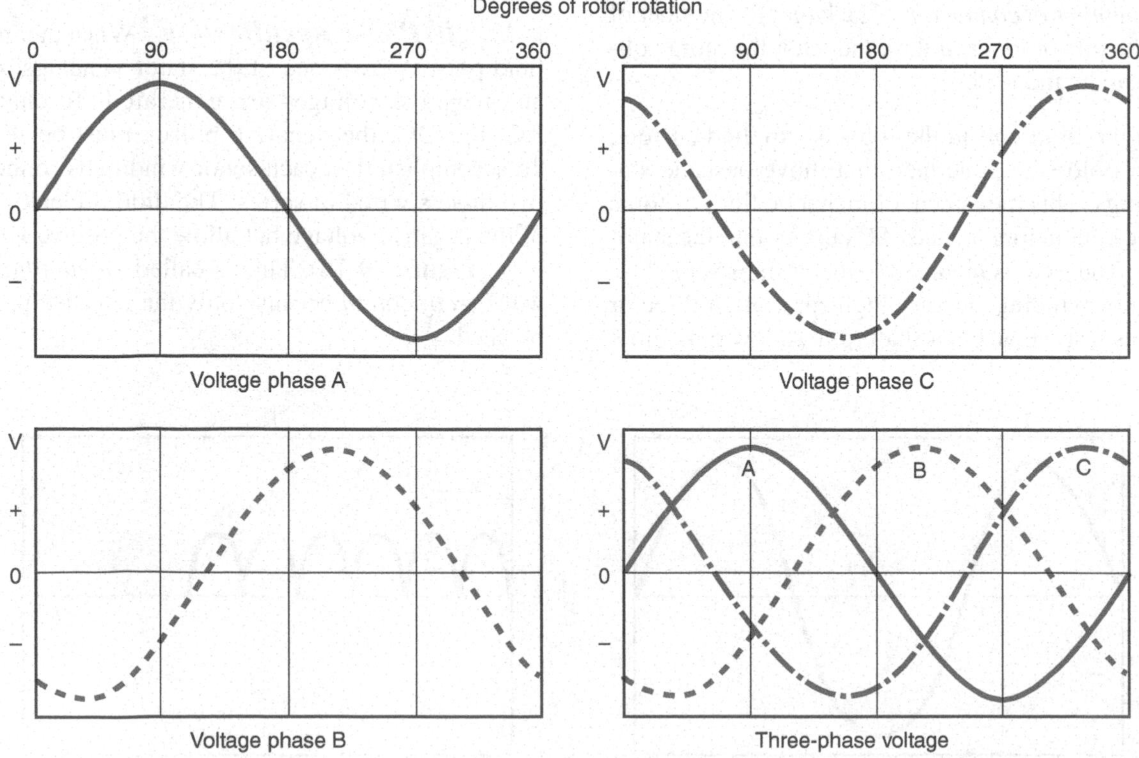

FIGURE 19-79 The three stator windings are staggered so that the output is also staggered. This produces a more even voltage output.

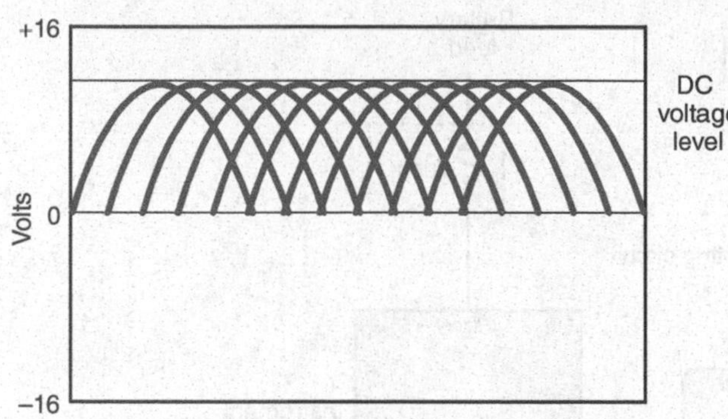

FIGURE 19-80 Rectified DC output on a scope shows the voltage from each stator winding.

Regulators are usually mounted in the generator and may include the brushes as part of the regulator's assembly (**Figure 19-85**).

Older systems use a variable resistor in the regulator for field control. When charging demand is low, current flow will be low. As demand increases, the regulator increases the current flowing through the coil. In most systems, actual current flow through the coil is less than 6 amps, even at full charging output. This is because the majority of generator output is derived by the speed and number of the magnetic lines used to induce current into the stator and not from the strength of the field produced by the rotor.

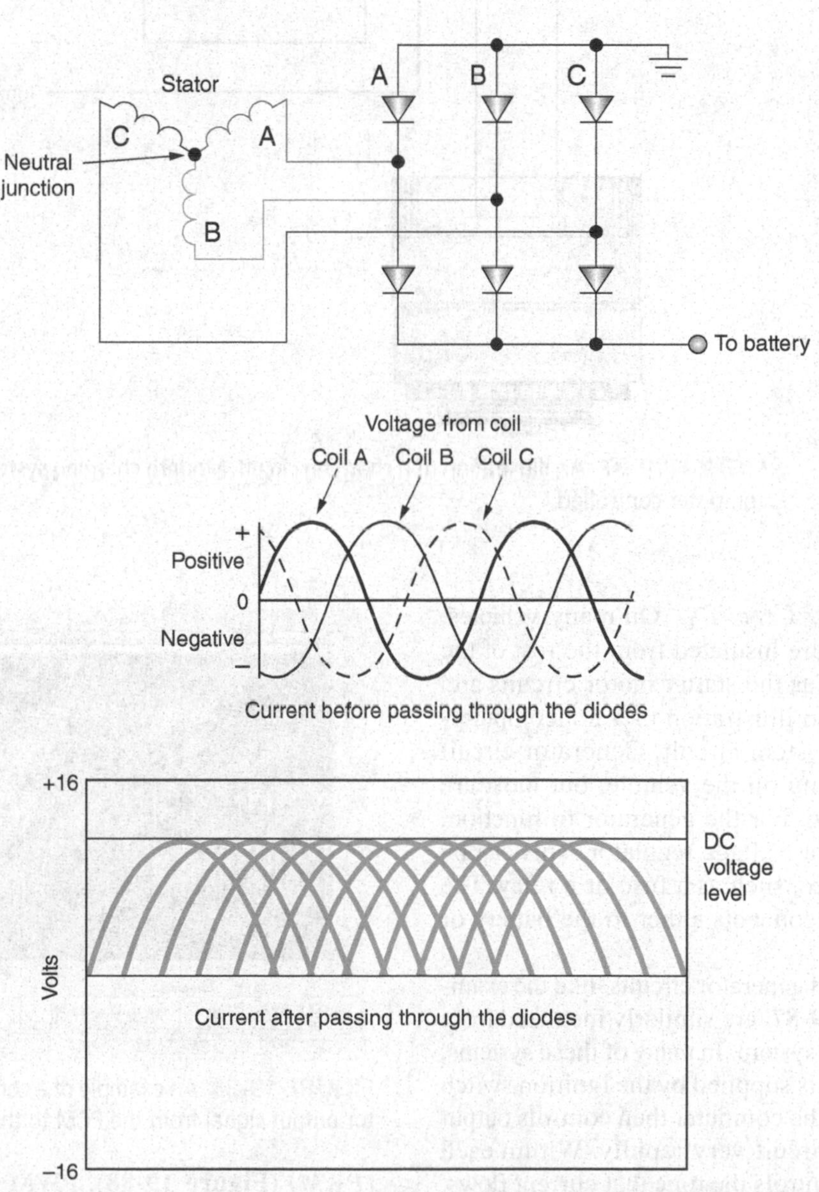

FIGURE 19-81 This shows how the diodes are wired to block negative AC voltage.

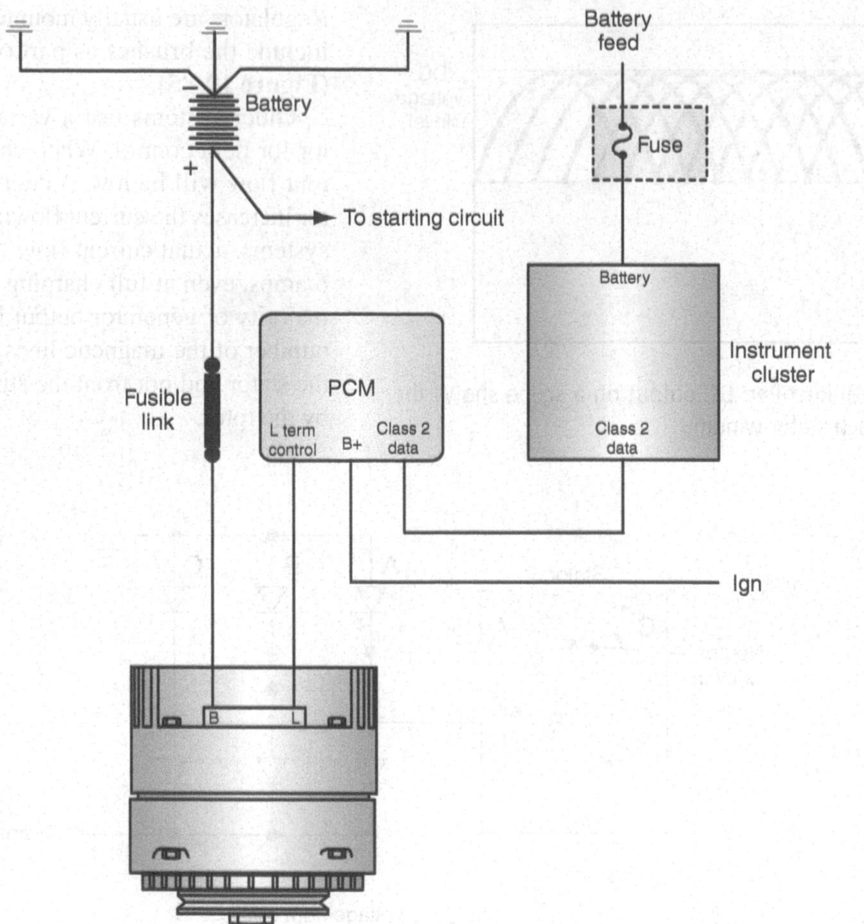

FIGURE 19-82 An illustration of a charging circuit. Modern charging systems are computer controlled.

■ ***Charging System Circuits.*** On many vehicles, the generator circuits are insulated from the rest of the electrical system, just as the starter motor circuits are. **Figure 19-86** shows an illustration of a noncomputer-controlled charging system circuit. Generator circuit wiring varies depending on the vehicle, but most are similar to this example. For the generator to function, power is supplied to the voltage regulator. This can be from an external source, such as a fuse or a relay. The generator output wire connects either to the battery or to a junction box.

Computer-controlled generator circuits, like the example shown in **Figure 19-87**, are similarly insulated from the rest of the electrical system. In many of these systems, power to the generator is supplied by the ignition switch and a dedicated fuse. The computer then controls output by pulsing the field circuit very rapidly. Within each pulse, the computer controls the time that current flows. This type of control is called *pulse-width modulation (PWM)* (**Figure 19-88**). PWM means the amount of time a component is on. As shown in Figure 19-88,

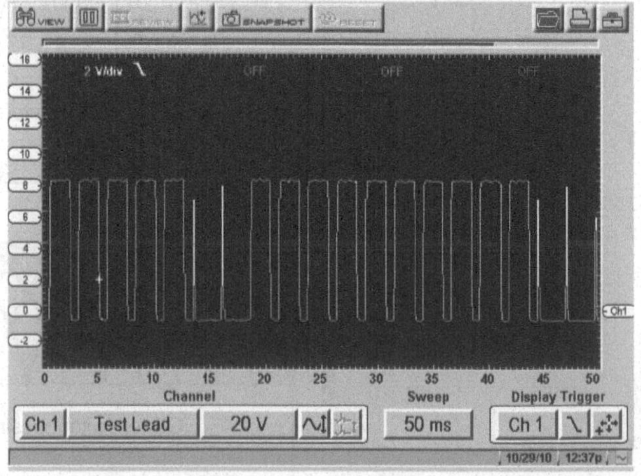

FIGURE 19-83 An example of a computer-controlled generator output signal from the PCM to the voltage regulator.

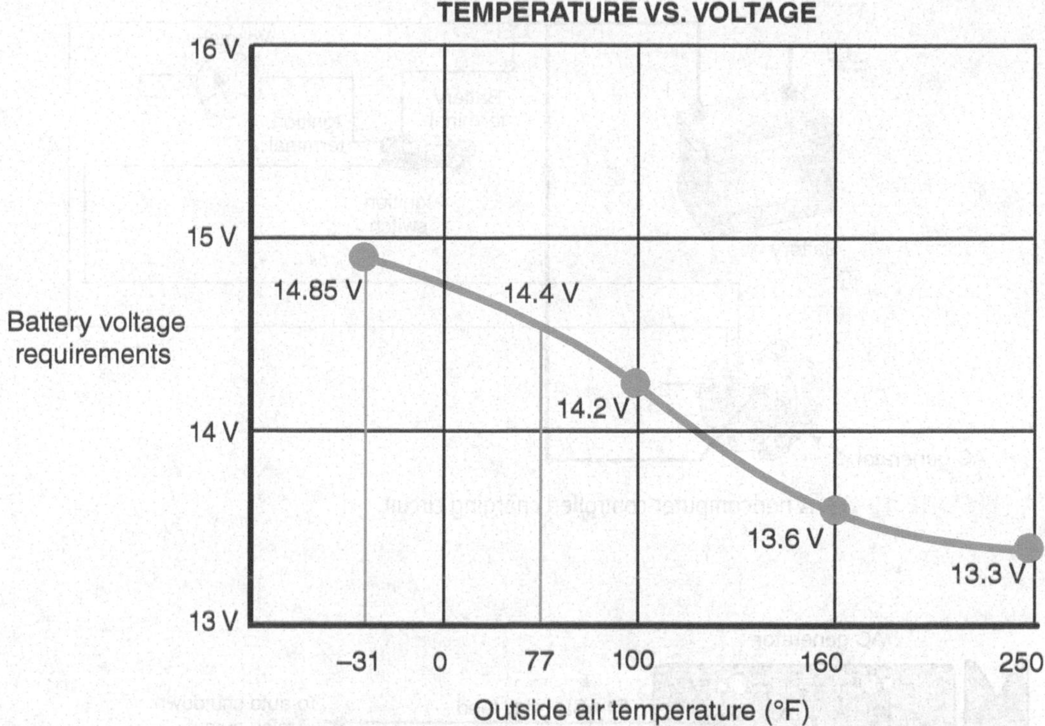

FIGURE 19-84 As battery temperature decreases, charging rate increases. This warms the battery and keeps it charged.

FIGURE 19-85 An example of a voltage regulator and brush assembly.

50% means that the device is on one-half of the time and off the other half. If a generator with a rating of 100 amps is running at a 50% pulse width, its charging output will be 50 amps. If increased charging output is needed, the pulse width can be increased, to 75%, for example. This increases generator output to 75 amps. The computer turns the field on and off very quickly, several hundred times per second.

Hybrid Charging Systems

Hybrid vehicles use the motor/generators not only to propel the vehicle but also to recharge the high-voltage batteries. This is accomplished by the electronics that control the motor/generator operation. The motor/generator can be used as an electric brake to slow the vehicle, which recaptures kinetic energy and converts it into electrical energy.

In a traditional AC generator, as output increases, the strength of the magnetic field inside the generator also increases. This increase in field strength slows the rotor and makes the generator driveshaft much more difficult to spin. Hybrids use this condition to their advantage. When the driver of the vehicle lifts his or her foot from the accelerator pedal and presses the brake pedal, the hybrid drive control system switches the motor/generator into power generation mode. This means that the energy of the moving vehicle is now used to turn the generator. The arrangement of the drivetrain and hybrid system is shown in **Figure 19-89**.

The drive wheels are mechanically connected to the motor/generators. When slowing or braking, they can generate electricity instead of using power. The electricity generated is then used to recharge the high-voltage

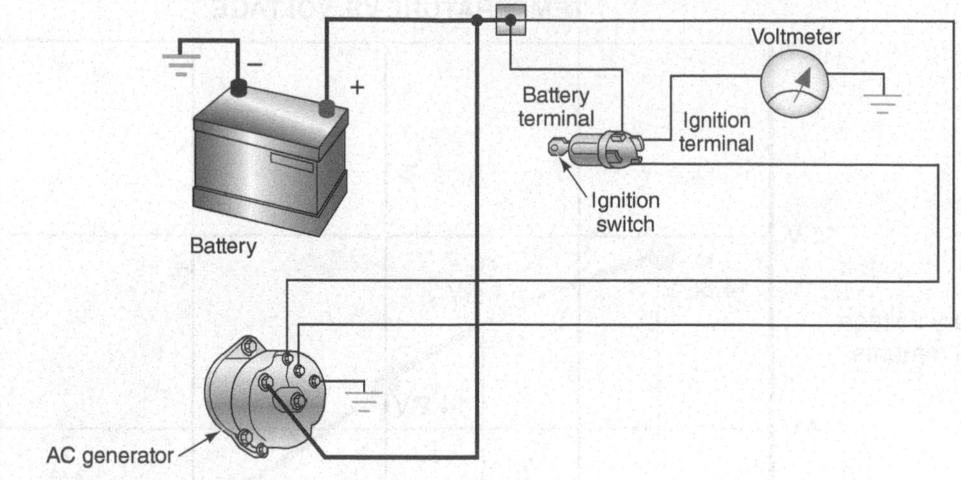

FIGURE 19-86 A noncomputer-controlled charging circuit.

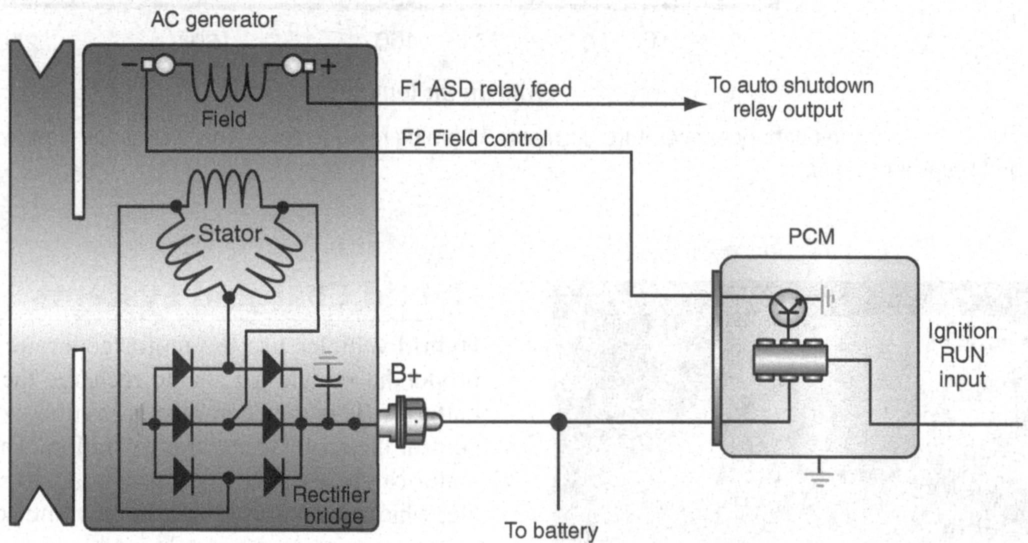

FIGURE 19-87 A computer-controlled charging circuit.

battery. The physical resistance of the motor/generator to spinning is used to slow the vehicle. This is called *regenerative braking*. In many cases, the motor/generator supplies most of the braking power necessary for normal driving. The service brakes are used at very low speeds, typically less than about 10 mph and when increased braking is needed, such as in a panic stop situation. The regenerative braking is so efficient that many hybrid vehicle owners have not worn out the service brake pads even after many years of driving.

To recharge the 12-volt battery, the high-voltage system has a 12-volt charging circuit that is part of the high-voltage battery control system. The function of the AC generator is contained within the hybrid system and is typically not repairable separately.

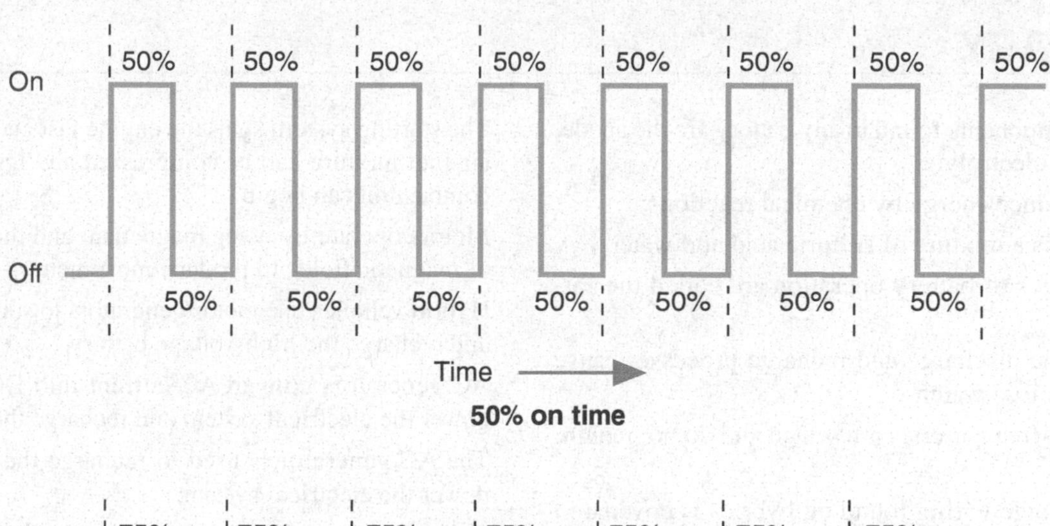

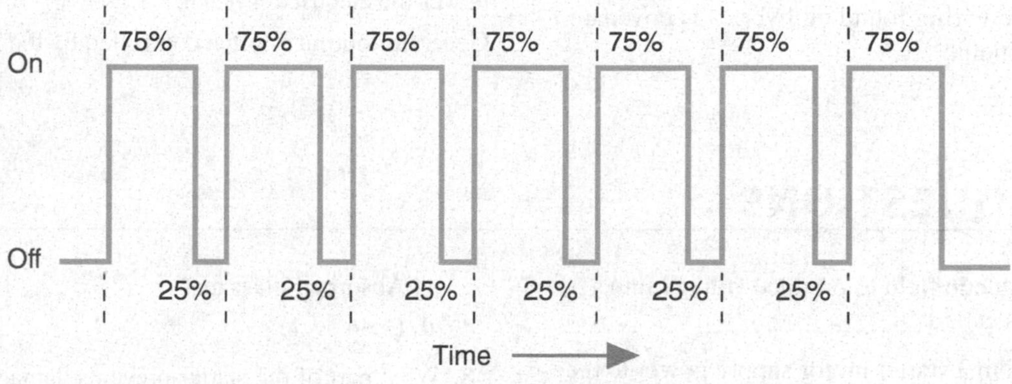

FIGURE 19-88 An example of a pulse-width modulated signal. The more "on" time, the higher the generator output will be.

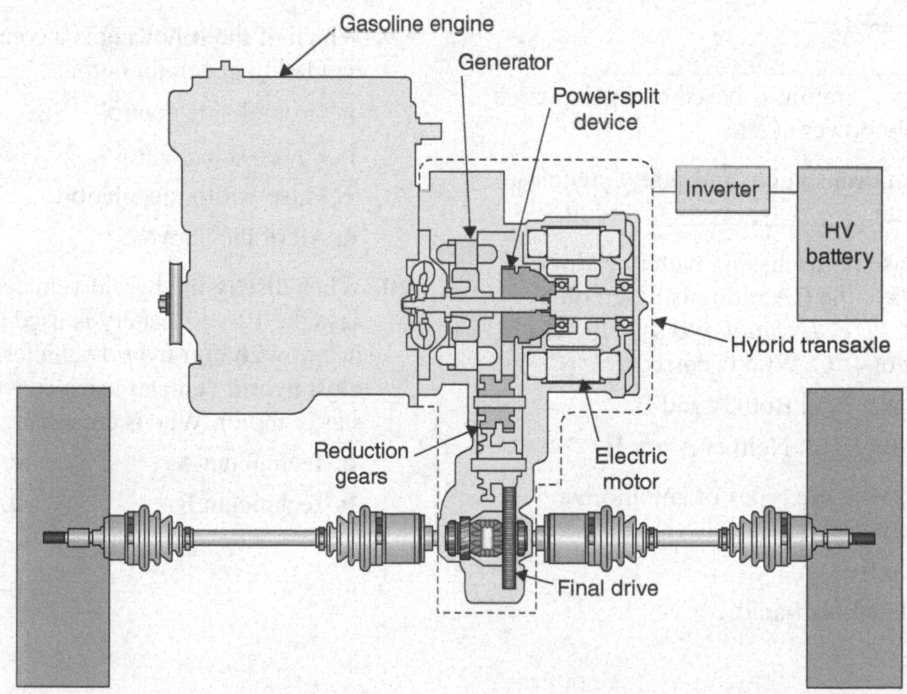

FIGURE 19-89 This hybrid drive system uses two motor/generators. The larger motor can be used to propel the vehicle and recharge the high-voltage battery when used as an electric brake.

SUMMARY

The three components found in any battery are the anode, cathode, and electrolyte.

Batteries produce energy by chemical reactions.

Battery acid is a mixture of sulfuric acid and water.

The four stages of battery operation are called the battery cycle.

Over time, the discharge and recharge processes cause the battery to lose water.

Maintenance-free batteries are sealed and do not require adding water.

The high-voltage wiring found on hybrids is covered in bright orange conduit.

The starting system spins the engine fast enough that the air-fuel mixture can be compressed and ignited so that combustion can begin.

Motors operate by using magnetism and the interaction of magnetic fields to produce movement.

Hybrid vehicles use motor/generators to start the engine and recharge the high-voltage battery.

AC generators convert AC current into DC current to power the electrical system and recharge the battery.

The AC generator is used to recharge the battery and power the electrical system.

Generator output is often controlled by the PCM.

REVIEW QUESTIONS

1. Using a magnetic field to produce voltage into a conductor is called ______________.
2. The brushes in a starter motor supply power to the ______________ segments.
3. In the generator, the ______________ coil produces magnetic fields within the ______________.
4. The ______________ ______________ ______________ rating is based on the battery's ability to supply power at 0°F.
5. The battery, ignition switch, and safety switch are part of the starter ______________ circuit.
6. Two technicians are discussing battery ratings. *Technician A* says the CA rating is based on a temperature of 32°F. *Technician B* says the CA rating is based on 0°C. Who is correct?
   a. Technician A
   b. Technician B
   c. Both A and B
   d. Neither A nor B
7. All of the following are types of automotive batteries except:
   a. Maintenance free
   b. Valve-regulated lead-acid
   c. Absorbed glass mat
   d. Deep cycle
8. What part of the starter prevents damage to the drive gear as the engine starts?
   a. Brush assembly
   b. Overrunning clutch
   c. Plunger
   d. Planetary gear
9. Which of the following is a common method of regulating generator output?
   a. Ground-side control
   b. Power-side control
   c. Pulse-width modulation
   d. All of the above
10. When discussing hybrid vehicles, *Technician A* says the 12-volt battery is used to crank the engine just as with non-hybrid vehicles. *Technician B* says most hybrid vehicles use a conventional 12-volt starter motor. Who is correct?
    a. Technician A
    b. Technician B
    c. Both A and B
    d. Neither A nor B

CHAPTER 20

Starting and Charging System Service

Chapter Objectives

At the conclusion of this chapter, you should be able to:

- Identify safe work practices for servicing the starting and charging system.
- Perform a battery inspection. (ASE Education Foundation MLR 6.B.1 & 6.B.4)
- Confirm proper battery capacity for vehicle application; perform battery capacity test; determine necessary action. (ASE Education Foundation MLR 6.B.2) Perform battery services such as charging, jump-starting, and maintaining vehicle memory circuits. (ASE Education Foundation MLR 6.B.1, 6.B.3, 6.B.5, & 6.B.6)
- Identify electronic modules, security systems, radios, and other accessories that require reinitialization or code entry after reconnecting vehicle battery. (ASE Education Foundation MLR 6.B.8)
- Diagnose starting system concerns such as no-crank no-start conditions.
- Test starter current draw and starting system voltage drops. (ASE Education Foundation MLR 6.C.1)
- Remove and replace a starter motor. (ASE Education Foundation MLR 6.C.4)
- Inspect the charging system.
- Test generator output. (ASE Education Foundation MLR 6.D.3)
- Perform charging system voltage drops. (ASE Education Foundation MLR 6.D.4)
- Remove and replace a generator. (ASE Education Foundation MLR 6.D.3)

KEY TERMS

battery corrosion	fast charging	starter cranking current
battery holddown	overcharging	three-minute charge test
battery load test	parasitic load	undercharge condition
conductance tester	slow charging	

Tools and Safety

To safely and effectively service the starting and charging systems, you need to understand the purpose and operation of a variety of tools. In addition, close attention to personal safety equipment is important because you will be working with batteries and with spinning pulleys and drive belts.

STARTING AND CHARGING SYSTEM TOOLS

Many of the tools used when servicing, diagnosing, and repairing the starting and charging systems are used for general electrical system service, such as the digital multimeter (DMM). Other tools as shown here are needed to measure the high amperage found in starting and charging circuits.

- A starting/charging system tester is used to measure voltage and current flow (**Figure 20-1**). This type of tester can be used to test battery capacity, starting and charging system performance.
- **Conductance testers** are used to test batteries, starters, and generator output (**Figure 20-2**).
- Current clamps can be used with a DMM or scope when you are testing the starting and charging systems (**Figure 20-3**).

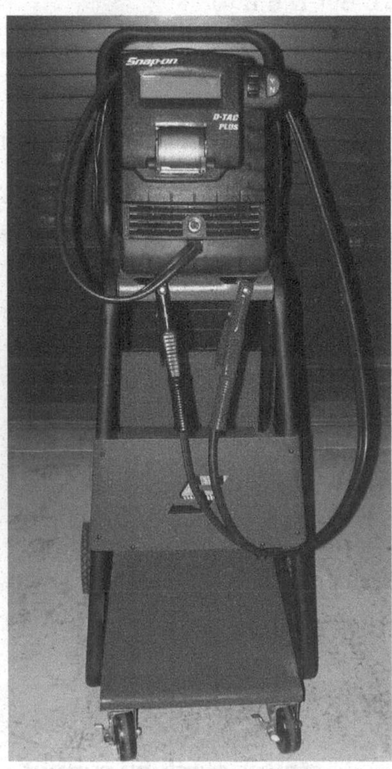

FIGURE 20-1 An example of a battery/starting/charging system tester. Often referred to as VAT tester for volt/amp tester.

FIGURE 20-2 A conductance tester is used to send a small AC current through the battery. This tester is small, easy to use, and inexpensive compared to the larger VAT-type testers.

FIGURE 20-3 An inductive current clamp is used to measure cranking and charging current.

- Digital scopes, such as the Snap-On Vantage Pro, can be used as meters or scopes (**Figure 20-4**). These tools are often used when you are diagnosing charging system concerns.
- Battery service tools, such as the post clamp pliers and terminal spreaders, are used when servicing or replacing a battery (**Figure 20-5**).
- Removing a very corroded battery terminal from a top-post battery may require a terminal puller (**Figure 20-6**). Prying on the terminal can break the post off the battery.

- Battery terminal brushes are used to clean the battery terminals and posts during service (**Figure 20-7**).
- Remote starter switches are used when performing starting system tests and bench testing starter motors (**Figure 20-8**).
- A memory saver is used to maintain battery power to the electrical system when the battery is disconnected. This prevents the loss of memories stored in the various modules. Several types of memory savers are available. The one shown in **Figure 20-9** plugs into a power outlet. Others plug into the DLC connector. Some shops use a boost pack.

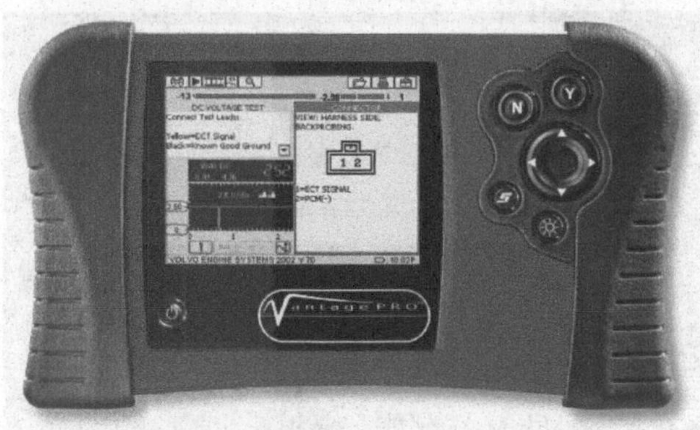

FIGURE 20-4 A lab scope can be used with an inductive clamp to test the starting and charging systems.

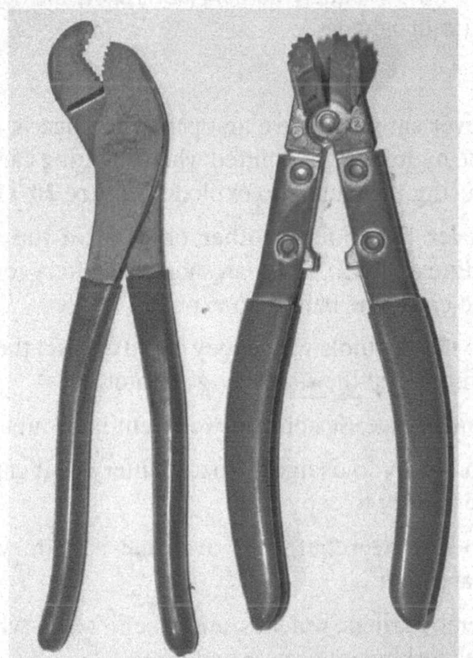

FIGURE 20-5 Battery pliers are used to remove and service top-post battery terminals.

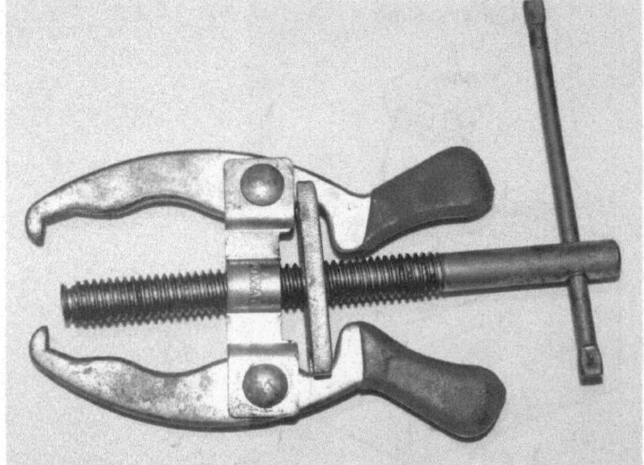

FIGURE 20-6 A terminal puller is used to remove stuck-on top-post terminals without damaging the battery.

FIGURE 20-7 Battery brushes are used to clean terminals and posts.

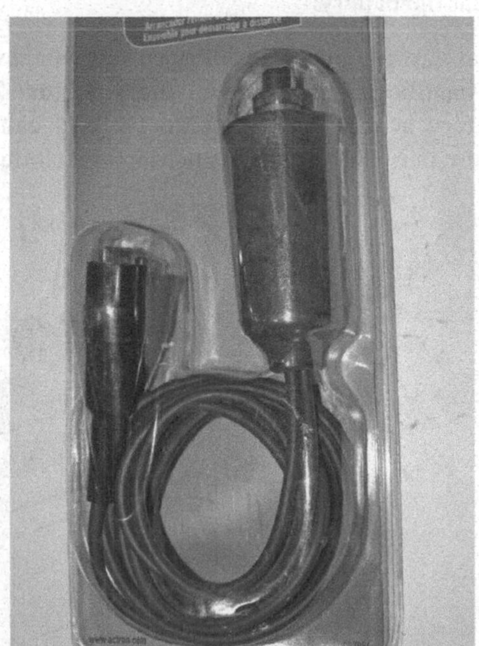

FIGURE 20-8 Remote starter switches are helpful when testing the starting circuit.

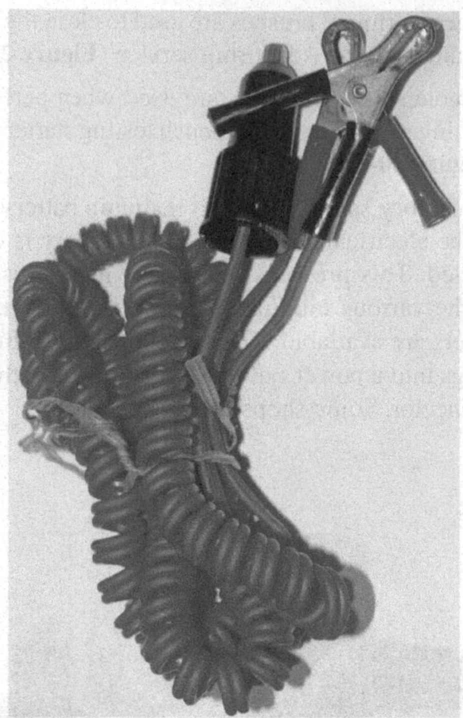

FIGURE 20-9 A memory-saving tool plugs into the lighter or auxiliary power outlet and connects to a second battery to maintain the different memories stored in vehicle systems.

SAFETY PRECAUTIONS FOR SERVICING THE STARTING AND CHARGING SYSTEMS

When working with batteries, extra safety precautions must be observed to prevent injury and damage to the vehicle. Follow these precautions when you are working on or near the battery:

- Wear gloves and use a battery carrier whenever you are handling or servicing a battery (**Figure 20-10**). Battery acid is highly corrosive and can cause severe burns to your skin and damage clothing.

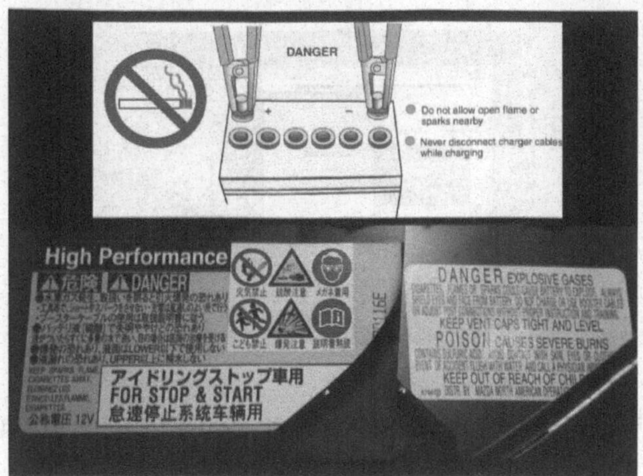

FIGURE 20-11 Battery warning labels show the dangers of working with batteries.

- Never smoke or have an open flame near a battery. The hydrogen gas emitted when charging can ignite, causing the battery to explode (**Figure 20-11**).
- Never lay tools or other objects on the battery (**Figure 20-12**). This can short the battery terminals and cause the battery to explode.
- Do not lay tools where they could contact the power terminals at the starter or generator.
- Charge batteries only in well-ventilated areas.
- Do not try to charge a frozen battery as it can cause the battery to explode.
- Do not overcharge or overheat a battery while charging it.
- Remove rings, watches, and other jewelry when you are working on or near batteries.
- Remove the negative cable first and reinstall the negative cable last.

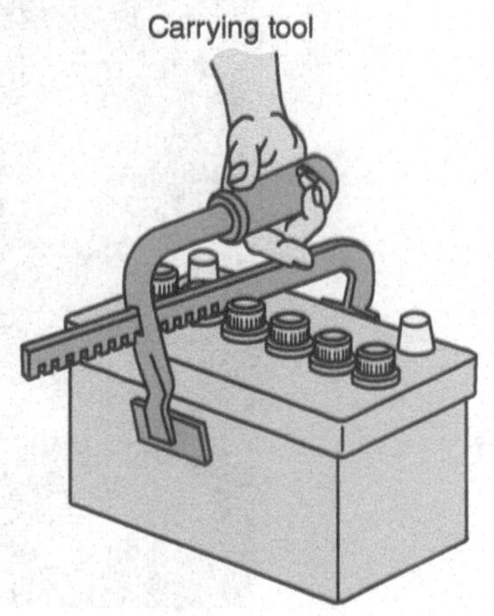

FIGURE 20-10 A battery carrying strap is used to prevent getting battery acid on you and your clothing.

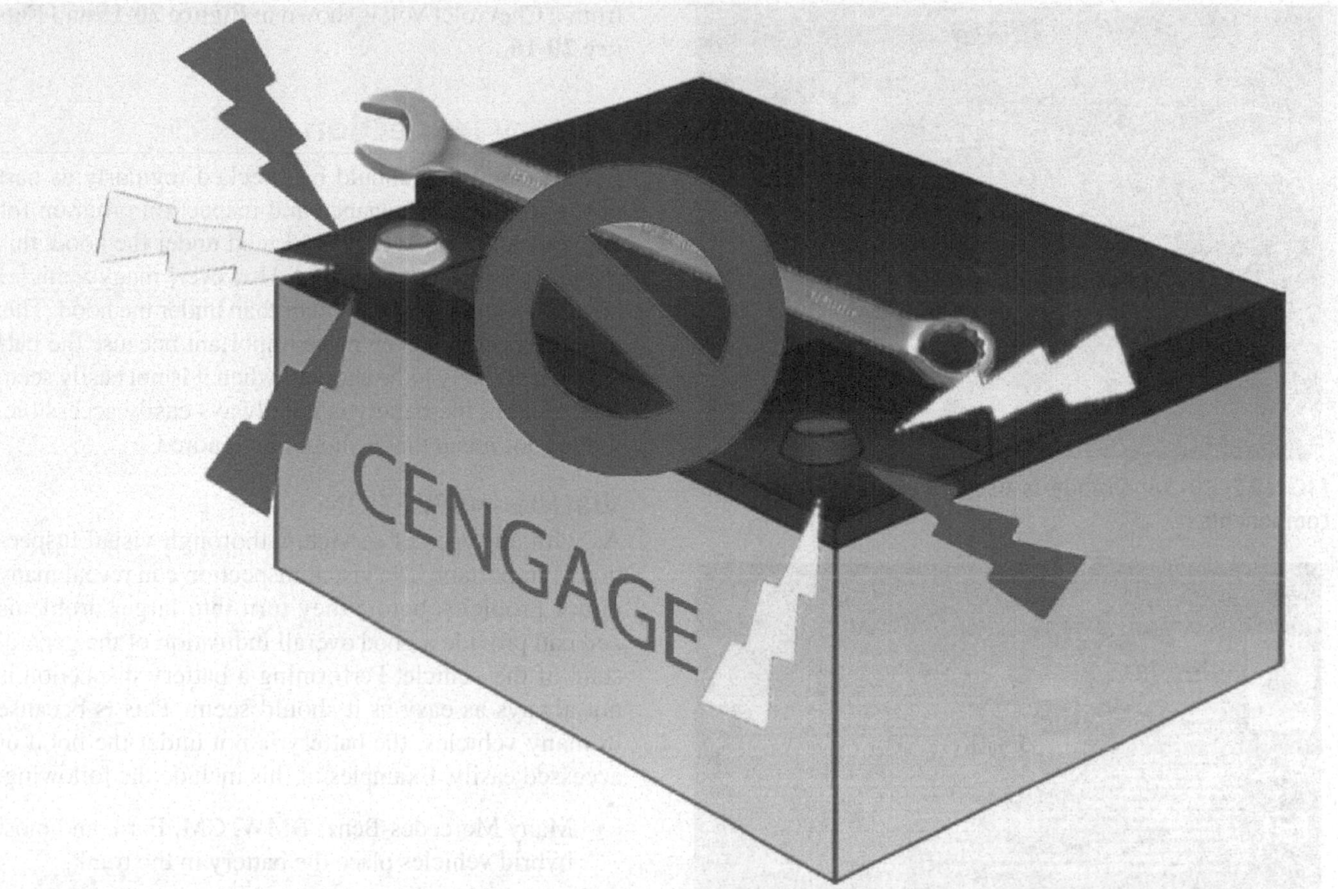

FIGURE 20-12 **Never lay tools on top of a battery. The tool could arc the terminals together, causing an explosion.**

- Batteries, starters, and battery cables can become hot if the starter is held in the cranking position for too long. A starter should never be cranked for more than 15 seconds at a time without at least a two-minute cool-off period. Continued cranking of the engine can overheat the starter and battery cables and can damage the starter.
- The generator can also become very hot when the output is high. Do not place your hands on the generator when the engine is running, especially if the charging system output tests have been performed.

FIGURE 20-13 **A warning decal on a Lexus hybrid vehicle. It is important to know battery location and type before working on the electrical system.**

■ ***Identify High-Voltage Circuits and Safety Precautions.*** When you are working on a hybrid vehicle, locate and read the service warnings and precautions published by the vehicle manufacturer before attempting any service or repairs (**Figure 20-13**). This applies even if you are not working directly on the high-voltage (HV) system because high-voltage components are located throughout the vehicle (**Figure 20-14**). Figure 20-14 shows the engine compartment of a Chevrolet Volt. Components and wiring for the high-voltage (HV) system are under the cowl, near the brake and coolant reservoirs, and throughout the compartment.

In most cases, you will not need to disable the HV system for regular vehicle service, including servicing the 12-volt battery. You should, however, be aware of the high-voltage components and what the orange wiring represents.

High-voltage wiring and components are identified by the orange conduit and connections. Never touch or place tools near the high-voltage components. Follow

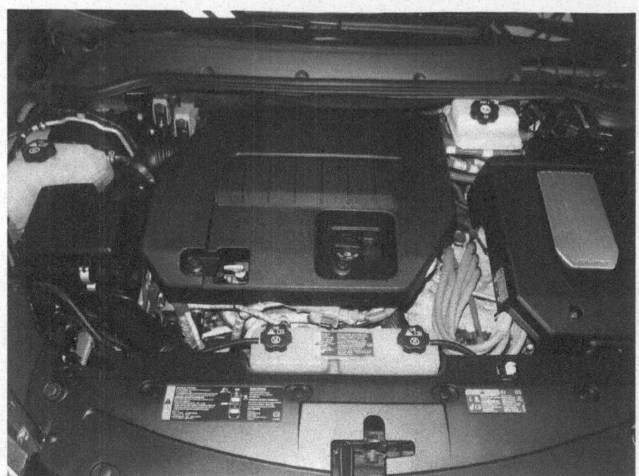

FIGURE 20-14 Orange is used to warn of high-voltage components.

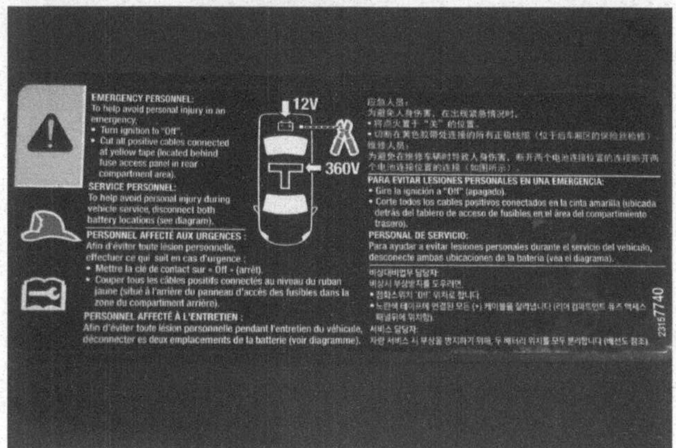

FIGURE 20-15 An example of a high-voltage warning decal. Never attempt to work on the high-voltage system without proper training and tools.

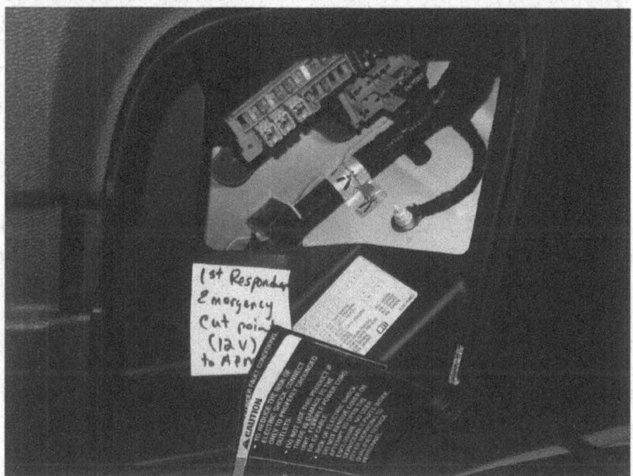

FIGURE 20-16 An example of an emergency high-voltage disconnect or cut point.

the procedures to disable the HV systems exactly if you need to do so.

In an emergency, hybrid and electric vehicles have methods of disconnecting the HV systems. An example from a Chevrolet Volt is shown in **Figure 20-15** and **Figure 20-16**.

Battery Inspection

Battery condition should be checked regularly as part of the routine maintenance and inspection program for the vehicle. If the battery is located under the hood, this is more easily accomplished. However, many vehicles have the battery in places other than under the hood. This makes inspection even more important because the battery is less likely to be checked when it is not easily seen. Just because the battery is not always easily accessible, it does not mean that it should be ignored.

VISUAL INSPECTION

As with all types of service, a thorough visual inspection is important. The visual inspection can reveal many minor problems before they turn into larger problems and can provide a good overall indication of the general state of the vehicle. Performing a battery inspection is not always as easy as it should seem. This is because in many vehicles, the battery is not under the hood or accessed easily. Examples of this include the following:

- Many Mercedes-Benz, BMW, GM, Ford, and most hybrid vehicles place the battery in the trunk.
- SUVs and minivans from General Motors and other manufacturers place the battery in a special compartment under the floor behind the front seats.
- Chrysler vehicles often have the battery located in the left front fender.
- Underhood batteries may be covered by engine control modules, junction boxes, or other types of covers.

Once you have located the battery or batteries, begin your inspection while following some basic safety precautions.

■ ***Battery Safety.*** As mentioned before, battery safety is of utmost concern to prevent personal injury and damage to the vehicle. Keep these precautions in mind when you are inspecting and servicing the battery:

- Never lay tools on the battery or near the battery connections.
- Never smoke or have an open flame near the battery.
- Do not overcharge or overheat the battery during charging.
- Never attempt to charge or jump-start a frozen battery.
- Wear gloves when you are handling the battery or working near battery acid or corrosion.

- Disconnect the battery negative connection first, then the positive. Reconnect the positive connection first, then the negative.
- Some vehicles, such as those with diesel engines and some Mercedes-Benz models, have two batteries. Service on these systems require disconnecting both batteries to prevent accidental arcing or damage.

■ ***Corrosion.*** Battery corrosion is a common problem (**Figure 20-17**). **Battery corrosion** forms from the release of battery acid. White corrosion is from the battery acid leaking out of the battery at the post. Corrosion can also form by a reaction between dissimilar metals at the terminals. Blue corrosion, as around the positive cable in **Figure 20-17**, indicates that the copper in the terminal is being eaten away.

Battery corrosion should be neutralized and removed. This can be done by applying a mixture of baking soda and water as a paste. The soda neutralizes the acid, making cleanup easier. Battery cleaning chemicals can also be used to neutralize the acid and aid in its cleanup. When you are working with battery acid, always wear chemical-resistant gloves to protect your hands.

If there is severe corrosion on and around the battery, remove the battery for cleaning. Whenever battery acid has leaked from a battery, the acid must be neutralized and cleaned. Battery acid will eat through metal battery trays and wiring, causing damage to the vehicle.

Carefully inspect the battery terminals and cables for corrosion buildup. Side-post batteries often hide corrosion between the battery terminal and cable terminal, and may require removing the cables to fully inspect their condition (**Figure 20-18**).

Not all battery corrosion look like that shown in Figure 20-17. A thin layer of corrosion can build between posts and terminals that can be hard to detect.

FIGURE 20-17 Battery corrosion can cause poor connections and eat through terminals and cables.

FIGURE 20-18 Corrosion on side-post connections can be very difficult to find unless you remove the terminal from the battery.

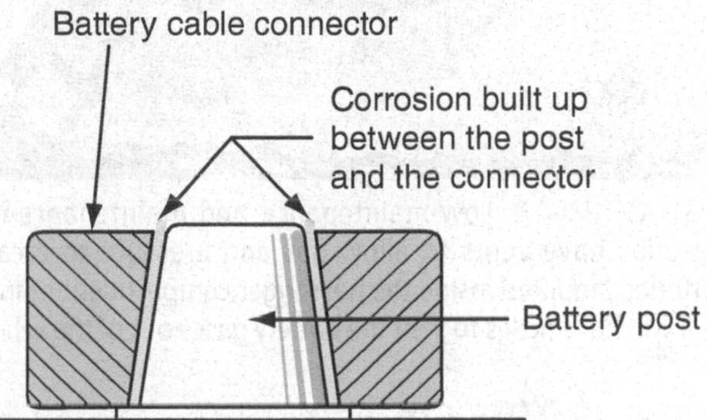

FIGURE 20-19 A very thin layer of corrosion can form between the post and the terminal, resulting in connection problems.

Figure 20-19 shows an illustration of how corrosion can build up around a battery post. This occurs from the galvanic reaction between the lead of the post and the copper of the terminal.

■ ***Battery Case.*** Inspect the battery case for signs of acid leaks, cracks, cap damage, and check vent condition. Acid leaks can be caused by damaged cell caps, improperly installed cell caps, and cracks in the plastic of the case. On side-post batteries, overtightening the terminal connections or using improper terminal bolts can damage the side-post connections and cause a leak. A battery with cracks in the case or leaks around terminals must be replaced.

Maintenance-free batteries are sealed, meaning that there is no method of checking the acid level or condition. Even though a battery may have what appears to be cell caps, these caps are not removable on maintenance-free batteries. Attempting to remove these caps will damage the battery and result in acid leaks. Maintenance-free batteries have vents on the sides. If these vents become plugged, pressure can build in the battery during charging. This can cause the battery case to swell and crack. Batteries mounted in locations other than the engine compartment have a vent tube (**Figure 20-20**). This allows the gas to vent outside of the vehicle. Inspect the vents and vent tubes to make sure the battery is venting properly.

FIGURE 20-20 Low-maintenance and maintenance-free batteries have vents to allow gas and pressure to escape. Batteries mounted inside the passenger compartment or trunk area will have hoses to vent the battery gases out of the vehicle.

FIGURE 20-22 A faulty battery terminal connection requires replacement of the terminal or cable. "Fixes" like this are not professional and are not a long-term fix.

■ ***Terminals and Posts.*** Inspect the terminals and posts for corrosion and make sure they are tight. Top-post batteries can have several types of cable terminals (**Figure 20-21**). If the terminal is loose on the post, intermittent connection problems can occur. Overtightening the terminal can damage the terminal and break the terminal bolts. Terminals that will not tighten securely require replacement (**Figure 20-22**).

Top-post batteries can be damaged if the terminals are pried from the posts. When removing a top-post battery connection, never pry on the terminal or cable. This can break the post off the top of the battery, ruining the battery. If the terminal is stuck to the post, use a terminal puller (**Figure 20-23**). This will allow you to safely remove the cable without damaging the battery.

FIGURE 20-23 Sometimes a terminal puller is necessary to remove a corroded and stuck terminal from the post. Do not pry on the terminal to remove it.

(a)

(b)

FIGURE 20-21 (a) An example of a top-post battery cable connection. (b) This battery connection bolts the cable to the terminal so the terminal can be replaced separately.

Side-post batteries can be damaged by overtightening the side-post connections. This can twist the terminal inside the battery, breaking it loose from the cell connections and causing an acid leak. Replacing the original battery cable bolt with a longer bolt can also damage the side-post terminals. The longer bolt threads too far into the terminal and breaks through the inside of the terminal into the case, and causes it to leak acid. In both of these situations, the battery must be replaced.

■ ***Battery Cables.*** Battery cables can become damaged by corrosion as the acid attacks the copper strands. Excessive corrosion can eat away the cables in addition to the terminals (**Figure 20-24**). When this happens, the entire cable needs to be replaced. Removing a damaged terminal and cutting back the cable and installing a new end is not a permanent repair (**Figure 20-25**). Side-post battery cables and ends are also subject to damage and improper repair (**Figure 20-26**). Battery cables should be checked for tightness and connection condition at all connection points.

FIGURE 20-24 Battery corrosion forms from gassing and leaking acid. Both liquid acid and corrosion are dangerous if allowed to contact skin.

FIGURE 20-25 Replacing the terminal is not a long-term repair.

FIGURE 20-26 An example of how not to make proper battery connections.

Battery cables can also be damaged by overheating. This can happen if the starter circuit is drawing excessive current, or if the starter is kept engaged until both the starter and cables overheat. Overheated battery cables often show damage to the insulation, which may appear melted or deformed.

Many battery cables have a smaller ground wire running to the body near the battery (**Figure 20-27**). This is a battery-to-body ground and must be in good condition or circuits using the body as ground can be affected.

■ ***Battery Holddowns and Covers.*** All new vehicles leave the factory with some type of **battery holddown** device (**Figure 20-28**). Check that the battery is properly secured. A loose battery can be damaged by bouncing around, which shortens the battery's life and may cause damage to the vehicle.

Common battery holddowns include a bolted-in wedge at the base of the battery or a metal strap that is secured across the top of the battery. Regardless of what type of holddown is used, it is important that it

FIGURE 20-27 An example of a battery-to-body ground wire.

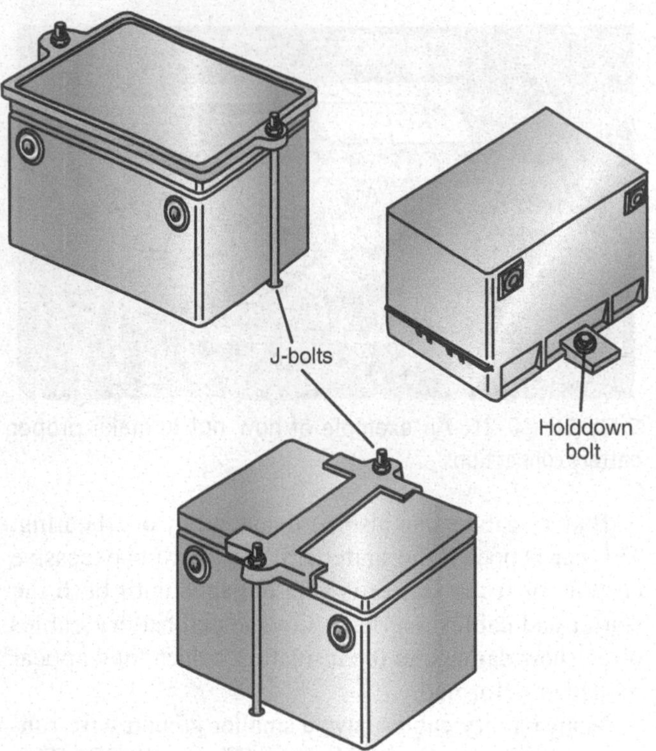

FIGURE 20-28 Common types of holddowns.

is in place and secures the battery. When you replace a battery, the holddown must be reinstalled with the new battery. If the holddown does not fit the new battery, double-check the battery application to make sure that it is correct for the vehicle. In some situations, any hold down is better than none at all (**Figure 20-29**).

Many batteries have some type of cover or shroud (**Figure 20-30**). These are used to help insulate the battery from high underhood temperatures. Some battery covers double as reservoirs for windshield washer fluid or as the coolant overflow tank. When you are replacing a battery, make sure that these covers are reinstalled correctly to protect the battery and prevent damage to the covers.

FIGURE 20-29 A customer fixes for the battery holddown after some collision damage.

FIGURE 20-30 Blankets are used to help insulate the battery from high underhood temperatures.

■ ***Built-In Hydrometer.*** A feature found on many batteries is the built-in hydrometer (**Figure 20-31**). The hydrometer gives an indication of the battery's state of charge based on the specific gravity of the electrolyte. Typically, the indicator displays green if the battery is fully charged, yellow if discharged, and black if the electrolyte is very low. A concern about using the built-in hydrometer for decisions about battery condition is that the hydrometer may only be showing the state of one cell out of the six cells.

REPLACING A BATTERY

■ ***Choosing the Correct Battery.*** While several batteries may physically fit into the spot for the battery, that does not mean that just any battery should be used in that vehicle. When you are selecting a replacement battery, refer to the BCI rating for the application. A replacement battery should be the same group number as specified by the vehicle manufacturer. Installing a battery that does not fit properly can cause clearance issues or prevent the holddown from keeping the battery in place correctly (**Figure 20-32**).

Next, determine the CCA rating of the vehicle. Installing a battery with too little CCA capacity may result in an engine that will not start in cold weather. The vehicle owner's manual and online service information often provides information about what CCA rating is needed.

Next, you should not have to make any modifications to the vehicle to make the battery fit. This includes things such as removing and not installing the battery holddown or covers, cutting terminals off the battery cable ends to install a different type of terminal, and creating

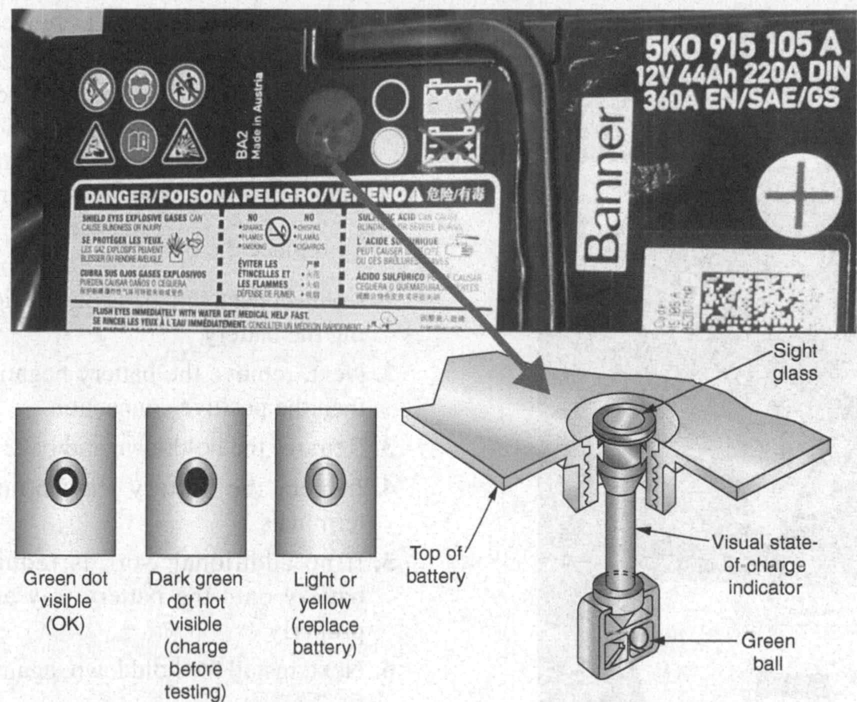

FIGURE 20-31 The built-in hydrometer can indicate battery condition based on the specific gravity of the electrolyte.

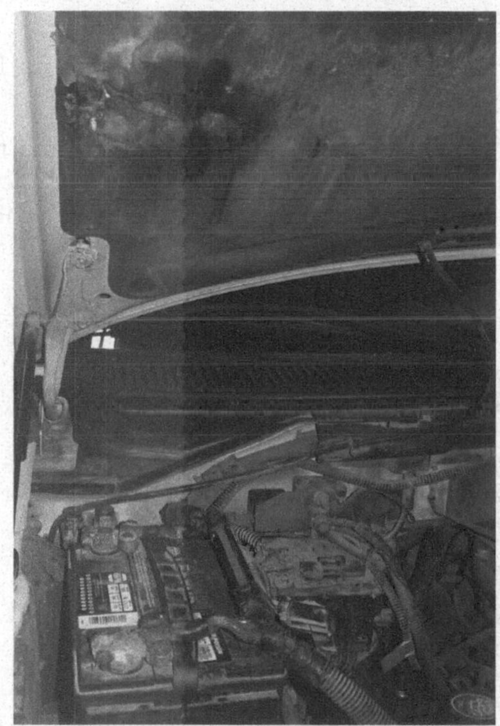

FIGURE 20-32 Installing the wrong battery can lead to problems such as the posts shorting out against the body.

additional clearance for the battery posts to keep them from interfering with other parts of the vehicle.

■ ***Identify and Maintain Memory Systems.*** Before you disconnect the battery on a vehicle, it is a good idea to install a memory-saving device. If you do not have a battery saver, a second battery with jumper cables or an auxiliary booster can be used. If you are using a second battery and jumper cables, connect the positive jumper cable lead to the second battery and then to the battery being disconnected. Next, connect the negative jumper cable clamp to the second battery and then to the battery being disconnected.

Another popular method is to use a power brick (**Figure 20-33**). These multipurpose jump start/power supplies are small, light, and easy to use in the shop. If you are using an auxiliary booster connect the positive lead to the battery positive cable first, then connect the negative lead to the battery negative connection (**Figure 20-34**). By installing a memory saver or a second battery, you can remove the battery cables without losing memory functions, such as clock settings, radio presets, and other functions stored throughout the various modules in the vehicle.

■ ***Replacing a Battery.*** Begin by identifying any systems that require reset or relearned when the battery is disconnected. While this step may not seem important, many systems are affected by disconnecting the battery. These include the following:

- Clocks.
- Radios with antitheft lockout codes.
- Some vehicles will trigger the antitheft system to inhibit engine starting, leaving you with a vehicle with a new battery that will not start.

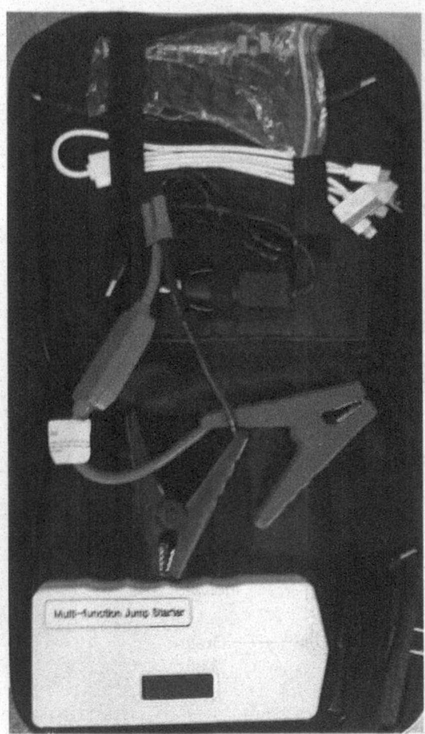

FIGURE 20-33 Power bricks or solid-state boosters like this are becoming popular for jump-starting a battery.

- Power windows with one-touch or express down/up functions.
- In some vehicles, disconnecting the battery can cause the airbag system to deactivate and set a trouble code.
- Memory settings such as power seats and mirrors.
- The engine and transmission control modules have learned adaptive strategies to compensate for wear and different operating conditions. The engine may not idle or shift correctly after memory loss until the operating parameters are relearned.

To replace the battery:

1. Connect the memory saving device before disconnecting the battery.
2. Next, remove the battery negative cable connection then the positive connection.
3. Remove the holddown and battery.
4. Inspect the battery tray, holddown, cables, and terminals.
5. If no additional work is required, place the new battery onto the battery tray and make sure it fits properly.
6. Next, install the holddown, again checking for proper fit.
7. Connect the positive cable connection to the battery then the negative cable.
8. Remove the memory saver or power source.
9. Start the engine and make sure no warning lights remain on.

Some vehicles require a new battery to be registered into the onboard computer system. This includes late-model BMW, Mini, Ford, and other vehicles.

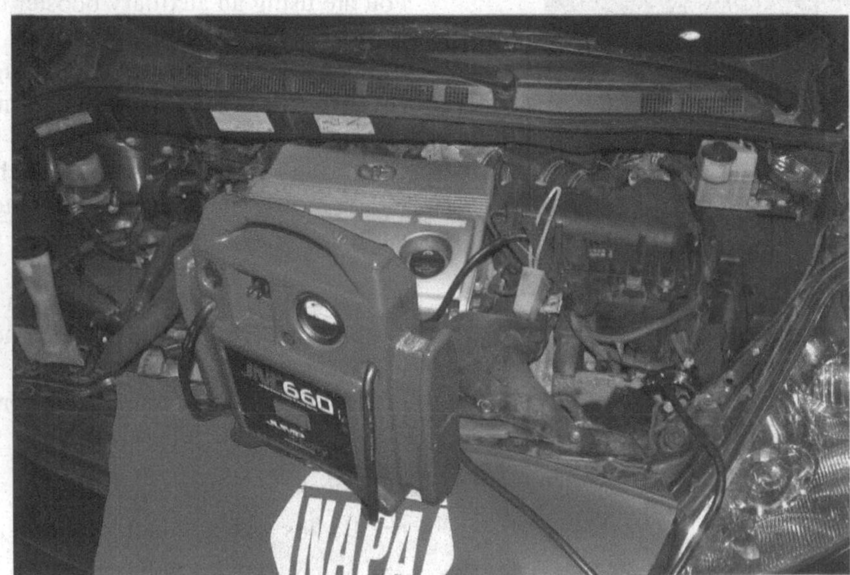

FIGURE 20-34 Using a booster battery pack to maintain electrical system power during a battery replacement.

Registration is necessary for proper charging system operation and other functions. To register a new battery, certain codes or details from the new battery are entered into the computer using a scan tool or battery reset tool. Failure to perform this step can result in improper charging and reduced life of the new battery. Refer to the vehicle manufacturer's service information for specific instructions.

IN-DEPTH INSPECTION

Once a visual inspection is performed, you may need to perform a more detailed inspection to fully determine the condition of the battery.

■ ***Battery Voltage Leak Test.*** Corrosion across the top of the battery can create a circuit that draws power from the battery, and cause it to prematurely discharge. This can be tested by using a voltmeter to measure the voltage across the top of the battery (**Figure 20-35**). Set the voltmeter to read VDC and place the negative lead on the negative battery cable. Next, place the positive lead on the top of the battery near the negative battery post, and note any voltage reading on the meter. Move the positive lead slowly away from the negative post and toward the positive post while you watch the voltmeter readings. If the meter displays voltage across the battery case, the battery should be cleaned and retested. The slight bit of voltage displayed by the meter can, over time, discharge the battery as the corrosion acts as a circuit.

■ ***Electrolyte Level.*** If the battery is not sealed, the acid level can be checked. Carefully remove the vent caps with a screwdriver and note the level of acid for each cell. The acid should be above the top of the plates but not completely filling the cell. There needs to be a small amount of space between the acid level and the caps to allow for expansion of the acid during charging. If the battery is overfilled, the acid will leak out of the caps as the battery is charged.

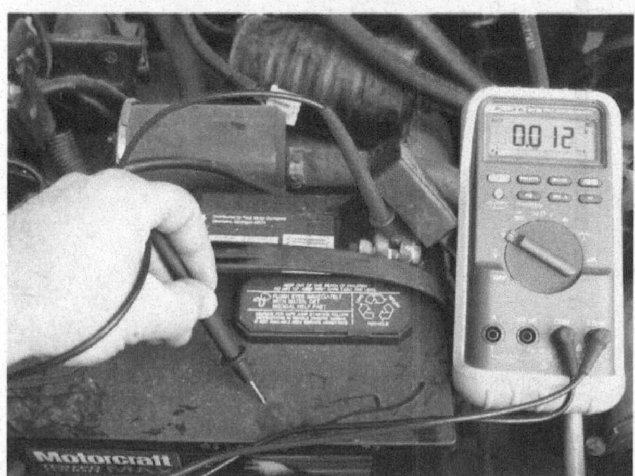

FIGURE 20-35 Checking for battery draw caused by corrosion across the top of the battery.

If one or more cells are low on electrolyte, this indicates a problem with the battery. Low acid level can be caused by a leak or by the acid being boiled out of the battery. Overcharging can boil the acid, leaving the cells exposed. If the acid level is low, add distilled water to the battery. Do not use ordinary tap water because it contains minerals and other contaminates that can react within the battery and shorten its life.

■ ***Battery State of Charge.*** The state-of-charge (SOC) test is used to determine the percentage of battery charge. A fully charged battery has 12.6 volts and is at 100% SOC. **Table 20-1** shows battery state of charge based on voltage readings. Test the battery voltage after the battery has set and the voltage has stabilized. A battery that has just been charged or is in a running vehicle shows greater than 12.6 volts due to what is called *surface charge*. To remove the surface charge of a battery in a vehicle, turn the headlights on for one minute, and then let the battery stabilize for two minutes before you check the voltage.

Checking the battery voltage with all the accessories off is called *the open circuit voltage test*. The voltage should be 12.6 volts if the battery is fully charged. A reading of less than 12.6 volts indicates the battery may not be able to maintain a full charge, and further testing is necessary.

State of charge can also be determined by testing the specific gravity of the electrolyte, however, this test is not as common because many batteries are sealed and testing the electrolyte is not possible. Remove the cell caps and check the acid level for each cell. The electrolyte should be clear like water. If it appears cloudy or has debris floating in the acid, the battery has been damaged by excessive vibration and should be replaced. Remove a few drops of acid and place on the refractometer (**Figure 20-36**). Look through the eyepiece and read the specific gravity on the scale (**Figure 20-37**). Each cell should read within a few points of each other.

TABLE 20-1

Open Circuit Voltage Table	
Open Circuit Voltage	**Charge Percentage**
11.7 volts or less	0
12.0 volts	25
12.2 volts	50
12.4 volts	75
12.6 volts or more	100

FIGURE 20-36 A refractometer can also be used to check battery electrolyte. Place a drop of electrolyte on the lens and look through the eyepiece. Read the state of charge on the correct scale.

Battery Service and Testing

Just as modern vehicles have changed the way many services and repairs are performed, the battery service has also changed due to new technologies and features. For example:

- Disconnecting the battery now means the loss of antitheft radio codes, personality presets, and adaptive strategies for the engine and transmission control modules.
- Just connecting a battery charger and turning it on can destroy AGM batteries if the charging rate is too high.
- New battery chargers that allow only very slight amounts or zero AC voltage are necessary when keeping battery voltage up during computer reprogramming. AC voltage from chargers can interfere with computer operation, cause programming errors, and even damage onboard modules.

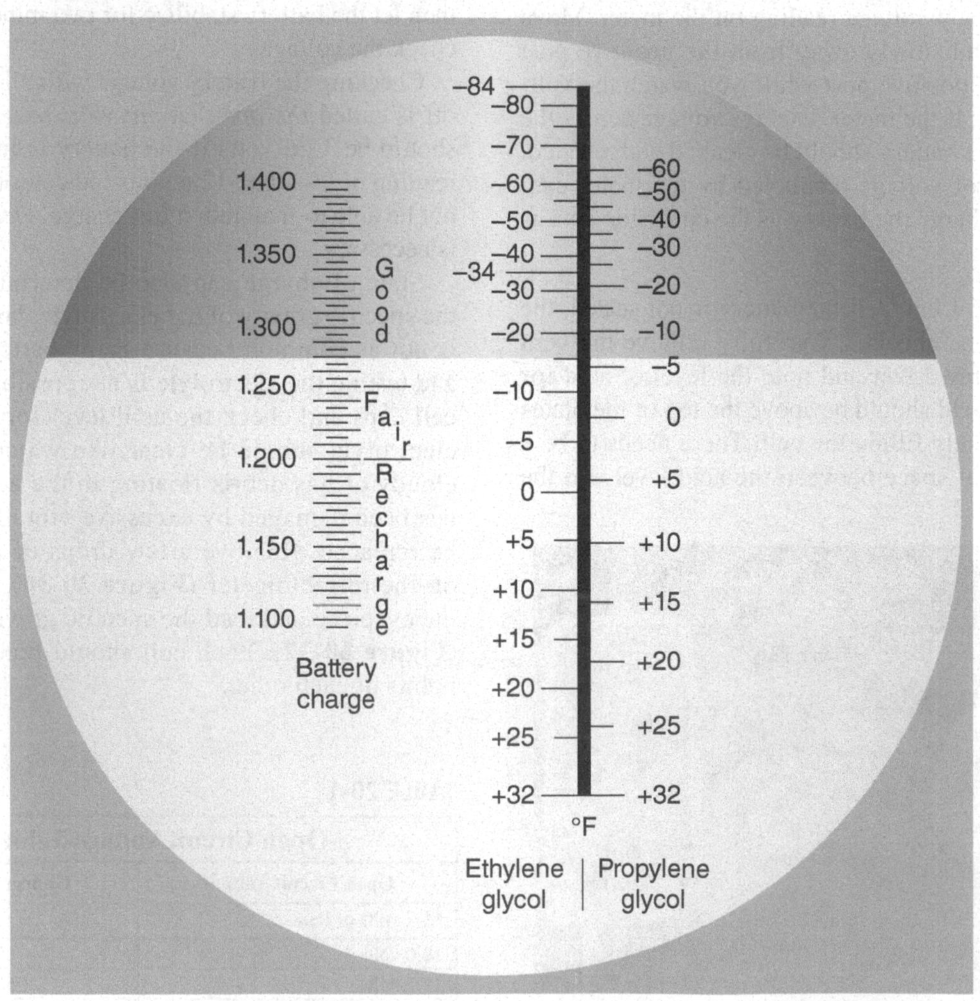

FIGURE 20-37 Checking battery acid specific gravity with a hydrometer. This illustrates how a refractometer displays battery state of charge.

BATTERY CHARGING

When you determine that a battery needs to be recharged, you also need to decide what is the best method to recharge the battery. There are two basic types of chargers: smart (automatic) chargers and manual chargers. An important aspect of battery service is understanding which charger to use and why.

■ ***Determine Battery Condition.*** Use the open circuit battery voltage or state of charge from hydrometer readings to determine the battery's condition. If the battery is low on electrolyte, add distilled water to the cells until the level is covering the plates, but do not overfill the battery. Only distilled water should be added to the battery. Distilled water does not have any impurities or minerals that are found in tap water. If the battery is only slightly discharged, with an open circuit voltage of 12 volts or more, you can use a fast charge to restore the battery. If the battery is severely discharged, a slow charger, also called *a trickle charger*, should be used.

■ ***Types of Charging.*** A fast charge is done at a high amperage rate for a short amount of time to quickly bring the battery back up to charge. The drawback is that a fast charge can overheat and damage a battery if too high of a charging rate is used or if the battery is severely sulfated. Slow charging supplies a low current, typically about 1 or 2 amps over a long period of time to slowly bring the battery back up to charge. Slow charging is the preferred method and should be used when time permits.

Most shops have manual battery chargers similar to the one shown in **Figure 20-38**. These chargers often have multiple charging rates and may have a booster rate that can be used to jump-start a dead battery. The charger may also have a timer mechanism so that charge rate and time can be set. Once the charge time expires, the charger shuts off. The drawbacks of these chargers is that they do not monitor battery voltage or current and can pass an excessive amount of AC current while charging.

Some chargers have built-in battery testers and can be programmed for different type of batteries and charging rates. **Figure 20-39** shows an example of a smart battery charger. These chargers can be used on lead-acid and AGM batteries, which require special charging procedures.

Other types of smart chargers can be set for standard, AGM, or gel-type batteries (**Figure 20-40**). These chargers monitor battery charge condition and will switch

FIGURE 20-38 A typical shop battery charger.

FIGURE 20-39 This type of smart charger can be used on lead-acid and AGM batteries. This charger will also test the battery before and after charging.

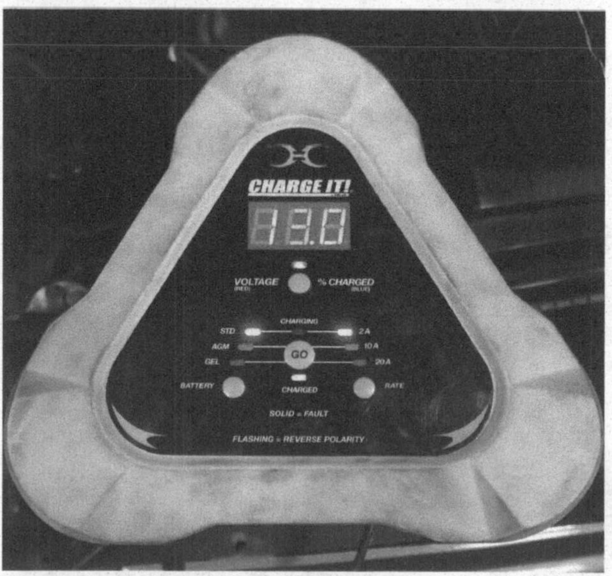

FIGURE 20-40 A smart charger that can charge standard batteries, AGM batteries, and gel batteries.

to a float or maintenance charge setting once the battery has reached 100% charge.

If keeping a battery charged while accessing the onboard computer system or when performing reprogramming of a module, clean chargers are necessary (**Figure 20-41**). These chargers have a very clean DC output, meaning very little AC voltage leaks through the charger. This is important when performing module programming to help prevent module failure during programming.

■ ***Charger Safety.*** Before you connect a battery charger and turn it on, there are several important safety precautions you need to follow to prevent injury and damage to the charger, battery, and vehicle.

- Do not attempt to charge a frozen battery. If there is the possibility that the battery could be frozen, remove the battery from the vehicle, and let it warm inside for 24 hours before attempting to test or charge it. Charging a frozen battery can cause a rapid expansion and heat build up inside of the battery and cause it to explode.
- Charge batteries only in well-ventilated areas so that the hydrogen gas can safely disperse.
- Check the condition of the charger before you operate it. Be sure the power cord has the ground lug attached and that the electric receptacle you are going to use is a grounded plug. Make sure the power cord insulation is intact and there is no exposed wiring. Check the charging cables and clamps. Make sure the cables are tightly attached to the clamps and the clamps spring tightly closed.
- Do not set batteries on top of the charger. During charging, the battery may leak acid, which could get into the charger and cause damage.

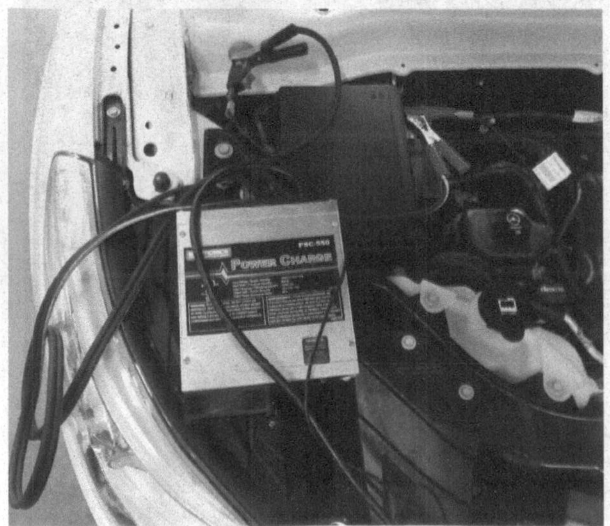

FIGURE 20-41 A clean charger is used when performing computer reprogramming.

- If you are charging a battery in the vehicle, make sure that the ignition and all accessories are off and the doors are closed.
- Be sure the charger is turned off and unplugged before you attach the charging cables.
- Attach the positive charging cable to the battery positive connection first then connect the charger negative lead to the battery negative connection. Do not connect the negative charging clamp to fuel lines or sheet metal body parts. Once the clamps have a solid connection to the battery, plug the charger into the power receptacle.
- Determine what type of charge rate is acceptable for the battery. Some AGM batteries require low-amperage charge rates or for voltage to remain below 14.8 volts.
- Set the charge rate and time. The charge rate and time are based on the battery's condition and temperature. Refer to the charging rates shown on the battery charger.
- Connect a voltmeter to the battery and note the charging voltage. If the voltage exceeds 15.5 volts, the battery is likely sulfated and needs to be replaced.

■ ***Battery Slow Charging.*** **Slow charging** is the best way to restore a discharged battery since a low charge rate does not cause excessive heat generation and allows the plates to more thoroughly charge throughout the plate volume. Slow or trickle charging is often done at a rate of 1 to 2 amps.

Before you use a trickle charger, make sure it is turned off and unplugged. Some trickle chargers do not have an on/off switch; they are on whenever they are plugged into the outlet, so check before you plug the charger in. As with larger battery chargers, first connect the battery cables from the charger to the battery positive and negative. Be sure to observe the correct polarity by connecting the red charger lead to the positive battery terminal and the black charger lead to the negative battery terminal. Next, plug the charger into an outlet. If the charger has an on/off switch, turn the charger on. Record the battery voltage using a voltmeter.

■ ***Fast Battery Charging.*** **Fast charging** can be used to quickly bring a discharged battery back up to charge, but can, if left unchecked, damage or destroy the battery. Fast charging may be done at 20 to 60 amps for short periods of time.

Refer to the battery charging table for the battery charger you are using. An example of charging rate table is shown in **Table 20-2**. This table will provide you with

TABLE 20-2

Open Circuit Voltage	Battery Specific Gravity	State of Charge (%)	Charging Time of Full Charge at 80°F (267°C)					
			AT 60 A	AT 50 A	AT 40 A	AT 30 A	AT 20 A	AT 10 A
12.6	1.265	100	Full charge					
12.4	1.225	75	15 min.	20 min.	27 min.	35 min.	48 min.	90 min.
12.2	1.190	50	35 min.	45 min.	55 min.	75 min.	95 min.	180 min.
12.0	1.155	25	50 min.	65 min.	85 min.	115 min.	145 min.	280 min.
11.8	1.120	0	65 min.	85 min.	110 min.	150 min.	150 min.	370 min.

the acceptable charge rate and time based on the type of battery being charged.

If the battery starts to leak acid or gets hot, stop the battery charger immediately and let the battery cool. If this occurs, the battery is extremely dangerous due to the gas being emitted and poses an explosion hazard.

Do not attempt to fast charge an AGM battery with a manual charger. Unregulated current and voltage can overheat and destroy the battery. Always read the warning labels on the battery before you attempt service.

BATTERY TESTS

Battery tests can be as basic as checking the battery voltage to more in-depth tests, such as load-testing the battery to check its ability to produce current to crank the engine. As a technician, you will need to be able to perform a variety of tests safely and accurately.

■ ***Battery State of Charge.*** Test the battery state of charge once the battery has had time to stabilize after charging. This may require the battery sitting for several hours before an accurate state of charge can be determined. Connect the leads of a voltmeter to the battery and measure the volts DC. A reading of 12.6 indicates a 100% charge, 12.45 indicates 75% charge, 12.25 indicates 50% charge, 12.06 indicates 25% charge, and 11.89 or below indicates a fully discharged battery. These numbers are for a battery at 80°F (30°C).

■ ***Conductance Test.*** The battery conductance test measures the ability of the battery to transmit current through the internal structure of the plate material and internal connections. Conductance testers are small, handheld testers that supply a small AC current through the battery. As the AC flows through the cells, it is changed slightly by the battery. The tester interprets this change and translates it into a reading of the battery's condition.

To use a conductance tester, connect the positive and negative test leads to the battery and enter the information asked on the display. Typically, you need to know the CCA or CA rating of the battery. Once the test is completed, the unit displays the results of the test (**Figure 20-42**). Some testers have built-in printers so test results can be given to the customer.

■ ***Sulfation Test.*** The sulfation test is also called the **three-minute charge test**. This test performs a high-rate charge (30–40 amps) for three minutes while the voltage is observed. Voltage should not exceed 15.5 volts. If voltage exceeds 15.5 volts, it indicates the battery plates are sulfated, and the battery is not accepting a charge. A 24-hour slow charge may be able to desulfate the plates and restore the battery. Recharging with a very low-amperage charger, such as a battery tender, may also restore a sulfated battery.

This test should be performed with the battery removed from the vehicle or with the negative battery cable disconnected. This is because a sulfated battery can pull 17 volts or more, which can cause damage to onboard systems. To perform this test, connect a battery charger with a 30 to 40-amp setting and a voltmeter to the battery posts. Start the charger and watch the meter readings. At the end of the three minutes, the voltage should not exceed 15.5 volts.

Battery sulfating occurs when a battery sits for a period of time and it is not fully charged. Sulfating also can occur if there is a charging system problem that prevents the battery from fully charging. Other causes include a

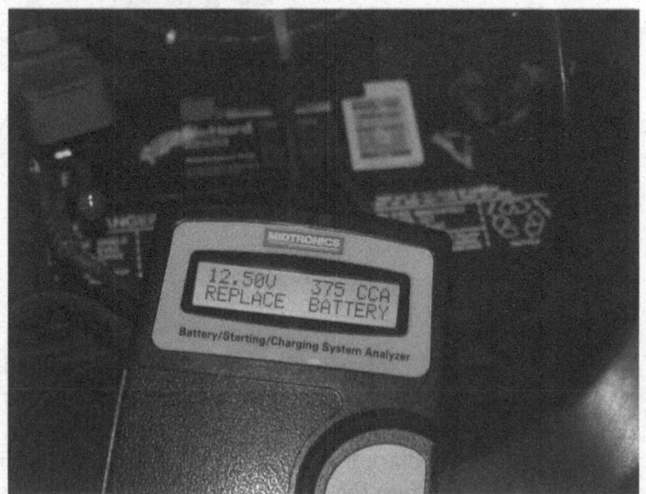

FIGURE 20-42 A conductance tester passes a small AC current through the battery.

loose generator drive belt, infrequent driving, and driving only in very short trips. Frequent discharging and rapidly recharging the battery, as often happens with batteries in school vehicles, can also cause the plates to sulfate.

A sulfated battery that does not accept a charge can also cause the generator to work too hard, which increases the temperature inside the generator. Over time, this can lead to reduced generator service life and even repeated generator failures.

■ ***Battery Load Test.*** The **battery load test** is also called *the high-capacity discharge test* or just *a load test*. The load test measures the battery's ability to produce current for 15 seconds and requires a battery load tester (Figure 20-1).

This test will load the battery by one-half of its CCA rating for 15 seconds, during which time the battery voltage should not drop below 9.6 volts. Refer to **Table 20-3** for temperature-voltage corrections for the load test. To perform the load test, the battery must be at least 75% charged. If the battery is less than 75%, it will probably not be able to pass the load test.

The load test applies one-half of the battery's CCA rating for 15 seconds. If the CCA rating of the battery is not known, you can substitute the equivalent of three times the amp/hour rating. If neither of these is known, use the following:

- Load amps for a four-cylinder engine equals twice the cubic inch displacement of the engine.
- Load amps for a six-cylinder engine equals one-and-a-half times cubic inch displacement of the engine.
- Load amps for an eight-cylinder engine equals the cubic inch displacement of the engine.

TABLE 20-3

Battery Voltage Chart

Electroylyte Temperature	Minimum Voltage under Load for 15 Seconds
70°F (21°C)	9.6
60°F (16°C)	9.5
50°F (10°C)	9.4
40°F (4°C)	9.3
30°F (−1°C)	9.1
20°F (−7°C)	8.9
10°F (−12°C)	8.7
0°F (−17.8°C)	8.5

Connect the load tester to the battery as illustrated in **Figure 20-43. Make sure that the test cables have a good connection to the battery.** If the connection is not clean and tight, the tester will not be able to properly load the battery, and it may cause sparks at the connection point. Place the arrow on the current clamp on one of the tester's battery cables with the arrow pointing in the direction of current flow, away from the battery on negative and toward the battery on the positive. It is important the you place the inductive clamp over one of the tester's cables. The clamp measures the current flowing from the battery, through the tester, and back to the battery.

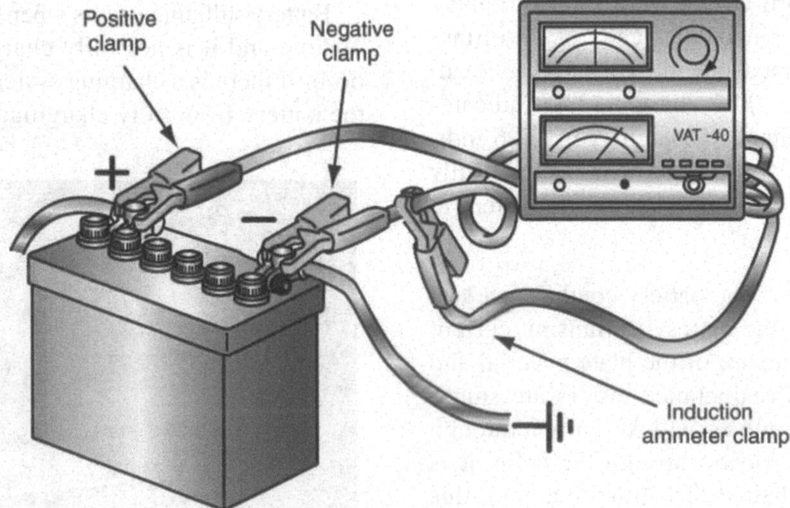

FIGURE 20-43 Using a VAT-type tester to load-test a battery.

1. Connect the positive and negative leads to the battery and the current clamp around either the positive or negative tester cable.
2. Place the clamp with the arrow in the direction of current flow.
3. Turn the load knob until the specified load amperage is shown on the amperage gauge.
4. Hold the load for 15 seconds and release the knob.
5. Let the battery sit for two minutes and record the voltage.

Before testing, note the battery voltage. If the battery is fully charged, 12.6 volts or slightly higher, you can perform the test. If the battery has recently been charged and the voltage is above 12.6, you need to remove the surface charge. This requires loading the battery one-quarter of its load capacity for 15 seconds. The battery must be at least 75% charged, with an open circuit voltage of 12.4 volts to conduct a load test. If the battery is less than 75% charged, it will probably not pass the load test.

To begin the test, turn the load knob (if equipped) and hold the load at the specified amperage for 15 seconds. If the tester you are using has a LOAD or TEST button, just press to begin. The battery voltage should not drop below 9.6 volts during the test. If battery is at room temperature and the voltage does drop below 9.6 volts, charge the battery and retest. If it fails a second time, the battery should be replaced. If the battery passes the first load test, allow the battery to rest for two minutes. Check the open circuit voltage; it should be close to 12.6 volts. If the voltage is close to 12.6 volts, the battery passes the test. If the voltage is less than 12.6 volts, charge and retest the battery.

Because the load test is actually drawing power from the battery, it is normal for the tester to get hot, smell, and possibly emit a small amount of smoke. This is because of the heat generated by the carbon resister inside the tester.

Do not apply excessive loads or apply loads for longer than 15 seconds to the battery. Because the tester is performing a rapid discharge of the battery, repeated tests, excessive current draw, or excessive test length can overheat and damage the tester and battery.

■ ***Parasitic Load Testing.*** If a battery in a vehicle has a low state of charge after sitting, a parasitic load could be the cause. **Parasitic load**, also called *parasitic draw* or *key-off battery drain*, occurs when a system or component continues to draw power even after the vehicle is shut down. All modern vehicles have some key-off draws (**Table 20-4**). These include things such as clock and radio memories, computer memories, antitheft and keyless entry systems. Typically, these draws go down over time as control modules enter a standby or sleep mode. If, however, a module does not shut down, it could cause a significant key-off draw on the battery. This can discharge the battery enough so that the vehicle may not start after sitting overnight. Determining if a parasitic load is present can be done in several ways.

Service Tip

If a battery tests good, it likely is good, and if a battery tests bad, it likely is bad. However, it is a good idea to perform more than one test on a battery before you decide whether it is good or bad.

Test Light

This is the least effective method of measuring parasitic draw. Disconnect the battery negative cable, and install a test light in series between the battery terminal and the battery cable connection (**Figure 20-44**). If the test light is brightly lit, a significant key-off draw is present. You may ask why a test light will light up when placed in series within the negative side of the circuit. If there is

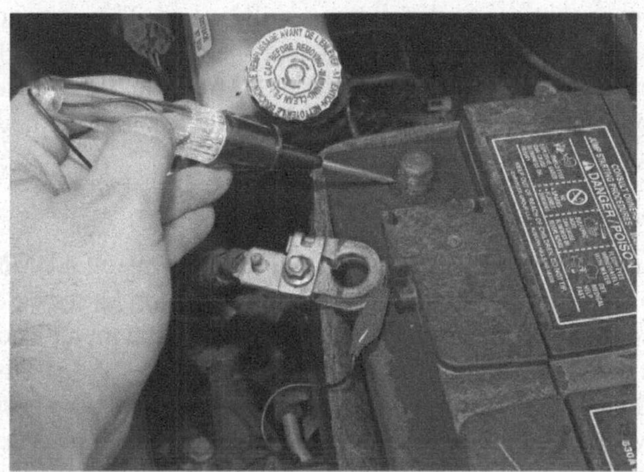

FIGURE 20-44 Testing parasitic draw with a test light. This method loses memories and does not indicate how much of a draw there is, only that a draw exists.

TABLE 20-4

Component or Module	Current Draw (mA) Typical	Current Draw (mA) Max
Electronic load control		1.0
Keyless entry system	2.0 0.3	2.0 0.4
Heated seat control modules		0.5
Instrument panel digital cluster	4.0	6.0
Light control module	0.5	1.0
Multi-function chime module (key removed from ignition)	1.0	1.0
Miscellanious modules		0.1
Pass key decoder module	.75	1.0
PCM	5.0	7.0
Entertainment network, communication system	7.0 1.8	8.5 3.5
RAC (theft deterrent) (illuminated entry) (auto door locks)		3.8

not a significant draw, there will not be enough current flow to power the light. If there is a sufficient draw, the test light will indicate current flow by lighting up. Placing the light in series allows the current flowing though the circuit to power the test light.

Leave the test light in place for several minutes and note the intensity of the light. If the light dims to being barely visible, the load has gone down. This can be caused by a system powering down, such as the keyless entry or interior lighting systems.

Using the test light for this test does not provide you with information about exactly how much current is being drawn from the battery, only that a draw is present. This is the least accurate method of testing parasitic load. Removing the battery cable to perform the test also means that memory functions will be lost. This may not please the owner of the vehicle, especially if the vehicle is equipped with an antitheft radio that requires a reset code. This method should only be used if other methods are not available.

Direct Measurement

Another way to check parasitic load is by using a DMM to measure the amperage flowing from the battery. Set the meter to measure DC amps and place the meter leads in the appropriate jacks in the meter (**Figure 20-45**). Disconnect the battery negative cable and connect the positive meter lead to the battery negative post and the meter negative lead to the negative battery cable connection. The amount of amperage draw is displayed on the meter (Figure 20-45). This method shows exactly how much current is being drawn from the battery but does require removing the battery cable, which results in lost memory functions.

Current Clamp

A similar way to measure parasitic draw is by using a DMM and a current probe or current clamp (**Figure 20-46**). Connect the current probe leads to the voltmeter positive and negative jack connections. Set the DMM on the DCV or DC millivolts scale. Turn the current probe on and zero the probe. The meter should display 0.0 volt with the probe zeroed. Locate the arrow on the current probe clamp. The arrow points in the direction of electron flow, so on the negative cable the arrow will point away from the battery. Clamp the probe around the negative battery cables, and note the reading on the meter.

The advantage of using the current clamp is that the battery does not need to be disconnected to perform the test. This means that memory functions are not lost and will not need to be reprogrammed, nor will computer adaptive strategies need to be relearned.

Using a graphing meter or digital scope is helpful when you perform this test. You can set the meter or scope to display the draw over a long period of time. This way you can then go on to other work while the meter or scope measures and graphs the draw. By using this method, you can easily see the graph as various modules and systems shut down. It is important to remember that if you open a door or wake the modules during the test, you need to start over. This is because some systems stay awake for an amount of time, perhaps an hour or more, before shutting down. You also may need to keep the door switches closed to allow the modules to shut down yet keep the doors open to allow access to fuse panels for testing.

FIGURE 20-45 Testing parasitic draw with a DMM placed in series and measuring the amperage provides you with an actual number which can be compared to specs. This method still requires disconnecting the battery.

FIGURE 20-46 Using a DMM and a current clamp will show how much of a draw is present, and the battery remains connected so memories are not lost.

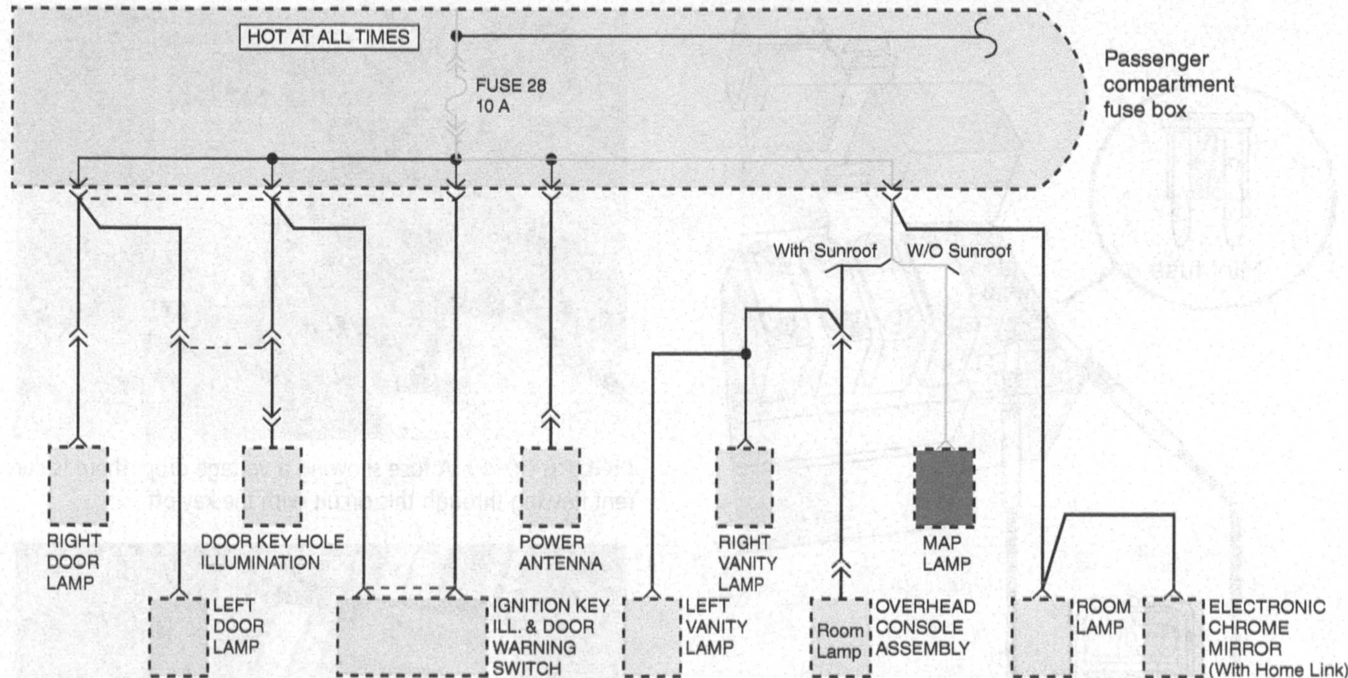

FIGURE 20-47 A sample wiring diagram. When diagnosing a parasitic draw, a wiring diagram will show you what components are connected to each fuse.

Voltage Drop Testing

Another way to check parasitic load is by measuring the voltage drop across each fuse. Each fuse is in series with the rest of the circuit it protects (**Figure 20-47**). Anytime there is current flowing through a circuit, a small but measurable voltage drop occurs across each component of the circuit. Any voltage drop across a fuse indicates that current is flowing through the circuit. The amount of voltage drop is related to the amount of current flow; more current flow equals a larger voltage drop. With the key and all accessories OFF, measure the voltage drop at each fuse (**Figure 20-48**). If a voltage reading is present (**Figure 20-49**), the circuit is drawing power and should be checked. Determine what circuits and components that fuse protects and unplug each one until the voltage drop disappears.

The allowable parasitic draw for the vehicle depends on the vehicle and what options may be installed. Refer to the manufacturer's service information for parasitic draw specs. A general rule is that 30 to 50 mA (0.030 to 0.050 amps) is usually acceptable. Some manufacturers may allow slightly more, while some may specify less. If a much higher draw is present, you may need to let the vehicle sit for an hour or two for various systems to shut down.

If the parasitic draw is excessive, you will need to determine the cause of the draw. Before you begin to test every circuit on the vehicle, double-check that no accessories were left on, such as a vanity mirror light, glove box light, or similar item. Eliminate any possibility that you somehow caused the draw.

If you are using a test light or direct current measurement to check the draw, remove each fuse, one by one, while monitoring the parasitic draw. If you remove a fuse and the draw suddenly disappears, reinstall the fuse and see if the draw returns. If it does, remove the fuse again and note the draw. If the draw is gone, locate what circuits are fed from that fuse and check each circuit for the draw. This is done by removing the load from each of the circuits one-by-one until the problem component is found. Common examples of components that can cause a parasitic draw are lights not shutting off, shorted generator diodes, and control modules not shutting down.

> **Service Tip**
>
> Once the cause of the draw is found and corrected, retest the parasitic draw. If the draw is still above specifications, another system or component is at fault. You will need to recheck each circuit again by removing the fuses one at a time. If the parasitic draw is within specifications, the problem is solved.

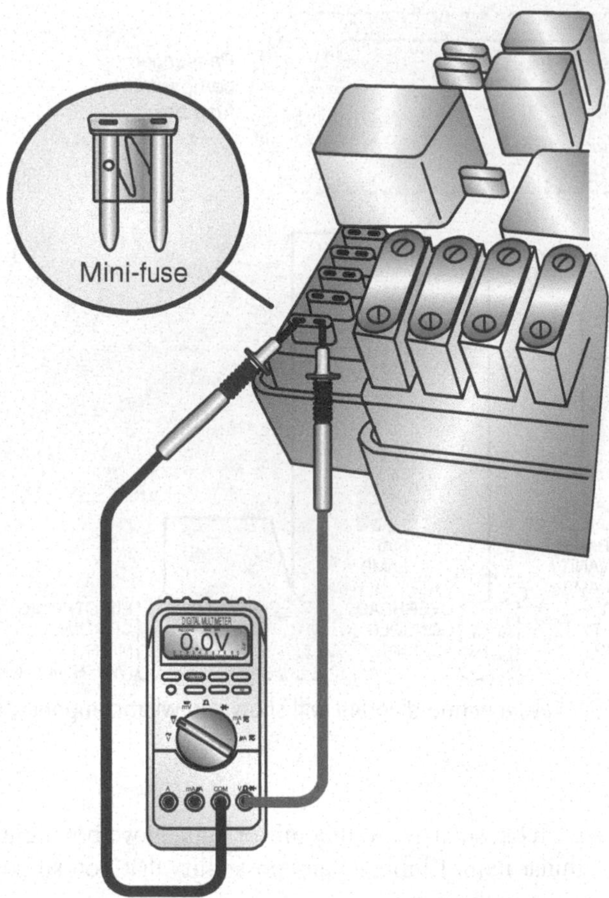

FIGURE 20-48 Measuring parasitic draw by checking the voltage drop across the fuse. A reading of 0.0 volt indicates that no current is flowing through the fuse.

FIGURE 20-49 A fuse showing a voltage drop. There is current flowing through this circuit with the key off.

FIGURE 20-50 Not a recommended repair to the battery cable.

BATTERY SERVICE

Battery service consists of performing a visual inspection of the battery, cables, terminals, holddown, and the area surrounding the battery. Service may also include cleaning and restoring the battery terminal connections, charging the battery, jump-starting, testing, and replacing the battery.

■ ***Battery Service.*** In many cases, a battery simply needs to be inspected and cleaned. Once you have performed a battery inspection and have determined that no significant problems are present, the battery and connections can often be cleaned in the vehicle. **Photo Sequence 11** shows how to clean a battery and the connections.

■ ***Replacing Battery Cables.*** If a battery terminal or cable becomes severely corroded or damaged, proper battery connections cannot be maintained. This requires that the cable be replaced. Even if only the terminal is corroded away or damaged, the entire cable may need to be replaced. It is not recommended that battery terminals be cut off and replaced (**Figure 20-50**). Cutting and installing a new end can lead to future corrosion and connection problems. Because positive battery cables on modern vehicles may contain the vehicle's main fuse (**Figure 20-51**), or be made of two separate cables for the starter and a power junction box, they can be very expensive. However, replacement is the best option for a complete repair.

FIGURE 20-51 A battery cable that contains the main fuse for the vehicle.

PHOTO SEQUENCE 11

Cleaning a Battery

PS11-1 Loosen the battery negative connection.

PS11-2 Use a battery terminal puller to prevent damage to the battery.

PS11-3 Loosen and remove the positive battery connection.

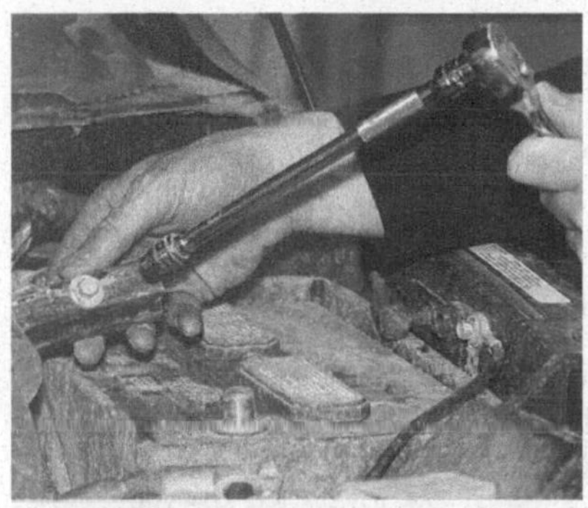

PS11-4 Remove the battery holddown and any shields.

PS11-5 Remove the battery and clean using a mixture of baking soda and water or a commercial battery cleaning solution or aerosol.

PS11-6 Rinse the battery with water.

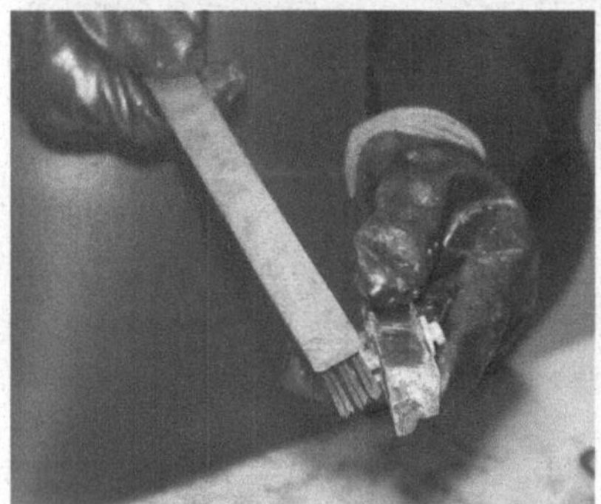

PS11-7 Clean the battery holddown hardware if necessary.

PS11-8 Clean the terminal clamps with a battery brush.

PS11-9 Clean the posts with a battery brush. Install the holddown and then reinstall the battery cables, positive first, then the negative.

When you replace a battery cable, remove the old cable and compare it to the replacement cable. Make sure that the new cable is the same length and diameter and has the same connections as the original cable. Be sure to clean and torque the cable connections when you are installing the new cable.

■ ***Jump-Starting with a Booster Pack.*** There may be times that you need to jump-start a vehicle. One quick way to do this is with a booster pack. There are two types of booster packs: AGM and charger/booster power packs. Booster packs that contain AGM batteries are charged by AC current. The booster has a built-in charger; all that is required is plugging it into a standard power outlet with an extension cord. Lithium power packs (**Figure 20-52**), are becoming popular because they are small, inexpensive, and can be used to recharge electronic devices and jump-start vehicles. However, these types of booster packs tend to have very short cables, which does not always allow the negative clamp to be attached to an engine bracket or other grounded location. Use caution if jump-starting a vehicle with this type of booster because both clamps may have to connect to the battery. Attach the positive clamp to the positive terminal first then the negative clamp to the negative terminal. Once the engine starts, quickly remove the booster from the battery to prevent damage to the pack.

FIGURE 20-52 Using a lithium battery or power brick to jump-start a vehicle.

Service Note

Jump-starting a dead battery, whether from a booster pack or a second vehicle, can cause voltage spikes in the electrical system. These spikes can damage generators, modules, and other electronic components. If possible, it is preferred that a dead battery be slowly recharged instead of jump-started.

Before using an AGM booster pack, check its charge level. This is often done by pressing a "test" button and looking at the voltage gauge. Inspect the positive and negative booster leads and clamps. Make sure the leads are in good condition and the clamps close tightly.

Connect the positive booster cable to the battery positive cable and then the negative cable to an engine ground, such as a bracket (**Figure 20-53**). This keeps any sparks away from the battery. Start the engine and remove the negative booster pack cable from the engine, then remove the positive cable at the battery.

■ ***Jump-Starting with Jumper Cables.*** When jump-starting a dead vehicle with another vehicle, observe the following safety precautions:

- Do not pull the vehicles so close together that they touch each other.
- Connect the jumper cables in this order: positive lead on the discharged battery, positive lead on the booster battery, negative lead on the booster battery, negative clamp on the engine of the dead vehicle (**Figure 20-54**).

FIGURE 20-53 Using a booster box to jump-start a vehicle.

Service Note

Jump-starting a dead battery from a second vehicle can cause voltage spikes in the electrical systems of both vehicles. These spikes can damage generators, modules, and other electronic components. If possible, it is preferred that a dead battery be slowly recharged instead of jump-started.

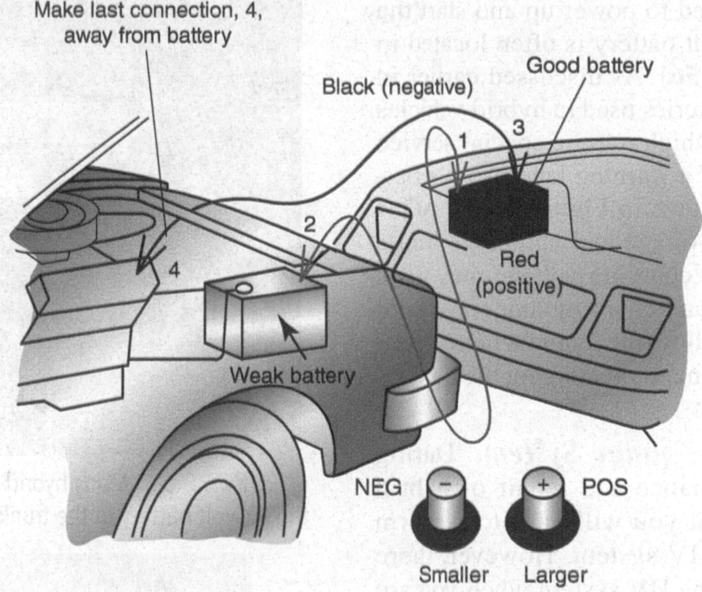

FIGURE 20-54 When using jumper cables, connect the cables in the order shown to prevent sparks and a possible battery explosion.

- Do not stand directly over either battery while you are making connections.
- Do not stand in between the vehicles.

Start the engine of the booster vehicle and let it idle for a moment, then crank the engine of the disabled vehicle. Once the engine starts, run it at a fast idle speed to help recharge the battery quickly. Carefully remove the negative jumper cable clamp from the engine ground on the jump-started vehicle. If the engine stays running, continue to remove the remaining jumper cable connections in reverse order of how you attached them.

Because not all vehicles have a battery that is accessible from under the hood, alternate jump-starting connection points are often provided. An example of remote jump-starting connections on a vehicle with an inaccessible battery is shown in **Figure 20-55**.

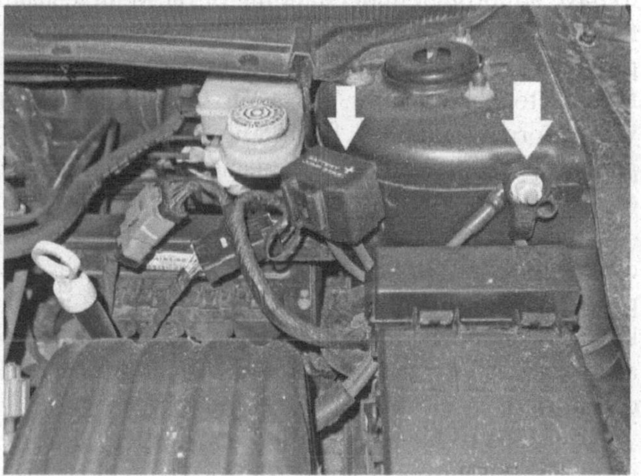

FIGURE 20-55 Vehicles with remote batteries provide special jump-start connections.

■ ***Diagnosing Battery Failure.*** Eventually batteries wear out and fail. The lifespan of the battery depends on the temperatures in which it operates. Batteries in very hot climates tend to have much shorter service lives than batteries in cooler climates. There are factors that can decrease the life of the battery and/or cause repeated battery failures. **Table 20-5** can be used to help determine the cause of premature battery failure.

TABLE 20-5

Symptom	Possible Cause
Battery is discharged or low	Excessive key-off draw, low generator output, poor battery connections, voltage drop in charging circuit
Repeated battery failure	Low generator output, poor battery connections, voltage drop in charging circuit, missing battery holddown, deep cycling of the battery
Corrosion returns after cleaning	Acid leaking through post openings, overcharging
Battery discharges when sitting	Excesseive key-off draw

■ ***Identify Hybrid Vehicle 12-Volt Battery Service Procedures.*** Even though a hybrid electric vehicle may operate with 115 to 600 volts in the hybrid system, a 12-volt battery is still used to power up and start the hybrid systems. The 12-volt battery is often located in the trunk area (**Figure 20-56**). As discussed earlier in this chapter, the 12-volt batteries used in hybrid vehicles are often AGM batteries, which require special service procedures. An example of a warning label from a battery in a Toyota Prius is shown in **Figure 20-57**. AGM batteries often have both voltage and amperage limits for charging. For example, charging voltage may need to be kept below 15 volts and charging amperage kept under 6 amps. Failure to follow the manufacturer's procedures can result in permanently damaging the battery.

FIGURE 20-56 Many hybrid vehicles (and nonhybrids) place the 12-volt battery in the trunk or cargo area.

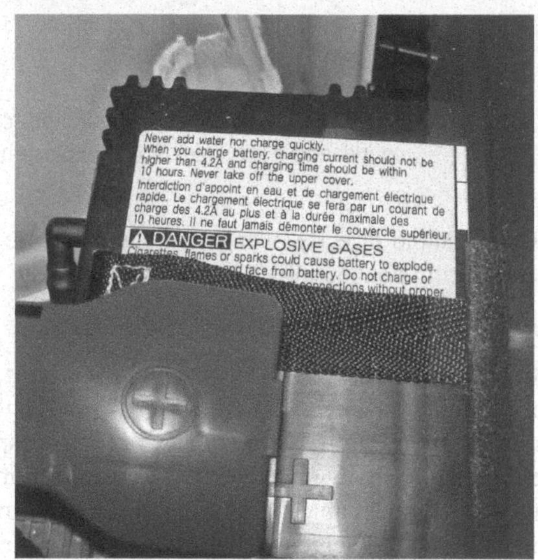

FIGURE 20-57 The AGM battery is used because it does not need to vent and because the majority of the power for the electrical system is provided by the HV battery.

■ ***Disabling the High-Voltage System.*** During course of normal maintenance and repair of hybrid vehicles, it is unlikely that you will have to perform any type of work on the HV system. However, there may be reason to disable the HV system when you are working on other parts or systems on the vehicle that are close to parts of the HV system. For safety purposes, you may need to disable the HV system. Before any attempt is made to disable the HV system, you need to have a thorough understanding of the vehicle manufacturer's procedures for shutting the system down. The procedure to disable the HV systems varies from manufacturer to manufacturer, so refer to the appropriate service information before beginning. The following is a general procedure and is not meant to be used in place of the manufacturer's specific procedures.

Remove the ignition key, smart key, or fob. If the vehicle has Push to Start, ensure that the vehicle has been properly shut down, the READY light is off, and the key fob is removed from the vehicle. Locate the 12-volt accessory battery and remove the 12-volt ground cable connection.

You need to have a certified and inspected pair of insulated HV gloves (**Figure 20-58**). These gloves need to be rated for at least 1,000 volts AC. Ensure that the gloves have been tested and certified within the last six months prior to use, as shown by the certification shown on the gloves in **Figure 20-59**. Examine the gloves for any signs of damage. One method to check for small cuts in the gloves is to inflate the glove by blowing into the hand opening and then roll the opening closed, trapping the air inside the glove. Continue to gently roll the closed glove toward the fingers as shown in **Figure 20-60**. If any air escapes, the gloves are damaged and cannot be used. If the gloves pass inspection, place a pair of leather work gloves over the HV gloves to protect the HV gloves from damage (**Figure 20-61**).

Next, locate the HV service disconnect plug or switch. This is located near the HV battery. An example of a HV

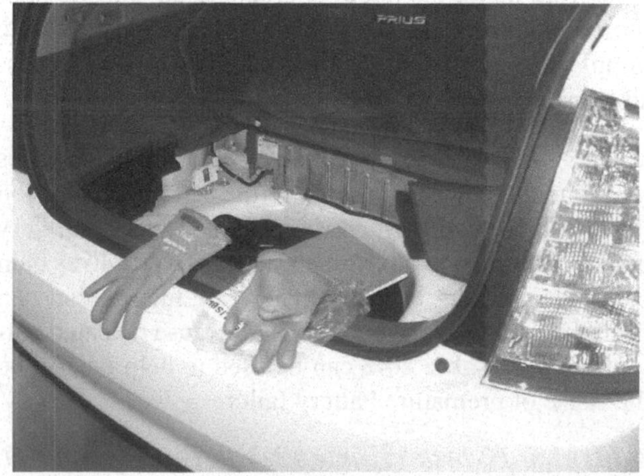

FIGURE 20-58 Special HV gloves are used when working on or near the high-voltage system.

FIGURE 20-59 HV gloves need to be inspected and certified regularly to prevent accidental electrocution.

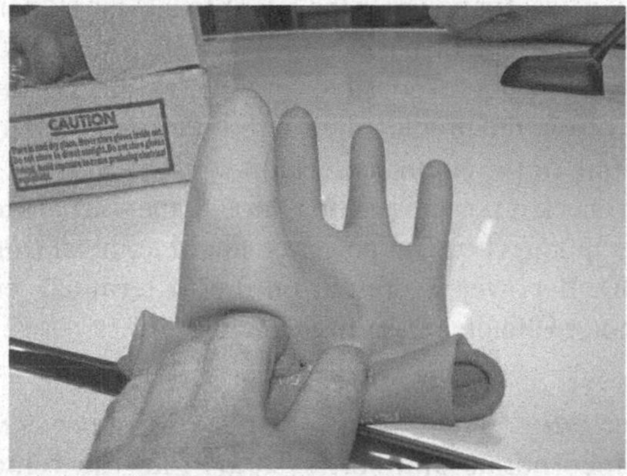

FIGURE 20-60 Checking an HV glove for leaks involves rolling it so trapped air is used to pressurize the glove. Leaks will be apparent as the air escapes.

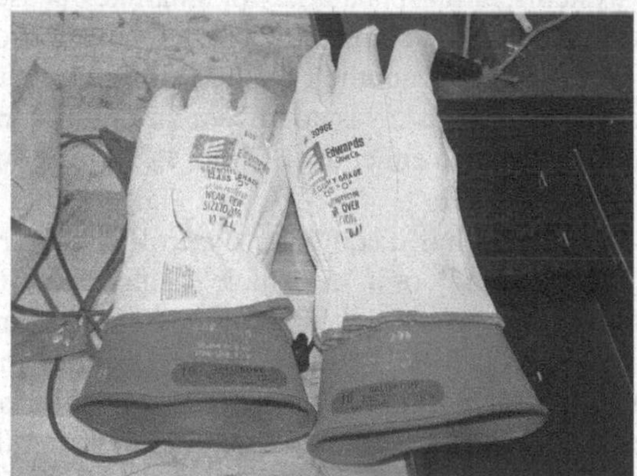

FIGURE 20-61 Because the HV gloves are not meant to be used for working on the actual components, heavy leather gloves are worn over the HV gloves to protect against cuts.

service disconnect is shown in **Figure 20-62**. Remove the plug or rotate the switch to disable the HV system. Place the service plug in your toolbox or other location away from the HV battery.

Removing the service plug separates the HV battery from the rest of the vehicle and components of the hybrid drive system. Having the plug removed also separates the HV battery in half electrically. This means that instead of having a 300-volt battery, there are two 150-volt batteries inside of the battery box. This is important because removing the service plug does not turn off or make the HV battery into a safe low-voltage battery. If the battery is removed for service or replacement, there is still plenty of voltage available to cause serious injury or death if the proper procedures are not followed.

To enable the HV system, install the service plug while you are wearing the leather-covered HV gloves. Reconnect the 12-volt battery, and verify that the vehicle is operational. There will probably be DTCs stored due to the loss of battery power. Install a scan tool and check and erase any DTCs stored by the battery and HV disconnect.

Starting System Testing and Service

On older vehicles, the starting system was a simple circuit with few components (**Figure 20-63**). Modern vehicles have complex starting circuits that include antitheft systems and remote start capability, but the basics of the starting circuit remain: the starter motor, control switches, and related wiring. Proper diagnosis of the starting system is now even more important because several control modules are used in the control of starter operation.

STARTER CIRCUITS

There are two starting systems circuits; the control circuit and the insulated power and ground circuits. Faults in the control circuit usually cause a no-crank concern.

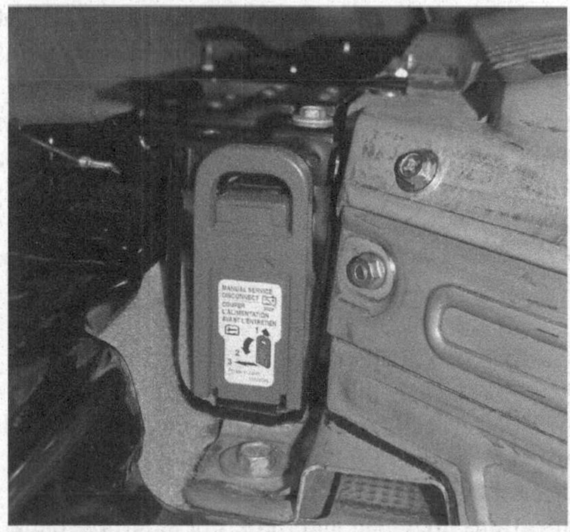

FIGURE 20-62 This HV disconnect is a common plug type, requiring the latch to be pulled up and then down and out to open the HV circuit.

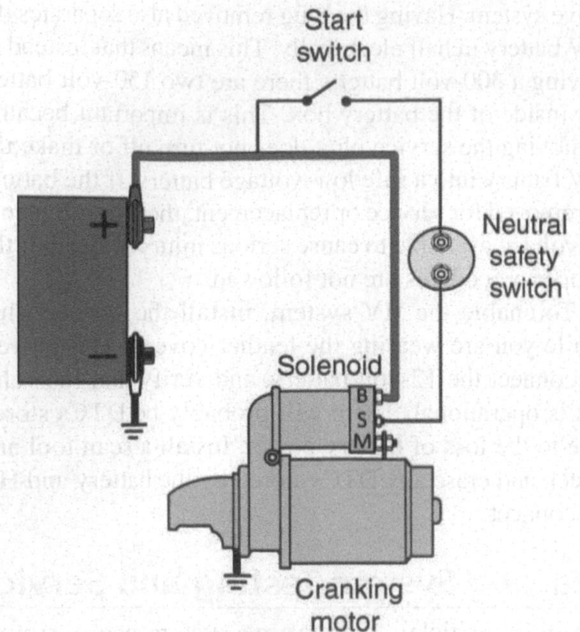

FIGURE 20-63 An illustration of a basic starting system circuit.

Problems in the insulated circuits can cause a no-crank or a slow-crank situation where the engine cranking speed is slower than normal. It is important to understand the operation of the system and to know how to diagnose these concerns.

The following sections are broken down into traditional starting systems, meaning those that are not electronically controlled by various modules and modern starting systems that incorporate antitheft, Push to Start, and onboard computers to operate.

■ ***Non-Computer Control No-Crank Testing.*** If an engine does not crank, first inspect and test the battery and battery connections. Too often battery problems are misdiagnosed and the starter gets replaced.

Once you have verified that battery and connections are good, check for power to the S terminal at the starter solenoid while trying to crank the engine

> **Service Note**
>
> As with all electrical circuits on the vehicle, everything begins and ends at the battery. First, make sure that the battery is fully charged and able to supply the current necessary to operate the starter. Inspect and verify that the battery connections are clean and tight. For side-post batteries, this may require removing the battery cables to perform a closer inspection.

(**Figure 20-64**). Either a DMM or test light can be used to check for power at the S terminal. Many technicians use a test light since it quickly can show if power is present when cranking. If there is power present at the S terminal and the starter does not crank, either the starter solenoid or the motor itself is at fault. If there is no power present, there is a problem in the control circuit. Test back through the circuit from the S terminal to the next component in the control circuit, typically the safety switch (**Figure 20-65**). Keep testing until power is found. For example, in Figure 20-65, there is no power until the supply side of the ignition switch. Because there is no power coming out of the ignition switch, the problem is in the switch itself.

If you test for power to the S terminal at the starter and power is present, then the control circuit is operating correctly and is not responsible for the no-crank condition. In this case, the problem lies in the starter motor circuit, solenoid, or the engine is seized.

To check if power is getting through the solenoid to the starter motor, check for power on the M terminal (**Figure 20-66**). If power is present on the M terminal, either the motor is faulty or the engine is seized.

■ ***Computer No-Crank Testing.*** Most newer vehicles incorporate antitheft systems into the vehicle's onboard computer systems. This often includes special keys, an antitheft module, the body control module (BCM), and powertrain control module (PCM). With these systems, a transponder key or smart key is used to authenticate that the correct key or fob is being used to start the engine. The antitheft module either approves or denies the key. If the key is approved, the antitheft module sends a signal to either the BCM or PCM to allow the engine to crank over.

If the vehicle has a computer-controlled starter circuit (**Figure 20-67**), or if the vehicle is equipped with an antitheft system, you need to determine if the system has enabled and locked out the starter motor. Begin by checking the dash for flashing security lights when the key is turned to ON and CRANK. An examples of an antitheft/security warning light is shown in **Figure 20-68**. If the security light flashes, this often indicates that the system is active and may be preventing the starter from operating. You will need to refer to the manufacturer's service information for further testing procedures.

Check the gear selector position, and, if the vehicle has push-to-start, check the brake pedal position sensor and START switch inputs on the scan tool (**Figure 20-69**). Push-to-start systems and some with a key require that the driver depress the brake pedal before the START circuit will close. If any of these inputs are

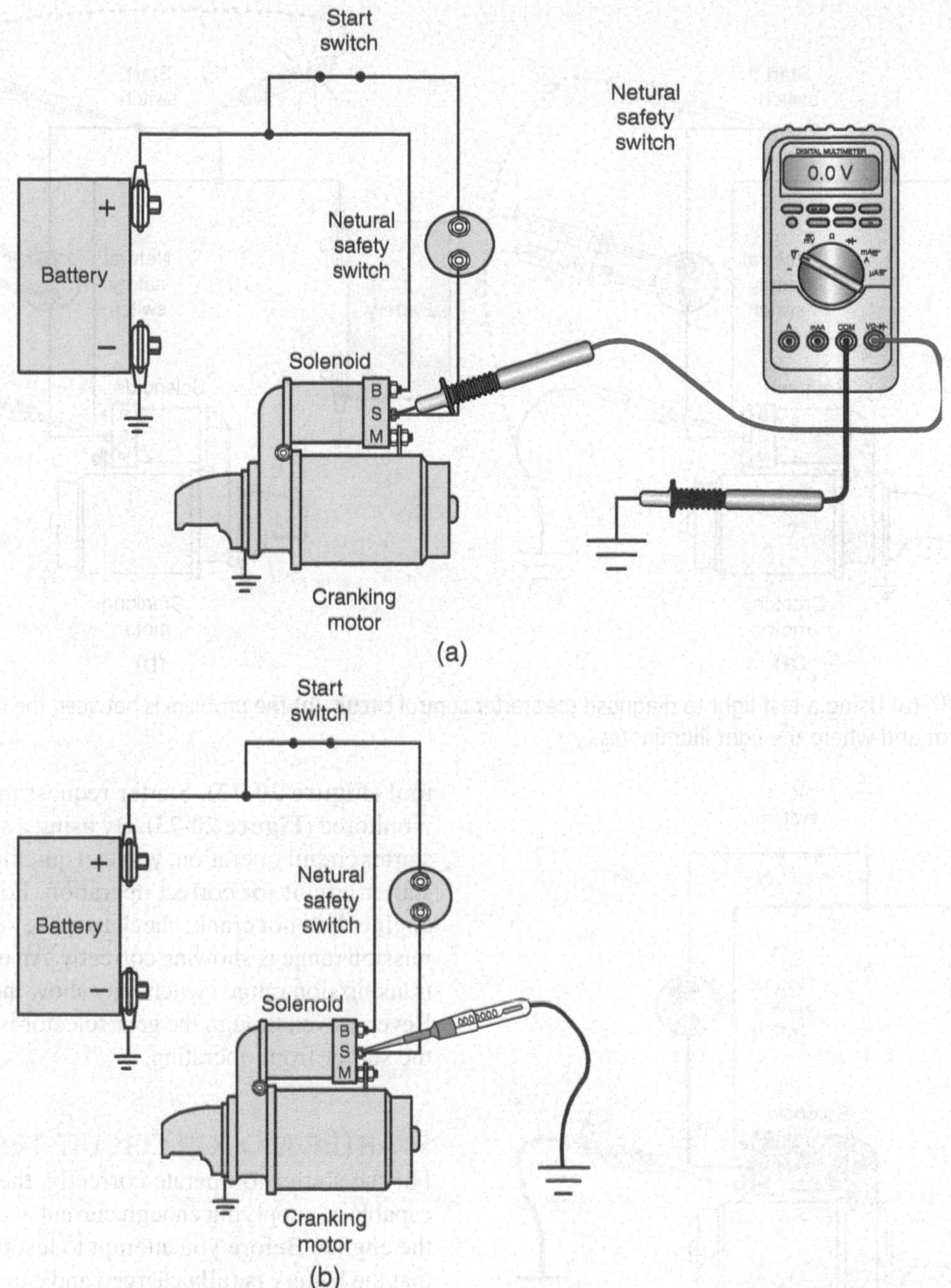

FIGURE 20-64 **(a) Checking for voltage with a DMM at the starter S terminal when cranking. No voltage means a problem in the control circuit. (b) Performing the same test with a nonpowered test light.**

missing, follow the manufacturer's diagnostic procedures to troubleshoot the circuit.

■ ***Control Circuit Component Testing.*** Park/Neutral switches, also called transmission range switches, allow the starter to operate only in the Park and Neutral positions. These switches can be tested by checking for voltage in and out of the switch (**Figure 20-70**). Voltage drop through the switch can also be checked (**Figure 20-71**).

Clutch safety switches are similar to brake light switches, normally open until the pedal is pressed. Check for power to the switch and out of the switch with the pedal pressed to the floor. The switch can be checked for voltage drop as well. Switches should have a voltage drop of 100 mV (0.100 volt) or less when closed.

To check total voltage drop of the control circuit, place the DMM positive lead on the battery positive terminal. Place the DMM negative lead on the S terminal

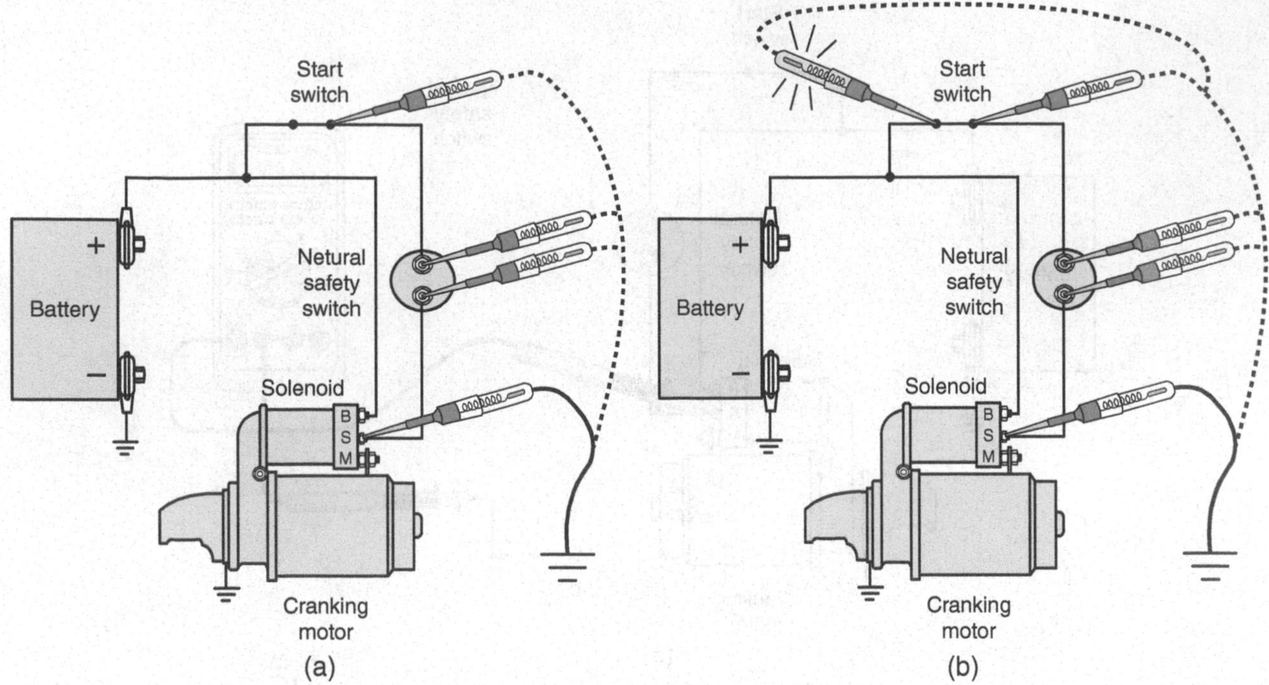

FIGURE 20-65 (a) Using a test light to diagnose the starter control circuit. (b) The problem is between the last location the light did not turn and where the light illuminates.

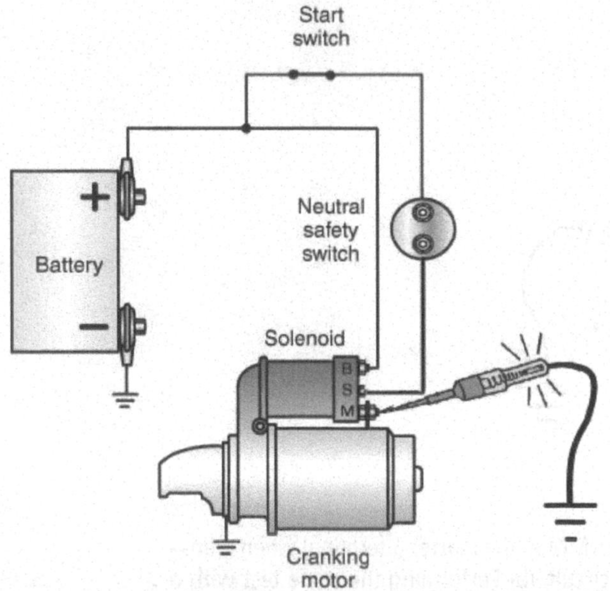

FIGURE 20-66 If the solenoid is working properly, there should be power at the M terminal when the key is in the Start position.

at the start solenoid or relay. Crank the engine for five seconds and note the reading. Typically, control circuit voltage drop should not exceed 200mV (0.200 volt). If the voltage drop is excessive, test each part of the circuit until the problem is found. Common causes of excessive voltage drop include corroded and damaged connections.

■ ***Computer Control Circuit Component Testing.*** On many vehicles, the operation of the transmission and clutch switches can be monitored on a scan tool (**Figure 20-72**). Starter request may be able to be monitored (**Figure 20-73**). By using a scan tool to check starter circuit operation, you can quickly test parts of the starter circuit for correct operation. For example, if the engine does not crank, check to make sure that the transmission range is showing correctly. An out-of-adjustment transmission range switch may show the transmission in Reverse even though the gear selector is in Park, keeping the starter from operating.

STARTER MOTOR CIRCUIT TESTING

For the starter to operate correctly, the battery must be capable of supplying enough current and voltage to crank the engine. Before you attempt to test the starter, ensure that the battery is fully charged and capable of producing the necessary amperage required by the starter.

■ ***Safety Precautions.*** Due to the high current draw of the starter circuit, the battery, battery cables, and starter can become very hot with prolonged cranking. Do not crank the engine for more than 15 seconds at a time, and allow at least two minutes for the starter to cool between tests.

■ ***Starter Cranking Current and Voltage Tests.*** To check the starter circuit operation, a voltmeter and inductive current meter are used. Testers such as the Snap-On AVR and Sun VAT series testers can measure cranking amps and volts at the same time. To test starter current draw, the tester's positive clamp attaches the battery positive connection and the negative clamp to the

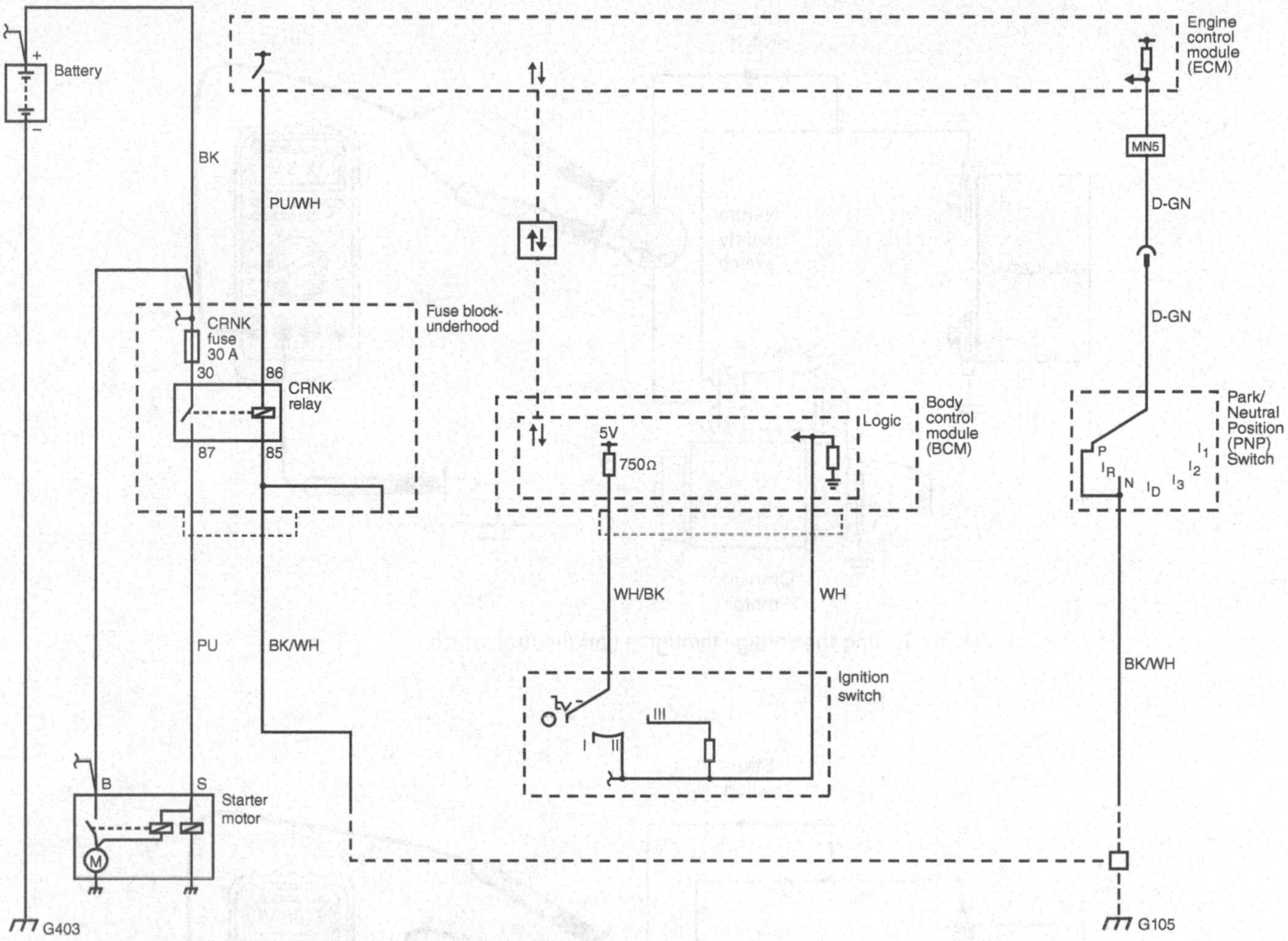

FIGURE 20-67 An example of a wiring diagram for a computer-controlled starting system.

FIGURE 20-68 A flashing security light along with an engine that will not crank indicates a problem with the antitheft system.

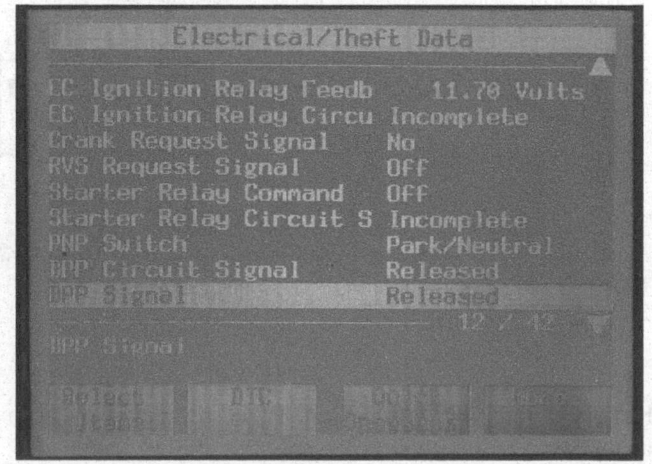

FIGURE 20-69 A scan tool can be used to check operation of the starting circuit. This example shows both the transmission gear and brake pedal position data. Both are important for modern starting systems.

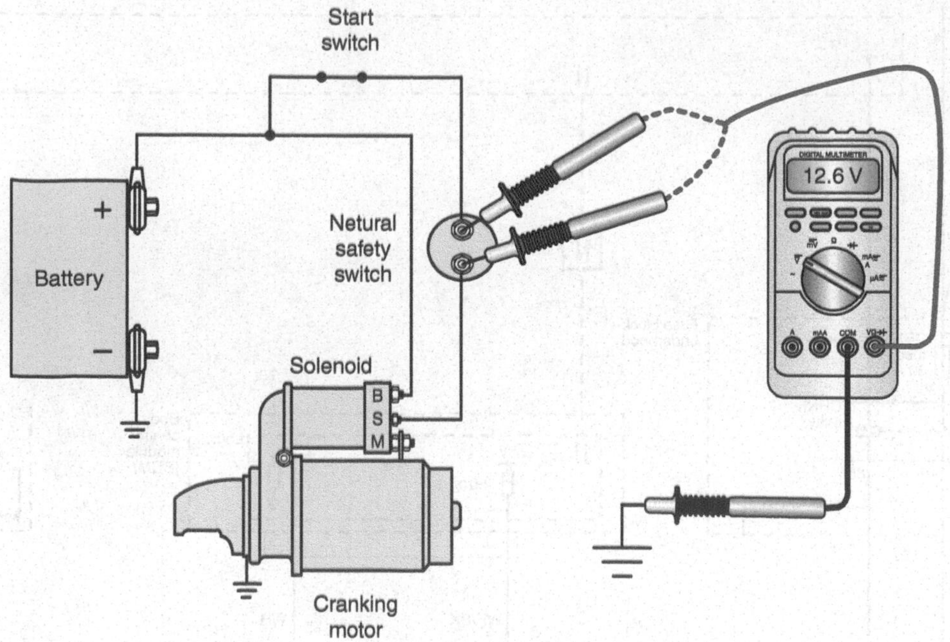

FIGURE 20-70 Testing the voltage through a park/neutral switch.

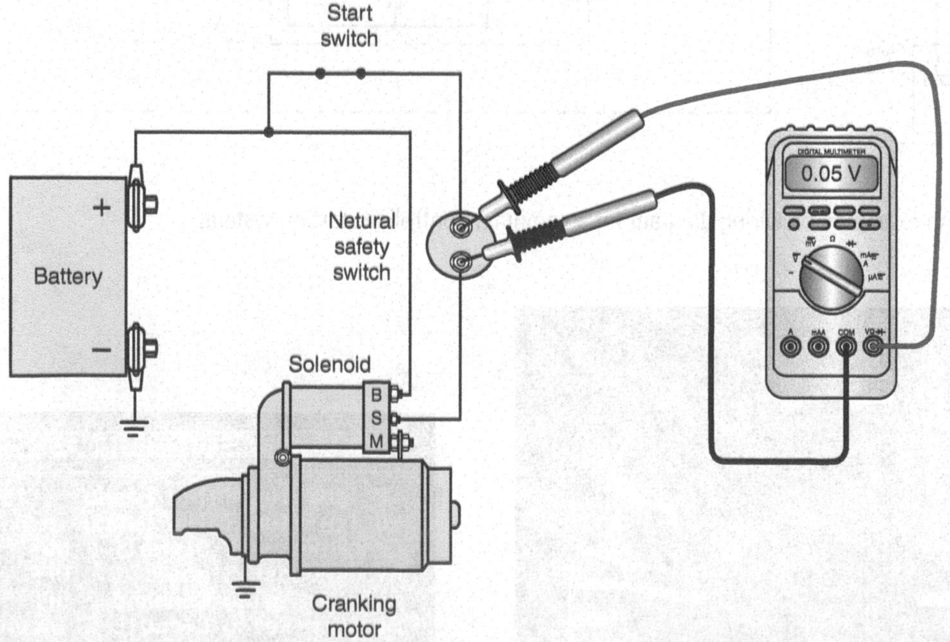

FIGURE 20-71 Testing voltage drop through the park/neutral switch. Voltage drop should be very low through a switch.

battery's negative connection. Place the current clamp around either the positive or negative battery cable connecting the battery to the vehicle (**Figure 20-74**).

Remove the fuel pump fuse or relay to prevent the engine from starting. Crank the engine for at least five seconds and note the cranking amperage and voltage. Refer to the manufacturer's specs for cranking amps. For most gasoline engines, cranking amps will be between 125 and 250 amps.

You may not find a cranking amp specification for testing on the car. Some manufacturers only provide starter no-load amperage specs. This is discussed later

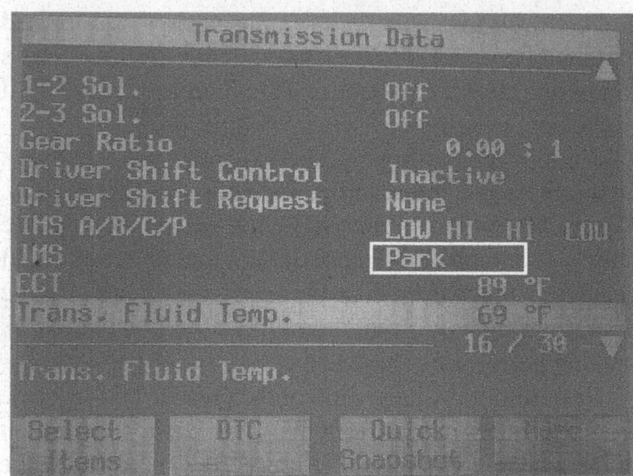

FIGURE 20-72 Checking the operation of the transmission range switch with a scan tool.

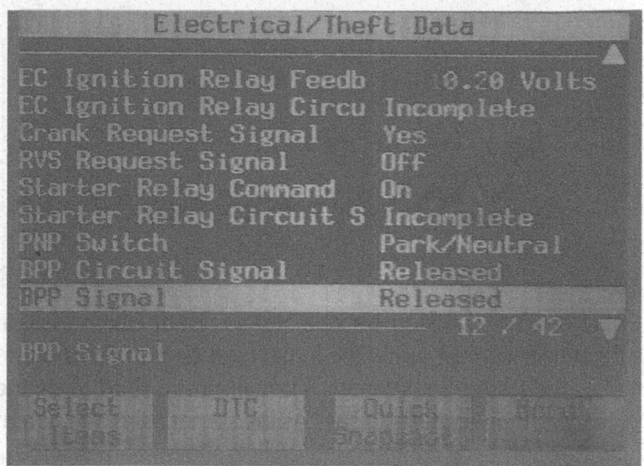

FIGURE 20-73 Checking starter or crank request with the scan tool.

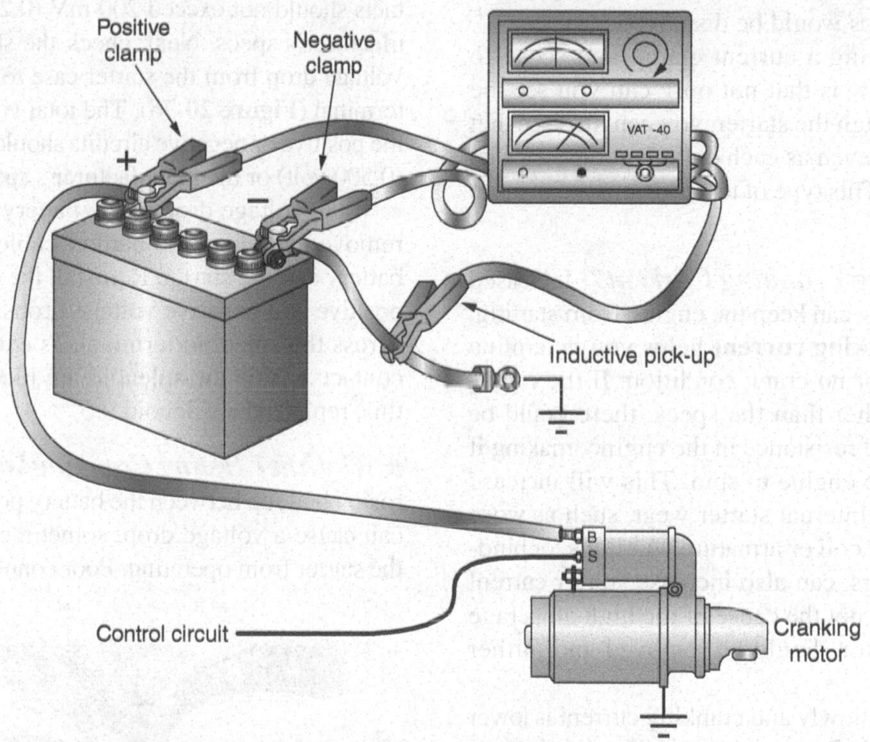

FIGURE 20-74 To test cranking volts and amps, connect the VAT to the battery positive and negative terminals and place the inductive clamp around either the positive cable from the battery to the starter motor or around the battery negative cable. In this example, the clamp is around the negative cable.

in this chapter. Cranking voltage should not fall below 9.6 volts. If voltage falls below 9.6 volts, charge and test the battery.

You may notice that cranking amperage is much higher as the starter first engages, and then the amperage drops and remains steady. This is normal and is caused by the initial current draw by the starter. Motor current is highest at zero armature rpm. Also, there is the additional resistance of getting the engine to start to rotate. The combination of these will cause the initial current flow to be high.

Figure 20-75 is a screen capture of an engine cranking and starting using a lab scope. The blue channel shows cranking current draw. The peak current draw, at the instant cranking begins, exceeds 600 amps. This is shown at the top left of the image. The scale values are given along the left edge of the image. Once the starter is spinning at normal cranking speed, current draw levels

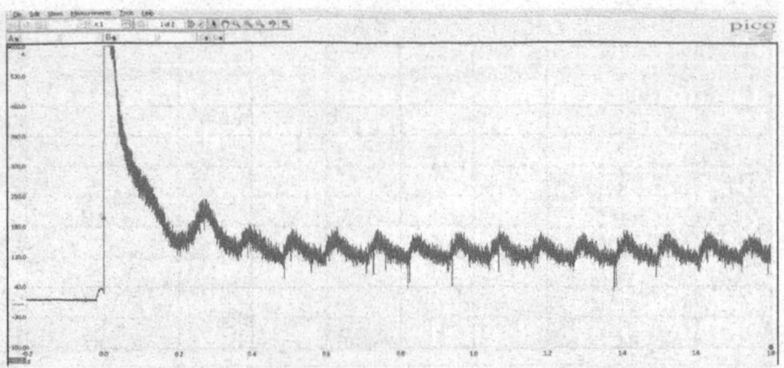

FIGURE 20-75 Using a scope and a current clamp to measure starter current draw. The initial high current draw is normal as the starter is stationary; it is really just a short to ground until the armature starts to spin. As the starter spins, the current flow through each compression cycle is shown on the scope.

out between approximately 110 and 180 amps. On the starting/charging tester shown in Figure 20-74, an average of about 145 amps would be displayed.

One benefit to using a current clamp and scope to check the starter draw is that not only can you see the current flowing through the starter, you can also see that the cranking draw is even as each cylinder completes its compression stroke. This type of test is discussed in more detail in Chapter 23.

■ ***Why Am I Testing Cranking Current?*** Increased cranking current draw can keep the engine from starting. Testing **starter cranking current** helps you determine the cause of a slow or no-crank condition. If the cranking amperage is higher than the specs, there could be increased mechanical resistance in the engine, making it more difficult for the engine to spin. This will increase starter current draw. Internal starter wear, such as worn brushes, shorted field coil or armature windings, or binding armature bushings, can also increase starter current draw. If the engine is not the cause of the high amperage draw, the starter motor should be removed and further testing performed.

If the starter spins slowly and cranking current is lower than the specs, check for poor connections at the battery and starter cables. A loose or corroded connection increases resistance in the circuit, creating an increased voltage drop. This takes power away from the starter. Check the battery ground cable connection on the engine. To accurately determine the condition of the battery and starter connections, you have to perform voltage drop tests.

■ ***Starter Motor Voltage Drop Testing.*** The starter motor circuit, because it is a high-amperage circuit, is typically allowed a slightly higher voltage drop than other circuits, up to 500 mV (0.500 volt). This higher specification is because of the large amount of current flow through the battery cables and starter. When cranking, a lot of electrons need to flow through the cables and connections. This rush of current flow and limited amount of wire for it to flow through increases the wire's temperature and resistance. This increases the circuit resistance, which increases the voltage drop.

Begin your testing by checking the voltage drops at the battery cables and posts on a top-post battery. **Figure 20-76** shows how to check voltage drop at the battery terminals. It is important to remember that to get accurate voltage drop measurements, the circuit must be operating, meaning the engine must be cranked while you are performing these tests.

Next, test the voltage drop of the positive battery cable from the battery positive terminal to the battery connection at the solenoid B terminal (**Figure 20-77**). To check the voltage drop through the solenoid, connect the DMM to the solenoid battery connection and the motor connection at the M terminal. Voltage drop through the solenoid contacts should not exceed 200 mV (0.200 volt) or the manufacturer's specs. Next, check the starter ground circuit voltage drop from the starter case to the battery negative terminal (**Figure 20-78**). The total voltage drop for either the positive or negative circuits should not exceed 500 mV (0.500 volt) or the manufacturer's specs.

If the voltage drop of the battery cables is excessive, remove and clean the battery cable connections at the battery and the starter. Reinstall the cables and retest the positive and negative voltage drops. If the voltage drop across the solenoid terminals is excessive, the internal contacts inside the solenoid are likely burnt. To correct this, replace the solenoid.

■ ***Why Am I Testing Cranking Voltage Drops?*** Corrosion buildup between the battery posts and the terminals can cause a voltage drop, sometimes enough to prevent the starter from operating. Poor connections at the battery,

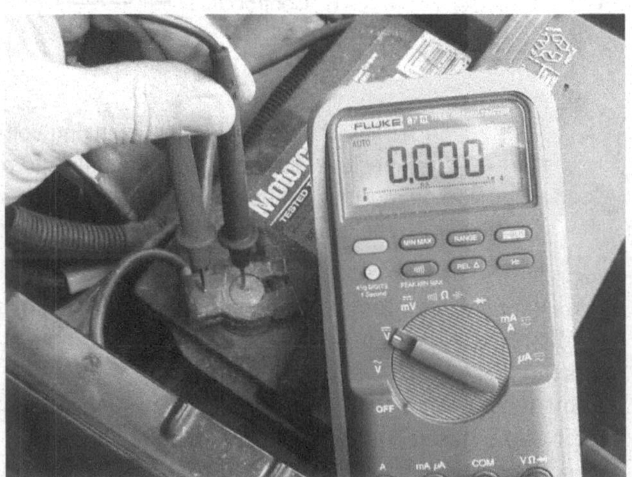

FIGURE 20-76 If the voltage supplied to the starter is low, check the voltage drop of the battery connections while cranking.

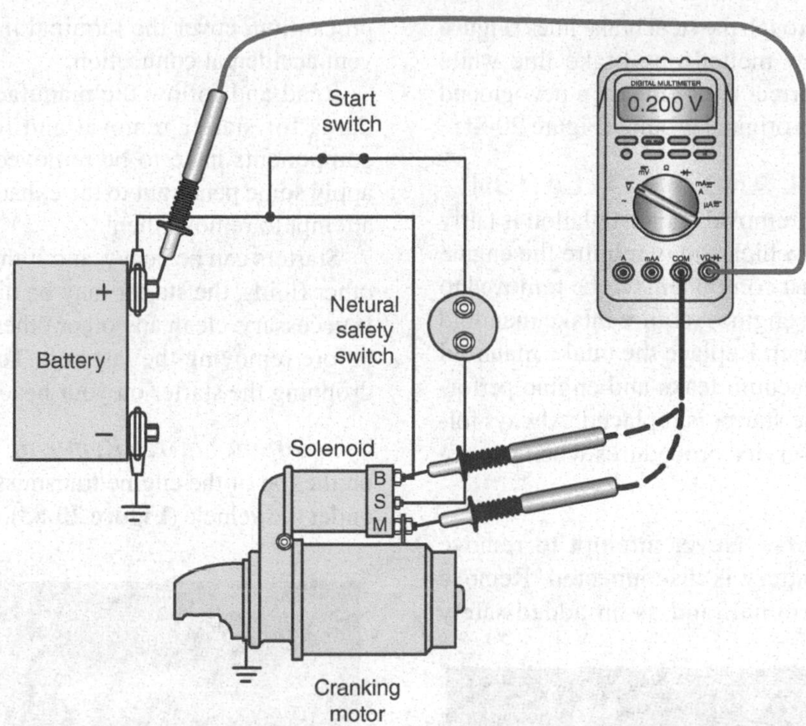

FIGURE 20-77 To test the positive starter motor circuit, place the voltmeter on the battery positive and the positive battery cable connection at the starter and crank the engine.

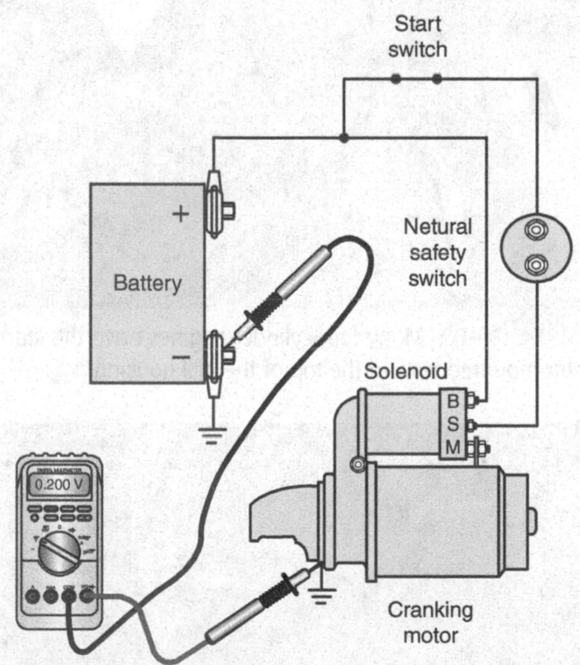

FIGURE 20-78 To check the starter ground circuit voltage drop, connect the voltmeter positive lead to the battery negative and the negative lead to the starter case and crank the engine.

starter battery cable connection, and the battery ground connections can easily be misdiagnosed as a battery or starter motor problem. Performing voltage drop tests

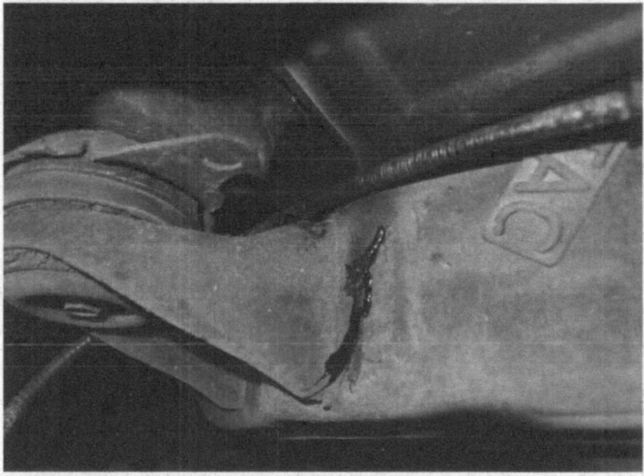

FIGURE 20-79 An example of what can happen when the proper grounds are faulty. This truck's starting circuit went to ground through the parking brake cable, eventually melting the cable.

before replacing parts helps make sure that the actual problem is found and corrected.

Excessive voltage drop in the starter circuit, because it is such a high load on the electrical system, can cause problems in other areas. Current flow will find a path possible, sometimes creating unwanted pathways to complete a circuit. **Figure 20-79** shows a melted parking brake cable from a truck with a faulty ground. The vehicle with the main ground connected to the stud shown in **Figure 20-80** has such a poor connection that when cranking the engine, the starter circuit grounded through the steel brake lines (**Figure 20-81**). The current flow melted one brake line while attempting to start. To correct the problem, a new ground had to be made beside the original ground (Figure 20-81).

STARTER REMOVAL AND INSTALLATION

On most vehicles, starter removal and installation is fairly simple. However, some vehicles may require the engine to be jacked up or exhaust components to be removed to remove the starter. Some engines require intake manifold removal to get to the starter. Replace the intake manifold gasket to help prevent vacuum leaks and engine performance problems once the starter is replaced. Always follow the manufacturer's service procedures when you are replacing the starter.

■ ***Safety Precautions.*** Never attempt to remove the starter unless the battery is disconnected. Remove the battery negative terminal, and as an added safety precaution, cover the terminal in electrical tape to prevent accidental connection.

FIGURE 20-80 A remote ground location that developed a poor connection over time and had to be relocated.

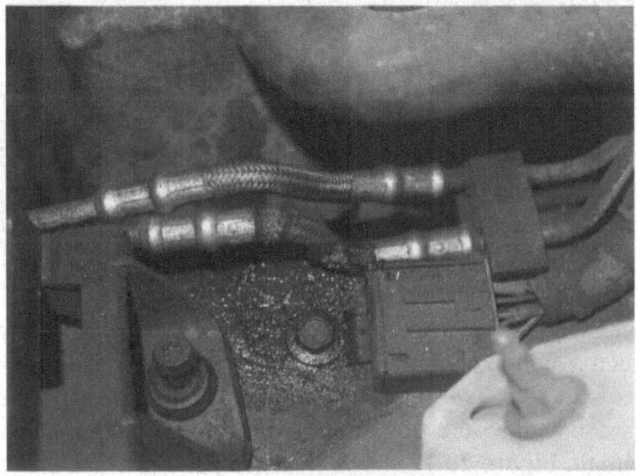

FIGURE 20-81 The starting circuit went to ground through the steel brake lines. This melted one of the lines and caused a brake fluid leak.

Read and follow the manufacturer's service procedures for starter removal and installation. If exhaust components have to be removed to access the starter, apply some penetrant to the exhaust fasteners before you attempt to remove them.

Starters can be heavy, and if they are covered in oil or other fluids, the starter may be difficult to keep hold of. If necessary, clean any oil or other liquid from the starter before removing the last bolt. This may keep you from dropping the starter on your head or feet.

■ ***Common Starter Removal.*** Starters can be located on the top of the engine/transmission (**Figure 20-82**), or under the vehicle (**Figure 20-83**). Regardless of location,

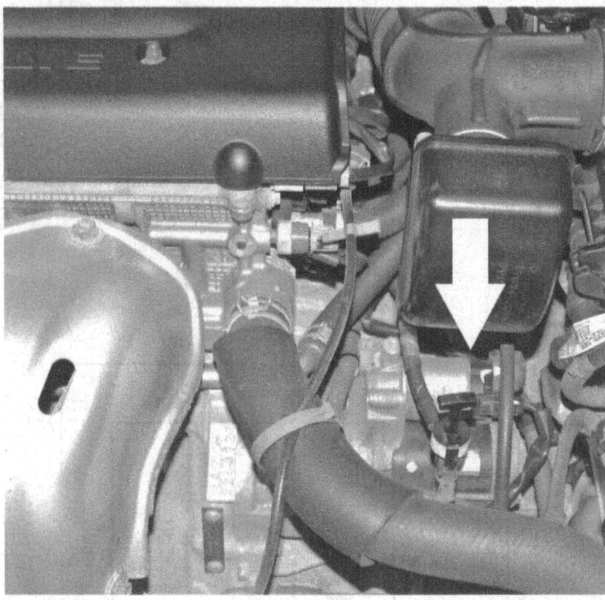

FIGURE 20-82 Many four-cylinder engines have the starter motor mounted toward the top of the bell housing.

FIGURE 20-83 Most V-type engines place the starter low on the engine.

remove the battery negative cable before you begin to remove the starter. For top-mounted starters, air intake ductwork may need to be removed to gain access to the starter. Install a fender cover to protect the vehicle, and carefully disconnect any electrical connectors, air intake hoses, and other items that may interfere with the starter removal. Next, disconnect the control circuit and positive battery connections to the solenoid. Remove the bolts securing the starter to the engine/transmission and remove the starter.

For starters mounted under the vehicle, remove the battery cable and control circuit wire first. Remove the starter bolts slowly, keeping the starter supported as the last bolt is removed. Do not let the starter hang by either the battery or control circuit wires. Once all the bolts are removed, carefully pull the starter out of the bell housing and place the starter on a workbench.

Inspect the battery cable and control circuit wiring for signs of corrosion or damage. Inspect the starter bolts and the threads in the engine/transmission for damage. If any threads are damaged, you need to either correct the damage or replace the fasteners before reinstalling the starter. If the starter bolt threads are very rusty, clean the threads with a wire brush. If necessary, use a tap to clean the threads in the engine or transmission before you reinstall the starter.

■ ***Inspect Flywheel/Ring Gear.*** Once the starter is removed, inspect the teeth of the flywheel or ring gear. Damaged teeth can cause incomplete starter engagement (**Figure 20-84**). This means the starter may spin but not actually crank the engine. Severely worn or damaged teeth cannot be repaired and require replacing the flywheel or ring gear.

Damaged flywheel/ring gear teeth can be caused by the driver engaging the starter after the engine is running or by incorrect starter installation. Some starters, such as on GM products, use a shim (**Figure 20-85**), to set the clearance between the starter drive gear and the flywheel/ring gear. If the clearance is not correct, damage to the starter drive and the flywheel/ring gear teeth can result. **Figure 20-86** shows how starter drive gear to flywheel/ring gear clearance is checked. If the clearance is too small, shims are added to increase the space between the starter drive gear and the flywheel. If the clearance is too large, shims are removed.

■ ***Starter Bench Testing.*** Vehicle manufacturers often only provide starter no-load current draw specs instead of on-vehicle cranking current specs. This is because starter current draw is affected by the battery, battery connections, battery cables, cable connections, and engine condition. A poor battery cable connection can affect the starter current draw, leading you to think that there is a problem with the starter when there actually is not. No-load current is checked with the starter removed from the vehicle. This is often called a starter bench test.

Place the starter in a bench vise and connect a fully charged battery or jumper pack and a remote starter switch to the starter (**Figure 20-87**). Connect an inductive ammeter or starting/charging system tester, and use the remote starter switch, engage the starter for five seconds. Note the cranking current draw and compare the reading to specs. No-load cranking current is often 60 to 100 amps, depending on the starter. If the current draw is excessive, suspect worn starter brushes or a binding armature. High current draw usually indicates that the starter needs to be replaced.

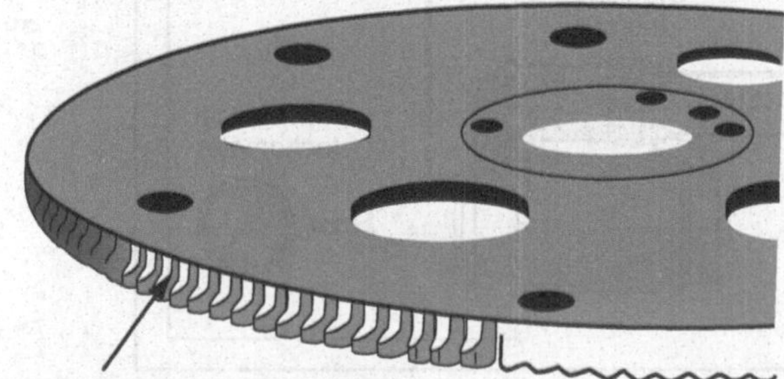

FIGURE 20-84 With the starter removed, inspect the teeth on the ring gear/flywheel for wear and damage. Damaged teeth can cause a crank no-start condition.

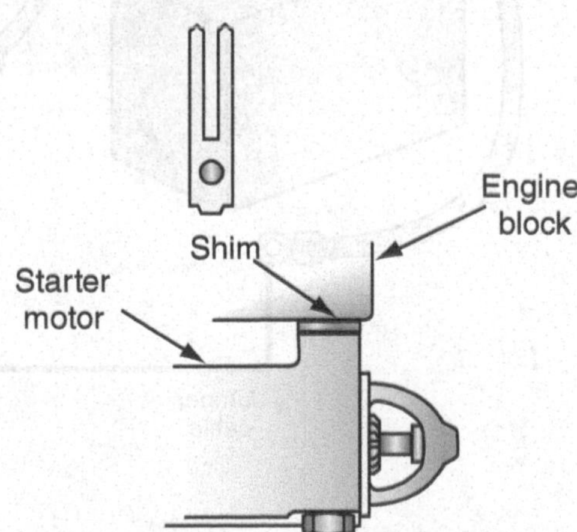

FIGURE 20-85 Many GM engines use a shim between the starter and the engine block to set the correct gear clearance. If a shim is in place when replacing a starter, reinstall the shim with the new starter and check gear clearance.

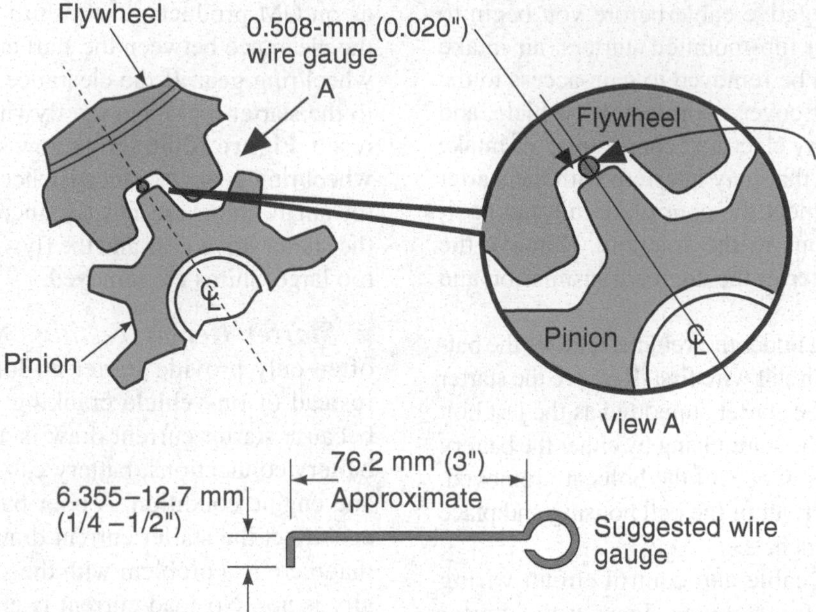

FIGURE 20-86 A special tool is used to measure the clearance between the starter drive gear and ring gear teeth, though many technicians use an Allen wrench instead.

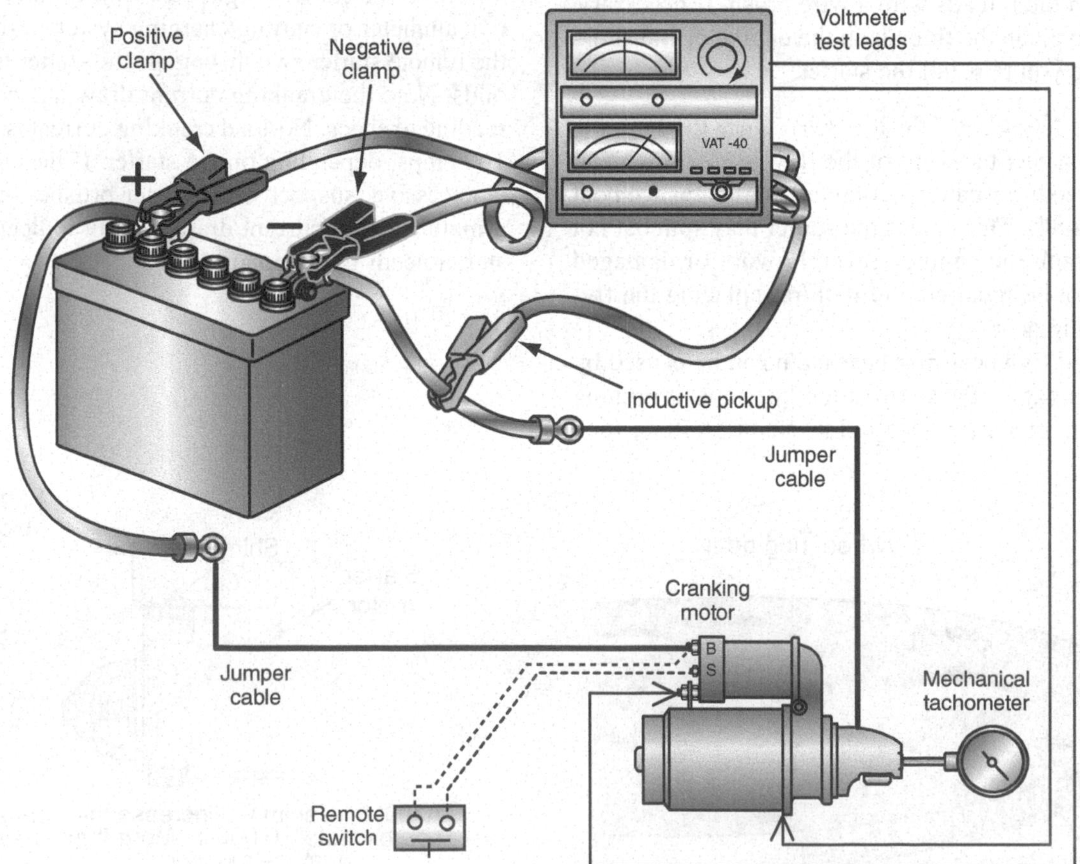

FIGURE 20-87 An illustration of how to connect a starter for bench testing. Many manufacturers only provide starter bench test or no-load testing specifications.

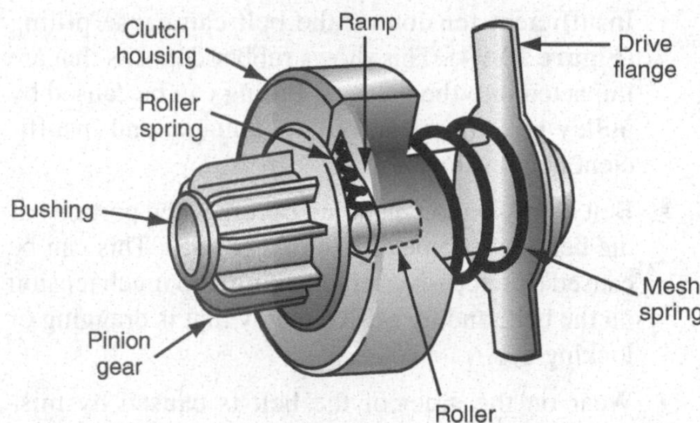

FIGURE 20-88 Overrunning clutches allow the drive gear to spin freely one way yet remain locked to the armature when spun the other way. A failed overrunning clutch can cause a starter to spin but not crank the engine.

■ ***Why Am I Bench Testing the Starter?*** This test more accurately checks the current used by the starter motor. For example, if a starter, while still installed on the engine, is drawing twice as much amperage as is normally seen for similar vehicles, this could lead you to think there is a problem with the starter motor. If you remove and bench test the starter and the current draw is well within specifications, then you need to examine why the draw is normal when it is tested off the vehicle. Most likely there is a problem with the engine, which is causing the starter to require more amperage to operate.

■ ***Inspect Starter Drive Gear.*** The starter drive, also called *the starter Bendix gear*, is mounted on the armature along a series of spiral grooves. As the drive gear moves out toward the starter's nose, along the grooves, it meshes with the flywheel/ring gear. The drive gear uses a one-way clutch to lock the gear to the armature shaft when it is spinning in one direction (**Figure 20-88**). This clutch allows the drive gear to rotate freely in the opposite direction. This is to prevent damage to the starter once the engine starts. The starter drives the flywheel/ring gear at approximately 400 rpm, but once the engine starts, the flywheel may spin 1,000 rpm or more. If the flywheel were allowed to drive the starter at such a high speed, the starter would be seriously damaged. The one-way clutch allows the drive gear to counter rotate freely, preventing damage.

Check the starter drive gear for signs of wear on the teeth. Turn the drive gear in both directions (**Figure 20-89**). In one direction, the gear should be locked to the armature. When spun in the opposite direction, the drive gear should spin freely without turning the armature. If the drive gear spins in both directions, the internal clutch has failed and the drive gear needs to be replaced.

■ ***Starter Installation.*** Carefully set the starter back into the opening in the bell housing and start the bolts by hand. If the starter requires shims, reuse the shims that were installed with the original starter. Locate the torque specs for the starter bolts and tighten to spec. It is important to torque the bolts to specs as both the starter and bell housing are often aluminum. Overtighting the bolts can strip the threads, damage the starter housing or even crack the bell housing. If the bolts are left loose, they will eventually vibrate loose. Reattach the battery cable and control circuit wiring. Do not overtighten the battery cable connection as the studs in the solenoid are easily broken.

For both topside and underside starters, make sure that the battery cable, control circuit wiring, and any other wires connected to the starter are not touching any metal or other components. Ensure that any wiring on the starter positive connection cannot touch the control circuit connection. This will cause the starter to engage as soon as the battery is reconnected.

Reconnect the battery negative cable and reconnect the starting system tester to the battery and battery cable. Perform several engine starts while watching the cranking voltage and amperage. Ensure that cranking amperage is within the manufacturer's specs and that the starter operates correctly.

Charging System Testing and Service

As with many systems, charging system diagnosis now often involves using a scan tool to check for diagnostic trouble codes (DTCs) and data parameters related to the charging circuit, as most generators are now controlled by the engine control module or PCM.

GENERATOR TESTING

Generator tests include charging current and voltage, but also performing voltage drop tests on the generator circuits and tests of the onboard computer system.

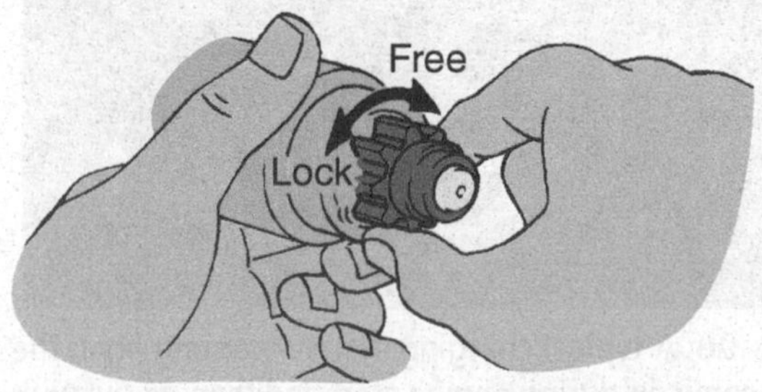

FIGURE 20-89 Checking a starter drive gear.

■ ***Visual Inspection.*** Begin the charging system tests with a visual inspection of the charge indicator or warning light on the instrument panel. The charge warning light should illuminate during the key-on bulb check and engine start and then go out. If the light remains on, a problem in the charging system is indicated. The charging system warning light is usually a picture of a battery (**Figure 20-90**). Some vehicles will display GEN, BAT, BATTERY, or even CHARGING SYSTEM FAULT on the dash. Vehicles with a volt gauge may turn on a CHECK GAUGES light. If a warning light is illuminated with the engine running and you are not sure of its meaning, refer to the vehicle owner's manual.

Perform a visual inspection of the belt, generator, and wiring. Inspect the battery and battery connections. Look for any aftermarket wiring and accessories, especially stereo systems. Improperly installed accessories can affect many aspects of the vehicle's operation, including the charging system. An aftermarket stereo, particularly with a high-wattage amplifier, can draw enough power to discharge the battery at engine idle speeds.

■ ***Generator Drive Belt.*** Because the generator is driven by an accessory drive belt, thoroughly inspect the belt for wear and damage. Some examples of types of belt wear are as follows:

- Severe cracking (**Figure 20-91**) is common on belts made of neoprene. As the belt ages and wears, cracks develop along the ribs. When cracks exceed four per inch per rib, the belt needs to be replaced.
- Newer belts are made of different materials and do not crack like neoprene belts. Instead, you have to look more closely at the belt to determine how much it is worn. As the belt wears, the ribs become narrower (**Figure 20-92**). To check the amount of wear on the belt, use a belt wear gauge (**Figure 20-93**). If the gauge fits flush with the ribs or below, the belt is worn and should be replaced.

FIGURE 20-90 A typical charging system warning light. The light should come on during engine start and then go out with the engine running.

- Insufficient tension on the belt can cause pilling (**Figure 20-94**). This shows rubber deposits that are impacted into the grooves. Pilling can be caused by pulley misalignment, pulley damage, and insufficient belt tension.
- Belt chunk-out or chunking occurs when portions of the belt have come off (**Figure 20-95**). This can be caused by excessive belt cracking, too much tension on the belt, and by a drive pulley that is dragging or locking up.
- Wear on the sides of the belt is caused by misalignment problems (**Figure 20-96**). Check for mismounted accessories that are forcing the belt sideways, as illustrated in **Figure 20-97**.

FIGURE 20-91 Older neoprene belts are prone to cracking. When cracking becomes excessive, as shown here, the belt needs to be replaced.

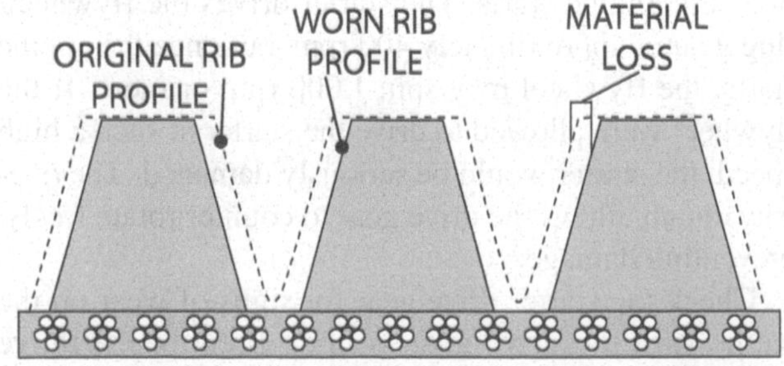

FIGURE 20-92 As the belt wears, the belt will ride lower into the pulley. This reduces the contact between the belt's ribs and the pulley.

FIGURE 20-93 Using a belt wear gauge to check a belt.

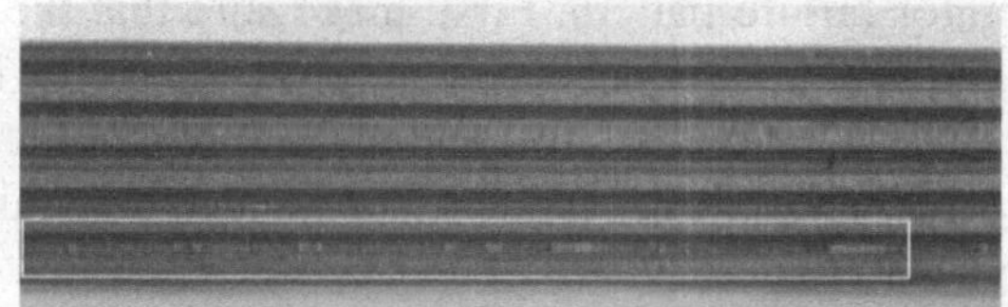

FIGURE 20-94 Pilling is the depositing of belt material in the grooves of the belt. This indicates one or more problems in the belt drive system.

FIGURE 20-95 Chunking occurs when large portions of belt material come off the belt.

When possible, check drive belt tension with a belt tension gauge (**Figure 20-98**). For this type of gauge, squeeze the handle and place the belt into the gauge (**Figure 20-99**). Release the handle and read the tension on the scale. The belt tension specifications vary depending on the size of the belt and whether it is a new or used belt. Refer to the vehicle manufacturer's service information for specifications.

FIGURE 20-96 Wear on the sides of the belt indicates an alignment problem.

If the tensioner cannot provide sufficient belt tension, the belt can slip during periods of high load. This includes periods requiring high generator output. A slipping belt reduces generator output and can lead to an undercharged battery. In addition, belt alignment should be checked. Even a slight amount of misalignment, as little as $\frac{1}{16}$ in., can result in belt noise and wear.

■ ***Battery Inspection.*** Just as when diagnosing a starting system complaint, perform a complete battery inspection when you are checking the charging system. Inspect battery cables and connections, check open circuit voltage, and perform a conductance test. It

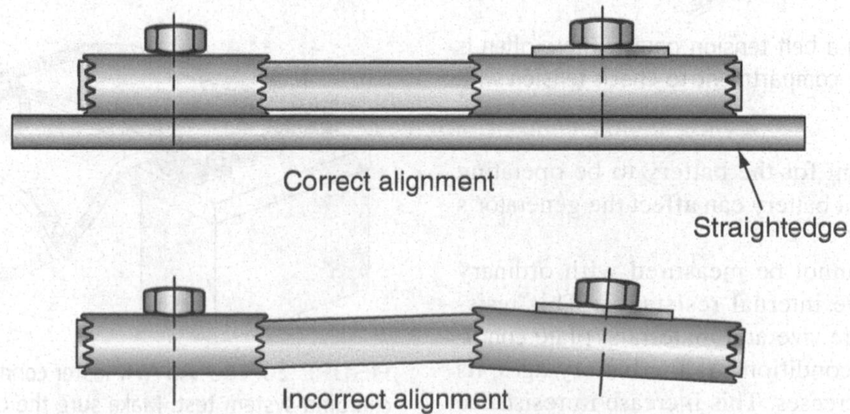

FIGURE 20-97 An illustration of belt misalignment.

FIGURE 20-98 An example of a belt tension gauge. Depending on the type of gauge, the reading will show how much tension is on the belt.

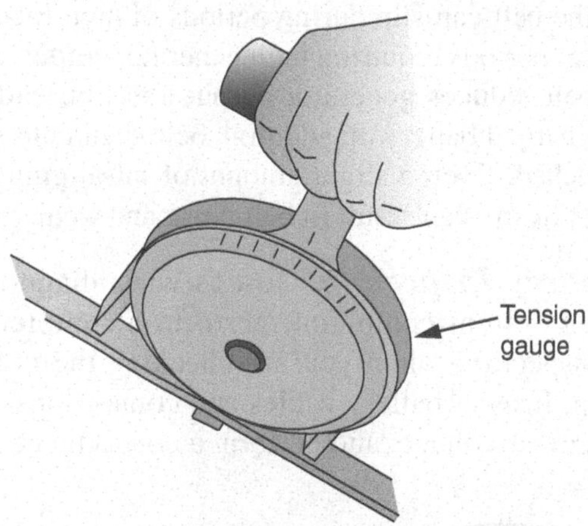

FIGURE 20-99 Using a belt tension gauge. There often is not room in the engine compartment to check tension with this type of gauge.

is extremely important for the battery to be operating properly, as a marginal battery can affect the generator's performance.

Even though it cannot be measured with ordinary meters, batteries have internal resistance. This resistance depends on plate size and materials, plate condition, and electrolyte condition. As the battery ages, its internal resistance increases. This increase in resistance can cause the generator to have to work harder to keep the battery charged. Over time, this will decrease the life of the generator. Diagnosing a battery resistance concern usually requires performing a conductance test and a three-minute charge test. If the battery internal resistance is high, the conductance tester will usually fail the battery. Unfortunately, this type of tester does not tell you why a battery fails, just if it does. The three-minute charge test is a good indicator of the internal battery condition, as discussed earlier in this chapter. Though not common, it is possible for an internal cell connection in a battery to intermittently open due to vibration. This can cause momentary drops in battery voltage, which causes the generator to increase output.

If you are presented with a vehicle with repeated generator failures and all of the other causes have been eliminated, you may need to replace the battery to cure the generator failure pattern. First, make sure that the battery and generator connections are good and do not have excessive voltage drop. Test and make sure that the battery passes a load test, conductance test, and three-minute charge test.

■ ***Charging System Output Voltage.*** Charging voltage varies depending on load, engine speed, and the vehicle being tested. Refer to the manufacturer's specifications for the charging system voltage output. Generally, charging voltage should be at least 0.5 volt above battery voltage and should not exceed 15.5 volts.

Connect a DMM or starting/charging system tester to the battery (**Figure 20-100**). If you are using a starting/charging system tester, place the inductive clamp around the positive battery cable to the starter with the arrow pointing toward the battery or around the negative cable with the arrow pointing away from the battery. Start by checking output voltage at idle with all accessories turned off. For most systems, voltage should be at least

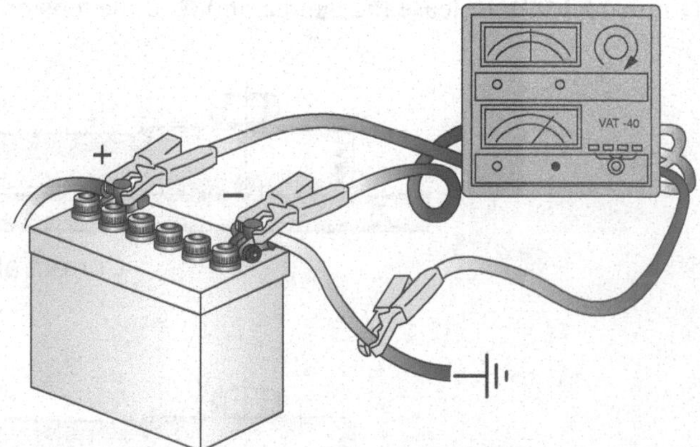

FIGURE 20-100 An AVR tester connected to a vehicle for a charging system test. Make sure the current clamp is around either a battery cable or the charging output wire on the generator.

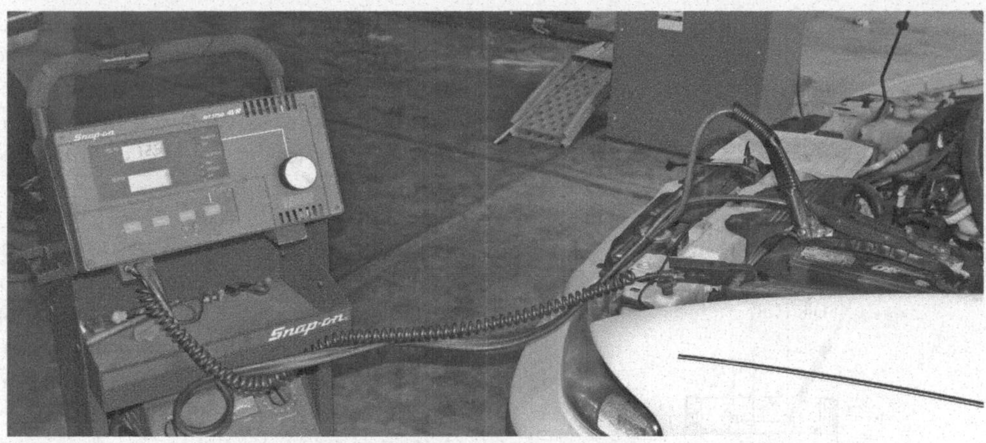

FIGURE 20-101 An AVR tester connected to a vehicle for a charging system test.

13.2 volts. Turn on the headlights and note the voltage. Bring the engine speed up to 2,000 rpm and note the voltage. Charging voltage should not have decreased with the headlights on and should increase as the rpm increases.

Lower-than-specified charging voltage can be caused by internal generator faults, such as open stator windings, a faulty voltage regulator, and faulty diodes. Excessive charging voltage can be caused by a faulty voltage regulator or a faulty field control circuit.

■ ***Charging System Output Amperage.*** Charging system output amperage also varies depending on load, engine speed, and the particular vehicle being tested. Refer to the manufacturer's specifications for charging system amperage output. An inductive ammeter or starting/charging tester is necessary to check output amperage.

As a rule, the generator should be able to produce 90% of its rated output, meaning a 100-amp generator should be able to produce 90 amps. Lower-than-specified charging amperage can be caused by internal generator faults, such as open stator windings, a faulty voltage regulator, and a faulty generator control circuit.

■ ***Output Testing with an AVR or VAT Tester.*** Starting/charging system testers, such as the Snap-On AVR and Sun VAT testers, can be used to load the generator and check both the voltage and the amperage output. Connect the tester as shown in **Figure 20-101** and note charging output at idle with accessories off. Refer to the manufacturer's service specs for the actual voltage and amperage output specs and at what engine speed output should be tested.

Generally, generator tests are done by bringing the engine speed up to 2,000 rpm and noting the output. Load the generator until the maximum amount of amperage is obtained and note the voltage. The generator should be able to produce 90% of its rated output without the voltage dropping below 12.6 volts. If the voltage drops below 12.6 volts during this test, the battery is either discharged or defective. In some cases, a battery that cannot maintain voltage affects the output of the generator. You need to replace the battery and retest the generator to accurately determine the condition of the charging system.

> **TECH INFO**
>
> Be aware that some charging systems turn the generator off at idle with little electrical load. This improves idle quality and saves fuel. Before condemning a generator because it does not charge it idle, be sure you understand how the charging system works.

■ ***Generator Voltage Drop Testing.*** Typically, the generator output and ground circuits are tested for voltage drop. To check the voltage drop of the output circuit, connect the positive lead of a DMM set to DC volts to the charging output terminal on the generator and the negative DMM lead on the battery positive terminal (**Figure 20-102a**). Check the voltage drop first at idle with all the accessories off. The voltage drop should be very low, less than 200 mV (0.200 volt). Next, turn on several accessories and note the voltage drop. The voltage drop may increase slightly as the charging system output increases. Finally, bring the engine speed up to 2,000 rpm and, using an AVR or VAT tester, load the charging system to its maximum output and note the voltage drop. The voltage drop should increase but should not exceed 500 mV (0.500 volt) or

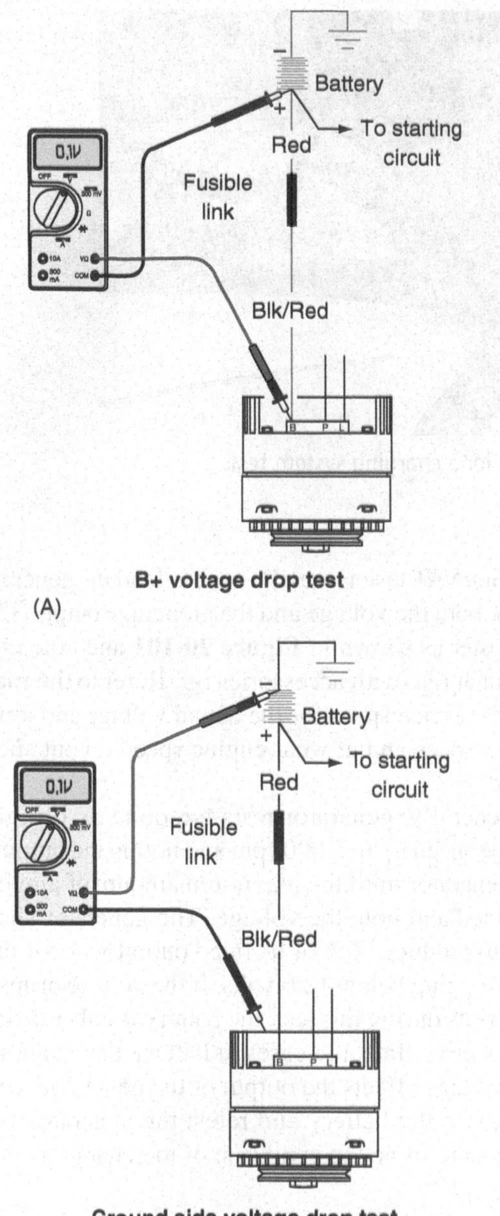

FIGURE 20-102 (a) An illustration of how to check charging system output circuit voltage drop. (b) How to check charging system ground circuit voltage drop.

the manufacturer's specs. If the voltage drop is excessive, check the battery positive connections at the battery and the generator output terminal. If the generator has a voltage-sensing circuit, check the voltage drop of this circuit under no-load and under load conditions. If excessive voltage drop is present in the sensing circuit, the generator output will be affected.

Generator ground circuit voltage drop is performed the same as the charging output circuit voltage drop, except the positive DMM lead is placed on the outside of the generator's case (**Figure 20-102b**). Check the ground circuit voltage drop at idle and under load at 2,000 rpm. The voltage drop should not exceed 500 mV (0.500 volt) or the manufacturer's specs. If the voltage drop is excessive, check the battery negative cable connections at the battery, engine ground, and body ground.

■ ***Why Am I Testing Generator Voltage Drop?*** A vehicle can have a no- or under-charge condition that can be misdiagnosed as a defective generator. This is due to excessive voltage drop in a connection. If the voltage drop is at the generator output terminal, overheating and damage to the terminal and cable can result. Excessive voltage drop at the field control or sensing circuit connections can cause charging output to be very low or high depending on the circuit.

■ ***AC Voltage Leak Testing.*** Another problem that should be tested for is AC voltage from the generator. A very small amount of AC voltage, less than 500 mV AC (0.500 volt AC) is usually acceptable. To test for AC, place the DMM on the VAC scale, and place the positive meter lead on the generator's output terminal and the negative meter lead on the battery negative terminal. Run the engine at idle and at 2,000 rpm and note the readings. Next, connect a VAT or AVR tester and repeat the idle and 2,000 rpm test but with a load applied to the charging system. Note the highest AC readings. A high AC voltage reading is caused by defective diodes in the generator.

■ ***Generator Control Circuit Testing.*** Generator control may be performed by an internal electronic voltage regulator or may be controlled by the engine control module.

Generator field control is often done by pulsing the field circuit off and on very quickly. Some systems, like those on General Motors vehicles, turn the field off and on 400 times per second. This is called a pulse-width modulated control signal. **Figure 20-103** shows the varying amount of off and on time of the pulses in a pulse-width modulated signal. The blue trace is the current output of the generator. When the signal is on, current flows through the field circuit. This generates the magnetic field in the coil and pole pieces, which results in AC being induced in the stator windings. The longer that the signal is on, the more current flows through the field. This results in more generator output.

This circuit control can be checked with a lab scope connected to the field control wire and ground (**Figure 20-104**). The red trace is the voltage signal to the generator from the PCM. As electrical demand increases, the on time of the signal should increase, allowing for increased generator output.

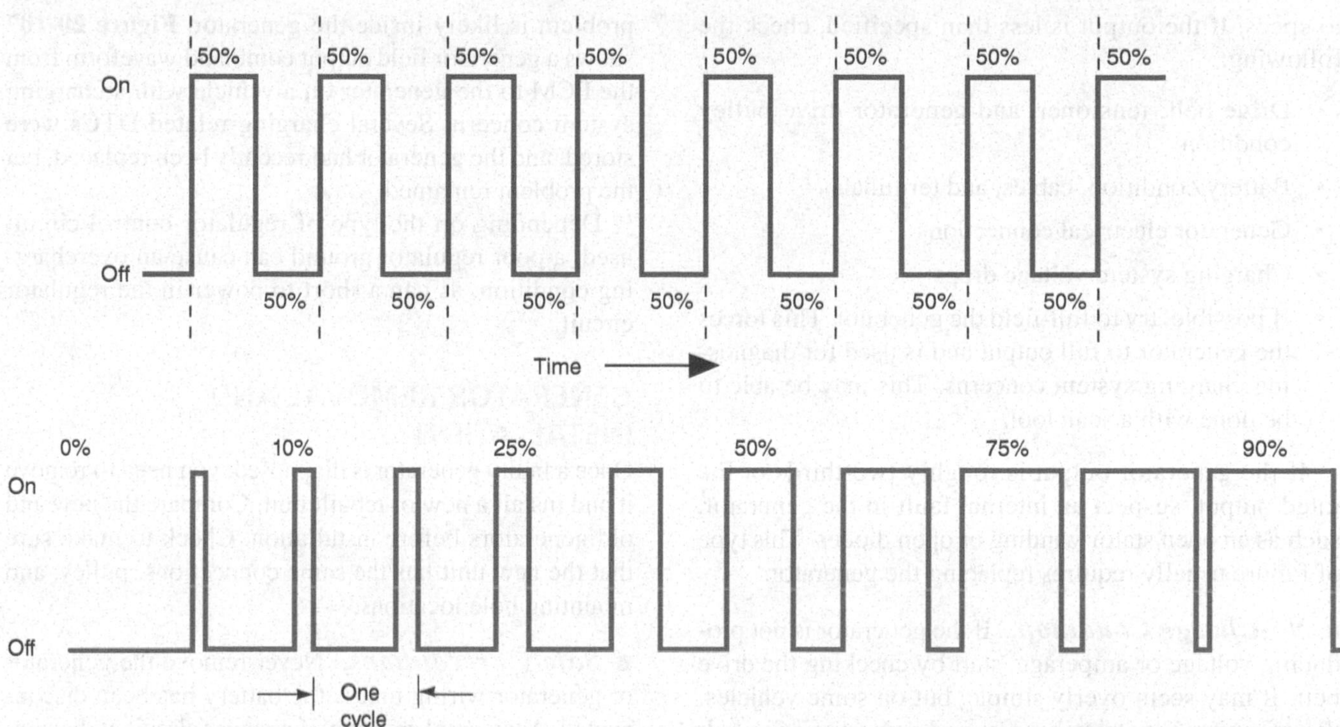

FIGURE 20-103 Pulse-width modulated signals vary the amount of time the pulse is on. In the upper image, the signal is on 50%. This would mean a generator capable of putting out 100 amps would be producing about 50 amps at 50% on time.

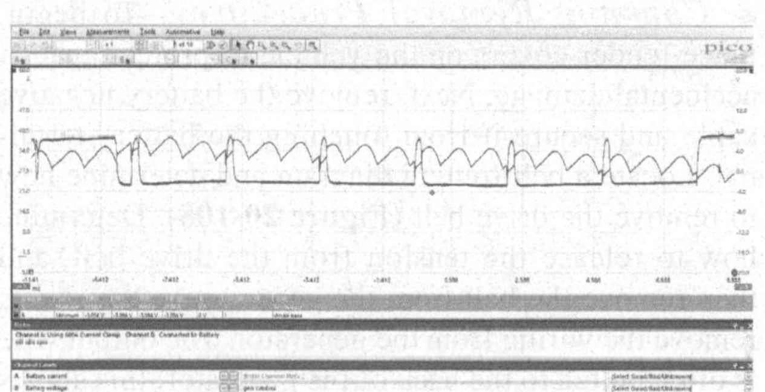

FIGURE 20-104 The red pattern is an example of a charging control signal from a computer to the generator. The blue pattern shows the actual output of the generator as tiny peaks of AC voltage. The overall result is an output of DC voltage.

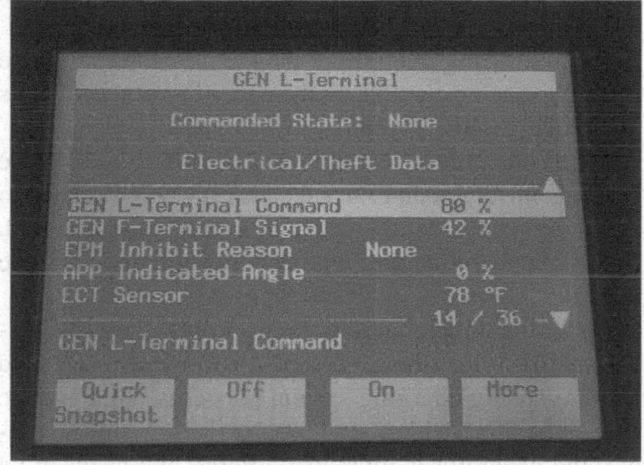

FIGURE 20-105 An example of using scan data to check charging system operation.

With most vehicles, the generator output can be monitored with a scan tool (**Figure 20-105**). In this example of scan data, the generator is commanded on at 80%, meaning the generator is commanded to produce about 80% of its output capacity.

DIAGNOSE UNDERCHARGE, NO-CHARGE, AND OVERCHARGE CONDITIONS

There are three possible charging system failure types: undercharging, no-charging, and overcharging. Of these, the first two are the most common concerns.

■ ***Undercharge Condition.*** An **undercharge condition** is one in which the charging system is producing voltage and amperage but at an insufficient rate to keep the battery charged. Begin by checking the charging system warning light on the dash or the voltage gauge as this can be an indicator of system performance. Next, perform a complete visual inspection of the battery, the charging system components, and look for any aftermarket accessories—especially stereos and amplifiers.

Check charging system output with an AVR or VAT tester and compare the voltage and amperage output to specs. If the output is less than specified, check the following:

- Drive belt, tensioner, and generator drive pulley condition
- Battery condition, cables, and terminals
- Generator electrical connections
- Charging system voltage drops
- If possible, try to full-field the generator. This forces the generator to full output and is used for diagnosing charging system concerns. This may be able to be done with a scan tool.

If the generator output is roughly two-thirds of the rated output, suspect an internal fault in the generator, such as an open stator winding or open diodes. This type of failure usually requires replacing the generator.

■ ***No-Charge Condition.*** If the generator is not producing voltage or amperage, start by checking the drive belt. It may seem overly simple, but on some vehicles, the generator is not easily seen, and a missing drive belt can be overlooked. Next, ensure that power is being supplied to the generator. Many vehicles use one or more fuses in the charging system. If power is not supplied to the generator, it will not be able to produce any output. Refer to a wiring diagram or the manufacturer's service procedures to ensure that all the necessary power circuits are intact (**Figure 20-106**).

If the generator is monitored or controlled by the PCM, connect a scan tool and check for DTCs related to the charging system. If a generator fault code is stored, follow the manufacturer's procedure for diagnosing the fault.

As in the undercharge tests, try to full-field the generator. This forces the generator to full output and is used for diagnosing charging system concerns. This may be able to be done with a scan tool. If there is still no output, the generator is most likely faulty and needs to be replaced.

■ ***Overcharge Condition.*** **Overcharging** occurs when the field control circuit does not limit the amount of current that is supplied to the field. This can cause overcharging of the battery and battery damage as well as damage to the electrical system components. Electronics are very sensitive to overvoltage situations and can easily be damaged if voltage rises above 18 volts.

If the generator is controlled by the PCM, connect a scan tool and check for DTCs related to the charging circuit. If the generator field can be controlled through the scan tool, try to turn the generator off. If the generator stops charging, the problem may be in the field control circuit. If the generator continues to charge, the problem is likely inside the generator. **Figure 20-107** shows a generator field output command waveform from the PCM to the generator on a vehicle with a charging system concern. Several charging-related DTCs were stored, and the generator had recently been replaced, but the problem remained.

Depending on the type of regulator control circuit used, a poor regulator ground can cause an overcharging condition, as can a short to power in the regulator circuit.

GENERATOR REMOVAL AND INSTALLATION

Once a faulty generator is diagnosed, you need to remove it and install a new or rebuilt unit. Compare the new and old generators before installation. Check to make sure that the new unit has the same connections, pulley, and mounting hole locations.

■ ***Safety Precautions.*** Never remove the generator or generator wiring unless the battery has been disconnected. Accidental shorting of a power circuit at the generator can cause a fuse or fuse link to blow. In addition, the generator has a high-amperage wiring connection that, if shorted to ground, can damage electronic components and could cause a fire.

■ ***Common Removal Procedures.*** To begin, place fender covers on the vehicle to protect against accidental damage. Next, remove the battery negative cable and secure it from touching the battery terminal. Locate a belt routing diagram and determine how to remove the drive belt (**Figure 20-108**). Determine how to release the tension from the drive belt, and then remove the belt from the generator pulley. Next, remove the wiring from the generator. The output wire is often bolted to the back of the generator; this allows a tight connection point and reduces the chance of a poor connection (**Figure 20-109**). Remove the fasteners securing the generator to the engine and remove the generator.

Depending on the vehicle and the generator that is installed, you may need to remove the drive pulley and reuse it on the replacement generator. Use an air impact wrench and impact socket to remove the drive pulley nut. Remove the pulley, washers, and fan if necessary. When you are installing the pulley on the new generator, install all the parts in the same order in which they were removed from the original generator, and torque the retaining nut to specs. Do not use the air impact wrench to tighten the drive pulley nut.

Some generators use an overrunning clutch, sometimes called a decoupler, in the drive pulley. This is done to increase belt and tensioner life, reduce noise, and

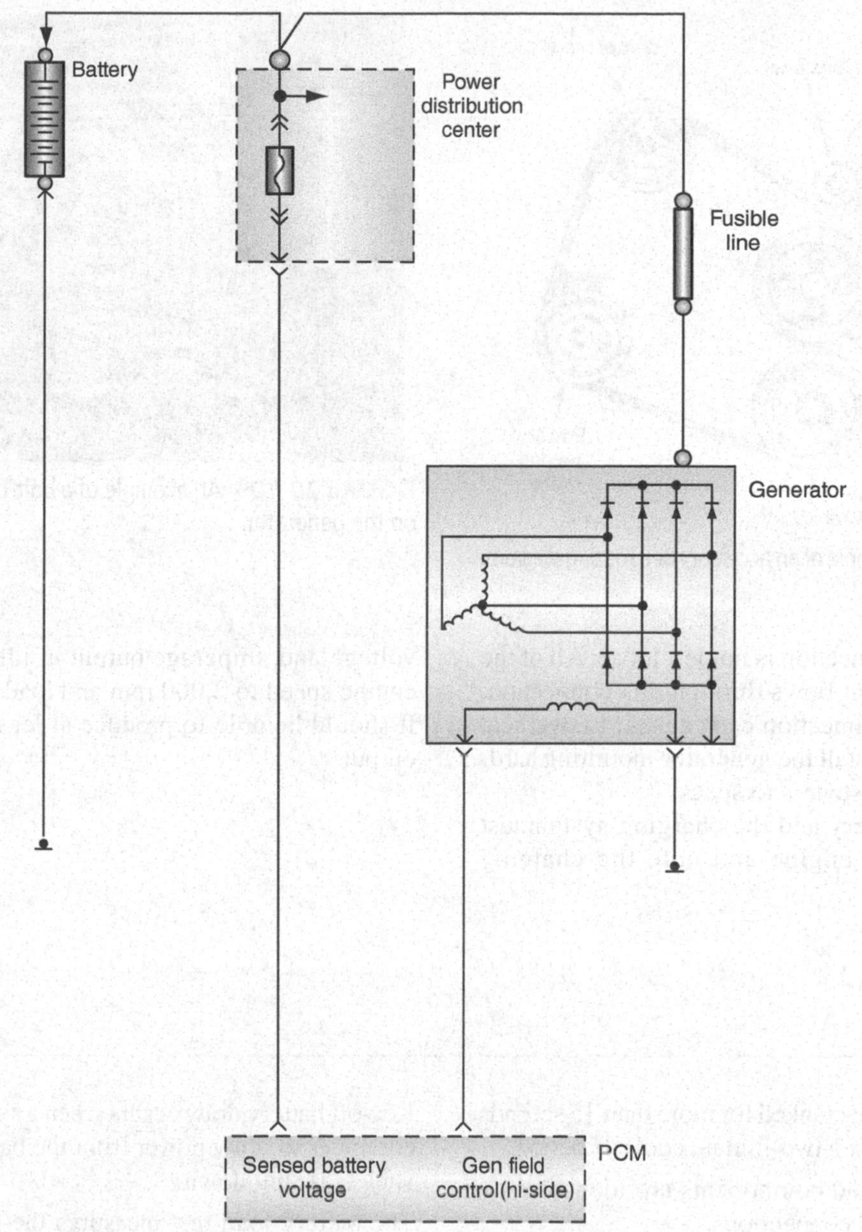

FIGURE 20-106 An example of a charging system wiring diagram. Before replacing a generator, make sure all power and ground circuits are intact.

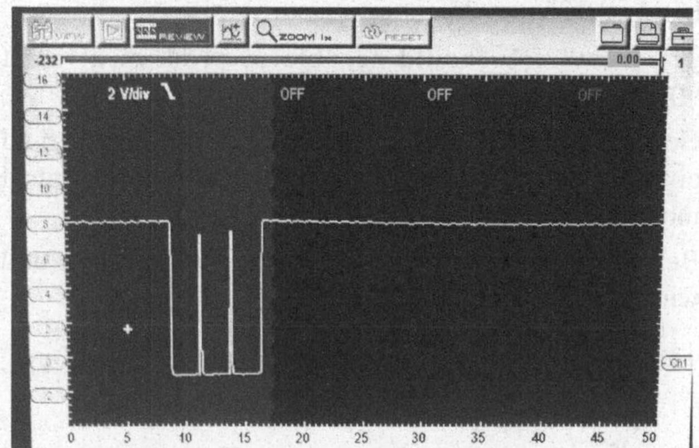

FIGURE 20-107 This waveform is from a vehicle with a charging system problem.

reduce surge loads on the generator. These clutch-type pulleys should be checked for proper operation before being installed on the replacement generator. Turn the decoupler by hand; it should free-wheel in one direction and have a spring-like feel while driving the generator shaft in the other direction. Special tools are required to remove and install decoupler pulleys.

■ ***Generator Installation.*** Visually check that the replacement generator matches the one that was removed. Check the electrical connections and location of connections to mounting holes or tabs. Verify that the drive belt is in good condition and is properly aligned on all drive pulleys. Reconnect the electrical connections and tighten the output stud fastener to specs. It is very

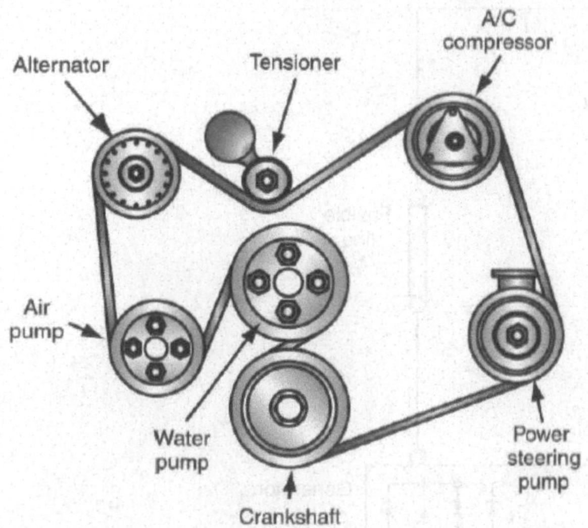

FIGURE 20-108 An example of an accessory belt routing diagram.

FIGURE 20-109 An example of a bolt-on battery connection on the generator.

important that this connection is not left loose. All of the charging system current flows through this connection, and a loose or poor connection can cause it to overheat and catch on fire. Reinstall the generator mounting hardware, and torque all fasteners to specs.

Reconnect the battery and the charging system test equipment. Start the engine and note the charging voltage and amperage output at idle. Next, bring the engine speed to 2,000 rpm and load-test the generator. It should be able to produce at least 90% of its rated output.

SUMMARY

A starter should never be cranked for more than 15 seconds at a time without at least a two-minute cool-off period.

High-voltage wiring and components are identified by the orange conduit and connections.

Inspect battery terminals and posts for corrosion and tight connections.

Side-post batteries can be damaged by overtightening the side-post connections.

Check that the battery is properly secured by the holddown.

A fully charged battery has 12.6 volts and is at 100% state of charge.

A battery fast charge is done at a high amperage rate for a short amount of time.

Slow battery charging supplies a low current over a long period of time.

Key-off battery draw occurs when a system or component continues to draw power from the battery even after the vehicle is shut down.

The battery load test measures the battery's ability to produce current for a short amount of time.

For most gasoline engines, cranking amps will be between 125 and 250 amps.

The generator should be able to produce at least 90% of its rated output.

Generator field control is often done by pulsing the field circuit off and on very quickly by the engine control module.

Before replacing a generator, make sure the belt and wiring are in good condition.

REVIEW QUESTIONS

1. A battery _______________ is used to keep the battery and place.
2. Excessive _______________ draw can drain the battery when the vehicle sits.
3. A severely worn or loose_______________ can cause low charging system output.
4. Starter current draw on a gas engine is typically between _______________and _______________ amps.
5. Before removing either the starter or generator, first _______________ the battery _______________ cable.
6. A vehicle has a slow cranking complaint. *Technician A* says corrosion at the battery cables could be the cause. *Technician B* says starter current draw and voltage drop tests should be performed. Who is correct?
   a. Technician A
   b. Technician B
   c. Both A and B
   d. Neither A nor B
7. Which is the preferred method of recharging a frozen battery?
   a. Fast charging
   b. Slow charging
   c. Clean charging
   d. None of the above are acceptable
8. All of the following are components of the starter control circuit except:
   a. Ignition switch
   b. Park-neutral switch
   c. Hold-in winding
   d. Antitheft module
9. A starter turns very slowly and has excessive cranking current draw. When removed from the vehicle and bench tested, the starter spins normally, and current draw is within specifications. *Technician A* says a weak battery may be the cause. *Technician B* says an internal engine problem may be the cause. Who is correct?
   a. Technician A
   b. Technician B
   c. Both A and B
   d. Neither A nor B
10. A vehicle has had the generator replaced three times in two months. *Technician A* says a poor generator ground connection may be the cause. *Technician B* says a faulty battery may be the cause. Who is correct?
    a. Technician A
    b. Technician B
    c. Both A and B
    d. Neither A nor B

CHAPTER 21

Lighting and Electrical Accessories

Chapter Objectives

At the conclusion of this chapter, you should be able to:

- Inspect and service interior and exterior lighting components. (ASE Education Foundation MLR 6.E.1)
- Aim headlights. (ASE Education Foundation MLR 6.E.2)
- Identify system voltage and safety precautions associated with high-intensity discharge headlights. (ASE Education Foundation MLR 6.E.3)
- Remove and reinstall a door panel. (ASE Education Foundation MLR 6.F.2)
- Describe the operation of keyless entry/remote-start systems. (ASE Education Foundation MLR 6.F.3)
- Diagnose concerns with power window and door lock systems.

KEY TERMS

CHMSL	**incandescent bulb**	**RKE**
dimmer switch	**indicator lights**	**warning lights**
halogen insert headlamp	**instrument panel**	**window regulator**
high-intensity discharge (HID) lamp	**maintenance reminders**	

Lighting Systems

Modern automotive lighting is used for both functional and decorative purposes. Exterior lights are required to meet governmental regulations for color and in brightness. Interior lighting is used to illuminate driver controls and to provide style and decoration inside the vehicle.

INTERIOR LIGHTING

Interior lighting for many years was not much more than a dome light and several bulbs inside the instrument panel that allowed the driver to see the dash gauges at night. Modern vehicles now have an array of interior lights, accent lights, vanity lights, puddle illumination lights, and dash displays that incorporate LEDs, fluorescent lighting, and even fiber optic lighting.

■ ***Convenience and Courtesy Lighting.*** This type of lighting, found in door panels, in the headliner, in and under dash panels, and throughout the passenger compartment, provides illuminations for safer vehicle entry and exit (**Figure 21-1**). Many cars and trucks activate the interior lights when the door handle is moved, a door is unlocked, or the remote approaches the vehicle or unlocks or locks the doors.

■ ***Components and Operation.*** The number of lights and how the systems operate vary depending on the vehicle. In general, interior lighting systems consist of several low-watt or light-emitting diodes (LEDs), door switches, an interior light switch, and either a lighting module or a body control module (BCM).

On older vehicles, the interior lights are often incandescent bulbs (**Figure 21-2**). An **incandescent bulb** provides light by heating a small wire, called a filament, until it glows. Many vehicles are now using LEDs for interior lighting. The benefits of using LEDs are reduced power consumption, negligible heat output, and a very long service life.

Door switches are used either to complete a light circuit or as an input for the BCM (**Figure 21-3**). Many vehicles have these switches built into the door latch assembly. On vehicles without lighting or body modules, the interior lights turn on when a door is opened and the switch completes the ground circuit for the lights (**Figure 21-4**). As you can see from the diagram, the lights can also be turned on and off from the courtesy lamp switch. On many vehicles, the BCM receives input from the door switches and then drives the lights either directly or by activating an interior light relay.

Vehicles that have interior lights that slowly dim and then go out use either the lighting or the body module to control the lights. To dim the light over time, the power to the bulb is controlled by using pulse width modulation. This means that the light is pulsed on and off rapidly. The longer the pulse is on, the brighter the light shines.

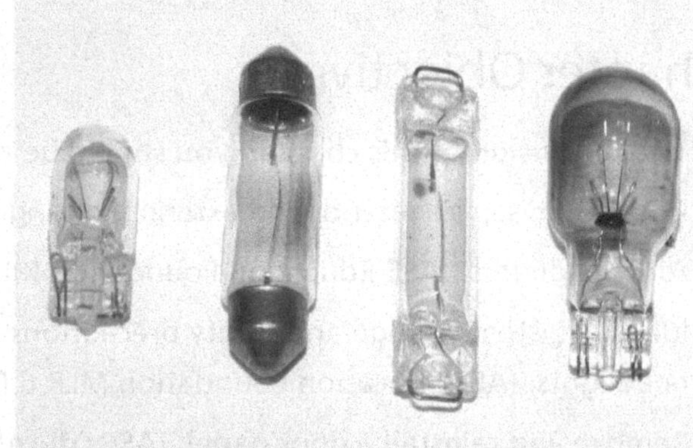

FIGURE 21-2 Various types of incandescent bulbs. These types of bulbs are still common in interior lighting.

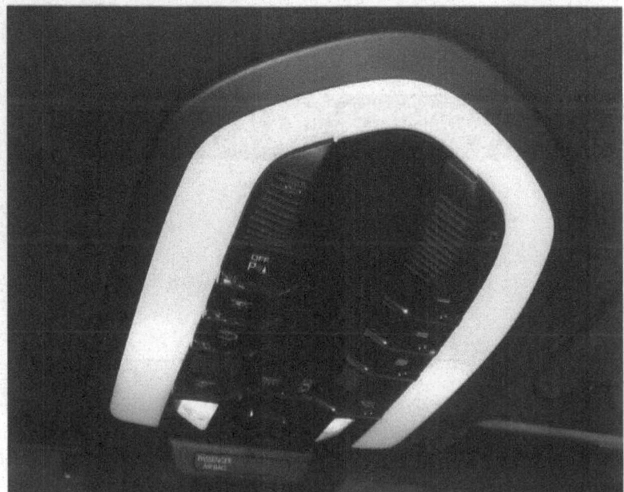

FIGURE 21-1 Interior illumination now includes many different types of lights.

FIGURE 21-3 An example of a door switch that is used for the interior lights and warning chime.

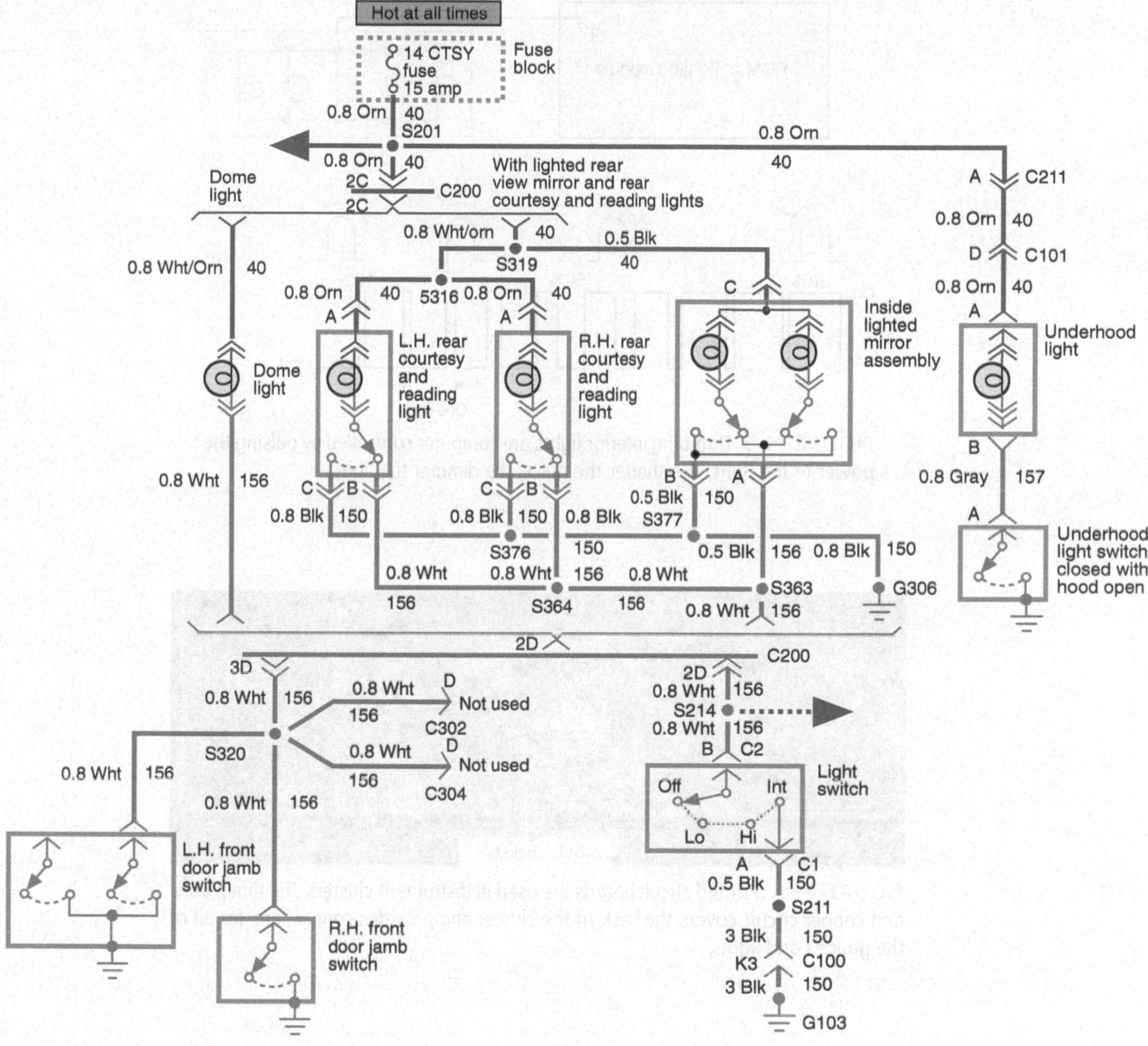

FIGURE 21-4 An example of an interior courtesy light circuit.

As the module decreases the "on" time of the pulse, the light dims. An example of a BCM-controlled pulse-width interior light control pattern is shown in **Figure 21-5**.

By monitoring or controlling the interior lights through the BCM, accidental battery discharge from an open door or a light left switched on can be averted. On many vehicles, if the lighting circuit remains on for an extended period, often 20 minutes, the BCM will turn the system off to prevent battery discharge.

■ ***Instrument Panel Illumination.*** Instrument panels, or IPs, contain the gauges, warning lights, and other driver information systems illuminated as part of the interior lighting circuit. IP illumination is controlled through the headlamp switch, and brightness is controlled through either a rheostat in the headlamp switch or a dimmer switch/sensor that supplies input to the instrument cluster or dash module. If incandescent bulbs are used, they are often placed into a flexible printed circuit panel (**Figure 21-6**). Each bulb is inserted into a socket that twists and locks to the printed circuit (**Figure 21-7**).

Many IPs do not use incandescent bulbs but rather liquid crystal displays (LCDs) or thin-film transistor (TFT) displays. Both LCDs and TFT displays are types of liquid crystal flat panel displays. These displays are typically

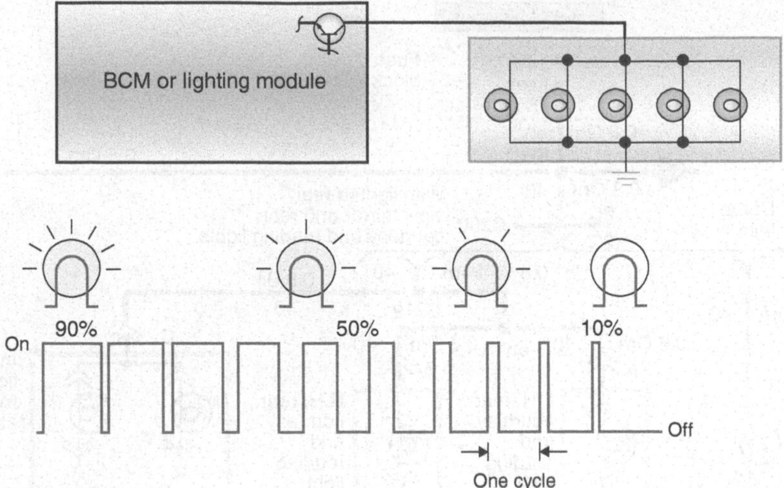

FIGURE 21-5 Dimming interior lights are computer controlled by pulsing the power to the light. The shorter the pulse, the dimmer the light.

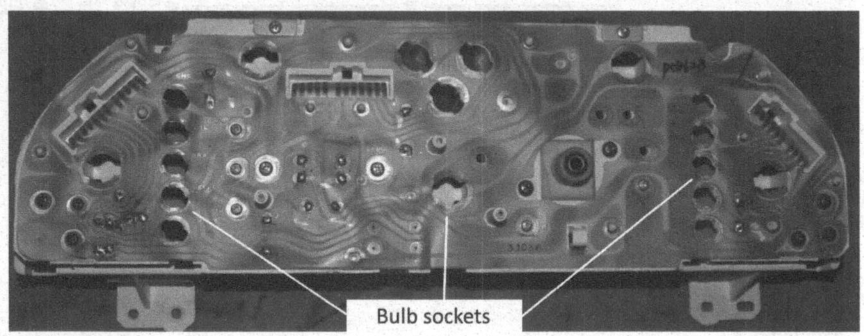

FIGURE 21-6 Printed circuit boards are used in instrument clusters. The thin plastic and copper circuit covers the back of the cluster and provides connections for all of the gauges and lights.

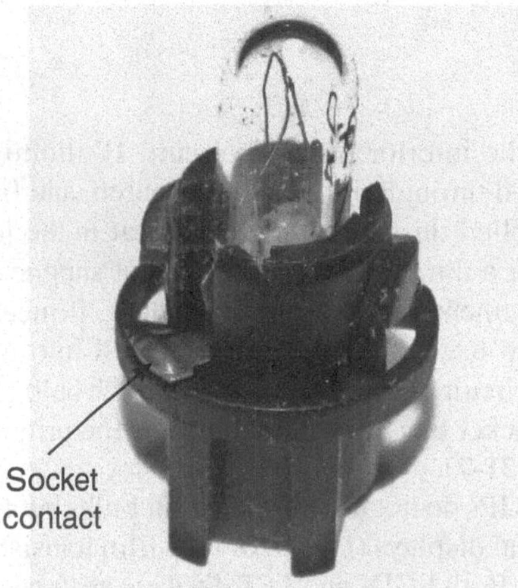

FIGURE 21-7 An example of an instrument panel illumination light bulb and socket.

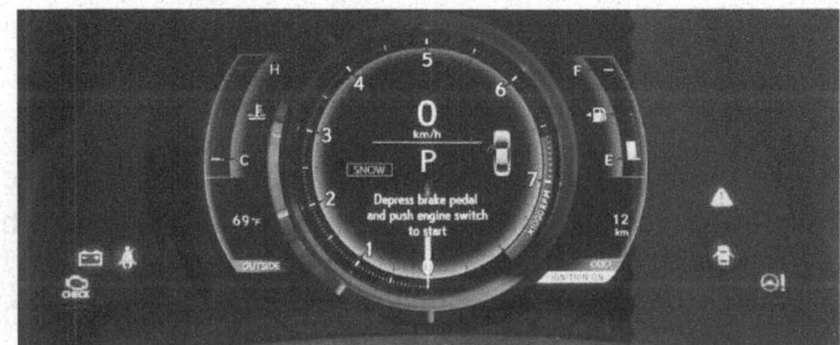

FIGURE 21-8 Many modern vehicles use electronic displays instead of fixed gauges. These displays are often customizable to suit the driver to show different vehicle parameters or navigation information.

configurable by the driver to show various elements of vehicle operation (**Figure 21-8**). These displays are completely electronic; they do not have separate illumination circuits and are not serviceable.

INTERIOR LIGHTING SERVICE

When you are beginning the diagnosis of a customer's concern, begin by verifying the complaint. Sometimes the customer may believe that a problem exists when in fact the vehicle is operating as designed. In this case, you need to educate the customer about how a system operates. On some vehicles, lighting and other functions have programmable options and the concern may be solved by changing various options in the on board computer program.

■ ***Bulb Testing.*** Incandescent bulbs can be checked visually and with an ohmmeter, although a visual inspection is not always 100% accurate as a break in a filament can sometimes be very hard to detect.

When you are inspecting a suspect bulb, remove the bulb from its socket and examine the connections and overall bulb condition. If the bulb has a cloudy or smoky appearance the bulb's seal has failed, allowing air to enter the bulb. When this happens, the bulb must be replaced (**Figure 21-9**).

To check the bulb with a digital multimeter (DMM), set the meter to read ohms and connect the leads to the bulb connections (**Figure 21-10**). Most bulbs have a resistance of 5 ohms or less. Sometimes a filament will break but remain in contact, causing an intermittent connection and intermittent bulb operation. To check this, while the ohmmeter is connected to the bulb connections, lightly tap the bulb while you watch the meter. If the reading changes or shows "OL," then the bulb likely has a broken filament.

Light emitting diodes are checked with a DMM either using ohms or, if the meter is able, using a diode test function. Use the resistance test if the meter you are using does not have the diode testing function. To check an LED by measuring resistance, connect the meter leads to the LED connection and note the reading. Then reverse the meter leads on the LED and remeasure. The LED should show low resistance one way and "OL" the other. If "OL" is obtained in both directions, the LED is open. If a low resistance reading is obtained in both directions, the LED is shorted.

A DMM with a diode test function can also be used to test LED lights. The DMM supplies a small current to the diode and measures the voltage drop between the meter leads. Connect the meter positive lead to the positive side of the LED and the meter negative lead to the LED negative lead. When testing using a DMM, the LED may light up. A good LED should show near 500 mV (0.500 volt). Reverse the meter leads on the LED connection and retest. The meter should read "OL."

■ ***Courtesy/Interior Lighting Diagnosis.*** Begin checking the operation of all interior lights by turning each light on by the master switch, then by opening each door one at a time, and lastly by turning each light on individually from its own switch. Narrow down which lights do not work at all regardless of the switch activation, then remove and test them. If the lights turn on from the master switch but not when a door is opened, you need to remove and inspect the door switches. Driver's door switches tend to wear out first since that is the most used door.

Switches can be tested for continuity, resistance, or voltage drop. Most DMMs have a continuity test function, which is a resistance test with an audible indicator of when continuity exists. Continuity means that there is a complete circuit or path between the meter's test leads. To test the continuity or resistance of a switch, remove the electrical connection from the switch, and connect the meter leads to switch terminals. Open and close the switch. On the continuity test function, the meter should beep as the switch is closed and stop beeping when the switch is opened.

FIGURE 21-9 Bulbs that appear smoky or discolored either are burned out or will be very soon.

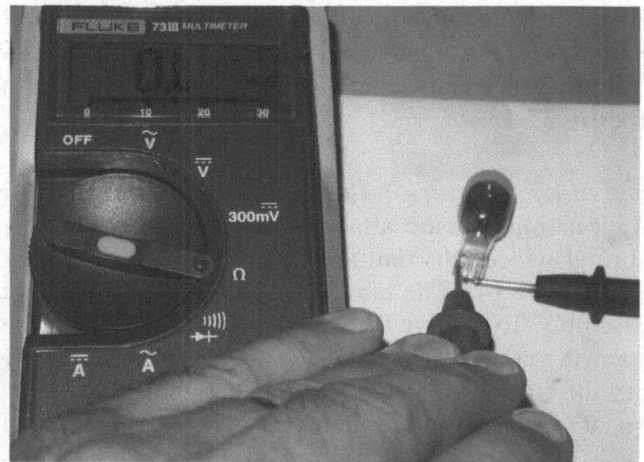

FIGURE 21-10 Testing a bulb with a DMM.

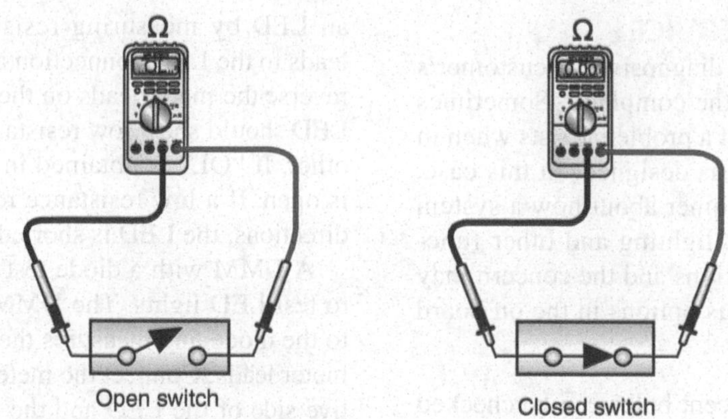

FIGURE 21-11 Testing the switch in the open position shows OL, which means there is no connection with the switch open.

When you are checking resistance, the meter should show a very low resistance reading with the switch closed and "OL" when it is open (**Figure 21-11**). While these tests provide a quick check of the switch's contacts, they do not test the switch under normal operating conditions. To perform a more accurate test, the switch should be voltage drop tested with the circuit in operation. Voltage drop across a switch should be low, no more than 200 mV (0.200 volt).

In many vehicles, the door switches complete a ground circuit though the switch itself (**Figure 21-12**). This type of switch can be used either to complete the ground path for a light assembly or to provide an input to a lighting or body module.

Many newer vehicles control interior lighting with a lighting module or the BCM. To diagnose a fault in one of these systems, connect a scan tool and check for trouble codes. If a diagnostic trouble code (DTC) is present, follow the diagnostic chart for the code. You can look at module input and output data for the system with a fault. Watch the switch input data on the scan tool while operating the switch. You may also be able to test the lights by using bidirectional active commands with the scan tool. An example of output control functions on a General Motors vehicle is shown in **Figure 21-13**.

■ ***Courtesy/Interior Light Replacement.*** Interior light assemblies are attached either by a screw or by spring-loaded tabs that are part of the light assembly (**Figure 21-14**). The tabs fit into openings in the light assembly. To remove the cover, place a small flat-blade screwdriver into the notch in the cover and then carefully pry the cover away from the light assembly. Care must be taken when you are removing a snap-fit light as the tabs are usually plastic and are easily broken. Once the cover is off, remove the bulb and replace it with the same bulb number. To reinstall the cover, align the tabs with the openings and gently push the cover back into place.

■ ***Bulb Trade Numbers.*** When you are replacing a bulb, whether it is used in the interior or exterior, it is important to replace it with the correct bulb. Each type of bulb has a trade number, which is used to classify the type of bulb and its wattage. The bulbs shown in **Figure 21-15** appear identical, but they are not. The 3057 and 3157 bulbs have the same shape and socket configuration, but the filaments have different resistances. **Figure 21-16** shows examples of common bulb trade numbers and their associated wattages.

FIGURE 21-12 A door switch is often simply a connection to ground with the switch closed. This then completes a light circuit or is used as an input for a module.

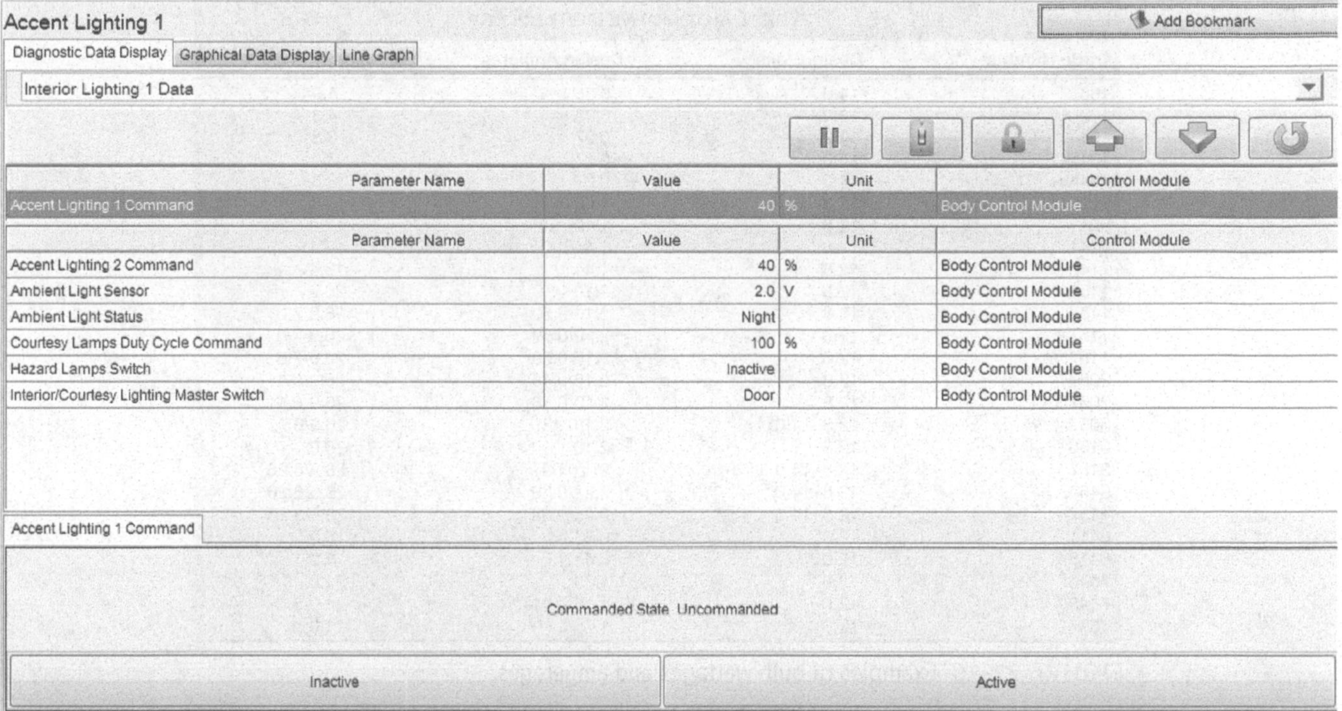

FIGURE 21-13 An example of using a scan tool to check lighting system operation.

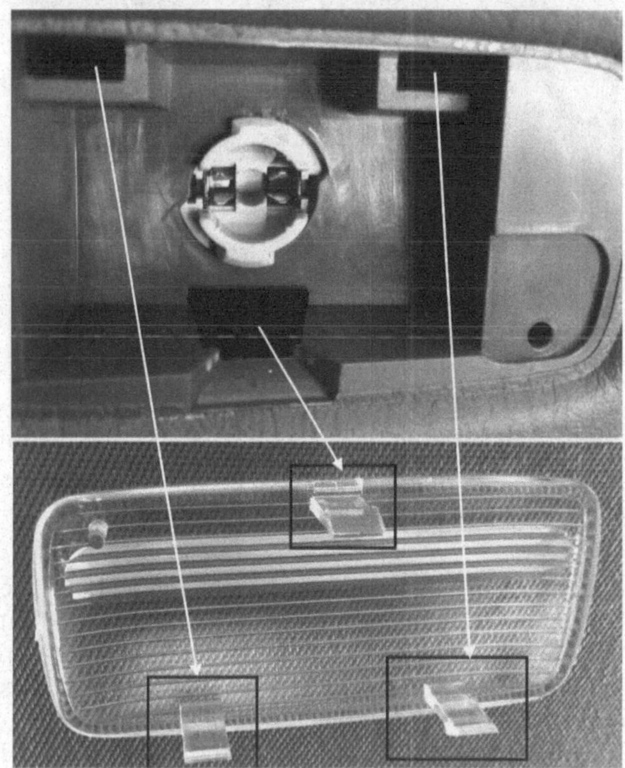

FIGURE 21-14 The tabs that hold light covers in place are small and easily broken if forced. Use care when removing and installing light covers to avoid damage. The tabs snap into place within the housing.

FIGURE 21-15 These bulbs are identical in appearance but note that the trade numbers are different. This indicates the bulbs are different wattages.

■ ***Instrument Panel Lighting Diagnosis.*** There are two common IP lighting concerns: either all bulbs are inoperative, or some bulbs work but others do not, leaving the IP partially illuminated.

If all IP illumination bulbs are inoperative, begin by checking the IP dimmer control. Though rare, it is possible that the dimmer has been turned all the way down and the customer has not checked it or does not know of its function. An example of two dimmer switches are shown in **Figure 21-17** and an illustration of the dimmer circuit is shown in **Figure 21-18**. If the dimmer is not turned off, then check the fuse for the IP lighting. On many vehicles, this fuse only has power supplied to it when the head-lamp switch is on. If power is present

TYPICAL AUTOMOTIVE LIGHT BULBS			
Trade Number	Design Volts	Design Amperes	Watts: P = A × V
168	14.0	0.35	4.9
192	13.0	0.33	4.3
194	14.0	0.27	3.8
194E-1	14.0	0.27	3.8
194NA	14.0	0.27	3.8
912	12.8	1.00	12.8
921	12.8	1.40	17.92
1141	12.8	1.44	18.4
1142	12.8	1.44	18.4
1156	12.8	2.10	26.9
1157	12.8	2.10/0.59	26.9/7.6
1157A	12.8	2.10/0.59	26.9/7.6
1157NA	12.8	2.10/0.59	26.9/7.6
2057	12.8	2.10/0.48	26.9/6.1
2057NA	12.8	2.10/0.48	26.9/6.1
3057	12.8–14.0	2.1/0.48	26.9/6.72
3156	12.8	2.10	26.9
3157	12.8–14.0	2.1/0.59	26.9/8.26
3457	12.8–14.0	2.23/0.59	28.5/8.26
4157	12.8–14.0	2.23/0.59	28.5/8.26
6411	12.0	0.833	10.0
6418	12.0	0.417	5.0
7440	12.0	1.75	21.0
7443	12.0	1.75/0.417	21.0/5.0
7507	12.0	1.75	21.0

FIGURE 21-16 Examples of bulb wattages and amperages.

on both sides of the fuse, you need a wiring diagram to examine the IP illumination circuit. In a dash lighting circuit, power runs through the dimmer, to the fuse, and then to the IP lights (**Figure 21-19**). Check for power in and out of the dimmer. If the dimmer is working correctly, the IP needs to be removed. Check for power and ground to the IP circuit. Inspect the connection for the IP circuit carefully. If voltage is present at both the power and ground terminals, there is a problem in the lighting ground circuit. It is also possible (though uncommon) that over time all the bulbs have burned out.

If only one or a few bulbs are inoperative, the most likely cause is that the bulbs have burned out. To replace the bulbs, the IP must be removed. Replace the bulbs with the same type of bulb since substituting a different bulb, with different resistance, will cause differences in the lighting of the IP. With the IP removed, advise the customer that it would be worth the extra cost to replace all the illumination bulbs at the same time.

FIGURE 21-17 The dimmer is used to control the brightness of the instrument panel lights.

EXTERIOR LIGHTING

Exterior lighting consists of the headlamps, turn signals, brake warning lamps, and marker or cornering lamps. Each of these types of lighting has specific functions and requirements. It is important for technicians to understand the operation of exterior lighting circuits and how to diagnose and fix problems related to them.

■ ***Headlamps.*** Headlamps and the headlamp housings are more than just a light source; they have become an integral part of vehicle styling and in some vehicles, part of the passive safety systems. Currently, there are three types of headlamps used on motor vehicles: the halogen insert bulb, the high-intensity discharge (HID) bulb, and LED lights (**Figures 21-20** through **21-22**). Many vehicles use a combination of HID and LED lights (**Figure 21-23**).

Halogen insert headlamp bulbs replaced sealed beam headlamps because their compact size allows for greater flexibility in the design of the headlamp assembly (**Figure 21-24**). Halogen inserts are secured into the headlamp lens assembly so that the light produced is captured and directed by the lens in such a way as to provide better overall visibility than what can

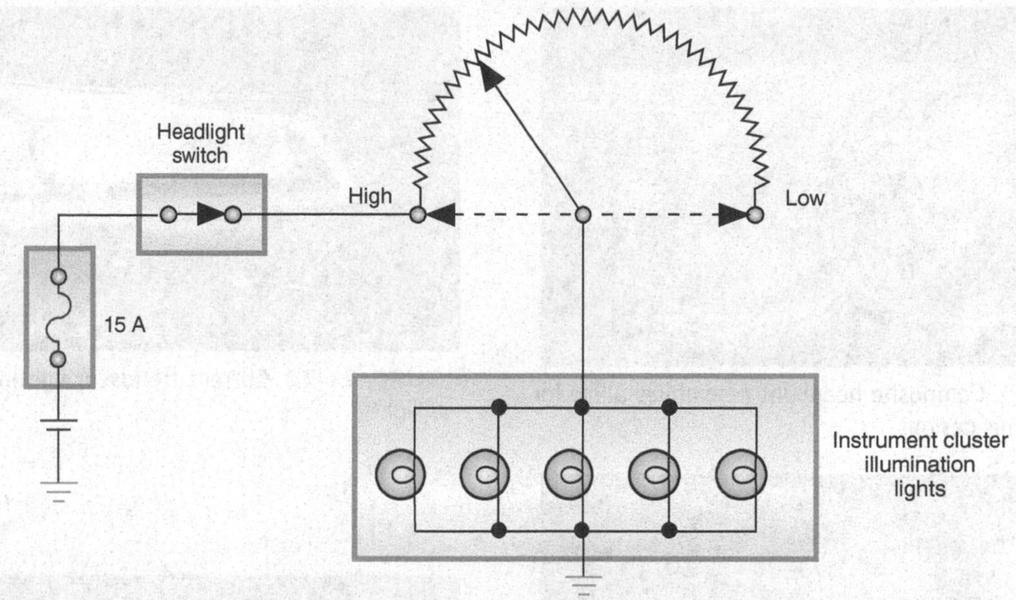

FIGURE 21-18 A rheostat may be used to control the brightness of the dash lights. As the wiper moves along the resistor, current flow to the bulbs increases or decreases.

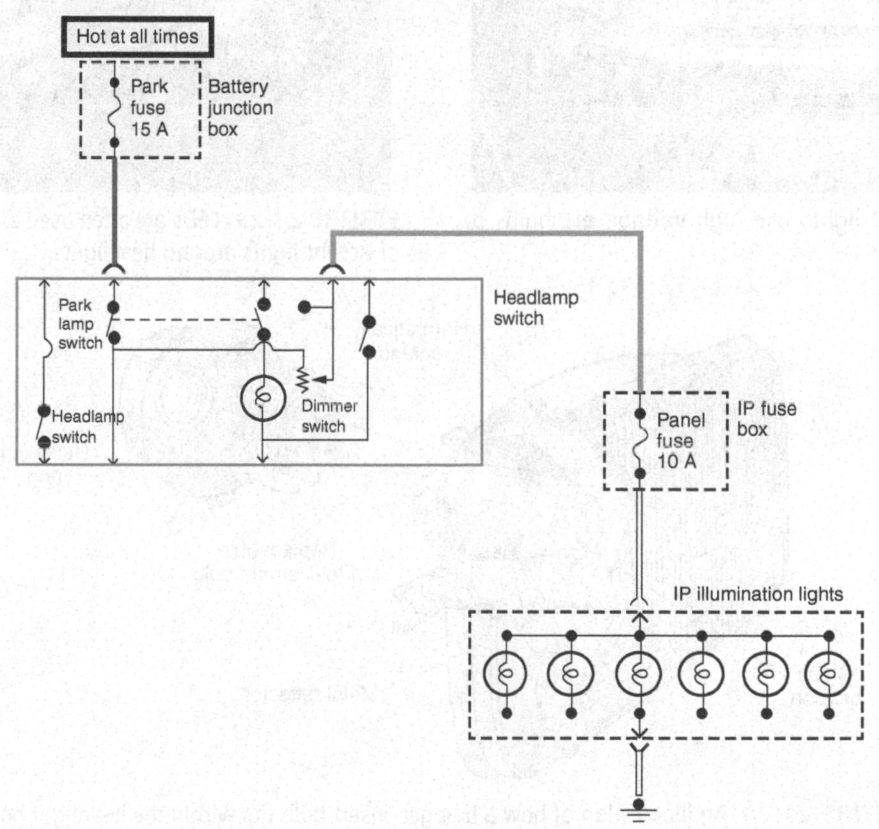

FIGURE 21-19 An illustration of a dash lighting circuit.

be obtained with a sealed beam lamp. Examples of halogen insert headlight bulbs are shown in **Figure 21-25**.

Many vehicles now use the **high-intensity discharge lamp** or HID lamp. These lamps provide light by using a high-voltage AC arc of current across two electrodes. The light from HID lamps looks blue and brighter than that produced by halogen lamps. Examples of an HID and halogen bulb are shown in **Figure 21-26.** Note the two filaments in the halogen bulb compared to the construction of the HID bulb. Current flow excites the gas

FIGURE 21-20 Composite headlight assemblies allow for more aerodynamic designs.

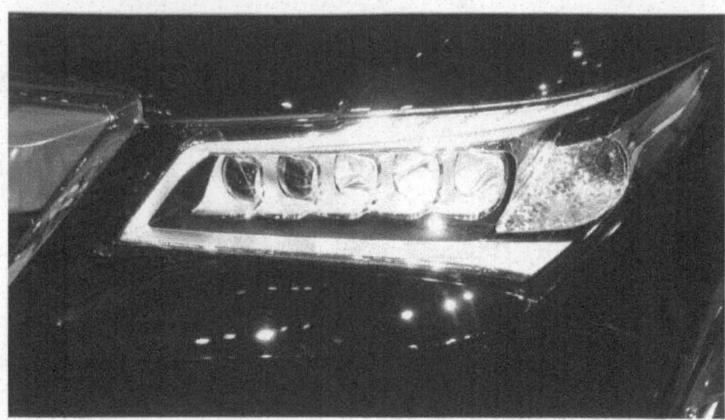

FIGURE 21-22 Current trends in lighting include LED headlights.

FIGURE 21-21 HID lights use high-voltage arc bulbs to generate light.

FIGURE 21-23 LEDs are often used as daytime running lights or accent lights around headlights.

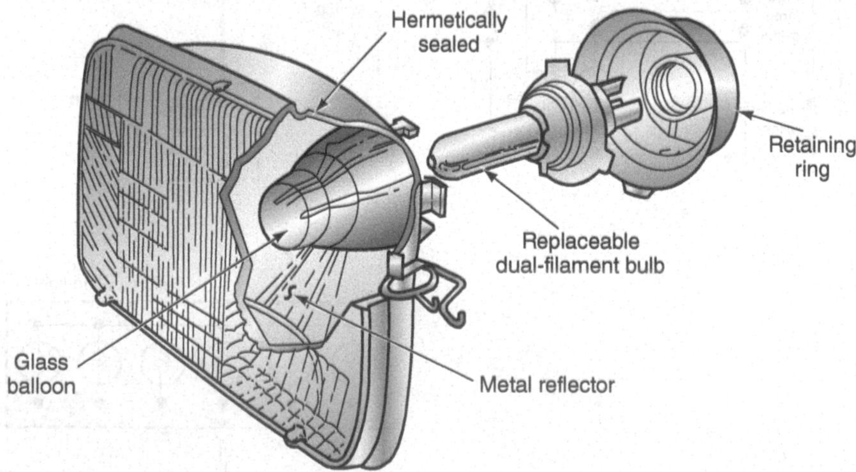

FIGURE 21-24 An illustration of how a halogen insert bulb fits within the headlight housing.

in the HID lamp and causes it to glow. To operate, HID lamps require a ballast and igniter since AC is used. **Figure 21-27** shows an HID lamp assembly. On the left is the ballast, which increases the voltage to bulb. An example of an HID lighting circuit is shown in **Figure 21-28**. HID systems require high voltage to operate, which, if you are not careful, could result in injury if proper service procedures are not followed.

HID bulbs can also contain mercury, which means be careful when handling the bulbs to prevent breaking the glass. Also, be aware that HID bulbs should not be disposed of in the regular trash because of the mercury

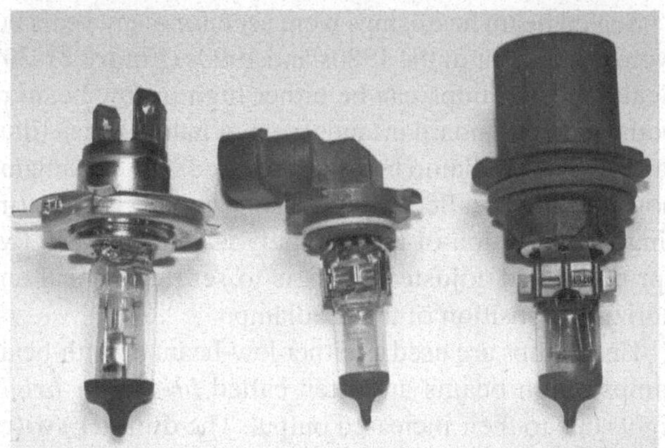

FIGURE 21-25 Halogen inserts come in several shapes and sizes, all of which attach to the headlight housing.

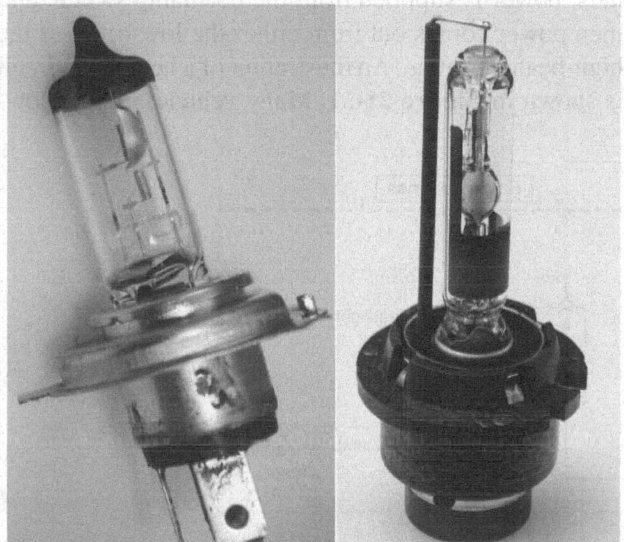

FIGURE 21-26 An example of a halogen insert bulb (left) and a HID bulb (right).

FIGURE 21-27 An example of a HID housing, ballast, and wiring.

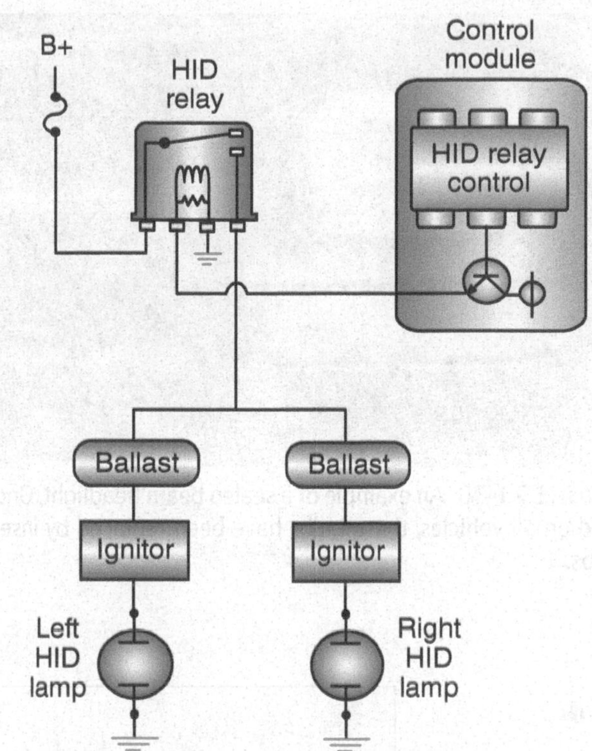

FIGURE 21-28 An example of a HID circuit.

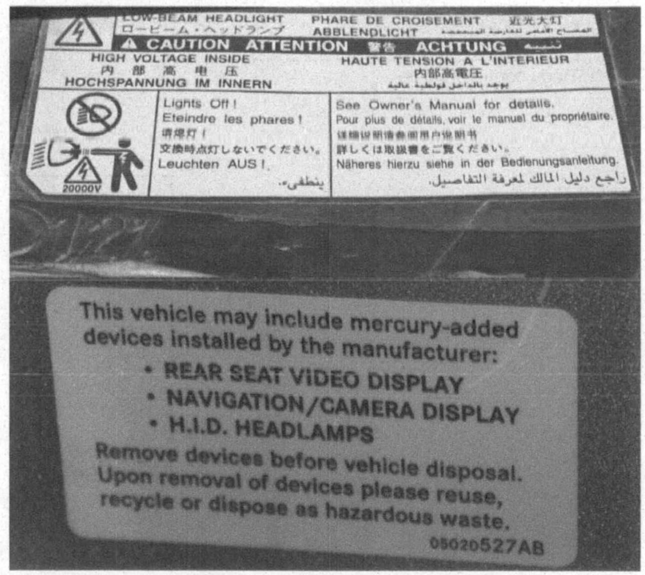

FIGURE 21-29 Do not attempt to service the HID lights without following the manufacturer's service procedures. These lights use several thousand volts upon startup and present a serious shock hazard. Like fluorescent bulbs, HID bulbs should be disposed of properly because they contain mercury.

content. An example of HID warning decals is shown in **Figure 21-29**.

Because true HID systems require ballasts and high voltage to operate, they are more expensive than other types of lighting and true HID systems cannot easily be retrofitted.

The latest trend in lighting is the use of LEDs. For years, LED lights were only used in a few applications due to cost and low relative brightness. Now many vehicles use LEDs as their primary sources of lighting, including as headlights (Figure 21-22).

FIGURE 21-30 An example of a sealed beam headlight. Once used on all vehicles, these bulbs have been replaced by insert bulbs.

Sealed beam headlamps were used for many years but were phased out in the 1980s and 1990s (**Figure 21-30**). Sealed beam lamps can be either high or low beam or both, and be standard incandescent or halogen gas–filled lamps. The headlamp bulbs are secured into a headlamp housing, often called *the headlamp bucket,* with a trim ring. The position of the assembly is adjusted with two spring-loaded adjusters. This is to set the vertical and horizontal position of the headlamp.

Headlamps are used as either low-beam or high-beam lamps. High beams are often called *brights* or *bright lights* due to their increased output. The **dimmer switch** controls the operation of high-beam lights. The dimmer switch is usually activated by pulling or pushing the turn signal stalk back or forward. The switch has three terminals; power is supplied from the headlamp switch, and then power comes out from either the low-beam or the high-beam terminal. An illustration of a headlight circuit is shown in **Figure 21-31**. Many vehicles control low-

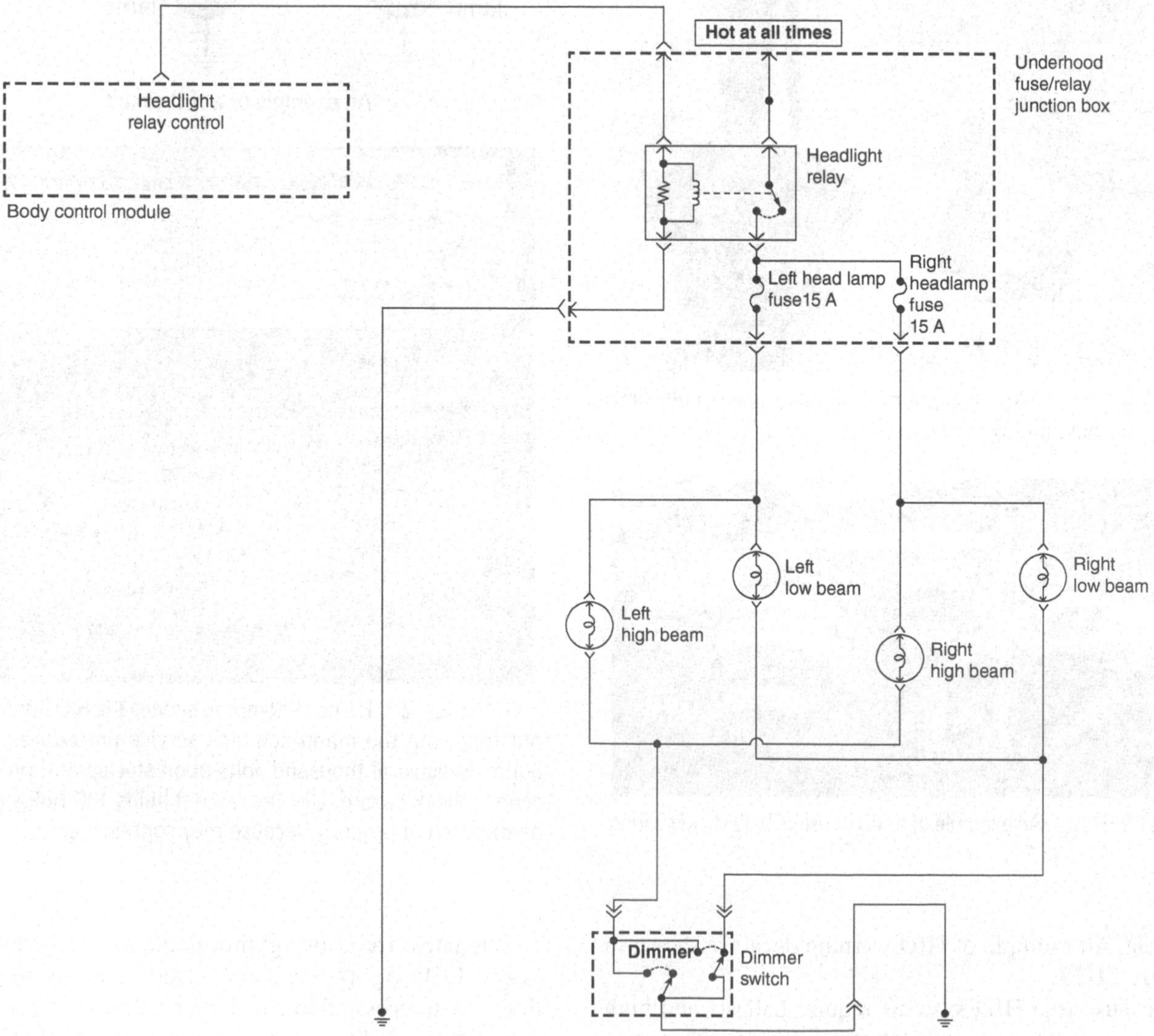

FIGURE 21-31 An example of a headlight circuit.

and high-beam operation through a lighting or body module. Instead of the **dimmer switch** controlling the headlights, it is used as an input for the module. Depending on which headlight mode is selected, the module activates either the low- or high-beam relay.

Some vehicles have automatic lighting control. Basic systems turned the headlights on as it gets dark outside. Advanced systems control the high-beam lights, turning them off and on based on driving conditions and the oncoming lights from other vehicles.

Another feature becoming more popular is adaptive lighting. Adaptive lighting systems control headlight motion, either left or right and/or up or down, based on driving conditions. These systems can turn the headlights based on steering inputs to help see around corners.

■ ***Daytime Running Lamps.*** Many countries have adopted laws that require motor vehicle headlamps to be on whenever the engine is running. This is to increase visibility and decrease accidents. Daytime running lamps (DRLs) operate whenever the ignition is on but turn off if the parking brake is applied. DRLs can use either the low- or the high-beam headlamps. The DRL circuit may use relays to switch the two headlamps into a series circuit, or a DRL module may control a separate DRL relay. The vehicle in **Figure 21-32** uses the low-beam headlights in series as the DRLs. Many newer vehicles are combining the DRLs and LED halo lights (**Figure 21-33**).

FIGURE 21-32 Daytime running lights (DRLs) typically keep the headlight bulbs on at about half power. Many newer vehicles use LED lights for the DRLs.

FIGURE 21-33 An example of LED daytime running lights.

■ ***Parking Lamps.*** The term parking lamps is used to describe any front, rear, or side marking lamps that are operated by the headlamp switch. **Figure 21-34** shows a typical parking lamp circuit. Power is supplied to the bulbs from the tail lamp relay with the headlight switch in both the Park and Headlamp positions. Some vehicles power the parking light circuit directly from the headlamp switch.

These lamps may have red or yellow lenses and may use a variety of incandescent bulbs. On some vehicles, the parking lamps are dual-filament bulbs; the other filament is used for the turning signal and brake warning lamps.

■ ***Turn Signals and Hazard Warning Lamps.*** The turn signal circuit is a key-on circuit that has its own set of switches to control the turn signal lamps. If the vehicle uses the same bulbs for the brake and turn signals, then the turn signal circuit connects to the brake light circuit. **Figure 21-35** shows an example of a turn signal and brake light circuit. Vehicles that use separate brake and turn signal bulbs do not interconnect these two circuits. When the turn signal switch is activated, voltage from the fuse flows through a flasher, through the switch contacts, and to the bulbs. Return springs in the switch, called *cancellation springs*, automatically cancel the turn signals when the steering wheel is rotated back to the straight ahead position.

There are two types of turn signal/hazard flasher units. Older vehicles use a current-sensitive flasher that operates similarly to a circuit breaker (**Figure 21-36**). As current flows through the turn signal or hazard circuits, it creates heat in the flasher. This heat causes a bimetallic strip to deform slightly. The movement opens the contacts in the flasher, opens the circuit, and causes the bulbs to go out. As the strip in the flasher cools, it returns to its original shape, allows the circuit to close, and turns the bulbs back on.

Since the flasher is current sensitive, an inoperative bulb reduces current flow in the circuit, which affects how the flasher operates. In addition, if bulbs of different resistances, and therefore wattages, are installed, the current flow in the circuit changes. This will affect the operation of the flasher, causing incorrect blink rates or the signals to not blink at all.

Modern vehicles can use either a hybrid electromechanical flasher or an entirely electronic flasher unit, called *a solid-state flasher* (**Figure 21-37**). The electromechanical flasher uses an electronic flasher control circuit to control a relay to open and close the light circuits. This type of

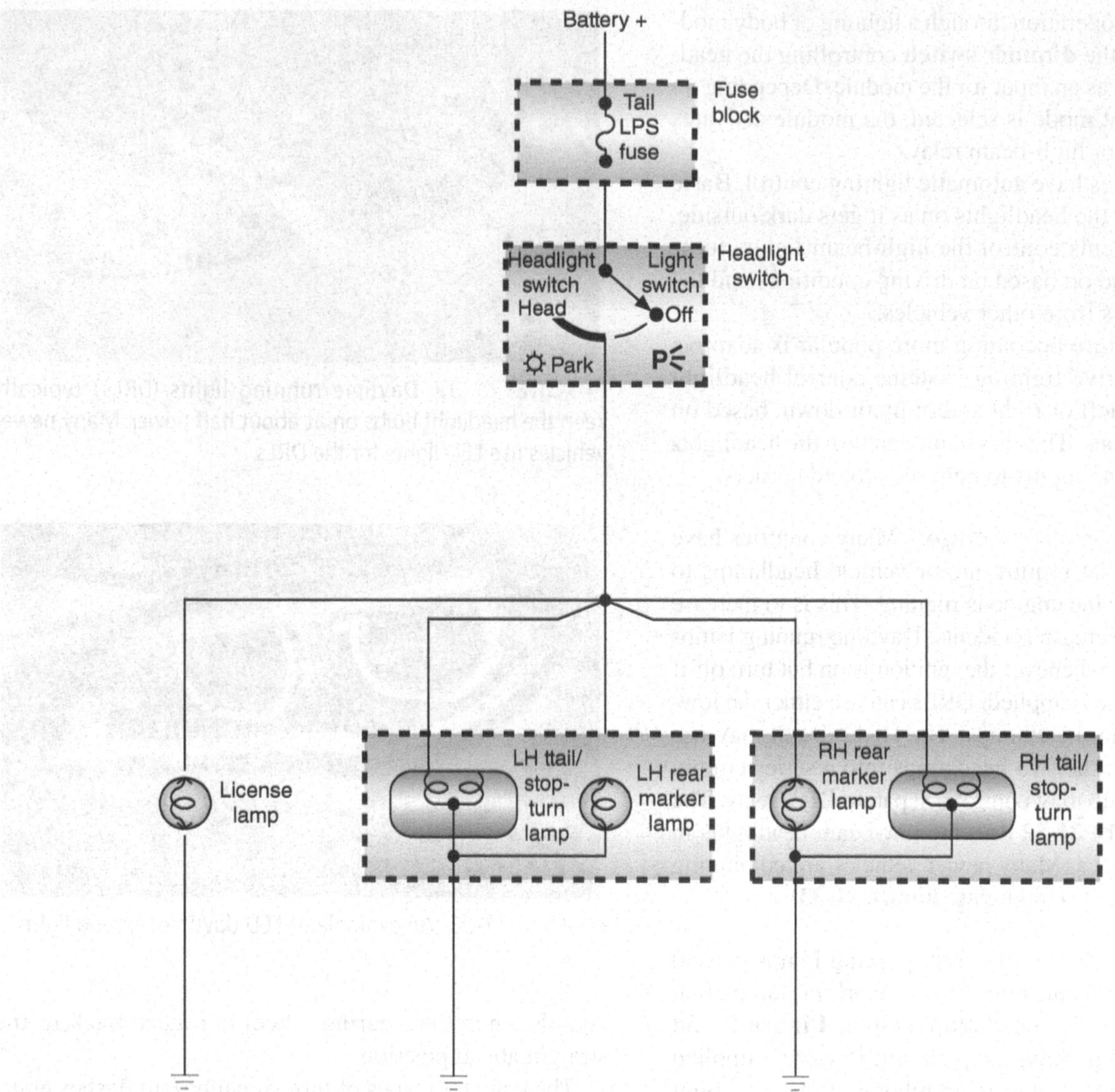

FIGURE 21-34 An illustration of a tail light circuit.

flasher can determine if a bulb is inoperative, and then blinks the turn signal at twice the standard rate to alert the driver. Solid-state flashers also monitor current flow in the circuit but use an electronic switch to open and close the turn signal or hazard lamp circuits. Both the electromechanical and solid-state flashers have a much longer service life than the older bimetallic flashers.

Many newer vehicles use a lighting module or the BCM to control turn signal operation. The module turns the front and rear signal lights on and off based on input from the turn signal switch. The turn signal "click" is supplied by the audio system.

The hazard warning lamps use either the brake or the turn signal bulbs and are a hot-at-all-times circuit. Activating the hazard lamps overrides the turn signal circuit. On older vehicles with bimetallic flashers, a separate hazard flasher is used since the current flow of all four turn signal bulbs is greater than that of the normal turn signal circuit.

■ ***Stop Lamps.*** The stop lamp circuit is often a hot-at-all-times circuit. **Figure 21-38** shows a conventual stop lamp circuit. When the driver presses the brake pedal, the brake lamp switch closes, allowing current to flow to the brake lamps.

On most modern vehicles, the brake pedal position sensor is an input monitored by the onboard computer system. This input is used to for brake and ABS functions, cruise control, transmission operation, as well as vehicle stability control operation.

Vehicles sold since the 1985 model year have been equipped with a center high-mounted stop lamp, or **CHMSL**. Also called a *third brake lamp*, it is often integrated into body design elements such as spoilers.

■ ***Reverse Lamp Circuit.*** Reverse lamps are the clear lens lights in the back of the vehicle. These lights are clear so that they are clearly distinguishable from brake or

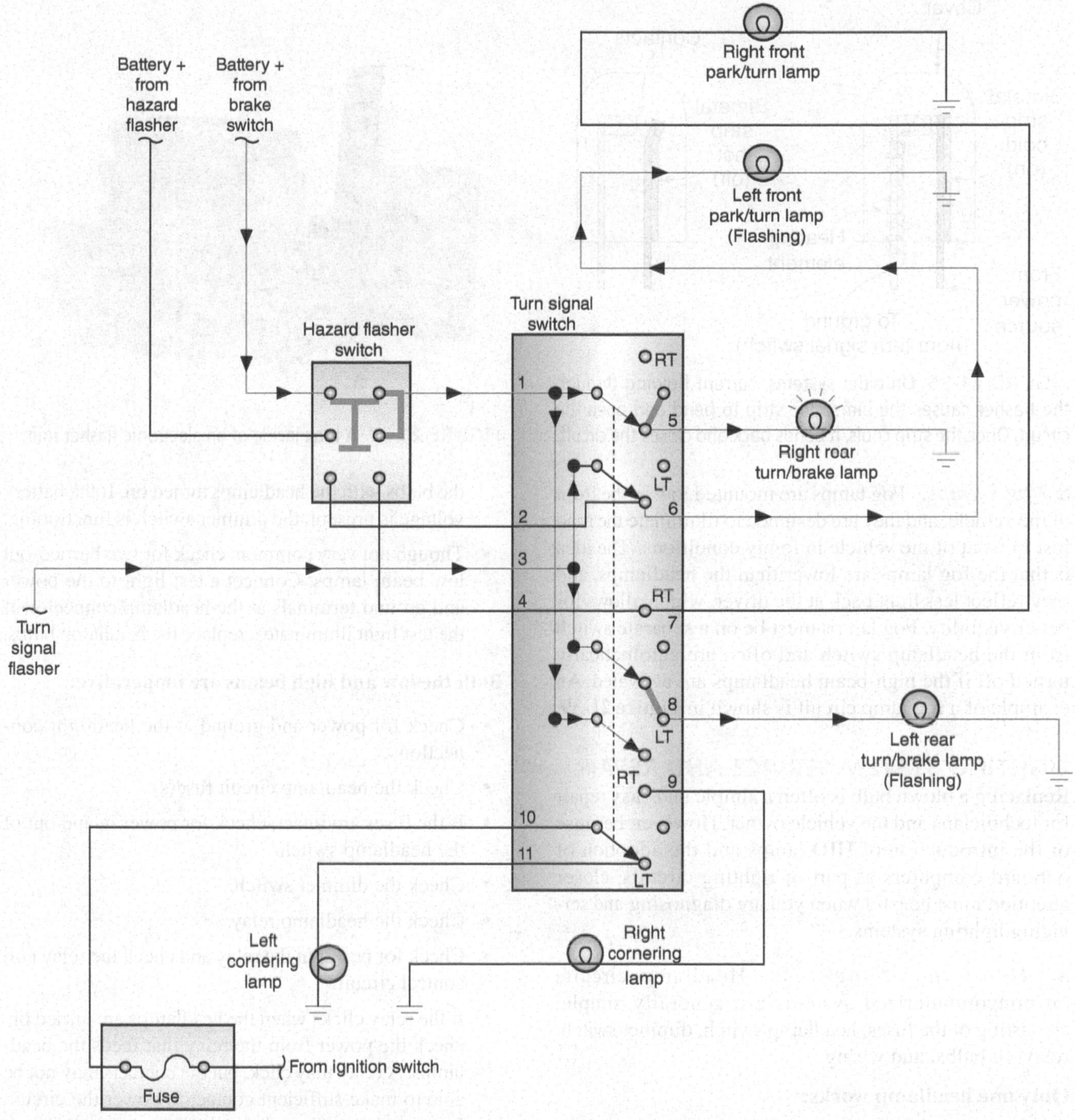

FIGURE 21-35 A simple turn signal circuit. Newer systems use the turn signal switch as an input for the BCM or lighting module to control turn signal light operation.

turning lamps. Clear lenses also provide better visibility as the light can illuminate the area behind the vehicle better.

The reverse light switch is often part of the Park/Neutral switch on automatic transmissions. When the transmission is shifted into Reverse, power flows from the fuse, through the reverse circuit terminals, and to the reverse lamps. On vehicles with manual transmissions, a separate reverse light switch is attached to the transmission case. When the transmission is placed in Reverse, the switch is closed, allowing current to flow to the reverse lamps.

On newer vehicles, the BCM or lighting module receives input from the transmission control or powertrain control module (PCM) when the vehicle is shifted into reverse. The BCM then commands the backup lights to turn on directly or commands the backup light relay on.

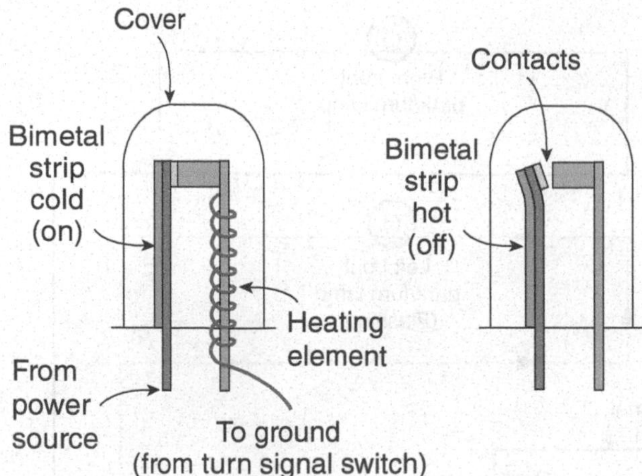

FIGURE 21-36 On older systems, current flowing through the flasher causes the bimetallic strip to bend and open the circuit. Once the strip cools, it bends back and closes the circuit.

FIGURE 21-37 A look inside of an electronic flasher unit.

■ ***Fog Lamps.*** Fog lamps are mounted low in the front of the vehicle, and they are designed to illuminate the road just in front of the vehicle in foggy conditions. The idea is that the fog lamps are lower than the headlamps, and they reflect less light back at the driver, which allows for better visibility. Fog lamps must be on a separate switch from the headlamp switch and often are automatically turned off if the high-beam headlamps are activated. An example of a fog lamp circuit is shown in **Figure 21-39**.

LIGHTING SYSTEM SERVICE AND REPAIR

Replacing a blown bulb is often a simple and easy repair for technicians and the vehicle owner. However, because of the introduction of HID lamps and the adoption of onboard computers as part of lighting circuits, closer attention must be paid when you are diagnosing and servicing lighting systems.

■ ***Headlamp Diagnosis.*** Headlamp circuits on noncomputerized systems are generally simple, consisting of the fuses, headlamp switch, dimmer switch, relay(s), bulbs, and wiring.

Only one headlamp works:

- Check if both work when the high beams are turned on.
- The most likely cause is a burned-out bulb.
- Some vehicles use two separate left and right headlight fuses. Before you assume the headlight is the fault, you should check for power and ground to the light with a meter.

Both low-beam headlamps are inoperative:

- Check high-beam operation.
- If both high beams operate correctly, the problem may be a faulty dimmer switch. Check the voltage at the bulbs with the headlamps turned on. If the battery voltage is present, the dimmer switch is functioning.
- Though not very common, check for two burned-out low-beam lamps. Connect a test light to the power and ground terminals at the headlamp connector. If the test light illuminates, replace the headlamp bulbs.

Both the low and high beams are inoperative:

- Check for power and ground at the headlight connections.
- Check the headlamp circuit fuse(s).
- If the fuses are intact, check for power in and out of the headlamp switch.
- Check the dimmer switch.
- Check the headlamp relay.
- Check for power at the relay and check the relay coil control circuit.
- If the relay clicks when the headlamps are turned on, check the power from the relay that feeds the headlamps. A relay may click, but the contacts may not be able to make sufficient contact to power the circuit. In most cases, if you do not have power to the headlights, you will need to refer to a wiring diagram.

One headlamp is significantly brighter than the other:

- Check that both bulbs are the same.
- Check the bulb connector. A corroded connection can cause enough of a voltage drop to make a bulb dim (**Figure 21-40**). A connection that is corroded may be able to be cleaned with contact cleaner and small wire brushes. If the corrosion has damaged the terminals or it cannot be removed, you will need to replace the connector.

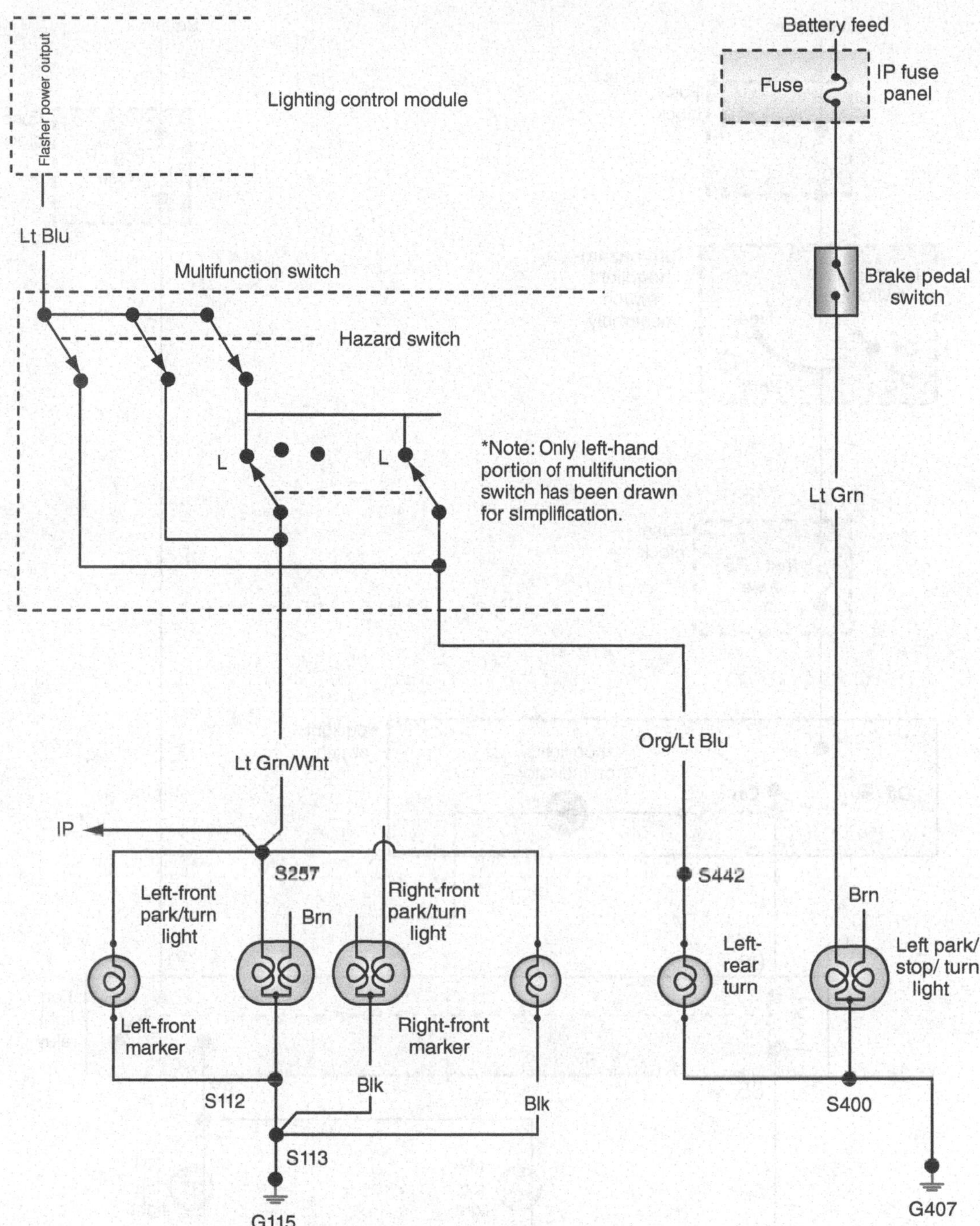

FIGURE 21-38 An illustration of a brake light circuit that does not use the same bulbs for the turn signals and brake lights.

Computerized headlight circuit diagnosis is similar except you will likely need a scan tool. First, check for DTCs and address any lighting control system codes first. You can use the scan tool to check for headlight and dimmer switch inputs (**Figure 21-41**). You may also be able to command the BCM or lighting module to turn on the lights to test circuit operation. An example of that type of function is shown in **Figure 21-42**.

When you are diagnosing a computerized lighting, HID or LED headlamp circuits, refer to the manufacturer's service information. HID systems use high voltage to power-up the bulbs. Incorrect testing procedures can lead to injury and damage to the vehicle.

HEADLAMP REPLACEMENT

In most cases, headlamp replacement is straightforward and usually simple. One area to be cautious of, however, is when you are installing certain aftermarket lights or lenses. For headlamps and lenses to be legal for street use, they must have the DOT stamped on them. Some

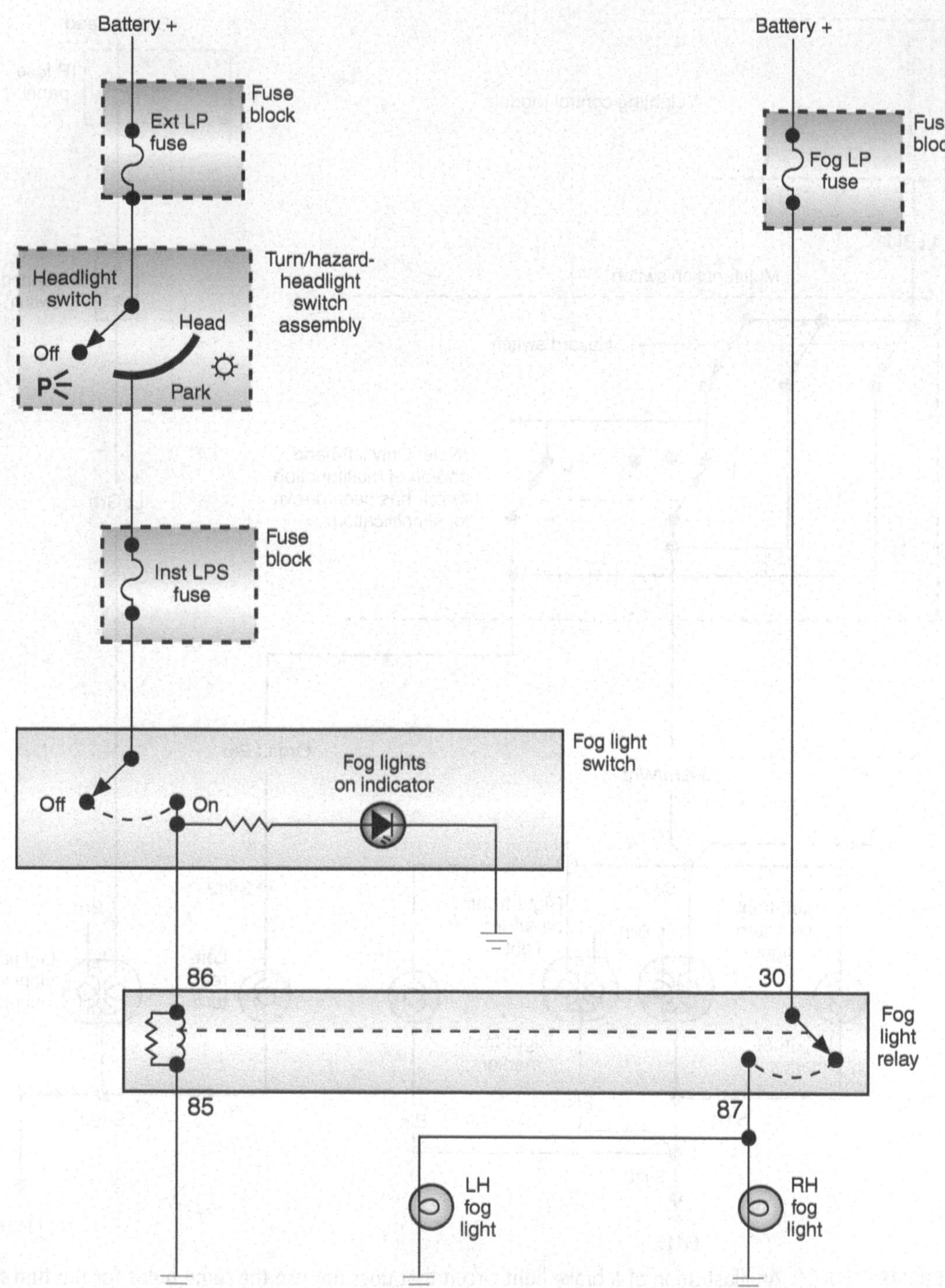

FIGURE 21-39 An example of a fog light circuit.

aftermarket lights and lenses, especially those that are tinted, may not be DOT certified, and therefore not legal.

■ ***Sealed Beam Headlamps.*** To replace most sealed beam headlamps, a trim ring or bezel must first be removed. To access the trim ring, you may have to remove a trim panel or even part of the grill to gain access to the light. Remove the screws that hold the bezel in place, and pull the headlamp out of the headlamp bucket. Disconnect the wiring and inspect the connector and terminals for corrosion or damage from excessive heat. Connect the wiring harness to the replacement bulb and set the bulb into place. Note that sealed beam bulbs are manufactured with the light beam pointing downward. If the bulb is installed upside down, the light will shine up instead of down. Make sure the brand name and other markings are right side up when

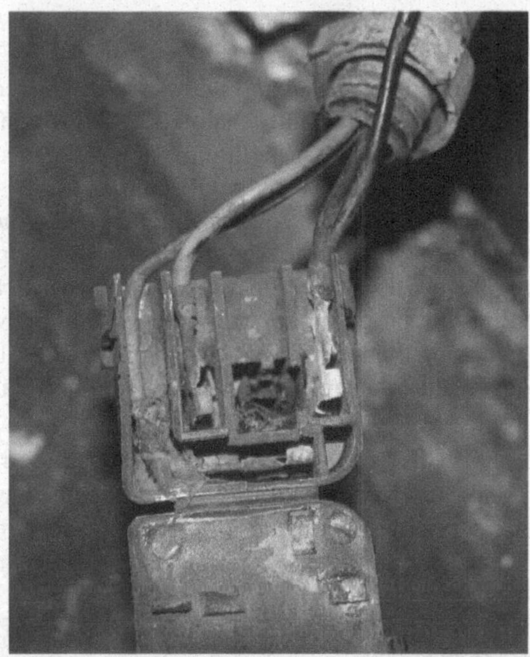

FIGURE 21-40 Corrosion in connectors, like this older, sealed beam connector, can cause the light to be dim or not work at all.

you install the bulb. Install the bezel and tighten the screws. Turn the headlamps on to make sure the bulb operates properly.

■ ***Halogen Inserts.*** Replacing the insert bulbs is typically not difficult, getting to them can be. Depending on the vehicle, the bulb may be accessible under the hood

```
Honda/Acura 1983-2012
**Lighting Data**
Headlight Switch(Auto)_____________OFF
Headlight Switch(Headlight)________OFF
Headlight Switch (High Beam)_______OFF
Headlight Switch(Off)_______________ON
Headlight Switch(Parking)__________OFF
Headlight Switch(Passing)__________OFF
Ignition Key Cylinder Light Command__OFF
Left Turn Signal Command___________OFF
Parking Light Command(Driver's)____OFF
Parking Light Command(Passenger's)___OFF
Passenger's Coutesy Light Command____OFF
1 2 3 4
```

FIGURE 21-41 Using a scan tool to monitor lighting system operation.

(**Figure 21-43**), through the fender (**Figure 21-44**), or you may have to remove items such as bumper covers.

Insert bulbs may twist and lock in place or use a locking ring or clip to secure the bulb to the lens assembly. The bulb shown in **Figure 21-45** uses a retaining ring that secures the bulb to the housing. Twist to loosen the ring, and then pull the bulb from the housing. The bulb shown in **Figure 21-46** uses a wire bail to hold the light in place. Push the highlighted end of the wire forward, and then pull up to release it from the latch. Swing the wire out from behind the bulb, and then remove the bulb. Bulbs like that shown in **Figure 21-47** twist and lock

Left Front Turn Signal Lamp

Add Bookmark

Diagnostic Data Display | Graphical Data Display | Line Graph

Exterior Lighting 1 Data

Parameter Name	Value	Unit	Control Module
Left Front Turn Signal/Hazard Lamp Command	Inactive		Body Control Module

Parameter Name	Value	Unit	Control Module
Automatic Headlamps Disable Switch	Inactive		Body Control Module
Headlamps On Switch	Inactive		Body Control Module
Park Lamps Switch	Inactive		Body Control Module
License Plate Lamps Command	Active		Body Control Module
Headlamps Flash Switch	Inactive		Body Control Module
High Beam Command	Inactive		Body Control Module
High Beam Select Switch	Inactive		Body Control Module
Front Fog Lamps Relay Command	Inactive		Body Control Module
Front Fog Lamps Switch	Inactive		Body Control Module
Backup Lamps Relay Command	Inactive		Body Control Module

Left Front Turn Signal/Hazard Lamp Command

Commanded State Uncommanded

Inactive | Active

FIGURE 21-42 Depending on the vehicle, you may be able to command lights through the scan tool to check operation.

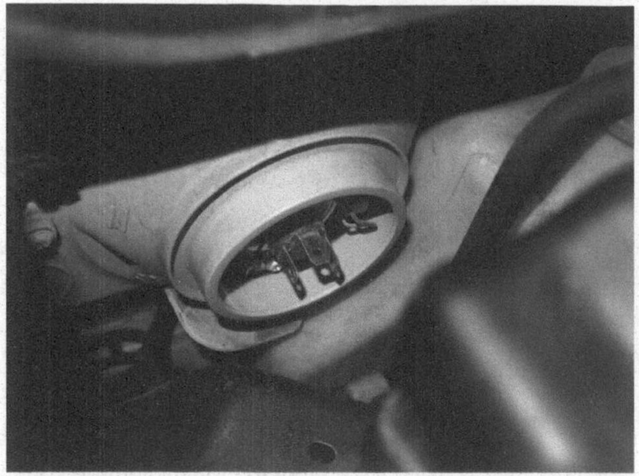

FIGURE 21-43 An example of a headlight bulb accessible from the engine compartment.

into place in the headlight housing; simply twist the bulb counterclockwise to release it and pull out.

With the bulb removed and the wiring disconnected, inspect the connector and terminals for corrosion or damage. When you are installing the new bulb, be careful not to touch the glass bulb. Oils and dirt from your skin can cause the glass to weaken and crack when it gets hot. During operation, these bulbs can reach temperatures of over 700°F, so proper installation is critical. If the bulb is not properly secured to the lens, it may work loose, melting the lens or even causing a fire. Install the bulb and reattach the retainer, then verify bulb operation.

■ ***HID Bulb Replacement.*** Be sure to read and follow all service precautions and procedures when you are servicing an HID system. Next, make sure the

(a)

(b)

FIGURE 21-44 (a) A headlight access door built into the inner fender. (b) Sometimes the inner fender liner must be removed to access the lights.

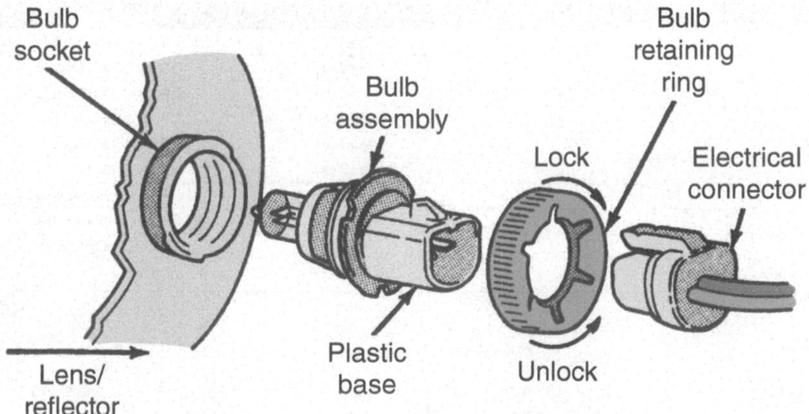

FIGURE 21-45 Some insert bulbs are held in place by a locking ring. Turn the ring counter clockwise to loosen and remove it from the housing.

FIGURE 21-46 Many insert bulbs are held in place with a wire bale. Simply press the end of the wire and unhook it from the tab holding it in place.

FIGURE 21-47 Some bulbs twist and lock into the headlight housing. Unplug the connector and rotate the bulb counterclockwise to remove. Align the tabs and rotate clockwise to install the new bulb.

lighting system is turned off. Some vehicles recommend disconnecting the 12-volt battery to prevent any chance of electric shock. Once the system is off, locate the service procedures for replacing the bulb. The typical replacement procedures include removing the cover over the HID bulb assembly, disconnecting the wiring connector, and then removing the bulb (**Figure 21-48**). Reinstall the new bulb, and secure it to the headlight housing. Turn the headlights on and verify their operation.

■ ***Headlamp Aiming.*** Proper aiming of the headlamps is important for safety reasons and so the vehicle complies with state and federal regulations. There are three methods of headlamp adjustment: headlamp aiming tools, built-in bubble levels, and automatically adjusted headlamps.

On many older vehicles, the headlamps are aimed by using either a headlamp aiming kit, or by using a special screen. Before you attempt either of these methods, several items need to be checked and adjusted. First, check the vehicle's tire pressure, and set the pressure to the manufacturer's specs. Check the vehicle's ride height to make sure that the vehicle is not sagging or uneven from side-to-side. Next, place the vehicle on a flat and level surface. If you are using an aiming kit, follow the manufacturer's instructions for installing and calibrating the aiming heads. If you are using an aiming screen, adjust the headlamp position until the light aligns with the appropriate indicators. Most headlamps use a spring-loaded adjustment screw (**Figure 21-49**). Turning the vertical adjustment screw moves the lamp up or down. The horizontal screw moves the lamp so that it points directly ahead of the vehicle.

Many vehicles have built-in aim indicators, like the bubble level (**Figure 21-50**). To adjust the headlamp position, turn the adjustment screw until the bubble level is centered.

Some vehicles have motorized headlamp assemblies that can automatically adjust for height changes, and on some vehicles, turn the assembly left and right based on steering wheel input. These systems are adjusted using special fixtures and through the on board computer system with a scan tool.

EXTERIOR LAMP SERVICE

Replacing exterior lights is a common task for an entry-level technician. Sometimes this process is very simple and sometimes it can be difficult. Regardless of the difficulty, it is important that the job is done correctly and that the right bulbs are used.

■ ***Why Does the Type and Wattage of the Bulb Matter?*** Many of today's cars and trucks use lighting modules or the BCM to monitor and control lighting functions. Ensuring the proper current flow through the lighting circuit is important to avoid damaging a module. When you are replacing a bulb, always replace it with the correct type and number. Installing a bulb that fits but does not have the correct wattage rating can cause the turn signals to flash at the wrong rate and can cause "light out" warning indicators to turn on. This is because the wattage of the bulb affects the current flow in the circuit.

■ ***Exterior Lamp Replacement.*** Parking lamps, turn signals, brake lamps, and backup lamps are often integrated into headlamp and tail lamp assemblies. Bulb replacement

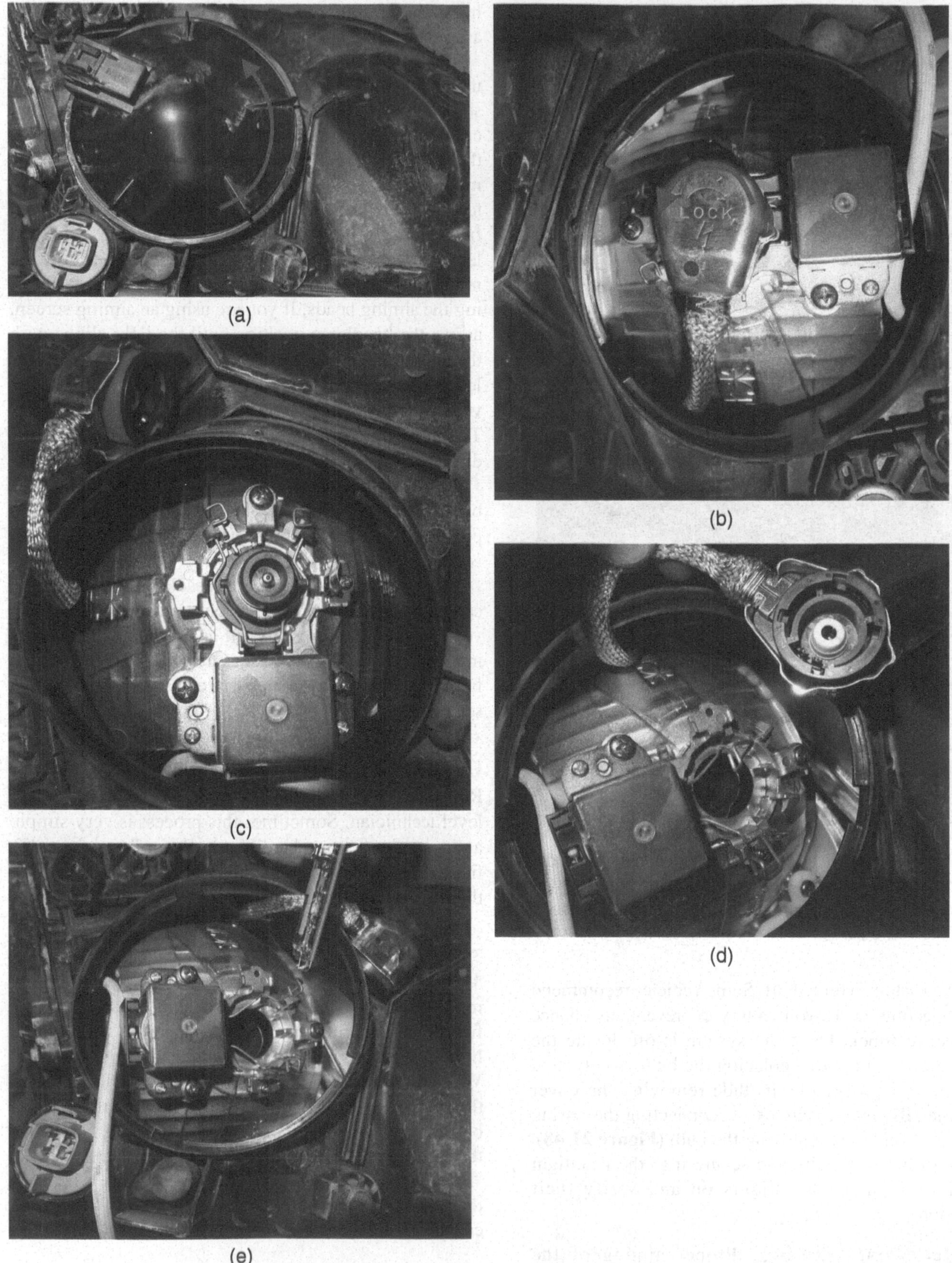

FIGURE 21-48 Replacing a HID bulb is similar to a standard bulb. (a) Remove the cover that protects the rear of the housing and the bulb. (b) Rotate the connector counterclockwise to unlock it from the bulb. (c) With the connector removed, unlatch the wire bale holding the bulb in place. (d) Inspect the connector and housing before installing the new bulb. (e) Install the new bulb while trying not to touch the glass.

FIGURE 21-49 An example of left and right headlight adjustment.

FIGURE 21-50 Many vehicles have built-in bubble levels for headlight aim adjustments.

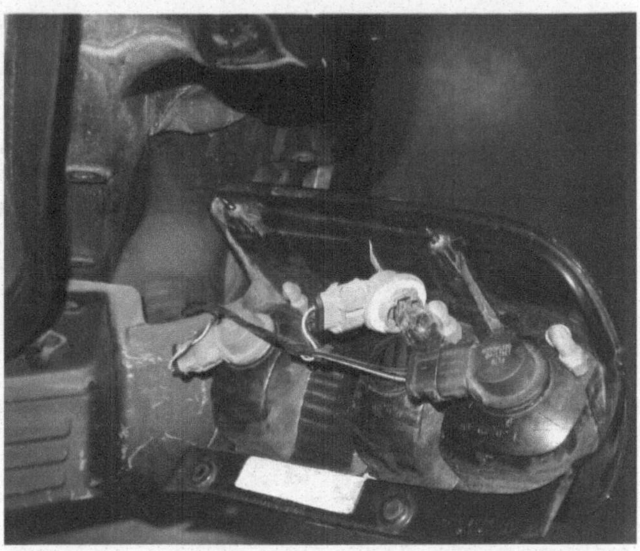

FIGURE 21-51 Removing a tail light housing to replace a bulb.

FIGURE 21-52 A bulb damaged by exposure to the outside elements.

usually requires the removal of the assembly (**Figure 21-51**). Many exterior lights use a twist-lock type of bulb socket. The socket is inserted into the light assembly, and then it is turned clockwise slightly until it locks into place. To remove, twist the socket counterclockwise until the socket pulls out of the assembly. Remove the bulb and check the terminals in the bulb connector for corrosion or damage. Some exterior lights are exposed to the outside elements, meaning that road dirt, snow, and salt can cover the wiring and connectors. This exposure, over time, can severely damage the wiring, connectors, and terminals, and bulbs (**Figure 21-52**).

Many exterior assemblies are secured to the body with one or more screws that must be removed to access the lights. The screws are typically either Phillips or Torx head and pass through the lens into a plastic retention nut. Be careful when you are removing these screws as the heads can easily strip if the screw is stuck in place. When you reinstall the lens, do not overtighten the screws as this can crack the lens housing.

■ ***Exterior Lamp Diagnosis.*** As with other diagnostic procedures, determine if a problem is with a particular piece of a circuit or is a fault of the entire circuit. If one lamp is out, it is likely a faulty bulb. If the entire circuit is inoperative, begin by checking for an open fuse or fuses. If the fuse is intact, you may need a scan tool and a wiring diagram. **Figure 21-53** shows the use of a scan tool to access BCM outputs for testing lighting concerns. Using a scan tool may allow you to monitor switch data or command the lights to turn on.

Faulty bulbs and poor connections cause most turn signal circuit faults. Incorrect turn signal blink rate is caused by either blown bulbs or incorrect bulbs. Since bulb resistance determines the current flow in the circuit, if you replace bulbs with other bulbs that have different resistances, it can cause the flasher unit to blink at a different rate. Many vehicles are showing up in shops with lighting problems, such as the turn signals not working properly. This is often caused by someone replacing the factory bulbs with LED bulbs. Talking to the vehicle

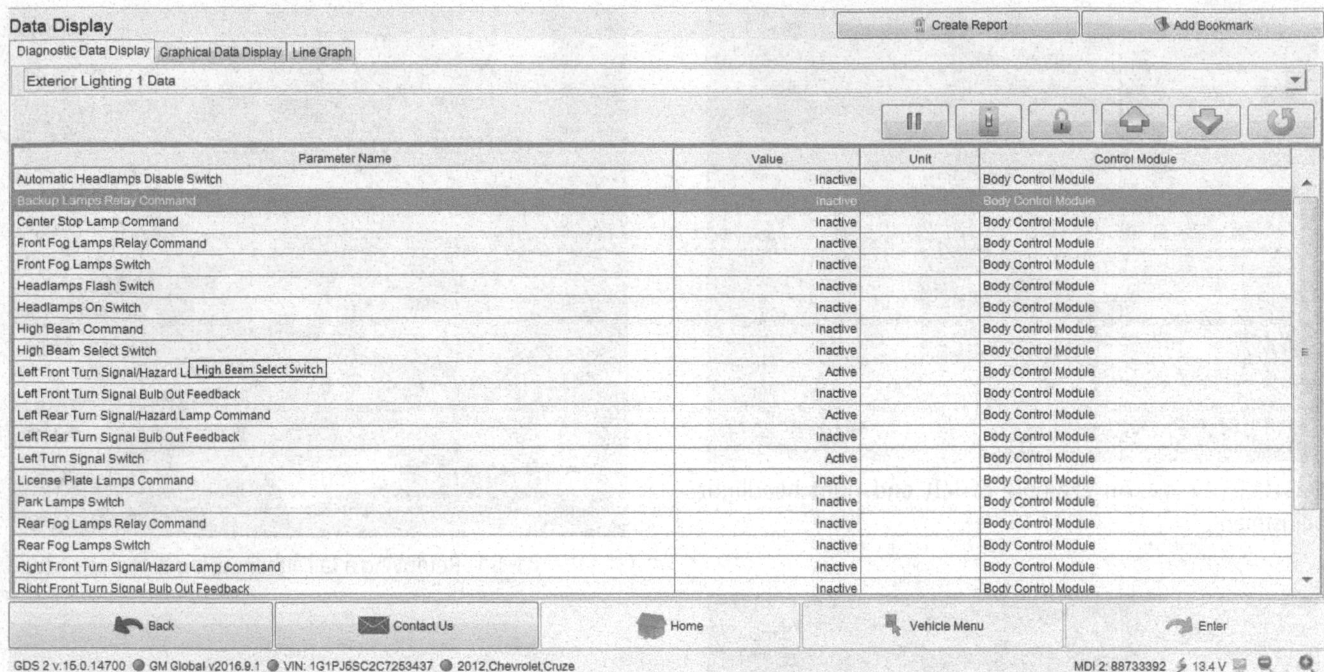

Parameter Name	Value	Unit	Control Module
Automatic Headlamps Disable Switch	Inactive		Body Control Module
Backup Lamps Relay Command	Inactive		Body Control Module
Center Stop Lamp Command	Inactive		Body Control Module
Front Fog Lamps Relay Command	Inactive		Body Control Module
Front Fog Lamps Switch	Inactive		Body Control Module
Headlamps Flash Switch	Inactive		Body Control Module
Headlamps On Switch	Inactive		Body Control Module
High Beam Command	Inactive		Body Control Module
High Beam Select Switch	Inactive		Body Control Module
Left Front Turn Signal/Hazard L…	Active		Body Control Module
Left Front Turn Signal Bulb Out Feedback	Inactive		Body Control Module
Left Rear Turn Signal/Hazard Lamp Command	Active		Body Control Module
Left Rear Turn Signal Bulb Out Feedback	Inactive		Body Control Module
Left Turn Signal Switch	Active		Body Control Module
License Plate Lamps Command	Inactive		Body Control Module
Park Lamps Switch	Inactive		Body Control Module
Rear Fog Lamps Relay Command	Inactive		Body Control Module
Rear Fog Lamps Switch	Inactive		Body Control Module
Right Front Turn Signal/Hazard Lamp Command	Inactive		Body Control Module
Right Front Turn Signal Bulb Out Feedback	Inactive		Body Control Module

FIGURE 21-53 Using a scan tool to test lighting circuits by monitoring if the circuit shows active when turned on.

owner and asking if LED replacement bulbs have been installed may save you some time and energy when diagnosing lighting concerns.

Even though electronic flashers are much more reliable than the older bimetallic units, that does not mean that they cannot fail. Electronic flashers contain a relay that is controlled by a small logic circuit. Both the electronics and the relay are subject to failure, which typically causes the turn signal circuit to stop operating.

Many vehicles have more than one set of switch contacts in the brake lamp switch. This is done as a redundancy for circuits such as cruise control. Be sure that when you test for power for the brake lamps that you also test the correct wires.

If the reverse lamps do not illuminate, check the fuse first. If the fuse is intact, locate the reverse lamp switch. For automatic transmission vehicles, the switch is often part of the Park/Neutral switch. Check for any adjustment procedure for the switch. Sometimes the Park/Neutral switch can be adjusted a slight amount. If there is no adjustment or if power is not present on the bulb side of the switch, the switch is likely faulty.

Refer to Table 21-1 for circuit faults and possible causes.

Driver Information Systems

Driver information, such as the speedometer, fuel gauge, engine temperature gauge, and system warning lights are usually part of the instrument panel (IP) display. Incorporated into the IP display, and often into navigation displays also, manufacturers are providing other types of information. These displays can be basic and provide minimal information, such as when a door is not shut or how many miles are left until the engine oil needs to be changed. Other systems provide maintenance, tire pressure, fuel economy, outside temperature, and many other pieces of information.

The **instrument panel** is part of the dash and contains the gauges, warning lights, and message centers that provide information to the driver about the operating condition of the vehicle. Instrument panels may have analog gauges (**Figure 21-54**), digital displays (**Figure 21-55**), or a combination of both. Older vehicles use basic analog display gauges that do not have any ability to be changed or configured by the driver. Newer vehicles often have electronic displays that can be changed to show different types of information depending on the driver's preferences.

SYSTEM INDICATOR/WARNING LIGHTS

Indicator lights, such as the high beam and turn signal indicators, are used to indicate when a circuit is on. **Warning lights** are used to alert the driver to a condition. For example, some vehicles have a "low traction" light, which is used to alert the driver to slippery road conditions. Other warning lights, like those discussed below, are illuminated when a fault is present so that the driver can have the vehicle inspected and repaired. Warning

FIGURE 21-54 An example of analog dash displays using gauges and pointers.

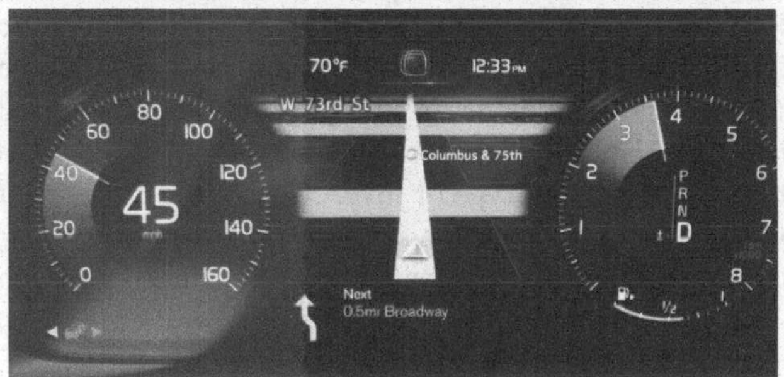

FIGURE 21-55 A digital dash display that replicates analog instruments.

FIGURE 21-56 All of the various warning lights should illuminate during the key on bulb check.

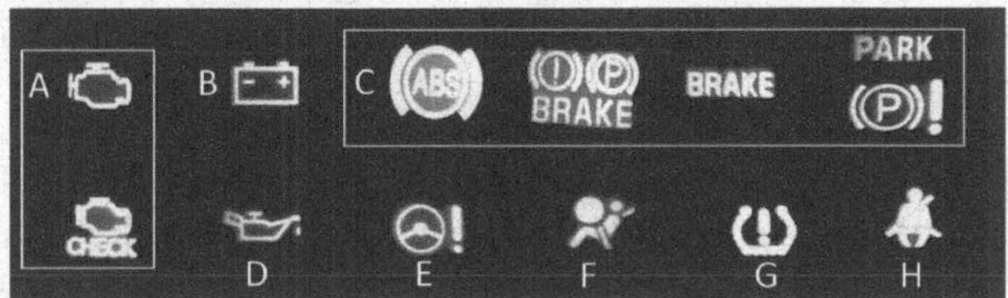

FIGURE 21-57 Examples of common warning lights.

lights are either yellow or red. Yellow lights indicate that faults exist and that the systems they indicate need to be inspected. Red lights are used to warn the driver of serious problems that should be looked at immediately.

When the ignition is turned on or the START button is pressed, the warning lights should all illuminate, perform a bulb check, and then go out once the engine has started. **Figure 21-56** shows an example of a dash during bulb check.

■ ***Service Engine Light.*** The service engine light has several names, including the service engine soon light (SES) and check engine light (CEL). This light is called *the malfunction indicator lamp (MIL)*. The MIL illuminates when the engine control module (ECM) determines that a fault exists in the emission control system that causes an increase in exhaust emissions. If the MIL is blinking during engine operation, this indicates that a misfire is occurring that is damaging the catalytic converter (**Figure 21-57a**).

■ ***Charging System Light.*** This light may display GEN, CHARGE, or a picture of a battery, it depends on the vehicle, and it illuminates if a fault is present in the charging system (**Figure 21-57b**).

■ ***Brake/ABS Warning Lights.*** The red BRAKE warning light can be triggered for several different reasons. First, this light should come on if the parking brake is applied. On some vehicles, this light may flash if the vehicle is driven and the parking brake is not released. Second, vehicles equipped with a brake fluid level sensor in the master cylinder reservoir illuminate the light when the brake fluid level drops below a certain point. Third, on vehicles equipped with a pressure differential valve in the brake system, the light turns on in the event of a loss of hydraulic brake pressure (**Figure 21-57c**).

The antilock brake system uses a yellow light indicating that a fault has occurred in the antilock brake system. If this light remains on, the ABS system is disabled. The service brakes should still function normally, but antilock operation is not available.

■ ***Oil Pressure Light.*** This light indicates that the engine oil pressure has fallen to a level that, if the engine continues to operate, may cause severe engine damage. An oil pressure sending unit completes the ground circuit for the light if the oil pressure falls below a preset limit, typically around 10 psi (**Figure 21-57d**).

■ ***Electric Power Steering.*** This light is used to indicate that a fault has occurred in the electric power steering assist system (**Figure 21-57e**).

■ ***Airbag/SRS/SIR.*** This yellow light, sometimes labeled SRS, SIR, or Inflatable Restraint, is used to indicate a fault in the airbag system, disabling the airbags. The airbag system light may also be used with an indicator for the side-impact airbags (**Figure 21-57f**).

The side-impact airbag light comes on if objects are placed in the passenger seat that press against the sides of the seat where the airbags are located.

■ ***Tire Pressure.*** Required on 2008 and newer vehicles, tire pressure monitoring systems must alert the driver if there is a significant drop in tire pressure. On some systems, this light may only come on after the vehicle has been driven for a short time, and after the ABS system has monitored wheel speed long enough to determine if one or more tires are underinflated. An example of the standard TPMS light is shown in **Figure 21-57g**. Some vehicles will also display the actual pressure or which tires are low on pressure.

■ ***Seat Belt.*** The seat belt light stays on until the front seat occupants have fastened their seat belts (**Figure 21-57h**).

■ ***Passenger or Side Airbag Off.*** Vehicles with occupant detection systems may turn the passenger side airbag off if less weight is detected on the passenger seat than is recommended for airbag deployment. Vehicles with side-impact airbags disarm the system if large packages are placed in the passenger seat. When this occurs, the side airbag light turns on (**Figure 21-58**).

■ ***Traction or Trac.*** This light, depending on the vehicle, may indicate slippery or reduced-traction road conditions. On vehicles with traction control, this light may indicate that a fault is present in the traction control system (**Figure 21-59**).

■ ***Electronic Throttle Control.*** Vehicles equipped with electronic throttle control (ETC) systems have a malfunction warning light that illuminates if a fault is present in the ETC system. This light usually comes on with the MIL and possibly a maintenance indicator. When a fault is present in the ETC system, throttle input is limited to a default or limp-in mode, and the engine rpm is reduced (**Figure 21-60**).

FIGURE 21-58 The red airbag light indicates a fault, the yellow indicates the side airbags are off.

FIGURE 21-59 This light indicates a fault in the traction control system.

MAINTENANCE REMINDERS

Maintenance reminders for many years were based solely on either time or mileage intervals. After a certain amount of time or miles were driven, a light on the IP would illuminate to remind the driver that maintenance, usually an engine oil change, was due. Many **maintenance reminders** are based on time, miles driven, how the vehicle is operated, and in what type of operating conditions. This allows the on board computer to determine a more realistic maintenance schedule based on real-world conditions.

Maintenance reminders may turn on a light that starts to blink when the vehicle starts. Many vehicles provide an oil life index. This may show either a percentage or how many miles before the next service is required (**Figure 21-61**). Other reminders display more information, such as what type of inspection is due at the next service interval. By more closely monitoring the vehicle's operation for maintenance needs, manufacturers can

FIGURE 21-60 This light indicates a fault in the electronic throttle control system.

FIGURE 21-61 Maintenance required lights are used to remind drivers that routine services are necessary.

prolong oil change intervals and reduce overall maintenance costs, which ultimately saves the consumer money.

■ ***Maintenance Reminder Reset.*** There are several methods of resetting the maintenance light or service index. A few of the common methods are described here.

- Reset button—this is used on older GM cars and trucks. A button in the fuse box is pressed and held, which resets the maintenance light.
- Press and hold the trip odometer—used on Toyota, Honda, Ford, and others, variations of holding the trip odometer button while turning the ignition from OFF to ON will reset the oil life index and/or maintenance light.
- Press the accelerator pedal to the floor five times in 10 seconds—also used on GM vehicles.
- Reset the maintenance interval in the driver information center.
- Reset using a scan tool.

To reset a maintenance light or service index, refer to the vehicle owner's manual or the service information for proper procedures.

COMMON ACCESSORIES

Many items that used to be options have now become standard equipment on many of today's cars and trucks. This is because of increased expectations by consumers and the decreasing cost of electronic components and small, powerful permanent magnets.

Accessories are often optional equipment, such as a keyless entry system. Depending on the vehicle's trim level, options may become standard equipment when they are part of an equipment package. In most cases, accessories are installed at the factory when the vehicle is built. If the customer wishes to add a feature, such as remote entry or remote start, the components may be able to be installed by the dealer.

FIGURE 21-62 This touchpad is a type of keyless entry system available on Ford products.

■ ***Remote Keyless Entry (RKE) and Remote Start.*** Also *called keyless entry*, **RKE** systems use a remote to activate the power door locks. On some vehicles, the keyless system also opens the rear hatches, operates the power-sliding side doors, and opens the trunk or rear hatch. Activating the locking feature often arms the vehicle's antitheft and/or alarm systems. A variation of remote entry used by Ford uses a keypad on the driver's door (**Figure 21-62**). In this system, the driver enters a five-digit code to unlock the door.

Most vehicles offer or have as standard equipment a remote key fob to lock, unlock, or open trunks, tailgates, or side doors. The fob may also be a smart key used with push-to-start systems. An example of this type of remote is are shown in **Figure 21-63**. These remotes are also part of the vehicle's antitheft system. The remotes disarm the security system when they are used to unlock the doors. In many vehicles, a remote control door lock receiver is mounted inside the driver's door. This receiver picks up the transmissions from the remote transmitter and sends the information across the on board network. If the remote is used for push-to-start, there are transmitters and receivers inside the passenger compartment to determine if the fob is present and allow the engine to start.

FIGURE 21-63 An example of a keyless entry remote.

Some vehicles have a remote start feature. This allows the driver to start the engine using the remote fob (**Figure 21-64**). To prevent theft, the vehicle locks the doors when the remote engine start is used. In addition, to drive the vehicle after using the remote start, the driver must unlock the door, use the key to unlock the steering, and then place the vehicle into gear. If the driver presses the brake pedal without the key in the ignition, the engine will shut off. These systems often have an automatic "time-out," meaning that the engine will shut off after a certain amount of time passes, usually around 10 minutes. This prevents the vehicle from being started and left running all day on accident.

Many newer vehicles use smart phone apps to locate the vehicle, lock and unlock the doors and start the engine. Vehicles equipped with OnStar and similar systems can be unlocked remotely in the event of the key being locked in the car on accident.

■ ***Power Windows.*** The power window system on most vehicles uses small DC motors mounted inside the door to move the window glass within the window track. The motor and track are commonly referred to as a **window regulator**.

Depending on the vehicle, when the power window switch is pushed, voltage is routed through the switch contacts and to one terminal of the motor. This causes the motor to spin in one direction. When the switch is pushed in the other direction, the voltage is reversed from the switch and is sent to the opposite motor terminal, which causes the motor to spin in the opposite direction. **Figure 21-65** shows a schematic of this type of circuit.

Newer vehicles use the BCM or a door module to control power window operation. When the up/down switch is activated, this input is received by the module, which then either raises or lowers the window. This provides the express up/down feature found on most vehicles. When the power window switch is pressed and held, or moved past its normal travel to a second position within the switch, the window automatically lowers or raises completely after the switch is released. Newer vehicles may also have an obstacle detection circuit for the power windows. It can stop and reverse a window that is going up if something either crosses the path of the window or contacts the window as it is raised.

FIGURE 21-64 This vehicle has remote engine start from the keyless entry remote.

■ ***Power Locks.*** Power locks can, depending on the vehicle, operate in several ways. The basic function is the ability to lock and unlock all the doors at the same time. Power door locks operate in a similar manner to power windows. A reversible DC motor is used to lock and unlock the door latch. An example of a power door lock circuit is shown in **Figure 21-66**.

Many vehicles incorporate power lock circuits into BCM functions. This allows for door lock/unlock functions from the remote control and integration with the vehicle's security system. In addition, some vehicles have an automatic locking function that locks the doors when the transmission is placed in Reverse or Drive, or after the vehicle begins to drive. These settings are often customizable through the driver information center.

POWER WINDOW AND DOOR LOCK SERVICE

Since increasing numbers of accessories are being installed in modern cars and trucks, there is an increasing need to be able to repair these items when they stop working. Among the most common accessory repairs are those for power window motors.

■ ***Non Computerized Power Windows.*** Power window circuits receive power from a fuse, which is often supplied to the power window master switch. The switch toggles the power and ground supplied to the window motor, meaning that polarity of the power and ground to the motor changes depending on whether the switch is placed in the up or down position. Since the motor is a brush-type DC motor, it spins clockwise or counterclockwise, depending on the polarity of the connections.

■ ***Computerized Power Windows.*** Some window circuits are monitored or controlled by the BCM or power window module. User request data from the window switch is sent to the BCM or window module. The module then activates the power window motor.

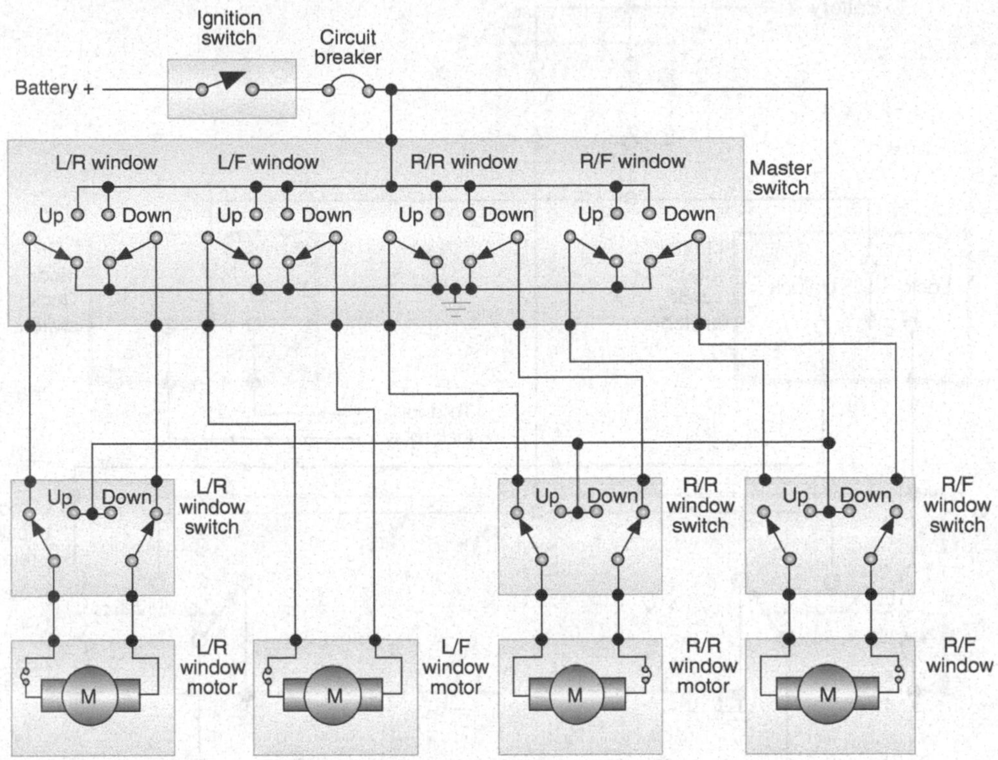

FIGURE 21-65 An illustration of a power widow circuit.

These are typically smart window systems, meaning that they monitor current flow through the motor for express up/down functions and obstruction detection.

■ ***Power Window Diagnosis and Repair.*** Begin your diagnosis by determining if all the power windows are affected or if just one window is inoperative. If all the power windows are inoperative, start by checking the fuse. Be aware that some vehicles use a fuse for each power window circuit so that one open fuse will not cause all the windows to stop working. If the fuse(s) are intact, obtain a wiring diagram for the circuit. Check for power and ground to the power window master switch. If either is missing, there may be an open in the wiring to the master switch. A common wiring failure is through the driver's door jamb. The movement of the door flexes the wiring to the point where eventually a wire or wires break. An example of this is shown in **Figure 21-67**.

If power and ground are present, test the switch by checking for power and ground through the switch when activating a window motor. This can be done by using a test light in place of the window motor. Unplug the window motor connection and connect a test light to the two terminals of the motor harness. Activate the window switch, and note the test light. The light should glow in both the up and down switch positions. If it does not, the switch is likely at fault. If the light glows in both directions, the window motor is faulty.

On a vehicle that uses a BCM or similar module for the power window circuit, connect a scan tool to the DLC and enter the body controller menu. The window switch activity can be checked via the scan tool and the scan tool can often be used to command motor operation (**Figure 21-68**). This allows you to test the switch, wiring, and motor all with the scan tool.

To replace a power window motor, the door panel must be removed. This requires careful handling and attention so that the plastic tabs and connections are not damaged during removal and installation. **Figure 21-69** illustrates how a door panel is attached to the door frame. For many vehicles, begin by removing any screws that attach door latch and armrests to the door. Check around the bottom of the door panel and behind any lights in the door for additional screws that secure the panel in place. Next, use a door panel screwdriver to pop the plastic tabs from the door frame (**Figure 21-70**). Once the door panel is loose, carefully lift the panel to release it from the window opening along the top of the door frame. Disconnect the interior door latch and any wiring to remove

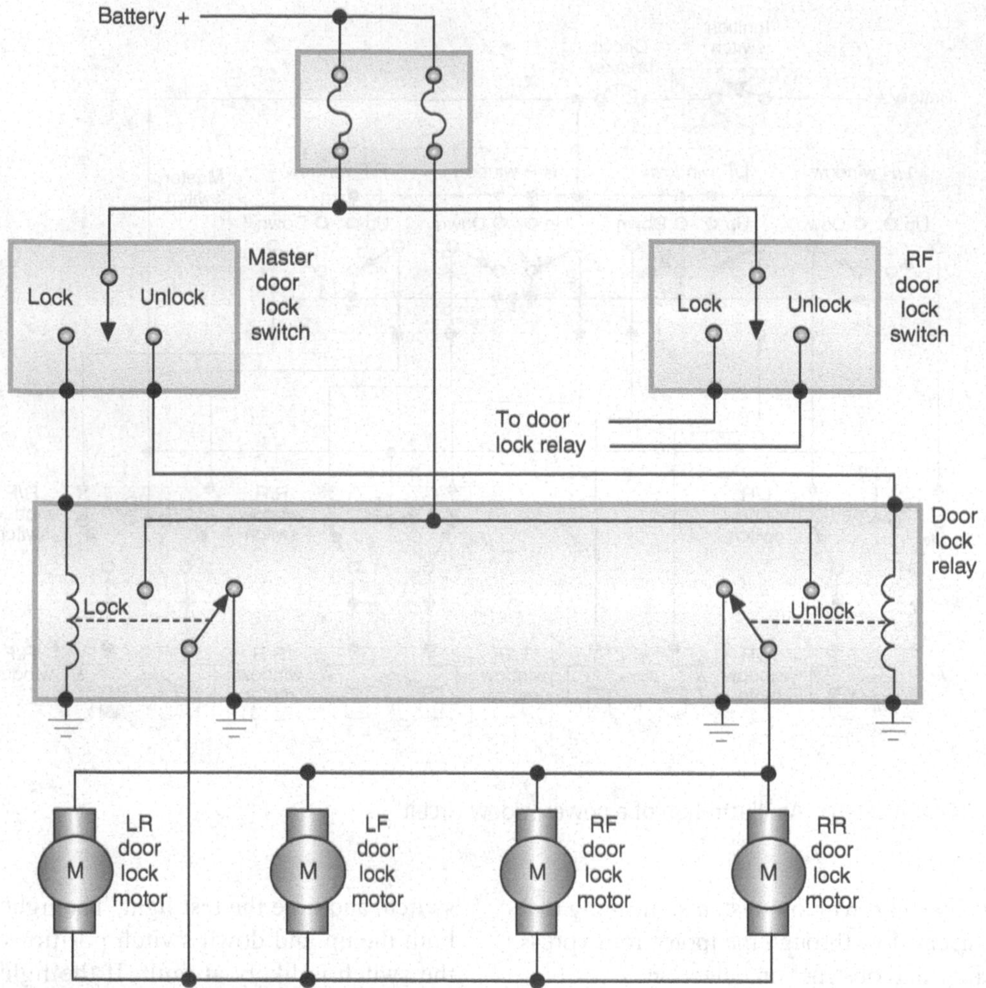

FIGURE 21-66 A power door lock circuit is very similar to a power window circuit.

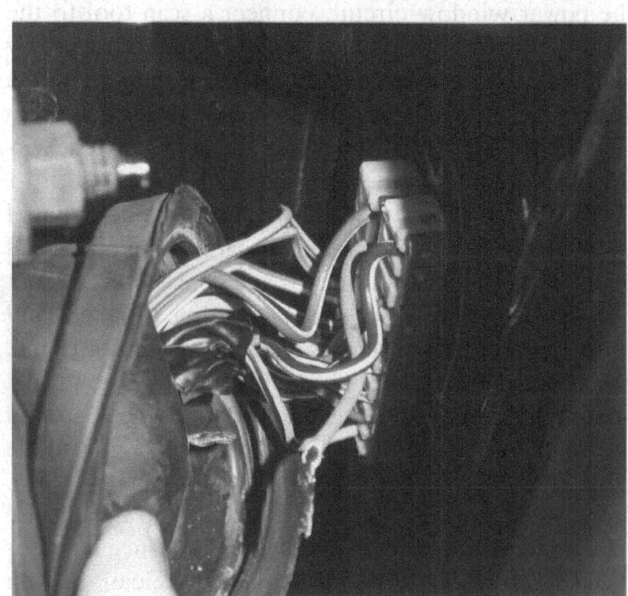

FIGURE 21-67 Wiring failures are common problems within the door jamb.

the panel. Once it is removed, place the panel in a safe and clean location, such as the back seat of the vehicle on which you are working (**Figure 21-71**).

To remove the power window motor, usually the window glass and regulator assembly must be removed. Locate the bolts securing the window glass to the regulator and remove. If possible, raise and secure the window in place in the window frame. Masking tape can be used to hold the glass from dropping. You may have to remove the glass. If so, lift the window carefully out of the window frame and set it aside in a safe location. Next, remove the bolts holding the regulator and motor to the door frame. Once the regulator is unbolted, you need to carefully maneuver it within the door until it can be removed from an opening in the frame.

Once the regulator is removed, install the new motor and regulator into the door and start each of the mounting bolts by hand. Make sure the regulator is properly positioned before you tighten the mounting bolts. Set the window back down into place and reattach the glass to the

File Function Setup TIS User Help

System Select | Stored Data | EMPS Live | **D-Door Motor Live**

2013 Prius V
2ZR-FXE

JTDZN3EU9D

Trouble Codes
Data List
Active Test

Parameter	Value
D Door P/W Auto SW	OFF
D Door P/W Up SW	ON
D Door P/W Down SW	OFF
Glass Position (Close-1/4)	OK
Glass Position (1/4-2/4)	OK
Glass Position (2/4-3/4)	OK
Glass Position (3/4-Open)	OK
Number of Trouble Codes	0

FIGURE 21-68 Using a scan tool command power window operation.

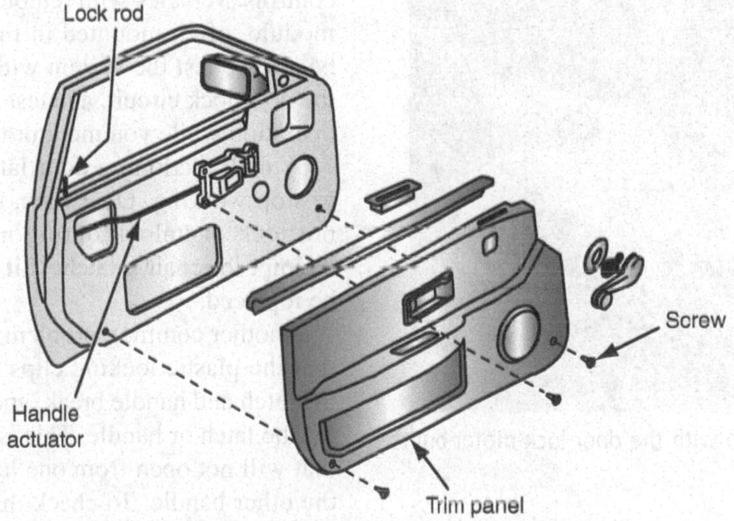

FIGURE 21-69 This illustrates how a door panel is attached.

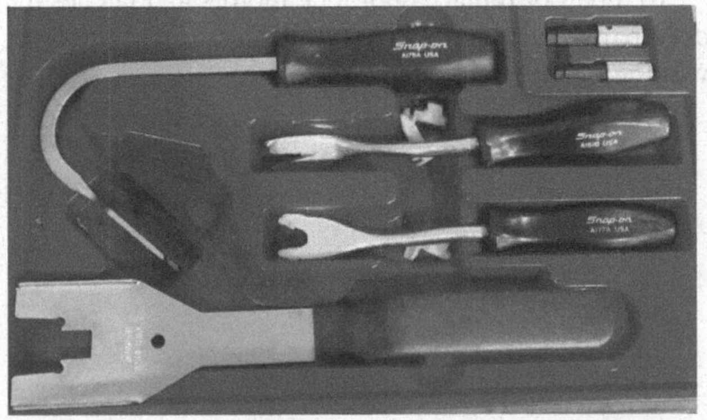

FIGURE 21-70 A door trim tool kit makes removing door panels easier.

regulator. Now, turn the key on and test the operation of the window. Make sure it moves up and down smoothly and that no binding or problems are present. If the operation is correct, reassemble the rest of the door panel.

■ ***Power Door Locks.*** Power door lock systems may use a small, remotely mounted motor or the motor may be attached directly to the latch (**Figure 21-72**). The motors are usually attached to the handles and latch with rods or cables, similar to the inside and outside door release handles.

■ ***Power Door Locks Diagnosis and Repair.*** Begin your diagnosis by checking to see if all the door locks are

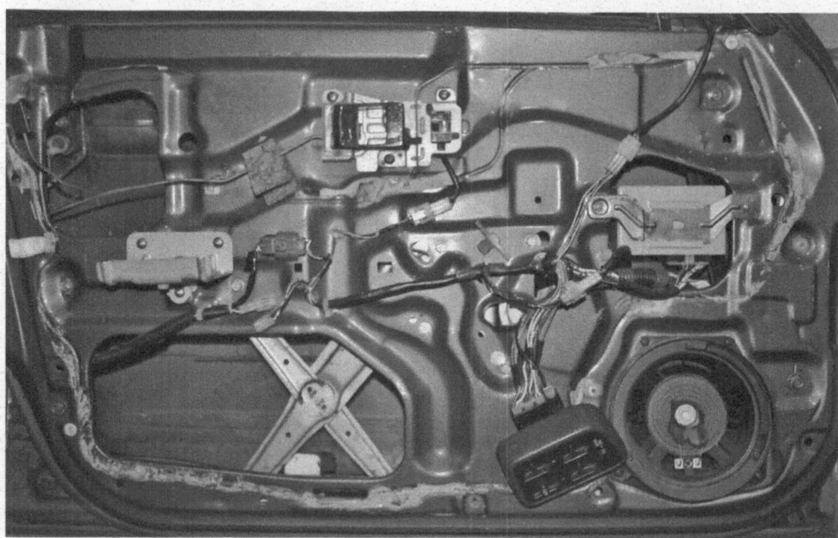

FIGURE 21-71 An example of a door with the trim panel removed.

FIGURE 21-72 A door latch with the door lock motor built into the latch.

inoperative or if only one door is affected. If all the locks are inoperative, begin by checking the circuit fuse. If the fuse is intact, obtain a wiring diagram to determine the next location for testing. If the system uses a remote, test the door lock operation using the remote and the interior controls. Vehicles with remote locks use a keyless entry module, often mounted in the driver's door. You may be able to test the system with a scan tool. If so, locate the door lock circuit, and test the switch and lock motor operation while you monitor the scan tool data.

A malfunctioning door latch can cause a door lock to stop working. Over time, the latches wear and may not lock or unlock during normal operation. Do not attempt to repair a latch; if it is faulty, the latch should be replaced.

Another common problem with door lock systems is that the plastic locking clips that hold the lock rods to the latch and handle break, and allow the lock rod to fall off the latch or handle. This is often indicated by a door that will not open from one handle but works fine from the other handle. To check this, remove the door panel and inspect the lock rods and clips. Replacement clips are usually available from the dealer.

To repair or replace door lock motors and latches the door panel must be removed as discussed in the power window service section.

SUMMARY

Courtesy lights provide illumination for safer vehicle entry and exit.

Door switches are used either to complete a light circuit or as an input for the BCM, which controls light operation.

Instrument panel illumination is controlled through the headlamp switch.

If all instrument panel illumination bulbs are inoperative, begin by checking the IP dimmer control.

HID lamps provide light by using a high-voltage AC arc of current across two electrodes.

High- or low-beam headlight activation is made by the dimmer switch.

Daytime running lamps operate whenever the ignition is on.

Solid-state flashers monitor current flow in the circuit but use an electronic switch to open and close the turn signal or hazard lamp circuits.

Reverse lamps are the clear lens lights in the back of the vehicle.

When you are installing the new bulb, be careful not to touch the glass bulb.

Many vehicles require that the headlamps be aimed by using a headlamp aiming kit or an aiming screen.

Some vehicles have built-in aim indicators that use a bubble level.

Power window motors are DC motors that spin clockwise or counterclockwise, depending on the polarity of the power and ground supplied to the motor.

Replacing a power window motor requires removing the door panel.

Power door lock motors may be separate motors, or they may be attached to the door latch.

Broken wiring in the driver's door jamb is a common failure that can cause various power window and door lock concerns.

REVIEW QUESTIONS

1. Bulbs that have a wire _______________ that glows when current flows through it are called incandescent bulbs.

2. LED lights can be checked with DMM using the _______________ test function.

3. The _______________ consumed by a bulb is based on the voltage and amperage that flow through the bulb.

4. A Smart key is used on vehicles with _______________ to _______________ systems to allow entry into the vehicle and to disarm the antitheft system.

5. The left front door has the _______________ window switch that controls all of the power windows.

6. All of the following are types of headlights except:

a. Halogen insert **c.** HID
b. TFT **d.** LED

7. A vehicle with power windows and door locks is experiencing intermittent window and lock operation. *Technician A* says a blown fuse may be the problem. *Technician B* says the wiring in the door jamb may be the cause. Who is correct?

a. Technician A **c.** Both A and B
b. Technician B **d.** Neither A nor B

8. *Technician A* says setting a package on the passenger seat may cause an illuminated yellow "side airbag off" light. *Technician B* says an illuminated yellow "side airbag off" light indicates a fault in the airbag system. Who is correct?

a. Technician A **c.** Both A and B
b. Technician B **d.** Neither A nor B

9. On a late-model car, the brake lights do not illuminate when the brake pedal is pressed. *Technician A* says a defective stop light switch may be the cause. *Technician B* says a scan tool may be used to check the brake light circuit. Who is correct?

a. Technician A **c.** Both A and B
b. Technician B **d.** Neither A nor B

10. *Technician A* says all bulbs with the same type of base are interchangeable. *Technician B* says installing a replacement turn signal bulb with the incorrect wattage can affect turn signal blink rate. Who is correct?

a. Technician A **c.** Both A and B
b. Technician B **d.** Neither A nor B

CHAPTER 22

Engine Performance Principles

Chapter Objectives

At the conclusion of this chapter, you should be able to:

- Explain the principles of engine operation.
- Identify components of the engine and its various subsystems.
- Describe four-cycle gasoline engine operation.
- Describe four-cycle diesel engine operation.
- Describe Atkinson cycle engine operation.

KEY TERMS

additives	**exhaust system**	**piston**
atkinson cycle	**horsepower**	**rotary engine**
atmospheric pressure	**hydrocarbons**	**throttle body**
biodiesel	**ignition system**	**torque**
camshaft	**intake stroke**	**vacuum**
compression ratio	**lubrication system**	**volumetric efficiency**
cooling system	**overhead camshaft**	
crankshaft	**overhead valve engine**	

The dominant engines in use today are the gasoline-powered Otto cycle engine and the diesel engine. The Otto cycle engine, named for German engineer Nicolaus Otto, has four strokes or cycles, meaning that each piston completes four up and down cycles to produce power. The air and fuel mixture is ignited by a high-voltage spark delivered to the combustion chamber by spark plugs.

The diesel engine, named for another German engineer, Rudolf Diesel, is the power behind ships, trains, trucks, heavy-duty equipment, and cars. Though similar in operation to Otto cycle engines—diesel engines also have four strokes—they do not use a spark to ignite the air–fuel mixture. Instead, diesel engines compress the air so much that the heat generated by this compression ignites the fuel that is injected into the cylinder.

Both gasoline- and diesel-powered engines operate in similar ways, drawing air into the cylinders, mixing the air and fuel together for combustion, and then exhausting the spent gases. Both engines also share many of the same basic components. Gasoline and diesel engines are classified as heat engines, meaning they convert thermal energy into mechanical output. **Figure 22-1** shows a modern gasoline engine and **Figure 22-2** shows a modern diesel engine. Both are very similar in appearance and construction.

Principles of Operation

Before you attempt to understand all of the engine's components and their operation, it is important to understand the fundamental principles that are involved in engine operation. Without this knowledge, it is difficult to understand how particular components and systems perform their functions and are affected by various operational problems.

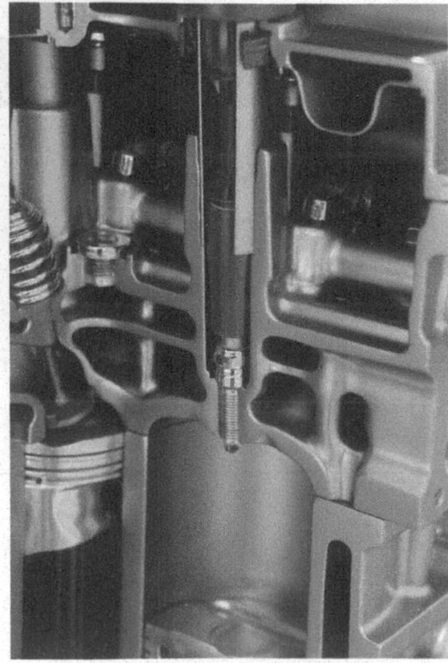

FIGURE 22-1 An inside look at a modern gasoline engine.

FIGURE 22-2 An example of a modern diesel engine.

AIR AND FUEL

Gasoline- and diesel-powered internal combustion engines (ICEs) are, at their most basic, air pumps. They pull large quantities of air in and pump it back out. The air is mixed with the fuel, which is ignited and burned. The process of burning the air and fuel generates power.

■ ***Air Pressure.*** The weight of the air that surrounds the earth is called **atmospheric pressure**. At sea level, this weight places approximately 14.7 pounds of pressure per square inch of surface area (**Figure 22-3**). The pressure

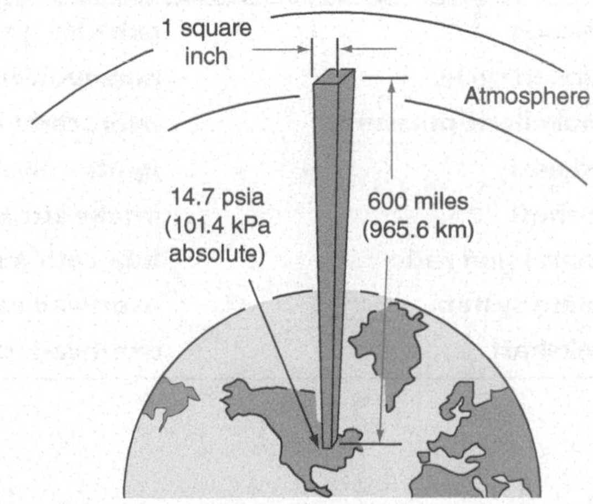

FIGURE 22-3 Air pressure is the result of gravity pulling down on the atmosphere and the weight of the nitrogen, oxygen, and other components of the air we breathe.

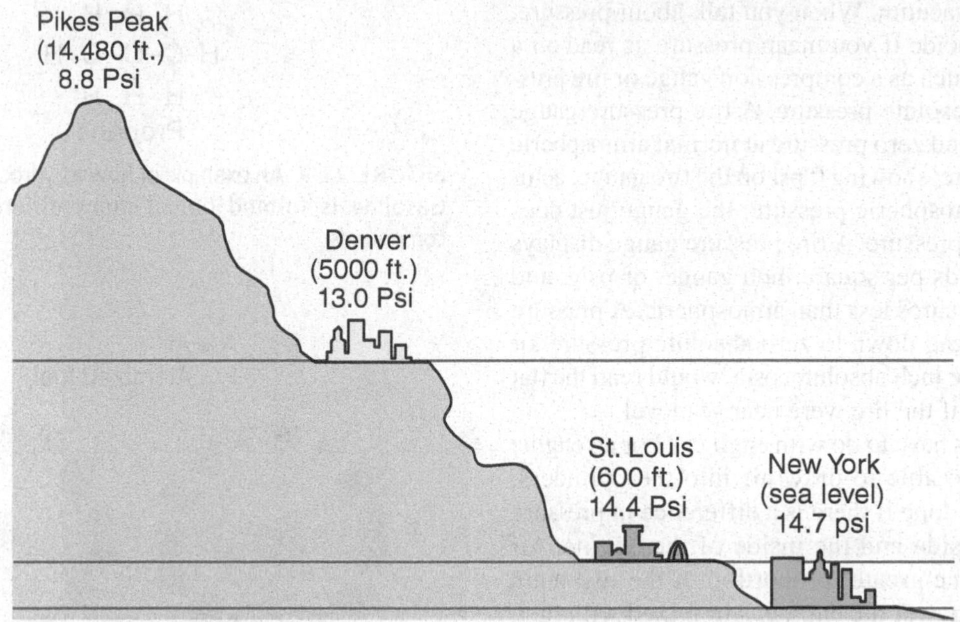

FIGURE 22-4 As elevation increases, there is less air above the ground. This results in lower air pressure as the elevation increases above sea level.

decreases as altitude increases; this is because as altitude increases there is less air above you, and so there is less pressure pushing down on you. **Figure 22-4** shows the differences in air pressure between sea level and other elevations. The difference in atmospheric pressure, such as from driving up into a mountainous area, can have a large impact on how an engine performs.

Inside the engine are cylinders, named literally because of their shape. Inside the cylinders are pistons (**Figure 22-5**). Above the cylinders is the cylinder head, which contains valves. The valves let air in and out of the cylinders. When the intake valve opens and the piston moves from the top of the cylinder toward the bottom of the cylinder, the volume of the cylinder increases and the pressure in the cylinder decreases.

Outside of the engine is atmospheric pressure. When the piston moves down, reducing the pressure in the cylinder, the air pressure outside of the engine forces air into the cylinder. This low pressure is called **vacuum**. Every time you drink with a straw you are experiencing air pressure and vacuum working together to cause movement.

Refer Chapter 4 in the Lab Manual for experiments with pressure and vacuum. Vacuum in the engine is not a complete absence of pressure; it is just pressure lower than atmospheric pressure. **Figure 22-6** shows a chart

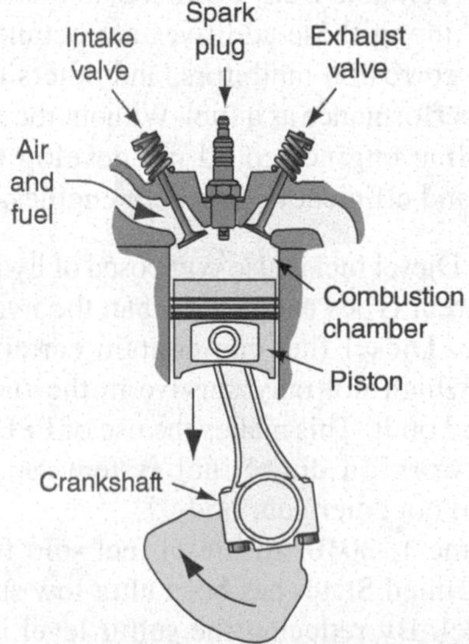

FIGURE 22-5 An illustration of the inside of an engine. Air enters the cylinder because outside air pressure forces air into the lower pressure in the cylinder as the piston moves downward.

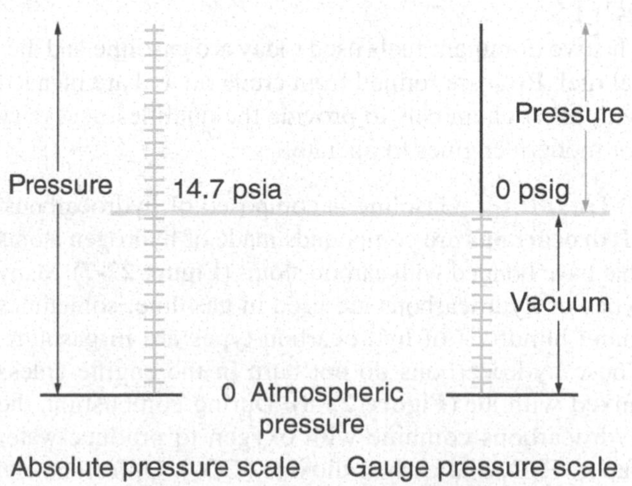

FIGURE 22-6 Pressure can be referenced as absolute or as gauge pressure and vacuum is pressure that is less than atmospheric pressure.

of pressure and vacuum. When you talk about pressure, you must first decide if you mean pressure as read on a pressure gauge, such as a compression gauge or tire pressure gauge, or absolute pressure. A tire pressure gauge is calibrated to read zero pressure at normal atmospheric pressure. A flat tire, showing 0 psi on the tire gauge, actually contains atmospheric pressure; the gauge just does not show us that pressure. A tire pressure gauge displays pressure in pounds per square inch gauge, or psig, and cannot show pressures less than atmospheric. A pressure gauge that can read down to zero absolute pressure, or pounds per square inch absolute, psia, would read the flat tire as 14.7 psia, if the tire were near sea level.

What does this have to do with engines? For an engine to run it must be able to draw air into the cylinders. This can only be done if there is a difference in pressure between the outside and the inside of the engine. Air also comprises the greater proportion of the two main ingredients that power the engine. Air, mixed with fuel, is burned in the cylinders. Much more air is used by the engine than fuel, roughly 15 times more air than gasoline. Unfortunately, not all the air drawn into the engine can be used to make power. The air we breathe is composed of nitrogen, oxygen, argon, and other trace gases. Roughly, nitrogen accounts for 78%, oxygen for 21%, and argon for slightly less than 1% of the total volume. Nitrogen, however, is generally inert, meaning it does not normally react with other substances. In the engine, nitrogen neither helps nor hurts combustion. At high temperatures, the nitrogen combines with oxygen to form oxides of nitrogen, which is an exhaust emission that causes smog, a form of air pollution. Oxygen, even though in a much smaller proportion of the overall volume of air entering the engine, is what allows the fuel to burn and the engine to produce power.

FUELS

The two dominant fuels used today are gasoline and diesel fuel. Both are refined from crude oil and are blended with other chemicals to provide the qualities necessary for modern engines to operate.

■ ***Gasoline.*** Gasoline is composed of hydrocarbons. **Hydrocarbons** are compounds made of hydrogen atoms that have bonded with carbon atoms (**Figure 22-7**). Many types of hydrocarbons are used in gasoline; sometimes many hundreds of hydrocarbon types are in gasoline. These hydrocarbons do not burn in the engine unless mixed with air (**Figure 22-8**). During combustion, the hydrocarbons combine with oxygen to produce water vapor, H_2O, and carbon dioxide (CO_2). Unfortunately, because the air, fuel, and combustion chamber do not allow for total combustion, other more harmful combinations also occur. These "leftovers" contribute to air pollution.

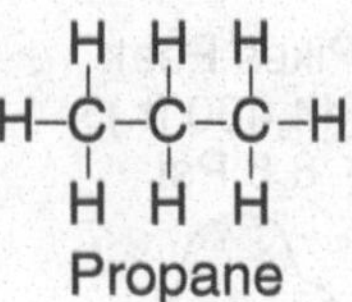

FIGURE 22-7 An example of how a hydrocarbon is arranged. Gasoline is formed out of many different hydrocarbon compounds.

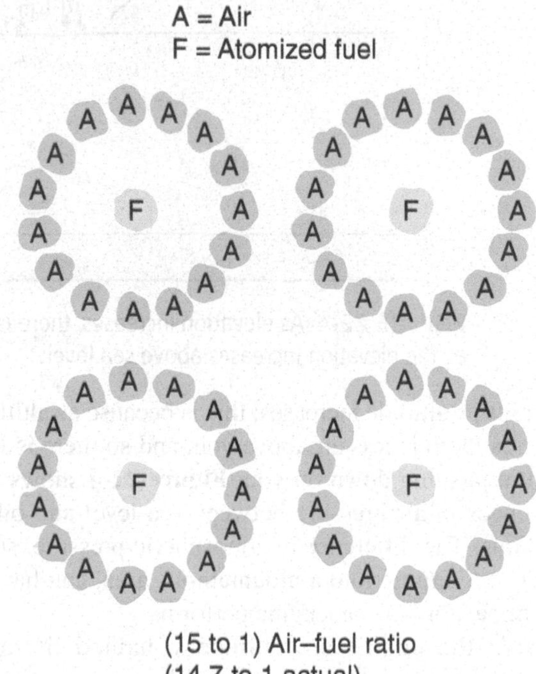

FIGURE 22-8 For combustion to be its most efficient, one part of fuel is combined with about 15 parts of air.

Gasoline contains many **additives** to make it useable as a fuel. Among these additives are octane boosters, detergents, corrosion inhibitors, and others to improve gasoline's performance as a fuel. Without the many additives, gasoline engines could not develop the power, reliability, and efficiency of modern engines.

■ ***Diesel.*** Diesel fuel is also composed of hydrocarbons, but of different types and blends than the hydrocarbons in gasoline. Diesel fuel can contain certain types of microbes, which not only survive in the fuel but also actually feed on it. This makes the use of PPE important when you work on diesel fuel systems so that these microbes do not enter your body.

Since June 1, 2010, all diesel fuel sold for on-road use in the United States has been ultra-low sulfur diesel (ULSD) fuel. By reducing the sulfur level in the fuel, diesel exhaust emissions are reduced, reducing air pollution, and acid rain.

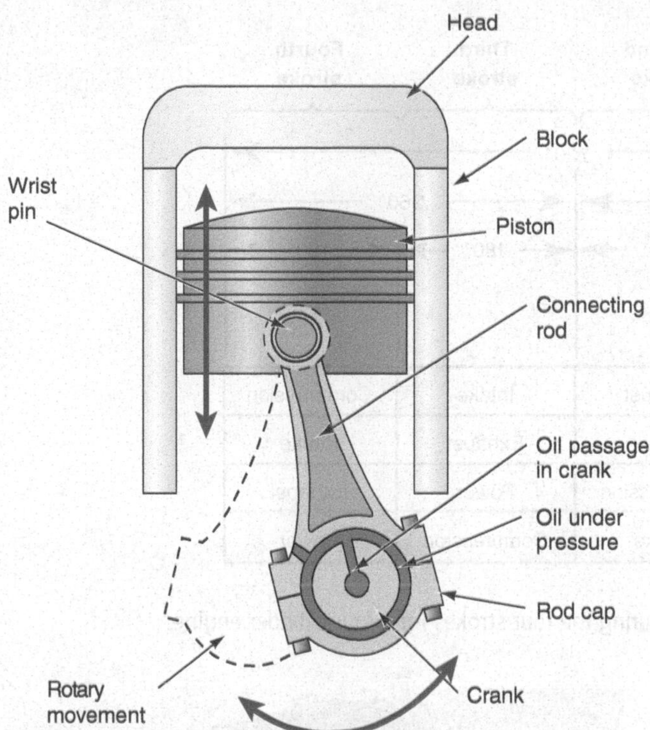

FIGURE 22-9 The piston moves up and down in the cylinder. This reciprocating or back-and-forth motion has to be converted into a circular or rotary motion.

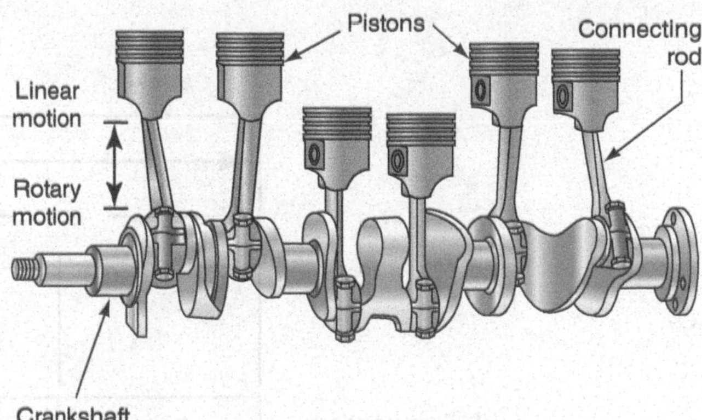

FIGURE 22-10 The movement of the connecting rods turns the crankshaft, which creates the rotary motion that eventually drives the wheels.

Biodiesel is a new trend in diesel fuel. **Biodiesel** is made from renewable resources, such as vegetable oils, animal fats used in cooking oils, and even algae. Pure biodiesel does not contain any petroleum products. Biodiesel can be blended and used with diesel fuel.

BASIC ENGINE OPERATION

Regardless of the type of ICE used, all are burning a mixture of a fuel and air to produce power. Gas and diesel engines are reciprocating engines, meaning that the power is produced by a reciprocating or back-and-forth motion (**Figure 22-9**). This motion must be converted into a rotary motion to drive the wheels. This is done by mounting the reciprocating piston to a rotating crankshaft via a connecting rod (**Figure 22-10**). As combustion takes place, the pressure increase in the combustion chamber forces the piston down in the cylinder. This downward motion is turned into rotary motion by the connecting rod attachment at the crankshaft. The crankshaft connects to the transmission and ultimately to the drive wheels. **Figure 22-11** shows a cutaway view of the engine and transmission to illustrate how power flows from the engine.

To produce the power to drive the crankshaft, gas and diesel ICEs have a series of four operating cycles. The four cycles, also called the four strokes, are intake, compression, combustion, and exhaust. You may notice that power is only produced during one of these four cycles, an inherent inefficiency of the four-cycle design. To complete the four cycles, the piston must travel up and down

FIGURE 22-11 Power flows from the crankshaft to the transmission and ultimately to the wheels.

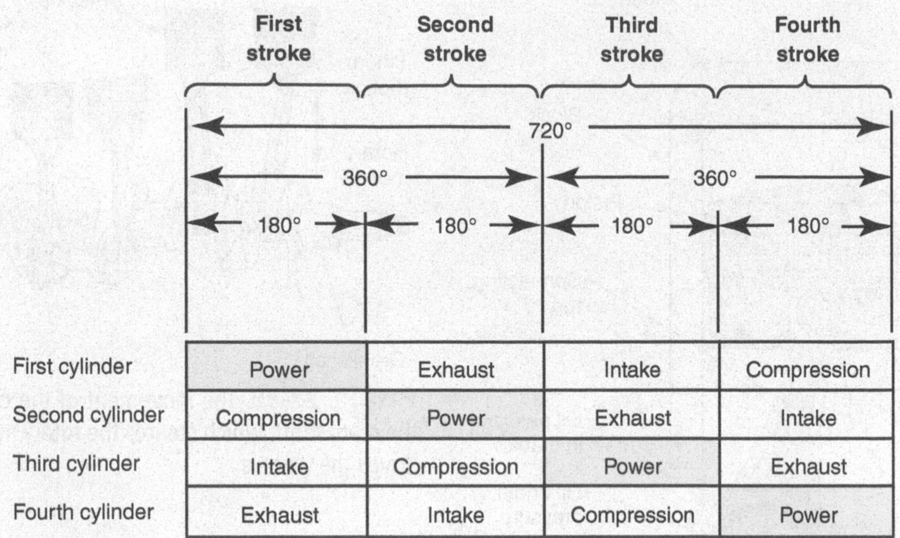

	First stroke	Second stroke	Third stroke	Fourth stroke
First cylinder	Power	Exhaust	Intake	Compression
Second cylinder	Compression	Power	Exhaust	Intake
Third cylinder	Intake	Compression	Power	Exhaust
Fourth cylinder	Exhaust	Intake	Compression	Power

FIGURE 22-12 This chart shows each cylinder during the four strokes for a four-cylinder engine.

the cylinder two complete times. Since power is only produced in one of the four cycles, more than one cylinder is used to produce enough power to propel the vehicle (**Figure 22-12**). The four cycles are discussed in more detail later in this chapter.

■ ***Gasoline Engines.*** Gasoline-powered engines have been in use since the beginning of internal combustion powered transportation. While the gas engine has improved greatly in performance, very little has actually changed in its basic operation.

All current production gasoline engines for automotive use in the United States are four-cycle, liquid-cooled, spark ignition engines. A typical gas engine consists of the engine block, cylinder head, intake and exhaust systems, fuel and ignition systems, and a network of sensors and a control module that monitors and adjusts fuel delivery (**Figure 22-13**).

FIGURE 22-13 An inside look at a modern four-cylinder gasoline engine.

■ ***Diesel engines.*** Diesel engines also date back to the earliest days of transportation and have also evolved into more powerful and efficient designs, while their basic operation has not changed in over 100 years.

Diesel engines tend to be larger and heavier than gasoline engines of similar size or displacement. Displacement means the amount of volume the pistons displace in the cylinders. When people refer to engine sizes, such as a Chevy 350 or 5.7, they are talking about the engine displacing 350 cubic inches or 5.7 liters of volume. Diesel engines are larger and heavier because the operating pressures in the combustion chamber of a diesel are much higher than those of a gas engine. This creates more stress on engine components, which requires larger and stronger components to handle the stress.

Diesel engines are also four-cycle, liquid-cooled engines. However, instead of using an ignition system to provide spark to ignite the air–fuel mixture, diesel engines use high compression to superheat the air in the cylinder. As the piston reaches the top of the cylinder during the compression stroke, fuel is injected. The superheated air ignites the fuel, and the resulting combustion forces the piston back down, turning the crankshaft. A typical diesel engine consists of the engine block, cylinder head, intake and exhaust systems, fuel system, and a network of sensors and a control module that monitors and adjusts fuel delivery (**Figure 22-14**).

■ ***Atkinson and Miller Cycle Engines.*** An **Atkinson cycle** engine is a modified gasoline four-cycle engine. The intake stroke is longer, until the piston is

FIGURE 22-14 An inside look at a modern four-cylinder diesel engine. A diesel engine, though very similar, operates differently from a gas engine. Most notably, diesel engines must use direct fuel injection, and there is no ignition system or spark plugs.

moving back up to compress the air–fuel mixture. This forces some of the air–fuel charge from the cylinder and back into the intake manifold. Once the intake cycle is over, the piston compresses the remaining air–fuel charge. The reduced charge generates less power, but it also uses less fuel and produces few exhaust emissions. This type of engine cycle is used in hybrid-electric vehicles like the Toyota Prius and others.

A Miller cycle engine operates in the same way as an Atkinson engine except that a turbo or supercharger is used. This increases the efficiency of the engine by forcing more air into the cylinders than what is normally drawn in.

■ ***Rotary Engines.*** For many years, Mazda was the only manufacturer selling a rotary engine in a production vehicle. Even though the Mazda RX-8 is no longer sold in the United States, there are many rotary engines still on the roads, and the engine may make a comeback as a generator in series-hybrid vehicles.

Unlike Otto and diesel engines, the **rotary engine** does not have any reciprocating parts, just rotating parts

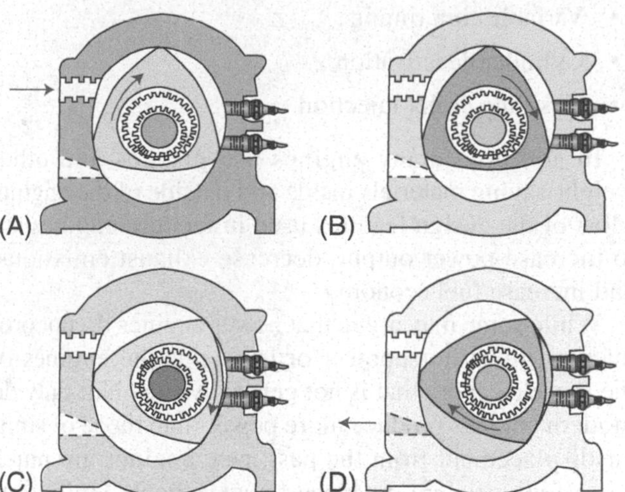

FIGURE 22-15 An illustration of a rotary engine. Rotary engines use a three-lobed rotor instead of reciprocating pistons.

(**Figure 22-15**). The three-sided rotor spins within the rotor housing on the eccentric shaft (egg or oval shaped). There is one combustion cycle for every revolution of the rotor, so there is one power stroke per revolution. This power cycle occurs more frequently than in a reciprocating gasoline or diesel engine, where two full crank revolutions are necessary for one power stroke. This means that in general, a rotary engine produces more power than a reciprocating gasoline engine of comparable displacement. Additionally, because there are no reciprocating motions, rotary engines can often reach higher maximum rpms compared to Otto cycle engines.

ENGINE DESIGN AND CONSTRUCTION

Engines have evolved greatly in efficiency and power production over the decades. Engine design has also evolved to utilize new materials and technologies to meet both consumer demand and legislative requirements to increase efficiency and reduce emissions.

■ ***Modern Engine Design.*** Modern engines, though mostly unchanged in basic operation, utilize design concepts that begin in the digital world long before they are ever produced as a physical product. This allows engineers and designers to test ideas and work out problems that arise before components go into production.

Many engines used in modern vehicles share some basic design qualities, these include:

- Overhead camshafts
- Four-valve cylinder heads
- Aluminum cylinder blocks and heads

- Variable cam timing
- Cylinder deactivation
- Gasoline direct injection

In addition, many engines use plastics and other weight-saving materials inside and outside of the engine. Most of the design features used in modern engines are to increase power output, decrease exhaust emissions, and increase fuel economy.

While some may argue that newer engines do not produce as much horsepower or torque as the engines of the muscle car era that is not generally true. Not only do modern engines produce more power than those of similar displacement from the past, new engines are much more fuel efficient, produce fewer exhaust emissions, and require less frequent maintenance.

■ ***Engine Construction.*** One reason that engines are lighter and more efficient are the materials from which they are constructed. Until the 1980s, most engines used cast iron engine blocks and cylinder heads. This made the engines strong, but heavy. Over time, aluminum began to be used in intake manifolds, then cylinder heads, and finally for the block.

Today's engines are mass-produced, typically by robots, and shipped from manufacturing plants to vehicle assembly plants all over the world. Many manufacturers have consolidated engine production so that a few engines can be used in many different vehicles sold in any market. This saves money and time in the building of the vehicle. In fact, some manufacturers collaborate on engine design and construction and use the same engine in many different models from different car companies.

■ ***Engine Lower End.*** The engine's lower end, also called *the bottom end* or *crankcase*, contains the crankshaft, pistons, connecting rods, oil pan, cylinder block, and oil pump (**Figure 22-16**).

The **crankshaft** is supported in the block by the main bearings. The crankshaft and main bearings are sandwiched between the main bearing caps or bedplate and the block (Figure 22-16). A bedplate is a section of the block that contains the lower main bearing halves. Bearings are supplied with engine oil to cool and lubricate the moving parts. The oil is pressurized by the oil pump and is fed through the lubrication system and forms a very thin oil film on the bearings and bearing journals. This film of oil, generally about 0.002 inches (0.05 mm) thick, separates the moving parts to prevent metal-on-metal contact and allows the crankshaft to spin smoothly on the bearings.

Attached to the crankshaft are the connecting rods. The smaller bearing journals on the crankshaft are for the connecting rods. Connecting rod bearings sit between the rod and the bearing journal on the crank (**Figure 22-17**). Oil is supplied to the connecting rod bearings just as for the crankshaft main bearings. The top of the connecting rod is attached to the piston with a piston or wrist pin. The wrist pin allows the piston to move on the connecting rod. The distance the piston travels is called *piston stroke*.

The **piston** forms the lower section of the combustion chamber, and the top of the piston is often shaped to aid in combustion. Valve reliefs are commonly cut into the top of the piston to prevent valve and piston damage when the piston is at the top of its stroke and the valves are open. The piston fits closely to the cylinder bore and is sealed by a set of compression rings (**Figure 22-18**). The top two rings are compression rings, which move across the cylinder wall on another very fine oil film, preventing metal-to-metal contact. The bottom ring is an oil control ring. The oil ring removes excess oil from the cylinder wall to prevent it from burning during the combustion stroke.

When combustion occurs, the pressure formed during combustion is exerted on the top of the piston, forcing it downward in the cylinder. This downward movement pushes on the connecting rod, which forces the crankshaft to spin. Momentum keeps the crankshaft spinning in between cylinder firings.

On some V-type engines, called **overhead valve engines**, the camshaft is mounted in the block (**Figure 22-19**). In this configuration, the camshaft (highlighted in red) is driven by the crankshaft by a timing chain or gearset. An example of a typical cam-in-block timing chain arrangement is shown in **Figure 22-20**. The **camshaft** has oblong-shaped lobes that push up on a valve lifter and pushrod (**Figure 22-21**). The cam lobes are positioned and shaped so that the valves open at the correct times for each cylinder's combustion cycle. The high part of the lobe opens the valve, and the valve is closed when the low part of the lobe is against the lifter. A pushrod pushes on the bottom of the rocker arm, which is mounted in the cylinder head. The camshaft on an overhead valve engine is also used to drive oil pump.

The engine oil pump is in the block on wet sump systems. A wet sump lubrication system is one in which the oil is kept inside the oil pan, beneath the block. When the engine is running, oil is drawn from the pan into the oil pump (**Figure 22-22**). The oil pan stores the oil that is not in circulation. The oil pump pickup or sump extends down into the lowest portion of the oil pan. This is the most common method of engine oiling.

Many high-performance cars use a dry sump lubrication system. A dry sump system does not store oil in the oil pan under the block; instead, oil is stored in a tank external to the engine, and two pumps are used to circulate oil. An example from a C7 Corvette is shown in

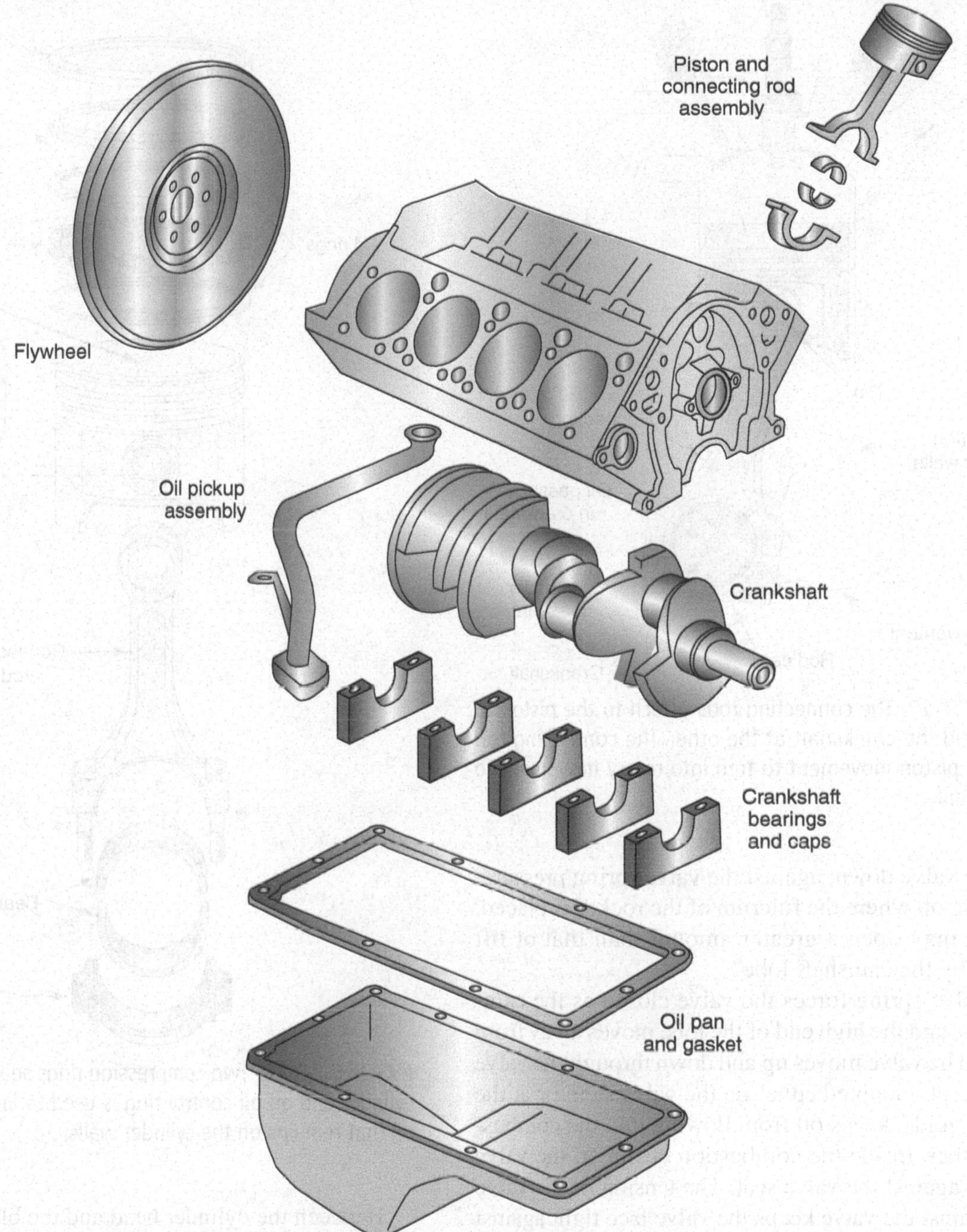

FIGURE 22-16 The engine lower or bottom end. The block holds the pistons, connecting rods, crankshaft, bearings, and oil pump. The oil pan bolts to the bottom of the block.

Figure 22-23. Dry sump systems allow the engine to sit lower since there is no oil pan storing the oil. This allows improved vehicle aerodynamics. Dry sump systems are not affected by braking, accelerating, and turning forces the way wet sump systems are.

■ ***Engine Top End.*** The cylinder head, intake manifold, exhaust manifold(s), camshaft(s), valves, and valve cover(s) form the engine top end or upper end (**Figure 22-24**). The cylinder head contains the intake and exhaust valves, which are used to let the air and fuel into the cylinder and the spent exhaust gases out. The term valve train is used to describe the valves and the components that open and close the valves. This includes the camshaft, rocker arms, lifters, pushrods, and valve springs.

■ ***Overhead Valve Engines.*** In overhead valve engines, the valves are opened by rocker arms (**Figure 22-25**). The pushrods, moved upward by the cam and lifter, push against the seat in the rocker arm. The rocker arm pivots on the rocker arm stud or bolt. As the rocker is pushed up by the pushrod, the opposite side of the rocker arm

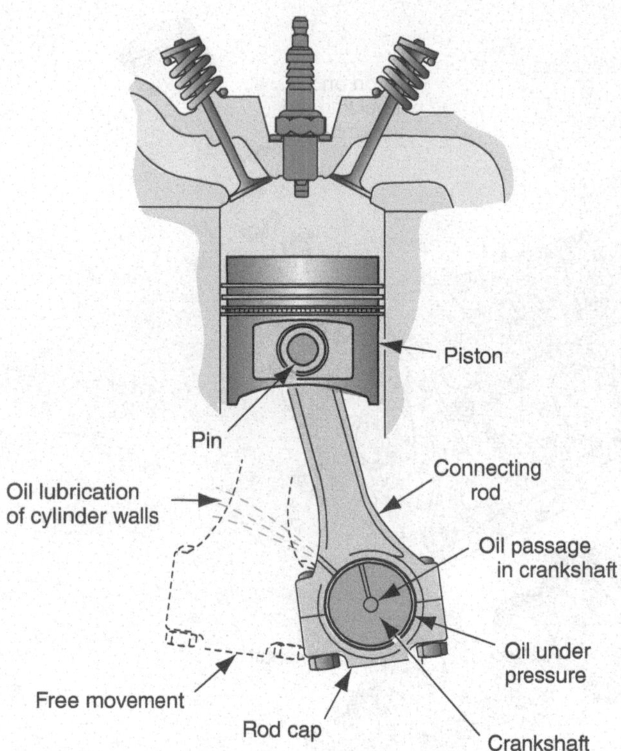

FIGURE 22-17 The connecting rods attach to the piston at one end and the crankshaft at the other. The connecting rod allows the piston movement to turn into rotary movement to turn the crank.

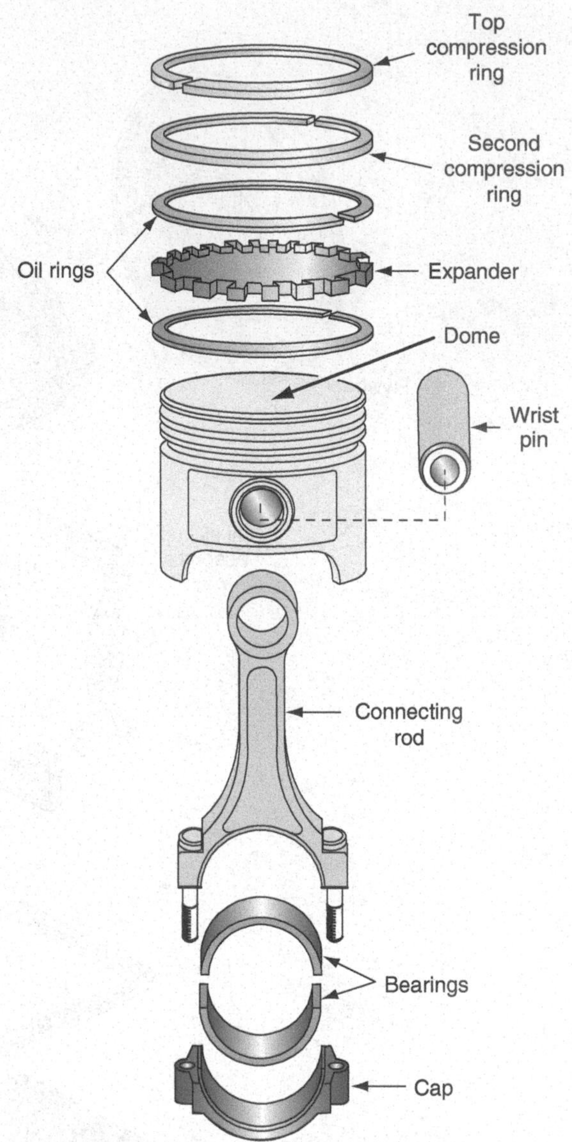

FIGURE 22-18 Two compression rings seal the piston to the cylinder and an oil control ring is used to limit the amount of oil that remains on the cylinder walls.

forces the valve down, against the valve spring pressure. Depending on where the fulcrum of the rocker is placed, the valve may open a greater amount than that of lift provided by the camshaft lobe.

The valve spring forces the valve closed as the camshaft turns, and the high end of the lobe moves away from the lifter. The valve moves up and down through the valve guide. A seal, mounted either on the valve stem or at the top of the guide, keeps oil from flowing into the combustion chamber. Inside the combustion chamber, the valve face seals against the valve seat. The tension of the valve spring against the valve keeps the valve face tight against the seat. A tight fit is necessary to prevent combustion loss between the valve face and valve seat.

■ *Overhead Cam Engines.* **Overhead camshaft (OHC) engines** have the camshaft located in the cylinder head (**Figure 22-26**). The cam rides in cam bearing journals formed in the head. This type of valve train may use one or two camshafts. Single overhead cam engines often open the valves using a rocker arm shaft and rocker arms. Engines with separate camshafts for the intake and exhaust valves are called *dual overhead cam (DOHC) engines* (**Figure 22-27**). This design typically reduces the number of components in the valve train as the cams sit on top of the valves, eliminating the need for rocker arms.

Between the cylinder head and the block is the cylinder head gasket (**Figure 22-28**). This gasket is typically a multilayer gasket with steel reinforcement rings around each cylinder bore. The head gasket seals vacuum, compression, combustion, oil, and coolant, making it one of the most important gaskets in the engine.

■ *Intake System.* Getting the air into the engine is the responsibility of the intake system (**Figure 22-29**). Modern vehicles use a network of ducts, hoses, chambers, and filters to route the airflow into the engine and to reduce operation noise.

Air enters the intake system at the inlet duct or snorkel, usually mounted in a fender or behind the front lights. An example of a fresh air intake system is shown

FIGURE 22-19 Camshaft location in an overhead valve engine. This design is used by many GM and Dodge V8 engines.

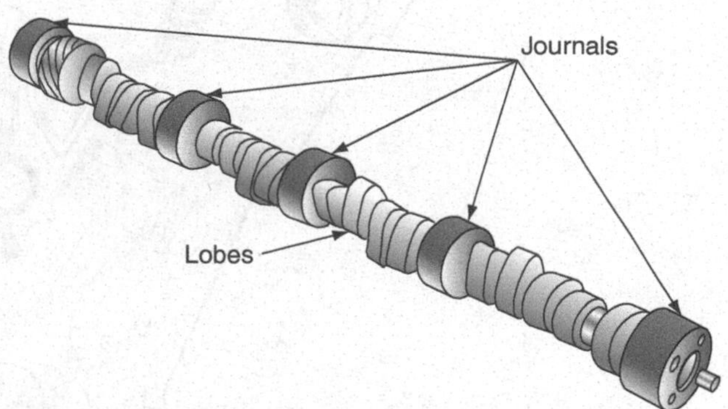

FIGURE 22-21 An illustration of a camshaft. Lobe shape determines when and how long the valves open.

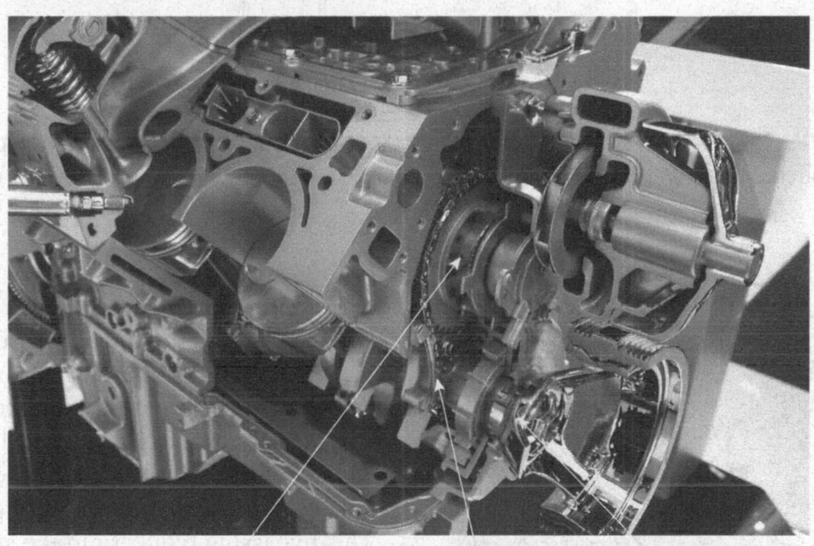

FIGURE 22-20 A timing chain and timing gears. The crank and cam must be kept in time or the engine will not run properly or could even suffer major damage.

in **Figure 22-30**. Air then enters the air filter housing, passing through the air filter element and into the intake hose to the throttle body.

Some new cars have hoses attached to the intake system that route engine sounds into the passenger compartment (**Figure 22-31**). These tubes are used to add to and enhance the engine sounds for the driver and passengers.

The **throttle body** is the valve that controls the amount of airflow into the engine (**Figure 22-32**). Older vehicles use a cable to connect the accelerator pedal to the throttle body. This mechanical throttle body has a coolant passage, used to prevent icing in cold weather.

Modern vehicles use a drive-by-wire throttle system (**Figure 22-33**). This means there is no direct physical connection between the accelerator pedal and the throttle body. The engine computer monitors the position of the accelerator pedal and then uses an electric motor to open and close the throttle.

The intake manifold routes the air to the intake port for each cylinder (**Figure 22-34**). Many engines use variable runner intake manifolds, meaning that the distance the air travels varies depending on engine rpm. During low-speed operation, long intake runners improve power by speeding up the airflow. At high rpms, short runners are used.

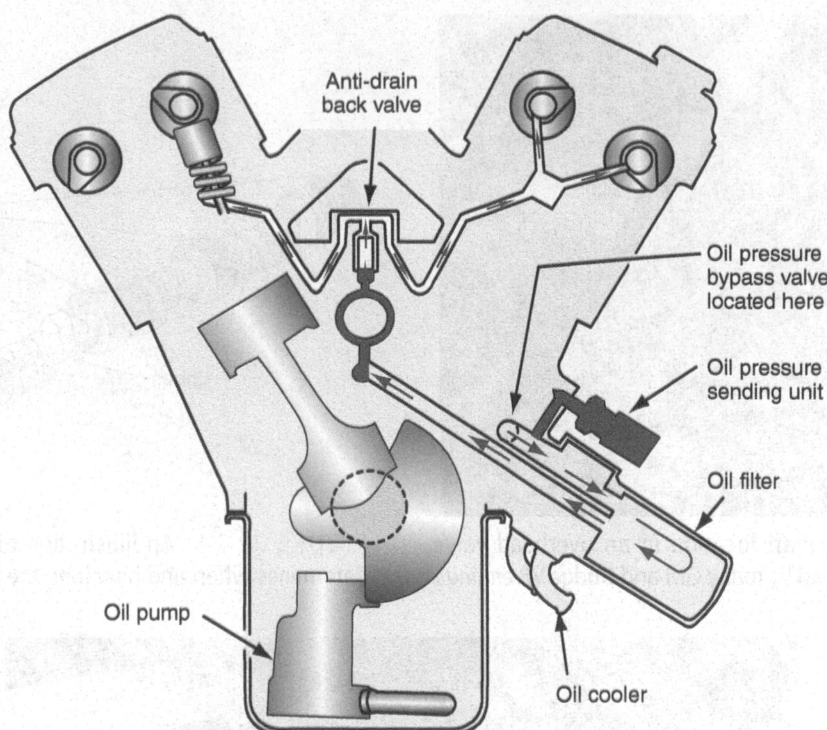

FIGURE 22-22 In a wet sump lubrication system, oil is pulled from the pan to the pump. It is then filtered and pumped to the rest of the engine.

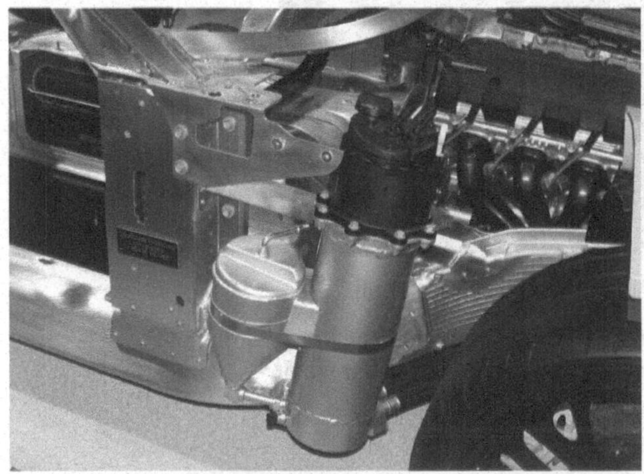

FIGURE 22-23 An example of a dry sump system. A large oil reservoir outside of the engine holds the oil, shown here located behind the right front tire.

On a turbocharged engine, the intake air is compressed and pumped from the outlet of the turbocharger. A turbocharger have two sections, the turbine and the compressor (**Figure 22-35**). Exhaust gases pass across the turbine wheel, causing it to spin. The turbine is connected through a shaft to the compressor. The compressor draws fresh air in and compresses it. The compressed air is then pumped into the intake manifold. The forcing of compressed air into the intake is called *boost*. Turbochargers generate more pressure as the speed of the turbine increases, so little air is pumped at low rpm. As engine speed increases, boost pressure also increases. Boost pressure must be controlled to prevent engine damage; this is done by the wastegate. The wastegate is a valve that opens to let exhaust gas bypass the turbine. This bypassing reduces turbine speed and decreases the amount of boost.

Some engines use a supercharger instead of a turbocharger (**Figure 22-36**). Superchargers, also called *blowers*, are belt driven by the engine (**Figure 22-37**). This means they are not dependent upon exhaust gases to develop boost pressure. A supercharger may use a set of lobes, a set of screws, or a set of scrolls that compress the air that is pumped into the engine.

The air that is pressurized, either by the turbo or supercharger, becomes hotter than normal intake air. This is because of the increase in pressure forcing the air molecules together more tightly, which increases their temperature. The heated air reduces performance since the air is less dense, meaning less oxygen will be present for a given volume of air. To offset this, turbocharged and supercharged engines use an intercooler. The intercooler is a type of heat exchanger, used to remove some of the heat from the air entering the engine.

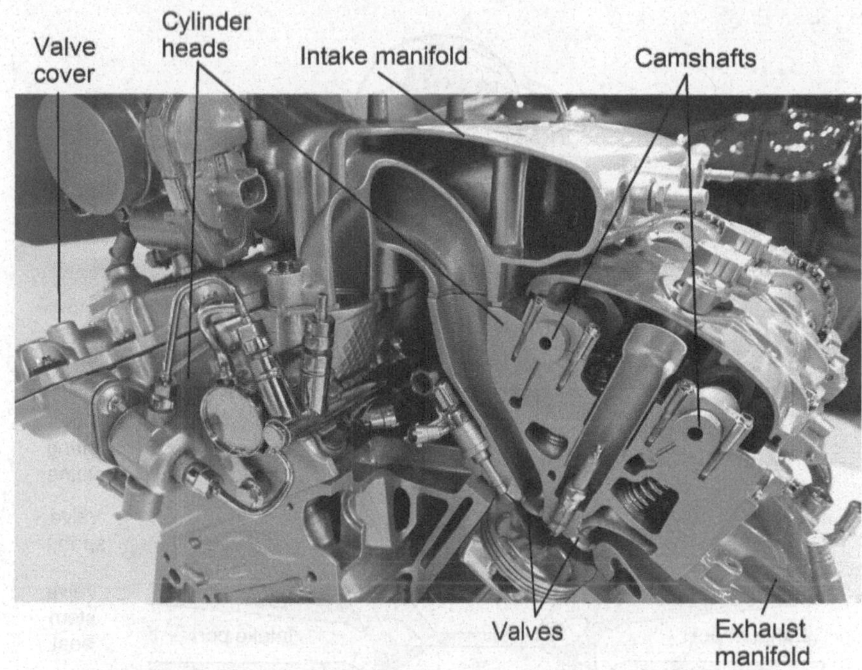

FIGURE 22-24 The upper part of a dual overhead cam engine.

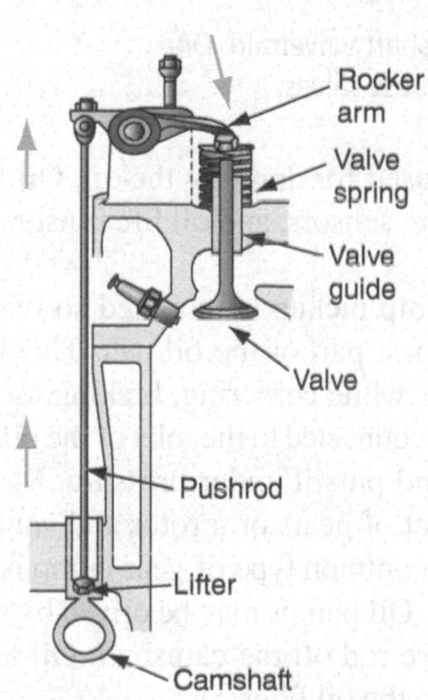

FIGURE 22-25 When the camshaft is located within the block, more valvetrain components are required to open the valve. Overhead valve (OHV) engines use lifters and pushrods to transmit the motion of the cam lobe to a rocker arm. The rocker arm pivots and pushes on the valve.

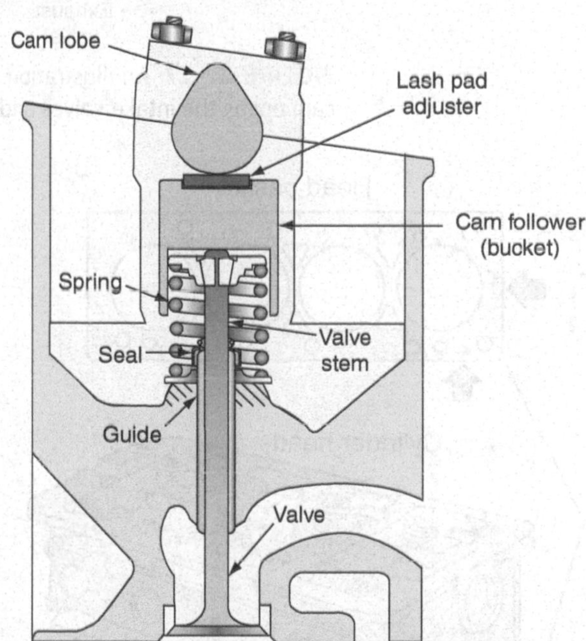

FIGURE 22-26 Overhead cam (OHC) engines often place the cam directly over the valve, eliminating a complex valvetrain. This arrangement is common with dual overhead cam engines.

ENGINE SUPPORT SYSTEMS

In addition to the engine itself, several systems are used to support its operation. Without these additional systems and components, the engine would not operate.

■ ***Lubrication System.*** The **lubrication system** supplies oil to the moving parts within the engine. In addition, oil removes heat from engine components, cleans the inside of the engine, and traps contaminants. Wet sump systems consist of the oil pan, oil pickup, oil pump, oil filter, and oil galleries or passages within the block. Oil flow in a typical engine is shown in **Figure 22-38.**

The oil pan is either stamped steel or aluminum and is bolted to the bottom of the engine block. A drain plug

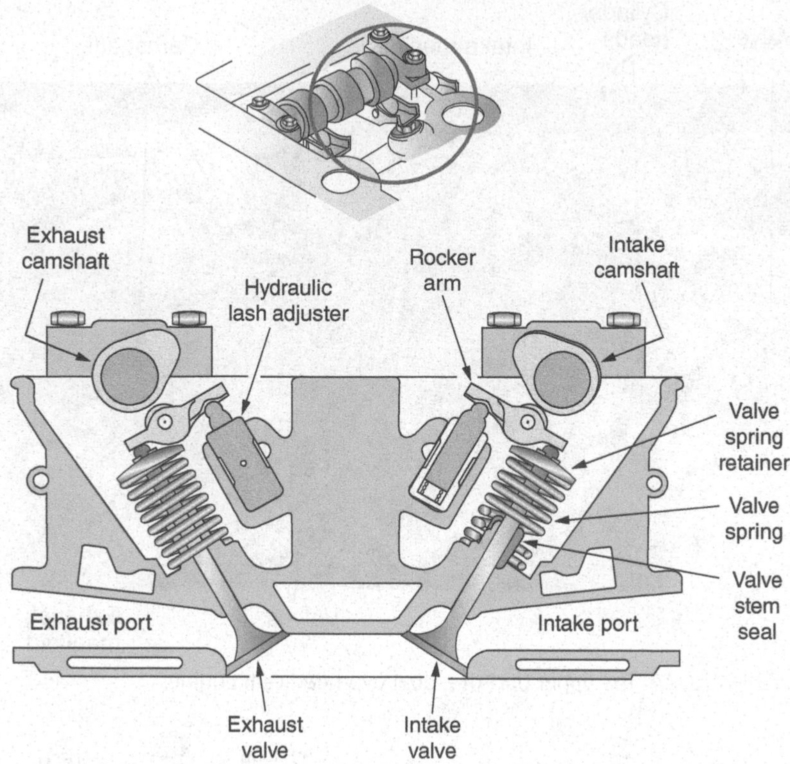

FIGURE 22-27 An illustration of a dual overhead camshaft valvetrain. One cam opens the intake valves and one cam opens the exhaust valves.

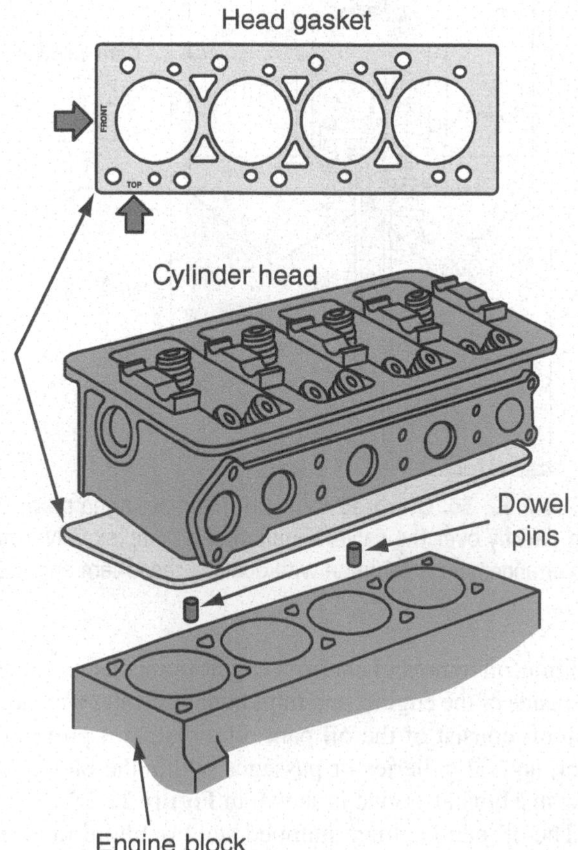

FIGURE 22-28 Sealing the cylinder head to the block is the job of cylinder head gasket.

in the pan is used for draining the oil. Oil level sensors, oil temperature sensors, and oil life sensors are mounted in the oil pan.

The oil pump pickup is mounted so that it draws oil from the deepest part of the oil pan. This helps prevent drawing in air while cornering, braking, or accelerating. The pickup is connected to the inlet of the oil pump, which pulls oil in and puts it under pressure. Most oil pumps use either a set of gears or a rotor and vanes to pressurize the oil. A common type of gear pump is illustrated in **Figure 22-39**. Oil pumps may be driven by the crankshaft or from a drive rod off the camshaft. Oil leaving the oil pump flows to the oil filter.

Oil filters remove contaminants and particles from the oil and are either cartridge or canister filters (**Figures 22-40** and **22-41**). Oil flows from the oil filter to the crankshaft main and rod bearings through passages called *oil galleries*.

Oil galleries are passages drilled or cast into the block and head for oil flow. **Figure 22-42** shows an example of oil flow through an engine. Oil is picked up from pan, goes to the pump, the oil filter, then through oil galleries that supplies the crankshaft. These passages also bring oil up to the camshaft and lifters in V-type engines and then up to the cylinder head. Drain holes in the head and block allow oil to drain back down into the oil pan.

Some diesel engines use high-pressure oil to open the fuel injectors. Known as hydraulic unit injector fuel systems, the engine control module (ECM) opens and closes an injector pressure regulator. The regulator delivers oil to the injector fuel rail, where a solenoid for each injector is controlled by the ECM, which supplies oil to the injector to open the injector and allow fuel flow.

■ ***Cooling System.*** The **cooling system** removes excess heat from the engine, maintains the engine's normal operating temperature, and provides heat for the passenger compartment. The components include the coolant, radiator, radiator cap, radiator fan, water jackets, thermostat, hoses, and heater core (**Figure 22-43**).

Coolant, also called *antifreeze*, is a mixture of chemicals and pure water and is used to absorb heat and transfer it away from the engine. The chemicals include ethylene glycol, corrosion inhibitors, dyes, and other

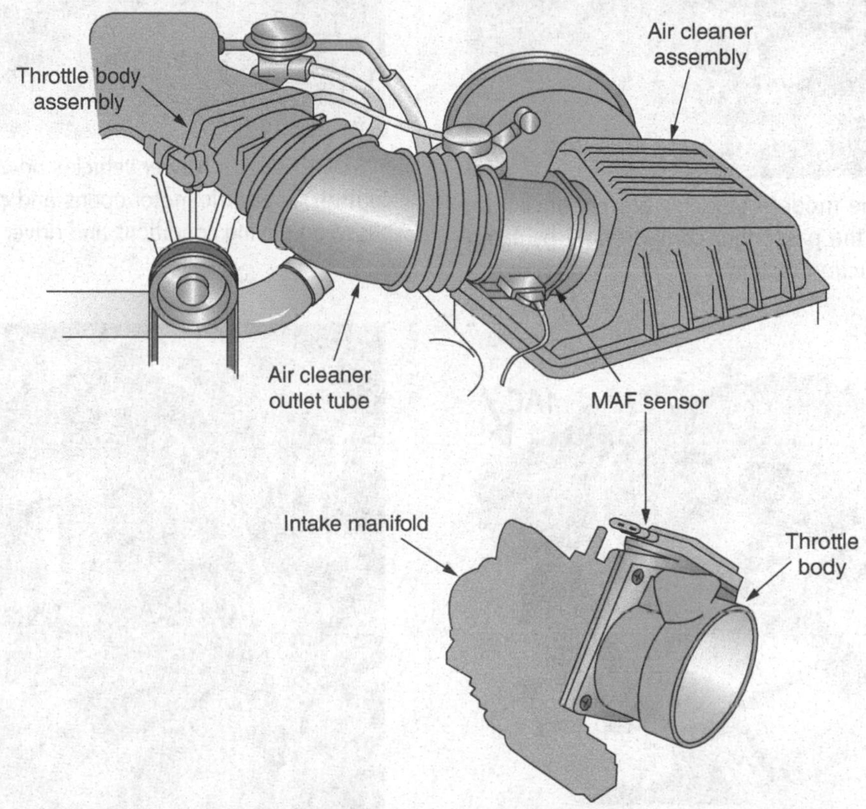

FIGURE 22-29 An illustration of a typical induction system.

FIGURE 22-30 This turbocharged engine has a large air intake on the right side of the engine compartment and an intercooler (on the top) to reduce incoming air temperature.

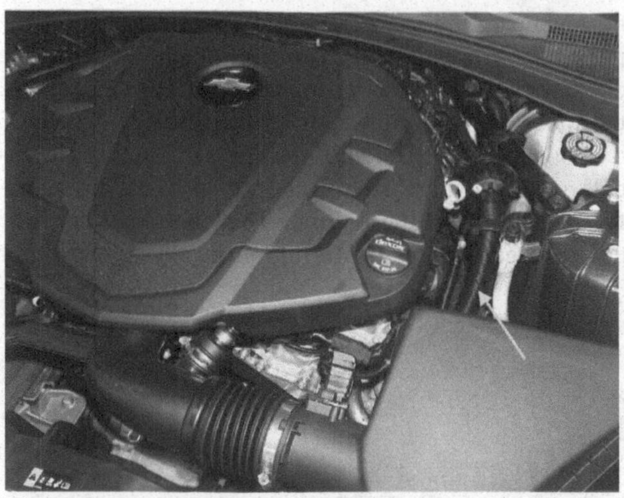

FIGURE 22-31 Some modern vehicles are enhancing the engine sounds within the passenger compartment by pumping in air from the induction system.

FIGURE 22-33 Newer vehicles now use electronic throttle control. An electric motor opens and closes the throttle plate based on driving conditions and driver input.

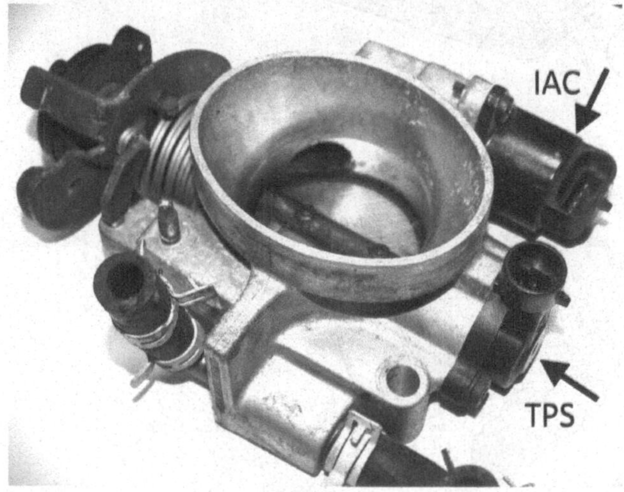

FIGURE 22-32 An example of an older, mechanical throttle body. Note the coolant hoses used to warm the throttle body and prevent icing in very cold weather.

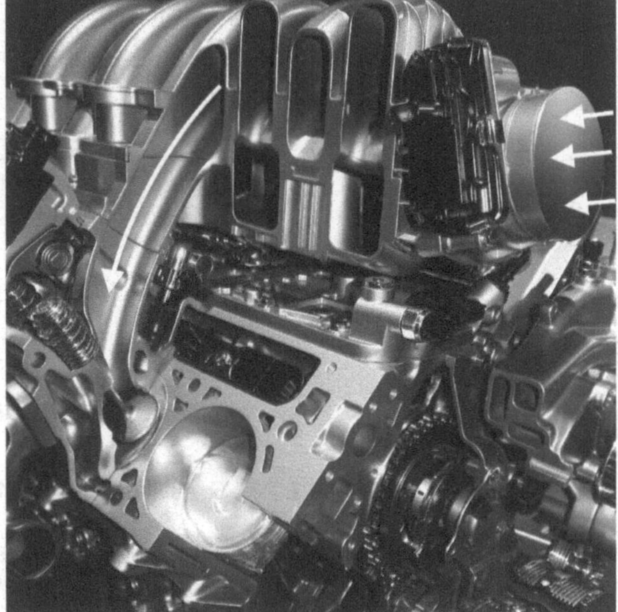

FIGURE 22-34 An example of an intake manifold. Air is funneled so that the best air flow is achieved to fill the combustion chamber properly when the intake valve opens.

ingredients. Coolant is available in many different varieties and colors, each with specific applications. However, color is not the main indicator of which type of coolant can be used in a particular vehicle. Only the specified coolant should be used as severe corrosion of engine components can occur if the incorrect coolant is installed.

The radiator is a heat exchanger, transferring heat from the coolant to the air surrounding the radiator. The radiator has a much larger surface area than it seems because of the number of fins attached to the coolant tubes.

The cooling system is sealed with a radiator cap or pressure tank cap. The cap allows pressure to build in the cooling system as the temperature increases. As pressure increases, so does the boiling point. For every pound of pressure in the system, the boiling point increases about 3°F. See **Table 22-1** for pressure and temperature correlations. An example of a cooling system cap and its pressure rating is shown in **Figure 22-44**. On systems with a radiator cap, the cap allows coolant to flow from the radiator and into an overflow bottle when the system is hot. This is because as the coolant heats up, it expands. As the coolant cools, it is pulled back into the radiator from the overflow tank. Most modern cooling systems use a pressure or surge tank, which has the cap and is the fill point for the system.

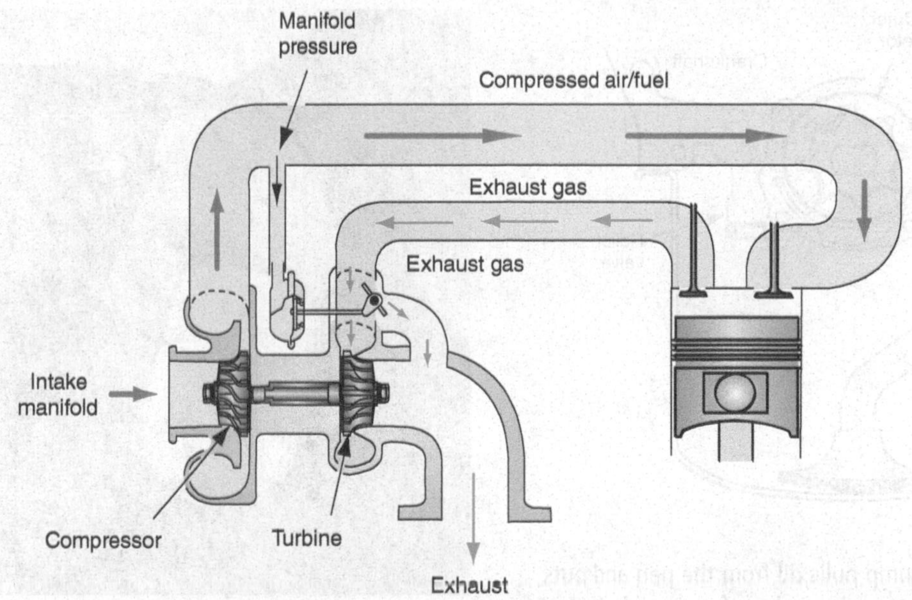

FIGURE 22-35 A turbocharger uses the flow of exhaust gases over a turbine to compress air going into the engine. Pushing in more air requires adding more fuel. This increases the performance of the engine.

FIGURE 22-36 A supercharger forces air into the engine but uses a belt-driven air pump.

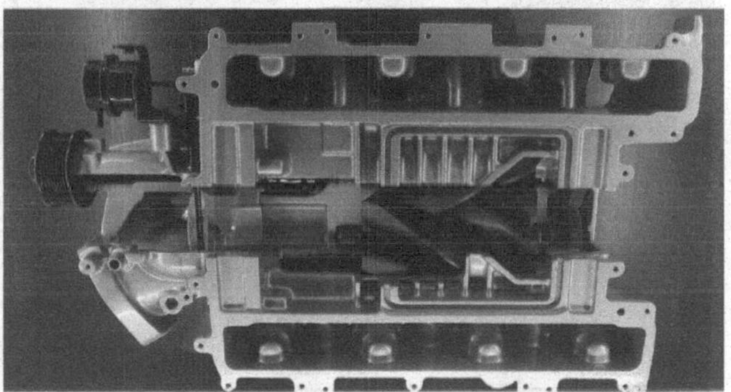

FIGURE 22-37 A view inside a supercharger and the three-lobed rotors the compress the incoming air.

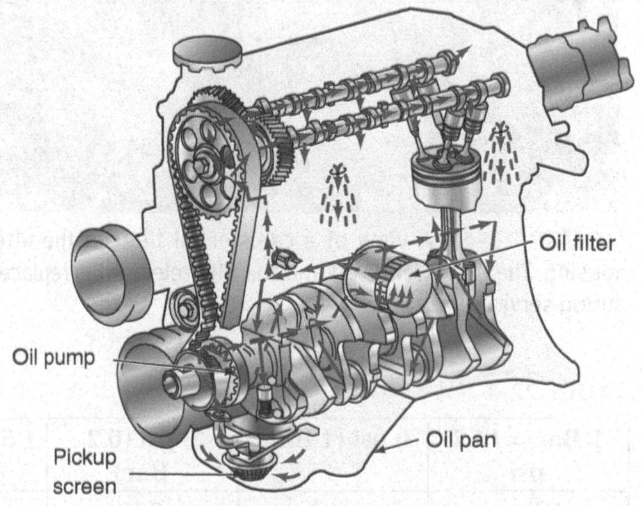

FIGURE 22-38 An illustration of how the oil circulates through the engine.

Attached to the radiator are the radiator fans. The fans pull air through the radiator to remove heat. Most vehicles use electric cooling fans because belt-driven fans consume engine power even when they are not needed, such as when the engine is cold. Electric fans are controlled by the engine control module based on input from a coolant temperature sensor. Fans may operate at either low or high speeds, depending on the heat load and on air conditioner use.

Some vehicles still use belt driven fans, though most cars and light trucks use electric fans. The accessory

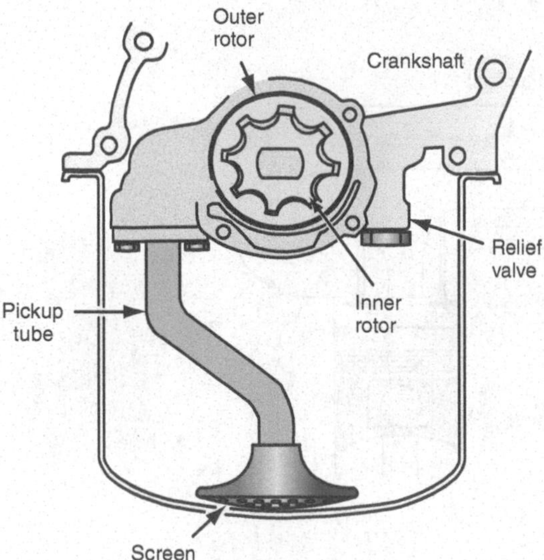

FIGURE 22-39 The oil pump pulls oil from the pan and puts it under pressure. The oil pump shown here forces oil between the teeth of the inner and outer gear to create pressure.

FIGURE 22-40 A view of a canister oil filter in the filter housing. The cap is removed and the filter element is replaced during service.

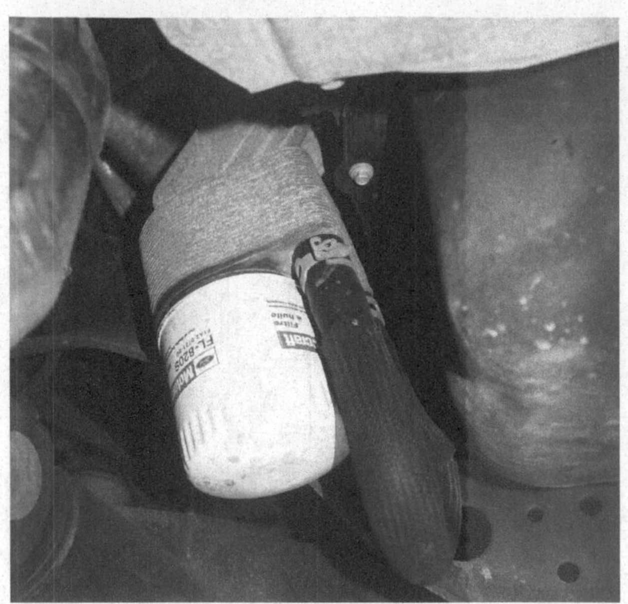

FIGURE 22-41 A cartridge oil filter. These filters thread onto a pipe connected to the engine. This example threads onto an oil cooler assembly.

drive belt is used to turn the fan, which is bolted to the water pump (**Figure 22-45**). When the engine is running, the crankshaft drives the belt, which turns the water pump and cooling fan. Belt driven fans are not used as much because spinning the fan is work for the engine, which reduces efficiency and increases fuel consumption. Electric can provide better thermal control and efficiency since they are only turned on when needed.

A few vehicles use hydraulic fans. In these systems, hydraulic fluid from the power steering system is used to drive the cooling fan. These systems regulate the flow of fluid so that fan speed can be increased or decreased depending on the temperature and cooling demands on the system.

Cast into the engine block and cylinder head are passages called water jackets. Coolant circulates through these passages, removing heat from the cylinders and valves. An illustration of these passages is shown in **Figure 22-46**.

The thermostat keeps coolant inside the engine during warmup and during cold weather operation. Once the coolant reaches a certain temperature, normally around 195°F (90°C) the thermostat fully opens, and lets hot coolant leave the engine and flow to the radiator. If the coolant

TABLE 22-1

1 Bar = 14.7 psi	0 psi (1 Bar)	3 psi (0.2 Bar)	5 psi (0.34 Bar)	10 psi (0.7 Bar)	15 psi (1 Bar)	20 psi (1.4 Bar)
Water	212	221	227	242	257	272
50/50	223	232	238	253	265	283

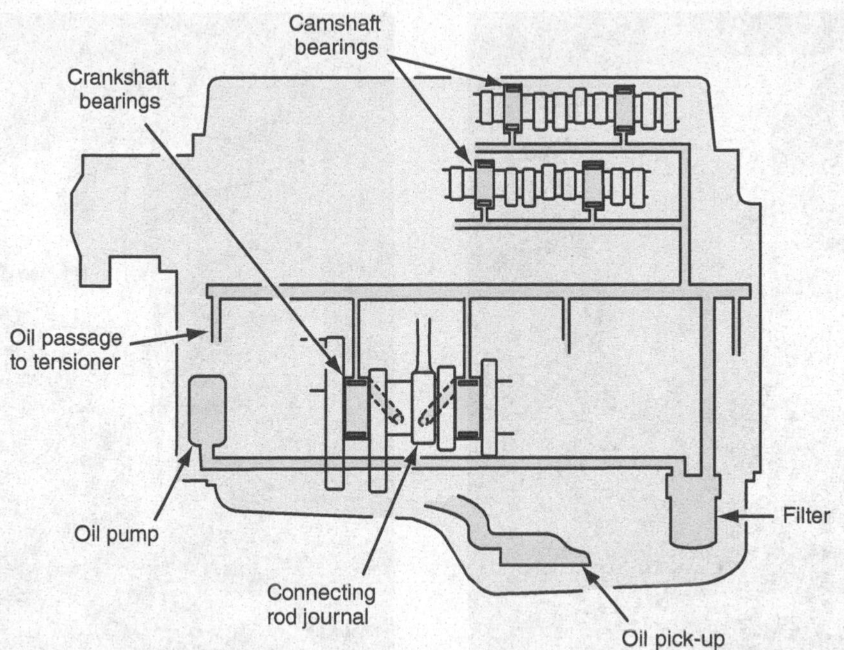

FIGURE 22-42 The oil galleries supply oil throughout the engine to moving parts.

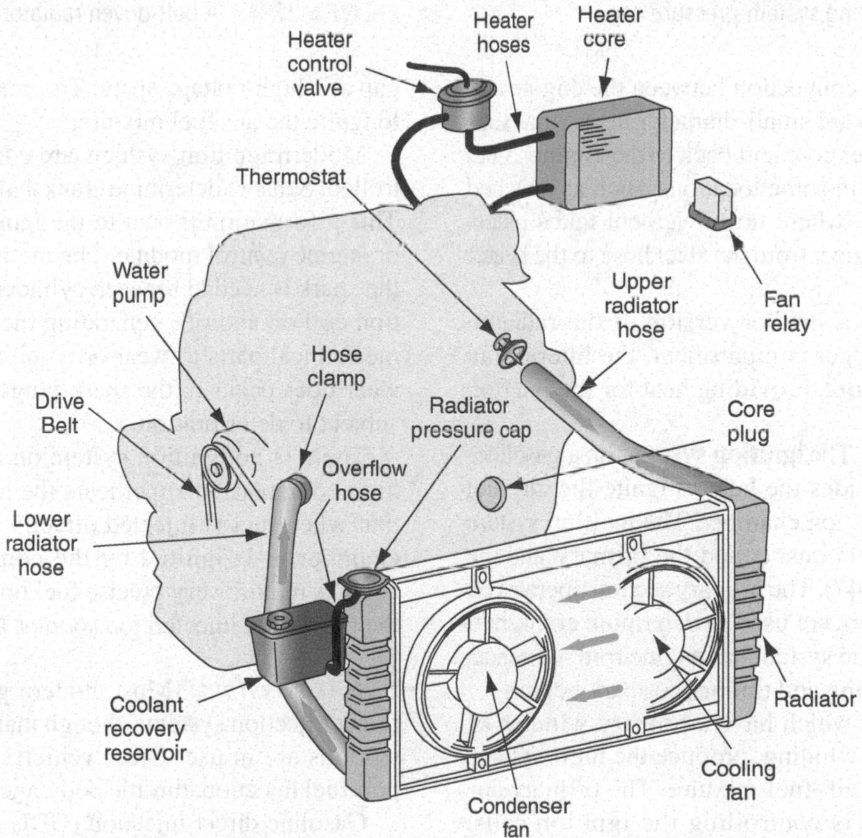

FIGURE 22-43 The cooling system removes excess heat from the engine and supplies heat for the passenger compartment.

temperature drops below the thermostat opening temperature, the thermostat closes. This provides hot coolant for the heater core and passenger compartment and regulates the engine's minimum operating temperature.

Coolant hoses connect the radiator to the engine, route coolant through the throttle body, and carry coolant to the heater core. Radiator hoses are large-diameter synthetic rubber hoses designed to handle underhood temperatures

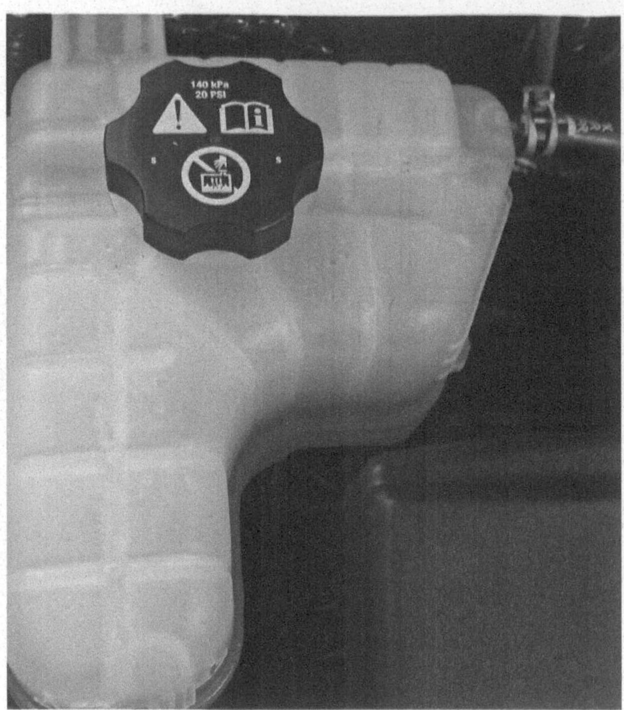

FIGURE 22-44 A cooling system pressure cap.

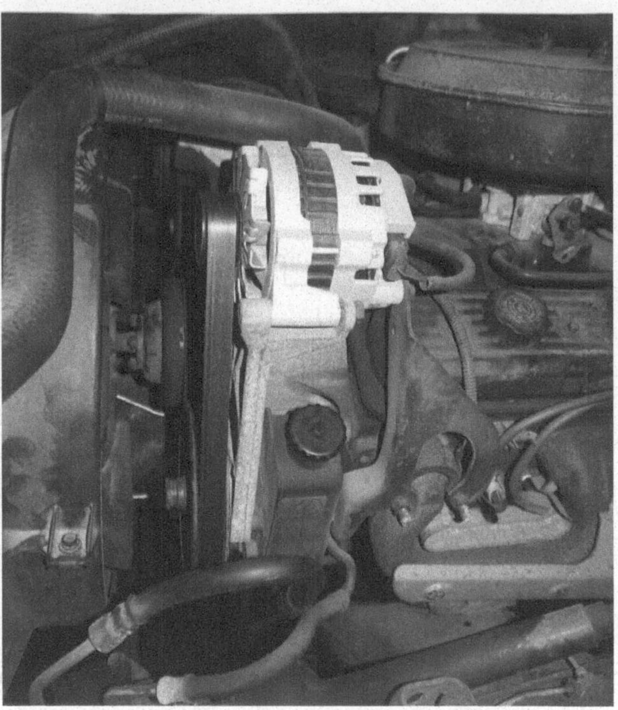

FIGURE 22-45 A belt-driven radiator fan.

and provide a flexible connection between the engine and radiator. Heater hoses are small-diameter hoses that supply coolant to the heater core and back to the engine. Steel coolant lines are used in some locations, such as attached to the engine or body where no movement takes place. Rubber hoses then connect from the steel hose to the heater core.

The heater core is a smaller version of the radiator. Located in the passenger compartment, the interior fan blows air across the core, providing heat for the interior.

■ *Ignition System.* The **ignition system**, on a gasoline-powered engine, provides the heat to ignite the air–fuel mixture in the combustion chamber. The ignition system is divided into two sections, called the primary and the secondary (**Figure 22-47**). The primary section operates on battery voltage. Sensors are used to determine crankshaft speed. Once the ignition system has engine rpm reference, the process of generating and timing the spark begins.

The ignition coils, which have a primary winding of wire and a secondary winding, produce the high-voltage spark that ignites the air–fuel mixture. The primary circuit's main function is controlling the ignition coils. When the coil's primary circuit turns off, magnetic fields in the coil's primary windings collapse. This causes the magnetic fields to collapse across the coil's secondary windings. The collapse induces voltage in the secondary windings, which creates a voltage spike. The spike is the spark, which is routed through the spark plug wires to the spark plugs. The voltage spike arcs across the spark plug gap as a high-voltage spark. The heat of the spark is used to ignite the air–fuel mixture.

Modern ignition systems are entirely computer controlled. Sensors determine crankshaft speed and position. This information is sent to the ignition control module or engine control module. The module determines when the spark is needed for each cylinder, then turns the ignition coil on and off, generating the spark. There are no mechanical parts to wear or require adjustment, though wear does occur to the spark plugs, and plug wires are subject to deterioration.

There is no ignition system on a diesel engine. The high compression superheats the air in the cylinder so that when fuel is injected directly into the combustion chamber, it is ignited by the compressed air. Diesel engines require very precise fuel injection timing so that the fuel is not injected too soon or too late.

■ *Fuel System.* Most modern gasoline engines use direct injection systems, though many port fuel injection systems are in use. Older vehicles can have a form of port fuel injection, throttle body injection or a carburetor.

Gasoline direct injection (GDI) systems use an electric supply pump and a mechanical high-pressure pump on the engine. The mechanical pump raises fuel pressure to 1,500–2,500 psi depending on the system. A high-pressure GDI pump is shown in **Figure 22-48**. Clearly visible is the back of the camshaft and the drive lobe for the pump. **Figure 22-49** shows a GDI injector in relation to the cylinder head and block.

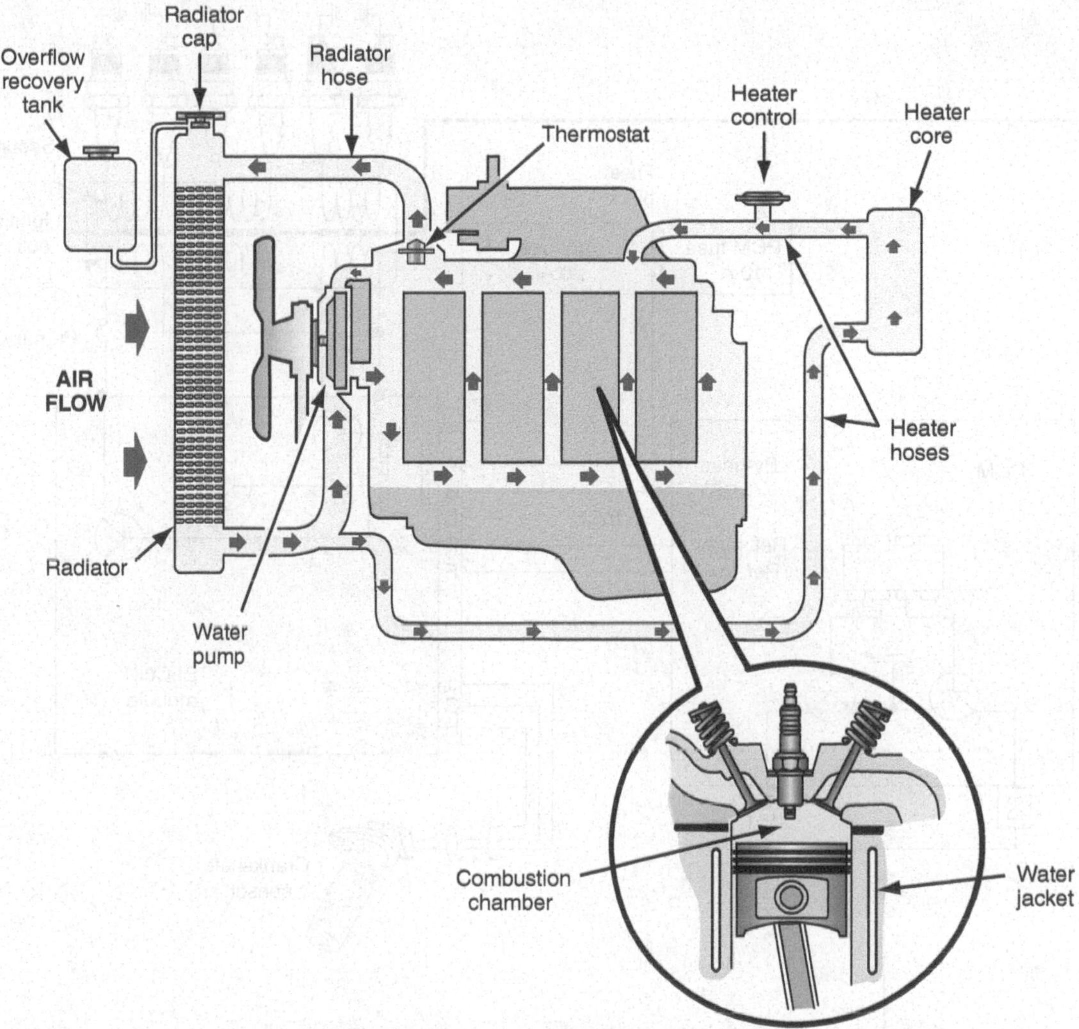

FIGURE 22-46 Passage formed into the block and head, called water jackets, allow coolant to pass near the combustion chambers and cylinder walls to pick up the heat created during combustion.

On port fuel injection systems, a fuel pump, usually mounted in the fuel tank, supplies fuel to the injectors (**Figure 22-50**). An inline fuel filter is used to remove contaminants before the fuel reaches the injectors. On port-injected engines, fuel pressure is usually between 40 and 50 psi.

Regardless of which type of injection system is used, on modern systems, the engine control module controls the opening of each injector based on input from crankshaft and camshaft position sensors and data from other sensors. The amount of time the injector is open is called the injector pulse width. The pulse width varies constantly, as changes in engine speed and load require more or less fuel delivery.

Diesel fuel systems have a low-pressure supply and high-pressure injection side. The low-pressure side uses a fuel transfer or lift pump, which supplies fuel to the injection pump. Inline between the fuel tank and injection pump is the fuel filter/fuel heater/water separator. The fuel filter removes contaminants, just as in a gasoline fuel system. Fuel heaters are used because at low temperatures diesel fuel can crystallize and plug fuel filters and injectors. Water separators remove water that may be present from the fuel storage tanks. Water can allow algae and bacteria growth in diesel fuel. The algae and bacterial growth can cause corrosion, which also can plug filters and injectors.

■ ***Exhaust System.*** The **exhaust system** performs several functions:

- It removes the spent exhaust gases from the engine.
- It helps transfer heat away from the head and exhaust valves.
- Quiets the noise of combustion.

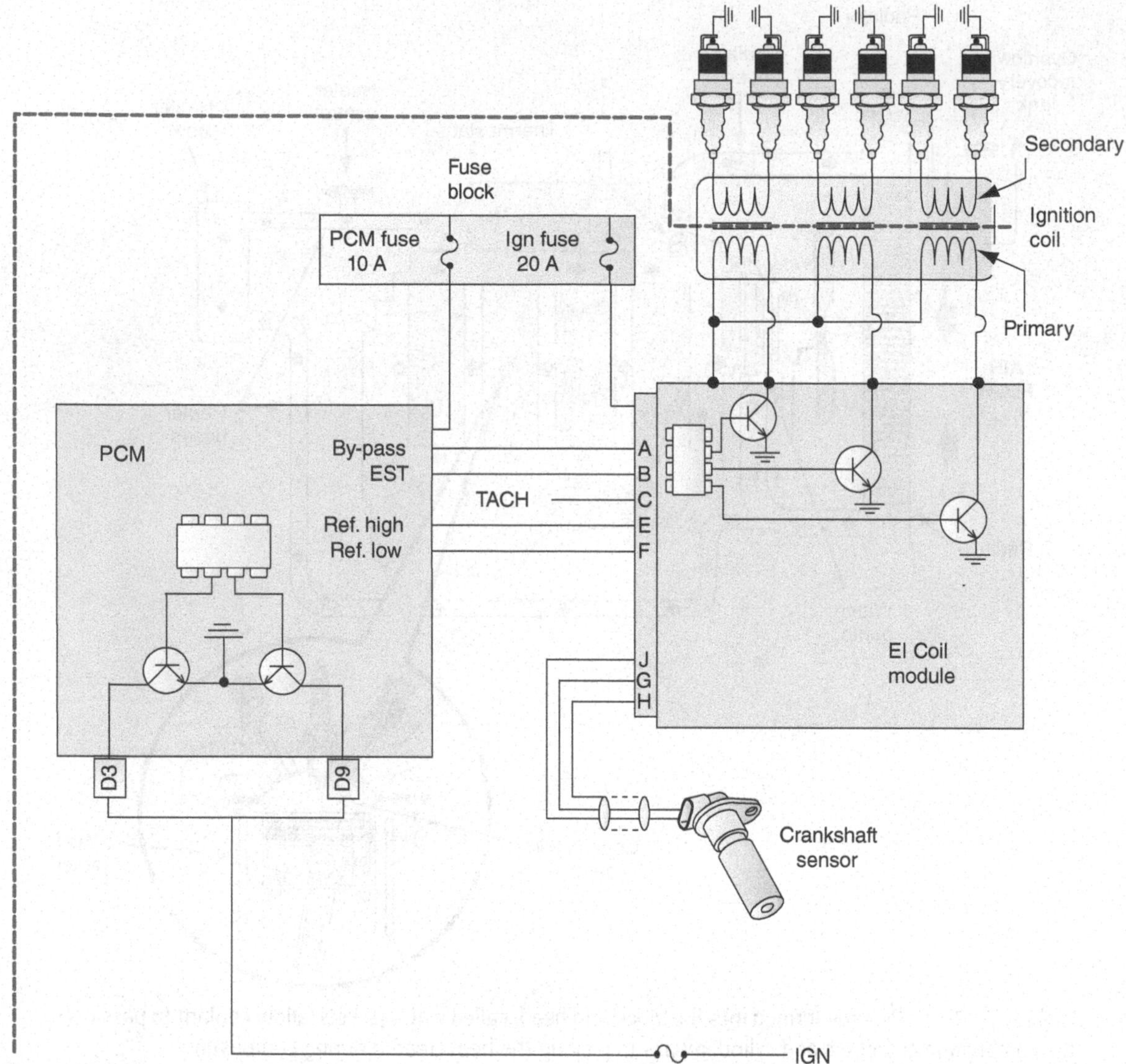

FIGURE 22-47 The ignition system generates the spark that ignites the air-fuel mixture for combustion.

- Routes harmful exhaust gases away from the passenger compartment.
- Uses platinum and other materials to convert harmful components of the exhaust gases into less harmful gases.

Attached to the cylinder head is the exhaust manifold or header (**Figure 22-51**). Each exhaust port in the head is matched to a port in the manifold. The manifold carries exhaust gases and heat away from the cylinder head. Many exhaust manifolds now contain small preconverters to help clean up the exhaust (Figure 22-51).

The catalytic converter converts harmful exhaust gases into less harmful and less polluting gases. A catalyst is something that aids in a chemical reaction without being used up during the process. The converter contains metals such as platinum, formed in a honeycomb structure. As the exhaust gas passes through the converter it reacts with the metals, losing and then gaining oxygen molecules, which alters the gases. After exiting the catalytic converter, the exhaust travels along a pipe to the muffler. Mufflers use chambers to break up the exhaust pulses, which quiet the noise of the engine.

BASIC ENGINE OPERATION

For an engine to operate correctly, a complex choreography of operations must take place. The engine's mechanical systems must be in good working condition to pull in and compress the air and fuel. The fuel and ignition systems must deliver at very exact times for maximum power to be developed. The entire combustion process only lasts for a few milliseconds—a few thousandths of a second. This requires all components and systems to operate in perfect unison. The combustion process is broken down into the four strokes, explained in detail below.

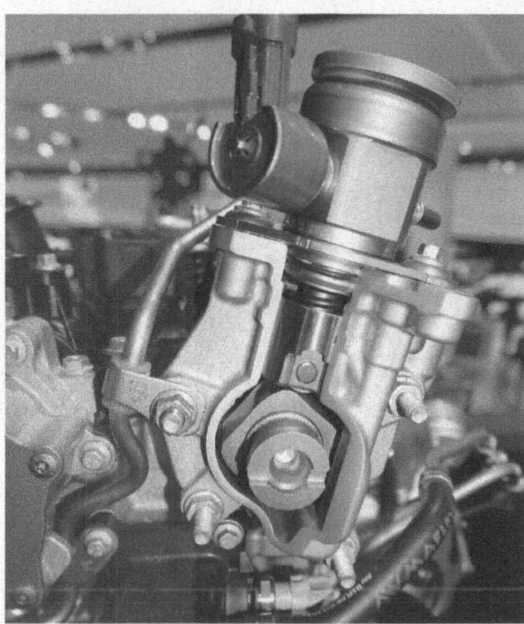

FIGURE 22-48 Many modern engines have gasoline direct injection. This is an inside look at the high-pressure fuel pump that supplies the injectors.

FIGURE 22-49 A view of a gasoline direction injection fuel injector. Fuel is delivered directly to the combustion chamber under high pressure.

■ *Intake Stroke.* The **intake stroke** begins when the intake valve opens and air and fuel begin to enter the cylinder (**Figure 22-52a**). If the engine has gasoline direct injection, the fuel does not enter the cylinder with the air. The intake stroke starts at the end of the exhaust stroke before the piston moves down in the cylinder. At the end of the exhaust stroke, the exhaust valve is still open, and the piston is near top dead center (TDC); this is called *valve overlap*. The cylinder is evacuating the exhaust gases through the open exhaust valve. The flow of the gases out of the cylinder help pull the intake air in. This effect is known as scavenging and is used to begin pulling the fresh air–fuel mixture into the cylinder before the piston is starting its downward movement. Just after the piston passes TDC, the exhaust valve closes, but the intake valve remains open.

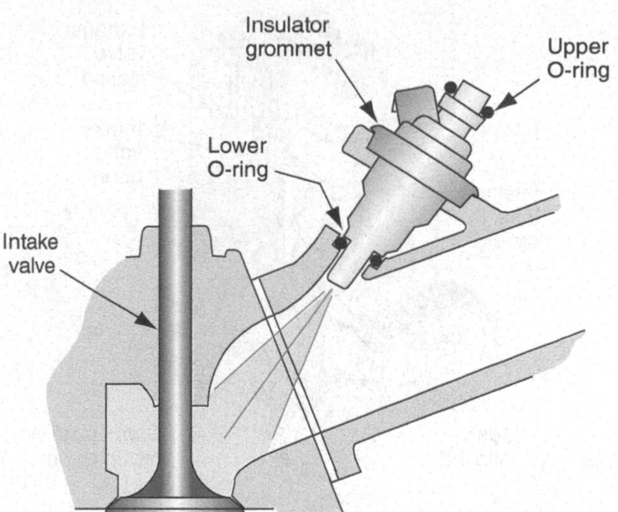

FIGURE 22-50 Port fuel injection systems spray fuel behind the intake valve. The fuel is then drawn into the cylinder with the incoming air when the valve opens.

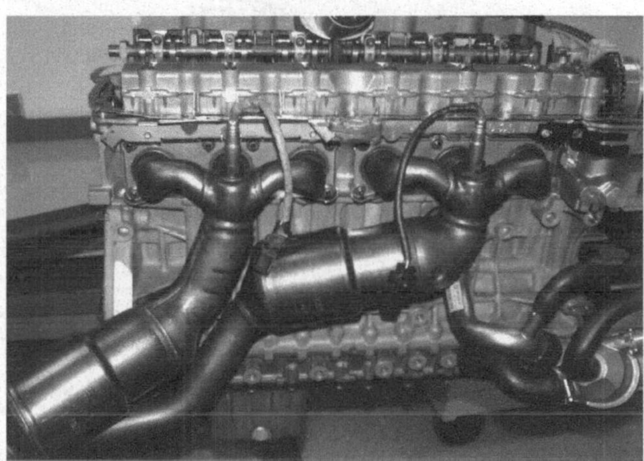

FIGURE 22-51 The exhaust system removes the burnt gases from the engine. This engine has two exhaust manifolds, each with a catalytic converter.

The downward piston movement and the restriction caused by the intake valve and throttle plate creates a low pressure above the piston. This low pressure or vacuum is at lower pressure than the atmospheric pressure above the throttle plates. Since pressure moves from high to low, air flows into the cylinder as the piston moves down. The amount of air that enters the cylinder depends on many factors, including outside pressure, pressure in the cylinder, throttle opening, valve opening, engine speed, intake runner design, and others.

Air and fuel continue to enter the cylinder until the intake valve closes. When the intake closes depends on the engine design, but it is generally after the piston reaches bottom dead center (BDC) and begins to move back upward in the cylinder.

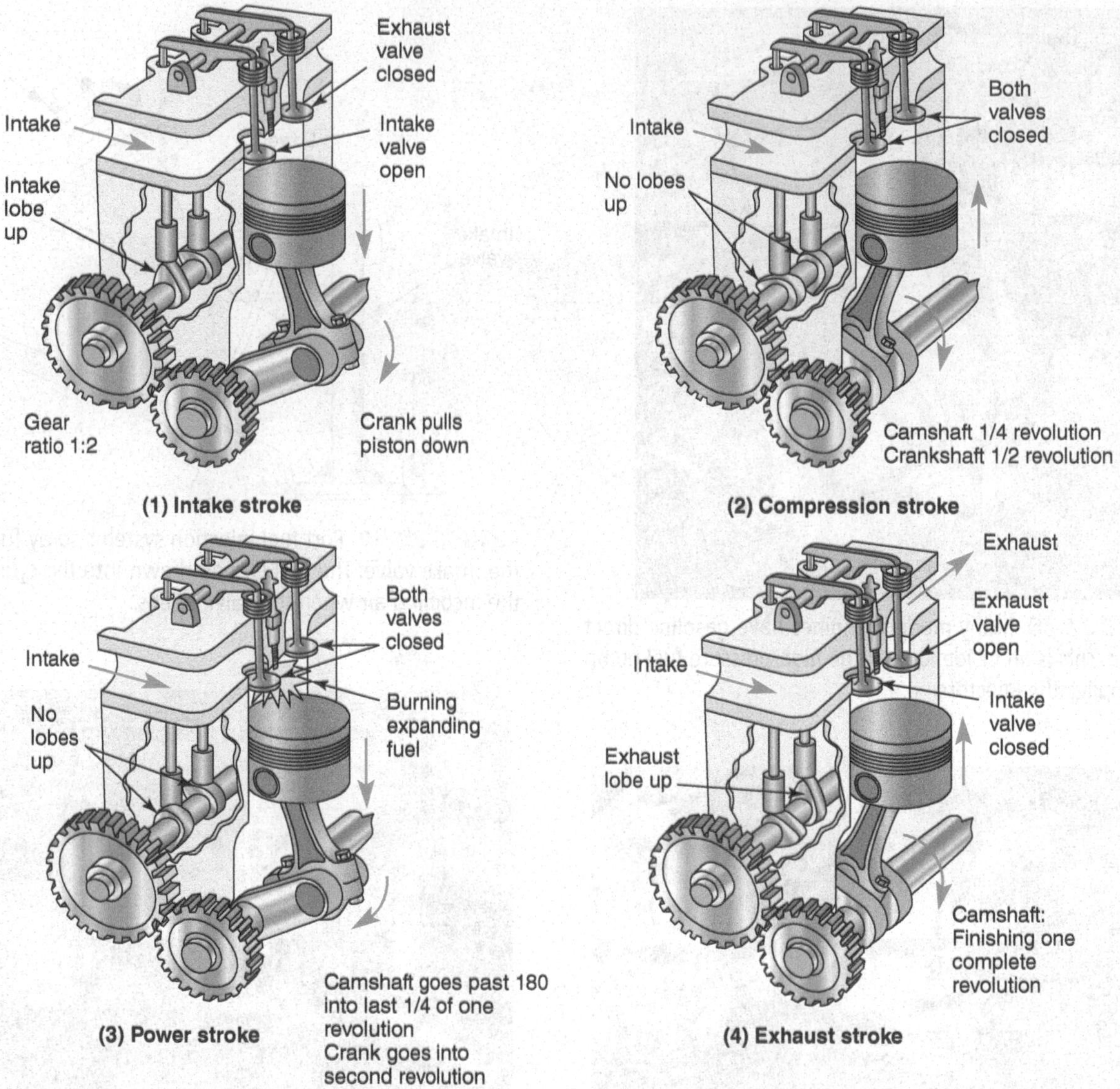

Exhaust
valve
closed
Intake
Intake
valve
open
Intake
lobe
up
Gear
ratio 1:2
Crank pulls
piston down
(1) Intake stroke
Intake
Both
valves
closed
No lobes
up
Camshaft 1/4 revolution
Crankshaft 1/2 revolution
(2) Compression stroke
Intake
Both
valves
closed
No
lobes
up
Burning
expanding
fuel
Camshaft goes past 180
into last 1/4 of one
revolution
Crank goes into
second revolution
(3) Power stroke
Exhaust
Exhaust
valve
open
Intake
Intake
valve
closed
Exhaust
lobe up
Camshaft:
Finishing one
complete
revolution
(4) Exhaust stroke

FIGURE 22-52 This illustrates the sequence of the four strokes during one complete cycle.

■ ***Diesel Intake Stroke.*** Air is drawn into the engine just as in a gasoline engine. Most diesel engines are turbocharged, so that air is forced into the cylinders during the intake stroke. Due to the longer stroke of the diesel engine, more air is drawn into the cylinders than during the intake stroke in a gasoline engine.

■ ***Compression Stroke.*** To be able to extract the chemical energy stored in gasoline, the air–fuel mixture, called the charge, must be compressed. An illustration of the compression stroke is shown in Figure 22-52b.

How much the air–fuel charge is compressed depends on the **compression ratio** of the engine (**Figure 22-53**). Compression ratios in a gasoline engine range from around 9:1 to 12:1. For many years, compression ratios had been relatively low, between about 9 and 9.5:1. This is because engines were "detuned," meaning they had lower compression ratios and reduced horsepower. With

Volume before
compression: 480 cc
Volume after
compression: 60 cc
TDC
BDC
Compression ratio: 8 to 1

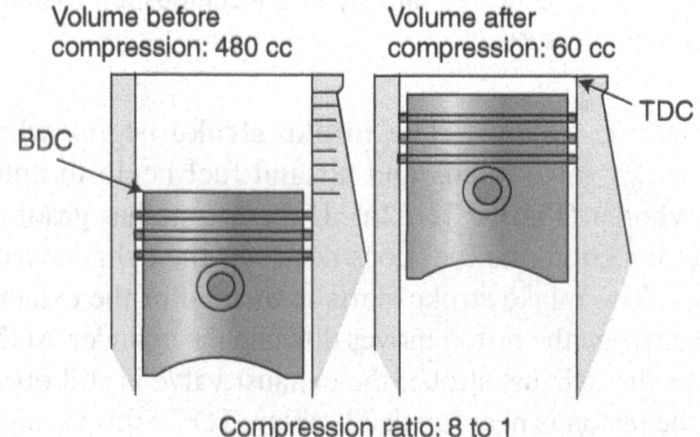

FIGURE 22-53 Compression ratio is based on the volume of the cylinder with the piston at bottom dead center compared to top dead center.

advances in engine design and gasoline direct injection, compression ratios are increasing. Many newer vehicles have compression ratios above 11:1. Increasing the compression ratio increases the amount of energy that can be extracted from the air–fuel mixture. Engines with turbo or superchargers typically have lower compression ratios, although that is also changing as improved management of the air and fuel systems are allowing higher compression ratios even on these engines.

As the mixture is compressed, its temperature increases as the air and fuel molecules are packed more tightly together in the decreasing volume of the cylinder. The fuel is present as a very fine mist of very small droplets. Small droplets can more easily attach to the surrounding air and burn more efficiently than a puddle of liquid gasoline. Compression of the mixture allows for a much higher release of energy when the mixture ignites. It also results in much more efficient combustion and more even heat distribution across the piston.

■ ***Diesel Compression Stroke.*** Compression ratios in a diesel are commonly between 15:1 and 22:1. The high compression heats the air to very high temperatures, as much as 1,000°F.

■ ***Combustion, Power Stroke, and Expansion Cycle.*** As the piston moves up during the compression stroke, the ignition system is beginning to generate the spark to ignite the compressed mixture. As the piston nears top dead center or TDC, the crankshaft and connecting rod begin to move past TDC, and the spark plug ignites the mixture. The high voltage is discharged across the plug gap, causing the fuel molecules to ignite and burn (Figure 22-52c).

The ignition creates a flame front that travels across the piston toward the cylinder walls, consuming the air–fuel mixture. This creates very high temperatures and greatly increases the pressure in the cylinder. The force of the expansion of the air–fuel mixture during combustion pushes down on the piston. This causes the connecting rod to turn the crankshaft. The power of the rotating crankshaft is used to drive the wheels.

Compressing the air–fuel mixture forces the molecules closer together and raises the temperature of the charge. When it is ignited, the air–fuel charge combusts very rapidly. This produces very high temperatures in the combustion chamber. When the temperature inside the combustion chamber increases, the pressure also increases. It is at this point that the chemical energy of the fuel is converted into heat energy. The heating and expansion of the combustion gases is converted into mechanical energy as the piston is forced down by the expansion of the gases during combustion.

Many people think that combustion is an uncontrolled explosion taking place within the engine. In reality, combustion is highly engineered so that the maximum amount of power is extracted most efficiently and with the fewest leftovers or emissions. Uncontrolled combustion, particularly spontaneous combustion, causes reduced power, poor performance, increased emissions, and in some cases, engine damage.

For most engines, the amount of piston travel during each stroke is equal. Once the fuel is compressed and ignited, power is produced during the expansion cycle, also called *the power* or *combustion stroke*.

■ ***Atkinson Engines and Expansion Cycles.*** Some modern engines, called Atkinson Cycle engines, use variable valve timing to increase efficiency by increasing the engine's expansion ratio. This means that the length of the expansion cycle is greater than the length of the intake cycle. By shortening the intake cycle compared to the expansion cycle, greater efficiency is achieved. This is done by keeping the intake valve open into the compression stroke (**Figure 22-54**). Some of the air–fuel mixture is displaced, reducing effective compression, but increasing the length of the expansion cycle. However, this efficiency also reduces power output and performance. For this reason, Atkinson Cycle engines are paired with electric motors to make up for the power loss. This arrangement is used in hybrid electric vehicles (HEVs) such as those from Toyota, Honda, and other manufactures.

■ ***Diesel Combustion.*** Once the air is compressed and cylinder temperature is high, fuel is injected into a precombustion chamber, which is often formed at the top of the piston. Combustion starts in the precombustion chamber, which then spreads across the top of the piston.

■ ***Exhaust Stroke.*** As the piston nears BDC on the combustion stroke, the exhaust valve begins to open. This allows the spent gases to exit the cylinder (Figure 22-52d). Most of the energy of the expanding gases has been used to turn the crankshaft, but pressure remains in the cylinder. When the exhaust valve opens, the higher pressure in the cylinder forces the gases out to the exhaust manifold. As the pressure of the gases equalize between the cylinder and the tailpipe, the momentum of the gas flow reduces pressure inside the cylinder. This helps extract all the exhaust gases. This low pressure also helps pull the fresh air–fuel mixture into the cylinder during valve overlap.

■ ***Diesel Exhaust Stroke.*** After combustion, exhaust gases vacate the cylinder via the exhaust valve, just as in a gasoline engine. As the exhaust gas passes the turbocharger, the turbine wheel is driven by the exhaust flow. The turbine drives the compressor, which pressurizes intake air and pumps it to the intake manifold, and starts the process over.

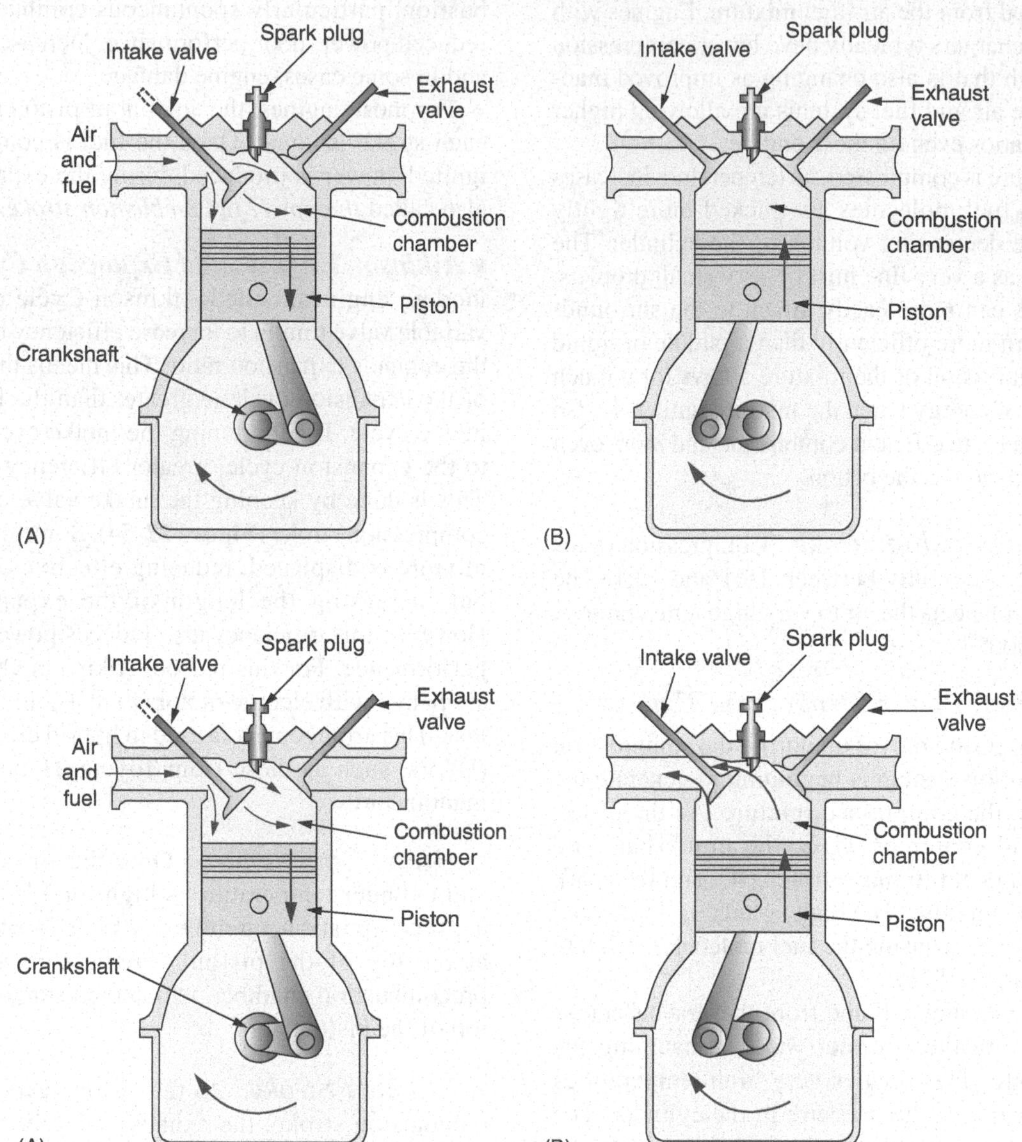

FIGURE 22-54 The Atkinson cycle engine uses a modified intake stroke to reduce the compression but increase expansion, making the engine more efficient.

■ ***Four Cycle Operation.*** During operation, a four-cylinder engine has each cylinder performing one of the four strokes at any given moment. At 900 rpm, an engine is running at 15 revolutions per second. A four-cylinder engine makes power every 180 degrees of crankshaft rotation, so 7.5 combustion cycles take place every second at 900 rpm. As another example, at 900 rpm there are 450 power strokes per cylinder. 450 power strokes per minute/60 seconds is 7.5 per second.

A six-cylinder engine produces power every 120 degrees of rotation and an eight-cylinder engine every 90 degrees of rotation. As the number of cylinders increases, power is generated more frequently, and results in smoother engine operation. For many years, luxury cars such as Jaguars used V12 engines with small displacements, often less than 5 liters. Power delivery is very smooth with very little vibration from the engine.

■ ***Horsepower and Torque.*** **Horsepower** is a rating of the amount of work performed in a specific amount of time. It was calculated by James Watt as the amount of weight a horse could move in one minute. Watt determined that a horse could lift 150 pounds 220 feet in one minute. Watt later revised his calculation and one horsepower is now standardized and referenced as the ability to pull 330 pounds 100 feet in one minute. This equals 33,000 ft.-lb. per minute or 550 ft.-lb. per second. This meant that, in Watt's time, a working horse could lift 550 pounds of coal one foot each second. Two horsepower could do the same work in one-half of the time. Most vehicle manufacturers now rate their combustion engine output in watts and HP. One HP is equal to about 746 watts. Some manufacturers are using watts as the primary engine power output rating. The traction motors used in hybrids are also rated in watts.

Measuring engine horsepower is not as easy as watching a horse pull a weight. This is because while the horse moves slowly enough to measure distance traveled, an engine is rotating very fast and is not moving in a straight line. To calculate HP, an engine is placed on an engine dynamometer, which measures the torque produced at the crankshaft. **Torque** is the measurement of the force that causes an object to rotate on its axis. Once the maximum torque output is determined, horsepower can be calculated. For example, an engine producing 300 ft.-lb. of torque at 4,000 rpm produces 228 HP at 4,000 rpm. Horsepower is equal to torque multiplied by engine speed and divided by 5,252. But why the formula and the 5,252? Refer to Watt's measurement of 150 lb. moved 220 feet in one second. Take the 150 lb., and apply it as a line from a curve (a tangent) of a one-foot-radius circle. The circle represents the flywheel at the back of the engine. This equals 150 foot-pounds of torque, 150 × 1 foot = 150 foot-pounds. Next, the 220 feet needs to be converted to rpm. The circumference of a one-foot-radius circle is 6.283185 feet, which when divided into 220 feet equals 35.014 rpm. So with the engine spinning at 35.014 rpm, the 150 ft.-lb. move 220 feet in one second. This equals 150 foot-pounds of at 35 rpm and one horsepower; which is the total amount of work performed in one minute. To find the constant used to calculate horsepower from torque, the constant x(x) = 150 ft.-lb. multiplied by 35.014 rpm divided by 1 HP, which equals 5,252.1. This means HP = torque times rpm divided by 5,252.

Now that we know where HP comes from, we can try to apply it to engine output. Unfortunately, HP measurement has changed over the years in many ways, leading to many different claims regarding engine power output. Some engines use Society of Automotive Engineers or SAE gross crankshaft HP, meaning the engine is tested with some belt-driven accessories but may not use the stock exhaust system. Other engines may use the SAE net crankshaft HP rating. Net HP is measured at the crankshaft but without the transmission or driveline losses and with all the belt-driven accessories and subsystems installed. SAE net HP is closer to the actual engine output as installed and operating in a production vehicle since all accessories and systems are used in the test. In 2005, SAE introduced a new rating called SAE certified crankshaft horsepower. This voluntary test requires an independent observer be present as the test is conducted. **Figure 22-55** shows an example of an engine rating using the SAE certification.

■ ***Efficiency.*** The power produced by the engine is determined, in part, by the overall engine efficiency. Increasing engine efficiency has been a prime factor in engine design in recent decades. Improvements in efficiency result in increased power output, decreased fuel consumption, and decreased exhaust emissions. **Figure 22-56** shows an example of typical losses of mechanical efficiency in the gas engine.

Overall gasoline engine efficiency is determined by the ratio of usable power at the crankshaft to the power supplied to the engine as the energy content of the fuel. **Figure 22-57** illustrates losses based on internal friction. Gasoline engines are about 20% efficient using this method. The remaining 80% of the energy is lost through heat and friction. Approximately 35% of the heat leaves with the exhaust. An additional 35% is lost

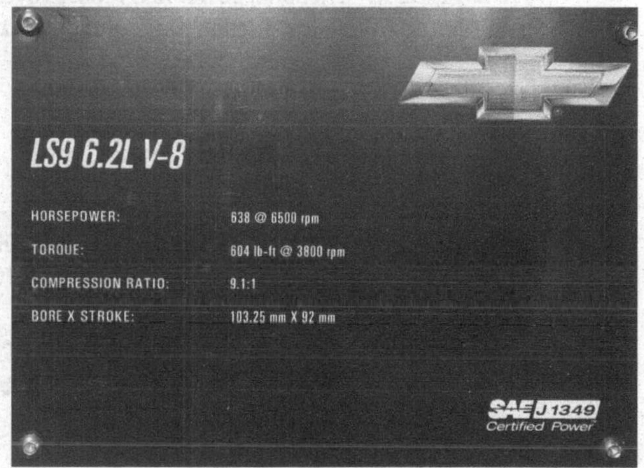

FIGURE 22-55 An example of an SAE engine horsepower rating.

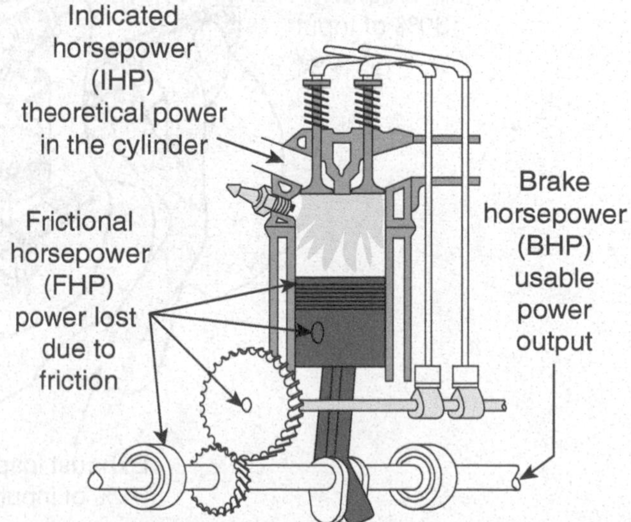

FIGURE 22-56 The actual power available to drive the wheels is less than the total power output of the engine.

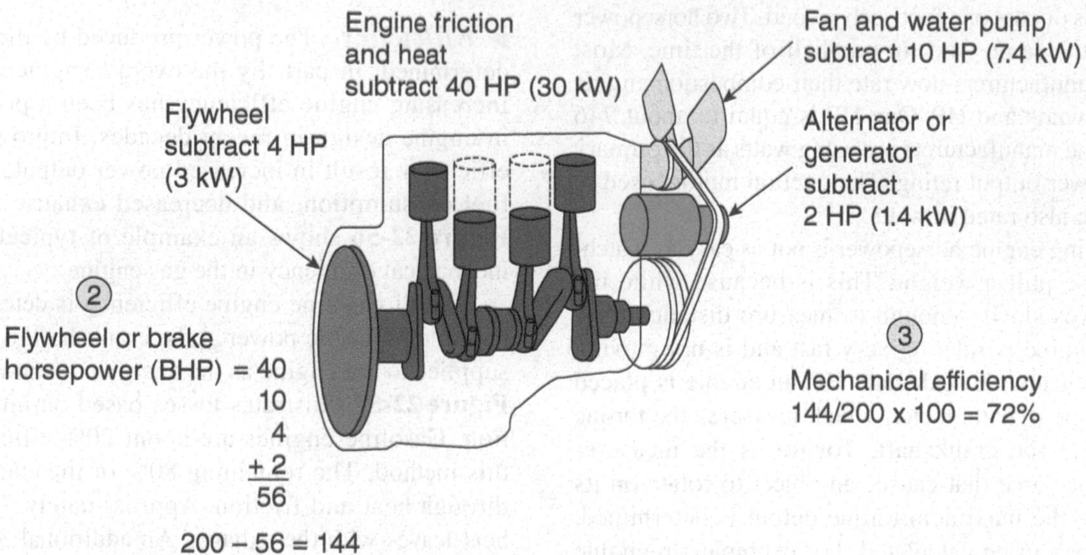

FIGURE 22-57 Frictional power loss reduces overall engine output. Modern engines use many new technologies, such as 0W20, low-friction rings, and others to reduce these losses.

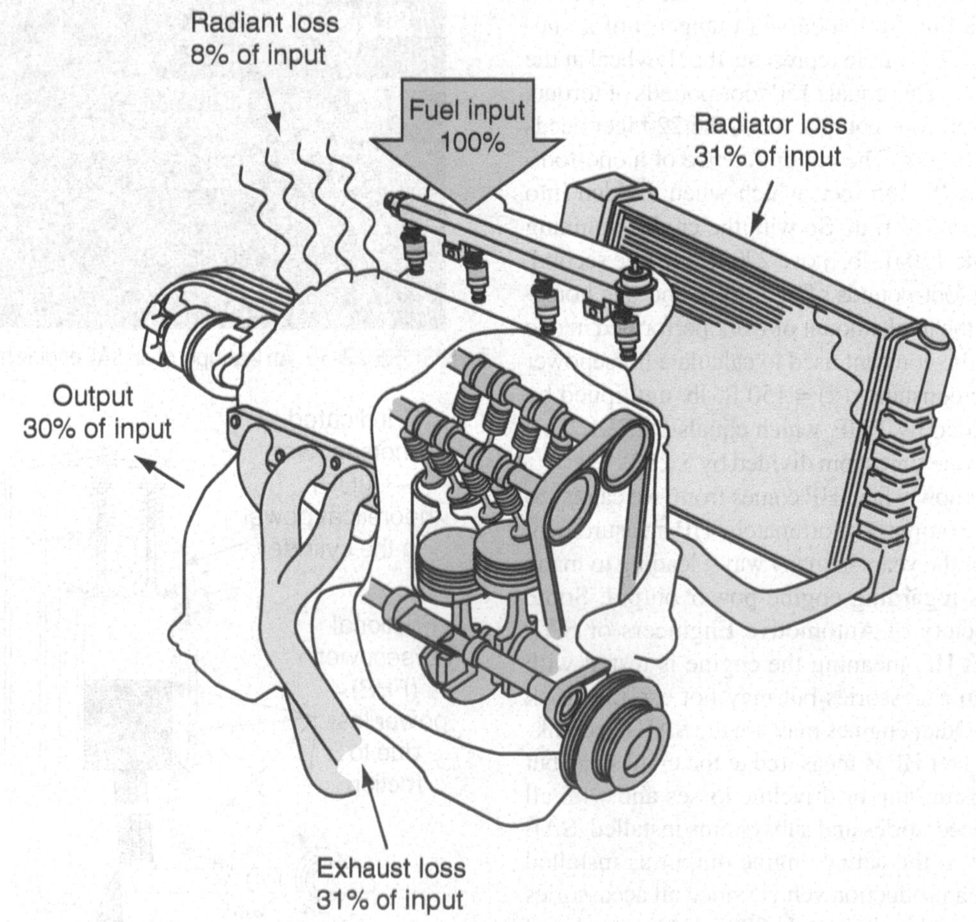

FIGURE 22-58 Much of the energy produced by the gas engine is lost as heat.

through the cooling system (**Figure 22-58**). The remaining 10% is lost through mechanical friction inside the motor, belt-driven accessories, and pumping loss inside the engine. Pumping loss is the result of the air not being able to enter efficiently into the cylinders at low rpms.

Diesel engines, because of the higher energy content of diesel fuel, higher compression ratios, and higher combustion temperatures, extract more power from combustion and are more efficient.

■ ***Volumetric efficiency.*** **Volumetric efficiency** is determined by the ratio of air drawn into the cylinder compared to the maximum air that could enter the cylinder (**Figure 22-59**). This is an indicator of how well the engine breathes. Most gas engines are about 80% to 90% volumetrically efficient. Restrictions in the intake and around the valves and ports restrict airflow, reducing efficiency.

One modern method of increasing efficiency on drive-by-wire engines is to keep the throttle plate fully open when the vehicle is cruising at a steady speed. This improves airflow through the engine by reducing pumping losses and turbulence in the intake manifold.

Modern engine designs attempt to reduce inefficiencies as much as possible. Some ways in which this is done are by using energy-conserving engine oils, using low-friction piston rings, low-mass pistons and connecting rods, hollow camshafts, and by extracting as much power from each combustion event as possible. Engines with gasoline direct injection can run at much leaner air–fuel mixtures, and save fuel and reduce emissions.

Engine efficiency is reduced as internal engine components wear. Valves do not seat as well nor do rings seal against the cylinder wall as tightly as the engine ages. Carbon deposits on the intake valves can absorb fuel and disturb the airflow into the cylinder. Not performing necessary maintenance, such as replacing spark plugs, also decreases how efficiently the engine operates.

HYBRID-ELECTRIC VEHICLE ENGINES

The gasoline engines used in many hybrid vehicles today incorporate the Atkinson cycle into engine operation. The Atkinson cycle extends the time that the intake valve is open well into the compression stroke. This forces some of the air–fuel mixture back up into the intake manifold. This reduces the compression ratio and power output, and saves fuel and decreases emissions.

The increase in efficiency comes from the extended intake stroke. As the piston moves upward, some of the air–fuel mixture is pushed back into the intake. This reduces the compression ratio and the displacement of the engine, but the expansion ratio, the distance the piston moves after combustion, remains the same. This extracts more power from combustion since the pressure generated by combustion is expanded over a greater volume than the volume of the cylinder during compression.

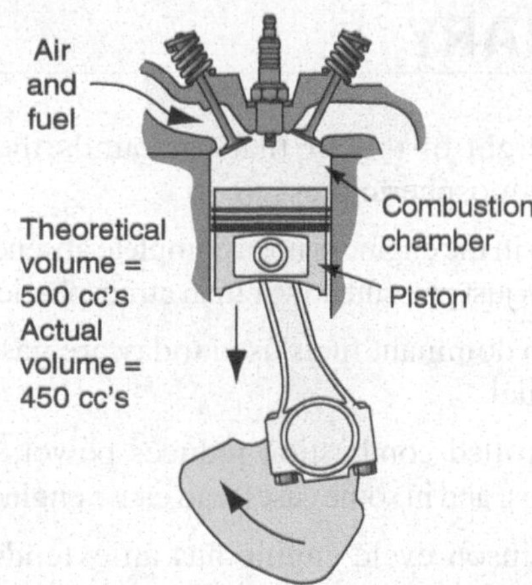

FIGURE 22-59 Volumetric efficiency is the difference between how much air a cylinder can theoretically hold compared to the actual amount.

FIGURE 22-60 A modern hybrid powertrain.

■ ***Hybrid Powertrains.*** Hybrid-electric vehicles (HEVs) combine ICEs and powerful electric motor/generators to propel the vehicle, recapture braking energy, increase fuel economy, and reduce exhaust emissions. An example of a modern Honda hybrid powertrain is shown in **Figure 22-60**. Note the large electric motor at the rear of the transaxle. Most currently available HEVs can be driven using just the electric motors before the ICE starts and is used for propulsion. The distance and speed the vehicle can operate in electric mode depends on the vehicle and the amount of charge in the high-voltage battery.

Because of the unique way in which HEVs combine two different power sources, their drivetrains are necessarily different from those of traditional vehicles. However, the basic engine operation remains the same

SUMMARY

- The weight of the air that surrounds the earth is called atmospheric pressure.
- Vacuum in the engine is not a complete absence of pressure; it is just pressure lower than atmospheric pressure.
- The two dominant fuels used today are gasoline and diesel fuel.
- Uncontrolled combustion reduces power, increases emissions, and in some cases, can cause engine damage.
- An Atkinson cycle engine has an extended intake stroke and longer expansion stroke.
- Rotary engines do not have any reciprocating parts, only rotating parts.
- The piston forms the lower section of the combustion chamber.
- The four strokes of internal combustion engines are the intake, compression, power, and exhaust strokes.
- When combustion occurs, the pressure pushes on the piston, forcing it downward in the cylinder.
- A wet sump lubrication system is one in which the oil is kept inside the oil pan.
- The throttle valve controls the amount of airflow into the engine.
- The exhaust system removes the spent exhaust gases from the engine.
- The cooling system removes excess heat from the engine.
- Coolant is a mixture of antifreeze and water.
- The ignition system on a gasoline-powered engine provides the heat to ignite the air–fuel mixture.
- In a diesel engine, the high compression superheats the air in the cylinder so that when fuel is injected directly into the combustion chamber, it is ignited.
- Horsepower is a rating of the amount of work performed in a specific amount of time.
- Gasoline engine efficiency is determined by the ratio of usable power at the crankshaft to the power supplied to the engine as the energy content of the fuel.

REVIEW QUESTIONS

1. Pressure that is less than atmospheric pressure is called ______________.

2. During the ______________ stroke, one valve is closed and the piston is moving downward in the cylinder.

3. The ______________ motion of the pistons is converted into the ______________ motion of the crankshaft.

4. Engines that use the Atkinson cycle have modified intake and ______________ strokes.

5. A ______________ is driven by the exhaust and is used to increase the amount of air delivered to the engine.

6. Which of the following is the correct order of the four strokes?

a. Exhaust, intake, ignition, compression
b. Compression, exhaust, intake, power
c. Intake, power, compression, exhaust
d. Power, exhaust, intake, compression

7. Engine design is being discussed. *Technician A* says the camshaft can be located either in the block or cylinder head(s). *Technician B* says all modern engines are either OHC or DOHC designs. Who is correct?

a. Technician A **b.** Technician B
c. Both A and B **d.** Neither A nor B

8. Horsepower is a rating of the engine's ability to:

a. Draw in and compress air
b. Produce torque
c. Perform work
d. None of the above

9. An Atkinson cycle engine ______________ the intake stroke.

a. Shortens **b.** Lengthens
c. Eliminates **d.** Inverts

10. *Technician A* says pressure greater than atmospheric pressure is measured as psia. *Technician B* says pressure that is less than atmospheric pressure is called vacuum. Who is correct?

a. Technician A
b. Technician B
c. Both A and B
d. Neither A nor B

CHAPTER 23

Engine Mechanical Testing and Service

Chapter Objectives

At the conclusion of this chapter, you should be able to:

- Inspect an engine for leaks. (ASE Education Foundation MLR 1.A.3)
- Perform engine mechanical tests, including:
  - Vacuum tests. (ASE Education Foundation MLR 8.A.2)
  - Cylinder power balance tests. (ASE Education Foundation MLR 8.A.3)
  - Cranking and running compression tests. (ASE Education Foundation MLR 8.A.4)
  - Cylinder leakage tests. (ASE Education Foundation MLR 8.A.5)
- Inspect the exhaust system for leaks and damage. (ASE Education Foundation MLR 8.C.3 & 8.C.4)
- Check and adjust valve lash. (ASE Education Foundation MLR 8.B.1)
- Inspect and replace a timing belt. (ASE Education Foundation MLR 1.A.5)
- Explain the differences between seals and gaskets.
- Install engine covers using gaskets, seals, and sealers as required. (ASE Education Foundation MLR 1.A.4)

KEY TERMS

cranking compression test	**oil pressure test**	**seal**
cylinder leakage test	**power balance test**	**UV dye**
gasket	**relative compression test**	**valve lash**
interference engines	**running compression test**	

The most advanced onboard computer systems in today's vehicles cannot correct for engine mechanical problems. While these systems do adapt and improve performance, they cannot compensate for problems such as low compression or incorrect valve timing. Therefore, before any attempt is made to correct a performance problem, the engine's mechanical condition must first be verified. This is done by performing some basic tests that check the engine's ability to draw in air, compress it, keep the combustion gases sealed in the combustion chamber, and expel the spent exhaust gases. These basic functions must occur for the engine to run properly.

Modern engine control systems provide the ability to monitor, control, and diagnose a wide variety of systems and components. The modern computerized engine achieves an incredible level of performance, while using less fuel and emitting fewer emissions than ever before. This is accomplished by using a network of sensors and control modules. However, these sophisticated computer systems cannot diagnose or compensate for every type of problem an engine may develop. Because of this, it is still necessary for the technician to be able to understand, diagnose, and correct mechanical problems.

FIGURE 23-2 An example of a vacuum/pressure gauge.

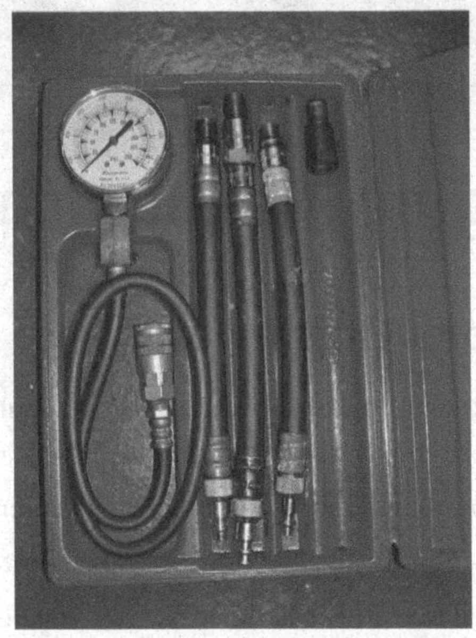

FIGURE 23-3 This type of compression has different adaptors that thread into the spark plug holes. The test gauge is connected to the adaptor hoses with a quick-disconnect fitting.

Engine Testing Tools

Many specialized tools are used when you test the engine. These tools are shown here with explanations. The use of these tools is explained within the context of the test being performed.

- An ultraviolet (UV) light is often used when you diagnose engine leaks. A kit is used with a special **UV dye**, a liquid additive that glows when exposed to UV light (**Figure 23-1**).
- A vacuum gauge is used to check engine vacuum (**Figure 23-2**). Some gauges can measure both vacuum and boost pressure.
- To test the engine's ability to draw in and compress the air–fuel mixture, a compression gauge is used. Mechanical gauges are being replaced by pressure transducers that are used with a scope (**Figure 23-3**).

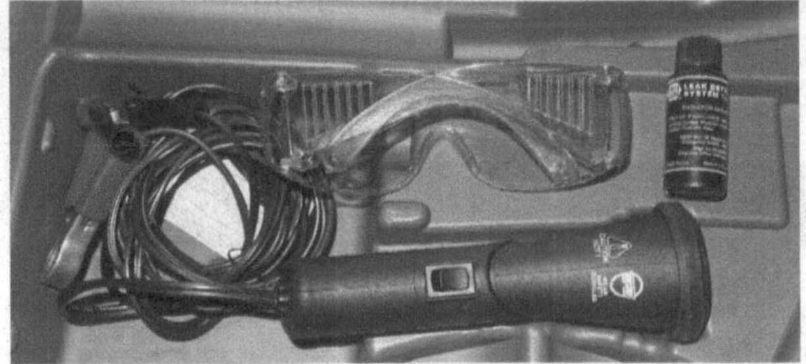

FIGURE 23-1 An example of a UV light kit that is used to help find different types of fluid leaks.

- If a cylinder shows low compression, a cylinder leak test is often performed. A cylinder leak tester measures the amount of air that passes through a cylinder (**Figure 23-4**).
- Locating engine noises can be easier when you use a stethoscope (**Figure 23-5**). Many shops use electronic stethoscopes which amplify noises, making them easier to identify.
- An exhaust backpressure tester is used to diagnose a restricted exhaust system (**Figure 23-6**).
- Lab scopes are used in a variety of testing situations (**Figure 23-7**). Scopes can display voltage and amperage over time, making them very useful for testing sensors and other components.

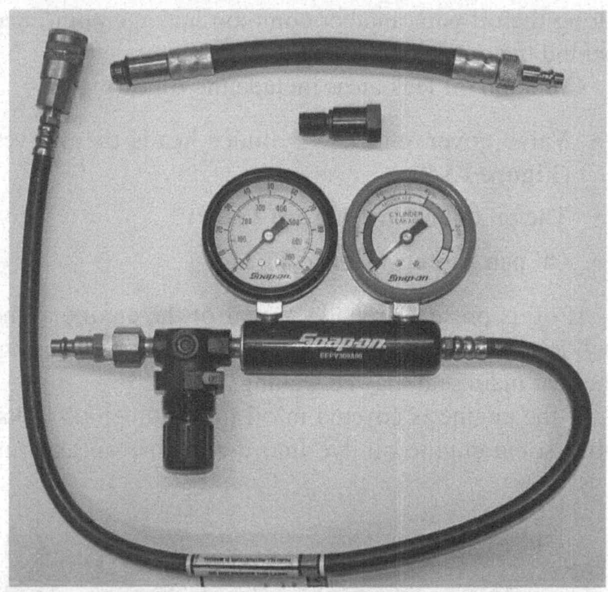

FIGURE 23-4 A cylinder leakage test kit. The top hose threads into the spark plug hole and connects to the hose on the tester. Air is supplied to the tester, and what passes through is displayed as leakage on the gauge.

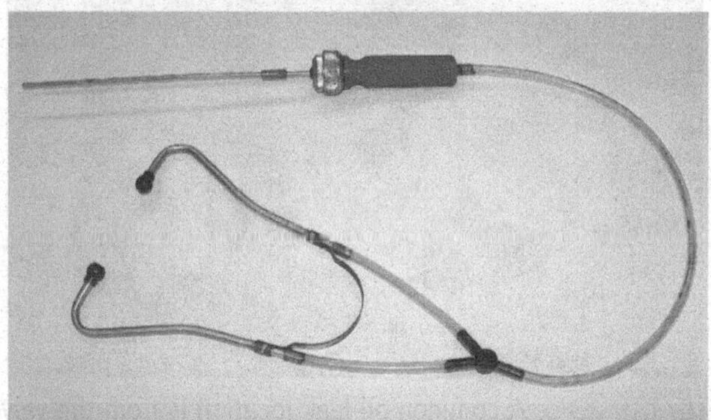

FIGURE 23-5 A stethoscope can be helpful when locating engine noises. Many shops use electronic stethoscopes.

FIGURE 23-6 An exhaust backpressure gauge is used to confirm a plugged exhaust system. This gauge threads into the oxygen sensor hole.

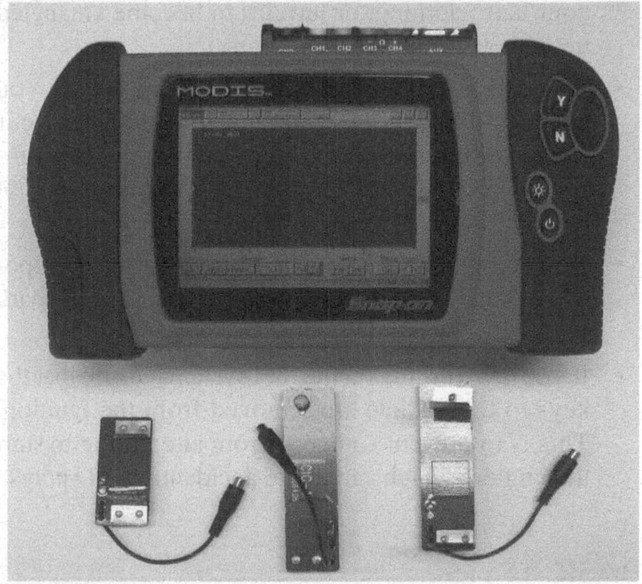

FIGURE 23-7 An example of a scanner/lab scope, used for many different types of voltage, resistance, amperage, pressure, and vacuum tests.

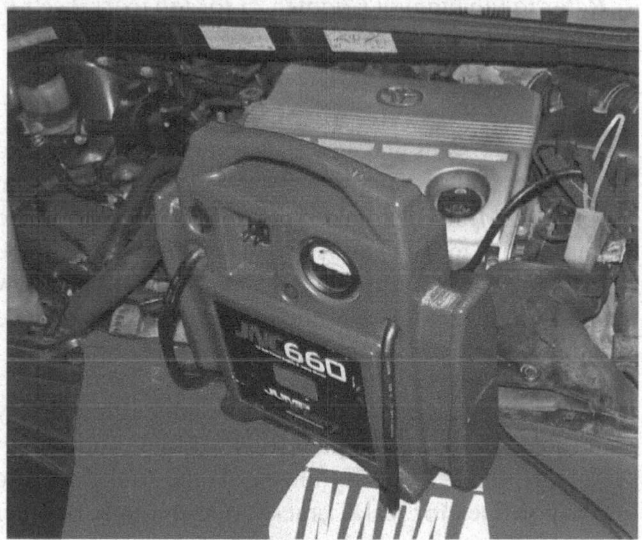

FIGURE 23-8 An example of using a fender cover to protect the vehicle during service.

Engine Testing Safety Precautions

As with all aspects of vehicle service, it is important to understand how to work safely to prevent injury and damage to tools, equipment, and the vehicle. The following are some general safety precautions:

- Whenever you are working on the engine or under the hood, be sure to use fender covers to protect the vehicle's finish (**Figure 23-8**).
- Be careful when you are working around rotating parts, such as drive belts and pulleys. Do not allow

your hair, clothing, or jewelry to become entangled in the belts or pulleys.

- If possible, allow the engine to cool before you perform any service. This not only reduces your chance of injury but also helps prevent engine damage. Removing fasteners and spark plugs from hot engines can easily cause damage to the threads.
- Do not place your hands or tools on or near components of the secondary ignition system, such as spark plug wires or ignition coils.
- If you are working on the fuel system, ensure that the keys to the vehicle are removed from the ignition. This is to prevent someone from attempting to start the engine, which can cause accidental fuel sprays.

BASIC TESTS

Before you make assumptions about an engine or its control systems, you must first evaluate the overall engine condition. This includes inspecting for leaks and verifying the internal components' integrity.

Refer to lab manual Chapter 23 for lab worksheets to perform the tests discussed in this section.

INSPECT THE ENGINE FOR LEAKS

Over time, the seals and gaskets in the engine harden and deteriorate, and eventually leaks develop. Inspecting an engine for a fluid leak is a common service performed by automotive technicians. First, determine what it is you are looking for. Usually the customer will let you know which fluid is leaking, but not always. You may only be told that something is leaking.

■ ***Fuel Leaks.*** Fuel leaks are usually immediately noticeable because of the fuel smell. If fuel is not spraying or dripping, carefully check all the fuel lines and fuel vapor lines for leaks, cracks, rust-through, or other damage. Do not use an electric shop light while you are looking for a fuel leak. If fuel drops onto a hot light, a fire is likely to result. Because leaking fuel will act as a cleaner, removing dirt, oil and undercoating, a clean spot near a fuel line may point toward the leak. To help you locate a small leak, a combustible gas detector or an exhaust gas analyzer can be used. These tools can detect small amounts of gas vapor, helping you narrow down the cause of the leak.

■ ***Engine Oil.*** The source of an engine oil leak, especially if it has been leaking for some time, can be difficult to locate. Start by inspecting the top of the engine around the valve covers and intake manifold gaskets, as these are common leak areas. Move down to check around the cylinder head gasket(s) and timing cover gasket. Inspect along the oil pan, another common leakage point, and behind the crankshaft pulley.

Common oil leak areas include the following:

- Valve covers and the cylinder heads themselves (**Figure 23-9**)
- The oil drain plug (**Figure 23-10**)
- Oil pan gaskets (**Figure 23-11**)

If oil is present around the rear of the engine at the bell housing, or is coming from inside the bell housing, the rear main seal may be leaking.

If the engine is covered in oil from numerous leaks, pour some engine oil dye into the engine oil and run

FIGURE 23-9 A common oil leak location is from the valve cover gasket. This gasket is leaking along the back side of the engine.

FIGURE 23-10 Oil pan drain plugs are common leak points. This vehicle has a replacement plug installed, possibly to try to fix a leak from the previous drain plug.

FIGURE 23-11 Oil filters and oil sending units are often places from which oil leaks.

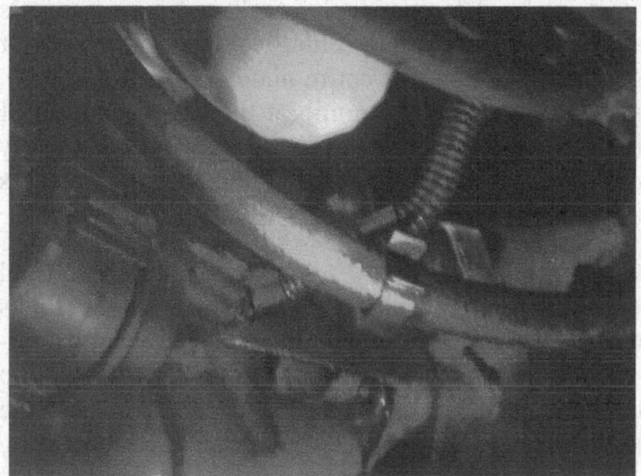

FIGURE 23-12 When several leaks are present, UV dye can help determine the exact cause of the leak. The dye will glow yellow under the UV light. **Be sure to wear the protective eyewear that comes with the UV light. Do not shine the light into your eyes.**

the engine. As the dye circulates throughout the engine and starts to leak, it will be visible under UV light (**Figure 23-12**). The dye will show as yellow when under the UV light rays. **Read and follow all the safety precautions and warnings when using a UV light kit. Do not expose your eyes or skin to direct UV light. Exposure to UV light can cause eye and skin damage.**

■ ***Coolant.*** Coolant leaks are often easy to distinguish due to the bright colors used to dye the coolant. In addition, coolant leaks may be accompanied by a sweet smell and steam. Begin by inspecting around the radiator and hoses.

- Weak or loose hose clamps can allow coolant to leak once the system is hot and has built up pressure. **Figure 23-13** shows an example of how coolant leaves a residue from a slow leak at a hose connection. Note the trail of corrosion below the hose indicating from where the leak is.

FIGURE 23-13 This coolant leak, while slow and slight, is leaving a trace of coolant down the side of the engine.

FIGURE 23-14 A slight coolant leak from the radiator shows the coolant starting to solidify.

- Check for leaks along the radiator core and side tanks. A rotten radiator may leak from several locations (**Figure 23-14**).
- Inspect around intake manifolds, core plugs, and where the cylinder head(s) bolt to the block. A leaking block core plug is shown in **Figure 23-15**.
- If coolant is dripping from the front of the engine, it is likely that the water pump is leaking (**Figure 23-16**).
- If the customer complains of steam from interior vents or wet carpeting, suspect the heater core is leaking (**Figure 23-17**).

Coolant can even leak through aluminum castings. Some engines may experience problems of porosity cracking, meaning that the aluminum casting of the head allows coolant to pass through the aluminum and leak into the oil.

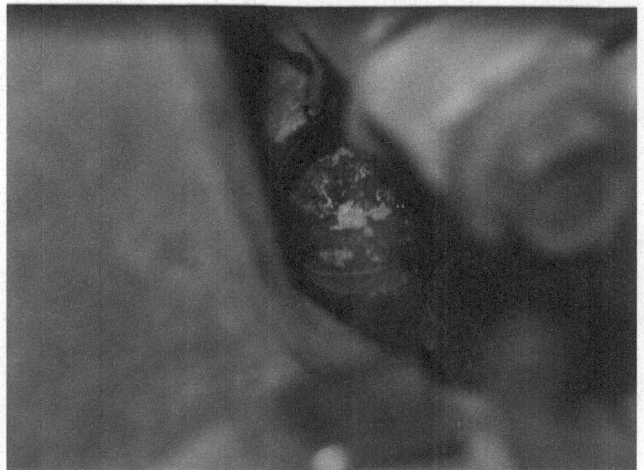

FIGURE 23-15 A rusted through and leaking core plug.

FIGURE 23-16 A leaking water pump allowing coolant to leak down the front of the engine and behind the crankshaft pulley.

■ ***Power Steering Fluid.*** Power steering fluid leaks, especially if the fluid is new and clear, may look like clean engine oil or brake fluid. Leaking power steering fluid from a faulty rack and pinion unit is shown in **Figure 23-18**. Fluid leaking from a hose or gearbox may coat the entire underside of the vehicle, making it difficult to pinpoint the exact location of the leak. A leak in the power steering system usually causes noise from the power steering pump as air enters the system.

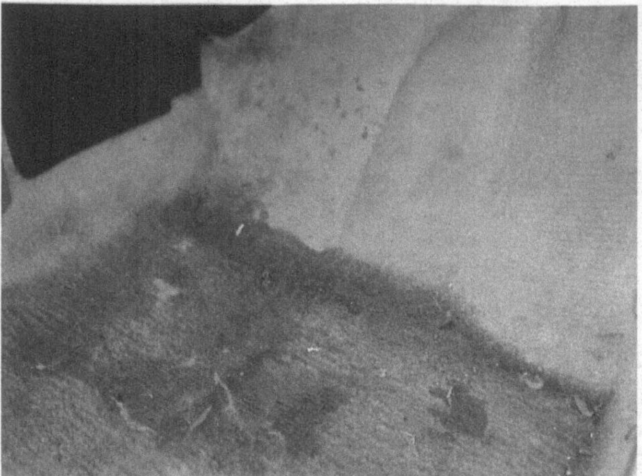

FIGURE 23-17 A leaking heater core that has soaked the carpet inside the passenger compartment.

■ ***Transmission Fluid.*** Leaking transmission fluid may look red, brown, or black depending on the fluid's age and condition. A common problem is a leaking transmission oil pan (**Figure 23-19**). Leaking transmission cooler lines may cause the leak to appear to be from the radiator or the engine, so pay careful attention during your inspection.

It may be necessary to clean the engine compartment and under the engine and transmission to verify all the fluid leaks. High-mileage vehicles may have leaks from several different systems and from multiple places just

FIGURE 23-18 Power steering fluid, depending on where the leak is from, can be confused with engine oil or brake fluid.

FIGURE 23-19 Automatic transmission fluid is red and easily identified as a leak. However, very old fluid may not appear red and may look like engine oil or power steering fluid.

from the engine. Clean the leaking areas with a degreaser to aid in locating the origins of the leaks.

Once the leak or leaks are identified, determine how to correct them. Leaking seals or gaskets can be replaced; however, you should try to figure out why the leak is occurring. For example, an engine may leak oil from one or more gaskets because of excessive pressure buildup inside the engine due to a faulty PCV system. Replacing the gaskets only fixes the leak temporarily. Inspecting and testing the PCV system is addressed in Chapter 24.

ENGINE MECHANICAL TESTING

Before any internal work or teardown of the engine begins, it is important to accurately diagnose a problem. The tests discussed in this section are used to help determine if a problem is internal to the engine or external. For example, an engine may have a misfire due to a fuel or ignition fault, a vacuum leak, incorrect valve adjustment, or more serious internal issues.

■ ***Oil Pressure Testing.*** An **oil pressure test** is performed if the oil pressure warning light remains on, if the oil pressure gauge reads low, or if unusual noises are heard from the engine. Low engine oil pressure can quickly cause major engine damage if it is not resolved.

Begin by checking the engine oil level, and add oil if it is low. Next, remove the oil pressure sending unit, and install the oil pressure gauge into the sending unit hole (**Figure 23-20**). The oil pressure sending unit is usually located close to the engine oil filter but can be anywhere on the engine block. Start the engine and note the oil pressure reading. Refer to the manufacturer's specs, but generally a warm engine should have at least 15 to 20 psi of oil pressure at idle speed, and it should increase as rpm increases. Worn main bearings, worn rod bearings, a clogged oil pickup, and a worn oil pump may cause low oil pressure. Unfortunately, once low oil pressure has been confirmed, at least partial engine disassembly

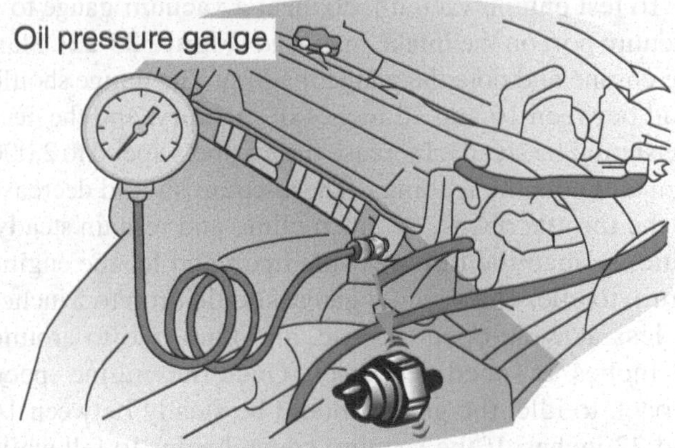

FIGURE 23-20 This illustrates how an oil pressure gauge is installed where the oil sending unit goes to check oil pressure.

is needed to determine the exact cause or causes. Low oil pressure can mean that major engine repairs are needed.

■ ***Vacuum Testing.*** Gasoline engines produce a vacuum during the intake stroke due to the restriction of airflow past the throttle plates and intake valves. This vacuum can be measured and used to diagnose a variety of engine problems. The better each cylinder seals, the more vacuum is produced. Ideally, each cylinder should produce an equal amount of vacuum.

Begin by listening for a vacuum leak. A leak from a vacuum hose will usually cause a whistling noise at idle speeds. However, the absence of an audible leak does not mean a leak is not present. Leaks around intake manifold gaskets and other locations often do not make noise.

Testing for vacuum leaks can be performed using a smoke machine (**Figure 23-21**). Connect the smoke machine to a vacuum hose on the intake manifold. Inject smoke into the engine and watch for signs of smoke coming out. The source of the smoke indicates the vacuum leak. Check the vacuum brake booster for leaks as well.

FIGURE 23-21 Using a smoke machine to look for a vacuum leak.

To test engine vacuum, connect a vacuum gauge to a vacuum port on the intake manifold (**Figure 23-22**). Start the engine and note the gauge reading. The gauge should read between 16 and 22 inches of mercury, and the needle should be steady. Increase the engine's speed to 2,000 rpm and note the readings. The vacuum should decrease as the throttle opens, and then clime and remain steady. Quickly snap the throttle wide open and let the engine return to idle. The vacuum gauge should drop to 5 inches or less at wide open throttle, and then rise to around 22 inches as speed decreases. Once the engine speed is back to idle, the gauge should be steady between 16 and 22 inches. If the vacuum gauge begins to fall while you are holding the engine at 2,000 rpm, the exhaust is restricted, preventing the flow of exhaust gas out of the engine. This will cause the exhaust gases to back up into the engine and prevent the air–fuel mixture from filling the cylinders properly.

FIGURE 23-22 Vacuum gauge readings can provide a great deal of information about engine operation.

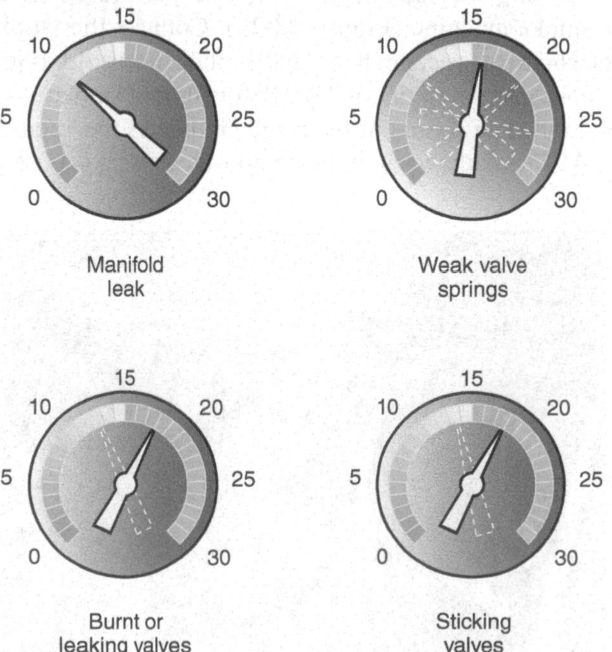

FIGURE 23-23 Often an overlooked test, the vacuum test can quickly identify a wide variety of engine problems.

This simple test is often overlooked by technicians, yet it can provide a great deal of information about the engine's mechanical condition. Some technicians now use a vacuum/pressure transducer to perform vacuum tests. The transducer connects to a vacuum hose and then to a lab scope. The reading from the transducer is displayed as a waveform. This method allows very close examination of each cylinder's vacuum signature and can be useful in locating valve sealing problems. The chart in **Figure 23-23** shows examples of common problems and how each can affect the vacuum gauge.

■ ***Power Balance Testing.*** The **power balance test** is used to determine if each cylinder is contributing power to the overall engine output. On many late-model vehicles, the on board computer can perform this test using a scan tool (**Figure 23-24**). During a power balance test, fuel injectors are disabled and the change in the engine's rpm is recorded. If a cylinder is not producing power, there will not be a drop in rpm when that cylinder's injector is disabled. This indicates the problem cylinder. All cylinders should drop rpm within about 10% of each other. **Figure 23-25** shows a power balance test being performed using the Ford IDS scan tool. Notice that cylinders 4 and 8 have negative numbers on the rpm scale on the left. This indicates that cylinders 4 and 8 are not producing power like the other cylinders.

The scan tool may allow you to select the cylinder(s) for testing (**Figure 23-26**). The cursor, shown highlighted in red, is used to select which cylinder to deactivate. By watching the scan data and listening to the engine, you can determine which cylinder, if any, is not producing power.

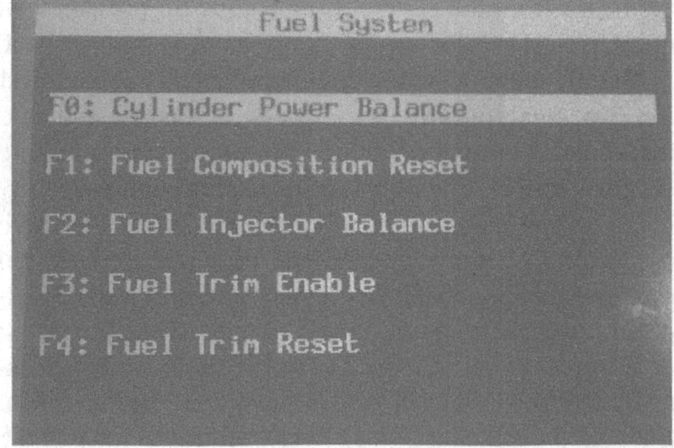

FIGURE 23-24 Using a Tech2 scan tool to perform a power balance test.

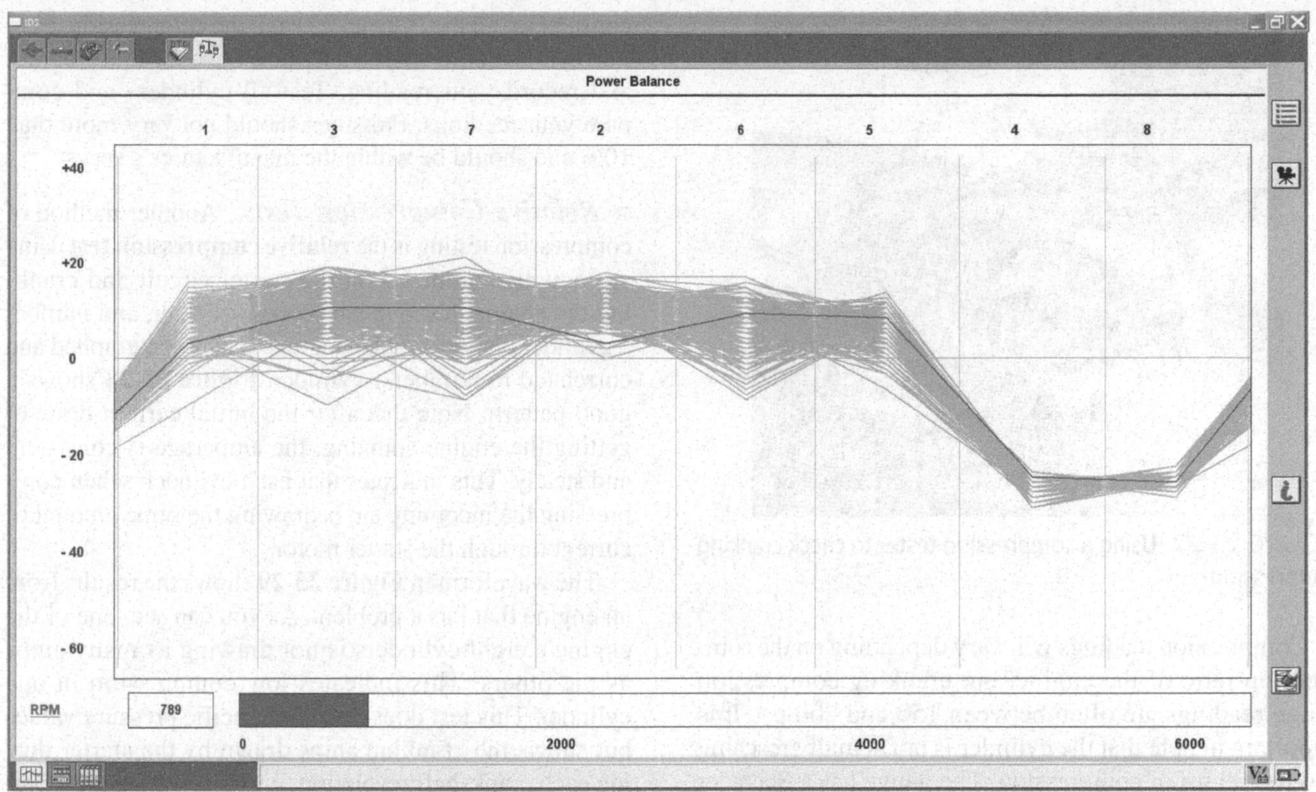

FIGURE 23-25 The power balance test results show that cylinders 4 and 8 are not producing power.

```
Asian 1983-2009
Power Balance Test
»#1  #2  #3  #4  #5   #6  Normal  Exit
     RPM            0  TPS(V)________0.58
ECT(°F)___________93  O2(mV)__________0
O2 MON__________LEAN  O2 B1-S2(mV)____0
O2 MON B1-S2____LEAN  MAF(V)_________0.24
INJ(mS)_________0.00  O2 HEATER______OFF
O2 HEATER B1-S2_OFF
```

FIGURE 23-26 Using a scan tool to turn off fuel injectors to perform a power balance test.

On older vehicles, an ignition analyzer connected to the primary side of the ignition coil can do this test. The analyzer interrupts the coil so that a cylinder does not spark, producing an rpm drop. The drawback of testing this way, by stopping the spark, is that unburned fuel passes through the cylinder and into the catalytic converter.

On vehicles with spark plug wires, some technicians insert a piece of vacuum hose or an adjustable spark tester in series with the spark plug wire, and use a test light to short the spark to ground. Done correctly, this will safely cancel or kill the cylinder. This allows the technician to test each cylinder without an analyzer. **Caution must be used when you perform this test to avoid being shocked by the ignition system. If the test light is not adequately grounded, the spark can easily jump to you to ground, causing you to be shocked and possibly injured.**

■ ***Cranking Compression Tests.*** A **cranking compression test** is used to check the ability of the engine to pull in air, compress the air, and keep it sealed in the cylinder. Usually a compression test is performed once a problem cylinder has been located. If a cylinder does not drop rpm during the power balance test, but has sufficient spark and fuel delivery, a compression test is often performed to check the cylinder's mechanical condition.

To test a specific cylinder, first allow the engine to cool and remove the spark plug from the cylinder to be tested. Next, thread the compression tester's hose into the spark plug hole and attach the gauge to the hose (**Figure 23-27**). Disable the fuel system and block the throttle wide open. This prevents fuel from washing the oil off the cylinder walls and allows for maximum airflow into the engine. Crank the engine over for five seconds. Note the needle reading of the first pulse during the test and the final reading on the gauge after five seconds. The first pulse should be at least 75% of the final reading.

FIGURE 23-27 Using a compression tester to check cranking compression.

Compression readings will vary depending on the compression ratio of the engine, but cranking compression gauge readings are often between 150 and 200 psi. It is important to note that the cylinder is not actually reaching 150 to 200 psi of compression. The gauge has a Schrader valve installed, just like a tire valve stem, which lets air into the gauge but not back out. This means that during the test, each time air is pulled into the cylinder and compressed, it is forced into the gauge, where the pressure reading increases. Compression pressure specs vary, so always refer to the manufacturer's service information. Worn rings, scored cylinder walls, burnt valves, damaged valve seats, worn cam lobes, or worn components that reduce valve opening can cause low compression readings.

If a cylinder indicates low compression, a wet compression test may be performed. A wet test involves squirting a small amount of engine oil into the cylinder and repeating the compression test. The oil will form a temporary seal between the piston rings and cylinder walls. If worn rings are causing the low compression, the oil will provide enough additional sealing to increase the compression reading. If the reading increases over the original by more than 10 percent, worn rings are likely the problem. If the reading does not increase, the low compression is likely caused by a leaking valve seat, cracked cylinder head, or leaking head gasket.

Cranking compression tests are also performed to determine overall engine condition. If an engine is not running properly, but the cause is not one or two specific cylinders, it may need a compression test on all cylinders. Start by removing all the spark plugs, setting them aside in order so that they can be reinstalled into the correct cylinders. Next, if possible, block the throttle plates open to allow maximum airflow into the cylinders. Connect a battery charger to the battery so that cranking speed remains constant for all the cylinders. Install the compression tester into cylinder number 1; crank the engine for five seconds, and record your reading. Test all cylinders and compare your readings. Pressures should not vary more than 10% and should be within the manufacturer's specs.

■ *Relative Compression Tests.* Another method of compression testing is the **relative compression test** using an inductive ammeter on the starter circuit and cranking the engine. Using a lab scope, ammeter, and number 1 cylinder pickup, cranking amperage can be graphed and correlated to number 1 cylinder. **Figure 23-28** shows a good pattern. Note that after the initial current draw of getting the engine spinning, the amperage is consistent and steady. This indicates that each cylinder, when compressing the incoming air, is drawing the same amount of current through the starter motor.

The waveform in **Figure 23-29** shows the results from an engine that has a problem. As you can see, one of the engine's eight cylinders is not drawing as many amps as the others. This indicates low compression in one cylinder. This test does not give specific pressure values but shows the cranking amps drawn by the starter during each crankshaft revolution. A cylinder with low compression draws fewer amps than a cylinder with higher compression, which shows on the waveform.

The same test is shown in **Figure 23-30** with the addition of a second test lead attached to the ignition system to determine which cylinder has low compression. The red spikes indicate the firing of cylinder number 1. Once cylinder 1 is identified, you can follow the engine's firing order to find out which cylinder has low compression. In this case, the firing order is 18436572, so the problem is with cylinder 2. This test can identify a weak cylinder quickly so that more in-depth testing can follow.

■ *Running Compression Tests.* A **running compression test** is done to check the breathing ability of the engine, specifically to see if the valve train is opening and closing the valves sufficiently or if the crank and cam are correctly timed. The difference between a cranking and running compression test is that a cranking test allows pressure to build up in the cylinder over several complete engine cycles. This can mask a problem such as worn cam lobes, which allow sufficient air to enter the cylinder at cranking rpm but do not open the valves enough at running rpm. Running compression tests can also pinpoint sticking valves and problems with variable valve timing (VVT) systems.

To perform a running compression test, remove the spark plug from the cylinder being tested. Next, disconnect the cylinder's fuel injector electrical connector if possible. If working on an older, distributor ignition system, disconnect and ground the spark plug lead for that

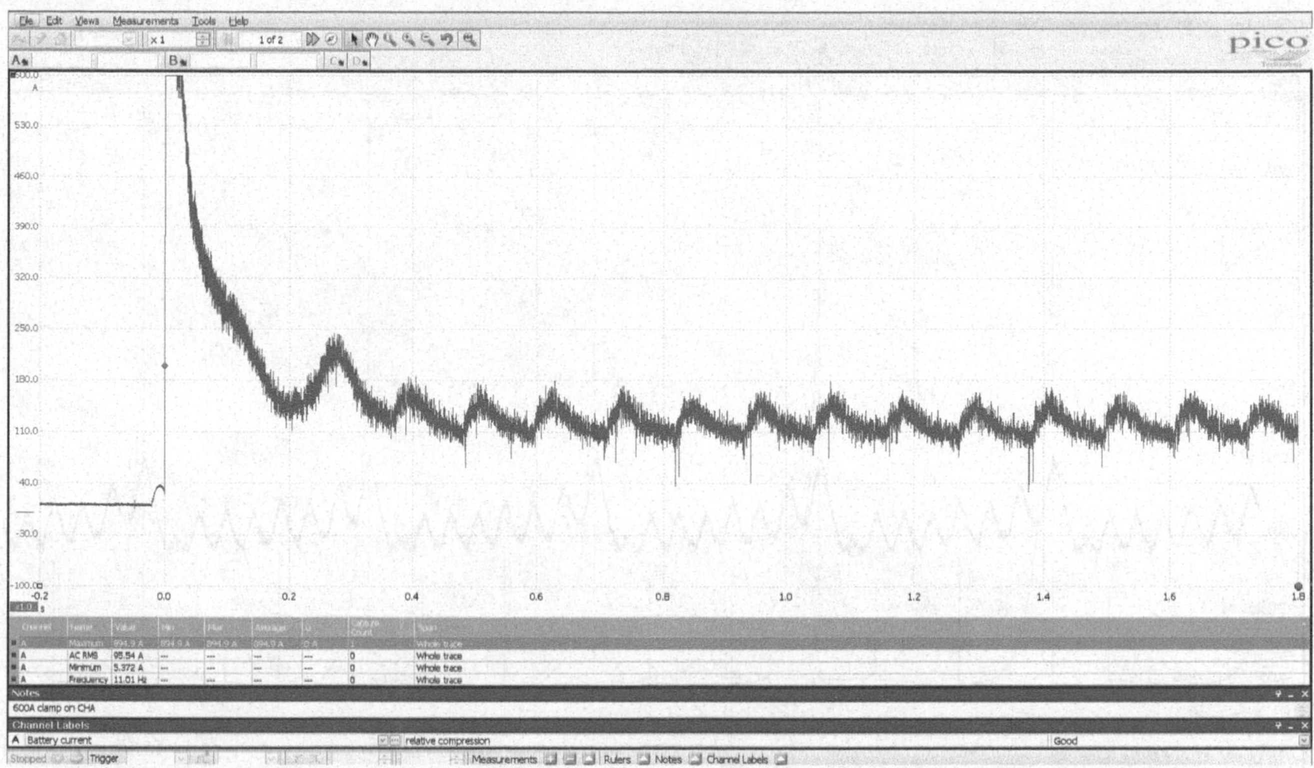

FIGURE 23-28 The results of using a current clamp and scope to perform a relative compression test. This is an example of an engine with good compression and a smooth, even compression pattern.

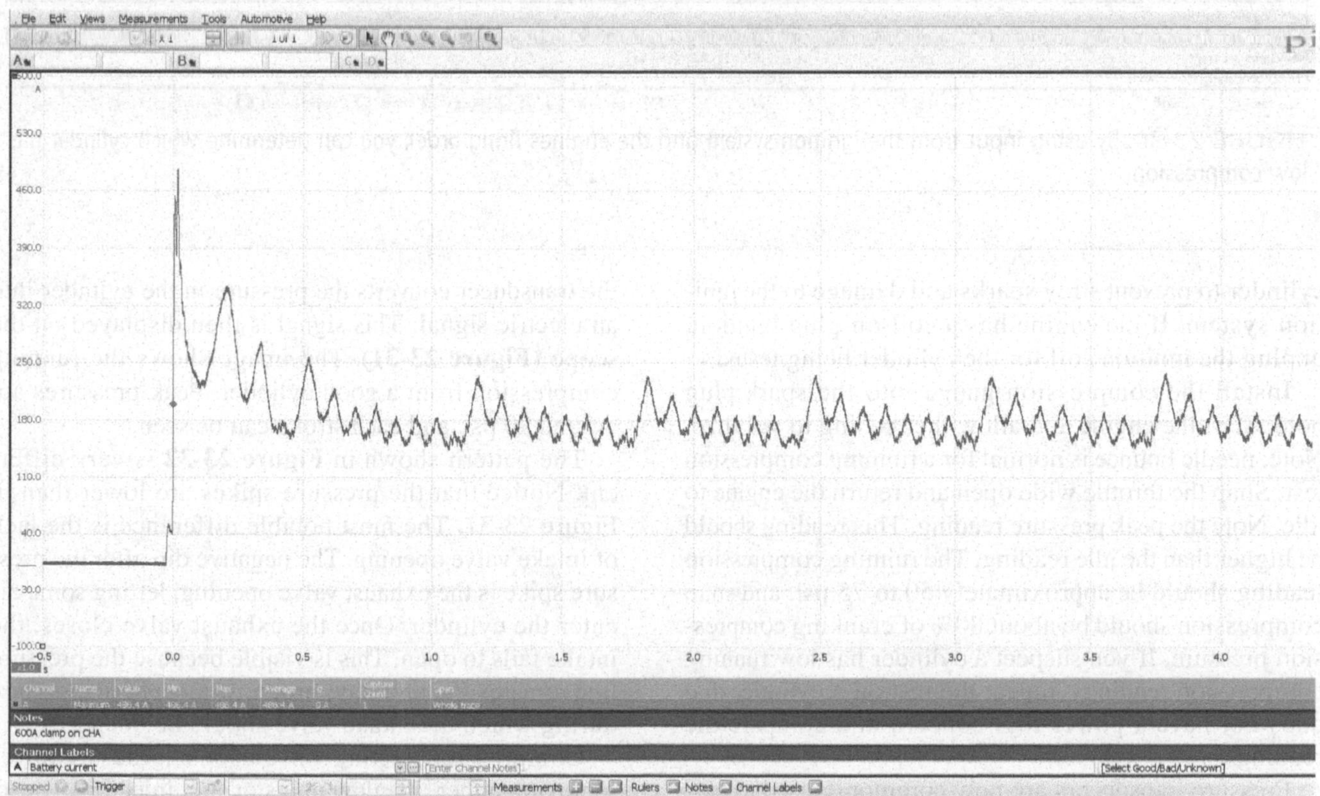

FIGURE 23-29 A relative compression test showing a cylinder that is not drawing as much current as the others. This indicates low compression on a cylinder.

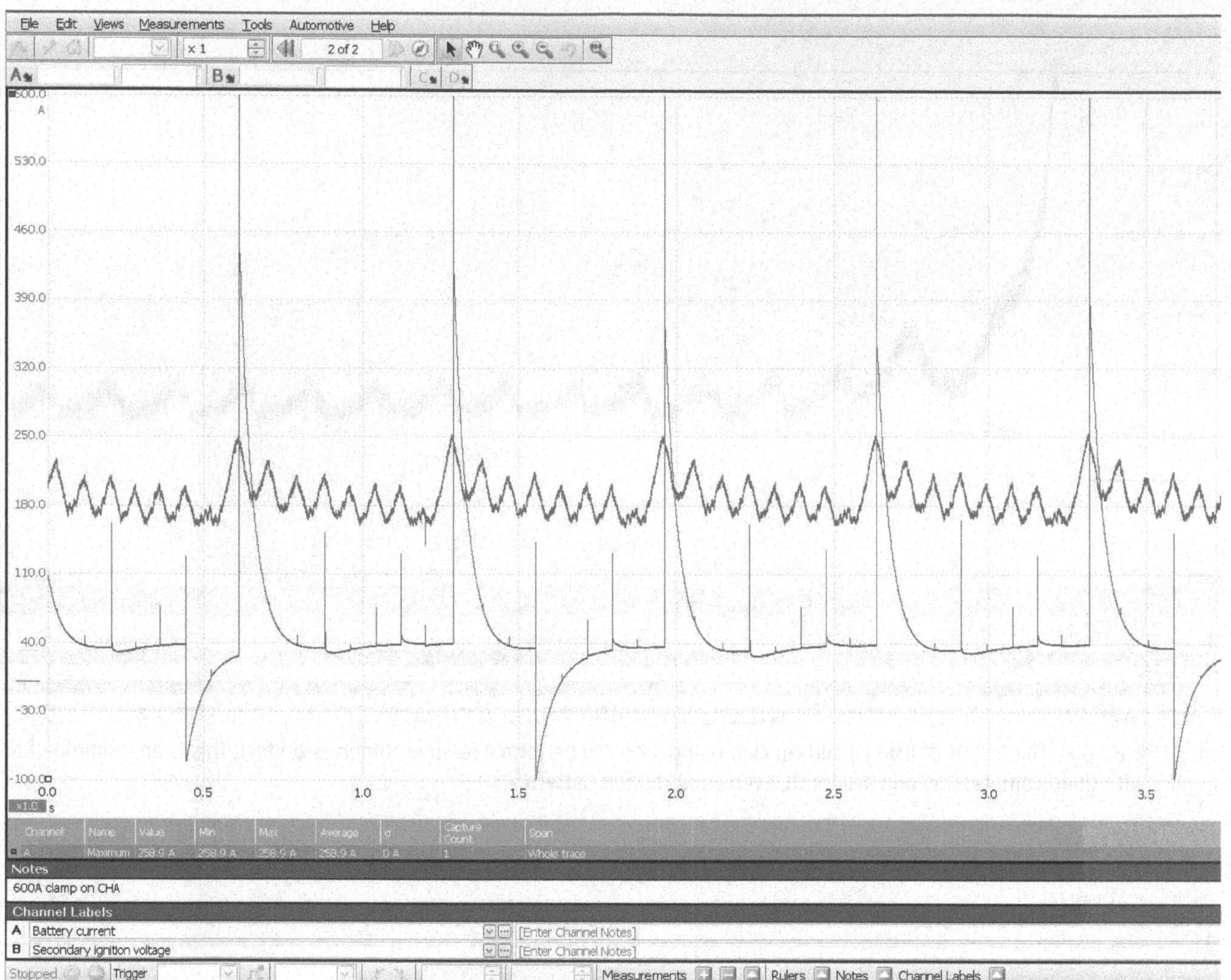

FIGURE 23-30 By using input from the ignition system and the engines firing order, you can determine which cylinder has low compression.

cylinder to prevent stray sparks and damage to the ignition system. If the engine has a coil-on-plug ignition, unplug the ignition coil for the cylinder being tested.

Install the compression gauge into the spark plug hole. Start the engine and allow the reading to stabilize. Note, needle bounce is normal for a running compression test. Snap the throttle wide open and return the engine to idle. Note the peak pressure reading. This reading should be higher than the idle reading. The running compression reading should be approximately 50 to 75 psi, and snap compression should be about 80% of cranking compression pressure. If you suspect a cylinder has low running compression readings, repeat the test on a cylinder that does not have a power loss concern and compare the readings.

Pressure transducers are now commonly used to perform compression tests. Connected to a lab scope and used in place of a conventional compression test gauge, the transducer converts the pressure in the cylinder into an electric signal. This signal is then displayed on the scope (**Figure 23-31**). The image shows the running compression from a good cylinder. Peak pressures are around 90 psi, and each stroke can be seen.

The pattern shown in **Figure 23-32** is very different. Notice that the pressure spikes are lower than in Figure 23-31. The most notable difference is the lack of intake valve opening. The negative dip after the pressure spike is the exhaust valve opening, letting some air enter the cylinder. Once the exhaust valve closes, the intake fails to open. This is visible because the pressure line remains flat and stays around atmospheric pressure during which the intake valve should be open.

Compression testing on diesel engines requires removing either the glow plugs or fuel injectors. Diesel compression gauges have a higher-pressure capability, but the testing is essentially the same.

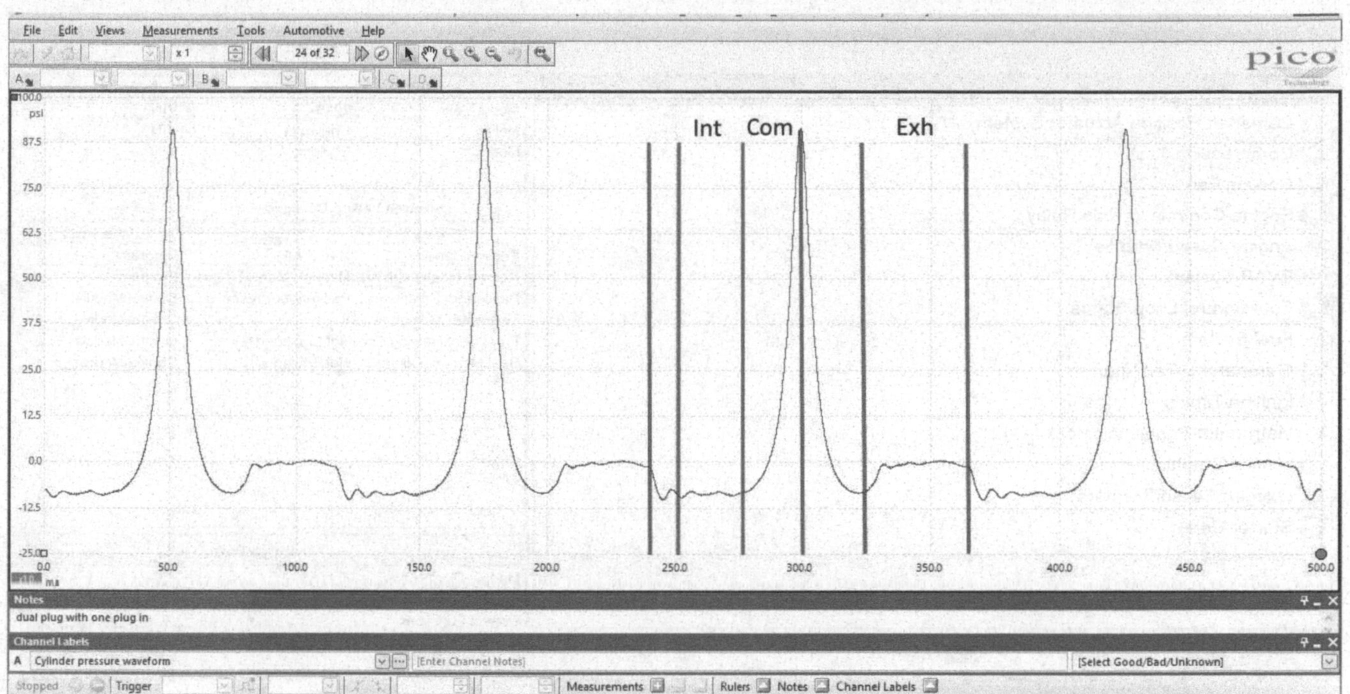

FIGURE 23-31 A running compression test using a pressure transducer installed in the spark plug hole. All parts of the four strokes are easily seen.

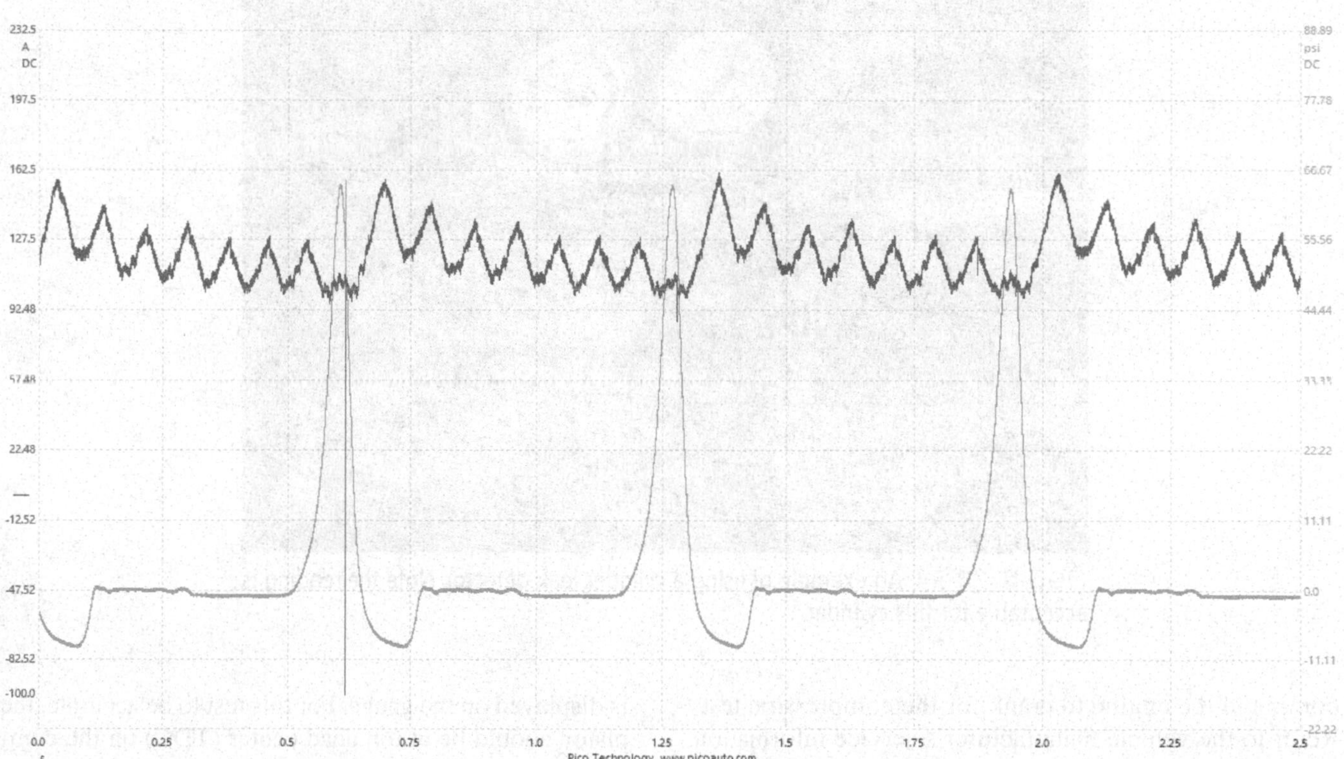

FIGURE 23-32 This screen shows the relative compression and the actual in-cylinder compression pressure using a pressure transducer.

Compression testing on modern vehicles, hybrids, and those with push-to-start require using a scan tool. Because hybrid vehicles do not use conventional 12-volt starter motors and often do not use traditional keys either, compression tests on these vehicles are performed differently. Shown in **Figure 23-33**, this test mode is used to

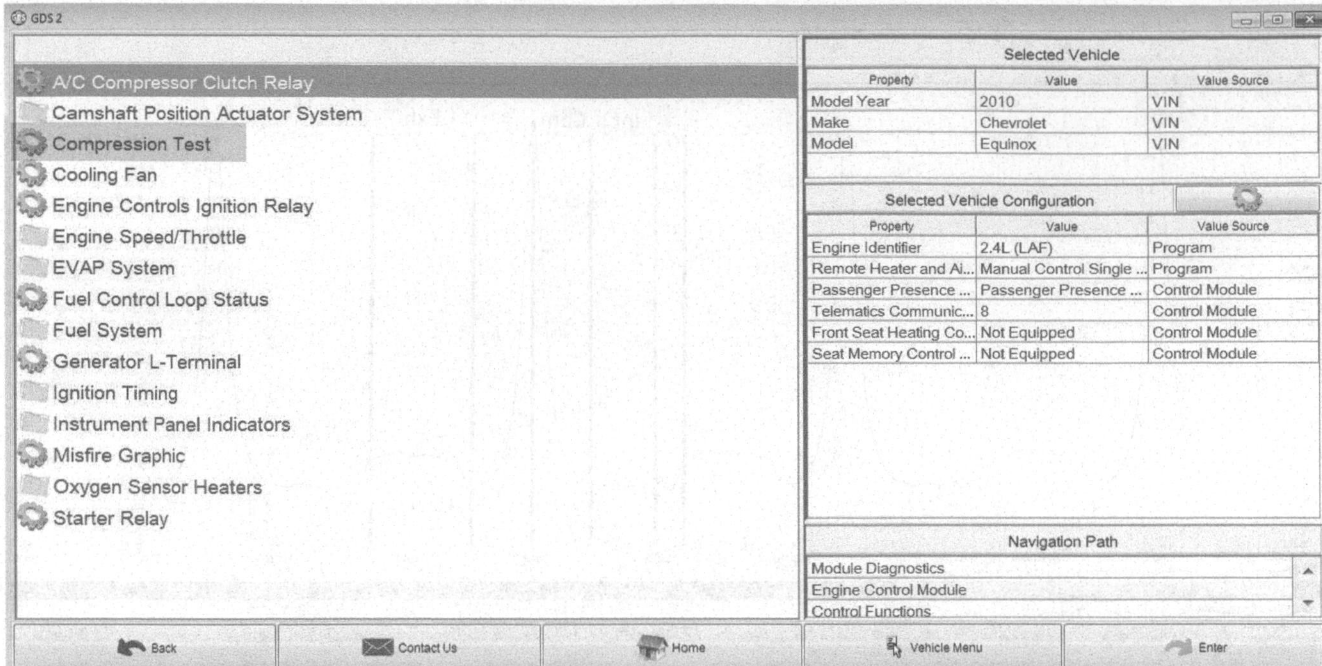

FIGURE 23-33 Using a scan tool for compression testing.

FIGURE 23-34 An example of using a cylinder leak detector. Note the reading is acceptable for this cylinder.

command the engine to crank for the compression test. Refer to the vehicle manufacturer's service information for procedures specific to the vehicle being tested.

■ ***Cylinder Leakage Tests.*** If a low-compression cylinder is identified, a **cylinder leakage test** can help determine where the pressure is going (**Figure 23-34**). The cylinder leak detector uses the shop air supply and a regulator to pressurize the cylinder. The rate of leakage is displayed on the gauge. For this test to be accurate, the piston should be at top dead center (TDC) on the compression stroke so both valves are closed. Generally, up to 20% leakage is considered normal for this test. The gauge shown in Figure 23-34 indicates that the leakage is acceptable.

If excessive leakage is present, listen for air coming from the exhaust tailpipe, throttle body, and oil fill cap. Remove the radiator cap and check for bubbles in the coolant. If an exhaust valve is burned or not seating, air will be heard escaping out of the exhaust pipe. An intake valve that is damaged or not seating will allow air to escape up through the intake, and it can be heard exiting the throttle body. If the air is leaking past the rings, air will be heard from the oil fill cap or dipstick tube. Bubbles in the coolant indicate a blown head gasket or cracked cylinder head.

An important note about leakage testing is that if the cylinder is not at TDC compression or if the piston moves when air is in the cylinder, a valve may open, allowing the air to escape. If this happens, the gauge will read a 100% leak rate, which can lead you down a false trail. Make sure the piston is at TDC compression and remains there. This may require using a socket and ratchet on the crankshaft pulley bolt to prevent the crank from rotating. Be careful when you are performing this test as the air entering the engine can cause the engine to spin, catching your fingers in belts or pulleys if you are not careful.

■ ***Exhaust Smoke.*** When examining a vehicle for engine problems, be sure to check what is coming out of the tail pipe. Excessive exhaust smoke is an indicator of several possible problems:

- Blue exhaust smoke means that engine oil is being burned in the combustion chamber. Worn valve stem seals, worn valve guides, and worn piston rings are possible causes.
- White smoke is usually caused by coolant entering the combustion chamber from a leaking head gasket or cracked head. White smoke can also be caused by automatic transmission fluid being burned. Many older vehicles use a vacuum modulator in the transmission. If the diaphragm ruptures, transmission fluid can be pulled into the engine through the vacuum port, and when burned it can cause white exhaust smoke.
- Black smoke is caused by excessive fuel being burned. A faulty fuel injector, excessive fuel pressure, or modifications to the engine and fuel system can cause this.

Modern diesel engines should emit only clear smoke.

- Black smoke indicates a rich fuel mixture, which can be caused by leaking injectors or clogged intake systems.
- Excessive black smoke can also be caused by a modified or "chipped" diesel engine control module.
- Blue smoke occurs when oil is getting into the combustion chamber, just as in a gasoline engine.
- White smoke from a diesel indicates a lean fuel mixture. This can be caused by low fuel pressure. A faulty glow plug circuit can cause white smoke on engine startup.

With both gasoline and diesel engines, check exhaust smoke at idle and under load, driving if possible. Some problems may not show up when idling in the shop but will when the engine is placed under load.

■ ***Engine Noise.*** All engines make noise, especially during warmup and cooldown periods, as the block and head(s) expand and contract. These noises typically sound like clicks or a creaking noise. Abnormal noises occur when there is a problem. Noises are usually classified as upper end, lower end, combustion related, or external to the engine.

Upper-end engine noises are valve train related. These noises are typically light tapping or ticking sounds that coincide with the opening and closing of the valves. Excessive valve clearance or lash, worn lifters, loose rocker arms, or a lack of proper lubrication to the valve train components can cause this. The components highlighted in **Figure 23-35** will often "tap" when loose or worn. To determine the cause of an upper-end noise, remove the valve cover and inspect the components. An engine with hydraulic lifters should not have any play between the rocker arms and the valves. If looseness is felt, check for a collapsed lifter, damaged rocker, or sticking valve. On engines with valve lash adjustments, check the clearance between the rocker arms and the valves. This procedure is discussed in Chapter 24.

Lower-end noises can be caused by worn main or rod bearings, which tend to make low-to-medium-pitch knocking sounds that get worse as engine speed increases. Piston slap is a hollow, bell-like noise most often occurring during cold engine warmup. Worn piston pins make a sharp metallic rap, most distinct at idle speed. Cracked flywheels or flex plates can mimic lower-end noises and can lead to misdiagnosis. Lower-end noises usually require removing the oil pan and flywheel inspection cover to determine the cause of the noise.

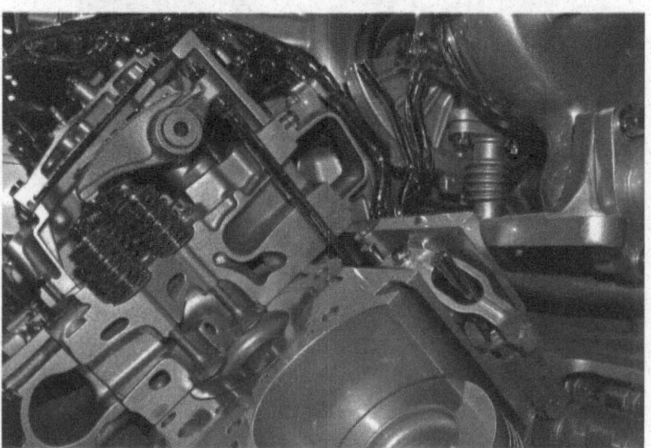

FIGURE 23-35 The components highlighted can, if worn or damaged, cause tapping noises.

Front-end engine noises can come from worn timing chain guides or weak tensioners (**Figure 23-36**). Damaged crankshaft balancers can cause knocking noises similar to a cracked flywheel (**Figure 23-37**).

Combustion noise is often due to detonation or spark-knock. This is a light pinging noise, like loose marbles under the hood. This can be caused by using fuel with too low an octane rating, advanced ignition timing, very lean mixtures, a malfunctioning EGR system, incorrect spark plugs being installed, or engine overheating.

Diesel engines can have combustion noises due to the use of fuel with an incorrect cetane rating. An illustration of fuel octane and cetane ratings is shown in **Figure 23-38**. A fuel with a low octane rating burns faster than a fuel with a high octane rating. Using a fuel with a lower octane number than specified by the vehicle's manufacturer, especially in older vehicles, can cause the fuel to spontaneously ignite in the combustion chamber before it is ignited by the spark plug. **Figure 23-39** shows an example of a fuel requirement decal. Many European manufactured cars and those with a turbo- or superchargers often require high-octane fuel. Make sure the customer understands if his or her vehicle has a high-octane fuel requirement.

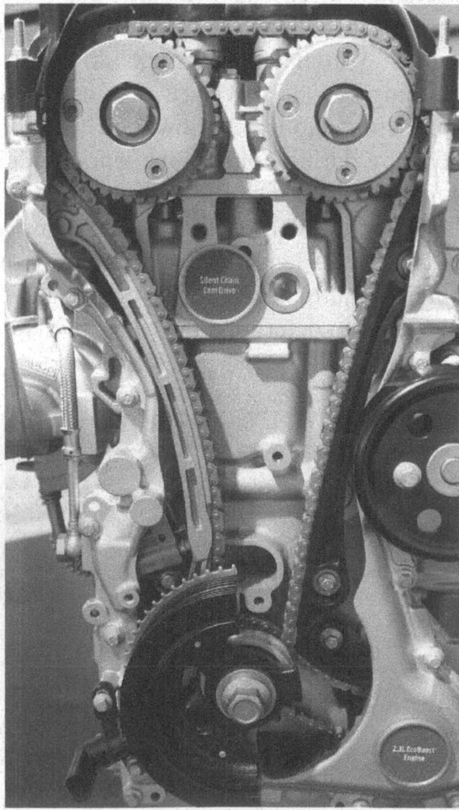

FIGURE 23-36 Problems with timing chains, chain guides, and tensioners can cause various noises from the front of the engine.

FIGURE 23-37 An example of a crankshaft pulley that is separating and cause noise.

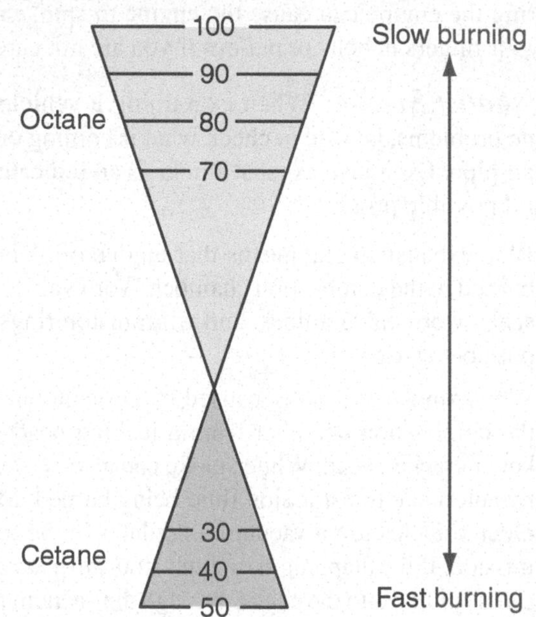

FIGURE 23-38 On older engines, using the incorrect octane fuel can cause detonation. A low-octane fuel ignites more readily than a high-octane fuel.

FIGURE 23-39 An example of a fuel octane requirement decal for a turbocharged VW.

If an engine is experiencing detonation, there may be a fault in the knock sensor circuit. Most late-model engines use a knock sensor to listen for detonation. If detonation is detected, the knock sensor signals the engine control module, which retards the ignition timing. If the sensor or circuit fails, detonation can go undetected.

Noises outside of the engine can be caused by induction and vacuum leaks. This usually causes a high-pitched hissing sound. An example of a leaking vacuum hose is shown in **Figure 23-40**. As the air is pulled through the cracks in the hose, it causes a whistling or hissing sound.

Chirping or squealing, especially upon engine startup or when the wheels are turned, is often caused by a loose or worn accessory drive belt. Cracked, chipped, or broken fan blades can create several types of noise, from clicking, whistling, grinding, or roaring sounds.

A clunk, coupled with a jerking or shifting feeling, may be caused by a worn or broken engine or transmission mount. To check for worn powertrain mounts, set the parking brake and block the drive wheels. Next, start the engine and place the transmission in reverse. Keeping your foot firmly on the brake, slowly increase engine speed while you watch for excessive engine movement. Repeat the process with the transmission in Drive. Normal engine movement is about an inch or less. If the engine moves more than an inch either way during the test, inspect the engine and transmission mounts for wear and damage.

The same problem can occur with sub-frame or cradle mounts (**Figure 23-41**). This vehicle had severe noises whenever shifted through the gears, accelerated, or decelerated as the entire sub-frame moved under the car.

Noises can be caused by worn, damaged, or broken components. As the component moves, it generates noise. Diagnosing noises can be difficult because sound can travel through the powertrain and body of the vehicle. To aid in diagnosing a noise, tools such as electronic stethoscopes are often used. One popular device is called *the Chassis Ear*. This tool has several microphones that are attached to the vehicle to help determine the cause of a noise.

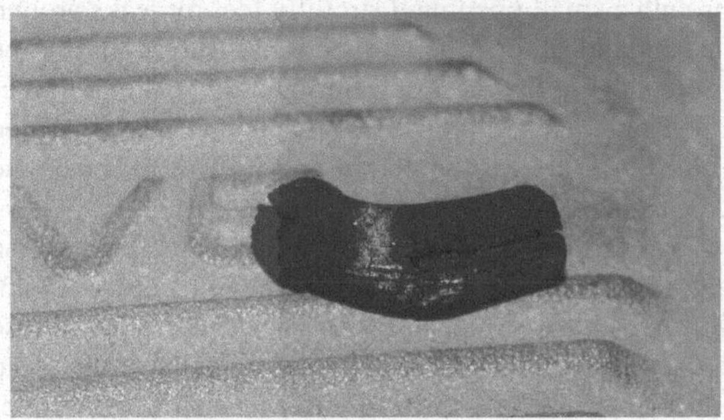

FIGURE 23-40 A broken vacuum hose will often cause a hissing noise when the engine is running.

■ ***Engine Vibration.*** Engine vibration can be caused by cylinder misfire, poor overall engine operation, an engine that is out of balance, and worn or broken engine mounts. A cylinder misfire disturbs power generation, causing the crankshaft to slow down during the cylinder's power stroke, and then speeding back up as the next cylinder fires. This is usually more noticeable in a four-cylinder engine than a six- or eight-cylinder engine because there are fewer cylinders producing power in a four-cylinder engine.

An engine that is producing power on all cylinders, but unevenly, will also vibrate. This can be caused by faults that affect all the cylinders, such as excessive EGR flow, low fuel pressure, or incorrect valve timing.

An out-of-balance condition can occur if the harmonic balancer or flywheel is replaced, or if there is a failure with a balance shaft. Externally balanced engines are balanced with the harmonic balancer and flywheel installed. The engine can become out of balance if one of those components is replaced. An internally balanced engine does not need to be rebalanced if either the harmonic balancer or flywheel is replaced.

Many engines use balance shafts to reduce vibrations. These shafts often counterrotate and cancel out engine vibrations. Balance shafts are timed with the crank and cam(s) and must be set correctly when replacing the timing belts or chains. Improper alignment of the balance shaft will result in a vibrating engine.

Engine mounts secure the engine and absorb vibration. Some mounts use a hard rubber damper (**Figure 23-42**).

FIGURE 23-41 Severely worn or missing cradle bushings can allow the powertrain to move and make noise.

FIGURE 23-42 An example of a powertrain mount. Worn mounts can allow the engine and/or transmission to move around and make noise.

Hard rubber mounts allow good engine support and some flexibility, but not all engine vibration dissipates through the mount; some is transferred to the body of the vehicle. Over time, the mount deteriorates and can separate, allowing the engine to vibrate freely. Many four-cylinder and V6-powered vehicles use hydraulic engine mounts, which are filled with either a hydraulic oil or glycol. This type of mount works well to reduce vibration until the mount starts to leak. Many newer vehicles use active motor mounts, meaning that the engine control module monitors and counteracts engine vibration. There are several types of active engine mounts being used today, including air, hydraulic, and magnetorheological mounts. Diagnosis of active mounts requires the use of a scan tool to monitor mount sensor data and output performance.

■ ***Exhaust System Inspection.*** An inspection of the exhaust system should be part of your diagnosis. Problems with the exhaust include restrictions, exhaust leaks, and mechanical damage. A restricted exhaust can cause reduced engine performance if the flow of exhaust gases prevents the normal induction of the air–fuel mixture. The exhaust can become restricted due to damage, such as a crushed pipe, or by a failed catalytic converter. To test for a restricted exhaust, a vacuum gauge can be used, as discussed earlier in this chapter, or by using an exhaust backpressure gauge (Figure 23-6 at the beginning of the chapter).

To use a backpressure gauge, unplug and remove the oxygen sensor located before the catalytic converter.

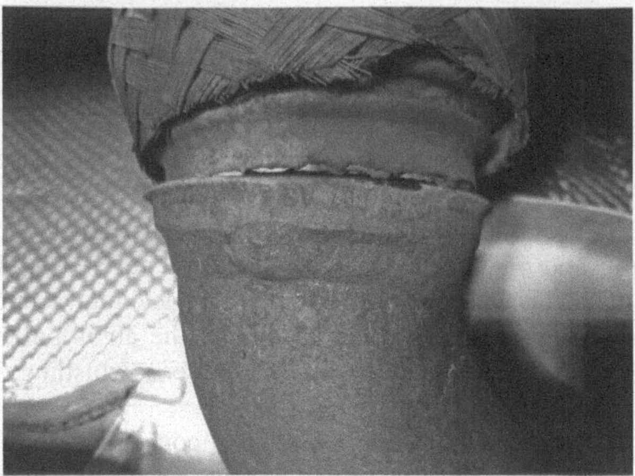

FIGURE 23-43 Exhaust leaks not only make noise but can be dangerous if fumes enter the passenger compartment.

Thread the gauge into the oxygen sensor fitting and start the engine. Check the backpressure at idle and at 2,000 rpm. A slight amount of backpressure is normal, less than 1–2 psi at idle and less than 3 psi at 2,000 rpm. If the pressure is excessive, check for damage to the exhaust system. If there are no signs of damage, tap the catalytic converter with a rubber mallet, and listen for rattling or knock from inside the converter. If noise is present, this indicates that the inside of the converter has failed and is broken up. This requires replacing the converter and possibly the muffler if pieces of the converter have broken loose and plugged the muffler. If no noise is present, it does not mean the converter is not the problem. You may need to separate the exhaust by disconnecting it where it goes into the converter and then retesting pressure. If the pressure is reduced, the restriction is either in the converter or after the converter.

Exhaust leaks can be caused by rust through of the pipes and muffler or by mechanical damage, such as driving over a curb. Check for leaks by listening at connections and along the pipes. Exhaust leaks often sound like a loud engine or like a hissing as exhaust gas is forced out through a small opening. Carefully feel for leaks around gaskets and other pipe connections. Replace any pipe or component that is rusted through and leaking. Sometimes an exhaust clamp will rust apart, allowing a pipe to loosen. Replace the clamp and tighten to repair the exhaust leak.

An example of a cracked front exhaust pipe at the engine manifold connection is shown in **Figure 23-43**. This leak had a loud, deep noise because it was close to the engine. **Figure 23-44** shows an exhaust manifold bolt that rusted off, causing a large exhaust leak under the hood. Flex pipe failures are common where the two sections are joined together (**Figure 23-45**).

FIGURE 23-44 A broken exhaust manifold bolt allows exhaust to escape, making a lot of noise.

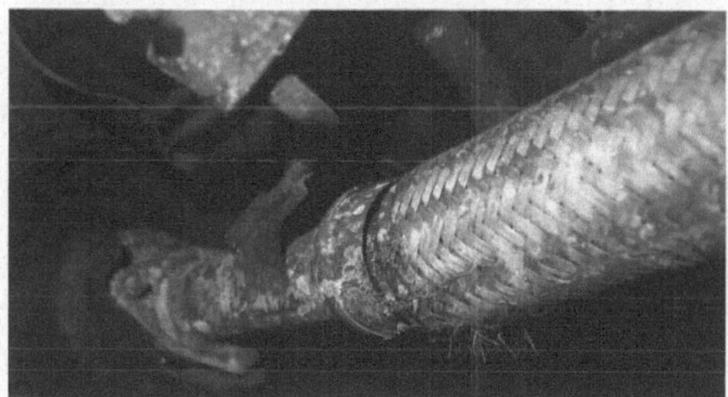

FIGURE 23-45 A common exhaust failure is in the flex pipes. These pipes fail from rust and from having to absorb the movements of the powertrain.

FIGURE 23-46 Check the hangers and other parts of the exhaust system. Broken hangers can allow pipes and mufflers to move and contact other components.

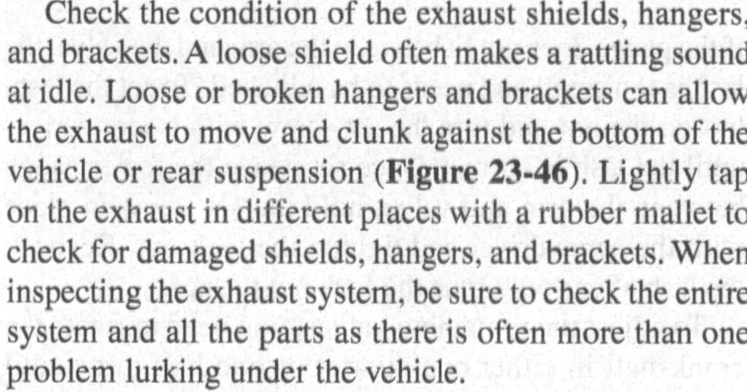

Check the condition of the exhaust shields, hangers, and brackets. A loose shield often makes a rattling sound at idle. Loose or broken hangers and brackets can allow the exhaust to move and clunk against the bottom of the vehicle or rear suspension (**Figure 23-46**). Lightly tap on the exhaust in different places with a rubber mallet to check for damaged shields, hangers, and brackets. When inspecting the exhaust system, be sure to check the entire system and all the parts as there is often more than one problem lurking under the vehicle.

Test Analysis

Test analysis is critical so that the true condition of the engine can be ascertained. Failure to perform testing or failure to understand test results can lead a technician to incorrect conclusions regarding engine condition. This is especially true for engine noises. Careful inspection and testing is necessary to prevent misdiagnosis and the selling of unneeded work. If you are unsure whether a problem with an engine is a major problem or a minor problem, get a second opinion. Ask a coworker or a more experienced technician to look at your test results before you decide to remove or replace an engine.

MAKING DECISIONS BASED ON TEST RESULTS

Once you have performed the tests necessary to determine what problems an engine has, you need to decide what the best options are for repair or replacement. Several factors determine whether attempts should be made to repair or replace the engine or its major components. These factors include the following:

- The age and mileage on the engine and on the vehicle
- The use and life expectancy of the vehicle
- The extent of the repairs required
- Customer expectations
- Warranty periods for parts and labor
- Warranty for a replacement engine
- Shop capabilities
- Access to machine shop services

■ ***Minor Engine Concerns.*** Some engine repairs, such as upper engine oil leaks, valve train noise, timing belts and chains, water pumps, and intake and exhaust leaks, and others are considered relatively minor and can be performed without removing the engine. Of course, there are exceptions to this, where some repairs are more easily performed with the engine out of the vehicle.

Many shops can provide repairs for these types of problems because very little dedicated engine-related equipment is needed, and these repairs usually do not require a lot of time to complete, unlike rebuilding an engine.

■ ***Major Engine Concerns.*** If an engine has lower-end noise, excessive ring wear, or damage to pistons, rods, or the crankshaft, removal is likely necessary. However, most shops do not rebuild engines in-house anymore due to the very large investment in tooling that is required. Engine rebuilding is done by specialty shops that have the machining tools necessary to restore the engine to its original condition. For most shops, the most economical approach is replacing a worn-out engine with a new or rebuilt engine from a trusted supplier. This reduces the time the vehicle is out of service and the time the technician spends on the repair, and it requires almost no cost to the shop for equipment or special tools.

ENGINE-RELATED SERVICES

Even as advanced as engine designs are today, all engines require some periodic services to remain in good working condition. Among the basic services still performed are valve adjustments, timing belt replacement, seal and gasket replacement, and cooling system repairs.

■ ***Valve Adjustments.*** Many overhead cam engines require periodic valve lash adjustment. These engines operate the rocker arm directly by the camshaft. The gap between the rocker arm and the top of the valve, called **valve lash**, is to allow for expansion of the components. If a gap were not present, the expansion of the rocker arm and the valve could cause the valve to hang open once both were hot. To prevent this, a slight gap is set between the rocker arm and the valve, which diminishes as the engine warms up.

A common lash adjustment method is shown in **Figure 23-47**. To adjust the lash, place cylinder 1 at TDC compression stroke. Make sure that both the intake and exhaust valves are closed. Locate the valve lash spec. In many cases, intake valve lash is set between 0.008 inch (0.2 mm) and 0.010 inch (0.25 mm) and exhaust valve lash is set between 0.010 inch (0.25 mm) and 0.012 inch (0.3 mm). Select the proper feeler gauge (**Figure 23-48**), and insert it between the rocker arm and the top of the valve (**Figure 23-49**). The feeler gauge should fit with a slight amount of friction or drag when moved between the rocker arm and the valve. If the rocker arm moves up and down with the gauge in place, the lash is excessive and needs to be reset.

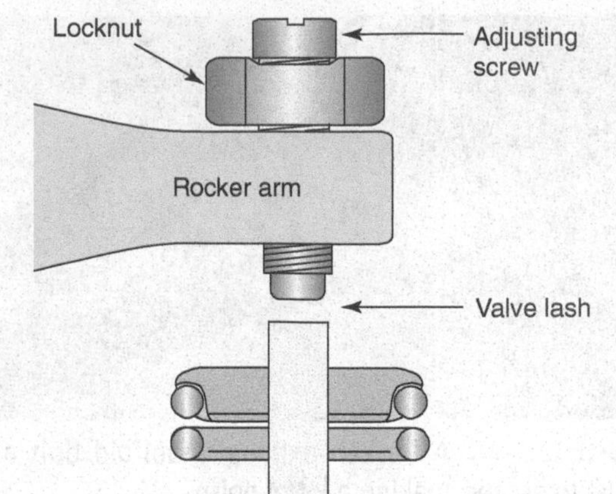

FIGURE 23-47 This illustrates valve lash, the small space or clearance between the rocker arm and the top of the valve.

FIGURE 23-48 A common feeler gauge set. Each strip is a different thickness.

If the gauge does not fit between the arm and the valve, the lash is too small and needs to be adjusted. To set the lash, loosen the nut and turn the set screw with a screwdriver until the lash is correct. Once set, leave the feeler gauge between the arm and valve, hold the set screw in place with the screwdriver and tighten the lock nut. Recheck the lash after you torque the lock nut to specs.

To adjust the remaining valves, you need to rotate the crankshaft in either one-quarter or one-half turns until all cylinders have been checked at TDC compression. The exact adjustment procedures vary depending on the engine. Always refer to the manufacturer's service procedures for specific steps for the vehicle being serviced.

Engines with camshafts that ride on top of the valves often require using a special shim to adjust the clearance between the cam lobe and the follower (**Figure 23-49**). To replace this type of shim, a special tool is required to push the valve down and remove and install the new shim.

■ ***Verify Camshaft Timing.*** One of the most critical aspects of engine operation is maintaining the correct correlation between the crankshaft and camshaft(s). Proper timing of the opening and closing of the valves in relation

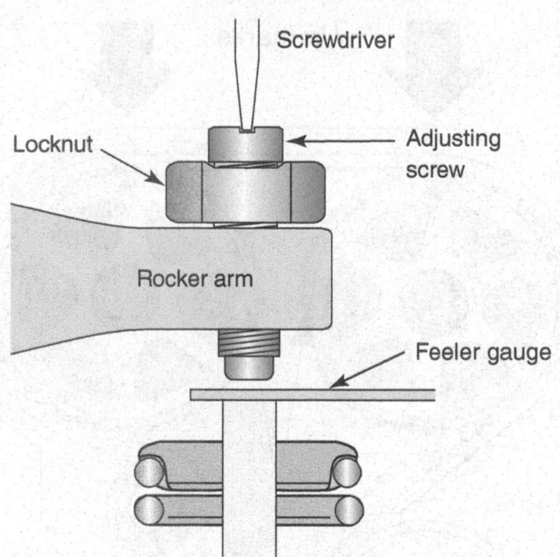

FIGURE 23-49 Using a feeler gauge to check valve lash. This is typically performed on a cold engine, but always check the manufacturer's service information.

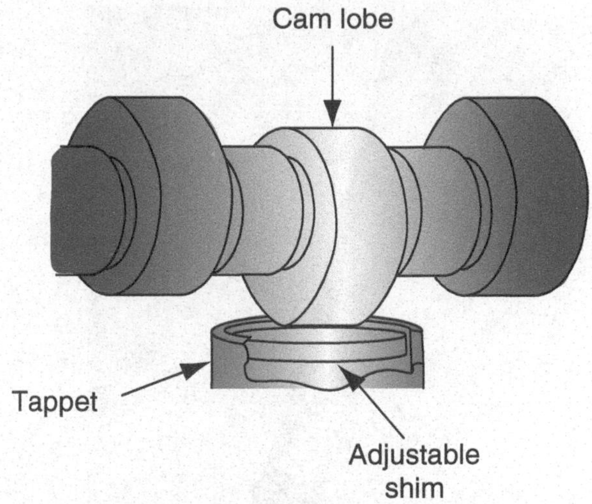

FIGURE 23-50 OHC engines may require a shim to adjust valve clearance. The old shim is removed and measured and the clearance measured. Adjusting the clearance usually requires replacing the shims with ones of different thicknesses.

to the pistons and crank is essential for the engine to run properly. As the engine ages, wear of the timing components, such as the timing belt, timing chain, tensioners, and guides, can allow changes in valve timing. These changes can range from slight, causing little or no noticeable impact on driving, to extreme. Many modern engines are **interference engines**, meaning that if the valve timing varies too much, the valves can contact the pistons. This often causes catastrophic engine damage, bending valves and even destroying pistons.

As shown in **Figure 23-51**, the clearance between the valves and pistons is already small, requiring valve relief cuts in the top of the piston. If the camshaft timing moves very much from wear on the timing components, contact between valves can occur. **Figure 23-52** shows an example of a valve bent due to contact with a piston.

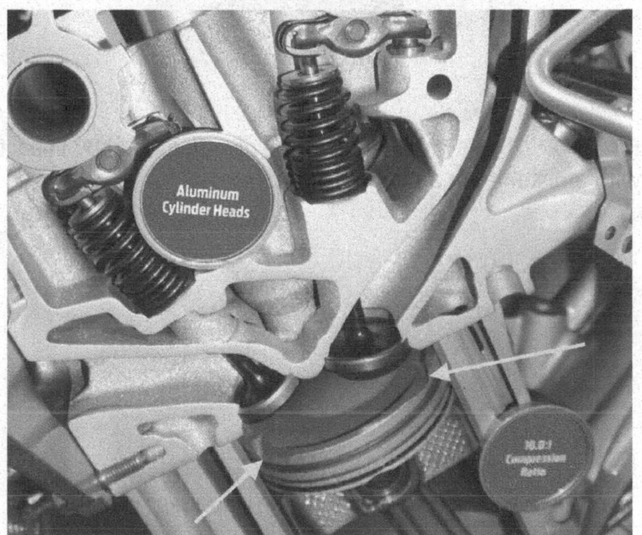

FIGURE 23-51 This shows the very limited space between the valves and the pistons. If engine timing changes, the pistons may hit the valves.

To physically check the timing, you will need to align a set of timing marks. Timing marks are placed on the crankshaft sprocket, cam sprockets, balance shaft sprockets, and others depending on the engine configuration. An illustration of timing marks for a timing belt engine is shown in **Figure 23-53**. Timing marks can be dots, dashes, arrows, or other visual markers that align with another mark (**Figure 23-54**). Begin by locating the diagrams for how the timing system is configured and how the marks align using the service information. This is a critical step because some engines have very simple timing arrangements and others can be quite complex.

For example, the engine illustrated in Figure 23-53 is a single overhead cam four-cylinder engine. To check camshaft timing, most engines like this one require setting the crankshaft so cylinder 1 is at TDC on the compression stroke. This means setting the crankshaft pulley to 0 degree or TDC on the timing scale (**Figure 23-55**). Once the crankshaft is set, remove the upper cam cover and check the alignment of the camshaft timing mark (**Figure 23-56**).

Multi-cam, V6, and V8 engines may not be as easy to check alignments due to having multiple chains and cams. Some engines require very precise procedures to check the alignment of each cam and chain. These steps may include rotating the crankshaft in several steps and checking various alignment points at each step. You will need to refer to the service information for specific procedures for the engine you are testing. Failure to follow

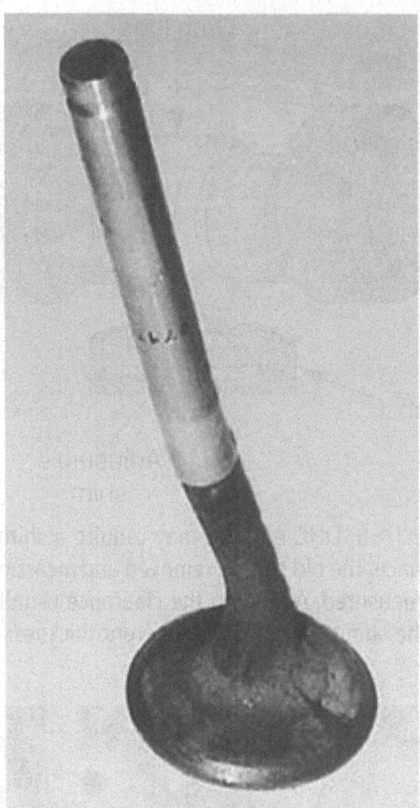

FIGURE 23-52 An example of a valve damaged by impact with a piston after a timing belt broke.

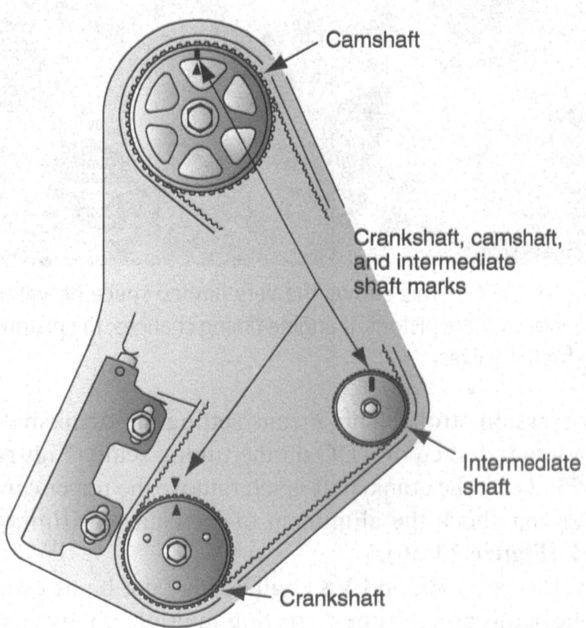

FIGURE 23-53 An example of timing mark alignments for a single overhead cam engine.

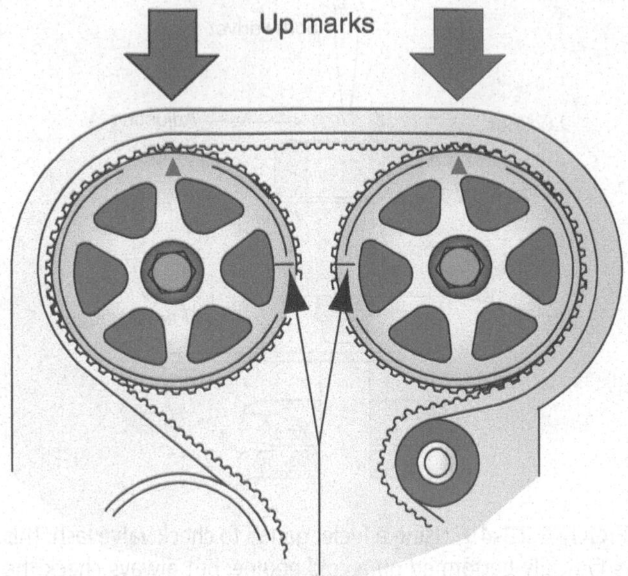

FIGURE 23-54 Timing marks can take many different forms and are not always easy to recognize during inspection.

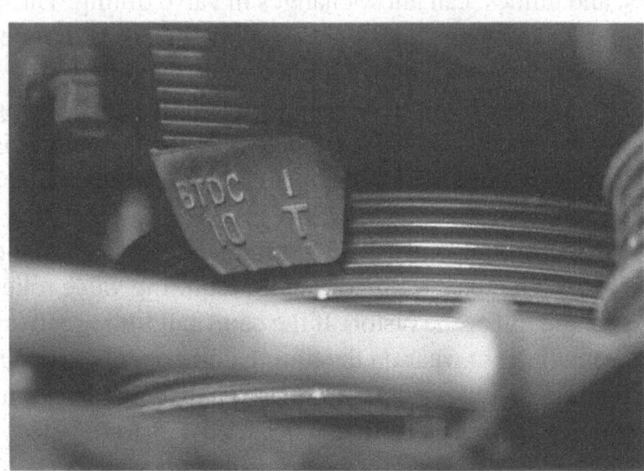

FIGURE 23-55 Checking valve timing requires setting the crankshaft to cylinder number 1 top dead center. How this looks will vary depending on the engine.

the service procedures, such as by rotating the engine in the wrong direction, can cause serious engine damage.

■ ***Gaskets and Seals.*** Gaskets and seals are used to attach various components together and to prevent leaks from occurring. For example, the oil pan bolts to the underside of the engine and must hold oil inside without leaking. The oil pan **gasket** provides the soft, malleable

FIGURE 23-56 Once the crankshaft is set to number 1 TDC, check the position of the timing mark(s) on the camshaft. If all marks are aligned properly, the engine is in time.

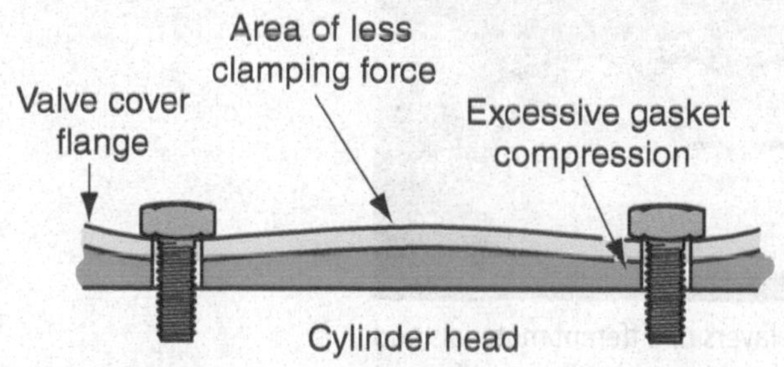

FIGURE 23-57 Gaskets are used between non moving parts to accommodate and fill in small surface differences between the parts.

FIGURE 23-58 Seals, such as this front crank seal, are used between a moving part and a non moving part.

FIGURE 23-59 Flexible gaskets are used in intake manifolds, valve covers, and oil pans.

interface between the engine block and oil pan that allows them to seal against each other and prevents the oil from leaking out (**Figure 23-57**). Gaskets are used between nonmoving parts, and **seals** are used between a moving part and a nonmoving part. Seals are used around the ends of the crankshaft where it extends out of the block and around the camshaft where it extends out of the cylinder head (**Figure 23-58**).

Oil leaks are a common complaint, especially on older, high-mileage engines. Valve covers, oil pan gaskets, and sending units are common causes of oil leaks. Camshaft seals behind the timing belt sprocket are also common leak points.

Gasket replacement procedures depend on where the gasket is located and what type of gasket is used. Gaskets can be flexible rubber, like some intake manifold gaskets (**Figure 23-59**), or hard, multilayer gaskets, such as a head gasket (**Figure 23-60**). Flexible gaskets are easily replaced—simply remove the gasket from the groove and install the new gasket, pressing it firmly into the groove. Reinstall the component and tighten the fasteners to specs. Hard gaskets often require more work to remove. One approach is to apply a gasket remover, a chemical that softens the gasket to make removal easier. Hard gaskets often need to be scraped from the surface to which they are stuck. Plastic or carbon scrapers can be used on aluminum parts, and steel scrapers can be used on steel and cast iron. Some technicians use an air-powered tool with a special disc that eats at the gasket. Caution is required when using this type of tool because debris from certain types of cleaning discs can cause severe engine damage if allowed to enter the cooling or lubrication systems. Only use approved cleaning discs and always cover holes in the engine that may allow gasket debris to enter.

Some engine components do not use a premade gasket; instead, a form-in-place gasket is used. Often called *silicone gasket maker*, this type of gasket is inexpensive and is often used on oil pans and differential covers. Several different types of gasket-making compounds are available, each with specific uses. When you are using a gasket-making product, make sure you are using the correct type for the application. Also, do not substitute gasket maker for a premade gasket. **Figure 23-61** shows how to apply gasket maker. Depending on the type of gasket maker being used, you may need to let the gasket set and form a skin before you install the component. Read the instructions that come with the product before you apply it to the component. Once the gasket is formed, install the component and torque the fasteners to specifications.

FIGURE 23-60 Head gaskets are made of layers of different materials and metals to seal the head to the block under extreme operating conditions.

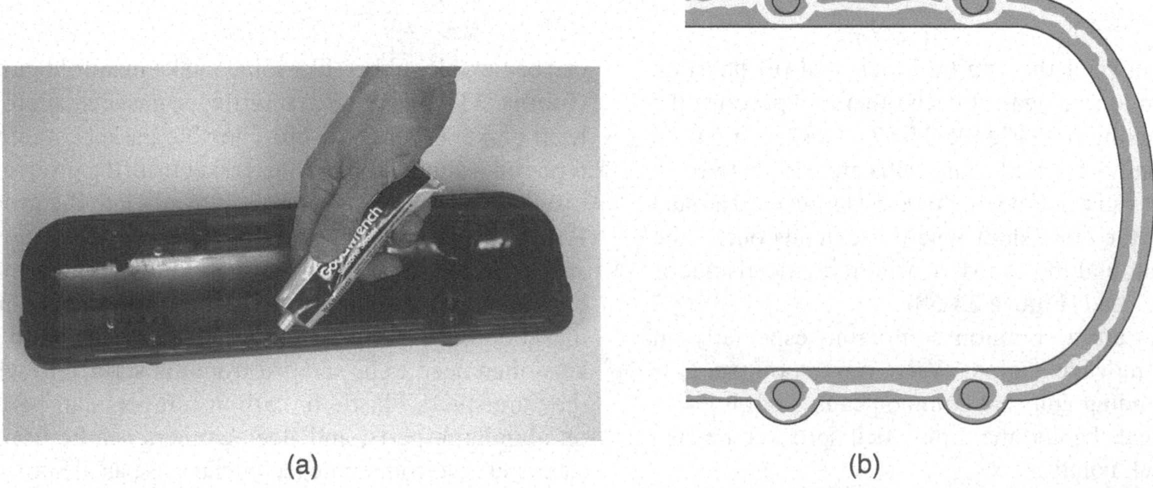

FIGURE 23-61 (a) Apply a thin bead of gasket maker around the sealing area of the component. (b) Be sure to place sealer around bolt holes but be careful not to apply too much.

SUMMARY

- Engine oil leaks can occur from gaskets and seals, and can be difficult to find if taking place over a long period of time. UV dye can be used to help find the source of the leak(s).
- Cleaning the underside of the engine and transmission may be necessary to find the source of a leak.
- Exhaust smoke color can indicate various internal engine problems. Blue smoke indicates oil burning, white is coolant burning in a gasoline engine, and black is excessive fuel.
- Abnormal engine noises may be caused by internal engine problems or by broken or damaged components external to the engine.
- The power balance test is used to determine if each cylinder is contributing power to overall engine output.
- Compression testing determines if an engine can draw in air and compress it within the cylinder.
- A running compression test is often used to locate problems with the valve train.
- Cylinder leak tests are performed to determine if a cylinder has a valve, ring, or head gasket problem.
- On some OHC engines, valve lash needs to be periodically checked and adjusted.

REVIEW QUESTIONS

1. A cylinder ______________________________ ______________________________ test is used to determine if a cylinder is contributing to engine output.

2. A cylinder ______________ test is used to isolate problems with rings, valves, and head gaskets.

3. A bouncing vacuum gauge needle indicates problems with the ______________.

4. An engine with excessive amounts of ______________ smoke from the exhaust may have a problem with piston sealing.

5. Many technicians perform a ______________ compression test using an inductive amp clamp and a scope.

6. An engine is not running well with a lack of power and a rough idle. *Technician A* says a relative compression test should be performed. *Technician B* says a cylinder power balance test should be performed. Who is correct?

a. Technician A
b. Technician B
c. Both A and B
d. Neither A nor B

7. A hissing sound can be heard in the engine compartment with the engine idling. Which is the most likely cause?

a. Incorrect valve timing
b. Vacuum leak
c. Excessive engine oil pressure
d. Exhaust leak

8. A gasoline engine is producing large amounts of white smoke from the exhaust. Which of the following is the most likely cause?

a. The fuel system is delivering too much fuel.
b. Oil pressure is too low.
c. The oil rings are worn.
d. Coolant is entering the combustion chamber.

9. An engine has a very noticeable tapping noise when running. *Technician A* says valve timing should be checked. *Technician B* says valve clearance should be checked. Who is correct?

a. Technician A
b. Technician B
c. Both A and B
d. Neither A nor B

10. *Technician A* says seals are used between two stationary parts and can be made of rubber, cork, or RTV. *Technician B* says seals are used between a moving part and a stationary part. Who is correct?

a. Technician A
b. Technician B
c. Both A and B
d. Neither A nor B

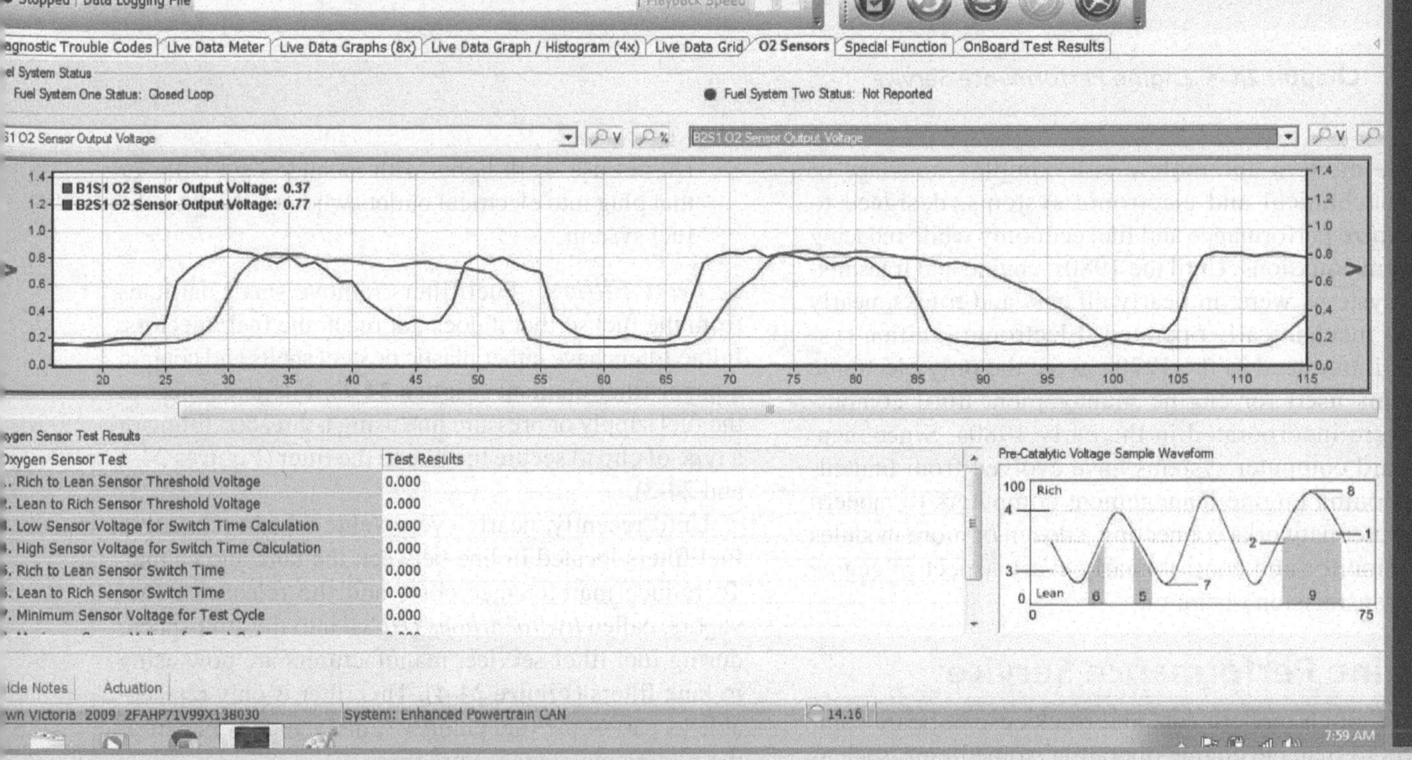

CHAPTER 24

Engine Performance Service

Chapter Objectives

At the conclusion of this chapter, you should be able to:

- Inspect and replace fuel filters and fuel lines. (ASE Education Foundation MLR 8.C.1)
- Remove and replace spark plugs. (ASE Education Foundation MLR 8.A.7)
- Inspect components of the positive crankcase ventilation (PCV) system. (ASE Education Foundation MLR 8.D.1)
- Explain OBD II operation.
- Explain OBD II monitors, their functions, and how to determine monitor status. (ASE Education Foundation MLR 8.B.2)
- Retrieve and understand OBD II diagnostic trouble codes. (ASE Education Foundation MLR 8.B.1)

KEY TERMS

active test
air filter
comprehensive component monitor (CCM)
continuous monitor
drive cycle
enhanced OBD (EOBD)
freeze frame
fuel trim
global OBD
history code
intrusive test
misfire monitor
monitor
noncontinuous monitors
oxygen sensor monitor
passive test
PCV valves
pending code
trip

The modern automobile is a complex marriage of mechanical and electronic systems, designed to maximize performance and fuel economy while reducing harmful emissions. Until the 1980s, engine and transmission systems were, in nearly all cars and trucks, nearly 100% mechanically operated. Electronic ignition systems, introduced in the 1970s, were the only electronic systems used for engine management until computers were incorporated in the early 1980s. Since then, onboard computer systems have evolved from limited-functioning engine management computers to modern computer networks connecting a dozen or more modules that monitor and control nearly every aspect of engine and transmission operation.

Engine Performance Service

Even though modern cars and trucks use onboard computers to manage engine operation, some items, such as filters, spark plugs, and other normal wear items, still require periodic inspection and replacement. In addition to being able to perform basic maintenance services on the engine, an understanding of the purpose and operation of the onboard computer systems is important to be able to communicate with customers about their vehicles and service requirements. In addition, an understanding of the onboard computer systems is vital for technicians to move beyond entry-level work into the more technical skills of a general technician.

It is not within the scope of this text to prepare you to become a general or diagnostic technician, but the information and procedures discussed here will allow you to perform certain necessary tasks expected from an entry-level technician in a shop.

FUEL SYSTEM SAFETY

The biggest safety precautions when you are working on the fuel system are related to the safe handling of gasoline. One of the properties that makes gasoline the dominant fuel for passenger cars, the flammability of the vapors, also makes working around gasoline hazardous. Whenever you work on the fuel system, observe the following safe work practices:

- Never smoke or allow an open flame near gasoline, diesel, or either fuel's vapors.
- Do not use tools that can create sparks near a fuel leak.
- If gasoline or diesel fuel contacts your skin, wash the area immediately.
- Store gasoline only in approved containers.
- Never use gasoline or diesel fuel as a cleaner or solvent.
- Do not use work lights with incandescent bulbs or that plug into electrical outlets when working on the fuel system.

■ ***Fuel Filters.*** Fuel filters remove small particles from the fuel so that it does not reach the fuel injectors. Inline filters have either plastic or steel shells and contain a paper filter element (**Figure 24-1**). Filters connect to the fuel supply or pressure line using a threaded fitting or a type of clip to secure the line to the filter (**Figures 24-2** and **24-3**).

Until recently, nearly every vehicle had serviceable fuel filters located in line between the tank and engine. To reduce maintenance costs and the release of fuel vapors, called *hydrocarbons (HCs)*, into the atmosphere during fuel filter service, manufacturers are now using in-tank filters (**Figure 24-4**). This filter is only replaceable as part of the fuel pump assembly and is not part of the routine maintenance program.

Vehicles with diesel engines often use a large two-stage fuel filter or a set of filters (**Figure 24-5**). The filter system also includes a water separator. Diesel engines

FIGURE 24-1 An example of a fuel filter. The paper element traps dirt as the fuel travels from the center to the outside of the element.

FIGURE 24-2 An example of a fuel filter with two different types of lines and connections.

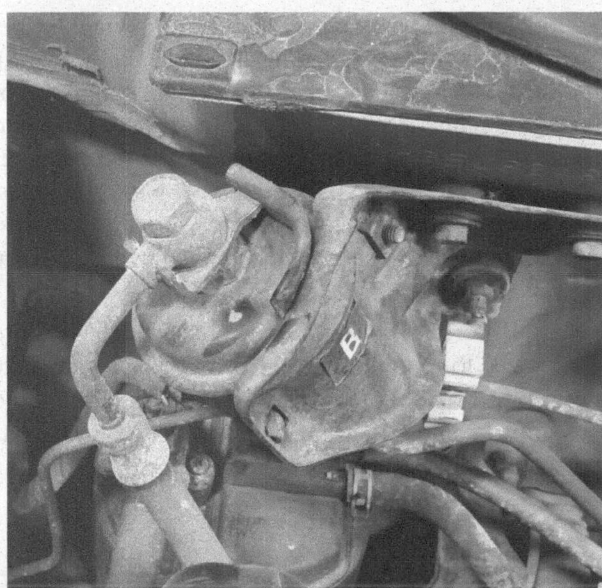

FIGURE 24-3 An example of a fuel filter that uses banjo fittings to attach the fuel lines.

FIGURE 24-4 Many manufacturers are placing the fuel filters in the fuel tank. This type of filter is not part of the normal maintenance schedule.

require periodic fuel filter service more frequently than gasoline engines because of the properties of diesel fuel that attract water and even bacterial growth in the fuel.

■ ***Fuel Tank and Line Inspection.*** Inspect fuel tanks and fuel lines for leaks and damage. Metal fuel tanks and lines (**Figure 24-6**), are prone to rust through. Inspect fuel lines for kinks, especially fuel return lines. A kinked pressure line can reduce fuel pressure and volume, while a kinked return line can increase fuel pressure. Check for leaks around the tank straps and fuel line retaining clips as moisture often collects in these areas.

Fuel lines, regardless of their construction material, use several types of connections. Special fuel line tools may be required to separate some of these types of connectors (**Figure 24-7**).

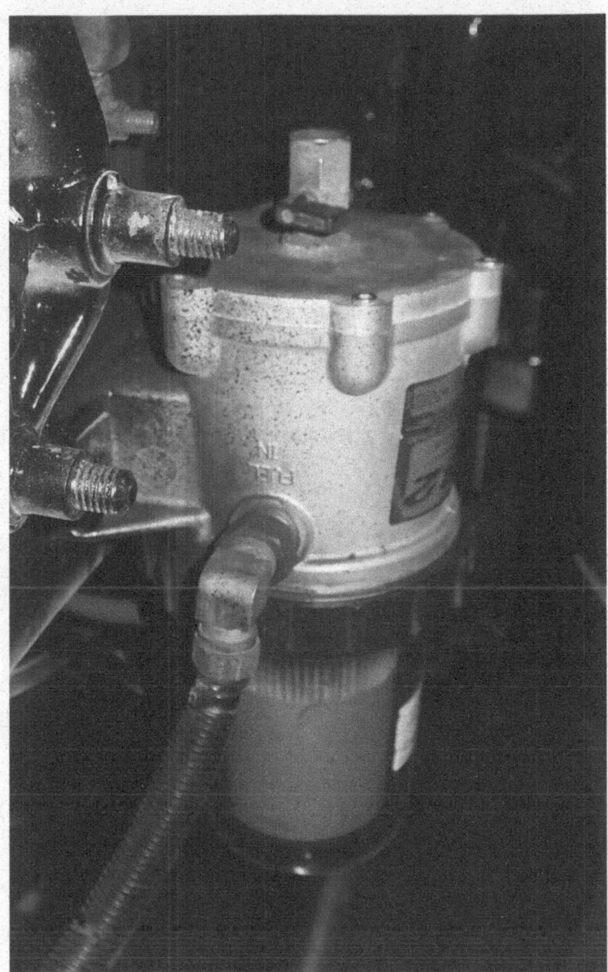

FIGURE 24-5 An example of a diesel fuel filter. Diesel engines use large filters to trap debris, water, and other contaminants.

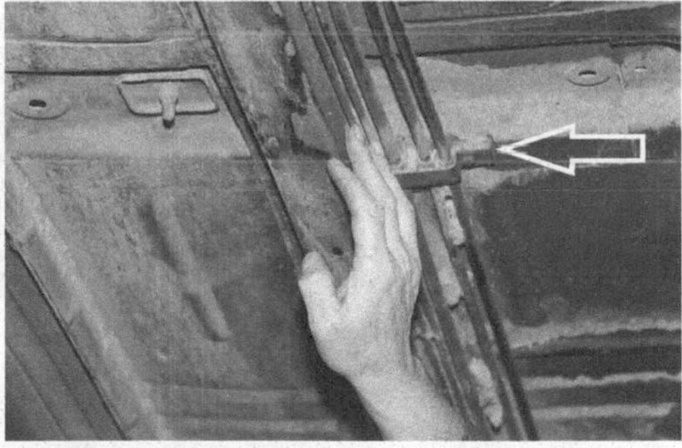

FIGURE 24-6 When inspecting fuel lines, check around the clamps that secure the line to the body as these are common rust points.

FUEL FILTER REPLACEMENT

1. Before you remove a fuel filter, **depressurize the fuel system.** This can be done by connecting a fuel pressure gauge to the fuel test port and releasing the pressure

FIGURE 24-7 Some fuel lines require special tools to disconnect the fittings.

through the fuel pressure tester or by removing the fuel pump fuse or relay and starting the engine (**Figure 24-8**). **Figure 24-9** shows an example of a fuel pump fuse or relay in a junction box. Remove the fuse or relay and start the engine, crank the engine several times after it stalls to reduce the pressure that is left in the lines. After the engine fails to start several times, fuel pressure should be low enough to begin removing the filter.

2. Next, locate the filter and inspect the connections. Many filters are located under the vehicle and are subjected to road dirt and corrosion, which can make removal difficult (**Figure 24-10**). You may need to apply some penetrant or even some brake parts cleaner to the connections to ease removal.

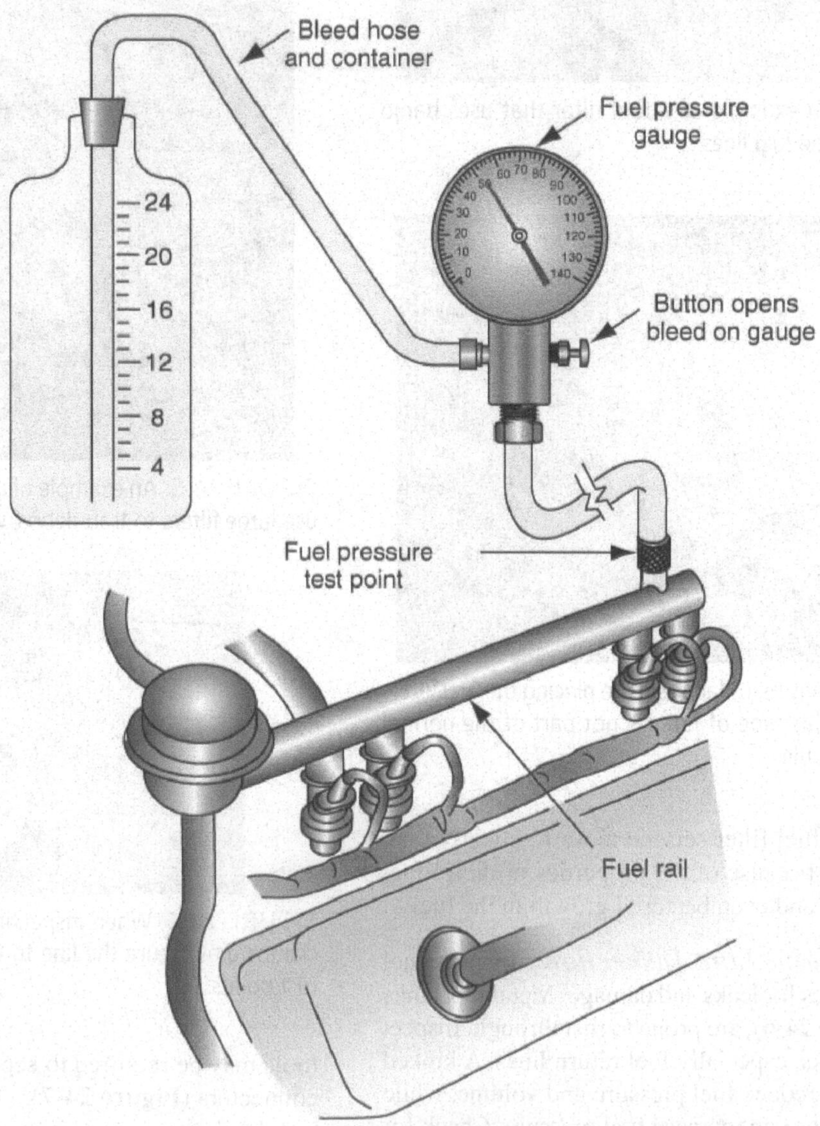

FIGURE 24-8 Replacing a fuel filter requires depressurizing the fuel system. This can be done using a fuel pressure gauge attached to the pressure test port. However, this often results in fuel spraying from the test port as the connection is made with the pressure tester, so this method should not be used on a hot engine. **Never open the fuel system until the fuel pressure is relieved.**

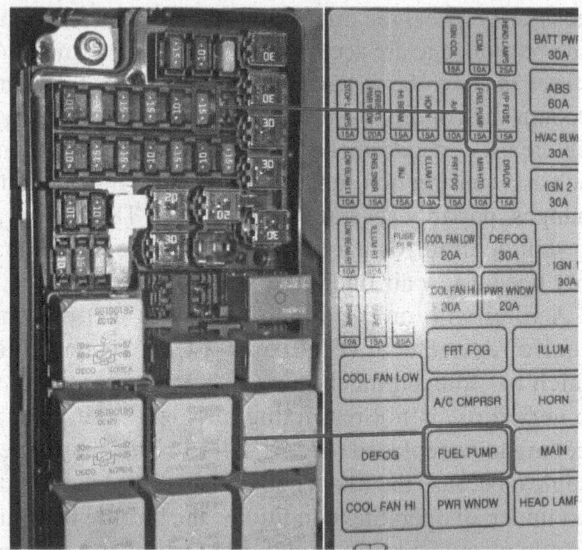

FIGURE 24-9 Removing a fuel pump fuse or relay and then running the engine until it stalls will remove most of the fuel pressure from the system. Continue to crank the engine about 5 seconds after it stalls to further reduce the fuel pressure in the lines.

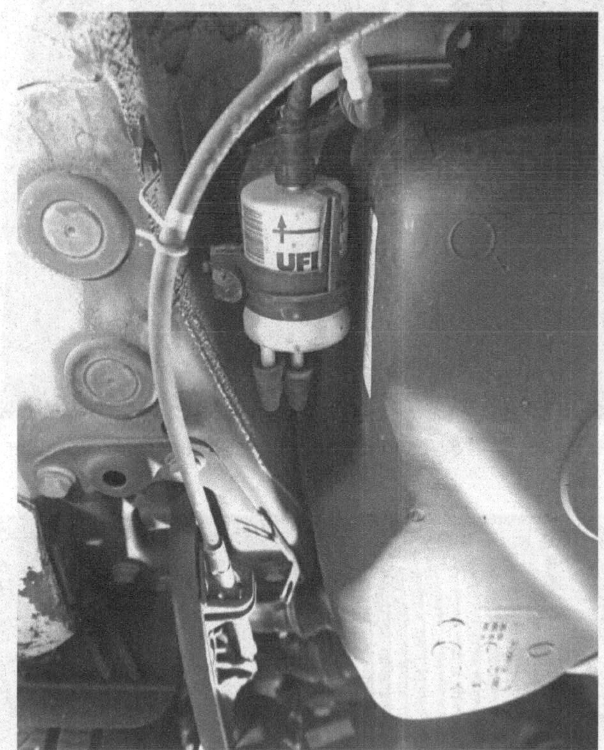

FIGURE 24-10 All fuel filters are directional and will have some method of indicating the correct way to install. Some, like this VW filter, cannot be installed backward.

3. Determine how to remove the fuel lines from the filter if necessary, then wrap a shop towel around the filter and the line you are removing first. Carefully remove the line and pull it loose from the filter. If

FIGURE 24-11 Some older GM and Ford filters can be installed either direction so look for arrows or other markings to indicate which way the filter is installed.

any residual pressure is in the fuel line, the towel will catch and absorb the leaking fuel.

4. Remove the second connection and remove the filter from its mounting bracket. Drain the fuel from the filter into a shop pan or gas can. Notice that the filter is directional in its installation, meaning that one side is the inlet and the other is the outlet (**Figure 24-11**). Make sure that the replacement filter matches the original filter in the direction of flow and that the line connections are the same.
5. Install the new filter so that the flow of fuel is correct. Reattach the fuel lines, making sure that each fully seats and is secure on the filter. If banjo fittings connect the lines to the filter, install the new gasket washers on the fittings and tighten the fittings to spec. Remount the filter to its bracket as necessary.
6. Reinstall the fuse or relay and turn the ignition on and off several times. This will turn the fuel pump on so that you can check for leaks. If no leaks are present, start the engine. The engine may cough, stall, and run rough for a second or two until the air in the fuel system passes and full fuel flow to the injectors is restored. Recheck again for leaks around the filter connections.

■ ***Air Cleaner.*** The air cleaner housing and **air filter** remove contaminants from the air before it enters the engine. Until port fuel injection systems became standard, air cleaners and filters were donut-shaped and mounted directly on top of the carburetor. This sometimes necessitated the ram-air hood scoops of years past. Modern vehicles mount the air cleaner lower, often behind the headlamp assembly. This allows for lower front-end designs and better vehicle aerodynamics.

The air filter is a pleated paper element (**Figure 24-12**). The weave of the filter traps small dirt particles and other debris. Over time, the filter becomes plugged with dirt, which can, if severely restricted, affect engine performance. Some vehicles are equipped with an airflow restriction gauge (**Figure 24-13**). As the air filter becomes clogged with dirt, the restriction increases. Eventually, the gauge will show red or otherwise indicate the filter needs to be changed.

■ ***Air Filter Inspection.*** Determining if an air filter needs replacement is not always easy. An airflow restriction gauge (Figure 24-13) is the best method of determining filter condition. A visual inspection does not necessarily reveal the true condition of the filter. Neither time nor mileage are accurate methods of determining if a filter needs to be replaced, because neither takes into account the conditions in which the vehicle operates. A vehicle used in extremely dusty and/or dirty conditions, such as a construction vehicle, may require frequent filter changes, possibly several times per year. The same type of vehicle used for normal day-to-day operation may only need a new air filter after several years.

A filter with only a slight amount of dirt does not require replacement (**Figure 24-14**). A second filter, which has trapped much more dirt, while appearing to be very dirty, is not restricting airflow to the point of needing to be replaced (**Figure 24-15**). This may appear illogical because the filter is obviously dirty, but consider that under all operating conditions, the throttle body and throttle plate(s) present the biggest restriction to airflow into the engine. For the air filter to restrict airflow to the point of causing a drivability concern, the filter must be restricted to the point where its surface area to pass

FIGURE 24-12 An example of a high-performance low-restriction air filter.

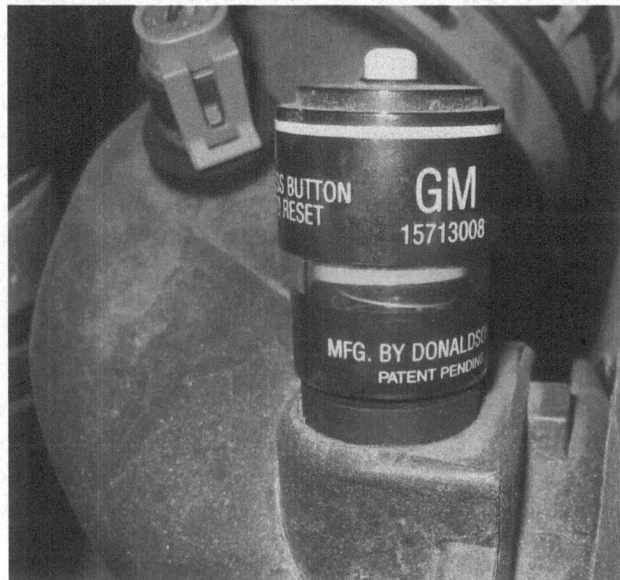

FIGURE 24-13 Restriction gauges are used to determine when the air filter should be replaced. Do not reset the gauge unless replacing the filter.

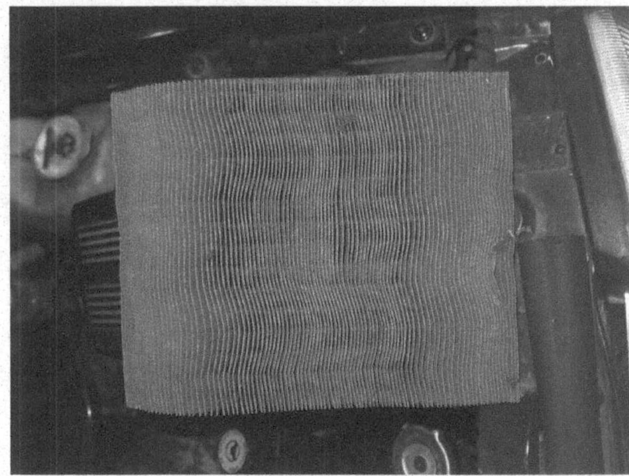

FIGURE 24-14 A slightly dirty but still serviceable air filter.

FIGURE 24-15 A dirtier but still serviceable air filter. The total amount of restriction by the dirt is still much less than the total amount of filter surface area.

air is less than the size of the throttle body. Considering the size of the area of the filter element, a filter needs to be almost completely clogged before it will affect drivability.

When you remove and inspect the filter, be careful not to damage the filter. Some filters are mounted in a way that makes removal difficult. Many technicians hold the filter up to a light to determine its condition. If light is visible through the pleats, it is good. If light does not penetrate the pleats, it needs to be replaced. As discussed previously, this test does not prove conclusively whether a filter needs to be replaced or not. Many times, the filter will have a very dirty portion where the incoming air is routed by the ductwork, while the remaining area of the filter is clean. This accumulation of dirt in one section does not necessitate replacement. Also, depending on the paper out of which the filter is made, light may or may not easily pass through.

Inspect the filter housing for dirt and for oil. Remove dirt or other debris from the filter housing. If oil is found in the housing, perform a thorough inspection of the positive crankcase ventilation (PCV) system. If the PCV is not functioning, oil can back up into the air cleaner housing. Once the cause of the problem has been corrected, clean all the oil from the induction components. On occasion, you may find a mouse nest in the air cleaner (**Figure 24-16**). Make sure the customer is informed of this condition, and remove the accumulated nesting materials.

When replacing the air filter, use either original equipment (OE) or high-quality aftermarket replacement filters. Inexpensive filters may use a lower-quality paper, which traps less dirt and allows small particles to pass through the filter and into the engine. Less expensive filters often do not fit exactly as they should, which can cause problems with some mass airflow (MAF) sensors mounted in the air cleaner housing. Make sure the replacement filter fits the housing correctly and provides an airtight fit when installed.

FIGURE 24-16 A rodent had taken up residence in the air cleaner housing. Advise the customer to remove food items and nesting materials from where the vehicle is parked to help keep mice out of the engine compartment.

Checking an air filter also means checking the other induction hoses and components (**Figure 24-17**). Many cars and trucks have several hoses that connect to the air cleaner housing or induction hose. These can be for the PCV system, discussed later in this chapter, or for noise enhancement. Check these components and make sure each is firmly connected.

Inspect the main induction hose itself. These hoses are prone to cracking and can allow unmetered (unaccounted for) air to enter the engine (**Figure 24-18**). Leaks in the induction hose can cause severe drivability concerns, such as stalling, surging, misfires, and may turn on the check engine light.

FIGURE 24-17 The highlighted hoses shown are part of the system that sends engine noise into the passenger compartment.

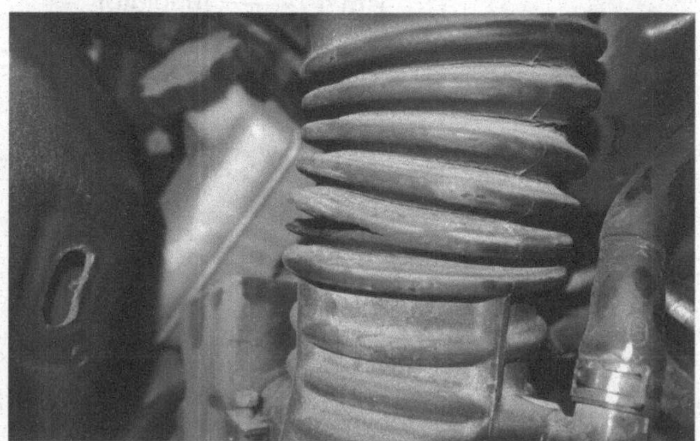

FIGURE 24-18 Inspect induction hoses for cracks and damage. Air leaks, like from the damaged hose shown here, can cause the engine to run very poorly.

SPARK PLUGS

Spark plugs are used to ignite the air–fuel mixture in the combustion chamber (**Figure 24-19**). Spark plugs are constructed of a steel shell, a porcelain insulator, an internal resistor, and the electrodes (**Figure 24-20**). Over time, the arcing of the spark across the terminals wears away at the terminals (**Figure 24-21**). This increases the air gap between the terminals, which affects the amount of voltage required to jump the gap.

FIGURE 24-19 An example of spark plug location. Most OHC engines place the spark plugs in the top of the combustion chamber as shown here.

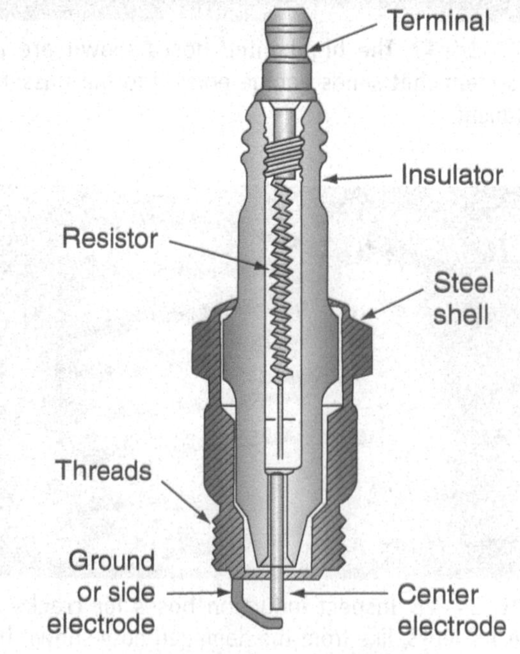

FIGURE 24-20 An inside view of a spark plug.

Even though spark plugs require changing much less frequently than ever before, they do still require service. Some vehicle manufacturers state that the spark plugs do not need to be changed until the engine has 100,000 miles of service, while others recommend inspection and changing plugs every 60,000 miles.

■ ***Replacing Spark Plugs.*** Before you remove the spark plugs, allow the cylinder head(s) to cool to the touch. **Removing plugs from hot aluminum heads can damage the threads inside the cylinder head.**

On engines with spark plug wires, twist the boots on the spark plugs before you attempt to pull the wire off (**Figure 24-22**). The boots shown in Figure 24-22 are located very close to the exhaust manifold. This proximity tends to bake the boots onto the plugs, which can make removal difficult. Note the damaged insulation

FIGURE 24-21 The spark jumping the air gap of the plug, over years of operation, wears away at the metal of the electrodes, which causes the gap to increase.

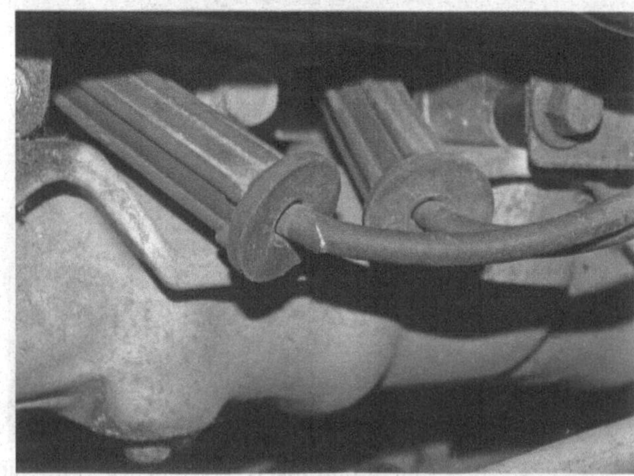

FIGURE 24-22 Inspect the plug wires and boots for damaged insulation. Broken insulation can allow the spark to jump from the wire and cause a misfire.

FIGURE 24-23 Plug wires that extend down into the cylinder head have very long boots to reach down to the plug.

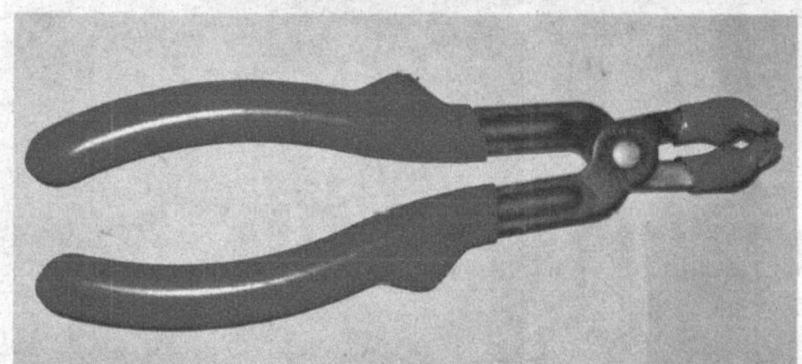
FIGURE 24-24 Plug wire pullers are helpful when the boots are stuck on the plugs and their location makes it difficult to get a good grip on the boots by hand.

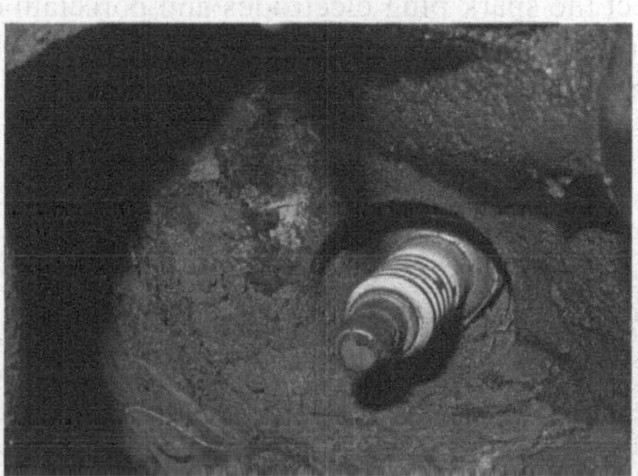
FIGURE 24-25 Before removing any spark plug, inspect and clean the area where the plug threads into the head. Failure to remove rust and dirt will allow it to fall down into the cylinder when the plug is removed.

on the plug wire in Figure 24-22. Inspect the plug wires and boots for damage. Broken insulation can allow the spark to arc from the plug wire to ground, causing a misfire. Broken insulation can also allow you to accidently get shocked if you touch the wire with the engine running.

On older overhead cam (OHC) and dual overhead cam (DOHC) engines, the plug wire boots pass through the valve cover(s) because the plugs are threaded into the top of the combustion chamber (**Figure 24-23**). If the boots are difficult to get to or are stuck on the plugs, a pair of plug wire boot pullers can help with removing the boots (**Figure 24-24**).

Once the plug boot can move easily, firmly pull the boot from the plug. It may be necessary to use an air blowgun to blow any dirt and debris from around the spark plug before you continue (**Figure 24-25**). Removing the rust prevents the rust and dirt from falling into the cylinder or getting into the threads of the combustion chamber when the plug is removed.

Once clean, install the spark plug socket over the plug, making sure that the socket fully engages the steel shell and the hex drive of the plug. Install a ratchet and turn counterclockwise. If the plug does not break loose under light to moderate force, stop. The plug may be seized in the head, requiring additional work to remove it. If the plug is stuck, try to turn the plug to ¼ of a turn counterclockwise (CCW), then spray a penetrating lubricant into the plug well and let it soak for 5 to 10 minutes. Next, try to turn the plug out; it may be necessary to turn the plug CCW then clockwise and back CCW several times until it is completely removed.

Service Warning

Service Warning: If you are removing plugs from late model Ford 3V V8 engines, special procedures are used to remove the plugs due to the possibility of the plug breaking off in the head. Follow the Ford service procedures.

For coil-on-plug (COP) engines, remove the fastener holding the coil in place over the spark plug (**Figure 24-26**). Gently twist the coil to break the seal between the plug boot and spark plug. Pull the coil up and off the plug. Inspect the boot for signs of arc-through. This sometimes shows as a white or gray residue on the plug boot (**Figure 24-27**). A coil that is arcing through the boot should be replaced. Install the appropriate spark plug socket and remove the plug.

When removing the plug wire or coil, inspect them for signs of oil leaking into the plug well. Many OHC engines have spark plug wells that pass through the valve cover (**Figure 24-28**). Seals are used to keep oil from getting into the plug well. These seals can leak and allow oil to get onto the plug boot or coil and plug (**Figure 24-29**).

Spark plugs can provide valuable information about what is happening inside the combustion chamber. Inspect the spark plug electrodes and porcelain extension under the positive electrode (**Figure 24-30**). The porcelain should be light tan or gray with little or no evidence of deposit formation.

FIGURE 24-26 An example of a coil-on-plug (COP) design. Remove the coil fasteners and then twist and gently pull the coils from the plugs.

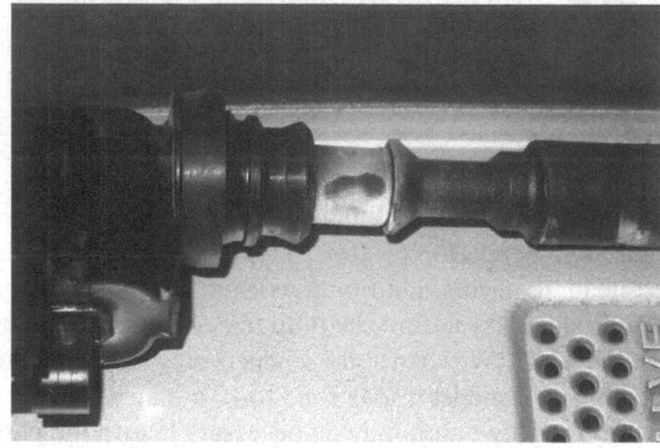

FIGURE 24-27 An example of a coil that has been arcing through the boot.

Inspect the upper porcelain insulation of the plug for signs of carbon tracking and cracks. Cracks in the insulation allow the spark to jump from the center resistor core and through the plug wire boot, causing a misfire. Make sure that the upper terminal connection is tightly screwed to the plug. The terminal can work loose and create a gap for the spark to arc across, which can affect the performance of the plug and affect the spark to the cylinder.

When you install the new plugs, first make sure that the replacement plugs match the old spark plugs. In most cases, only the OE spark plugs should be installed. Using

FIGURE 24-28 Plug wire boots on OHC engines are often hidden inside the valve cover. If the boots come out wet with oil, the valve cover seals around the plug tube need to be replaced. The oil will eventually destroy the plug wires.

FIGURE 24-29 A spark plug from an engine that had leaking plug seals in the valve cover.

FIGURE 24-30 A spark plug with normal wear and coloring.

replacement spark plugs of a different brand and/or type can result in engine misfires and performance concerns. Next, determine if antiseize should be used on the spark plug threads. Refer to the manufacturer's service information regarding the use of antiseize and to locate the plug's torque specification. If antiseize is to be used, apply a very light coat to the plug threads (**Figure 24-31**). **Do not apply too much antiseize to the plug—this can cause improper seating in the cylinder head and cause misfires.** If you replace the plugs on a Ford 3V V8 engine, coat the lower ground shield with antiseize before installation.

Depending on the type of plug, the gap needs to be checked and possibly adjusted. To check the gap between the positive and ground electrodes, locate the spark plug gap spec. This spec is typically between 0.035 and 0.060 inch (0.9 to 1.5 mm). Using a plug gap gauge (**Figure 24-32**), carefully insert the correct gauge into the plug gap (**Figure 24-33**). If the gap is too small, the tool can be used to open the gap by bending the ground electrode away from the positive electrode. If the gap is too large, use the tool to carefully bend the ground electrode down toward the positive electrode. Always remeasure the gap after making any adjustment. Caution must be used when you are checking and gaping plugs with platinum and iridium electrodes as these are easily damaged. An example of a platinum tip plug is shown in **Figure 24-34**.

To install the spark plugs, start the plug into the threads using only the plug socket, a piece of rubber hose, or an old plug wire boot. This ensures that the plug starts easily and does not cross-thread. **Never start a plug using a ratchet or an air tool.** Once the plug is fully threaded in and seated against the head, torque it to specifications. This step is important to obtain the proper seating of the plug to the head. A plug that is left loose will allow combustion leaks and arcing in the ground path, and it can cause damage to the head and plug due to inadequate heat dissipation. Overtightening the plug can damage the plug and cylinder head and strip the plug threads. Spark plug torque is usually about 13 to 18 foot-pounds. Once it is torqued, reinstall the plug wire, making sure that the wire terminal fully seats over the plug terminal.

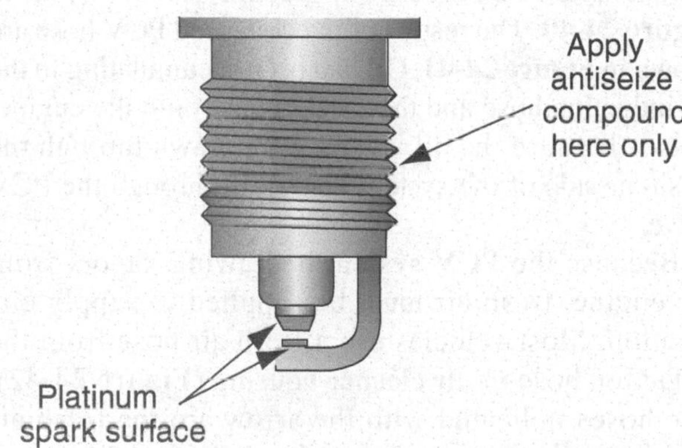

FIGURE 24-31 Do not apply antiseize to the threads unless directed to by the service information. Do not apply too much antiseize or misfires may result.

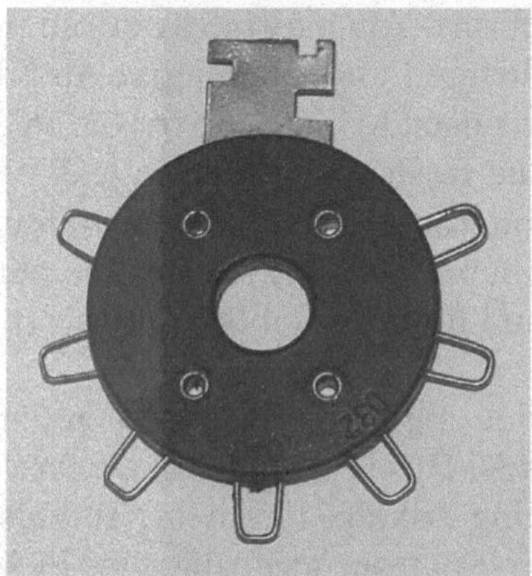

FIGURE 24-32 An example of a plug gap tool.

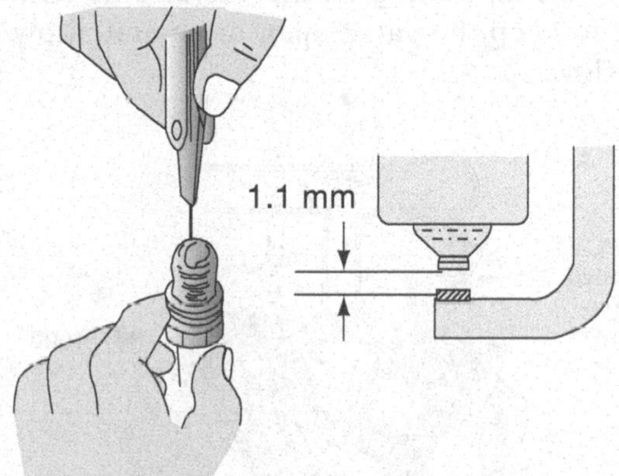

FIGURE 24-33 Checking the spark plug gap. Be careful not to bend or damage the electrodes with the tool. Be extra careful when working on platinum or iridium plugs as the electrodes can be damaged by the gap tool.

FIGURE 24-34 A platinum tip spark plug.

POSITIVE CRANKCASE VENTILATION SYSTEMS

The positive crankcase ventilation (PCV) system was the first emission control device installed in automotive applications (**Figure 24-35**). The vapors in the crankcase are a result of blowby. Combustion gases that get past the piston rings and build up in the lower part of the engine are called *blowby*. If left unchecked, the vapors will build up pressure, causing oil leaks. The PCV system uses vacuum to pull these vapors out of the crankcase and back into the intake for burning. Before the introduction of the PCV valve, crankcase vapors were simply allowed to flow out of the engine.

An example of a PCV valve and hose is shown in **Figure 24-36**. PCV operation is based on the amount of vacuum supplied to the valve (**Figure 24-37**). At idle, vacuum is high, which causes the valve to be almost completely seated; this results in very little vapor flow. As engine speed and throttle opening increase, there is less vacuum pulling on the valve. This allows the spring to keep the valve open more and allow more vapor flow.

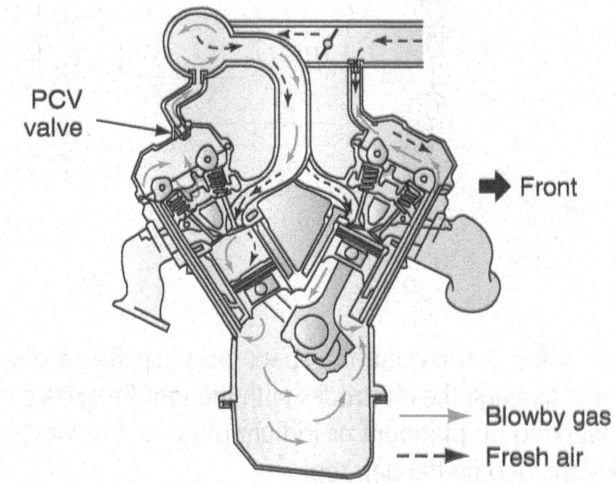

FIGURE 24-35 An illustration of the flow through the engine from the PCV system.

FIGURE 24-36 An example of a PCV valve and hose. Notice the retention clamp to hold the valve firmly in the valve cover.

Some engines use a fixed orifice instead of a PCV valve. The orifice meters the amount of air circulated through the engine as a PCV valve does but does not contain a spring-loaded check valve. Other engines use a centrifugal device, usually located on the rear of a camshaft, to separate oil vapor and liquid oil before the vapors are removed from the engine (**Figure 24-38**).

■ ***PCV System Inspection.*** **PCV valves** are a controlled vacuum leak, pulling blowby gases from the crankcase and returning them to the intake. Typically, the PCV valve is attached to a large-diameter vacuum hose. The vacuum source for the PCV is usually second in size only to the vacuum source for the power brake booster (**Figure 24-39**). Inspect the vacuum hose and valve for signs of cracks and vacuum leaks. With the engine running, remove the valve from the valve cover and place your thumb over the inlet of the valve. The valve should snap, and vacuum should be present at the inlet. If the valve does not snap and/or no vacuum is felt, check for vacuum supply to the valve. A blocked vacuum supply hose will prevent the valve from drawing the vapors from the engine. If vacuum is present to the valve but not at the valve inlet, suspect a plugged valve. An example of a failed PCV vacuum hose is shown in **Figure 24-40**. The results of the collapsed PCV hose are shown in **Figure 24-41**. Oil had been accumulating in the air induction hose and then leaking out onto the engine. This is because the oil vapors were drawn through the fresh-air side of the system instead of through the PCV valve.

Because the PCV system is drawing vapors from the engine, fresh air must be supplied to supply circulation. Most vehicles use a fresh air hose from the induction hose or air cleaner housing (**Figure 24-42**). The hoses indicated with the arrow are the fresh air supply hoses. Without a supply of fresh air to circulate through the system, the vacuum drawn by the PCV valve causes a vacuum in the engine crankcase. This can cause liquid oil to be drawn out by the PCV valve. Inspect the fresh air hose for cracks, rotting, and for tight connections. You may have noticed that the fresh air hose is located after the engine air filter. This is so that the air used to circulate for the PCV system is filtered. If the fresh air hose has a leak or comes off, unfiltered underhood air will be drawn into the engine instead of filtered air. This negates the purpose of the air cleaner and filter assembly by letting dirty air enter the engine.

Late-model engines have a PCV retention device (**Figure 24-43**). This is to eliminate the possibility of the valve becoming dislodged from the. This also prevents a large vacuum leak from occurring if the PCV valve were not installed properly or fell out.

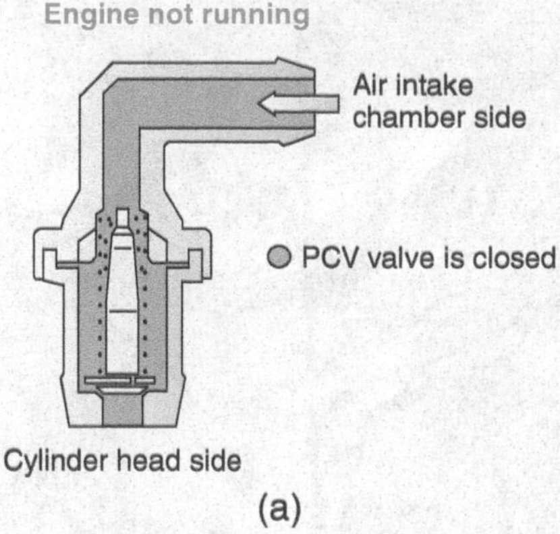

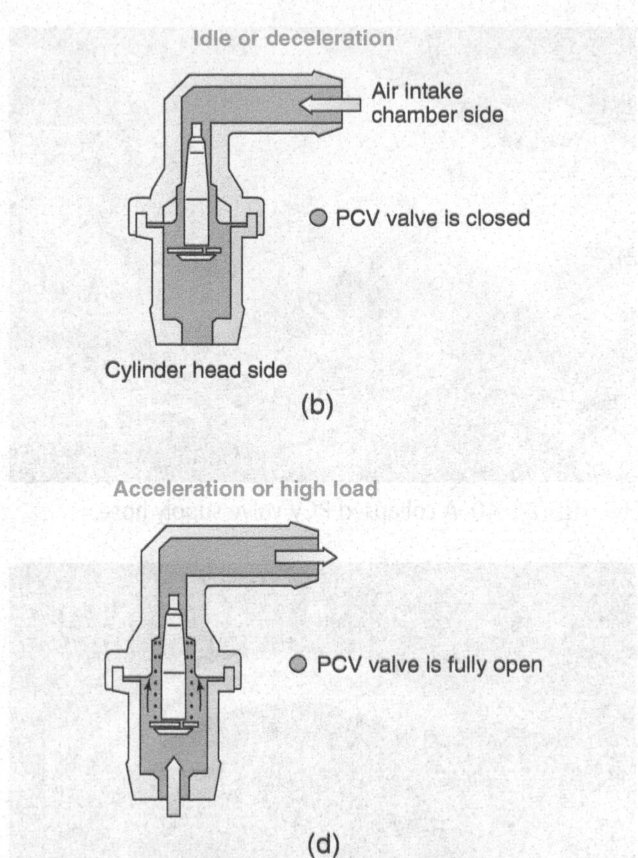

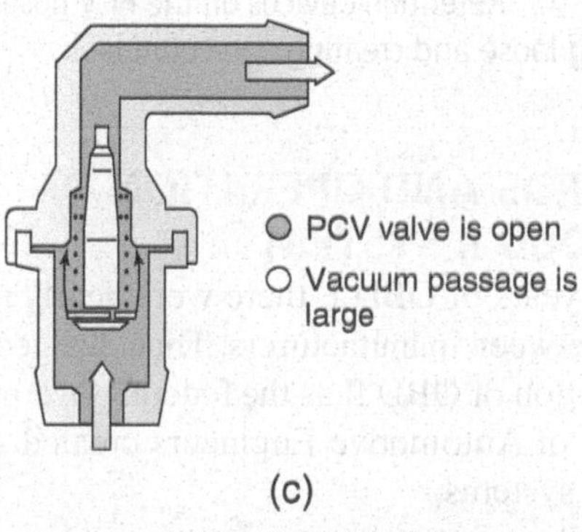

FIGURE 24-37 (a) With the engine off, the internal spring keeps the valve seated. (b) At idle, the high vacuum pulls the valve off the seat but very little flow is achieved. (c) Under partial throttle opening, the valve allows vapor to flow from the engine to the intake. (d) High throttle openings mean low vacuum, which allows increased flow through the valve.

FIGURE 24-38 An oil/vapor separator driven by the camshaft helps to keep oil from being pulled through the PCV system.

FIGURE 24-39 PCV valves have retainers to prevent a large vacuum leak if the valve were dislodged. The electrical connection is a heater, used to prevent freezing of the valve in cold weather.

Onboard Diagnostic Systems

The first onboard computer systems began to appear in the late 1970s on a small number of vehicles. It was not until the early 1980s that the larger auto manufacturers began to install computerized engine management systems. This early system, called onboard diagnostics I or OBD I, was used until the 1996 model year. Since 1996, passenger cars and light-duty vehicles sold in the United States have been required to have OBD II (meaning generation two) for onboard emission control monitoring.

FIGURE 24-40 A collapsed PCV valve supply hose.

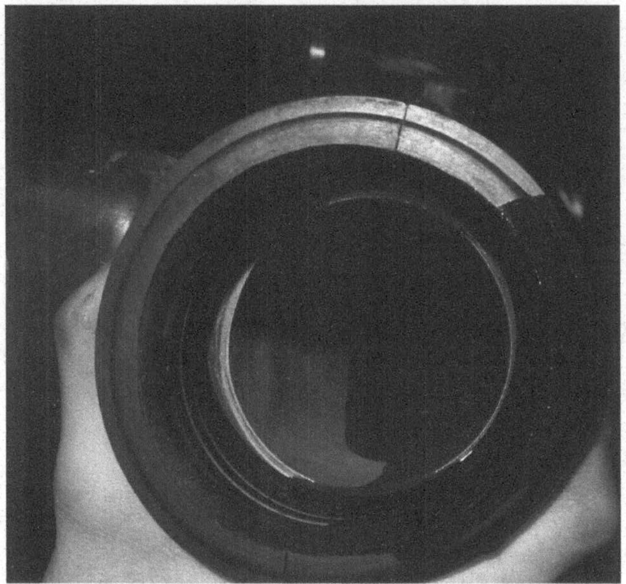

FIGURE 24-41 The collapsed PCV valve hose prevented the valve from operating correctly. This led to oil backing up into the intake hose and creating an oil leak.

FIGURE 24-42 The PCV system requires a fresh-air supply to allow proper circulation through the engine.

FIGURE 24-43 Retention devices on the PCV hose prevent it from working loose and creating a vacuum leak.

STANDARDS AND OPERATION OF THE OBD II SYSTEM

During the years of OBD I, there were nearly no shared standards between manufacturers. That changed with the implementation of OBD II as the federal government and the Society of Automotive Engineers created standards for OBD II systems.

■ ***OBD II Standards.*** There are many standards for OBD II systems, most of which specify operating conditions and aspects of the systems' function. For the technician, standards related to accessing the OBD II system include:

- Size, shape, location, and pin configuration of the data link connector (DLC) (**Figure 24-44**).
- Basic scan tool functions, such as automatically being able to determine the type of communication protocol used, displaying the status of onboard test results, and other functions described later in this chapter.
- Standardization of generic trouble codes and code descriptions.

Beyond the standards for communication and scan tool functions, OBD II also prescribes onboard testing modes, which are discussed later this chapter.

OBD II systems are required to provide comprehensive monitoring of exhaust emissions. The major requirement for OBD II is to turn on the malfunction indicator light (MIL) if emissions increase 150% over the federal test parameter (FTP) for that vehicle. Along with increased monitoring, OBD II systems must have self-diagnostic capabilities. This means the system must be able to detect problems within itself, such electrical faults and for conflicting sensor data.

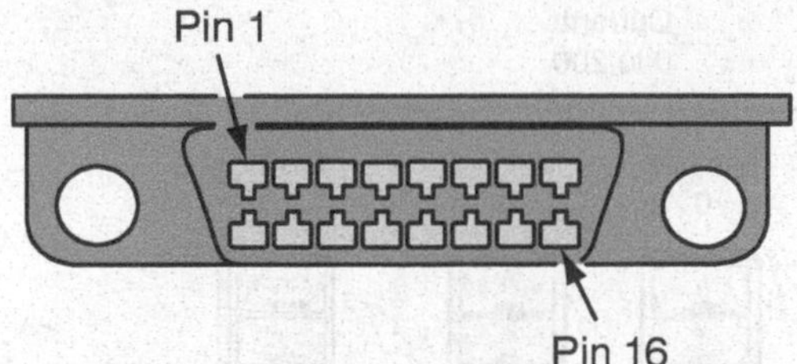

Pin 1	Secondary UART 8192 baud serial data Class B 160 baud serial data
Pin 2	J1850 Bus + line on 2 wire systems, or single wire (class 2)
Pin 3	Ride control diagnostic enable
Pin 4	Chassis ground pin
Pin 5	Signal ground pin
Pin 6	PCM/VCM Diagnostic enable
Pin 7	K line for International Standards Organization (ISO) application
Pin 8	Keyless entry enable, or MRD theft diagnostic enable
Pin 9	Primary UART
Pin 10	J1850 Bus-line for J1850-2 wire applications
Pin 11	Electronic Variable Orifice (EVO) steering
Pin 12	ABS diagnostic, or CCM diagnostic enable
Pin 13	SIR diagnostic enable
Pin 14	E&C bus
Pin 15	L line for International Standards Organization (ISO) application
Pin 16	Battery power from vehicle unswitched (4 amps max)

FIGURE 24-44 The standard OBD II data link connector. Not every vehicle will have a terminal in each of the 16 pin locations.

■ ***Functionality.*** OBD I systems were only required to monitor a few emission-related systems and inputs to the engine control module or ECM. Items monitored included the fuel system through oxygen sensor input and the exhaust gas recirculation (EGR) system. OBD II requires much more specific and comprehensive monitoring of components and systems. This includes monitoring:

- Catalytic converter operation
- Engine misfires
- Fuel vapor evaporation collection, storage, and use
- Secondary air injection into the exhaust
- Fuel deliver control
- Oxygen sensor operation and performance
- Exhaust gas recirculation
- Positive crankcase ventilation
- Engine cooling system temperature
- Air conditioning system pressure

■ ***Self-Diagnostic Capability.*** The main purpose of OBD II is for detecting and testing of malfunctions that can affect emissions. In addition to there being more systems and components monitored, OBD II systems must also monitor each component that is an input or is an output of the ECM. What this means is that the ECM must be able to determine if a fault occurs in the network of sensors, outputs, or wiring. OBD II codes can indicate if voltage in the circuit is too high or too low, or if the circuit is open or shorted.

In addition, OBD II must be able to determine if a misfire is occurring, determine how severe the misfire is, and alert the driver when the engine is misfiring (**Figure 24-45**). The ECM uses data from the crankshaft (CKP) and camshaft (CMP) position sensors to determine if the engine is misfiring. This capability did not exist in OBD I systems.

OBD II MODES OF OPERATION

OBD I systems had, at most, three modes of operation. The first mode was displaying a trouble code. This was often done by flashing a light in sequence, such as "flash, pause, flash-flash." This would indicate a code 12. Many vehicles' computer system provided only this function. The second mode was the reading of serial data. This allowed the technician to read data provided by the ECM. The third, and least common mode, was bidirectional control. This provided the technician a way to control a component or system with the scan tool. This was so that a specific test or function could be utilized. OBD II systems have 10 modes of operation, which include and expand the functions of OBD I.

■ ***Global (Generic) and Enhanced (Vehicle Manufacturer) Modes.*** In the ECM, there are two different and separate sections, **global OBD** and *enhanced OBD*. The global section provides codes, data, and emissions-related information to anyone who can connect a scan tool or code reader to the DLC. This

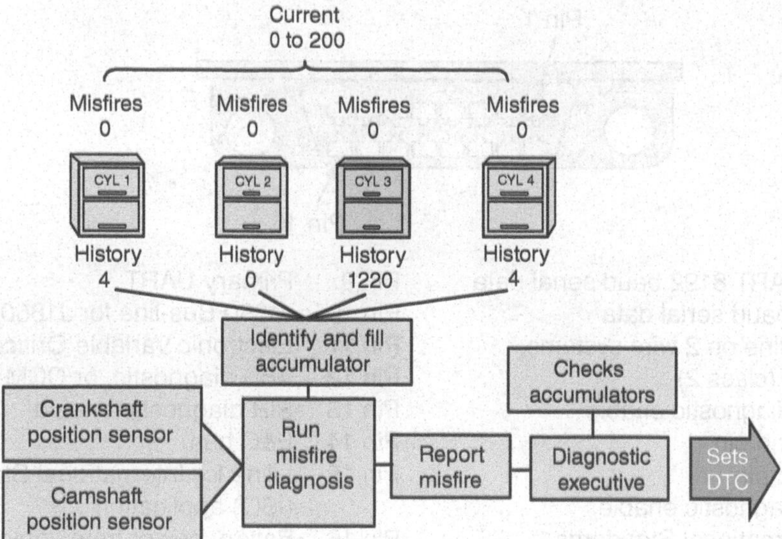

FIGURE 24-45 An illustration of how the PCM determines if a misfire is occurring.

allows those in the aftermarket to retrieve basic information to help diagnose the vehicle. Additionally, the data provided by the global function is raw data, meaning that it cannot contain substitute values. Substitute values are used by the ECM if a fault exists, and the input data is thought to be incorrect. The global programming controls the operation of the MIL and performs the onboard testing functions. Because of the mandate for access to basic information through the data link connector, anyone with a scan tool or code reader can retrieve basic data. This is part of Global or Generic OBD II mandate. **Figure 24-46** shows the various modes available in Global OBD II with a scan tool.

In addition to providing access for generic scan tools and code readers, the generic OBD II mode is being used to collect data about vehicle use and driving habits. Several companies sell data recording devices that plug into the DLC and record the operational information. This data can be used by parents to keep track of a young driver or by some insurance companies to provide discounts for safe driving practices. An example of an insurance data monitor is shown in **Figure 24-47**.

In addition to global functions, there is access to what is called **enhanced OBD (EOBD)** or the manufacturer's section. The enhanced section is VIN specific. This means that the vehicle's identification number or VIN must be entered to configure the scan tool to the system. The enhanced mode provides information that the vehicle manufacturer decides to include, which means the amount and type of data will vary depending on the vehicle. Often there is much more data and more functions available from enhanced mode compared to generic mode. This includes features such as controlling onboard

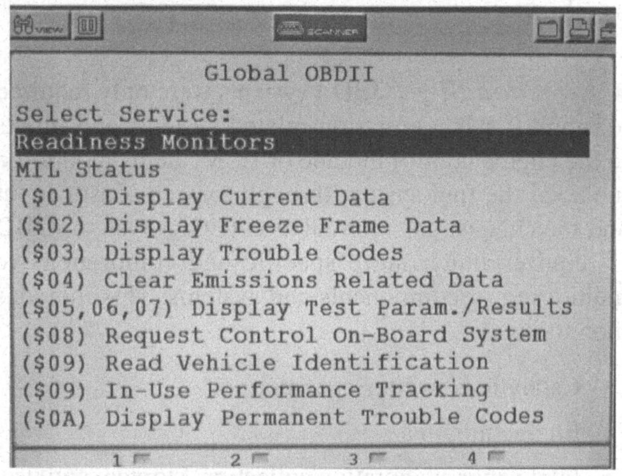

FIGURE 24-46 The list of current monitors in the OBD II system.

FIGURE 24-47 Generic OBD II functions provide data for use by anyone, including insurance companies to track driving habits.

systems. An example of features available in enhanced mode is shown in **Figure 24-48**.

When accessing an OBD II system, many technicians begin by checking for codes and data using the global function before switching over to the EOBD system. In many cases, the information kept in the global system is sufficient to diagnose the vehicle.

It is important to note that all aspects of OBD II, global, enhanced, communications, and so on, apply only to emissions-related systems. The vehicle manufacturer is not obligated to provide access to other systems such as ABS, traction control, body, or any others under Global OBD II mandates.

MONITORS

One of the major changes from OBD I to OBD II systems is the introduction of monitors. A *monitor* is a method the onboard computer uses to determine if a system or component is operating properly.

■ ***Principles of OBD II Monitors.*** A **monitor** is a test run by the ECM on components and systems to determine that certain operating conditions have been met. There are two types of OBD II monitors, continuous and noncontinuous. As the names imply, a continuous monitor is one that operates pretty much whenever the engine is running. A noncontinuous monitor only operates when certain operational conditions have been met.

The purpose of the monitors is to enable the OBD II system to be self-monitoring. This reduces the need for external emissions testing because the ECM is now responsible for testing every emissions-related component and system. There are currently 13 monitors being used, eight of which are common to all OBD II vehicles since 1996. The remaining five monitors are used as necessary on certain vehicles. An example of monitor test status is shown in **Figure 24-49**.

There are three ways in which a monitor can perform a test: passive, active, and intrusive test strategies.

A **passive test** is a "wait and see" strategy used by the ECM for detecting obvious faults, such as open or short circuits. An example of this is the comprehensive component monitor or CCM. The CCM monitors circuits for opens, shorts, and invalid input.

An **active test** is performed by the ECM by taking direct control of the component to determine its functionality. In this case, the ECM may command a component into operation to determine if the component responds to a command. An example of this test would be when the ECM commands air pump operation to test if the oxygen sensor(s) show a corresponding change in exhaust oxygen content.

An **intrusive test** is like an active test except that the result of the test may have an impact on drivability. For example, if the ECM is not receiving information regarding exhaust gas recirculation (EGR) flow, it can command the EGR valve to open during operating conditions in which the EGR flow is normally not used, such as at idle or low-speed driving. If the EGR valve is working, the exhaust flowing into the intake air stream will have an effect on performance and can be noted by the manifold absolute pressure (MAP) and oxygen (O_2) sensors. In this way, the ECM can determine if the EGR is working and if the problem is in the EGR flow sensor.

■ ***Using Monitors as Repair Verification.*** Because monitors are the way the PCM performs self-tests, they can also be used by the technician to verify if a repair is complete. For example, say a vehicle has a code set for insufficient EGR flow. You remove the EGR valve and clean the EGR passages that are plugged with carbon and clear the DTC. To be sure the problem is solved, test drive the vehicle so that the EGR monitor will run. If the monitor runs and passes, the problem has been corrected.

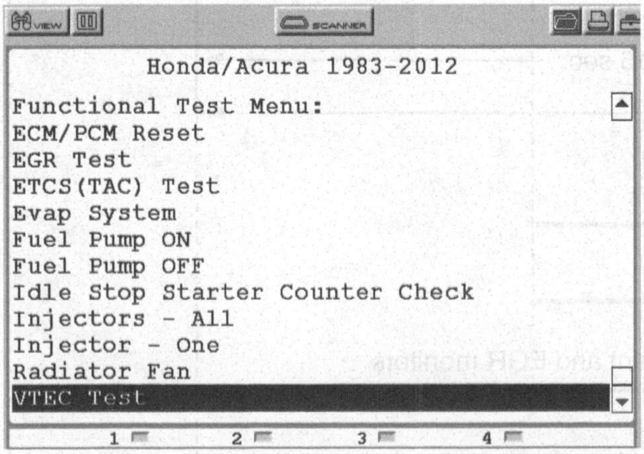

VIEW SCANNER

Honda/Acura 1983-2012

Functional Test Menu:
ECM/PCM Reset
EGR Test
ETCS(TAC) Test
Evap System
Fuel Pump ON
Fuel Pump OFF
Idle Stop Starter Counter Check
Injectors - All
Injector - One
Radiator Fan
VTEC Test

1 2 3 4

FIGURE 24-48 Functional tests allow you to take command of certain components or systems for diagnostic purposes.

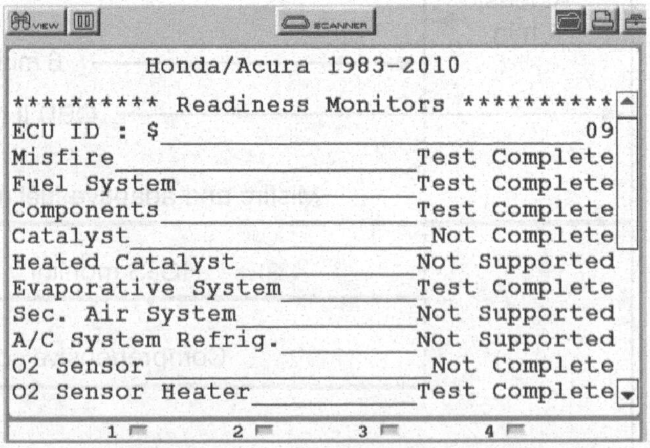

VIEW SCANNER

Honda/Acura 1983-2010

********** Readiness Monitors **********

ECU ID : $	09
Misfire	Test Complete
Fuel System	Test Complete
Components	Test Complete
Catalyst	Not Complete
Heated Catalyst	Not Supported
Evaporative System	Test Complete
Sec. Air System	Not Supported
A/C System Refrig.	Not Supported
O2 Sensor	Not Complete
O2 Sensor Heater	Test Complete

1 2 3 4

FIGURE 24-49 An example of a monitor status report. Note that this vehicle did not support all of the possible tests listed.

■ ***Do Monitors Affect Each Other?*** Yes, because some monitors, such as the catalytic converter monitor, depend on the operation of other components. For example, for the catalytic converter monitor to run, the heated oxygen sensor (HO_2S) monitor must run and pass. This is because the catalytic converter monitor uses the oxygen sensors as part of its test.

When one monitor prevents another monitor from running it is called a conflict. In this example, if the HO_2S monitor is not able to pass, the catalyst monitor will be placed on hold pending successful completion of the HO_2S testing. This means that a failed component can lead to a group of tests that are incomplete because of a fault (Figure 24-49). **Figure 24-50** shows how various monitors run based on vehicle operation. This is important to note because, in many cases, very specific operating conditions must be met for some monitors to run. If these conditions are not met—for example, due to a change in driving habits or climate conditions—the failure of one monitor to run can then also keep other monitors from running.

■ ***Are Some Monitors More Important Than Others?*** Since all monitors are used for emissions monitoring, no single monitor is more important than any other. In addition, all OBD II vehicles use the mandated eight monitors listed in the next section. There are five additional monitors that can be used as determined by the manufacturer. While not less important, these five are not always used because a vehicle may not need the system or component that is the subject of the monitor.

CONTINUOUS MONITORS

There are three **continuous monitors** that are active at almost all times when the engine is running: these are the comprehensive component monitor or CCM, fuel control, and misfire monitors. These monitors are continuous because a failure in the systems watched by a continuous monitor can have a significant effect on both emissions and the catalytic converter.

■ ***Comprehensive Component Monitor.*** The **comprehensive component monitor (CCM)** is a diagnostic test that is designed to check all input and output circuits for the ECM. This means monitoring for open and short circuits, functionality, and the rationality of input signals. Most components monitored by the CCM are not tested by other monitors nor are they parts of monitors.

For example, the engine coolant temperature (ECT) sensor circuit is monitored by the CCM. Five volts are supplied to the ECT sensor, which, based on its temperature, will vary the voltage drop in the circuit. The ECM uses this voltage drop to determine engine coolant temperature. If, however, the ECT connector is damaged and the circuit opens, the circuit resistance becomes infinite, and no voltage will return to the ECM. The ECM then can determine that the sensor return voltage is absent and set a diagnostic trouble code (DTC) for the ECT circuit.

Functionality tests are performed by the ECM commanding an actuator to respond to a command. An example would be if the ECM commanded the electronic throttle control motor to open the throttle slightly. This action is then verified by data from the throttle position sensors and

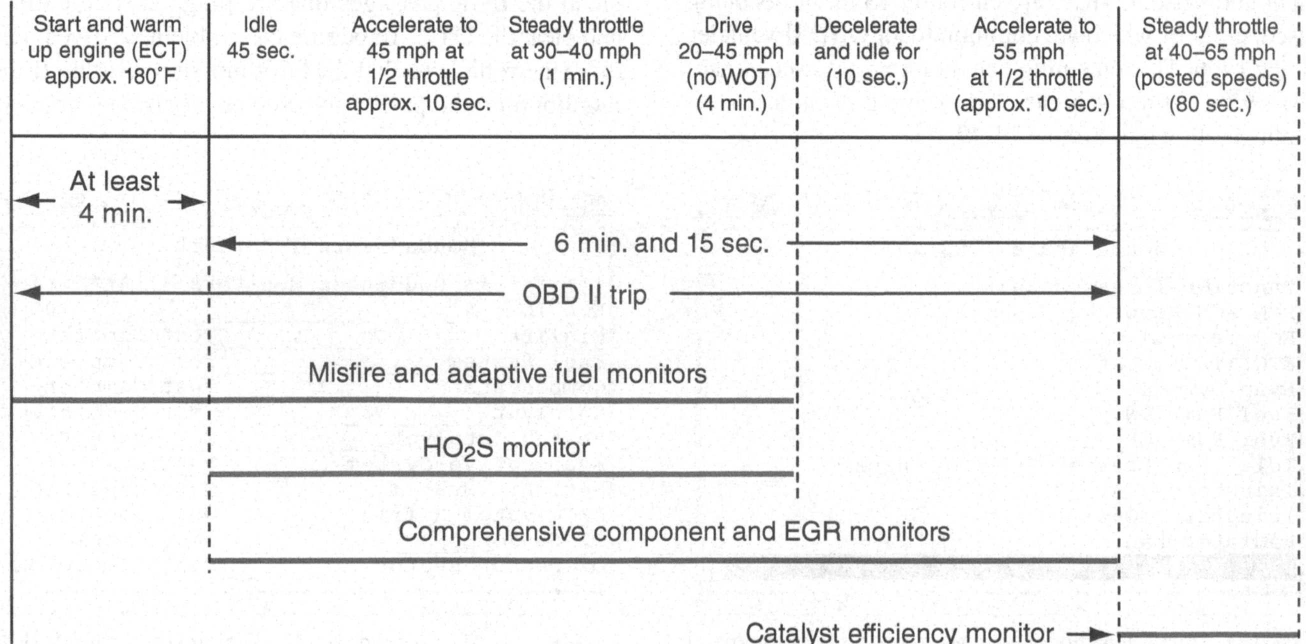

FIGURE 24-50 A trip allows the PCM to run a specific monitor. A drive cycle should allow all monitors to run.

crankshaft position sensor. The throttle position sensors would report an increase in the throttle opening, and the crankshaft sensor would report an increase in engine rpm.

Rationality testing is used to determine if input signals make sense. Using the ECT circuit again as an example, an open in the ECT circuit will cause the ECM to see an input temperature of approximately −40°F (−40°C). By comparing the ECT data to the input from the intake air temperature (IAT) sensor, the ECM can determine if the ECT reading is correct or out-of-range. If the IAT signal indicates that the temperature is 60°F (15.5°C), the ECM can determine that the ECT data is incorrect and use a substitute value to continue operation while setting a code and turning on the MIL.

■ ***Misfire Monitor.*** The **misfire monitor** is a test used to evaluate combustion efficiency. This monitor runs continuously at idle, under acceleration, and at cruise. The principle behind the misfire monitor is that the crankshaft position (CKP) sensor can detect the changes in the rotational speed of the crankshaft as it fails to accelerate correctly following a misfire. **Figure 24-51** shows a graphic illustration of how this is accomplished. Because there is not an increase in crankshaft rpm when a cylinder does not produce power, the ECM is able to determine that a misfire has occurred. By examining both CKP and camshaft position (CMP) signals, the ECM can then determine which cylinder is experiencing the misfire. There are 13 misfire DTCs in OBD II, ranging from P0300 for a random cylinder misfire to P0301 to P0312, which are used to specify which cylinder is misfiring. A P0301 is used for cylinder number 1, P0302 for number 2 on up to P0312 for cylinder 12.

Additionally, OBD II requires a distinction between three types of misfires. They are:

- Type A misfires. These are catalyst-damaging misfires, which must be detected within 200 to 1,000 crankshaft revolutions. Type A misfires cause the MIL to blink twice per second.
- Type B misfires. These are emissions-threatening misfires, meaning that emissions will exceed 150% of the federal test procedure or FTP standards. Type B misfires are detected within 1,000 to 4,000 crankshaft revolutions. If the ECM detects a type B misfire, a pending code is stored but the MIL is not illuminated. If the misfire occurs on the second trip, a DTC is stored and the MIL is commanded on.
- Type C misfires. This type of misfire causes emissions to exceed California's standards and uses the same two-trip detection as a type B misfire.

Rough road conditions can cause the ECM to believe that a misfire is occurring when in fact it is not. This is because the wheels and tires, responding to rough road conditions, apply torque to the engine, causing changes in engine speed. To address this issue, manufacturers use rough road detection software in the ECM to help filter out false misfire events from actual events. Many vehicles use input from the ABS wheel speed sensors to determine if a rough road condition is present. Another approach is to unlock the torque converter so that the engine and transmission are isolated from each other. If the drive wheels are experiencing a rough road, unlocking the torque converter clutch will prevent many of the speed changes from being transferred to the crankshaft.

■ ***Fuel Control Monitor.*** OBD II mandates that the fuel delivery system be monitored constantly for the ability to maintain fuel control and emission standards. The system is malfunctioning if emissions exceed 150% of the federal test procedure standards or FTP for that vehicle. The FTP is the standard the vehicle must meet for exhaust and fuel emissions, which vary depending on

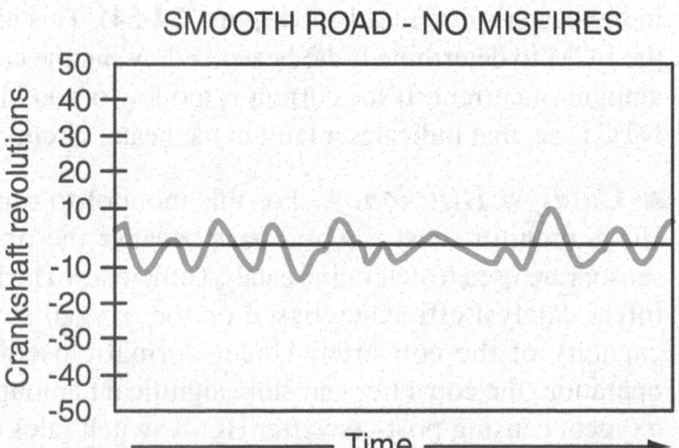

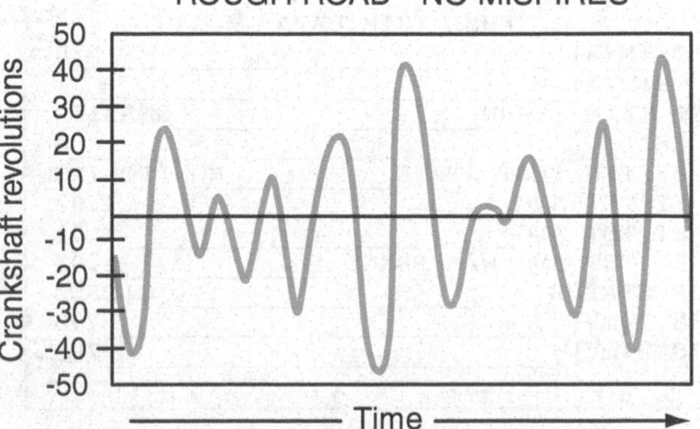

FIGURE 24-51 Using crankshaft and camshaft position sensors, the ECM can determine if a misfire is occurring. The crankshaft sensor provides the rpm data and the relationship from the camshaft sensor allows the ECM to determine which cylinder or cylinders are affected.

the model year, type of vehicle, and the engine installed. The fuel control monitor begins as soon as the vehicle enters closed-loop operation. Closed-loop means that the ECM is basing fuel delivery requirements on the data provided by the oxygen sensors. Closed-loop operation begins once the oxygen sensors warm up and provide feedback on exhaust oxygen content.

Fuel control is represented by short-term fuel trim (STFT) and long-term fuel trim (LTFT) on the scan tool (**Figure 24-52**). **Fuel trim** means the making of slight adjustments to fuel delivery to keep the engine operating efficiently. If the fuel system requires no correction, then ST and LT fuel trims will read 0%. Any positive number represents the ECM adding fuel, while any negative number indicates fuel delivery is being reduced. Most manufacturers allow a change of up to 10%, either positive or negative, as allowable fuel trim correction. Fuel correction is based on input from the oxygen sensors; if the O_2 signal is above 450 mV, the exhaust is considered rich, and the ECM shortens injector pulse. The reduced injector pulse causes the exhaust gases to contain more oxygen, which the O_2 sensor reads as lean, which causes the fuel trim to lengthen injector pulse again. This process continues, switching between rich and lean as long as the system is in closed-loop operation.

The term stoichiometric is used to describe when the engine is running at the perfect air–fuel ratio. While the OBD II system can keep the engine running very close to stoichiometric, minor changes are necessary to adapt to operating conditions.

■ ***Why Switch Between Rich and Lean?*** Many people ask if it would be easier if the ECM just kept the fuel trim as close to zero correction as possible instead of switching back and forth between rich and lean constantly. The short answer is the catalytic converter needs the constantly switching rich and lean exhaust gases to function properly.

Chevrolet 1981-2010

FUEL TRIM DATA

ST TRIM(%)	0
LT TRIM(%)	1
FUEL TRIM LEARN	ENABLED
FT CELL	6
FT SYSTEM TEST STATE	MONITORING
ST FT TST AVG	1.02
LT FT TST AVG	1.01
LT FT TST AVG w/o PURGE	1.03
LOOP STATUS	CLOSED
HO2S 1(mV)	795
HO2S 2(mV)	785

FIGURE 24-52 An example of fuel trim data. This vehicle has very little fuel trim correction at this engine speed.

NONCONTINUOUS MONITORS

Noncontinuous monitors require certain enable criteria to be met before the monitor can run. These criteria will vary slightly from manufacturer to manufacturer and for each monitor. This section describes the noncontinuous monitors in general. For information specific to a vehicle, you will need to refer the manufacturer's service information.

■ ***Oxygen Sensor Monitor.*** The **oxygen sensor monitor** tests for the sensor's time to reach activity, for signal voltage, and for the signal response rate. The time-to-activity test is performed on cold engine starts and compares how long it takes the sensor to become active to a programmed parameter. For example, if the programmed time is 60 seconds, the ECM will expect the sensor to become active and provide data within 60 seconds of engine startup. If the time is longer than the programmed time, an oxygen sensor DTC will be stored. The response time test examines how long the transitions between rich to lean and lean to rich take (**Figure 24-53**). The signal response rate test is used to determine if the sensor is reaching the proper voltage switch levels (Figure 24-53). This test can also determine if the sensor is shorted to ground or to voltage based on the voltage signal.

Rear oxygen sensors are monitored for similar conditions. The exception is that rear sensors, those after the catalytic converter, should not indicate a rapidly switching rich/lean condition.

■ ***Oxygen Sensor Heater Circuits.*** The O_2 heater circuits are tested for proper operation by monitoring the heater circuit driver in the ECM for voltage changes when the heaters are switched on and off. The term driver is used for the output control circuit of the ECM. The output control circuit "drives" the output component, commanding it to turn on or off. The ECM may also monitor current flow in the heater's circuit driver (**Figure 24-54**). This allows the ECM to determine if the heater is drawing the correct amount of current. If the current is too low or too high, a DTC is set that indicates a fault in the heater circuit.

■ ***Catalyst Efficiency.*** For this monitor to run, the HO_2S monitor must run and pass because the oxygen sensors are used to determine catalyst efficiency. The ECM infers catalyst efficiency based on the oxygen storage capacity of the converter. Under normal closed-loop operation, the converter can store significant amounts of oxygen, causing post-converter HO_2S switch rates to be much slower and of decreased amplitude compared to pre-converter sensors. **Figure 24-55** shows an example

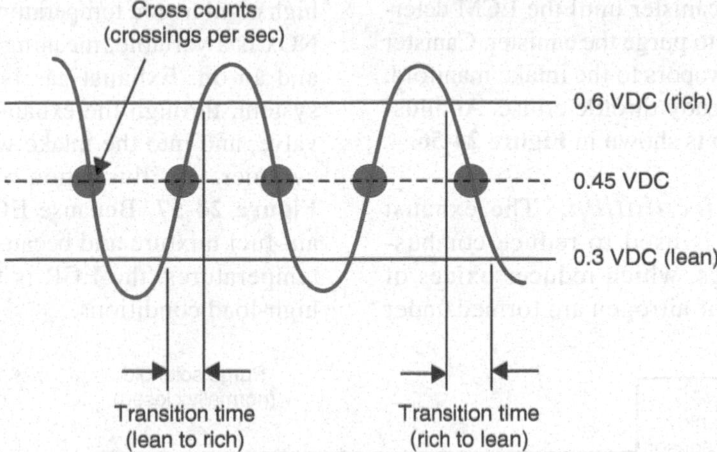

FIGURE 24-53 The amount of voltage produced by the O_2 sensor is examined by the ECM to determine if the sensor is working properly. Oxygen sensor activity is also examined for the time between certain points or levels of voltage change. The PCM can determine if the sensor is responding too slowly based on this data.

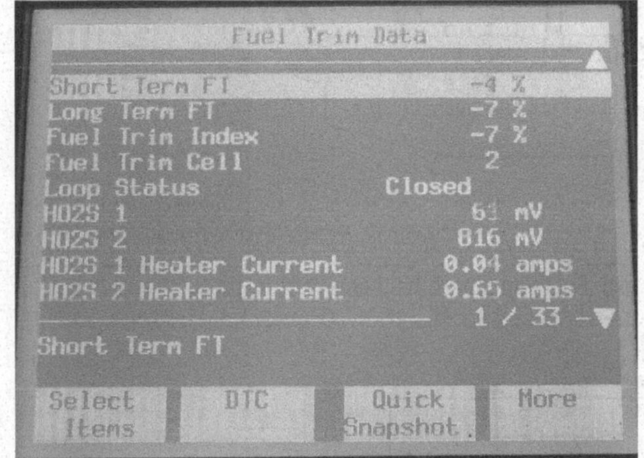

FIGURE 24-54 An example of heated oxygen sensor current data. This information is used to evaluate heater function.

of both pre- and post-converter HO_2S signals for a good catalytic converter. As catalyst efficiency deteriorates over time, the switch rate of the post-converter sensors increases, as does the amplitude of the signal. When the switch rates of the post-catalyst HO_2S sensor reach a predetermined value, the ECM turns on the malfunction indicator lamp and stores a catalyst efficiency code.

■ ***Evaporative Emissions.*** The evaporative (EVAP) system collects fuel vapors from the fuel tank, stores the vapors in the charcoal canister and then routes the vapors to the intake manifold to be burned in the engine. These vapors are collected to reduce the number of hydrocarbons (HCs) that are released into the atmosphere. Fuel vapor is collected during the refueling process. This is called *onboard refueling vapor recovery (ORVR)*. Fuel vapors

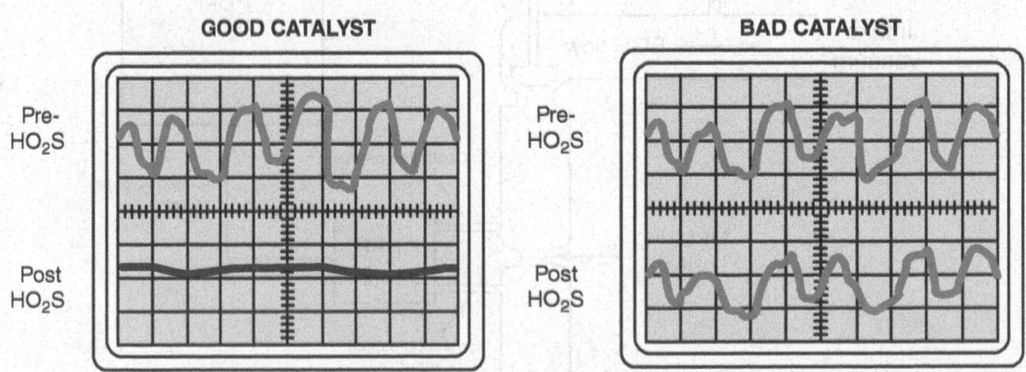

FIGURE 24-55 Pre-and post-oxygen sensor data is used to determine the efficiency of the catalytic converter.

are stored in the charcoal canister until the ECM determines the appropriate time to purge the canister. Canister purge, the releasing of the vapors to the intake manifold, generally occurs during steady throttle cruise. An illustration of the EVAP system is shown in **Figure 24-56**.

■ ***Exhaust Gas Recirculation.*** The exhaust gas recirculation (EGR) is used to reduce combustion chamber temperatures, which reduces oxides of nitrogen or NOx. Oxides of nitrogen are formed under high combustion temperatures, above 2,500°F. The *x* in NOx is a variable, meaning the gas may be NO_1, NO_2, and so on. Exhaust gas is routed from the exhaust system, through the exhaust gas recirculation (EGR) valve, and into the intake where it is supplied to each cylinder. An illustration of this system is shown in **Figure 24-57**. Because EGR displaces some of the air–fuel mixture and because NOx only forms at high temperatures, the EGR is typically only used under high-load conditions.

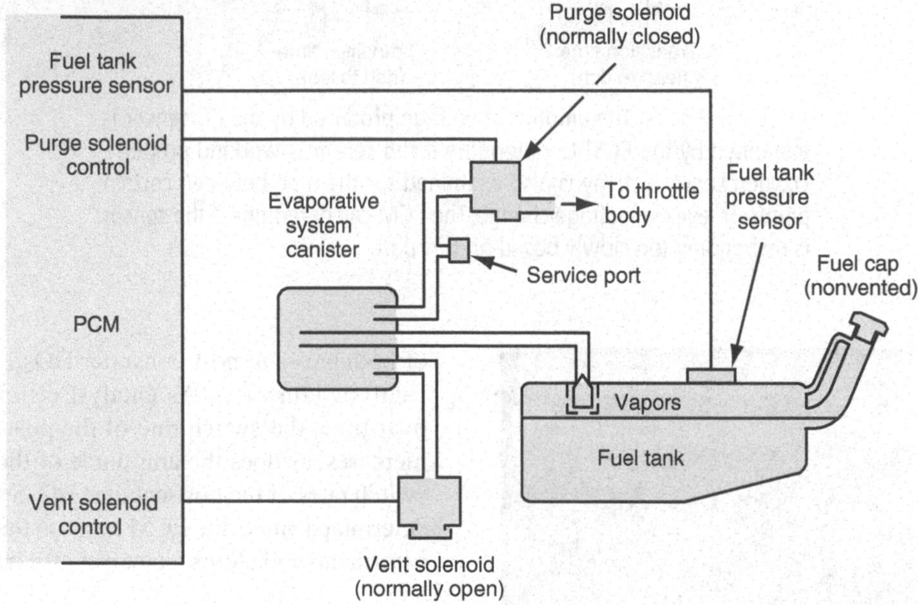

FIGURE 24-56 An illustration of the EVAP system. Fuel vapors are held and sent to the engine during steady cruising conditions.

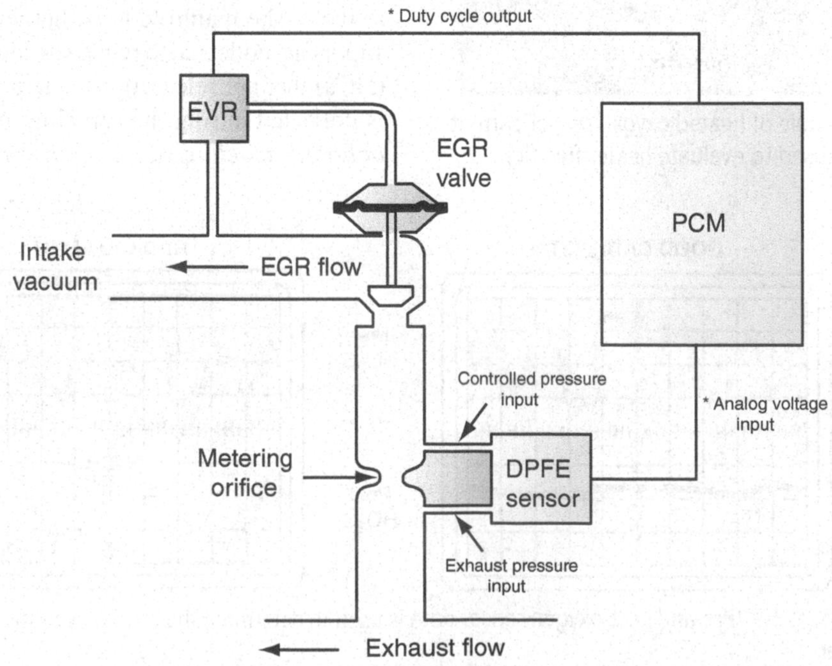

FIGURE 24-57 An illustration of an EGR system. The recirculated exhaust gas is used to lower combustion chamber temperatures and reduce exhaust emissions.

Many systems monitor EGR flow is by using a manifold absolute pressure (MAP) sensor. During engine closed-throttle deceleration or during steady-throttle cruise, the ECM commands EGR flow. The flow of exhaust gases increases the pressure in the intake, which is read by the MAP sensor.

■ ***Positive Crankcase Ventilation.*** The positive crankcase ventilation (PCV) valve uses engine vacuum to draw blowby crankcase vapors out of the engine and return them to the intake manifold for inclusion in the intake air stream for burning. Newer OBD II vehicles have a PCV monitor that checks for PCV flow once per drive cycle. A drive cycle is discussed in the next section. Vehicles with the monitor have some type of PCV valve retention lock that prevents accidental disconnection of the valve from the intake or valve cover. The ECM monitors PCV flow to check for valve operation and can detect excessive flow and restricted flow.

■ ***Heated Catalyst.*** This monitors the electrically heated catalyst found on a few vehicles. This heated catalyst is for quicker engine warmup and to get into closed-loop operation more quickly to reduce emissions.

■ ***Secondary Air.*** Some vehicles use a secondary air injection system that pumps fresh air into the exhaust so that any unburned hydrocarbons can burn. This system, used on cold engine starts, reduces hydrocarbon and carbon monoxide emissions. On many OBD I systems, the air injection reaction (AIR) system pumped fresh air into the exhaust manifold or into the catalytic converter (**Figure 24-58**). Most OBD II systems supply air to the converter only.

■ ***Air Conditioning.*** Used to monitor pressures in the air conditioning system to check for refrigerant leaks. If the pressure drops below a specified amount, an A/C system code is set.

■ ***Thermostat.*** The thermostat monitor is used to detect if engine operation temperature is too low. The ECM monitors ECT, IAT, and engine operating conditions such as load and rpm and determines if engine coolant temperature has increased to a preset limit. There may also be a time limit, how long the engine operates, before the thermostat is considered faulty.

Trips and Drive Cycles

Trips and drive cycles were new for OBD II as part of the system of monitors and self-diagnostics. To enable the monitoring of specific components and systems, the OBD system requires parameters for when certain operations can take place; these parameters are called **trips**. A trip is used to allow a monitor to run for a specific component or system. A **drive cycle** is a series of trips designed to allow all monitors to run.

TRIPS AND DRIVE CYCLES

For the ECM to run a self-diagnostic on a system or component, enabling criteria must be met. These criteria, such as range of coolant temperature, throttle opening, and engine load, when met form a trip. A trip consists of all the enabling criteria being met to run a monitor.

> **Word Wall**
> *Enabling Criteria*—This means a specific number and type of operating conditions that must be met before an onboard test will begin.

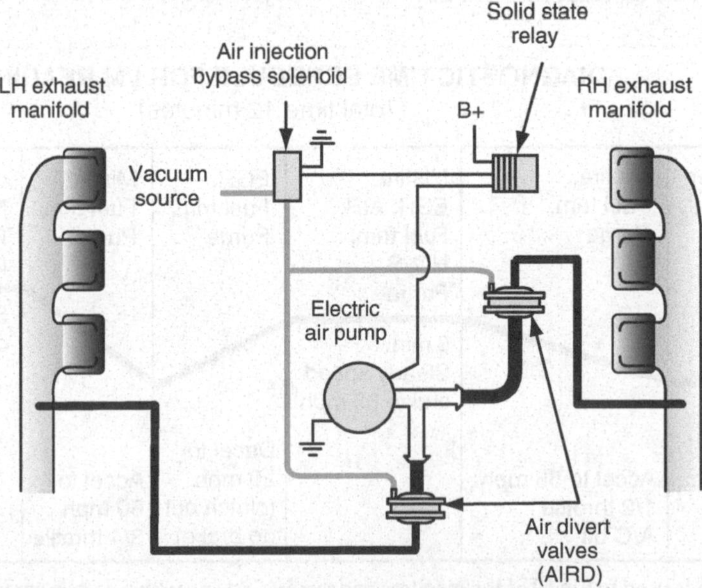

FIGURE 24-58 Not used on all vehicles, the air injection system pumps fresh air into the exhaust to burn any leftover fuel.

■ ***Trips Run Specific Monitors.*** Each monitor has its own trip and enabling criteria. For example, for the EVAP monitor to run, a vehicle may require the following conditions:

EVAP 0.040" Leak Detection Monitor		
Condition for Monitor to Begin	**MIN**	**MAX**
Engine OFF time	4 hours	
Time since engine start	360 seconds	2,500 seconds
Intake air temperature	40°F	95°F
BARO pressure	75 kPa	
Engine load	20%	75%
Vehicle speed	42 mph	85 mph
Purge amount	75%	100%
Purge flow	0.05 lbs/min	0.10 lbs/min
Fuel level	15%	80%
Fuel tank pressure	−17 H2O	1.5 H2O
Battery voltage	11.5 volts	16 volts

FIGURE 24-59 An example of enable criteria for a monitor to run. Each system has its own set of enable criteria.

This is just an example of what criteria may be used. Information regarding what is necessary to satisfy both trips and drive cycles is available in the manufacturer's service information and from Prodemand and AllData.

Until the specific criteria for a certain monitor are met for a trip, a DTC will not set unless the fault triggers a separate monitor, such as the CCM, to recognize the problem. For example, a fault with an EGR valve causing insufficient exhaust gas flow will not set a DTC until the EGR trip has run and the ECM determines there is insufficient EGR flow. If the vehicle is not driven in a way for the EGR monitor criteria to begin, the monitor will not run, and the DTC will not set. However, if the EGR fault is due to a damaged EGR electrical connector, it may be detected as a circuit fault by the CCM but will not set an EGR flow DTC.

■ ***Drive Cycles.*** Completing a drive cycle allows all criteria for all monitors to be met (**Figure 24-60**). Once the criteria are met, the monitors run, and the system either passes or fails.

Once a repair has been made, the vehicle will need to be driven to satisfy the enable criteria for its trip. After a successful completion of the trip, further driving may be necessary to complete a drive cycle so that all monitors can run. Typically, all monitors must be completed and passed to satisfy state and local emission testing requirements.

■ ***Monitors and Emission Tests.*** Many states have adopted checking the OBD II readiness monitors as part of their inspection and maintenance (I/M) programs. By plugging into the DLC, the examining station can obtain data on DTCs and monitor status. If one or more monitors are incomplete or failed, the vehicle will not pass the I/M test and will need to be repaired and retested.

CODES: CURRENT, PENDING, AND HISTORY

DTCs, often just called *codes*, are used to help diagnose faults in the onboard diagnostic system. Codes generally do not specify replacing any particular component; instead they provide the technician with a starting point to begin diagnosis.

■ ***What Is a Code?*** Simply stated, a code is a method of organizing faults that can occur in the onboard computer systems. A code is a means of communication between the technician and the computer.

■ ***OBD I Codes.*** Most OBD I codes were composed of one or two numbers, such as 6, 12, or 45. Numbers were used because the use of flash codes, those read by

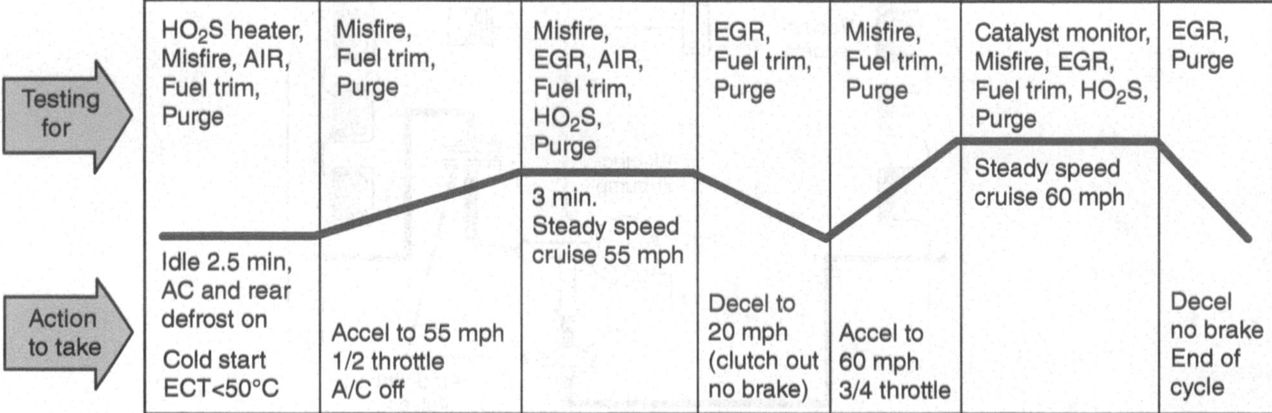

FIGURE 24-60 A drive cycle is used to run all of the monitors and requires driving within certain parameters for the test to complete.

watching the number of times the check engine light or other light blinked in a sequence, was a common method of retrieving codes. It would not have been easy to use flash codes to express letters of the alphabet.

One of the problems with OBD I was the lack of any standardization of the codes. A code 12 from a GM vehicle did not mean the same thing on a Ford, Chrysler, or any other vehicle.

■ ***ODB II Codes.*** Diagnostic trouble codes under OBD II mandates are much more standardized than before. All codes have a prefix that indicates to which major group the code belongs. Powertrain codes start with P, body codes start with B, chassis codes start with C, and network communication codes begin with U. This immediately allows the technician to determine what types of codes are stored. A description of how a DTC is categorized is shown in **Figure 24-61**.

Four types of DTCs are used in OBD II: type A, B, C, and D codes. Type A DTCs are emissions related and are detected during the first trip. Examples of this type of DTC are misfire, fuel control, and CCM test codes. Type B DTCs are also emissions related but require two trips to set a hard code. During the first detection, a pending code is stored; on the second trip a current or hard DTC is set, the MIL is requested on, and a **freeze frame** is stored. A freeze frame is a snapshot of pieces of data stored at the time the code set. The information contained in the freeze frame will vary depending on the code and the vehicle. Type C DTCs are not emissions related and do not store a freeze frame. Type C DTCs can cause a service light to illuminate on the instrument panel but not the MIL. Type D DTCs are not emissions related, do not store freeze frames, and do not generally cause any warning lamp to illuminate.

■ ***Current Codes.*** A current code, sometimes called *a hard code*, can be a one-trip or a two-trip code and can be set by any of the monitored systems. Common one-trip codes are set by a cylinder misfire, such as a P030x,

The SAE J2012 standards specify that all DTCs will have a five-digit alphanumeric numbering and lettering system. The following prefixes indicate the general area to which the DTC belongs:

1. P — power train
2. B — body
3. C — chassis
4. U — network codes

The first number in the DTC indicates who is responsible for the DTC definition.

1. 0 — SAE
2. 1 — manufacturer

The third digit in the DTC indicates the subgroup to which the DTC belongs. The possible subgroups are:

0 — Total system
1 — Fuel-air control
2 — Fuel-air control
3 — Ignition system misfire
4 — Auxiliary emission controls
5 — Idle speed control
6 — PCM and I/O
7 — Transmission
8 — Non-EEC powertrain

The fourth and fifth digits indicate the specific area where the trouble exists. Code P1711 has this interpretation:

P — Powertrain DTC
1 — Manufacturer-defined code
7 — Transmission subgroup
11 — Transmission oil temperature (TOT) sensor and related circuit

FIGURE 24-61 A breakdown of what each part of a DTC means.

where x is the specific cylinder experiencing the misfire. For example, a P0303 indicates a misfire detected for cylinder 3. An example of current codes set in a PCM are shown in **Figure 24-62**.

A current two-trip code is set after the ECM detects a problem that occurred either in two successive trips or in two nonsuccessive trips that had almost exactly the same operating conditions.

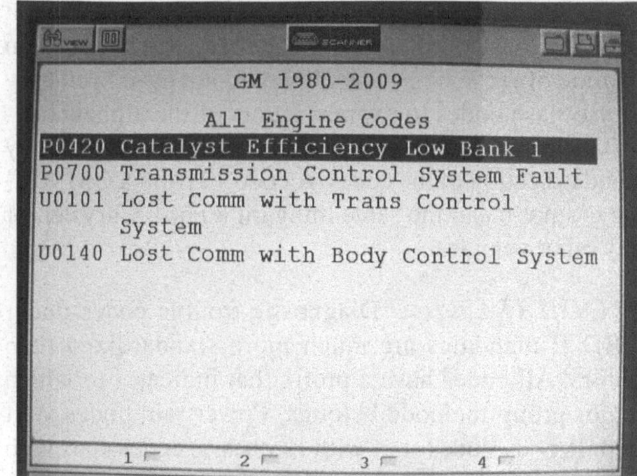

FIGURE 24-62 Examples of current DTCs.

■ ***Freeze Frame or Failure Records.*** When a DTC sets a freeze frame will also be recorded. A freeze frame is a snapshot of specific pieces of data that the PCM records when the DTC sets in memory (**Figure 24-63**). This data is valuable for being able to operate the vehicle under the conditions in which the DTC set recrate the conditions for confirming the complaint and the repair once finished.

■ ***Pending Codes.*** A **pending code** means that the ECM has detected an issue, has stored a pending code in memory, and is waiting for a second trip to take place so that the code can be confirmed.

■ ***History Codes.*** A **history code** is one that remains in the ECM memory after a fault has been corrected. After 40 consecutive warmup cycles with no further

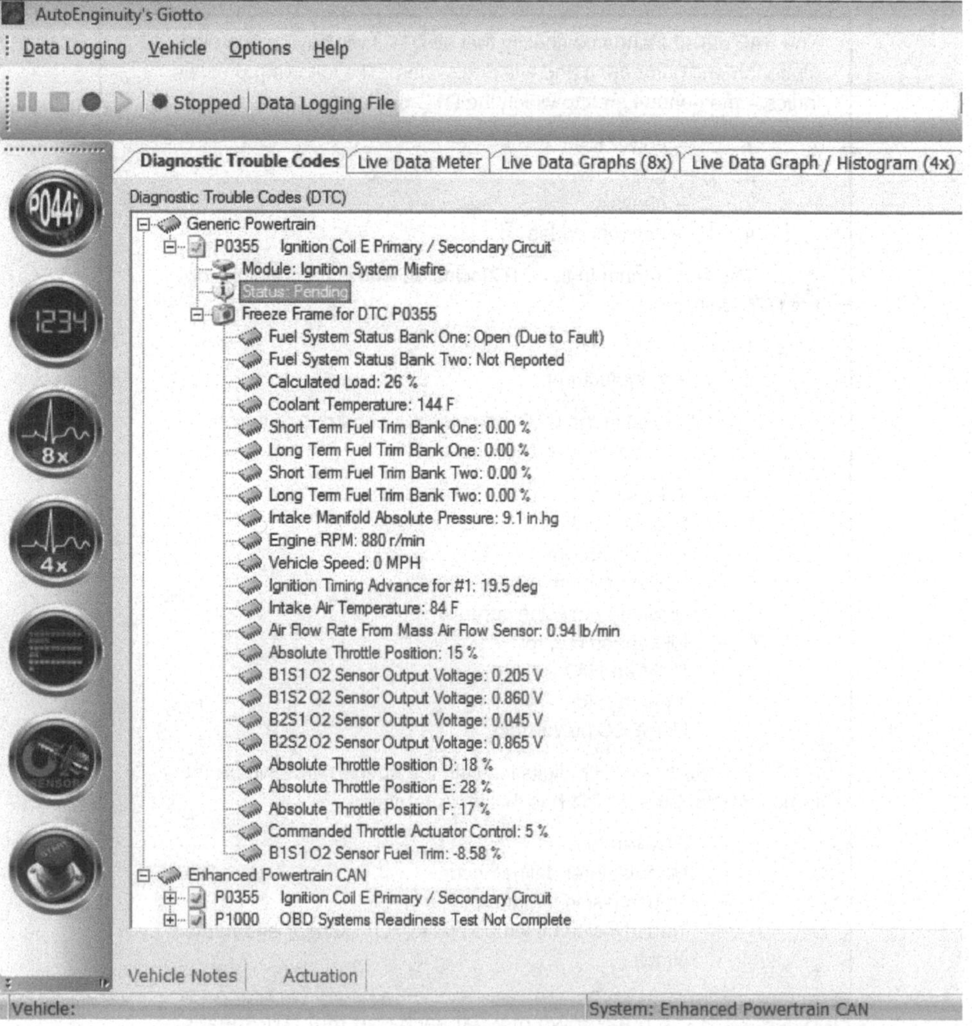

FIGURE 24-63 A pending code is set in memory and the PCM is waiting for a second occurrence before making it a current DTC.

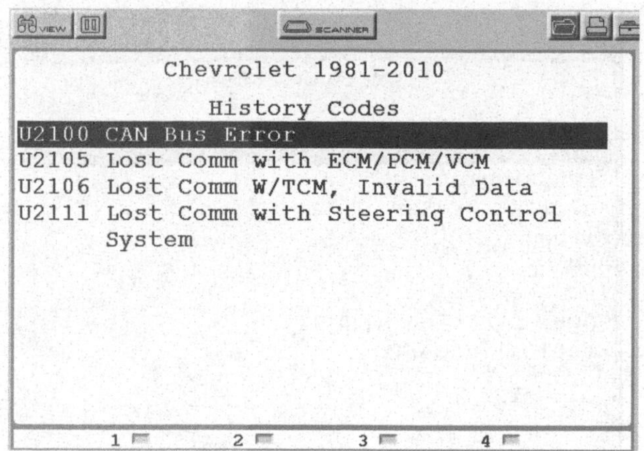

FIGURE 24-64 Examples of history codes. These DTCs were set due to a blown fuse.

FIGURE 24-65 A common type of generic scan tool.

faults detected, the DTC will clear from the ECM memory. Type C and D codes are also stored as history codes (**Figure 24-64**).

USING A SCAN TOOL

Using a scan tool has become another basic skill for today's technicians. This is due to the increasing use of computers and networks to monitor and control all types of vehicle functions.

When first introduced to scan tools, some people are overwhelmed and intimidated by what often seems to be an extremely complex array of connectors, cables, tool interfaces, and incomprehensible data on a screen. Most, after a little time to get familiar with a particular tool, become very comfortable in its use in very little time.

There are many different scan tools on the market, from basic code readers to complex and expensive scan tools, but there are only two types of scan tools, OE and aftermarket. Examples of OE tools include the General Motors Tech2 and MDI/GDS, Ford IDS and WDS, Chrysler wiTech, Toyota Techstream, and many others. Companies such as Snap-On Tools, OTC Tools, Auto Enginutiy, and others make aftermarket tools. The difference is that the OE tools provide the same functionality for anyone who uses the tool, whether employed by the manufacturer in a dealership or by a technician working in an independent shop. Aftermarket tools are designed to be used on many different makes and models of vehicles and are not specific to any one brand of car or truck. These tools allow a technician to access some of the systems and data on vehicles but often do not provide the depth of access available with the OE tools.

■ ***Snap-On Solus, Solus Pro, Modis.*** Many shops have scan tools made by Snap-On Tools (**Figure 24-65**). Still in common use are the Solus family and the Modis. Newer models, such as the Verus, Verdict, Solus and Modis Ultra are very similar and operate in basically the same manner. The latest models use wireless connections, Wi-Fi to access service information, and have touch-screen interfaces.

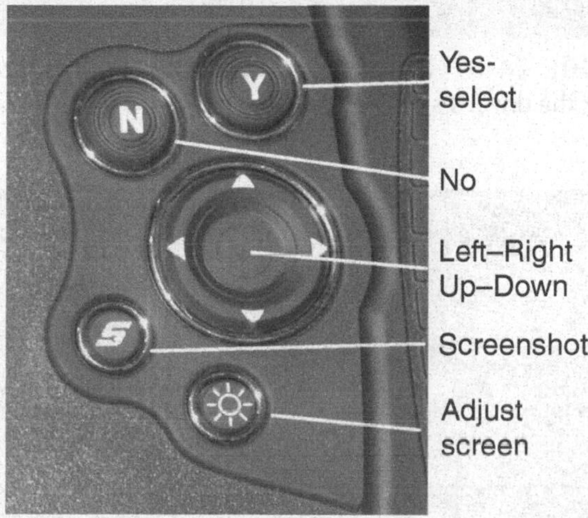

FIGURE 24-66 The navigation buttons for some of Snap-On's scan tools.

The Solus and Modis have a very basic user interface, containing just a few buttons. The Yes, No, and four-way pad are the major controls for navigating the menus. The functions of the buttons are shown in **Figure 24-66**.

To begin, locate the OBD II data link connector (DLC) on the vehicle; this is under the driver's side of the dash (**Figure 24-67**). Connect the OBD II connector to the DLC. Press the power button, and the scan tool will boot. The initial screen allows you to select vehicles by U.S. domestic, Asian, or European manufacturer or to use Global OBD II mode. This allows you to use either enhanced onboard diagnostics (EOBD) or Global/Generic OBD II functions.

When using EOBD, follow the VIN entry steps to program the vehicle information into the scan tool. Once entered, the tool will direct you to install a certain

FIGURE 24-67 The data link connector (DLC) is located under the driver's side of the dash.

FIGURE 24-68 Older Snap-On scan tools use personality keys to program the tool to the specific vehicle.

personality key (**Figure 24-68**). Once connected to the network, you will have a list of systems from which you can select (**Figure 24-69**).

Once you have finished using the scan tool, press the power button until the screen displays the Turn Off box. Use the Yes button to turn the tool off. This process allows the tool to shut down properly.

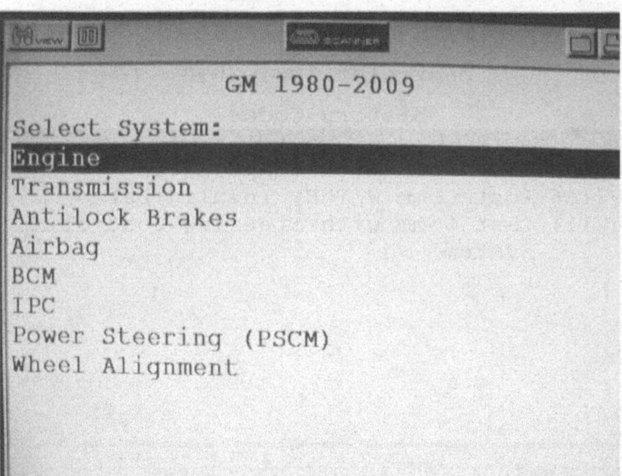

FIGURE 24-69 The systems available will vary depending on the year, make, and model vehicle you are working on.

■ ***OE Scan Tools.*** Many shops have OE or manufacturer scan tools such as the GM MDI/GDS, Toyota Techstream, Chrysler wiTech, and Ford IDS/WDS systems. These are PC-based scan tools, meaning that the software runs from a laptop. These systems typically allow for wireless data transfer between the PC and the vehicle interface.

While there are similarities between using an aftermarket and an OE scan tool, you will find that the amount of data and other diagnostic functions available with an OE tool are much greater. The advantage of using an aftermarket scan tool is that the tool is basically the same no matter what vehicle you are using it on, so you only have to learn how to use one tool. All the OE scan tools are different, and they often require a greater learning curve to become proficient in using their full capabilities. Examples of a GM GDS, Toyota Techstream, and Ford IDS interface are shown in **Figure 24-70**, **Figure 24-71**, and **Figure 24-72**.

Refer to Chapter 24 in the Lab Manual for exercises on using various types of scan tools.

■ ***OBD III.*** The successes of OBD II include a much more robust onboard monitoring system and increased availability of information for technicians. However, just as with OBD I systems, there is still an issue regarding the delay between detection of an emissions fault and its actual repair. In places with mandated emissions testing, a vehicle with an illuminated MIL and emissions-related faults will eventually be tested, and likely repaired. In parts of the country that do not require emissions testing for vehicle registration, there is no incentive or requirement that any fault be repaired.

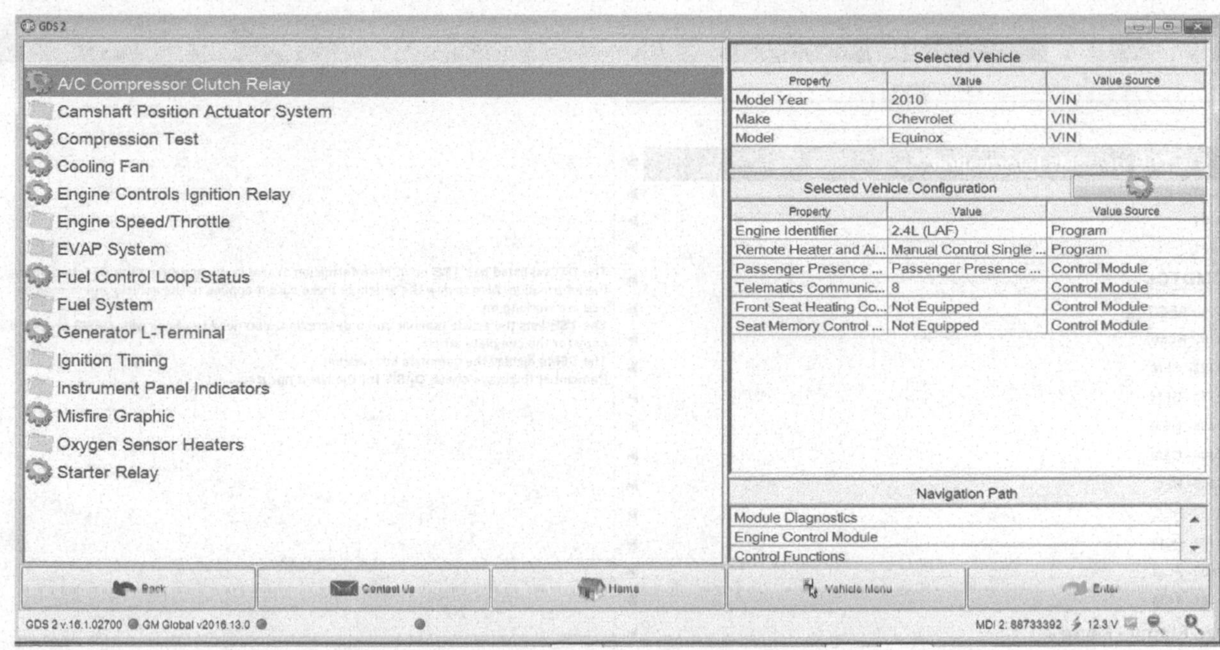

FIGURE 24-70 A screen from the General Motors Global Diagnostic System (GDS) scan tool interface. This screen shows control functions available for the PCM.

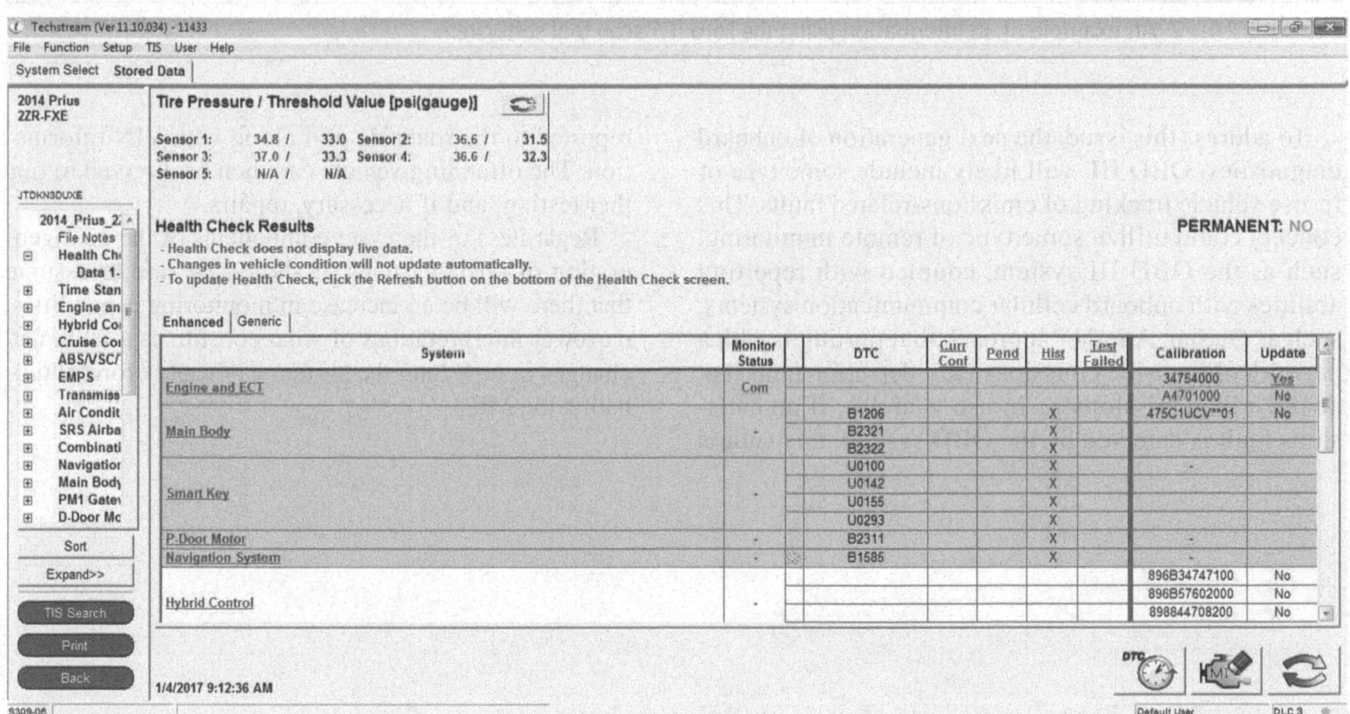

FIGURE 24-71 A screen from the Toyota Techstream scan tool interface. This is the initial screen once the scan tool begins communicating with the network.

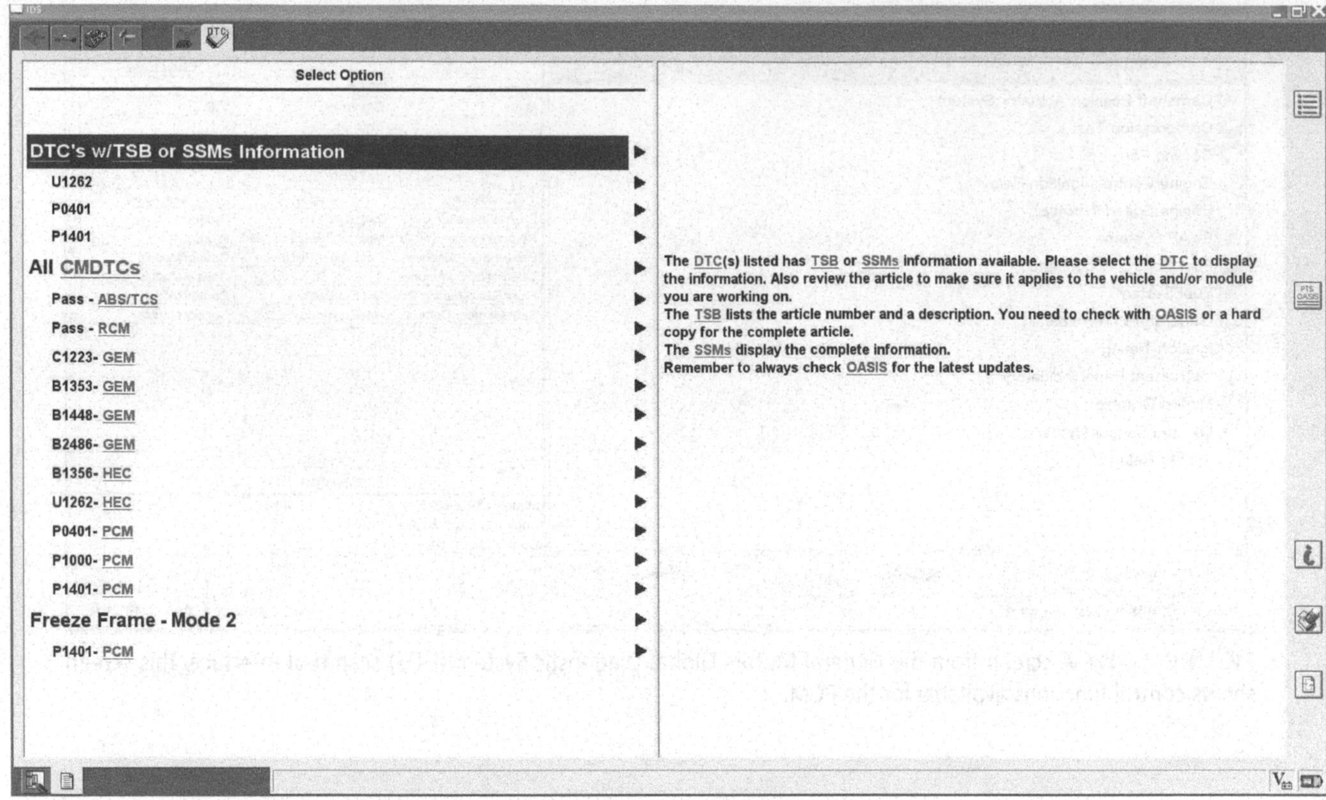

FIGURE 24-72 An example of an information using the Ford IDS scan tool software.

To address this issue, the next generation of onboard diagnostics, OBD III, will likely include some type of in-use vehicle tracking of emissions-related faults. This concept could utilize some type of remote monitoring, such as the OBD III system, coupled with reporting abilities with onboard cellular communication systems, such as OnStar. Another approach to reporting is with a network of roadside emissions fault detection units that analyze the emissions of in-use vehicles. If an emissions fault is detected by the OBD system, this fault is reported to the roadside unit along with VIN information. The offending vehicle can then be directed to further testing, and if necessary, repairs.

Regardless of the exact components of the next generation of OBD systems, it should be safe to assume that there will be an increase in monitoring capabilities, narrower interpretations of what constitutes a fault, and changes in how long the system can wait before illuminating the MIL.

SUMMARY

Many vehicles no longer have fuel filters that are replaced as part of the maintenance schedule.

Depressurized the fuel system before removing the fuel filter to prevent fuel from spraying out.

Let the engine cool down before removing the spark plugs to prevent damage to the cylinder head.

Clean rust and debris from around the spark plugs holes before removing the plugs.

Check all positive crankcase ventilation (PCV) system hoses to see if they are collapsed and replace if necessary.

The first generation of onboard diagnostic systems varied greatly from manufacturer to manufacturer.

The malfunction indicator light (MIL) comes on if emissions increase 150% over the federal test parameter.

OBD II must be able to determine if a misfire is occurring, determine the severity of the misfire, and alert the driver.

Global or Generic OBD II data provides codes, data, and emissions-related information accessible by a generic scan tool.

The enhanced data is VIN specific and provides information based on the vehicle manufacturer.

Type A misfires are catalyst-damaging misfires.

Type B misfires are emissions-threatening misfires.

Fuel control is represented by short-term fuel trim (STFT) and long-term fuel trim (LTFT).

Noncontinuous monitors require certain enable criteria to be met before the monitor can complete.

DTCs can be current codes, pending codes, or history codes.

REVIEW QUESTIONS

1. A _______________ is a driving pattern that allows all of the monitors to run.

2. When the ECM first detects a fault, a _______________ code is set.

3. Always start spark plugs into the cylinder head by _______________ to prevent damage to the threads.

4. The _______________ _______________ ventilation system removes blowby gases from the engine and burns them.

5. The two modes of scan tool operation are _______________ and _______________ OBD.

6. When discussing OBD systems. *Technician A* says an engine misfire code would have a P prefix. *Technician B* says P codes indicate a problem with a powertrain system. Who is correct?

a. Technician A **c.** Both A and B
b. Technician B **d.** Neither A nor B

7. *Technician A* says antiseize should be applied to the entire threaded section of a spark plug before installing the plug. *Technician B* says excessive application of antiseize could cause a misfire. Who is correct?

a. Technician A **c.** Both A and B
b. Technician B **d.** Neither A nor B

8. Which of the following is not one of the OBD II system monitors?

a. EVAP **c.** Cruise control
b. EGR **d.** Fuel control

9. *Technician A* says that even if a misfire code is set, all the monitors will still run. *Technician B* says that a misfire code may cause the MIL to flash. Who is correct?

a. Technician A **c.** Both A and B
b. Technician B **d.** Neither A nor B

10. A vehicle equipped with OBD II has an illuminated MIL. *Technician A* says the code can be checked using generic mode. *Technician B* says only generic codes will display using a generic or Global OBD II scan tool. Who is correct?

a. Technician A **c.** Both A and B
b. Technician B **d.** Neither A nor B

Drivetrains and Transmissions

Chapter Objectives

At the conclusion of this chapter, you should be able to:

- Describe automatic and manual transmission operation. (ASE Education Foundation MLR 2.A.1, 3.A.1, & 3.C.1)
- Drain and replace transmission fluid and filters. (ASE Education Foundation MLR 2.B.4)
- Identify, check, and adjust transmission fluid levels. (ASE Education Foundation MLR 8.C.1)
- Inspect for fluid leaks. (ASE Education Foundation MLR 2.A.4 & 3.A.2)
- Inspect and adjust transmission linkage. (ASE Education Foundation MLR 2.B.1)
- Inspect powertrain mounts. (ASE Education Foundation MLR 2.B.3)
- Describe continuously variable transmission and hybrid transmission operation. (ASE Education Foundation MLR 2.C.1 & 2.C.2)
- Drain and refill manual transmissions and final drive units. (ASE Education Foundation MLR 3.A.2)
- Inspect clutch fluid level, check for leaks, and bleed a hydraulic clutch system. (ASE Education Foundation MLR 3.B.1 & 3.B.2)
- Describe the operation of electronically controlled manual transmissions. (ASE Education Foundation MLR 3.C.1)
- Inspect and service front-wheel drive half shafts. (ASE Education Foundation MLR 3.D.2)
- Check locking hub assemblies. (ASE Education Foundation MLR 3.D.3)
- Inspect the differential and refill with the correct lubricant. (ASE Education Foundation MLR 3.E.1, 3.E.2 & 3.E.3)

KEY TERMS

automatic transmission fluid (ATF)
clutch disc
constant velocity joints
continuously variable transmission (CVT)
differential
dual-clutch transmission (DCT)
limited-slip differential (LSD)
planetary gearset
power-split device
pressure plate
release bearing
shift solenoid
torque converter

To make the power produced by the engine usable, the rotary motion (and torque) of the engine's crankshaft travels through a transmission and drivetrain to the drive wheels. The word powertrain is typically used to talk about the engine and transmission, and drivetrain refers to the configuration or layout of the powertrain and how the wheels are driven.

Drivetrain Types

The job of the drivetrain is to take the power produced by the engine and transfer it to the driving wheels. The drivetrain generally includes components such as the transmission, differential, and driveshaft(s) or axles. There are two basic drivetrain configurations: rear-wheel drive (RWD) and front-wheel drive (FWD). These may be combined into either an all-wheel drive (AWD) or a four-wheel drive (4WD) arrangement as well.

DRIVETRAIN CONFIGURATIONS

How a vehicle's drivetrain is laid out defines the type and use of that vehicle. Though some cars and SUVs are available as AWD models, the basic layout of how the wheels are driven is based on if the vehicle will be RWD or FWD. This is because the placement of components such as differentials and axles require space under the vehicle. Space that is taken up by drivetrain components reduces space for passengers, cargo, and other items.

■ ***Rear-Wheel Drive.*** Though not quite as popular as it once was, RWD is still used in modern vehicle production (**Figure 25-1**). Nearly all full-size trucks and SUVs use either a RWD or a 4WD configuration. Many sports cars, such as the Camaro, Mustang, Challenger, and performance sedans are RWD.

Rear-wheel drive, also called a *longitudinal drivetrain,* places the driving components along the centerline of the vehicle (**Figure 25-2**). Power flows from the rear of the engine to the transmission, which is typically mounted directly to the back of the engine. A driveshaft connects the rear of the transmission to the differential (**Figure 25-3**).

The job of the differential is to drive the rear wheels. To do this, the power must make a 90-degree turn (**Figure 25-4**). The **differential** also allows the drive wheels to turn at different speeds while turning corners (**Figure 25-5**).

■ ***Front-Wheel Drive.*** The majority of modern cars and small SUVs are FWD models. This is because FWD provides better poor-weather traction than traditional RWD systems and allows for a lower and more spacious rear seating area due to not having a rear differential taking up space. An illustration of a FWD drivetrain is shown in **Figure 25-6**.

Front-wheel drive, also called *a transverse drivetrain,* incorporates the transmission and differential into one unit, called *a transaxle.* The transaxle bolts directly to the rear of the engine just as in a RWD configuration. However, in a transaxle, power from the transmission section is transmitted to the differential through either a gear-to-gear or a chain drive (**Figure 25-7**).

To get power from the transaxle to the wheels, two independent driveshafts, often called *half shafts* or *constant velocity (CV)* shafts are used (**Figure 25-8**). In

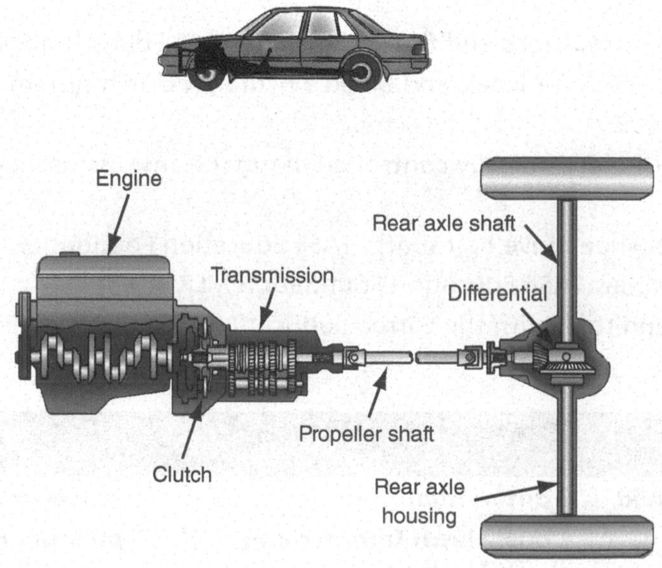

FIGURE 25-1 This illustrates how a rear-wheel drivetrain is configured.

FIGURE 25-2 An example of a RWD drivetrain.

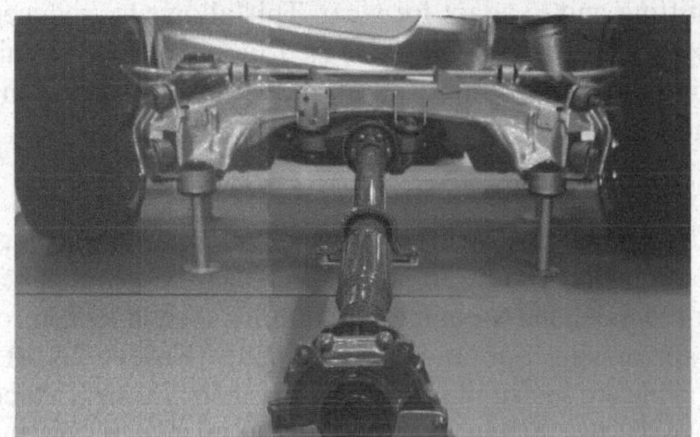

FIGURE 25-3 The driveshaft connects the transmission to the rear differential.

addition to driving the front wheels, CV shafts also must allow for left/right steering and up and down suspension movements (**Figure 25-9**). Unlike U-joints used in rear wheel drivetrains, **constant velocity joints** and shafts allow for the transfer of power and torque without any change in rotational speed.

Transmissions

The job of any type of transmission is to allow the engine to run at its most efficient speeds so that power is not wasted. This is accomplished by changing the output speed of the transmission compared to the input speed

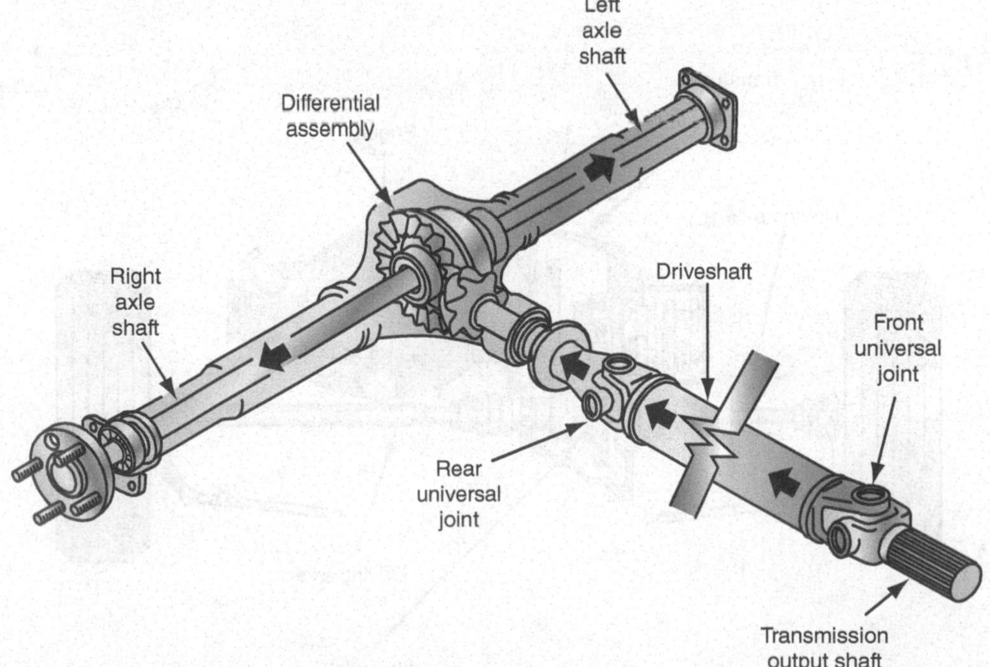

FIGURE 25-4 This illustrates the operation of a differential. In straight-line driving, the power from the driveshaft splits and drives each rear wheel.

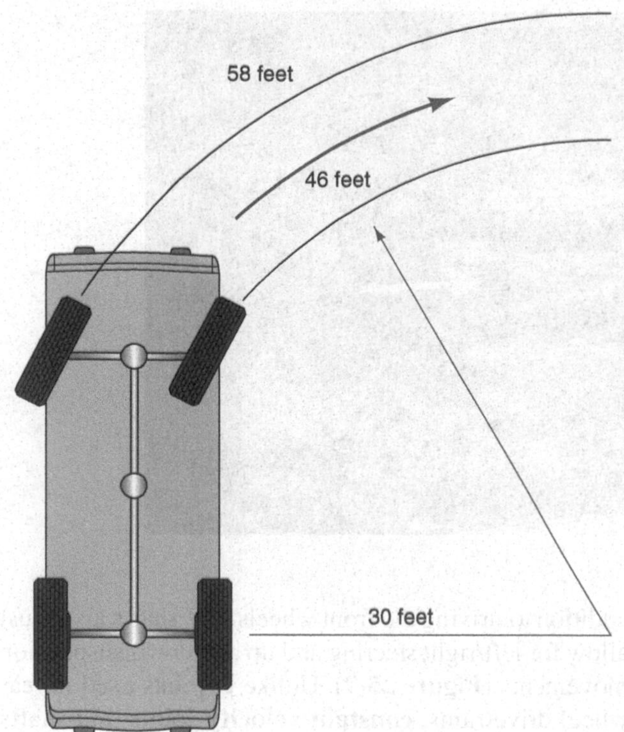

FIGURE 25-5 To be able to smoothly go around corners, the differential must allow the wheels to spin at different speeds.

from the engine's crankshaft. By changing gear ratios and speeds, the transmission allows the engine to use its power more efficiently.

How the transmission accomplishes this task depends on whether it is an automatic, continuously variable, manual, or electronically controlled manual transmission, such as the dual-clutch transmissions now in popular use.

Automatic Transmissions

Automatic transmissions were not widely available or used in automobiles until the mid-1950s. Though modern automatic transmissions are more advanced than ever before, the principles and basic components have not changed since the first automatic transmissions.

Until the 1980s, automatic transmissions really were automatic, meaning that their operation was controlled within the transmission itself. As onboard computer systems became standard, the automatic transmission also began to be controlled by a powertrain control module or PCM. The PCM uses inputs such as throttle position, vehicle speed, engine coolant and transmission oil temperature, and other information to control transmission shifting and to adapt for wear. This allows for reduced exhaust emissions and increased fuel economy by tailoring engine and transmission operation based on the driving conditions.

PRINCIPLES OF OPERATION

The major components of an automatic transmission are a torque converter, oil pump, planetary gears, clutches and bands, and a complex hydraulic system to control transmission operation and shifting (**Figure 25-10**). Visible are sets of planetary gearsets and clutches, which provide the different gears of the transmission. At the front of the transmission, where it connects to the engine, is the torque converter.

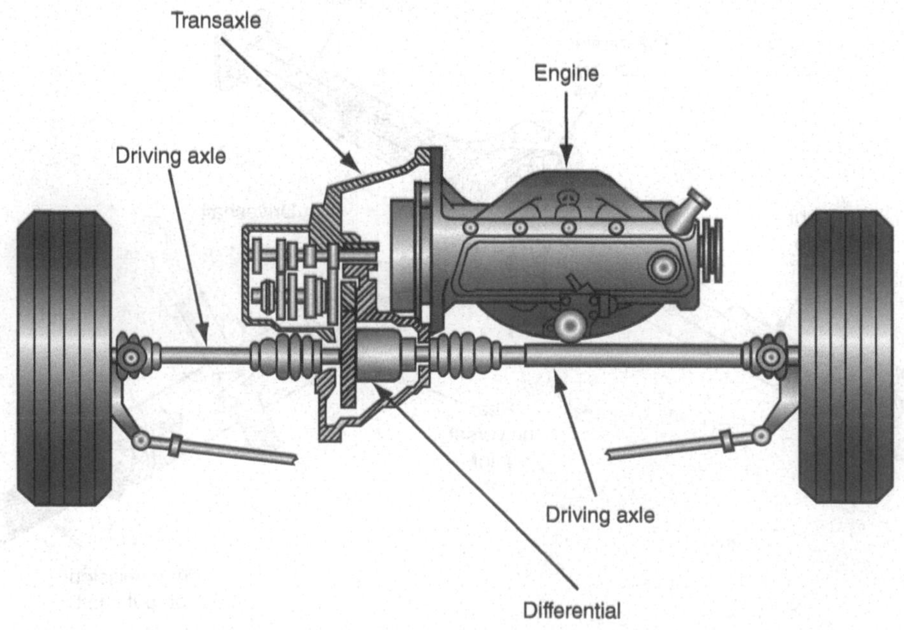

FIGURE 25-6 This illustrates how a front-wheel drivetrain is configured.

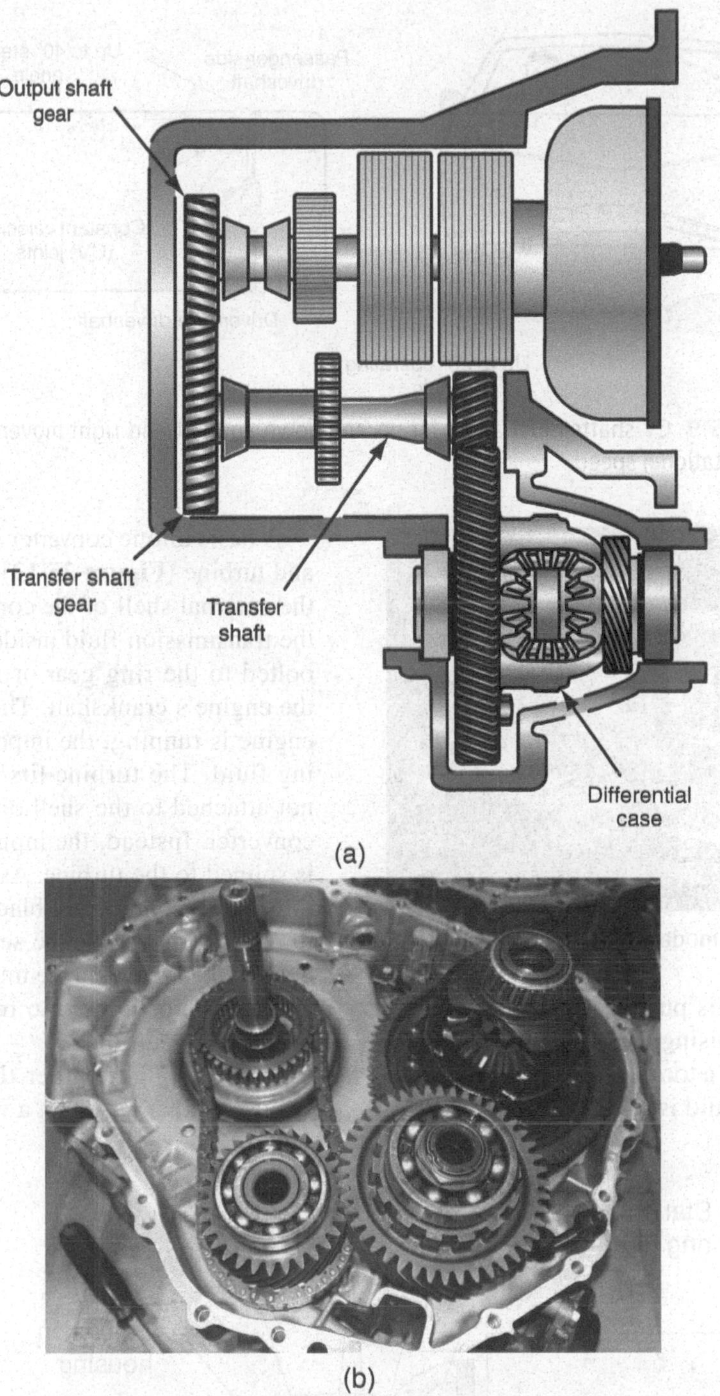

FIGURE 25-7 (a) This illustrates how power flows out of the transmission and through the differential to drive the front wheels. (b) An example of a differential from a FWD car.

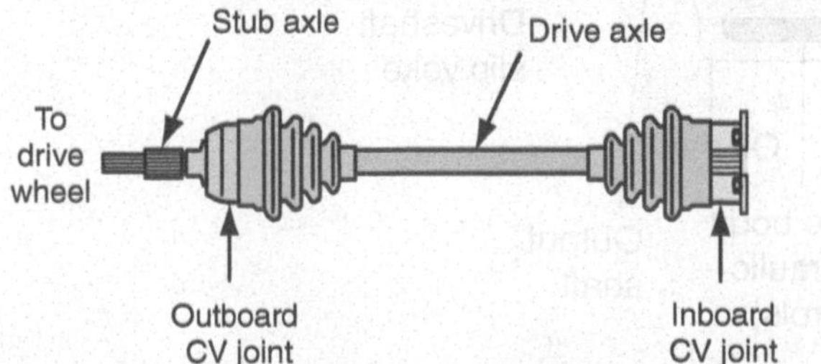

FIGURE 25-8 A CV shaft contains two types of joints and an axle shaft to connect the transaxle to the wheels.

■ ***Torque Converters.*** A **torque converter** is used to join the engine's crankshaft to the input shaft of an automatic transmission (**Figure 25-11**). A torque converter is a fluid coupling, meaning that it couples or joins the engine and transmission together by use of a fluid, in this case, automatic transmission fluid. The easiest way to visualize the action of a torque converter is by examining the action of two fans facing toward each other (**Figure 25-12**). The fan on the left is driving the fan on the right. The left fan represents the engine. The

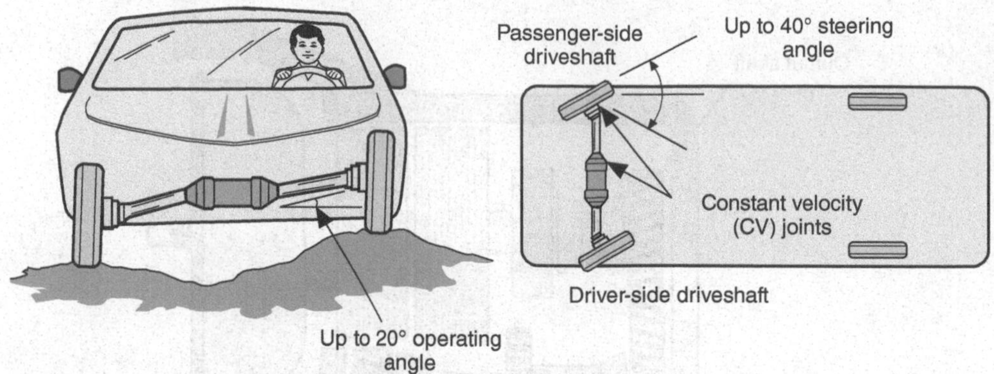

FIGURE 25-9 CV shafts must allow for up and down and left and right movements without changing rotational speed.

FIGURE 25-10 Inside of a modern automatic transmission.

air moved by the left fan is pushing against the blades of the fan on the right, causing them to spin. This same process takes place inside a torque converter except that automatic transmission fluid is used instead of air.

A basic torque converter includes the impeller, stator, and turbine (**Figure 25-13**). The impeller is made into the external shell of the converter and is used to pump the transmission fluid inside the converter. The shell is bolted to the ring gear or flex plate and is driven by the engine's crankshaft. This means that whenever the engine is running, the impeller is spinning and pumping fluid. The turbine fits inside the converter but is not attached to the shell or directly to any part of the converter. Instead, the input shaft of the transmission is splined to the turbine. As the fluid is pumped by the impeller, it contacts the blades of the turbine, causing it to rotate. This is like the second fan mentioned above. Sandwiched between the impeller and turbine is the stator. The stator is used to redirect fluid flow based on operating conditions.

If you thought earlier that using a fan to push air against another fan is not a very efficient way to spin the

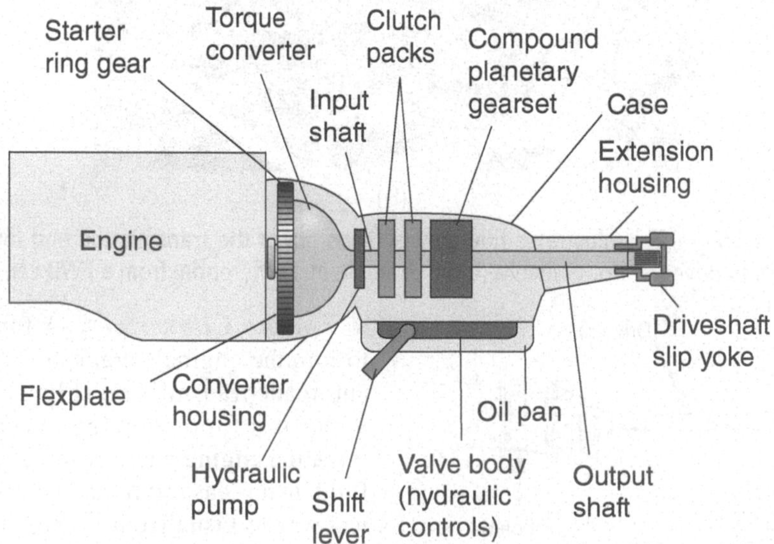

FIGURE 25-11 The torque converter is used to join the engine's crankshaft to the input shaft of an automatic transmission and allow the engine to idle with the transmission still in gear.

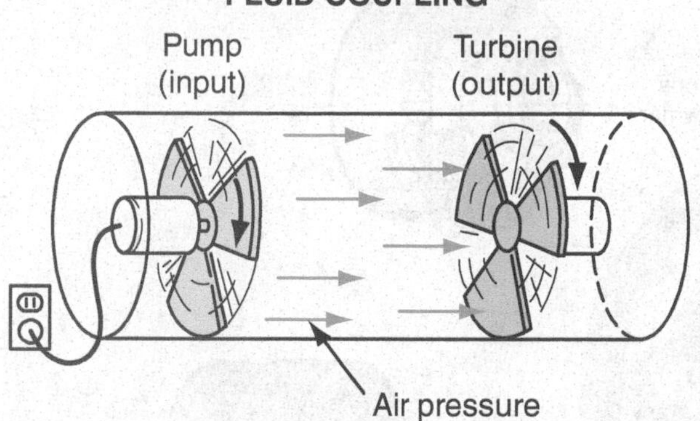

FIGURE 25-12 A torque converter uses a pump to push fluid to drive the input shaft of the transmission.

second fan, you are correct. The rotational speed of the second fan will not equal that of the first fan because of loss of force through the air. The same problem occurs in a torque converter. Losses from friction and heat result in about a 90% efficient transfer within the converter. This means that an impeller spinning at 1,000 rpm would move enough fluid for the turbine to spin at about 900 rpm. To correct this loss, modern converters use a fourth component, called *a torque converter lock-up clutch*. When applied, this clutch locks the impeller and turbine together creating an overdrive operating condition. During steady-throttle cruising, this allows the transmission to lock to the engine, reducing engine rpm and saving fuel.

The rear of the shell of the torque converter installs into the front transmission oil pump. Because the shell spins when the engine is running, the torque converter is driving the transmission oil pump also.

■ ***Planetary Gears.*** A **planetary gearset** is made of three gears meshed together, the ring gear, the sun gear, and planetary gears (**Figure 25-14**). In operation, one gear is the driving gear, one gear is the driven gear, and one gear is held in place. To allow for multiple gear ranges, planetary gears are combined into compound gearsets.

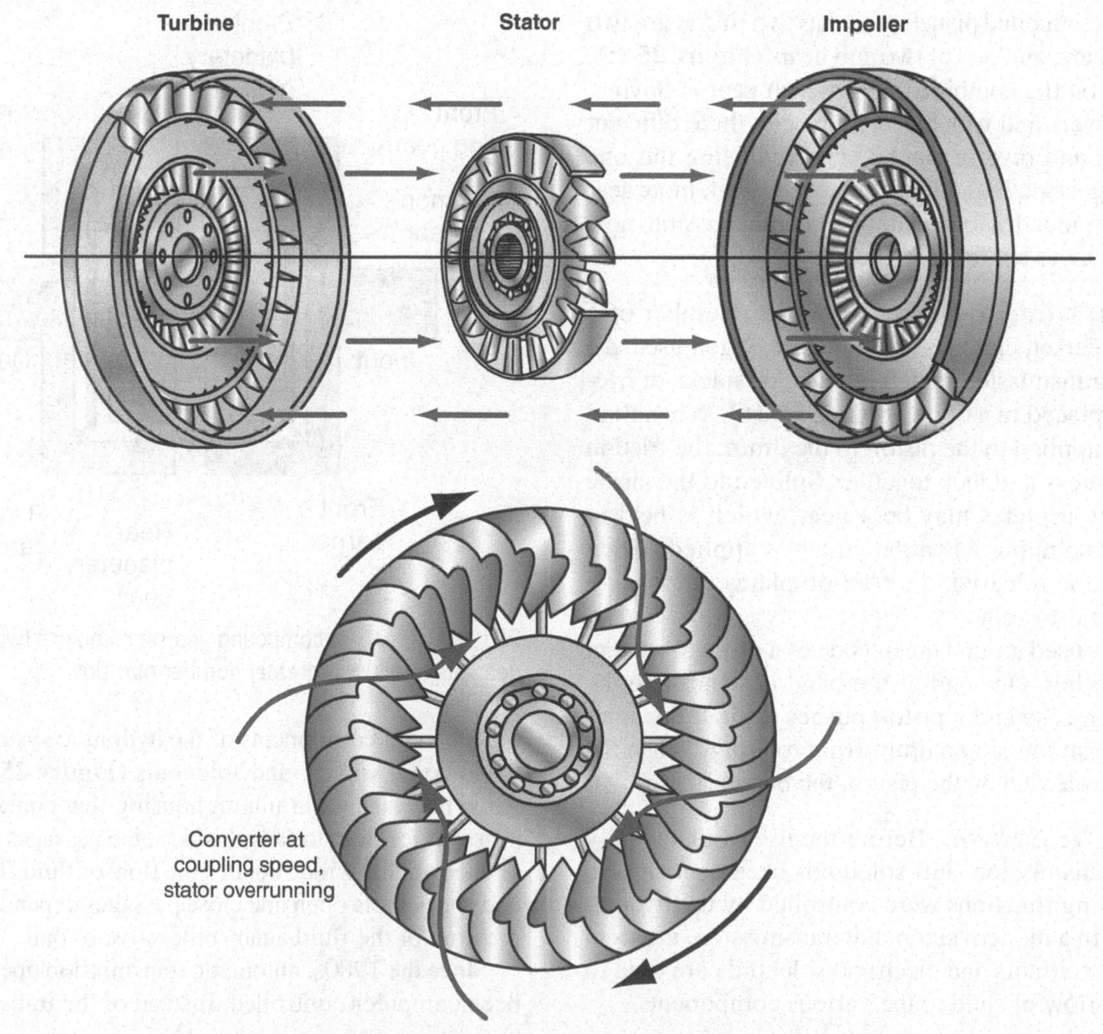

FIGURE 25-13 The fluid pumped by the impeller turns the turbine. The turbine drives the input shaft of the transmission. The stator is used to increase the efficiency of the torque converter.

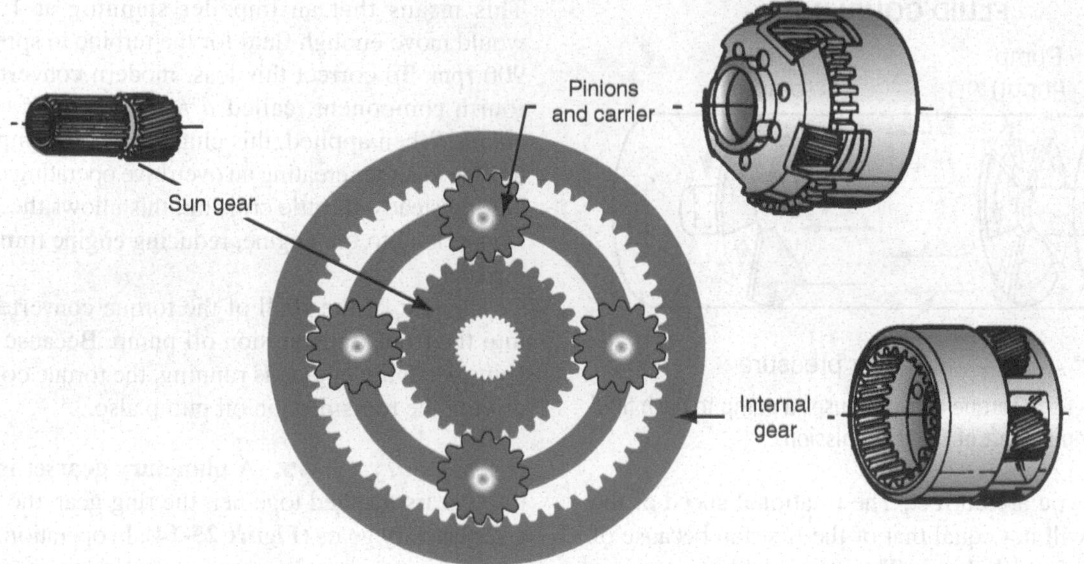

FIGURE 25-14 A simple planetary gearset. One part is held, one drives and the other is driven. Depending on which component does what, various speeds can be achieved.

A common compound planetary set has two ring gears, two planetary gears, and one or two sun gears (**Figure 25-15**). Depending on the combination of which gear is driving, which is driven, and which gears are held, three different gear ranges and reverse can be achieved using this one compound gearset. By adding a second gearset, more gear ratios can be added. Modern automatic transmissions now range from six to ten forward speeds.

■ ***Clutches and Bands.*** To hold a member of a planetary gearset, clutches and bands are often used. An automatic transmission clutch is made of stacks of friction plates placed in a drum (**Figure 25-16**). When fluid pressure is applied to the piston in the drum, the friction plates compress and lock together. Splined to the inside of the friction plates may be a gear, which is held in place from spinning when the clutch is applied. When the pressure is released, the friction plates unlock and allow the gear to spin.

A band is used around the outside of a drum to prevent it from rotating. One end of the band is secured to the transmission case and a piston pushes against the other end of the band to stop a drum from rotating. When the pressure is released on the piston, the band releases.

■ ***Hydraulic System.*** Before the use of electrically operated transmission shift solenoids became standard, all the shifting functions were controlled by the hydraulic system. In a modern automatic transmission, a system of hydraulic circuits and electrical solenoids are used to control the flow of fluid to the various components.

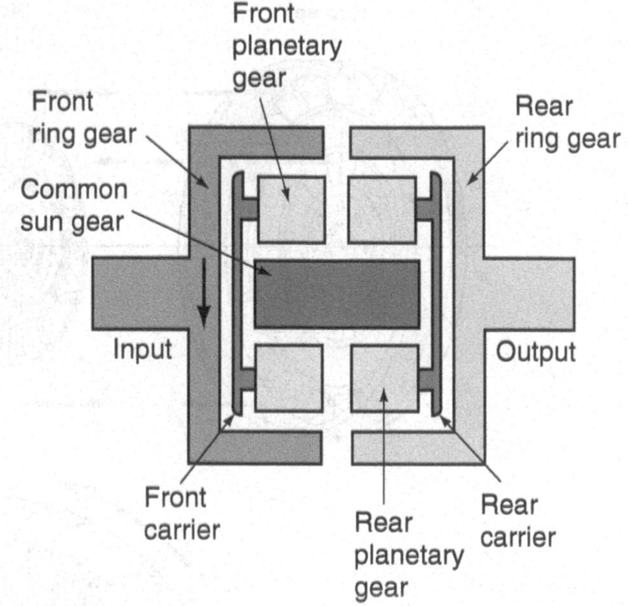

FIGURE 25-15 A compound gearset combines two planetary gearsets to allow a greater number of ratios.

The main components of the hydraulic system are the valve body, valves, and solenoids (**Figure 25-17**). The valve body is an aluminum housing that contains many passages for fluid to flow. Inside some passages are valves or check balls, which control the flow of fluid. The valves and check balls open and close passages depending on the pressure of the fluid against the valve or ball.

Since the 1980s, automatic transmission operation has been computer controlled. Instead of the transmission's

FIGURE 25-16 A clutch in a modern automatic transmission. When applied, the clutch is used to hold a member of a planetary gearset.

hydraulic system controlling upshifts and downshifts, the PCM commands a collection of shift solenoids to control shifting. A **shift solenoid** is electrically operated, like a starter solenoid. Located in the valve body, solenoids are used to open and close hydraulic passages. When a solenoid allows fluid to pass, the fluid can be used to activate a clutch or band. The PCM uses data from the various engine and transmission sensors to fine-tune transmission operation for increased performance and fuel economy.

Also part of the hydraulic system is the transmission cooler and cooler lines, illustrated in **Figure 25-18**. The transmission cooler is often located in the engine radiator, though separate external coolers are also popular. Steel lines and rubber hoses connect the transmission to the cooler. Just like engine oil, if it is overheated, transmission fluid loses its lubricating properties and begins to bake onto internal components. Coolers are especially important on cars, trucks, and SUVs that perform any type of towing or hauling of heavy loads as the additional weight increases the operating temperature of the transmission.

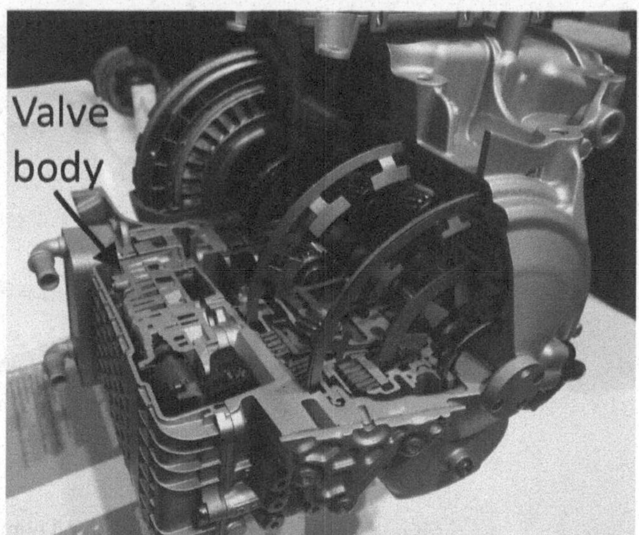

FIGURE 25-17 The valve body controls fluid flow inside the transmission.

CONTINUOUSLY VARIABLE TRANSMISSIONS

A **continuously variable transmission (CVT)** is a type of automatic transmission, but instead of having gears, clutches, and bands, it uses two variable-diameter pulleys and a belt.

■ ***Principles of Operation.*** The idea behind a CVT transmission is that by using two pulleys of varying diameters, the transmission can constantly vary the gear ratio so engine efficiency can be increased. For example, if the most efficient rpm range for an engine to operate in is from 2,500 to 3,000 rpm, an automatic transmission must shift gears to keep the engine running within that range. A CVT transmission varies the diameter of the input and output pulleys so that the engine speed can remain within that rpm range as much as possible. This eliminates shifting and keeps the engine running at its most efficient speed. An inside look at a CVT is shown in **Figure 25-19**.

The input (driving) and output (driven) pulleys can change diameter, which changes the "gear" ratio between them (**Figure 25-20**). The driving pulley is driven by the engine crankshaft, and the driven pulley is attached to the output shaft and drive wheels. A steel link chain or belt is used with the pulleys (**Figure 25-21**). Depending on the design, a CVT may or may not have a torque converter.

One negative of the CVT is that, because it does not shift, the familiar feeling of shifting gears is absent. For some people, this is hard to get used to. To solve this concern, some manufacturers have programmed "shift points" into their transmission control systems to

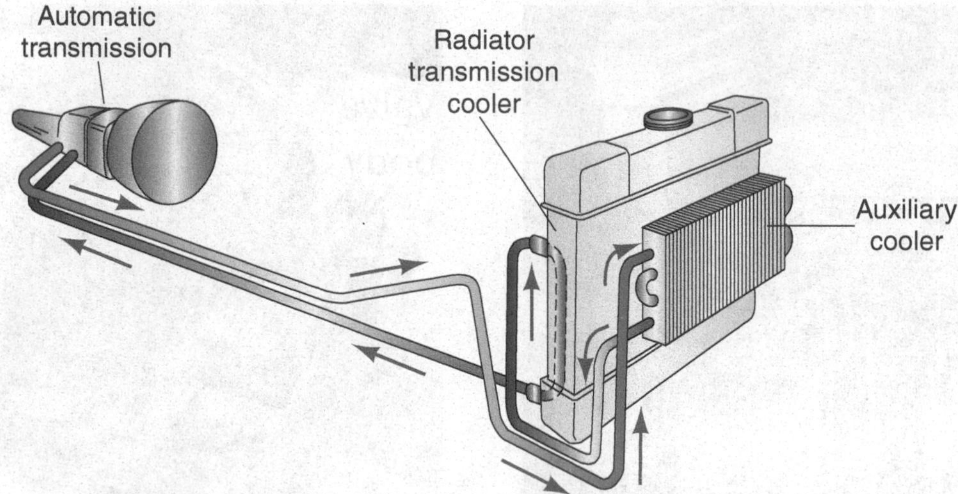

FIGURE 25-18 The transmission oil pump sends fluid to the cooler, usually located in or in front of the radiator. Keeping the transmission fluid from overheating is crucial to the service life of the transmission.

replicate the feeling of changing gears. Other manufacturers offer a "manual" mode of shifting the CVT for a sportier feel.

HYBRID POWERTRAINS

Depending on the vehicle, a hybrid vehicle may have a manual transmission, an automatic transmission, a CVT, or a power-split device, which is not a transmission like any of the other three mentioned.

■ ***Hybrid Power-Split Device.*** A **power-split device** is a form of transmission, in that it transmits power from the engine to the drive wheels, but it also receives power input from the electric motors. **Figure 25-22** shows an illustration of the Toyota system used in the Prius and other hybrid electric vehicles (HEVs). You will notice that the core of the power-split device is a planetary gearset, discussed earlier in this chapter. This gearset is connected to the gasoline engine by the planetary gears, motor/generator 1 (MG1) by the sun gear, and motor/generator 2 (MG2) by the ring gear. The ring gear is the output to the drive wheels through the differential gears. With this arrangement, any of the three can be the input in the system, depending on operating conditions.

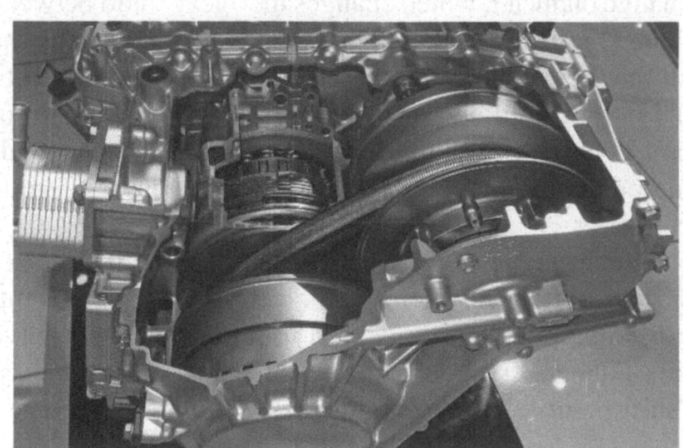

FIGURE 25-19 The inside of the "gears" in a CVT.

For example, when taking off from a stop under electric power only, MG2 is the input and is powering the drive wheels. During hard acceleration and cruising conditions, the gasoline engine is the input via the planetary gears. During deceleration and regenerative braking, the wheels drive MG2 as a generator to produce electricity to recharge the battery. MG1 is not used to propel the car; instead it is used to generate electrical power.

■ ***Honda IMA Systems.*** Another version of a hybrid transmission is the Integrated Motor Assist or IMA system used by Honda. In this arrangement, an electric motor is placed between the engine and transmission (**Figure 25-23**). The IMA configuration used the electric motor as an assist unit, providing additional power to drive the wheels and for engine stop/start capability. The transmissions used on IMA vehicles were either traditional manual or CVT units.

Honda has recently switched to an electric transaxle in their hybrid vehicles, similar to what is used by Toyota and other manufacturers. An example of a Honda Earth Dreams hybrid powertrain is shown in **Figure 25-24**.

Manual Transmissions

For many years, the only transmissions available in cars and trucks were manual transmissions. Over time, the manual transmission has mostly been replaced by the

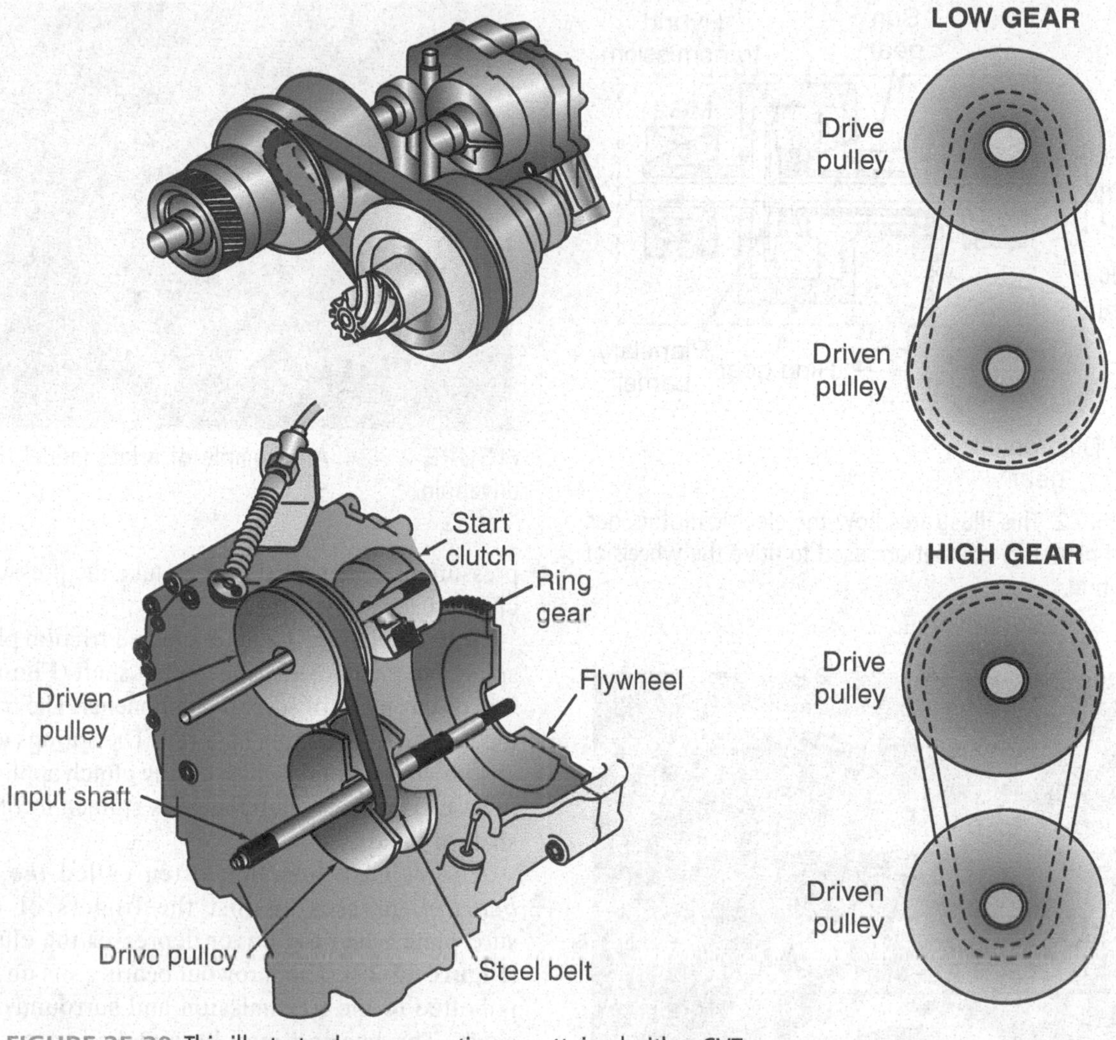

FIGURE 25-20 This illustrates how gear ratios are attained with a CVT.

FIGURE 25-21 The drive belt in a CVT.

automatic transmission, with a few exceptions. Many drivers, however, prefer a manual to an automatic, especially in sports cars.

With a manual transmission, a clutch connects the engine to the transmission and transmits the power from the crankshaft to the transmission. Inside a manual, several sets of gears create the different gear ratios necessary to use the power from the engine efficiently, and synchronizers allow the gear changes to be smooth and quiet.

CLUTCHES

A clutch assembly has a pressure plate, clutch disc, release bearing, cable or hydraulic system, and the flywheel (**Figure 25-25**). When the driver presses the clutch pedal, the clutch disengages. This unlocks the clutch assembly, and the clutch disc is no longer in contact with the flywheel and is not driving the input shaft of the transmission. When the clutch pedal is up, the clutch is applied. This means that the clutch assembly is locked together, and the flywheel is driving the input shaft of the transmission through the clutch disc.

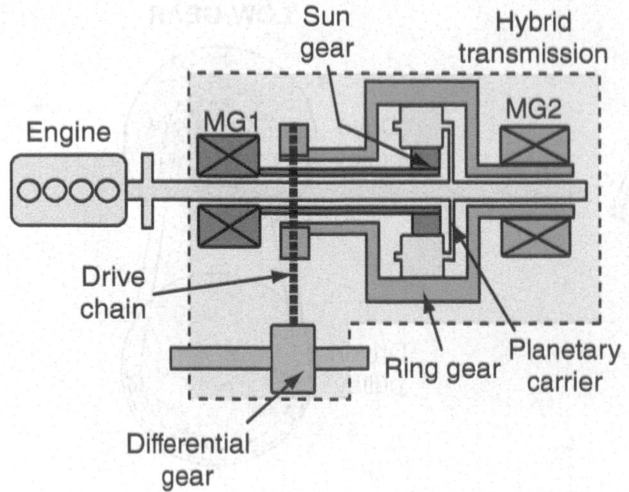

FIGURE 25-22 This illustrates how the electric motors, gas engine, and planetary gearset are used to drive the wheels of a Toyota hybrid.

FIGURE 25-23 This shows the rotor (center) and stator of a Honda IMA assist motor.

The **pressure plate** bolts to the flywheel and holds the clutch disc between the plate and the flywheel (**Figure 25-26**). The job of the pressure plate is to keep the clutch disc tight against the flywheel when the clutch is applied. When the driver presses the clutch pedal, the pressure plate is pulled back to take the pressure off the clutch disc (**Figure 25-27**).

FIGURE 25-24 An example of a late-model Honda HEV drivetrain.

The **clutch disc** is a double-sided friction plate that is splined to the transmission's input shaft (**Figure 25-28**). The disc is made of several components. The friction surface is like brake lining material. The springs are used to absorb vibration in the disc during clutch application and release. The center hub flange is splined to fit the input shaft of the transmission.

The **release bearing**, often called *the throwout bearing*, presses against the fingers of the pressure plate when the driver depresses the clutch pedal (**Figure 25-29**). The throwout bearing sits on a hub that is bolted to the transmission and surrounds the input shaft. The bearing may be attached to a shift fork, which is used to move the bearing against the pressure plate (**Figure 25-30**). In some vehicles, the throwout bearing and slave cylinder are both attached to the hub (**Figure 25-31**).

To activate the throwout bearing, the driver depresses the clutch pedal. The pedal may connect to a cable, called *the clutch cable*, which connects at the other end at the shift fork. Most vehicles use a hydraulic clutch system, consisting of a small master cylinder and a slave cylinder (**Figure 25-32**). The advantages of a hydraulic clutch are that no adjustments are necessary, and there is no cable to stretch. A disadvantage is that if a leak develops in the system, pressure is lost and the clutch cannot be released.

The flywheel is bolted to the crankshaft and the pressure plate bolts to the flywheel. Typically, a flywheel is heavy, about 20 pounds or more, and an inch or more in thickness. The reason for the mass is because the flywheel needs to be strong to hold the clutch and pressure plate without distorting and to serve as rotational mass to help keep the engine running at low rpm.

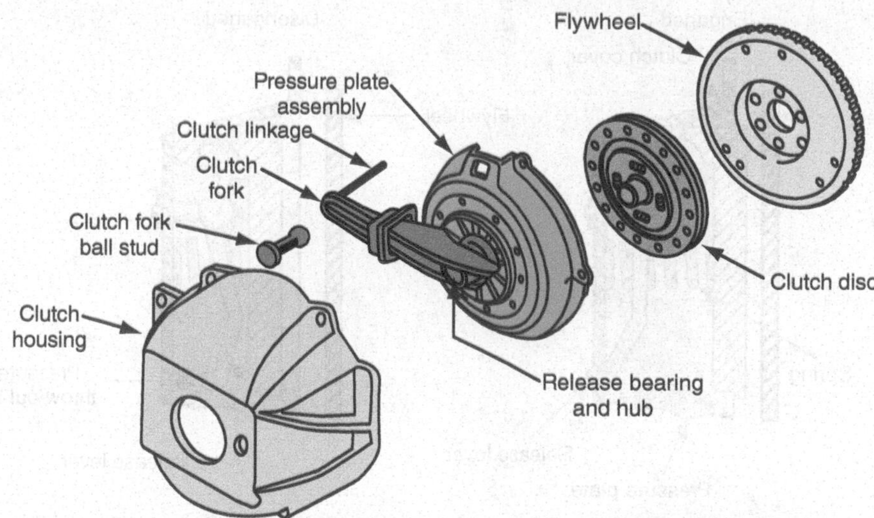

FIGURE 25-25 This illustration shows the components of the clutch assembly.

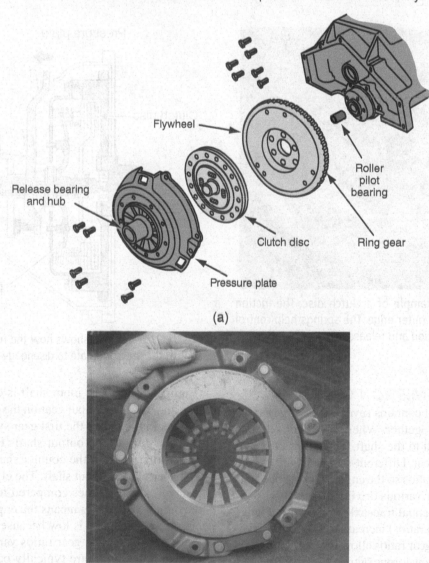

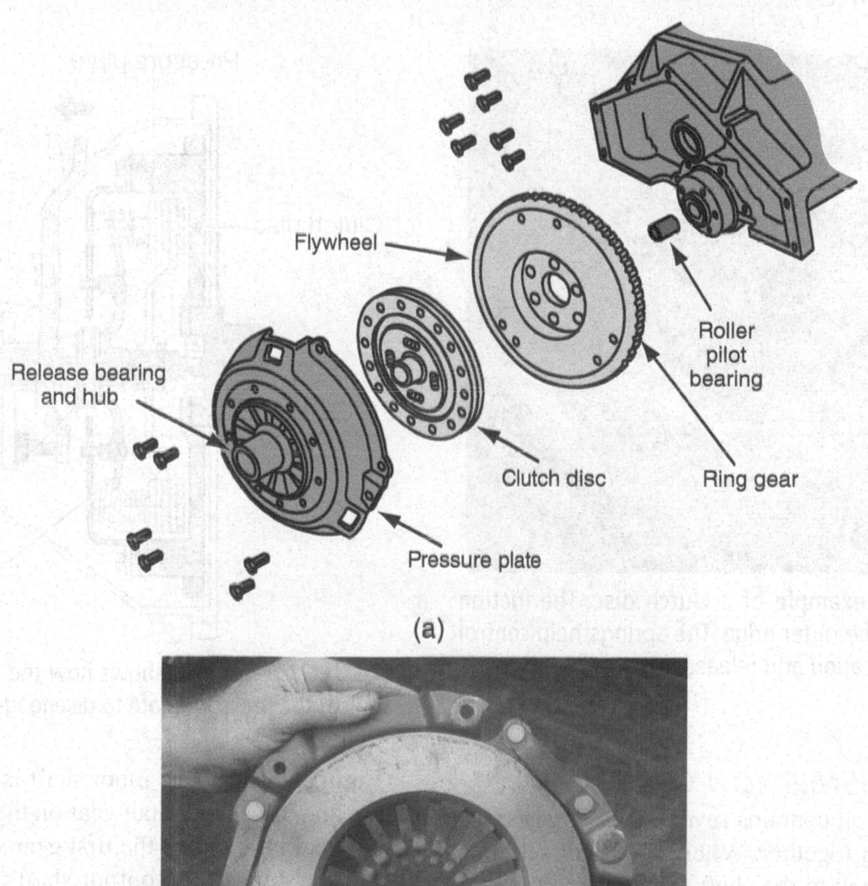

FIGURE 25-26 (a) The pressure plate holds the clutch disc tight against the flywheel. (b) The release bearing presses against the center of the pressure plate, which has fingers that attach to the friction plate.

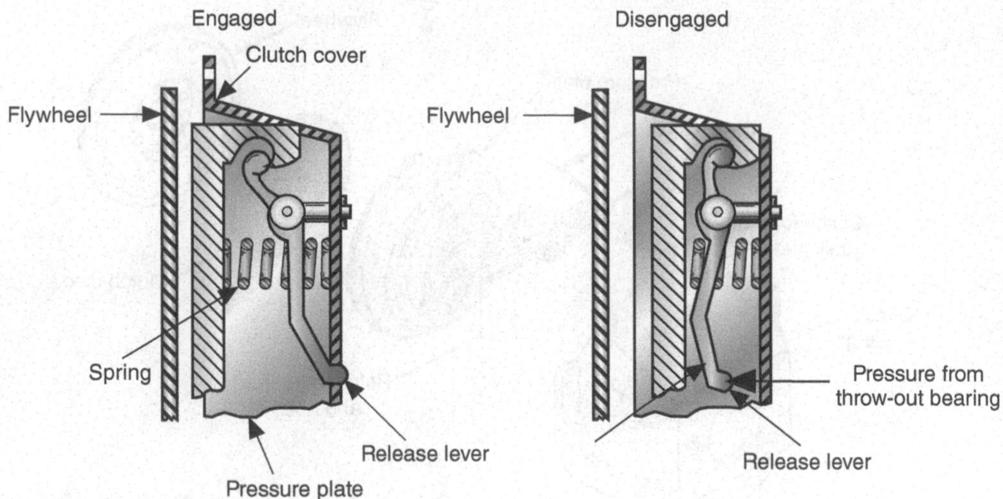

FIGURE 25-27 An illustration of how the pressure plate applies against and releases the clutch disc.

FIGURE 25-28 An example of a clutch disc. The friction surfaces are around the outer edge. The springs help control vibration during application and release.

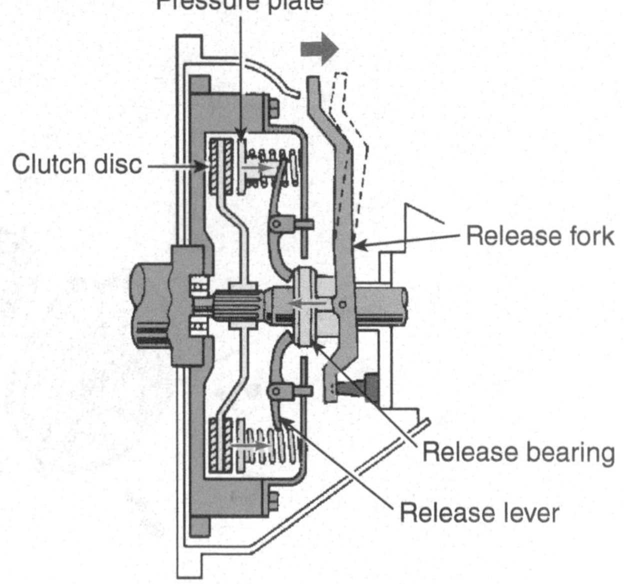

FIGURE 25-29 This shows how the release bearing applies against the pressure plate to disengage the clutch.

MANUAL TRANSMISSION OPERATION

A manual transmission contains several sets of gears that are in constant mesh together. When the driver selects a gear, the gear is locked to the shaft. This allows the gear to then drive another gear. Different-size gears are used to provide different gear ratios so the engine can operate at its most efficient rpms for various driving speeds. An example of the gears in a manual transaxle is shown in **Figure 25-33**. Low gears (high ratios) increase torque and accelerate quickly, while low gear ratios allow lower engine speed and economy but with low torque output. Most modern cars and light trucks have six forward gears and one reverse gear.

A manual transmission consists of an input shaft, output shaft, counter shaft, gears, and synchronizers (**Figure 25-34**). The input shaft is driven by the clutch disc and drives the input gear on the counter shaft. When first gear is selected, the first gear synchronizer and hub lock the gear to the output shaft (**Figure 25-35**). This allows first gear on the counter shaft to drive the large first gear on the output shaft. The effect is that the small gear turns many times compared to the larger gear on the output shaft. This means the engine revs up quickly but the output speed is low because of the difference in the gear ratios. First gear ratios vary depending on the type of vehicle but are typically between 3:1 and 4:1. This means that it takes three to four engine revolutions to make one revolution of the output shaft. This is why when in first gear the car has a lot of torque to accelerate,

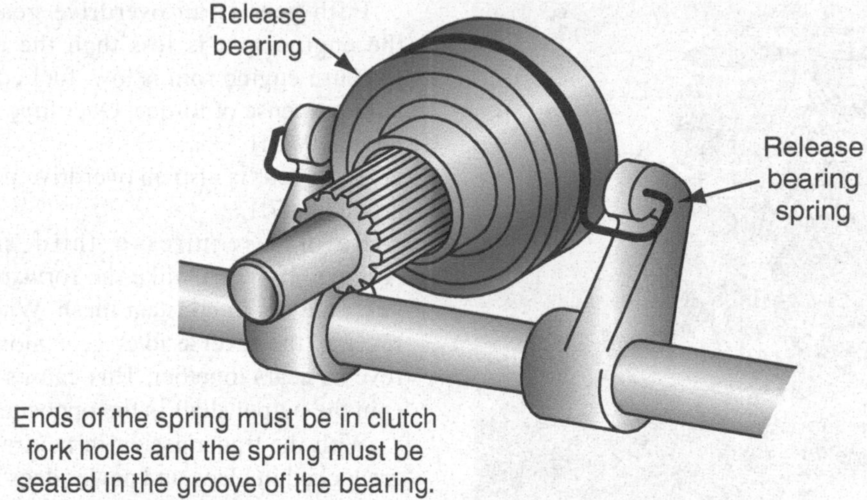

FIGURE 25-30 An example of a common release bearing configuration.

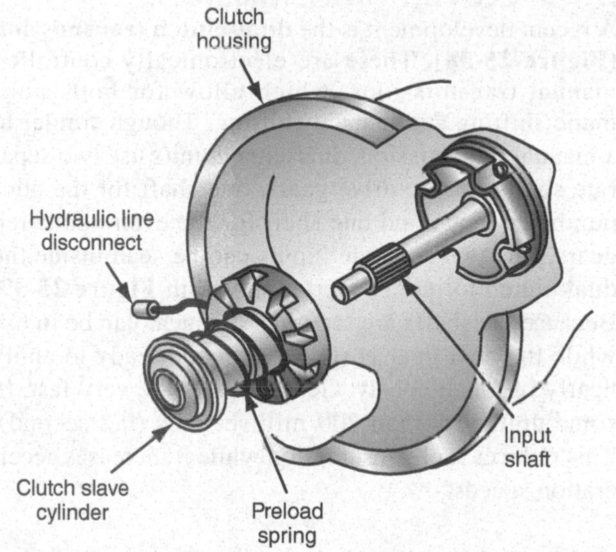

FIGURE 25-31 Some designs integrate the release bearing and slave cylinder into one unit, located inside the bell housing.

but does not reach a very high speed. How gear ratio affects vehicle speed is illustrated in **Figure 25-36**.

Synchronizers are used to allow gear shifting without the clash or grinding of gears. In modern transmissions, all forward gears are synchronized. A synchronizer allows smooth gear changes by matching the speed of the gear and the shift collar (Figure 25-35). As the driver shifts gears, the cone shape of the blocking ring provides friction against the cone on the gear. This friction causes the two parts to match speed. Once they are spinning at the same speed, the lock sleeve can slide over the teeth of the blocking rings and lock the gear to the shaft.

When second gear is selected, first gear is unlocked and second gear is locked. Second gear ratios are less than those of first, typically around 2:1. This still provides increased torque, but less than first gear. Second gear is for acceleration and allows for higher vehicle speed than first gear.

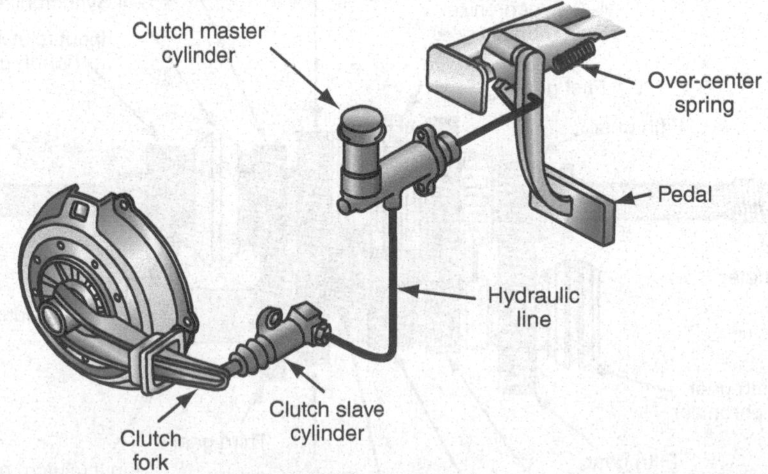

FIGURE 25-32 Most vehicles use a hydraulic clutch system. Brake fluid is used to transmit force from the clutch master cylinder to the clutch slave cylinder. This eliminates the adjustment required with clutch cables.

FIGURE 25-33 This shows how the gears are in constant mesh in a manual transmission/transaxle.

Third gear, has a lower ratio than second but still has some torque multiplication, yet it allows for a wider range of engine and vehicle speed. A typical third gear ratio would be around 1.3:1.

Fourth gear is often direct drive, meaning none of the gears are used. The input and output shafts are locked together, and engine rpm passes through the transmission without being changed. Direct drive has a gear ratio of 1:1.

Fifth gear is an overdrive gear. This means that the engine rpm is less than the output shaft speed. Because engine rpm is low, fuel economy is increased at the expense of torque. Overdrive gear ratios are often around 0.85:1.

Sixth gear is also an overdrive gear with ratios often around 0.57:1.

Reverse requires a third shaft and bearing (**Figure 25-37**). Unlike the forward gears, the reverse gears are not in constant mesh. When the driver selects reverse, the reverse idler gear moves to mesh the two reverse gears together. This causes the counter shaft to spin the output shaft in the opposite direction.

With the transmission in neutral, none of the gears are locked in place and power does not pass through the transmission.

DUAL-CLUTCH TRANSMISSIONS

A recent development is the **dual-clutch transmission** (**Figure 25-38**). These are electronically controlled manual transmissions, which allow for both automatic shifting and manual shifting. Though similar to a manual transmission, dual-clutch units use two separate shafts for the drive gears, one shaft for the odd-numbered gears and one shaft for the even-numbered gears. The two separate inputs can be seen inside the dual-clutch torque converter shown in **Figure 25-39**. Because the shafts are separate, one gear can be in use while the next gear is staged or made ready to apply nearly instantaneously. Gear changes are very fast, in some units less than 200 milliseconds (0.2 second). This reduces fuel consumption while it increases acceleration speeds.

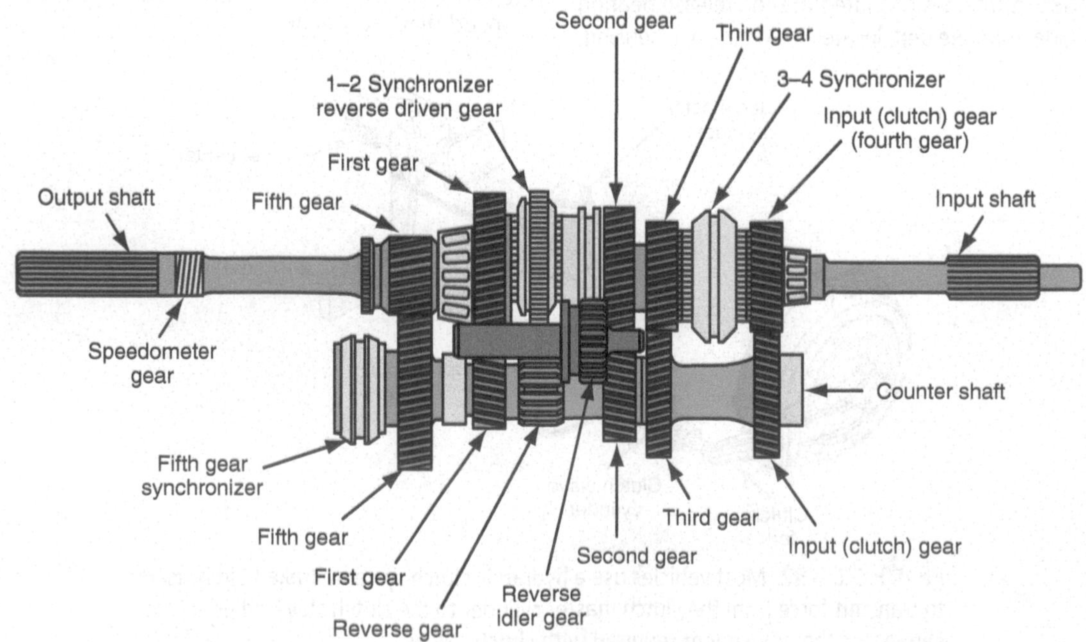

FIGURE 25-34 The components of a five-speed manual transmission.

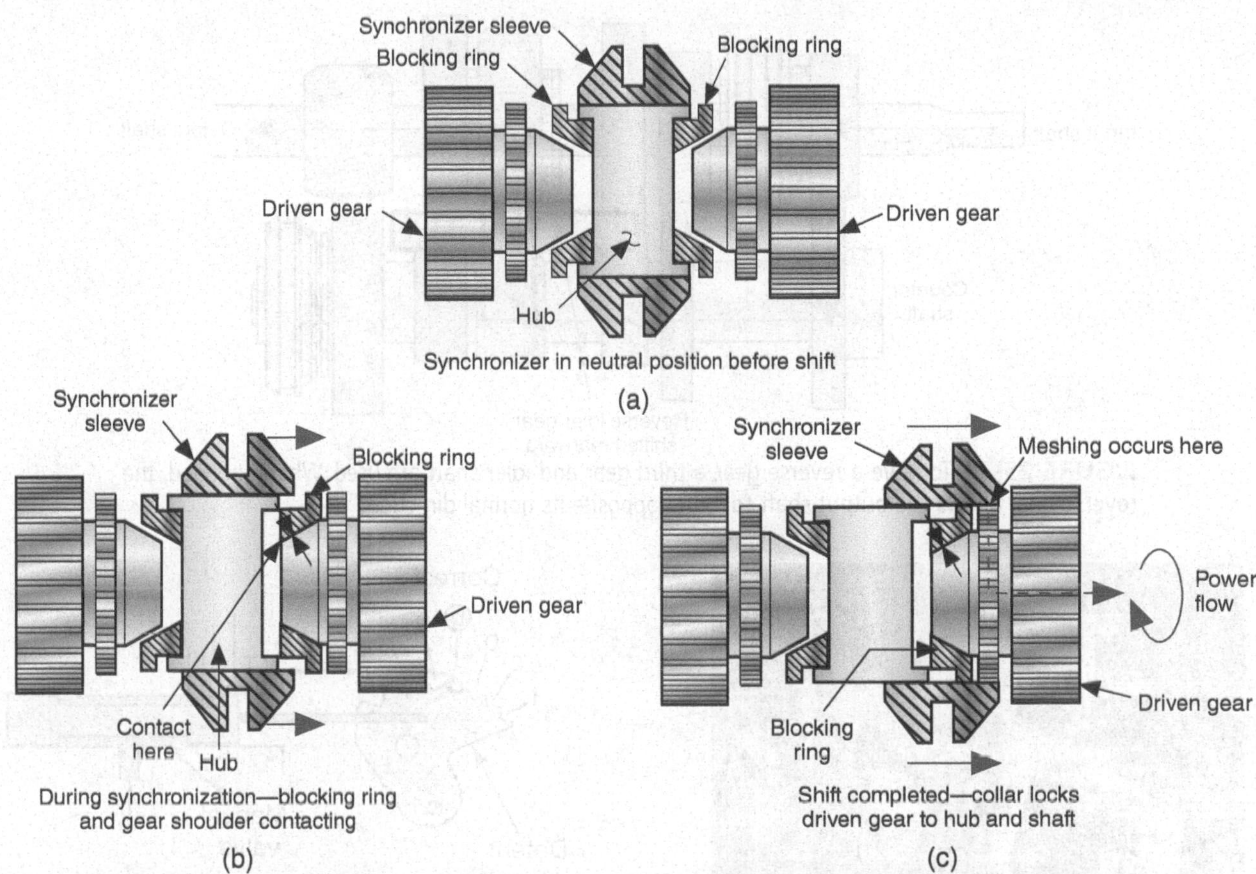

FIGURE 25-35 (a) A synchronizer and hub assembly are used to allow the smooth changes from one gear to another. (b) When shifting gears, the sleeve moves from the hub toward the blocking ring of gear being selected. (c) As the sleeve slides over the blocking ring, the two speeds match and the gear is locked to the shaft.

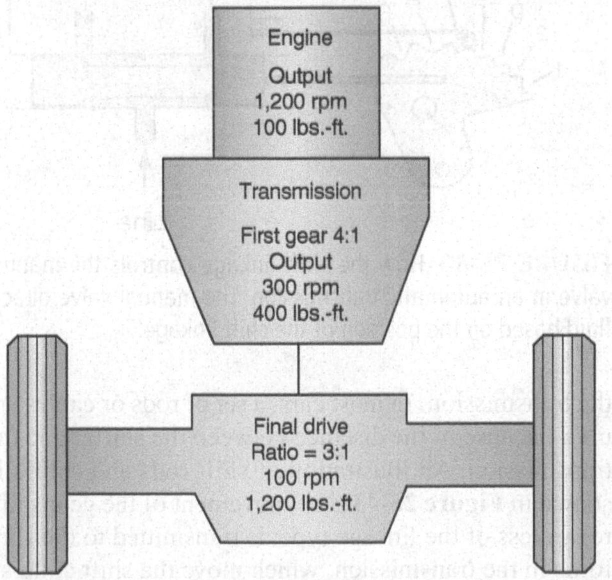

FIGURE 25-36 Power from the engine flows through the transmission and final drive gears. The final gear ratio depends on the gear ratios in the transmission and in the differential.

SHIFT LINKAGE

The shift linkage, whether mechanical or electrical, connects the driver to the transmission. Modern automatic transmissions can use mechanical linkage, cables, or no direct linkage at all; instead, the onboard computer system is the linkage. Manual transmissions can use cables, rods, or mount the gear shifter directly to the transmission.

■ ***Automatic Transmission Linkage.*** Many automatic transmissions use a shift cable. The cable is attached to the gear shifter at one end and to the transmission shift lever at the other. Inside the transmission, the shift lever is connected to the manual shift valve (**Figure 25-40**). The manual shift valve directs fluid through the valve body based on the position of the shift lever.

Many newer automatic transmissions do not use any linkage; instead, the PCM or transmission control module (TCM) is used. With this system, the gear shifter is an input device to the computer and is not physically connected to the transmission. The computer monitors shifter position and shifts the transmission accordingly. This is common on vehicles with manual shifting modes or paddle shifters (**Figure 25-41**).

■ ***Manual Transmission Linkage.*** Depending on the location of the transmission, the shift linkage may directly connect the shifter to the shift rods in the transmission, which is common in trucks. **Figure 25-42** shows an example of a gear shifter that connects directly into

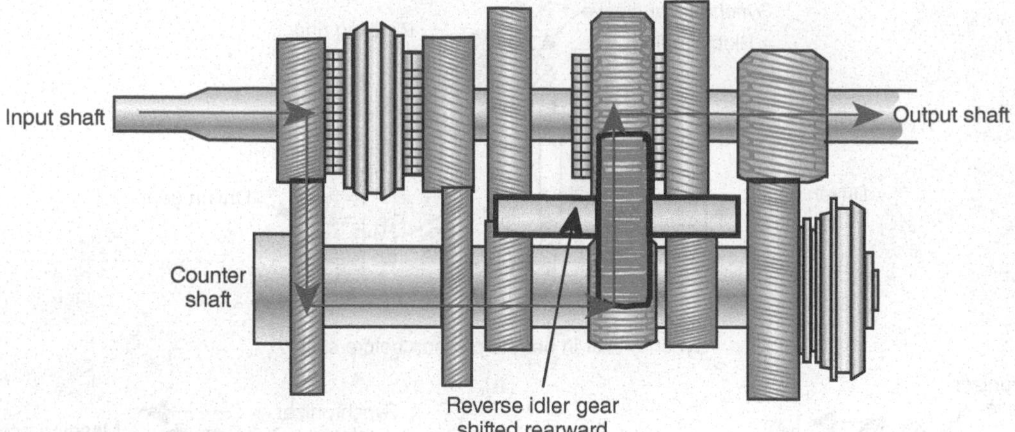

FIGURE 25-37 To have a reverse gear, a third gear and idler shaft are used. When engaged, the reverse gear allows the output shaft to rotate opposite its normal direction.

FIGURE 25-38 The inside of a dual-clutch transmission looks similar to inside a manual transmission.

FIGURE 25-39 The dual input clutches of a dual-clutch transmission. Each clutch is splined to its own gear shaft.

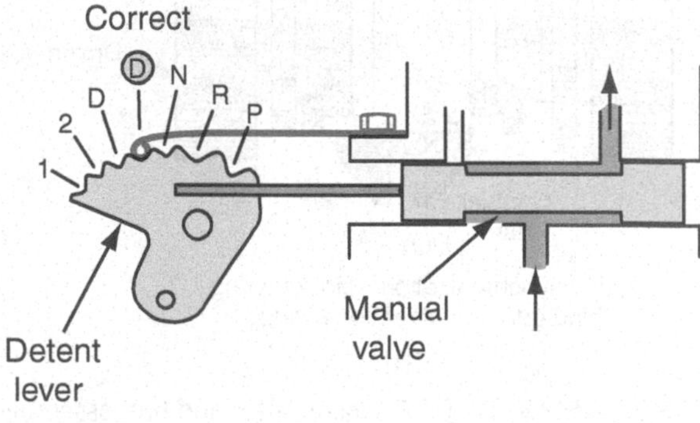

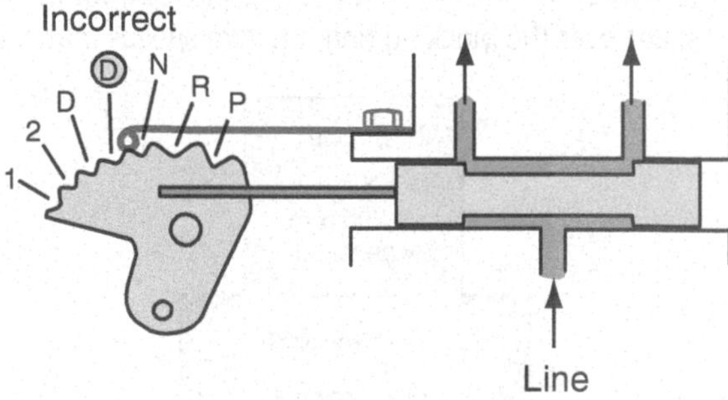

FIGURE 25-40 How the shift linkage controls the manual valve in an automatic transmission. The manual valve directs fluid based on the position of the shift linkage.

the transmission. In most cars, a set of rods or cables are used because of the distance between the shifter and the transmission. An illustration of shift rods and cables is shown in **Figure 25-43**. The movement of the gear shift, regardless of the linkage type, is transmitted to the shift forks in the transmission, which move the shift collars.

Transmission Inspection and Service

Even though modern transmissions are much more reliable than ever before, periodic inspection and service is still required to keep the transmission operating

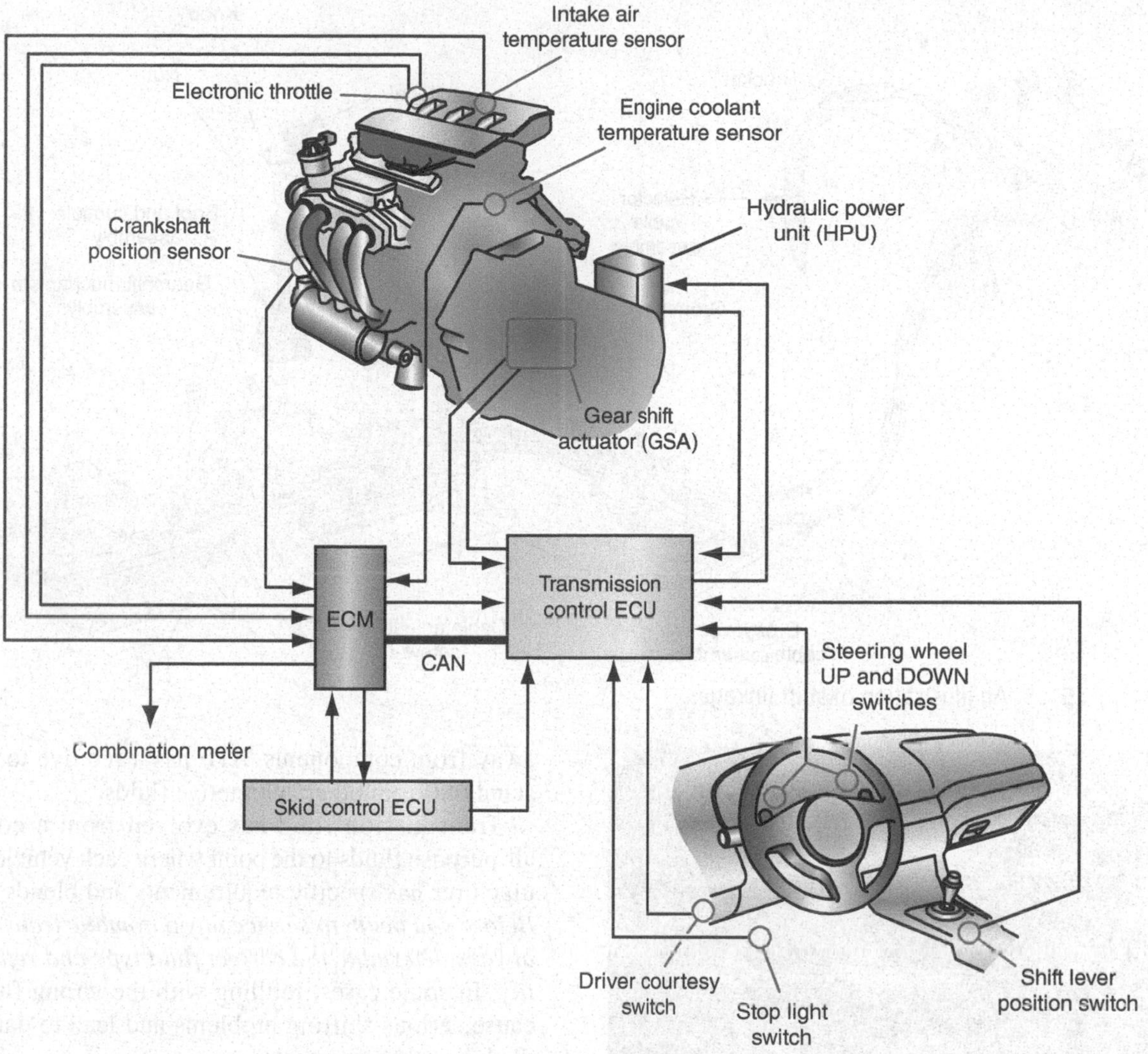

FIGURE 25-41 An illustration of a modern computer-controlled transmission. No physical connections, such as linkages, are used.

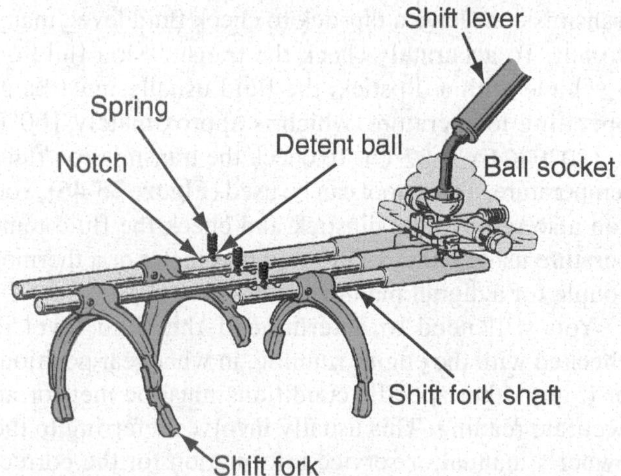

FIGURE 25-42 An example of manual transmission shift linkage.

correctly. This usually includes inspecting for fluid leaks, adjusting fluid levels, replacing the fluid and filter, adjusting the shift linkage, and checking and replacing transmission mounts.

AUTOMATIC TRANSMISSION INSPECTION AND SERVICE

Inspection and service of an automatic transmission involves checking for fluid leaks, changing the fluid, and checking and adjusting shift linkage.

■ ***Automatic Transmission Inspection.*** Because the automatic transmission relies on fluid to transfer power from the engine, an average transmission may contain 10 to 14 quarts of transmission fluid. Over time, seals and gaskets can leak, and the fluid itself can become contaminated with metal and clutch material.

Inspect the transmission for signs of fluid leaks. Leaks are common around pan gaskets (**Figure 25-44**) and from axle seals (**Figure 25-45**). Check the transmission cooler lines for leaks as the rubber hoses become cracked with age.

■ ***Automatic Transmission Fluid.*** **Automatic transmission fluid (ATF)** is a specially blended, high-detergent hydraulic oil designed specifically for use in

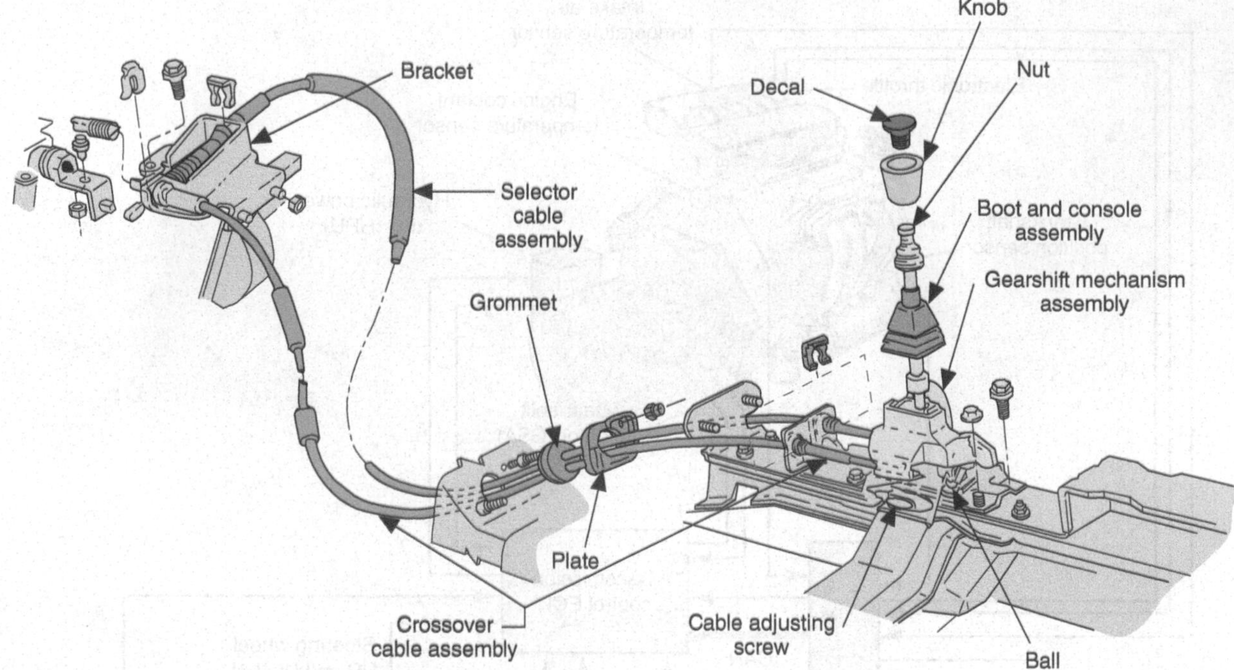

FIGURE 25-43 An illustration of shift linkage.

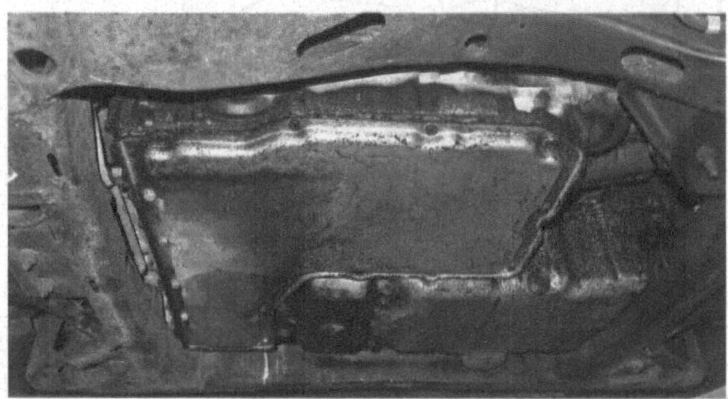

FIGURE 25-44 A leaking transmission pan.

FIGURE 25-45 Inspect around axle seals for leaks.

automatic transmissions. Transmission fluid contains special friction modifiers that allow it to work with the clutches and other components in the transmission. Like engine oil, ATF helps clean internal parts and carries heat away from components. ATF has a red dye to make it stand out from other automotive fluids.

Transmission fluid has evolved from a couple of all-purpose fluids to the point where each vehicle's manufacturer has specific requirements and blends of ATF. *Before you begin to service any automatic transmission, always determine the correct fluid type and refill quantity.* In some cases, refilling with the wrong fluid may cause serious shifting problems and lead to damage of the internal components.

■ ***Checking ATF.*** While many modern automatic transmissions have a dipstick to check fluid level, many do not. To accurately check the transmission fluid on a vehicle with a dipstick, the fluid usually must be at operating temperature, which is approximately 160°F to 180°F (71°C – 82°C). To check the transmission fluid temperature, a scan tool can be used (**Figure 25-46**). You can also remove the dipstick and check the fluid temperature using either an infrared pyrometer or a thermocouple for a digital multimeter.

You will need to determine if the fluid level is checked with the engine running, in what gear position, or if any other specific conditions must be met for an accurate reading. This usually involves referring to the owner's manual or service information for the correct procedure. Next, locate the dipstick, which can sometimes be a challenge. **Figure 25-47** shows an example of an engine compartment and transmission dipstick. The dipstick may be labeled for the transmission, or it may not be. Be sure you are checking the correct dipstick.

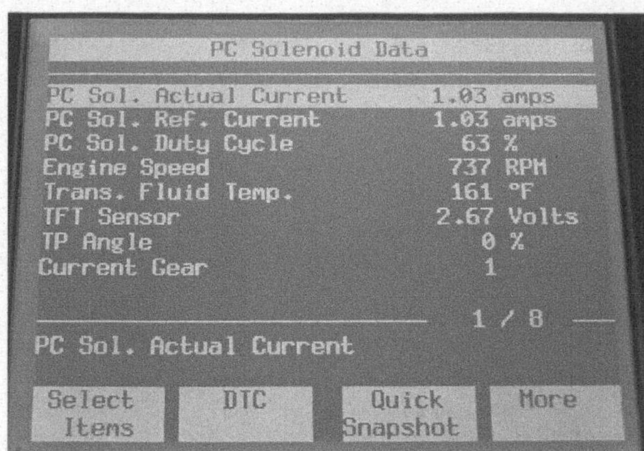

FIGURE 25-46 Modern transmissions are more sensitive to under- or over-filling. Use a scan tool to check transmission fluid temperature when checking and adjusting fluid level.

FIGURE 25-47 The rare and disappearing transmission dipstick.

Once it is located, remove the dipstick from the transmission, clean the entire dipstick with a clean shop towel, reinstall the dipstick, and remove once more. Hold the dipstick horizontally and check the fluid level against the marks on the lower end of the dipstick (**Figure 25-48**). Check the fluid for color, smell, and contaminates. ATF is typically red, though it does turn dark, often brown or even black over time. It should not have a burnt smell or have a gritty feel or particles visible in the fluid or on the dipstick. Check the dipstick for signs that the fluid has burned or varnished on the stick. This indicates that the fluid has overheated.

As with other vital fluids, there are many different types of transmission fluids in use. Do not add fluid to the transmission unless you are certain that it is low and that the correct fluid is being used. The recommended fluid type may be stamped on the dipstick; if not, refer to either the owner's manual or the service information to locate the correct fluid type.

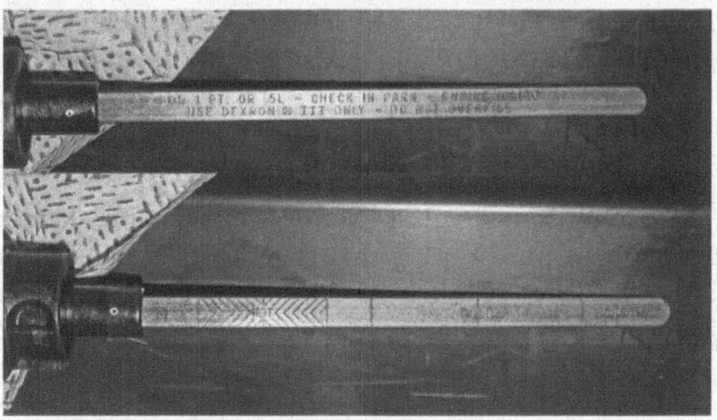

FIGURE 25-48 The dipstick may have cold and hot markings or simple low and full marks. You will need to determine what the marks on the dipstick mean.

If fluid needs to be added and the correct fluid is available, place a funnel in the transmission filler opening or dipstick tube (**Figure 25-49**). Make sure the funnel is clean and dry before use. Next, slowly pour fluid into the funnel. Depending on how much is needed, start by adding one-half of a quart, remove the funnel and recheck the fluid level. Add more fluid as needed. After you have added fluid, you need to recheck the fluid level. Recheck it a few times because sometimes the fluid sticks to the inside of the dipstick tube and makes it hard to get an accurate reading.

■ ***Where's the Dipstick?*** Many modern vehicles do not have a transmission dipstick. There are several designs and ways to check these transmission, including:

- A dipstick tube without a dipstick. Some vehicles provide a dipstick tube that is capped. To check the ATF level, the transmission must be within a specified operating temperature, typically 150°F–200°F (65.5°C–93.3°C). Remove the cap

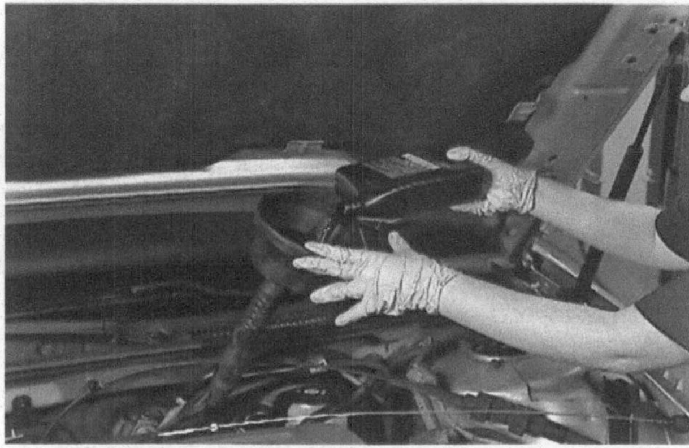

FIGURE 25-49 Adding fluid through the dipstick tube.

and insert a special tool (a dipstick) of certain length. Remove the tool and measure how far up the fluid is on the stick. A variation of this requires a special tool to be inserted into a special opening in the transmission case to check fluid level.

- Check bolt or plug. Many transmissions have a bolt or plastic plug that is removed to check the ATF level (**Figure 25-50**). The small bolt in the center of the plug is removed to check fluid level. Again, the transmission fluid must be within a specified operating temperature, typically 150°F–200°F (65.5°C–93.3°C). Remove the bolt or plug. If fluid streams out, the level is correct. If only a dribble comes out or no fluid comes out, the level is low or the fluid temperature is too low (**Figure 25-51**).

The important part of checking the ATF is that the fluid is within the correct operating temperature. Too low of a temperature will give a low fluid reading. Be sure to read and follow the vehicle manufacturer's service information for specific requirements and procedures to check the fluid.

FIGURE 25-50 A dipstickless transmission with a check plug in the pan.

■ ***Automatic Transmission Fluid Service.*** There are a couple of different ways in which ATF can be serviced, each of which depends on the vehicle. Many vehicles have a transmission pan that is removed to access the transmission filter. A large drain pan is placed under the transmission, and the bolts securing

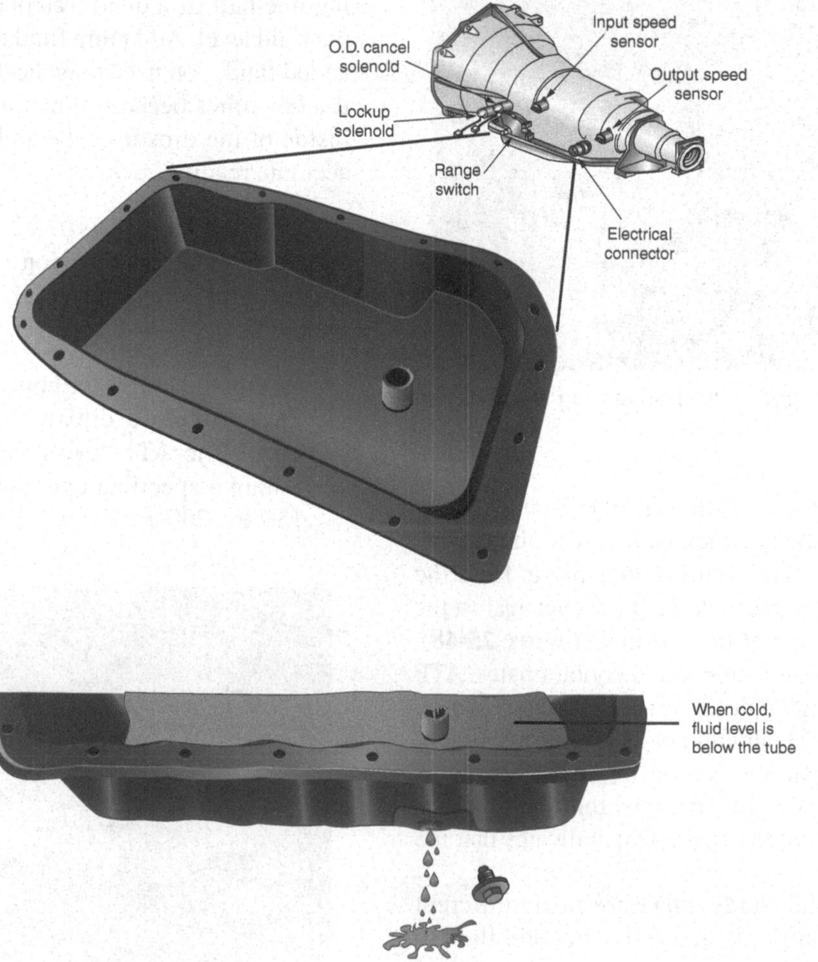

FIGURE 25-51 This illustrates how a check plug is used to check the fluid level.

the pan in place are removed from all but one corner of the pan. This allows the pan to tilt downward and makes draining easier and less messy (**Figure 25-52**). Once the fluid is drained, the remaining bolts are removed and the pan is taken off. Next, the filter is removed, the pan is cleaned, and the gasket is replaced. Once the new filter is installed, the pan is then reinstalled and the bolts are torqued. After the pan is secured, the fluid is replaced. This type of service is still done on many domestically produced FWD and RWD vehicles.

Some vehicles have adopted the use of an external, threaded transmission filter, just like an engine oil filter (**Figure 25-53**). On these transmissions, replace the filter when you are draining and refilling the transmission. This type of filter is much faster to replace, and the service tends to be less expensive. Some vehicles do not have a replaceable filter, so the transmission fluid is drained from a drain plug and then refilled with new fluid (**Figure 25-54**).

FIGURE 25-52 Removing the transmission pan to change the fluid and filter.

FIGURE 25-53 Some vehicles have threaded transmission filters to make service easier.

FIGURE 25-54 To drain the transmission fluid, remove the drain plug.

Some shops use a transmission fluid flushing system that connects to the transmission cooler lines. The flush unit then pumps the old fluid out while it pumps the new fluid into the system. The main concern with this service is that because there are so many different types of fluids in use, contaminating a system with the wrong fluid is possible unless all fluid is removed from the flush machine after each use. Refer to the service information before using a transmission fluid machine because some vehicle manufacturers state not to perform this type of service on their vehicles.

■ ***Transmission Linkage.*** Over time, the transmission linkage can wear, and allow a slight amount of play in the linkage to develop. This may be felt as play in the shifter or by the transmission not engaging into the correct gear range as selected by the driver. To check the linkage, begin by shifting the transmission through its gear ranges while noting any play in the shifter. If play is felt, you need to determine if the play is at the shifter or at the transmission. If the vehicle has a column-mounted shifter, inspect under the dash at the shift jacket on the steering column. If a console shifter is used, you probably need to remove the console to inspect the shifter assembly.

If the source of the play is in the linkage at the transmission, you may be able to adjust the linkage to remove the play. On mechanical linkages loosen the adjustment nut and move the linkage to remove the play (**Figure 25-55**). When you adjust the linkage, follow the manufacturer's service procedures to ensure that the adjustment is correct. If the linkage cannot be properly adjusted, there is likely a problem with the cable or other linkage component, which requires replacement.

■ ***Transmission Range Switch.*** Often called *the TR switch* or *sensor*, the transmission range switch is used to provide information to the computer about

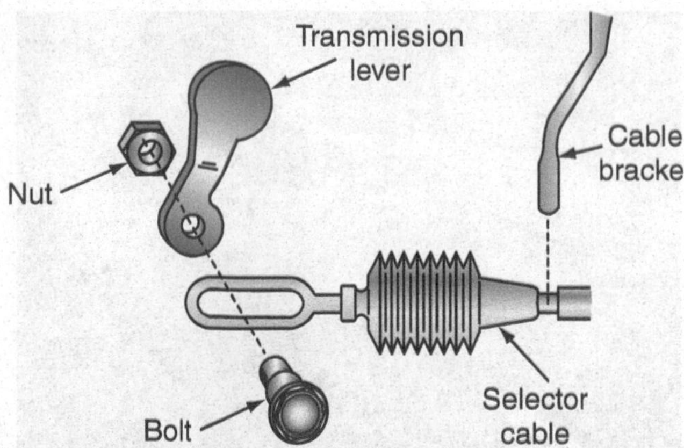

FIGURE 25-55 For transmission with mechanical linkage, adjustment may be necessary if shifting or starter concerns arise.

FIGURE 25-57 An example of a powertrain mount.

which gear or range the transmission is placed in by the driver. The computer uses this information for shifting. A faulty switch can cause the transmission to not upshift, not stay in gear, delay gear engagement, and even allow the engine to start in positions other than Park and Neutral.

To check the operation of the TR sensor, connect a scan tool and locate the transmission data. Make sure the sensor is displaying the correct gear range for each of the gear selector positions. An example of this type of data is shown in **Figure 25-56**. Depending on the type of sensor, you may need a digital multimeter (DMM) to check the resistance through each position of the switch. Always refer to the manufacturer's service procedures for proper testing information.

■ ***Transmission Mounts.*** Various mounts are used to keep the powertrain—the engine and transmission—in place and reduce vibrations transmitted to the passenger compartment (**Figure 25-57**). Over time, these

Transmission Data

Engine Torque	-9 ft-lbs
Calc. Throttle Position	0 %
Engine Speed	776 RPM
Transmission ISS	671 RPM
Transmission OSS	0 RPM
Vehicle Speed	0 mph
Commanded Gear	1
1-2 Sol.	On
2-3 Sol.	On

1 / 29

Engine Torque

Select Items | DTC | Quick Snapshot | More

FIGURE 25-56 Checking indicated gear position compared to what the gear shifter indicator shows.

mounts wear from constant movement, heat, and rot. As the mounts wear, the powertrain can develop excessive amounts of movement, which can cause noises, clunks, vibration, and binding in the drive axles. For these reasons, it is important to inspect the mounts during routine service.

To check the mounts for excessive movement, begin by blocking the drive wheels with wheel chocks. Next, start the engine, set the parking brake, firmly hold the service brake, and place the transmission in Reverse. Increase engine speed while a helper watches the movement of the engine and transmission. Repeat with the transmission in Drive. If the engine or transmission moves excessively, meaning more than an inch in any direction, the mounts may be worn. Take a close look at the mounts and look for signs of cracking, missing rubber, oil contamination, and oil leaking from the mount as many powertrain mounts are hydraulic. If any mount is excessively worn or damaged, it needs to be replaced.

MANUAL TRANSMISSION SERVICE

Because of the overall simplicity of the manual transmission, there is little for the technician to inspect and service under normal operating conditions. Inspect for signs of fluid loss during routine service. Fluid leaks can result from leaking gaskets and front and rear seals. Whenever a leak occurs, check the transmission vent to make sure it is not plugged. A plugged vent allows pressure to increase in the transmission case, which can result in fluid being forced out of seals and gaskets that otherwise would not leak.

Even though the manual transmission is a sealed unit, proper maintenance, such as inspecting and replacing the transmission lubricant, is still part of a routine maintenance schedule.

■ ***Manual Transmission Fluid Service.*** As with all vehicle fluids, the manual transmission fluid requires periodic maintenance. The fluid picks up metal shavings, and over time, the additives wear out. This necessitates periodic fluid replacement. First, determine the type of fluid used. This can vary from selected engine oils to automatic transmission fluids as the correct fluid. To check fluid level, locate the service plug in the side of the transmission (**Figure 25-58**). Remove the plug and check to see if the fluid is up to the bottom of the hole. In most cases, the fluid level should just reach the bottom of the check plug hole. If it is low, top it off with the correct fluid, reinstall the plug, and torque it to specifications.

To replace the transmission fluid, place a pan under the drain plug, remove the fill plug. and then remove the drain plug. Once it is empty, reinstall the drain plug and refill the transmission. Tighten both plugs to specifications.

■ ***Check and Adjust Hydraulic Clutch Fluid.*** Clutch hydraulic fluid should be checked as part of the normal maintenance and inspection program. Most vehicles use brake fluid in the clutch system, though check the cap or service information is checked before adding fluid if low. Many vehicles use the brake fluid to fill the clutch master cylinder (**Figure 25-59**).

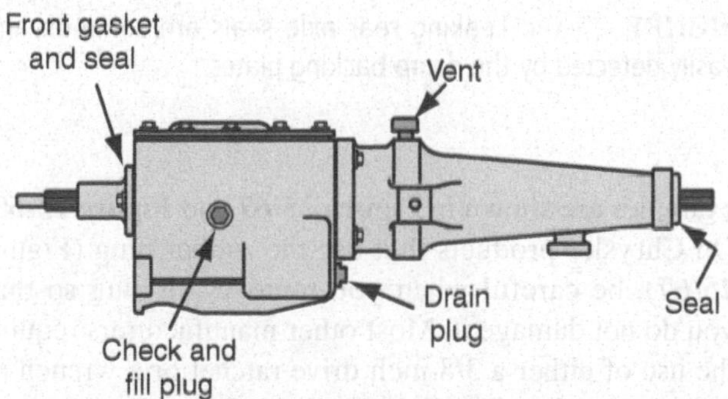

FIGURE 25-58 To check fluid level, remove the check plug on the side of the transmission. Be sure the plug is the correct plug before removing. Some transmissions have plugs for other purposes, which should not be removed.

FIGURE 25-59 An example of a hydraulic clutch master cylinder and shared fluid reservoir.

Like most brake master cylinders, the clutch fluid can usually be checked without removing the cap. However, in some cases the fluid cannot be seen and the cap must be removed (**Figure 25-60**).

If the fluid level is low, if there is a complaint of a soft clutch pedal, or the clutch does not disengage properly, inspect the clutch hydraulic system for leaks. Leaks can occur at the master cylinder, often allowing fluid to travel down the inside of the firewall. Look up under the dash and check where the clutch pedal pushrod enters the back of the clutch master cylinder. If fluid is present, the master cylinder is leaking and should be replaced (**Figure 25-61**).

Leaks from the slave cylinder on or in the transmission are also common. The slave cylinder shown in **Figure 25-62** was leaking and fluid was collecting inside the bell housing of the transmission. Check the fluid line from the master cylinder down to the slave cylinder for leaks also. Both steel and rubber fluid lines are susceptible to damage and leaks.

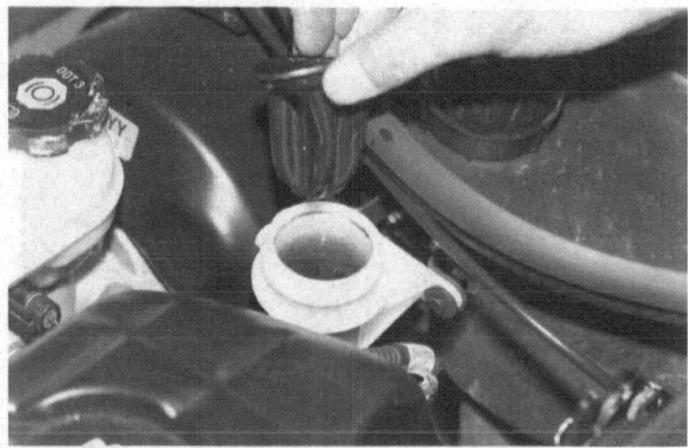

FIGURE 25-60 Checking hydraulic clutch fluid level.

FIGURE 25-61 If the hydraulic clutch fluid is low, check under the dash for signs of fluid leaking from the back of the master cylinder.

■ ***Differential Inspection.*** Inspecting a FWD differential for leaks means checking around the axle shaft seals, as shown earlier in this chapter. Because the differential is made into the transaxle housing, there typically are not any other areas for leaks to occur.

Differentials on RWD vehicles are prone to leaks at the pinion shaft seal (**Figure 25-63**). Leaks from the axle shaft seals often show up as dampness on the backing plate of the rear brakes (**Figure 25-64**). The differential covers and gaskets are also common places for leaks to occur (**Figure 25-65**). Whenever a leak is present, check the differential's vent hose for blockage or damage (**Figure 25-66**). The vent allows the pressure in the differential to dissipate as the fluid heats up during driving. A blocked vent will allow pressure to increase, which can push fluid past seals and even damage the seals.

■ ***Differential and Transfer Case Lubricant.*** Often, differential lubricant is an overlooked maintenance item. To check the differential fluid, locate the plug. Two examples are shown in **Figure 25-67** and **Figure 25-68**. On Chrysler products that use the rubber plug (Figure 25-67), be careful when you remove the plug so that you do not damage it. Most other manufacturers require the use of either a 3/8-inch drive ratchet or a wrench to remove the plug. The fluid level should be up to the bottom of the fill plug hole.

FIGURE 25-62 A leaking concentric clutch slave cylinder.

FIGURE 25-63 An example of a leaking pinion seal in a rear differential.

FIGURE 25-64 Leaking rear axle seals are common and easily detected by the damp backing plates.

If fluid is needed, refer to the owner's manual or service information for the correct lubricant. Limited-slip differentials (LSDs) generally require a specific type of lubricant be used, so do not assume that any gear lubricant can be used. **Limited-slip differentials** often use small clutch packs, like those used in automatic transmissions. For the clutch pack to work properly, a special lubricant is required. If it is needed, an additive of special LSD friction modifiers can be used. Failure to use the correct lubricant can cause failure of the limited-slip components, damage to the gears and bearings, and even failure of the differential unit.

Most manufacturers specify periodic replacement of the differential lubricant, similar to transmission fluid. On most differentials, the cover must be removed to completely drain all the old fluid. Place a pan under the differential and remove the bolts holding the cover in place and then remove the cover (**Figure 25-69**). This also requires replacing the cover gasket (**Figure 25-70**).

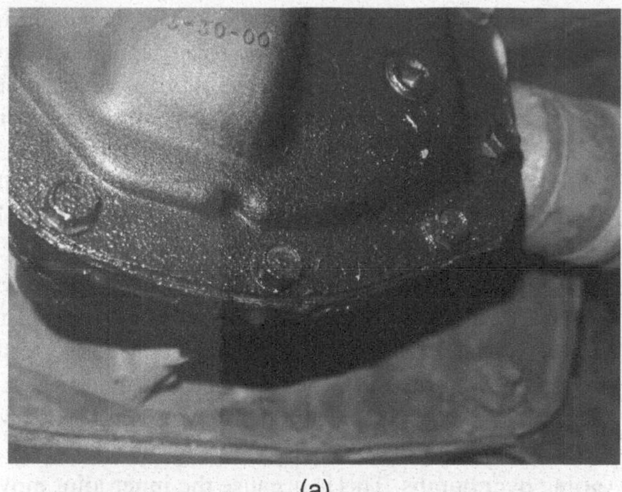

(a)

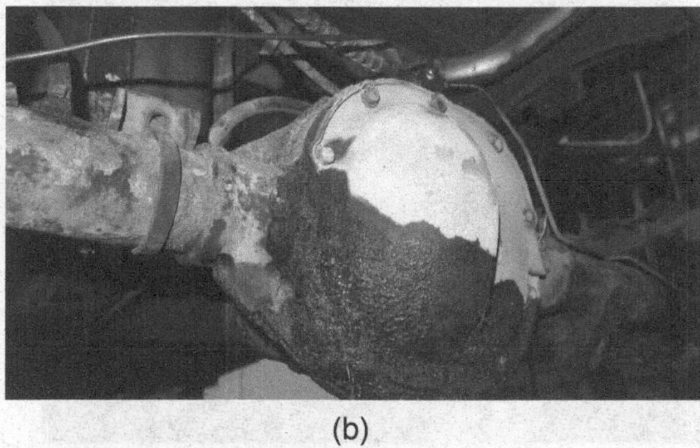

(b)

FIGURE 25-65 (a) A leaking rear differential cover. (b) This plastic differential cover was cracked and leaking.

FIGURE 25-66 Always check the differential vent when fixing a leak. A plugged vent will allow pressure to build and cause the leak to return if not corrected.

FIGURE 25-68 Many differentials use a 3/8-inch drive plug for checking and filling the unit.

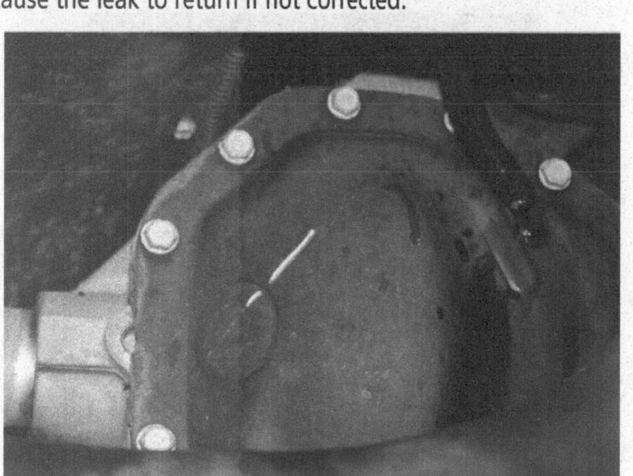

FIGURE 25-67 An example of a differential plug on a Chrysler product. Be careful not to damage the plug when removing it.

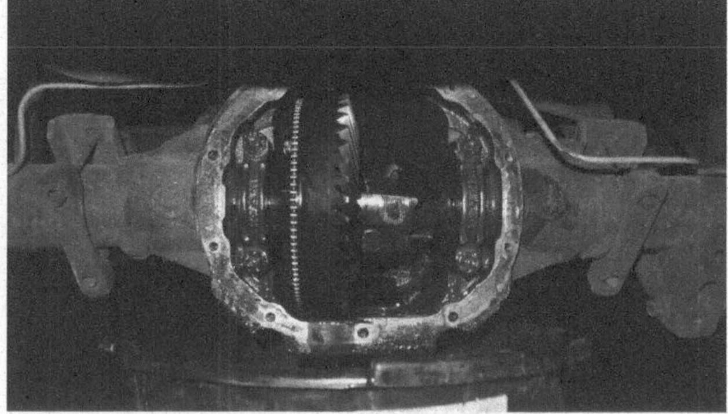

FIGURE 25-69 A differential with the cover removed for fluid service and leak repair.

Some shops use a fluid extraction pump to remove the fluid through the fill plug. This method removes most of the fluid and does not require replacing the cover gasket.

Some differential housings have a drain plug. Remove the plug to drain the fluid, then reinstall the plug, and fill with the correct type of lubricant.

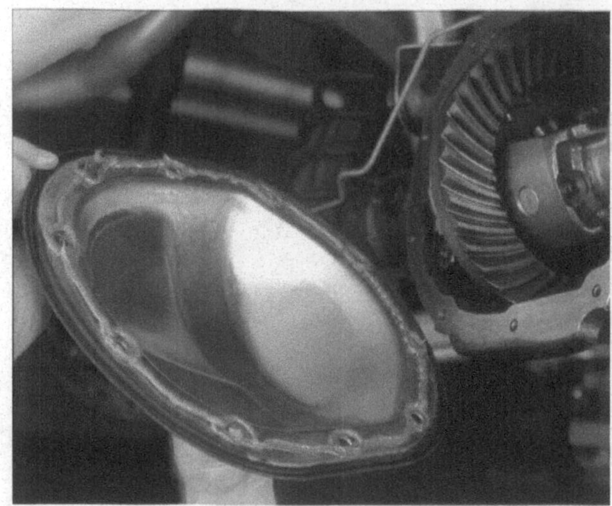

FIGURE 25-70 A replacement gasket may not be available so a form-in-place gasket is used to seal the cover.

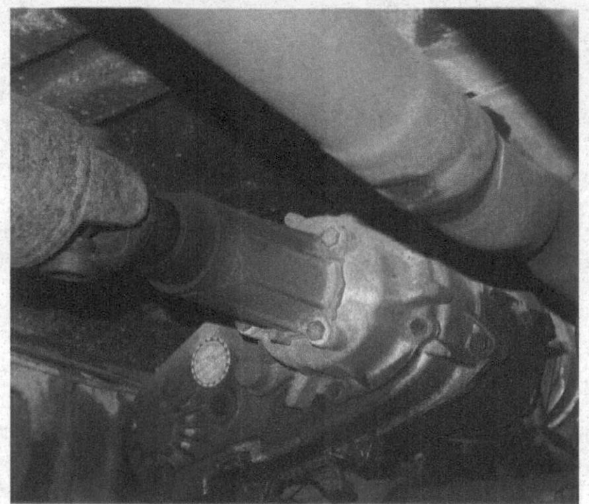

FIGURE 25-71 Inspect transfer cases for leaks as they are prone to the same issues as differentials and transmissions.

Inspecting a transfer case is similar to checking the transmission and differential. Begin by inspecting for leaks around the axle seals, joints between case sections, and where motors, actuators, or electrical connections pass though the case. An example of a case with several leaks is shown in **Figure 25-71**.

To check the transfer case lubricant, locate and remove the check plug, similar to that in a differential. If the fluid level is low, determine the correct fluid for the application. As with most fluids, the transfer case may require a special lubricant. Check the service information before topping off the fluid.

Axle Shafts

Axle shaft service, replacing CV boots or the shaft itself, is a common repair procedure. The axles are subject to wear and damage from road debris. Inspect the boots and axle shafts as part of a normal inspection program and during routine services. **Figure 25-72** shows an example of a split CV boot. The boots become brittle with age and eventually split. This allows the grease to come out and for dirt, water, and other debris to get into the boot and the CV joint. Once the joint is open to the environment, it usually does not take long for the contaminants to damage the joint beyond repair.

Typically, when an outer CV joint is worn excessively, it will cause a loud clicking sound during turns. This is caused by the ball bearings being loose in the cage. An example of CV joint construction is shown in **Figure 25-73**. An inner joint that is worn and loose will clunk when the driver is accelerating, decelerating, or going over bumps. This is because the inner joint moves in and out as the wheel moves up and down. When this occurs, either the joint or the entire shaft will need to be replaced. Whether you should replace the joint or the axle is often dependent upon the type of vehicle and the availability of parts.

■ ***Axle Shaft Removal and Installation.*** For most FWD vehicles, axle shaft removal is much the same. Begin by removing the wheel and tire assembly. Next, remove the axle nut. This can be done with an air impact driver or by hand tools if a helper is available to press and hold the brake to keep the axle from turning. Once the axle nut is removed, disconnect the lower ball joint-to-steering knuckle connection (**Figure 25-74**). This may require removing a pinch bolt or a nut from the ball joint stud. If the ball joint nut is a friction nut, do not reuse the old nut; replace it with a new nut.

FIGURE 25-72 Close inspection of the CV boot reveals that it is split and leaking grease.

FIGURE 25-73 Inside of a CV joint.

FIGURE 25-74 To remove the CV shaft, unbolt the strut or ball joint to allow the steering knuckle to move.

FIGURE 25-75 An axle puller may be required to separate the CV shaft from the wheel bearing.

FIGURE 25-76 Pry the inner joint away from the transaxle.

Separate the ball joint from the steering knuckle so that the knuckle can move freely. In some cases, the axle's stub shaft slips easily from the hub, allowing the hub to be moved out of the way. Many times, however, the stub shaft requires a special tool to push the shaft out of the hub. Tighten the tool to force the stub shaft out of the hub (**Figure 25-75**). Once the hub is free from the axle, use a wire or bungee cord to hold the hub away from the axle. Next, place a pry bar between the inner CV joint housing and the outer edge of the transaxle case (**Figure 25-76**). Pry against the CV joint housing to remove it from the transaxle. Before completely removing the axle shaft from the transaxle, place a fluid catch pan under the transaxle to catch any transmission fluid that leaks out.

The axle shafts on some vehicles are bolted to a flange at the transaxle. To remove this type of axle you must remove the retaining bolts holding the axle to the flange.

If you are replacing only the damaged CV joint, mount the axle in a bench vice by clamping the steel shaft in the vice's jaws. Next, cut the boot clamps from the joint and pull the boot back from the joint (**Figure 25-77**). Then, using a brass hammer, tap the joint from the shaft. In some cases, you need to use a pair of snap ring pliers to spread a snap ring apart to remove the joint. Once the joint is removed, pull the boot from the shaft and discard it. Clean the shaft of grease, and slide the new boot and small clamp onto the axle shaft. Open the new bag of CV grease and pack the grease into the new joint. Install the new retaining ring or snap ring on the axle and then install the joint onto the axle, making sure it seats fully and is locked into place. Apply the remaining grease to the joint and into the boot. Slide the boot over the CV joint and install the new clamp. Crimp both clamps onto the boot and joint. This usually requires a special tool (**Figure 25-78**). Once the clamps are secured, clean up any mess and prepare to reinstall the axle.

To reinstall the axle shaft, first apply a light coat of grease to the slip ring on the inner CV joint shaft. Use the grease to center the ring around the shaft; this eases installation by keeping the ring from catching around

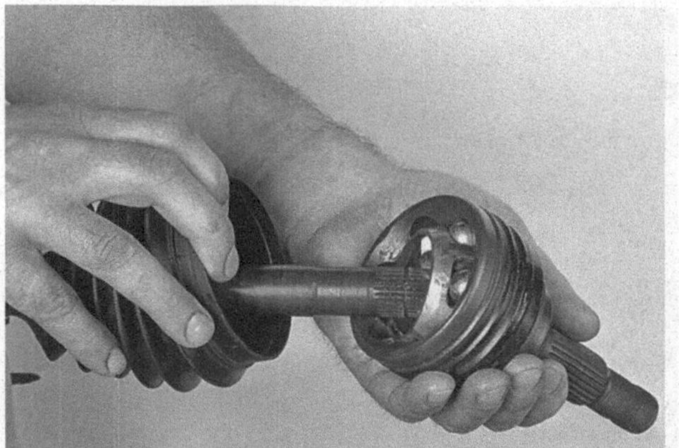

FIGURE 25-77 Remove the old boot and joint. Thoroughly clean the joint and dry it before packing in the new grease.

FIGURE 25-78 Installing new boot clamps often requires a special tool to crimp the clamp tightly in place.

the bottom of the shaft. Next, carefully install the inner joint straight into the transaxle, keeping the axle centered. After the axle is started into the differential gear inside the transaxle, use a rubber mallet to fully seat the axle. Once the axle is seated, reinstall the outer stub shaft into the hub. When the threaded section of the stub shaft is protruding from the hub, install the new axle nut and begin to tighten. This pulls the axle into place in the hub. Once the axle is fully seated, reconnect the lower ball joint to the steering knuckle. Install the pinch bolt or nut and tighten to specs. To tighten the axle nut, either have a helper apply the brakes or install the wheel and tire and lower the vehicle to place a light amount of weight on the tire. Tighten the axle nut to specs. It is very important that the axle nuts be properly torqued as it is also setting the preload on the wheel bearings. Failure to properly torque the axle nuts can cause premature wheel bearing failure.

■ ***RWD U-Joint and Driveshaft Inspection.*** The driveshaft, also called *a prop* or *propeller shaft*, connects the transmission to the rear differential. Universal joints, or U-joints, are used to allow the shaft to move as the differential moves up and down (**Figure 25-79**).

In many cases, worn U-joints will make noise or cause a vibration as they rotate. Inspect the U-joints for looseness or excessive wear (**Figure 25-80**). To fully check a U-joint, you may need to remove the driveshaft to check for play or binding in the joint.

To replace a U-joint, first remove the driveshaft from the vehicle. If the vehicle is on a drive-on lift, be sure to block the wheels to keep the vehicle from rolling. Place the transmission in neutral and make index marks on the flanges (if necessary) so that the parts can be reinstalled as they came apart (**Figure 25-81**). Next, remove the nuts, bolts, and clamps holding the rear U-joint to the differential. Use a prybar to remove the rear joint from the pinion flange and carefully remove the driveshaft from the transmission.

Make index marks on the driveshaft and flanges before removing the U-joints. This is so each part is reinstalled as it was to reduce vibration during operation. Next, remove the clips holding the U-joints in place (**Figure 25-82**). To remove the U-joints, either a joint press or sockets and a bench vice can be used. A joint press is similar to a ball joint press and is used to press the caps out and reinstall them. If a press is not available, you can use a couple of impact sockets to press the caps out. Place a socket that is close to the same size as the U-joint cap against one cap and another, larger

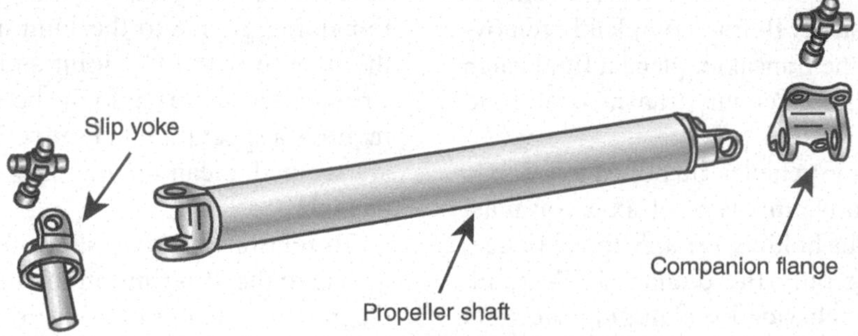

FIGURE 25-79 An illustration of a driveshaft and U-joints.

(a)

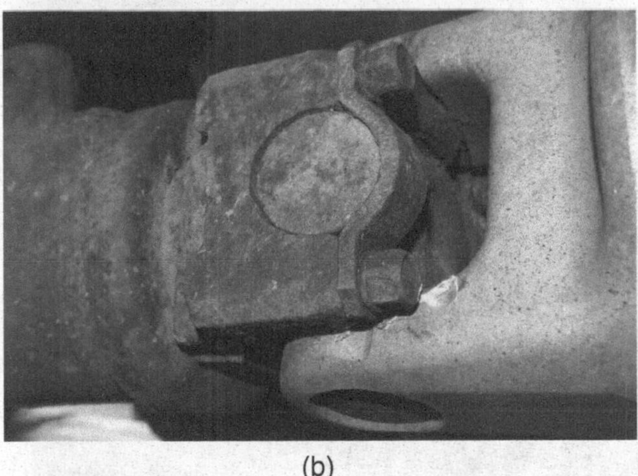

(b)

FIGURE 25-80 (a) An example of a worn out and failing U-joint. (b) This joint was allowing the driveshaft to dig into the clamp bolt on the pinion shaft.

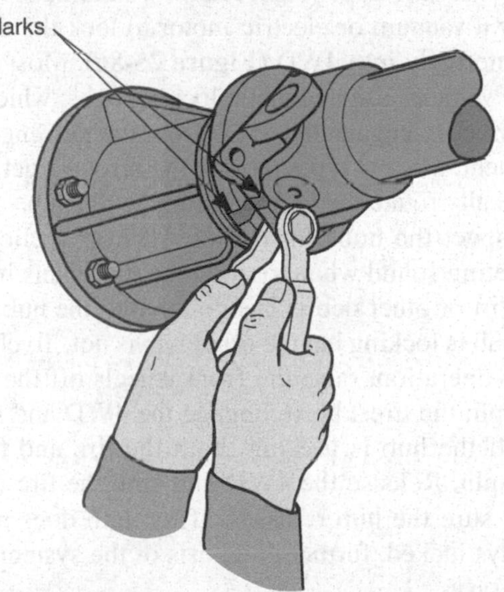

FIGURE 25-81 Mark witness marks on the components so that they can be reinstalled with the same alignment.

FIGURE 25-82 Remove the clips that secure the caps in the flanges. Discard the clips since the new joint comes with new clips.

socket, over the cap on the other side (**Figure 25-83**). Tighten the driveshaft and sockets in a bench vise. The smaller socket will push the U-joint cap into the other socket. Press the remaining caps out and install the new joint (**Figure 25-84**). Install the new clips and check for smooth movement of the joints.

Reinstall the driveshaft by sliding the front yoke onto the transmission's output shaft. Next, realign the marks on the rear flange and install the driveshaft. Torque all fasteners to specs.

Replacement U-joints typically have grease fittings. Once the driveshaft is installed, inject grease into each joint until a slight amount of grease escapes from the caps. Wipe off any excess grease that comes out. Sending the vehicle out without greasing the joints will cause very rapid wear and failure of the new U-joints.

FIGURE 25-83 Use a small and a large socket to press the caps from the joint.

FIGURE 25-84 Carefully install the new joint and press the caps into place. Use the sockets to finish pressing the caps fully into position. Install the new clips and grease fitting.

FIGURE 25-85 Not as common as they once were, manually locking hubs require getting out of the vehicle and turning the hub selector to the Lock position.

■ ***Locking Hub Inspection.*** Some 4WD trucks use manual locking front hub assemblies (**Figure 25-85**). This type of hub is used to lock the front wheels when the driver is using the 4WD, and to unlock the wheels when the 4WD is not in use. Locking hubs are either manual or automatic (Figure 25-85). Automatic hubs use either a vacuum or electric motor to lock the differential internally into 4WD (**Figure 25-86**). Most modern 4WD vehicles use automatic locking hubs, which allow the driver to engage the 4WD from the passenger compartment. Manual types require the driver to get out and physically rotate the lock assembly at the hub.

FIGURE 25-86 An example of a front differential hub locking motor.

Inspect the hubs for noises, such as a clicking or ratcheting sound when rotating the tires. This indicates that dirt or other debris has gotten into the hub, or that one hub is locking but the other hub is not. To check the hub's operation, raise the front wheels off the ground and spin the tires. Next, engage the 4WD and spin the tire. If the hub is locking, both the tire and the axle will spin. Release the 4WD and spin the tire again to make sure the hub releases. If the hub does not lock or stays locked, further diagnosis of the system will be necessary.

Checking automatic hubs is similar. With the wheels off the ground, engage the 4WD and turn a front wheel. If the 4WD is working, both front wheels should be locked together and turn in the same direction. If the 4WD does not engage, you will need to inspect the system to determine the cause of the fault. Common 4WD failures include damages to vacuum or wiring to the differential lock and faulty actuators. Follow the manufacturer's service information for specific diagnostic procedures.

SUMMARY

The job of any type of transmission is to take the power produced by the engine and transfer it to the driving wheels.

Modern automatic transmissions are controlled by the powertrain control module or PCM.

A torque converter is used to join the engine's crankshaft to the input shaft of an automatic transmission and is a type of fluid coupling.

A planetary gearset is made of three gears meshed together, the ring gear, the sun gear, and planetary gears.

The PCM commands a collection of shift solenoids to control shifting in automatic transmissions.

A CVT transmission varies the diameter of the input and output pulleys so that the engine speed can remain within a certain rpm range as much as possible.

With a manual transmission, a clutch connects the engine to the transmission and transmits the power from the crankshaft to the transmission.

Low gear ratios increase torque and accelerate quickly, while high gear ratios allow lower engine speed and economy but with low torque output.

Synchronizers are used to allow gear shifting without the clash of gears.

Refilling a transmission with the wrong fluid may cause serious shifting problems and lead to damage of internal components.

The transmission range switch or sensor is used to provide information to the computer about which gear or range the transmission is placed in by the driver.

LSDs generally require a specific type of lubricant be used that contains an additive of special friction modifiers.

Most modern 4WD vehicles use automatic locking hubs, which allow the driver to engage the 4WD from the passenger compartment.

When replacing CV joints, clean all the old grease out of the joint before packing in the new grease.

Before replacing a U-joint, mark the relationship between the driveshaft and flanges to help keep the shaft in balance.

REVIEW QUESTIONS

1. The component in the torque converter that drives the transmission input shaft is the _______________.
2. A planetary gearset consists of the sun gear in the center and _______________ gear around the outside.
3. A continuously variable transmission uses two changeable _______________ and a belt.
4. A hydraulic clutch system uses a _______________ cylinder and a _______________ cylinder to apply the clutch.
5. Special _______________ modifiers may need to be added when refilling a limited-slip differential.
6. When discussing a FWD car: A loud clunk is heard from the right front wheel when accelerating. *Technician A* says a worn CV joint may be the cause. *Technician B* says a worn engine or transmission mount could be the cause. Who is correct?
   a. Technician A
   b. Technician B
   c. Both A and B
   d. Neither A nor B
7. Which of the following is most likely to be a fifth or sixth gear ratio?
   a. 0.85:1
   b. 1:1
   c. 2.25:1
   d. 3.9:1
8. Which of these may be necessary when checking automatic transmission fluid level?
   a. A scan tool
   b. A special dipstick tool
   c. A thermometer
   d. All of these
9. *Technician A* uses a bench vice and sockets to remove and install a new U-joint. *Technician B* drives the old U-joint caps out and the new caps in with a hammer. Who is correct?
   a. Technician A
   b. Technician B
   c. Both A and B
   d. Neither A nor B
10. A dual-clutch transmission is based on which type of transmission?
    a. CVT
    b. Automatic
    c. Manual
    d. Hybrid

CHAPTER 26

Heating and Air Conditioning

Chapter Objectives

At the conclusion of this chapter, you should be able to:

- Describe the components and operation of the heating and air conditioning systems. (ASE Education Foundation MLR 7.A.1)
- Inspect air conditioning drive belts. (ASE Education Foundation MLR 7.B.1)
- Identify hybrid vehicle high-voltage air conditioning components. (ASE Education Foundation MLR 7.B.2)
- Inspect the air conditioning condenser for airflow restrictions. (ASE Education Foundation MLR 7.B.3)
- Inspect heating and air conditioning ducts. (ASE Education Foundation MLR 7.D.1)
- Replace a cabin air filter. (ASE Education Foundation MLR 7.D.1)
- Identify sources of odors from the air conditioning system. (ASE Education Foundation MLR 7.D.2)
- Inspect and service cooling system hoses. (ASE Education Foundation MLR 7.C.1)
- Inspect and replace the engine thermostat. (ASE Education Foundation MLR 1.C.3)
- Determine the correct coolant type. (ASE Education Foundation MLR 1.C.4)
- Drain, flush, refill, and bleed the cooling system. (ASE Education Foundation MLR 1.C.4)

KEY TERMS

air bleed valve	**evaporator**	**radiator fan**
blower motor	**heater core**	**refractometer**
cabin air filter	**HVAC system**	**refrigerant**
compressor	**pressure tank**	**thermostat**
condenser	**radiator**	**water pump**

The heating, ventilation, and air conditioning system, HVAC for short, is responsible for maintaining engine temperature and passenger comfort. Heating and defrosting systems are required in cars and trucks and until the adoption of electronics in the automobile, remained basically unchanged. For many years, air conditioning, simply called *A/C*, was an expensive option for the customer when ordering a new car or truck. However, for nearly all modern vehicles, A/C is now standard equipment. In addition, while the basic components and operation of the A/C have not changed significantly in many years, the refrigerants used in the system have. This is because of environmental concerns about the chemicals used to make the refrigerants.

The HVAC System

The **HVAC system** incorporates the engine cooling system, the passenger compartment heating system, the components in the dash that move and circulate air within the passenger compartment, and the air conditioning system. Each of these systems and the components are separate yet also connected.

THE COOLING SYSTEM

The main function of the cooling system is to remove excess heat from the engine and to maintain the proper engine operating temperature. As part of this, heat absorbed by the cooling system is used to warm the passenger compartment as desired by the occupants.

A basic cooling system contains the engine, coolant, water pump, radiator, radiator fan, thermostat, heater core, and coolant hoses (**Figure 26-1**).

■ ***The Engine.*** All new cars and trucks sold in the United States have liquid-cooled engines. This means that a liquid is used to transfer heat away from the engine to the surrounding air. During combustion, temperatures can reach 2,500°F. Some of the heat and energy is used to turn the crankshaft and drive the wheels, but most, about two-thirds, is wasted and must be removed.

In the engine, surrounding the cylinders and combustion chambers, are coolant passages (**Figure 26-2**). These passages allow coolant to circulate and absorb heat. As the coolant is pumped through the system, heat is carried away and dissipated to the outside air.

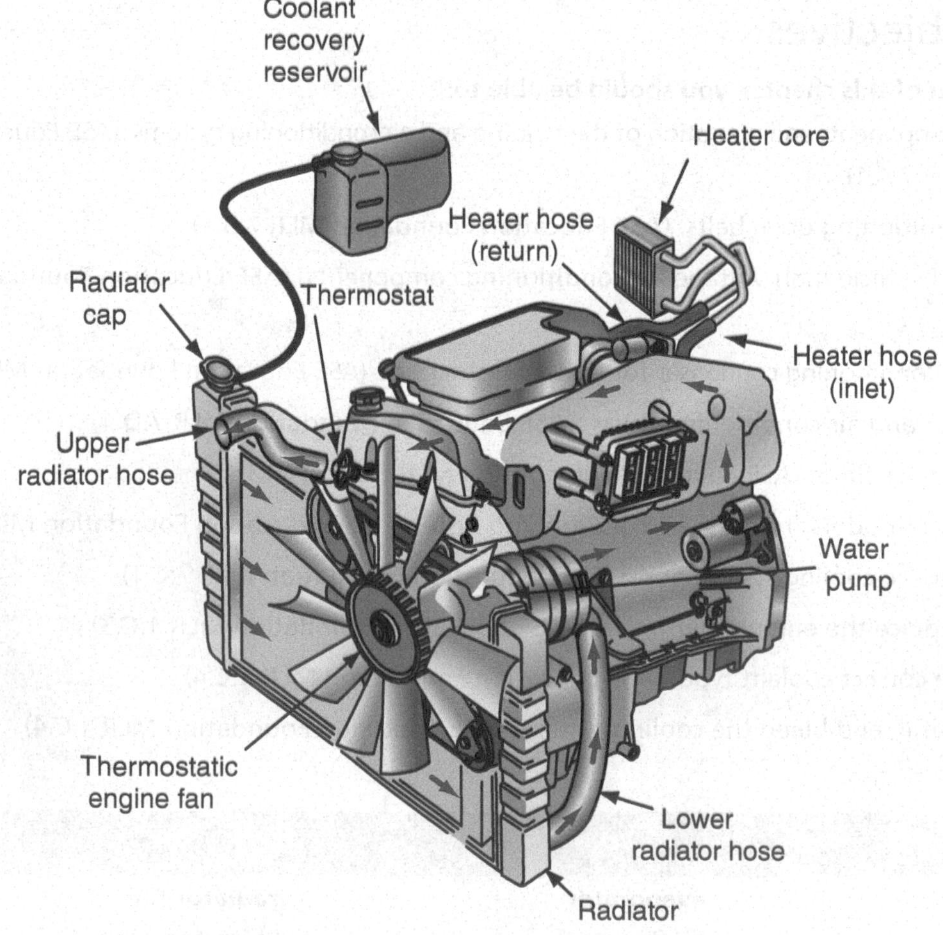

FIGURE 26-1 This illustrates the flow of coolant through the system.

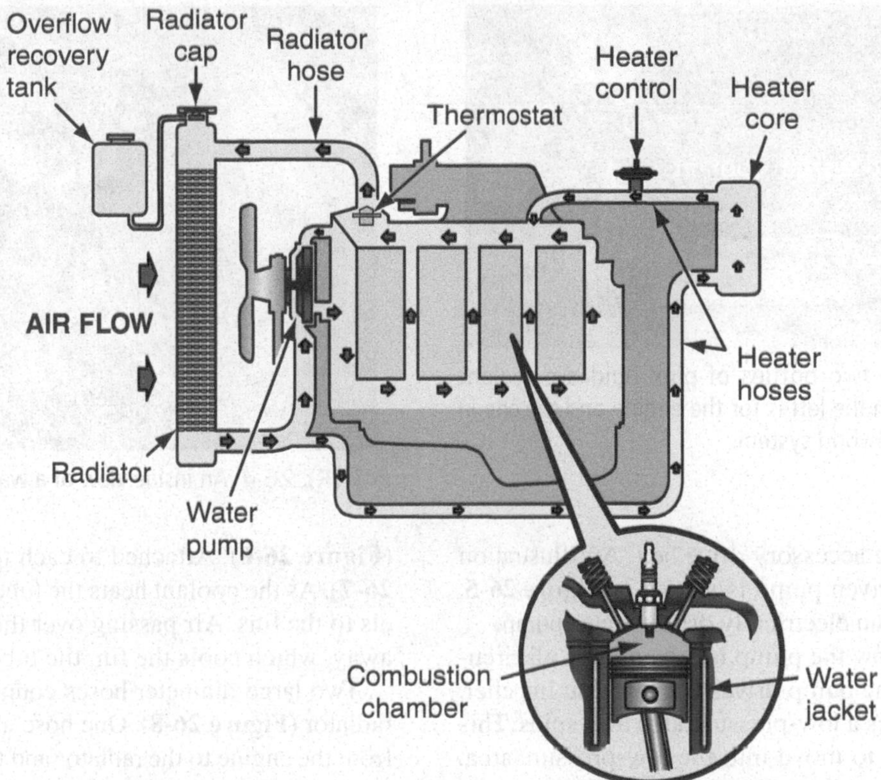

FIGURE 26-2 An example of a basic cooling system. Coolant flows around the cylinders, where it picks up heat. The coolant then flows to the radiator, where heat is removed before it flows back into the engine.

■ ***Coolant.*** Coolant, also referred to as antifreeze, is a mixture of ethylene glycol, other special chemicals, and pure water. Coolant is the medium or go-between by which heat transfer takes place in the cooling system. Coolant can be purchased as full-strength coolant, meaning it must be mixed with water, or it can also be purchased premixed. Premixed coolant is sold as a 50/50 mixture of coolant and distilled or pure water.

Modern vehicles have different requirements for coolant, and all vehicle manufacturers have specific coolant specifications. This means that before you add to or service the cooling system, you need to determine exactly which coolant is required for the application. Do not rely on the color of the coolant to decide which coolant to use. When you select a coolant, refer to the owner's manual or service information. Even though all current factory coolants are ethylene glycol based, they are not all the same. There are many varieties of coolant, including the following:

- Dex-Cool, which is used by General Motors and is made with organic acid technology (OAT).
- Coolant that uses hybrid acid organic technology (HOAT), used by Ford and other manufacturers.
- Inorganic acid technology (IAT) coolants, which are used in the traditional green coolant.
- Coolants that do not contain silicates, borates, and other chemicals and coolants that do contain one or more of these additives.

Beyond these differences, some vehicles, such as hybrid electric vehicles (HEVs), have two or more different coolants, one for the engine and one (or more) for the hybrid drive systems (**Figure 26-3**). The coolant in the bottle on the left is for the engine, and the coolant in the bottle on the middle is for the high-voltage system. Note that both coolants look the same but each is different, and they are not interchangeable. In addition, supercharged engines may have a separate coolant type just for the intercooler. Because of all the variations in coolant specifications, it is critical that you identify the correct coolant for the application.

■ ***Water Pump.*** To circulate the coolant through the engine and cooling system, a **water pump** is used. A cutaway view of a water pump is shown in **Figure 26-4**. The water pump on most engines is driven either by the timing

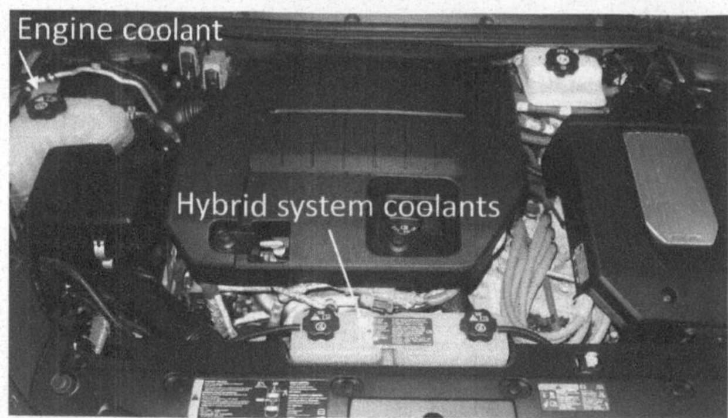

FIGURE 26-3 The two bottles of pink fluid are coolant reservoirs; the one on the left is for the engine and the one in the middle is for the hybrid system.

FIGURE 26-4 An inside view of a water pump.

belt/chain or by the accessory drive belt. An illustration of a timing belt–driven pump is shown in **Figure 26-5**. A few engines use an electrically driven water pump.

Regardless of how the pump is driven, they all circulate the coolant. The pump driveshaft turns an impeller. The impeller creates a low-pressure area as it spins. This causes the coolant to move into the low-pressure area. Opposite the low-pressure side of the impeller, the coolant exits and flows through the system.

■ ***Radiator.*** The **radiator** transfers the heat absorbed by the coolant to the outside air. As hot coolant flows into the radiator, it travels along a series of tubes (**Figure 26-6**). Attached to each tube are fins (**Figure 26-7**). As the coolant heats the tubes, the heat also travels to the fins. Air passing over the fins draws the heat away, which cools the fin, the tubes, and the coolant.

Two large-diameter hoses connect the engine to the radiator (**Figure 26-8**). One hose allows coolant to flow from the engine to the radiator and the other hose returns the coolant to the engine at or near the water pump. Rubber hoses are used to allow for engine movement during operation.

The radiator may have a radiator cap (**Figure 26-9**). This cap seals the radiator fill neck and is used to allow the pressure to increase in the cooling system up to a

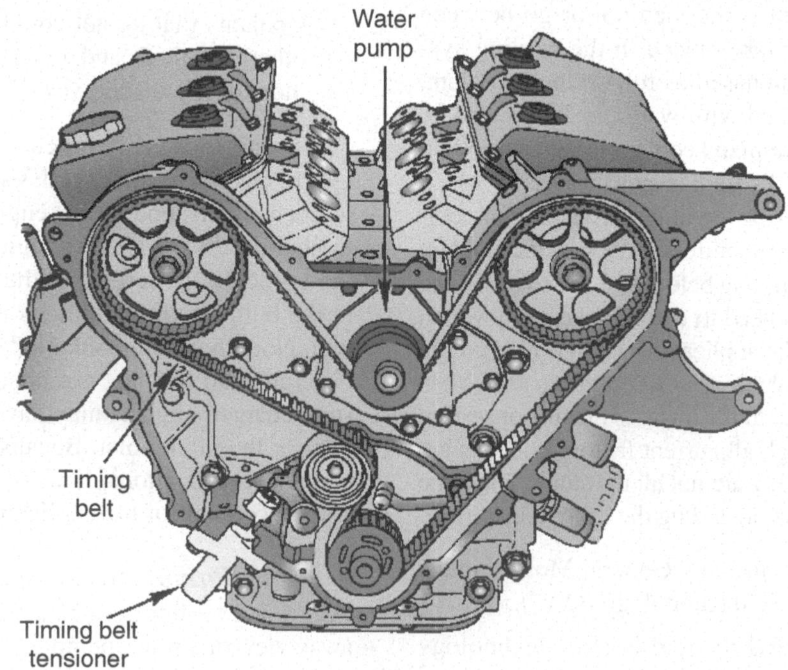

FIGURE 26-5 An example of an engine that drives the water pump with the timing belt.

certain point, typically between 14 and 18 pounds per square inch (psi). As the coolant picks up the heat from combustion it expands. This causes the pressure on the coolant to increase. As the pressure increases, the boiling point of the coolant also increases, about 3°F for every pound of pressure. This is important because modern engines often operate at temperatures of 225 to 235°F

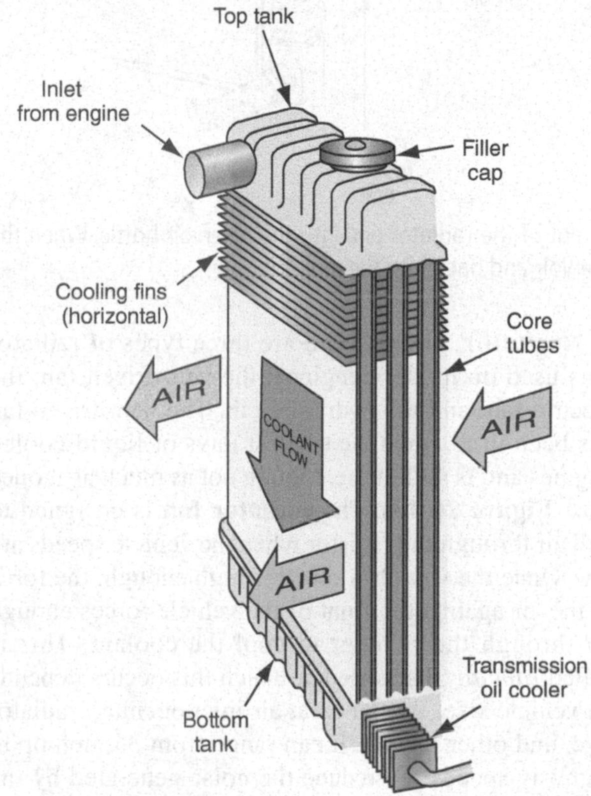

FIGURE 26-6 Radiators are either crossflow or downflow, depending on system design. This illustrates a downflow radiator.

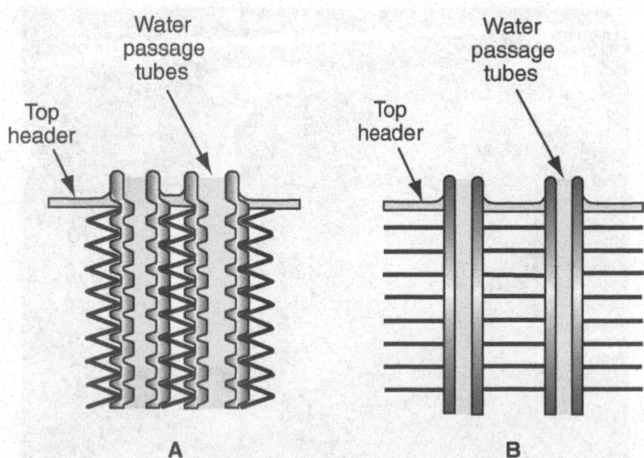

FIGURE 26-7 The fins surrounding the coolant tubes allow for a much greater surface area for the air to pass over and remove heat.

or higher. This type of cap has a pressure relief valve (**Figure 26-10**). If the pressure in the system reaches the predetermined pressure limit, the valve opens and allows coolant to flow out of the radiator and into an overflow or reservoir bottle (Figure 26-10). Once the coolant temperature and pressure decrease as the engine cools down, the coolant flows out of the reservoir and back into the engine.

This type of cap has a safety release position, used to allow pressure to be released when the cap is removed. Turning the cap to the release point allows coolant to flow from the radiator to the reservoir. Turning the cap to the second position releases the cap from the radiator fill neck (**Figure 26-11**).

Many cooling systems use a surge or **pressure tank**, also called *a degass bottle* (**Figure 26-12**). On these systems, the pressure cap is on the tank instead of on the radiator. In many cases, this type of cap is threaded to the tank. Turning the cap counterclockwise releases pressure in the system and allows the cap to be removed.

FIGURE 26-8 An example of a radiator hose connecting to the engine.

FIGURE 26-9 This is a common type of two-position radiator cap.

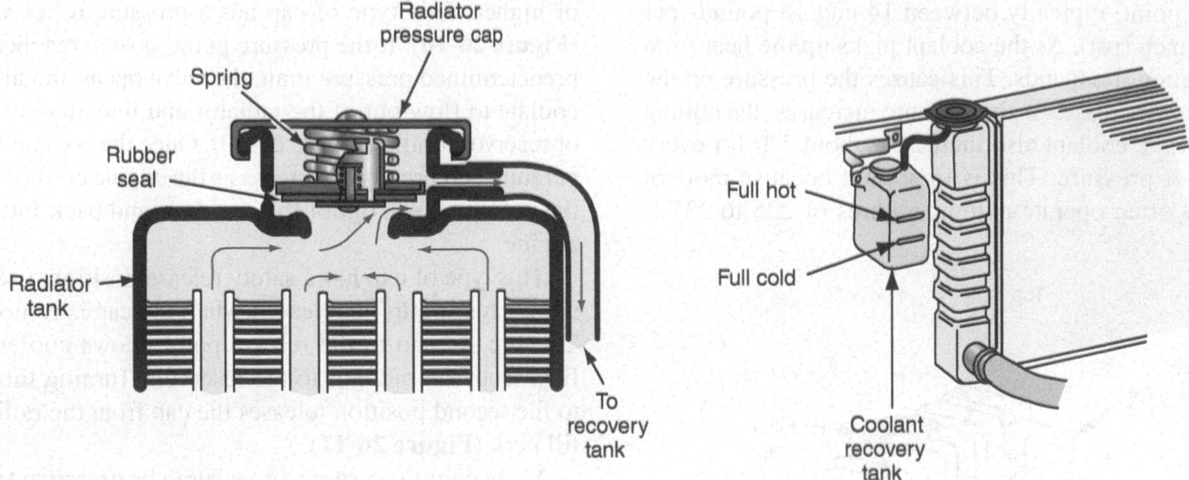

FIGURE 26-10 When the valve in the cap opens, the coolant passes out of the radiator and into the reservoir bottle. When the engine cools, the decrease in pressure draws the coolant out of the reservoir and back into the radiator.

This tank is the fill point and coolant reservoir for the system. Marks on the tank indicate if the coolant level is low, and also show a maximum fill line. The space in the tank above the maximum fill line allows for coolant expansion when it gets hot.

Never remove the radiator/cooling system cap on a hot engine. Hot, pressurized coolant can erupt from the system and cause severe burns.

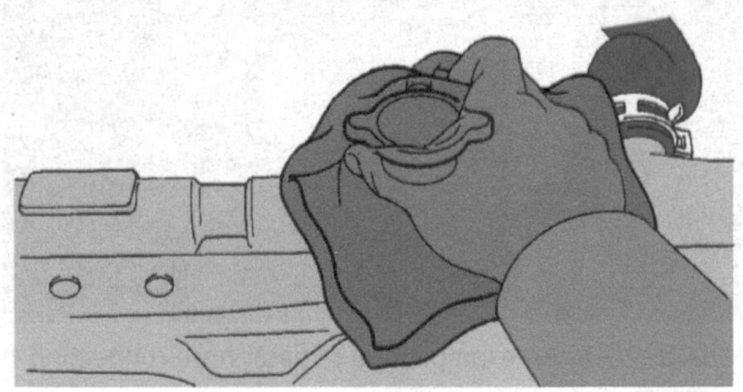

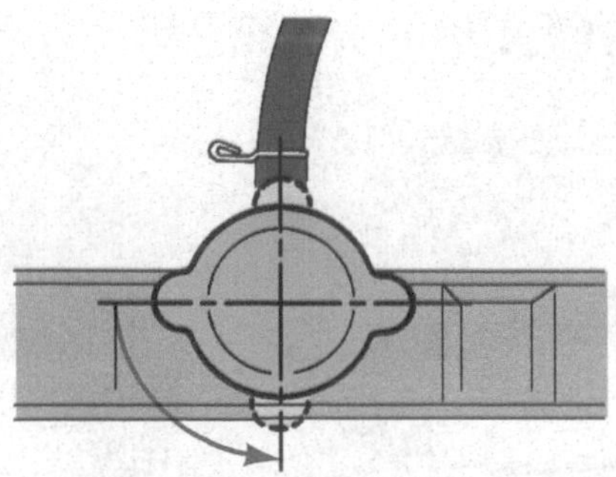

FIGURE 26-11 When removing a two-position safety cap, first turn the cap slowly counterclockwise to release the pressure. Once the pressure is released, press down and turn the cap counterclockwise until it stops, and remove the cap.

■ ***Radiator Fan.*** There are three types of radiator fans used on modern engines: the belt-driven fan, the electric fan, and the hydraulic fan. The belt-driven fan has been used since the earliest days of liquid-cooled engines and is still in use, though not as much as it once was (**Figure 26-13**). The **radiator fan** is designed to pull air through the radiator when the vehicle speeds are low. Once the vehicle's speed is high enough, the force of the air against the front of the vehicle forces enough air through the radiator to cool the coolant. This is called *ram air*. The speed at which this occurs depends on vehicle size, frontal area, air inlet opening, radiator size, and other factors. It can range from 35 mph up to highway speeds. To reduce the noise generated by the fan and to reduce vibration, fan blades are commonly unequally spaced.

A belt, driven by the engine crankshaft, drives the water pump pulley to which the fan is attached. This means that

FIGURE 26-12 An example of a pressure or surge tank. The caps on these are typically threaded and do not have a pressure release position like two-position caps.

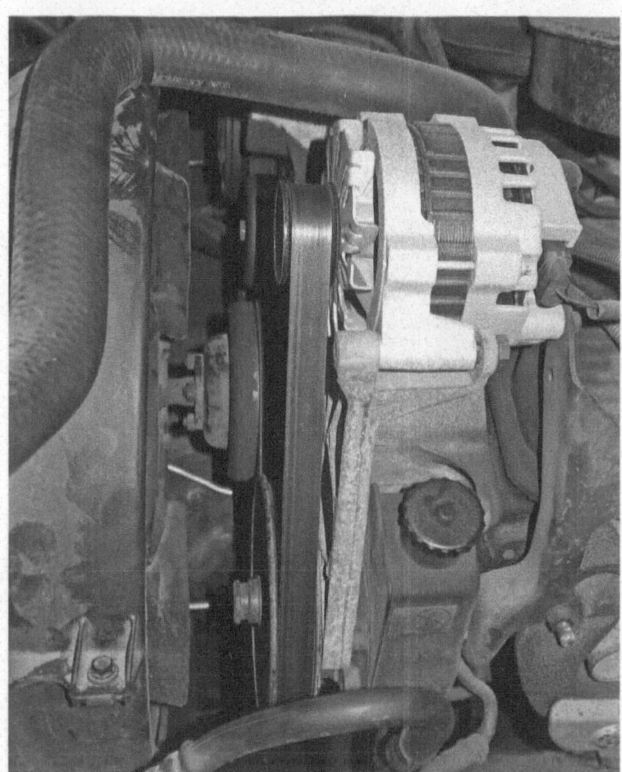

FIGURE 26-13 An example of a belt-driven fan.

when the engine is running, the fan is spinning and moving air. This is a waste of energy because a cold engine does not require air to be pulled through the radiator. Also, once the ram air effect is sufficient, the radiator fan is no longer needed, but it still spins.

To reduce the power loss that results from driving the fan all the time, a fan clutch is used. A fan clutch allows the fan to uncouple from the drive pulley when it is not needed (**Figure 26-14**). A thermostatic spring in the clutch is used to sense the temperature of the air passing through the radiator. If the air temperature is low, the spring unlocks the clutch, allowing the fan to freewheel. When the air temperature increases to a set point, the spring moves and locks the clutch and fan together.

Enclosing the fan area behind the radiator is a fan shroud (**Figure 26-15**). This plastic shroud is used to direct the airflow from the fan through the radiator. Without the shroud in place, the effectiveness of the fan is reduced. This is because the combination of the fan and the shroud together increase the speed of the air passing across the radiator. This increases cooling capacity as it directs airflow through the entire core section of the radiator. Removing the shroud eliminates the channeling of the airflow and reduces the efficiency of the cooling system.

Most vehicles now use electric cooling fans (**Figure 26-16**). The fan or fans mount directly to the radiator in a fan shroud. The engine control module (ECM) controls

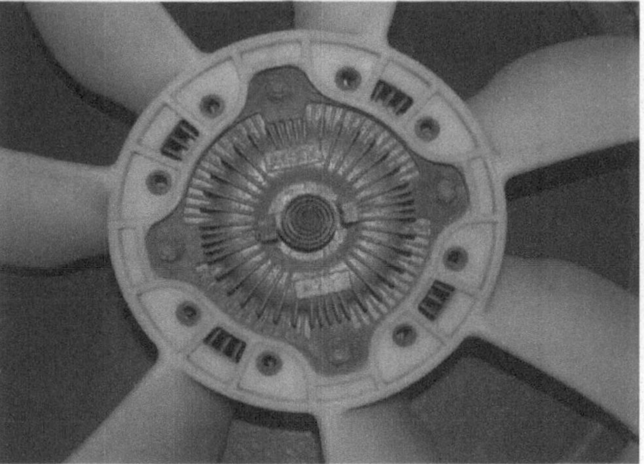

FIGURE 26-14 An example of a fan clutch. The spring in the middle is the thermostatic spring, used to lock and unlock the clutch based on temperature.

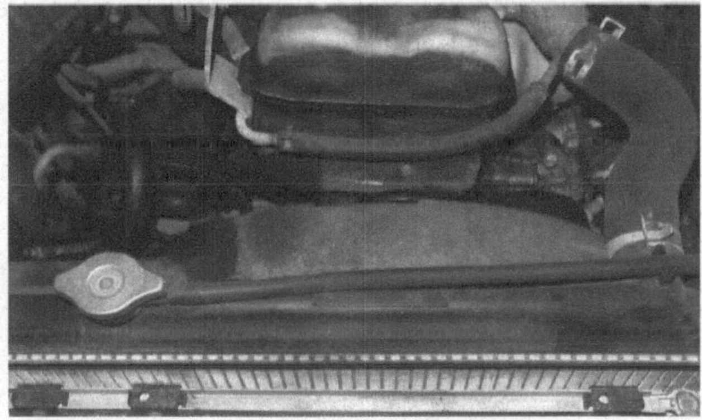

FIGURE 26-15 An example of a fan shroud. The shroud is important for directing air through the radiator for proper cooling.

the operation of the fan. The ECM monitors engine coolant temperature through a sensor located in the engine, typically near the thermostat. When the coolant temperature reaches a certain point, often around 220 to 230°F, the ECM grounds the fan relay, which turns on the fan. An example of this circuit is shown in **Figure 26-17**. Because the fan only runs when it is needed, engine temperature is more precisely controlled and there is less wasted energy than with a belt-driven fan. Older vehicles may have a sensor or switch mounted in the radiator that is used to control cooling fan operation.

Some vehicles are using hydraulically operated cooling fans. In the system used on Jeep models, high-pressure fluid from the power steering system passes through a fan drive motor. The fluid drives the fan and then either passes on to the steering gearbox or is routed back to the power steering fluid reservoir. Fan speed is controlled by a fan control module and is based on engine coolant temperature, air conditioning system pressure, and transmission fluid temperature.

FIGURE 26-16 An example of an electric fan. These fans are enclosed in a combination housing and shroud.

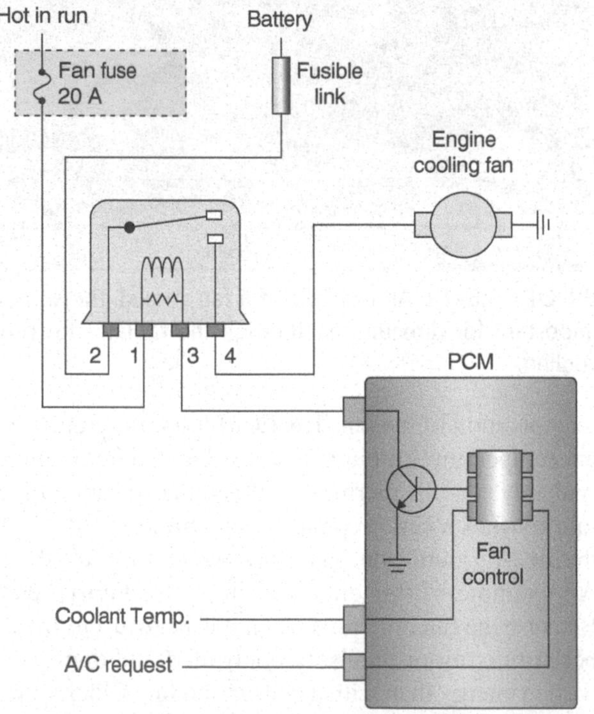

FIGURE 26-17 An example of a basic cooling fan circuit.

■ ***Thermostat.*** The **thermostat**, usually located in the intake manifold, prevents coolant flow from the engine to the radiator until the coolant temperature reaches a certain point (**Figure 26-18**). Keeping the coolant in the engine allows the coolant to pick up heat and supply heated coolant for warming the passenger compartment. Without a thermostat, the coolant circulates constantly, and the heat absorbed by the coolant is constantly transferred at the radiator. While this may sound like a good idea, it is not because for the engine to operate efficiently, it must operate at high temperatures. Constantly flowing coolant, especially in cool or cold weather, keeps engine temperature too low. In general, the thermostat begins to open around 170°F, and it is fully open by about 195°F. An engine without a thermostat or one that is stuck open may operate at only 140 to 160°F.

FIGURE 26-18 An example of a thermostat.

The thermostat also acts as a restrictor to coolant flow. Even when fully open, the thermostat restricts the flow of coolant from the engine to the radiator. This is to allow sufficient time for the coolant to be able to pick up heat from inside the engine. Without the thermostat, the coolant could pass through the engine so quickly that it would not remove enough heat to effectively manage engine temperature.

Inside the thermostat is a wax pellet and a piston (**Figure 26-19**). When the coolant temperature is high enough, the wax begins to melt and flow. This moves the piston upward, which opens the flange and allows coolant to pass through. When coolant temperature decreases, the spring overcomes the pressure of the cooled wax and closes the flange.

■ ***Heater Core.*** The **heater core** is a mini radiator, used to supply heat to the passenger compartment (**Figure 26-20**). Small-diameter heater hoses connect the

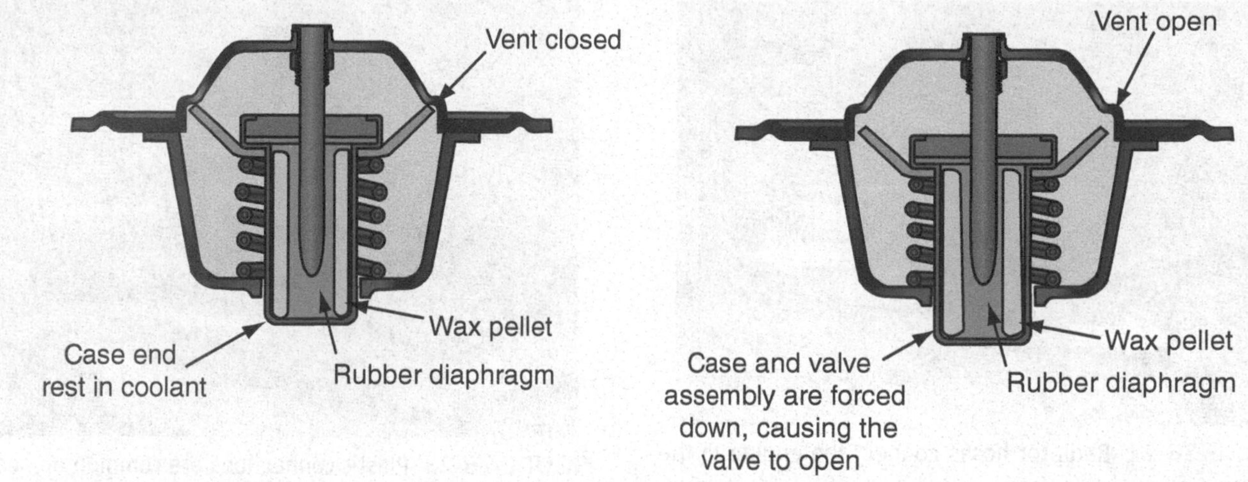

FIGURE 26-19 As the wax heats, it moves and opens the thermostat. The spring is used to close the thermostat when coolant temperature drops.

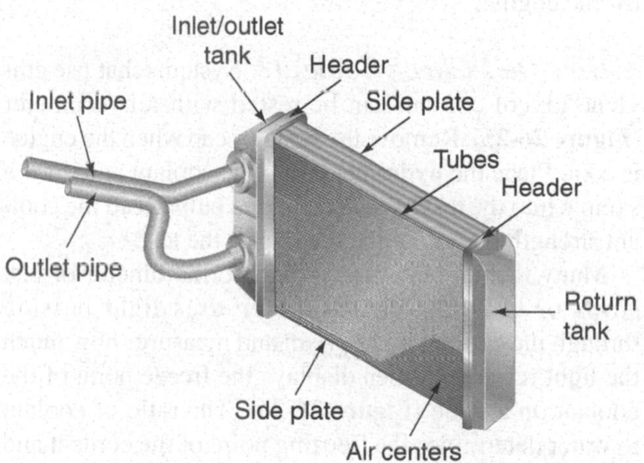

FIGURE 26-20 An example of a heater core. Many cores are aluminum and use plastic pipes.

FIGURE 26-21 A heater control valve shuts off the flow of coolant when cooler air is selected by the passengers.

heater core to the engine. Coolant flow to the core may be controlled by a heater control valve (**Figure 26-21**). The control valve is operated by the heater control head in the passenger compartment and is used to block the flow of hot coolant when it is not wanted, such as during warm weather. Engines without a heater control valve have coolant constantly circulating through the core.

As the engine and coolant warm up, the heater core temperature also increases. The passenger compartment fan blows air across the core, warming the air coming from the vents.

It is important to note that cooling system components, especially the radiator and heater core, should be electrically insulated from the engine and body so that current flow through these components does not occur. Current flowing through cooling system components can lead to metal damage and coolant leaks.

■ ***Hoses.*** Rubber hoses and steel lines are used to connect the various cooling system components together. **Figure 26-22** shows an example of a rubber upper radiator hose. Rubber hoses are used to allow for engine movement during operation. Many vehicles have plastic connections between hoses and lines (**Figure 26-23**).

COOLING SYSTEM SERVICE

The cooling system requires periodic service to keep the engine and other cooling system components protected against corrosion. Over time, the additives in the coolant

FIGURE 26-22 Radiator hoses connect the engine to the radiator.

FIGURE 26-23 Plastic connections are common on heater hoses.

break down, and the ability of the coolant to provide rust and corrosion protection is reduced. Cooling system hoses and other components are subject to high and very low temperatures and go through thousands of pressure cycles. In addition, the rubber in the hoses and metals used on other parts will eventually break down from exposure to chemicals and oxygen.

For the cooling system to operate correctly, it must be sealed and pressure tight. As the coolant heats and expands, its volume increases, and requires an overflow or expansion tank to hold the coolant. If a leak develops in the system, instead of the coolant moving from the radiator into the overflow and back as it cools, coolant will leak out and air will be drawn back in as the system cools (**Figure 26-24**).

Air entering the system can lead to corrosion as the oxygen reacts in the cooling system. In addition, leaks cause air pockets in the system. Air pockets can affect heater output and more importantly, cause uneven cooling of components, such as the cylinder head, which can lead to overheating and engine damage. Inspection and maintenance of the cooling system is of vital importance for the engine.

■ ***How Do I Check Coolant?*** Systems that use ethylene glycol coolant can be tested with a hydrometer (**Figure 26-25**). Remove the radiator cap when the engine is cool. Place the hydrometer into the coolant and draw a sample into the test by squeezing the bulb. Read the coolant strength by noting the reading in the test.

Many technicians use either a refractometer or test strips or both. A **refractometer** uses light passing through the coolant and a prism and measures how much the light refracts. It then displays the freeze point of the coolant on a scale (**Figure 26-26**). The ratio of coolant to water determines the freezing point of the coolant and affects the boiling point. Remove the radiator cap when the engine is cool and place few drops of coolant on the refractometer. Hold the tool so you can look through the lens and note the reading on the scale.

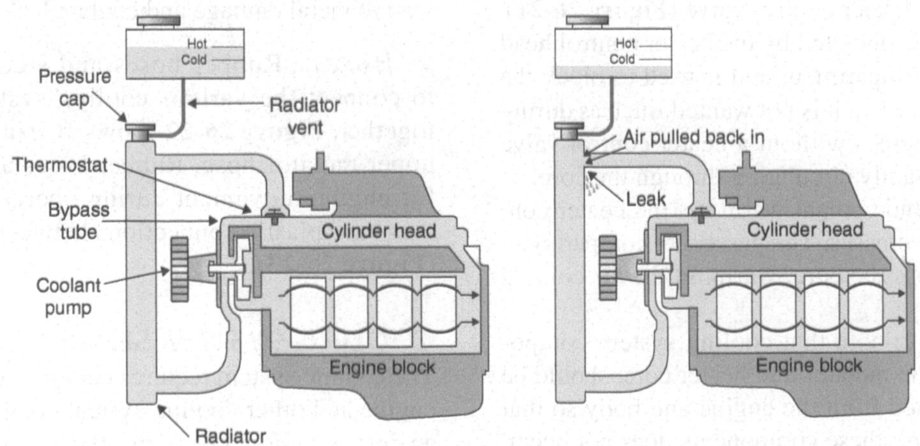

FIGURE 26-24 When a leak occurs, air is drawn back into the system.

Coolant condition can also be checked by using test strips (**Figure 26-27**). These strips can determine the coolant protection factor and pH levels. To use, insert a test strip into the coolant for about one second and remove. Compare the color of the test pads on the strip to the chart on the strip's container. Coolant pH is typically between 8 and 11. As the pH drops, the coolant becomes acidic, which can be an indicator that the additives in the coolant are depleted.

■ ***What Coolant Should Be Used?*** Before performing any work on the cooling system, first determine what type of coolant is required. There are many different types of coolant in use. Color alone is not the way to tell which coolant to use. Refer to the vehicle owner's manual or service information. As a general rule, domestic vehicles made before the mid-1990s use inorganic acid technology (IAT) green ethylene glycol coolant. Since the 1990s, most manufacturers have been using organic acid technology (OAT) or hybrid organic acid technology (HOAT) extended or long-life coolants (**Figure 26-28**).

FIGURE 26-25 Using a coolant hydrometer to check the condition of the coolant.

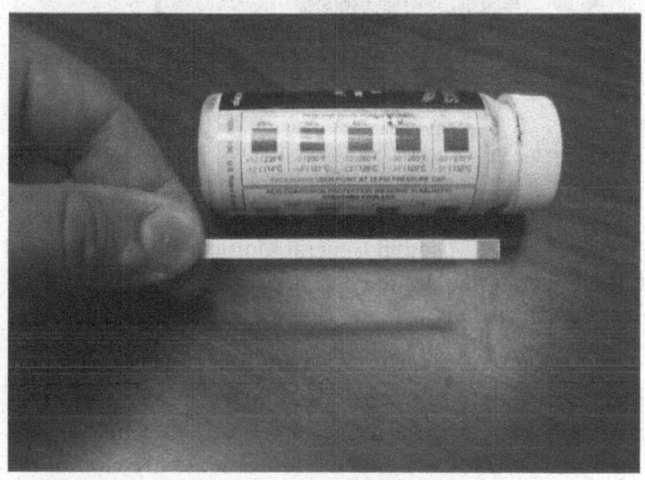

FIGURE 26-27 An example of a coolant test strip.

1. Place a few drops of the sample fluid on the measuring prism and close the cover.

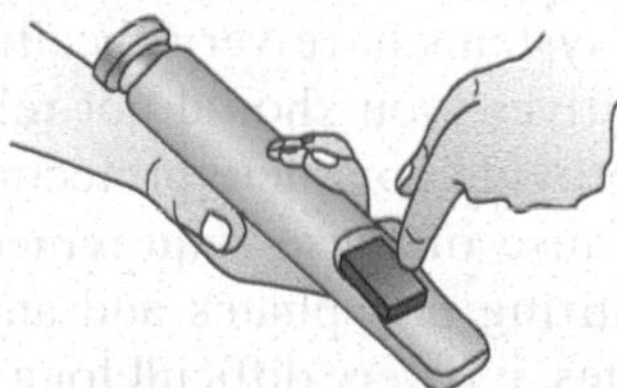

2. Hold up to a light and read the scale.

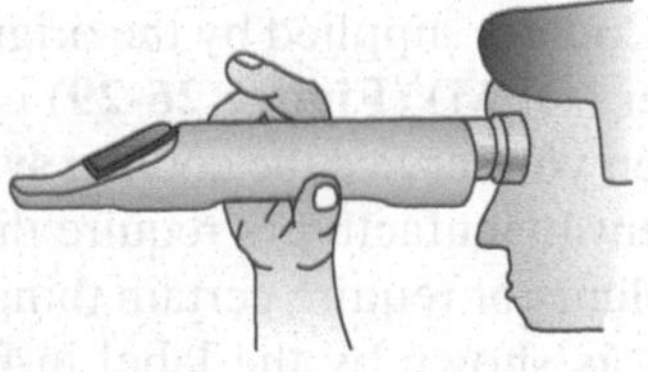

FIGURE 26-26 This illustrates using a refractometer to determine coolant condition.

FIGURE 26-28 Used in General Motors vehicles, Dex-Cool long-life coolant is orange and is not compatible with traditional green coolant.

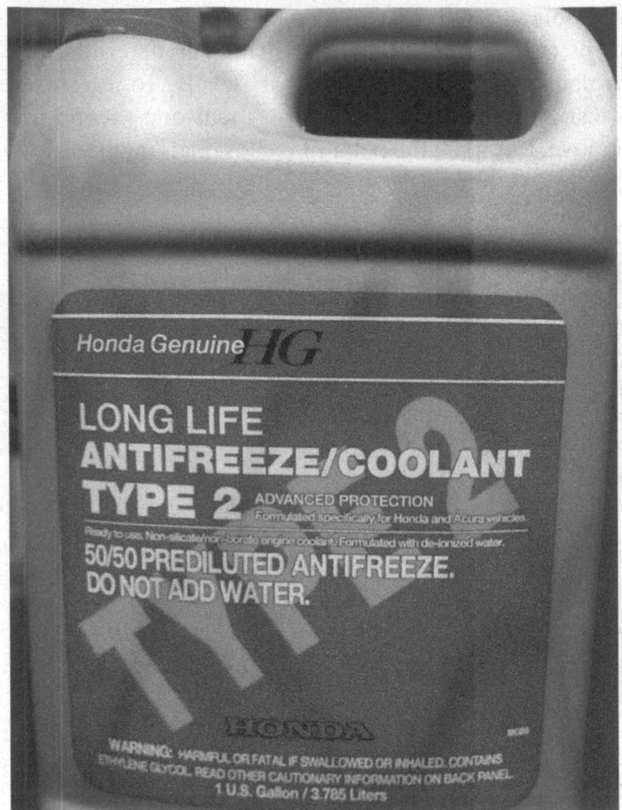

FIGURE 26-29 An example of original equipment (OE) coolant.

These coolants are not compatible with traditional IAT green coolant and cannot be mixed together in the cooling system. Even extended life coolants are not all the same and cannot be mixed together.

Universal coolants are available in full strength and premixed 50/50 solutions. However, because some cooling systems have very specific requirements regarding additives, you should not rely on a universal coolant to provide complete protection for any given vehicle. Because of these requirements, such as one vehicle requiring phosphates and another requiring no phosphates, it is very difficult for a universal coolant to either have or not have a specific additive for the vehicle being serviced. The best solution for the long-term health of the cooling system is to use the coolant specified by the vehicle manufacturer.

Coolant supplied by the original equipment manufacturer (OEM) (**Figure 26-29**) is usually the best choice when you service the cooling system. This is because different manufacturers require different additives in their coolants or require certain things to *not* be in their coolant, as shown by the label in **Figure 26-30**. An example of another manufacturer's concern about using the correct coolant is shown in **Figure 26-31**. It is important that when you top-off or refill the system, the mix of pure water and coolant remain close to 50/50.

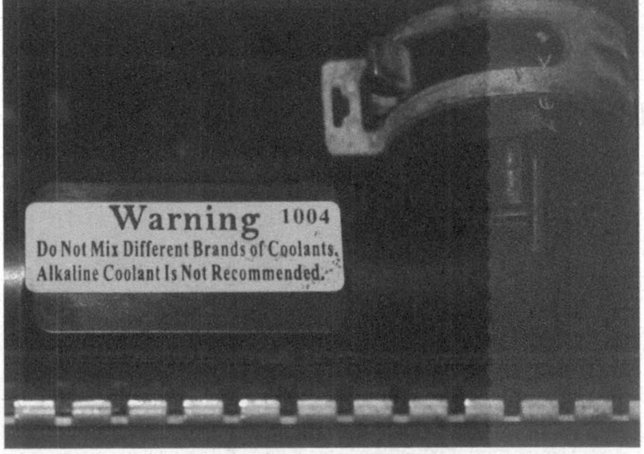

FIGURE 26-30 An underhood decal warning what type of coolant should not be used.

■ ***How Do I Find a Coolant Leak?*** To locate a leak in the cooling system, you may need to perform a pressure test. This test uses a cooling system pressure

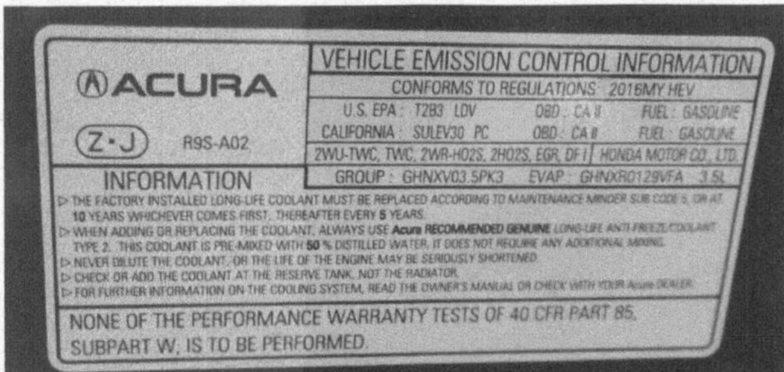

FIGURE 26-31 This label contains important information about using the correct coolant.

FIGURE 26-32 Using a manual pressure tester to pressurize the cooling system to check for leaks.

tester to apply pressure to the system for inspection. Testers can be manual (**Figure 26-32**), meaning you have to pump the tester to apply pressure to the system, or they can use shop air pressure (**Figure 26-33**). Regardless of the type, adapters are used to connect the tester to the radiator or pressure tank.

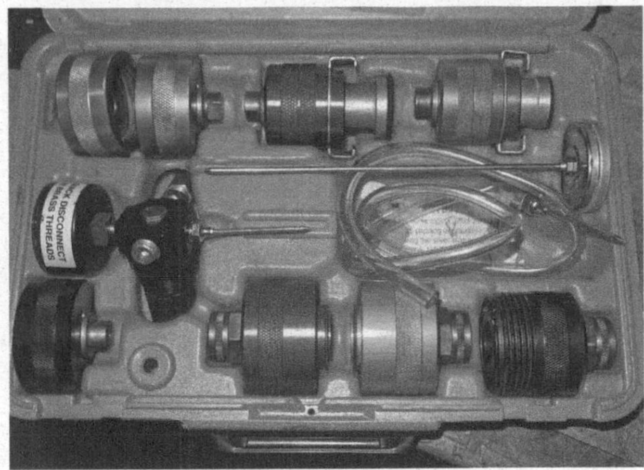

FIGURE 26-33 An air-powered cooling system pressure tester.

With the engine cool, carefully remove the radiator cap. Inspect the underside of the cap and radiator fill neck for damage. Next, select the appropriate adapter and attach it to the fill neck. Check the radiator cap for the maximum pressure for the cooling system (**Figure 26-34**). Some caps, like the cap on the right in Figure 26-34, are rated in bar pressure instead of psi. One bar is equivalent to atmospheric pressure at 14.7 psi. A cap rated at 0.9 bar maintains pressure at about 13 psi. The pressure on the cap is the maximum amount of pressure the system can safely maintain.

Once the correct adaptor is installed, begin to apply pressure to the cooling system; **do not exceed the pressure shown on the cap.** Exceeding maximum system pressure can damage components and cause leaks. Watch the pressure gauge on the tester. If it remains steady, it means there is not a large leak. Examine the components of the cooling system and look for evidence of leaks and traces of coolant (**Figure 26-35**). If the needle drops quickly or pressure cannot be maintained at all, look

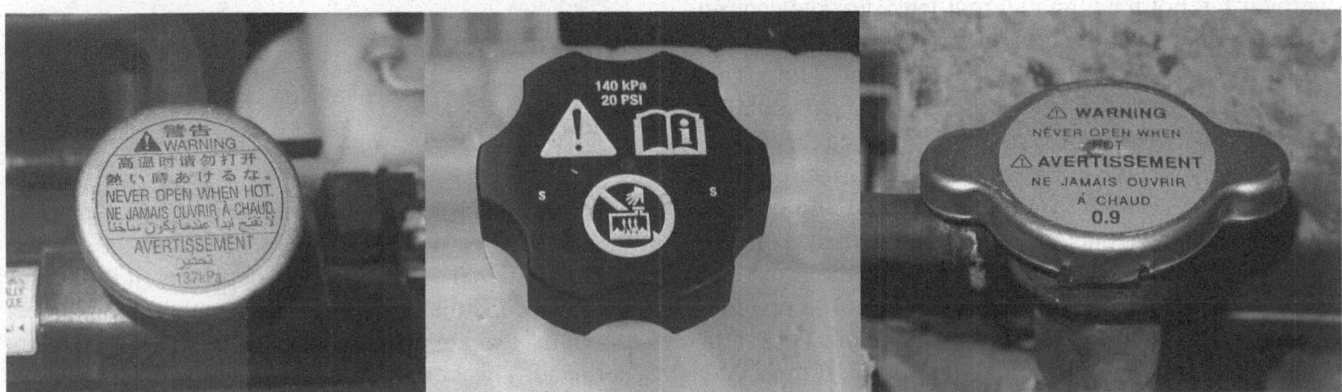

FIGURE 26-34 The cooling system pressure can be found on the cap on most vehicles.

FIGURE 26-35 An example of a small coolant leak leaving a trace on the engine.

for a large leak. Listen for the hissing sound of pressure escaping or for bubbling. Look around the engine compartment and under the vehicle for dripping coolant. If no sign of a leak is visible from the engine compartment yet the system will not hold pressure, check inside the passenger compartment for coolant leaking from the floor vents. A leaking heater core may leak coolant onto the floor or allow coolant vapor to combine with the air from the interior fan. This often looks like steam coming from the vents with the engine running and the heater on. An example of a leaking heater core is shown in **Figure 26-36**.

A small leak, depending on its location, can be difficult to locate as the coolant moves across and down various components. To help locate the source, an ultraviolet (UV) dye can be added to the cooling system. Once dye is added, run the engine until it reaches normal operating temperature to allow the dye to circulate completely. Next, using a special UV light and glasses, shine the light over the engine compartment while you are looking for the glow of the dye.

A faulty pressure cap may cause a cooling system to overheat but not have an external leak. For the cooling system to operate at the correct temperature, the system pressure is raised, typically between 13 and 20 psi. This is to increase the boiling point of the coolant. The ability to increase and maintain system pressure is a function of the cap. A cap that cannot maintain the correct pressure can cause the cooling system to overheat and boil. To check the cap, a special adapter is used with the pressure tester (**Figure 26-37**). Attach the cap to the tester and apply pressure. The cap should hold pressure up to its maximum rating and then bleed pressure off. For example, a cap rated at 15 psi should hold 15 psi. If more than 15 psi is applied to the cap, the pressure should release

FIGURE 26-36 Coolant coming out of the floor air vents from a leaking heater core.

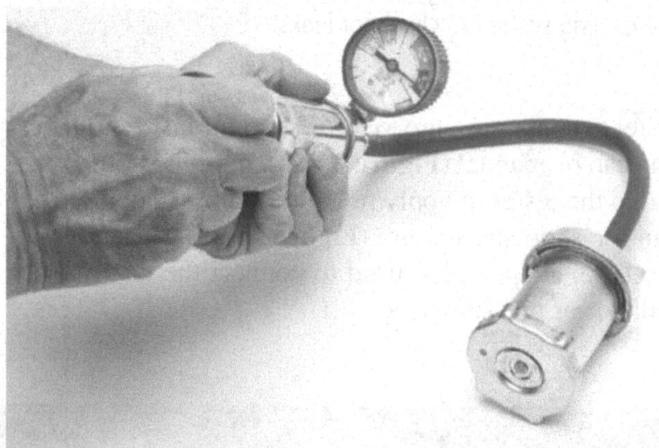

FIGURE 26-37 Testing a radiator cap to see if it can hold system pressure.

and drop to 15 psi and hold. A cap that does not maintain the correct pressure needs to be replaced.

If an external coolant leak is not present, it is possible that the coolant is leaking into the combustion chamber through a leaking cylinder head gasket or cracked cylinder head. This condition often causes excessive smoke from the exhaust but not in all cases. A combustion leak increases the pressure and temperature in the cooling system and usually causes an overheating condition. To check for this problem, connect a cooling system pressure tester to the radiator or pressure tank. Start the engine and run it at 2,000 rpm, and note the pressure on the tester's gauge. If there is a rapid increase in pressure, a combustion leak is likely. Another test involves using a special test kit called a *block tester*. This test draws vapor from the cooling system into a special fluid which changes color if combustion gases are present.

■ ***Inspect and Service Cooling System Hoses.*** Cooling system hoses are made from reinforced rubber and polymers that provide flexibility and durability, typically for many years of vehicle operation. Eventually, though, the hoses and the clamps that secure them will weaken and fail, causing coolant leaks.

Inspect the hoses and connections for signs of coolant loss (**Figure 26-38**). This type of leak, while slight, not only allows coolant to leak out of the system, but also allows air to be pulled in. A ruptured hose can cause a small leak to become a major leak and lead to total loss of coolant. This can cause severe engine damage or destruction. Also during your inspection, squeeze the hose to check its rigidity. A hose may look fine from the outside, but actually be weak and near failure on the inside (**Figure 26-39**). A hose should be pliable but not soft while being firm but not hard or brittle. **Figure 26-40** shows examples of common hose problems.

Hose clamps secure the hoses in place (**Figure 26-41**). Over time, the clamps lose their tension, which can allow coolant to seep through the connection. Some hose connections, such as those used where a rubber heater hose connects to a steel pipe, may use a compression clamp to secure the hose. This type of clamp must be cut off and a new hose clamp installed when you replace the hose.

To replace a damaged or leaking hose, first drain the coolant into a clean shop pan. Most radiators have a drain (**Figure 26-42**). Remove the radiator cap, place a pan under the radiator, and open the drain. To help avoid making a mess, attach a length of rubber hose to the drain spout, and place the other end in the drain pan. This greatly reduces the amount of coolant that splatters on the shop floor. Once the coolant is drained, remove the hose clamps and secure the hose in place. The hose may be corroded in place, making it difficult to remove. In this case, use a utility knife to cut the hose and peel it back. Do not use a screwdriver or other tool to pry the hose off its connection. Many cooling system parts are plastic and will snap if you pry on them.

Once the hose is removed, inspect its connection for corrosion. The outlet housing that is shown in **Figure 26-43**

FIGURE 26-38 An example of a small leak around a hose connection.

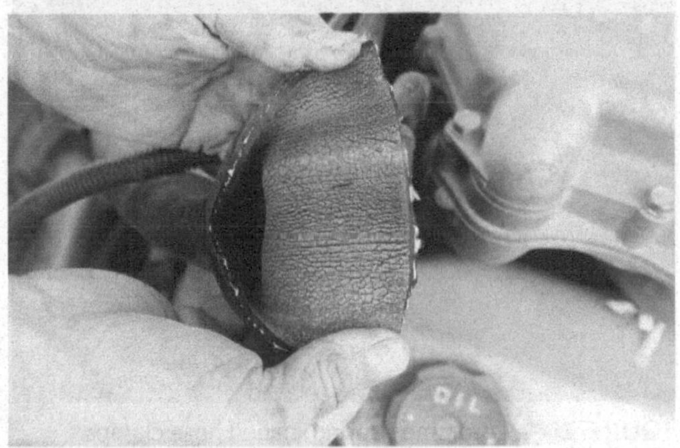

FIGURE 26-39 Hose degradation and failure from the inside.

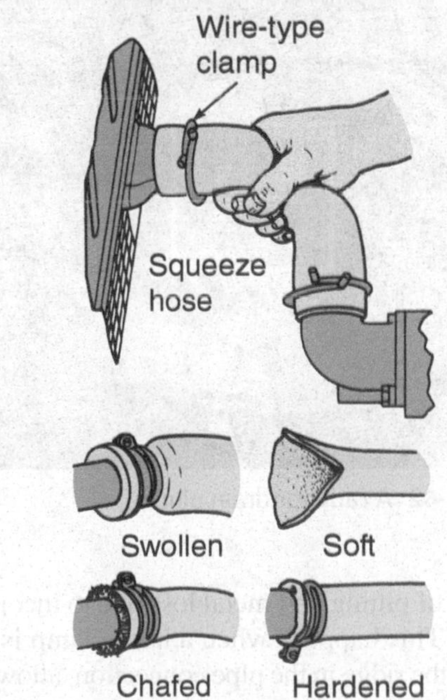

FIGURE 26-40 Check hoses for problems by visual and hands-on inspection.

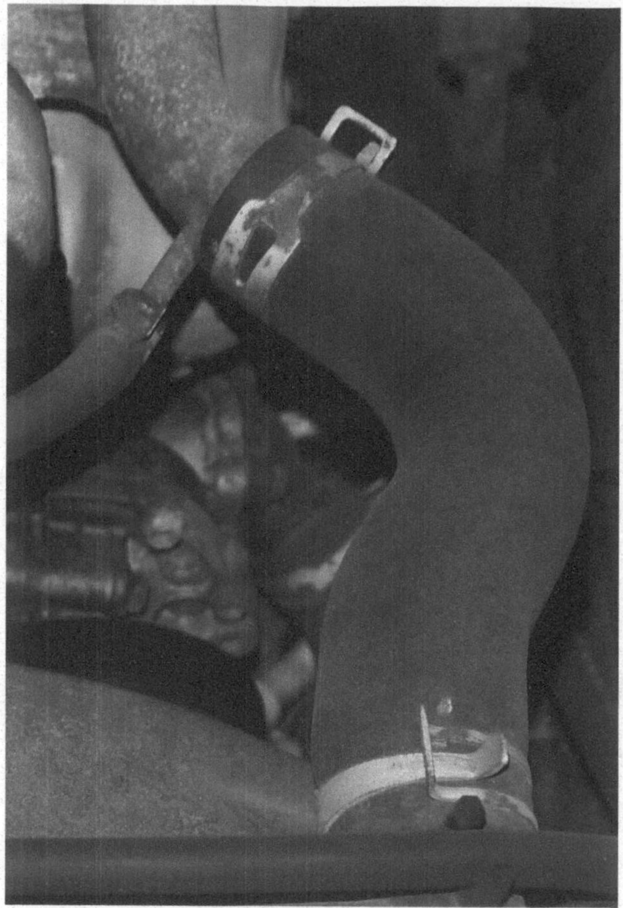

FIGURE 26-41 Common spring-loaded hose clamps.

FIGURE 26-42 A radiator drain plug.

FIGURE 26-43 Erosion of the housing is due to coolant staying trapped against the metal from improper hose clamp position.

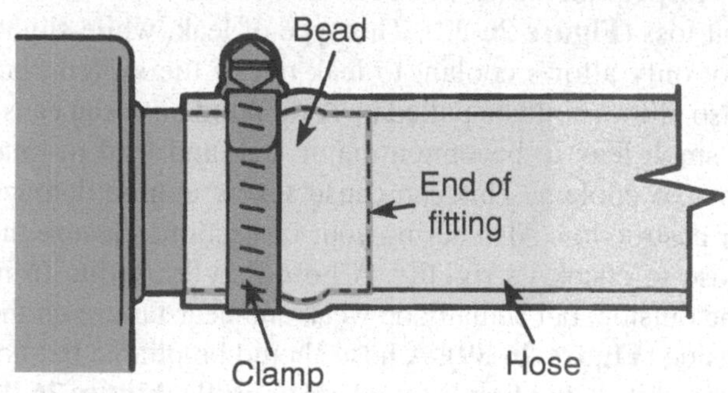

FIGURE 26-44 How to properly place the hose clamp in relation to the pipe.

shows signs of pitting and metal loss due to incorrect hose installation. This happens when a hose clamp is installed too far from the ridge in the pipe connection, allowing coolant to accumulate in the gap between the hose clamp and the end of the pipe (**Figure 26-44**). When you install a hose, make sure the clamp is placed as close to the ridge as possible.

Before you install the new hose, double-check to confirm that it is the correct shape and size for the application. Installing a hose that is not the correct shape may allow it to contact other components, which can damage the hose. Next, install the hose clamps over the hose and fit the hose into place. Depending on the type of clamp, you may need to tighten the clamp with a screwdriver or nut driver. Do not overtighten the clamp. This will damage the hose and could even crack the component to which the hose is attached. If the hose is secured with a spring clamp, a special tool may be needed to spread and hold the clamp during installation (**Figure 26-45**).

Once the hose is installed, tighten the radiator drain plug, and refill the cooling system with the correct type of coolant. You may need to bleed air from the cooling system after hose replacement. This is covered later in this chapter.

■ ***Inspect and Replace the Engine Thermostat and Verify Engine Operating Temperature.*** A malfunctioning thermostat can cause two different concerns, low engine temperature and a lack of heat or poor

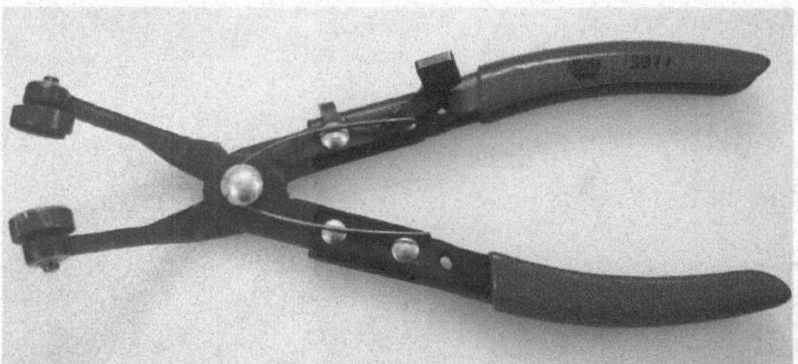

FIGURE 26-45 An example of hose clamp plyers for spring-loaded clamps.

heater performance or higher-than-normal engine temperature and overheating. A thermostat that is stuck open will cause low engine operating temperature and poor heater performance (**Figure 26-46**). This is because instead of keeping the coolant inside the engine during warmup, the coolant circulates and dissipates heat as it flows through the radiator. In addition to poor heater performance, this can cause an increase in fuel consumption and exhaust emissions and cause the malfunction indicator light (MIL) to come on. The illuminated MIL will be caused by a diagnostic trouble code (DTC) set for the engine temperature being too low. A vehicle with a stuck-open thermostat typically has poor heat output temperature, and a temperature gauge that does not rise into the normal range.

If the fill neck for the cooling system is at the radiator or other location where coolant flow can be observed, you may be able to see if the thermostat is stuck open by observing the coolant flow. When the engine is cold, remove the cap, start the engine, and look down into the radiator. If coolant can be seen circulating, the thermostat is likely stuck open.

A thermostat that is stuck closed will cause the engine to overheat because the coolant will not be able to exit the engine and flow through the radiator. This usually causes the engine temperature to increase quickly, to the point where the temperature gauge is well above normal. If left running, the engine will overheat and boil over. This is dangerous, as extremely hot coolant and steam spews from the engine and is likely to cause engine damage, such as blowing a head gasket. To check for a stuck closed thermostat, run the engine to normal operating temperature. If possible, connect a scan tool to monitor the engine coolant temperature sensor data. If the radiator hoses remain cool to warm and the engine temperature is reading normal or hot, the thermostat is likely stuck. Shut the engine off, and let it cool before you replace the thermostat.

Once you have determined a thermostat is faulty, it will need to be replaced. First, determine the location of the thermostat. Locations vary based on the type of

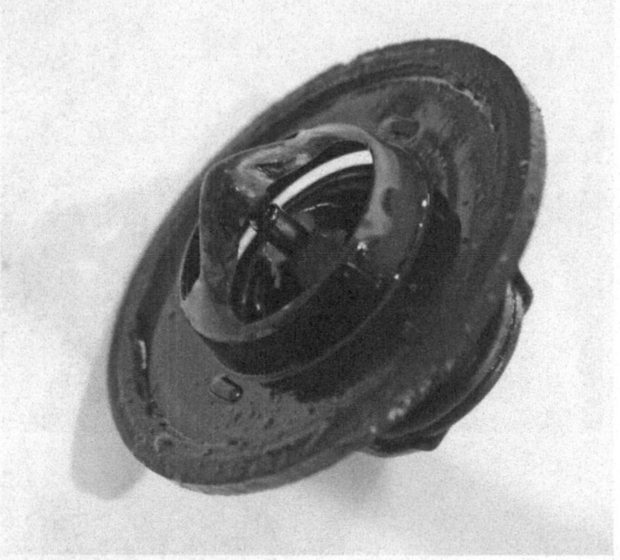

FIGURE 26-46 This thermostat is stuck open, preventing the vehicle from reaching operating temperature and causing very poor heater output.

engine and whether the cooling system is standard or reverse flow. The two most common locations for the thermostat are at the upper radiator hose connection to the engine and at the lower radiator hose connection to the engine. Locations vary greatly from engine to engine, as does the work required to replace the thermostat. If you are not sure where the thermostat is located, refer to the vehicle's service information.

To replace the thermostat, let the engine cool and drain the coolant. In general, remove the bolts that hold the thermostat housing to the engine, and then remove the thermostat and seal (**Figure 26-47**). Most housings use a rubber seal (Figure 26-46), and others use a paper gasket between the housing and the engine. To remove the gasket from an aluminum surface, use a carbon or plastic scraper so the aluminum is not damaged. If necessary, apply a gasket removal chemical.

Once the gasket surfaces are clean, install the thermostat. Be sure the temperature-sensing pellet is placed into the engine correctly. Though rare, a thermostat can sometimes be installed upside down. This will cause the engine to overheat. Reinstall the thermostat housing and tighten the bolts to specifications. Refill the system with the correct coolant and start the engine. Place the heater on "hot" and turn on the interior fan.

Check for leaks as the engine warms up. To monitor the engine temperature, watch the coolant temperature gauge on the dash or install a scan tool and check engine coolant temperature sensor data (**Figure 26-48**). As the coolant temperature approaches 170°F (76.6°C), the thermostat should begin to open. This will likely cause

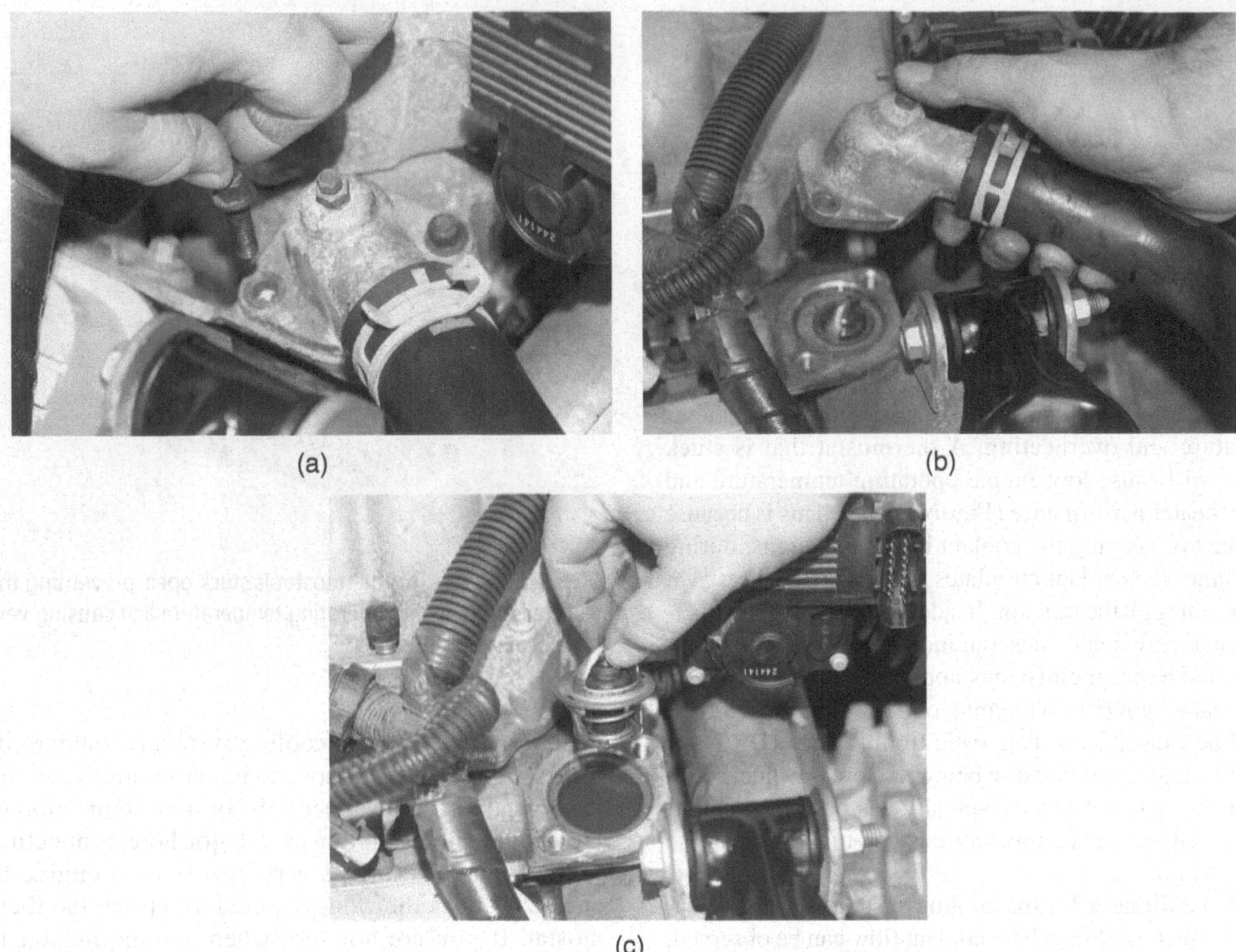

FIGURE 26-47 (a) Locate and remove the thermostat housing bolts once the coolant has drained. (b) Remove the housing from the engine. (c) Remove the thermostat.

VIEW SCANNER

Chevrolet 1981-2010

FUEL TRIM DATA

CAT PROTECTION ACTIVE	NO
POWER ENRICH	NO
DCEL FUEL C/OFF	NO
EVAP PURGE(%)	13
ENGINE SPEED	674
DESIRED IDLE	672
COOLANT(°C)	93
INTAKE AIR(°C)	36
MAF(gm/Sec)	2.80
AIRFLOW(gm/Sec)	2.80
ENGINE LOAD(%)	31

1 2 3 4

FIGURE 26-48 An example of scan tool data showing engine coolant temperature.

air bubbles to rise in the filler neck. Once the thermostat opens completely, typically by 195°F (90.5°C) you will likely need to add coolant to completely refill the system. When it is full, install the radiator cap and continue to allow the engine to warm up while you monitor the temperature on the scanner. If the system uses an electric radiator fan, expect the fan to come on between 220°F and 230°F (104.4°C and 110°C). Once you have verified that the fans are working properly and that there are no leaks, turn off the engine and disconnect the scanner. Top off the coolant in the overflow if necessary.

■ ***Drain, Flush, Refill, and Bleed the Cooling System.*** As discussed earlier in this chapter, draining the coolant is performed by opening the drain on the radiator. Some vehicles may also have a drain on the engine block. Some vehicles, particularly larger vans with rear heating, may have a separate drain for the rear heater core. Because radiator drains are typically made of plastic, many technicians do not attempt to open them because of the chance of the drain breaking or not closing completely. For this reason and because some vehicles do not have drains installed in the radiator, removing the lower radiator hose to drain the coolant is an option.

Flushing the cooling system removes the old coolant and circulates clean water through the system to remove rust and corrosion. A cooling system flush should be performed anytime the system has experienced a failure, such as from a leaking radiator, heater core, intake manifold gaskets, or other leak that has compromised the coolant. This includes if the system has been topped off with universal coolant or even tap water, as both can change the chemical balance of the coolant and cause corrosion problems. A leak lets air into the system and air can cause reactions in the coolant, which can lead to other failures. **Figure 26-49** shows an example of coolant that has been exposed to air from a leak in the system.

There are several ways a cooling system can be flushed; which method is used will usually depend on the type of vehicle and what equipment the shop has.

- A flush machine is connected to the system, and the old coolant is pumped out as new coolant is pumped in.
- A "T" adapter is installed in a heater hose that allows you to connect a garden hose to the system to flush out the old coolant.
- The thermostat is removed once the system is drained and the system refilled with water. Then the engine is run to allow the water to circulate and is then drained. The thermostat is reinstalled and the system is filled with fresh coolant.
- Refilling the drained system with water, and then running the engine to normal temperature, then draining the water and refilling with coolant.
- A pulsating flush gun, connected to a water hose and a shop air hose, is used to flush the system by injecting pulsating water.

FIGURE 26-49 Sludge in the cooling system can be from air getting in from a leak.

These techniques listed above are general and are not meant to replace the recommended service procedures specified by the vehicle manufacture. Refer to the service information for specific service instructions. Improper cooling system service can lead to severe engine damage and failure of many other cooling system components. Failure to correct the cause of the problem, such as incorrect coolant or iron or aluminum oxide residue can lead to repeat failures of cooling system components.

Many shops use a suction tool or air lift to refill an empty cooling system (**Figure 26-50**). The tool connects to the cooling system filler neck and to the shop's compressed air system. Using vacuum created by the shop air moving through the tool, air is sucked from the cooling system. Once a certain level of vacuum is reached, a valve is opened that allows coolant to pass through the tool and into the cooling system. This rapidly fills the system and removes any trapped air.

Because air trapped in the cooling system is a serious concern, many cooling systems have **air bleed valves** installed (**Figure 26-51**). In general, the air bleed valves should be open while you refill the cooling system to allow air to escape. Once coolant flows out of the valve, the valve is closed. Even with bleed valves, air can remain trapped in the cooling system. While you run the engine, grasp a heater or radiator hose and feel for pulsing or bubbling within the hose. If this is felt, air is trapped in the system and it must be removed. To remove the air often requires running the engine at various speeds. You may need to jack the corner of the vehicle where the fill neck is located. Open the air bleed valves. Then shut off and restart the engine until the air pocket moves to the filler neck where it can dissipate.

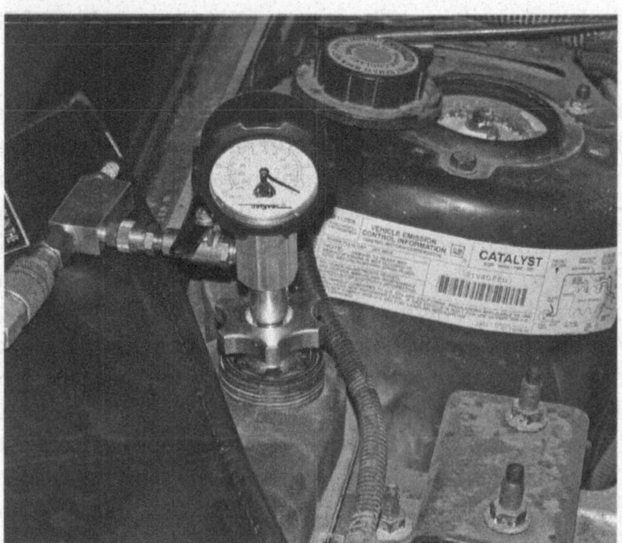

FIGURE 26-50 A vacuum or air-lift tool makes refilling the cooling system faster and it helps eliminate air pockets.

FIGURE 26-51 Air bleed valves are used to help trapped air escape when filling the system.

Make sure that all air is removed from the system before you release a vehicle. Trapped air can cause low heater output or even prevent the thermostat from opening. This can cause the engine to overheat and suffer severe damage. Because of the problems of removing trapped air, many technicians use the vacuum fill method described earlier. Once the system is refilled, check the coolant percentage and protection factor.

THE VENTILATION SYSTEM

The ventilation system works with both the heating and air conditioning systems to route airflow for the passenger compartment. Most of the components of the ventilation system are hidden beneath the dash and are seldom seen by the driver or passengers. The components of the ventilation system include the heater core/evaporator housing, interior fan motor, HVAC controls, and the ductwork.

■ ***Heater Core/Evaporator Case.*** This housing, also called *the HVAC case*, contains the heater core and air conditioning evaporator, blend and mode doors, and blower motor. The case is usually located behind the dash, though on some trucks it is under the hood and bolted to the firewall (**Figure 26-52**). Located near the core and evaporator is the blower motor (Figure 26-52). Air temperature is controlled by the blend door. The door directs air either at the heater core or at the evaporator. The blend door position is determined by the temperature selection on the HVAC control panel. Mode selection, where the air blows out from, and whether fresh or recirculated air is used, is controlled by separate doors in the case. At the bottom of this housing is a drain, used to allow water from the evaporator to drain out under the vehicle when the A/C is operating.

■ ***Cabin Air Filters.*** Common on many cars and trucks, **cabin air filters** are often mounted in the heater core/evaporator case near the blower motor. They can also be located at the fresh air inlet to the HVAC system. These filters are used to trap dirt, dust, and other contaminants from entering the HVAC system and passenger compartment (**Figure 26-53**). Some vehicles use filters that can trap pollen and other allergens.

■ ***Interior Fan Motor.*** The interior fan motor, or **blower motor**, circulates the air through the HVAC ductwork and passenger compartment. An example of a blower motor and its location in the HVAC case is shown in **Figure 26-54**. The speed of the fan is controlled by the HVAC control panel.

A common method of controlling blower speed is by using a stepped resistor block. When the driver selects the low fan speed, several resistors, usually three, are placed in series in the fan motor circuit (**Figure 26-55**). The resistors cause a voltage drop in the fan circuit, reducing the voltage and current available for the fan. Selecting the medium-low speed places two resistors in series, while medium-high places one resistor in series. When high speed is selected, no resistors are used; full voltage and current flow through the blower circuit, and the fan runs at full speed.

■ ***HVAC Controls.*** The HVAC controls, sometimes called *the control head*, are used to control the temperature of the air circulating in the passenger compartment, the amount of airflow by the controlling blower speed, and the direction of air flowing out of the vents. Interior air temperature is controlled by moving the temperature selector, which in turn changes the position of the blend door. The way this is done depends on the vehicle. The most common methods of blend door control are by a cable, vacuum diaphragm, and by electric motor. The HVAC case shown in **Figure 26-56** shows where both vacuum and cable connections are used to control operation. When the driver changes the temperature on the control head, the blend door moves to allow more or less airflow across either the heater core or the A/C evaporator.

In addition to the blend door, a fresh air/recirculated air option is part of the HVAC controls. This allows the driver to select air from the outside of the vehicle to pass through the heater core/evaporator housing or to have the air inside the passenger compartment recycle through the system.

Another function of the HVAC head is the selection of where the air is discharged. The most common options are defrost mode, which blows air onto the inside of the windshield, the panel vents, which direct airflow toward the passengers, and the floor vents for keeping feet warm. Combinations of each are common, such as sending air to both the defrost and floor vents (**Figure 26-57**).

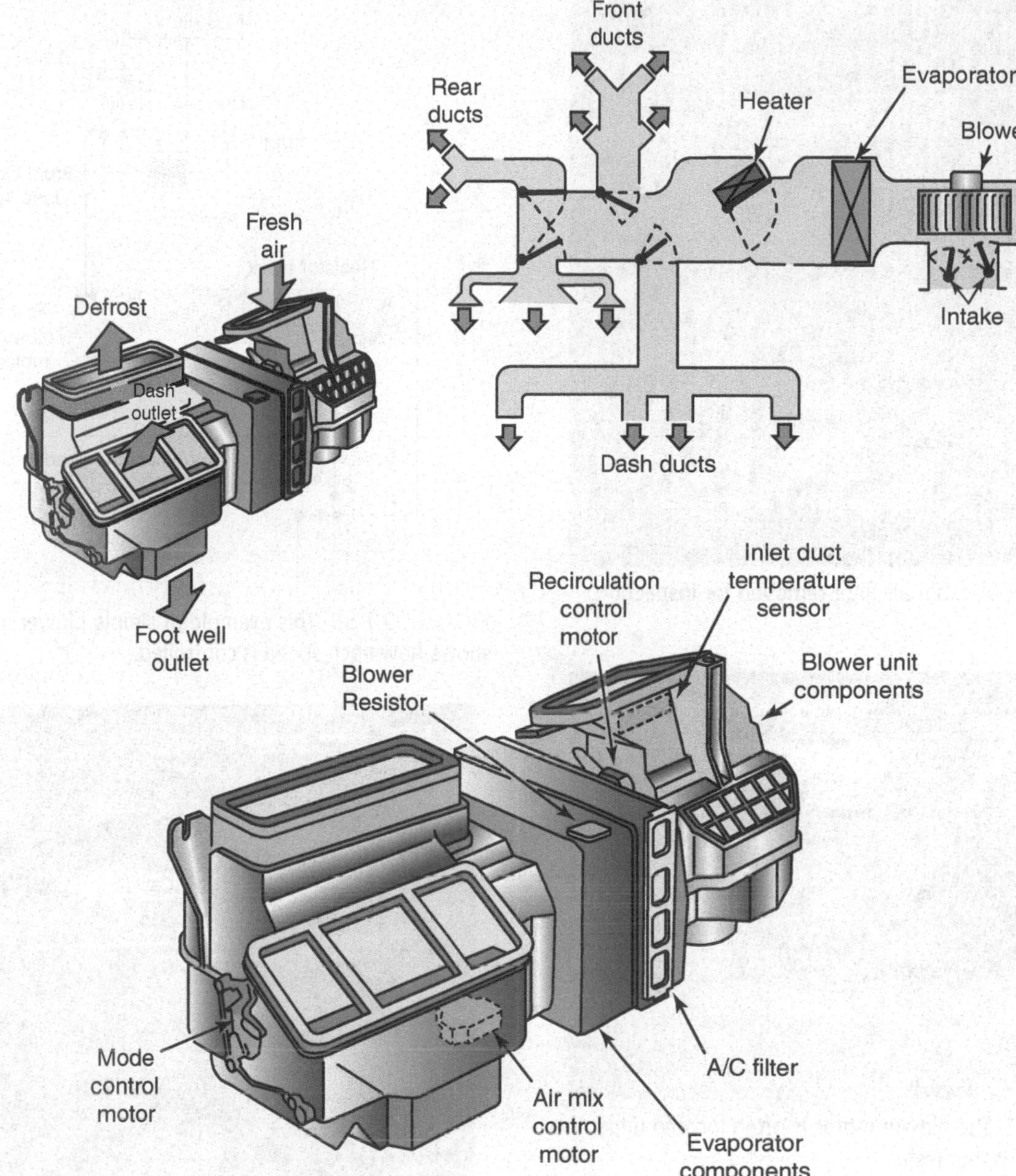

FIGURE 26-52 An illustration of the HVAC or heater case.

The HVAC control head also provides a method of manually turning on or off the A/C compressor.

■ ***HVAC Ductwork.*** Connecting the numerous vents, blower motor, and other HVAC components is the ductwork. Made of plastic, the ductwork routes the air through the system to the vents. In most cases, the ductwork is rigid and is not subject to disassembly unless you are removing major sections for repairs. However, some vehicles use flexible sections of ductwork to connect to vents mounted along the lower dash trim panels.

Most vents have slats that can be closed to stop or greatly reduce the flow of air from the vent. The vents are mounted so that each can pivot to direct airflow as desired by the operator.

AIR CONDITIONING SYSTEM

The air conditioning or A/C system uses principles of heat transfer to cool the passenger compartment. The components of the A/C system are the refrigerant, compressor, accumulator or receiver-dryer, evaporator, condenser, and either an orifice tube or an expansion valve.

Before discussing the components of the system, it is important to have a basic understanding of how the A/C works. The system is divided into two sections, the high-pressure and low-pressure sides. The high-pressure side

FIGURE 26-53 A cabin air filter removed for inspection.

FIGURE 26-54 The blower motor is often located under the passenger side of the dash.

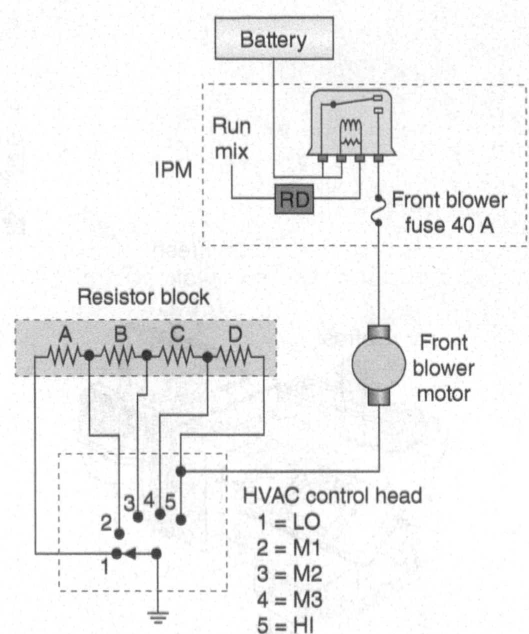

FIGURE 26-55 This example of simple blower motor circuit shows how each speed is controlled.

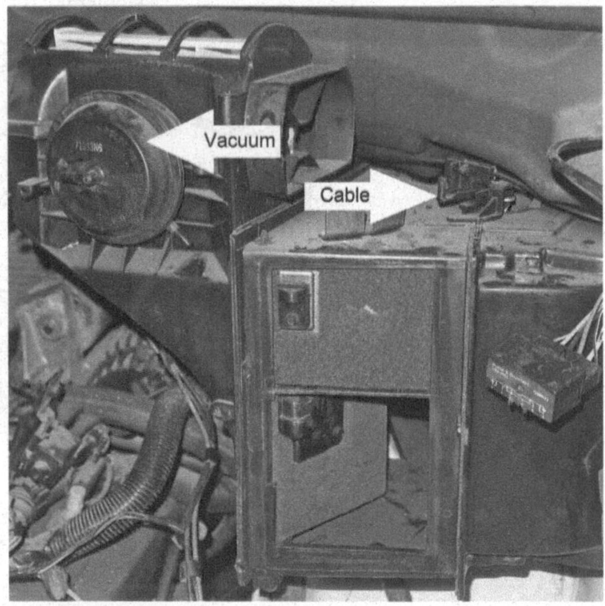

FIGURE 26-56 Older vehicles often use vacuum and cables to control the modes of operation.

contains the high side of the compressor, the condenser, accumulator, and lines and hoses between them and up to the expansion valve or orifice tube. The low-pressure side contains the evaporator, low side of the compressor, receiver-dryer, and the lines and hoses that connect them (**Figure 26-58**).

A **refrigerant** is a chemical that is used to transfer heat in the air conditioning system. The most common refrigerant currently in use is R-134a. Under atmospheric pressure, R-134a has a boiling point of around −16.4°F. To prevent the refrigerant from boiling before we want it to, it is kept under pressure in the system. Just as increasing the pressure in the cooling system raises the boiling point of the coolant, increasing the pressure in the A/C system increases the boiling temperature of the refrigerant (**Figure 26-59**). This is important because when the refrigerant changes from a liquid to a vapor, it absorbs heat. As the refrigerant passes through the evaporator in the passenger compartment, it picks up heat and evaporates. This gas then travels out to the condenser, where it loses heat to the air passing over the condenser. As the heat is lost, the refrigerant condenses from a vapor back to a liquid.

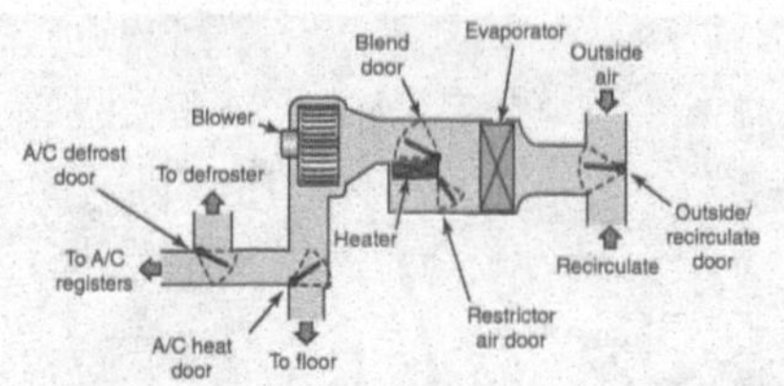

FIGURE 26-57 Airflow is routed to the different vents by moving doors inside the HVAC case.

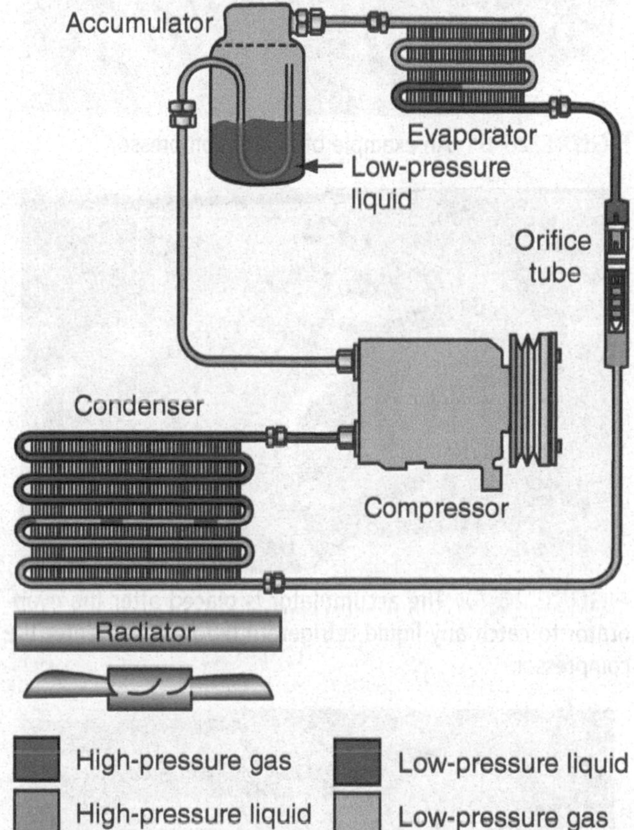

FIGURE 26-58 The A/C system is divided into low- and high-pressure sides.

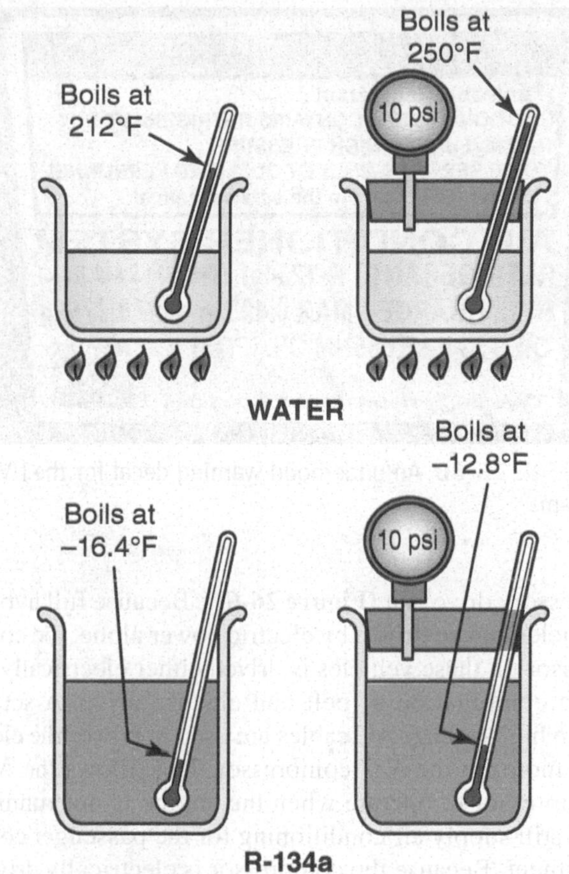

FIGURE 26-59 Increasing pressure on the refrigerant increases its boiling point.

■ ***Refrigerant.*** From 1995 through 2017, new cars sold in the United States were equipped with R-134a as the refrigerant in the A/C system. R-134a replaced R-12, which was in use for decades, but was phased out due to its damaging effects on the ozone layer of the atmosphere.

Manufacturers began phasing in a new refrigerant, R-1234yf, with the 2013 model year vehicles. There is some controversy regarding the change in refrigerant. In 2012, Mercedes-Benz recalled vehicles equipped with R-1234yf and replaced the refrigerant with R-134a. This is because under certain situations, released R-1234yf could ignite and catch fire. An example of an R1234yf underhood warning decal is shown in **Figure 26-60**. R-1234yf is currently being installed in some new vehicles as of the 2016 model year.

The new refrigerant has a reduced environmental impact compared to R-134a and offers a slight increase in cooling capacity, meaning a slightly lower temperature can be reached. However, just as when R-134a replaced R-12, R-1234yf is not backward compatible, meaning it cannot be used with R-12 or R-134a systems. This means shops need to purchase new equipment to service R-1234yf systems.

■ ***Air Conditioning Compressor.*** The A/C **compressor** separates the low- and the high-pressure sides of the system. The intake of the compressor, called *the suction side*, draws evaporated refrigerant from the passenger compartment and pumps it under high pressure to the condenser, located near the radiator in the front of the vehicle.

On all but electric and hybrid electric vehicles, the A/C compressor is driven by the crankshaft through the

FIGURE 26-60 An underhood warning decal for the HVAC system.

accessory drive belt (**Figure 26-61**). Because full hybrid vehicles can be driven by electric power alone, the compressor on these vehicles is driven either electrically or by a combination of belt and electric drive. A set of three high-voltage AC cables are used to power the electric motor in the A/C compressor. This allows the A/C compressor to operate when the engine is not running and still supply air conditioning for the passenger compartment. Because the compressor is electrically driven on hybrids, special refrigerant oils and service procedures are required. Do not attempt to service the A/C system in a hybrid vehicle without training on the system. The high-voltage cables are in bright orange conduit just as other parts of the high-voltage system on hybrid vehicles.

■ ***Accumulator/Receiver-Dryer.*** These two components are similar in appearance—see **Figure 26-62** and **Figure 26-63**—and are often confused. An accumulator is placed in the low side of the system, after the evaporator and before the compressor, and is used to trap liquid refrigerant before it can enter the compressor. This is to prevent damage to the compressor as it can only pump refrigerant vapor. A receiver-dryer is placed in the high-pressure side between the condenser and the expansion valve and is used to store refrigerant until it is needed by the system.

■ ***Evaporator.*** The **evaporator** is in the passenger compartment (**Figure 26-64**). As the low-temperature liquid refrigerant enters the evaporator, the warmer air of the passenger compartment heats and evaporates the liquid. As the refrigerant evaporates, it absorbs heat, which cools the air passing over the evaporator. This air, circulated by the interior fan, is what cools the passenger compartment.

FIGURE 26-61 An example of an A/C compressor.

FIGURE 26-62 The accumulator is placed after the evaporator to catch any liquid refrigerant before it can enter the compressor.

FIGURE 26-63 A receiver-dryer is placed between the condenser and the expansion valve.

■ ***Condenser.*** Once the refrigerant has absorbed heat and has turned into a vapor, it is pumped by the compressor to the **condenser**, which is typically located in front

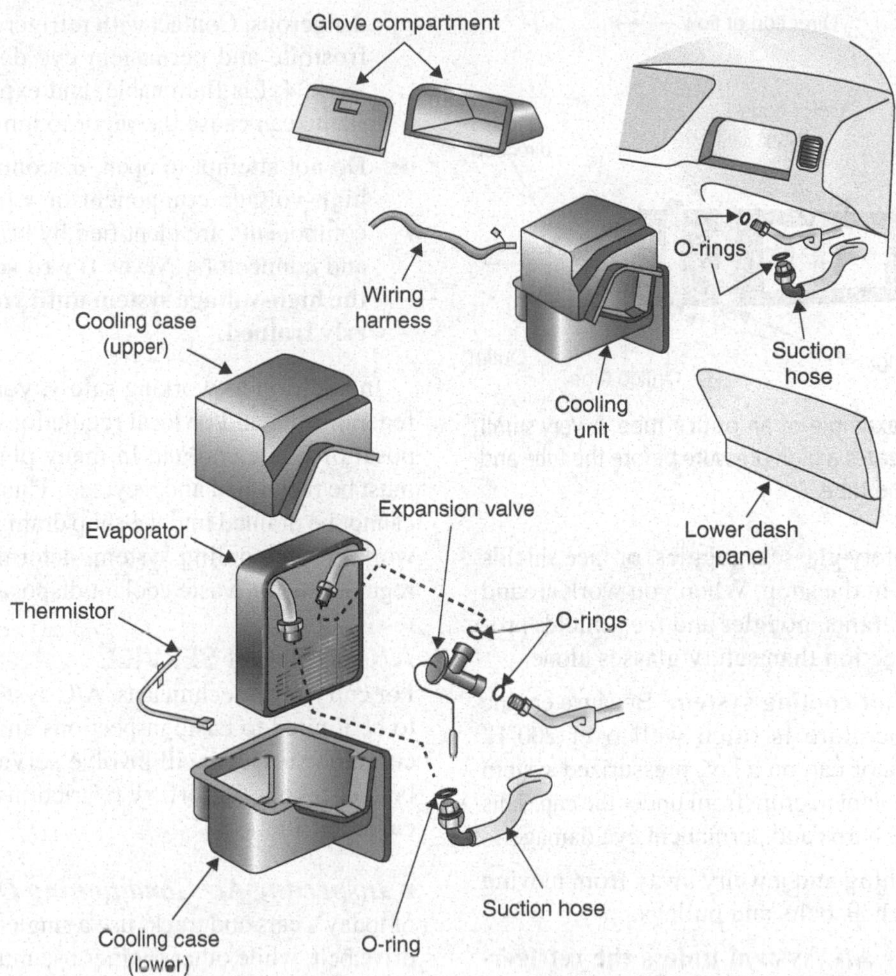

FIGURE 26-64 The evaporator is often located deep in the HVAC case in the dash.

of the engine radiator. The air flows over the condenser and cools the refrigerant in the same way heat is removed from the engine coolant at the radiator (**Figure 26-65**).

■ ***Orifice Tube/Expansion Valve.*** To drop the pressure of the refrigerant before it enters the evaporator, either an orifice tube or expansion valve is used. An example of an orifice tube is shown in **Figure 26-66** and an illustration of an expansion valve is shown in **Figure 26-67**. As the refrigerant passes the orifice tube or valve, its pressure drops. The decrease in pressure also reduces the refrigerant's temperature. Dropping the pressure of the liquid before it enters the passenger compartment allows it to boil or change state more easily and allows for the absorption of more heat from the passenger compartment.

FIGURE 26-65 This truck has a separate A/C condenser mounted beside the radiator. Note the small dents in the cooling fins from road debris.

SERVICING THE HVAC SYSTEM

It is not within the scope of this text to provide detailed diagnostic and service procedures for the entire HVAC system. Instead, the most common concerns and services are addressed.

Before you attempt to work on the HVAC system, it is important to understand how to do so safely. The following safety precautions should be followed to prevent personal injury and damage to the vehicle. **Figure 26-68** shows examples of warning labels used for the air conditioning system.

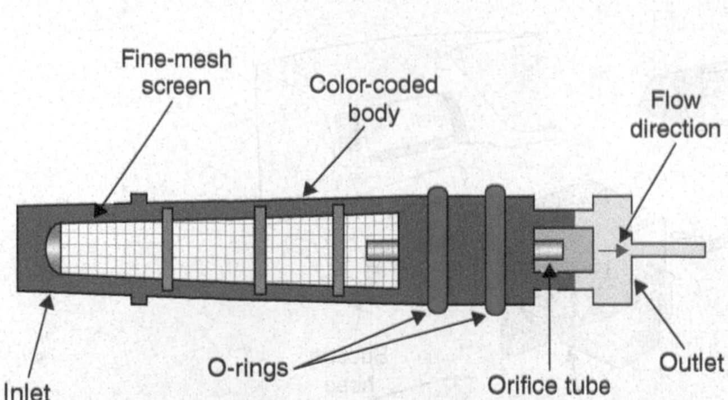

FIGURE 26-66 An example of an orifice tube. A very small opening in the tube creates a high pressure before the tube and a low pressure after the tube.

- Always wear safety glasses, goggles, or face shields when you work in the shop. When you work around liquids or refrigerants, goggles and face shields provide better protection than safety glasses alone.
- **Never open a hot cooling system.** Because engine operating temperature is often well over 200°F, removing a radiator cap on a hot, pressurized system will allow the coolant to erupt from under the cap. This can cause severe burns and permanent eye damage.
- Keep loose clothing and jewelry away from moving components such as belts and pulleys.
- **Never open the A/C system unless the refrigerant has been reclaimed from the system.** Release of refrigerant into the atmosphere is illegal and dangerous. Contact with refrigerant vapor can cause frostbite and permanent eye damage. Refrigerant R-1234yf is flammable, and exposure to a spark or flame can cause the vapor to ignite.
- Do not attempt to open, disconnect, or service any high-voltage component or wiring. High-voltage components are identified by bright orange conduit and connectors. **Never try to service any part of the high-voltage system until you have been properly trained.**

In addition to working safely, you also must follow federal, state and/or local regulations regarding the disposal of waste coolant. In many places, waste coolant must be reclaimed and recycled. This means that coolant cannot be drained into the shop drain system. Before you work on the cooling system, determine what the local regulations for waste coolant disposal are.

A/C SYSTEM SERVICE

For entry-level technicians, A/C system service is likely to be limited to basic inspections and repairs. The most common services will involve servicing the A/C drive belt, checking for airflow restrictions, and servicing the cabin air filters.

■ ***Inspecting Air Conditioning Drive Belts.*** Many of today's cars and trucks use a single multi-rib accessory drive belt, while other engines use multiple belts (**Figure 26-69**). Regardless of the number, belts require periodic inspection and replacement. Belts should be inspected

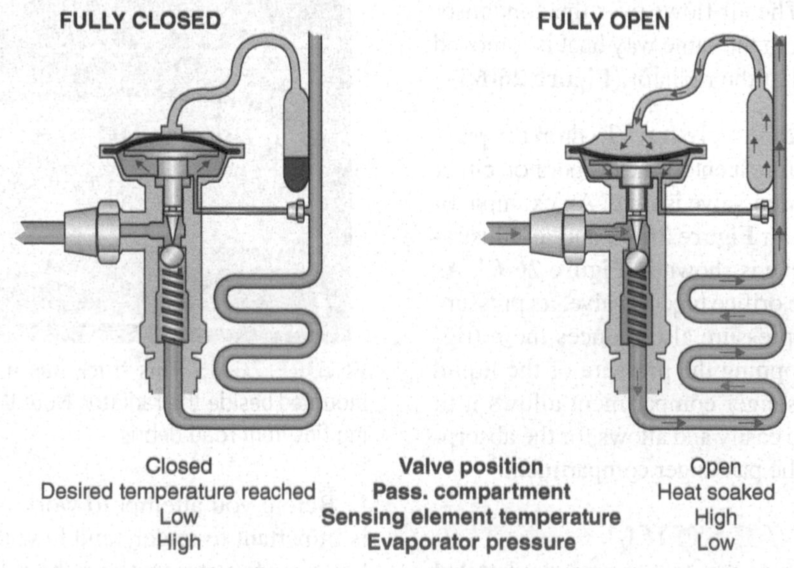

FIGURE 26-67 This illustrates an expansion valve. The valve controls the amount of refrigerant flowing through the evaporator.

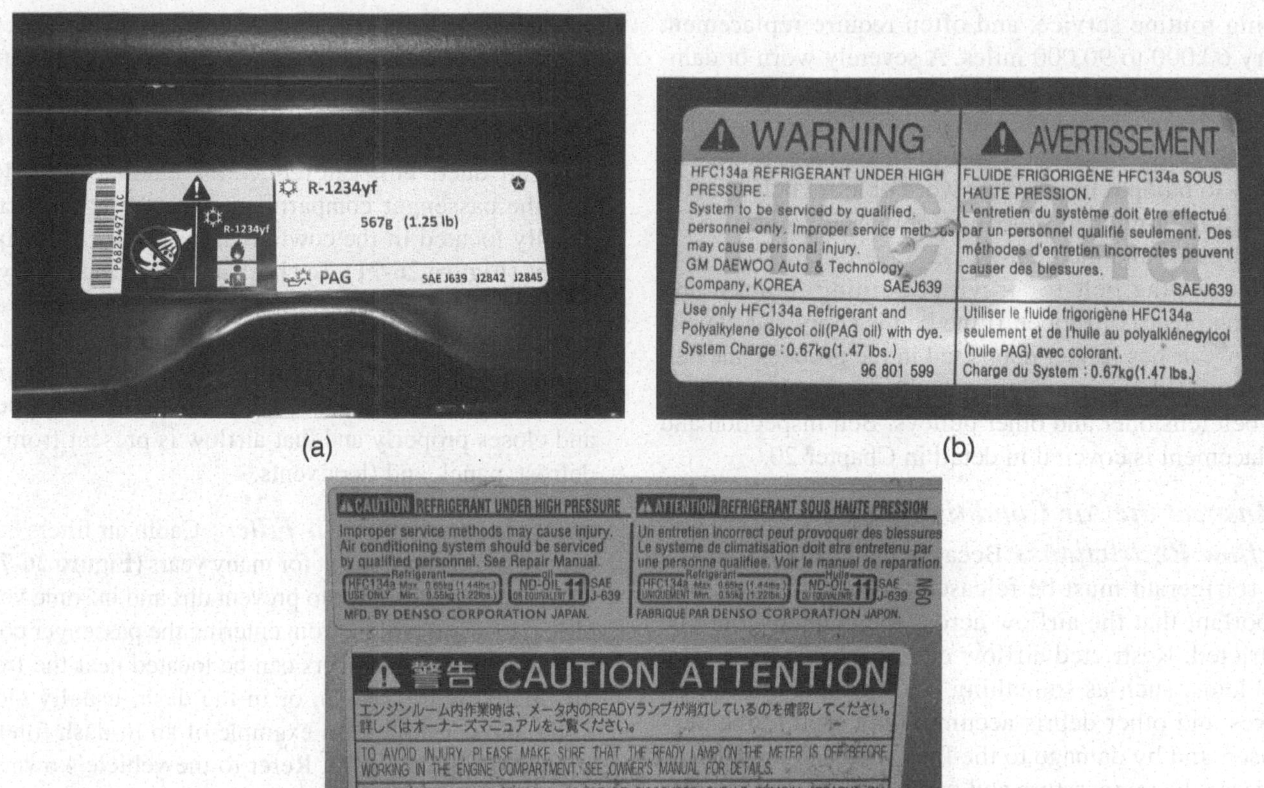

FIGURE 26-68 (a) An HVAC identification label for R-1234yf. (b) An HVAC identification label for R-134a. (c) An HVAC identification label for a hybrid electric vehicle. Note the same R-134a refrigerant but different refrigerant oil.

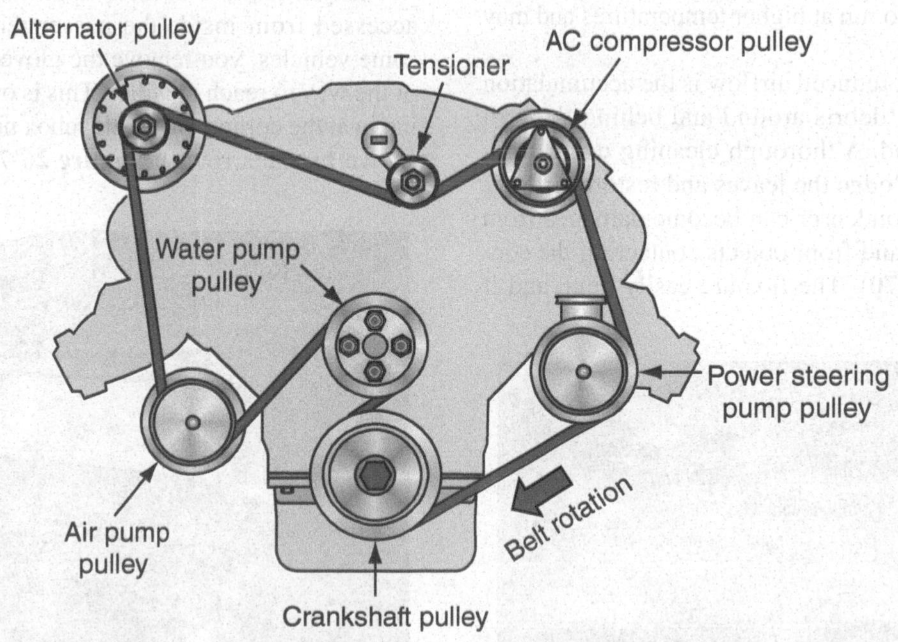

FIGURE 26-69 An illustration of an accessory belt routing.

during routine service, and often require replacement every 60,000 to 90,000 miles. A severely worn or damaged A/C compressor belt can cause poor A/C performance due to the compressor not being driven properly. This can allow the compressor to slip, which reduces its ability to pump refrigerant. A worn or loose belt can also cause a high-pitched squealing or chirping noise as the belt slips around the pulleys.

Inspect the belt for fraying, burning, cracks, and missing pieces of rubber. If the belt is shiny, has chunks missing, or has rubber deposited in the grooves, the belt tension is likely insufficient and you should examine the belt tensioner and other pulleys. Belt inspection and replacement is covered in detail in Chapter 20.

■ ***Inspect the Air Conditioning Condenser for Airflow Restrictions.*** Because the heat absorbed by the refrigerant must be released at the condenser, it is important that the airflow across the condenser not be restricted. Restricted airflow can be caused by several problems, such as something blocking the condenser, leaves and other debris accumulating around the condenser, and by damage to the fins. Reduced airflow can increase the temperature and pressure in the system and reduce A/C performance.

In cold climates, people will sometimes place a piece of cardboard in front of the condenser and radiator to improve heater performance. Unfortunately, sometimes this obstruction does not get removed and causes problems as the weather warms up. If the condenser and radiator are blocked, not only will the A/C system suffer, but the engine is likely to run at higher temperatures and may overheat.

Another cause of reduced airflow is the accumulation of leaves and other debris around and behind the grill area of the front end. A thorough cleaning using compressed air may dislodge the leaves and restore airflow.

The fins of the condenser can become damaged from improper handling and from objects contacting the condenser (**Figure 26-70**). The fins are easily bent, and if enough are bent, the cooling capacity of the condenser will be reduced. Fins may be straightened using a special comb.

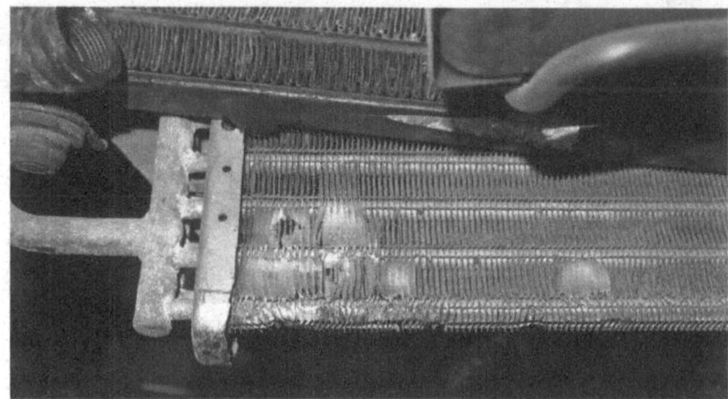

FIGURE 26-70 Improper handling during service can damage the cooling fins, as shown here.

■ ***Inspect Heating and Air Conditioning Ducts.*** Over time, dirt and debris can collect in the fresh air duct. This can reduce the amount of airflow into the passenger compartment. The fresh air duct is usually located in the cowl area, just below the wiper blades (**Figure 26-71**). To check and clean the inlet, you may need to remove the cowl. Once it is accessible, use a shop vacuum to clean the area.

The interior ducts are usually trouble-free. As part of an HVAC inspection, make sure each vent opens and closes properly and that airflow is present from all defrost, panel, and floor vents.

■ ***Replace a Cabin Air Filter.*** Cabin air filters have been common equipment for many years (**Figure 26-72**). The filters are designed to prevent dirt and in some vehicles, odors and pollen, from entering the passenger compartment. Cabin air filters can be located near the fresh air inlet (Figure 26-71), or in the dash, usually close to the blower motor. An example of an in-dash filter is shown in **Figure 26-73**. Refer to the vehicle's owner's manual or service information for the location and service procedure. The steps listed below are general and not specific to any particular car or truck.

Filters located near the fresh air inlet are usually easy to replace. Locate the filter cover and remove. Next, pull the filter from the opening. Clean any dirt and debris from around the inlet and replace the filter. Reinstall the filter cover.

Many filters are in the evaporator case and must be accessed from inside the passenger compartment. In some vehicles, you remove the glovebox or move it out of the way to reach the filter. This is often done by pressing in at the corners of the glovebox near the travel stops, shown by the arrows in **Figure 26-74**. This allows the

FIGURE 26-71 Some cabin air filters are located at the fresh air inlet and are very easy to change.

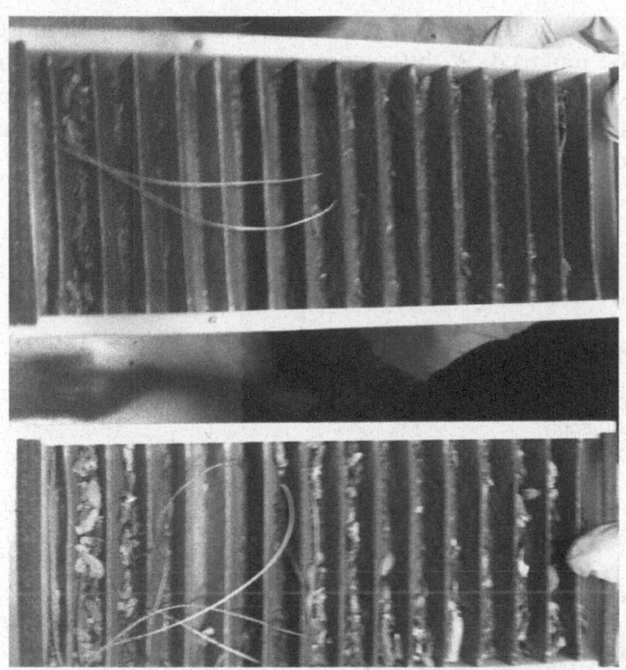

FIGURE 26-72 An example of a cabin air filter after a couple of years in service.

FIGURE 26-73 Some filters can only be accessed by removing various dash components.

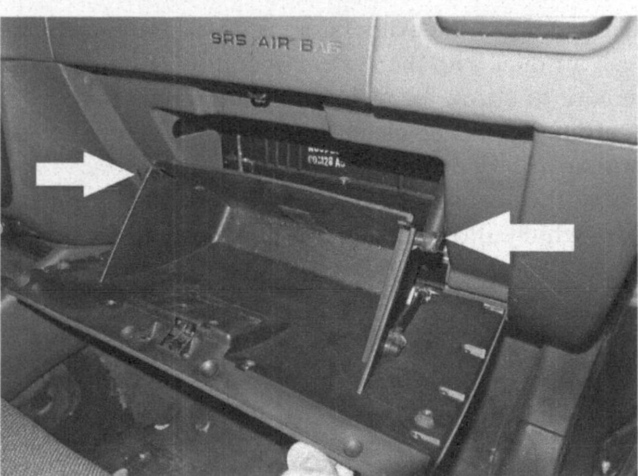

FIGURE 26-74 To remove some cabin filters, the glovebox is dropped out of the way by pressing in on the guides/stops. Make sure you are following the correct service procedure before you begin service to prevent damage to components.

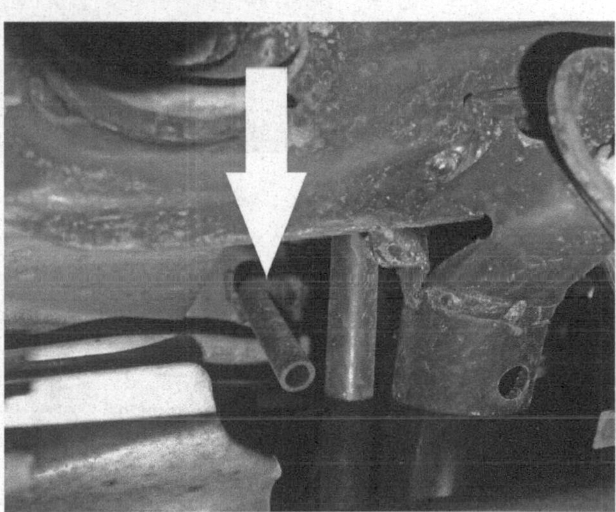

FIGURE 26-75 An example of an evaporator drain.

glovebox to swing down out of the way. Next, remove the filter cover and filter. Clean any dirt and debris from the filter housing and replace the filter. Reinstall the cover and set the glovebox back into place.

Some vehicles require partial disassembly of the dash to access the filter. To remove the filter, the lower dash trim is removed followed by several braces. Follow the service information for details about how to properly service this type of filter.

■ ***Identify Sources of Odors from the Air Conditioning System.*** During A/C operation, the absorption of heat by the refrigerant at the evaporator causes condensation to form on the evaporator core. This is the same effect as when a glass or can with a cold drink sits out in a warm room. The heat transfer causes the moisture in the air to condense into water at the colder surface. Due to the water condensing on the evaporator, the case has a drain, where the water drips out under the vehicle. Typically, the drain protrudes out from the firewall on the passenger side (**Figure 26-75**). Be sure that the evaporator drain is clear by turning the A/C on and checking for water dripping out under the passenger side of the vehicle. If some moisture remains on the evaporator, it can cause mildew and odors from the HVAC system. Eliminating these odors typically involves either applying a disinfectant to the evaporator core or spraying it into a fresh air duct with the system operating (**Figure 26-76**). Be sure to check for technical service bulletins (TSBs) regarding odors and cleaning procedures before you perform any service to the system.

Another cause of HVAC odors is a leaking heater core, which allows coolant to leak out and accumulate in the heater case. If the heater core is leaking, it must be replaced. With the core removed, you will need to clean as much of the case and ductwork as possible to prevent the odors from remaining after the repair is completed.

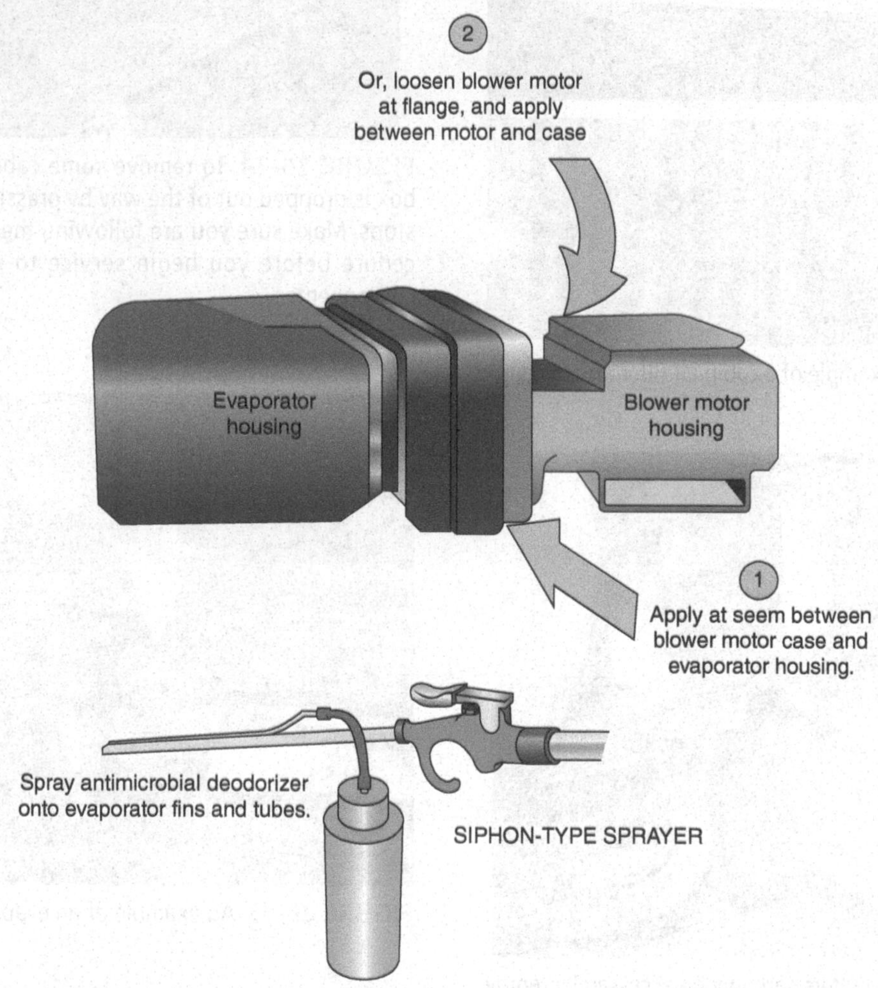

FIGURE 26-76 An illustration of applying a cleaning agent to reduce A/C system odors.

SUMMARY

The main function of the cooling system is to remove excess heat from the engine and to maintain the proper engine operating temperature.

All new cars and trucks sold in the United States have liquid-cooled engines.

This radiator cap seals the radiator fill neck and is used to allow the pressure to increase in the cooling system.

Never remove the cap from a hot cooling system.

Three types of radiator fans used on modern engines are the belt-driven fan, the electric fan, and the hydraulic fan.

A cooling system pressure tester is used to help locate leaks in the cooling system.

A malfunctioning thermostat can cause two different concerns, low engine temperature and a lack of heat or poor heater performance or higher-than-normal engine temperature and overheating.

Flushing the cooling system removes the old coolant and circulates clean water through the system to remove rust and corrosion.

Cabin air filters are designed to prevent dirt, odors, and pollen from entering the passenger compartment.

The blend door may be controlled by a cable, a vacuum diaphragm, or an electric motor.

There are currently two refrigerants in use, R-134a and R-1234yf. Neither are compatible with each other.

The A/C compressor separates the low- and the high-pressure sides of the system.

The condenser cools the refrigerant in the same way in which the radiator removes heat from the coolant.

Mildew and odors from the HVAC system can be caused by moisture remaining on the evaporator core.

REVIEW QUESTIONS

1. The acronym HVAC is used for the heating, _______________, and air conditioning system.
2. The component that allows pressure to increase in the cooling system is the _______________.
3. A thermostat that is stuck_______________can cause poor heater performance.
4. The _______________ removes heat from the refrigerant as air passes over it.
5. The _______________ separates the high- and low-pressure sides of the A/C system.
6. Which of the following is the most common type of cooling fan used on cars and light trucks?
   a. Belt drive
   b. Direct drive
   c. Hydraulic
   d. Electric
7. A car's engine overheats within minutes of starting. *Technician A* says the thermostat may be faulty. *Technician B* says the cooling fan may not be working. Who is correct?
   a. Technician A
   b. Technician B
   c. Both A and B
   d. Neither A nor B
8. A vehicle has very little airflow coming from the vents with the fan set on high speed. *Technician A* says the cabin air filter may be plugged. *Technician B* says the blend door motor may be faulty. Who is correct?
   a. Technician A
   b. Technician B
   c. Both A and B
   d. Neither A nor B
9. An engine overheats when driving on the highway but not when driven in traffic. Which of the following is the most likely cause?
   a. Defective thermostat
   b. Air trapped in the system
   c. Radiator fan not working
   d. Restricted radiator flow
10. *Technician A* says all OAT are similar and can be used interchangeably. *Technician B* says traditional green coolant can be mixed with long-life coolants. Who is correct?
    a. Technician A
    b. Technician B
    c. Both A and B
    d. Neither A nor B

CHAPTER 27

Vehicle Maintenance

Chapter Objectives

At the conclusion of this chapter, you should be able to:

- Explain how maintenance differs from repair.
- Explain why periodic maintenance is important for modern vehicles.
- Interpret vehicle maintenance schedules and indicator displays.
- Review vehicle service history. (ASE Education Foundation RST Service #4)
- Locate technical service bulletins (TSBs) and other service information. (ASE Education Foundation MLR 1-8.A.1)
- Perform engine oil change. (ASE Education Foundation MLR 1.C.5)
- Inspect and service engine coolant. (ASE Education Foundation MLR 1.C.4)
- Inspect automatic transmission fluid. (ASE Education Foundation MLR 2.A.2, 2.A.3, & 2.A.4)
- Inspect manual transmission fluid. (ASE Education Foundation MLR 3.A.2 & 3.A.3)
- Inspect and service engine air filters. (ASE Education Foundation MLR 8.C.2)

KEY TERMS

ACEA standards
air filter
American Petroleum Institute (API)
diesel exhaust fluid
hybrid organic acid technology (HOAT)
ILSAC
maintenance
maintenance reminder
multiviscosity oils
organic acid technology (OAT)
severe service
Society of Automotive Engineers

Maintenance

Before we go into depth regarding how to perform maintenance on a vehicle, you should understand what maintenance is and how it applies to the vehicle, the technician, and the shop.

WHAT IS MAINTENANCE?

Maintenance is the act of keeping something in a state of good operating condition. This usually means to inspect and replace items before a failure occurs and a repair is necessary. For example, replacing the drive belt as a maintenance item before it breaks and leaves the vehicle broken down on the side of the road. For the automobile, maintenance is performed to ensure the safety, reliability, drivability, comfort, and longevity of the vehicle. The maintenance needs of a vehicle depend on the year, make, model, and driving conditions in which the car operates.

■ ***Maintenance versus Repair*** Maintenance usually requires performing certain services, such as changing the engine oil and filter. This is done periodically to ensure that the dirt and contaminants that dilute the oil are removed from the engine, reducing wear on parts and prolonging the life of the engine. A repair is a service to correct a problem, such as replacing a leaking gasket. The gasket may have failed due to age and the constant thermal cycling of engine operation, allowing engine oil to leak from the engine. Once the gasket has been replaced, the engine oil may need to be changed also. Replacing the gasket is a repair to correct a problem, while the oil change is done to maintain the operation of the engine.

■ ***Why Is Maintenance Important?*** Maintenance is important for the vehicle, the technician, and for the shop. Even though modern vehicles are more advanced, more efficient, and require less maintenance than ever before, no vehicle in the market yet requires no maintenance. Even electric vehicles, such as Tesla's require tire rotations, brake services, and periodic inspections and services. The following is a list of common maintenance items for most vehicles:

- Washing and cleaning the interior and exterior
- Replacing the engine oil and filter
- Replacing the transmission fluid and filter
- Inspecting and replacing engine coolant
- Inspecting and replacing brake fluid
- Inspecting and replacing spark plugs and ignition wires
- Inspecting and replacing timing and accessory drive belts
- Inspecting, cleaning, and tightening battery terminals
- Replacing air and fuel filters
- Replacing cabin air filters
- Replacing wiper blades and adding washer fluid
- Inspecting and adjusting tire pressure
- Rotating and balancing tires
- Checking and adjusting wheel alignment
- Inspecting components of the brake and suspension systems
- Inspecting and lubricating steering and suspension joints, U-joints, and body hinges

These are just examples of common maintenance items. This list does not cover all makes and models of vehicles. Some vehicles have fewer items while others have more. Taking care of the items that need regular inspection and service provides opportunities for the technician and the shop. Periodic maintenance is important to the technician because it provides paid work. It allows you to inspect the vehicle so that additional problems can be corrected. This often means educating the customer about the importance of regular maintenance. By educating the customer about maintenance and the potential to avoid costly repairs, you can gain customer loyalty and trust, ultimately increasing your business.

Over time, vehicle manufacturers have greatly improved the reliability of cars and trucks. This has led to a decrease in the number of vehicles that return to the service department for repairs. Due to advancements in design and the need to decrease the cost of ownership, vehicles require less maintenance than ever before. These two factors alone—reduced maintenance and fewer repairs—can make it difficult for the service department to bring the customer back in. When the vehicle does return for maintenance work, it is important for the shop to make sure that all services are performed correctly, on time, and at a value to the customer. By making vehicle owners aware of the maintenance requirements of their vehicles, the shop can better provide these services and earn the trust of the customer.

One way to build this relationship with the customer is to review the vehicle's maintenance schedule with the customer. This schedule is included with the owner's manual and provides information about what types of maintenance are necessary and how frequently.

CONSEQUENCES OF IMPROPER MAINTENANCE

Failure to perform routine maintenance, especially fluid changes, can cause serious damage to the vehicle. Engine oil change intervals may range from a few thousand miles to up to 15,000 miles on some vehicles. If these intervals are ignored, rapid breakdown of the engine oil can cause increased wear, damage to internal components, and even engine failure. Other consequences of ignoring maintenance include the following:

- Brake system damage or brake failure
- Loose steering
- Poor ride and handling
- Suspension component failure, which can cause an accident
- Broken timing belts, which can cause serious engine damage

■ ***What Maintenance Is Required and When?*** The vehicle owner's manual usually contains information about what periodic maintenance should be performed and when. Many vehicles use a time/mileage schedule (**Figure 27-1**). This table outlines what items should be checked, serviced, or replaced at each interval. The schedule is also divided into normal and severe operating conditions. **Severe service** conditions typically are when *any* of the following conditions apply:

- Prolonged idling
- Use in dusty conditions
- Frequent starts and stops
- Towing or hauling
- Use in very cold or very hot climates
- Other factors, such as fleet use

If the vehicle is operated in any of these conditions, then the severe schedule should be followed. This is an important conversation to have with the customer. Many people do not understand that their everyday driving habits are harder on the vehicle than what is considered normal operation. This usually means the vehicle should be serviced twice as frequently when following the severe service operation schedule.

Many of today's cars and trucks use some type of electronic **maintenance reminder** system. At the most basic, these systems illuminate a light on the dash to remind the driver to have the engine oil changed (**Figure 27-2**). Systems that are more sophisticated calculate oil life based on driving and operating conditions. Some vehicles, in addition to notifying the driver when oil changes are needed, also provide information about what types of inspections

Maintenance Schedule for Severe Conditions

Service at the indicated distance or time–whichever comes first.	miles × 1,000	15	30	45	60	75	90	105	120
	km × 1,000	24	48	72	96	120	144	168	192
	months	12	24	36	48	60	72	84	96
Replace engine oil and oil filter		Replace every 3,750 miles (6,000 km) or 6 months							
Check engine oil and coolant		Check oil and coolant at each fuel stop							
Clean (○) or replace (●) air cleaner element Use normal schedule except in dusty conditions		○	●	○	●	○	●	○	●
Inspect valve clearance		Adjust only if noisy						●	
Replace spark plugs								●	
*Replace timing belt*1, balancer belt *1, *2, and inspect water pump								●	
Inspect and adjust drive belts			●		●		●		●
Inspect idle speed								●	
Replace engine coolant		120,000 miles (192,000 km) or 10 years, then every 60,000 miles (96,000 km) or 5 years							
Replace transmission fluid	MT	Every 60,000 miles (96,000 km) or 3 years							
	AT	60,000 miles (96,000 km) or 3 years, then every 30,000 miles (48,000 km) or 2 years							
Inspect front and rear brakes		Every 7,500 miles (12,000 km) or 6 months							
Replace brake fluid		Every 3 years (independent of mileage)							
Check parking brake adjustment		●	●	●	●	●	●	●	●
Replace dust and pollen filter*3			●		●		●		●
Lubricate all hinges, locks and latches		●	●	●	●	●	●	●	●
Rotate tires (Check tire inflation and condition at least once per month)		Rotate tires every 7,500 miles (12,000 km)							
Visually inspect the following items:									
Tie rod ends, steering gear box, and boots Suspension components, driveshaft boots		Every 7,500 miles (12,000 km) or 6 months							
Brake hoses and lines (including ABS) All fluid levels and conditions of fluids Cooling system hoses and connections #Exhaust system, #fuel lines and connections Lights and controls, vehicle underbody		●	●	●	●	●	●	●	●

*1 : See timing belt on page **283** to determine need for replacement.
*2 : Balancer belt applies to 4-cylinder models only.
*3 : See Dust and Pollen Filter on page **282** for replacement information under special driving conditions.

U.S. Owners
Follow the Severe Conditions Maintenance Schedule if you drive your vehicle *MAINLY* under one or more of the following conditions:

- Driving less than 5 miles (8 km) per trip or, in freezing temperatures, driving less than 10 miles (16 km) per trip.
- Driving in extremely hot [over 90°F (32°C)] conditions.
- Extensive idling or long periods of stop-and-go driving.
- Trailer towing, driving with a car top carrier, or driving in mountainous conditions.
- Driving on muddy, dusty, or de-iced roads.

Canadian Owners
Follow the Maintenance Schedule for Severe Conditions.

:See information on maintenance and emissions warrantly, last column, page **239**.

242 Maintenance

FIGURE 27-1 An example of a maintenance schedule. It defines when certain items should be checked and replaced based on time and mileage.

or services should be performed (**Figure 27-3**). This vehicle display indicates not only that the oil needs to be changed soon, but also that maintenance service B1 is due. To determine what service B1 means, the customer or shop needs to refer to the vehicle owner's manual or maintenance schedule.

Once the specified maintenance has been performed, the maintenance timer is reset. The reset procedure is typically very simple but varies from manufacturer to manufacturer and even between different models from the same manufacturer. A few common methods include the following:

- Holding the trip odometer reset button down and turning the ignition switch to the on position until the maintenance light goes out
- Pressing the accelerator pedal all the way down three times within five seconds after turning on the ignition switch

FIGURE 27-2 A maintenance light or message alerts the driver that some type of maintenance is required.

FIGURE 27-3 This vehicle uses a wrench, a service code, and an oil life index. The wrench indicates that the maintenance is due, the code correlates to the maintenance schedule in the owner's manual, and the oil life index alerts the driver to how soon the engine oil needs to be changed.

- Pressing a reset button located in a fuse panel or under the dash
- Resetting through a maintenance menu in the driver information center

The preceding methods certainly do not cover every car and truck on the market, so refer to the owner's manual or service information to determine the correct method for the vehicle being serviced. It is important for the reminder to be reset so that the onboard computer resets and begins to monitor the process from the beginning. This allows more accurate performance tracking by the computer to calculate when the next maintenance should occur.

Maintenance Services

Maintaining the relationship with the customer is an important part of maintaining vehicles. In many cases, the owner or driver of a vehicle simply does not pay much attention to maintenance schedules. Because of this, it is important for you, as the professional, to keep the customer aware of the need for ongoing maintenance, and to stay up to date with changes in services, fluids, and procedures.

While the customer may not want to know every detail about a particular service, such as an engine oil change, it is important for you to know the details about the vehicle being serviced. When a vehicle is in for service, look at previous service records to see what work has been done in the past and when. It is never good to try to sell the customer a long list of maintenance items only to find out that the services have already been performed.

Oil change intervals are often determined by miles driven and by the conditions under which the vehicle operates. Knowing how your customer drives their vehicle is important for you to make recommendations for how often services need to be performed. In addition, you should regularly check for updated parts and service information released as technical service bulletins (TSBs). TSBs are available from the vehicle manufacturers, and from aftermarket service information systems such as AllData, ProDemand, and Identifix. These bulletins can be very important for determining the correct fluids to use as new fluids are released and older types are phased out.

COMMON MAINTENANCE CHECKS

The following are general items that are often checked when a vehicle is in for an oil change. Taking the time to check these not only tells customers that your shop is trying to keep them safe, it can also help you find additional services or repairs that need to be addressed, which can earn the shop and you more money.

■ ***Brake Operation.*** **Before you move any vehicle for service, always check the brake pedal feel and travel before you begin to move the vehicle.** Make sure that the brake pedal is firm and does not go to the floor. If the brakes pass this first test, place the vehicle in gear and slowly start to pull away. Give the brakes a quick check once you are moving just to make sure they are working properly.

If the vehicle has a standard transmission and the parking brake is set when you get in, it should be safe to use the parking brake when you park the vehicle again. If the vehicle has an automatic transmission and the parking brake is not set, do not try to use the parking brake unless you confirm with the vehicle's owner that it works correctly. Because the parking brake on automatic transmission vehicles tends to be used less frequently, it can stick if it is applied and may not fully release.

■ ***Check Tire Pressure.*** Use a known good tire gauge, such as those that meet ANSI gauge standards. If this type of gauge is not available, use one that has been checked against other gauges and on many tires. Check the pressure of all the tires on the vehicle, including the spare. Begin by removing the valve core cap and placing the gauge on the valve stem firmly so that the pin inside the gauge depresses the valve core (**Figure 27-4**). Check the pressure reading against the tire decal information. If the tires are reading a little higher than the specified pressure, it may be due to the vehicle having been driven recently, causing the tires to heat up and the pressure to increase. Do not let air out of the tires if the tires are not being checked cold, that is, after sitting overnight or at least six hours. If a tire is significantly lower than the rest, look in the tread for a nail or screw puncture. If a tire is very low and no immediate cause is visible, be sure to tell the customer about the low tire and recommend a more thorough inspection to prevent a possible flat tire or a blowout. When you check the spare tire, remember that most temporary mini-spares require inflation to 60 psi.

When you check the tire pressure, look at the wear on the tires. Look for signs of damage to the tires, wear bar protrusion, cuts, dry rotting, date codes, and other problems that can shorten the life of the tire. If a tire or tires are worn excessively on one side, you should recommend that the wheel alignment be checked. Tire inspection and service is discussed in greater detail in Chapter 5.

■ ***Strut and Shock Absorber Inspection.*** When the vehicle is up on a lift, inspect the struts and shocks for signs of oil leaks. A shock or strut that is wet needs to be replaced (**Figure 27-5**). Inspect mounting bushings for dry rot. With the vehicle on the shop floor, bounce each corner a few times and count the number of times the vehicle rebounds. Generally, good shocks and struts will rebound one and a half to two times. More than two bounces can indicate that the shocks and struts are weak. Further information about shock, strut, and suspension inspection can be found in Chapter 7.

■ ***Inspect Wiper Blades and Washers.*** Some shops, as part of the service program, wash each vehicle that comes in for service. This is a good time to check the operation of the wiper and washer system. If your shop does not wash customer vehicles, look at the windshield and note any streaks or areas where the wiper blades do not clean the glass. Test the washer fluid spray and note wiper effectiveness. The wiper blades should move smoothly across the windshield without noise, streaks, or leaving sections untouched. The washer fluid should spray with enough pressure to cover most of the windshield. Check both front and rear windshields as applicable. If the blades leave streaks, inspect the blade for damage. Sometimes cleaning the blades will restore much of their effectiveness. Use a towel that is wet with windshield washer fluid to clean the length of the blade (**Figure 27-6**).

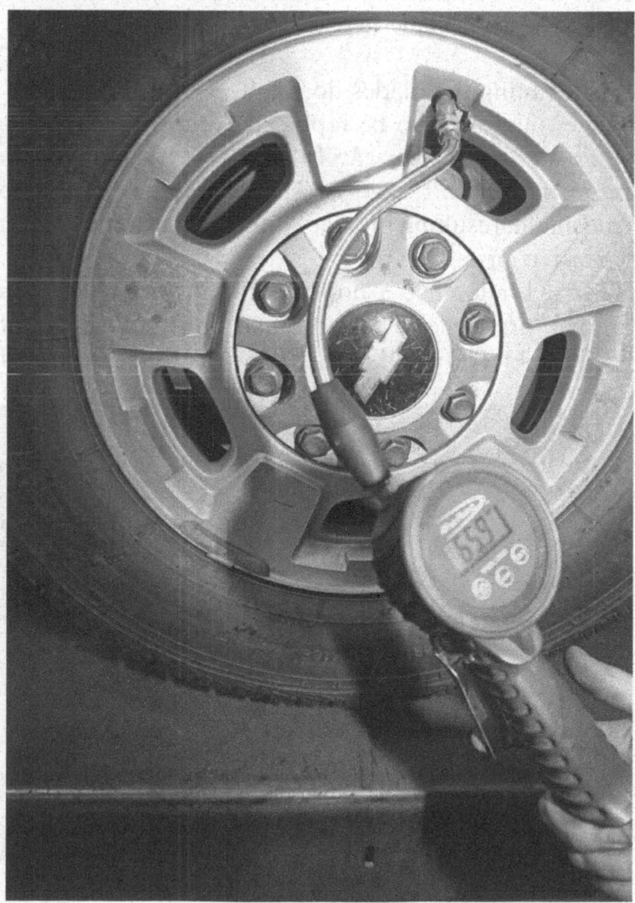

FIGURE 27-4 Checking tire pressure is part of any routine maintenance inspection.

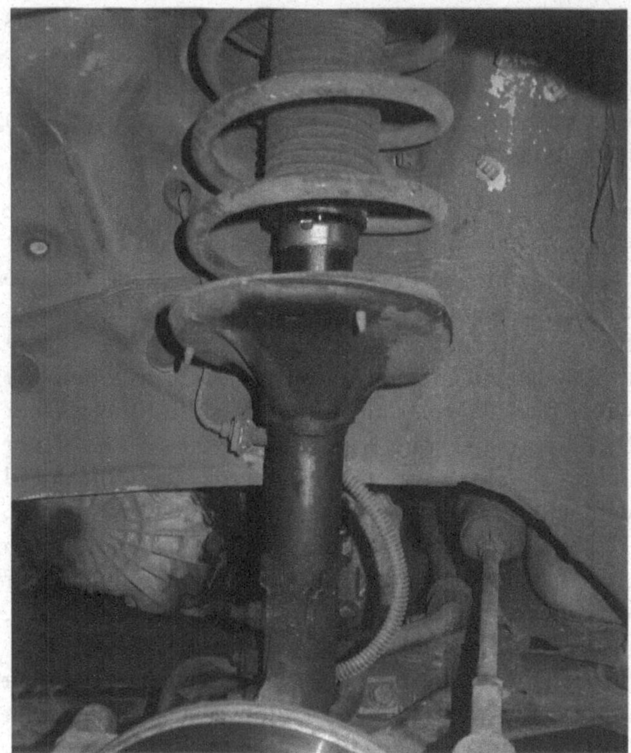

FIGURE 27-5 Leaking shocks and struts are an opportunity for the technician and shop and replacing them improves the ride and handling of the vehicle.

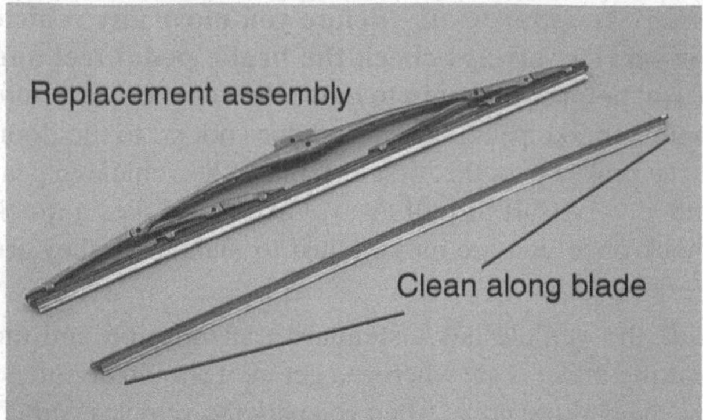

FIGURE 27-6 If the wiper blades are not performing well, try cleaning the blades and retesting. If performance does not improve, suggest replacing the blades.

If cleaning the blades does not restore their performance, they need to be replaced. When you replace wiper blades, be very careful not to let the wiper arm snap down onto the glass. This can chip or crack the windshield, resulting in a very unhappy customer and an expensive mistake to fix.

Most blades use a hook latch (**Figure 27-7**). To remove, press the release tab and pull the arm out of the J hook. To install, place the new blade so that it slides into the J hook and latches into place. An example of how a blade is removed and installed is shown in **Figure 27-7b**. Once the new blades are installed, turn on the washers and make sure the blades clean the windshield.

■ ***Air Filter Inspection.*** The air cleaner housing and **air filter** remove contaminants from the air before it enters the engine. The air filter is a pleated paper element (**Figure 27-8**). The weave of the filter traps small dirt particles and other debris. Over time, the filter becomes plugged with dirt, which can, if severely restricted, affect engine performance.

With the filter removed, inspect the filter housing for dirt and for oil. Remove dirt or other debris from the filter housing. If oil is found in the housing, perform a thorough inspection of the PCV system. If the PCV is not functioning, oil can back up into the air cleaner housing. Once the cause of the problem has been corrected, clean all the oil from the induction components. This is discussed in more detail in Chapter 24.

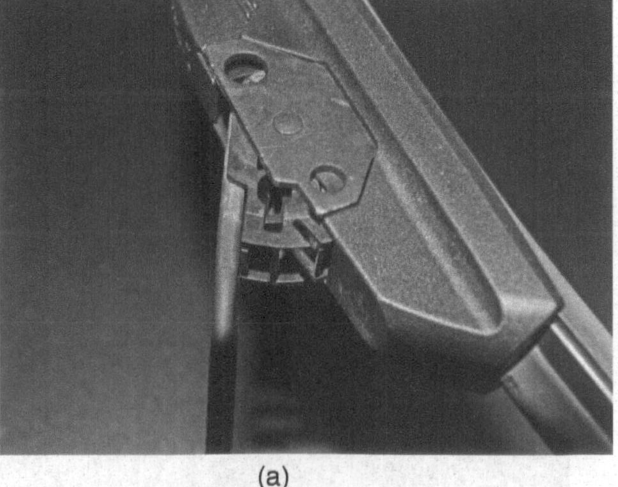

(a)

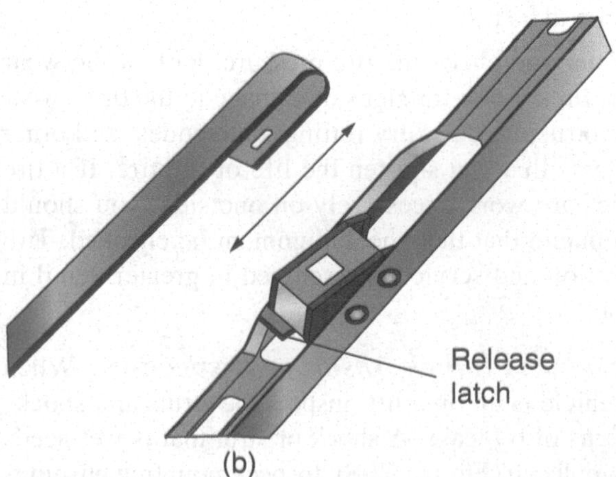

(b)

FIGURE 27-7 (a) A common type of latch securing the blade to the wiper arm. (b) Remove the blade by unlocking the latch and pulling it from the arm. Reinstall the blade over the hook and then lock it into place. It should click as it seats fully in the hook.

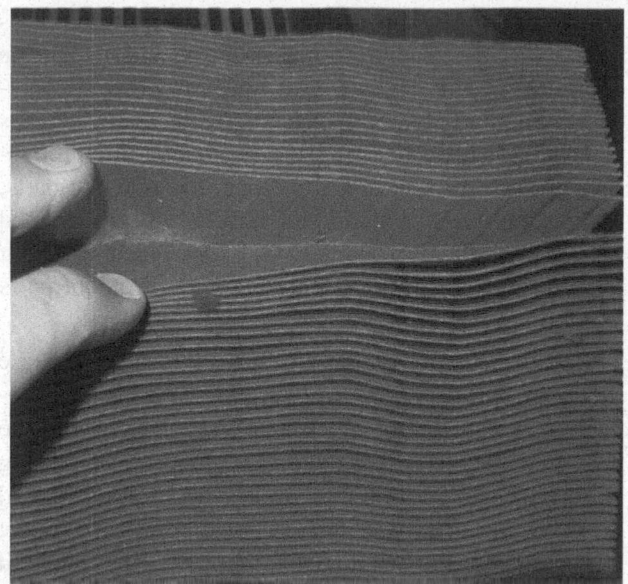

FIGURE 27-8 An example of an air filter with minimal dirt.

When you replace the air filter, use either original equipment (OE) or high-quality aftermarket replacement filters. Inexpensive filters may use a lower-quality paper, which traps less dirt and allows small particles to pass through the filter and into the engine. Less expensive filters often do not fit exactly as they should, which can cause problems with some mass airflow (MAF) sensors mounted in the air cleaner housing. Make sure the replacement filter fits the housing correctly and provides an airtight fit when installed.

Some vehicles have performance air filters installed that require periodic cleaning and an application of a dirt trapping oil. Care should be used when servicing these filters so that excessive oil is not applied. Over-oiling can cause oil to be pulled from the filter and into the engine. This can damage MAF sensors and cause drivability concerns.

■ ***Drive Belt Checks.*** Many modern cars and trucks use one multirib accessory drive belt, often called *a serpentine belt*. However, there are vehicles that use multiple belts, not all of which are mounted to the front of the engine so do not assume that an engine has only one belt.

Multirib belts should be inspected at each service and often require replacement every 60,000 to 90,000 miles. A worn or damaged belt can cause erratic power steering assist, reduced charging system output, and a high-pitched screeching or squealing noise as the belt slips around the pulleys.

Types of belt wear to look for include the following:

- Severe cracking. As the belt ages and wears, cracks develop along the ribs.
- Belt chunk-out or chunking occurs when portions of the belt have come off.
- Wear on the sides of the belt caused by misalignment problems.

Belt inspection and service is covered in greater detail in Chapter 20.

Stretch fit belts are used on several models of cars and light trucks. Instead of the belt having an automatic or manual tensioning device, stretch fit belts are stretched onto the pulleys during installation and cut off when replaced. Special tools are used to install the new belt to prevent damage to the belt and pulleys. Inspect these belts for wear and problems like ordinary drive belts.

■ ***Seat Belts.*** Perform a quick inspection of the seat belts and latches when you move the vehicle in and out of the shop. There should be no cuts in the seat belts, and the latches should fasten securely and release easily. Inspect the belts themselves for signs of wear or damage (**Figure 27-9**).

■ ***Warning Lights.*** When the ignition is first placed in the Run position, all the dash warning lights should illuminate; this is called *bulb check*. Once the engine starts, the lights should go out. Some warning lights, such as the ABS or security, may flash several times once the engine is started, but these should also go out after a few seconds. If a yellow or red light remains on, it usually indicates that a problem is present in the system indicated by the light. **Figure 27-10** shows an example of common dash warning lights. Chapter 21 provides additional information about specific dash warning lights and the possible causes for them to stay on after the engine starts.

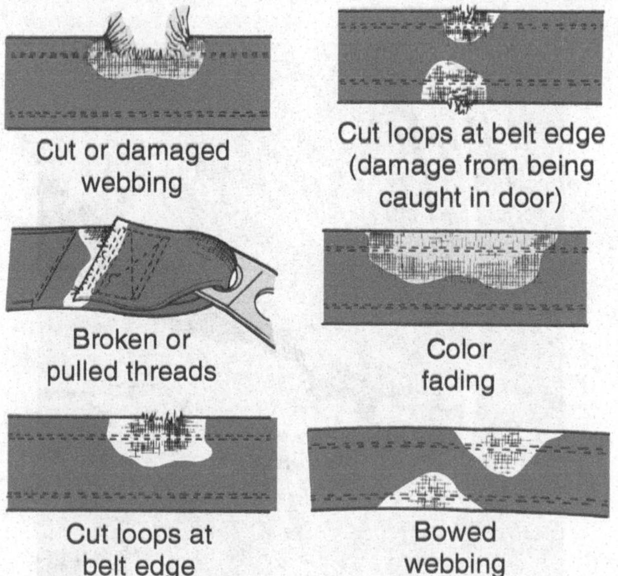

FIGURE 27-9 Inspect seat belts for abnormal wear and tear.

■ ***Cabin Air Filters.*** Most newer vehicles have a cabin air filter that requires periodic replacement. Cabin filters trap pollen, dirt, and odors from the air entering the passenger compartment. Filters can be located in the cowl area in front of the windshield or in the passenger compartment.

Checking and replacing a cabin air filter can be as simple as removing a cover and a filter. Other vehicles require removing interior trim panels to access the filter. Replacing cabin filters is covered in detail in Chapter 26. Filters will, over time, clog with dirt, reducing the amount of airflow from the interior fan, which may prompt a customer complaint of inadequate heating or cooling. Many customers do not even know that their vehicle is equipped with a cabin filter, so it is important for you to educate your customers about maintenance issues such as these.

■ ***Chassis Lubrication.*** Although not as common as it once was, the need to lubricate steering and suspension components and U-joints is still present on some vehicles, particularly larger trucks and SUVs. Some vehicles have ball joints and steering sockets that require lubrication (**Figure 27-11**). In many cases, replacement parts, such as ball joints and tie rods, have grease fittings and require periodic lubrication. Even if the vehicle was built without grease fittings, it may have replacement parts that require attention.

Before you connect the grease gun, clean the grease fitting off with a shop towel (**Figure 27-12**). This prevents dirt from being injected into the socket with the new grease. Next, place the grease gun onto the fitting (**Figure 27-13**). Pump new grease into the joint until grease begins to expand the boot. Be careful not to inject too much grease; some boots are sealed to the joint and

FIGURE 27-10 During the key-on bulb test, a number of lights should illuminate.

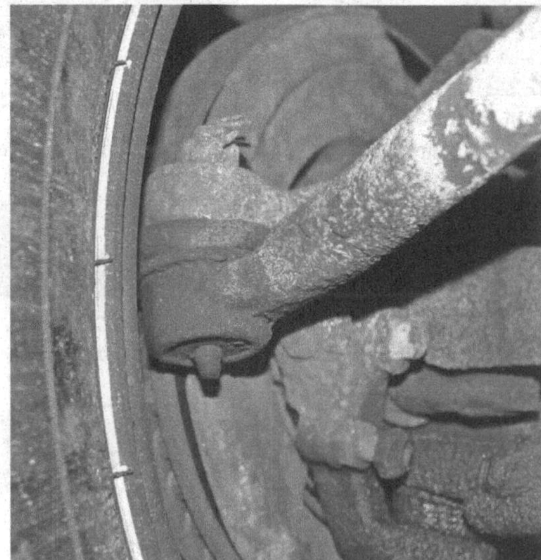

FIGURE 27-11 Inspect the steering and suspension components, both front and rear, for grease fittings.

FIGURE 27-12 Before injecting new grease, use a shop rag to clean the fittings.

FIGURE 27-13 Using a grease gun to lubricate a steering joint.

will expand until they rupture if too much grease is injected. Once all the joints have been lubricated, wipe off any excess grease from the fittings and around the boots.

Check the driveshaft for grease fittings while under the vehicle. Many four-wheel drive (4WD) trucks and SUVs have grease fittings on the shafts and U-joints (**Figure 27-14**). A special adaptor may be needed to get grease into the U-joint fitting due to the limited amount of space in the joint.

Even though most vehicles do not have greaseable steering joints, you should still inspect tie rods, ball joints, and other steering and suspension components. Problems like damage grease boots will eventually lead to loose parts (**Figure 27-15**). A thorough inspection and discussion with the customer can help prevent small problems from becoming larger ones down the road.

■ ***Constant Velocity Boot Inspection.*** Front-wheel drive (FWD), all-wheel drive (AWD), 4WD, and some rear-wheel drive (RWD) vehicles use small drive axles, called *half-shafts*. Over time, the boots that protect the constant velocity joints at the ends of the axles deteriorate and fail, allowing the constant velocity (CV) grease to leak from the joint (**Figure 27-16**). Once the boot is open to the environment, it does not take long for damage to the joint to occur. This usually requires joint or axle shaft replacement. Sometimes, the split boot is not as visible as that shown in Figure 27-16. Pull back on the boots to inspect in between the folds of the boots (**Figure 27-17**).

■ ***Exterior Lights and Horn.*** When you perform an oil change, it is easy to make a quick check of the exterior lights. Check the low- and high-beam headlamps, turn signal, brake, and backup lights. Turn signal lights should operate at the front and rear of the vehicle and should have equal blink rates from side-to-side. Also, check the four-way hazard operation. Brake and backup lights can often be checked while sitting in the vehicle and using the reflections behind you to see if the lights are working.

FIGURE 27-14 Many trucks still have grease fittings and components that require regular lubrication. Check the vehicle carefully for grease points.

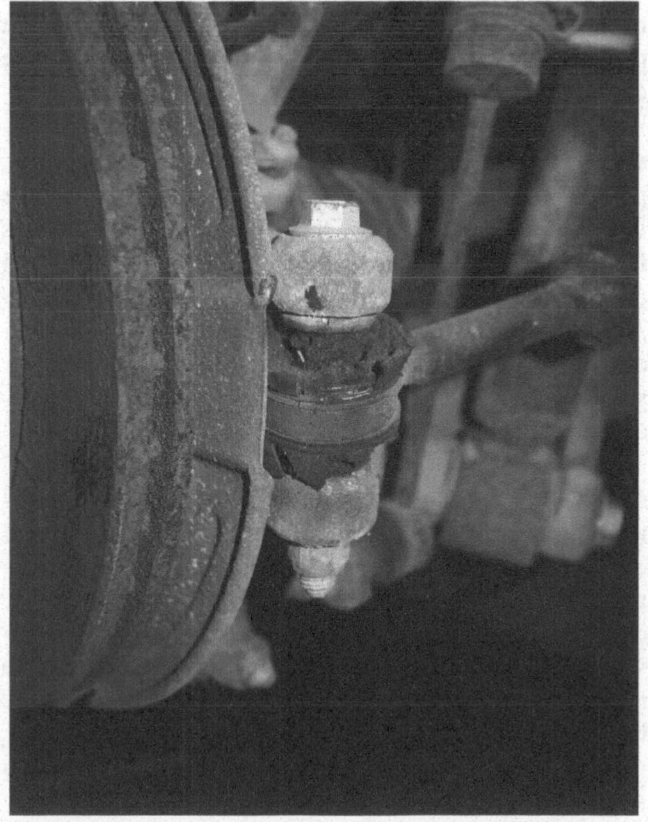

FIGURE 27-15 An example of a damage tie rod boot is an opportunity.

Perform a quick tap on the horn pad as you are exiting the shop to make sure that the horn operates correctly.

Fluid Checks and Services

All vehicles on the road today have fluids that require inspection and service. Even electric vehicles have brake fluid that requires occasional service. Fluid service has become a major part of routine maintenance and much can be said about how and when services should be performed. As a technician, you should be familiar with what the vehicle manufacturer requires and recommends based on the operating conditions of the car or truck.

COMMON FLUID CHECKS AND SERVICES

When you perform an engine oil change, it is common to check the level and condition of the other fluids, such as the power steering, brake, coolant, and transmission fluids. If the vehicle needs a fluid added, it is important that you use the proper fluid to top off the system. The following are general guidelines for common fluid services. Always refer to the vehicle's service information for specific requirements and procedures.

■ ***Power Steering Fluid.*** Power steering fluid may be in a reservoir mounted on the power steering pump (**Figure 27-18**), or in a remotely mounted reservoir (**Figure 27-19**). If the power steering fluid is low, make a note on the service order. Do not top off the power steering fluid unless you are sure you have the correct fluid for the vehicle.

■ ***Power Steering Fluid Service.*** Until recently, power steering fluid was generally not a maintenance item. Some manufacturers specify replacement intervals for the power steering fluid. While most vehicles can use generic power steering fluid, some, such as Honda/Acura, require special fluid, available from their dealers. Use of non-OE fluid can cause a failure of the power steering system. Flushing the power steering fluid is discussed in Chapter 9.

■ ***Brake Fluid.*** Brake fluid is in the brake fluid master cylinder reservoir, generally located in the left rear corner of the engine compartment. Most vehicles have a plastic reservoir so that the fluid level can be easily checked (**Figure 27-20**). If you must remove the reservoir cap, clean the cap and the surrounding area before removing it. This helps prevent dirt from entering the brake system.

If topping off brake fluid, use only the correct brake fluid from a sealed container. **Using anything other than brake fluid can cause complete brake system failure. Be sure to use only the specified brake fluid.** If the brake fluid is at the MINIMUM level, you should suggest a brake inspection. It is likely that the brake pads and rotors are sufficiently worn and may be due for replacement.

■ ***Brake Fluid Service.*** Due to the tight tolerances and the high-speed operation of the valves used in ABS hydraulic units, brake fluid flushing, the removal and replacement of the fluid, has become a part of the routine maintenance

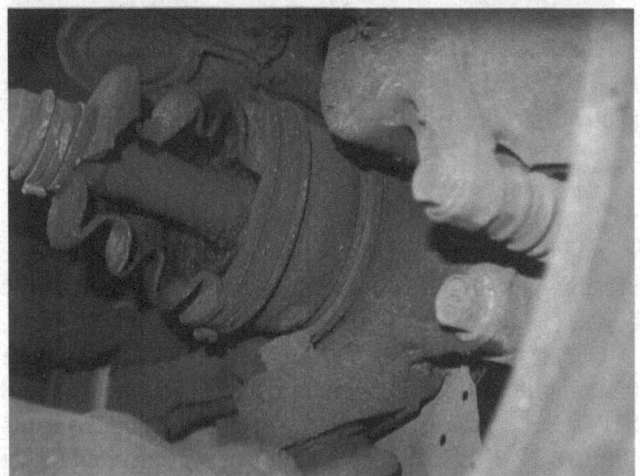

FIGURE 27-16 An example of a damaged CV boot and joint.

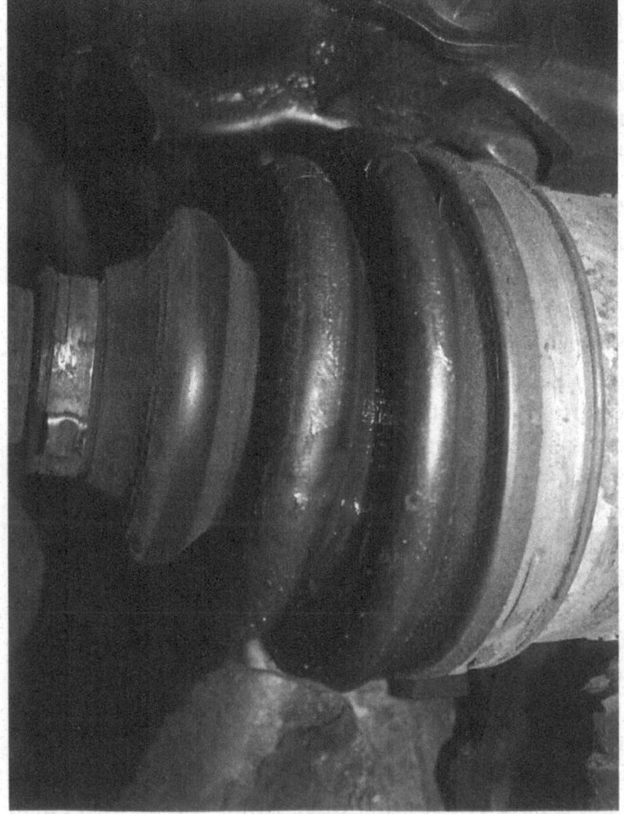

FIGURE 27-17 Look closely at CV boots for leaking grease.

schedule. Because brake fluid is hygroscopic, the accumulation of moisture in the system leads to rust, corrosion, and copper ion contamination. The old fluid should be flushed every two to three years to remove the contaminants and restore the brake fluid boiling point. Brake fluid service is covered in more detail in Chapter 11.

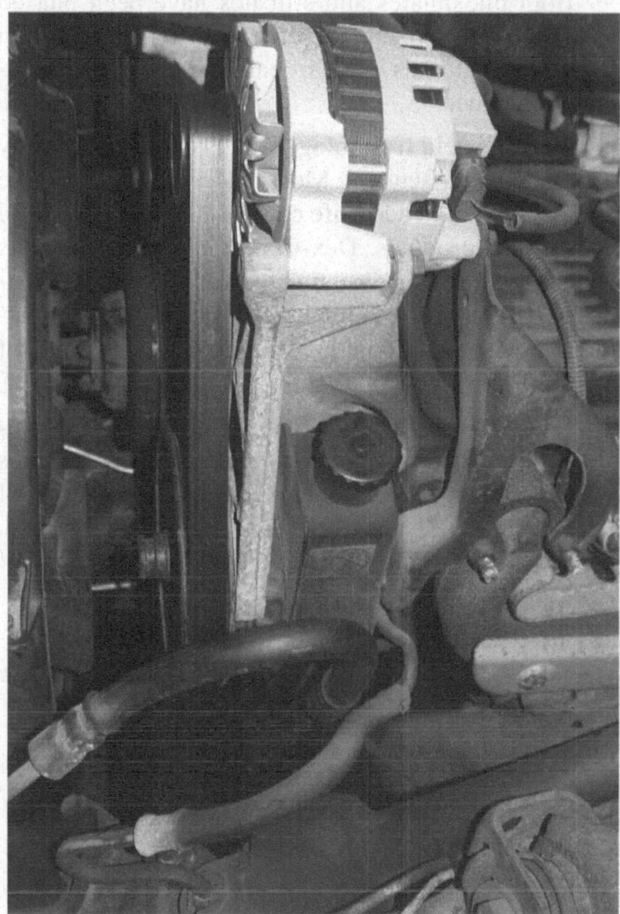

FIGURE 27-18 An example of a power steering reservoir attached to the power steering pump.

FIGURE 27-19 An example of a remote power steering reservoir.

■ ***Coolant.*** Coolant level can often be checked by looking at the expansion tank (**Figure 27-21**). If the vehicle does not have a tank, radiator cap removal may be necessary to check the coolant level and condition.

Some vehicles, mostly hybrids, have two or more separate coolant types, one coolant for the engine and one or more for the hybrid drive system (**Figure 27-22**). Pay careful attention to which coolant you are checking so that only the right coolant is used if adding some is necessary.

For many years, one coolant was used in most everything. Today, there are many different types of coolants in use, most of which are incompatible with traditional green coolant. To make things more confusing, color is not the way to tell which coolant is right for a vehicle. A few of the common types of coolant available today include the following:

- Traditional green, ethylene glycol–based coolant, which uses inorganic acid technology (IAT) and

FIGURE 27-20 Checking brake fluid level is easy with plastic reservoirs. Check for MIN and MAX lines on the reservoir.

FIGURE 27-21 An example of a coolant pressure tank. Coolant level can easily be seen and compared to the marks on the tank.

contains silicates and phosphates. Ethylene glycol is the base product for all OE coolants right now, but this type of coolant is not used as the factory fill in modern vehicles. Details about IAT coolant is shown in **Figure 27-23**. Look closely at the contents of the coolant and you will notice that it contains phosphates and silicates, which were common coolant ingredients for many years. The phosphates and silicates were used as corrosion inhibitors. These additives are rapidly consumed, requiring the coolant to be changed about every two to four years. Both phosphates and silicates have been phased out of use because of their incompatibility with newer materials used in seals and gaskets in modern engines.

FIGURE 27-22 Hybrids and some supercharged engines have two different cooling systems. Be aware of which coolant you are checking.

- **Organic acid technology (OAT)** coolant is used by many manufacturers. Most modern engines require the use of extended-life coolants, such as Dex-Cool for GM products. Dex-Cool and similar coolants use organic acids and do not contain nitrates, phosphates, or silicates, making this coolant incompatible with traditional green coolant. An example of this type of coolant is shown in **Figure 27-24**.
- **Hybrid organic acid technology (HOAT)** coolants are used by Ford, Chrysler, and Mercedes and others. HOAT coolant combines properties of IAT and OAT coolants. This coolant may be red, pink, yellow, or blue (**Figure 27-25**). This type of

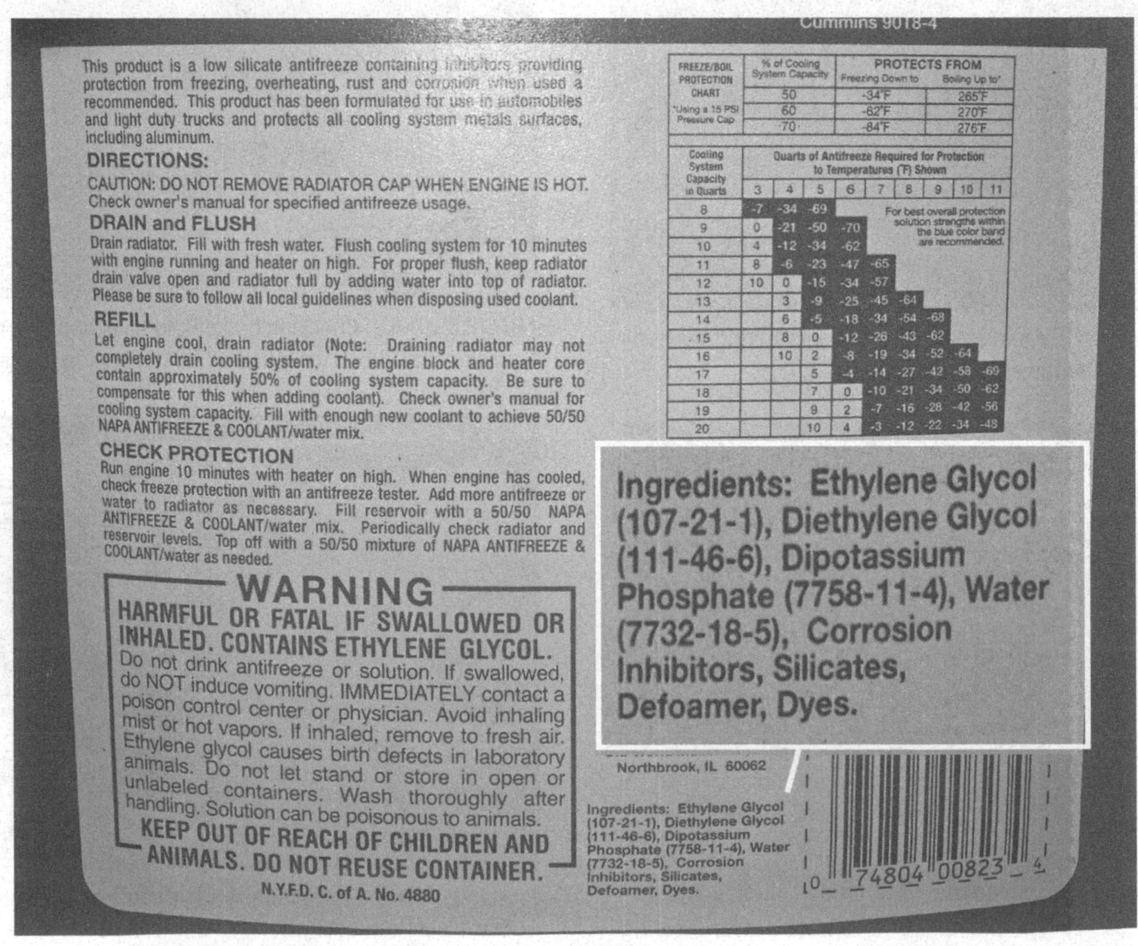

FIGURE 27-23 Traditional green inorganic acid technology coolant.

coolant is not compatible with either traditional green or extended-life organic acid coolants like Dex-Cool.

- HOAT coolants that are silicate free but contain phosphates are used in some Asian vehicles.
- G48 ethylene glycol coolant, is low-silicate and phosphate free, and is used by BMW. This coolant is like the G-05 coolant approved by Mercedes-Benz.
- Universal coolants are available in full strength and premixed 50/50 solutions. However, because some cooling systems have very specific requirements regarding additives, you should not rely on a universal coolant to provide complete protection for any given vehicle. The best solution for the long-term health of the cooling system is to use the coolant specified by the vehicle manufacturer.

■ ***How Do I Check Coolant?*** If the coolant is low and the correct coolant is not available, note it on the service order and recommend that a complete cooling inspection be performed. When you inspect the coolant, first ensure that the cooling system has cooled and is not under pressure. **Never remove a radiator cap from a hot cooling system. The release of hot coolant can cause severe burns.**

Carefully remove the radiator or pressure tank cap. Inspect the cap and the fill neck for signs of corrosion or sludge (**Figure 27-26**). If the coolant level is acceptable, test the coolant condition with a hydrometer, refractometer, or pH test strips. Testing the coolant is discussed in more detail in Chapter 26.

Service Note

Caution! Do not remove a radiator or pressure tank cap from a hot cooling system. This can cause severe burns and injury.

■ ***How to Choose the Correct Coolant?*** So how do you know which coolant is right for a vehicle? Begin by checking the vehicle owner's manual for information about coolant types. This can usually be found by looking in the manual's index under "Fluids." You can also check the radiator cap or on a decal on or near the expansion tank (**Figure 27-27**). If no information is found on

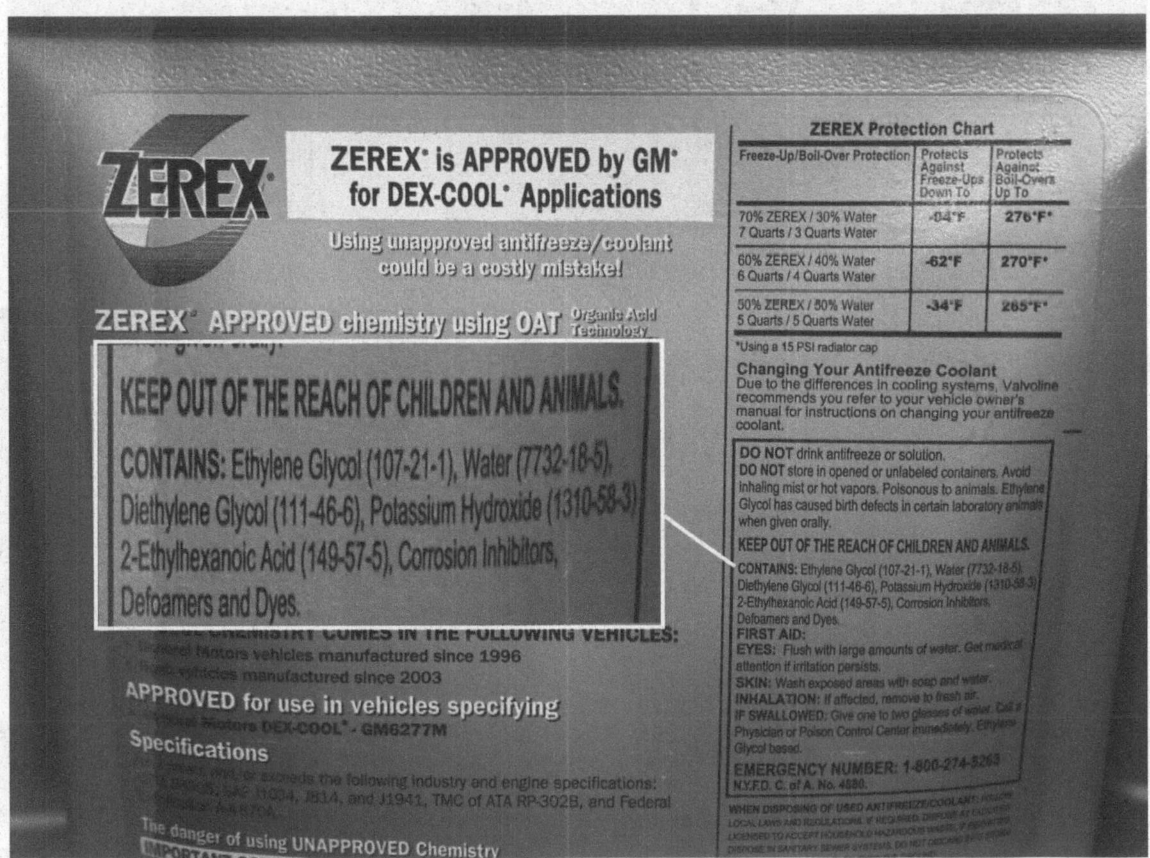

FIGURE 27-24 The ingredients of an OAT coolant.

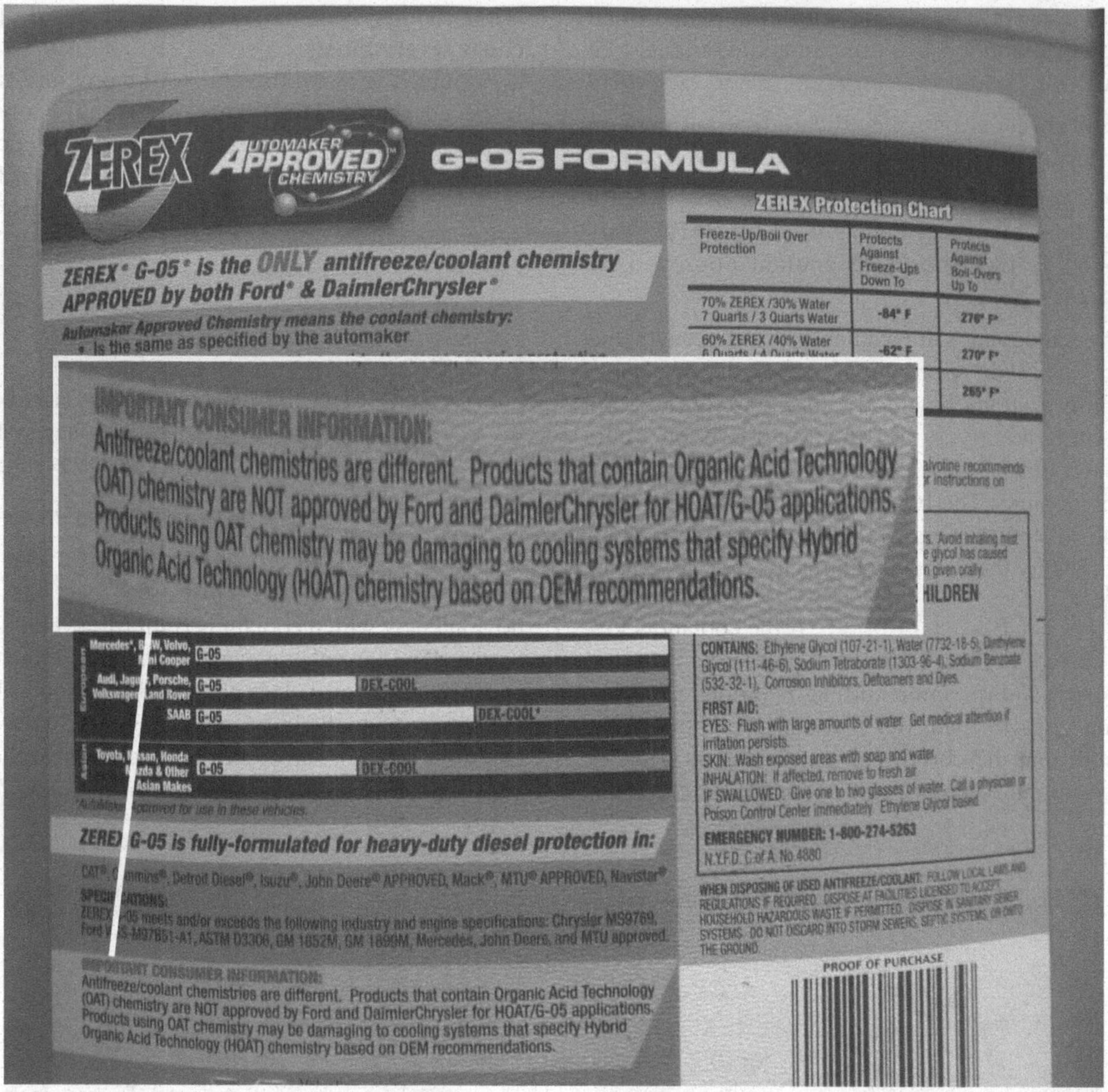

FIGURE 27-25 The ingredients of a HOAT coolant.

FIGURE 27-26 Coolant sludge indicates cooling system problems or that the wrong coolants have been used.

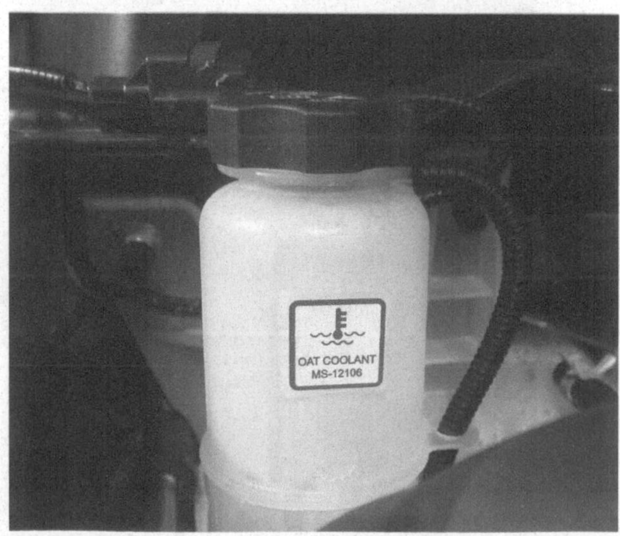

FIGURE 27-27 Check the coolant tank for information about what coolant should be used.

TABLE 27-1

A sample of different coolants in use by a few manufacturers.

Manufacturer	Specification	Color	Acid	Silicates	Amine	Nitric	Borate	Phosphate
Ford	ESE-M97B44-A	Diesel additive						
	WSS-M97B51-A1	Gold	HOAT	low	no			no
	WSS-M97B44-D / D2	Orange	HOAT					
	WSS-M97B55-A	Dark green	HOAT					
	ESE-M97B44-A	Green	IAT	yes				yes
General Motors	GM 6277M	Orange	OAT	no				no
Chrysler	MS-7170	Pink	HOAT					
	MS-9769	Pink	HOAT					
	MS-12106	Purple	OAT					
Honda	Long Life Type 2	Blue	OAT	no			no	
Toyota	Super Long Life Coolant	Pink/Red	HOAT	no	no	no	no	

the vehicle, check the service information. Most manufacturers publish fluid specifications, such as those shown in **Table 27-1**.

■ ***Automatic Transmission Fluid.*** Many modern automatic transmissions have a dipstick to check fluid level. However, many do not. To accurately check the automatic transmission fluid (ATF), is typically must be at operating temperature. This is often around 150°F – 200°F (65.5°C–93°C). To check the transmission fluid temperature, a scan tool can be used. You can also remove the dipstick and check the fluid temperature using either an infrared pyrometer or a thermocouple for a digital multimeter. ATF inspection is covered in detail in Chapter 25.

■ ***Manual Transmission Fluid.*** A manual transmission requires periodic inspection and maintenance. The fluid picks up metal shavings, and over time, the additives wear out. This necessitates periodic fluid replacement. To check fluid level, locate the service plug in the side of the transmission. Remove the plug and check to see if the fluid is up to the bottom of the hole. In most cases, the fluid level should just reach the bottom of the check plug hole. If it is low, top it off with the correct fluid, reinstall the plug, and torque it to specifications.

■ ***Differential and Transfer Case Lubricants.*** Both differentials and transfer cases are an often-overlooked maintenance item. To check the differential fluid, locate the check and fill plug (**Figure 27-28**). On Chrysler products that use the rubber plug, be careful when you remove the plug so that you do not damage it. Most other manufacturers require the use of either a 3/8-inch drive ratchet or a wrench to remove the plug. The fluid level should be up to the bottom of the fill plug hole. If fluid is needed, refer to the owner's manual or service information for the correct lubricant.

Checking a transfer case is similar. Locate the check plug and remove (**Figure 27-29**). The fluid should be up to the hole. If the fluid is low, determine the correct type of fluid to top off the level.

■ ***Washer Fluid.*** Inspect the fluid level in the washer bottle and top it off as necessary. Some vehicles, such as some minivans and SUVs, use two separate washer fluid reservoirs, one for the front and one for the rear washers. Always check both reservoirs. The washer fluid reservoir often has a blue cap to help identify it as the washer fluid (**Figure 27-30**). Many reservoirs do not make it easy to determine the fluid level. Sometimes the reservoir cap will have a built-in dipstick to determine fluid level.

■ ***Diesel Exhaust Fluid.*** A new item on the maintenance list is refilling the diesel exhaust fluid (DEF) reservoir. Newer diesel engine vehicles, to meet diesel exhaust emission standards, have a fluid tank that must be refilled. **DEF** is a chemically pure urea and deionized water solution, approximately 33% urea and 67% pure water, that is injected in a special diesel catalytic converter to reduce NOx emissions (**Figure 27-31**). The ammonia in the urea reacts with the NOx to produce nitrogen, oxygen, and carbon dioxide.

Vehicles equipped with DEF tanks require periodic refills. Time between refills varies with mileage and driving conditions but five gallons to every 5,000–8,000 miles

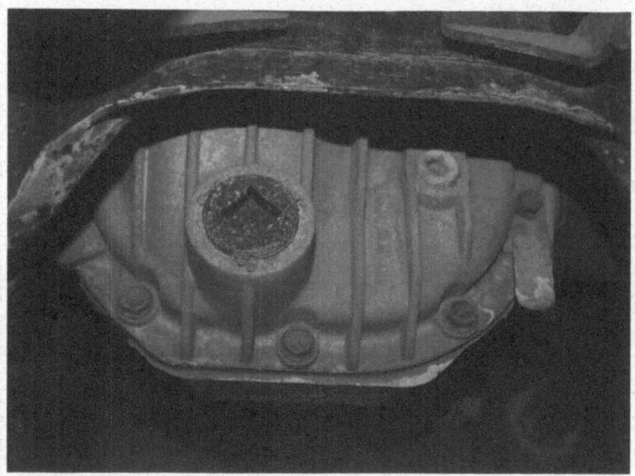
FIGURE 27-28 An example of a rear differential fluid check plug.

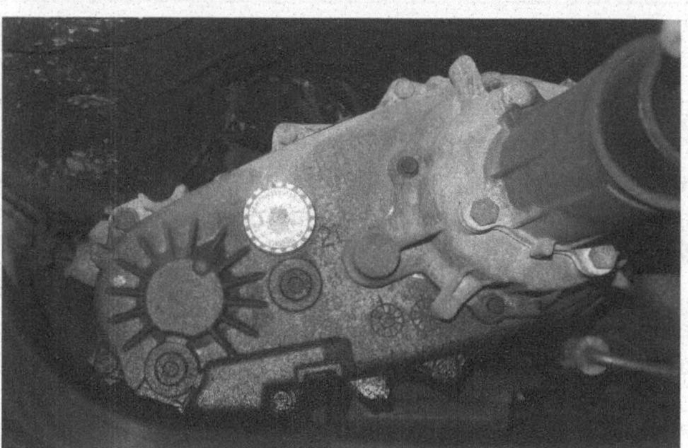
FIGURE 27-29 Transfer case fluid check and drain plugs.

driven is common. If the system runs empty, the OBD system will eventually place the engine in a reduced power mode until the system is refilled and the cleaning cycle initiated. DEF is available in one and five gallon jugs and from the vehicle manufacturer, aftermarket parts stores, and at gas stations (**Figure 27-32**).

Once the driver information center displays the "Exhaust fluid service" or similar indicator, the system is low enough to need fluid added. Locate the DEF reservoir and fill cap, which may be located under the hood (**Figure 27-33**). Note the capacity and warning decal for the DEF (**Figure 27-34**). DEF fill locations are often next to the fuel filler (**Figure 27-35**). Remove the cap and fill the system. Depending on the vehicle, the tank may hold five or more gallons.

While DEF is not toxic or considered hazardous, it will leave residues, that if left, can damage painted surfaces. Be sure to thoroughly clean up any spilled fluid.

FIGURE 27-30 Windshield washer fluid caps are typically blue for easy identification.

■ ***Fluid Leaks.*** When you are under the vehicle, look for signs of fluid leaks. As vehicles age it is not uncommon for small leaks to develop. A slight bit of seepage is not a concern. However, there should not be engine oil, transmission fluid, coolant, or other liquids coating or dripping from the underside of the vehicle (**Figure 27-36**). Tracing down the source of a leak can be difficult and often requires more time and effort than is available during an oil change. Note the presence of the leak on the service order and alert the customer. Leak inspection is discussed in more detail in Chapter 23.

■ ***Exhaust Leaks.*** In many cases, an exhaust leak can be heard when you move the vehicle. An exhaust leak, like that caused by the broken pipe shown in **Figure 27-37**, can pose a danger from the carbon monoxide in the exhaust. Look at the exhaust system components when you are under the vehicle. Exhaust pipes should be intact and free of leaks. Check the exhaust hangers to ensure that the system is not contacting the underbody and rear suspension components. Exhaust inspection is discussed in more detail in Chapter 23.

ENGINE OIL SERVICE

Even though modern vehicles require less maintenance than ever, all the fluids require periodic service. One of the most common services performed by entry-level technicians is the oil change. Even though the job is usually not complex, performing an oil change on modern vehicles can often require more than just removing a drain plug and pouring in new oil.

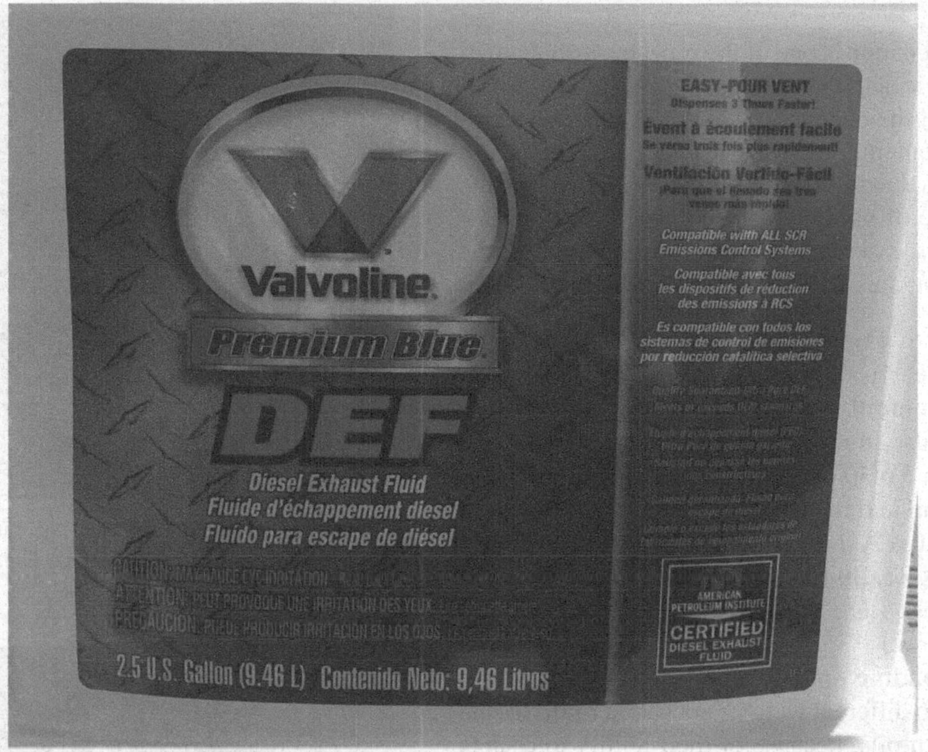

FIGURE 27-31 DEF is required for vehicles to meet diesel exhaust emission standards.

ENGINE OIL

The engine oil change is one of the most basic and often-performed services in the world of vehicle maintenance. And because of the changes in engine design and maintenance requirements, it is now less understood than in the past. This is due to several factors, which include the widespread use of variable valve timing (VVT) and/or lift mechanisms, extended service requirements, and the necessity of not causing damage to the catalyst.

■ ***Oil Basics.*** For many years, a few types of oil were used in nearly all applications. The vehicle manufacturers commonly specified oils such as 10W30 and 5W30 for use in most engines. These designations mean that for a 5W30 oil, the oil has the cold viscosity of a 5-weight oil and the warm viscosity of a 30-weight oil. The W in 10W30 means winter, so you can think of the oil being 5-weight for winter (cold) and 30-weight for summer (hot). Oil viscosity ratings are based on tests ranging from −35°F to more than 300°F (−37°C to 150°C) depending on the weight of the oil.

Oils with two different weights are called **multiviscosity oils**. Modern oils, because of their ability to change viscosity at different temperatures, provide much better protection during cold starts than conventional single-weight oils or even multiviscosity oils from 30 years ago. The ability of oil to flow easily when it is cold is vital to prevent cold-start engine wear. In addition, oil needs to remain consistent at higher temperatures so that it will remain a barrier between moving parts (**Figure 27-38**). Oil also performs other functions, such as a detergent, a heat exchanger, and as a corrosion inhibitor. Modern engine oils are complex mixtures of different additives and base oil stocks and are designed to provide very specific qualities necessary for engine operation.

■ ***Oil Ratings.*** Many modern engines require oils such as 0W20, 5W20, 5W30, up to 5W40, and others. Weights are not the only way in which engine oil is rated;

FIGURE 27-33 Some vehicles place the DEF tank under the hood. Be sure which bottle you are filling whenever replacing or topping off a fluid.

FIGURE 27-34 An example of a DEF bottle and warning label.

FIGURE 27-32 DEF at the pump is common at larger filling stations and truck stops.

FIGURE 27-35 Many vehicles place the DEF fill point near the fuel filler.

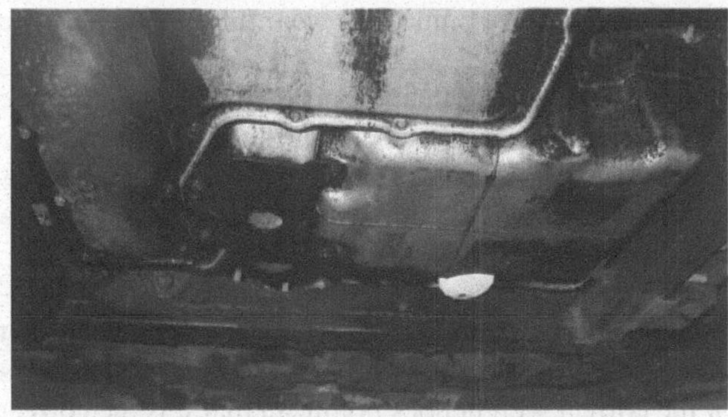

FIGURE 27-36 Older and high-mileage vehicles often have a fluid leak. This vehicle had both engine oil and transmission fluid leaking.

FIGURE 27-37 Inspect the exhaust for damage and leaks. An exhaust leak can be dangerous for the drive and occupants and the customer made aware of any problems.

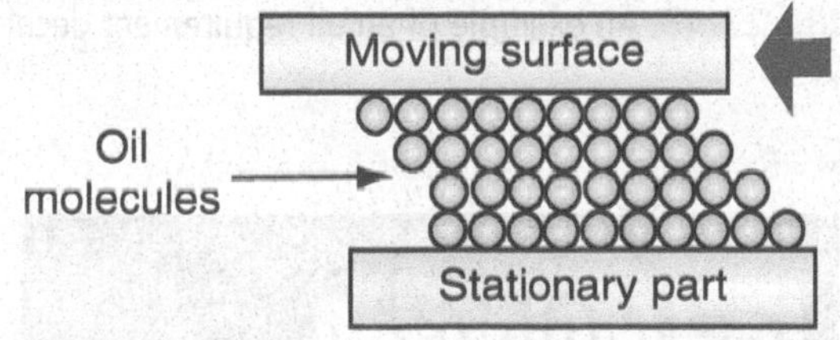

FIGURE 27-38 Oil acts as a barrier and as a lubricant between moving parts.

all vehicle manufacturers have specific oil requirements for many if not all their engines. Beyond the classification of engine oil by its weight, several other methods and approvals are also used. Technicians and consumers may be familiar with oils having both an SAE and API rating. The **Society of Automotive Engineers**, SAE, oversees the testing and certification of the viscosity ratings. The **American Petroleum Institute (API)**, develops the basic oil specifications, which are updated periodically. Oil with an API approval has a starburst symbol (**Figure 27-39**) displayed on the bottle. API oil ratings, the most current being SN in 2012, coincides with the latest ILSAC GF-5 rating discussed below.

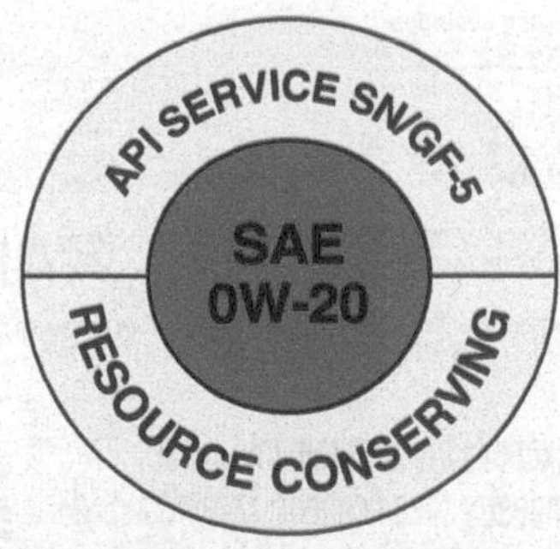

FIGURE 27-39 An example of the API service "donut."

As vehicle manufacturers develop new engine technologies, reduce internal friction, improve fuel economy, and reduce exhaust emissions, new engine oils must be developed as well. SN-compatible oils are backward compatible for older engines that require SM, SL, SJ, and earlier oils. Further information about API engine oil certifications can be found at API.org. It is important to note that while newer oils are backward compatible, older oil types cannot be used in newer engines requiring the latest oil rating. Do not use oils such as an SL-rated oil (2001 model year) in a 2010 engine as the SL oil will not have the properties needed for the newer engine.

In addition to SAE and API ratings, oils may also receive certifications from the International Lubricant Standardization and Approval Committee, or **ILSAC**. This committee represents both domestic and Japanese auto manufacturers. The most common ILSAC standards for modern engines is the GF-5 ratings. GF-5 was new for the 2011 model year vehicles and has reduced amounts of phosphorus, which can damage catalytic converters. GF-5 has improved seal and gasket compatibility to reduce oil leaks, increased corrosion protection, and increased energy-conserving qualities to reduce fuel consumption. Do not use an older oil, such as a GF-4 oil, in an engine that requires GF-5 as the previous oil will not have the correct qualities required by the newer engine. An example of an oil bottle with the ILSAC rating is shown in **Figure 27-40**.

Oils used in vehicles from European manufacturers often need to meet **ACEA standards**. ACEA stands for the Association of Constructors of European Automobiles. These ratings are for both gasoline- and diesel-powered engines that have specific requirements for sulphated ash, phosphorus, sulfur (SAPS), and high temperature/high shear (HTHS) viscosity ratings. ACEA ratings are listed as A1, A3, A5, B1, B3, B4, B5, and C1 through C4 (**Figure 27-41**).

The vehicle manufacturers themselves also have very specific oil ratings. Examples of these are the Ford WSS-M2C945-A, WSS-M2C946-A specs, GM 4718M, GM 6094M, and Dexos rating (**Figure 27-42**). An example from a VW is shown in **Figure 27-43**. Nearly every vehicle manufacturer has its own oil ratings for its engines, and putting the wrong oil into an engine

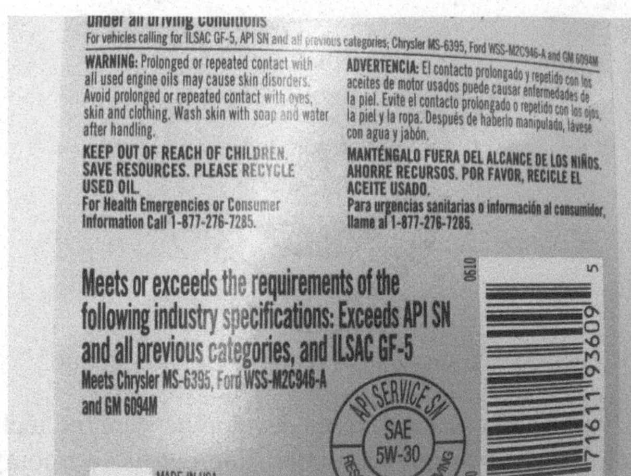

FIGURE 27-40 An example of an oil that meets the ILSAC GF-5 standards.

FIGURE 27-41 An example of an oil that meets the ACEA A1/B1 standards.

FIGURE 27-42 Many vehicles, like this General Motors car, indicate what oil is required on the oil fill cap.

can lead to rapid component wear and serious engine damage.

Look carefully at the oil certifications found on the labels. Because nearly all modern engines have some type of oil requirement, you need to be able to locate this information and determine what oil is appropriate for a vehicle (**Figures 27-44** and **27-45**).

Due to changes in engine technology and longer maintenance intervals, it is vitally important that the correct oil is used. Just because an oil has an API starburst and is of the correct weight does not mean that it is the correct oil for the application. As a professional, it is up to you to determine what the oil requirements for your customers' vehicles are and to locate and use the correct oil. The consequences for using the wrong oil in the short term can include an illuminated malfunction indicator lamp (MIL) and the setting of variable valve timing DTCs.

FIGURE 27-43 An example of an oil requirement decal on a VW.

FIGURE 27-44 This oil meets Ford WSS-M2C945-A oil standard. Checking the oil bottles for these types of certifications is important when selecting a product.

The longer-term effects include oil sludge, rapid internal component wear, and engine failure. Several manufacturers, including Toyota and Mercedes Benz, have been sued in class-action lawsuits regarding engine oil problems and the resulting engine failures.

■ ***Engine Oil Change.*** Before you begin changing the oil, take time to check and document the engine oil level and condition. **Figure 27-46** shows an example of checking the oil before performing the oil change. Note that the oil level is low, at the Min (minimum) line on the dipstick. Also, note if there are signs of oil baking onto the dipstick (**Figure 27-47**). Oil varnish, like that shown here, can indicate poor maintenance practices, such as going too long between oil changes.

Next, check the vehicle owner's manual (if available) for the recommended oil. If the manual is not with the vehicle, refer to either the manufacturer's service information or a similar resource, such as a lubrication guide, to ensure that the correct oil is being used. The oil cap may indicate what weight of oil is required but usually does not specify any additional requirements like those discussed in this section (**Figure 27-48**). The owner's manual may only provide basic information, such as use an APR approved oil. If this is the case, refer to the service information to check whether there are specific oil requirements for the engine. Once these two items are done, proceed with the oil change by placing fender covers on the vehicle to protect its finish. Next, remove the oil fill cap and set it aside. Removing the cap will allow for easier draining of the crankcase. Some technicians place the oil fill cap on the hood latch. This is a way to ensure that the cap gets reinstalled because the hood will not close if the oil cap is in the way of the latch.

Raise the vehicle and position the oil drain under the engine, near the oil drain plug. Position the drain so that

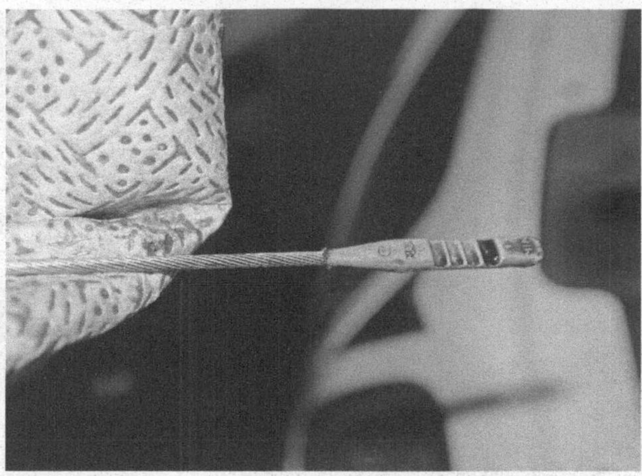

FIGURE 27-46 Check and record the oil level before draining the oil. Documentation of neglect can save the shop if a dispute ever occurs.

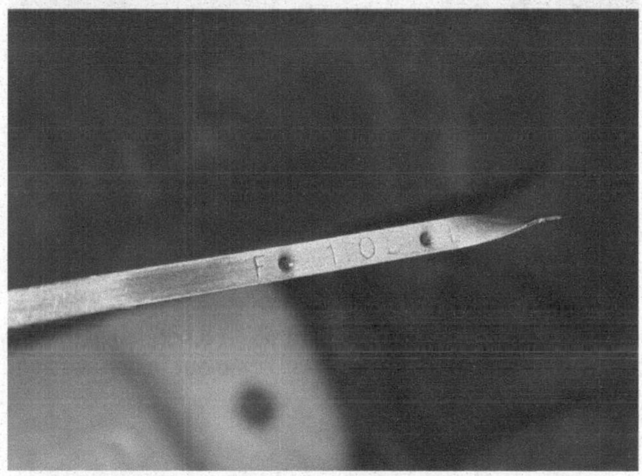

FIGURE 27-47 Note any signs of oil varnish on the dipstick. This can indicate improper maintenance, which can damage the engine.

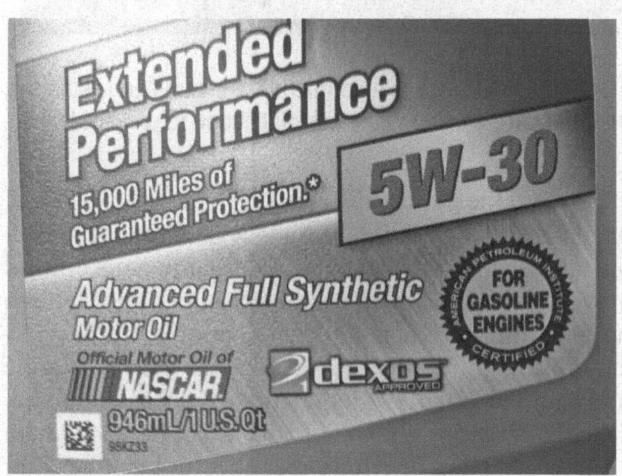

FIGURE 27-45 Dexos approved oils will have the Dexos logo on the bottle.

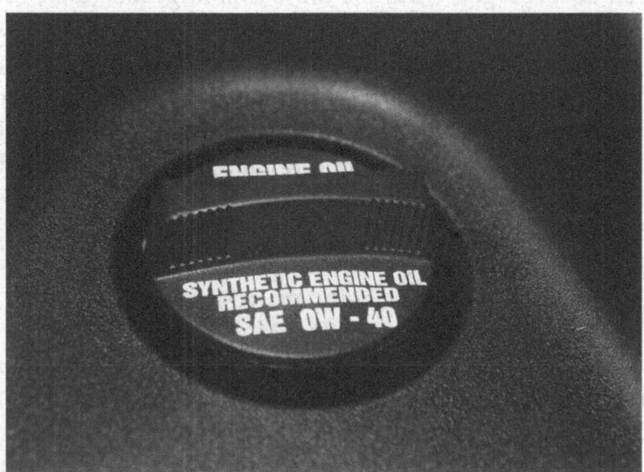

FIGURE 27-48 The oil cap may only provide what weight oil should be used.

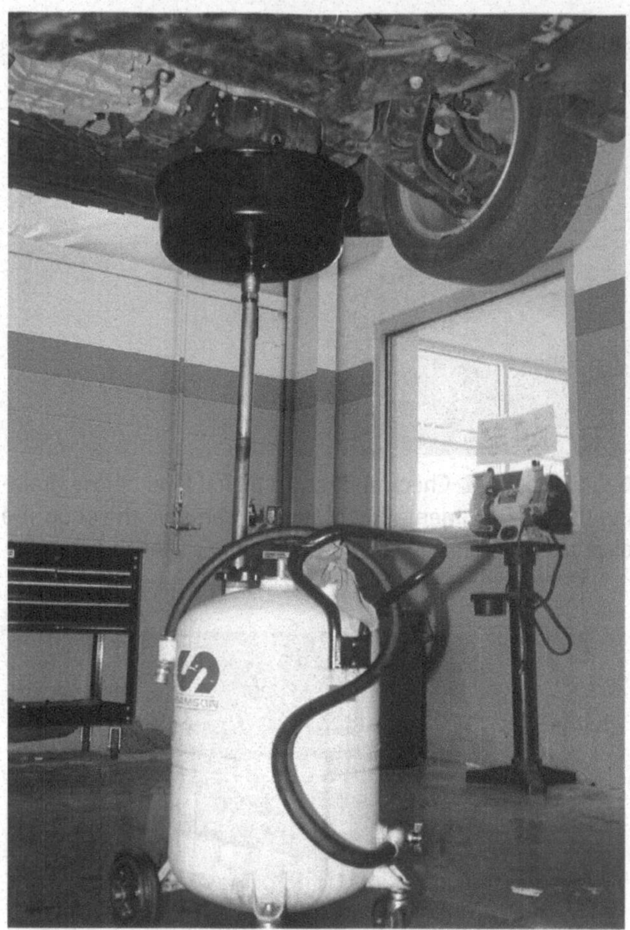

FIGURE 27-49 Place the drain pan so it can catch the oil coming out. Remember, hot oil will come out faster and at a different angle than when cold.

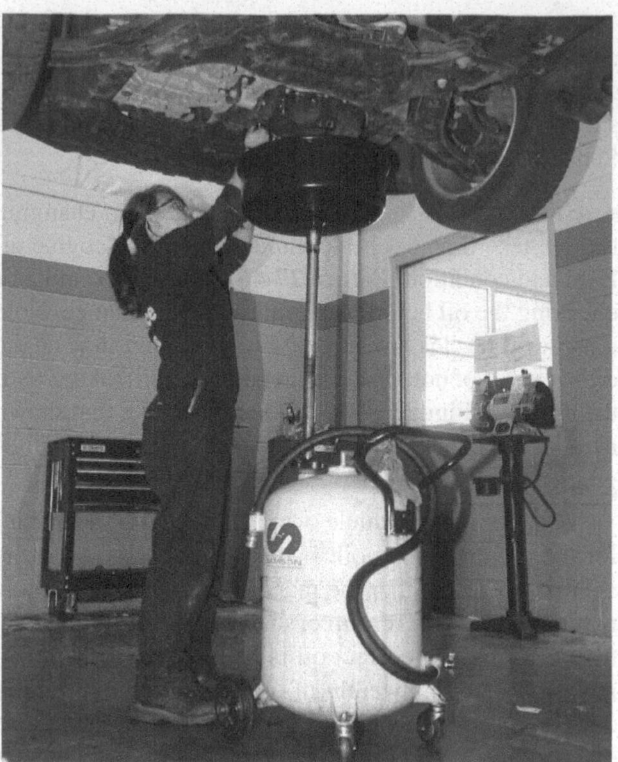

FIGURE 27-50 You may be able to replace the filter while the pan is still catching oil draining from the pan.

FIGURE 27-51 Many manufacturers recommend replacing the drain plug washer with each oil change to prevent leaks.

it will be able to catch the oil flowing out of the plug hole (**Figure 27-49**). If the engine is hot, the oil will flow faster and in more of a horizontal direction if the drain plug is on the side of the oil pan, so position the drain pan accordingly. Once the oil is draining, locate the oil filter and try to loosen it by hand. Do not completely loosen the filter unless it is located very close to the drain pan (**Figure 27-50**). If the filter is tight, use an oil filter wrench, to loosen the filter. Once the oil has drained and is dripping slowly from the oil pan, prepare to reinstall the drain plug. Many manufacturers recommend that a new oil drain plug washer be installed at each oil change. If necessary, install a new washer on the plug and hand-tighten the plug into the oil pan, using a torque wrench, tighten it to specs (**Figure 27-51**).

Remove the oil filter if you have not already done so. Inspect the filter mounting pipe and gasket surface. Clean the gasket surface with a shop towel. Before installing, verify that the new filter matches the old filter where the seal is installed. Make sure that the old filter's gasket is not stuck to the engine. If it is, remove and discard the gasket. Installing the new filter on top of the old gasket will cause the oil to leak, which can result in engine damage or failure. If the filter mounts upward or even on its side, pour a small amount of new oil into the new filter and apply a light coat of oil to the gasket. Prefilling the filter helps reduce the amount of time the engine runs without lubrication when it is restarted. Next, thread the new filter onto the filter pipe and tighten it per specs. This generally means to tighten the filter one-half to three-quarters of a turn after the seal contacts the filter base on the engine. Do not use a filter wrench to tighten the filter unless you cannot get the filter tight enough using your hands.

Many new vehicles use cartridge oil filters (**Figure 27-52**). To replace this type of filter, first remove the cover, and then the filter (**Figure 27-53**). Special sockets are often needed to remove these housings (**Figures 27-54** and **27-55**). The opening for this filter is shown in **Figure 27-56**. Remove the filter from the housing, clean the housing with a shop rag, remove the O-ring seal, and replace with the new O-ring. Place the new filter into the housing and install it back in the engine (**Figure 27-57**). Torque the housing to specs to prevent damage and leaks.

Using a funnel, pour the correct amount of new oil into the engine. Most modern engines with wet-sump oil systems hold between four and seven quarts of oil. However, many smaller four-cylinder engines require slightly less than four quarts to refill, so check the refill capacity specs before you accidentally overfill the system. Engines with dry-sump oil systems, external oil coolers, and diesel trucks usually require more oil, up to 12 quarts to fill the system. Always check the service information for the refill capacity so that you do not over- or underfill the engine. Some shops have oil dispenser, which can be set for a certain amount of oil to be installed (**Figure 27-58**).

FIGURE 27-52 New and old cartridge filters. Always install new O-ring seals on the filter housing to prevent leaks.

FIGURE 27-53 Some cartridge filter housings are easily accessible, others are not. Most require a special tool to remove.

Once the oil is poured in, check under the engine for signs of leaks. If no leaks are found, reinstall the oil fill cap and start the engine. Watch the oil pressure gauge or oil pressure warning light on the dash when the engine is started. The oil gauge should rise quickly, or if an oil light is used, it should go out a couple of seconds after the engine starts. If the gauge does not rise or the light stays on, shut the engine off immediately and check for an oil

FIGURE 27-54 A selection of different oil filter wrenches.

FIGURE 27-55 These sockets are used on cartridge-style oil filter housings.

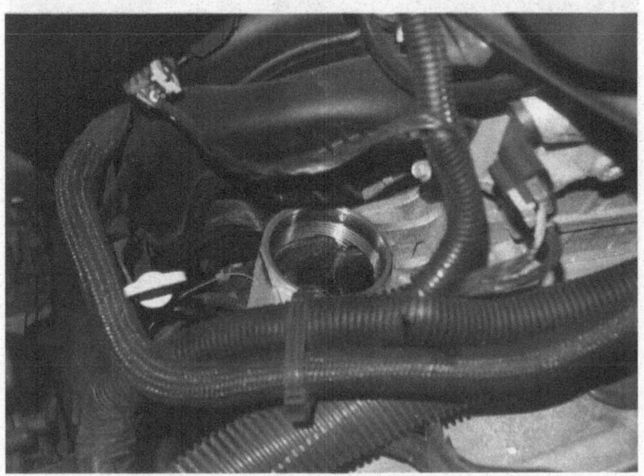

FIGURE 27-56 The engine with the oil filter removed. Be careful to not drop anything down there.

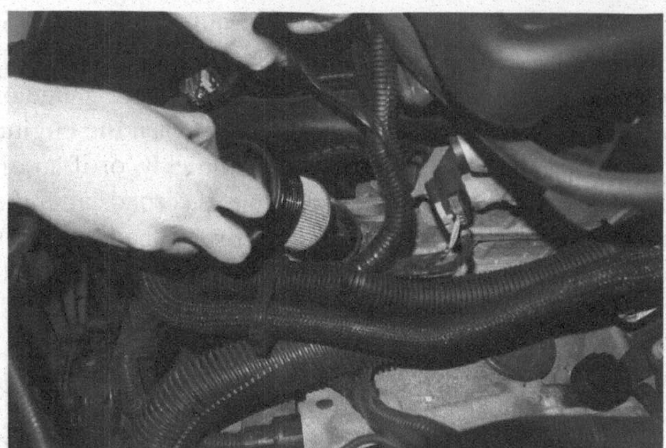

FIGURE 27-57 Reinstalling the housing and new filter. Torque the housing to specifications.

FIGURE 27-58 Refill the oil with the correct type and amount.

FIGURE 27-59 Once filled, start the engine and check for leaks. Shut the engine off and recheck the oil level, add if necessary.

FIGURE 27-60 An example of a filter crusher.

leak. Once the engine has run for about 10 seconds, turn the engine off and check again for any signs of a leak.

Let the engine sit for a couple of minutes to allow oil to drain back into the oil pan, then check the oil level on the dipstick. **Figure 27-59** shows the oil level on the dipstick after the oil change. Add oil if necessary and then recheck your work. If the vehicle has an oil life monitor or maintenance required indicator, reset it for the customer. Clean up your tools and work area, and return the car or truck to the customer.

Once your shift is over, it is time to clean up and take care of your tools and the shop equipment. Your shop may have an oil filter crusher (**Figure 27-60**). These are used to get waste oil out of the filter before disposal of the filter. Place the filter into the crusher compartment and close the cover. Some crushers, like that pictured here, are air powered while others require manual application of force. Once the filter is crushed, dispose of it following local hazardous waste regulations.

SUMMARY

Regardless of the age or type of vehicle, there remains a need for periodic maintenance. Being able to take care of these needs for your customers is important both for you and for your customer. Educating your customers about their vehicles' maintenance requirements, and how preventive maintenance can save on future repair costs, can build trust and loyalty between the customers and the shop.

Maintenance is the act of keeping something in a state of good operating condition.

A repair is performed to correct a problem.

Failing to perform maintenance often leads to repairs that are more expensive.

The vehicle's owner's manual contains information about when and what periodic maintenance should be performed.

Many of today's cars and trucks use some type of electronic maintenance reminder system.

Inspecting steering and suspension and other under-car components should be performed during routine maintenance work.

Because of the changes in engine technology and the increased time between maintenance intervals, it is vitally important that the correct oil is used.

Before beginning to service any automatic transmission, always determine the correct fluid type and refill quantity.

There are many different types of coolants in use today, and color is not the recommended way to tell one coolant from another.

REVIEW QUESTIONS

1. Maintenance schedules classify needed services based on whether the vehicle is operated in normal or _______________ conditions.
2. After performing an oil change on a vehicle, you need to reset the_______________reminder.
3. After the vehicle is driven, tire pressure _______________, so do not let air out if the readings are slightly higher than the recommended pressure.
4. A customer's service _______________ is a record of what services have been performed and when.
5. For engine oils, such as 5W20, the W means_____ __________.
6. Which of the following is *not* part of a maintenance routine?
   - **a.** Wheel balancing and alignment
   - **b.** Cooling system flush
   - **c.** Tire puncture repair
   - **d.** Inspecting and replacing drive belts
7. Use of incorrect engine oil can cause which of the following concerns:
   - **a.** Variable valve timing DTCs
   - **b.** Rapid and excessive engine wear
   - **c.** Sludge accumulation
   - **d.** All of the above
8. Updated parts and service information is released by the vehicle manufacturer as:
   - **a.** Maintenance bulletins
   - **b.** Technical service bulletins
   - **c.** Repair order updates
   - **d.** None of the above
9. A vehicle is having routine maintenance performed. *Technician A* says all orange coolants are OAT based and therefore interchangeable. *Technician B* says all ethylene glycol–based coolants are interchangeable. Who is correct?
   - **a.** Technician A
   - **b.** Technician B
   - **c.** Both A and B
   - **d.** Neither A nor B
10. *Technician A* says the automatic transmission fluid should be checked when cold and the vehicle has sat overnight. *Technician B* says special tools may be needed to check the transmission fluid level. Who is correct?
    - **a.** Technician A
    - **b.** Technician B
    - **c.** Both A and B
    - **d.** Neither A nor B

APPENDIX A

MILLI-METERS	INCHES		MILLI-METERS	INCHES	
	DECIMALS	FRACTIONS		DECIMALS	FRACTIONS
.1	.00394		11.9063	.46875	15/32
.2	.00787		12.00	.47244	
.3	.01181		12.3031	.484375	31/64
.3969	.015625	1/64	12.70	.5000	1/2
.4	.01575		13.00	.51181	
.5	.01969		13.0969	.515625	33/64
.6	.02362		13.4938	.53125	17/32
.7	.02756		13.8907	.546875	35/64
.7938	.03125	1/32	14.00	.55118	
.8	.0315		14.2875	.5625	9/16
.9	.03543		14.6844	.578125	37/64
1.00	.03937		15.00	.59055	
1.1906	.046875	3/64	15.0813	.59375	19/32
1.5875	.0625	1/16	15.4782	.609375	39/64
1.9844	.078125	5/64	15.875	.625	5/8
2.00	.07874		16.00	.62992	
2.3813	.09375	3/32	16.2719	.640625	41/64
2.7781	.109375	7/64	16.6688	.65625	21/32
3.00	.11811		17.00	.66929	
3.175	.125	1/8	17.0657	.671875	43/64
3.5719	.140625	9/64	17.4625	.6875	11/16
3.9688	.15625	5/32	17.8594	.703125	45/64
4.00	.15748		18.00	.70866	
4.3656	.171875	11/64	18.2563	.71875	23/32
4.7625	.1875	3/16	18.6532	.734375	47/64
5.00	.19685		19.00	.74803	
5.1594	.203125	13/64	19.05	.7500	3/4
5.5563	.21875	7/32	19.4469	.765625	49/64
5.9531	.234375	15/64	19.8438	.78125	25/32
6.00	.23622		20.00	.7874	
6.35	.2500	1/4	20.2407	.796875	51/64
6.7469	.265625	17/64	20.6375	.8125	13/16
7.00	.27559		21.00	.82677	
7.1438	.28125	9/32	21.0344	.828125	53/64
7.5406	.296875	19/64	21.4313	.84375	27/32
7.9375	.3125	5/16	21.8282	.859375	55/64
8.00	.31496		22.00	.86614	
8.3344	.328125	21/64	22.225	.875	7/8
8.7313	.34375	11/32	22.6219	.890625	57/64
9.00	.35433		23.00	.90551	
9.1281	.359375	23/64	23.0188	.90625	29/32
9.525	.375	3/8	23.4157	.921875	59/64
9.9219	.390625	25/64	23.8125	.9375	15/16
10.00	.3937		24.00	.94488	
10.3188	.40625	13/32	24.2094	.953125	61/64
10.7156	.421875	27/64	24.6063	.96875	31/32
11.00	.43307		25.00	.98425	
11.1125	.4375	7/16	25.0032	.984375	63/64
11.5094	.453125	29/64	25.4001	1.0000	1

Table title: **Metric, Decimal and Fraction Conversions**

APPENDIX B

English-Metric Conversion Chart

CONVERSION FACTORS

Unit	To Unit	Multiply By
LENGTH		
Millimeters	Inches	.03937
Inches	Millimeters	25.4
Meters	Feet	3.28084
Feet	Meters	.3048
Kilometers	Miles	.62137
Miles	Kilometers	1.60935
AREA		
Square Centimeters	Square Inches	.155
Square Inches	Square Centimeters	6.45159
VOLUME		
Cubic Centimeters	Cubic Inches	.06103
Cubic Inches	Cubic Centimeters	16.38703
Liters	Cubic Inches	61.025
Cubic Inches	Liters	.01639
Liters	Quarts	1.05672
Quarts	Liters	.94633
Liters	Pints	2.11344
Pints	Liters	.47317
Liters	Ounces	33.81497
Ounces	Liters	.02957
WEIGHT		
Grams	Ounces	.03527
Ounces	Grams	28.34953
Kilograms	Pounds	2.20462
Pounds	Kilograms	.45359
WORK		
Centimeter Kilograms	Inch Pounds	.8676
Inch Pounds	Centimeter Kilograms	1.15262
Meter Kilograms	Foot Pounds	7.23301
Foot Pounds	Newton Meters	1.3558
PRESSURE		
Kilograms/ Sq. Centimeter	Pounds/Sq. Inch	14.22334
Pounds/Sq. Inch	Kilograms/Sq. Centimeter	.07031
Bar	Pounds/Sq. Inch	14.504
Pounds/Sq. Inch	Bar	.06895
Atmosphere	Pounds/Sq. Inch	14.696
Pounds/Sq. Inch	Atmosphere	.06805
TEMPERATURE		
Centigrade Degrees	Fahrenheit Degrees	(C° × $^{9}/_{5}$) + 32
Fahrenheit Degrees	Centigrade Degrees	(F° − 32) ×$^{S}/_{g}$

Inches	Decimals	MM
1/64	.016	.397
1/32	.031	.794
3/64	.047	1.191
1/16	.063	1.588
5/64	.078	1.984
3/32	.094	2.381
7/64	.109	2.778
1/8	.125	3.175
9/64	.141	3.572
5/32	.156	3.969
11/64	.172	4.366
3/16	.188	4.763
13/64	.203	5.159
7/32	.219	5.556
15/64	.234	5.953
1/4	.250	6.350
17/64	.266	6.747
9/32	.281	7.144
19/64	.297	7.541
5/16	.313	7.938
21/64	.328	8.334
11/32	.344	8.731
23/64	.359	9.128
3/8	.375	9.525
25/64	.391	9.922
13/32	.406	10.319
27/64	.422	10.716
7/16	.438	11.113
29/64	.453	11.509
15/32	.469	11.906
31/64	.464	12.303
1/2	.500	12.700
33/64	.516	13.097
17/32	.531	13.494
35/64	.547	13.891
9/16	.563	14.288
37/64	.578	14.684
19/32	.594	15.081
39/64	.609	15.478
5/8	.625	15.875
41/64	.641	16.272
21/32	.656	16.669
43/64	.672	17.066
11/16	.687	17.463
45/64	.703	17.859
23/32	.719	18.256
47/64	.734	18.653
3/4	.750	19.050
49/64	.766	19.447
25/32	.781	19.844
51/64	.797	20.241
13/16	.813	20.638
53/64	.828	21.034
27/32	.844	21.431
55/64	.859	21.828
7/8	.875	22.225
57/64	.891	22.622
29/32	.906	23.019
59/64	.922	23.416
15/16	.938	23.813
61/64	.953	24.209
31/32	.969	24.606
63/64	.984	25.003

Courtesy of Snap-on.

Metric Conversion Chart

METRIC CONVERSION: lb. ft. to N·m

The chart below can be used to convert pound foot to newton metre. The left hand column lists pound foot in multiples of 10 and the numbers at the top of the columns list the second digit. Thus 36 pound foot is found by following the 30 pound foot line to the right to "6" and the conversion is 49 N·m.

lb. ft.	0 N·m	1 N·m	2 N·m	3 N·m	4 N·m	5 N·m	6 N·m	7 N·m	8 N·m	9 N·m
0	0	1.36	2.7	4.1	5.4	6.8	8.1	9.5	10.9	12.2
10	13.6	14.9	16.3	17.6	19	20.3	21.7	23.1	24.4	25.8
20*	27	28	30	31	33	34	35	37	38	39
30	41	42	43	45	46	47	49	50	52	53
40	54	56	57	58	60	61	62	64	65	66
50	68	69	71	72	73	75	76	77	79	80
60	81	83	84	85	87	88	90	91	92	94
70	95	96	98	99	100	102	103	104	106	107
80	109	110	111	113	114	115	117	118	119	121
90	122	123	125	126	127	129	130	132	133	134
100	136	137	138	140	141	142	144	145	146	148

*Above 20 lb. ft. the converted N·m readings are rounded to the nearest N·m

METRIC CONVERSION: kg·cm to N·m

The chart below can be used to convert kilogram centimetre to newton metre. The left hand column lists kg·cm in multiples of 10 and the numbers at the top of the columns list the second digit. Thus 72 kg·cm is found by following the 70 kg·cm line to the right to "2" and the conversion is 7.1 N·m.

kg· cm.	0 N·m	1 N·m	2 N·m	3 N·m	4 N·m	5 N·m	6 N·m	7 N·m	8 N·m	9 N·m
0	0	.098	.20	.29	.39	.49	.59	.69	.78	.88
10	.98	1.08	1.18	1.27	1.37	1.47	1.57	1.67	1.76	1.86
20	2.0	2.1	2.2	2.3	2.4	2.5	2.6	2.7	2.8	2.8
30	2.9	3.0	3.1	3.2	3.3	3.4	3.5	3.6	3.7	3.8
40	3.9	4.0	4.1	4.2	4.3	4.4	4.5	4.6	4.7	4.8
50	4.9	5.0	5.1	5.2	5.3	5.4	5.5	5.6	5.7	5.8
60	5.9	6.0	6.1	6.2	6.3	6.4	6.5	6.6	6.7	6.8
70	6.9	7.0	7.1	7.2	7.3	7.4	7.5	7.6	7.7	7.8
80	7.9	7.9	8.0	8.1	8.2	8.3	8.4	8.5	8.6	8.7
90	8.8	8.9	9.0	9.1	9.2	9.3	9.4	9.5	9.6	9.7
100	9.8	9.9	10.0	10.1	10.2	10.3	10.4	10.5	10.6	10.7

One oz. in. = 28.35 gms. in.
One lb. in. = 1.152 kg.cm
One lb. ft. = .138 kg.m

One kg·cm = .8679 lb. in.
One kg·cm = 7.233 lb. ft.
One N·cm = .0885 lb. in.
One N·m = .7375 lb. ft.

Courtesy of Snap-on Tools

APPENDIX D

A Comparison of Various Pressure Standards

Pressure vs. Vacuum			
PSIG	**PSI (abs)**	**BAR**	**IN·Hg**
2	16	1.14	–
1	15	1.07	–
0	14.2	1.00	0
–1	13	0.94	2
–2	12	0.87	4
–3	11	0.80	6
–4	10	0.72	9
–5	9	0.64	11
–6	8	0.57	13
–7	7	0.50	15
–8	6	0.44	18
–9	5	0.37	20
–10	4	0.30	22
–11	3	0.22	24
–12	2	0.14	26
–13	1	0.07	28
–14	0	0.00	30

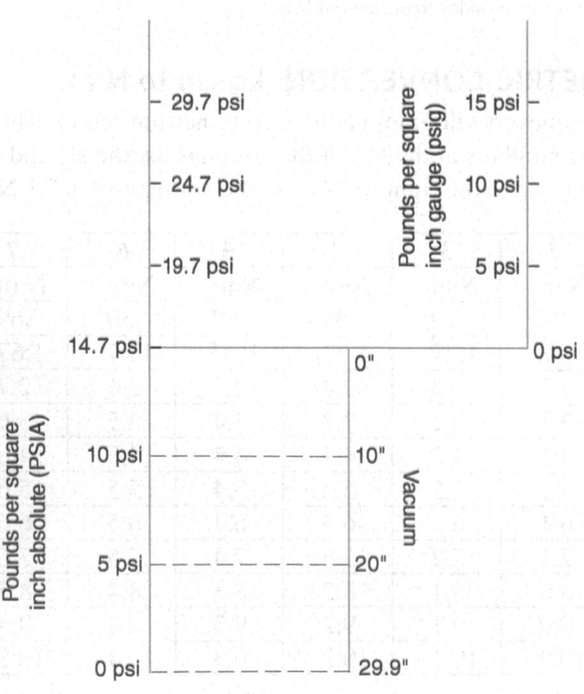

APPENDIX E

Common Acronyms

A/C	air conditioning
A/F sensor	air/fuel ratio sensor
A/T	automatic transmission/transaxle
M/T	manual transmission/transaxle
AAT sensor	ambient air temperature sensor
APP sensor	accelerator pedal position sensor
IAT sensor	intake air temperature sensor
FT	fuel trim
MAF sensor	mass airflow sensor
IAC	idle air control
DLC	data link connector
GEN	generator
BARO	barometric air pressure sensor
LTFT	long term fuel trim
STFT	short term fuel trim
BPP sensor	brake pedal position sensor
CMP	camshaft position sensor
CKP	crankshaft position sensor
EVAP	evaporative emission control system
TCC	torque converter clutch
TBI	throttle body injection
MIL	malfunction indicator light/lamp
CL	closed loop
OL	open loop
CVT	continuously variable transmission
ECT	engine coolant temperature sensor
RPM	revolutions per minute
KS	knock sensor
GDI	gasoline direct injection
EFI	electronic fuel injection
SEFI	sequential electronic fuel injection
PFI	port fuel injection
CMFI	central multiport fuel injection
DTC	diagnostic trouble code
DPFE	differential/delta pressure feedback EGR
EGR	exhaust gas recirculation
ICM	ignition control module
VSS	vehicle speed sensor
WSS	wheel speed sensor
ABS	antilock brake system
TC	traction control
VSC	vehicle stability control
TWC	three way catalyst
TWC + OC	three way catalyst + oxidation catalyst
PCM	powertrain control module
ECM	electronic control module
O2S	oxygen sensor
HO2S	heated oxygen sensor
FF	flex fuel
HEV	hybrid electric vehicle
EV	electric vehicle
EREV	extended range electric vehicle
4WD	four wheel drive
FWD	front wheel drive
RWD	rear wheel drive
AWD	all wheel drive
FF	freeze frame
FRP	fuel rail pressure
TB	throttle body
WOT	wide open throttle
GND	ground
g/s	grams per second
ATF	automatic transmission fluid
PS fluid	power steering fluid
OBD I	onboard diagnostic system 1st generation
OBD II	onboard diagnostic system 2nd generation
PSP	power steering pressure
PID	parameter identification
PCV	positive crankcase ventilation
TR	transmission range
PWM	pulse width modulation
TAC	throttle actuator control
TOT	transmission oil temperature
VCM	vehicle control module
B+	battery positive voltage
IMRC	intake manifold runner control
TCM	transmission control module

GLOSSARY

A

above-ground lift. A type of automotive lift in which the structure sits entirely above floor level. The main types are the swing arm-type and the drive-on type.

accumulator. A storage structure for pressurized brake fluid in many ABS systems and the Powermaster brake booster system.

ACEA standards. Standards created by the Association of Constructors of European Automobiles.

active (direct) TPMS. A tire pressure monitoring system that uses sensors placed in each wheel to directly measure and transmit the pressure information to the vehicle's on-board computer system.

active test. An approach to component testing that involves the ECM taking direct control of the component in order to determine its functionality.

additives. Substances added to gasoline to make it more effective as a fuel in modern engines. Additives are also used in lubricants such as engine oil, transmission fluid, and others to create the specific qualities needed for each fluid.

aftermarket parts. Repair parts that are purchased from a source other than the manufacturer of the original parts.

AGM battery. An absorbed glass mat battery that contains acid in an absorbent mat, eliminating acid leaks. The hydrogen given off during charging remains inside the battery; thus AGM batteries are also called recombination batteries because the hydrogen and oxygen recombine inside the battery. AGM batteries require special charging procedures.

air bleed valve. A valve used to remove air from the cooling system.

air filter. is responsible for removing dirt and debris from entering with the fresh air that is drawn into the engine.

air hammer. An air-powered tool fitted with a chisel or cutting bit to separate bushings or suspension components.

air pressure. The amount of force the air in a container such as a tire exerts against the inner surface area of the tire.

air springs. A thick, tough bag filled with air that acts as a spring. Air springs are used on most large commercial semi trucks and trailers. Some systems can adjust the air pressure to control ride height.

Allen wrench. A simple, L-shaped wrench with a hexagonal cross section, used to turn fasteners with a six-sided socket in the head.

alternating current. The form in which electrical current is generated at power stations and transmitted along power lines to customers. Alternating current is created by causing electrons to flow in an alternating positive and negative manner.

American Petroleum Institute. A trade association whose activities include developing technical specifications and ratings for oils used in the automotive industry.

amperage. The amount of electrical current flow that is occurring through a conductor.

amp-hour (AH) rating. A battery rating that describes how many amp-hours of current the battery can supply for 20 hours before voltage falls below 10.5 volts.

anchor. In a drum brake, a fixture used to minimize twisting forces on the brake shoes as the brakes are applied.

anode. An electrode or electrical conductor into which current flows.

asbestos. A hazardous compound that, due to its high heat resistance and insulating qualities, has been used in many automotive applications, especially brake pads and clutch discs.

asymmetrical tires. A tire whose tread pattern differs from the inside to the outside of the tread.

Atkinson cycle. A modified four-cycle gasoline engine. Compared to the conventional model, the intake stroke in the Atkinson cycle is allowed to continue longer, until the piston is moving back up to compress the air-fuel mixture. This forces some of the air-fuel charge from the cylinder and back into the intake. The reduced charge generates less power but also uses less fuel and produces fewer emissions.

atmospheric pressure. The weight of the air that surrounds the earth. At sea level, this weight exerts 14.7 pounds of pressure per square inch of surface area.

automotive machinist. A machine shop worker who specializes in such engine repair tasks as reboring cylinder blocks, replacing cylinder liners or sleeves, fitting pistons and connecting rods, and repairing cracks.

automatic transmission fluid (ATF). The lubricant used by an automatic transmission.

axle puller. A tool used to free the axle from the bearing in the steering knuckle.

B

backing plate. The stamped steel plate, attached to the axle assembly, that holds the components of a drum brake assembly.

ball-peen hammer. A hammer with a ball-shaped surface on one side of the head; often used when servicing ball joints and other suspension parts.

battery charger. A device used to recharge a battery that has become discharged or to charge newly filled batteries.

battery corrosion. A buildup of material on battery terminals, either from hydrogen and oxygen condensing back onto the battery or from galvanic reactions between dissimilar metals at the terminals. Corrosion creates electrical resistance at the terminals.

battery cycle. The stages of battery operation: charged, discharging, discharged, and recharging.

battery holddown. A device that prevents the battery from moving in its mounting when the vehicle is in motion.

battery load test. A test that measures the battery's ability to produce current by loading it at one-half of its CCA rating for 15 seconds, during which time the battery voltage should not drop below 9.6 volts. Special equipment is required. Also called a high-capacity discharge test.

battery reserve capacity. A rating, in minutes, of how long a battery at 80°F can supply 25 amps to the electrical system before voltage drops below 10.5 volts in the event that the charging system fails.

bearing preload. A thrust load placed against bearings to prevent axial or side-to-side movement and to eliminate end play.

bench bleeding. The prefilling and purging of air from a master cylinder before installing it on the vehicle. This reduces the amount of bleeding that needs to be done on the vehicle.

bench grinders. A motor-driven grinding wheel or wire wheel, used for such tasks as reshaping metal, sharpening tools, and removing rust from parts.

biodiesel. Diesel fuel that is made from renewable resources, such as vegetable oils, animal fats used in cooking oils, and even algae. Pure biodiesel does not contain any petroleum products.

blow guns. An attachment for a shop air hose that is used to blow away dirt and debris and to dry components using a high-pressure stream of air.

blower motor. The interior fan motor, used to circulate air in the passenger compartment.

body language. Communication by means of expression and posture.

brake drum. A cast iron or composite aluminum shell with an internal friction surface against which the shoes to rub in a drum brake.

brake fluid. A specially formulated, nonmineral oil-based fluid designed specifically for the demands of the brake system.

brake fluid flushing. The replacement of old fluid with new fluid in the brake hydraulic system.

brake hardware. In a drum brake, the springs and related parts of the brake assembly. In disc brake systems, brake hardware can include caliper guides, shims, bushings, and caliper pin bolts.

brake light switches. A switch mounted to the brake pedal bracket that activates the brake lights when the pedal is pressed. Brake light switches are also inputs for the onboard computer system.

brake pads. In a disc brake, the brake pad is the friction material or pad lining and the backing plate for the lining that clamps down on the brake rotor to slow the wheel.

brake pedal height. The distance from the brake pedal to the floor with the pedal depressed.

brake rotors. The disc, most commonly made of cast iron, against which the caliper clamps the brake pads to slow the vehicle.

brake shoes. The metal backing on which the lining material is attached in a drum brake.

bypassing. The passage of brake fluid around the master cylinder pistons due to seal failure inside the master cylinder.

C

cabin air filter. A filter used to remove dust and other contaminants from the air circulating into the passenger compartment.

calibration decal. A decal found on the engine or transmission of some vehicles that provides specific information about emissions system devices and calibrations.

caliper pistons. That part of a disc brake caliper that applies pressure against the brake pad. Pistons are made of steel, plastic, or aluminum.

camshaft. A crankshaft-driven shaft containing oblong-shaped lobes that push on valve lifters to open cylinder head valves.

carbon monoxide. A colorless, odorless, toxic gas that is found in the exhaust of a running engine.

catalyst efficiency. A test run by the oxygen sensors in an OBD II system to determine whether the catalytic converter is working properly.

cathode. An electrode or conductor from which current flows out.

CCM. Comprehensive component monitor. This is a diagnostic test designed to check the integrity of all input and output circuits for the engine control module in an OBD II vehicle.

centerbore. A large hole in the center of a wheel where it mounts to a hub on the vehicle.

chain of command. The formal hierarchy or structure of supervision in a businesss.

chemical reaction. A change in the bonds between the atoms of different chemical elements or compounds.

CHMSL. center high-mounted stop lamp. All vehicles sold in the United States since the 1985 model year have been equipped with a CHMSL.

circuit. In electricity, a path for electrons to flow from a source of higher electrical potential to a source of lower potential.

circuit breakers. A circuit protection device that acts like a switch to open the circuit when current flow becomes excessive.

circuit faults. A condition that can cause a circuit to not function properly or not function at all.

circuit protection. Devices such as fuses, fuse links, and circuit breakers in an electrical circuit that stop the flow of electricity in the event that a fault occurs in the circuit. These help to protect people from injury and prevent damage to the vehicle's electrical system.

clearance gauge. For drum brakes, a tool used to measure the drum diameter and then transfer that dimension to the shoes in order to set the shoe-to-drum clearance.

coefficient of friction (CoF). A number that expresses the ratio of force required to move an object divided by the mass of the object.

coil spring. A length of steel wound into a coil shape. The shape of the coil creates a spring action. Coil springs are found in front and rear suspension systems.

cold cranking amps. A rating based on a battery's ability to deliver current for 30 seconds at 0°F before the total voltage drops to 7.2 volts.

collapsible steering column. A safety feature that allows a steering column to telescope in the event of a front-end collision, to prevent it from being driven into the driver's chest.

collision technician. A technician who specializes in repairing body or structural damage to vehicles.

combination wrench. A common repair tool on which one end is a box-end wrench for loosening or tightening a specific size of fastener, and the other end has an open wrench to quickly remove the loosened fastener.

composite rotor. A disc brake rotor that combines a steel hub with a cast iron friction surface. This type of rotor offers a slight weight reduction over a cast iron rotor.

comprehensive monitoring. Term used to describe the level of exhaust emissions monitoring and self-diagnostics of OBD II systems.

compression stroke. The stage of engine operation during which the air-fuel mixture is compressed in the cylinder by the upward motion of the piston.

condenser. A radiator-like component that removes heat from the air conditioning refrigerant.

conductance tester. A small, hand-held tester that supplies a small AC current through the battery. As the AC flows through the cells, it is changed slightly by the battery. The tester interprets this change and translates it into a reading of the battery's condition.

conductor. Wires, terminals, and connectors that carry electricity in circuits.

connectors. Fittings used to connect wiring to components and to other wires.

contact patch. That part of a tire's tread that makes physical contact with the road and where the tire supports the weight of the car.

continuous monitor. An OBD II emissions monitor that is active at almost all times the engine is running. Continuous monitors include the CCM, fuel control, and misfire monitors.

continuously variable transmission (CVT). A type of automatic transmission that uses two variable-diameter pulleys and a belt instead of gears, clutches, and bands.

coolant. A mixture of water, ethylene glycol, and additives that carries heat away from the engine and transfers it to the radiator and outside air.

coolant exchanger. A device for extracting and replacing the coolant in a vehicle and bleeding any trapped air from the system.

cooling fan. A belt, electric, or hydraulically driven fan used to pull air through the radiator to remove heat from the coolant.

cooling system. The system that removes excess heat from an engine and provides heat for the passenger compartment. Components include the coolant, radiator, radiator cap, radiator fan, water jackets, thermostat, hoses, and heater core.

cranking amps. A rating is similar to cold cranking amps, except that it is measured at 32°F instead of 0°F.

cranking compression test. A test conducted by turning the engine over briefly with a compression tester attached to determine a cylinder's ability to pull in air, compress the air, and remain sealed.

crankshaft. The shaft to which the pistons' connecting rods are attached and through which power is transmitted from the engine to the transmission and ultimately to the drive wheels.

creepers. A small rolling platform used by technicians to move around under a vehicle that is only raised off the floor a slight amount.

cylinder leakage test. A test performed on a suspect cylinder to determine where the pressure is going. The cylinder leak detector uses the shop air supply and a regulator to pressurize the cylinder, and the rate of leakage is displayed on the gauge.

D

dead (rigid) axle. A solid rear axle that does not drive the rear wheels.

dead blow hammer. A hammer whose head is filled with lead shot in order to reduce elastic rebound when striking a surface.

dependent suspension. A method of suspending the wheels on a vehicle so that the movement of one wheel affects the opposite wheel. This method sacrifices some ride quality for added strength.

digital multimeter. A handheld electrical testing device that can measure AC and DC voltage, resistance, amps, and other measurements.

diode. A simple semiconductor used as a one-way electrical check valve.

direct (active) TPMS. A tire pressure monitoring system that uses sensors placed in each wheel to directly measure and transmit the pressure information to the vehicle's onboard computer system.

direct current. Electrical current that flows only from zero to positive and does not alternate negative and positive as AC does; the type of current created by the use of a battery.

directional tires. A tire whose tread pattern requires it to be used to rotate mainly in one direction, meaning that right- and left-side tires are not interchangeable.

disc brake micrometer. A clamp-type instrument for making very precise measurements of disc brake rotor thickness.

dissolved copper. Copper from the internal parts of the brake system that is absorbed by brake fluid as the corrosion inhibitors in the brake fluid are depleted.

driver. A driver is part of a computer or electronic module that activates another component by supplying either power or ground to the circuit.

drive belt tensioner. A device, typically an adjustable or spring-loaded pulley, used to maintain proper tension on a drive belt.

drive cycle. A period of vehicle operation that allows all criteria for all monitors to be met.

drum brake micrometer. A tool used to measure brake drum diameter and to check for out-of-round.

drum in hat. A miniature drum brake assembly housed within the hat of a rear disc brake rotor to serve as the parking brake.

dry-park check. A method of checking a steering linkage that involves placing the vehicle on a lift and, with the engine off, having a helper rock the steering wheel back and forth while a technician beneath the vehicle watches and listens to the steering linkage.

dual-clutch transmission. An electronically controlled manual transmission that permits both automatic shifting and manual shifting.

duo-servo. or servo drum brakes use the motion of the drum and leverage to apply additional force against the secondary brake shoe.

duo-servo brake. A type of drum brake that uses leverage to increase brake application force. This design places the shoe anchor at the top of the brake assembly; when the brakes are applied, forward or primary shoe twists with the drum rotation, pushing against the rear or secondary shoe and placing more force on the secondary shoe than was originally applied by the wheel cylinder alone. Also called a self-energizing brake.

dust boot. In a disc brake caliper, an accordion-type seal that protects the outside of the piston and the piston bore from debris. In a drum brake wheel cylinder, two dust boots protect the pistons.

dynamic balance. The distribution of weight around the axis of rotation of a rotating wheel and tire.

E

EBCM (electronic brake control module). The control unit for an antilock brake system.

electrical shock. The passage of an electric current through a person's body. Electric shock can cause burns and can overcome normal nerve impulses.

electricity. The flow of electrons from a source of higher electrical potential to a source of lower electrical potential through a conductor.

electrohydraulic unit. In an ABS system, this unit usually contains electric motors, solenoids, and valves that control the flow of brake fluid to the wheel brakes based on input from the EBCM.

electrolyte. A substance, such as the sulfuric acid and water mix used in a battery, that contains free ions that make the substance electrically conductive.

electromagnetism. The term used to discuss the whole scope of electricity and magnetism.

electronic stethoscope. An electronic listening tool used to isolate the source of problematic noises.

electronics. Electrical components that contain integrated circuits, computer chips, and similar components that operate with very small voltages and amperage, are typically not serviceable, and are replaced as a unit. These types of devices are often called solid-state components.

electrons. Subatomic particles whose activity creates electric currents.

emissions decal. An under-hood decal that contains such information as the emission year for which the vehicle is certified, installed emission control devices, and what emission standards the vehicle meets. Other information may also be included.

employability skills. The so-called soft skills, such as initiative and work ethic, that make one a good employee.

employment plan. For a worker, a self-assessment of personal strengths and weaknesses and short- and long-term goals, and an action plan for achieving the goals.

end play. Side-to-side movement at the end of an axle.

end-takeoff rack. A form of rack-and-pinion steering in which a socket on the inner tie rod end is connected directly to the end of the rack.

engine hoists. A lifting device used to remove engines from vehicles.

enhanced OBD. Data and functions available from an OBD II system through a scan tool that can be programmed to a vehicle's VIN.

entry-level employee. A beginning employee in the repair shop who often must complete an initial probationary period.

entry-level technician. A beginning technician capable of making simple vehicle repairs, whose typical assignments include basic inspections and maintenance services.

ethylene glycol coolant. The traditional inorganic- acid-based green coolant that contains both

exhaust system. The structure that removes spent exhaust gases from the engine, quiets the noise of combustion, routes harmful exhaust gases away from the passenger compartment, and converts harmful components of the exhaust gases into less harmful gases.

eyewash stations. An emergency station in the workplace that enables technicians to quickly irrigate their eyes when exposed to irritants.

F

fade. A decrease in the effectiveness of a brake that occurs when the friction surfaces overheat, typically as a result of continuous heavy braking.

fast charging. Charging a battery at a high amperage rate for a short amount of time; not the best method of charging because it risks overheating the battery.

fender cover. A protective cover placed over the vehicle body during service to protect the finish from damage from tools, corrosive materials, and other hazards.

fixed caliper. Disc brake calipers that are mounted directly to the steering knuckle. Each piston receives equal fluid pressure and pushes on a brake pad. This type of caliper is used mainly in high-performance applications.

flasher units. A control unit for the vehicle's turn signals.

flat-rate guide. A written guide to information about original equipment (OE) parts cost, aftermarket parts cost (some software), and estimated labor times for specific repairs.

flexible coupler. An intermediate coupling between a steering gearbox and the steering column, designed to accommodate variations in alignment between the two.

floating calipers. A disc brake caliper that "floats," or moves back and forth, on bolts in its mounting hardware when the brakes are applied.

floating drum. The most common type of brake drum, which slips over the hub and is held in place by the wheel fasteners. Often, accumulated rust makes removal difficult.

floor jack. A lifting device that uses a hydraulic cylinder to convert pumping actions on a handle into a smaller upward movement of the arm.

frame. That portion of a body-over-frame vehicle on which the body is mounted. The frame is often ladder-shaped, with two long rails that run the length of the vehicle and several crossmembers attached to the rails.

free play. At the brake pedal, the slight amount of pedal movement at the released position before the pushrod begins to move into the booster and master cylinder.

freeze frame. An element of data storage related to a Type A diagnostic trouble code in an OBD II system.

frequency. In springs, the number of times a spring oscillates or bounces before returning to its rest point.

fuel trim. An adjustment to fuel delivery by the engine control module, based on input from the oxygen sensors.

fuse. A common circuit protection device that works by allowing a current-carrying element in the fuse to melt, and thus to open the circuit, when the current flow generates excessive heat.

fuse link. A circuit protection device that features a smaller-gauge wire than is used in the rest of the circuit. If excessive current flow occurs, the smaller wire melts, opening the circuit.

G

gasket. A soft, malleable interface between nonmoving parts, such as that between the engine block and the oil pan, that allows them to seal against each other and prevent leaks.

gearbox. In a steering mechanism, the use of gears to increase the mechanical advantage and decrease driver effort in turning the steering wheel while converting the wheel's motion into motion that moves the vehicle's front wheels.

generic (global) OBD II. Codes and data available to anyone who connects a scan tool to the data link connector of an OBD II system.

grabbing. A brake problem in which the brake applies too quickly or with too much force, which causes the wheel to lock.

grease fittings. A small, nipple-like fitting threaded into a component for the injection of grease. A spring-loaded check ball inside the fitting allows grease in but keeps it from being pushed back out.

gear ratio. The difference between input speed and output speed created by a gear train in a transmission.

H

halogen insert headlamp. A headlamp assembly that features a replaceable halogen bulb.

hardware. For computers, physical parts such as hard drives, DVD drives, fans, processors, memory, and power supplies. Hardware also applies to the springs, guides, and other pieces of drum and disc brake assemblies.

hazardous wastes. Any substance that can affect public health or damage the environment. Nearly every liquid, chemical, and solvent used by the vehicle and for service and repair is considered a hazardous waste.

heater core. A small, radiator-like component used to supply heat for the passenger compartment.

high-intensity discharge lamp. A headlamp that provides light by using a high-voltage AC arc of current across two electrodes, which excites gas in the lamp and causes it to glow.

history code. A diagnostic trouble code that remains in the ECM memory after a fault has been corrected. After 40 consecutive warmup cycles with no further faults detected, the history code will clear from the ECM memory.

hold mode. One of three modes of operation for a wheel brake in an ABS system; used to limit any further pressure increase in a wheel brake circuit. Also called isolate mode.

holddown spring tool. A tool used to remove coil-type holddown springs in a drum brake.

holddown springs and pins. Springs and pins that hold the shoes to the backing plate and keep the shoes in position on raised pads in a drum brake.

horsepower. A rating of the amount of work performed by an engine. Originally based on the amount of weight a horse could move in one minute. One HP is equal to about 746 watts.

hub cap. A decorative cap secured to the outside of a wheel. in one of three ways: by spring-type mounting clips around the circumference of the hubcap, by the lug nuts against the hubcap, or by false lug nuts that thread onto the actual lug nuts, keeping the hubcap tight against the rim.

HVAC. Acronym for a vehicle's heating, ventilation, and air conditioning system.

hybrid organic acid technology. Technology that forms the basis for some long-lasting modern coolants. It is represented as a hybrid of inorganic and organic acid coolant technology.

hydraulic assist. An external power brake assist system that uses hydraulics, typically from the power steering system to a hydraulic power brake booster.

hydraulics. The science of the mechanical properties of fluids.

hydrocarbons. Compounds made of hydrogen atoms that have bonded with carbon atoms. Gasoline is composed of hydrocarbons.

hygroscopic. A characteristic of brake fluid by which it absorbs moisture. Over time, this can damage the hydraulic system components.

I

I-beam. A beam with an I-shaped cross section; in an I-beam suspension system, I-beams are mounted to the crossmember at one end with a bushing; at the other end the beam connects to upper and lower ball joints and to a radius arm, which limits forward and backward movement of the suspension.

ignition system. That part of a gasoline-powered engine that provides the heat to ignite the air-fuel mixture in the combustion chamber by sending a high-voltage spark through a spark plug gap.

impact wrench. An air-powered tool for loosening lug nuts and other tight fasteners.

incandescent bulb. A bulb that produces light by passing an electric current through a filament, which heats and glows.

independent suspension. A method of suspending the wheels on a vehicle that permits each wheel to respond individually to varying road conditions. This is the optimal approach for ride quality.

indicator lights. A light on the instrument panel that indicates when a circuit is on.

indirect (passive) TPMS. A system that uses the vehicle's antilock brake system to monitor tire pressure based on wheel speed.

induction. The production of an electric current by moving a conductor through a magnetic field.

inductive current clamp. A device that measures current flow by detecting the strength of the magnetic field created by current flowing through a conductor.

in-ground lift. A type of automotive lift in which the hydraulics are installed beneath the concrete shop floor, and the controls are operated above ground level.

initiative. A behavior that causes a person to start to do something or to help out in some way without having been asked to do so.

ILSAC. The International Lubricant Standardization and Approval Committee, representing both domestic and Japanese auto manufacturers, which creates standards and ratings for oils used in the automotive industry.

inner tie rod socket. A special tool for removing and reconnecting an inner tie rod end to a rack gear.

instrument panel. That part of the vehicle's dash that contains the gauges, warning lights, and message centers to provide information to the driver about the operating condition of the vehicle.

insulated circuit. An electrical circuit to which no additional circuits are attached. The starter motor circuit, because of its high load, is a prime example.

intake stroke. The stage of engine operation during which the intake valve opens and air and fuel begin to enter the cylinder.

intrusive test. An approach to component testing that is similar to active testing except that the result of the test may have an impact on driveability. *See also* active test.

J

jack. A tool used to raise a vehicle off the ground.

jounce. Upward movement of a wheel that compresses the spring.

K

kV (kilovolts). Used when measuring thousands of volts.

kPa (kilopascals). A metric unit of air pressure measurement.

L

lead-acid battery. The standard battery type used in motor vehicles that contains six cells, each composed of several lead-based plates submerged in a mixture of sulfuric acid and water.

leading-trailing brakes. A nonservo type of drum brake in which the shoe anchor is at the bottom of the backing plate; when the brakes are applied, no additional force from the leading shoe can be applied to the trailing shoe.

leaf spring. A long, semi-elliptical piece of flattened steel used on the rear of many vehicles. The spring is attached to the frame through a shackle assembly that permits changes in the effective length of the spring as it is compressed. Several such leaves can be stacked together for additional load-carrying capacity.

LED headlamp. A headlamp whose light is provided fully or partially by light-emitting diodes.

lifelong learning. Continuing to pursue training and other learning opportunities in order to grow in knowledge and skill throughout one's career.

limited-slip differential (LSD). A limited-slip differential uses clutches or other methods to reduce the amount of power to a drive wheel with reduced traction.

line technician. An experienced mid-level technician with ASE and/or manufacturer certification. Capable of efficiently repairing most systems on modern vehicles. Often involved in training entry-level employees.

linear measurement. A measurement from point A to point B, such as the wheelbase of a vehicle or the amount of free play in a pedal. Sometimes calculated by taking two measurements and subtracting one from the other, as when determining if an object such as a brake drum is out-of-round.

live axle. A depended style rear axle that drives the wheels. A four-wheel drive vehicle may have a front live axle.

load. In electrical applications, anything that consumes electrical power.

load-carrying ball joint. A ball joint that supports the vehicle weight carried by the springs.

load-sensing proportioning valve. A type of combination brake system proportioning valve that varies the amount of pressure passing to the rear brakes based on vehicle load, which is gauged by the position of a lever attached to the valve. *See also* proportioning valves.

low-maintenance battery. A heavy-duty lead-acid battery that requires less frequent addition of water than a standard lead-acid battery.

lubrication system. The system that supplies oil to the moving parts within an engine.

M

MacPherson strut. A combined coil spring, shock absorber, and bearing plate used as a front suspension component in many front wheel drive cars.

maintenance. The act of keeping something in a state of good operating condition.

maintenance reminder. An instrument panel signal that alerts the driver to maintenance items that need attention.

maintenance-free battery. A sealed lead-acid battery to which water cannot be added. They use slightly different plate materials than standard batteries and release almost no gas.

manual bleeding. Typically a two-person procedure in which a helper pumps the brakes while a technician opens and closes the bleeder screws.

master cylinder. The part of a hydraulic brake system that receives input force through a pushrod when the driver presses on the brake pedal. The pistons inside the master cylinder move forward, pressurizing fluid in the brake lines, which connect to the output pistons at the wheel brakes. Hydraulic clutch systems use a single piston master cylinder to activate the slave cylinder.

material safety data sheet (MSDS). Detailed information about a hazardous substance, including: its chemical components; flammability, reactivity, toxicity, and corrosiveness; and first aid instructions for accidental exposure.

mechanical safety latch. A safety device on a lift that engages whenever the lift is lowered; it prevents the lift from falling in the event of a sudden loss of hydraulic pressure.

metering valve. A type of combination brake system valve that introduces a slight delay in the application of the front disc brakes in relation to the rear drum brakes. This allows more balanced braking and better vehicle control. *See also* proportioning valves.

metric measurement. Measurement based on the meter and portions of the meter, which are all based on units of 10.

misfire monitor. A test used by OBD II systems to evaluate combustion efficiency by detecting the changes in the rotational speed of the crankshaft as it slows down following a misfire.

modified strut. A strut-style shock absorber that has the spring separate from the strut.

motivation. The intrinsic and extrinsic factors that direct a person to seek and succeed in a career and other activities.

multilink suspension. A front suspension system in which the steering knuckle pivots on upper and lower ball joints and the strut is mounted rigidly to the body at the upper strut mount.

multimeter. A device capable of measuring AC and DC volts, amperage, resistance, duty cycle, frequency, and temperature in an electric circuit.

multi-rib belts. A type of drive belt that features multiple longitudinal ribs along its length.

multiviscosity oils. An oil that can change viscosity as temperature varies. This is important for effective flow at cold temperatures.

N

National Automotive Technician Education Foundation. A national organization that certifies high school and post-secondary education programs for automotive service technicians.

National Institute for Automotive Service Excellence. A national nonprofit organization formed in 1972 to provide testing and certification for auto mechanics, to improve the quality of service, and to improve the image of the repair industry. ASE currently certifies automotive technicians in nine areas and offers advanced certifications for engine performance specialists.

no-crank condition. A situation in which the starter system does not turn the engine over upon attempted startup.

noncontinuous monitors. An OBD II emissions monitor that requires certain enabling criteria to be met before the monitor can run. These criteria will vary slightly from manufacturer to manufacturer and for each monitor.

nonverbal communication. Communication with others by means other than words, such as body language, gestures, appearance, and overall demeanor.

O

OE parts. Also know as original equipment. Parts that are purchased from the vehicle's original manufacturer.

offset. The position of the mounting surface on a wheel compared to the center of wheel depth. Wheel offset can be zero, positive, or negative.

Ohm's law. A scientific law that states that one volt can push one amp through one ohm of resistance. Commonly expressed as $E = I \cdot R$ where E is voltage, I is amperage, and R is resistance.

ohm. The standard measure of electrical resistance, represented by the Greek symbol Ω (omega) on electrical test equipment.

oil pressure test. A test of the pressure generated in an engine's lubrication system; a warm engine typically should have at least 15 to 20 psi of oil pressure at idle.

on-car lathe. A tool for machining a brake rotor without removing it from the vehicle. The purpose of machining the rotor on the vehicle is to eliminate runout problems caused by stacked tolerances.

open circuit. An electrical circuit in which the path for current flow is broken, causing the circuit not to function.

organic acid technology. Technology that forms the basis for some long-lasting modern coolants. Organic acid coolant typically contains nitrates and phosphates but no silicates.

overcharging. A condition in which the field control circuit in the battery charging system does not limit the amount of current that is supplied to the field, creating a potential for damage to the battery and electronics.

overhead valve engine. A valve train configuration in which the camshaft is mounted in the engine block and the valves are operated by valve lifters, pushrods, and rocker arms.

oversteer. A driving condition in which the rear tires reach their cornering limit before the front tires.

oxygen sensor. A sensor located in the exhaust system used to determine the content of the exhaust gases for fuel control by the ECM.

oxygen sensor monitor. An OBD II emissions monitor that tests for an oxygen sensor's time to reach activity, for signal voltage, and for the signal response rate.

P

$P = F/A$. The input pressure equation in hydraulics, where pressure P is found by dividing force F by piston size A.

parallel circuit. An electrical circuit that contains individual branches for each load.

parallelism. The extent to which the friction surface of a brake disc varies in thickness. Slight variations in rotor thickness result in brake pedal pulsation as the pads must move in and out to compensate. Also called taper or thickness variation.

parallelogram linkage. A linkage in the form of a parallelogram that connects a recirculating ball gearbox to the steering knuckle.

parasitic load. A system or component that continues to draw power from the battery even after the vehicle is shut down. All modern vehicles have some key-off drains, which are monitored by control modules, but a malfunction can result in excessive drain.

parking brake. A mechanical system of holding the brakes to prevent vehicle movement.

parts technician. A person who works directly with customers to find repair and replacement parts for their vehicles.

parts washer. A special tub in which a solvent is pumped through to facilitate the removal of dirt, oil, grease, rust, or other substances from parts.

passive (indirect) TMPS. A system that uses the vehicle's antilock brake system to monitor tire pressure based on wheel speed.

passive test. An approach to component testing by the ECM that monitors circuits for faults such as opens and shorts.

PCV valve. A spring-loaded valve through which blowby combustion gases are drawn out of the crankcase and into the intake system.

pending code. A diagnostic trouble code that is stored in memory, but not displayed, the first time the ECM detects a potential fault. A second trip is required to confirm the fault and display a hard code.

percentage. Meaning the part of the whole; used to calculate invoice items such as sales tax, fees, and parts markups, and used in the shop to calculate ratios such as the amount of antifreeze to mix with water.

permanent magnet motor. An electric motor that uses permanent magnets instead of electromagnets as the field coils. They draw less power than motors with electromagnetic field coils and can be made smaller as a result.

personal hygiene. Standards of personal cleanliness and grooming that are important for presenting a professional image at the workplace.

personal protective equipment (PPE). Clothing and other protective items worn by technicians to minimize their exposure to personal hazards in the workplace.

piston. An engine part that forms the lower section of the combustion chamber. Pressure is exerted on the top of the piston, forcing it down in the cylinder. This movement is transmitted through the connecting rod to the crankshaft. The term piston also refers to pistons used in hydraulic components, such as master cylinders, wheel cylinders, and calipers. The pistons in hydraulic components are used to apply pressure or force.

Pitman arm puller. A puller tool used to separate a Pitman arm from the Pitman shaft of a steering gearbox.

planetary gearset. A planetary gearset contains a ring gear, planetary gears, and a sun gear in constant mesh.

pliers. Any of the various types of simple hand tools used to grasp objects on the automobile.

pounds per square inch. or (psi) is used to express the amount of force or pressure exerted on an object's surface area.

power balance test. A test used to determine if each cylinder is contributing power to overall engine output. During the test, fuel injectors are disabled and the change in engine rpm is recorded. If a cylinder is not producing power, there will not be an rpm drop when that cylinder's injector is disabled, alerting you to the problem cylinder.

power-split device. A power-split device is used by Toyota hybrid vehicles in place of traditional automatic or manual transmission.

power source. A source of high electrical potential, typically a battery in automotive applications.

power steering fluid. A lubricant used in a power steering system. Modern vehicles often need specific proprietary fluids.

power steering fluid flush. The withdrawal of old fluid from a power steering system and replacement with new fluid in order to remove debris and extend the life of the system.

power steering pump. The gear- or belt-driven pump that supplies pressurized fluid to the steering gearbox.

prealignment inspection. Inspecting steering and suspension components prior to performing a wheel alignment.

pressure bleeding. A method of brake system bleeding that uses pressure to force the air out and fluid through the brake system. A special adapter must be installed over the master cylinder reservoir.

pressure differential. The principle behind the operation of a vacuum brake booster. An assist is provided by the action of atmospheric pressure against the vacuum in the system.

pressure differential valve. A valve in a dual-circuit hydraulic brake system that closes off one circuit in response to a leak in order to maintain hydraulic pressure in the other circuit. Preserves brake pressure at two of the four wheels.

pressure increase mode (abs). One of three modes of operation for a wheel brake in an ABS system; used to reapply pressure to the brake circuit to slow the wheel again.

pressure plate. The pressure plate holds the clutch disc against the flywheel. When the clutch is released, the pressure plate releases the force against the clutch disc.

pressure tank. Also called a surge or degas tank, used to store coolant and provide a way to check and fill the cooling system.

proportioning valves. Hydraulic valves commonly used to help control brake application on vehicles with combination brake systems, meaning front disc brakes and rear drum brakes. The proportioning valve limits pressure to the rear brakes to prevent wheel lockup.

psi (pounds per square inch). An English unit of pressure measurement.

psia. Pounds per square inch absolute. Zero on a psia gauge means zero pressure, unlike the typical scale on a tire gauge, for which zero on the gauge means atmospheric pressure (14.7 psi).

psig. Pounds per square inch gauge. Air pressure units displayed on a tire pressure gauge, which cannot show pressures less than atmospheric.

puller. A tool designed for a particular type of removal or for removing a specific part.

pulsation. A shudder or pulsing of the vehicle and/or brake pedal during stopping, commonly caused in drum brakes by an out-of-round brake drum and in disc brakes either by variations in brake rotor thickness (parallelism) or side-to-side rotor movement (runout).

punches. A simple impact tool used to remove pins and to create indentations in preparation for drilling.

R

rack and pinion. A type of steering assembly that features a rack gear that moves the front wheels in response to the motions of a pinion gear attached to the steering column.

radiator. Used to remove heat from engine coolant. As the coolant passes through the tubes inside the radiator, air flowing over the fins and tubes removes heat from the coolant.

ratio. A method of expressing a relationship of two values relative to each other—for example, a 10:1 compression ratio.

rebound. The release of stored energy from a spring. When a wheel and tire move downward over uneven road conditions it is called rebound.

recirculating ball gearbox. An older style of steering gearbox whose main components are a worm gear on the steering shaft, a ball nut that moves on ball bearings up and down the shaft as the worm gear turns, and a sector gear that engages with the ball nut and transfers steering inputs to the steering linkage.

refractometer. A tool used to measure the specific gravity of a liquid, such as coolant or battery acid. Light passing through the tool refracts based on the liquid the light passes through. Results are shown on a scale.

relative compression test. A quick way to identify a weak cylinder by cranking the engine with an inductive ammeter on the starter circuit. With a lab scope, cranking amperage can be graphed and correlated to a particular cylinder. A cylinder with low compression will draw fewer amps than a cylinder with higher compression, which will show on the graph.

relay. A device that allows a small amount of current to control a large amount of current. A coil of very fine wire is wound many times around an iron core; when current flows through the coil, it acts as an electromagnet to close a contact switch.

release bearing. The release bearing applies against the pressure plate to disengage the clutch.

release mode (abs). One of three modes of operation for a wheel brake in an ABS system; used when the EBCM senses that even after holding pressure to a wheel brake, the wheel is still slowing too rapidly. The EBCM commands the pressure in the circuit be released so that the wheel can begin to rotate again. Also called pressure decay or dump mode.

respirators. Any of several types of breathing apparatus designed to protect the technician's airways from dust and various contaminants.

résumé. A document that showcases your experience, education, and career goals and allows potential employers to quickly learn about you.

return path. In an electrical circuit, the negative or ground path back to the battery or negative part of a circuit.

return spring tool. A tool used to remove and install the high-tension return springs in a drum brake.

return springs. Small tension springs attached to drum brake shoes that pull the shoes back away from the drum when the brakes are released.

ride height. The height at which the vehicle should rest on its suspension. Front-to-back and side-to-side variations should be minimal.

RKE. Remote keyless entry; the use of a remote to activate the power door locks to lock, unlock, and on some vehicles, open rear hatches or the trunk.

rotary engine. An engine that features a spinning three-sided rotor instead of reciprocating pistons. There is one combustion cycle for every revolution of the rotor, so there is one power stroke per revolution.

rough cut. The initial cut when a brake disc is machined, intended to remove surface defects and restore the disc surfaces to parallel.

rubber mallet. A mallet with a solid rubber head, used for such jobs as installing hub caps, when the surface could be marred by a harder hammer.

run-flat tires. A tire that can support itself and the vehicle even if air pressure in the tire is lost.

running compression test. A compression test done at running rpm to check the breathing ability of the engine, specifically to see if the valve train is opening and closing the valves sufficiently.

runout. A defect that causes side-to-side or lateral movement in a spinning object. In wheels it is commonly caused by imperfections in the wheel's shape; in disc brakes it can be caused by a distorted rotor, hub flange distortion, or worn wheel bearings.

S

schematics. Electrical wiring diagrams that use standard symbols to identify circuits and their components.

screwdriver. A simple tool used to loosen and tighten screws. In auto repair, the most common screw head styles are the standard or straight, the Phillips, and the Torx.

seal. A snug interface between a moving part and a nonmoving part that prevents leaks. For example, a crankshaft seal is used where the shaft extends out of the engine block.

sealed beam headlamp. A type of headlamp in which the filament, reflector, and lens are permanently sealed as a unit; used in vehicles for many years but phased out in favor of replaceable-bulb lamps.

self-adjuster assembly. In drum brakes, a mechanism that expands the shoes slightly when the vehicle is driven in reverse and the brakes are applied. This compensates for increases in shoe-to-drum distance as the brake wears.

self-energizing. *See* duo-servo brake.

semiconductor. Device made from a material whose electrical properties are midway between those of a conductor and those of an insulator; their conductivity increases as temperature increases. Their use is the foundation of modern electronics.

semi-independent. A system that uses a fixed rear axle that twists slightly under loads.

serial data. Vehicle operation data provided by the ECM to a technician via a scan tool.

series circuit. An electrical circuit in which all of the parts of the circuit occur in a single path.

series-parallel circuit. An electrical circuit that combines the features of series and parallel circuits.

service advisor. A person who takes service orders from customers and communicates with each customer about the technician's diagnosis and estimated repair costs.

service order. Also called a repair order (RO). This is a legal contract between the customer and the shop regarding the services to be performed. It contains pertinent customer information, vehicle information, and a record of services and repairs.

shift solenoid. A shift solenoid is an electrically operated solenoid that either allows or blocks the flow of transmission fluid.

shims. When attached to the backing of a disc brake pad, helps reduce noise.

shimmy. A side-to-side shaking motion of the steering wheel caused by a wheel/tire that is dynamically out-of-balance.

shock absorber. A suspension component that dampens the oscillations (bounce) of a spring. The most common type is the direct double-acting hydraulic shock absorber.

short circuits. An unwanted connection between two different electrical circuits. A short to power causes a circuit to operate when another, unrelated circuit is used. A short to ground occurs when the power supply of a circuit touches or connects to ground.

sliding caliper. A type of floating disc brake caliper that is mounted on sliding keyways.

slow charging. Charging a battery at a relatively low rate, commonly 1 to 2 amps, in order to minimize heat generation and to allow the plates to more thoroughly charge throughout the plate volume.

Society of Automotive Engineers. An organization founded to create common engineering and design standards for the automotive industry; oversees such tasks as the testing and certification of oil viscosity ratings.

sockets. A fittings that is used with a ratchet handle to quickly and safely loosen and tighten a specific size of fastener. Sockets come in many sizes and configurations.

specific gravity. A measurement of the density of a substance compared to water. Specific gravity tests are performed on battery acid and coolant.

spring compressors. A tool used to safely compress coil springs for removal and installation.

spring rate. The amount of force needed to compress or twist a spring a certain amount. Springs can have either linear or variable rates.

sprung weight. Vehicle weight that is carried by the springs.

square seal. The seal in a disc brake caliper that seals each piston in the bore. The seal also acts as a return spring for the piston.

squealer. A built-in wear indicating device on a disc brake pad that is named after the noise it makes when it comes into contact with the rotor.

stabilizer bar. A steel bar attached to the lower control arms or axle assembly and to the body or frame; designed to minimize body roll when the vehicle is cornering. Also called sway bar or anti-roll bar.

starter current draw. An electrical test of the amperage used by the starter circuit during engine startup. For most gasoline engines, cranking amps will be between 125 and 250 amps.

starter motor. An electric motor powered by DC current from the battery to spin the engine for initial startup.

state of charge. The percentage of battery charge. A fully charged battery has 12.6 volts and is at 100 percent SOC.

static balance. The distribution of weight around the axis of rotation of a stationary wheel and tire.

static electricity. The accumulation of an electrical charge by a substance not normally conductive to electricity.

stator. The stationary windings of wire in an AC generator. Stator windings occur in two different configurations, the wye winding and the delta winding.

steering gear. or gearbox is the component that changes the rotary motion of the steering wheel into the linear side-to-side motion to turn the front wheels.

steering ratio. For a steering mechanism, the ratio of steering wheel movement to front wheel and tire movement; found by dividing the total number of degrees the steering wheel turns by the total number of degrees of front wheel movement.

stranded wire. is a conductor formed by combining several small wires called strands into a larger wire.

strut spring compressor. A tool used to compress the spring in a MacPherson-type strut for disassembly and assembly.

T

taper. *See* parallelism.

temperature rating. A component of uniform tire quality grade information that indicates the tire's resistance to heat and its ability to dissipate heat. Graded as A, B, or C.

terminals. Electrical pins and sockets within connectors that carry current.

test light. A simple, unpowered electrical tester for making basic power and ground circuit tests. When the clip is attached to a good ground and the probe end is touched to a power source, the light bulb will light.

thermal cycling. A pattern of heating and cooling that, over time, can cause resistance to build up in electrical connections.

thermostat. A valve used to keep coolant inside the engine to allow for engine warmup and to provide heated coolant for the heater core.

thickness variation. *See* parallelism.

three-minute charge test. A test in which a high-rate charge (30–40 amps) is applied to a battery for three minutes, while voltage is observed. Voltage should not exceed 15.5 volts; excessive voltage indicates that the battery plates are sulfated and the battery is not accepting a charge.

tire rotation. Periodically removing the wheels and tires from their current location and moving them to another corner of the vehicle in order to maximize tread life.

tire slip. The amount of traction between the tire and the road. A freely rolling wheel has zero tire slip, while a locked wheel moving over the pavement has 100 percent slip.

torque. A measurement of twisting force.

torque converter. The torque converter connects the engine's crankshaft to the input shaft of an automatic transmission.

torsion bar. A length of steel bar fastened to a control arm at one point and the vehicle's frame at another. Movement of the control arm imparts twist on the torsion bar. The absorption of the twist is the spring action of the torsion bar.

Torx drives. A driver bit with a six-sided star shape.

track bar. A bar used in many rear suspensions to limit front and rear movement of the rear axle.

traction rating. A component of uniform tire quality grade information that is based on the tire's friction during a wet skid test. Traction is rated AA, A, B, or C.

transistors. A solid-state switch containing three layers of semiconducting material. A transistor consists of a base, a collector, and an emitter. When voltage is applied to the base, current can flow from the collector to the emitter and to the load in the circuit.

treadwear rating. A component of uniform tire quality grade information that is used to compare one tire to another based on expected service life of the tire.

trip. A set of parameters that define when a specific OBD II monitor is enabled.

U

ultra-low sulfur diesel. Diesel fuel that contains reduced levels of sulfur as an antipollution measure. As of June 1, 2010, all diesel fuel sold in the United States had to be of this type.

undercharge condition. A condition in which the charging system produces voltage and amperage but at an insufficient rate to keep the battery charged.

understeer. A driving condition in which the front tires reach their cornering limit before the rear tires.

unibody. The space frame or unitized body construction approach used for most modern cars, in which there is no separate frame.

unsprung weight. Vehicle weight that is suspended beneath the springs and therefore is not carried by the springs. Examples include the weight of the wheels and tires, brake components, control arms, steering knuckles, and some rear axles.

unwanted resistance. Excessive resistance in an electrical circuit that reduces current flow, causing the intended load in the circuit to operate improperly or not at all.

V

vacuum. In an engine, air pressure that is lower than atmospheric pressure.

vacuum assist. An external power brake assist system that uses the vacuum created by the operation of a gasoline-powered engine.

vacuum bleeding. The use of a vacuum evacuation tool to pull fluid through the brake system to remove air and to flush out and replace the old fluid.

vacuum check valve. A valve in a vacuum-type power brake assist system that prevents vacuum from being released back to the engine. This allows the power brake booster to store a vacuum reserve.

valve lash. The gap between a rocker arm and the top of the valve stem. The gap allows for expansion of the components.

vise. A bench-mounted device with jaws used to clamp and hold an object.

VIN. Also known as Vehicle Identification Number is a 17-digit serial number that uniquely identifies a vehicle.

visual inspection. is performed by the technician to look for obvious concerns, such as fluid leaks or severely worn tires.

voltage. An amount of electrical potential, that is, the capacity to do work.

voltage drop. The measurement of voltage lost through wiring, components, and connections due to electrical resistance.

voltage regulator. A device that controls the amount of current that is supplied to the field winding in an AC generator.

volumetric efficiency. The ratio of air drawn into a cylinder compared to the maximum air that could enter the cylinder. Most gas engines are about 80 to 90 percent volumetrically efficient.

W

warning lights. A light on the instrument panel that alerts the driver to a condition or to a fault.

warranty time. Estimated labor time to perform specific repairs that are covered under the manufacturer's warranty.

web browser. Software for locating and displaying websites.

wet compression test. A compression test performed on a cylinder with low compression to determine the cause of the problem. A wet test involves squirting a small amount of engine oil into the cylinder and retesting; if the oil causes compression to increase, worn rings are suspected.

wheel. The metal rim on which an automotive tire is mounted.

wheel bearing. Any of several types of bearings that reduce friction at the interface of the axle and the hub.

wheel cylinder. An output cylinder in a drum brake assembly. The wheel cylinder contains pistons that move the brake shoes against the drum when the brakes are applied.

wheel speed sensors. An ABS system sensor that enables the EBCM to determine wheel speed.

window regulator. In a power window system, a small DC motor and track mounted inside the door to move the window glass.

wire gauge. The cross-sectional size of a wire.

wiring harness. A bundle of wires and connectors. A harness typically has many connections to components and is used for an entire system or section of the vehicle.

work ethic. A commitment to the job that is demonstrated by such actions as consistent on-time attendance, dependability, and a positive attitude.

Z

Z.87 standard. An ANSI standard for the impact resistance of industrial eyewear.

INDEX

C

E

I

T

U

V

W

Y

Z